Tectonic evolution of northwestern México and the southwestern USA

Edited by

Scott E. Johnson
Department of Earth Sciences
5790 Bryand Global Sciences Center
University of Maine
Orono, Maine 04469-5790
USA

Scott R. Paterson
Department of Earth Sciences
University of Southern California
Los Angeles, CA 90831-0740
USA

John M. Fletcher
Departamento de Geología
CICESE
P.O. Box 434843
San Diego, California 92143
USA

Gary H. Girty
Department of Geological Sciences
San Diego State University
San Diego, California 92182-1020
USA

David L. Kimbrough
Department of Geological Sciences
San Diego State University
San Diego, California 92182-1020
USA

and
Arturo Martín-Barajas
Departamento de Geología
CICESE
P.O. Box 434843
San Diego, California 92143
USA

THE
GEOLOGICAL
SOCIETY
OF AMERICA

Special Paper 374

3300 Penrose Place, P.O. Box 9140 ▪ Boulder, Colorado 80301-9140 USA

2003

Published by The Geological Society of America, Inc.
3300 Penrose Place, P.O. Box 9140, Boulder, Colorado 80301
www.geosociety.org

Printed in U.S.A.

GSA Books Science Editor: Abhijit Basu

Library of Congress Cataloging-in-Publication Data

Tectonic evolution of northwestern México and the southwestern USA / edited by Scott E. Johnson …
[et al.].
    p. cm. — (Special paper ; 374)
    Includes bibliographic references, index.
    ISBN 0-8137-2374-4 (softcover)
    1. Geology, Structural—Mexico—Baja California (Peninsula) 2. Geology, Structural—California,
Southern. 3. Geology, Stratigraphic—Mesozoic. 4. Geology, Stratigraphic—Tertiary. I. Johnson, Scott E.
Special papers (Geological Society of America) ; 374.

QE629.5.B35T43 2003
551.8'0972'2—dc22

                                                                    2003049445

Cover: Cover: Stromatic migmatites with folded, transposed biotite-rich layers and several generations of leucosome. The outcrop occurs within the transitional section of the batholith, in the hanging wall of the Main Mártir thrust, northern Sierra San Pedro Mártir, Baja California, México. These rocks and the migmatization predate ca. 127 Ma granitoids (now orthogneisses) that intruded them. Photo by Erwin A. Melis.

10 9 8 7 6 5 4 3 2 1

Contents

Preface

R. Gordon Gastil

This volume is dedicated to R. Gordon Gastil, Professor Emeritus at San Diego State University, in recognition of his outstanding contributions to the geosciences through geologic mapping and interpretation Earth's history. Together with students and colleagues, Gordon mapped thousands of square kilometers in Arizona, California, and México, including the entire state of Baja California! He was awarded the prestigious Dibblee Medal on August 21, 2002, in recognition of his extraordinary accomplishments in field geology and geologic mapping. Gordon's insatiable curiosity and geologic insight served as a model for many hundreds of San Diego State University (SDSU) students across five decades who learned the art of geologic mapping under his direction.

Although well known for his mapping efforts in Baja California, Gordon's broad range of interests and remarkable ability to break new ground across scientific disciplines is truly the hallmark of his rich scientific career. His breadth and originality are perhaps best illustrated by his 1960 American Journal of Science paper titled "The Distribution of Mineral Dates in Time and Space." This highly innovative contribution presented the first global synthesis of continental crustal evolution based on radiometric dating and distribution of rock units. His approach of weighing mineral dates against outcrop areas validated basic concepts of "orogenic periodicity" and established the basic pattern of global-scale episodic continental growth still recognized today. Many of the questions and problems posed in Gordon's 1960 paper endure as central topics of research in the earth sciences today.

Russell Gordon Gastil was born in San Diego, California, June 25, 1928, to Francis and Russell Gastil. By the time he reached high school, he was already attracted to geology as a profession because of the possibilities for exploration and discovery. He enrolled at San Diego State Teachers College (now San Diego State University) where he came under the influence of Professor Baylor Brooks, who later established the Department of Geological Sciences at SDSU. At the time, no geology degree was offered at SDSU, so

Professor Brooks directed Gordon to the University of California at Berkeley, where he completed a B.Sc. in geology. He then continued on in the same department to complete a Ph.D., which he received in 1953. Gordon's doctoral thesis at Berkeley was a study of Proterozoic crystalline basement rocks in south-central Arizona, for which he had fellowship support from the Oak Ridge National Laboratory. He mapped and used a portable scintillometer to measure in situ radioactivity of rocks as part of this work; this experience helped establish his early interest in the rapidly developing techniques of radiometric dating.

Upon graduation and following a short stint working for Shell Oil Company, Gordon was drafted into the U.S. Army and served 22 months in southern Alaska and the Aleutians where he did geologic mapping with John Reed. Along the way, he collected granite samples for lead-alpha dating by Esper Larson at Harvard and subsequently did zircon separation work at the Naval Gun factory in Washington, D.C.

Following Army service, Gordon spent the next three summers doing mineral exploration and mapping in Labrador for a mining company (Canadian Javelin), spending the intervening winters in Montreal and at Harvard. He then returned to California, where he taught for three semesters at the University of California at Los Angeles before Baylor Brooks brought Gordon onto the San Diego State College faculty as a full-time faculty member in 1959.

Gordon received his first National Science Foundation grant at SDSU—for Pb-alpha dating—in 1961. In 1963, he and Ned Allison secured National Science Foundation funding to begin producing a reconnaissance geologic map of the State of Baja California. They hired six undergraduate students that first year (including two from Universidad Autónoma de Baja California in Ensenada), ordered air photographs, and the quest was on! Most of the mapping was done over the next four years but continued until it was compiled at a scale of 1:250,000 in 1971. The map subsequently was published in 1975 as part of Geological Society of America Memoir 140, Reconnaissance Geology of the State of Baja California. What is remarkable about this mapping feat is not only that it was accomplished with undergraduate students, but that the general inaccessibility of large areas made the going extremely tough (at the time, there were no reliable topographic base maps and virtually no paved or graded roads). Along the way, Gordon mapped a large section of the State of Sonora, including Tiburon Island on the mainland edge of the Gulf of California. The Baja California map, published in three separate sheets and now out of print, remains as a primary regional mapping contribution still highly sought after by all Baja California geologists. It is certainly appropriate that these maps are being made available again in this volume as PDF files on the volume's accompanying CD-ROM.

Gordon's mapping experience in Baja California and surrounding areas formed the foundation of his research at SDSU. He has always freely shared his enthusiasm and ideas with students, colleagues, and anyone else interested in Earth. He ultimately supervised 58 master's theses at SDSU along with many dozens of undergraduate research projects known as senior reports. Most of this material has been published in some form, and a visit to the SDSU library to review thesis material is now almost a standard pilgrimage undertaken by anyone initiating new research or mining projects in Baja California.

Gordon's research on the tectonic and geologic history of Peninsular California and adjacent México has established his position in the upper echelon of Cordilleran geologists. Highlights are almost too numerous to mention. He was the first to recognize the strong transverse asymmetry across the Peninsular Ranges batholith and its natural separation into distinct western and eastern provinces (a feature independently recognized by Lee Silver and co-workers at the California Institute of Technology). Building on work of Richard Mirriam and Esper Larsen, he also recognized contrasting patterns of pluton zonation within the Peninsular Ranges batholith, including ring dikes and cone sheet structures in the western province. Following the arrival of Daniel Krummenacher at SDSU in 1969, Daniel and Gordon initiated two major projects in K-Ar dating. The first, across the northern third of the Peninsular Ranges batholith, documented the strongly asymmetrical west-to-east Cretaceous uplift and cooling history of the batholith; the second involved dating of Cenozoic volcanic rocks in the circum–Gulf of California and relating this history to the evolving North American–Farallon–Pacific plate boundaries. Current knowledge of the pre-batholithic stratigraphy of Peninsular California is largely based on the work of Gordon and his students; much of this work is published in Geological Society of America Special Paper 279, edited by Gordon and Rick Miller.

Starting in 1973, with support from PEMEX, Gordon and his students initiated major mapping projects in the Vizcaíno Peninsula region of west-central Baja California that established much of the basic geology of the region, including documentation of ophiolite complexes. He did early work on the "elevated erosion surfaces" of the Peninsular Ranges, speculating on their tectonic significance. Together with students and using a magnetic susceptibility meter, he established the ilmenite-magnetite line within the Peninsular Ranges batholith and pioneered its use as a provenance tool in sedimentary successions.

Inasmuch as Gordon and his students have published frequently through his career in Geological Society of America journals, as well as his major contributing efforts to Memoir 140 and Special Paper 279, it is certainly appropriate that this volume be included in the Geological Society of America Special Paper series. The 17 papers published in this volume reflect both the diversity of the geology in Baja California and surrounding regions as well as the diversity of contributions made by Gordon and his colleagues. We briefly introduce these papers here. We also draw attention to the accompanying CD-ROM, which contains a rich collection of resources, including: (1) the three out-of-print geologic maps of Baja California introduced above, which were originally published by the Geological Society of America; (2) an out-of-print geologic map of coastal Sonora, México, prepared by Gastil and Krummenacher and originally published by the Geological Society of America; (3) an out-of-print geologic map of southern Sinaloa that accompanies the paper by Henry et al., and which was originally published by the Geological Society of America; (3) a color magnetic map of Baja California to accompany the overview paper by Sedlock; (4) three extensive compilations of geochronologic data which accompany the two papers by Grove et al. and the paper by Ortega-Rivera; (5) sample locations, geochronologic data and a color geologic map of the Vizcaíno Peninsula that accompany the paper by Kimbrough and Moore; (6) color geologic maps to accompany the Vizcaíno Peninsula paper by Sedlock; and (7) compiled geochemical and geochronological data from southern California that accompanies the papers by Todd et al. and Shaw et al.

The papers in this volume largely focus on the Mesozoic and Tertiary history of Baja California and nearby regions of mainland México. The papers are organized by scale, geologic age, and topic. The volume begins with an extensive tectonic overview by Sedlock with the goals of summarizing our present understanding of the geology of Baja California, addressing several controversial issues, and providing a framework of Mesozoic and Tertiary histories discussed in greater detail in the remaining papers. This paper is followed by three other Mesozoic tectonics papers. The first two papers, one by Kimbrough and Moore and the other by Sedlock, examine Triassic and Jurassic ophiolites and the accompanying deformation, metamorphism, and magmatism in the Vizcaíno Peninsula. These suprasubduction zone ophiolite–arc sequences are compared to others in the Cordillera. In the following paper, Wetmore et al. examine characteristics of two Mesozoic arcs and juxtaposed metasedimentary packages exposed in Baja California and southern California. They conclude that the Santiago Peak and Alisitos arcs exposed north and south of the Aqua Blanca Fault, respectively, have distinct histories, with the Santiago Peak representing a continental margin arc and the Alisitos representing a collided oceanic arc.

In the following paper, Umhoefer attempts to link Cretaceous events in Baja California to events elsewhere in the Cordillera and concludes that major collision (125–120 Ma) in Baja California led to northward escape of the central and northern Cordillera. One of the controversial topics introduced in the Umhoefer and Sedlock overview papers is the question of lateral transport of Baja terranes. In the next paper, Symons et al. examine paleomagnetic and geobarometric data from the La Posta pluton, California, and conclude that this pluton intrudes across a suture across which the western terrane was displaced 2700 + 1000 km relative to both the eastern terrane and stable North America.

Five papers focus on the timing, characteristics, and source regions of magmatism in the Peninsular Ranges batholith. Shaw et al. examine a suite of strongly deformed, Jurassic transitional and peraluminous magmas in southern California, which are now known to also exist along the axial zone of the Peninsular Ranges batholith in Baja California. Source regions proposed include young mantle-derived magmas, evolved basinal fill metasediments, or metavolcanics in an arc environment. Todd et al. present field, petrographic, and geochemical characteristics of a suite of I-type, Cretaceous plutons in an east-to-west transect across the Peninsular Ranges batholith in southern California and discuss the implications for its evolution. Henry et al. examine the southern continuation of the Peninsular Ranges batholith in Sinaloa, México and discuss both similarities and differences with the Peninsular Ranges batholith in Baja California. Tulloch and Kimbrough compare the Peninsular Ranges batholith to the Median batholith of New Zealand and suggest similarities to Archean trondhjemite-tonalite-granodiorite granitoids. They suggest that both the Peninsular Ranges batholith and Median batholith magmas were generated by melting of mafic lower-arc crust.

The following four papers present extensive summaries of geochronologic or geobarometric data across and along the Peninsular Ranges batholith and examine their implications for batholith cooling, uplift, and regional tectonics. Ortega-Rivera discusses the eastward decrease in both U/Pb ages and various cooling ages, notes that geochrons patterns are consistent across Baja and into mainland Mexico and concludes that no large-scale pre-rifting displacement occurred. Rothstein and Manning present pressure-temperature and metamorphic data for the Peninsular Ranges batholith and reexamine thermal models and the tectonic implications of likely thermal gradients in arcs. Grove, Lovera, and Harrison examine Late Cretaceous cooling of the east-central Peninsular Ranges batholith and its relationship to La Posta pluton emplacement, Laramide shallow subduction, and forearc sedimentation. They speculate that pluton emplacement drove denudation between 91 and 86 Ma and removal of lower crust and mantle lithosphere triggered erosional denudation between 78 and 68 Ma. Grove, Jacobson, Barth, and Vucic present 850 detrital zircon ages from high–pressure-temperature schists (Pelona, Orocopia, Rand, Portal Ridge, Sierra de Salinas) underplated in the Late Cretaceous–early Tertiary and discuss two distinct source terranes for sediments in these units.

The final three papers examine sedimentation and volcanism in the Tertiary. Johnson examines Pliocene faulting, sedimentation, and volcanism near Loreto Baja California Sur and the implications for transtensional tectonics. Oskin and Stock examine Cenozoic volcanism and tectonics of the continental margins of the Upper Delfin basin in northeastern Baja California and western Sonora. They conclude that extension and volcanism were closely linked and acted as a coupled system to localize Pacific–North American plate motion in the Gulf. Paz Moreno et al. discuss tholeiitic to alkalic basalts in the Quaternary Moctezuma volcanic field of the central Sonoran Basin and Range Province and the implications for enclosed xenoliths.

We are excited about this volume and think it represents well Gordon Gastil's broad interests and love of geology. We thank Gordon for his long years of friendship as well as his efforts in understanding this complex region. We also thank the authors for their diligence in helping with this collection of papers.

Scott E. Johnson
Scott R. Paterson
John M. Fletcher
Gary H. Girty
David L. Kimbrough
Arturo Martín-Barajas

Geological Society of America
Special Paper 374
2003

Geology and tectonics of the Baja California peninsula and adjacent areas

Richard L. Sedlock*

Department of Geology, San José State University, San José, California 95192-0102, USA

INTRODUCTION

This volume honors Russell "Gordon" Gastil, emeritus professor at San Diego State University's Department of Geological Sciences, who served as professor there from 1959 to 1993. Gordon mapped and supervised the mapping of tens of thousands of square kilometers of México, California, and Arizona. One of his many contributions was GSA Memoir 140, which included the first reconnaissance map of the northern half of the Baja California peninsula (Gastil et al., 1975). This long out-of-print opus probably is the most-cited work in all geo-literature pertaining to the peninsula and is jealously guarded by those who own a copy. Gordon and his students continued their studies in the region over the next two decades, and many of these analyses were collected in GSA Special Paper 279 (Gastil and Miller, 1993).

Gordon's early interest in the geology and tectonics of the Peninsular Ranges in southern California and northern Baja California (e.g., Bushee et al., 1963; Gastil, 1968) expanded to include a wide variety of geologic and tectonic problems throughout northwestern México. The regional focus of papers in this volume reflects his seminal work in the Peninsular Ranges, other parts of Baja, islands in the Gulf of California, and the Mexican states of Sonora, Sinaloa, Nayarit, and Jalisco (Fig. 1).

The goals of this introductory paper are to (1) summarize our present understanding of the geology and tectonics of southernmost California and Baja California, the Gulf of California, and western mainland México; (2) identify outstanding problems and suggest avenues of research that might address those problems; and (3) provide context for the other papers in this volume. The emphasis on tectonics-oriented issues reflects Gordon's interests as well as my own. Coverage of other topics, such as paleontology and modern sedimentary environments, is incomplete or absent.

This paper first describes the salient aspects of the geography, modern tectonic setting, seismicity, and geophysical attributes of the region. It then summarizes the geology in terms of specific terranes and describes major batholiths throughout the region. The geology of western mainland México is not covered systematically, though some components of it are addressed in later sections of the paper.

The bulk of this paper is an analysis of major events in the Mesozoic and Cenozoic history of the Baja California peninsula and nearby regions. Three topics receive the most attention, reflecting the scope of other papers in this volume and the focus of tectonics-oriented research in the region in the last few decades: (1) the nature, origin, and development of the Peninsular Ranges batholith and its host rocks in southernmost and Baja California; (2) possible northward translation of part, most, or all of southernmost and Baja California in the Late Cretaceous, Paleogene, or both; and (3) late Cenozoic transtension, rifting, and formation of oceanic crust in the Gulf of California. Other tectonics-related issues are addressed, but in less detail.

Keywords: Baja California, Gulf of California, tectonic history, paleomagnetism.

* sedlock@geosun.sjsu.edu

Sedlock, R.L., 2003, Geology and tectonics of the Baja California peninsula and adjacent areas, *in* Johnson, S.E., Paterson, S.R., Fletcher, J.M., Girty, G.H., Kimbrough, D.L., and Martín-Barajas, A., eds., Tectonic evolution of northwestern México and the southwestern USA: Boulder, Colorado, Geological Society of America Special Paper 374, p. 1–42. For permission to copy, contact editing@geosociety.org. © 2003 Geological Society of America.

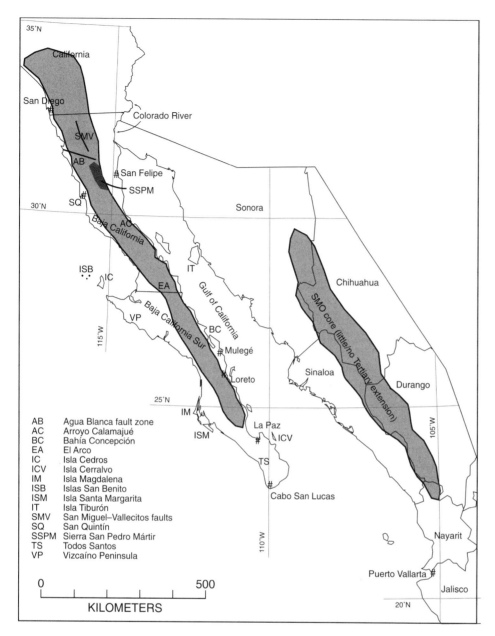

Figure 1. Approximate outlines of the Peninsular Ranges and unextended core of Sierra Madre Occidental (shaded). Positions of Agua Blanca and San Miguel–Vallecitos fault zones are approximate and do not include probable northward continuations. Outline of Sierra Madre Occidental (SMO) core modified from Henry and Aranda (2000) and Ferrari et al. (2002).

AB   Agua Blanca fault zone
AC   Arroyo Calamajué
BC   Bahía Concepción
EA   El Arco
IC   Isla Cedros
ICV   Isla Cerralvo
IM   Isla Magdalena
ISB   Islas San Benito
ISM   Isla Santa Margarita
IT   Isla Tiburón
SMV   San Miguel–Vallecitos faults
SQ   San Quintín
SSPM Sierra San Pedro Mártir
TS   Todos Santos
VP   Vizcaíno Peninsula

GEOGRAPHIC BACKGROUND

The Baja California Peninsula

The long (1300 km), narrow (45–240 km) Baja California peninsula is separated from mainland México by the similarly shaped Gulf of California (also known as the Sea of Cortez) (Fig. 1). The peninsula comprises two states of the Republic of México: Baja California, which is often incorrectly called Baja California Norte, extends from the U.S. border south to latitude 28°N, and Baja California Sur extends from latitude 28°N south to Cabo San Lucas (Fig. 1). To avoid ambiguity about the meaning of the term "Baja California," in places

I use the informal terms "Baja" and "the peninsula" to refer to the entire peninsula and reserve "Baja California" for the northern state.

Baja's physical geography is dominated by the Peninsular Ranges, a spine-like topographic high that continues northward into southern California (Fig. 1). The maximum elevation of this geographic backbone ranges from about 1800 m to 3078 m in the north (31°–32.5°N) to about 300 m near 24°N, where it merges into low plains near La Paz. East of 111°W, the "Los Cabos block" includes ranges up to 2000 m high. Low mountains of the Vizcaíno Peninsula between 27°N and 28°N are separated from the Peninsular Ranges by the Vizcaíno Plain, which probably hosted shallow seas earlier in the Quaternary.

The peninsula's remote location and difficult terrain rendered it fairly inaccessible until the 1970s, when a paved transpeninsular road was completed. Much of the peninsula still is roadless, and most of the other parts of it are accessible only via four-wheel drive vehicles, cooperative weather and road conditions, and luck.

The Gulf of California

The Gulf of California, formerly known as the Sea of Cortez, is an elongate basin that extends from its head near the mouth of the Colorado River at 32°N to a broad transition zone to the Pacific Ocean at 23°N to 20.5°N (Fig. 1). Prior to opening of the gulf in the late Cenozoic, the Baja California peninsula probably lay farther southeast such that its southern tip was adjacent to Nayarit and Jalisco on the mainland (Stock and Hodges, 1989; Lonsdale, 1991; Lyle and Ness, 1991). Gulf islands include blocks of thinned continental crust that were stranded during late Cenozoic transtensional opening of the gulf and volcanic edifices constructed during opening of the gulf.

Western Mainland México

From north to south, the Gulf of California is bounded on the east by the Mexican states of Sonora, Sinaloa, Nayarit, and Jalisco (Fig. 1). The dominant geographic and geologic feature of western México is the Sierra Madre Occidental, an elongate plateau of Cenozoic volcanic rocks that trends north-northwest–south-southeast from about 30°N in eastern Sonora to 20°N near the Nayarit–Jalisco border. The narrow core of the Sierra Madre Occidental (Fig. 1) consists of weakly extended to unextended Tertiary strata that attain an average maximum elevation of about 2000 m. Much of Sonora exhibits basin-and-range morphology, with older basement rocks exposed in northwest-trending ranges separated by alluviated basins. Between 28°N and 24°N, pre–Sierra Madre Occidental rocks crop out in a thin band between the Sierra Madre Occidental and a flat coastal plain that borders the gulf. Between 24°N and 21°N, Cenozoic volcanic rocks completely obscure older rocks as far east as 103°W.

MODERN TECTONIC SETTING, SEISMICITY, AND GEOPHYSICAL DATA

Northwestern México hosts the transtensional Pacific–North America plate boundary. Most plate motion is accommodated by structures within the Gulf of California itself and along its western margin (i.e., the eastern edge of the Baja California peninsula), as indicated by modern seismicity (USGS–MIDAS, 1998). Western mainland México is tectonically inactive, as is most of the peninsula. However, north of 31°N, the San Miguel–Vallecitos and Agua Blanca fault zones transfer dextral slip from the northern Gulf across the peninsula into the Pacific borderland. The San Miguel–Vallecitos fault system has a maximum geologic slip rate of 0.55 mm/yr (Hirabayashi et al., 1996) and a geodetic slip rate of 1.2 ± 0.6 mm/yr (Dixon et al., 2003) and

has generated several M>6 earthquakes in historical time. The Agua Blanca fault zone has a late Quaternary to Holocene geologic slip rate of 3–7 mm/yr (Rockwell et al., 1993; Dixon et al., 2003) and a geodetic slip rate of 6.2 ± 1.0 mm/yr (Dixon et al., 2003). The most recent rupture on the Agua Blanca fault zone, in A.D. 1640 ± 160 (Rockwell et al., 1993), may have been part of an earthquake sequence that also affected the Rose Canyon and Newport-Inglewood faults along the southern California coast (Grant and Rockwell, 2002).

At 24.5°N to 23°N, the extensional component of Pacific–North America plate motion appears to be taken up along a set of north-northwest–striking normal faults that transect the southernmost part of the peninsula (e.g., Fletcher and Munguía, 2000). The average rate of normal slip on the major fault of this system (San José del Cabo) is 0.4–0.7 mm/yr (Fletcher et al., 2000). Focal mechanisms of historical earthquakes in the Gulf of California east of La Paz show nearly pure normal slip that is interpreted as kinematically distinct from deformation along the spreading centers and transforms in the gulf axis (Fletcher and Munguía, 2000).

Seismic and gravity data have been used to estimate the thickness and structure of the crust in northwestern México. Offshore northwestern Baja, the crust thickens from ~12 km in the outer borderland to ~17 km in the inner borderland (Couch et al., 1991). Thickness estimates of the northern Peninsular Ranges include a maximum of 28 km (Couch et al., 1991) and a maximum of 40 km, thinning to about 33 km at the Pacific coast (Lewis et al., 2001). Estimates of maximum thickness of the Peninsular Ranges in the central and southern peninsula include 22 km (Couch et al., 1991) and 33 km (Romo, 2002). East of the Peninsular Ranges, the thickness of the crust is ~21 km between the Salton Trough and head of the gulf (Couch et al., 1991), at least 13 km (Couch et al., 1991) or 15–18 km (Lewis et al., 2001) in the northern gulf, ~10 km in the central gulf, and ~8 km in the southern gulf (Couch et al., 1991). In the northern and central gulf, the crust is thinner near the peninsula than near mainland México; crust in the southern gulf lacks such asymmetry (Couch et al., 1991). Geochemical data from Quaternary basalts in northeastern Sonora are interpreted to indicate that the lithosphere has thinned greatly since the Oligocene but do not permit quantitative determination of past or current thickness (Paz Moreno et al., this volume).

Electrical conductivity, mass density, and magnetic susceptibility data collected along a northeast-southwest transect through the northern Vizcaíno Peninsula led Romo et al. (2001) to infer the presence of a conducting surface that dips 15° from the west coast 100 km eastward and then 45° to the east coast of Baja. Romo (2002) infers that this slab represents the abandoned Farallon plate that is still present beneath the Vizcaíno Peninsula.

The new North American Magnetic Anomaly Map (NAMAG, 2002) includes data from all of México plus adjacent oceanic regions. The most remarkable feature in northwestern México is the steep gradient that runs nearly continuously through the peninsula from 31°N (the latitude of the Agua Blanca fault) to 23.5°N, where it terminates west of the Los Cabos block (Fig. 2). This gradient separates a 75-km-wide band of weak magnetic anomalies on the east from slightly more variable, but consistently strong, magnetic anomalies on the west. As discussed in a later section, it roughly coincides with a major terrane boundary.

TERRANES OF THE BAJA CALIFORNIA PENINSULA

Previous Geologic Maps

No detailed geologic map of the entire Baja California peninsula has been published that incorporates recent mapping. Gastil et al. (1975) compiled a three-map set of the peninsula north of 28°N (excluding Pacific islands) at 1:250,000 using data published through 1970. Fenby and Gastil (1991) compiled a ~1: 1,500,000 map of the entire peninsula and surrounding marine regions in which they grouped rocks by age and genesis; rocks older than 37.5 Ma were not differentiated. The fifth edition of the geologic map of México, the most recent (Ortega-Gutiérrez

et al., 1992), was published at 1:2,000,000 and shows generalized map units for the entire peninsula.

Terrane Terminology

In this paper, I assign Mesozoic and older rocks around the Gulf of California to terranes (Fig. 3). The terrane concept is not a fundamental necessity: my basic premise is that the region can be discussed usefully in terms of assemblages or associations (i.e., terranes) that differ from one another. The key Mesozoic and older components of each of the terranes are listed in Table 1 and described in more detail in the following sections.

Several terrane-naming schemes have been implemented by other workers (Table 1). I use the names shown at the top of Table 1 and in Figure 3 for the following reasons. *Caborca* has precedence in the literature and is much better established in the literature than Cortes or Serí. The name "Guerrero," especially

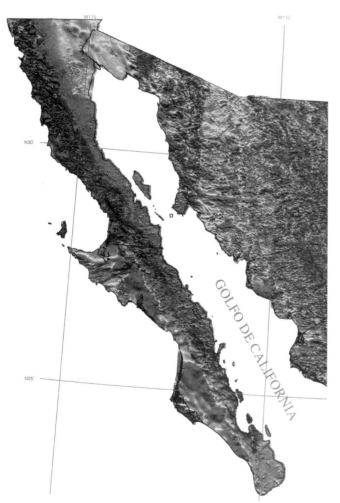

Figure 2. Map of total magnetic field of Baja and surrounding areas from North American Magnetic Anomaly Map (NAMAG, 2002). Color magnetic map is available on the CD-ROM accompanying this volume.

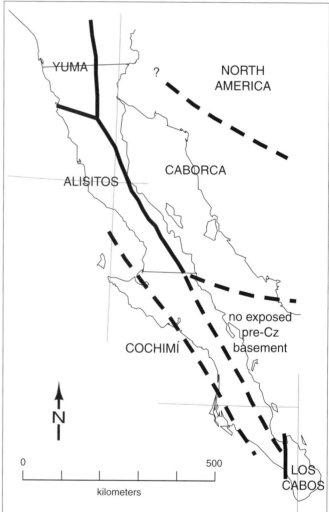

Figure 3. Terrane nomenclature as used in this paper. See text and Table 1 for explanation.

TABLE 1. TERRANE NOMENCLATURE FOR BAJA

This paper	Caborca	Yuma	Alisitos	Los Cabos block	Cochimí
Primary constituents	Proterozoic to Early Paleozoic shelf strata Jurassic to Early Cretaceous arc	Triassic(?)–Jurassic flysch (Bedford Canyon Complex) Early Cretaceous arc (Santiago Peak Volcanics)	Early K island arc (Alisitos)	Prebatholithic rocks (undated) K plutonic rocks	Mesozoic island arcs, ophiolites, subduction complexes (blueschists)
Previous workers					
Howell et al., 1985	Caborca	Guerrero	Guerrero	La Paz	Vizcaíno
Coney and Campa, 1987; Silberling et al., 1992	Cortes	Santa Ana	Santa Ana	La Paz	Vizcaíno, Magdalena
Sedlock et al., 1993	Serí	Yuma	Yuma	Pericú	Cochimí
Dickinson and Lawton, 2001	Caborca	Guerrero	Guerrero	Guerrero	Guerrero

as used by Dickinson and Lawton (2001), covers so many different rocks over such a large area that it is of little use for regional analyses like the one in this paper. In addition, as discussed later, the nature and history inferred for the "Guerrero superterrane" by Dickinson and Lawton (2001) may not adequately apply to rocks in the Baja California peninsula. The *Alisitos* terrane, as defined in this paper, was subsumed within the *Yuma* terrane as defined by Sedlock et al. (1993). The *Los Cabos terrane* is a more accurate term than the La Paz terrane because La Paz probably lies outside the boundaries of the block, and it reflects the common usage of the term "Los Cabos block" by many workers. *Cochimí* is used instead of Vizcaíno or Viscaíno because the latter name also has been applied to part of the central California borderland (e.g., McCulloch, 1987), and because the terrane covers a much larger area than just the Vizcaíno Peninsula.

Note that this set of terrane names refers only to rocks in the Baja California peninsula and, in the case of the Caborca terrane, northern Sonora. In this paper, I only briefly treat other rocks on the Mexican mainland (i.e., south and east of the Caborca terrane) and thus do not apply terrane terminology there. Pre-Cenozoic rocks are not exposed along the Gulf margin of the southern part of the peninsula (Fig. 3), so no terrane assignment is made. The southeastern part of the peninsula may be underlain by Mesozoic or older rocks similar to those in Sinaloa, but no direct or indirect evidence constrains the nature or age of the basement there.

Caborca Terrane

The oldest rocks in the peninsula are greenschist- to amphibolite-facies metamorphic rocks that crop out in isolated ranges within a low-lying strip adjacent to the northern Gulf of California. The stratigraphy and provenance of these rocks can be deduced fairly clearly despite subsequent metamorphism and strain, and many workers have postulated a correlation with unmetamorphosed late Proterozoic and Paleozoic rocks of the Caborca terrane in central Sonora (Gastil and Krummenacher,

1977b; Gastil and Miller, 1981; Miller and Dockum, 1983; Gastil, 1985; Gastil et al., 1991b; Gastil, 1993). The metamorphic rocks in northeastern Baja California include distinct shelf and basinal assemblages separated by a buried contact. The boundary is correlated with a thrust that emplaced basinal rocks atop shelfal strata in central Sonora during Permo-Triassic time (Gastil and Miller, 1984; Stewart, 1990; Gastil, 1993). The shelf-basin boundary is inferred to cross Isla Tiburón into eastern Baja California and curve northward or northeastward into southeastern California (Stewart, 1990; Gastil, 1993).

The basement of the metamorphic rocks in northeastern Baja California is unknown, but Proterozoic–early Paleozoic shelf strata in Sonora are underlain by crystalline basement rocks of the North American craton that range in age from 1.75 Ga to 1.1 Ga (Anderson and Silver, 1981). The southern limit of Precambrian basement is interpreted to be near latitude 28°N (Valencia-Moreno et al., 2001, and see Figure 1 in Henry et al., this volume).

The southern boundary of the Caborca terrane generally is depicted as a roughly east-west line near 27°N (e.g., Sedlock et al., 1993; Dickinson and Lawton, 2001), but it may lie even farther south. Paleozoic basinal strata in southern Sonora and adjacent Sinaloa (Mullan, 1978; Henry and Fredrikson, 1987; Gastil et al., 1991b) are similar to and may be correlative with the basinal strata of the Caborca terrane in Sonora.

In Baja California near 30°N, Early Permian to Early Triassic rocks record a transition from deep-water basinal rocks to shallow-water shelfal strata similar to shelfal strata in the southern Great Basin (Buch and Delattre, 1993).

In the Middle Jurassic, the western margin of the Caborca terrane south of the Agua Blanca fault zone was the site of arc magmatism presumably related to subduction of oceanic lithosphere beneath North America. This magmatic episode has been recognized only recently by Schmidt and Paterson (2002), who obtained a pair of 164 Ma U-Pb ages from widespread orthogneiss near 30.5°N. The geographic extent and temporal range of this arc have not been established, but arc magmatism either persisted into or was reestablished in the Early Cretaceous, based on 135–127 Ma U-Pb ages from granite and orthogneiss in the northern Sierra San Pedro Mártir (Johnson et al., 1999) and a 132 Ma U-Pb age from gabbro in the southern Sierra San Pedro Mártir (Schmidt and Paterson, 2002). Given the lack of detailed mapping and precise dating elsewhere, Jurassic and Early Cretaceous (meta)plutonic rocks probably are more widespread than generally thought.

Yuma Terrane

The Yuma terrane consists of Mesozoic basinal and volcanic-arc rocks that accumulated along the western edge of North America. These rocks are assigned to a separate terrane rather than to the Caborca terrane or to North America because part or all of them may have undergone margin-parallel translation (see later section). This definition of the Yuma terrane differs from that of Sedlock et al. (1993) by excluding the Alisitos arc, which here is considered a separate terrane. The Yuma is faulted against the Alisitos at the Agua Blanca fault zone.

Basinal Assemblage (Bedford Canyon Complex)

Prebatholithic basinal strata consist chiefly of pervasively deformed terrigenous turbidites of probable Triassic and Jurassic age. In southern California, different exposures of these rocks are assigned to the Bedford Canyon Formation, French Valley Formation, and Julian Schist. Limestone clasts in the Bedford Canyon Formation contain Bajocian to Callovian fossils (Moran, 1976), including ammonite species that have affinities southward rather than northward (Imlay, 1963, 1965). The Bedford Canyon Formation yielded Rb/Sr whole-rock isochron ages of about 230 Ma and 175 Ma (Criscione et al., 1978). Volcaniclastic rocks that contain Tithonian *Buchia* and a flow or dike that yielded a U-Pb zircon age of about 152 Ma originally were assigned to the Santiago Peak Volcanics (Balch et al., 1984), but probably are correlative with the Bedford Canyon Formation (Wetmore et al., this volume). Detrital zircons suggest Late Proterozoic and Paleozoic to early Mesozoic sources and may have been derived from igneous rocks of these ages in Arizona and northwestern México (Gastil and Girty, 1993).

The French Valley Formation and Julian Schist contain rare fossils of possible Late Triassic age (Hudson, 1922; Lamb, 1970), and both units have yielded Jurassic U-Pb ages related to later intrusion or metamorphism (Davis and Gastil, 1993; Girty et al., 1993). In Baja California north of the Agua Blanca fault zone, the Vallecitos Formation and related rocks lack fossils and have yielded Rb/Sr whole-rock isochron ages of about 206 Ma (Davis and Gastil, 1993) and about 167 Ma (Chadwick, 1987).

Wetmore et al. (this volume) propose that all the units described above be grouped into a single tectonic unit, which they term the Bedford Canyon Complex, on the basis of similar age, provenance, lithology, submarine-fan depositional environments, and deformation. I follow their recommended usage in this paper.

The Bedford Canyon Complex is intruded by strongly deformed S-type orthogneiss that is pre-Cretaceous in age. Early radiometric studies suggested Triassic(?) ages with large uncertainties (Todd et al., 1991; Thomson and Girty, 1994), but recent U-Pb work indicates intrusion ages of 170 ± 2 Ma to 161 ± 2 Ma (Shaw et al., this volume). The Bedford Canyon Complex is intruded by Early Cretaceous plutons of the western Peninsular Ranges batholith and mid-Cretaceous plutons of the eastern Peninsular Ranges batholith.

Arc Assemblage (Santiago Peak Volcanics)

Early Cretaceous volcanic arc strata in the Yuma terrane are assigned to the Santiago Peak Volcanics (Larsen, 1948; Tanaka et al., 1984; Carrasco et al., 1993; numerous student theses cited in Wetmore et al., this volume). This unit consists of subaerially erupted flows, breccias, tuffs, and rare epiclastic deposits that were emplaced from at least as early 128 Ma to at least as late as 112 Ma. Major element, trace element, and isotopic data indicate that the Santiago Peak Volcanics formed in a continental-margin arc rather than in an island arc (Wetmore et al., this volume).

The contact between the Santiago Peak Volcanics and underlying basinal assemblage (Bedford Canyon Complex) is an angular unconformity (Larsen, 1948; Wetmore et al., this volume) rather than a fault, as has been generally inferred (Criscione et al., 1978; Gastil et al., 1981; Todd et al., 1988; Dickinson and Lawton, 2001). Evidence includes a basal conglomerate, a deformation discontinuity at the contact, and hypabyssal intrusions in the Bedford Canyon Complex that feed flows in the overlying Santiago Peak Volcanics.

Alisitos Terrane

The Alisitos terrane consists chiefly of the Alisitos Formation, a thick (>6 km) sequence of Early Cretaceous volcanic flows, volcanic breccia, and volcaniclastic rocks with a thin but laterally persistent intercalation of reef limestone (Santillán and Barrera, 1930; Allison, 1955, 1974). Sedimentary rocks are more abundant in the upper part of the section. Most of the formation accumulated in submarine environments. Most of the Alisitos is undated, but the limestone contains Albian fauna and volcanic rocks have yielded U-Pb ages of 116 ± 2 Ma (Carrasco et al., 1995) and 114.8 ± 1.5 (Johnson et al., 2003). The Alisitos Formation crops out as far south as El Arco at 28°N, and similar rocks of unknown age crop out near Loreto (McLean, 1988).

The Alisitos Formation probably accumulated in an island arc rather than in a continental arc (Silver and Chappell, 1988; Johnson et al., 1999; Wetmore et al., this volume). Mafic and intermediate volcanic rocks are more abundant than silicic rocks. Precambrian zircons have been identified in the volcanic rocks, though geochemical and isotopic data suggest derivation from a depleted mantle source without contamination by continentally derived materials (Silver and Chappell, 1988; Johnson et al., 1999; Tate et al., 1999; Tate and Johnson, 2000). A lithic arenite from the younger part of the Alisitos contains Proterozoic and Archean detrital zircon (P. Wetmore, 2003, personal commun.).

Contact relations of the Alisitos Formation have been studied most closely in the Sierra San Pedro Mártir near 31°N (Fig. 1). The Alisitos strata are overthrust by Mesozoic basinal strata along the Main Mártir thrust, a ductile shear zone that was active between ca. 115 Ma and 105 Ma (Johnson et al., 1999, 2003; Schmidt and Paterson, 2002). Shortening strain in the Alisitos dies out westward away from the contact (Johnson et al., 1999; Wetmore et al., 2002, this volume). Northward from the Sierra San Pedro Mártir, the thrust has a strike nearly parallel to the Agua Blanca fault zone, where it juxtaposes the Alisitos with the Santiago Peak Volcanics (Wetmore et al., 2002, this volume).

Roughly similar relationships are present near El Arco (28°N), where metamorphosed volcanic and volcaniclastic rocks assigned to the Alisitos terrane crop out in a discontinuous belt that trends about N70W (Fig. 4). Alisitos strata show a southwest-to-northeast increase in metamorphic grade (incipient to greenschist facies) and an increase in the intensity of a steep, northwest-striking foliation with steep mineral lineation (Sedlock, 1997). To the northeast, the Alisitos is probably faulted against amphibolite-facies metasedimentary rocks of uncertain origin (see next section) along a buried or intruded contact similar to the Main Mártir thrust. Limited K-Ar dating (Barthelmy, 1979) of deformed and undeformed granitoids in the region suggests that shortening is latest Early Cretaceous. About 5–8 km southeast of the contact with the metasedimentary rocks, the Alisitos is faulted against an assemblage of greenschist-facies gabbro, serpentinized ultramafic rocks, and minor diorite and pillow basalt of unknown age (Barthelmy, 1979; Sedlock, 1997). Rangin (1978) and Radelli

(1989) speculated that these rocks represent a disrupted ophiolite, but an alternative explanation is that they are fault-bounded fragments of the western Peninsular Ranges batholith.

Unassigned Strata of the "Transition Zone"

Metamorphosed prebatholithic strata crop out in fault-bounded structural horses at several places along the boundary between the Alisitos and Caborca terranes. Strata in these horses consist of flysch with abundant interbedded volcanic rocks. These rocks probably are not correlative with the volcanic-poor flysch of the Bedford Canyon Complex north of the Agua Blanca fault zone (Schmidt et al., 2002; Wetmore et al., this volume). The origin and tectonic affinity of these rocks are unknown, so they are not assigned to either terrane. Schmidt et al. (2002) suggested that these rocks form a distinct unit along the "transition zone" between the eastern and western parts of the Peninsular Ranges batholith. This hypothesis requires additional detailed mapping to explain the apparent discontinuity of rocks in each of their outcrop areas. For more detailed descriptions of the following regions, see the cited sources or the extensive discussion in Schmidt et al. (2002).

In the Sierra San Pedro Mártir near 31°N, metamorphosed basinal rocks are interbedded with metavolcanic rocks and marble that were derived from basalt, silicic tuff, and tuffaceous sandstone (Schmidt et al., 2002; Schmidt and Paterson, 2002). The ages of the protoliths are poorly constrained, though Wetmore et al. (this volume) suggest a possible Middle to Late Jurassic age.

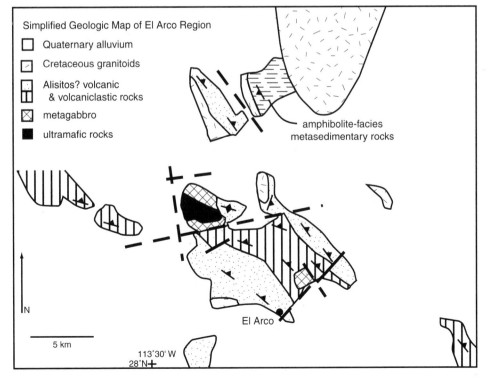

Figure 4. Geologic map of El Arco region near 28°N.

Near El Marmol (about 30°N), a 6-km-thick greenschist-to amphibolite-facies flysch unit that includes rare quartzite conglomerate and andesite contains probable Aptian-Albian fauna (Phillips, 1993). The ages of the thin-bedded middle member (5000 m) and nonmarine upper member (850 m) have not been determined with fossils, but accumulation of 6 km of mostly thin-bedded material could easily have lasted into the late Albian or Cenomanian. After deposition, the entire sequence was shortened into isoclinal folds and metamorphosed to greenschist facies (Phillips, 1993).

At Arroyo Calamajué, near 29.6°N (Fig. 1), broken formation derived from metamorphosed flysch of unknown age may be correlative with the Cretaceous flysch farther north (Griffith and Hoobs, 1993). North of El Arco (28°N), upper-greenschist facies to lower-amphibolite facies quartzite and pelitic schist crop out northeast of probable Alisitos rocks (Barthelmy, 1979; Radelli, 1989; Sedlock, 1997; Fig. 4).

The Los Cabos Terrane

Prebatholithic metasedimentary and minor metaigneous rocks in the Los Cabos terrane at the tip of the Baja California peninsula (Ortega-Gutiérrez, 1982; Gastil and Miller, 1983; Aranda-Gómez and Pérez-Venzor, 1989) were metamorphosed to amphibolite facies during late Early Cretaceous plutonism and have unknown protolith ages and depositional environments. Metasedimentary and plutonic rocks were penetratively deformed in the mid-Cretaceous (Aranda-Gómez and Pérez-Venzor, 1989). Strongly mylonitized garnetiferous gneiss near Todos Santos (Fig. 1) may be a pre-Late Cretaceous metamorphic core complex (Carrillo-Chávez, 1992). Enigmatic late Paleozoic and Triassic K-Ar dates obtained from the deformed plutonic rocks may indicate that the mid-Cretaceous deformation and metamorphism overprinted older fabrics and assemblages, but these dates are considered to be invalid by some workers (G. Gastil, 1991, personal commun.). The strain and metamorphism thwart attempts to correlate these rocks with other assemblages in western México, but their inferred protoliths resemble both the enigmatic Paleozoic rocks of Sinaloa and southern Sonora (see Caborca terrane) and the Bedford Canyon Complex of the Yuma terrane.

The western edge of the Los Cabos terrane has been inferred to be the La Paz fault (e.g., Hausback, 1984; Sedlock et al., 1993), but recent stratigraphic studies show that this fault probably has not accommodated significant exhumation of Mesozoic rocks since the early Miocene, i.e., the Mesozoic rocks were exposed to near-surface conditions by the early Miocene (Fletcher et al., 2000; Fletcher and Munguía, 2000). In this paper, I consider the western boundary of the Los Cabos terrane to be the western boundary of the "Los Cabos block" which, though not rigorously defined, is generally considered to be the western limit of exposed Mesozoic rocks. Such Mesozoic rocks lie east and west of the "La Paz fault" as defined by Hausback (1984), so the contact must lie within fairly low-lying terrane west of that fault and south of La Paz (Fletcher et al., 2000). Undisrupted Miocene rocks and geomorphologic features in this region suggest the terrane boundary has been inactive since at least the early Miocene (Fletcher et al., 2000). Pre-Miocene tectonism probably must have occurred to explain the elevation of the Cretaceous rocks of the Los Cabos terrane with respect to most of Baja California Sur, but otherwise the nature and history of this inferred terrane boundary are poorly understood.

The modern eastern boundary of the Los Cabos block is interpreted to be the San José del Cabo fault and its northward projection into the gulf, east of Isla Cerralvo (Fletcher et al., 2000; Fletcher and Munguía, 2000). The fault system has accommodated nearly pure normal slip and over 5 km of surface uplift in the Neogene, with average uplift rates of 0.4–0.7 mm/yr (Fletcher et al., 2000).

Cochimí Terrane

The Cochimí terrane consists of Mesozoic blueschist, ophiolite, and island-arc terranes overlain by Cretaceous forearc basin strata (Rangin, 1978; Kimbrough, 1985; Moore, 1985; Sedlock, 1988b, 1999; Kimbrough and Moore, this volume). These oceanic rocks crop out on Isla Cedros, Islas San Benito, and the Vizcaíno Peninsula near 28°N and on Isla Magdalena and Isla Santa Margarita near 24°30'N (Fig. 1). Similar rocks probably underlie most of the continental shelf between 30°N and 23°N, judging from gravity and magnetic anomalies (Couch et al., 1991).

The Cochimí terrane is divided into three structural units (Sedlock, 1988b, 1999). The structurally lowest unit (western Baja terrane of Sedlock, 1988b) comprises a fairly intact subduction complex that crops out only on Islas San Benito, Isla Cedros, and Isla Santa Margarita. Blueschist protoliths included Triassic oceanic crust, Triassic to Cretaceous pelagic sedimentary rocks, and Cretaceous terrigenous turbidites that were deposited as the oceanic rocks neared North America (Sedlock and Isozaki, 1990). The blueschists were metamorphosed under high-pressure, low-temperature conditions (Sedlock, 1988a) during mid-Cretaceous subduction (Baldwin and Harrison, 1989) and were exhumed while subduction continued in the Late Cretaceous to Paleogene (Sedlock, 1988a, 1996, 1999).

The structurally highest unit, informally referred to as the "upper plate," includes Mesozoic island arcs, ophiolites, and volcanogenic to terrigenous cover strata that crop out in all areas except Islas San Benito (Kimbrough, 1985; Moore, 1985; Sedlock, 1988b; Kimbrough and Moore, this volume). Component arcs and ophiolites have varying metamorphic and structural histories that suggest independent tectonic development, but all were transferred from the oceanic realm to the continent in the Jurassic to Early Cretaceous. Kimbrough and Moore (this volume) conclude that Triassic and Jurassic ophiolite in the Vizcaíno-Cedros region formed in suprasubduction zone settings. They infer that the ophiolites, and Middle Jurassic to Early Cretaceous arc plutonic rocks by which they are intruded, formed in the forearc of the North American subduction margin. Sedlock (this volume) describes four stages of deformation in a Triassic ophiolite on the Vizcaíno Peninsula, including extension during Triassic emplacement, Jurassic shortening, strike slip in the Cretaceous forearc, and synsubduction extension (see below). The accreted arc and ophiolite terranes were overlapped by forearc basin turbidites that range in age from lower Aptian (R. Sedlock, unpubl. data) to middle Eocene (Helenes, 1984; Smith and Busby, 1993a; Kimbrough et al., 2001). Most upper-plate rocks have been confined to shallow crustal levels throughout their history. Metamorphism typically ranges from absent to low-grade, and primary igneous and sedimentary structures typically are well-preserved. All upper-plate units were affected by brittle extension that constitutes the youngest or only strain event.

The structurally intermediate unit is a serpentinite-matrix mélange that occupies fault zones up to 500 m thick at all contacts between the blueschists and the upper-plate rocks. Most blocks are serpentinized ultramafic rocks, and the serpentinite matrix probably formed by shearing of similar rocks. About half of the other blocks are fragments of hanging-wall or footwall rocks, up to 250 m across, which typically crop out within a few kilometers of their sources. The remaining blocks comprise diverse metamorphic rocks that are exotic, i.e., derived from source terranes not exposed elsewhere in the Cochimí terrane. Exotic blocks include eclogite, coarse-grained blueschist, and amphibolite (Moore, 1986; Sedlock, 1988a, 1999) and yield $^{40}Ar/^{39}Ar$ ages ranging from 170 Ma to 94 Ma (Baldwin and Harrison, 1989, 1992; Bonini and Baldwin, 1994, 1998).

In summary, the three structural units of the Cochimí terrane form a sandwich with brittlely distended upper-plate rocks juxtaposed against deep-level blueschists along fault zones marked by the serpentinite-matrix mélange. The fault zones are interpreted as shallowly dipping crustal-scale normal faults that have accommodated synsubduction exhumation of the footwall blueschists (Sedlock, 1988b, 1996, 1999).

MESOZOIC PLUTONIC ROCKS

Plutonic rocks, chiefly of Cretaceous age, are widespread in the Peninsular Ranges batholith of southern and Baja California. The following discussion focuses on the Peninsular Ranges batholith and briefly addresses similar rocks that are distributed over large parts of Sonora, Sinaloa, Jalisco, and the Los Cabos terrane (see Fig. 1 of Henry et al., this volume). Henry et al. (this volume) summarize a large volume of new and previously published work from the Sinaloa batholith, compare these rocks with those of other batholiths around the gulf region, and recommend additional studies that would help develop an integrated view of all the batholiths.

Peninsular Ranges Batholith

The Peninsular Ranges batholith crops out continuously from 34°N to 28°N. Isolated outcrops are present on the east coast of Baja California Sur as far south as 26°N, and diorite and gneissic xenoliths have been collected in Tertiary basalt at 26.4°N (Demant, 1981; Hausback, 1984; López-Ramos, 1985; McLean, 1988). Gravity anomalies indicate that the batholith probably continues in the subsurface at least as far south as 26°N (Couch et al., 1991), and the strikingly continuous, narrow magnetic low (Fig. 2) suggests that batholithic rocks may extend south to the Los Cabos block.

Plutonic rocks in the Peninsular Ranges batholith display differences in age, petrology, geochemistry, and isotopic signatures that traditionally have been used to subdivide the batholith into western and eastern zones, here referred to as the western Peninsular Ranges batholith and eastern Peninsular Ranges batholith. Nomenclature and interpretation of these zones in different parts of the elongate batholith is confusing and controversial; as discussed in a later section, such problems can be attributed to the different tectonic settings and histories of rocks on opposite sides of the Agua Blanca fault zone. In the following summary, I sometimes refer to "western Peninsular Ranges batholith" and "eastern Peninsular Ranges batholith" because earlier publications have used such terminology. However, evidence for the independence of the Santiago Peak and Alisitos arcs (see later section) strongly argues for separate treatment of the plutonic rocks associated with them. Thus, where data permit such distinction, the following discussion refers to northwestern, southwestern, and eastern components of the batholith, with the Agua Blanca fault zone as the north-south divider.

Most studies of the Peninsular Ranges batholith have been undertaken in the more easily accessible outcrops in southern California. In this region, the Peninsular Ranges batholith has been divided into western and eastern zones (typically the western Peninsular Ranges batholith and eastern Peninsular Ranges batholith; the northwestern Peninsular Ranges batholith and eastern Peninsular Ranges batholith of this paper) on the basis of age, geochemical, and geophysical gradients and discontinuities (e.g., Gastil et al., 1981; Gastil, 1983; Todd et al., this volume, and references therein). The northwestern Peninsular Ranges batholith intrudes both the Santiago Peak Volcanics and Bedford Canyon Complex subterranes of the Yuma terrane; the eastern Peninsular Ranges batholith north of the Agua Blanca fault zone intrudes some plutons of the northwestern Peninsular Ranges batholith, both Yuma subterranes, and metamorphic rocks of the Caborca terrane. Proposed explanations of these differences, which are discussed at length in later sections, include (1) changing plate motions (Walawender et al., 1990), (2) development as two separate magmatic arcs (e.g., Johnson et al., 1999), and (3) eastward migration of a single magmatic arc (e.g., Todd et al., 1988, this volume).

Todd et al. (this volume) synthesized a large body of work from the Peninsular Ranges batholith in southern California that includes detailed maps and abundant geochemical, compositional, and paleobarometric data. They assign about 60 distinct plutons to 10 granitic suites on the basis of lithology, mineralogy, and texture. Most of the suites were emplaced in the western part of the Peninsular Ranges (i.e., the northwestern Peninsular Ranges batholith of this paper) between 120 Ma (or possibly earlier) and as late as ca. 105 Ma. Two of the suites (Granite Mountain and La Posta) were emplaced between ca. 104 Ma and 94 Ma chiefly in the eastern part of the Peninsular Ranges (eastern Peninsular Ranges batholith), but they also intrude slightly older plutonic suites in the western region.

Ortega-Rivera (this volume) compiled previously published and newly presented U/Pb, $^{40}Ar/^{39}Ar$, and fission-track data from the Peninsular Ranges batholith north of ~30.5°N. The compilation supports uninterrupted eastward-migrating plutonism, at least for the Peninsular Ranges batholith north of the Agua Blanca fault zone, where about 80% of the samples were collected.

Western Peninsular Ranges Batholith

Cretaceous plutons in the western Peninsular Ranges batholith were emplaced between ca. 140 Ma and 102 Ma (Silver and Chappell, 1988; Johnson et al., 1999). Compositions range from gabbro to rare granite and include low-K quartz gabbro to quartz diorite. Ring complexes display ample evidence of magma mixing and mingling (Johnson et al., 1999; Tate et al., 1999; Tate and Johnson, 2000; Johnson et al., 2002; Todd et al., this volume). Chemical and isotopic data imply that the plutons of the western Peninsular Ranges batholith—both north and south of the Agua Blanca fault zone—were emplaced in oceanic crust (Silver and Chappell, 1988), and at least some of them were generated at shallow depths (Tate et al., 1999).

The northwestern Peninsular Ranges batholith includes gneissic granites in a belt that extends at least 150 km north from the international border and that is at least 45 km wide (Shaw et al., this volume). These deformed plutonic rocks intrude the Julian Schist and are concentrated along the Cuyamaca–Laguna Mountains shear zone (Todd et al., 1988; Thomson and Girty, 1994). Near 32.8°, the Harper Creek and Cuyamaca Reservoir gneisses have yielded 10 zircon U-Pb ages between 179 Ma and 149 Ma (Middle and Late Jurassic) plus a single 245 ± 39 Ma (Triassic) age (Thomson and Girty, 1994; Shaw et al., this volume). Shaw et al. (this volume) use geochemical data to infer that these gneisses were emplaced as S-type and transitional I- to S-type granitoids derived at least partly from the Julian Schist and conclude that post-magmatic foliation and isoclinal folding occurred from the Late Jurassic (or earlier) until the mid-Cretaceous.

Pre-Cretaceous plutonic and metaplutonic rocks have not been reported from the southwestern Peninsular Ranges batholith (i.e., the Alisitos terrane).

Eastern Peninsular Ranges Batholith

The eastern Peninsular Ranges batholith includes mid-Cretaceous plutons that were intruded between ca. 105 Ma and ca. 90 Ma (Silver and Chappell, 1988; Todd et al., this volume, and references therein). The eastern Peninsular Ranges batholith ranges in composition from gabbro to rare granite but is dominated by tonalite-trondhjemite-granodiorite plutons, including abundant "La Posta–type" hornblende or biotite tonalite (Gastil et al., 1975; Todd et al., 1988; Walawender et al., 1990; Gastil et al., 1991a). This "La Posta tonalite–trondjemite–granodiorite suite" has been

studied chiefly in southern California, but similar rocks probably crop out semi-continuously as far south as 26°N (Gastil et al., 1975; Kimbrough et al., 2001). Depths of emplacement were probably 5–20 km in areas north of 30°N and 5–15 km in areas south of 30°N (Rothstein and Manning, this volume). The La Posta tonalite–trondjemite–granodiorite suite intruded between 98 Ma and 93 Ma, implying a flux rate similar to that of the Columbia River basalt province (Kimbrough et al., 2001), and effectively concluded volumetrically significant intrusion in the Peninsular Ranges batholith (Grove et al., this volume, Chapter 13).

The La Posta pluton itself, which straddles the international border, contains a linear north–south screen of Julian Schist that may form a major structural discontinuity. Geothermobarometry of plagioclase and amphiboles indicates that the west side of the pluton equilibrated at pressures of 2–3 kb, while the west side of the pluton equilibrated at pressures of 4–5 kb (Symons et al., this volume). Because paleomagnetic data are inconsistent with tilting of the batholith, these authors infer that a fault or faults within the screen of Julian Schist accommodated up to 8 km of differential uplift.

As noted earlier, plutons of the eastern Peninsular Ranges batholith intrude Jurassic–Early Cretaceous granitoids and orthogneiss that have been documented in the Sierra San Pedro Mártir and that may be present in much of the Peninsular Ranges. These important new findings pose significant constraints on tectonic models of the Mesozoic development of Baja, as discussed in a later section.

Boundary Between Western and Eastern Zones of Peninsular Ranges Batholith

The location, nature, and significance of the boundary between western and eastern zones of the Peninsular Ranges batholith have been points of confusion and disagreement for many years. An important step toward resolving these issues is recognizing that the independence of the Santiago Peak and Alisitos arcs implies the independence of the western Peninsular Ranges batholith/eastern Peninsular Ranges batholith boundary north and south of the Agua Blanca fault zone. Possible interpretations of these boundaries are explored in more detail in a later section.

Part of the confusion surrounding the boundary between eastern and western zones of the Peninsular Ranges batholith is that its location varies slightly depending on the criterion used to identify it. Criteria that have been used include the differences or changes in the nature of prebatholithic rocks, the ages of plutons, the composition of plutons, isotope geochemistry, metamorphic assemblage, geophysical parameters, and cooling history. Schmidt et al. (2002) cogently summarized the situation and suggested that the boundary be considered a transition zone that incorporates most or all of these changes rather than a sharp boundary. These authors consider this transition zone to be a distinct belt with its own stratigraphy, magmatism, and structures. While I do not formally follow their approach, I concur that the eastern part of the Peninsular Ranges batholith—i.e., the plutons and host rocks east of this transition zone—appears to form a continuous belt along the entire Peninsular Ranges, uninterrupted by the Agua Blanca fault zone.

The Northwestern Peninsular Ranges Batholith/Eastern Peninsular Ranges Batholith Boundary

In southern California and northern Baja California (north of the Agua Blanca fault zone), the boundary between the northwestern Peninsular Ranges batholith and eastern Peninsular Ranges batholith is a 5- to 25-km-wide zone that hosts several geochemical and geophysical gradients or discontinuities, including age of pluton emplacement, $\delta^{18}O$, initial Sr, magnetite versus ilmenite, I-type versus S-type, gravity anomalies, and magnetic anomalies (summaries in Todd et al., 1988, this volume; Gastil, 1993). The boundary zone also hosts Cretaceous shortening structures that differ from structures in rocks on either side of it.

Between 33.2°N and 32.7°N, the boundary between the western Peninsular Ranges batholith and eastern Peninsular Ranges batholith coincides with the Cuyamaca–Laguna Mountains shear zone, which hosted east-over-west shortening between ca. 118 Ma and ca. 115 Ma (Thomson and Girty, 1994). Todd et al. (1988) suggested that these features lie 10–30 km east of an inferred, unexposed suture between the Bedford Canyon Complex and Santiago Peak Volcanics, but a suture does not seem to be a viable candidate given the evidence for an unconformable contact between these units (see earlier section).

The Southwestern Peninsular Ranges Batholith/Eastern Peninsular Ranges Batholith Boundary

South of the Agua Blanca fault zone, it is more difficult to recognize most of the gradients and discontinuities that mark the internal Peninsular Ranges batholith boundary to the north. The magnetite-ilmenite line has been delineated as far south as 28.5°N, but most other features have not been traced south of about 31°N.

In the Sierra San Pedro Mártir, between 32°N and 30.5°N, the boundary between the southwestern Peninsular Ranges batholith and eastern Peninsular Ranges batholith lies near the Main Mártir thrust. This east-dipping structure places metamorphic rocks of continental affinity over the Alisitos, with a metamorphic break of at least 3 kb (Johnson et al., 1999; Schmidt and Paterson, 2002; Wetmore et al., 2002). The thrust is intruded by plutons of the western and eastern zones of the batholith (Johnson et al., 1999; Schmidt and Paterson, 2002).

The Main Mártir thrust forms part of the western zone of a 20-km-wide doubly vergent structural fan that is fully exposed only in the southern part of the sierra (Schmidt and Paterson, 2002). In the northern Sierra San Pedro Mártir, the central and eastern zones of the fan may have been obliterated by Late Cretaceous plutons of the eastern Peninsular Ranges batholith (Johnson et al., 1999).

The timing of slip on the Main Mártir thrust appears to differ in the northern and southern parts of the range. In the north, the thrust has orientation and kinematics similar to those of the Cuyamaca–Laguna Mountains shear zone but hosted shortening slightly later (115–105 Ma; Johnson et al., 1999, 2003). In the

south, the boundary accommodated shortening over a ≥ 40-m.y.-period that started prior to 132 Ma (Schmidt and Paterson, 2002).

At Arroyo Calamajué (29.6°N), Alisitos strata are tectonically interleaved with Paleozoic(?) metasedimentary rocks of the Caborca terrane within a 5 to 7-km-wide fault zone that was active ca. 103–100 Ma (Griffith and Hoobs, 1993). However, insufficient information is available from surrounding plutons to ascertain the position of the boundary between the western Peninsular Ranges batholith and eastern Peninsular Ranges batholith.

Near El Arco (28°N), the Alisitos probably is faulted against metasedimentary rocks of North American affinity (Fig. 4; Sedlock, 1997), but insufficient data are available from plutonic rocks to evaluate the position of the western Peninsular Ranges batholith–eastern Peninsular Ranges batholith boundary in this region.

Plutonism in Sonora and Sinaloa

Cretaceous plutonic rocks form a composite batholith in Sinaloa, across the Gulf of California from Baja California Sur (Henry et al., this volume). Early layered gabbros have hornblende K-Ar ages of 139 Ma and 134 Ma, though it is unclear whether these dates record emplacement or metamorphic cooling ages or have been affected by excess argon. Mafic tonalites and granodiorites were intruded before or during a major strain episode at ca. 101 Ma that may have commenced by 110 Ma or earlier and that may have persisted until ca. 90 Ma. Post-tectonic granodiorites and granites were emplaced in the Late Cretaceous and early Tertiary and show progressive eastward younging at a rate of ~1–1.5 km/Ma. Posttectonic intrusions probably cooled rapidly at depths of less than 3 km. Rare earth element (REE) patterns are typical of rocks underlain by Precambrian crust, but initial Sr ratios for eight samples range from 0.7030 to 0.7062 (Henry et al., this volume, and sources therein), suggesting that the region is not underlain by Precambrian basement.

Middle to Late Jurassic plutonic rocks in Sonora are interpreted as part of the Jurassic arc of southwestern North America (Anderson and Silver, 1979; Tosdal et al., 1990). A varied suite of Cretaceous to Paleogene granitoids crops out in central Sonora; K-Ar ages range from 90 Ma to 40 Ma (Damon et al., 1983; Roldán-Quintana, 1991). Few Early Cretaceous ages have been reported from the Sonora batholith, which contrasts with abundant Early Cretaceous granitoids in the Peninsular Ranges batholith, Sinaloa, and the Los Cabos terrane (see below). East-west geochemical and isotopic trends are obscured by strong north-south gradients attributed to the batholith's emplacement across the southern margin of North America (Valencia-Moreno et al., 2001).

Plutonism in Jalisco and the Los Cabos Terrane

The zoned Puerto Vallarta batholith in western Jalisco (southeastern corner of Fig. 1) is interpreted to have intruded between ca. 103 Ma and ca. 91 Ma, based on U-Pb zircon and Rb-Sr whole-rock isochron ages; biotite cooling ages range from ca. 84 Ma to ca. 75 Ma (Schaaf et al., 1995). Limited petrologic and geochemical data suggest that the batholith is older and more mafic to the west.

The Los Cabos terrane contains deformed granitoids ranging in composition from granite to hornblende gabbro, and undeformed granites, tonalites, and a 75 km² body of gabbronorite (Schaaf et al., 2000). Undeformed plutons in the northern and southern part of the Los Cabos terrane are interpreted to have intrusion ages of 129 ± 15 Ma and 115 ± 4 Ma, respectively, based on Rb-Sr isochron diagrams. Biotite cooling ages for the same plutons are 116 ± 2 Ma and 90 ± 2 Ma, respectively. Inconclusive field relations and isotopic data prevent determination of the age of the deformed plutons relative to the undeformed ones; deformation may have occurred prior to intrusion of the undeformed suite or as a result of it (Schaaf et al., 2000).

Schaaf et al. (2000) conclude that similarities in geochemical and isotopic data and cooling history indicate a "common magmatic evolution" of the undeformed granitoids of the Los Cabos terrane and the Puerto Vallarta batholith. However, examination of the isotopic and geochronologic data for each (their Table 5) raises questions about such "magmatic consanguinity." While granitoids in the southern part of the Los Cabos terrane have initial Sr and initial Nd ratios similar to those of the northern part of the Puerto Vallarta batholith, they differ significantly from those of the south-central and western parts of the Puerto Vallarta batholith. They also have intrusion and cooling ages 10–20 m.y. older than those anywhere in the Puerto Vallarta batholith. Furthermore, undeformed granitoids in the northern Los Cabos terrane differ significantly from granitoids in the Puerto Vallarta batholith—and from those in the southern Los Cabos terrane—in intrusion and cooling age, initial Sr, and initial Nd. A close relationship between the two regions has been postulated in tectonic reconstructions of México (e.g., Sedlock et al., 1993), but much more work is needed to determine whether granitoids of the Los Cabos terrane and the Puerto Vallarta batholith were once part of a contiguous, comagmatic suite.

Comparison of Batholiths

The batholiths in the Peninsular Ranges, Sinaloa, Sonora, Jalisco, and Los Cabos terrane show similarities and differences. All areas contain similar types of intrusions. All areas except Jalisco show a sequence from early gabbro to syntectonic to posttectonic rocks; Jalisco lacks syntectonic rocks. The Sinaloa batholith is dominated by potassic granodiorites, whereas the Peninsular Ranges batholith is dominated by calcic tonalites. All areas except the Los Cabos terrane show an eastward decrease in the age of magmatism. Syn-intrusive deformation ceased by ca. 105 Ma in the Peninsular Ranges batholith, by 89 Ma and possibly by 101 Ma in Sinaloa, and by about 98 Ma in the Los Cabos terrane. Posttectonic magmatism migrated eastward at 10 km/Ma in the Peninsular Ranges, Sonora, and Jalisco, but only at 1.0–1.5 km/Ma in Sinaloa.

Although the similarities among these batholiths permit matching them across the Gulf of California, the rocks contain no diagnostic piercing points that require such a restoration. Indeed, as noted by Henry et al. (this volume), similar rocks crop out along the entire length of the Cordillera of the Americas.

Syn-Plutonic Metamorphism

Intrusion of the Peninsular Ranges batholith and other mid-Cretaceous plutonic rocks resulted in contact metamorphism of local host rocks, yielding widespread development of high-temperature, low- to moderate-pressure metamorphic assemblages. Rothstein and Manning (this volume) studied the effects of metamorphism by the eastern Peninsular Ranges batholith in northern Baja California, where metamorphic effects are relatively unobscured by later deformation. Thermobarometric calculations for eight pelitic schists indicate peak metamorphic temperatures of 475–720 °C and pressures of 3–6 kb. Rothstein and Manning combine their results with data from the eastern Peninsular Ranges batholith in southern California, the Sierra Nevada batholith, and the Ryoke metamorphic belt in Japan to define a maximum background geotherm (i.e., a regional value not strongly affected by proximity to a particular pluton) of ~40 °C/km between 0 and 10 km depth and ~22 °C/km between depths of 10 km and 25 km. The latter is significantly lower than the 35–50 °C/km gradients that had been inferred previously and may require modification of models of seismic velocity, denudation history, and mechanical evolution (Rothstein and Manning, this volume).

In the Sierra San Pedro Mártir, country rocks on either side of the southwestern Peninsular Ranges batholith–eastern Peninsular Ranges batholith boundary zone were metamorphosed under markedly different conditions. West of the boundary, the Alisitos is lower greenschist to subgreenschist grade (Gastil et al., 1975; Johnson et al., 1999; Schmidt and Paterson, 2002). In the western boundary zone, wall rocks contain biotite, garnet, and andalusite, suggesting temperatures <550 °C and pressures <4 kb (Schmidt and Paterson, 2002). In the west-central, central, and eastern parts of the boundary zone, pressures exceeded 5 kb and temperatures reached the granite solidus, based on the presence of sillimanite, fibrolite, and migmatite (Schmidt and Paterson, 2002). Available data do not support interpretations of regional tilt of the Peninsular Ranges or a gradual west-to-east increase in paleodepth.

A roof pendant of sillimanite-grade metasedimentary rocks of the Bedford Canyon Complex near the U.S.-México border is enclosed within the 94 Ma La Posta pluton. Gem-bearing pegmatite dikes that cut the pendant probably formed from early water-undersaturated melts produced by the breakdown of muscovite in these metamorphosed strata (Walawender, this volume). Gem-bearing pegmatites throughout southern California may have formed by the intrusion of other La Posta–type plutons into the pre-batholithic metasedimentary rocks.

MESOZOIC TECTONICS

In this section, I examine new papers in this volume and analyze previously published works that relate to key aspects of the Mesozoic tectonic history of the Baja California peninsula, southernmost California, and northwestern mainland México. Time, space, and lack of familiarity preclude detailed analysis of some aspects of mainland tectonic evolution, particularly for pre-Cretaceous time. For example, this paper does not address the Mojave–Sonora Megashear or Triassic–Jurassic sedimentary rocks in Sonora. Such issues are discussed at some length in Sedlock et al. (1993), but new developments have arisen in the past decade. Some post-batholithic (post ~85 Ma) events of Cretaceous age are discussed in the section on Cenozoic tectonics.

The Peninsular Ranges Batholith Arc/Arcs

Models that have been proposed to explain the nature and significance of the boundary between the western Peninsular Ranges batholith and eastern Peninsular Ranges batholith differ in interpreting the Peninsular Ranges batholith as a single arc or a composite arc. Models that propose a composite arc further differ on whether the western component was a locally constructed fringing arc or a remotely constructed island arc.

A Single Arc

Walawender et al. (1991) proposed that the Peninsular Ranges batholith in southern California formed above a single subduction zone that shallowed progressively beneath western North America starting ca. 105 Ma. Thomson and Girty (1994) proposed that the western Peninsular Ranges batholith–eastern Peninsular Ranges batholith boundary in southern California is a pre-Triassic crustal boundary, possibly the rifted western margin of North America that was reactivated as an intra-arc structure in the Early Cretaceous. Todd et al. (this volume) used the distribution of rock types, mineralogical and chemical trends, and geophysical gradients in southern California to conclude that Peninsular Ranges batholith plutonism migrated eastward across a pre-existing lithospheric boundary such as a continental margin, though they infer that the boundary was Late Jurassic–earliest Cretaceous rather than pre-Triassic.

Tulloch and Kimbrough (this volume) note petrologic and geochemical similarities between the Peninsular Ranges batholith and the Median batholith of New Zealand and propose that both formed as single, zoned batholiths emplaced along their respective continental margins. They suggest that the western Peninsular Ranges batholith was generated by partial melts from underplated lower crust of basaltic composition and that this underplated crust later was transferred deeper and farther inboard in the forearc to generate magmatic rocks of the eastern Peninsular Ranges batholith.

Ortega-Rivera (this volume) interpreted geochronologic data to indicate that the entire width and length of the Peninsular Ranges batholith accumulated in place, with strong indicators of eastward migration since at least 130 Ma. She bases her interpretation on radiometric data compiled north of ~30.5°N, with much denser sampling north of the Agua Blanca fault zone than south of it.

Dickinson and Lawton (2001) assigned the Yuma and Alisitos terranes to their Guerrero superterrane, which they infer was accreted to North America in the Early Cretaceous; they envision accretion of the Baja component of the superterrane by 121 Ma. They consider the Alisitos a "post-accretion arc assemblage," requiring that much or most of the Peninsular Ranges batholith was constructed on the continental margin. However, their interpretation cannot be reconciled with the clear evidence summarized in this paper that the Alisitos is not post-accretion and instead was accreted to the continent between ca. 115 Ma and ca. 105 Ma. Rocks in the Baja California peninsula probably should be considered independently of the Guerrero terrane of mainland México (e.g., Tardy et al., 1994) rather than grouped with them as in Dickinson and Lawton's superterrane model (see later section).

Composite with Fringing Arc

In the most commonly cited model, the western Peninsular Ranges batholith–eastern Peninsular Ranges batholith boundary is interpreted as the suture zone produced by accretion of a fringing arc (Santiago Peak and Alisitos rocks) to the western edge of North America in the Early Cretaceous (e.g., Gastil et al., 1975; Gastil et al., 1981; Todd et al., 1988; Griffith and Hoobs, 1993; Busby et al., 1998). The origin of this fringing arc typically is attributed to back-arc extension that started in the Jurassic. Gastil et al. (1975, 1981) suggested that the northern fringing arc (Santiago Peak) was sutured to the North American margin in the Early Cretaceous, whereas the southern arc (Alisitos) was accreted in the middle to Late Cretaceous.

Composite with Exotic Island Arc

Johnson et al. (1999), working in the northern Sierra San Pedro Mártir, interpreted the southwestern Peninsular Ranges batholith–eastern Peninsular Ranges batholith boundary as the site of amalgamation of an exotic island arc terrane (Alisitos Formation intruded by the older plutons of the southwestern Peninsular Ranges batholith) with the western edge of North America (North America and Yuma and Caborca terranes intruded by the eastern Peninsular Ranges batholith) between ca. 115 Ma and 108 Ma. Recent geochronologic work (Johnson et al., 2003) has adjusted the younger age bracket to 105 Ma. Johnson et al. suggest that the Alisitos arc formed above an east-dipping subduction zone and that a second subduction zone dipped east beneath the margin of North America.

A similar interpretation is presented by Symons et al. (this volume) based on new and previously published paleomagnetic data. They suggest that the Alisitos arc, Santiago Peak arc, and western Peninsular Ranges batholith formed a substantial, though unquantified, distance from North America in the Pacific basin and were accreted to North America, including the eastern Peninsular Ranges batholith, in the mid-Cretaceous.

Hybrid Model

None of the models described above can adequately explain all the data that have been collected along the length of the boundary in the Peninsular Ranges. However, a hybrid model proposed by Wetmore et al. (2002, this volume) and Schmidt et al. (2002) reconciles most of the fundamental aspects of other proposals. These workers present convincing evidence that the Agua Blanca fault zone marks a major Mesozoic discontinuity, and they adopt Johnson et al.'s (1999) suggestion that the Alisitos arc south of the fault zone was an exotic island arc terrane that accreted to North America at 115–105 Ma. Strain along the Main Mártir Thrust was generated by this accretionary episode. An ancestral Agua Blanca fault zone accommodated left-reverse slip when the Alisitos underthrust the Santiago Peak Volcanics, which represented a continental arc built on the western margin of North America (Schmidt et al., 2002; Wetmore et al., 2002, this volume). A somewhat similar Mesozoic history of the Agua Blanca fault zone was inferred by Gastil et al. (1975, 1981). Wetmore et al. (this volume) suggest that Triassic–Jurassic rocks that would have comprised the North American accretionary prism prior to accretion of the Alisitos arc were removed by subduction beneath North America rather than by lateral translation.

The hybrid model incorporates the accreted-arc model of Johnson et al. (1999) and is consistent with the interpretations of Thomson and Girty (1994) and Todd et al. (this volume) for southern California. Like the model of Schmidt and Paterson (2002), the hybrid model requires little or no lateral translation along the western Peninsular Ranges batholith–eastern Peninsular Ranges batholith boundary. Finally, aspects of the "fringing-arc collision model" would still apply south of the Agua Blanca fault zone, but Wetmore et al. note several points that they suggest are misinterpreted in other models: the interpretation of the boundary in the Calamajué area; the correlation of "flysch-like" strata deposited in a Triassic-Cretaceous backarc basin (e.g., Gastil et al., 1981; Busby et al., 1998); and structural and lithologic data used to infer backarc rifting (Busby et al., 1998).

Another intriguing line of support for an exotic origin of the Alisitos terrane is the striking magnetic lineament that roughly bisects the peninsula along its length between the Los Cabos terrane and the Agua Blanca fault zone (Fig. 2). Higher magnetic values west of the lineament are consistent with Alisitos oceanic basement, and lower values to the east are consistent with continental crust. The sharp, linear feature—probably the most pronounced lineament on the entire magnetic map of North America—may correspond to the suture between an oceanic island arc and the continental margin. In any case, it closely tracks the boundary between the eastern and western parts of the Peninsular Ranges batholith from the Agua Blanca fault zone at least as far south as 28°N.

Although the hybrid model explains many observations, incompatibilities persist. (1) Johnson et al. (1999) and Schmidt and Paterson (2002) worked only about 50 km apart in the Sierra San Pedro Mártir, yet they arrived at different conclusions regarding the timing of cessation of slip on the Main Mártir thrust. (2) Wetmore et al. (this volume) conclude that the Santiago Peak arc, plus the western Peninsular Ranges batholith plutons that intrude it, developed on the western edge of North America, but Symons et al. (this volume) infer that the Santiago Peak arc developed in the Pacific realm and was accreted along with the Alisitos arc.

(3) The early (125 Ma or earlier) onset of eastward-migrating Peninsular Ranges batholith plutonism inferred by Ortega-Rivera (this volume) is incompatible with attachment of the Alisitos terrane between 115 Ma and 105 Ma; eastward migration of the Peninsular Ranges batholith north of the Agua Blanca fault zone does seem to be well-supported. Obviously, satisfactory resolution of these and other issues requires additional mapping, structural analysis, and geochemical and geochronologic study along the entire length of the Peninsular Ranges batholith, especially in less-studied parts of the peninsula.

Exhumation and Denudation

The arc associated with the Peninsular Ranges batholith was exhumed and denuded from ca. 100 Ma to ca. 75 Ma. The driving force for these processes was long assumed to be the accretion of the western Peninsular Ranges batholith, but this cause is unlikely for two reasons. First, accretion occurred much earlier (roughly 115–105 Ma). Second, if the hybrid model is correct, arc collision did not even occur north of the Agua Blanca fault zone. Alternative driving forces include increased plate coupling (Schmidt et al., 2002) and thermal-mechanical effects related to emplacement of the voluminous suite of tonalite, trondhjemite, and granite plutons (Kimbrough et al., 2001; Grove et al., this volume, Chapter 13). Near 33°N, a two-phase cooling history has been delineated (Lovera et al., 1999; Grove et al., this volume, Chapter 13). An initial, short-lived stage peaked at ca. 90 Ma and a longer period of cooling began at ca. 78 Ma and peaked at ca. 74 Ma.

Westward tilting of the Peninsular Ranges batholith has been proposed by many workers (Ague and Brimhall, 1989; Butler et al., 1991; Ague and Brandon, 1992; Dickinson and Butler, 1998) to account for the deeper structural levels exposed in the eastern Peninsular Ranges batholith. However, a simple model of westward tilt is untenable if the plutonic rocks in Sonora and Sinaloa are considered to have originally been east of and contiguous with the Peninsular Ranges batholith, i.e., the actual "eastern Peninsular Ranges batholith" prior to post–90 Ma separation. Grove et al. (this volume, Chapter 13) bypass this problem by assuming symmetrical erosional denudation centered on the eastern Peninsular Ranges batholith, but there are no reports of eastward tilt in the Sonora-Sinaloa region.

Northward Transport?

One of the most contentious questions in geoscientific studies of the Baja California peninsula is whether some of its constituent rocks underwent significant northward transport with respect to North America in the Cretaceous, Paleogene, or both. The problem commonly is portrayed or perceived in terms of two mutually exclusive hypotheses: (1) rocks in Baja and adjacent southern California were transported ≥10° northward after the mid-Cretaceous and before late Cenozoic opening of the Gulf of California, and (2) rocks in Baja and southern California have undergone northward translation only due to Gulf opening.

In the following discussion, the term northward transport refers to northward displacement (i.e., across latitude, not necessarily margin-parallel) that occurred prior to the opening of the Gulf of California in the late Cenozoic. Northward transport, reported in degrees of latitude with 95% confidence limits, is the difference between the paleolatitude determined paleomagnetically and the paleolatitude calculated from a paleomagnetic reference pole. All values of northward transport in this discussion are for pre-Gulf times, with 2° of transport subtracted to account for 255 km of post–6 Ma, northwest-southeast–directed dextral slip within the Gulf (Oskin et al., 2001; Oskin and Stock, this volume). Note that northward transport does not require that rocks of Baja or southern California were translated along or parallel to the western margin of North America. Displacements across longitude lines cannot be evaluated with paleomagnetic data.

Clarifying the Hypothesis

Since Teissere and Beck (1973) first recognized anomalously shallow inclinations in rocks of the Peninsular Ranges batholith in southern California, dozens of paleomagnetic and geologic studies have addressed the hypothesis of northward transport of Mesozoic and Cenozoic rocks in Baja and southern California. However, the hypothesis "rocks in Baja and southern California have undergone northward transport prior to the opening of the Gulf of California" groups rocks of Jurassic to Eocene age and may be too general to be tested effectively. In its place, I propose a group of related hypotheses that can be tested individually. Each hypothesis takes the form "X has undergone northward transport prior to the opening of the Gulf of California," where X is a combination of the seven distinct rock assemblages listed in the top row of Table 2 ranging from none (hypothesis H) to all seven (hypothesis A). Hypotheses in higher rows in the table are more restrictive, limiting their scope to rock assemblages that are progressively farther west. Hypothesized transport of a particular rock assemblage (e.g., Alisitos) requires transport of rock units farther west (for this example, Cochimí basement). To avoid redundancy, each rock assemblage is discussed below only for the hypothesis that it distinguishes from the hypothesis in the row above (e.g., Caborca is discussed only in hypothesis C, though its transport is implied in hypotheses A and B).

Hypothesis A: All Neogene and Older Rocks Underwent Northward Transport Prior to the Opening of the Gulf of California

Paleomagnetic data from Miocene volcanic and sedimentary rocks indicate negligible late Cenozoic northward transport (Hagstrum et al., 1987), and the tectonic history of the peninsula is sufficiently well-constrained from marine magnetic anomalies in the eastern Pacific to render this hypothesis untenable. No published evidence or models support hypothesis A.

Hypothesis B: All Pre-Neogene Rocks Underwent Northward Transport Prior to the Opening of the Gulf of California

Paleomagnetic data from the Cretaceous San Bartolo pluton in the Los Cabos terrane indicate insignificant northward transport and rotation (Hagstrum et al., 1985). Geologic links to Jalisco (e.g., Schaaf et al., 2000) suggest that the Los Cabos terrane was west of Jalisco prior to the opening of the gulf, but, as noted in an earlier section, the validity of the geologic links has not been clearly established. Based on these sparse data, I conclude that hypothesis B lacks supporting evidence.

Hypothesis C: All Pre-Neogene Rocks Except the Los Cabos Terrane Underwent Northward Transport Prior to the Opening of the Gulf of California

No paleomagnetic results have been reported from Paleozoic and early Mesozoic rocks in the narrow band of Caborca terrane in northeastern Baja California. This is understandable given the high levels of metamorphism and strain that most of these rocks have experienced. As discussed at the end of this section, the similarity of these rocks to Paleozoic rocks in Sonora is widely interpreted to indicate negligible northward transport of these rocks, but the correlation and interpretation are equivocal.

TABLE 2. ROCK COMBINATIONS FOR NORTHWARD TRANSPORT HYPOTHESES A–H

	Cochimí basement	Alisitos arc + wPRb	Santiago Peak arc + wPRb	Late K marine strata	Caborca Pz-Mz rocks	Los Cabos block	Neogene strata
H							
G	√						
F	√	√					
E	√	√	√				
D	√	√	√	√			
C	√	√	√	√	√		
B	√	√	√	√	√	√	
A	√	√	√	√	√	√	√

Note: wPRb—western Peninsular Ranges batholith; Pz-Mz—Paleozoic-Mesozoic.

Hypothesis D: All Pre-Neogene Rocks West of the Caborca Terrane Underwent Northward Transport Prior to the Opening of the Gulf of California

Paleomagnetic studies of Late Cretaceous to Eocene strata in Baja and southern California have yielded anomalously shallow inclinations (i.e., low paleolatitudes) that could be explained by northward transport (Table 3). In general, more pre-Gulf transport is indicated for older rocks, with sites in the Valle Group averaging about 12° and rocks younger than 72 Ma averaging about 5°. Negligible transport is indicated by some studies, particularly for rocks younger than about 58 Ma (late Paleocene).

Any northward transport that was experienced by sedimentary rocks in Table 3 must also have affected the rocks on which these strata lie. Strata younger than 82 Ma overlap the western Peninsular Ranges batholith, Alisitos Formation, or Santiago Peak Volcanics. The depositional basement of the Valle Group is rarely exposed, but it includes Late Jurassic–Early Cretaceous volcanogenic rocks in the northern Vizcaíno Peninsula and (locally) Early Cretaceous sedimentary rocks in the southern Vizcaíno. The Valle Group was derived from Alisitos-like and Peninsular Ranges batholith–like sources, and it structurally overlies ophiolitic and arc rocks of the Cochimí terrane (Moore, 1985; Sedlock, 1988b, this volume).

TABLE 3. PALEOMAGNETIC DATA FROM LATE CRETACEOUS TO EOCENE SEDIMENTARY ROCKS IN BAJA AND SOUTHERN CALIFORNIA

Unit studied	Age (Ma)	Measured paleolatitude (°N)	Expected paleolatitude (°N)	Northward transport (°)	Rotation (°)	Source
Poway Group	42 ± 1	29 ± 5	33 ± 5	4 ± 4	5 ± 7	1
La Jolla Group	48 ± 2	27 ± 3	33 ± 3	6 ± 3	17 ± 5	1
Las Tetas de Cabra Formation	55 ± 1	31 ± 4	28 ± 3	−3 ± 4	−5 ± 6	2
Bateque Formation	55 ± 1	23 ± 6	28 ± 3	5 ± 5	−19 ± 9	2
Silverado Formation	60 ± 2	25 ± 7	37 ± 3	12 ± 6	−18 ± 10	3
Silverado Formation	62 ± 2	26 ± 6	37 ± 3	11 ± 5	−20 ± 8	1
Punta Baja Formation	70 ± 3	29 ± 13	34 ± 5	5 ± 11	10 ± 19	4
Point Loma Formation (normal only)	72 ± 2	22 ± 4	37 ± 5	14 ± 5	35 ± 8	5
Point Loma Formation (reversed only)	72 ± 2	20 ± 12	37 ± 5	17 ± 10	42 ± 17	5
Rosario Formation (P Baja)	74 ± 6	26 ± 7	34 ± 5	8 ± 7	11 ± 11	3
Rosario Formation (P San Jose)	77 ± 3	25 ± 2	36 ± 5	11 ± 5	11 ± 6	4
Ladd Formation and Williams Formation	82 ± 8	27 ± 5	38 ± 5	11 ± 6	−29 ± 9	3
Valle Formation, N Vizcaíno 1	85 ± 1	22 ± 8	36 ± 4	13 ± 8	25 ± 13	6
Valle Formation, N Vizcaíno 2	87 ± 1	20 ± 5	36 ± 4	16 ± 5	33 ± 9	6
Valle Formation, N Vizcaíno 3	90 ± 2	25 ± 4	36 ± 4	11 ± 5	32 ± 7	6
Valle Formation, Cedros	90 ± 2	22 ± 5	37 ± 4	15 ± 5	6 ± 8	6
Valle Formation, S Vizcaíno 1	90 ± 2	25 ± 2	35 ± 4	9 ± 4	11 ± 5	6
Valle Formation, N Vizcaíno 4	90 ± 2	32 ± 6	36 ± 4	4 ± 6	−22 ± 9	6
Valle Formation, S Vizcaíno 2	94 ± 2	20 ± 1	35 ± 4	14 ± 3	14 ± 4	6
Valle Formation, Vizcaíno sites	94 ± 8	24 ± 12	36 ± 4	12 ± 10	28 ± 9	7
Valle Formation, Cedros sites	95 ± 5	21 ± 3	37 ± 4	16 ± 4	14 ± 6	8

Note: Abridged after Table 2 in Dickinson and Butler (1998). Sources: 1—Lund and Bottjer, 1991; Lund et al., 1991; 2—Flynn et al., 1989; 3—Morris et al., 1986; 4—Filmer and Kirschvink, 1989; 5—Bannon et al., 1989; 6—Patterson, 1984; 7—Hagstrum et al., 1985; 8—Smith and Busby, 1993.

Hypothesis E. Only the Cochimí, Alisitos, and Santiago Peak Rocks Underwent Northward Transport Prior to the Opening of the Gulf of California

Little is known about the paleomagnetism of the Santiago Peak Volcanics. An abstract by Yule and Herzig (1994) reported paleomagnetic data from volcanic flows that correspond to a 3° ± 8° anomaly, as calculated by Dickinson and Butler (1998). Paleomagnetic studies of western Peninsular Ranges batholith rocks in southern California that are coeval and partly comagmatic with the Santiago Peak Volcanics have yielded anomalously low inclinations (Teissere and Beck, 1973; Hagstrum et al., 1985; Symons et al., this volume). Symons et al. (this volume) derived a combined paleopole from the western Peninsular Ranges batholith and Santiago Peak Volcanics that they interpret to indicate 11° ± 4° (1210 ± 440 km) of northward transport and 20° ± 6° of clockwise rotation. In contrast, their data from the La Posta pluton, a component of the eastern Peninsular Ranges batholith that intrudes the Julian Schist, indicate nonsignificant northward transport and rotation since ca. 100 Ma.

Hypothesis F. Only the Cochimí and Alisitos Terranes Underwent Northward Transport Prior to the Opening of the Gulf of California

No paleomagnetic results have been reported from the Alisitos Formation. Two studies of overlapping Late Cretaceous (roughly 80–70 Ma) strata determined moderately low inclinations, corresponding to northward transport of 8° ± 7° and 11° ± 5° (Table 3). Two studies of overlapping earliest Eocene (55 Ma) strata determined small to negligible inclination anomalies (5° ± 5° and –3° ± 4°, Table 3). This small data set could be explained by northward transport that ceased in the latest Cretaceous or Paleocene or by minor compaction shallowing.

As discussed in an earlier section, Johnson et al. (1999), Wetmore et al. (2002, this volume), and Schmidt et al. (2002) have presented strong evidence that the Alisitos arc developed west of, and was not influenced by, North America. They cannot quantify the intervening distance, and paleomagnetic study can shed no light on cross-longitudinal transport, so the preaccretion origin and trajectory of the Alisitos are unknown. However, the absence of continental components in the volcanogenic part of the terrane makes it unlikely that the Alisitos arc was translated northward while attached to the western margin of North America.

Hypothesis G. Only the Cochimí Terrane Underwent Northward Transport Prior to the Opening of the Gulf of California

Paleomagnetic data have been collected from two locations in the oceanic "basement" (pre–Valle Group) rocks of the Cochimí terrane. Triassic chert, sandstone, limestone, and pillow basalt in an ophiolite on the Vizcaíno Peninsula yielded inclination anomalies of 18° ± 11.3° (Hagstrum et al., 1985). The ophiolite probably lay outboard of the North American continental margin at least as early as the Late Jurassic to earliest Cretaceous (see discussions in Kimbrough and Moore, this volume, and Sedlock, this volume), but the paleolatitude of accretion is not known.

Paleomagnetic data from red ribbon chert of Early Jurassic to Early Cretaceous age (Sedlock and Isozaki, 1990) are interpreted to indicate about 25° of northward transport and 55° clockwise rotation since ~100 Ma (Hagstrum and Sedlock, 1990, 1992). The subduction complex was added to North America near the peak of metamorphism at 110–100 Ma (Baldwin and Harrison, 1989) and was juxtaposed with ophiolitic and island-arc rocks across crustal-scale normal faults during Late Cretaceous to Paleogene exhumation (Sedlock, 1996, 1999). Thus, the ophiolite and subduction complex may have followed independent trajectories and rates of northward transport prior to their juxtaposition (Sedlock et al., 1993).

Hypothesis H. No Rocks in Baja or Southern California Underwent Northward Transport Prior to the Opening of the Gulf of California

Anomalously low paleomagnetic inclinations have been measured in ophiolite and subduction-complex rocks of the Cochimí terrane, suggesting that the Cochimí has undergone significant margin-parallel dextral slip. Many aspects of this inferred displacement are indeterminate, but the paleomagnetic data and profound lithologic differences across the Cochimí/Yuma terrane boundary (see further discussion in later section) suggest that at least part of the Cochimí terrane underwent pre-Gulf northward transport.

Rebuttals to Northward Transport

Proposed northward transport of rocks in southern and Baja California has been countered by rebuttals that include reinterpretation of the significance of the paleomagnetic data and enlistment of various geological arguments. Below, I examine these arguments and conclude that none provide sufficient grounds for rejecting all hypotheses of northward transport.

Compaction shallowing. The shallow inclinations of the Cretaceous–Eocene strata shown in Table 3 have been attributed to compaction shallowing by Butler et al. (1991) and Dickinson and Butler (1998). During compaction shallowing, sedimentary particles are compacted, flattened, or rotated toward horizontal due to the pressure of overlying sediment; magnetic particles are sufficiently moved or modified to result in shallower paleomagnetic inclinations.

The only study of sediment compaction of rocks (distinct from disaggregated, redeposited slurries of such rocks; see below) from Baja or southern California is that of Smith and Busby (1993b), who found that sandstones and shales in the Valle Group have similar inclinations independent of grain size. Because strata with a higher percentage of mud grains will compact more readily, Smith and Busby concluded that the Valle did not experience compaction shallowing. Butler and Dickinson (1995) and Dickinson and Butler (1998) disagreed, noting that it is not known whether mudstones undergo more compaction than sandstones after acquisition of post-depositional remanent magnetism. In their arguments against northward transport, Butler and Dickinson assume that mudstones do not undergo more compaction than sandstones after acquisition of post-depositional remanent magnetism, but such a relation has not been demonstrated.

Tan and Kodama (1998) experimentally tested Late Cretaceous (Campanian-Turonian, ca. 80–70 Ma) strata in southern California by disaggregating and redepositing them in a few hours, imparting post-depositional remanent magnetism, and compacting the deposit. They concluded that compaction produced the shallow inclinations that had been measured paleomagnetically. A similar result was generated by Kodama and Davi (1995) from the Pigeon Point Formation and was used to reinterpret paleomagnetic data from those strata. Li et al. (2001) have begun similar testing of the Valle Group, with disaggregation/redeposition of a single sample and inclination correction of 11 other samples using the magnetic particle anisotropy model of Jackson et al. (1991). They concluded that burial compaction accounts for about 6° of inclination shallowing in coarse-grained sandstones.

It is unclear how the results of these experiments relate to processes that produce sedimentary rocks. Obviously, the experiments cannot replicate the complexity of geologic variables such as changes in temperature and pressure, changes in pressure and temperature gradients, changes in the geochemistry of pore fluids, and other dynamic variables that real sediments encounter during the hundreds, thousands, or millions of years needed for lithification to run its course.

Butler and Dickinson (1995) assume that the 10–30° of inclination shallowing in young Pacific sediments (Gordon, 1990; Tarduno, 1990) was due to compaction. Such was the preferred interpretation of both original authors, but each expressed caveats about other possible causes of the shallowing. Even if the shallowing in these samples was due to compaction, it does not logically follow that all sediments are compacted 10–30° after acquisition of post-depositional remanent magnetism or even that all sediments are compacted at all. Preliminary work of Li et al. (2001) recognizes a smaller amount of compaction in the Valle Group than was postulated by Butler and Dickinson. In a paleomagnetic study of Late Cretaceous strata in the Salinia terrane, Whidden et al. (1998) obtained identical paleolatitudes from red beds with chemical remanence (hematite) and forearc-basin marine strata with detrital remanence, and concluded that the marine strata did not undergo significant inclination shallowing due to compaction.

The following points summarize the controversial interpretation of paleomagnetic data from marine sedimentary rocks. (1) Most clastic sediments, and all that consist chiefly of mud, undergo compaction and volume loss prior to lithification. (2) If remanent magnetism is imparted shortly after deposition and before much compaction occurs, inclination shallowing will result. (3) If remanent magnetism is imparted after compaction has started, inclination shallowing will be less, with a zero value in cases where magnetism is imparted after compaction has ceased. (4) Currently, there is no way to determine the degree to which a given rock was compacted prior to magnetization. (5) Compaction in artificially redeposited aggregates occurs under controlled conditions and in a much shorter time frame than natural compaction. (6) At least some marine strata in the Cordillera have not undergone significant compaction shallowing. (7) Studies of compaction shallowing have focused on rocks suspected of later tectonic motion; it is difficult to envision valid application of compaction-shallowing studies without a body of data from a sizeable control group of well-known tectonic history. Thus, studies of compaction shallowing currently cannot confirm or disconfirm hypotheses of northward transport.

Tilting of the Peninsular Ranges batholith. Anomalous inclinations from plutonic rocks can also be produced by post-magnetization tilt (Beck, 1980, 1992). Butler et al. (1991) suggested that the low inclinations from the western Peninsular Ranges batholith could have been produced by east-side-up rotation of the Peninsular Ranges batholith around a northwest-trending axis. Ague and Brandon (1992) conducted a paleobarometric analysis of the Peninsular Ranges batholith in southern California. For the western Peninsular Ranges batholith, they calculated 19° of east-side-up rotation around a N20W-trending axis, with uncertainties of ~5.5° in both the rotation and the axial trend. Ague and Brandon concluded that the paleomagnetic data from the western Peninsular Ranges batholith is explained best by 700 ± 450 km of northward transport. Uncertainties in the tilt-axis azimuth and tilt amount result in a larger (roughly 7–8.5°) confidence oval that grazes the confidence circle of the expected direction. Dickinson and Butler (1998) argued that Ague and Brandon's result could be attributed to the combined uncertainties of the paleomagnetic and paleobarometric calculations.

Dickinson and Butler's (1998) supporting arguments for their hypothesis of horizontal-axis tilting of the western Peninsular Ranges batholith depend on their treatment of statistical uncertainties. Their hypothesis requires the statistically extremely improbable case in which each of four uncertainties (expected direction, tilt-corrected direction, tilt-axis azimuth, and tilt amount) is resolved in such a way as to *minimize* the angular distance between the tilt-corrected and expected directions. They have selected one of innumerable possible combinations of uncertainties and used this preferred but statistically unlikely combination to argue that "paleomagnetic data do not require northward transport of terranes" (p. 1278). Obviously, the same uncertainty data also permit many other interpretations, most of which are statistically more likely. For example, the data could be used to *maximize* the angular distance between the tilt-corrected and expected directions, consistent with at least 11° of northward transport.

Rigid-block tilting of the Peninsular Ranges batholith, whether en masse or in domains, has proven difficult to evaluate. The search for paleohorizontal by geobarometric means is not straightforward, as the preceding discussion of the Ague and Brandon data set indicates. Each of the following studies relies to some degree on assumptions and interpretive preference of geobarometric data, structural data, or both to estimate paleohorizontal.

Ortega-Rivera et al. (1997) presented detailed $^{40}Ar/^{39}Ar$ data from the Sierra San Pedro Mártir pluton at 31°N that showed eastward decreases in the cooling ages of hornblende and biotite. They described four ways to explain these patterns, two of which require eastward tilting of the pluton and two of which do not.

Geobarometric calculations restrict tilting to ≤15° in an indeterminate direction. Ortega-Rivera (this volume) expanded on this study and suggested that results from several plutons indicate minor east-side-up tilt. However, it is unclear whether such patterns are typical of most plutons, and her data do not allow determination of the size or boundaries of tilt domains, if any.

Symons et al. (this volume) noted several reasons for rejecting the hypothesis of batholith tilt in southern California. Bedding in the Santiago Peak Volcanics, which would have been tilted along with the underlying plutonic rocks, is subhorizontal. Likewise, bedding in the sedimentary rocks of the Bedford Canyon Complex (usage of Wetmore et al., this volume) is variable but lacks a preferred orientation that would favor the tilt model. Finally, the western Peninsular Ranges batholith and eastern Peninsular Ranges batholith have similar geobarometric trends (Ague and Brandon, 1992), and paleomagnetic results were interpreted by Symons et al. to indicate that the La Posta pluton (eastern Peninsular Ranges batholith) has not been tilted.

Emplacement depths of Peninsular Ranges batholith plutons in southern California and northern Baja California generally increase from west to east (Silver et al., 1979; Gastil, 1983; Todd et al., this volume, and references therein). Some of this difference has been attributed to late Cenozoic uplift and tilting that accompanied the opening of the Gulf of California (Gastil et al., 1975; Silver et al., 1979; Kerr and Kidwell, 1991; Axen et al., 2000). Todd et al. (this volume) interpret paleobarometric data from 39 samples in southern California to indicate a general west-to-east increase in equilibration pressure. However, the map pattern shows much variability that cannot be explained by simple east-side-up tilt of a single crustal block.

Near San Marcos (32.2°N), north of the Agua Blanca fault zone, Böhnel et al. (2002) conducted a paleomagnetic study of a 120-Ma dike swarm intruded by a younger (110–100 Ma) pluton. A combined paleopole for the two units yielded northward transport of 6° ± 5°. Böhnel et al. applied a tilt correction to change the mean dip of the dike swarm from 79° to vertical, reducing northward transport to zero. They assume original verticality of the dikes on theoretical grounds but did not describe field evidence that the dikes (which display some attitudinal variation) actually were emplaced statistically vertically.

South of the Agua Blanca fault zone, thermobarometric analyses of the Peninsular Ranges batholith and metamorphic assemblages in its host rocks do not show a west-to-east increase in metamorphic pressure or emplacement depth (Schmidt and Paterson, 2002; Rothstein and Manning, this volume), so significant regional tilting of the Peninsular Ranges seems unlikely.

Böhnel and Delgado-Argote (2000) suggested that the San Telmo pluton, a 100–90 Ma suite of granitic to gabbroic intrusives on the western edge of the western Peninsular Ranges batholith at 31°N, has not been tilted appreciably. They base their interpretation on the <20° dip of bedding of metamorphosed Alisitos strata in a few small roof pendants. Because they do not present dip directions in the Alisitos, tilting cannot be discounted in their study area, nor can it be precisely determined.

Geologic correlations. Another frequently raised objection to the northward-transport hypothesis is that correlatives of transported rocks in the Baja California peninsula cannot be identified in southwestern coastal México, to which the peninsula would have been adjacent after restoration of 1000–1500 km of margin-parallel slip. Dickinson and Lawton (2001) dismiss northward transport by reasoning that (1) plutonic rocks in southern México and the peninsula have different ages and chemical character, and (2) the Chortis block (nuclear Central America) lay adjacent to southern México in the Late Cretaceous to early Tertiary. Whether or not these statements are valid, their application to the question of northward transport of Baja is not. First, part of the peninsula may have lain outboard of the Chortis block, or Baja and Chortis may have formed a contiguous section of the Late Jurassic–Early Cretaceous continental arc in southern México (Sedlock et al., 1993). Second, western Baja California, comprising the Alisitos arc and western Peninsular Ranges batholith intrusives, and possibly the Cochimí terrane, probably was not near the continental margin during much of its independent existence. As argued by Johnson et al. (1999) and Wetmore et al. (this volume), the Alisitos/western Peninsular Ranges batholith probably formed in an island arc that followed an unknown trajectory across the eastern Pacific basin prior to late Early Cretaceous accretion to North America. Thus, following the usage and logic of Cowan et al. (1997), I conclude that an "absence of correlatives" does not disconfirm the northward-transport hypothesis because other reasonable explanations have not been ruled out.

Late Precambrian to Paleozoic rocks in Sonora are widely held to be correlative with similar, though deformed and metamorphosed, rocks in northeastern Baja California (both groups of rocks are assigned to the Caborca terrane). Nevertheless, robust supporting evidence for such a correlation is sparse. In fact, the most detailed stratigraphic study—of metamorphic rocks in the San Felipe area by Anderson (1993)—reveals correlations that are "tentative" and "provisional" at best. Anderson "tentatively correlated" his units D and G with Late Proterozoic to Early Cambrian basalt and quartzite in northwestern Sonora. However, his units C and H differ significantly from rocks in the Sonoran section, and he cautioned that, in general, "the stratigraphic correlation of the remaining area is unclear." The San Felipe section probably accumulated in a shelfal environment, like the Sonoran section, but the stratigraphic differences are significant considering that the two sections are only ~200 km apart (after restoration of late Cenozoic opening of the Gulf of California).

Gehrels et al. (2002) determined the ages of detrital zircons in metasedimentary rocks of presumed Cambrian and Early Ordovician age near San Felipe and San Marcos, respectively. They interpret their results to indicate that the fragment of the Caborca terrane in northeastern Baja California received detritus from currently adjacent parts of North America, and thus that those strata underwent negligible net northward transport prior to late Cenozoic opening of the Gulf of California.

Where are the faults? An argument commonly invoked against the northward-transport hypothesis is that geologists have not identified the strike-slip faults, fault zones, and shear zones along which large displacement would have occurred. Umhoefer (2000) summarized four reasons that such structures appear to be absent from active continental margins. First, strike-slip faults may be destroyed or reactivated by later tectonism, uplift, and erosion. Corollaries not explicitly stated by Umhoefer include burial by later sedimentary or volcanic units and obliteration by intrusives. Second, slip may have been distributed on numerous anastomosing faults, none of which may have accommodated a large magnitude of slip. Third, faults may have formed in narrow ocean-floored basins, perhaps due to transtensional strain, where they would have a very low preservation potential. Fourth, strike-slip faults are subordinate to dip-slip faults at all but the most highly oblique plate boundaries; strike slip must be accommodated in some way, but not necessarily on discrete strike-slip faults.

Thus, the search for sites of accommodation of northward transport of rocks in southern and Baja California should consider options not limited to through-going strike-slip faults. To account for the possibility of differential northward-transport of different rock assemblages, such a search should include the following boundaries: Cochimí versus Alisitos, Alisitos versus Caborca-Yuma, Yuma versus Caborca, and all Mesozoic rocks in Baja (perhaps excluding the Los Cabos terrane) versus mainland México.

The boundary between the Cochimí and Alisitos terranes is completely covered by the Pacific Ocean or by late Cenozoic sedimentary rocks, but its subsurface position can be estimated from the distribution of outcrops and subcrops and from distinctive gravity anomalies (Sedlock et al., 1993). Its inferred position roughly corresponds to the break in slope of a conducting surface near 28°N that was reported by Romo et al. (2001). Geologic differences between the Cochimí and Alisitos terranes imply that the boundary is a northwest-striking fault or fault zone, but the nature and dip of that structure are unknown. South of about 30°N, a north-northwest–striking positive gravity anomaly attributed to "gabbro intrusions" and "Franciscan-like rocks" coincides almost exactly with the mapped extent of the Cochimí terrane (Couch et al., 1991). Faults inferred between en echelon positive anomalies on the continental shelf and slope between 26° and 23 °N (Couch et al., 1991) strike more westerly than and may be truncated by splays of the Tosco–Abreojos fault zone to the west, which partly accommodated the tangential component of Pacific–North America relative motion in the Miocene (see later section). These faults may have accommodated margin-parallel northward transport of outboard slices of Cochimí, Franciscan, and related rocks driven by the tangential component of Late Cretaceous to Paleogene right-oblique convergence (Sedlock et al., 1993). Strike-slip faulting in the Late Cretaceous forearc has been inferred by Busby et al. (1998) to explain distinctive patterns in forearc-basin strata deposited between Cochimí and Alisitos basement rocks.

The boundary between the Alisitos terrane and the Caborca and Yuma terranes is interpreted by Johnson et al. (1999) and Wetmore et al. (2002, this volume) to be the site of arc-continent collision in the late Early Cretaceous. The striking linear gradient in the magnetic data (Fig. 2) may indicate that this boundary is a fault contact that extends from the Agua Blanca fault zone to the Los Cabos block. If the Alisitos has undergone more northward transport than rocks to the east and north, the effects may be recorded by suitable structures or strain patterns associated with the boundary. Schmidt and Paterson (2002) used kinematic data and the continuity of map units to infer that little, if any, strike-slip had occurred along this segment of the boundary. However, following the reasoning of Cowan et al. (1997) for similar arguments related to the Baja B.C. controversy, large-scale strike-slip cannot be formally disconfirmed until other reasonable interpretations are ruled out. Much fundamental work on other parts of this boundary is required before its tectonic history comes into focus.

As discussed in an earlier section, the boundary between the Yuma terrane and the Caborca terrane roughly corresponds to the boundary between the northwestern Peninsular Ranges batholith and eastern Peninsular Ranges batholith. Two interpretations of the origin of the boundary are (1) the western Peninsular Ranges batholith represents a fringing arc that collapsed against the continent (e.g., Gastil et al., 1981), and (2) the entire Peninsular Ranges batholith developed at a single, progressively eastward-migrating arc (e.g., Todd et al., this volume). Each interpretation could be consistent with northward translation of the Yuma terrane inferred from paleomagnetic data (hypothesis E). Such translation requires faults or distributed strain that have not been identified in southern California, so either the features remain unrecognized or such translation did not occur. In the latter case, the Yuma terrane should be subsumed within the Caborca terrane, for no significant structural boundary would separate them. An important caveat for "boundary hunters" is that faults and distributed strain need not be spatially coincident with any of the older structures or boundaries in the region, including the Late Jurassic–earliest Cretaceous lithospheric boundary inferred by Todd et al. (this volume), any lithologic boundary between prebatholithic rocks, and the boundary between the northwestern Peninsular Ranges batholith and eastern Peninsular Ranges batholith. Additional geologic and paleomagnetic studies are needed to help ascertain the location and nature of the Yuma–Caborca boundary.

The structural boundary between the Baja California peninsula and mainland México—currently marked by the late Cenozoic transtensional plate boundary beneath the gulf—was suggested by Gastil (1991) as the most likely of several candidate sites for a major strike-slip fault zone along which Baja rocks may have been transported northward. Thus, he was addressing only hypothesis C of this paper. The preservation potential of Late Cretaceous–Paleogene strike-slip faults in this region is exceedingly low. In the Oligocene to middle Miocene, the crust in northwestern México hosted arc magmatism that generated widespread volcanic rocks. Since about 12 Ma, crustal extension has caused foundering of thinned crust and formation of oceanic crust in the Gulf of California. A through-going zone of weakness such as a fossil dextral fault system could have focused arc magmatism and later extensional strain.

Gastil (1991) noted that the apparent correlation of the Caborca terrane between Sonora and northern Baja California precludes significant pre-Gulf dextral slip between mainland México and the peninsula. However, the Caborca terrane's long-term proximity to an active continental margin exposed it to numerous potential margin-parallel displacements. Thus, it is important to distinguish between northward transport and net northward transport of the Caborca rocks in northeastern Baja California. If correlations between Sonora and Baja are valid (see reservations above), net northward transport must be negligible. However, few data pertain to the pre-Cretaceous history of this margin. Beck (1989) pointed out that an earlier episode of southward transport may have been followed by slip of the opposite sense but similar magnitude. Sedlock et al. (1993) suggested that Caborca rocks in Baja California were translated southward during Jurassic left-oblique convergence and that most of the peninsula was translated northward during Late Cretaceous right-oblique convergence. In this hypothesis, accumulation of part or most of the Bedford Canyon Complex, growth of the Santiago Peak arc, growth of the Jurassic–Early Cretaceous arc on the western margin of the Caborca terrane, and accretion of the Alisitos terrane would have occurred at the latitude of southern México. This model requires translation on faults that subsequently were obliterated by Cenozoic arc magmatism, transtension, and ocean-crust formation near what is now the Gulf of California. This hypothesis is inconsistent with the paleomagnetic evidence for negligible northward transport of the La Posta pluton of the eastern Peninsular Ranges batholith in southern California (Symons et al., this volume) and depends on the paleomagnetic data from Late Cretaceous–Paleogene turbidites that are suspected of compaction shallowing.

Status of Northward-Transport Hypotheses

As described above, Butler et al. (1991) and Dickinson and Butler (1998) have proposed an alternative to hypotheses of northward transport such as those outlined in this paper. Their hypothesis requires that plutons of the western Peninsular Ranges batholith are tilted to the southwest and by the same amount; that the inclinations of layered sedimentary rocks were strongly flattened, and flattened by the same amount, throughout southern and Baja California, whatever their lithology, age, and depositional environment; and that layered sedimentary rocks underwent local clockwise rotations of similar magnitude. Thus, these processes, though acting independently and at different times and places, must have individually reoriented each of an originally tightly grouped set of paleopoles to generate a similarly tight grouping ~12° away. They tacitly assume that this remarkably high level of coincidence was attained, and on that basis they reject hypotheses of northward transport. However, as detailed in this paper, the evidence for compaction-shallowing and batholith tilt is equivocal, and geologic arguments against northward transport are not clear, simple, unambiguous, or model-independent.

I conclude that northward transport hypotheses G (Cochimí), F (add Alisitos and western Peninsular Ranges batholith), E (add Santiago Peak and western Peninsular Ranges batholith), and

D (add Late Cretaceous strata) remain valid, though unproven, because no well-established evidence precludes pre-Gulf northward transport of any of these rock assemblages. Hypothesis C, though more speculative than D–G, has not been invalidated by crucial tests. Hypotheses A, B, and H are considered improbable. The Dickinson and Butler (1998) hypothesis also remains valid despite problems discussed in this paper. Like hypotheses C–G, its day of reckoning awaits crucial tests that make predictions and a search for evidence that can eliminate it. Needed are more results from plutonic and sedimentary rocks, more tests of tilting, and a better understanding of compaction phenomena. If future paleomagnetic studies continue to indicate low inclinations, inferred tilting or compaction for each would approach an exorbitant level of coincidence. In such circumstances, the Dickinson and Butler hypothesis reasonably could be rejected.

Cowan et al. (1997), describing a similar controversy surrounding the Baja B.C. hypothesis in the northwestern Cordillera, noted that many, and perhaps most, geologists accord hypotheses of large-scale transport a low "prior probability," i.e., they have passed collective judgment that such hypotheses are unlikely to be true, essentially placing a higher burden of proof on them than on hypotheses of negligible transport. This situation is surprising given (1) non-rigid accommodation of distributed strain in western North America hundreds or thousands of kilometers from the continental margin since at least the late Paleozoic, (2) the evidence for large-scale displacement along active plate margins, and (3) the recognition of mechanisms capable of causing such displacements. It seems more appropriate to demand similar or a greater burden of proof for the autochthoneity of tectonic entities, particularly older tectonic entities in outboard positions along mature plate margins.

Cretaceous Tectonism Unrelated to Alisitos Accretion

Laramide Shortening

In the Late Cretaceous to early Tertiary, the loci of arc magmatism and crustal shortening shifted eastward away from the continental margin. The Laramide Orogeny is widely interpreted to have been triggered by the shallowing of the angle of subduction beneath North America (e.g., Coney and Reynolds, 1977). In northern México, Laramide shortening was accommodated as far east as 100°W (sources in Sedlock et al., 1993) and is recorded across all of Sonora, though in many areas this deformation is masked by Tertiary volcanic rocks and extensional features. King (1939) mapped north-south–striking folds and thrust faults of probable Late Cretaceous age throughout central Sonora, and Hardy (1981) inferred east-vergent thrusting prior to the Late Cretaceous–Paleogene dike intrusion in several ranges south of Caborca.

Southern California

The Pelona, Orocopia, and Rand Schists and similar rocks in southern California are generally inferred to have accreted beneath North America during Laramide orogenesis (e.g., Burchfiel and Davis, 1981). The age and significance of the Pelona, Orocopia, and Rand Schists have been the subject of much debate, and several models of their origin and development have been proposed (see summary in Grove et al., this volume, Chapter 14). Ion microprobe U-Pb dating of >700 detrital zircons from schists of sedimentary protolith throughout this belt indicates a Late Cretaceous to earliest Tertiary depositional age (Grove et al., this volume, Chapter 14). These results contradict earlier interpretations that various Pelona, Orocopia, and Rand Schists are intruded by Late Jurassic (Mukasa et al., 1984), Early Cretaceous (James, 1986), and Late Cretaceous (Ross, 1977; Silver and Nourse, 1986) plutonic rocks.

Ar/Ar cooling ages indicate that the sedimentary protoliths of the Pelona, Orocopia, and Rand Schists were eroded from their source, deposited, buried, accreted, and metamorphosed within less than 10 m.y., with accretion ages ranging from 91 Ma in the northwest to 48 Ma in the southeast (Grove et al., this volume, Chapter 14). These results and other factors led Grove et al. to favor a hypothesis in which the schists formed in the shallowly dipping Laramide subduction zone, but they note that their results also are consistent with a model in which the schists formed by underthrusting of forearc-basin strata (Barth and Schneiderman, 1996).

Los Cabos Terrane

Metasedimentary and plutonic rocks of the Los Cabos terrane were penetratively deformed in the mid-Cretaceous (Aranda-Gómez and Pérez-Venzor, 1989). This strain was roughly synchronous with late Early Cretaceous accretion of the Alisitos terrane, but no kinematic or other links have yet been established between the two regions.

Regional Perspective

Plate Motions

Marine magnetic anomalies of Late Jurassic to mid-Cretaceous age are poorly preserved to unpreserved in the Pacific basin, so it is impossible to ascertain the number, distribution, and relative motion of oceanic plates in the eastern Pacific in this time interval. Linear velocities predicted by global plate–hotspot circuits suggest a sinistral tangential component of convergence at Mexican latitudes until about 119 Ma, though the uncertainties are large (Engebretson et al., 1985). Terrestrial data from western North America have been used to infer the sense, angle, and age of changes of the obliquity of oceanic plates to the west, but it is unclear how accurate or sensitive these proxies may be. For the Late Jurassic and Early Cretaceous, such data generally are interpreted to indicate sinistrally oblique plate convergence (e.g., Avé Lallemant and Oldow, 1988; see discussion in Umhoefer, this volume). Two papers in this volume propose sinistral translation in Baja in the Early Cretaceous. Wetmore et al. interpret mapping and kinematic data on the ancestral Agua Blanca fault zone to indicate that it hosted left-oblique slip during accretion of the Alisitos terrane. The reverse component of slip suggests that the Alisitos was emplaced from the southwest relative to North America rather than from a more westerly direction or along the

strike of an ancestral Agua Blanca transform (cf., Gastil et al., 1975; 1981; Schmidt et al., 2002). Sedlock speculated that a fault zone in the Vizcaíno Peninsula (Cochimí terrane) is a sinistral structure that formed in the Early Cretaceous North American forearc at an unspecified latitude.

By about 85 Ma, the Farallon and newly formed Kula plates converged with North America with a dextral tangential component (Engebretson et al., 1985). Relative motions among the Kula, Pacific, and North America plates cannot be reconstructed precisely, but right-oblique relative motion at Mexican latitudes probably included a margin-parallel component of 30–100 km/Ma for a period of 10–25 m.y. (Engebretson et al., 1985). The margin-parallel component may have been expressed, at least in part, by dextral-slip fault systems in the North American forearc (e.g., Page and Engebretson, 1984; Sedlock et al., 1993). Such dextral slip could have accommodated northward transport of rocks in southern and Baja California interpreted from paleomagnetic data.

Accretion of "Guerrero Superterrane"

Dickinson and Lawton (2001) grouped many distinct arc terranes into the Guerrero superterrane, expanding the territorial limits of and thereby superseding the Guerrero terrane of Coney and Campa (1987). Embracing a hypothesis proposed by Lapierre et al. (1992) and Tardy et al. (1994), they inferred that the Guerrero superterrane consisted chiefly of Early Cretaceous oceanic island arcs above a west-dipping subduction zone or zones that consumed oceanic lithosphere prior to accretion of the superterrane to North America by ca. 120 Ma. The subducted oceanic realm was termed the Arperos basin by Lapierre, Tardy, and colleagues (e.g., Lapierre et al., 1992; Tardy et al., 1994; Freydier et al., 1996; Freydier et al., 2000) and the Mezcalera plate by Dickinson and Lawton (2001).

A thorough analysis of these interpretations is beyond the scope of this paper, but the Guerrero superterrane/Arperos basin model is contested by many geologists working in southern México. Elías-Herrera and Ortega-Gutiérrez (1998) described field evidence that is difficult to reconcile with the model and presented new geochemical data that conflict with the geochemical story on which Lapierre, Tardy, and Freydier base their tectonic interpretations. Elías-Herrera et al. (2000) present new geologic and radiometric data that they interpret to indicate multiple accretion events in southern México, and they recommend reexamination of the idea that the Guerrero volcanic rocks formed as island arcs atop Mesozoic oceanic lithosphere. Salinas-Prieto et al. (2000) present structural data indicating that accretion-related shortening occurred as late as the Paleocene. Cabral-Cano et al. (2000) concluded that key geologic relationships can be interpreted to indicate that Guerrero arcs and Arperos basin were fringing arcs and intervening backarc basins, respectively.

Thus, the Guerrero superterrane hypothesis proposed by Dickinson and Lawton (2001) should be viewed with caution until much more progress is made toward understanding the geology of southwestern México. As discussed earlier in this paper, geologic and radiometric data from the Alisitos Formation, Santiago Peak Volcanics, and other rocks in the Baja California region are difficult to reconcile with assignment of these rocks to the Guerrero superterrane.

Speculative Models of Western North America in the Early Cretaceous

The following key premises could be used to formulate models of the Early Cretaceous tectonic development of the Baja California peninsula. (1) Paleozoic rocks of the Caborca terrane in northeastern Baja California and Sonora are correlative, with net offset equal to that caused by opening of the Gulf of California. (2) The western margin of the Caborca terrane hosted a Late Jurassic–Early Cretaceous continental arc above an east-dipping subduction zone. (3) The Alisitos Formation formed as an oceanic island arc at an unknown but significant distance west of North America. (4) The Alisitos terrane accreted to the Caborca and Yuma terranes between 115 Ma and 105 Ma. (5) An ancestral Agua Blanca fault zone hosted left-reverse slip during Alisitos accretion. (6) Parts of the peninsula underwent northward transport in the Late Cretaceous and Paleogene. (7) Data from the southern Sierra San Pedro Mártir prohibit significant strike slip during or after Alisitos accretion. Premise 6 is particularly complex, comprising several "subpremises" or discrete hypotheses described earlier in this paper.

Premise 2 merits further discussion before we consider large-scale Mesozoic reconstructions involving the peninsula. The presence of a Late Jurassic–Early Cretaceous continental arc on the western margin of the Caborca terrane (Schmidt and Paterson, 2002) requires the existence of two separate arcs at this time. This "west Caborcan arc" clearly was emplaced into the continental margin, whereas the Alisitos arc developed in the oceanic realm, out of range of continental influence (premise 3). This arc pair indicates that Dickinson and Lawton's (2001) hypothesis for approach and accretion of the Guerrero superterrane (their Fig. 8) is inconsistent with the geology of northern Baja California. The "west Caborcan arc" only recently has been recognized in northern Baja California despite decades of work in the region. Thus, it is possible that rocks of a Late Jurassic to earliest Cretaceous continental arc are present in other parts of western México but are unrecognized or unexposed. Sparse earliest Cretaceous granitoids in Sinaloa (Henry et al., this volume) could be parts of such an arc. The existence of such an arc would require rethinking regional tectonic models (e.g., Sedlock et al., 1993; Dickinson and Lawton, 2001) and would neutralize some geologic arguments against northward transport of the Peninsular Range's batholith and Alisitos terranes.

The seven premises listed above can be combined in many ways to construct different tectonic models, but uncertainties prevent formulation of a unique explanation. Nevertheless, I present two schematic reconstructions of the Late Jurassic to Late Cretaceous history of the peninsula (Figs. 5 and 6) that demonstrate the viability of different interpretations of premise 6 (northward transport) and that illuminate some of the many unknowns and

uncertainties that persist. Rigorous testing of these and other models requires many more data than currently are available.

Figure 5 shows a speculative reconstruction that satisfies premises 1–5 and 7 and does not prescribe or require pre-Gulf northward transport of any rocks in the Baja California peninsula. In the Late Jurassic to early Early Cretaceous (Fig. 5A), a magmatic arc developed along the western margin of southern California and northwestern México. Magmatism appears to have been less voluminous than before or after this period, possibly due to the marked obliquity of convergence between oceanic lithosphere in the Pacific with North America. Nevertheless, magmatic rocks of this age are present in central California, southern California (Yuma terrane), the "western Caborcan arc" of northeastern Baja California, and possibly Sinaloa (Schmidt et al., 2002; Wetmore et al., this volume; Henry et al., this volume). By about 120 Ma (Fig. 5B), the Alisitos island arc in the Pacific was nearing the western margin of México, though the trajectory of its path in the Pacific is unconstrained. Figure 5B shows two of many possibilities: in option A, the Alisitos moved east or east-southeast with respect to North America, possibly along transform faults parallel to the ancestral Agua Blanca fault zone; in option B, the Alisitos moved north or northeast with respect to North America, possibly on an unidentified plate other than the Farallon. In either case, plutons of the western Peninsular Ranges batholith intruded the Yuma terrane as early as 140 Ma. By 110 Ma, the Alisitos terrane was accreting to the Yuma (Santiago Peak Volcanics) and Caborca terranes in northern Baja California (Fig. 5C); if the linear magnetic gradient (Fig. 2) corresponds to the accretionary front of the Alisitos, then the latter was accreted to crust of unknown nature and age near what is now Sinaloa. Accretion of the Alisitos terrane started at about 115 Ma and ceased by 105 Ma (Johnson et al., 1999, 2003). Plutons of western Peninsular Ranges batholith intruded the Alisitos during accretion and continued to be emplaced into the Yuma terrane. Earliest Late Cretaceous (100–90 Ma) posttectonic plutons of the eastern Peninsular Ranges batholith stitched the Alisitos, Yuma, and Caborca terrane in northeastern Baja California (Fig. 5D).

Figure 6 shows a second speculative reconstruction that satisfies premises 1–7 and invokes not only northward transport in the Late Cretaceous but also southward transport in the Late Jurassic to early Early Cretaceous. By the beginning of the Late Jurassic (Fig. 6A), the southwestern margin of North America was undergoing left-oblique convergence that probably generated left-lateral strike-slip faults in the forearc and arc near the continental margin. Such faults displaced the part of the Caborca terrane in Baja California and the older parts of the Yuma terrane (Bedford Canyon Complex and intrusive rocks) southward to the latitude of central México by about 125 Ma (Fig. 6B). Tectonism ascribed to accretion of the Guerrero superterrane (e.g., Dickinson and Lawton, 2001) would have occurred east of the southernmost position of the displaced rocks. Accretion of oceanic island arcs (e.g., Lapierre et al. (1992), Tardy et al. (1994), Dickinson and Lawton, 2001) is difficult to reconcile with the reconstruction in Figure 6. Alternatively, Guerrero tectonism may have resulted

R.L. Sedlock

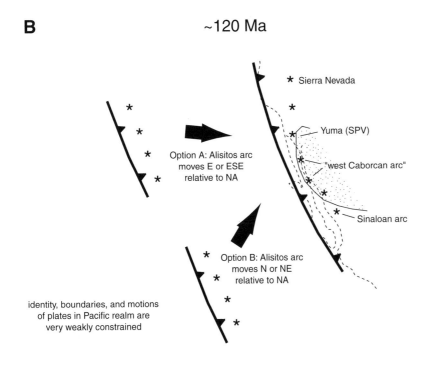

A **160–125 Ma**

inferred motion relative
to North America

identity, boundaries, and motions
of plates in Pacific realm are
very weakly constrained

sinistral faults in forearc

★ Sierra Nevada

★

★ — Yuma (Santiago Peak Volcanics)

★ — "west Caborcan arc"

★

★ — Sinaloan arc?

B **~120 Ma**

★ Sierra Nevada

★

★ — Yuma (SPV)

★ — "west Caborcan arc"

★

★ — Sinaloan arc

★

★

★

Option A: Alisitos arc
moves E or ESE
relative to NA

★

★

Option B: Alisitos arc
moves N or NE
relative to NA

★

★

identity, boundaries, and motions
of plates in Pacific realm are
very weakly constrained

Figure 5. Speculative reconstruction of Baja California and surroundings in Late Jurassic to Late Cretaceous that eschews northward translation. Dotted pattern indicates Caborca terrane (Paleozoic strata) of northwestern México. Stars represent active magmatic arcs. Light dashed line indicates approximate pre-Gulf shoreline (~6 Ma). NA—North America; SPV—Santiago Peak Volcanics; wPRB—western Peninsular Ranges batholith; ePRB—eastern Peninsular Ranges batholith. A: 160–125 Ma. B: ~120 Ma. C: ~110 Ma. D: ~95 Ma. See text for explanation.

from closure of a backarc basin behind a magmatic arc that lay east of the southernmost position of the Caborcan rocks (Cabral-Cano et al., 2000). By 120 Ma, the Alisitos oceanic island arc approached the margin of southwestern México (Fig. 6C) and accreted to the displaced Yuma and Caborca rocks there between 115 Ma and 105 Ma (Fig. 6D). Plutons of the western Peninsular Ranges batholith intruded the Alisitos during accretion and continued to be emplaced into the Yuma terrane (Fig. 6D). Earliest Late Cretaceous (100–90 Ma) post-tectonic plutons of the eastern Peninsular Ranges batholith stitched together the Alisitos, Yuma, and Caborca terrane in northeastern Baja California (Fig. 6E). In the Late Cretaceous to Paleocene, the Mesozoic rocks of the peninsula were transported northward due to the tangential component of right-oblique convergence between North America and oceanic lithosphere (Kula, Farallon, or both) (Fig. 6F). Northward transport dating to this age has been proposed by many workers, including Teissere and Beck (1973), Hagstrum et al. (1985), and Lund and Bottjer (1991).

C ~110 Ma

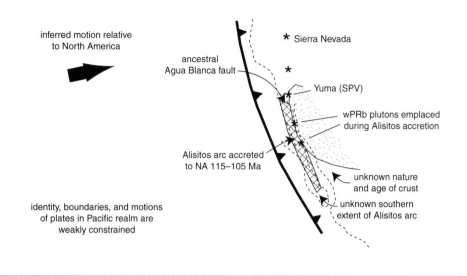

inferred motion relative
to North America

ancestral
Agua Blanca fault

* Sierra Nevada

*

Yuma (SPV)

wPRb plutons emplaced
during Alisitos accretion

Alisitos arc accreted
to NA 115–105 Ma

unknown nature
and age of crust

unknown southern
extent of Alisitos arc

identity, boundaries, and motions
of plates in Pacific realm are
weakly constrained

Figure 5 *(continued)*.

D ~95 Ma

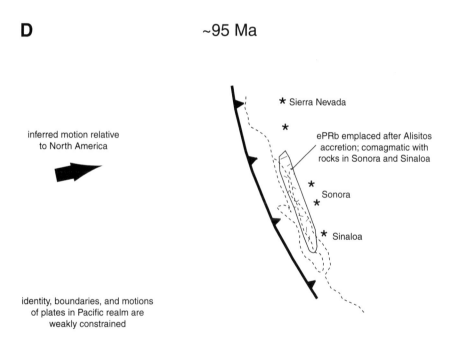

inferred motion relative
to North America

* Sierra Nevada

*

ePRb emplaced after Alisitos
accretion; comagmatic with
rocks in Sonora and Sinaloa

*
Sonora
*

* Sinaloa

identity, boundaries, and motions
of plates in Pacific realm are
weakly constrained

Other viable scenarios can be developed that incorporate northward transport of parts of the Baja California peninsula. Hypothesis G of this paper probably is compatible with all seven premises outlined above because the Cochimí terrane is likely to be structurally independent of the rest of the peninsula. Hypotheses C through F all involve northward transport of the Alisitos terrane. If premises 4 and 5 above are valid (accretion of the Alisitos to Caborca + Yuma at 115–105 Ma, with left-reverse slip on the Agua Blanca fault zone), and if hypothesis C is valid (transport of Caborca rocks in northeastern Baja California), then hypotheses D, E, and F must also be true (transport of Alisitos, Santiago Peak, and overlying sedimentary rocks). In other words, because the three terranes were stitched by plutons of the Peninsular Ranges batholith, the entire assemblage must have been transported if any part of it can be shown to have been transported after intrusion of the Peninsular Ranges batholith. Alternatively, if hypothesis C is valid and hypotheses D, E, and F are invalid, then premise 7 and probably premise 5 are not valid. In this case, the Alisitos and

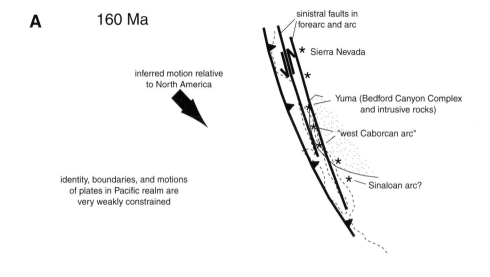

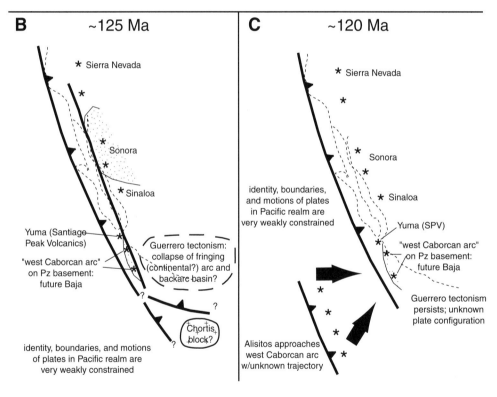

Figure 6. Speculative reconstruction of Baja California and surroundings in Late Jurassic to Paleocene that incorporates northward translation. Dotted pattern indicates Caborca terrane (Paleozoic strata) of northwestern México. Stars represent active magmatic arcs. Light dashed line indicates approximate pre-Gulf shoreline (~6 Ma). NA—North America; Pz—Paleozoic; PRb—Peninsular Ranges batholith; wPRb—western Peninsular Ranges batholith; SPV—Santiago Peak Volcanics. A: 160 Ma. B: ~125 Ma. C: ~120 Ma. D: ~110 Ma. E: ~95 Ma. F: ~60 Ma. See text for explanation.

southwestern Peninsular Range's batholith would have developed an unspecified distance south and west of the continental margin prior to northward transport, but other parts of the peninsula would have developed more or less in situ.

Umhoefer (this volume) proposes a speculative reconstruction of the Cretaceous tectonic evolution of western North America that incorporates several controversial or speculative hypotheses proposed by other workers; he candidly points out that his choices exclude other hypotheses a priori. By adopting the Early Cretaceous triple junction postulated by Dickinson and

Lawton (2001) offshore of the modern México-California boundary, Umhoefer necessarily excludes most hypotheses that involve Cretaceous northward transport of rocks in southern and Baja California (premise 6; hypotheses C–F of this paper). South of this triple junction, he accepts the Dickinson and Lawton model of the approach and accretion of the Guerrero superterrane at 125–115 Ma, despite the evidence for premise 4 (accretion at 115–105 Ma) presented by Johnson et al. (1999), Wetmore et al. (2002, this volume) and Schmidt et al. (2002). Umhoefer's model tacitly embraces premises 1, 3, and 7 and does not address premise 2 or 5.

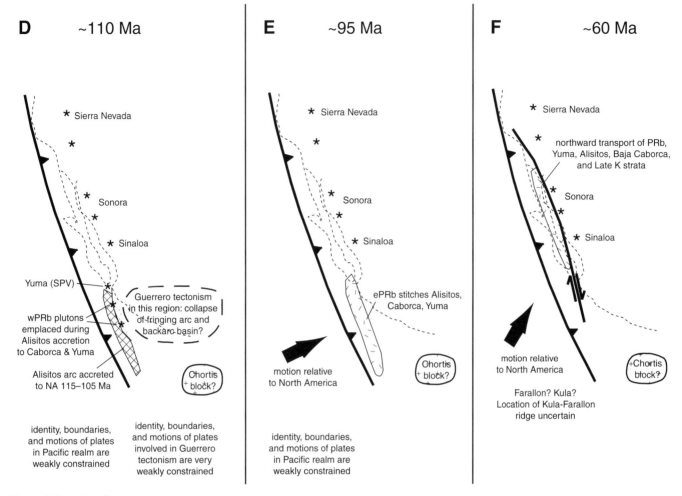

Figure 6. (*continued*)

He also addresses numerous aspects of Cordilleran geology in the United States and Canada using a tectonic-escape model to explain complex margin-parallel strike slip that bears upon the Baja B.C. controversy and various aspects of the age and distribution of Cretaceous batholiths in the western United States.

Another observation that should be incorporated in reconstructions of the Baja California peninsula in the Mesozoic is that the entire Peninsular Ranges batholith underwent increased deformation, exhumation, and magmatism starting in the earliest Late Cretaceous (about 100–95 Ma). These processes affected rocks north and south of the Agua Blanca fault zone, so they probably did not stem from arc accretion but instead resulted from some larger-scale process such as increased coupling in the subduction zone (Schmidt et al., 2002).

CENOZOIC TECTONICS

In this section I examine new papers in this volume and analyze previously published works that relate to key aspects of the Cenozoic tectonic history of the Baja California peninsula, southernmost California, and northwestern México. Post-batholithic latest Cretaceous events also are included in this section. This section focuses on processes of and products that ensued from the birth and development of the Pacific–North America plate boundary, particularly the partitioning of dextral slip and extension during the transition from continental rifting to sea-floor spreading in the Gulf of California.

Late Cretaceous to Late Oligocene (85–30 Ma)

The Peninsular Ranges Batholith and Other Batholiths

The locus of arc magmatism had migrated eastward out of Baja by about 90 Ma (Gastil, 1983; Silver and Chappell, 1988). From about 90 Ma to 65 Ma, the eastern part of the Peninsular Ranges batholith in southern California and northern Baja California (between roughly 33.5°N and 32°N) was exhumed 12–20 km along east-side-up structures such as the Cuyamaca–Laguna Mountains shear zone (Todd et al., 1988; Lovera et al., 1997).

Grove et al. (this volume, Chapter 13) investigated thermochronolonogic relationships using their own data and data of

Ortega-Rivera (this volume) from the eastern Peninsular Ranges batholith in southern California. Like earlier workers, they recognize two phases of rapid cooling of the eastern Peninsular Ranges batholith. The earlier episode, recorded chiefly in rocks west of a biotite 85 Ma K-Ar age contour, lasted from ca. 91 Ma to 86 Ma and probably was caused by rapid erosion and transient heating due to emplacement of the La Posta tonalite–trondjemite–granodiorite suite. The later episode, recorded chiefly in rocks east of the biotite age contour, lasted from ca. 78 Ma to ca. 68 Ma and may have been caused by shallowing of the Laramide subduction zone. The thermal history deduced by Grove et al. (this volume, Chapter 13) is very similar to the thermal history independently calculated from forearc strata in the same region (Lovera et al., 1999). Grove et al. (this volume, Chapter 13) simulate denudation of the Peninsular Ranges batholith via symmetric erosional denudation centered upon the most heavily denuded east-central batholith.

Late Cretaceous and early Tertiary volcanogenic and plutonic rocks indicate that a continental arc was active in Sonora and Sinaloa (Henry et al., this volume). These rocks are difficult to study because they are so extensively covered by middle Tertiary volcanogenic rocks of the Sierra Madre Occidental (see below).

Post-batholithic Rocks of the Baja California Peninsula

This section describes major components of a post-batholithic assemblage of Late Cretaceous and Cenozoic sedimentary and volcanic rocks that probably accumulated in southern California and the Baja California Peninsula. Excluding probable Cenozoic slip on the western boundary of the Los Cabos block (Fletcher et al., 2000), Baja may have been assembled into a fairly stable configuration by the end of the Late Cretaceous or earliest Tertiary.

Along the west coast of Baja California, postbatholithic marine clastic strata rest nonconformably on, and probably were derived from, the Alisitos Formation and Peninsular Ranges batholith (Gastil et al., 1975; Bottjer and Link, 1984). The Upper Cretaceous Rosario Formation contains abundant *Coralliochama Orcutti*, a warm-water rudist absent from Cretaceous rocks east of the San Andreas fault in California (Saul, 1986). Stable-isotope data from nannofossils in the Rosario Formation near 31°N support the idea that the tropics in the Cretaceous were at least as warm as today (Maestas et al., 2003). Additional work on these superbly preserved assemblages may contribute to the northward-transport debate. Johnson et al. (1996) documented the persistence of a Late Cretaceous (77–70 Ma) rocky shoreline along the Pacific coast at 32°N, demonstrating that not all postbatholithic sedimentation on the west flank of the Alisitos arc occurred in marine forearc basins. Biofacies distribution around a nearshore island along this Late Cretaceous rocky shoreline shows that dominant wave attack was from the south (Johnson and Hayes, 1993). Application of the modern north Pacific circulation model to Maastrichtian time (Patzkowsky et al., 1991) suggests that the paleo-island lay sufficiently far south that it was controlled by tropical summer cyclones rather than the north-to-south component of the north Pacific gyre. Some postbatholithic rocks in Baja California were deposited in tectonically active

basins of uncertain tectonic setting (Boehlke and Abbott, 1986; Cunningham and Abbott, 1986).

In the late Paleocene to middle Eocene, an incised paleodrainage transported distinctive rhyodacite (Poway) clasts from Sonora westward across the tectonically quiet Peninsular Ranges to the Pacific coastal plain in southern California (Bartling and Abbott, 1984; Axen et al., 2000). Uplift of the Peninsular Ranges between ca. 45 Ma and ca. 33 Ma disrupted this deposystem, and the Peninsular Ranges batholith replaced mainland México as the dominant source of sediment (Walsh and Demere, 1991). Middle Miocene basalts, tuffs, and fluvial and shallow marine clastic rocks along the coast between Ensenada and San Diego were derived from western volcanic and Franciscan sources that subsequently were submerged or displaced (Ashby, 1989).

In Baja California Sur, postbatholithic Late Cretaceous and Paleogene clastic rocks probably were derived from the southward continuation of the Peninsular Ranges batholith, which crops out locally on the eastern coast. Mid-Cretaceous to Paleogene marine clastic rocks of the Valle Formation crop out on Isla Cedros and the western Vizcaíno Peninsula, and similar strata are present on Isla Magdalena and Isla Santa Margarita (Blake et al., 1984; Sedlock, 1993; Smith and Busby, 1993a; Kimbrough et al., 2001). Latest Cretaceous and Paleogene marine clastic rocks containing benthic and planktonic forams and minor tuff (e.g., Bateque and Tepetate Formations) crop out in the eastern Vizcaíno Peninsula and along the Pacific coast east of Isla Magdalena (Heim, 1922; Mina, 1957; Hausback, 1984; López-Ramos, 1985; McLean et al., 1987). Wells in these regions and north of the Vizcaíno Peninsula penetrated Late Cretaceous to Eocene marine clastic rocks and minor tuff (Mina, 1957; López-Ramos, 1985). Eocene nonmarine rocks near Loreto probably are more proximal lateral equivalents of the shallow marine strata (McLean, 1988).

Late Oligocene to Middle Miocene (30–12 Ma)

Plate-Tectonic Developments

In the late Oligocene, plate motions caused two major changes that had far-reaching impacts on the geology of northwestern México. (1) At about 28 Ma, the Farallon–North America subduction margin along the western margin of North America encountered the Pacific–Farallon ridge-transform system at the latitude of the modern international border, forming the new Pacific–North America transform boundary (e.g., Atwater, 1989). (2) Starting at about 25 Ma, the locus of active subduction-related magmatism jumped trenchward from central México to a relatively narrow band near the Pacific coast (e.g., Severinghaus and Atwater, 1990).

As North America continued to move westward relative to the Pacific–Farallon boundary in the late Oligocene to middle Miocene, the Mendocino and Rivera triple junctions moved north and south, respectively, lengthening the Pacific–North America transform boundary. The Rivera triple junction probably lay offshore central Baja by ca. 16 Ma and offshore the tip of Baja by ca. 10 Ma (Lonsdale, 1991; Atwater and Stock,

1998). Each southward jump of the triple junction extinguished magmatism progressively farther south in the short-lived late Oligocene–Miocene arc. By the late Miocene, all of northwestern México had been transferred from a subduction margin to a transform margin. The resulting coupling of North America to the Pacific plate caused wrenching at the plate boundary and within the North American plate (Stock and Hodges, 1989; Bohannon and Parsons, 1995).

Sedimentary Strata in the Baja California Peninsula

Late Oligocene to early Miocene sedimentary rocks are only sparsely distributed in the peninsula north of ~27°N. Helenes and Carreño (1999) attributed this stratigraphic hiatus (ca. 30 Ma to ca. 20 Ma) to emergence and surface uplift caused by attempted subduction of young, buoyant Farallon lithosphere.

In contrast, late Oligocene to Miocene shallow-water strata are preserved in Baja California Sur from 27°N to 24°N. Near 26°N, late Oligocene to early Miocene shallow marine clastic rocks and subordinate tuffs and flows contain commercially viable phosphorite deposits (Hausback, 1984; Grimm, 1999). Near La Paz (24°N), lower Miocene volcaniclastic strata were deposited in lagoonal and alluvial-fan environments (Schwennicke et al., 1996). The middle to upper Miocene Salada Formation, which crops out in the Pacific coastal plain near 25°N, is a thin (<100 m) sequence of richly fossiliferous marine strata that were deposited in a shallow embayment (Smith, 1992).

Pre-Gulf Volcanism

Western México hosted the emplacement of a thick pile of middle Tertiary rhyolitic ignimbrites in the Sierra Madre Occidental (Fig. 1). This thick, areally extensive assemblage comprises the world's largest silicic volcanic province. Most of the Sierra Madre Occidental volcanics were emplaced in distinct Oligocene (32–28 Ma) and early Miocene (24–20 Ma) pulses (McDowell and Clabaugh, 1979; Cochemé and Demant, 1991; Ferrari et al., 2002).

In the early to middle Miocene, a terrestrial calc-alkaline arc was centered on what is now roughly the axis of the Gulf of California. Part of the arc probably foundered during gulf-related extension (see below), but remnants of a chain of 25–11 Ma stratovolcanoes crop out continuously in eastern Baja California Sur and southeastern Baja California (Gastil et al., 1979; Sawlan, 1991). I follow the recommended stratigraphic revisions of Umhoefer et al. (2001) in assigning all these rocks to the Comondú Group. Vent-facies rocks of the axial core of this calc-alkalic arc are exposed on the eastern margin of the peninsula from 29° to 25°N (Hausback, 1984). The waning stage of orogenic magmatism was contemporaneous with alkalic and tholeiitic volcanism that started by ca. 13 Ma in eastern Baja and within the developing Gulf of California rift (Hausback, 1984; Sawlan and Smith, 1984; Sawlan, 1991).

In contrast to the laterally continuous exposures of the Comondú Group in Baja California Sur, Miocene arc rocks in Baja California are preserved in isolated volcanic provinces (Gastil et al., 1975). In the Puertecitos Volcanic Province near 30.5°N, Miocene (21–12 Ma) volcanic rocks include andesitic calc-alkaline arc rocks as young as 16–15 Ma, a less-abundant basaltic facies, and the widespread 12.6 Ma rhyolitic Tuff of San Felipe (Gastil et al., 1975; Nagy et al., 1999; Stock et al., 1999; Oskin and Stock, this volume and references therein). The Puertecitos rocks have been offset 255 ± 10 km across the Gulf of California from correlative rocks on Isla Tiburón and in adjacent Sonora (Oskin et al., 2001). In the San Luis Gonzaga volcanic field near 30°N, Martín et al. (2000) described andesites, dacites, and rhyolites of middle Miocene age and noted that the age range and style of volcanism of volcanic rocks in Baja California differ from those of the Comondú in Baja California Sur.

Basin-and-Range Extension

Starting in the middle Oligocene, western North America underwent significant extension that produced the Basin and Range province. The causes of this dramatic extensional episode are still uncertain, but possible contributors include gravitational collapse of overthickened crust (Coney and Harms, 1984; Axen et al., 1993) and rollback and foundering of the shallowly dipping slab of previously subducted Farallon lithosphere (Severinghaus and Atwater, 1990; Nieto-Samaniego et al., 1999). Ferrari et al. (2002) propose that slab breakoff controlled the timing and location of late Oligocene extension and volcanism in the Sierra Madre Occidental.

Basin-and-Range extension affected much of México north of 20°N (Fig. 1; Henry and Aranda-Gómez, 1992). Between about 30°N and 23°N, the province forms two belts separated by the relatively unextended linear core of the Sierra Madre Occidental, hereafter referred to as the Sierra Madre Occidental core (Figs. 1 and 7).

In the eastern belt, extension began in the late Oligocene (Henry and Aranda-Gómez, 1992; Aguirre-Díaz and McDowell, 1993; Gans, 1997; Ferrari et al., 1999; Ferrari et al., 2002) and was generally directed northeast-southwest. In Sonora, extensional basins contain coarse-grained alluvial deposits up to several hundred meters thick that are generally but imprecisely assigned to the Baucarit Formation (King, 1939; McDowell et al., 1997). These rocks contain few fossils, but basal and interbedded basalts range in age from 27 Ma to 16 Ma (Bartolini et al., 1994; Gans, 1997; McDowell et al., 1997). This episode of extension also generated metamorphic core complexes (25–18 Ma) in north-central Sonora (Nourse et al., 1994). In Sinaloa, north-northwest–striking normal faults and dikes formed as early as 32 Ma (Henry, 1989), but most faulting and tilting of Cenozoic strata in this region probably began after ca. 12 Ma (Henry and Aranda-Gómez, 2000). In the western belt, extension began in the late Oligocene in Sonora but in the middle to late Miocene in Sinaloa and Nayarit (Gans, 1997; McDowell et al., 1997; Henry and Aranda-Gómez, 2000).

Middle Miocene to Present (since 12 Ma)

Latest Middle to late Miocene rifting of continental crust in western México ultimately led to seafloor spreading in the Gulf

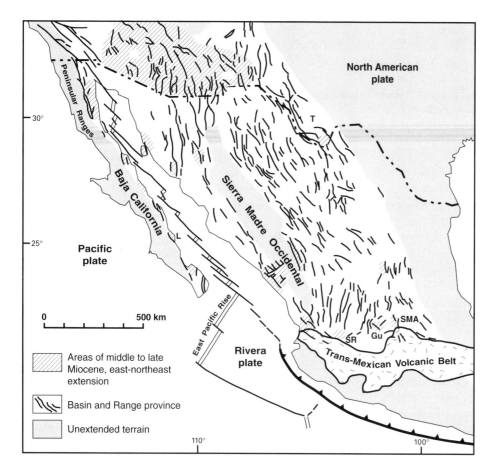

Figure 7. Extent of middle to late Miocene Basin-and-Range extension in México, from Henry and Aranda-Gómez (2000). Gu—Guanajuato; L—Loreto; SR—Santa Rosa; SMA—San Miguel de Allende; T—Trans-Pecos, Texas.

of California. However, the early rifting stage produced limited volumes of volcanic rocks that accumulated locally rather than along the length of the rift (Gastil and Krummenacher, 1977b; Oskin and Stock, this volume, and references therein). Late Miocene volcanic centers such as those near Puertecitos formed at accommodation zones in the developing rift; regions that had been weakened by early to middle Miocene arc volcanism became the site of concentrated extension during the late Miocene development of this "proto-Gulf" rift (Axen, 1995; Oskin and Stock, this volume, and references therein).

Plate-Tectonic Developments

At ca. 12 Ma, Pacific motion relative to North America increased from ~33 mm/yr to ~52 mm/yr along an azimuth of about N60W (Atwater and Stock, 1998). At ca. 8 Ma, the azimuth of Pacific motion shifted to about N37W at a rate of ~52 mm/yr. The rate increase at 12 Ma is coeval with and may have triggered widespread extension in the gulf region, but the azimuth change at 8 Ma is not easily related to tectonic developments in the region; as described below, changes of regional extent in the direction of extension did not occur until ca. 6 Ma. Latest middle to late Miocene transtension along the Pacific–North America transform boundary was partitioned into margin-parallel dextral slip on

offshore faults like the Tosco-Abreojos (Spencer and Normark, 1979; Normark et al., 1987) and margin-normal extension on normal faults in what is commonly termed the Gulf Extensional Province (see below). This episode of northeast-southwest to east-west "proto-Gulf" extension lasted from ca. 12 Ma to ca. 6 Ma.

At ca. 6 Ma, most Pacific–North America motion was concentrated in the future Gulf of California and was transferred northward to the San Andreas system in California (e.g., Lonsdale, 1989). By about 5 Ma, extension within the transtensional plate boundary had rotated to a direction 30–50° oblique to the margin, resulting in the rifting of the Baja California peninsula from North America and development of the San Andreas fault system sensu stricto in southern California; by 3.5 Ma, seafloor spreading had begun in the newly formed Gulf of California (Lonsdale, 1989, 1991; DeMets, 1995). Offset of correlative tuffs in eastern Baja California, on Isla Tiburón, and in coastal Sonora dated at 12 Ma and 6.3 Ma (Oskin et al., 2001, this volume) indicates that the gulf accommodated a maximum of 255 ± 10 km of post–6 Ma dextral slip and precludes larger offsets hypothesized by Dickinson (1996) and Gans (1997). Lonsdale (1989) suggested that the transfer of Pacific–North America slip from the Pacific margin of the peninsula to the Gulf of California was a protracted transition that lasted until 3.5 Ma, but Oskin et al. (2001) conclude that nearly all Pacific–North America transform boundary transform motion was transferred from the borderland to the gulf region and San Andreas system between 6.3 Ma and 4.7 Ma.

Proto-Gulf Extension

Latest middle to late Miocene (roughly 12–6 Ma) extension oriented generally normal to the Pacific–North America transform boundary-affected parts of Baja, Sonora, Sinaloa, and Nayarit, and all gulf islands—an area commonly referred to as the Gulf Extensional Province (Gastil et al., 1975; Stock and Hodges, 1989). Henry and Aranda-Gómez (2000) noted that extension of similar age and orientation affected much of México east of the Sierra Madre Occidental core and inferred that the Gulf Extensional Province simply represents the western branch of Basin-and-Range extension in México (Fig. 7).

Eastern Baja California peninsula. Proto-gulf extension has been documented along the full length of the eastern margin of the peninsula. Tilting and normal faulting due to east-west to east-northeast–west-southwest extension began by 12 Ma in eastern Baja California (Dokka and Merriam, 1982; Stock and Hodges, 1990; Lee et al., 1996, and sources in their Table 3). In the Sierra El Mayor at 32°N, east-west extension on high-angle normal faults that sole into an exposed listric detachment fault began ca. 12 Ma and continued into the Pleistocene (Siem and Gastil, 1994; Axen and Fletcher, 1998; Axen, 1999; Axen et al., 2000). A northwest-striking accommodation zone near 30.5°N apparently separated a zone of margin-normal extension to the north from the undeformed Puertecitos Volcanic Province to the south (Dokka and Merriam, 1982; Stock and Hodges, 1990; Axen, 1995) and may have linked north-northwest–striking normal faults along the early Main Gulf Escarpment (Dokka and Merriam, 1982; Stock and Hodges, 1990; Axen, 1995; Nagy et al., 1999; Nagy, 2000; Stock, 2000). The Puertecitos region also features an early syn-rift sequence of volcanic and nonmarine sedimentary rocks that formed between ca. 12 Ma and 6 Ma, with offset equivalents on Isla Tiburón and in central Sonora to the southeast (Stock et al., 1999; Oskin and Stock, this volume, and references therein). Near Bahía Concepción at 29°N, Miocene extension is associated with basaltic volcanism (Delgado-Argote and García-Abdeslem, 1999).

The gulf margin in the central and southern peninsula can be divided into segments separated by accommodation zones (Axen, 1995), but detailed work has been reported only for the Loreto region near 26°N. There, Umhoefer et al. (2002) recognized late Miocene extension oriented roughly N65E–S65W on the Loreto segment. Structural and topographic characteristics distinguish the Loreto segment from inferred segments to the north and south, but more data are needed to evaluate these differences. The Bahía Concepción–Mulegé region (26.7°N) probably hosted an accommodation zone in the late Miocene (Ledesma-Vázquez and Johnson, 2001). In the La Paz–Los Cabos region (24°–23°N), extension on north-northwest–striking normal faults probably began ca.12 Ma (Fletcher et al., 2000).

Nayarit. In Nayarit, extension along northwestern-striking normal faults probably began ca. 12 Ma (Gastil et al., 1978; Gastil et al., 1979; Nieto-Samaniego et al., 1999) or perhaps earlier in the middle Miocene (Ferrari et al., 2002). Data from several localities indicate that extension persisted until at least 10 Ma and that net proto-Gulf extension was <20% (sources in Henry and Aranda-Gómez, 2000).

Sinaloa. In southern Sinaloa, extension along north-north-west–striking, moderately dipping normal faults produced 20–50% extension of middle Tertiary volcanic rocks (Henry and Fredrikson, 1987; Henry, 1989). Extension was underway by 11 Ma and probably began by 12 Ma or perhaps 13 Ma (Henry and Aranda-Gómez, 2000). Faults can be grouped into vergence domains separated by transfer zones, similar to the geometry described above for the Baja California peninsula. In a given domain, parallel faults probably merge downward into a master fault of uncertain geometry. Normal separation on individual splays ranges up to 7 km, and domain geometry and spacing resembles that near Loreto (Umhoefer et al., 2002), which lay adjacent to southern Sinaloa prior to gulf opening.

Sonora. Proto-Gulf extension was documented in coastal Sonora on Isla Tiburón by Gordon Gastil and colleagues (Gastil and Krummenacher, 1977a, 1977b; Neuhaus et al., 1988). In central Sonora, late Cenozoic volcanic rocks were tilted during normal faulting that probably started ca. 12.5 Ma (McDowell et al., 1997; Mora-Alvarez and McDowell, 2000). Latest middle to late Miocene extension in Sonora probably was <20% (Henry and Aranda-Gómez, 2000). Early syn-rift volcanic and nonmarine rocks of latest middle to late Miocene age on Isla Tiburón and the adjacent Sonoran mainland are correlative with similar rocks in the Puertecitos region of eastern Baja California (see above). A northwest-striking fault on southern Isla Tiburón is correlated with an accommodation zone in the Puertecitos region (Oskin et al., 2001). The conjugate Sonora and Baja margins show extensional vergence domains of opposing sense (Oskin and Stock, this volume), suggesting that the Gulf obscures or obliterated transfer or accommodation zones in a complex rift system.

Eastern Basin and Range of México. Latest middle to late Miocene extension of proto-Gulf age also has been documented in several parts of the eastern branch of the Basin and Range province. At the latitude of Nayarit, these regions include Santa Rosa near Guadalajara (Moore et al., 1994; Ferrari, 1995) and Guanajuato and San Miguel de Allende (Nieto-Samaniego et al., 1999). Proto-Gulf extension in the Guanajuato area was <20%. In western Durango, a 175-km-long graben at the eastern edge of the Sierra Madre Occidental started forming at 13–12 Ma (Henry and Aranda-Gómez, 2000). In trans-Pecos Texas, minor (<10%) extension and basin-filling strata are roughly 11 Ma to 9 Ma (Dickerson and Muehlberger, 1994). In San Luis Potosí state, 13–10 Ma hawaiitic lavas and volcanic necks probably were emplaced along preexisting fracture systems in the late Miocene (Henry and Aranda-Gómez, 2000). Henry and Aranda-Gómez (2000) suggested that these localities are representative of the late Cenozoic strain history throughout north-central México and that the latest middle to late Miocene extension episode affected a broad region from the southern United States south to about 21°N and as far east as 100°W.

Was the Proto-Gulf a Marine Basin?

A proxy for the age of extension and gulf formation is the age of marine sedimentary rocks, which accumulated in basins that presumably formed due to tension and surface foundering. This proxy does not indicate the amount of extension, and in some areas it simply may record when a previously subaerial volcanic arc sank beneath sea level and was flooded by a marine incursion. Late Cenozoic marine strata crop out in many areas around and within the gulf, and their significance recently has been reinterpreted using new data.

In the northern Salton Trough, fossil and radiometric data from strata of the Fish Creek Gypsum and Imperial Formation are interpreted to indicate marine incursion at 9.5–8 Ma (Dean, 1996) and 7.4–6 Ma (McDougall et al., 1999), respectively. Paleoecologic arguments further restrict the age of marine incursion recorded in the Imperial Formation to 6.5–6.3 Ma (McDougall et al., 1999). In the southern Salton Trough, marine microfossils in the Imperial Formation are consistent with an earliest Pliocene or latest Miocene (<6 Ma) onset of marine incursion (Martín-Barajas et al., 2001).

Gastil et al. (1999) interpreted mapping, paleontologic, and radiometric data from southwestern Isla Tiburón to indicate marine incursion by at least 11 Ma (Fig. 8, top). They reported marine conglomerate with abundant late middle to late Miocene megafossils that is capped by an ignimbrite that yielded an 11.2 ± 1.3 Ma K-Ar date. The conglomerate overlies marine sandstone that contains microfossil assemblages interpreted as late Miocene–Pliocene (6.4–4 Ma). The sandstone overlies sedimentary breccia cut by a 9.02 ± 1.18 Ma (^{40}Ar/^{39}Ar) dike, and the breccia overlies volcanic rocks that yield ages of 12.9 ± 0.4 Ma (K-Ar) and 11.4 ± 2.6 Ma (^{40}Ar/^{39}Ar). To resolve the discrepancies among these ages, they disregarded the microfossil ages from the marine sandstone and concluded that the entire assemblage formed between ca. 13 Ma and 11 Ma.

Using new mapping and radiometric dates, Oskin (2002) and Oskin and Stock (2003) concluded that the interpretation of Gastil et al. (1999) is untenable and that marine incursion occurred no earlier than ca. 6.5 Ma (Fig. 8, bottom). Key aspects of the revision are (1) abandonment of the 11.2-Ma date on a capping rhyodacite flow that had been mapped as an ignimbrite by Gastil et al.; (2) reassignment of a 12.9-Ma date on volcanic breccia to a bed that underlies, rather than is interbedded with, marine strata; and (3) acceptance of the late Miocene–Pliocene ages on microfossils in the marine strata. Marine incursion to the Tiburón region thus occurred no earlier than 6.5 Ma (Oskin and Stock, 2003), diminishing the likelihood that a transpeninsular seaway connected the Pacific with the gulf region by the middle Miocene (Helenes and Carreño, 1999).

Late Miocene marine strata near the mouth of the Gulf of California (Carreño, 1985; McCloy et al., 1988; Smith, 1991) indicate that the Pacific Ocean had access to the southern part of the proto-Gulf by ca. 9–8 Ma, possibly due to locally large extensional strains and crustal thinning (Stock and Hodges, 1989).

Development of the Gulf of California

Complete rifting of North American continental crust occurred in the southernmost gulf by ca. 3.6 Ma (DeMets, 1995). Between 27.5°N and 23.5°N, graben, half graben, islands, and submarine highs within about 50 km of the Baja coastline comprise borderland-like relief (Nava-Sánchez et al., 2001). Deep basins in the central Gulf are not magnetically lineated but probably are underlain by oceanic lithosphere (Lonsdale, 1991). In the northern gulf and the Salton trough, basins are probably underlain by transitional crust (Fuis et al., 1984). Oskin and Stock (this volume) use geologic correlations to argue that the conjugate Sonora and Baja margins rifted apart cleanly, with subsidence of continental crust limited to narrow strips about 5–10 km wide.

In Pliocene–Quaternary time, strain within the gulf has been accommodated by a complex of normal, dextral, and oblique-slip faults, vertical-axis rotations, and spreading centers (Lonsdale, 1989; Fenby and Gastil, 1991; Lonsdale, 1991; Stock, 2000). Some modern transform faults in the gulf axis may be reactivated normal faults from the late Miocene "proto-Gulf" stage (Karig and Jensky, 1972).

The rift system in the gulf is highly oblique but shares many features with orthogonal rift systems, including segmentation, accommodation zones between segments, and listric master normal faults (e.g., Dokka and Merriam, 1982; Axen, 1995; Umhoefer et al., 2002). Axen (1995) suggested that late Miocene segmentation in the "proto-Gulf" had little bearing on the development of the Pliocene–Quaternary spreading system in the gulf, based on differences in geometry, relative ages, and extension directions of late Miocene and Pliocene–Quaternary structures. However, Stock (2000) and Nagy (2000) related accommodation zones in eastern Baja California to the Tiburón and Guaymas transforms.

In some areas around the periphery of the gulf, the extension direction changed from east-northeast–west-southwest to roughly west-northwest–east-southeast ca. 6 Ma, roughly coeval with the initiation or acceleration of transtensional opening of the modern gulf (Gastil and Krummenacher, 1977b; Henry, 1989; Umhoefer et al., 2002). In other areas, the Pliocene–Quaternary extension direction is unchanged from the late Miocene direction (Stock and Hodges, 1990; Lewis and Stock, 1998b; Fletcher et al., 2000).

Stratigraphic relations and fission-track thermochronology indicate that Mesozoic granitoids of the Los Cabos terrane were exposed at the surface by the mid-Cenozoic, and possibly as early as the Late Cretaceous, and that the modern topographic high is due to 5.2–6.5 km of uplift in the footwall of the San José del Cabo normal fault (Fletcher et al., 2000; Fletcher and Munguía, 2000). The San José del Cabo is the easternmost of a system of down-to-the-east, north- to north-northwest–striking normal faults that transect the southern peninsula between 24.5°N and 23°N. Fletcher and colleagues find no evidence for a shift from northeast-southwest to northwest-southeast extension

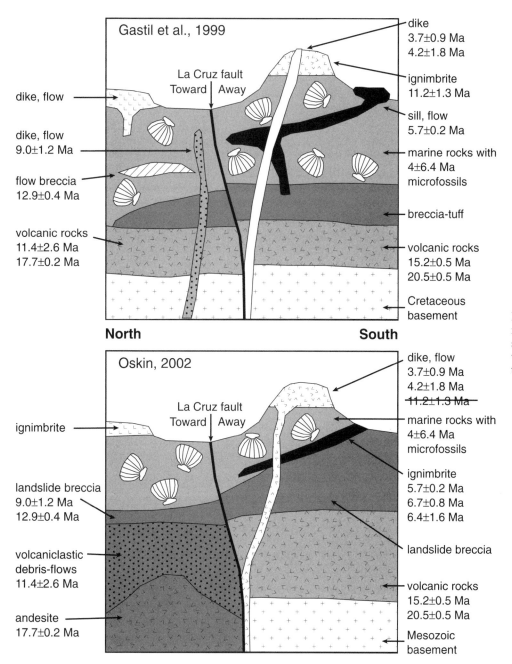

Gastil et al., 1999

La Cruz fault
Toward | Away

dike, flow

dike, flow
9.0±1.2 Ma

flow breccia
12.9±0.4 Ma

volcanic rocks
11.4±2.6 Ma
17.7±0.2 Ma

North   **South**

dike
3.7±0.9 Ma
4.2±1.8 Ma

ignimbrite
11.2±1.3 Ma

sill, flow
5.7±0.2 Ma

marine rocks with
4±6.4 Ma
microfossils

breccia-tuff

volcanic rocks
15.2±0.5 Ma
20.5±0.5 Ma

Cretaceous
basement

Oskin, 2002

La Cruz fault
Toward | Away

ignimbrite

landslide breccia
9.0±1.2 Ma
12.9±0.4 Ma

volcaniclastic
debris-flows
11.4±2.6 Ma

andesite
17.7±0.2 Ma

dike, flow
3.7±0.9 Ma
4.2±1.8 Ma
11.2±1.3 Ma

marine rocks with
4±6.4 Ma
microfossils

ignimbrite
5.7±0.2 Ma
6.7±0.8 Ma
6.4±1.6 Ma

landslide breccia

volcanic rocks
15.2±0.5 Ma
20.5±0.5 Ma

Mesozoic
basement

Figure 8. Comparison of stratigraphic and structural interpretations of southwestern Isla Tiburón, modified very slightly from Oskin (2002). See text for explanation.

in the southern peninsula during the late Cenozoic, and conclude that the original partitioning of Pacific–North America transform boundary relative motion into strike slip along gulf-axis transforms and extension across gulf-margin normal faults still applies in this region.

Near Loreto (26°N), tilting and faulting of Pliocene fan-delta and coastal ramp deposits in the southern Loreto basin (Umhoefer and Dorsey, 1997) occurred on dextral and normal faults. The extension direction of S80E–N80W is rotated about 35° clockwise from the direction of proto-Gulf extension in this region (Umhoefer et al., 2002). Pleistocene marine terraces yield uplift rates that are two to three times higher than in the Bahía

Concepción region to the north (Mayer and Vincent, 1999), indicating that the southern Loreto basin continues to actively accommodate a component of Pacific–North America transform boundary motion. On Isla Carmen offshore Loreto, marine sedimentary rocks overlie a 3.5- to 3.1-Ma unconformity (Dorsey et al., 2001). The northern Loreto basin is bounded on its north side by several faults that have been active in Pliocene–Quaternary time (Zanchi, 1994), possibly including an extension of a gulf fracture zone (Johnson et al., this volume). The Pliocene–Pleistocene Mencenares Volcanic Complex in this region may have been emplaced in an accommodation zone in rift-related normal faults (Johnson et al., this volume).

Near 27°N, undisturbed Pliocene strata indicate that the locus of extension shifted east of Bahía Concepción by the early Pliocene, relegating the region to tectonic quiescence that persisted through the Pleistocene (Johnson and Ledesma-Vazquez, 2001).

Heat flow, gravity, and seismic data indicate that oceanic lithosphere is not present in the Gulf of California north of about 30.4°N. Nagy and Stock (2000) infer that the Pacific–North America transform boundary in this region is a zone of distributed transtensional strain that links the San Andreas system and Main Gulf Escarpment to the north with spreading centers in the Delfín and Tiburón basins at 30°–29°N. This distributed strain also affected eastern Baja California north of about 30°N, where Nagy (2000) estimated ≤4% east-northeast-west-northwest–directed extension on normal faults that strike roughly north-south. Structural, paleomagnetic, and sedimentologic data have been used to infer that a diffuse late Miocene accommodation zone in this region has persisted to the Holocene (Martín-Barajas et al., 1997; Lewis and Stock, 1998a, 1998b).

In the ranges just north of the head of the gulf, east-west extension occurred on high-angle and listric normal fault systems that were active by ca. 12 Ma and have persisted into the Quaternary (Siem and Gastil, 1994; Axen and Fletcher, 1998).

Pliocene–Recent Volcanism

Late Cenozoic extension resulted in only minor volcanism east of the gulf. The largest of several small Quaternary basalt fields scattered throughout Sonora is the Pinacate, which covers about 1500 km^2 just east of the Colorado Delta at the head of the Gulf. Basalts from the Pinacate have yielded Ar/Ar ages of 38 ± 8 ka, 27 ± 6 ka, and 12 ± 4 ka (Gutmann et al., 2000). Paz Moreno et al. (this volume) report on the Moctezuma volcanic field in northeastern Sonora, which consists of tholeiitic basalts that erupted from fissure vents at 1.7 ± 0.74 Ma, and alkaline basalts that erupted from individual cones at 0.53 ± 0.2 Ma. Geochemical data indicate that the Quaternary lithosphere beneath the Moctezuma field is significantly thinner than it was during eruption of early Miocene basalts in the region.

Latest Cenozoic volcanism in the Baja California peninsula produced several large volcanic centers along the Gulf margin and rare outcrops on the Pacific side. Below, I highlight some of the key features, but more thorough summaries are provided by Saunders et al. (1987) and Sawlan (1991).

Basaltic to rhyolitic lavas and pyroclastic rocks of Holocene age constitute Isla San Luis along the eastern margin of the peninsula near 30°N (Paz Moreno and Demant, 1999; Hausback et al., 2003). In the Jaraguay and San Borja volcanic fields between 30°N and 28.5°N, Saunders et al. (1987) proposed the term "bajaite" to describe volcanic rocks with high Mg, Ni, Cr, and Sr, low Rb, and high K/Rb and LaYb. They attributed these traits to attempted ridge subduction, but Sawlan (1991) attributed them to partial melting of subduction-related metasomatic rocks. Near 27.5°N, latest Cenozoic volcanic rocks include the Pliocene–Quaternary La Reforma caldera and Quaternary Las Tres Virgenes volcanic field on the eastern margin of the peninsula

and Quaternary basalts and andesites on Isla Tortuga at 27.5°N (Batiza, 1978; Hausback et al., 2000). Submarine highs in the same region are inferred to consist of Pliocene to Quaternary volcanic rocks (Fabriol et al., 1999). At 26.3°N, intermediate to silicic calc-alkalic volcanic rocks were erupted in the Pliocene and possibly Pleistocene to form the Mencenares volcanic complex (Bigioggero et al., 1995).

Late Cenozoic volcanic rocks in the western part of the peninsula are rare. At San Quintín on the Pacific coast (30°N), 10 distinct Quaternary volcanic complexes consist chiefly of alkali basalt that have yielded Ar/Ar ages of 126 ± 8 Ma and 90 ± 10 Ma on older units (Luhr et al., 1995) and ^{3}He dates ranging from 165 ± 13 ka to 22 ± 5 ka (Williams, 1999). The basalts contain a variety of ultramafic and deep crustal xenoliths (e.g., Cabanes and Mercier, 1988, and references therein). Geochemical trends of the basalts are consistent with progressive partial melting of spinel lherzolite at unusually shallow levels in the mantle (Luhr et al., 1995). On northwestern Isla Santa Margarita near 24.5°N, Mesozoic oceanic rocks are intruded by hypabyssal rhyodacite that yielded Ar/Ar ages of 6.2 ± 0.1 Ma and 4.9 ± 0.1 Ma and may have formed by partial melting of a young subducted slab (Bonini and Baldwin, 1998).

CONCLUSIONS

This paper has provided geologic and tectonic context for the other contributions in this volume, concentrating on major tectonic developments such as the nature and origin of the Peninsular Ranges batholith, possible northward transport of parts of the Baja California peninsula in the Late Cretaceous and Paleogene, and the late Cenozoic transition to a transtensional plate boundary. Remarkable progress on these and other aspects of the geology of the peninsula and its surroundings has been made since the pioneering studies of Gordon Gastil and his colleagues and students (Gastil et al., 1975). Nevertheless, many uncertainties, disagreements, and controversies persist; in fact, many of these issues could not have been raised prior to the significant advances of the last three decades. Below I summarize some of the salient issues and try to identify the types of information that could aid in their resolution.

1. The nature of the boundary between the western and eastern zones of the Peninsular Ranges batholith. This boundary probably had independent histories on opposite sides of the Agua Blanca fault zone prior to Alisitos accretion by 105 Ma. Many questions remain about the location, nature, and development of the Peninsular Ranges batholith in both regions, particularly south of the fault zone. Field mapping, petrologic and isotopic studies, and radiometric dating are needed to answer these questions.

2. The trajectory of the Alisitos island arc prior to its accretion to North America. Recent studies have established the Alisitos as an island arc that accreted to North America in the late Early Cretaceous, but its birthplace and subsequent trajectory in the Pacific realm are very weakly constrained. The absence of continental influence until late in its history appears to pre-

clude simple translation along the continental margin, but little else can be said about its pre-accretion history. Studies of Early Cretaceous plate motions are unlikely to help reconstruct motion of the Alisitos arc, given the minimal amount of ocean floor of that age. Paleomagnetic studies may contribute to determining whether the peninsula was transported northward (see item 3 below), but such data cannot resolve translation across longitude. Isotopic, geochronologic, and provenance studies of the Alisitos may provide some insight but are certain to face obstacles in data collection and interpretation.

3. Postulated northward transport of not only the Alisitos, but of most other Mesozoic rocks of the peninsula. The *Northward Transport* section of this paper outlined the several hypotheses of northward transport that should be considered independently. Many fundamental questions remain unsolved, and resolution of each hypothesis will be challenging. Needed are more tests of plutonic rocks elsewhere in the peninsula, tests of possible tilting, more tests of sedimentary rocks, and a better understanding of compaction phenomena. In addition, interpretation of paleomagnetic results requires sufficient mapping, dating, and structural analysis, which have not yet been completed for many parts of the peninsula.

4. Reconstruction of the Late Jurassic to Early Cretaceous continental arc in northwestern México. Did the Baja California peninsula lie outboard of Sonora and Sinaloa during this entire time interval, or did it lie elsewhere for at least part of the time? The key unknown is whether Sonora and Sinaloa hosted a continental arc in the Late Jurassic and Early Cretaceous. Although few arc rocks of appropriate age have been recognized in these areas, it does not follow that such rocks are absent. First, the region has been affected by repeated events of younger tectonism, plutonism, and volcanism that could have overprinted or obscured older arc rocks. Second, the Late Jurassic–Early Cretaceous "west Caborcan arc" only recently has been identified in Baja California, in an area that had been studied more thoroughly than much of Sonora and Sinaloa. Unless a thorough search for such arc rocks in Sonora and Sinaloa comes up empty, the existence of this arc should remain an open question.

5. Development of a Miocene (post–12 Ma) pre-Gulf tectonic depression and marine incursion into that depression. Although much progress has been made on this issue in the last 15 years, details of the structural and stratigraphic development of the "proto-Gulf" are lacking in many areas and are difficult to synthesize regionally even where they have been established. Additional mapping, dating, and structural and stratigraphic analysis are likely to continue to clarify our understanding of this rift.

6. Transfer of strike slip from the Pacific borderlands to the southern gulf region. When and how was dextral slip of the Pacific–North America transform boundary transferred to the southern gulf region from the continental margin to the west? The transform boundary was firmly established within the gulf by the end of the Pliocene, but the late Miocene and Pliocene slip budgets in this region are poorly understood. Fundamental unknowns include the timing and distribution of dextral slip on faults in the Pacific borderlands such as the Tosco-Abreojos, on faults in the extended regions on either side of the Gulf of California, and on faults within the gulf itself. Some progress has been made on these problems, but evaluation of competing explanations requires careful mapping and structural analysis throughout the region.

Many other important geologic and tectonic questions about the Baja California peninsula, the Gulf of California, and western mainland México also are unresolved. Continued international collaboration undoubtedly will not only answer some of these questions but also reveal exciting new avenues of research in this remarkable region.

ACKNOWLEDGMENTS

This paper summarizes some of the major aspects of the geology and tectonics of the Baja California region, but it does not provide complete coverage of the large number of contributions in the literature. I regret the number of important contributions that were omitted (knowingly or not) and apologize for any errors committed while interpreting and synthesizing the many works cited here.

I thank many colleagues for conversations, reprints, preprints, abstracts, e-mail missives, diagrams, and other useful information, including Helge Alsleben, Myrl Beck, Tim Dixon, Becky Dorsey, Luca Ferrari, John Fletcher, Marty Grove, Jon Hagstrum, Brian Hausback, Chris Henry, Jim Ingle, Alex Iriondo, Markes Johnson, Ken MacLeod, Harold Magistrale, Kristin McDougall, Mike Oskin, Jaime Roldán-Quintana, José Romo, LouElla Saul, Doug Smith, Judy Terry Smith, David Symons, Paul Umhoefer, Paul Wetmore, and Wendi Williams. Chris Henry and Paul Umhoefer provided digital files that formed the templates for Figure 1 and Figures 5 and 6, respectively, and Chris Henry and Mike Oskin sent their original digital files that, with very slight modifications, became Figures 7 and 8, respectively. Andy Barth, Jon Hagstrum, Scott Johnson, Steve Lund, and Scott Paterson provided thoughtful reviews of early versions of this manuscript. I thank the volume editors, especially Scott and Scott, for their invitation, encouragement, and patience.

Jorge Ledesma-Vázquez and his colleagues at Ciencias Marinas (UNAM) and CICESE in Ensenada and at UABCS in La Paz have organized biennial meetings of the Peninsular Geological Society (Sociedad Geológica Peninsular) in Ensenada, La Paz, and Loreto that have served as welcome opportunities to share findings with other workers. These meetings and field trips have resulted in numerous advances in our understanding of the region's geology and tectonics and helped inspire the 1996 Penrose Conference on the Gulf of California (Umhoefer et al., 1996).

Finally, I thank the many colleagues with whom I've investigated various aspects (not only geological) of the peninsula, including José Jorge Aranda-Gómez, Suzanne Baldwin, Jennifer Bonini, Alejandro Carrillo-Chavez, John Fletcher, Jon Hagstrum, Dave Kimbrough, Jorge Ledesma-Vazquez, Tom Moore, José Antonio Pérez-Venzor, Doug Smith, and, of course, Gordon Gastil.

REFERENCES CITED

Ague, J.J., and Brandon, M.T., 1992, Tilt and northward offset of Cordilleran batholiths resolved using igneous bathymetry: Nature, v. 360, p. 146–149.

Ague, J.J., and Brimhall, G.H., 1989, Magmatic arc asymmetry and distribution of anomalous plutonic belts in the batholiths of California: Effects of assimilation, crustal thickness, and depth of crystallization: Geological Society of America Bulletin, v. 100, p. 912–927.

Aguirre-Díaz, G.J., and McDowell, F.W., 1993, Nature and timing of faulting and synextensional magmatism in the southern Basin and Range, central-eastern Durango, México: Geological Society of America Bulletin, v. 105, p. 1435–1444.

Allison, E.C., 1955, Middle Cretaceous gastropoda from Punta China, Baja California, México: Journal of Paleontology, v. 29, p. 400–432.

Allison, E.C., 1974, The type Alisitos Formation (Cretaceous, Aptian-Albian) of Baja California and its bivalve fauna, Geology of Peninsular California, Pacific Sections, American Association of Petroleum Geologists, Society of Economic Paleontologists and Minerologists, and Society of Economic Geologists, p. 20–59.

Anderson, P.V., 1993, Prebatholithic stratigraphy of the San Felipe area, Baja California norte, México, *in* Miller, R.H., ed., The prebatholithic stratigraphy of Peninsular California: Boulder, Colorado, Geological Society of America Special Paper 279, p. 1–10.

Anderson, T.H., and Silver, L.T., 1979, The role of the Mojave–Sonora Megashear in the tectonic evolution of northern Sonora, *in* Roldán-Quintana, J., ed., Geology of northern Sonora: Geological Society of America Field Trip Guidebook: Institute of Geology, Universidad Nacional Autónoma de México, Hermosillo, Sonora, Mexico, and University of Pittsburgh, Pittsburgh, Geological Society of America Annual Meeting, p. 59–68.

Anderson, T.H., and Silver, L.T., 1981, An overview of Precambrian rocks in Sonora, México: Revista del Instituto de Geología, Universidad Nacional Autónoma de México, v. 5, p. 131–139.

Aranda-Gómez, J.J., and Pérez-Venzor, J.A., 1989, Estratigrafía del complejo cristalino de la región de Todos Santos, Estado de Baja California Sur: Revista del Instituto de Geología, Universidad Nacional Autónoma de México, v. 8, p. 149–170.

Ashby, J.R., 1989, A résumé of the Miocene stratigraphic history of the Rosarito Beach basin, northwestern Baja California, México, *in* Abbott, P.L., ed., Geologic studies in Baja California: Los Angeles, California, Pacific Section, Society of Economic Paleontologists and Mineralogists, Book 63, p. 27–36.

Atwater, T., 1989, Plate tectonic history of the northeast Pacific and western North America, *in* Decker, R.W., ed., Decade of North American geology: Boulder, Colorado, Geological Society of America, p. 21–72.

Atwater, T., and Stock, J., 1998, Implications of recent refinements in Pacific–North America plate tectonic reconstructions: Eos (Transactions, American Geophysical Union), v. 79, p. 206.

Avé Lallemant, H.G., and Oldow, J.S., 1988, Early Mesozoic southward migration of Cordilleran transpressional terranes: Tectonics, v. 7, p. 1057–1075.

Axen, G., 1995, Extensional segmentation of the Main Gulf Escarpment, México and United States: Geology, v. 23, p. 515–518.

Axen, G.J., 1999, Strain partitioning among low-angle normal faults, strike-slip faults, and incipient spreading centers, northern Gulf of California, Mexico and Salton Trough, California [abs.]: Eos (Transactions, American Geophysical Union), v. 80, p. S340.

Axen, G.J., and Fletcher, J.M., 1998, Late Miocene–Pleistocene extensional faulting, northern Gulf of California, Mexico and Salton Trough, California: International Geology Review, v. 40, p. 217–244.

Axen, G.J., Grove, M., Stockli, D., Lovera, O.M., Rothstein, D.A., Fletcher, J.M., Farley, K., and Abbott, P.L., 2000, Thermal evolution of Monte Blanco dome: Low-angle normal faulting during Gulf of California rifting and late Eocene denudation of the eastern Peninsular Ranges: Tectonics, v. 19, p. 197–212.

Axen, G.J., Taylor, W.J., and Bartley, J.M., 1993, Space-time patterns and tectonic controls of Tertiary extension and magmatism in the Great Basin of the western United States: Geological Society of America Bulletin, v. 105, p. 56–76.

Balch, D.C., Bartling, S.H., and Abbott, P.L., 1984, Volcaniclastic strata of the Upper Jurassic Santiago Peak Volcanics, San Diego, California, *in* Bachman, S.B., ed., Tectonics and sedimentation along the California margin: Los Angeles, California, Pacific Section, Society of Economic Paleontologists and Mineralogists, p. 157–170.

Baldwin, S.L., and Harrison, T.M., 1989, Geochronology of blueschists from west-central Baja California and the timing of uplift of subduction complexes: Journal of Geology, v. 97, p. 149–163.

Baldwin, S.L., and Harrison, T.M., 1992, The P-T-t history of blocks in serpentinite-matrix mélange, west-central Baja California: Geological Society of America Bulletin, v. 104, p. 18–31.

Bannon, J.L., Bottjer, D.J., Lund, S.P., and Saul, L.R., 1989, Campanian/Maastrichtian stage boundary in southern California: Resolution and implications for large-scale depositional patterns: Geology, v. 17, p. 80–83.

Barth, A.P., and Schneiderman, J.S., 1996, A comparison of structures in the Andean orogen of northern Chile and exhumation structures in southern California: An analogy in tectonic style: International Geology Review, v. 38, p. 1075–1085.

Barthelmy, D.A., 1979, Regional geology of the El Arco porphyry copper deposit, Baja California, *in* Gastil, R.G., ed., Baja California geology: San Diego, Department of Geological Sciences, San Diego State University, p. 127–138.

Bartling, W.A., and Abbott, P.L., 1984, Upper Cretaceous sedimentation and tectonics with reference to the Eocene, San Miguel Island and San Diego area, California, *in* Abbott, P.L., ed., Upper Cretaceous depositional systems, southern California–Northern Baja California: Los Angeles, Pacific Section, Society of Economic Paleontologists and Mineralogists, Book 36, p. 85–102.

Bartolini, C., Damon, P., Shafiqullah, M., and Morales, M., 1994, Geochronologic contributions to the Tertiary sedimentary–volcanic sequences ("Baucarit Formation") in Sonora, Mexico: Geofisica Internacional, v. 33, p. 67–77.

Batiza, R., 1978, Geology, petrology and geochemistry of Isla Tortuga, a recent tholeiitic island in the Gulf of California: Geological Society of America Bulletin, v. 89, p. 1309–1324.

Beck, M.E., Jr., 1980, Paleomagnetic record of plate-margin tectonic processes along the western edge of North America: Journal of Geophysical Research, v. 85, p. 7115–7131.

Beck, M.E., Jr., 1989, Paleomagnetism of continental North America: Implications for displacement of crustal blocks within the western Cordillera, Baja California to British Columbia, *in* Mooney, W.D., ed., Geophysical framework of the continental United States: Boulder, Colorado, Geological Society of America Memoir 172, p. 471–492.

Beck, M.E., Jr., 1992, Some thermal and paleomagnetic consequences of tilting a batholith: Tectonics, v. 11, p. 297–303.

Bigioggero, B., Chiesa, S., Zanchi, A., Montrasio, A., and Vezzoli, L., 1995, The Cerro Mencenares volcanic center, Baja California Sur: Source and tectonic control on postsubduction magmatism within the gulf rift: Geological Society of America Bulletin, v. 107, p. 1108–1122.

Blake, M.C., Jr., Jayko, A.S., Moore, T.E., Chavez, V., Saleeby, J.B., and Seel, K., 1984, Tectonostratigraphic terranes of Magdalena Island, Baja California Sur, *in* Frizzell, V.A., Jr., ed., Geology of the Baja California peninsula: Los Angeles, California, Pacific Section, Society of Economic Paleontologists and Mineralogists, p. 183–191.

Boehlke, J.E., and Abbott, P.L., 1986, Punta Baja Formation, a Campanian submarine canyon fill, Baja California, México, *in* Abbott, P.L., ed., Cretaceous stratigraphy western North America: Los Angeles, Pacific Section, Society of Economic Paleontologists and Mineralogists, Book 46, p. 91–101.

Bohannon, R.G., and Parsons, T., 1995, Tectonic implications of post-30 Ma Pacific and North American relative plate motions: Geological Society of America Bulletin, v. 107, p. 937–959.

Böhnel, H., and Delgado-Argote, L.A., 2000, Paleomagnetic data from northern Baja California (Mexico): Results from the Cretaceous San Telmo batholith, *in* Stock, J.M., ed., Cenozoic tectonics and volcanism of Mexico: Boulder, Colorado, Geological Society of America Special Paper 334, p. 157–165.

Böhnel, H., Delgado-Argote, L.A., and Kimbrough, D.L., 2002, Discordant paleomagnetic data for middle-Cretaceous intrusive rocks from northern Baja California: Latitude displacement, tilt, or vertical axis rotation?: Tectonics, v. 21, no. 5, DOI 10.1029/2001TC001298.

Bonini, J.A., and Baldwin, S.L., 1994, ^{40}Ar/^{39}Ar geochronology of accreted terranes from southwestern Baja California Sur, *in* U.S. Geological Survey Circular 1107, p. 34.

Bonini, J.A., and Baldwin, S.L., 1998, Mesozoic metamorphic and middle to late Tertiary magmatic events on Magdalena and Santa Margarita Islands, Baja California Sur, Mexico: Implications for the tectonic evolution of the Baja California continental borderland: Geological Society of America Bulletin, v. 110, p. 1094–1104.

Bottjer, D.J., and Link, M.H., 1984, A synthesis of Late Cretaceous southern California and northern Baja California paleogeography, *in* Bachman, S.B., ed., Tectonics and sedimentation along the California margin: Los Angeles, California, Pacific Section, Society of Economic Paleontologists and Mineralogists, p. 171–188.

Buch, I.P., and Delattre, M.P., 1993, Permian and Lower Triassic stratigraphy along the 30th parallel, eastern Baja California Norte, México, *in* Miller, R.H., ed., The prebatholithic stratigraphy of Peninsular California: Boulder, Colorado, Geological Society of America Special Paper 279, p. 77–90.

Burchfiel, B.C., and Davis, G.A., 1981, Mojave Desert and environs, *in* Ernst, W.G., ed., The geotectonic development of California (Rubey Volume I): Englewood Cliffs, N.J., Prentice-Hall, p. 217–252.

Busby, C., Smith, D., Morris, W., and Fackler-Adams, B., 1998, Evolutionary model for convergent margins facing large ocean basins: Mesozoic Baja California, México: Geology, v. 26, p. 227–230.

Bushee, J., Holden, J., Geyer, B., and Gastil, R.G., 1963, Lead-alpha dates for some basement rocks of southwestern California: Geological Society of America Bulletin, v. 74, p. 803–806.

Butler, R.F., and Dickinson, W.R., 1995, Comment: "Shallow magnetic inclinations in the Cretaceous Valle Group, Baja California: remagnetization, compaction, or terrane translation?": Tectonics, v. 14, p. 218–222.

Butler, R.F., Dickinson, W.R., and Gehrels, G.E., 1991, Paleomagnetism of coastal California and Baja California: Alternatives to large-scale northward transport: Tectonics, v. 10, p. 561–576.

Cabanes, N., and Mercier, J.-C.C., 1988, Insight into the upper mantle beneath an active extensional zone: The spinel-peridotite xenoliths from San Quintín (Baja California, Mexico): Contributions to Mineralogy and Petrology, v. 100, p. 374–382.

Cabral-Cano, E., Lang, H.R., and Harrison, C.G.A., 2000, Stratigraphic assessment of the Arcelia–Teloloapan area, southern Mexico: Implications for southern Mexico's post-Neocomian tectonic evolution: Journal of South American Earth Sciences, v. 13, p. 443–457.

Carrasco, A.P., Kimbrough, D.L., Herzig, C.T., and Meeth, G.L., 1993, Discovery of accretionary lapilli in the Santiago Peak Volcanics of southern and Baja California, *in* Rendina, M.A., ed., Geologic investigations in Baja California: Santa Ana, California, South Coat Geological Society, p. 145.

Carrasco, A.P., Kimbrough, D.L., and Herzig, C.T., 1995, Cretaceous arc-volcanic strata of the western Peninsular Ranges: Comparison of the Santiago Peak Volcanics and Alisitos Group, *in* III International Meeting on Geology of the Baja California Peninsula: La Paz, México, Peninsular Geological Society, p. 19.

Carreño, A.L., 1985, Biostratigraphy of the late Miocene to Pliocene on the Pacific Island Maria Madre, Mexico: Micropaleontology, v. 31, p. 139–166.

Carrillo-Chávez, A., 1992, San Pedro mylonitic gneiss, southern Baja California peninsula: The southernmost and oldest of the Cordilleran metamorphic core complexes?: Geological Society of America Abstracts with Programs, v. 24, no. 5, p. 13.

Chadwick, B., 1987, The geology, petrography, geochemistry, and geochronology of the Tres Hermanos–Santa Clara region, Baja California, Mexico [M.S. thesis]: San Diego State University, 601 p.

Cocheme, J.-J., and Demant, A., 1991, Geology of the Yécora area, northern Sierra Madre Occidental, México, *in* Jacques-Ayala, C., ed., Studies of Sonoran geology: Boulder, Colorado, Geological Society of America Special Paper 254, p. 81–94.

Coney, P.J., and Campa, M.F., 1987, Lithotectonic terrane map of Mexico (west of the 91st meridian): U.S. Geological Survey Miscellaneous Field Studies Map MF-1874-D, scale 1:2,500,000, 1 sheet.

Coney, P.J., and Harms, T.A., 1984, Cordilleran metamorphic core complexes: Cenozoic extensional relics of Mesozoic compression: Geology, v. 12, p. 550–554.

Coney, P.J., and Reynolds, S.J., 1977, Cordilleran Benioff zones: Nature, v. 270, p. 403–406.

Couch, R.W., Ness, G.E., Sanchez-Zamora, O., Calderón-Riveroll, G., Doguin, P., Plawman, T., Coperude, S., Huehn, B., and Gumma, W., 1991, Gravity anomalies and crustal structure of the Gulf and Peninsular Province of the Californias, *in* Simoneit, B.R.T., ed., The gulf and peninsular province of the Californias: American Association of Petroleum Geologists Memoir 47, p. 25–45.

Cowan, D.S., Brandon, M.T., and Garver, J.I., 1997, Geologic tests of hypotheses for large coastwise displacements—A critique illustrated by the Baja British Columbia controversy: American Journal of Science, v. 297, p. 117–173.

Criscione, J.J., Davis, T.E., and Ehlig, P., 1978, The age of sedimentation/diagenesis for the Bedford Canyon Formation and the Santa Monica Formation in southern California: A Rb/Sr evaluation, *in* McDougall, K., ed., Mesozoic paleogeography of the western United States, Pacific Coast Paleogeography Symposium 2: Los Angeles, Pacific Section, Society of Economic Mineralogists and Paleontologists, p. 385–396.

Cunningham, A.B., and Abbott, P.L., 1986, Sedimentology and provenance of the Upper Cretaceous Rosario Formation south of Ensenada, Baja California, México, *in* Abbott, P.L., ed., Cretaceous stratigraphy western North America: Los Angeles, Pacific Section, Society of Economic Paleontologists and Mineralogists, p. 103–118.

Damon, P.E., Shafiqullah, M., and Clark, K.F., 1983, Geochronology of the porphyry copper deposits and related mineralization of México: Canadian Journal of Earth Sciences, v. 20, p. 1052–1071.

Davis, T.E., and Gastil, R.G., 1993, Whole-rock Rb/Sr isochrons of fine-grained clastic rocks in Peninsular California, *in* Miller, R.H., ed., The prebatholithic stratigraphy of peninsular California: Boulder, Colorado, Geological Society of America Special Paper 279, p. 157–158.

Dean, M.A., 1996, Neogene Fish Creek Gypsum and associated stratigraphy and paleontology, southwestern Salton Trough, California, *in* Seymour, D.C., ed., Annual Field Trip Guidebook 24: San Diego, California, South Coast Geological Society, p. 123–148.

Delgado-Argote, L.A., and García-Abdeslem, J., 1999, Shallow Miocene basaltic magma reservoirs in the Bahí de los Angeles basin, Baja California, Mexico: Journal of Volcanology and Geothermal Research, v. 88, p. 29–46.

Demant, A., 1981, Plio-Quaternary volcanism of the Santa Rosalía area, Baja California, México, *in* Roldán-Quintana, J., ed., Geology of northwestern México and southern Arizona: Instituto de Geología, Universidad Nacional Autónoma de México, Estación Regional del Noreste, p. 295–307.

DeMets, C., 1995, A reappraisal of seafloor spreading lineations in the Gulf of California: Implications for the transfer of Baja California to the Pacific plate and estimates of Pacific–North America motion: Geophysical Research Letters, v. 22, p. 3545–3548.

Dickinson, W.R., 1996, Kinematics of transrotational tectonism in the California Transverse Ranges and its contribution to cumulative slip along the San Andreas transform fault system: Boulder, Colorado, Geological Society of America Special Paper 305, 46 p.

Dickinson, W.R., and Butler, R.F., 1998, Coastal and Baja California paleomagnetism reconsidered: Geological Society of America Bulletin, v. 110, p. 1268–1280.

Dickinson, W.R., and Lawton, T.F., 2001, Carboniferous to Cretaceous assembly and fragmentation of Mexico: Geological Society of America Bulletin, v. 113, p. 1142–1160.

Dickerson, P.W., and Muehlberger, W.R., 1994, Basins in the Big Bend segment of the Rio Grande rift, *in* Kather, S.M., ed., Basins of the Rio Grande rift: Structure, stratigraphy, and tectonic setting: Boulder, Colorado, Geological Society of America Special Paper 291, p. 283–297.

Dixon, T., Decaix, J., Farina, F., Furlong, K., Malservisi, R., Bennett, R., Suarez-Vidal, F., Fletcher, J., and Lee, J., 2003, Seismic cycle and rheological effects on estimation of present-day slip rates for the Agua Blanca and San Miguel–Vallecitos faults, northern Baja California, Mexico: Journal of Geophysical Research, DOI 10.1029/2000JB000099.

Dokka, R.K., and Merriam, R.H., 1982, Late Cenozoic extension of northeastern Baja California: Geological Society of America Bulletin, v. 93, p. 371–378.

Dorsey, R.J., Umhoefer, P.J., Ingle, J., and Mayer, L., 2001, Late Miocene to Pliocene stratigraphic evolution of northeast Carmen Island, Gulf of California: Implications for oblique-rifting tectonics: Sedimentary Geology, v. 144, p. 97–123.

Elías-Herrera, M., and Ortega-Gutiérrez, F., 1998, The Early Cretaceous Arperos oceanic basin (western Mexico). Geochemical evidence for an aseismic ridge formed near a spreading center—Comment: Tectonophysics, v. 292, p. 321–326.

Elías-Herrera, M., Sánchez-Zavala, J.L., and Macias-Romo, C., 2000, Geologic and geochronologic data from the Guerrero terrane in the Tejupilco area, southern Mexico: New constraints on its tectonic interpretation: Journal of South American Earth Sciences, v. 13, p. 355–375.

Engebretson, D.C., Cox, A., and Gordon, R.G., 1985, Relative motions between oceanic and continental plates in the Pacific Basin: Boulder, Colorado, Geological Society of America Special Paper 206, p. 64 p.

Fabriol, H., Delgado-Argote, L.A., Dañobeitia, J.J., Córdoba, D., González, A., García-Abdeslem, J., Bartolomé, R., Martín-Ateinza, B., and Frias-

Camacho, V., 1999, Backscattering and geophysical features of volcanic ridges offshore Santa Rosalia, Baja California Sur, Gulf of California, Mexico: Journal of Volcanology and Geothermal Research, v. 93, p. 75–92.

Fenby, S.S., and Gastil, R.G., 1991, Geologic-tectonic map of the Gulf of California and surrounding areas, *in* Simoneit, B.R.T., ed., The Gulf and Peninsular Province of the Californias: American Association of Petroleum Geologists Memoir 47, p. 79–83.

Ferrari, L., 1995, Miocene shearing along the northern boundary of the Jalisco block and the opening of the southern Gulf of California: Geology, v. 23, p. 751–754.

Ferrari, L., López-Martínez, M., Aguirre-Díaz, G., and Carrasco-Nuñez, G., 1999, Space-time patterns of Cenozoic arc volcanism in central México: From the Sierra Madre Occidental to the Mexican Volcanic Belt: Geology, v. 27, p. 303–306.

Ferrari, L., López-Martínez, M., and Rosas-Elguera, J., 2002, Ignimbrite flare-up and deformation in the southern Sierra Madre Occidental, western Mexico: Implications for the late subduction history of the Farallon plate: Tectonics, v. 21, no. 4, DOI 10.1029/2001TC001302.

Filmer, P.E., and Kirschvink, J.L., 1989, A paleomagnetic constraint on the Late Cretaceous paleoposition of northwestern Baja California, México: Journal of Geophysical Research, v. 94, p. 7332–7342.

Fletcher, J.M., and Munguía, L., 2000, Active continental rifting in southern Baja California, Mexico: Implications for plate motion partitioning and the transition to seafloor spreading in the Gulf of California: Tectonics, v. 19, p. 1107–1123.

Fletcher, J.M., Kohn, B.P., Foster, D.A., and Gleadon, A.J.W., 2000, Heterogeneous Neogene cooling and exhumation of the Los Cabos block, southern Baja California: Evidence from fission-track thermochronology: Geology, v. 28, p. 107–110.

Flynn, J.J., Cipolletti, R.M., and Novacek, M.J., 1989, Chronology of early Eocene marine and terrestrial strata, Baja California, México: Geological Society of America Bulletin, v. 101, p. 1182–1196.

Freydier, C., Martínez-R, J., Lapierre, H., Tardy, M., and Coulon, C., 1996, The Early Cretaceous Arperos oceanic basin (western México): Geochemical evidence for an aseismic ridge formed near a spreading center: Tectonophysics, v. 259, p. 343–367.

Freydier, C., Lapierre, H., Tardy, M., Martínez-R, J., and Coulon, C., 2000, The Early Cretaceous Arperos basin: An oceanic domain dividing the Guerrero arc from nuclear Mexico evidenced by the geochemistry of the lavas and sediments: Journal of South American Earth Sciences, v. 13, p. 325–336.

Fuis, G.S., Mooney, W.D., Healy, J.H., McMechan, G.A., and Lutter, W.J., 1984, A seismic refraction survey of the Imperial Valley region, California: Journal of Geophysical Research, v. 89, p. 1165–1189.

Gans, P.B., 1997, Large-magnitude Oligo-Miocene extension in southern Sonora: Implications for the tectonic evolution of western México: Tectonics, v. 16, p. 388–408.

Gastil, R.G., 1968, Fault systems in northern Baja California and their relation to the origin of the Gulf of California: Stanford, Stanford University Publications in Geological Sciences, v. 11, p. 283–286.

Gastil, R.G., 1983, Mesozoic and Cenozoic granitic rocks of southern California and western México: Boulder, Colorado, Geological Society of America Memoir 159, p. 265–275.

Gastil, R.G., 1985, Terranes of peninsular California and adjacent Sonora, *in* Howell, D.G., ed., Tectonostratigraphic terranes of the Circum-Pacific region: Houston, Texas, Circum-Pacific Council for Energy and Mineral Resources, Earth Science Series 1, p. 273–283.

Gastil, G., 1991, Is there a Oaxaca-California megashear? Conflict between paleomagnetic data and other elements of geology: Geology, v. 19, p. 502–505.

Gastil, R.G., 1993, Prebatholithic history of Peninsular California, *in* Miller, R.H., ed., The prebatholithic stratigraphy of Peninsular California: Boulder, Colorado, Geological Society of America Special Paper 279, p. 145–156.

Gastil, R.G., and Girty, M.S., 1993, A reconnaissance U-Pb study of detrital zircon in sandstones of Peninsular California and adjacent areas, *in* Miller, R.H., ed., The prebatholithic stratigraphy of Peninsular California: Boulder, Colorado, Geological Society of America Special Paper 279, p. 135–144.

Gastil, R.G., and Krummenacher, D., 1977a, Reconnaissance geological map of coastal Sonora between Puerto Lobos and Bahía Kino: Boulder, Colorado, Geological Society of America Map and Chart Series MC-16, scale 1:150,000, 1 sheet.

Gastil, R.G., and Krummenacher, D., 1977b, Reconnaissance geology of coastal Sonora between Puerto Lobos and Bahía Kino: Geological Society of America Bulletin, v. 88, p. 189–198.

Gastil, R.G., and Miller, R.H., 1981, Lower Paleozoic strata on the Pacific plate of North America: Nature, v. 292, p. 828–830.

Gastil, R.G., and Miller, R.H., 1983, Pre-batholithic terranes of southern and peninsular California, U.S.A. and México: Status report, *in* Stevens, C.H., ed., Pre-Jurassic rocks in western North American suspect terranes: Los Angeles, California, Pacific Section, Society of Economic Paleontologists and Mineralogists, p. 49–61.

Gastil, R.G., and Miller, R.H, 1984, Prebatholithic paleogeography of peninsular California and adjacent México, *in* Frizzell, V.A., Jr., ed., Geology of the Baja California peninsula: Pacific Section, Society of Economic Paleontologists and Mineralogists, p. 9–16.

Gastil, R.G., and Miller, R.H., editors, 1993, The prebatholithic stratigraphy of Peninsular California: Boulder, Colorado, Geological Society of America Special Paper 279, 163 p.

Gastil, R.G., Phillips, R.P., and Allison, E.C., 1975, Reconnaissance geology of the state of Baja California: Boulder, Colorado, Geological Society of America Memoir 140, 170 p.

Gastil, G., Krummenacher, D., and Jensky, W.A., II, 1978, Reconnaissance geology of west-central Nayarit, México: Boulder, Colorado, Geological Society of America Map and Chart Series MC-24, scale 1:200,000, 1 sheet, 8 p.

Gastil, R.G., Krummenacher, D., and Minch, J., 1979, The record of Cenozoic volcanism around the Gulf of California: Geological Society of America Bulletin, v. 90, p. 839–857.

Gastil, R.G., Morgan, G.J., and Krummenacher, D., 1981, The tectonic history of peninsular California and adjacent México, *in* Ernst, W.G., ed., The geotectonic development of California (Rubey Volume I): Englewood Cliffs, N.J., Prentice-Hall, p. 284–306.

Gastil, G., Kimbrough, J., Tainosho, Y., Shimizu, M., and Gunn, S., 1991a, Plutons of the eastern Peninsular Ranges, southern California, USA, and Baja California, México, *in* Hanan, B.B., ed., Geological excursions in southern California and México, *in* Guidebook for the 1991 Annual Meeting of the Geological Society of America: San Diego, California, Geological Society of America, p. 319–331.

Gastil, R.G., Miller, R., Anderson, P., Crocker, J., Campbell, M., Buch, P., Lothringer, C., Leier Englehardt, P., DeLattre, M., Hoobs, J., and Roldán-Quintana, J., 1991b, The relation between the Paleozoic strata on opposite sides of the Gulf of California, *in* Jacques-Ayala, C., ed., Studies of Sonoran geology: Boulder, Colorado, Geological Society of America Special Paper 254, p. 7–18.

Gastil, R.G., Neuhaus, J., Cassidy, M., Smith, J.T., Ingle, J.C., and Krummenacher, D., 1999, Geology and paleontology of southwestern Isla Tiburón, Sonora, Mexico: Revista Mexicana de Ciencias Geológicas, v. 16, p. 1–34.

Gehrels, G.E., Stewart, J.H., and Kenter, K.B., 2002, Cordilleran-margin quartzites in Baja California—Implications for tectonic transport: Earth and Planetary Science Letters, v. 199, no. 1–2, p. 201–210.

Girty, G.H., Thomson, C.N., Girty, M.S., Miller, J., and Bracchi, K., 1993, The Cuyamaca-Laguna Mountains shear zone, Late Jurassic plutonic rocks and Early Cretaceous extension, Peninsular Ranges, southern California, *in* Rendina, M.A., ed., Geologic investigations in Baja California, México, Annual Field Trip Guidebook 21: Santa Ana, California, South Coast Geological Society, p. 173–180.

Gordon, R.G., 1990, Test for bias in paleomagnetically determined paleolatitudes from Pacific plate Deep Sea Drilling Project sediments: Journal of Geophysical Research, v. 95, p. 8397–8404.

Grant, L.B., and Rockwell, T.K., 2002, A northward-propagating earthquake sequence in coastal southern California?: Seismological Research Letters, v. 73, p. 461–469.

Griffith, R., and Hoobs, J., 1993, Geology of the southern Sierra Calamajué, Baja California Norte, México, *in* Miller, R.H., ed., The prebatholithic stratigraphy of peninsular California: Boulder, Colorado, Geological Society of America Special Paper 279, p. 43–60.

Grimm, K.A., 1999, Stratigraphic condensation and the redeposition of economic phosphorite: Allostratigraphy of Oligo-Miocene shelfal sediments, Baja California Sur, Mexico, *in* Lucas, J., ed., Marine authigenesis: From microbial to global: Society for Sedimentary Geology (SEPM) Special Publication 64, p. 325–347.

Gutmann, J.T., Turrin, B.D., and Dohrenwend, J.C., 2000, Basalt rock from the Pinacate volcanic field yields notably young $^{40}Ar/^{39}Ar$ ages: Eos (Transactions, American Geophysical Union), v. 81, p. 33.

Hagstrum, J.T., and Sedlock, R.L., 1990, Remagnetization and northward translation of Mesozoic red chert from Cedros Island and the San Benito

Islands, Baja California, México: Geological Society of America Bulletin, v. 102, p. 983–991.

Hagstrum, J.T., and Sedlock, R.L., 1992, Paleomagnetism of Mesozoic red chert from Cedros Island and the San Benito Islands, Baja California, Mexico, revisited: Geophysical Research Letters, v. 19, p. 329–332.

Hagstrum, J.T., McWilliams, M., Howell, D.G., and Gromme, C.S., 1985, Mesozoic paleomagnetism and northward translation of the Baja California peninsula: Geological Society of America Bulletin, v. 96, p. 1077–1090.

Hagstrum, J.T., Sawlan, M.G., Hausback, B.P., Smith, J.G., and Gromme, C.S., 1987, Miocene paleomagnetism and tectonic setting of the Baja California peninsula, México: Journal of Geophysical Research, v. 92, p. 2627–2640.

Hardy, L.R., 1981, Geology of the central Sierra de Santa Rosa, Sonora Mexico, *in* Roldán-Quintana, J., ed., Geology of northwestern México and southern Arizona: Instituto de Geología, Universidad Nacional Autónoma de México, Estación Regional del Noreste, p. 73–98.

Hausback, B.P., 1984, Cenozoic volcanic and tectonic evolution of Baja California Sur, México, *in* Frizzell, V.A., Jr., ed., Geology of the Baja California peninsula: Los Angeles, California, Pacific Section, Society of Economic Paleontologists and Mineralogists, p. 219–236.

Hausback, B.P., Stock, J.M., Dmochowski, J.E., Farrar, C.D., Fowler, S.J., Sutter, K., Verke, P., and Winant, C.D., 2000, To be or not to be a caldera—La Reforma caldera, Baja California Sur, Mexico: Geological Society of America Abstracts with Programs, v. 32, no. 7, p. A502.

Hausback, B.P., Cook, A., Farrar, C.D., Giambastiani, M., Martin, A., Paz-Moreno, F., Stock, J., and Dmochowski, J.E., 2003, Isla San Luis volcano, Baja California, Mexico—Late Holocene eruptions: Geological Society of America Abstracts with Programs, v. 35, no. 4, p. 29.

Heim, A., 1922, Notes on the Tertiary of southern Lower California: Geological Magazine, v. 59, p. 529–547.

Helenes, J., 1984, Dinoflagellates from Cretaceous to Early Tertiary rocks of the Sebastian Vizcaíno basin, Baja California, Mexico, *in* Frizzell, V.A., Jr., ed., Geology of the Baja California peninsula: Los Angeles, California, Pacific Section, Society of Economic Paleontologists and Mineralogists, p. 89–106.

Helenes, J., and Carreño, A.L., 1999, Neogene sedimentary evolution of Baja California in relation to regional tectonics: Journal of South American Earth Sciences, v. 12, p. 589–605.

Henry, C.D., 1989, Late Cenozoic Basin and Range structure in western México adjacent to the Gulf of California: Geological Society of America Bulletin, v. 101, p. 1147–1156.

Henry, C.D., and Aranda-Gómez, J.J., 1992, The real southern Basin and Range: Mid- to late Cenozoic extension in México: Geology, v. 20, p. 701–704.

Henry, C.D., and Aranda-Gómez, J., 2000, Plate interactions control middle–late Miocene, proto-Gulf and Basin and Range extension in the southern Basin and Range: Tectonophysics, v. 318, p. 1–26.

Henry, C.D., and Fredrikson, G., 1987, Geology of part of southern Sinaloa, México adjacent to the Gulf of California: Boulder, Colorado, Geological Society of America Map and Chart Series MCH063, scale 1:250,000, 1 sheet, 14 p.

Hirabayashi, C.K., Rockwell, T.K., Wesnousky, S.G., Stirling, M.W., and Suarez-Vidal, F., 1996, A neotectonic study of the San Miguel-Vallecitos fault, Baja California, México: Bulletin of the Seismological Society of America, v. 86, p. 1770–1783.

Howell, D.G., Jones, D.L., and Schermer, E.R., 1985, Tectonostratigraphic terranes of the Circum-Pacific region, *in* Howell, D.G., ed., Tectonostratigraphic terranes of the Circum-Pacific region: Houston, Texas, Circum-Pacific Council for Energy and Mineral Resources, Earth Science Series 1, p. 3–30.

Hudson, F.S., 1922, Geology of the Cuyamaca region of California: California University Department Geological Science Bulletin, v. 13, p. 175–223.

Imlay, R.W., 1963, Jurassic fossils from southern California: Journal of Paleontology, v. 37, p. 97–107.

Imlay, R.W., 1965, Jurassic marine faunal differentiation in North America: Journal of Paleontology, v. 39, p. 1023–1038.

Jackson, M.J., Banerjee, S.K., Marvin, J.A., Lu, R., and Gruber, W., 1991, Detrital remanence inclination errors and anhysteretic remanence anisotropy: Quantitative model and experimental results: Geophysical Journal International, v. 104, p. 95–103.

James, E.W., 1986, Pre-Tertiary palegeography along the northern San Andreas fault: Eos (Transactions, American Geophysical Union), v. 67, p. 1215.

Johnson, M.E., and Hayes, M.L., 1993, Dichotomous facies on a Late Cretaceous rocky island as related to wind and wave patterns (Baja California, México): Palaios, v. 8, p. 385–395.

Johnson, M.E., and Ledesma-Vazquez, J., 2001, Pliocene–Pleistocene rocky shorelines trace coastal development of Bahía Concepción, gulf coast of Baja California Sur (Mexico): Palaeogeography, Palaeoclimatology, Palaeoecology, v. 166, p. 65–88.

Johnson, M.E., Ledesma-Vázquez, J., Clark, H.C., and Zwiebel, J.A., 1996, Coastal evolution of Late Cretaceous and Pleistocene rocky shores: Pacific rim of Baja California, México: Geological Society of America Bulletin, v. 108, p. 708–721.

Johnson, S.E., Tate, M.C., and Fanning, C.M., 1999, New geologic mapping and SHRIMP U-Pb zircon data in the Peninsular Ranges batholith, Baja California, Mexico: Evidence for a suture?: Geology, v. 27, p. 743–746.

Johnson, S.E., Schmidt, K.L., and Tate, M.C., 2002, Ring complexes in the Peninsular Ranges Batholith, Mexico and the USA: Magma plumbing systems in the middle and upper crust: Lithos, v. 61, p. 187–208.

Johnson, S.E., Fletcher, J.M., Fanning, C.M., Vernon, R.H., Paterson, S.R., and Tate, M.C., 2003, Structure, emplacement, and lateral expansion of the San Jose tonalite pluton, Peninsular Ranges batholith, Baja California, Mexico: Journal of Structural Geology, v. 25, p. 1933–1957.

Karig, D.E., and Jensky, W., 1972, The Protogulf of California: Earth and Planetary Science Letters, v. 17, p. 169–174.

Kerr, D.R., and Kidwell, S.M., 1991, Late Cenozoic sedimentation and tectonics, western Salton Trough, California, *in* Hanan, B.B., ed., Geological excursions in southern California and Mexico, Geological Society of America, National Meeting Guidebook: San Diego, California, Department of Geological Sciences, San Diego State University, p. 397–416.

Kimbrough, D.L., 1985, Tectonostratigraphic terranes of the Vizcaíno Peninsula and Cedros and San Benito Islands, Baja California, México, *in* Howell, D.G., ed., Tectonostratigraphic terranes of the Circum-Pacific region: Houston, Texas, Circum-Pacific Council for Energy and Mineral Resources, Earth Science Series 1, p. 285–298.

Kimbrough, D.L., Smith, D.P., Mahoney, J.B., Moore, T.E., Grove, M., Gastil, R.G., Ortega-Rivera, A., and Fanning, C.M., 2001, Forearc-basin sedimentary response to rapid Late Cretaceous batholith emplacement in the Peninsular Ranges of southern and Baja California: Geology, v. 29, p. 491–494.

King, R.E., 1939, Geological reconnaissance in northern Sierra Madre Occidental of México: Geological Society of America Bulletin, v. 50, p. 1625–1722.

Kodama, K.P., and Davi, J.M., 1995, A compaction correction for the paleomagnetism of the Cretaceous Pigeon Point Formation of California: Tectonics, v. 14, p. 1153–1164.

Lamb, T.N., 1970, Fossiliferous Triassic metasedimentary rocks near Sun City, Riverside County, California: Geological Society of America Abstracts with Programs, v. 2, no. 110-111.

Lapierre, H., Ortiz, L.E., Abouchami, W., Monod, O., Coulon, C., and Zimmerman, J.-L., 1992, A crustal section of an intra-oceanic island arc: The Late Jurassic–Early Cretaceous Guanajuato magmatic sequence, central México: Earth and Planetary Science Letters, v. 108, p. 61–77.

Larsen, E.S., Jr., 1948, Batholith and associated rocks of Corona, Elsinore, and San Luis Rey quadrangles, southern California: Boulder, Colorado, Geological Society of America Memoir 29, 182 p.

Ledesma-Vázquez, J., and Johnson, M.E., 2001, Miocene–Pleistocene tectonosedimentary evolution of Bahía Concepción region, Baja California Sur (México): Sedimentary Geology, v. 144, p. 83–96.

Lee, J., Miller, M.M., Crippen, R., Hacker, B., and Vázquez, J.L., 1996, Middle Miocene extension in the Gulf Extensional Province, Baja California: Evidence from the southern Sierra Juárez: Geological Society of America Bulletin, v. 108, p. 505–525.

Lewis, C.J., and Stock, J.M., 1998a, Late Miocene to Recent transtensional tectonics in the Sierra San Fermín, northeastern Baja California, Mexico: Journal of Structural Geology, v. 20, p. 1043–1063.

Lewis, C.J., and Stock, J.M., 1998b, Paleomagnetic evidence of localized vertical axis rotation during Neogene extension, Sierra San Fermín, northeastern Baja California, México: Journal of Geophysical Research, v. 103, p. 2455–2470.

Lewis, J.L., Day, S.M., Magistrale, H., Castro, R.R., Astiz, L., Rebollar, C., Eakins, J., Vernon, F.L., and Brune, J.N., 2001, Crustal thickness of the Peninsular Ranges and Gulf Extensional Province in the Californias: Journal of Geophysical Research, v. 106, p. 13,599–13,611.

Li, Y., Kodama, K.P., and Smith, D.P., 2001, A compaction-corrected inclination for the middle Cretaceous Valle Group in Vizcaíno terrane, Baja California, Mexico [abs.]: Preliminary results: Eos (Transactions, American Geophysical Union), v. 82, F339.

Lonsdale, P., 1989, Geology and tectonic history of the Gulf of California, *in* Decker, R.W., ed., Decade of North American geology: Boulder, Colorado, Geological Society of America, p. 499–521.

Lonsdale, P., 1991, Structural patterns of the Pacific floor offshore of peninsular California, *in* Simoneit, B.R.T., ed., The Gulf and peninsular province of the Californias: American Association of Petroleum Geologists Memoir 47, p. 87–125.

López-Ramos, E., 1985, Geología de México, Tomo II, edición 3a, primera reimpresión: México, D.F. (published privately), 453 p.

Lovera, O.M., Grove, M., Kimbrough, D.L., and Abbott, P.L., 1999, A method for evaluating basement exhumation histories from closure age distributions of detrital minerals: Journal of Geophysical Research, v. 104, p. 29,419–29,438.

Luhr, J.F., Aranda-Gómez, J.J., and Housh, T.B., 1995, San Quintín Volcanic Field, Baja California Norte, México: Geology, petrology, and geochemistry: Journal of Geophysical Research, v. 100, p. 10,353–10,380.

Lund, S.P., and Bottjer, D.J., 1991, Paleomagnetic evidence for microplate tectonic development of southern and Baja California, *in* Simoneit, B.R.T., ed., The gulf and peninsular Province of the Californias: American Association of Petroleum Geologists Memoir 47, p. 231–248.

Lund, S.P., Bottjer, D.J., Whidden, K.J., Powers, J.E., and Steele, M.C., 1991, Paleomagnetic evidence for Paleogene terrane displacements and accretion in southern California, *in* May, J.A., ed., Eocene geologic history, San Diego region: Los Angeles, California, Pacific Section, Society for Sedimentary Geology (SEPM), p. 99–106.

Lyle, M., and Ness, G.E., 1991, The opening of the southern Gulf of California, *in* Simoneit, B.R.T., ed., The gulf and peninsular province of the Californias: American Association of Petroleum Geologists Memoir 47, p. 403–423.

Maestas, Y., MacLeod, K., Douglas, R., Self-Trail, J., and Ward, P., 2003, Late Cretaceous foraminiferal paleoenvironment and paleoceanography of the Rosario Formation, San Antonio del Mar, Baja California, Mexico: Journal of Foraminiferal Research, v. 33 (in press).

Martín-Barajas, A., Téllez-Duarte, M., and Stock, J.M., 1997, The Puertecitos Formation: Pliocene volcaniclastic sedimentation along an accommodation zone in northeastern Baja California, *in* Ledesma-Vázquez, J., ed., Pliocene carbonate and related facies flanking the Gulf of California, Baja California, Mexico: Boulder, Colorado, Geological Society of America Special Paper 318, p. 1–24.

Martín, A., Fletcher, J.M., López-Martínez, M., and Mendoza-Borunda, R., 2000, Waning Miocene subduction and arc volcanism in Baja California: The San Luis Gonzaga volcanic field: Tectonophysics, v. 318, p. 27–51.

Mayer, L., and Vincent, K.R., 1999, Active tectonics of the Loreto area, Baja California Sur, Mexico: Geomorphology, v. 27, p. 243–255.

McCloy, C., Ingle, J.C., and Barron, J.A., 1988, Neogene stratigraphy, diatoms, and depositional history of Maria Madre Island, Mexico: Evidence of early Neogene marine conditions in the southern Gulf of California: Marine Micropaleontology, v. 13, p. 193–212.

McCulloch, D.S., 1987, Regional geology and hydrocarbon potential of offshore central California, *in* Vedder, J.G., ed., Geology and resource potential of the continental margin of western North America and adjacent ocean basins—Beaufort Sea to Baja California: Houston, Texas, Circum-Pacific Council for Energy and Mineral Resources Earth Science Series, no. 6, p. 353–401.

McDougall, K., Poore, R.Z., and Matti, J., 1999, Age and paleoenvironment of the Imperial Formation near San Gorgonio Pass, southern California: Journal of Foraminiferal Research, v. 29, p. 4–25.

McDowell, F.W., and Clabaugh, S., 1979, Ignimbrites of the Sierra Madre Occidental and their relation to the tectonic history of western México: Boulder, Colorado, Geological Society of America Special Paper 180, p. 113–124.

McDowell, F.W., Roldán-Quintana, J., and Amaya-Martínez, R., 1997, Interrelationship of sedimentary and volcanic deposits associated with Tertiary extension in Sonora, México: Geological Society of America Bulletin, v. 109, p. 1349–1360.

McLean, H., 1988, Reconnaissance geologic map of the Loreto and part of the San Javier quadrangles, Baja California Sur, México: U.S. Geological Survey Miscellaneous Field Studies Map MF-2000, scale 1:50,000, 1 sheet.

McLean, H., Hausback, B.P., and Knapp, J.H., 1987, The geology of west-central Baja California Sur, México: U.S. Geological Survey Bulletin, v. 1579, p. 1–16.

Miller, R.H., and Dockum, M.S., 1983, Ordovician conodonts from metamorphosed carbonates of the Salton Trough, California: Geology, v. 11, p. 410–412.

Mina, F., 1957, Bosquejo geológico del Territorio Sur de la Baja California: Asociación Mexicana de Geólogos Petroleros Boletín, v. 9, p. 139–270.

Moore, G., Marone, C., Carmichael, I.S.E., and Renne, P., 1994, Basaltic volcanism and extension near the intersection of the Sierra Madre volcanic province and the Mexican Volcanic Belt: Geological Society of America Bulletin, v. 106, p. 383–394.

Moore, T.E., 1985, Stratigraphy and tectonic significance of the Mesozoic tectonostratigraphic terranes of the Vizcaíno Peninsula, Baja California Sur, México, *in* Howell, D.G., ed., Tectonostratigraphic terranes of the Circum-Pacific region: Houston, Texas, Circum-Pacific Council for Energy and Mineral Resources, Earth Science Series 1, p. 315–329.

Moore, T.E., 1986, Petrology and tectonic implications of the blueschist-bearing Puerto Nuevo mélange complex, Vizcaíno Peninsula, Baja California Sur, México, *in* Brown, E.H., ed., Blueschists and Eclogites: Boulder, Colorado, Geological Society of America Memoir 164, p. 43–58.

Mora-Alvarez, G., and McDowell, F.W., 2000, Miocene volcanism during late subduction and early rifting in the Sierra Santa Ursula of western Sonora, Mexico, *in* Stock, J.M., ed., Cenozoic tectonics and volcanism of Mexico: Boulder, Colorado, Geological Society of America Special Paper 334, p. 123–141.

Moran, A.I., 1976, Allochthonous carbonate debris in Mesozoic flysch deposits in Santa Ana Mountains, California: American Association of Petroleum Geologists Bulletin, v. 60, p. 2038–2043.

Morris, L.K., Lund, S.P., and Bottjer, D.J., 1986, Paleolatitude drift history of displaced terranes in southern and Baja California: Nature, v. 321, p. 844–847.

Mukasa, S.B., Dillon, J.T., and Tosdal, R.M., 1984, A late Jurassic minimum age for the Pelona-Orocopia Schist protolith, southern California: Geological Society of America Abstracts with Programs, v. 16, p. 323.

Mullan, H.S., 1978, Evolution of the Nevadan orogen in northwestern México: Geological Society of America Bulletin, v. 89, p. 1175–1188.

Nagy, E.A., 2000, Extensional deformation and paleomagnetism at the western margin of the Gulf extensional province, Puertecitos Volcanic Province, northeastern Baja California, Mexico: Geological Society of America Bulletin, v. 112, p. 857–870.

Nagy, E.A., and Stock, J.N., 2000, Structural controls on the continent-ocean transition in the northern Gulf of California: Journal of Geophysical Research, v. 105, p. 16,251–16,269.

Nagy, E.A., Grove, M., and Stock, J.M., 1999, Age and stratigraphic relationships of pre- and syn-rift volcanic deposits in the northern Puertecitos Volcanic Province, Baja California, Mexico: Journal of Volcanology and Geothermal Research, v. 93, p. 1–30.

NAMAG, 2002, North American Magnetic Anomaly Map: U.S. Geological Survey.

Nava-Sánchez, E.H., Gorsline, D.S., and Molina-Cruz, A., 2001, The Baja California peninsula borderland: Structural and sedimentological characteristics: Sedimentary Geology, v. 144, p. 63–82.

Neuhaus, J.R., Cassidy, M., Krummenacher, D., and Gastil, R.G., 1988, Timing of protogulf extension and transtensional rifting through volcanic/sedimentary stratigraphy of SW Isla Tiburón, Gulf of California, Sonora, México: Geological Society of America Abstracts with Programs, v. 20, p. 218.

Nieto-Samaniego, A.F., Ferrari, L., Alaniz-Alvarez, S.A., Labarthe-Hernández, G., and Rosas-Elguera, J., 1999, Variation of Cenozoic extension and volcanism across the southern Sierra Madre Occidental volcanic province, México: Geological Society of America Bulletin, v. 111, p. 347–363.

Normark, W.R., Spencer, J.E., and Ingle, J.C., 1987, Geology and Neogene history of the Pacific continental margin of Baja California Sur, México, *in* Vedder, J.G., ed., Geology and resource potential of the continental margin of western North America and adjacent ocean basins—Beaufort Sea to Baja California: Houston, Texas, Circum-Pacific Council for Energy and Mineral Resources Earth Science Series, no. 6, p. 449–472.

Nourse, J.A., Anderson, T.H., and Silver, L.T., 1994, Tertiary metamorphic core complexes in Sonora, northwestern México: Tectonics, v. 13, p. 1161–1182.

Ortega-Gutiérrez, F., 1982, Evolución magmática y metamórfica del complejo cristalino de La Paz, Baja California Sur: México, D.F., Sociedad Geológica Mexicana, VI Convención Nacional, Libro de Resúmenes, p. 90.

Ortega-Rivera, A., Farrar, E., Hanes, J.A., Archibald, D.A., Gastil, R.G., Kimbrough, D.L., Zentilli, M., López-Martínez, M., Féraud, G., and Ruffet, G., 1997, Chronological constraints on the thermal and tilting history of the Sierra San Pedro Mártir pluton, Baja California, México, from U/Pb, ⁴⁰Ar/³⁹Ar, and fission-track geochronology: Geological Society of America Bulletin, v. 109, p. 728–745.

Oskin, M., and Stock, J., 2003, Marine incursion synchronous with plate-boundary localization in the Gulf of California: Geology, v. 31, p. 23–26.

Oskin, M., Stock, J., and Martín-Barajas, A., 2001, Rapid localization of Pacific–North America plate motion in the Gulf of California: Geology, v. 29, p. 459–462.

Oskin, M.E., 2002, Part I. Tectonic evolution of the northern Gulf of California, Mexico, deduced from conjugate rifted margins of the Upper Delfin basin. Part II. Active folding and seismic hazard in central Los Angeles, California [Ph.D. thesis]: Pasadena, California, California Institute of Technology, 481 p.

Page, B.M., and Engebretson, D.C., 1984, Correlation between the geologic record and computed plate motions for central California: Tectonics, v. 3, p. 133–155.

Patterson, D.L., 1984, Paleomagnetism of the Valle Formation and the Late Cretaceous paleogeography of the Vizcaíno Basin, Baja California, México, *in* Frizzell, V.A., Jr., ed., Geology of the Baja California peninsula: Los Angeles, California, Pacific Section, Society of Economic Paleontologists and Mineralogists, p. 173–182.

Patzkowsky, M.E., Smith, L.H., Markwick, P.J., Engberts, C.J., and Gyllenhaal, E.D., 1991, Application of the Fujita-Ziegler paleoclimate model: Early Permian and Late Cretaceous examples: Palaeogeography, Palaeoclimatology, Palaeoecology, v. 86, p. 67–85.

Paz Moreno, F.A., and Demant, A., 1999, The Recent Isla San Luis volcanic centre: Petrology of a rift-related volcanic suite in the northern Gulf of California, Mexico: Journal of Volcanology and Geothermal Research, v. 93, p. 31–52.

Phillips, J.R., 1993, Stratigraphy and structural setting of the mid-Cretaceous Olvidada Formation, Baja California Norte, Mexico, *in* Miller, R.H., ed., The prebatholithic stratigraphy of peninsular California: Boulder, Colorado, Geological Society of America Special Paper 279, p. 97–106.

Radelli, L., 1989, The ophiolites of Calmalli (El Arco) and the Olvidada nappe of northern Baja California and west-central Sonora, México, *in* Abbott, P.L., ed., Geologic studies in Baja California: Los Angeles, California, Pacific Section, Society of Economic Paleontologists and Mineralogists, p. 79–85.

Rangin, C., 1978, Speculative model of Mesozoic geodynamics, central Baja California to northeastern Sonora (México), *in* McDougall, K.A., ed., Mesozoic paleogeography of the Western United States: Los Angeles, Pacific Section, Society of Economic Paleontologists and Mineralogists, p. 85–106.

Rockwell, T.K., Schug, D.L., and Hatch, M.E., 1993, Late Quaternary slip rates along the Agua Blanca fault, Baja California, México, *in* Rendina, M.A., ed., Geologic investigations in Baja California, México: Santa Ana, California, South Coast Geological Society, Annual Field Trip Guidebook No. 21, p. 53–91.

Roldán-Quintana, J., 1991, Geology and chemical composition of the Jaralito and Aconchi batholiths in east-central Sonora, México, *in* Jacques-Ayala, C., ed., Studies of Sonoran geology: Boulder, Colorado, Geological Society of America Special Paper 254, p. 69–80.

Romo, J.M., 2002, Conductividad eléctrica de la litósfera de Baja California en la región de Vizcaíno, B.C.S. [Ph.D. thesis]: Ensenada, Baja California, México, CICESE, 151 p.

Romo, J.M., Gómez-Treviño, E., Perez-Flores, M.A., García-Abdeslem, J., Esparza, F.M., and Flores-Luna, C.F., 2001, Electrical structure of the Baja California lithosphere beneath Vizcaíno region: Eos (Transactions, American Geophysical Union), v. 82, no. 47, p. F321.

Ross, D.C., 1977, Pre-intrusive metasedimentary rocks of the Salinian block, California—A paleotectonic dilemma, *in* Fritsche, A.E., ed., Paleozoic Paleogeography of the western United States, Pacific Coast Paleogeography Symposium 1: Los Angeles, California, Pacific Coast Section, Society of Economic Paleontologists and Mineralogists, p. 371–380.

Salinas-Prieto, J.C., Monod, O., and Faure, M., 2000, Ductile deformations of opposite vergence in the eastern part of the Guerrero terrane (SW Mexico): Journal of South American Earth Sciences, v. 13, p. 389–402.

Santillán, M., and Barrera, T., 1930, Las posibilidades petroliferas en la costa occidental de la Baja California, entre los paralelos 30 y 32 de latitud norte: Anales del Instituto de Geología, Universidad Nacional Autónoma de México, v. 5, p. 1–37.

Saul, L.R., 1986, Pacific west coast Cretaceous molluscan faunas: Time and aspect of changes, *in* Abbott, P.L., ed., Cretaceous stratigraphy western North America: Los Angeles, Pacific Section, Society of Economic Paleontologists and Mineralogists, p. 131–136.

Saunders, A.D., Rogers, G., Marriner, G.F., Terrell, D.J., and Verma, S.P., 1987, Geochemistry of Cenozoic volcanic rocks, Baja California, México: Implications for the petrogenesis of post-subduction magmas: Journal of Volcanology and Geothermal Research, v. 32, p. 223–245.

Sawlan, M.G., 1991, Magmatic evolution of the Gulf of California rift, *in* Simoneit, B.R.T., ed., The gulf and peninsular province of the Californias, American Association of Petroleum Geologists Memoir 47, p. 301–369.

Sawlan, M.G., and Smith, J.G., 1984, Petrologic characteristics, age and tectonic setting of Neogene volcanic rocks in northern Baja California Sur, México, *in* Frizzell, V.A., Jr., ed., Geology of the Baja California peninsula: Los Angeles, California, Pacific Section, Society of Economic Paleontologists and Mineralogists, Book 39, p. 237–251.

Schaaf, P., Morán-Zenteno, D., Hernández-Bernal, M.S., Solís-Pichardo, G., Tolson, G., and Kohler, H., 1995, Paleogene continental margin truncation in southwestern Mexico: Geochronological evidence: Tectonics, v. 14, p. 1339–1350.

Schaaf, P., Böhnel, H., and Pérez-Venzor, J.A., 2000, Pre-Miocene paleogeography of the Los Cabos Block, Baja California Sur: Geochronological and palaeomagnetic implications: Tectonophysics, v. 318, p. 53–69.

Schmidt, K.L., and Paterson, S.R., 2002, A doubly vergent fan structure in the Peninsular Ranges batholith: Transpression or local complex flow around a continental margin buttress?: Tectonics, v. 21, no. 5, DOI 10.1029/2001TC001353.

Schmidt, K.L., Wetmore, P.H., Johnson, S.E., and Paterson, S.R., 2002, Controls on orogenesis along an ocean-continent margin transition in the Jura-Cretaceous Peninsular Ranges batholith, *in* Barth, A., ed., Contributions to crustal evolution of the southwestern United States: Boulder, Colorado, Geological Society of America, p. 49–71.

Schwennicke, T., González-Barba, G., and De Anda Franco, N., 1996, Lower Miocene marine and fluvial beds at Rancho La Palma, Baja California Sur, Mexico: Boletín del Departamento de Geología, Uni-Son, v. 13, no. 1, p. 1–14.

Sedlock, R.L., 1988a, Metamorphic petrology of a high-pressure, low-temperature subduction complex in west-central Baja California, México: Journal of Metamorphic Geology, v. 5, p. 205–233.

Sedlock, R.L., 1988b, Tectonic setting of blueschist and island-arc terranes of west-central Baja California, México: Geology, v. 16, p. 623–626.

Sedlock, R.L., 1993, Mesozoic geology and tectonics of blueschist and associated oceanic terranes in the Cedros-Vizcaíno-San Benito and Magdalena-Santa Margarita regions, Baja California, México, *in* McDougall, K., ed., Mesozoic paleogeography of the western United States-II: Los Angeles, California, Pacific Section, Society of Economic Paleontologists and Mineralogists, Book 71, p. 113–125.

Sedlock, R.L., 1996, Syn-subduction forearc extension and blueschist exhumation in Baja California, México, *in* Platt, J.P., ed., Dynamics of subduction: American Geophysical Union Monograph 96, p. 155–162.

Sedlock, R.L., 1997, Field-based investigations of Mesozoic rocks near El Arco, central Baja California, México, *in* IV International meeting on geology of Baja California peninsula, Ensenada, B.C., Mexico, Peninsular Geological Society.

Sedlock, R.L., 1999, Evaluation of exhumation mechanisms for coherent blueschists in western Baja California, México, *in* Willett, S.D., ed., Exhumation processes: Normal faulting, ductile flow, and erosion: London, Geological Society [London], p. 29–54.

Sedlock, R.L., and Isozaki, Y., 1990, Lithology and biostratigraphy of Franciscan-like chert and associated rocks, west-central Baja California, México: Geological Society of America Bulletin, v. 102, p. 852–864.

Sedlock, R.L., Ortega-Gutiérrez, F., and Speed, R.C., 1993, Tectonostratigraphic terranes and tectonic evolution of México: Boulder, Colorado, Geological Society of America Special Paper 278, p. 153 p.

Severinghaus, J., and Atwater, T.M., 1990, Cenozoic geometry and thermal condition of the subducting slabs beneath western North America: Boulder, Colorado, Geological Society of America, Memoir 176, p. 1–22.

Siem, M.E., and Gastil, R.G., 1994, Mid-Tertiary to Holocene extension associated with the development of the Sierra El Mayor metamorphic core complex, northeastern Baja California, México, *in* Ross, T.M., ed., Geological investigations of an active margin: Geological Society of America, Cordilleran Section Guidebook, p. 107–119.

Silberling, N.J., Jones, D.L., Monger, J.W.H., and Coney, P.J., 1992, Lithotectonic terrane map of the North American Cordillera, Map I-2176: U.S. Geological Survey, Miscellaneous Investigation Series, scale 1:5,000,000, 2 sheets.

Silver, L.T., and Chappell, B.W., 1988, The Peninsular Ranges batholith: An insight into the evolution of the Cordilleran batholiths of southwestern North America: Transactions of the Royal Society of Edinburgh: Earth Sciences, v. 79, p. 105–121.

Silver, L.T., and Nourse, J.A., 1986, The Rand Mountains "thrust" complex in comparison with the Vincent thrust-Pelona schist relationship, southern California: Geological Society of America Abstracts with Programs, v. 18, p. 185.

Silver, L.T., Taylor, H.P., Jr., and Chappell, B., 1979, Some petrological, geochemical, and geochronological observations of the Peninsular Ranges batholith near the international border of the U.S.A. and México, *in* Todd, V.R., ed., Mesozoic crystalline rocks: San Diego, California, Department of Geological Sciences, San Diego State University, p. 83–110.

Smith, J.T., 1991, Cenozoic marine mollusks and paleogeography of the Gulf of California, *in* Simoneit, B.R.T., ed., The gulf and peninsular province of the Californias: American Association of Petroleum Geologists Memoir 47, p. 637–666.

Smith, J.T., 1992, The Salada Formation of Baja California Sur, México [abs]: La Paz, Baja California Sur, Peninsular Geological Society, First International Meeting on Geology of the Baja California Peninsula Memoir, p. 23–32.

Smith, D.P., and Busby, C.J., 1993a, Mid-Cretaceous crustal extension recorded in deep-marine half-graben fill, Cedros Island, Mexico: Geological Society of America Bulletin, v. 105, p. 547–562.

Smith, D.P., and Busby, C.J., 1993b, Shallow magnetic inclinations in the Cretaceous Valle Group, Baja California: Remagnetization, compaction, or terrane translation?: Tectonics, v. 12, p. 1258–1266.

Spencer, J.E., and Normark, W.R., 1979, Tosco-Abreojos fault zone: A Neogene transform plate boundary within the Pacific margin of southern Baja California, México: Geology, v. 7, p. 554–557.

Stewart, J.H., 1990, Position of Paleozoic continental margin in northwestern México: Present knowledge and speculations: Geological Society of America Abstracts with Programs, v. 22, p. 86–87.

Stock, J.M., 2000, Relation of the Puertecitos Volcanic Province, Baja California, Mexico, to development of the plate boundary in the Gulf of California, *in* Stock, J.M., ed., Cenozoic tectonics and volcanism of Mexico: Boulder, Colorado, Geological Society of America Special Paper 334, p. 143–156.

Stock, J.M., and Hodges, K.V., 1989, Pre-Pliocene extension around the Gulf of California, and the transfer of Baja California to the Pacific plate: Tectonics, v. 8, p. 99–116.

Stock, J.M., and Hodges, K.V., 1990, Miocene to Recent structural development of an extensional accommodation zone, northeastern Baja California, México: Journal of Structural Geology, v. 12, p. 315–328.

Stock, J.M., Lewis, C.J., and Nagy, E.A., 1999, The Tuff of San Felipe: An extensive middle Miocene pyroclastic flow deposit in Baja California, Mexico: Journal of Volcanology and Geothermal Research, v. 93, p. 53–74.

Tan, X., and Kodama, K.P., 1998, Compaction-corrected inclinations from southern California Cretaceous marine sedimentary rocks indicate no paleolatitudinal offset for the Peninsular Ranges terrane: Journal of Geophysical Research, v. 103, p. 27,168–27,192.

Tanaka, H., Smith, T.E., and Huang, C.H., 1984, The Santiago Peak volcanic rocks of the Peninsular Ranges batholith, southern California: Volcanic rocks associated with coeval gabbros: Bulletin Volcanologique, v. 47, p. 153–171.

Tarduno, J.A., 1990, Absolute inclination values from deep-sea sediments: A reexamination of the Cretaceous Pacific record: Geophysical Research Letters, v. 17, p. 101–104.

Tardy, M., Lapierre, H., Freydier, C., Coulon, C., Gill, J.-B., Mercier de Lépinay, B., Beck, C., Martínez-R, J., Talavera-Mendoza, O., Ortiz, H.E., Stein, G., Bourdier, J.-L., and Yta, M., 1994, The Guerrero suspect terrane (western Mexico) and coeval arc terranes (the Greater Antilles, and the Western Cordillera of Columbia): A late Mesozoic intra-oceanic arc accreted to cratonal America during the Cretaceous: Tectonophysics, v. 230, p. 49–74.

Tate, M.C., and Johnson, S.E., 2000, Subvolcanic and deep-crustal tonalite genesis beneath the Mexican Peninsular Ranges: Journal of Geology, v. 108, p. 721–728.

Tate, M.C., Norman, M.D., Johnson, S.E., Fanning, C.M., and Anderson, J.L., 1999, Generation of tonalite and trondhjemite by subvolcanic fractionation and partial melting in the Zarza Intrusive Complex, western Peninsular Ranges batholith, northwestern Mexico: Journal of Petrology, v. 40, p. 983–1010.

Teissere, R.F., and Beck, M.E., Jr., 1973, Divergent Cretaceous paleomagnetic pole position for the Southern California batholith, U.S.A.: Earth and Planetary Science Letters, v. 18, p. 296–300.

Thomson, C.N., and Girty, G.H., 1994, Early Cretaceous intra-arc ductile strain in Triassic-Jurassic and Cretaceous continental margin arc rocks, Peninsular Ranges, California: Tectonics, v. 13, p. 1108–1119.

Todd, V.R., Erskine, B.G., and Morton, D.M., 1988, Metamorphic and tectonic evolution of the northern Peninsular Ranges batholith, southern California, *in* Ernst, W.G., ed., Metamorphism and crustal evolution of the western United States (Rubey Volume VII): Englewood Cliffs, N.J., Prentice Hall, p. 894–937.

Todd, V.R., Shaw, S.E., Girty, G.H., and Jachens, R.C., 1991, A probable Jurassic plutonic arc of continental affinity in the Peninsular Ranges batholith, southern California: Tectonic implications: GSA Abstracts with Programs, v. 23, p. 104.

Tosdal, R.M., Haxel, G.B., and Wright, J.E., 1990, Jurassic geology of the Sonoran Desert region, southern Arizona, southeastern California, and northernmost Sonora, *in* Reynolds, S.J., ed., Geologic evolution of Arizona: Arizona Geological Society, Digest 17, p. 397–434.

Umhoefer, P.J., 2000, Where are the missing faults in translated terranes?: Tectonophysics, v. 326, p. 23–35.

Umhoefer, P.J., and Dorsey, R.J., 1997, Translation of terranes: Lessons from central Baja California, México: Geology, v. 25, p. 1007–1010.

Umhoefer, P.J., Stock, J.M., and Martín-Barajas, A., 1996, Tectonic evolution of the Gulf of California and its margins (Penrose Conference Report): GSA Today, v. 6, no. 8, p. 16–17.

Umhoefer, P.J., Dorsey, R.J., Willsey, S., Mayer, L., and Renne, P., 2001, Stratigraphy and geochronology of the Comondú Group near Loreto, Baja California Sur, Mexico: Sedimentary Geology, v. 144, p. 125–147.

Umhoefer, P.J., Mayer, L., and Dorsey, R.J., 2002, Evolution of the margin of the Gulf of California near Loreto, Baja California peninsula, Mexico:, v. 114, p. 849–868.

U.S. Geological Survey—Middle America Seismograph, 1998, Caribbean seismicity 1900–1994: U.S. Geological Survey—Middle America Seismograph (MIDAS Consortium), scale 1:6,500,000, 1 sheet.

Valencia-Moreno, M., Ruiz, J., Barton, M.D., Patchett, P.J., Zürcher, L., Hodkinson, D.G., and Roldán-Quintana, J., 2001, A chemical and isotopic study of the Laramide granitic belt of northwestern Mexico: Identification of the southern edge of the North American Precambrian basement: Geological Society of America Bulletin, v. 113, p. 1409–1422.

Walawender, M.J., Gastil, R.G., Clinkenbeard, J.P., McCormick, W.V., Eastman, B.G., Wernicke, R.S., Wardlaw, M.S., Gunn, S.H., and Smith, B.M., 1990, Origin and evolution of La Posta-type plutons, eastern Peninsular Ranges batholith, southern and Baja California, *in* Anderson, J.L., ed., The nature and origin of Cordilleran magmatism: Boulder, Colorado, Geological Society of America, p. 1–18.

Walawender, M.J., Girty, G.H., Lombardi, M.R., Kimbrough, D., Girty, M.S., and Anderson, C., 1991, A synthesis of recent work in the Peninsular Ranges batholith, *in* Hanan, B.B., ed., Geological excursions in southern California and México: Geological Society of America, 1991 Annual Meeting, Guidebook, p. 297–318.

Walsh, S.L., and Demere, T.A., 1991, Age and stratigraphy of the Sweetwater and Otay formations, San Diego County, California, *in* May, J.A., ed., Eocene geologic history, San Diego region: Los Angeles, Pacific Section, Society for Sedimentary Geology (SEPM), p. 131–148.

Wetmore, P.H., Schmidt, K.L., Paterson, S.R., and Herzig, C., 2002, Tectonic implications for the along-strike variation of the Peninsular Ranges batholith, southern and Baja California: Geology, v. 30, p. 247–250.

Whidden, K.J., Lund, S.P., Bottjer, D.J., Champion, D., and Howell, D.G., 1998, Paleomagnetic evidence that the central block of Salinia (California) is not a far-traveled terrane: Tectonics, v. 17, p. 329–343.

Williams, W.J.W., 1999, Evolution of Quaternary intraplate mafic lavas detailed using ^{3}He surface exposure and ^{40}Ar/^{39}Ar dating, and elemental and He, Sr, Nd, and Pb isotopic signatures: Potrillo Volcanic Field, New Mexico, U.S.A., and San Quintin Volcanic Field, Baja California Norte, Mexico [Ph.D. thesis]: El Paso, Texas, University of Texas, 193 p.

Yule, J.D., and Herzig, C.T., 1994, Paleomagnetic evidence for no large-scale northward translation of the southern California Peninsular Ranges: GSA Abstracts with Programs, v. 26, no. 7, p. A461.

Zanchi, A., 1994, The opening of the Gulf of California near Loreto, Baja California, México from basin and range extension to transtensional tectonics: Journal of Structural Geology, v. 16, p. 1619–1639.

MANUSCRIPT ACCEPTED BY THE SOCIETY JUNE 2, 2003

Geological Society of America
Special paper 374
2003

Ophiolite and volcanic arc assemblages on the Vizcaíno Peninsula and Cedros Island, Baja California Sur, México: Mesozoic forearc lithosphere of the Cordilleran magmatic arc

David L. Kimbrough

Department of Geological Sciences, San Diego State University, San Diego, California 92182-1020, USA

Thomas E. Moore

U.S. Geological Survey, M.S. 901, 345 Middlefield Road, Menlo Park, California 94025, USA

ABSTRACT

Mesozoic ophiolites in the Vizcaíno Peninsula and Cedros Island region of Baja California Sur are suprasubduction zone Cordilleran-type ophiolites structurally juxtaposed with underlying high pressure-temperature subduction complex assemblages. The region is divided into three separate tectonostratigraphic terranes, but here we recognize stratigraphic, intrusive, and petrologic links between these terranes and interpret the evolution of the entire region within the same Late Triassic to Early Cretaceous tectonic framework. Several phases of extension are recognized, including two major phases that resulted in development of distinct ophiolite assemblages. The Late Triassic Vizcaíno Peninsula Ophiolite (221 ± 2 Ma) represents the earliest stage of this history and comprises a complete spreading center sequence with depleted upper mantle and mafic crustal rocks, including sheeted dike complex. Jurassic arc magmatic rocks with low-Ti arc tholeiite and boninite geochemical affinities were intruded through and constructed on the Triassic ophiolite basement. Ultra-depleted arc-ankaramites on Cedros Island may represent an initial phase of arc rifting that was followed by major Middle Jurassic extension and production of the Cedros Island Ophiolite (173 ± 2 Ma). The Late Jurassic–Early Cretaceous Coloradito and Eugenia Formations contain mudflows and olistostrome blocks intercalated with arc volcanogenic sediment and rift-related pillow lavas; these units record extension and/or transtension and provide the earliest definite evidence of arc-continent interaction in the region. Middle Jurassic to Early Cretaceous arc plutonic rocks (ca. 165–135 Ma) were shallowly intruded into low greenschist-facies ophiolite and arc volcanic basement. Plutonic rocks range in composition from gabbro to granodiorite, but tonalite dominates. These intrusions are typical I-type Cordilleran batholithic rocks with relatively primitive arc geochemical affinities (initial $^{87}Sr/^{86}Sr$ range from ~0.704 to 0.706), but they are distinctly calcic in nature, a feature common to the adjacent Cretaceous Peninsular Ranges batholith.

The Vizcaíno-Cedros region correlates to ophiolitic terranes of the western Sierra-Klamath belt and Coast Ranges of California and Oregon that were constructed in part across the North American margin. Age, stratigraphic, and petrochemical data from the Vizcaíno-Cedros region support previously proposed forearc rifting models developed for the U.S. sector of the Cordilleran orogen that interpret the ophiolite assemblages as autochthonous or parautochthonous forearc lithosphere constructed outboard of the Mesozoic continental margin arc.

Keywords: ophiolite, volcanic arc, Baja California, Mesozoic, stratigraphy, petrology, tectonics.

Kimbrough, D.L., and Moore, T.E., 2003, Ophiolite and volcanic arc assemblages on the Vizcaíno Peninsula and Cedros Island, Baja California Sur, México: Mesozoic forearc lithosphere of the Cordilleran magmatic arc, *in* Johnson, S.E., Paterson, S.R., Fletcher, J.M., Girty, G.H., Kimbrough, D.L., and Martín-Barajas, A., eds., Tectonic evolution of northwestern México and the southwestern USA: Boulder, Colorado, Geological Society of America Special Paper 374, p. 43–71.

INTRODUCTION

Understanding the Mesozoic tectonic evolution of the Cordillera of western North American requires correct interpretation of ophiolite and volcanic arc terranes along the outboard edge of the continental margin (Saleeby, 1992). These rocks comprise a broad accretionary belt that stretches along the western North American Cordillera from Alaska to México. Following their initial recognition along the California margin as on-land fragments of oceanic lithosphere (Bailey et al., 1970), ophiolites, in particular, have played a central role in reconstruction of ancient plate boundaries and as a result have been the focus of wide-ranging and in-depth multidisciplinary studies aimed at understanding their tectonic significance (e.g., Dilek et al., 2000).

Despite this concerted research effort, ophiolites continue to pose first-order problems. This is well illustrated by the controversy over the origin of the Middle Jurassic Coast Range Ophiolite of California, for which three distinct models lead to remarkably different interpretations of regional tectonic relations (Dickinson et al., 1996). Though most investigators now favor a suprasubduction zone origin for the Coast Range Ophiolite, its site of origin has persisted as a central unresolved issue. One model interprets the ophiolite as an allochthonous, east-facing intraoceanic island arc that collided and accreted with the Sierran continental margin arc as part of the Nevadan orogeny (e.g., Schweickert and Cowen, 1975; Moores and Day, 1984; Dickinson et al., 1996; Ingersoll, 2000), while a second model interprets the ophiolite as autochthonous or parautochthonous forearc lithosphere of the Cordilleran Mesozoic continental arc (Saleeby, 1981, 1992; Stern and Bloomer, 1992).

This paper presents an updated view of stratigraphic relationships and the tectonic evolution of ophiolite and volcanic arc assemblages in the Vizcaíno Peninsula and Cedros Island region of west-central Baja California, ~600 km south of the U.S.-México border (Fig. 1). Here, Mesozoic ophiolite, volcanic arc, and subduction complex assemblages are exposed in a ~250 × 40 km northwest-trending belt within a localized region of uplift in the mostly submerged borderland of Baja California (Kilmer, 1977; Gastil et al., 1978; Rangin, 1978; Barnes, 1984; Kimbrough, 1985; Moore, 1985, 1986; Sedlock et al., 1993; Busby et al., 1998). Similar but less well-known units are present ~300 km south of the Vizcaíno Peninsula on Magdalena and Santa Margarita Islands (Rangin, 1978; Blake et al., 1984).

South of the Coast Range Ophiolite of California, the Vizcaíno-Cedros region contains by far the best exposure of paleoceanic assemblages within a ~2000-km-long segment at the southern end of the southwestern Cordillera orogen. Preservation of thick, intact, well-exposed, and low metamorphic-grade stratigraphic sequences overlying ophiolite and volcanic-arc basement provides an exceptional record of Mesozoic stratigraphy and tectonics (Fig. 2). On the Vizcaíno Peninsula, Late Triassic ophiolite basement and overlying strata comprise a superb Cordilleran ophiolite exposure (Moore, 1983, 1985). Paleontologic analysis of overlying ribbon chert of the San Hipolito Formation provided the first comprehensive analysis of a Triassic radiolarian fauna from the Western Hemisphere (Pessagno et al., 1979). On Cedros Island, Middle Jurassic rifted-arc and -ophiolite basement assemblages with overlying volcaniclastic rocks comprise an extraordinarily clear record of supra-subduction zone ophiolite generation (Kimbrough, 1984, 1985) and thus provide insight into processes operating in modern (mostly submerged) volcanic arc settings (e.g., Busby-Spera, 1988a; Critelli et al., 2002). In both areas, these assemblages structurally overlie blueschist-bearing subduction assemblages that are correlative with the Franciscan Complex of California. The similarity of Vizcaíno-Cedros basement to California–Oregon Mesozoic assemblages in the Klamath Mountains, western Sierra Nevada foothills, and California Coast Range has long been recognized (e.g., Suppe and Armstrong, 1972; Jones et al., 1976; Rangin, 1978; Kimbrough, 1985; 1989), but details of correlations and tectonic implications have remained uncertain.

This paper provides a reevaluation of ophiolite and volcanic arc basement terranes in the Vizcaíno-Cedros region (Kimbrough, 1985; Moore, 1985) and links the history of the terranes together. Zircon U-Pb and hornblende ^{40}Ar/^{39}Ar data are presented that provide critical geochronological constraints. A new extensional basin model for Late Jurassic–Early Cretaceous tectonics and sedimentation is presented. We also provide field and petrologic characterization of Jurassic and Cretaceous arc plutonic rocks in the region. Kimbrough (1985) noted the similarity of Vizcaíno-Cedros assemblages to the western Sierra-Klamath belt and Coast Range Ophiolite; here these similarities and regional implications are examined in more detail.

The new data and analysis support forearc-rifting models based on analogs from western Pacific arcs (Saleeby, 1981, 1992; Stern and Bloomer, 1992) to explain the origin of the Vizcaíno-Cedros ophiolite assemblages. We conclude that these rocks likely formed in the forearc region of the Mesozoic Cordilleran continental arc.

TERRANE ANALYSIS

Regional mapping by Kilmer (1977, 1979, 1984), Robinson (1975, 1979), Moore (1977, 1983, 1985), and Rangin (1978), supported by stratigraphic and topical studies of many other investigators (e.g., references in Frizzel, 1984, and Sedlock et al., 1993), provide the basis for terrane analysis of Mesozoic rock on the Vizcaíno Peninsula and Cedros Island. Moore (1985) and Kimbrough (1985) applied terrane concepts as outlined by Coney et al. (1980) and Jones et al. (1983) and recognized three discrete tectonostratigraphic terranes in "upper plate" Mesozoic ophiolite and volcanoplutonic arc rocks of the region; the Choyal, Vizcaíno Norte, and Vizcaíno Sur terranes. The Choyal terrane encompasses Cedros Island; the Vizcaíno Norte and Vizcaíno Sur terranes are separated by the Sierra Placer mélange (Fig. 1). These terranes are depositional basement for Cretaceous forearc basin turbidites of the Valle Group (e.g., Patterson, 1984; Busby-Spera and Boles, 1986; Smith and Busby, 1993; Kimbrough et al., 2001). This terrane subdivision was adopted by subsequent investigators (Sedlock, 1988b, Baldwin and Harrison, 1992) and is incorporated into the Cochimi composite terrane of Sedlock et al. (1993).

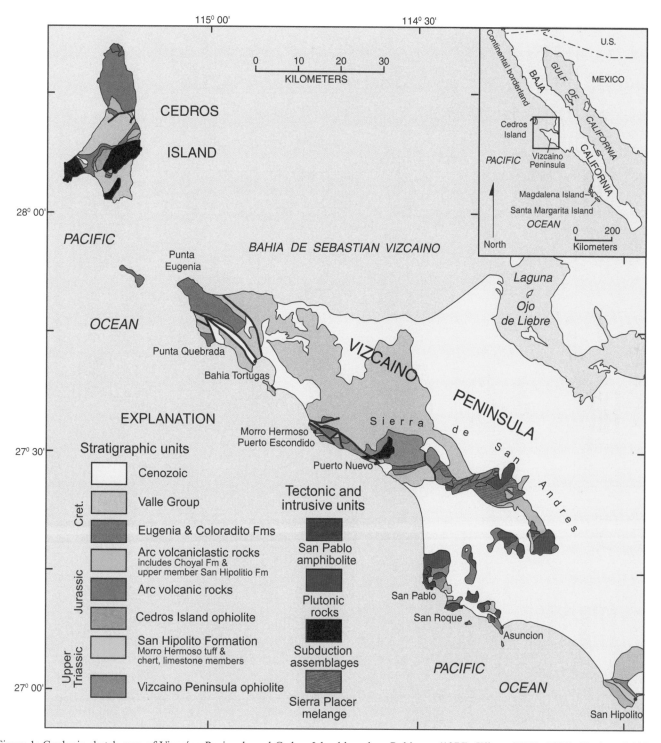

Figure 1: Geologic sketch map of Vizcaíno Peninsula and Cedros Island based on Robinson (1975), Kilmer (1977, 1984), Finch and Abbott (1977), and Moore (1983). Fm—formation.

The application of terrane analysis has been useful in identifying the major stratigraphic and structural boundaries in the Cedros-Vizcaíno region and has successively elucidated the geologic histories of its constituent stratigraphic domains. Terrane subdivision is a conservative approach that allows for the possibility that subduction-accretion processes assembled crustal elements that originated in widely disparate tectonic settings. In the Cedros-Vizcaíno region, however, there is no compelling evidence that any of the terranes are far-traveled with respect to one another and no evidence of accretionary structures or

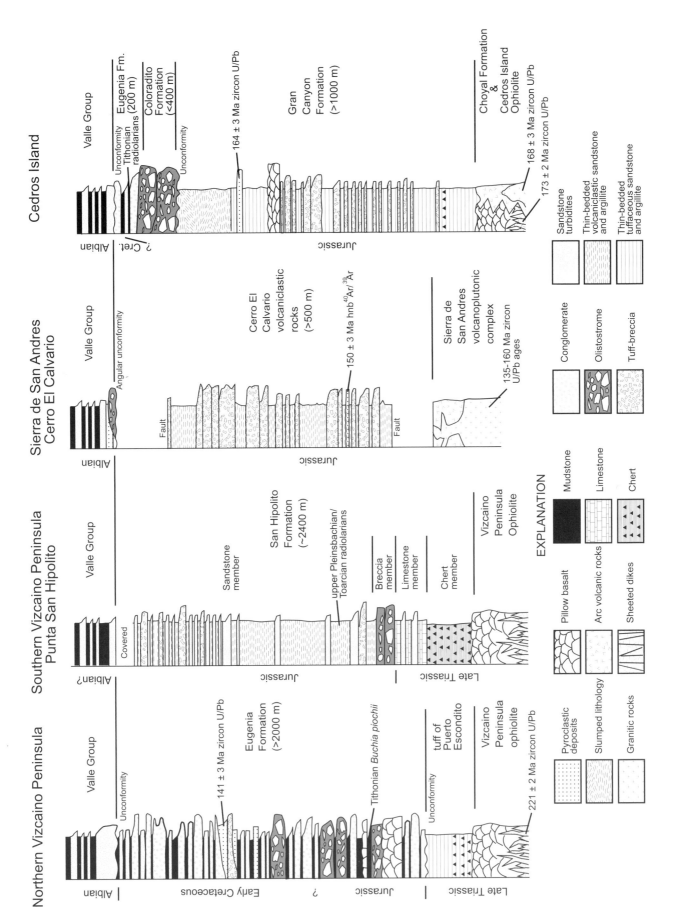

Figure 2: Generalized stratigraphic columns from four principal geologic domains in the Vizcaíno Peninsula and Cedros Island region.

metamorphism. Instead, the terranes occupy a structural position above subduction assemblages where they compose the substrate for mid-Cretaceous forearc basin deposits. Although the terranes contain stratigraphic discontinuities, they all appear to have originated in similar ensimatic arc environments that can be interpreted as a single coherent basement framework to the Valle Group forearc basin.

In this paper we postulate geographic continuity of these terranes in the same paleogeographic setting beginning in Late Triassic time. In this view, stratigraphic variations between terranes are consistent with those observed along and across strike in facies sequences associated with oceanic volcanic arcs (e.g., Fisher and Smith, 1991; Busby and Ingersoll, 1995; Taylor, 1995). We believe that this interpretation of the geology provides a more robust picture of the geologic evolution of the region.

GEOLOGIC FRAMEWORK

Vizcaíno Peninsula Ophiolite—Late Triassic ocean crust and overlying sedimentary rocks

Remnants of Upper Triassic ophiolite are exposed discontinuously in a northwest-trending belt from Punta Quebrada to Punta San Hipolito along the southwestern edge of the Vizcaíno Peninsula (Moore, 1977, 1983; Pessagno et al., 1979; Barnes, 1984; Fig. 1). The Late Triassic sedimentary cover of this ophiolite belt is assigned to the San Hipolito Formation. Everywhere they appear these strata are siliceous and pelagic in aspect, but they display differences in thickness and in content of volcanogenic detritus.

A ~60-km-long, east-southeast–trending mountainous region referred to here as Sierra de San Andres is the major topographic feature of the Vizcaíno Peninsula. Upper Triassic ophiolite in the west half of the range (the Sierra de San Andres ophiolite) is separated from a pre–Upper Jurassic volcanic arc assemblage in the eastern half of the range by the steeply dipping Sierra Placer serpentinite-matrix mélange (Moore, 1983, 1985; Fig. 2). Upper Triassic ophiolite assemblages further south and east of the Sierra San Andres exposed along the coast from Punta San Pablo to Punta San Hipolito were referred to as the La Costa ophiolite (Moore, 1985). Here, we introduce the name "Vizcaíno Peninsula Ophiolite" to collectively refer to all of the Late Triassic ophiolite remnants on the Vizcaíno Peninsula. Principal exposures, referred to individually by geographic location, are briefly reviewed below beginning at the northwestern end of the belt.

Punta Quebrada

Coastal exposures between Punta Quebrada and Punta Rompiente reveal a well-preserved section of pillow lava (~100 m thick) that contains interpillow limestone with Carnian to upper-middle Norian radiolarian assemblages (Barnes, 1984). Tectonized metagabbroic rocks and minor serpentinite containing diabase dikes lie in apparent stratigraphic continuity beneath the pillow lava. A thin sequence of discontinuous pink dolomite and recrystallized red chert conformably overlies the pillow lava.

Sierra de San Andres

The Sierra de San Andres ophiolite remnant is the most intact exposure of the Vizcaíno Peninsula Ophiolite (Fig. 1). The ophiolite here comprises a large structural antiform that plunges to the west beneath the Eugenia and Valle Formations. It is underlain by a blueschist-bearing serpentinite-matrix mélange (the Puerto Nuevo mélange) exposed in a structural window in the core of the antiform (Moore, 1986). From structural base to top, the ophiolite section includes serpentinized harzburgite-dunite tectonite, cumulate clinopyroxene + plagioclase ± olivine ± orthopyroxene gabbro, noncumulate isotropic high-level clinopyroxene gabbro, dike and/or sill complex, and extrusive pillow lava and breccia. The sequence has an estimated minimum thickness of 3000 m but is internally disrupted by numerous low-angle faults that cut out section, and more locally by late Cenozoic northwest-trending strike-slip faults. Everywhere the low-angle faults display the characteristics of normal faults, indicating that substantial extensional tectonism occurred following formation of the ophiolite.

Plagiogranite and albitite are present as small, isolated stocks and anastomosing dikes in noncumulate gabbro of the ophiolite. The plagiogranite is medium-grained with sparse green-brown hornblende and contains amphibolitized xenoliths of the noncumulate gabbro host. Both the plagiogranite and gabbro host contain miarolitic cavities and have undergone static greenschist-facies recrystallization. Barnes and Mattinson (1981) reported that a small mass of albitite collected from high-level noncumulus gabbro and dike complex yielded a sphene U-Pb age of 220 ± 2 Ma.

Near Morro Hermoso, the ophiolite remnant is depositionally overlain by 200–300 m of thin-bedded vitric tuff with minor interstratified tuffaceous chert, fine volcaniclastic sandstone, and minor pink limestone. Interpillow carbonate and a thin black chert in tuff ~15 m above the pillow lava both contain Carnian(?) to upper-middle Norian radiolarian assemblages. Upper Triassic *Monotis* sp. and conodonts also are present in this section. Barnes (1982, 1984) correlated this section with the Upper Triassic and Jurassic San Hipolito Formation (see below).

Punta San Pablo–Asuncion Headlands

A great mass of pillow lava cut by dikes and irregular bodies of microgabbro crops out at Punta Asuncion. Westward and inland along the coast from Punta Asuncion these rocks are underlain by sheeted dike complex and noncumulus gabbro. Similar rocks are present on Isla San Roque and Isla Asuncion just offshore. A strongly deformed assemblage of sheared ultramafic and cumulate gabbroic rocks is exposed in the Punta San Pablo headlands. These rocks are interpreted to comprise a southeast-striking ophiolite sequence that obliquely intersects the coast and has cumulative thickness estimated at 5000 m (Moore, 1983, 1985).

Punta San Hipolito

The top of the San Pablo–Asuncion ophiolite section is interpreted to be exposed 30 km to the south of Asuncion at Punta San Hipolito (Moore, 1983), where pillow basalt is depositionally overlain by Late Triassic strata of the San Hipolito Formation (Finch and Abbott, 1977; Pessagno et al., 1979). The San Hipolito Formation comprises a 2400-m-thick, homoclinally dipping section of Upper Triassic and Jurassic marine sedimentary rocks deposited on pillow lava (Fig. 2).

The basal portion of the San Hipolito section is a 450-m-thick sequence of tuffaceous chert, limestone, and interbedded tuff. Exceptionally rich, well-preserved radiolarian assemblages in the chert member as well as interpillow areas in underlying basalt indicate a Late Triassic age (late Carnian to late Norian). The upper limit of the *Capnodoce* radiolarian zone, widespread in the western Cordillera, is located within the chert member. Above this, reef-associated limestone and volcanic blocks in the Breccia member suggest debris-flows were shed from a nearby shallow volcanic region in a warm, tropical setting.

Jurassic–Cretaceous Volcanoplutonic Arc Magmatism; the San Andres–Cedros Complex

Arc volcanic and volcaniclastic rocks intruded by granitoids and lesser gabbroic/dioritic plutonic rocks crop out widely in the northern part of Cedros Island (Choyal Formation of Kilmer, 1979) and in the central to southeastern part of the Sierra de San Andres on the Vizcaíno Peninsula (Fig. 1). Jones et al. (1976) and Rangin (1978) correlated these exposures and collectively referred to them as the San Andres–Cedros volcanic-plutonic complex.

Arc Volcanic Rocks on Vizcaíno Peninsula

Arc volcanic rock in central to southeastern Sierra de San Andres consists mainly of andesite flows and breccias and mafic pillow lava interbedded locally with volcaniclastic sandstone, conglomerate, and breccia. Thick sequences of pre–Upper Jurassic volcaniclastic turbidites cut by andesite dikes and sills are in local depositional contact on this volcanic basement. These rocks are all metamorphosed to greenschist facies mineral assemblages and are structurally disrupted in the central part of the range by the steeply dipping Sierra Placer serpentinite matrix mélange. Moore (1984) described a particularly well-bedded and relatively intact sequence of volcaniclastic turbidites at Cerro El Calvario (Fig. 2) in the central part of the range and interpreted it as a proximal coarse-grained debris apron deposited along a steep volcanic slope. The Sierra de San Andres volcanic rocks have island-arc tholeiites and boninite geochemical affinities (Moore, 1983).

The upper part of the San Hipolito Formation at Punta San Hipolito is the ~1840-m-thick Sandstone Member, which consists of volcanogenic sandstone, siltstone, and tuffs with minor limestone and conglomerate (Finch and Abbott, 1977). Radiolarians recovered from thin limestone beds in the lower part of the sandstone member are upper Pliensbachian and/or Toarcian in age (Whalen and Pessagno, 1984), but the age of the upper part of the member is not constrained by fossil or isotopic data. Moore (1984) correlated the Sandstone Member of the San Hipolito Formation to the Cerro El Calvario sequence on the basis of compositional and sedimentological similarity.

Arc Volcanic Rocks on Cedros Island—Choyal Formation and Cedros Island Ophiolite

Volcanic arc rocks are widely exposed north of Arroyo El Choyal in the northern part of Cedros Island and were mapped as the Choyal Formation by Kilmer (1979). These rocks comprise a gently southward-dipping pile of basaltic-andesitic-dacitic submarine flows, volcaniclastic rocks and dikes with an estimated minimum stratigraphic thickness of 5 km. The deepest part of the pile at Punta Norte is intruded by several small granodiorite and tonalite bodies (Kilmer, 1979). A voluminous andesite dike and sill complex is well developed in the upper part of the pile. The Choyal is depositionally overlain by Middle Jurassic arc volcanogenic strata of the Gran Cañon Formation along the northern flank of the Pinos syncline (Kilmer, 1979; Kimbrough, 1984; Busby-Spera, 1988a; Fig. 2). The volcanic rocks are a compositionally diverse suite of subalkaline lavas with tholeiitic, boninitic, and calc-alkaline affinities. Primitive arc ankaramites with extreme subalkaline chrome diopside phenocryst compositions present in the Punta Norte area (Kimbrough, 1982) may record a pre–Middle Jurassic phase of arc rifting.

In the southern half of Cedros Island, rocks representing most levels of an idealized ophiolite section are present in fault-bounded blocks (Jones et al., 1976; Kimbrough, 1982). The ophiolite comprises three distinct members that are interpreted as different levels of an originally intact oceanic crust-mantle sequence. From base to top, these include: (1) dunite-harzburgite tectonite mixed with cumulate ultramafics and pyroxenite, (2) cumulate and noncumulate gabbroic rocks cut by mafic dikes, and (3) a volcanic unit of pillow lava, hyaloclastites, and massive flows cut by dikes. Tuffaceous radiolarian chert and crystal-lithic tuff of the Gran Cañon Formation rest depositionally on top of the volcanic section.

The Cedros Island Ophiolite is separated from the Choyal Formation by a northeast-southwest–trending, graben-like structure that obliquely intersects the east coast of the island between Gran Cañon and Arroyo Choyal (the Pinos syncline of Kilmer, 1979). Correlative sections of the Gran Cañon Formation are in depositional contact on these two basement assemblages on opposing flanks of the Pinos syncline. Based on stratigraphic, geochemical, and age relationships, the Cedros Island Ophiolite is interpreted to reflect major extension and rifting in an active inter- or backarc-basin setting comparable to rift basins in the modern western Pacific ocean (Kimbrough, 1982, 1984, 1985). The close spatial association and unequivocal stratigraphic linkage of the ophiolite to the rifted edge of older arc basement makes this an extraordinary example of a supra-subduction zone ophiolite. This model was developed in greater detail by Busby-Spera (1988a) and Critelli et al. (2002).

Kilmer (1979, 1984) defined the Gran Cañon Formation and collected fossils from a white tuff bed near the middle of the formation in Gran Cañon. This bed yielded the bivalve *Bositra buchi* (Roemer) as well as the ammonoids *Oppelia* sp., *Lytoceras* sp., and *Phylloceras* sp., and various plant fragments. *Bositra buchi* (Roemer) is considered Bajocian to Callovian (176–159 Ma) by Imlay (1968, personal commun.), and also has been found in the Bedford Canyon Formation in the Santa Ana Mountains, California, as well as Jurassic beds in Oregon and Alaska.

Coloradito and Eugenia Formations

These units consist of coarse-grained volcanogenic sediment-gravity flow deposits and represent a distinctive ca. Tithonian-Valanginian phase of proximal arc volcanism on Cedros Island and the Vizcaíno Peninsula that overlaps partly in age with intrusion of plutons in the southeastern Sierra de San Andres (Boles, 1978; Hickey, 1984; Barnes, 1984; Boles and Landis, 1984; Kimbrough et al., 1987). An outstanding feature is the presence of mudflows and continentally derived olistostrome blocks that record proximity to the continental margin (Fig. 2).

Arc Plutonic Complex on Cedros Island

Granitoid and dioritic intrusive rocks crop out in several small areas around the steep Punta Norte headlands on Cedros Island where they intrude volcanic and hypabyssal wallrock of the Choyal Formation volcanic arc assemblage (Fig. 1). The total surface outcrop area of these intrusions is ~4 km^2. Kilmer (1979) mapped two east-west–trending faults in this area that coincide with major canyons. The southern canyon contains economic gold and copper mineralization that has been mined intermittently since ~1890.

The Punta Norte intrusions span a broad compositional range from hornblende-pyroxene diorite to granodiorite or granite (Kimbrough, 1982). The largest volume at present erosion level is relatively homogeneous hornblende tonalite and pink hornblende-biotite granodiorite. Mafic intrusives are exposed in sea cliff exposures around Campo Punta Norte. Here, hornblende-clinopyroxene quartz diorite and quartz monzodiorite, as well as other less voluminous gabbroic and dioritic rocks, are intruded by the hornblende tonalite. The tonalite contains abundant angular and sub-angular equidimensional stope blocks (mostly 0.5–3 m across) that comprise 50% or more of the outcrop locally. Stope blocks of volcanic wallrock and dikes are also present in the tonalite.

The three main intrusions around the north tip of the island have near vertical and subparallel intrusive contacts with up to 500 m of relief, which imparts a sheeted geometry to the intrusions. Contacts with volcanic wallrock are sharp; stope blocks of wallrock are present in strongly foliated border zones with mafic schlieren smeared out parallel to contacts. Dike-like or irregular apophyses of granitoids locally splay outward into surrounding volcanic host rocks. These border zones grade rapidly inward to non-foliated and less mafic tonalite or granodiorite.

The texture of larger tonalite and granodiorite masses varies from hypidiomorphic granular to near 100% granophyric intergrowths of plagioclase, quartz, and lesser K-feldspar. The rock around the Punta Norte lighthouse is a spectacular granophyric-textured leucogranodiorite with epidote-filled miarolytic cavities. Miarolytic cavities in the granophyric-textured rocks, along with the low/sub-greenschist metamorphic grade of surrounding Choyal Formation wallrock, indicate a shallow depth of emplacement.

Punta Norte intrusives all show evidence of intense alteration by late hydrothermal fluids. Plagioclase is albitized, biotite completely replaced by chlorite, and secondary epidote and sphene is common. Pyroxene in dioritic rock is partially replaced by green hornblende and/or fibrous amphibole. The general abundance of late magmatic K-feldspar, even in relatively mafic pyroxene-bearing dioritic rocks, reflects the generally potassic nature of this suite. Anomalously K-rich mid-Mesozoic intrusive suites have also been documented from Jurassic volcanic arc belts in the Klamath Mountains and western Sierra Nevada foothills of California (Snoke et al., 1982).

Traces of copper mineralization are widespread in the Choyal Formation. The economic gold and copper mineralization at Punta Norte is present in hydrothermally altered volcanic rock along the trace of one of the east-west–trending faults mapped by Kilmer (1979). Ore mineralization is disseminated irregularly in areas of extensively fractured rock. Thin veins filled with secondary copper sulfide minerals cut the volcanics on hillsides in the area of the mine.

Arc Plutonic Complex on the Vizcaíno Peninsula

Arc plutonic rocks form a northeast-southwest–trending belt that extends from the headlands between Asuncion and San Pablo to the eastern Sierra de San Andres highlands (Fig. 1). In the coastal region between Asuncion and the San Pablo headland, these plutons intrude the San Pablo amphibolite and hypabyssal and gabbroic part of the Vizcaíno Peninsula Ophiolite. In the southeastern Sierra de San Andres they intrude Jurassic volcanic and volcaniclastic rocks. Intrusions vary in scale from small stocks and anatomizing dikes up to plutons 40 km^2 in area in the southeastern Sierra de San Andres. Intrusive relationships into the volcanic and volcaniclastic rocks are demonstrated by the presence of abundant tonalitic dikes and by prominent contact metamorphic aureoles containing hornblende hornfels mineral assemblages along the pluton margins. The presence of miarolytic cavities in the plutons and narrow contact aureoles indicate that the plutons were emplaced at high structural levels.

The plutons that intrude ophiolite basement are exposed over areas as large as 15 km^2. Intrusive relationships are demonstrated by the presence of abundant dikes intruding the amphibolite and ophiolitic sections, xenoliths of the amphibolite and ophiolitic rocks in the plutons, and contact metamorphic overprints on cumulate peridotite and gabbro of the ophiolite that are as high in grade as hornblende hornfels adjacent to pluton margins. Both the plutons and ophiolitic country rocks are strongly tectonized under brittle extensional conditions, especially in the San Pablo headlands.

Plutons range from gabbro to granodiorite in composition, but tonalite is most common. Hornblende, biotite, and, rarely, muscovite are present as essential minerals and display medium- to coarse-grained hypidiomorphic textures and, locally, seriate textures. Plagioclase is commonly strongly oscillatory zoned and forms megacrysts in some plutons (Table 1).

The plutons and their country rocks are depositionally overlain by the Aptian–Albian Asuncion Formation (Barnes, 1984).

TABLE 1 DESCRIPTION OF U/PB AND ^{40}AR/^{39}AR DATING SAMPLES

Sample	Latitude (north)	Longitude (west)	Lithology	Petrography	Field setting
ASCZ-1	27°28' 41.9"	114°32' 24.0"	plagiogranite	hypd gr; f-m grd; access op, sph, ap, zr; bt replaced by chl; pl replaced by ep + ab; contains small amphibolite inclusions	Two-m-thick dike from ~0.25 km² area of irregular stock-like intrusion w/ dikes and veins into altered noncumulate hb-cpx gb of SSAO. Located in Arroyo San Cristóbal, 6 km east of Puerto Nuevo. Contains abundant amphibolitized xenoliths of wall rock.
MHFZ-1	27° 31'50"	114° 43'06"	albitite (plagiogranite)	hypd gr; m grd; 70–80% euh-sub albite w/ interstitial patches of phrenite, ep, cc; access sph and sph+op	Isolated (2 x 5 m) lens-shaped intrusive mass in high-level gabbro & dike complex of the VPO to S of Puerto Escondito & W of Morro Hermosa fault; 219.9 ± 2 Ma sphene U/Pb age reported by Barnes and Mattinson (1981).
80-AS-3B	27°08' 56.7"	114°19' 35.1"	granodiorite	hypd gr; m grd; access op, ap, sph, zr; osc-zoned pl; k-spar and locally protoclastic qz are interstial; k-spar commonly rims pl	Coastal exposure of stock intruding dike and pillow complex of VPO, 3.4 km west of Asuncion. From 12 km² area of irregular stocks and dikes of SACC containing abundant amphibolized xenoliths.
81-SRq-7	27°1046.8"	114°2348.9"	hb-bt tonalite	hypd gr; m grd; QAP = 25:1:74 (581); C.I. = 16 access op, sph, ap, zr; strongly osc-zoned pl	Coastal exposure of 15 km² tonalitic pluton of SAAC at Pueblo San Roque. Intrudes and metamorphoses cumulate gabbro of VPO. Contains scarce mafic xenoliths.
81-TA-26	27°17' 22.6"	114°15' 06.0"	hb-bt tonalite	hypd gr; m-c grd; QAP = 31:3:66 (523); C.I. = 13; access op, sph, ap, zr; strongly osc-zoned pl with hb inclusions	Isolated hilltop exposure west of Tres Amigos pluton, 17 km north-northeast of Asuncion. Contains sparse mafic xenoliths and 1 cm pl and hb phenocrysts; bordered by wide zone of amphibolitized mafic volcanic and volcaniclastic rocks of SAAC.
CN 95	27°18'34.5"	114° 1013.6"	hb-bt tonalite	hypd gr; m/c grained / weakly foliated, scattered sparse mafic inclusions, 1–2 m wide pink leucocratic cross-cutting aplite dikes	Interior of largest granitoid pluton (~40 km²) in Vizcaino Peninsula–Tres Amigos pluton of Barnes (1982).
79-AR-53	27°26' 04.1"	114°25' 30.3"	gneissic hb tonalite	granoblastic; m grd; access ap, sph, zr; qz polygonized; pl replaced by ep + wm + ab; bt replaced by chl; hb rimmed by act; strong foliation	Prominently foliated ~2 km² pluton, 12 km SSE of San Jose de Castro. Contains sparse mafic xenoliths; bordered by amphibolitized mafic volcanic rocks of SAAC. Pluton and wall rocks are tectonized and surrounded by serpentinite-matrix melange of Sierra Placer melange.
322-12	27°20' 35.3"	114°18' 34.6"	hb tonalite	seriate, f-m grd; access op, zr; interstial myrmikitic intergrowths; miarolitic cavities filled with chl + qz + ep + sph + act; strongly osc-zoned pl	2 m thick dike intruding mafic volcanic rocks of SAAC, 23 km north of Asuncion. Dike is located 0.25 km north of amphibolitized border zone of 8 km² pluton of SAAC and is probably related to that body.
322-5	27°14' 11.5"	114°26' 15.7"	bt-hb tonalite	hypd gr; m grd; QAP = 33:4:63 (722); CI = 17; access op, ap, sph, zr; bt entirely replaced by chl; veins filled with qz + ab + ep + cc; mild cataclastic texture	2 m thick dike intruding partly amphibolitized cumulate gabbro and peridotite of VPO, 3.5 km NE of San Pablo town. Dike is apparently related to pluton of sample 322-10, located 1 km to north of the sample locality. Dike and wall rocks are strongly tectonized.

(continued)

TABLE 1 DESCRIPTION OF U/PB AND ^{40}AR/^{39}AR DATING SAMPLES (continued)

Sample	Latitude (north)	Longitude (west)	Lithology	Petrography	Field setting
322-11	27°14' 23.8"	114°28' 34.2"	bt-hb tonalite	hypd gr; m gr; access op, ap, sph, zr, ep; osc-zoned pl; wm + ab +cc + ep replaces pl cores; some by replaced by chl; qz is strongly interstitial	Collected from ~0.25 km² stock in southern marginal zone of 10 km² pluton of SAAC, 2.5 km NNW of Pueblo San Pablo. Pluton discordantly intrudes San Pablo amphibolite body and displays a complex 1 km wide marginal zone of abundant dikes and xenoliths.
CI-1	28°06' 54"	115°15' 22"	plagiogranite	m gr; qtz and altered pl; QAP = 42:0:58 (1020); CI = 5; hb ~4%; pl to ab + ep + clay; vein-filled fractures w/ phrenite	Plagiogranite intruded as irregular dikes and veins over an area of ~5 x 30 m into hydrothermally altered noncumulus hnb-cpx gabbro, qtz diorite and mafic dikes; plutonic section of Cedros Island Ophiolite on SW side of Monte Cedros.
925-13	28°11' 56"	115°10' 29"	hb-pl xtal lithic lapilli tuff	m gr; xtal lithic tuff w/zoned pl; euh grn hb; 3–4 mm angular volc frags; zeolite + minor cc matrix	5 m thick tabular tuff bed in Gran Cañon Fm ~900 m above dep contact on ophiolite basement—just above interbedded pillow lava horizon and interbedded w/ more thinly bedded volcanogenic ss, slts sequence—collected from floor of Gran Cañon.
829-1	28°22' 06"	115°13' 01"	granophyric leuco-tonalite	m gr; hypd gr; ab + qz intergrowths; bt to chl; patches of chl + ep + clay; access op, zr, ap	Largest plutonic unit in Punta Norte headland area at hill 1714'; sharp vertical intrusive contact w/ ~1500' of relief; volcanic + hypabyssal wallrock; block stoping and schieleran parallel to contacts.
829-8	28°22' 06"	115°12' 30"	granodiorite	m gr; hypd gr; subh pl and qz w/ interstitial k-spar, secondary chl, ab, ep, cc, clays; access zr, ap, op	Sheeted intrusive unit to east of 829-1; similar contact relations.
830-9	28°21' 48"	115°11' 40"	leuco-granodiorite	granophyric intergrowths of ab+qz in circular sprays and as cuniform intergrowths; intergrown qz+kspar; hnb replaced by op + clay; QAP = 37:25:38 (572)	Plutonic unit surrounding the Punta Norte lighthouse ~1 km N of Campo Punta Norte.
CAC-1	28°13' 53"	115°12' 50"	granophyric leuco-tonalite	m gr; hypd gr; ab + qz intergrowths; mafic minerals to cc + clay; access zr, op, ap	~100 x 200 m olistolith slide block(?) in Coloradito Formation near base of formation in Arroyo Coloradito—mapped by Kilmer as an intrusion.
80EC-55G	27° 24' 37"	114° 16' 12"	hb-pl xtal lithic tuff breccia	monolithic hb-pl andesite tuff breccia w/clasts up to 7 cm; 30% euhedral plag, 10% euhedral green-brown hb, 1% opaques, & 60% groundmass; pervasive static greenschist facies recrystallization, hb has alteration rims of leucoxene	Lower eastern flank of Cerro El Calvario from the lower part of a well-bedded volcaniclastic turbidite section ~600 m thick described by Moore (1984).

Note: hypd gr—hypidiomorphic granular; f grd—fine-grained; m grd—medium-grained; c grd—coarse-grained; QAP—normalized modal percentages of quartz, alkali feldspar & plagioclase from thin section point counts; (n)—total no. of point counts; C.I.—color index; osc—oscillatory; grndms—groundmass; access—accessory mineral; op—opaques; ap—apatite; zr—zircon; sph—sphene; hb—hornblende; pl—plagioclase; k-spar= potassium feldspar; qz—quartz; bt—biotite; ms—muscovite; cpx—clinopyroxene; ol—olivine; ep—epidote; chl—chlorite; ab—albite; act—actinolite; cc—calcite; wm—white mica; euh—euhedral; subh—subhedral; VPO—Vizcaíno Peninsula Ophiolite; SAAC—San Andres–Cedros arc complex.

Troughton (1974) reported a hornblende K-Ar age determination of 143.3 ± 2.7 Ma from tonalite intruding cumulate gabbro and peridotite in the San Pablo headlands, as well as hornblende and biotite K-Ar ages of 154.1 ± 1.0 and 122.9 ± 1.8 Ma, respectively, from a pluton in the Sierra de San Andres. Concordant U-Pb ages on zircon of 145 and 153 Ma were determined from two other plutons by J. Wright (reported in Barnes, 1982, 1984).

ISOTOPIC AGE DATA

Zircon U/Pb

Zircon U/Pb analyses are reported from 14 samples (Fig. 3). Locations, map units, petrography, and field settings for each of these samples are summarized in Table 1. Analytical data are presented in Table 2 and plotted on concordia diagrams in Figure 4.

Zircon was separated by conventional techniques using a Wilfley Table, heavy liquids, and a Frantz magnetic separator. The least magnetic zircons from each sample were split into size fractions and then handpicked to remove any contaminating grains. Zircon dissolution and ion exchange chemistry for separation of uranium and lead followed procedures modified from Krogh (1973). Isotope ratios were measured by thermal ionization mass spectrometry (TIMS) with the Finnegan MAT 261 multicollector and the AVCO 35-cm single collector instruments at the University of California at Santa Barbara and the Fisons VG Sector 54 multicollector instrument at San Diego State University. Analytical uncertainties, blanks, and common lead corrections are outlined in Table 2. Most of the samples yield concordant to near-concordant U/Pb dates that are interpreted to closely approximate crystallization ages. The relatively simple systematics for these samples are interpreted to reflect the low metamorphic grade of the samples and negligible or absent inherited components of radiogenic lead.

Much of the data presented here were originally reported by Kimbrough (1982). Subsequent to this publication, two of the mixed ^{235}U/^{208}Pb spikes used to determine uranium and lead concentrations were recalibrated, which had the effect of raising the U/Pb ages by 1%–2%, thus making the ^{206}Pb*/^{238}U and ^{207}Pb*/^{206}Pb* dates from individual fractions more nearly concordant. The recalculated U/Pb ages presented here supersede data reported in Kimbrough (1982). It is important to note that despite usage of the newer spike calibrations here, the interpreted ages are little changed because the original dates were based on high-precision ^{207}Pb*/^{206}Pb* dates rather than ^{206}Pb*/^{238}U dates as discussed below.

Sample ASCZ-1 is a plagiogranite from the Vizcaíno Peninsula Ophiolite collected in Arroyo San Cristóbal. The plagiogranite zircons have low uranium concentrations, ranging from 121 to 145.5 ppm. Each of the three fractions yielded slightly discordant results with ^{206}Pb*/^{238}U dates ranging from 215.5 to 217.2 Ma. Well-determined ^{207}Pb*/^{206}Pb* dates from each of the fractions are slightly older and in agreement with one another within analytical uncertainty at 221.0 ± 3.8 Ma, 219.9 ± 3.4 Ma, and 221.8 ± 3.6 Ma.

The small but consistent differences in ^{206}Pb*/^{238}U and ^{207}Pb*/^{206}Pb* dates is attributable to slight, relatively recent lead loss from the zircon. This interpretation is further supported by the negative correlation between uranium concentration and ^{206}Pb*/^{238}U dates. The interpreted crystallization age for this sample is 221 ± 2 Ma based on the weighted mean of the ^{207}Pb*/^{206}Pb* dates. The zircon systematics of ASCZ-1 and reliance upon high-precision ^{207}Pb*/^{206}Pb* ages to establish plagiogranite crystallization ages from TIMS data is similar to data from the Coast Range Ophiolite of California (Mattinson and Hopson, 1992) as well as data from the Dun Mountain Ophiolite of New Zealand (Kimbrough et al., 1992).

Sample CI-1 is a plagiogranite from the Cedros Island Ophiolite that yielded concordant results from all four analyzed fractions. The best results are from the two relatively low U fractions (~286 ppm) that also had the lowest common Pb contents as indicated by their relatively high ^{206}Pb/^{204}Pb ratios (>9000). The interpreted crystallization age for this sample is 173 ± 2 Ma based on the weighted mean of the ^{206}Pb*/^{238}U dates. This age is in good agreement with the higher precision ^{207}Pb*/^{206}Pb* dates from the two low uranium fractions.

Six fractions from three separate Punta Norte tonalite and granodiorite intrusions (829-1, 829-8, 830-9) yield similar concordant or near-concordant ages. The ^{206}Pb*/^{238}U dates from these intrusions all range from 163.0 Ma to 167.5 Ma. Three of the fractions have poorly determined ^{207}Pb*/^{206}Pb* dates due mainly to high common Pb contents. The three fractions with relatively well-determined ^{207}Pb*/^{206}Pb* dates, however, are in good agreement with the ^{206}Pb*/^{238}U dates. The interpreted age of 166 ± 3 Ma for the Punta Norte granitoids is based mainly on the ^{206}Pb*/^{238}U dates.

A single fraction of low-uranium zircon (38.7 ppm) from sample 925-7 (Gran Cañon Formation dacite tuff bed from near the top of the formation) yielded a perfectly concordant age of 164 Ma with a well-determined ^{207}Pb*/^{206}Pb* date.

Ten fractions from seven different samples of San Andres arc plutonic rocks yielded concordant to variably discordant results with ^{206}*Pb/^{238}U dates ranging from 134.5 Ma to 155.6 Ma. Although these results are consistent with previously reported ages from other workers, none of the results here provide high precision crystallization ages. Sample 81-Srq-7 is the youngest pluton dated at ca. 135 Ma; the interpreted age of 322-11 is ~160 based on the well-determined ^{207}Pb*/^{206}Pb* date and consideration of hornblende ^{40}Ar/^{39}Ar data from the same sample (see below). A well-determined ^{207}Pb*/^{206}Pb* of 191.3 for one of the two fractions analyzed from sample 322–5 SP may indicate a small component of inherited radiogenic lead in this sample.

Three fractions from the Santa Eugenia tuff-breccia of Hickey (1984) (Eugenia Formation) yield near-concordant results that indicate a crystallization age of 141.5 ± 3 Ma. This is consistent with a Tithonian age call from fossiliferous strata in the Eugenia Formation ~600 m stratigraphically beneath the tuff bed (Hickey, 1984).

Five fractions from a leucotonalite olistostrome slide block within the Coloradito Formation on Cedros Island yield ^{206}*Pb/^{238}U dates that range widely from 129.8 Ma to 170.7 Ma. Ages

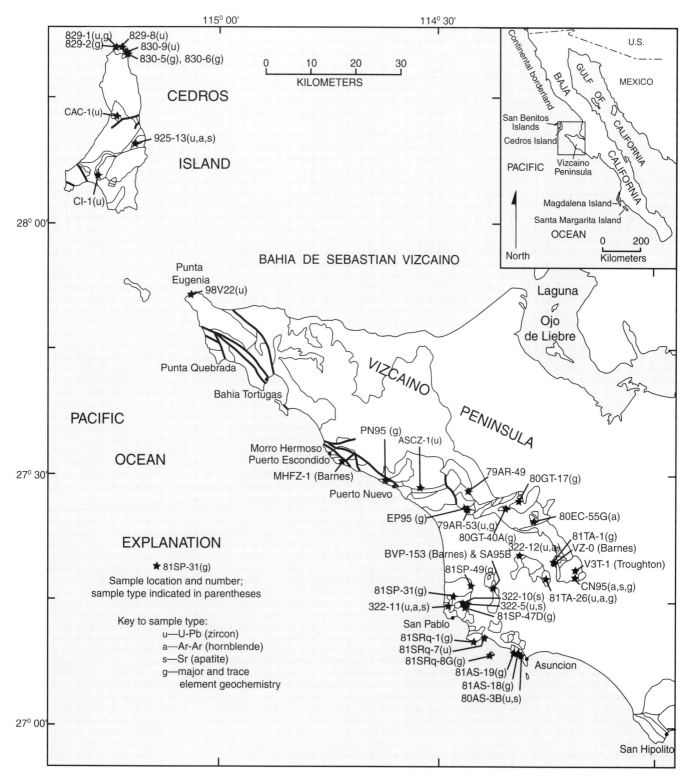

Figure 3: Sample locality map.

TABLE 2. ZIRCON U-PB ISOTOPIC DATA

Sample	Fraction	Weight (g)	Pb (ppm)	U (ppm)	Pb isotopic compositions			Ages (m.y.)			Error
					206/208	206/207	206/204	206*/238	207*/235	207*/235	
Vizcaíno Peninsula ophiolite plagiogranite—Arroyo San Cristóbal											
ASCZ-1	<200	0.0188	4.553	121.0	4.652	18.292	3576.5	217.2	217.5	221.0	±3.8
ASCZ-1	100-200	0.0288	5.423	145.5	4.689	18.288	3546.1	215.5	215.8	219.9	±3.4
ASCZ-1	>100	0.0160	4.969	132.9	4.681	18.152	3257.3	215.9	216.4	221.8	±3.6
San Andres magmatic arc granitoids											
81-TA-26	fg	0.0252	6.509	287.5	7.604	17.694	1898.1	139.9	139.7	136.4	±1.8
81-TA-26	cg	0.0344	3.097	134.7	8.251	18.440	2739.7	144.0	143.9	141.1	±2.9
81-SRq-7	<100	0.0364	6.098	280.7	7.272	18.749	3179.7	134.5	134.4	133.8	±3.9
81-SRq-7	>100	0.0340	3.026	129.6	6.519	15.995	1059.3	139.8	139.2	129.7	±14
80-AS-3B	b	0.0204	5.736	271.9	6.538	18.419	2924.0	132.7	133.8	162.0	±9
79AR-53	b,205	0.0600	220	7252	2.338	16.681	1373.6	148.9	149.6	161.0	±8
322-5 SP	b	0.0027	34.252	1258.9	4.144	19.683	10695.2	155.6	156.4	168.3	±2.8
322-5 SP	b	0.0170	32.044	1215.0	4.125	19.667	15873.0	150.8	153.2	191.3	±1.9
322-11	b	0.0376	7.585	324.0	6.479	18.005	2331.0	142.2	143.2	158.8	±3.5
322-12	b	0.0314	1.671	63.3	5.291	13.668	609.8	150.1	150.1	149.5	±8.2
Eugenia Formation–Santa Eugenia tuff											
98V22	140-200a	0.0025	4.964	224.2	7.615	19.332	5295.0	138.7	139.2	146.5	±2.7
98V22	<400a	0.0025	4.347	195.0	7.350	19.915	17250.2	139.4	140.9	168.0	±3
98V22	200-325a	0.0089	7.066	313.8	7.737	20.093	14471.1	141.5	141.1	135.9	±3.5
Cedros Island ophiolite plagiogranite—West La Lena											
CI-1	B	0.0067	18.275	614.9	4.908	19.384	6666.7	174.5	173.9	166.1	±8.9
CI-1	B	0.0036	10.591	358.7	4.857	18.616	3364.7	172.5	171.9	164.3	±9.3
CI-1	B	0.0184	8.449	286.7	4.936	19.577	9901.0	173.4	173.6	176.2	±4
CI-1	b, 205	0.0061	8.396	287.5	4.867	19.602	9090.9	171.8	171.6	170.0	±3
Punta Norte granitoids											
829-1	b	0.0120	19.482	603.4	3.478	12.659	469.7	167.5	161.9	80.3	±34
829-1	fg	0.0159	19.821	689.0	4.479	17.336	1826.2	164.3	165.2	177.9	±8.3
829-1	b	0.0132	18.305	658.1	5.847	18.957	4364.9	167.4	167.3	166.0	±5.1
829-8	B	0.0046	10.129	354.7	4.348	17.882	2259.9	163.0	163.3	167.6	±7
829-8	B	0.0058	14.169	368.6	2.509	7.723	185.6	166.8	169.8	211.1	±63
830-9	b, 205	0.0020	4.507	126.3	2.242	7.371	168.4	166.1	164.2	137.0	±29
Gran Cañon Formation tuff											
925-13	Cg	0.0020	1.089	38.7	5.249	15.253	904.2	164.1	164.1	164.0	±3
Coloradito Formation leucotonalite											
CAC-1	>100	0.0025	38.317	1334.5	3.281	17.556	1986.5	154.5	155.8	174.4	±2.8
CAC-1	<200	0.0090	15.768	552.6	4.205	16.535	1343.2	160.3	161.1	173.1	±3.8
CAC-1	<200	0.0214	15.544	516.5	4.343	17.259	1661.1	170.7	169.4	151.9	±5.6
CAC-1	>100	0.0071	28.010	994.7	3.220	17.164	1724.1	150.6	152.5	182.7	±13
CAC-1	>100, L	0.0064	21.126	928.9	4.100	19.296	7290.6	129.8	132.8	186.1	±5.3

Note: Fractions: 100, 140, 200, 325, 400 = mesh sizes; fg & cg = fine & coarse; b = bulk; a = air abrasion; L = HF leach Separation of U and Pb was done using HCl column chemistry. Concentrations were determined using mixed ^{208}Pb/^{235}U and ^{205}Pb/^{235}U spikes. Lead isotopic compositions corrected for ~0.10% ± 0.05% per mass unit mass fractionation. Ages calculated with following decay constants: 1.55125E-10 = ^{238}U and 9.8485E-10 = ^{235}U. Present day ^{238}U/^{235}U = 137.88. Common lead corrections made using Stacey and Kramers (1975) model lead isotopic compositions. Total lead blanks averaged c. 25 picograms.

are not correlated to uranium content of zircon, which ranges from ~500 to 1300 ppm. The ^{207}Pb*/^{206}Pb* dates are more consistent with four of the five fractions ranging between ca. 173 Ma to 186 Ma. Although the complicated systematics make interpretation difficult, the likely crystallization age of the sample based on this data is Early to Middle Jurassic (~170–185 Ma).

Hornblende ^{40}Ar/^{39}Ar

New hornblende ^{40}Ar/^{39}Ar step-heating data are reported here from five granitic rocks and two tuff beds (Fig. 3). Argon analyses were done at the University of California at Los Angeles using techniques discussed in Quidelleur et al. (1997). Loca-

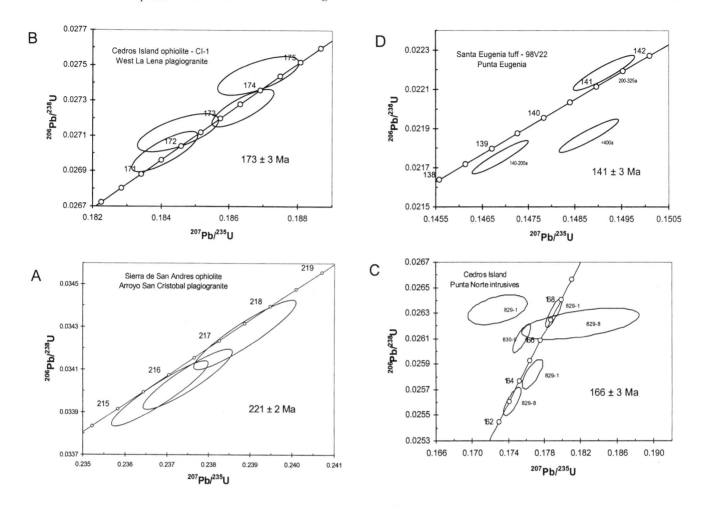

Figure 4: Zircon U-Pb concordia diagrams.

tions, map units, petrography and field settings for each of these samples are summarized in Table 1. Note that four of the samples also have zircon U/Pb analyses reported here. A summary of the ^{40}Ar/^{39}Ar data is presented in Figure 5. Complete analytical results are available from the GSA Data Repository.[1]

Results from all seven samples exhibit variably "disturbed" age and Ca/K spectra (Fig. 5). The irregular nature of the age spectra is most likely attributable to two main factors: (1) minor impurity of analyzed mineral separates (i.e., amphibole exsolution effects or pyroxene cores in hornblende and/or intergrown fine-grained mica that is hard to detect), and (2) later thermal disturbance/low-grade metamorphic overprinting of the rocks that produced loss of radiogenic argon (cf. Gaber et al., 1988; Baldwin

et al., 1990). In general, apparent ages increase over the first two to three heating steps to attain maximum apparent ages for the middle half of the age spectrum and then decrease over the last few steps. Ca/K exhibits a similar behavior but tends to never define a consistent plateau. The very high Ca/K ratios in some steps are consistent with the presence of pyroxene or effects of amphibole exsolution in some samples (Harrison and Fitz Gerald, 1986). None of the samples exhibit the characteristic "U-shaped" profiles indicative of samples containing excess radiogenic ^{40}Ar (McDougall and Harrison, 1999). Because of this, we interpret the data as providing reliable minimum age constraints on the crystallization ages of individual samples. In some instances, there is evidence that the ages obtained may closely approximate crystallization ages. For example, for the Gran Cañon tuff (925–13), the maximum ^{40}Ar/^{39}Ar age step of 160 ± 2 Ma is in good agreement with the zircon U/Pb age of 164.0 ± 3 Ma. The weighted-mean preferred ^{40}Ar/^{39}Ar age of 160 ± 2 Ma for sample 322–11, a Sierra San Andres tonalite, is in good agreement with a well-determined ^{207}Pb*/^{206}Pb* age from the same sample, suggesting this is a good approximation of the crystallization age.

[1]GSA Data Repository Item 2003172, Hornblende ^{40}Ar/^{39}Ar analytical techniques and data, and whole rock geochemistry sample localities and descriptions, is available on request from Documents Secretary, GSA, P.O. Box 9140, Boulder, CO 80301-9140, USA, editing@geosociety.org, at www.geosociety.org/pubs/ft2003.htm, or on the CD-ROM accompanying this volume.

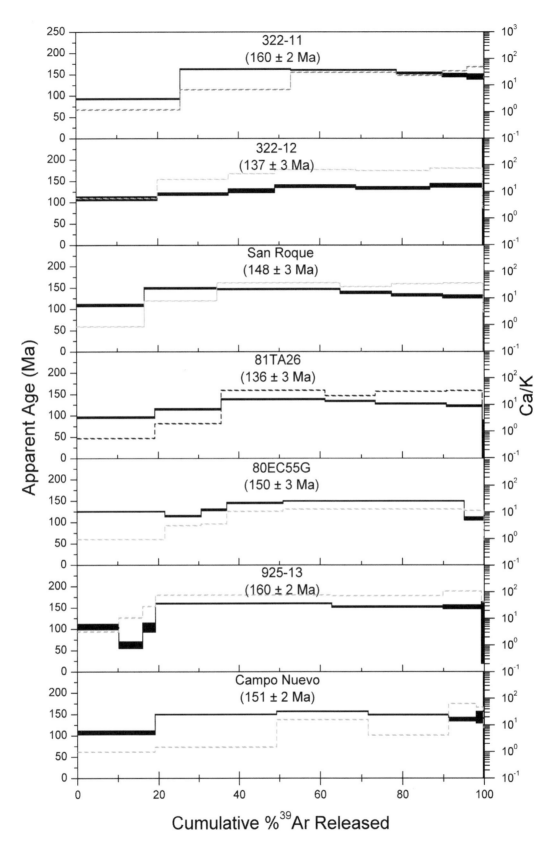

Figure 5: Hornblende $^{40}Ar/^{39}Ar$ age spectra and Ca/K ratios. Ages and uncertainties shown for individual samples are interpreted minimum crystallization ages (see text).

GEOCHEMISTRY OF ARC PLUTONIC ROCKS

Whole Rock Major and Trace Element Chemistry

Whole-rock powders were prepared from 21 fresh samples of plutonic rock in a tungsten carbide shatter box for major and trace element analysis (Tables 3 and 4). Sample locations are shown in Figure 3 and details are provided in the GSA Data Repository (see footnote 1). Seventeen of the rock powders were analyzed by Bondar Clegg & Company Ltd., by X-ray fluorescence spectrometry (XRF) (major elements + Ba, Nb, Rb, Sr, Zr, Y). Four samples were analyzed by Activation Laboratories by inductively-coupled-plasma atomic-emission spectrometry (ICP/AES) (major elements) and inductively-coupled-plasma mass spectroscopy (ICP/MS) (trace elements).

The Vizcaíno and Cedros plutons vary widely in composition as reflected by SiO_2 contents that vary from ~54 to 74 wt percent (Table 3). Tonalite and granodiorite dominate the suite, however, as demonstrated by normative feldspar composition plots (Fig. 6). The normative compositions correspond well to modal compositions determined by point counting from several of the samples. These rocks are strongly calcic as illustrated by the classical Peacock diagram that indicates a Peacock Index (value of SiO_2 where $Na_2O + K_2O = CaO$) for the suite of 65% (Fig. 7A). This is comparable to the Peacock Index of Cretaceous plutonic rocks from the western Peninsular Ranges batholith, which stands out as one of the most calcic batholiths of the North American Cordillera (Silver and Chappell, 1988). The alumina saturation index of the suite is ~1, indicating metaluminous to peraluminous compositions typical of I-type granitoids (Fig. 7B). On the Pearce et al. (1984) Rb/Y+Nb tectonic discrimination diagram, all samples fall well within the "volcanic arc granitoid" field (Fig. 8A). Chrondrite normalized REE abundance patterns for the Tres Amigos and Puerto Nuevo tonalite are comparable to patterns determined from the western zone of the Peninsular Ranges batholith (Gromet and Silver, 1987), except that these rocks have no Eu anomaly (Fig. 8B). Mid to heavy REE are essentially unfractionated.

Apatite $^{87}Sr/^{86}Sr$

Apatite $^{87}Sr/^{86}Sr$ isotopic ratios are reported here for nine arc plutonic granitoid samples from the Vizcaíno Peninsula and one sample of crystal-vitric tuff from the Gran Cañon Formation on Cedros Island (Table 5). Strontium isotopic analyses were performed by TIMS using the VG Sector 54 multicollector mass spectrometer at San Diego State University. Strontium was separated from apatite using 0.25 ml columns packed with Eichrom Sr-specific column (Sr•Spec®) resin. Isotope ratios were measured using a dynamic multicollector measurement routine. These ratios provide an approximation of the initial $^{87}Sr/^{86}Sr$ ratios of the rocks because 1) the rocks have relatively low Rb/Sr ratios and, 2) apatite is a Rb-free phase with much lower Rb/Sr ratios relative to whole rock Rb/Sr ratios.

The measured $^{87}Sr/^{86}Sr$ ratios range from 0.70359 to 0.70642. These "initial" ratios may reflect minor accumulation of radiogenic ^{87}Sr since magmatic crystallization of these rocks at ca. 150 Ma. Nevertheless, the overall low values are consistent with generation of the Vizcaíno granitoids from mantle sources with little or no involvement of ancient continental crust.

DISCUSSION

Repeated episodes of extensional deformation are a characteristic feature of Mesozoic ophiolite and arc basement in the Vizcaíno Peninsula and Cedros Island region of west-central Baja California and exerted a primary control on patterns of magmatism and sedimentation. The Late Triassic Vizcaíno Peninsula Ophiolite, which records a major phase of magmatism induced by mantle upwelling linked to extension and seafloor spreading, represents the earliest phase of this history.

In the discussion that follows, we first consider aspects of Vizcaíno-Cedros geology that bear on tectonic environments of formation and possible relationships between the three Mesozoic terranes in "upper plate" rocks originally defined by Moore (1985) and Kimbrough (1985); the Choyal, Vizcaíno Norte, and Vizcaíno Sur terranes. Next we discuss regional correlations of this geology to plate edge assemblages along strike to the north in California and Oregon. Finally we apply previously proposed forearc rifting models (Saleeby, 1981, 1992; Stern and Bloomer, 1992) to the Vizcaíno-Cedros region that we conclude best explain the geologic history.

Suprasubduction Zone Origin of the Late Triassic Vizcaíno Peninsula Ophiolite

Based on petrography and field relationships, the Vizcaíno Peninsula Ophiolite plagiogranites are interpreted as minor siliceous differentiate of tholeiitic magmas that produced the main plutonic-volcanic suites preserved in the ophiolite crustal section (Moore, 1983). The zircon U/Pb age of 221 ± 2 Ma reported here for plagiogranite in Arroyo San Carlos on the southern flank of the Sierra de San Andres antiform is in tight agreement with the sphene U-Pb age of 220 ± 2 Ma reported from albitite on the opposing flank of this structure (Barnes and Mattinson, 1981). Paleontologic ages from interpillow sediment at top of the ophiolite crustal section at three widely spaced localities (Punta San Hipolito, Sierra de San Andres, and Punta Quebrada) all yield nearly identical Carnian to Norian radiolarian age calls that are supported by the occurrence of *Halobia* sp. and *Monotis* sp. in two of the three sections. The Carnian-Norian boundary in the Palmer and Geissman (1999) geologic time scale is assigned an absolute age of 221 ± 9 Ma, in good agreement with the zircon and sphene radiometric ages. The 221 ± 2 Ma zircon age may, in fact, be useful in refinement of geologic time scale calibration since it represents a maximum age for the upper limit of the *Capnodoce* radiolarian zone (Carnian-Norian), which is located within the chert member of the San Hipolito Formation at Punta San Hipolito.

TABLE 3. WHOLE-ROCK MAJOR AND TRACE ELEMENT ANALYSES FOR ARC PLUTONIC ROCKS FROM THE SAN ANDRES–CEDROS COMPLEX

	Punta Norte granitoids, Cedros				Tres Amigos pluton, Sierra de San Andres				San Pablo headlands		
	PN830-6	PN830-5	PN829-2	PN829-1	$CN95	81TA-26	81TA-1	81TA-1d	81SP-47D	81SP-49	81SP-31
Major oxides (wt%)											
SiO_2	53.83	62.55	72.97	73.58	66.82	67.07	68.29	68.78	57.15	66.11	74.68
TiO_2	0.49	0.72	0.45	0.38	0.39	0.4	0.36	0.36	0.37	0.36	0.15
Al_2O_3	15.64	15.23	13.03	12.53	16.12	16.37	14.42	14.58	15.2	15.17	13.18
Fe_2O_3	2.83	3.22	1.57	1.33	1.31	1.86	2.77	2.81	2.09	2.07	0.51
FeO	5.02	3.09	1.8	1.7	1.60	1.29	1.67	1.61	4.24	2.83	1.02
Fe_2O_3*	8.41	6.66	3.57	3.22	3.09	3.29	4.62	4.6	6.8	5.21	1.65
MnO	0.19	0.15	0.08	0.06	0.05	0.06	0.11	0.11	0.12	0.11	0.02
MgO	6.02	2.19	1.08	0.69	1.72	1.92	1.77	1.79	5.89	2.1	0.85
CaO	9.12	4.64	2.09	1.91	4.02	4.76	4.73	4.78	9.15	5.34	1.41
Na_2O	3.18	4.01	4.84	4.08	4.46	4.13	3.27	3.38	1.95	2.94	3.76
K_2O	0.58	0.43	0.21	2.03	1.36	1.14	0.3	0.26	0.93	1.46	3.47
P_2O_5	0.03	0.31	0.03	0.03	0.13	0.03	0.06	0.03	0.22	0.03	0.03
LOI	1.67	1.74	1.55	0.96	0.44	0.59	1.76	1.7	1.38	1.23	0.93
Total	99.16	98.63	99.9	99.47	98.60	99.76	99.69	100.37	99.16	100.06	100.13
Trace elements (ppm)											
Ba	179	64	62	291	495.8	399	299	286	469	535	766
Nb	<5	<5	5	<5	5.1	5	<5	<5	<5	<5	<5
Rb	10	8	6	35	25.5	26	11	12	19	30	52
Sr	187	251	198	157	349.3	372	343	340	314	255	104
Zr	61	100	167	149	107.7	113	66	67	52	69	67
Y	18	23	30	34	11	15	20	19	13	18	24

Note: Major and trace element analyses by XRF from Bondar Clegg & Company Ltd.

*Major element analyses by ICP from Activation Laboratories; Ba, Nb, Rb, Sr, Zr, Y by ICP/MS from Lithogeochem (provided through Activation Laboratories).

#Major element analyses by XRF from Bondar Clegg & Company Ltd.

(continued)

TABLE 3. WHOLE-ROCK MAJOR AND TRACE ELEMENT ANALYSES FOR ARC PLUTONIC ROCKS FROM THE SAN ANDRES–CEDROS COMPLEX (continued)

	San Roque headlands				Asuncion and central Sierra de San Andres					Puerto Nuevo
	81SRQ-1	81SRQ-8G	79AR-53	#El Placer 95	#San Andres 95B	81AS-18	81AS-19	80GT-17	80GT-40A	#PN95
Major oxides (wt%)										
SiO$_2$	63.87	73.44	59.55	59.96	55.72	71.48	68.97	45.74	51.74	64.33
TiO$_2$	0.44	0.37	0.28	0.29	0.42	0.32	0.4	0.65	0.54	0.30
Al$_2$O$_3$	16.8	13.87	15.35	15.42	14.25	14.57	14.77	17.53	17.3	17.80
Fe$_2$O$_3$	1.81	1.89	2.87	1.98	1.62	1.61	2.48	6.77	4.46	1.59
FeO	1.74	0.71	4.44	5.10	4.70	1.29	1.42	6.24	5.4	1.20
Fe$_2$O$_3$*	3.74	2.68	7.8	7.64	6.84	3.04	4.06	13.7	10.46	2.92
MnO	0.07	0.02	0.14	0.14	0.14	0.02	0.06	0.22	0.17	0.07
MgO	2.34	0.54	3.59	3.39	7.43	0.96	1.23	6.02	4.48	1.99
CaO	4.87	2.12	7.98	6.32	11.23	4.14	4.53	12.63	9.8	4.72
Na$_2$O	4.2	5.17	1.52	1.03	2.32	3.73	3.45	1.31	1.9	4.73
K$_2$O	1.43	1	1.87	3.25	0.24	0.15	0.73	0.63	1.49	1.09
P$_2$O$_5$	0.27	0.03	0.03	0.27	0.04	0.15	0.03	0.15	0.03	0.14
LOI	0.94	0.41	2	1.98	1.02	0.96	0.95	0.73	1.45	1.84
Total	98.97	99.65	100.11	99.69	99.65	99.52	99.18	99.31	99.36	99.93
Trace elements (ppm)										
Ba	444	482	668	912.9	118.6	189	426	475	867	336.1
Nb	<5	6	<5	0.0	0.6	<5	<5	<5	<5	2.1
Rb	37	32	36	58.6	4.0	5	16	15	25	25.9
Sr	330	121	204	160.9	254.6	320	256	451	381	587.2
Zr	93	214	15	7.2	26.6	46	54	12	28	72.8
Y	16	30	13	11	11	12	17	13	16	7

Note: Major and trace element analyses by XRF from Bondar Clegg & Company Ltd.
#Major element analyses by ICP from Activation Laboratories; Ba, Nb, Rb, Sr, Zr, Y by ICP/MS from Lithogeochem (provided through Activation Laboratories).

TABLE 4. WHOLE ROCK TRACE ELEMENT
ANALYSES OF ARC PLUTONIC ROCKS

	#CN95	#EP95	#SA95B	#PN95
*Sc	7	44	49	7
*Be	0	1	0	0
V	58	285	209	55
Cr	18	36	243	20
Co	10.6	22.9	33.2	7.6
Ni	69	40	103	45
Cu	6	50	22	10
Zn	68	68	73	67
Ga	16	14	13	15
Ge	1.2	1.6	1.5	0.9
As	0	90	6	11
Rb	25.5	58.6	4.0	25.9
Sr	349.3	160.9	254.6	587.2
Y	11	11	11	7
Zr	107.7	7.2	26.6	72.8
Nb	5.1	0.0	0.6	2.1
Mo	2.3	2.3	3.1	3.0
Ag	0.8	0.0	0.6	0.0
In	0.0	0.0	0.0	0.0
Sn	2	3	2	2
Sb	0.6	2.9	0.0	2.2
Cs	0.5	1.4	0.2	1.2
Ba	495.8	912.9	118.6	336.1
La	10.4	1.4	4.3	5.9
Ce	21.6	2.8	9.4	12.1
Pr	2.05	0.31	0.98	1.21
Nd	9.1	1.9	5.0	5.3
Sm	1.9	0.6	1.3	1.2
Eu	0.61	0.27	0.46	0.43
Gd	1.9	1.0	1.6	1.0
Tb	0.3	0.2	0.3	0.2
Sy	1.6	1.6	1.7	1.0
Ho	0.4	0.4	0.4	0.2
Er	1.1	1.1	1.3	0.7
Tm	0.16	0.20	0.19	0.10
Yb	1.1	1.4	1.1	0.7
Lu	0.19	0.21	0.17	0.12
Hf	2.7	0.3	0.8	0.2
Ta	0.44	0.00	0.00	0.24
W	1.2	1.6	1.2	1.2
Tl	0.3	0.6	0.0	0.3
Pb	0	0	0	6
Bi	0.5	0.3	0.0	0.0
Tn	2.43	0.78	1.20	1.37
U	0.60	0.16	0.22	0.31

#Trace element analyses by ICP-MS from Lithogeochem
provided through Activation Laboratories.
*Sc & *Be by ICP-AES from Activation Laboratories.

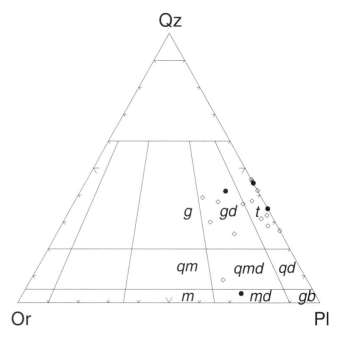

Figure 6: Normative quartz-orthoclase-plagioclase (Qz-Or-Pl) whole rock composition plot of Vizcaíno (open diamonds) and Cedros (filled circles) arc plutonic samples; classification after Streckeisen (1976). Other abbreviations: t—tonalite; gd—granodiorite; g—granite; qd—quartz diorite; qmd—quartz monzodiorite or quartz monzogabbro; qm—quartz monzonite; gb—diorite or gabbro; md—monzodiorite or monzogabbro; m—monzonite.

These relationships are consistent with the absence of stratigraphic and sedimentologic evidence for any significant hiatus between the igneous generation of the ophiolite and depositional onlap by basal strata of the San Hipolito Formation. The agreement of radiometric and biostratigraphic ages coupled with the evidence of proximal arc volcanic activity from the Upper Triassic tuff at Morro Hermoso (Barnes, 1982, 1984) suggests that the ophiolite was generated close to a volcanic arc. This—with a variety of petrologic characteristics from the ophiolite, including geochemical affinities of volcanic rock and local presence of cumulate orthopyroxene in gabbro, the local presence of keratophyre and quartz keratophyres in volcanic sections, and abundant vescicles in pillow rims locally up to 2 cm across that are indicative of eruption in shallow water—suggests that the Vizcaíno Peninsula Ophiolite was formed in a suprasubduction zone setting rather than at a mid-ocean ridge spreading center (Moore, 1983, 1985).

Stratigraphic variations in Late Triassic ophiolite remnants were one basis for distinguishing the Vizcaíno Norte and Vizcaíno Sur terranes (Moore, 1985). Late Triassic sedimentary cover on ophiolite basement at Punta San Hipolito, for example, is dominated by radiolarian chert with minor tuff, while the sequence 90 km away at Morro Hermoso is much more volcanogenic in nature (Fig. 2). Ophiolite basement in these two areas

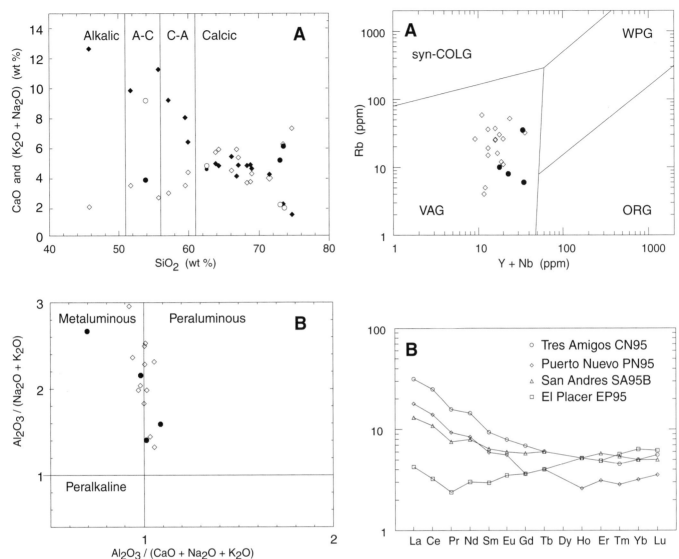

Figure 7: Vizcaíno (open diamonds) and Cedros (filled circles) arc plutonic whole rock composition plots. (A) Peacock alkali-lime plot; (B) Alumina saturation index diagram.

Figure 8: Whole rock trace element plots. (A) Rb/Y+Nb tectonic discrimination diagram plot of Pearce et al. (1984) for Vizcaíno (open diamonds) and Cedros (filled circles) arc plutonic rocks. Abbreviations: VAG—volcanic arc granitoid; syn-COLG—syn-collisional granitoids; WPG—within plate granitoids; and ORG—oceanic ridge granitoids. (B) Chrondrite normalized REE abundance patterns for selected Vizcaíno arc plutonic rocks.

also exhibits a number of differences (Moore, 1983). Relative to the Sierra de San Andres ophiolite remnant (Vizcaíno Norte), the ophiolite sequence around the Punta San Pablo–Asuncion headlands (Vizcaíno Sur) contains a significant fraction of keratophyre and quartz keratophyre in the volcanic section, more orthopyroxene in its basal cumulate section, and displays more variable geochemical characteristics. These features are interpreted to represent slow, disorganized spreading in a forearc or interarc basin that is perhaps closer to active arc volcanoes; the volcanic section of the Sierra de San Andres remnant, in contrast, has a more uniform basaltic composition with arc tholeiite geochemical affinity perhaps indicative of better-organized spreading at greater distance from arc volcanoes.

None of the differences between Vizcaíno Norte versus Vizcaíno Sur ophiolite remnants outlined above require that these age-equivalent Late Triassic ophiolite remnants originated in different tectonic settings. The variations are consistent with their development as parts of the same rifted volcanic arc (cf. Taylor, 1995). The situation is analogous to significant stratigraphic variation between remnants of the Coast Range ophiolite over similar distances (Hopson et al., 1981; Robertson, 1989). Although some displacements may have occurred between different remnants of the Vizcaíno Peninsula Ophiolite

TABLE 5. APATITE ^{87}SR/^{86}SR "INITIAL SR"

Sample	^{87}Sr/^{86}Sr
San Roque	0.70359 ± 07
322-10	0.70366 ± 08
11-79-1	0.70419 ± 05
CN95	0.70431 ± 12
AR 43B	0.70460 ± 11
322-14	0.70472 ± 27
925-13	0.70587 ± 09
322-11	0.70594 ± 27
322-5	0.70635 ± 35
80-AS-3B	0.70642 ± 10

Note: Quoted uncertainties are at one-sigma level. NBS 987 Sr standard ^{87}Sr/^{86}Sr = 0.71021 ± 2.

across the Sierra Placer mélange zone, we conclude that there is nothing in the geology that requires these remnants to be exotic with respect to one another.

Jurassic Arc Magmatism and Rifting Superimposed on Late Triassic Ophiolite Basement

The Vizcaíno Peninsula Ophiolite was the host for Jurassic ensimatic arc rocks that intrude and stratigraphically overlie ophiolite basement. This relationship is clearly demonstrated around the Punta San Pablo/Asuncion headlands, where ophiolite basement is intruded by arc-related granitoid plutons, and at Punta San Hipolito and the western Sierra de San Andres, where Jurassic volcaniclastic strata overlie pillow lava and sedimentary cover of the Triassic ophiolite (Figs. 1 and 2).

An important issue is the relationship of Jurassic arc rocks on Cedros Island versus the Vizcaíno Peninsula. Extrusive volcanic rocks in both areas are mainly andesite flows and breccias and mafic pillow lava that have low-Ti arc tholeiite and boninite geochemical affinities and are interpreted as vent facies of oceanic arc volcanic centers (Rangin, 1978; Kimbrough, 1979, 1982; Moore, 1983; Rangin et al., 1983).

Similarity of K/Ar ages from Punta Norte tonalite intrusions on Cedros Island (145 ± 6 Ma, Suppe and Armstrong, 1972) and the Vizcaíno Peninsula (154 ± 1 Ma, 143 ± 3 Ma, Troughton, 1974) was cited as support for this correlation by Jones et al. (1976) and Rangin (1978). Our zircon U/Pb age of 166 ± 3 Ma for the Punta Norte intrusions is substantially older than the age reported by Suppe and Armstrong (1972) and indicates that the Punta Norte intrusions are slightly older that the ca. 160–135 Ma Sierra de San Andres plutons. The plutons in both areas, however, share broad petrologic and geochemical similarities with relatively primitive arc geochemical affinities (initial ^{87}Sr/^{86}Sr range from ~0.704 to 0.706) and distinctly calcic compositions.

Jurassic marine turbiditic volcanogenic sequences on Cedros Island (Gran Cañon Formation) and the Vizcaíno Peninsula (Sierra de San Andres) are lithologically and sedimentologically similar, representing epiclastic and primary volcaniclastics derived from arc volcanoes (Moore, 1984; Kimbrough, 1984; Busby-Spera, 1988a; Fig. 2). Compositionally homogeneous pyroclasts in both areas, including monolithologic tuff breccias and devitrified glass with little or no textural modification, document proximal volcanic activity. The Gran Cañon Formation contains intercalated pillow lava in the upper part of the section; in the Sierra de San Andres, numerous dikes and sills of porphyritic pyroxene andesite intrude the volcaniclastic rocks. The 148 ± 3 Ma hornblende ^{40}Ar/^{39}Ar minimum age reported here from Cerro El Calvario is the only direct age data available from volcaniclastic strata in the central and southeastern Sierra de San Andres (Fig. 1). However, the ca. 160 Ma age of the oldest crosscutting plutons in the Sierra San Andres–San Pablo area implies pre–160 Ma ages for at least some of these strata. Middle Jurassic strata of the Gran Cañon Formation on Cedros (ca. 173–164 Ma) have much better age constraints, as well as more intact basement depositional framework. Although the Sierra de San Andres sections studied by Moore (1984) may contain a higher proportion of epiclastic versus primary volcaniclastic rock compared to the Gran Cañon Formation (Busby-Spera, 1988a), and details of the rift-basin stratigraphy documented from the Gran Cañon Formation (Busby-Spera, 1988a; Critelli et al., 2002) are not recognized in the Sierra de San Andres, we nevertheless conclude that vent facies volcanic and volcaniclastic sequences in both areas are broadly correlative proximal volcanic arc sequences.

The ~1840-m-thick sandstone member of the San Hipolito Formation at Punta San Hipolito is comprised of turbiditic volcanogenic sandstone, siltstone, and conglomerate, with minor tuff and limestone (Finch and Abbott, 1977; Pessagno et al., 1979; Fig. 2). Upper Pliensbachian and/or Toarcian (Early Jurassic) radiolarians from thin limestone beds near the base of the section (Whalen and Pessagno, 1984) indicate an absolute age of ca.195–180 Ma (cf. Palmer and Geissman, 1999). The overall thickening and coarsening-upward character of the section appears to record submarine fan progradation. The Early Jurassic age of the San Hipolito section suggests that it is partly or entirely older than the Cerro El Calvario sequence. If this is the case, the sandstone member may represent part of an earlier forearc basin sequence that predates inception of proximal Middle Jurassic magmatism in the area.

We conclude that there is not a strong basis for distinguishing Jurassic basement on Cedros Island from that on the Vizcaíno Peninsula. Distinctive, Late Jurassic strata of the Eugenia Formation, which depositionally overlaps basement on Cedros Island and the northern Vizcaíno Peninsula, supports this conclusion. The Eugenia Formation cover sequence rules out significant post–Late Jurassic displacements between Cedros Island and the Vizcaíno Peninsula, and similarity of pre–Eugenia Jurassic assemblages as outlined above is compatible with development of these areas as parts of the same Jurassic arc. An important implication of this correlation is that the Late Triassic Vizcaíno Peninsula Ophiolite is part of the rifted basement framework to the Middle Jurassic Cedros Island Ophiolite.

Cedros Island

The record of Jurassic arc magmatism on Cedros Island is particularly clear. Arc-ankaramites and boninitic affinity volcanic rocks at the base of the ~5-km-thick Choyal Formation arc complex record an initial Early Jurassic(?) phase of upper plate extension and renewed arc volcanism within the framework of Late Triassic ophiolite basement. These rocks are intruded by shallow-level Punta Norte granitoids that yield zircon U-Pb ages of 166 ± 3 Ma presented here and which must also predate major Middle Jurassic extension and formation of the Cedros Island Ophiolite at 173 ± 2 Ma. The ankaramites are characterized as extreme subalkaline, low Al_2O_3, and low TiO_2 chrome-diopside phenocryst compositions compatible with high degrees of melting of a sub-arc mantle source (Kimbrough, 1982). Ankaramites of similar composition occupy a forearc position in the Solomon Islands (Stanton and Bell, 1972), suggesting the Choyal Formation may indicate forearc, as opposed to backarc, rifting.

The Choyal Formation volcanic pile appears to record an early stage of arc rifting characterized by eruption of primitive ankaramites, boninites, low-K arc tholeiites, and calc-alkaline series magmas that culminated with major rifting and formation of the Cedros Island Ophiolite. The plagiogranite zircon U/Pb age of 173 ± 2 Ma is interpreted to be the time of formation of the main plutonic-volcanic crustal section of the ophiolite. This age is in good agreement with the Bajocian-Callovian paleontologic age from the middle part of the overlying Gran Cañon Formation based mainly on the occurrence here of bivalve *Bositra buchi* (Roemer) discovered by Frank Kilmer.

Critelli et al. (2002) inferred a phase of Middle Jurassic arc extension and rifting recorded by basaltic lavas interstratified with dacitic pyroclastic rocks near the top of the Gran Cañon Formation. There, caldera-forming events associated with arc-rifting typically occur within 10–15 m.y. after the birth of a backarc basin. The zircon U-Pb age of 164.0 ± 3 Ma from dacite tuff in the Gran Cañon Formation directly dates the time of inferred rifting (Fig. 2). In the alternative forearc model proposed here, the change in magmatic composition with time reflects transition from an early phase of forearc rifting and rapid crust production reflected by Choyal arc volcanism and ophiolite formation to more a more stabilized post-boninintic phase of arc evolution. Rheological studies suggest that areas of thick crust and high heat flow in active volcanic arcs are mechanically weak zones that focus lithospheric extension (Kusznir and Park, 1987). We speculate that the initial phase of forearc arc magmatism represented by the Choyal Formation acted as a guide to the location of rifting and extension that produced the Cedros Island Ophiolite.

Late Jurassic to Early Cretaceous Tectonics and Sedimentation—The Coloradito and Eugenia Formations

The Coloradito and Eugenia Formations on Cedros Island and the northern Vizcaíno Peninsula record a spectacular and distinctive episode of tectonics and sedimentation in the Vizcaíno-Cedros region (Kilmer, 1979; Moore, 1983; Barnes, 1984; Boles and Landis, 1984; Hickey, 1984, Figs. 1 and 2). These widespread units are characterized by marine volcanogenic strata intercalated with local pillow lavas and major sedimentary mélange/olistostrome intervals with megablocks of Triassic chert, Paleozoic sandstone, and continental detritus. Coloradito and Eugenia strata are in clear depositional contact on Gran Cañon Formation strata on Cedros Island and on Late Triassic Vizcaíno Peninsula Ophiolite basement and its sedimentary cover in the Punta Quebrada and Morro Hermoso areas. The 141.5 ± 3 Ma crystallization age for the Santa Eugenia tuff (Hickey, 1984) from the Eugenia Formation reported here overlaps the age range of arc plutons in the Sierra de San Andres (160–135 Ma) and supports the correlation of earlier investigators that volcanogenic strata of the Eugenia Formation were derived from arc volcanoes of the Cedros–San Andres arc complex (Rangin, 1978; Barnes, 1984).

The Coloradito Formation has been mapped only on Cedros Island, where it is characterized by a sheared black argillite matrix with abundant olistostrome slide blocks (Kilmer, 1979; Boles and Landis, 1984), whereas the Eugenia is recognized in both Cedros Island and the Vizcaíno Peninsula. We emphasize here broad similarities in depositional style and composition of mudflows and olistostrome blocks in both the Eugenia and Coloradito Formations that suggest to us these are part of the sample tectonic/stratigraphic unit. Tithonian to Neocomian ages from the Eugenia Formation on Cedros and the Vizcaíno (Hickey, 1984; Boles and Landis, 1984) provide the best basis for stratigraphic age assignment.

The Coloradito and Eugenia Formations provide the earliest definite record of arc-continent interaction in the Vizcaíno-Cedros area and were related to an arc-continent collision event (e.g., Kimbrough, 1982; Boles and Landis, 1984). However, contractional structures related to such an event have not been identified; regional thrusting and nappe structures related to the Late Jurassic Nevadan orogeny by Rangin (1978) have been reinterpreted as low-angle normal faults and/or conformable stratigraphic sequences (e.g., Moore, 1983; Sedlock, 1988b).

The dramatic shift in depositional environment represented by Coloradito-Eugenia required the initiation of a high-energy transport system with steep topographic gradients. The cause of uplift requires a tectonic explanation. We suggest that the Coloradito-Eugenia Formations were deposited in an extensional or transtentional depositional basin that collected active arc volcaniclastic debris as well as basement blocks from topographic highs on the edges of the basin. This interpretation is supported by the nonorogenic geochemical characteristics of pillow lavas that are intercalated in the Eugenia Formation (Rangin et al., 1983). The Sierra Placer serpentinite mélange represents a regional scale structure on the margin of the Coloradito-Eugenia depositional basin that potentially could have accommodated some of the displacement associated with formation of this basin.

Correlation of Vizcaíno-Cedros Region to California–Oregon Plate Edge Assemblages

Late Paleozoic to Early Triassic tectonic truncation and accretion processes along the southwestern margin of the Cordillera by strike-slip faulting was followed by establishment of a new, active continental margin arc that recorded eastward subduction of oceanic crust beneath North America (Hamilton, 1969; Burchfiel and Davis, 1972; Saleeby, 1982; Fig. 9). Fringing terranes that constitute much of the western Sierra-Klamath belt, the basement of the Cretaceous Great Valley forearc basin, and the California Coast Range ophiolite occupy the forearc region of the continental margin arc (Saleeby and Busby-Spera, 1992).

These fringing terranes are characterized by two distinct episodes of rifting and magmatism that produced ophiolite assemblages; the first occurred in the Late Triassic to Early Jurassic, and the second occurred in the Middle Jurassic (Saleeby and Busby-Spera, 1992). This history closely parallels events in the adjacent continental margin arc.

Schweickert and Cowan (1975) interpreted the fringing terranes as exotic island arc terranes that collided with North America near the end of the Jurassic to produce the Nevadan orogeny. Subsequent work, however, has documented a wide variety of stratigraphic, sedimentary provenance, pluton stitching, geochemical, and paleomagnetic relations that indicate that the fringing terranes, particularly those elements within the western Sierra-Klamath belt, must have formed as part of the North America plate margin (Saleeby, 1981; Saleeby et al., 1982; Harper and Wright, 1984; Wyld and Wright, 1988; Edelman et al., 1989). The western Sierra-Klamath belt in this view was constructed in large part across older ensimatic basement assemblages previously accreted to the Cordilleran plate edge and formed the framework for the later Middle Jurassic episode of rifting that led to the formation of the Josephine and Smartville Ophiolites and Coast Range Ophiolite (Saleeby, 1981, 1992).

Saleeby and Busby-Spera (1992) related initiation of widespread extension and suprasubduction zone magmatism in the Late Triassic to a major change in absolute plate motion of North America indicated by the J1 cusp in the apparent polar wander path (May and Butler, 1986). Uncertainties in timing of the J1 cusp permit it to have occurred as early as ca. 220 Ma, which corresponds closely to the age of the Vizcaíno Peninsula Ophiolite. Late Triassic–Early Jurassic rocks in California that broadly correlate to the Vizcaíno Peninsula Ophiolite include the Rattlesnake Creek terrane of the western Klamath Mountains (Wright, 1982) and the Fiddle Creek complex (Edelmen et al., 1989) and assemblages in the lower American River and lower Kaweah River areas (Saleeby, 1982) in the western Sierra Nevada (Fig. 9).

The geologic evolution of Vizcaíno-Cedros Mesozoic assemblages matches closely that of the fringing terranes of California-Oregon along strike to the north (cf. Jones et al., 1976). Both regions display the two main phases of extensional rifting and

ophiolite generation, have similar petrologic affinities, were overprinted by Late Jurassic–Early Cretaceous arc magmatism, and form basement to Cretaceous forearc basin sequences. Two distinct episodes of extension that embedded a Middle Jurassic ophiolite sequence within the framework of a Late Triassic ophiolite is a key relationship that links the history of the Vizcaíno-Cedros region to the western Sierra-Klamath belt. In addition, the western Sierra-Klamath belt was intruded by plutonic suites of broadly similar age (~165–135 Ma) and chemistry (e.g., Bateman, 1992; Saleeby et al., 1989) as those of the Vizcaíno-Cedros region. The close resemblance in so many respects leaves little doubt that these areas are parts of the same tectonic regime.

Despite the close parallels, there are at least three significant differences between the Vizcaíno-Cedros region and the western Sierra-Klamath belt: (1) the western Sierra de San Andres remnant of the Vizcaíno Peninsula Ophiolite represents a complete spreading center ophiolite sequence with depleted upper mantle and mafic crustal rocks, including a well-developed sheeted dike complex. Substantial tracts of juvenile oceanic crust such as this are not recognized in western Sierra-Klamath belt, although an extensional framework is clearly implied by sheeted dikes and basinal sequences possibly reflecting diffuse spreading (Saleeby, 1992). (2) Primary relationships with older late Paleozoic North American basement assemblages are not directly preserved in the Vizcaíno-Cedros region as they are in the western Sierra-Klamath belt. The absence of such relationships may in part reflect the more organized spreading and development of juvenile Late Triassic oceanic crustal sequences in the Vizcaíno-Cedros region, as well as the comparatively limited area of exposure. The possibility of zircon inheritance in one of the fractions of the San Pablo tonalite reported here (sample 322–5), as well as $^{87}Sr/^{86}Sr$ initials that range up to ~0.706 in several samples including 322–5 may, however, indicate that components of continental basement are present in the source region of these magmas. (3) The Vizcaíno-Cedros region did not experience the strong contractional deformation characteristic of the western Sierra-Klamath belt that overlapped and post-dated Middle Jurassic magmatism (Fig. 9). This deformation is expressed in the Klamath Mountains as dramatic telescoping via a major west-vergent thrust system that imbricated terranes. This deformation, referred to as the Siskiyou event by Coleman et al. (1988), is represented in the western Sierras by a variety of structures, including southwest vergent thrusting, tight folds, and steeply dipping axial surface schistosities (Sharp, 1988). The deformation may have been an early phase of Late Jurassic Nevadan orogeny thrusting (Saleeby and Busby-Spera, 1992).

We suggest that the Coloradito-Eugenia Formations are alongstrike correlatives of *Buchia Piochii*-bearing, Late Jurassic volcanogenic strata in the western Peninsular Ranges near San Diego (Fife et al., 1967) from Penasquitos Canyon north to the San Dieguito River (Fig. 9). The timing of Late Jurassic Coloradito–Eugenia basin initiation further suggests it may be a more southerly expression of the sinistral-sense transtentional-transpressional environment documented for the Owens Mountain dike swarm of the western Sierra Nevada foothills by Wolf and Saleeby (1995), who related this deformation to abrupt large-magnitude changes in the

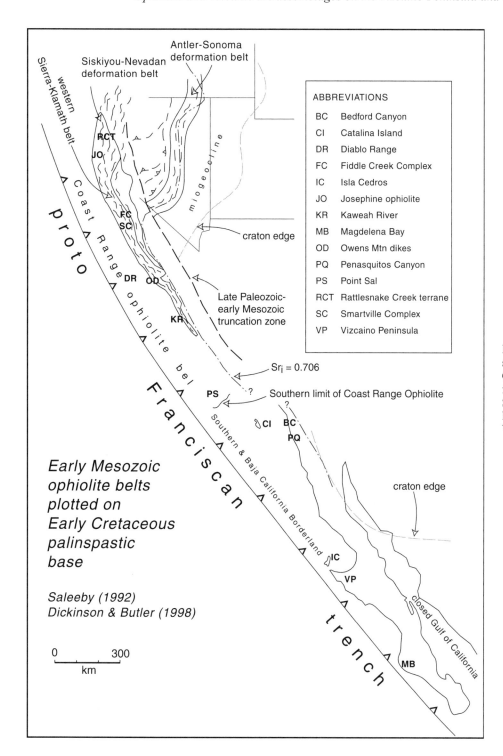

Figure 9: Regional ophiolite belts and selected assemblages of southwestern Cordillera plotted on palinspastic base for Early Cretaceous time (adapted from Saleeby, 1992, and Dickinson and Butler, 1998). Craton edge in México from Valencia-Moreno et al. (2001).

direction of North America plate motion at the J2 cusp in the apparent polar wander path for North America (ca. 150 Ma; May and Butler, 1986). This Late Jurassic belt of deformation, however, is inferred to extend southward from the Sierra Nevada across the Mojave Desert and into northern Sonora, México, well inboard of the borderland position occupied by the Coloradito-Eugenia basin.

Middle Jurassic Ophiolites

The Middle Jurassic Cedros Island Ophiolite is considered here to be part of the same regime of intra-arc/forearc rifting and basin formation represented by the Smartville, Josephine, and Coast Range Ophiolites of California. All of these ophiolites are interpreted as having originated in suprasubduction zone setting(s) based on age, geochemical data, and arc volcanic and volcaniclastic strata that depositionally overlie ophiolite basement (e.g., Evarts, 1977; Menzies et al., 1980; Harper, 1984; Shervais and Kimbrough, 1985; Kimbrough, 1985; Robertson, 1989, 1990; Shervais, 1990; Saleeby, 1992; Ingersoll, 2000; Harper, 2002). The Cedros Island, Coast Range, and Josephine Ophiolites include igneous rocks with boninitic affinity (Kimbrough, 1982; Harper, 1984). The Smartville Ophiolite, however, includes no boninites and apparently was derived from less depleted mantle sources.

The Cedros Island Ophiolite zircon U/Pb plagiogranite age of 173 ± 2 Ma overlaps the 173–165 Ma range of plagiogranite zircon crystallization ages from the Coast Range (Mattinson and Hopson, 1992). Note that the Coast Range Ophiolite ages quoted here are distinctly older than the 164–156 Ma range reported earlier by Hopson et al. (1981). These younger ages were reevaluated by Mattinson and Hopson (1992) based on newer, high-precision $^{207}Pb*/^{206}Pb*$ data that indicate older ages than originally thought for igneous generation of the Coast Range Ophiolite. Zircon U/Pb data indicate ages of ca. 164–162 Ma for the Josephine Ophiolite (Harper et al., 1994, 2002) and ca. 164–160 Ma for the Smartville Complex (Saleeby et al., 1989; Edelman et al., 1989), which suggests these assemblages may be up to ca. 10 m.y. younger.

The Cedros Island Ophiolite was blanketed shortly after its formation by proximally derived arc volcanic sediments (Kimbrough, 1984; Busby-Spera, 1988a). In this respect it bears closest similarity to the Josephine and Smartville ophiolites, as well as the Diablo Range segment of the Coast Range Ophiolite that is overlain by the Lotta Creek tuff (Evarts, 1977; Robertson, 1989). The southern Coast Range ophiolite (e.g., Point Sal), in contrast, is overlain by a distal volcanopelagic succession that is interpreted to track progressive northward transport of ophiolite basement from its site of origin during Middle to Late Jurassic time (Hopson et al., 1981; Pessagno et al., 2000).

The Cedros Island and Coast Range Ophiolites were underthrust and attenuated by high P/T subduction complex assemblages of the Franciscan Complex (Platt, 1975; Jones et al., 1976; Jayko et al., 1987; Sedlock, 1988a, 1988b; Sedlock and Isozaki, 1990; Baldwin and Harrison, 1992). Dating of high-grade eclogitic and amphibolite mélange blocks yield a spread of apparent ages of ca. 170–160 Ma from Cedros Island (Baldwin and Harrison, 1992) and ca. 165–150 Ma in Coast Ranges of California (Ross and Sharp, 1986, 1988; Mattinson, 1988); thermal modeling suggests these ages reflect initiation of subduction beneath hot suboceanic mantle of the hanging wall (e.g., Peacock, 1992). Ages of ca. 115–100 Ma for high-grade blueschist on Cedros Island (Baldwin and Harrison, 1992), and ca. 120–115 Ma on Catalina Island (Grove and Bebout, 1995) evidently record a separate subduction initiation event in the Southern California borderland.

Application of Forearc Spreading Models to the Vizcaíno-Cedros Region

A wide variety of plate tectonic reconstructions have been proposed for the Vizcaíno-Cedros region, all of which interpret parts or all of the Mesozoic ophiolite complexes as allochthonous terranes that were accreted to the North American plate margin by Late Jurassic (e.g., Rangin, 1978; Barnes, 1984). The Cedros Island Ophiolite, for example, has been interpreted as a backarc basin ophiolite formed behind an east-facing arc that collided and accreted with the North American margin (Landis and Boles, 1984; Busby-Spera, 1988a; Critelli et al., 2002; cf. Schweickert and Cowan, 1975). There are several problems with this interpretation (cf., Saleeby, 1992). (1) The entire frontal arc and accretionary prism associated with the postulated east-facing arc is missing. (2) The predicted crustal suture that must lie to the east has not been recognized, and there is no structural or stratigraphic evidence for a major collision and crustal shortening at this time. (3) The ophiolite has boninitic affinities characteristic of rifted forearc settings that are unlike mid-oceanic ridge basalt and backarc basin basalts typical of Cenozoic Western Pacific backarc basin ocean crust sequences (Taylor, 1995). (4) Correlative ophiolite and arc assemblages of the western Sierra-Klamath belt were constructed in part across North American affinity basement assemblages as discussed above.

We conclude that forearc rifting models developed for Middle Jurassic California ophiolite belts best explain coeval rifting and magmatism in the Vizcaíno-Cedros region. Two different forearc rifting models have been proposed. The Saleeby (1981, 1992) model calls for strongly oblique forearc rifting along the continental margin driven by oblique subduction and slab rollback of Panthalassan lithosphere; the analog is the active Andaman Sea, where strongly oblique subduction of Indian Ocean lithosphere beneath the Sumatra drives forearc rifting and dextral transcurrent faulting. In this model, Middle Jurassic ophiolite fragments underwent margin-parallel transport and are interpreted as parautochthonous assemblages. Stern and Bloomer (1992) offered an alternative model based on analogy to the Eocene Izu-Bonin-Mariana arc system. This model, the "subduction zone infancy model," suggests that generation of California ophiolite basement occurred in situ and was related to subduction initiation and slab rollback that drove strong extension oriented generally perpendicular to the plate margin. In both forearc-rifting models, dynamics of the subducting plate is the critical factor initiating forearc spreading and magmatism.

Stern and Bloomer (1992) emphasized high magma production rates in the "infant arc" phase of evolution; they estimated a production rate of 120–180 km³/km/Ma of new crust for the Izu-Bonin-Mariana arc sustained over a 10 m.y. episode, although unaccounted for remnants of Philippine Sea plate embedded in the forearc region would reduce this estimate (e.g., Debari et al., 1999). Gravity and magnetic highs over the Central Valley of California were interpreted by Cady (1975) as a signature of Coast Range Ophiolite basement beneath the Great Valley Group forearc strata. If this is correct, the extent of Middle Jurassic ophiolite basement along the Alta and Baja California margin with its ca. 10 m.y. duration of emplacement is consistent with high magma production rates as estimated by Stern and Bloomer (1992) for the Izu-Bonin-Mariana infant arc phase of evolution.

Boninites (53–56% SiO_2, MgO > 8%, TiO_2 < 0.5%), arc ankaramites, and low-Ti arc tholeiites in Jurassic rocks on Cedros Island and in the Sierra de San Andres indicate extensive melting of depleted mantle sources. Boninites are produced by melting of harzburgitic mantle (olivine + orthopyroxene) in the forearc region of intraoceanic island arcs and are characteristic of the earliest stages of subduction (Pearce et al., 1992). Experimental studies indicate that boninites form by melting at shallow mantle depths and that deeper melting to produce associated arc-tholeiites is expected (van der Laan et al., 1989). Oceanic arcs are the type example for boninite magmatism, but at least one example from a continental margin arc–backarc system is recognized (Piercey et al., 2001). Although geochemical affinities and magma flux arguments alone cannot indisputably distinguish between backarc, intra-arc, or forearc settings for magma generation (e.g., Taylor, 1995), the geochemical evidence from Mesozoic rocks of west-central Baja California favors a forearc setting. Pillow lavas and sheeted dikes of the Vizcaíno Peninsula Ophiolite are mostly mid-oceanic-ridge basalt and low-Ti arc-tholeiite affinity but also contain a few boninites dikes and extrusive rocks and locally abundant orthopyroxene in cumulate sequences that may indicate a boninite affinity.

Comparison of ophiolite formation age versus the age of Franciscan high-grade blocks that mark subduction initiation to the west along their outboard edge provides a test of alternative forearc versus backarc models of ophiolite generation; the backarc model predicts that initiation of Franciscan subduction should postdate accretion of the backarc and associated features by arc-arc collision along the margin, whereas the forearc model predicts that subduction initiation should predate forearc extension and ophiolite generation. The oldest metamorphic ages from high-grade Franciscan complex rocks on Cedros Island and California appear to just overlap igneous generation of the Jurassic ophiolites. Wakabayashi and Dilek (2000) concluded that the evidence favors a backarc collision/accretion model, while Dickinson et al. (1996) noted that the overlap in ages might present a problem for the backarc collision/accretion model. We believe that it is difficult to reconcile the relative timing of these events with certainty because (1) it involves comparison of different dating methods: U/Pb sphene, K/Ar, and $^{40}Ar/^{39}Ar$ hornblende and white micas from Franciscan metamorphic rock versus U/Pb zircon from ophiolite plagiogranite; and (2) there is no certainty that the ages of the high-grade Franciscan blocks at this crustal level represent the precise onset of widespread lithospheric subduction along the margin. Nevertheless, the close coincidence of these events is compatible with Franciscan subduction initiation driving forearc rifting and magmatism to produce the Middle Jurassic ophiolites.

It is hard to distinguish between the margin-normal (Saleeby, 1992), versus margin-parallel (Stern and Bloomer, 1992), forearc spreading models for Jurassic California ophiolites. The ophiolite assemblages are generally too disrupted to reliably reconstruct spreading directions from sheeted dikes on a regional basis. Paleomagnetic data and radiolarian faunal successions from the southern Coast Range Ophiolite, however, may indicate that ophiolite remnants here originated far to the south near the paleo-equator and were transported rapidly north before accreting to North America in the Late Jurassic (Hopson et al., 1981; Pessagno et al., 2000); this evidence supports the Saleeby (1992) model. Stern and Bloomer (1992), however, cited conflicting paleomagnetic studies (e.g., Mankinen et al., 1991) and similar faunal successions from presumably in situ strata in the western Klamath Mountains and concluded that the Coast Range Ophiolite is not necessarily far-traveled. It is relevant to note that paleomagnetic data from Cretaceous Valle forearc strata that suggested 12° of northward displacement of the basin relative to North America (Smith and Busby, 1993) must now be questioned because possible effects of inclination shallowing in sandstones have not been accounted for (Tan and Kodama, 1998). Recent work on Cretaceous Valle Group strata also indicates Cenomanian-Turonian provenance linkage to the adjacent Peninsular Ranges batholith history (Kimbrough et al., 2001). Additionally, the strongly calcic nature of arc-plutonic rocks in the Vizcaíno-Cedros region (165–135 Ma) noted here is a feature in common with the adjacent Peninsular Ranges batholith (140–80 Ma), providing circumstantial evidence for in situ formation of Vizcaíno-Cedros as forearc basement to the Peninsular Ranges. There is, however, no apparent Peninsular Ranges source for peraluminous Late Jurassic granite in Albian-age Valle Group strata, a relationship that may be consistent with margin-parallel transport (Kimbrough et al., 1987).

If Triassic and Jurassic ophiolites of the Vizcaíno-Cedros region were generated in the forearc of the Cordilleran continental arc, then why isn't there more continentally derived detritus in overlying forearc strata? A possible answer to this question comes from Busby-Spera (1988b), who provided evidence that the early Mesozoic continental arc was strongly extensional and formed "a continuous graben depression, more than 1000 km long, similar to the modern extensional arc of Central America…. This depression may have acted as a long-lived trap (more than 40 m.y.) for craton-derived quartz sand."

Another potential problem with the Stern and Bloomer (1992) infant-arc model applied to the Jurassic Cordillera is that subduction was apparently initiated along the margin in the Triassic following strike-slip truncation of the margin (cf. Ingersoll, 2000). This problem may in part reflect incomplete knowledge of tectonic environments in the early Mesozoic. Armstrong and Ward (1993) noted that early Mesozoic arc magmatism that tracks from the outer part of the miogeocline southward into the continental interior in southern California, Arizona, and México was characterized by separate Late Triassic and Middle to Late Jurassic culminations, coeval with the two episodes of ophiolite generation in the Vizcaíno-Cedros region. We speculate that the earlier subduction regime was followed by a major plate boundary reorganization in the Middle Jurassic associated with initiation of Franciscan subduction that triggered voluminous forearc spreading. This event corresponds closely in time with the onset of spreading in the central Atlantic at ca. 172 Ma. If subduction was initiated along an old fracture zone (cf. Izu-Bonin-Marianas), it would have required Late Triassic to Early Jurassic spreading centers to have been oriented at high angles to the margin. Although we also favor a forearc origin for the Vizcaíno Peninsula Ophiolite, Late Triassic crustal production rates appear to have been lower than for Middle Jurassic time. MacPherson and Hall (2001) noted that Eocene boninite magmatism of the IBM forearc may require additional tectonic or thermal factors, perhaps a plume-related thermal anomaly, to explain this high-flux magmatic event.

A series of recent papers (e.g., Tardy et al., 1994; Johnson et al., 1999; Dickinson and Lawton, 2001; Wetmore et al., 2002) proposed that Jura-Cretaceous intraoceanic arc terranes were accreted to Mexican segments of the western North America margin from late Early Cretaceous to Late Cretaceous time. Accreted oceanic arc assemblages in these models include the Guerrero terrane and the Alisitos arc of Baja California. The forearc generation model favored here presents a difficulty for these models, since the Vizcaíno-Cedros region sits outboard of the postulated exotic assemblages. Independent of the forearc model, the presence of continentally derived sediment in Late Jurassic strata of the Coloradito and Eugenia Formations must be accounted for.

CONCLUSIONS

The Late Triassic to Early Cretaceous history of magmatism and tectonics in the Vizcaíno-Cedros region suggests immature magmatic arc and associated basinal environments. Stratigraphic and intrusive relations between different elements of this system are consistent with geographic continuity in the same paleogeographic setting beginning in Late Triassic time. Two major phases of rifting, Late Triassic and Middle Jurassic, produced distinct suprasubduction-zone ophiolite assemblages. Subsequent Late Jurassic to Early Cretaceous arc magmatism intrudes and caps rifted screens of older ophiolite assemblages.

Application of a forearc generation model to this setting is based on (1) boninitic affinities of ophiolite and arc rocks; (2) consideration of relationships to coeval ophiolite and arc rocks of the western Klamath Mountains, Sierra Nevada foothills, and California-Oregon Coast Ranges; (3) the viability of actualistic models based on western Pacific arc systems; and (4) the apparent absence of a major suture belt and associated forearc and subduction complex assemblages predicted by collision-accretion models. The forearc model suggests that the Vizcaíno-Cedros region must constitute autochthonous or parautochthonous assemblages that were constructed outboard of the early Mesozoic continental arc. Although there are unresolved questions, we believe this interpretation best explains local relationships in the context of regional plate tectonic evolution of the Cordilleran margin.

The transition from boninitic and primitive arc tholeiitic magmatism associated with subduction initiation, forearc extension, and rapid crustal production to an established arc characterized by a more well-defined magmatic front behind cooler, stabilized forearc lithosphere is an outstanding feature of subduction zone evolution (Stern and Bloomer, 1992). The Vizcaíno-Cedros region may provide important insight into Cenozoic western Pacific arcs representative of this process.

ACKNOWLEDGMENTS

Financial support was provided by the Geological Society of America, National Science Foundation, and Department of Geological Sciences at San Diego State University. We are grateful to Gordon Gastil and Frank Kilmer for invaluable encouragement in the early phase of our work in Baja California. Marty Grove assisted with $^{40}Ar/^{39}Ar$ analyses and interpretation. Stan Keith provided whole-rock chemistry. Tony Carrasco provided able assistance in figure preparation. John Shervais, Brian Mahoney, and Arturo Martin reviewed the manuscript and made helpful suggestions.

REFERENCES CITED

Armstrong, R.L., and Ward, P.D., 1993, Late Triassic to earliest Eocene magmatism in the North American Cordillera: Implications for the Western Interior Basin, *in* Caldwell, W.G.E., and Kauffman, E.G., eds., Evolution of the Western Interior Basin: Geological Association of Canada Special Paper 39, p. 49–72.

Bailey, E.H., Blake, M.C., Jr., and Jones, D.L., 1970, On-land Mesozoic ocean crust in California Coast Ranges: U.S. Geological Survey Professional Paper 700-C, p. C70-C81.

Baldwin, S.L., Harrison, T.M., and Fitz Gerald, J.D., 1990, Diffusion of ^{40}Ar in metamorphic hornblende: Contributions to Mineralogy and Petrology, v. 105, p. 691–703.

Baldwin, S.L., and Harrison, T.M., 1992, The P-T-t history of blocks in serpentinite-matrix mélange, west-central Baja California: Geological Society of America Bulletin, v. 104, p. 18–31.

Barnes, D.A., 1982, Basin analysis of volcanic arc-derived, Jura-Cretaceous sedimentary rocks, Vizcaíno Peninsula, Baja California Sur, Mexico [Ph.D. thesis]: Santa Barbara, University of California, 240 p.

Barnes, D.A., 1984, Volcanic arc derived, Mesozoic sedimentary rocks, Vizcaíno Peninsula, Baja California Sur, Mexico, *in* Frizzell, V.A., Jr., ed., Geology of the Baja California peninsula: Field Trip Guidebook—Pacific Section, Society of Economic Paleontologists and Mineralogists Pacific Section, v. 39, p. 119–130.

Barnes, D.A., and Mattinson, J.M., 1981, Late Triassic–Early Cretaceous age of eugeosynclinal terranes, western Vizcaíno Peninsula, Baja California Sur, Mexico: Geological Society America Abstracts with Programs, v. 13, p. 43.

Bateman, P.C., 1992, Plutonism in the central part of the Sierra Nevada Batholith, California: U.S. Geological Survey Professional Paper 1483, 186 p.

Blake, M.C., Jr., Jayko, A.S., Moore, T.E., Chavez, V., Saleeby, J., and Seel, K., 1984, Tectonostratigraphic terranes of Magdalena Island, Baja California Sur, *in* Frizzell, V.A., Jr., ed., Geology of the Baja California peninsula: Field Trip Guidebook—Pacific Section, Society of Economic Paleontologists and Mineralogists Pacific Section, v. 39, p. 183–191.

Boles, J.R., 1978, Basin analysis of the Eugenia Formation (Late Jurassic), Punta Eugenia area, Baja, California, *in* Howell, D.G., and McDougall, K.A., eds., Mesozoic paleogeography of the western United States, Pacific Coast Paleogeography Symposium 2: Sacramento, California, Society of Economic Paleontologists and Mineralogists, p. 493–498.

Boles, J.R., and Landis, C.A., 1984, Jurassic sedimentary mélange and associated facies, Baja California, Mexico: Geological Society of America Bulletin, v. 95, p. 513–521.

Burchfiel, B.C., and Davis, G.A., 1972, Structural framework and evolution of the southern part of the Cordilleran orogen, western United States: American Journal of Science, v. 272, p. 97–118.

Busby, C.J., and Ingersoll, R.V., 1995, Tectonics of sedimentary basins: Cambridge, Massachusetts, Blackwell Science, p. 579.

Busby, C.J., Smith, D., Morris, W., and Fackler-Adams, B.N., 1998, Evolutionary model for convergent margins facing large ocean basins; Mesozoic Baja California, Mexico: Geology, v. 26, p. 227–230.

Busby-Spera, C.J., 1988a, Evolution of a Middle Jurassic back-arc basin, Cedros Island, Baja California: Evidence from a marine volcaniclastic apron: Geologic Society of America Bulletin, v. 100, p. 218–233.

Busby-Spera, C.J., 1988b, Speculative tectonic model for the early Mesozoic arc of the southwest Cordilleran United States: Geology, v. 16, p. 1121–1125.

Busby-Spera, C.J., and Boles, J.R., 1986, Evolution of subsidence styles in forearc basin; example from Cretaceous of southern Vizcaíno Peninsula, Baja California, Mexico: American Association of Petroleum Geologists Bulletin, v. 70, p. 463–464.

Cady, J.W., 1975, Magnetic and gravity anomalies in the Great Valley and Western Sierra Nevada Metamorphic Belt, California: Boulder, Colorado, Geological Society of America Special Paper 168, p. 56.

Coleman, R.G., Manning, C.E., Mortimer, N., Donato, M.M., and Hill, L.B., 1988, Tectonic and regional metamorphic framework of the Klamath Mountains and adjacent coast ranges, California and Oregon, *in* Ernst, W.G., ed., Metamorphism and crustal evolution of the western United States, Rubey Volume VII: Englewood Cliffs, New Jersey, Prentice Hall, p. 1061–1097.

Coney, P.J., Jones, D.L., and Monger, J.W.H., 1980, Cordilleran suspect terranes: Nature, v. 288, p. 329–333.

Critelli, S., Marsaglia, K.M., and Busby, C.J., 2002, Tectonic history of a Jurassic backarc-basin sequence (the Gran Cañon Formation, Cedros Island, Mexico), based on compositional modes of tuffaceous deposits: Geological Society of America Bulletin, v. 114, p. 515–527.

Debari, S.M., Taylor, B., Spencer, K., and Fujioka, K., 1999, A trapped Philippine Sea Plate origin for MORB from the inner slope of the Izu-Bonin Trench: Earth and Planetary Science Letters, v. 174, p. 183–197.

Dickinson, W.R., and Butler, R.F., 1998, Coastal and Baja California paleomagnetism reconsidered: Geological Society of America Bulletin, v. 110, p. 1268–1280.

Dickinson, W.R., and Lawton, T.F., 2001, Carboniferous to Cretaceous assembly and fragmentation of Mexico: Geological Society of America Bulletin, v. 113, p. 1142–1160.

Dickinson, W.R., Hopson, C.A., and Saleeby, J., 1996, Alternate origins of the Coast Range Ophiolite (California): Introduction and implications: GSA Today, v. 6, no. 2, p. 1–10.

Dilek, Y., Moores, E.M., Elthon, D., and Nicolas, A., 2000, Ophiolites and oceanic crust: New insights from field studies and the Ocean Drilling Program: Boulder, Colorado, Geological Society of America Special Paper 349, 552 p.

Edelman, S.H., Day, H.W., and Bickford, M.E., 1989, Implications of U-Pb zircon ages for the tectonic setting of the Smartville and Slate Creek complexes, northern Sierra Nevada: Geology, v. 17, p. 1032–1035.

Evarts, R.C., 1977, The geology and petrology of the Del Puerto ophiolite, Diablo Range, central California Coast Ranges, *in* Coleman, R.G., and Irwin, W.P., eds., North American ophiolites: Oregon Department of Geology and Mineral Industries Bulletin, v. 95, p. 121–139.

Fife, D.L., Minch, J.A., and Crampton, P.J., 1967, Late Jurassic age of the Santiago Peak Volcanics, California: Geological Society of America Bulletin, v. 78, p. 299–303.

Finch, J.W., and Abbott, P.L., 1977, Petrology of a Triassic marine section, Vizcaíno Peninsula, Baja California Sur, Mexico: Sedimentary Geology, v. 19, p. 253–273.

Fisher, R.V., and Smith, A.G., 1991, Sedimentation in volcanic settings: Society of Economic Paleontologists and Mineralogists Special Publication 45, p. 257.

Frizzell, V.A., Jr., ed., 1984, Geology of the Baja California peninsula: Field Trip Guidebook—Pacific Section, Society of Economic Paleontologists and Mineralogists Pacific Section, v. 39, 273 p.

Gaber, L.J., Foland, K.A., and Corbato, C.E., 1988, On the significance of argon release from biotite and amphibole during $^{40}Ar/^{39}Ar$ vacuum heating: Geochimica et Cosmochimica Acta, v. 52, p. 2457–2465.

Gastil, G., Morgan, G.J., and Krummenacher, D., 1978, Mesozoic history of peninsular California and related areas east of the Gulf of California, *in* Howell, D.G., and McDougall, K.A., eds., Mesozoic paleogeography of the western United States, Pacific Coast Paleogeography Symposium 2: Sacramento, California, Society of Economic Paleontologists and Mineralogists, p. 107–115.

Gromet, L.P., and Silver, L.T., 1987, REE variations across the Peninsular Ranges batholith: Implications for batholithic petrogenesis and crustal growth in magmatic arcs: Journal of Petrology, v. 28, p. 77–125.

Grove, M., and Bebout, G.E., 1995, Jurassic and Cretaceous tectonic evolution of coastal southern California: Insights from the Catalina Schist: Tectonics, v. 14, p. 1290–1308.

Hamilton, W.B., 1969, Mesozoic California and the underflow of Pacific mantle: Geological Society of America Bulletin, v. 80, p. 2409–2430.

Harper, G.D., 1984, The Josephine ophiolite, northwestern California: Geological Society of America Bulletin, v. 95, p. 1009–1026.

Harper, G.D., 2002, A Lau Basin model for the formation of the Josephine Ophiolite and related rift facies, Klamath Mountains, Oregon-California: Geological Society of America Abstracts with Programs, v. 34, no. 5, p. A-22.

Harper, G.D., and Wright, J.E., 1984, Middle to Late Jurassic tectonic evolution of the Klamath Mountains, California-Oregon: Tectonics, v. 3, p. 759–772.

Harper, G.D., Saleeby, J.B., and Heizler, M., 1994, Formation and emplacement of the Josephine ophiolite and the Nevadan orogeny in the Klamath Mountains, California-Oregon: U/Pb zircon and $^{40}Ar/^{39}Ar$ geochronology: Journal of Geophysical Research, v. 99, p. 4293–4321.

Harrison, T.M., and Fitz Gerald, J.D., 1986, Exsolution in hornblende and its consequences for $^{40}Ar/^{39}Ar$ age spectra and closure temperature: Geochimica et Cosmochimica Acta, v. 50, p. 247–253.

Hickey, J., 1984, Stratigraphy and composition of a Jura-Cretaceous volcanic arc apron, Punta Eugenia, Baja California Sur, Mexico, *in* Frizzell, V.A., Jr., ed., Geology of the Baja California peninsula: Field Trip Guidebook—Pacific Section, Society of Economic Paleontologists and Mineralogists Pacific Section, v. 39, p. 149–160.

Hopson, C.A., Mattinson, J.M., and Pessagno, E.A., Jr., 1981, Coast Range Ophiolite, western California, *in* Ernst, W.G., ed., The geotectonic development of California, Rubey Volume I: Englewood Cliffs, New Jersey, Prentice-Hall, p. 418–510.

Ingersoll, R.V., 2000, Models for origin and emplacement of Jurassic ophiolites of northern California, *in* Dilek, Y., Moores, E., Elthon, D., and Nicolas, A., eds., Ophiolites and oceanic crust: New insights from field studies and the Ocean Drilling Program: Boulder, Colorado, Geological Society of America Special Paper 349, p. 395–402.

Jayko, A.S., Blake, M.C., Jr., and Harms, T., 1987, Attenuation of the Coast Range ophiolite by extensional faulting, and the nature of the Coast Range "thrust": Tectonics, v. 6, p. 475–488.

Johnson, S.E., Tate, M.C., and Fanning, C.M., 1999, New geologic mapping and SHRIMP U-Pb zircon data in the Peninsular Ranges Batholith, Baja California, Mexico: Evidence for a suture?: Geology, v. 27, p. 743–746.

Jones, D.L., Blake, M.C., Jr., and Rangin, C., 1976, The four Jurassic belts of northern California and their significance to the geology of the southern California borderland, *in* Howell, D.G., ed., Aspects of the geologic history of the California continental borderland: American Association of Petroleum Geologists Pacific Section Miscellaneous Publication 24, p. 343–362.

Jones, D.L., Howell, D. G., Coney, Peter J., Monger, James W. H., 1983, Recognition, character, and analysis of tectonostratigraphic terranes in western North America, *in* Hashimoto, M., and Uyeda, S., eds., Accretion tectonics in the circum-Pacific regions: Tokyo, Terra Scientific Publishing Company, p. 21–35.

Kilmer, F.H., 1977, Reconnaissance geology of Cedros Island, Baja California, Mexico: Southern California Academy of Sciences Bulletin, v. 76, p. 91–98.

Kilmer, F.H., 1979, A geological sketch of Cedros Island, Baja California, Mexico, *in* Abbott, P.L., and Gastil, R.G., eds., Baja California geology, field guides and papers for Geological Society of America Annual Meeting: San Diego, San Diego State University, Department of Geological Sciences, p. 11–28.

Kilmer, F.H., 1984, Geology of Cedros Island, Baja California, Mexico: Humboldt, California, Humboldt State University, 69 p.

Kimbrough, D.L., 1979, Deformation and emplacement of mid-Jurassic oceanic crust; Cedros Island, Baja Sur California: Geological Society of America Abstracts with Programs, v. 11, no. 7, p. 458.

Kimbrough, D.L., 1982, Structure, petrology and geochronology of Mesozoic paleooceanic basement terranes on Cedros Island and the Vizcaíno Peninsula [Ph.D. thesis], University of California, Santa Barbara, 395 p.

Kimbrough, D.L., 1984, Paleogeographic significance of the Middle Jurassic Gran Cañon Formation, Cedros Island, Baja California Sur, *in* Frizzell, V.A., Jr., ed., Geology of the Baja California peninsula: Field Trip Guidebook—Pacific Section, Society of Economic Paleontologists and Mineralogists Pacific Section, v. 39, p. 107–118.

Kimbrough, D.L., 1985, Tectonostratigraphic terranes of the Vizcaíno Peninsula and Cedros and San Benito Islands, Baja California, Mexico, *in* Howell, D.G., ed., Tectonostratigraphic terranes of the circum-Pacific region: Houston, Texas, Circum-Pacific Council for Energy and Mineral Resources, p. 285–298.

Kimbrough, D.L., 1989, Franciscan Complex rocks on Cedros Island, Baja California Sur, Mexico: Radiolarian biostratigraphic ages from a chert mélange block and petrographic observations on metasandstone, *in* Abbott, P., ed., Geologic studies in Baja California: Los Angeles, California, Society of Economic Paleontologists and Mineralogists Pacific Section, v. 63, p. 103–108.

Kimbrough, D.L., Hickey, J., and Tosdal, R., 1987, U-Pb ages of granitoid clasts in upper Mesozoic arc-derived strata of the Vizcaíno Peninsula, Baja California, Mexico: Geology, v. 15, p. 26–29.

Kimbrough, D.L., Mattinson, J.M., Coombs, D.S., Landis, C.A., and Johnston, M.R., 1992, Uranium-lead ages from the Dun Mountain ophiolite belt and Brook Street Terrane, South Island, New Zealand: Geological Society of America Bulletin, v. 104, p. 429–443.

Kimbrough, D.L., Smith, D.P., Mahoney, J.B., Moore, T.E., Gastil, R.G., Ortega Rivera, M.A., and Fanning, C.M., 2001, Forearc basin sedimentary response to rapid Late Cretaceous batholith emplacement in the Peninsular Ranges of southern and Baja California: Geology, v. 29, p. 491–494.

Krogh, T.E., 1973, A low contamination method for hydrothermal decomposition of zircon and extraction of U and Pb for isotopic age determination: Geochimica et Cosmochimica Acta, v. 37, p. 485–494.

Kusznir, N.J., and Park, R.G., 1987, The extensional strength of the continental lithosphere; its dependence on geothermal gradient, and crustal composition and thickness, *in* Coward, M.P., Dewey, J.F., and Hancock, P.L., eds., Continental extensional tectonics: Geological Society of London Special Publication 28, p. 35–53.

Macpherson, C.G., and Hall, R., 2001, Tectonic setting of Eocene boninite magmatism in the Izu-Bonin-Mariana forearc: Earth and Planetary Science Letters, v. 186, p. 215–230.

Mankinen, E.A., Gromme, C.S., and Williams, K.M., 1991, Concordant paleolatitudes from ophiolite sequences in the northern California Coast Ranges: Tectonophysics, v. 198, p. 1–21.

Mattinson, J.M., 1988, Constraints on the timing of Franciscan metamorphism: Geochronological approaches and their limitations, *in* Ernst, W.G., ed., Metamorphism and crustal evolution of the western United States, Rubey Volume VII: Englewood Cliffs, New Jersey, Prentice Hall, p. 1023–1034.

Mattinson, J.M., and Hopson, C.A., 1992, U/Pb ages of the Coast Range Ophiolite: A critical reevaluation based on new high-precision Pb/Pb ages: American Association of Petroleum Geologists Bulletin, v. 76, p. 425.

May, S.R., and Butler, R.F., 1986, North American Jurassic apparent polar wander: Implications for plate motion, paleogeography and Cordilleran tectonics: Journal of Geophysical Research, v. 91, p. 11519–11544.

McDougall, I., and Harrison, T.M., 1999, Geochronology and thermochronology by the ^{40}Ar/^{39}Ar method: New York, Oxford University Press, 282 p.

Menzies, M., Blanchard, D., and Xenophontos, C., 1980, Genesis of the Smartville arc-ophiolite, Sierra Nevada foothills, California: American Journal of Science, v. 280-A, p. 329–344.

Moore, T.E., 1977, Structure and petrology of the Sierra de San Andres Ophiolite, Vizcaíno Peninsula, Baja California Sur, Mexico [M.Sc. thesis]: San Diego, San Diego State University, 83 p.

Moore, T.E., 1983, Geology, petrology, and tectonic significance of the Mesozoic paleooceanic terranes of the Vizcaíno Peninsula, Baja California Sur, Mexico [Ph.D. thesis]: Stanford, California, Stanford University, 376 p.

Moore, T.E., 1984, Sedimentary facies and composition of Jurassic volcaniclastic turbidites at Cerro El Calvario, Vizcaíno Peninsula, Baja California Sur, Mexico, *in* Frizzell, V.A., Jr., ed., Geology of the Baja California peninsula: Field Trip Guidebook—Pacific Section, Society of Economic Paleontologists and Mineralogists Pacific Section, v. 39, p. 131–148.

Moore, T.E., 1985, Stratigraphic and tectonic significance of the Mesozoic tectonostratigraphic terranes of the Vizcaíno Peninsula, Baja California Sur, Mexico, *in* Howell, D.G., ed., Tectonostratigraphic terranes of the circum-Pacific region: Houston, Texas, Circum-Pacific Council for Energy and Mineral Resources, p. 315–329.

Moore, T.E., 1986, Petrology and tectonic implications of the blueschist-bearing Puerto Nuevo mélange complex, Vizcaíno Peninsula, Baja California Sur, Mexico, *in* Evans, B.W., and Brown, E.H., eds., Blueschists and eclogites: Boulder, Colorado, Geological Society of America Memoir 164, p. 43–58.

Moores, E.M., and Day, H.W., 1984, Overthrust model for Sierra Nevada: Geology, v. 12, p. 416–419.

Palmer, A.R., and Geissman, J., 1999, Geologic time scale: Boulder, Colorado, Geological Society of America.

Patterson, D.L., 1984, Los Chapunes and Valle sandstones: Cretaceous petrofacies of the Vizcaíno basin, Baja California, Mexico, *in* Frizzell, V.A., Jr., ed., Geology of the Baja California peninsula: Field Trip Guidebook—Pacific Section, Society of Economic Paleontologists and Mineralogists Pacific Section, v. 39, p. 161–172.

Peacock, S.M., 1992, Blueschist-facies metamorphism, shear heating, and P-T-t paths in subduction zones: Journal of Geophysical Research, v. 97, p. 17693–17707.

Pearce, J.A., van der Laan, S.R., Arculus, R.J., Murton, B.J., Ishii, T., Peate, D.W., and Parkinson, I.J., 1992, Boninite and harzburgite from Leg 125 (Bonin-Mariana forearc): A case study of magma genesis during the initial stages of subduction, *in* Fryer, P.e.a., ed., Proceedings of the Ocean Drilling Program, scientific results, Sites 778–786, Bonin/Mariana Region: College Station, Texas, Ocean Drilling Program, p. 623–659.

Pessagno, E.A., Jr., Finch, J.W., and Abbott, P.L., 1979, Upper Triassic radiolaria from the San Hipolito Formation, Baja California: Micropaleontology, v. 25, p. 160–197.

Pessagno, E.A., Jr., Hull, D.M., and Hopson, C.A., 2000, Tectonostratigraphic significance of sedimentary strata occurring within and above the Coast Range ophiolite (California Coast Ranges) and the Josephine ophiolite (Klamath Mountains), northwestern California, *in* Dilek, Y., Moores, E., Elthon, D., and Nicolas, A., eds., Ophiolites and oceanic crust: New insights from field studies and the Ocean Drilling Program: Boulder, Colorado, Geological Society of America Special Paper 349, p. 383–394.

Piercey, S.J., Murphy, D.C., Mortensen, J.K., and Paradis, S., 2001, Boninitic magmatism in a continental margin setting, Yukon-Tanana terrane, southeastern Yukon, Canada: Geology, v. 29, p. 731–734.

Platt, J.P., 1975, Metamorphic and deformational processes in the Franciscan Complex, California; some insights from the Catalina Schist terrane: Geological Society of America Bulletin, v. 86, p. 1337–1347.

Quidelleur, X., Grove, M., Lovera, O.M., Harrison, T.M., Yin, A., and Ryerson, F.J., 1997, Thermal evolution and slip history of the Renbu Zedong Thrust, southeastern Tibet: Journal of Geophysical Research, v. 102, p. 2659–2679.

Rangin, C., 1978, Speculative model of Mesozoic geodynamics, central Baja California to northeastern Sonora (Mexico), *in* Howell, D.G., and McDougall, K.A., eds., Mesozoic Paleogeography of the western United States, Pacific Coast Paleogeography Symposium 2: Tulsa, Oklahoma, Society of Economic Paleontologists and Mineralogists, p. 85–106.

Rangin, C., Girard, D., and Maury, R., 1983, Geodynamic significance of Late Triassic to Early Cretaceous volcanic sequences of Vizcaíno Peninsula and Cedros Island, Baja California, Mexico: Geology, v. 11, p. 552–556.

Robertson, A.H.F., 1989, Palaeoceanography and tectonic setting of the Coast Range ophiolite, central California: Evidence from the extrusive rocks and the volcaniclastic sediment cover: Marine and Petroleum Geology, v. 6, p. 194–220.

Robertson, A.H.F., 1990, Sedimentology and tectonic implications of ophiolite-derived clastics overlying the Jurassic Coast Range Ophiolite, California: American Journal of Science, v. 290, p. 109–163.

Robinson, J.W., 1975, Reconnaissance geology of the northern Vizcaíno Peninsula, Baja California Sur, Mexico [M.S. thesis]: San Diego, San Diego State University, 114 p.

Robinson, J.W., 1979, Reconnaissance geologic map of the northern Vizcaíno Peninsula, Baja California Sur, Mexico, *in* Abbott, P.L., and Gastil, R.G., eds., Baja California Geology: Field Guides and Papers for 1979 Geological Society of America Annual Meeting: San Diego, San Diego State University, Department of Geological Sciences, p. 5–7.

Ross, J.A., and Sharp, W.D., 1986, ^{40}Ar/^{39}Ar and Sm/Nd dating of garnet amphibolite in the Coast Ranges, California: Eos (Transactions, American Geophysical Union), v. 67, p. 1249.

Ross, J.A., and Sharp, W.D., 1988, The effects of sub-blocking temperature metamorphism on the K/Ar systematics of hornblende: ^{40}Ar/^{39}Ar dating of polymetamorphic garnet amphibolite from the Franciscan Complex, California: Contributions to Mineralogy and Petrology, v. 100, p. 213–221.

Saleeby, J., 1981, Ocean floor accretion and volcanoplutonic arc evolution of the Mesozoic Sierra Nevada, *in* Ernst, W.G., ed., The geotectonic development of California, (Rubey Volume I): Englewood Cliffs, New Jersey, Prentice-Hall, p. 133–181.

Saleeby, J., 1982, Polygenetic ophiolite belt of the California Sierra Nevada: Geochronological and tectonostratigraphic development: Journal of Geophysical Research, v. 87, p. 1803–1824.

Saleeby, J., 1992, Petrotectonic and paleogeographic settings of U.S., Cordilleran ophiolites, *in* Burchfiel, B.C., Lipman, P.W., and Zoback, M.L., eds., The Cordilleran orogen: Conterminous U.S.: Boulder, Colorado, Geological Society of America, Geology of North America, v. G-3, p. 653–682.

Saleeby, J., and Busby-Spera, 1992, Early Mesozoic tectonic evolution of the western U.S. Cordillera, *in* Burchfiel, B.C., Lipman, P.W., and Zoback, M.L., eds., The Cordilleran orogen: Conterminous U.S.: Boulder, Colorado, Geological Society of America: The Geology of North America, v. G-3, p. 107–168.

Saleeby, J., Harper, G.D., Snoke, A.W., and Sharp, W.D., 1982, Time relations and structural-stratigraphic patterns in ophiolite accretion, west-central Klamath Mountains, California: Journal of Geophysical Research, v. 87, p. 3831–3848.

Saleeby, J., Shaw, H.F., Niemeyer, S., Moores, E.M., and Edelman, S.H., 1989, U/Pb, Sm/Nd, and Rb/Sr geochronological and isotopic study of the northern Sierra Nevada ophiolitic assemblages: Contributions to Mineralogy and Petrology, v. 102, p. 205–220.

Schweickert, R.A., and Cowan, D.S., 1975, Early Mesozoic tectonic evolution of the western Sierra Nevada, California: Geologic Society of America Bulletin, v. 86, p. 1329–1336.

Sedlock, R.L., 1988a, Metamorphic petrology of a high-pressure, low-temperature subduction complex in west-central Baja California Mexico: Journal of Metamorphic Geology, v. 6, p. 205–233.

Sedlock, R.L., 1988b, Tectonic setting of blueschist and island-arc terranes of west-central Baja California, Mexico: Geology, v. 16, p. 623–626.

Sedlock, R.L., and Isozaki, Y., 1990, Lithology and biostratigraphy of Franciscan-like chert and associated rocks in west-central Baja California, Mexico: Geological Society of America Bulletin, v. 102, p. 852–864.

Sedlock, R.L., Ortega-Guitierrz, F., and Speed, R.C., 1993, Tectonostratigraphic terranes and the tectonic evolution of Mexico: Boulder, Colorado, Geological Society of America Special Paper 278, 153 p.

Sharp, W.D., 1988, Pre-Cretaceous crustal evolution in the Sierra Nevada region, California, *in* Ernst, W.G., ed., Metamorphism and crustal evolution of the western United States, Rubey Volume VII: Englewood Cliffs, New Jersey, Prentice-Hall, p. 823–864.

Shervais, J.W., 1990, Island arc and ocean crust ophiolites: Contrasts in the petrology, geochemistry and tectonic style of ophiolite assemblages in the California Coast Ranges, *in* Malpas, J., et al., eds., Ophiolites, oceanic crustal analogues: Nicosia, Cyprus, Geological Survey Department, p. 507–520.

Shervais, J.W., and Kimbrough, D.L., 1985, Geochemical evidence for the tectonic setting of the Coast Range ophiolite: Geology, v. 13, p. 35–38.

Silver, L.T., and Chappell, B., 1988, The Peninsular Ranges batholith: An insight into the Cordilleran batholiths of southwestern North America: Transactions of the Royal Society of Edinburgh, Earth Sciences, v. 79, p. 105–121.

Smith, D.P., and Busby, C.J., 1993, Mid-Cretaceous crust extension recorded in deep-marine half-graben fill, Cedros Island, Mexico: Geological Society of America Bulletin, v. 105, p. 547–562.

Snoke, A.W., Sharp, W.D., Wright, J.E., and Saleeby, J., 1982, Significance of mid-Mesozoic peridotitic to dioritic intrusive complexes, Klamath Mountains—western Sierra Nevada, California: Geology, v. 10, p. 160–166.

Stacey, J.S., and Kramers, J.D., 1975, Approximation of terrestrial lead isotope evolution by a two-stage model: Earth and Planetary Science Letters, v. 26, p. 207–221.

Stanton, R.L., and Bell, J.D., 1972, Volcanic and associated rocks of the New Georgia group, British Solomon Islands Protectorate: Overseas Geology and Mineral Resources, Great Britain, v. 10, p. 113–145.

Stern, R.J., and Bloomer, S.H., 1992, Subduction zone infancy: Examples from the Eocene Izu-Bonin-Mariana and Jurassic California arcs: Geologic Society of America Bulletin, v. 104, p. 1621–1636.

Streckeisen, A.L., 1976, To each plutonic rock its proper name: Earth Science Reviews, v. 12, p. 1–33.

Suppe, J., and Armstrong, R.L., 1972, Potassium-argon dating of Franciscan metamorphic rocks: American Journal of Science, v. 272, p. 217–233.

Tan, X., and Kodama, K.P., 1998, Compaction-corrected inclinations from Southern California Cretaceous marine sedimentary rocks indicate no paleolatitudinal offset for the Peninsular Ranges terrane: Journal of Geophysical Research, v. 103, p. 27,169–27,192.

Tardy, M., Lapierre, H., Freydier, C., Coulon, C., Gill, J.B., Mercier de Lepinay, B., Beck, C., Martinez, R.J., Talavera M.O., Ortiz H.E., Stein, G., Bourdier, J.L., and Yta, M., 1994, The Guerrero suspect terrane (western Mexico) and coeval arc terranes (the Greater Antilles and the Western Cordillera of Colombia); a late Mesozoic intra-oceanic arc accreted to cratonal America during the Cretaceous: Tectonophysics, v. 230, p. 49–73.

Taylor, B., 1995, Backarc basins: Tectonics and magmatism: New York, Plenum Press, p. 524.

Troughton, G.H., 1974, Stratigraphy of the Vizcaíno Peninsula near Asuncion Bay, Territorio de Baja California, Mexico [M.Sc. thesis]: San Diego, San Diego State University, 83 p.

Valencia-Moreno, M., Ruiz, J., Barton, M.D., Patchett, P.J., Zuercher, L., Hodkinson, D.G., and Roldan-Quintana, J., 2001, A chemical and isotopic study of the Laramide granitic belt of northwestern Mexico; identification of the southern edge of the North American Precambrian basement: Geological Society of America Bulletin, v. 113, p. 1409–1422.

van der Laan, S.R., Flower, J.F.J., and Koster van Gros, A.F., 1989, Experimental evidence for the origin of boninites: Near-liquidus phase relations to 7.5 kbar, *in* Crawford, A.J., ed., Boninites: London, England, Unwin Hyman, p. 112–147.

Wakabayashi, J., and Dilek, Y., 2000, Spatial and temporal relationships between ophiolites and their metamorphic soles: A test of models of forearc ophiolite genesis, *in* Dilek, Y., Moores, E.M., Elthon, D., and Nicolas, A., eds., Ophiolites and oceanic crust: Boulder, Colorado, Geological Society of America Special Paper 349, p. 53–64.

Wetmore, P.H., Schmidt, K.L., Paterson, S.R., and Herzig, C.T., 2002, Tectonic implications for the along-strike variation of the Peninsular Ranges batholith, southern and Baja California: Geology, v. 30, p. 247–250.

Whalen, P.A., and Pessagno, E.A., Jr., 1984, Lower Jurassic radiolaria, San Hipolito Formation, Vizcaíno Peninsula, Baja California Sur, *in* Frizzell, V.A., Jr., ed., Geology of the Baja California peninsula: Society of Economic Paleontologists and Mineralogists, Pacific Section, v. 39, p. 53–66.

Wolf, M.B., and Saleeby, J., 1995, Late Jurassic dike swarms in the southeastern Sierra Nevada Foothills Terrane, California: Implications for the timing and controls of Jurassic Orogenesis, *in* Miller, D.M., and Busby, C., eds., Jurassic magmatism and tectonics of the North American Cordillera: Boulder, Colorado, Geological Society of America Special Paper 299, p. 203–228.

Wright, J.E., 1982, Permo-Triassic accretionary subduction complex, southwestern Klamath Mountains, northern California: Journal of Geophysical Research, v. 87, p. 3805–3818.

Wyld, S.J., and Wright, J.E., 1988, The Devil's Elbow ophiolite remnant and overlying Galice Formation: New constraints on the Middle to Late Jurassic evolution of the Klamath Mountains, California: Geological Society of America Bulletin, v. 100, p. 29–44.

MANUSCRIPT ACCEPTED BY THE SOCIETY JUNE 2, 2003

Geological Society of America
Special Paper 374
2003

Four phases of Mesozoic deformation in the Sierra de San Andres ophiolite, Vizcaíno Peninsula, west-central Baja California, México

Richard L. Sedlock*

Department of Geology, San José State University, San José, California 95192-0102, USA

ABSTRACT

New field and structural data from superb exposures on the Vizcaíno Peninsula of west-central Baja California, México, help delineate four phases of Mesozoic deformation that occurred in disparate tectonic environments.

Phase 1 extension, which affected only the Sierra de San Andres ophiolite, occurred after Late Triassic formation of the ophiolite but prior to Late Jurassic to earliest Cretaceous shortening. Extension may have occurred at or near the host spreading center shortly after ophiolite formation or during Jurassic plate-margin tectonism of unclear type or cause. Detachment faults are present, but they did not produce the extreme thinning seen at some modern spreading centers.

Phase 2 shortening caused thrusting and folding of Phase 1 structures. If the ophiolite formed in a suprasubduction-zone environment outboard of the North American plate, then Phase 2 may have been caused by accretion of the ophiolite to North America. Alternatively, Phase 2 may reflect shortening of the ophiolite in a suprasubduction-zone environment within the North American plate. Thrusting and folding are not widely or penetratively developed, suggesting that total shortening was minor.

Phase 3 produced a subvertical, roughly east-west–striking fault zone that cuts Phase 1 and Phase 2 structures. The current distribution of key units suggests that this fault zone may have accommodated sinistral slip in the Early Cretaceous forearc, consistent with plate-motion studies that infer left-oblique convergence between North America and oceanic lithosphere in the eastern Pacific at this time.

Phase 4 extension affected units as young as early Late Cretaceous in the study area and is correlated with syn-subduction forearc extension recognized in Mesozoic oceanic rocks throughout western Baja California. Extension began during the Early Cretaceous, coeval with Phase 3 sinistral faulting, and persisted throughout the Cretaceous into the Paleogene.

Keywords: Baja California, Vizcaíno Peninsula, ophiolite, extension.

INTRODUCTION

Ophiolites are remnants of oceanic lithosphere that are exposed at Earth's surface (e.g., Coleman, 1977). Most ophiolites are tectonically disrupted, and structures in many of them have been attributed to extensional deformation at or near the spreading center (e.g., Dilek et al., 1998). Ophiolites are also subject to dismemberment during accretion at plate margins and during whatever post-accretion tectonism they undergo thereafter.

In this paper I present field and structural data from a superbly exposed Late Triassic ophiolite in the Vizcaíno peninsula of west-central Baja California, México. A complex suite of structures is interpreted to record several distinct phases of deformation in the tectonic history of the region.

*sedlock@geosun.sjsu.edu

Sedlock, R.L., 2003, Four phases of Mesozoic deformation in the Sierra de San Andres ophiolite, Vizcaíno Peninsula, west-central Baja California, México, *in* Johnson, S.E., Paterson, S.R., Fletcher, J.M., Girty, G.H., Kimbrough, D.L., and Martín-Barajas, A., eds., Tectonic evolution of northwestern México and the southwestern USA: Boulder, Colorado, Geological Society of America Special Paper 374, p. 73–92. For permission to copy, contact editing@geosociety.org. © 2003 Geological Society of America.

GEOLOGY OF WESTERN BAJA CALIFORNIA

Regional Setting

Mesozoic rocks of Baja California can be divided into those of continental or Mesozoic arc affinity that underlie most of Baja and those of oceanic affinity that underlie the Vizcaíno peninsula and offshore islands. Continental and arc rocks of "mainland Baja" include Paleozoic and Mesozoic metasedimentary rocks of western North America, the Early Cretaceous Alisitos–Santiago Peak volcanic arc, and the Cretaceous (130–90 Ma) Peninsular Ranges batholith (Gastil et al., 1975, Silver and Chappell, 1988). The western flank of the continental assemblage is onlapped by Late Cretaceous to Paleogene siliciclastic marine strata, chiefly turbidites, which were probably deposited in a forearc basin (Gastil et al., 1975).

Oceanic rocks in western Baja consist of Mesozoic blueschist, ophiolite, and island arc terranes overlain by Cretaceous forearc basin strata (Rangin, 1978; Kimbrough, 1985; Moore, 1985; Sedlock, 1988a, 1999). These rocks have been assigned to the Vizcaíno composite terrane by Coney and Campa (1987) and to the Cochimí composite terrane by Sedlock et al. (1993). Oceanic rocks crop out on Isla Cedros, Islas San Benitos, and the Vizcaíno peninsula near 28°N and on Isla Magdalena and Isla Santa Margarita near 24.5°N (Fig. 1). Similar rocks probably underlie most of the continental shelf between 30°N and 23°N, judging from gravity and magnetic anomalies (Couch et al., 1991).

Mesozoic Oceanic Rocks in Western Baja

Mesozoic oceanic rocks that crop out in western Baja are divided into three structural units (Sedlock, 1988a, 1999). The structurally lowest unit (western Baja terrane of Sedlock, 1988a) comprises a fairly intact subduction complex that crops out only on Islas San Benito, Isla Cedros, and Isla Santa Margarita. Blueschist protoliths included Triassic oceanic crust, Triassic to Cretaceous pelagic sedimentary rocks, and Cretaceous terrigenous turbidites that were deposited as the oceanic rocks neared North America (Sedlock and Isozaki, 1990). The blueschists were metamorphosed under high-pressure, low-temperature conditions (Sedlock, 1988b) during mid-Cretaceous subduction (Baldwin and Harrison, 1989) and were exhumed while subduction continued in the Late Cretaceous to Paleogene (Sedlock, 1988b, 1996, 1999).

The structurally highest unit, informally referred to as the "upper plate," includes Mesozoic island arcs, ophiolites, and volcanogenic to terrigenous cover strata that crop out in all areas except the Islas San Benito (Kimbrough, 1985; Moore, 1985; Sedlock, 1988a, 1993). Island arc and ophiolite terranes were incorporated into the western edge of North America by the end of the Jurassic (Boles and Landis, 1984; Hickey, 1984). The accreted terranes were overlapped by forearc basin turbidites that range in age from lower Aptian (R. Sedlock, unpubl. data) to early Eocene (Smith and Busby, 1993; Kimbrough et al., 2001). Most upper-plate rocks have been confined to shallow crustal levels throughout their history; metamorphism typically ranges from absent to very low-grade, and primary igneous and sedimentary structures typically are well-preserved. All upper-plate units were affected by brittle extensional tectonism that constitutes the youngest or only strain event.

The structurally intermediate unit is a serpentinite-matrix mélange that occupies fault zones up to 500 m thick at all contacts between the blueschists and the upper-plate rocks. Most blocks are serpentinized ultramafic rocks, and the serpentinite

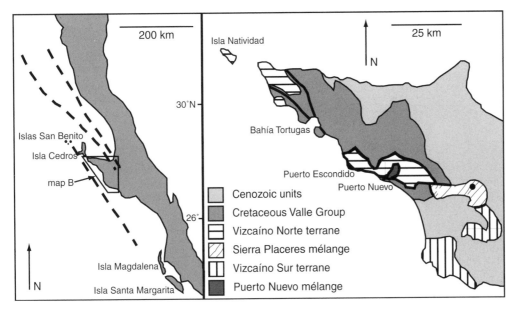

Figure 1. A: Map showing location of map B and key islands and offshore fault zones referred to in text. B: Generalized geology and terrane nomenclature of Vizcaíno peninsula. Solid circle denotes block of Triassic sedimentary and volcanic rocks in Sierra Placeres mélange as mapped by Moore (1983, 1985); see discussion.

matrix probably formed by shearing of similar rocks. About half of the other blocks are fragments of hanging-wall or footwall rocks, up to 250 m across, which typically crop out within a few kilometers of their sources. The remaining blocks comprise diverse metamorphic rocks that are exotic, i.e., derived from source terranes not exposed in western Baja. Exotic blocks include eclogite, coarse-grained blueschist, and amphibolite (Moore, 1986; Sedlock, 1988a, 1999) and yield $^{40}Ar/^{39}Ar$ ages ranging from 170 Ma to 94 Ma (Baldwin and Harrison, 1989, 1992; Bonini, 1994; Bonini and Baldwin, 1994).

In summary, the three structural units form a sandwich with brittlely distended upper-plate rocks juxtaposed against deep-level blueschists along fault zones marked by the serpentinite-matrix mélange. The fault zones are interpreted as shallowly dipping, crustal-scale normal faults that have accommodated synsubduction exhumation of the footwall blueschists (Sedlock, 1988a, 1996, 1999).

STUDY AREA

The study area is an elevated region that includes named and unnamed ranges northwest, north, and east of the small fishing villages of Puerto Nuevo and San Cristóbal (Fig. 2). According to Mexican topographic maps (1:50,000, published by Secretaria de programación y presupuesto [SPP] in 1982), the study area includes the Sierra Puerto Nuevo, Sierra Los Ajos, and Sierra El Chorrito, but these ill-defined entities probably do not completely include all areas shown in Figure 2. The study area comprises the western part of the Sierra de San Andres, as delimited by Moore (1983), but current maps do not apply this geographic term to the study area.

My work builds on the foundation supplied by Moore (1983, 1985, 1986), who mapped Mesozoic rocks throughout much of the Vizcaíno peninsula. The study area described herein is a subset of the area studied by Moore. Petrologic and geochemical data summarized below come mostly from his work and are augmented somewhat by my own observations. The geologic map and sections presented here resemble Moore's in a general way but differ significantly in detail; combined with new geometric and kinematic structural data, these revisions lead to tectonic interpretations that, in several respects, differ markedly from Moore's.

MESOZOIC MAP UNITS

Mesozoic map units within the study area include six stratigraphic units and two tectonic units (Figs. 2 and 3). In terms of the regional structural units described above, all of the stratigraphic units are part of the upper plate. The ophiolitic rocks and Eugenia Formation are part of the Vizcaíno Norte terrane (Moore, 1983, 1985), while the Valle Group constitutes an overlap assemblage. The tectonic units are the Puerto Nuevo mélange, which was named by Moore (1983) but is reinterpreted here, and the Tangaliote mélange, which is newly named and described here. In terms of the regional structural units described above, the Tangaliote fault zone is contained within the upper plate, while the

Puerto Nuevo mélange comprises the intermediate unit. Lower-plate rocks do not crop out on the Vizcaíno peninsula.

Sierra de San Andres Ophiolite

Triassic Harzburgite and Other Ultramafic Rocks (Trh)

Ultramafic rocks of this unit consist of 85–100% serpentinite (chrysotile, chlorite, and magnetite), with 0–15% relict primary phases, sparse preservation of relict textures, and compositional layering that is fairly well-defined when viewed from a distance (Fig. 4) but difficult to recognize at the outcrop. The protolith of most of these rocks probably was harzburgite with ~80% olivine. Other protoliths include dunite, wehrlite, and orthopyroxenite (Moore, 1983). Coarse-grained gabbro forms isolated horizons up to 4 m thick in a few places. Chromite forms chromitite nodules up to 10 m long and disseminated layers up to 15 cm thick. Serpentinite pseudomorphs of orthopyroxene (bastite) are common. Moore (1983) reported a tectonic fabric formed by alignment of such grains, but this fabric is difficult to recognize at most outcrops.

Unit Trh is cut by rodingitized diabase intrusions that range in thickness from a few centimeters to >5 m (Moore, 1983). Diabase intrusions east of Arroyo Puerto Nuevo, and some of those west of the arroyo, are parallel or subparallel to compositional layering and are interpreted as sills (Fig. 5, A and B). Other diabase intrusions west of Arroyo Puerto Nuevo are roughly orthogonal to the compositional layering and the sills (Fig. 5A) and are interpreted as dikes. Diabase bodies are somewhat less strained than the surrounding host rocks, indicating a higher degree of competency than the surrounding serpentinite. Because the sills and dikes are so visually and rheologically distinct from their serpentinized hosts, I used their lateral continuity as a rough, but useful, indicator of brittle extension. Individual dikes were traceable for up to 250 m along strike in areas of low strain but for no more than a few meters in areas of high strain.

Triassic Gabbro (Trg)

Cumulate gabbros make up >90% of this unit in the study area, though Moore (1983) reported greater compositional variety elsewhere in the western Vizcaíno peninsula. Compositional layering is common; its orientation varies over distances of several to tens of meters but roughly defines a bimodal distribution (Fig. 5C). The unit also includes minor noncumulate gabbro and lenses, dikes, and sills of olivine gabbro, gabbronorite, norite, and plagiogranite. The gabbro unit is locally cut by 10–50-cm-thick intrusions of diabase. A plagiogranite stock yielded a 220 U-Pb age on zircon (Kimbrough, 1982), the chief radiometric evidence for a Triassic age of the ophiolite. The gabbros probably underwent hydrothermal metamorphism at greenschist- to amphibolite-facies conditions (Moore, 1983).

Triassic Hypabyssal and Volcanic Rocks (Trv)

This map unit includes dikes and sills, chiefly diabase with minor basalt, and basaltic pillow lavas and volcanic breccias.

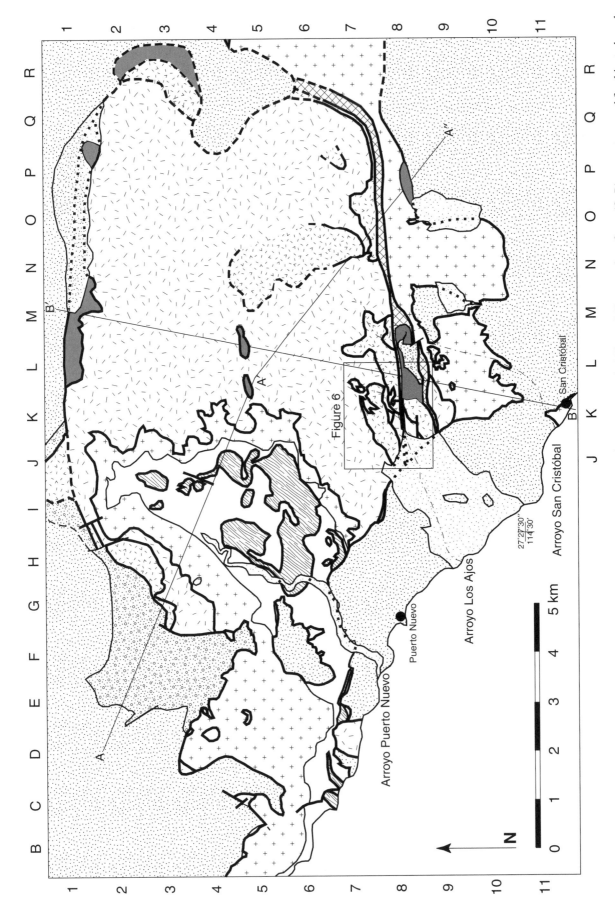

Figure 2. Geologic map of study area, showing locations of structure sections A–A″ and B–B′ (see Fig. 3) and of Figure 6. Eastern edge of area (Q, R) was not mapped for this study and is modified slightly from Moore (1983). A color version of this figure is on the CD-ROM accompanying this volume.

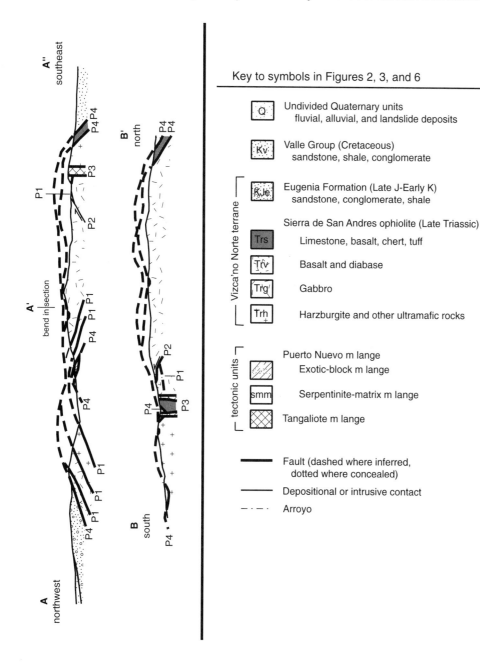

Key to symbols in Figures 2, 3, and 6

Q — Undivided Quaternary units
fluvial, alluvial, and landslide deposits

Kv — Valle Group (Cretaceous)
sandstone, shale, conglomerate

KJe — Eugenia Formation (Late J-Early K)
sandstone, conglomerate, shale

Sierra de San Andres ophiolite (Late Triassic)

Trs — Limestone, basalt, chert, tuff

Trv — Basalt and diabase

Trg — Gabbro

Trh — Harzburgite and other ultramafic rocks

Puerto Nuevo m lange

— Exotic-block m lange

smm — Serpentinite-matrix m lange

— Tangaliote m lange

━━━ Fault (dashed where inferred, dotted where concealed)

─── Depositional or intrusive contact

─ · ─ · Arroyo

Figure 3. Structure sections A–A″ and B–B′; see Figure 2 for location. Note bend at A′ in A–A″. P1, P2, P3, P4 refer to deformation phase of indicated fault. A color version of this figure is on the CD-ROM accompanying this volume.

The hypabyssal and volcanic rocks are grouped in a single map unit because (1) they are difficult to distinguish in weathered outcrops, and (2) they are intimately intermixed along primary igneous contacts in many areas. In these intermixed areas, I use the relative abundance of the diabasic intrusions to infer the direction of stratigraphic younging (lower abundance at the top).

Diabase intrusions are 20 cm to 1.5 m thick, have chilled margins up to 2 cm thick, locally contain xenoliths of older diabase, and generally are sheeted. Volcanic rocks include pillow basalt, pillow breccia, and other volcanic breccia. Interpillow limestone is present in a few areas. Pillows range from 25 cm up to 2 m across. These rocks contain assemblages indicative of

prehnite-pumpellyite-, greenschist-, and possibly lower amphibolite-facies metamorphism (Moore, 1983), but igneous textures are well preserved. Trace-element geochemistry of diabase and basalt is consistent with, but does not prove, formation in a supra-subduction-zone setting (Moore, 1985).

Much of the area mapped as unit Trv consists of fault-bounded slices of sheeted diabase intrusions that cannot be identified as dikes or sills because they lack adjacent gabbro or basalt to indicate paleohorizontal. However, primary igneous contacts are preserved within a 1 km² fault-bounded slice in Arroyo Los Ajos (Fig. 6). Sheeted dikes root downward at a high (65–90°) angle into an underlying transitional contact with chiefly isotro-

R.L. Sedlock

Figure 4. View of ultramafic unit of ophiolite (unit Trh). Compositional layering forms alternating lighter and darker bands that dip ~30° into the slope. Lines outside photo are parallel to apparent dip of layering. Thin white streaks near base and top of slope are diabase sills roughly parallel to layering. Slope is ~200 m high.

pic gabbro of unit Trg and intrude overlying pillow basalts. The dike complex also contains much rarer sills that range up to 5 m thick. This fault-bounded slice contains the upper 100 m of the underlying gabbro unit, a 50–100-m-thick transition from gabbro to diabase dikes, a 275-m-thick complex of sheeted dikes cross-cut at high angles by rare sills, and the basal 5 m of the transition zone between pillow basalts and diabase dikes. Based on the relations in Arroyo Los Ajos and on a few other fault slices that contain relatively intact stratigraphy, I interpret other outcrops of sheeted diabase in unit Trv as dikes with subordinate sills.

Triassic Sedimentary Rocks (Trs)

This unit crops out (1) in the Tangaliote fault zone (Fig. 2, K,L-8), where it overlies volcanic rocks of the ophiolite in

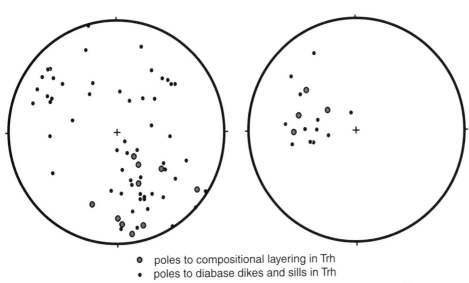

- ● poles to compositional layering in Trh
- • poles to diabase dikes and sills in Trh

A. West of Arroyo Puerto Nuevo   B. East of Arroyo Puerto Nuevo

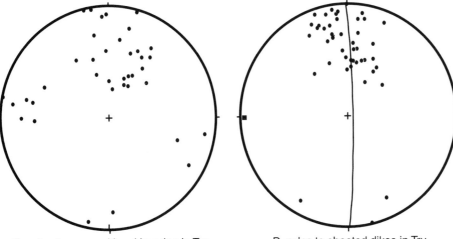

C. poles to compositional layering in Trg   D. poles to sheeted dikes in Trv

Figure 5. Lower-hemisphere plots of poles to planar elements in ophiolitic units. A, B: Poles to compositional layering (gray circles) and diabase dikes and sills (dots) in ultramafic unit, west and east of Arroyo Puerto Nuevo, respectively. C: Poles to compositional layering in gabbro unit east of Arroyo Puerto Nuevo. D: Poles to sheeted dikes in Unit Trv near Arroyo Los Ajos.

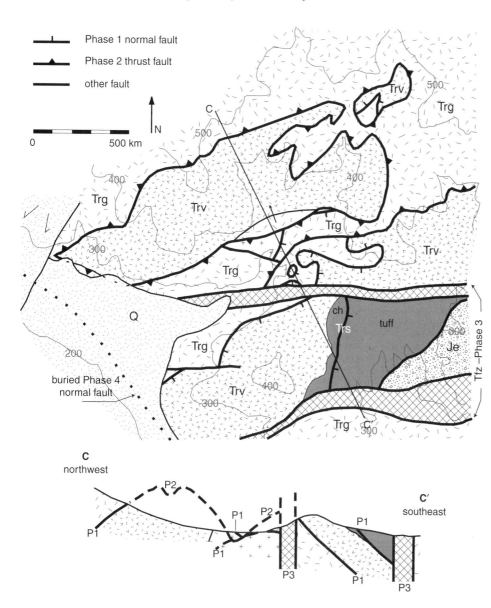

Figure 6. Geologic map and structure section C–C′ of lower Arroyo Los Ajos; see Figure 2 for location. Symbols as in Figures 2 and 3. Triassic sedimentary unit Trs consists of limestone-basalt-chert unit (ch) and tuff unit. Tfz = Tangaliote fault zone. Arrowed contact near center of map indicates gradation from underlying gabbro to sheeted dikes (see text). A color version of this figure is on the CD-ROM accompanying this volume.

conformable, possibly depositional, contact, and (2) in tectonic horses within fault zones in the eastern half of the study area (L,M-1; P-1; R-3; R-4; K,L,M-5; O,P-8). In the Tangaliote fault zone and in the horse at O,P-8, this unit displays the same stratigraphic sequence: a 10-m-thick basal unit of pink limestone, a 15-m-thick middle unit of basalt and local red radiolarian chert, and an upper unit of multicolored tuff that is at least 150 m thick. Individual tuff beds are 1–50 cm thick, contain devitrified pumice fragments, are finely laminated, and are locally cross-bedded. The tuff unit also contains rare interbeds of red radiolarian chert and tuffaceous chert.

Barnes (1982) and Moore (1983) reported similar rocks from elsewhere in the Vizcaíno that contain Carnian-Norian fossils and that locally are termed the San Hipolito Formation. Barnes (1982) reported mineral assemblages that indicate these rocks experienced zeolite facies metamorphism.

Upper Jurassic–Lower Cretaceous Eugenia Formation (KJe)

The Eugenia Formation consists chiefly of volcanogenic sedimentary rocks that Hickey (1984) divided into two members. The Tithonian-Neocomian lower member, consisting chiefly of coarse-grained sandstone, conglomerate, and minor shale, was deposited as a deep-water debris apron along the edge of a volcanic high. The Aptian-Albian upper member, consisting of rhythmically interbedded sandstone and shale, was deposited in a submarine fan. The upper member of the Eugenia is difficult to distinguish from the Cretaceous Valle Group.

The Eugenia Formation crops out in two parts of the study area. The lower member overlies Triassic sedimentary rocks of the ophiolite along a normal fault within the Tangaliote fault zone (Fig. 6). Moore (1983) also recognized this outcrop of Eugenia, which he mapped as a klippe atop his Sierra Placeres mélange.

Both members of the Eugenia Formation also crop out over much of the western part of the study area, where Moore mapped them as Valle Group. Bedding in these outcrops dips gently to moderately north to west (Fig. 7). The Eugenia is distinguishable from the Valle by its darker color, which enables the two to be differentiated despite mediocre exposure and little relief. Sandstones in the Eugenia typically are thicker-bedded, more poorly sorted, and more massive than those in the Valle, and conglomerates in the Eugenia generally are more poorly sorted and poorly rounded than their Valle counterparts. In the western outcrop area, the Valle depositionally overlies the Eugenia with slight (up to 20°) angular discordance. Because exposures are poor to mediocre in this low-relief area, I cannot discount the possibility that the mapped Eugenia also includes minor erosional remnants of overlying Valle strata.

Cretaceous Valle Group (Kv)

The Valle Group consists of deep-water sandstone, shale, and conglomerate ranging in age from Albian to Eocene. It is exposed throughout the Vizcaíno peninsula and on nearby Isla Cedros (Smith and Busby, 1993), and probable correlatives crop out on Isla Magdalena and Isla Santa Margarita to the south (Blake et al., 1984). Valle strata have not been dated in the study area, but stratigraphic similarities to the Los Indios section ~50 km northwest (Kimbrough et al., 2001) suggest that the section in the study area is chiefly Cenomanian.

The Valle Group mantles the older Mesozoic rocks that crop out in the uplifted core of the study area (Fig. 2). The Valle also crops out in several klippe within the southern part of the core (F-6, K-9, L-9). The Valle is in fault contact with all ophiolite units and depositionally overlies the Eugenia Formation. Beds within the Valle dip gently to moderately (Fig. 7); strike variations are examined in the Discussion section. The Valle was not subjected to regional metamorphism, but limited (<1 km²) alteration zones adjacent to major fault zones may have resulted from the circulation of hydrothermal or corrosive fluids.

Valle strata display a wide range of well-developed sedimentary structures that indicate deposition from sediment gravity flows at neritic to bathyal depths (Smith et al., 1993). Most rocks are easily interpreted in terms of submarine-fan facies, ranging from channel-fill conglomerates to basin-plain and interchannel mudstones. The Valle Group is interpreted as fill of the forearc basin of the active Peninsular Ranges batholith to the east (Kimbrough et al., 2001).

Previously unrecognized chaotic deposits have been mapped in roughly continuous belts in two locations: (1) along the coast near San Cristóbal (J-10,11), and (2) in a ~1-km-wide swath in the eastern part of the study area (B-4 to E-1). The disrupted strata form stratigraphic units within mid-fan turbidites and are interpreted as slump deposits, possibly due to collapse of the walls of a submarine channel. Exposures lie roughly equal distances from, and on either side of, the antiform centered on Arroyo Puerto Nuevo, raising the possibility that they are part of a single stratigraphic unit breached by the rising antiform. Ongoing studies of these rocks will help clarify their stratigraphic and tectonic significance.

Puerto Nuevo Mélange

A tectonic window into an antiform in Arroyo Puerto Nuevo exposes a complex mélange unit that tectonically underlies the

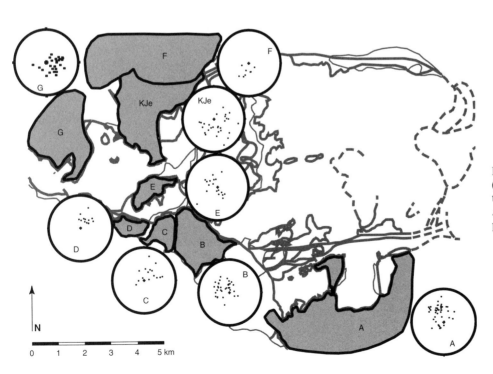

Figure 7. Poles to bedding in domains (shaded regions) of the Eugenia Formation (KJe) and Valle Group (domains A–F). Equal-area, lower-hemisphere projections.

Sierra de San Andres ophiolite. Moore (1983, 1986) termed this unit the Puerto Nuevo mélange and subdivided it into upper and lower "serpentinite breccia" members that bound an intervening "exotic block" member. The upper and lower members are indistinguishable and display block-in-matrix fabric of structural origin, so I map them as a single unit of serpentinite-matrix mélange (smm). I map the intervening member as exotic-block mélange (ebm).

The serpentinite-matrix mélange consists of resistant blocks of serpentinized ultramafic rocks and minor diabase immersed in powdery, light-colored serpentinite dominated by chrysotile and magnesite. Block size and block:matrix ratio increase upward toward the base of overlying ultramafic rocks (unit Trh). I interpret these variations as a downward-increasing strain gradient, and I concur with Moore (1986) that the serpentinite-matrix mélange formed by tectonic abrasion of unit Trh. The gradational contact between unit Trh and underlying serpentinite-matrix mélange was mapped at the top of matrix-supported blocks, corresponding roughly to a 2:1 ratio of unsheared to sheared rock. In many areas, this change also corresponds with a break in slope (steeper slopes in unit Trh) and a slight color change (red in strongly oxidized unit Trh, softer pastels in serpentinite-matrix mélange). The structural base of the serpentinite-matrix mélange unit is not exposed.

The exotic-block mélange contains a variety of metamorphic blocks, with representatives from prehnite-pumpellyite, greenschist, blueschist, epidote-amphibolite, and eclogite facies (Moore, 1986). Many blocks were derived from sources that do not crop out in the Vizcaíno peninsula. Mélange matrix consists of strongly sheared, scaly serpentinite dominated by antigorite. The upper and lower contacts of the exotic-block mélange are slickensided surfaces. The exotic-block mélange is laterally discontinuous, juxtaposing the overlying and underlying serpentinite-matrix mélange in the southwest (D-6,7) and south (I-6,7) and near isolated exotic blocks in the north (I-3, J-3).

Tangaliote mélange

Serpentinite-matrix mélange forms thin, mappable zones associated with a high-angle fault zone that strikes eastward across the study area (J-8 to Q-7). Moore (1985) considered this unit to be part of his Sierra Placeres mélange; however, as explained in the Structural History section, I consider this fault zone and mélange to be distinct from the Sierra Placeres and have named both after a small rancho in lower Arroyo Los Ajos.

Between Arroyo Los Ajos and Arroyo San Cristóbal, the Tangaliote mélange forms marginal belts of the 500-m-wide Tangaliote fault zone. These marginal belts range from 5 m to 150 m in thickness but generally are 40–75 m thick. Mélange blocks range in size from a few centimeters to over 100 m across and were derived chiefly from the ophiolitic units Trh, Trg, and Trv. A single amphibolite block of enigmatic protolith crops out near Arroyo San Cristóbal. Mélange matrix consists of moderately resistant, scaly, dark brown to black serpentinite.

A few hundred meters east of Arroyo San Cristóbal (M-8), the Tangaliote fault zone narrows and the marginal mélange belts

merge to form a single belt of serpentinite-matrix mélange that continues eastward, where it exhibits a 60° left bend that is at least 1 km long (Q-7).

STRUCTURAL HISTORY OF THE OPHIOLITE

The Sierra de San Andres ophiolite near Puerto Nuevo exhibits evidence of four discrete phases of deformation of Mesozoic age. Structures were assigned to Phase 1, 2, 3, or 4 on the basis of cross-cutting relationships and the age of affected map units. From oldest to youngest, the four deformation phases are (1) extension of pre-Late Jurassic, possibly Triassic, age; (2) shortening of Late Jurassic–earliest Cretaceous age; (3) strike-slip faulting of Early Cretaceous age; and (4) extension of Early Cretaceous to Paleogene age.

Most structures were assigned with confidence to a particular deformation phase, but ambiguity arose in the case of normal faults that do not cut rocks younger than the Triassic ophiolite, i.e., such faults could be associated with Phase 1 or Phase 4. In these and a few other problematic cases, structures were interpreted using structural style, kinematics, orientation, and geographic distribution. Figure 8 shows the distribution of faults assigned to each phase of deformation.

Phase 1: Pre-Late Jurassic Extension

Phase 1 normal faults cut the Triassic ophiolite but not the Eugenia Formation or Valle Group, and they are truncated by structures of Phases 2–4. I interpret similar faults that are not cut by younger structures as Phase 1 faults because their structural style, kinematics, or both are typical of unequivocal Phase 1 faults, or because their orientation, location, or both are more characteristic of Phase 1 than of Phase 4 faults. Phase 1 faults dip shallowly to moderately.

Trg/Trh

A laterally continuous normal fault that places the gabbro unit (Trg) atop the ultramafic unit (Trh) forms a laterally continuous, sinuous arc around the antiformal uplift centered near Arroyo Puerto Nuevo (Figs. 2 and 9A). The fault is truncated at a high angle at both its southeast (H-7) and southwest (F-4) terminations by Phase 4 faults. The contact generally dips 20°–40° and is subplanar, but locally it exhibits 10°–70° undulations with amplitudes and wavelengths up to ~10 m. For much of its extent, the contact is marked by a 3–15-m-thick zone of tectonically interleaved, 0.5–2-m-thick slices of serpentinite and shattered gabbro (Fig. 9B). In the footwall, the uppermost 5–20 m of the ultramafic unit consists of sheared serpentinite, and the spacing and intensity of foliation decrease farther from the contact. In the hanging wall, the lowermost 1–5 m of gabbro displays pervasive fractures and greenschist-facies alteration that die out farther from the contact. In some areas, the Trg/Trh contact is comparatively sharp, with no tectonic mixing of hanging and footwall, and wall-rock damage zones as thin as 1 m. Fault-striae

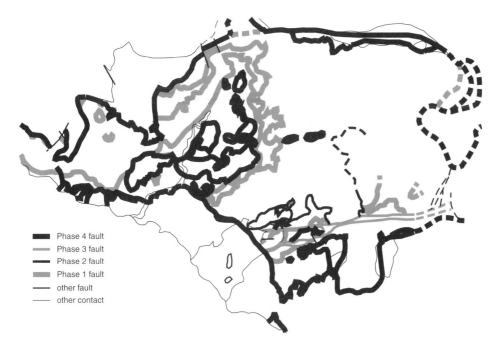

Figure 8. Distribution of faults assigned to specific deformation phases. A color version of this figure is on the CD-ROM accompanying this volume.

- ███ Phase 4 fault
- ███ Phase 3 fault
- ▬▬ Phase 2 fault
- ▬▬ Phase 1 fault
- ─── other fault
- ─── other contact

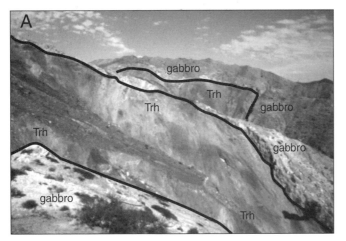

Figure 9. Phase 1 normal faults that place gabbro (unit Trg) above ultramafic rocks (unit Trh) of the ophiolite. A: Shallowly dipping, slightly undulatory contact in lower Arroyo Puerto Nuevo. B: Tectonically interleaved serpentinite (darker color) and gabbro (lighter color) along Trg/Trh contact in mid-Arroyo Puerto Nuevo. Note extension of layers. Hiking pole is ~1.2 m long.

data (Fig. 10) do not yield well-defined kinematic axes (very low eigenvalues). Phase 1 normal faults make 35–67° angles with compositional layering in overlying gabbros.

Isolated gabbro klippe west of the southwest termination of the main Trg/Trh surface (D-4, E-5) are interpreted as erosional remnants of the main Trg unit, and the underlying surface is correlated with the main Phase 1 Trg/Trh fault because of similarities in structural style, orientation, and geographic position. The Trg/Trh contact is also preserved along the margins of several gabbro bodies that form structural horses in Phase 4 faults in this same area (B-5, C-5, D-3, E-4, F-4). I correlate these with the

Phase 1 Trg/Trh contact but cannot discount the possibility that some or all slip occurred during Phase 4 deformation.

An isolated body of gabbro overlies the ultramafic unit east of Arroyo Los Ajos (K-9). The contact strongly resembles the continuous contact in Arroyo Puerto Nuevo and is assigned with confidence to Phase 1 because it is truncated by the high-angle Phase 3 fault to the northeast.

Trv/Trg

Several discontinuous faults place diabase, basalt, or both above gabbro around the margins of the uplifted core of the study

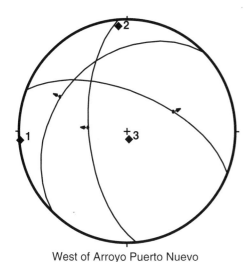

West of Arroyo Puerto Nuevo

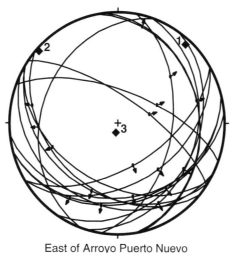

East of Arroyo Puerto Nuevo

Figure 10. Fault-striae data from Phase 1 fault that places gabbro unit (Trg) over ultramafic unit (Trh). Arrows show slip direction of hanging wall. 1, 2, 3: extension, intermediate, and shortening axes of unweighted moment tensor.

area. The best exposures of these faults are in the south-central part of the region (Figs. 6 and 8), where they are cut by the high-angle Phase 3 fault to the south (L-8, P-7) and by Phase 2 thrusts (K-8, N-7). The contacts are subplanar but locally form 10°–30° undulations with wavelengths up to ~20 m and amplitudes of several meters. The contact between Trv and Trg is generally sharp, with little interleaving of adjacent wall rocks, and wall-rock damage zones are generally ≤5 m thick. Phase 1 normal faults make 41°–48° angles with sheeted diabase dikes in overlying unit Trv.

In the Arroyo Los Ajos region, the Trv/Trg contact is also present in a klippe (K, L-7), where it was passively translated in the hanging wall of a Phase 2 fault, and within a large horse in the high-angle Phase 3 fault zone (K-8).

Near the head of Arroyo Puerto Nuevo, the Trv/Trg contact crops out continuously for ~2.5 km, roughly parallel to and structurally above the underlying Trg/Trh contact (G-3 to H-2). Phase 1 Trv/Trg contacts are inferred in low-relief outcrops along the northern margins of the study area (J-1, R-3).

Phase 2: Late Jurassic–Earliest Cretaceous Shortening

Trg/Trv

Thrusts of probable Late Jurassic to earliest Cretaceous age place gabbro above diabase and basalt in the south-central part of the study area (J-8 to M-7, N-7 to M-5; Fig. 8). These faults cut Phase 1 normal faults and are in turn truncated to the south by the high-angle Phase 3 fault. The easternmost thrust is not as well-exposed or accessible and has not been traced fully, but the western fault is spectacularly exposed at Arroyo Los Ajos (J-8), where it has been folded into a west-plunging antiform (Figs. 6 and 11A).

The thrust surface lacks the irregular undulations typical of Phase 1 faults but is locally folded into coaxial parasitic folds (Fig. 11B). The contact is sharp (Fig. 11C), lacking the tectonic interleaving characteristic of Phase 1 normal faults. In the foot-wall, fault-produced foliation is closely spaced within 50–100 cm of the contact and hackly out to 2–8 m from the contact. In the hanging wall, cataclastic foliation dies out 1–5 m from the contact. Uncommon fault striae (five measurements) plunge 20°–60° west to west-northwest.

Minor structures in the footwall of the Phase 2 thrust in Arroyo Los Ajos are also attributed to Phase 2 deformation. Sheeted dikes are warped into broad, open folds (wavelength ~200 m) around a subhorizontal east-west axis (Fig. 5D) that is subparallel to that of the antiform. Rare quartz-epidote veins, found only within ~10 m of the thrust, are openly to tightly folded (wavelength ~10 cm). Similar structures have not been observed elsewhere in the field area, and their proximity to the thrust strongly suggests they are synkinematic.

Puerto Nuevo Mélange

The contacts between the exotic-block and serpentinite-matrix members of the Puerto Nuevo mélange were interpreted as the margins of a zone of thrust faulting by Moore (1986) and thus may have formed during Phase 2 shortening. Below, I present evidence that slip on the faults that bound the exotic-block mélange occurred during Phase 4 extension. Nevertheless, the formation of the exotic-block mélange itself, i.e., the juxtaposition of diverse exotic blocks, may indeed have formed during Phase 2 shortening, possibly during overthrusting by the ophiolite as suggested by Moore (1986). Many exotic blocks display evidence of strong folding, consistent with shortening that caused thrusting. Such deformation was imparted before the blocks were incorporated into the exotic-block mélange.

Phase 3: Early Cretaceous Strike-Slip

Phase 3 deformation consists of a subvertical, generally east-west–striking fault zone from lower Arroyo Los Ajos (J-8) eastward to the eastern edge of the mapped area, where it forms

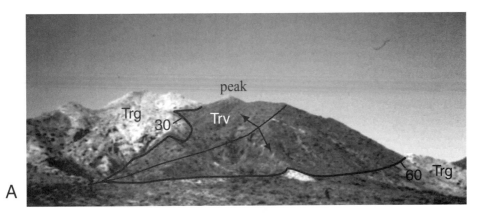

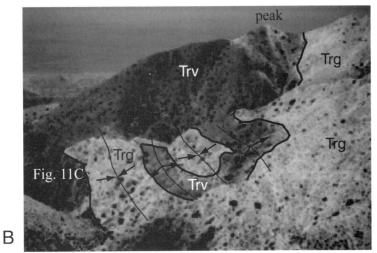

Figure 11. Phase 2 deformation. A: Antiformally folded Phase 2 thrust places ophiolitic gabbro (unit Trg) above sheeted dikes and pillow basalt (unit Trv) at J-8. View to east-southeast. South limb dips ~60° south, north limb dips ~30° north. Antiform nose at front left. Foreground covered by alluvium. Peak is same as that in B. B: Antiformally folded Phase 2 thrust at K-7, J-8. View to west-southwest. Peak is same as that in A. Undulatory outcrop pattern of thrust is due to parasitic folds (note "hook" in outcrop pattern of Trv). Location of C is shown at lower left. C: Truncation of sheeted dikes in Unit Trv (orientation shown by solid lines) by gabbro (light color) in hanging wall of Phase 2 thrust. Outcrop is ~20 m high. See B for location.

a 60° left bend (Q-7). Moore (1983) grouped this fault zone with the Sierra Placeres mélange, which crops out over a much larger area farther to the southeast (Fig. 1B). However, I suggest that this fault zone is tectonically distinct from the Sierra Placeres mélange and name it the Tangaliote fault zone after a small rancho in lower Arroyo Los Ajos.

The Tangaliote fault zone differs from the Sierra Placeres mélange in the following respects. (1) The Sierra Placeres mélange is up to 15 km wide; the Tangaliote fault zone is no more than 0.6 km across. (2) Blocks in the Sierra Placeres mélange are up to 8 km across; the maximum block area in the Tangaliote fault zone is ~1 km². (3) The Tangaliote fault zone has parallel, nearly straight-line contacts, indicating that it is approximately vertical and tabular; in contrast, the Sierra Placeres mélange has curved, irregular contacts. (4) Most blocks in the Sierra Placeres mélange are derived from adjacent island-arc rocks; such blocks are absent from the Tangaliote fault zone.

Alternatively, the Tangaliote fault zone may comprise the northernmost branch of the Sierra Placeres mélange, as suggested by Moore (1983, 1985). Block size and mélange thickness in the Tangaliote fault zone fall within the range of such values in the Sierra Placeres mélange, and incomplete mixing of blocks within the mélange may account for the absence of arc-derived blocks in the Tangaliote. The origin and significance of the Tangaliote fault zone are addressed more fully in the Discussion section.

The thickness of the Tangaliote fault zone and the nature of its tectonic inclusions vary along strike (Fig. 2). Between Arroyo Los Ajos and Arroyo San Cristóbal, the fault zone is 400–750 m wide and is bounded on both margins by 5- to 50-m-thick belts of the Tangaliote mélange (Fig. 12). The interior of the fault zone between these two arroyos consists of a faulted stratigraphic section that is informally referred to as "the Orphan" (K-9 to M-8). From bottom to top, this orphaned stratigraphic section includes Triassic gabbro, sheeted dikes, and basalts of the ophiolite; Triassic limestone, basalt, chert, and tuffs of unit Trs; and sandstones of the lower member of the Eugenia Formation.

The section within "the Orphan" is thinned by numerous normal faults. I assign faults between ophiolitic units to Phase 1. The normal fault that separates the Eugenia from underlying Trs must be younger than Phase 1, however. I infer that it formed during early Phase 4 extension (see below).

"The Orphan" terminates a few hundred meters east of Arroyo San Cristóbal. From there eastward to the edge of the mapped area, the fault zone consists solely of Tangaliote mélange. The 100–200 m

thickness of the mélange roughly approximates the combined thickness of the two marginal belts of the Tangaliote mélange west of Arroyo San Cristóbal. Thus, I interpret "the Orphan" as a lenticular inclusion that locally widened the fault zone.

Striae are abundant on foliation planes within the serpentinite matrix of the Tangaliote mélange. Rakes within the foliation planes range from 0° (horizontal) to 50°, suggesting that the latest phase of slip on this straight, subvertical fault zone was dominantly strike-slip.

At its western limit, the Tangaliote fault zone is truncated by a through-going, northwest-striking Phase 4 normal fault, though the intersection is covered by Quaternary alluvium (Fig. 6). To the east, the Tangaliote fault zone forms a 60° left bend (Q-7) that is truncated by a Phase 4 normal fault (R-6). Along its mapped length, the Tangaliote fault zone truncates structures that developed during Phase 1 and Phase 2.

Phase 4: Cretaceous Extension

Phase 4 extension produced normal fault and vein systems that cut strata of the Valle Group and Eugenia Formation and synkinematic normal faults that place these strata atop Triassic rocks of the Sierra de San Andres ophiolite. Phase 4 also produced structural horses of units Trv, Trs, and KJe, and faults within the Puerto Nuevo mélange.

Phase 4 normal faults cut faults of Phase 3 (beneath alluvium at J-8) and faults of Phase 1 (F-4, H-7, K-9), and thus must also be younger than faults of Phase 2. Phase 4 faulting must have been active after deposition of the Valle Group (probably Cenomanian), but field evidence described below indicates that it began in the Early Cretaceous while Phase 3 strike-slip faulting was still active. Phase 4 faulting is regionally correlated with mid-Cretaceous to Paleogene normal faulting recognized in all Mesozoic rock units in western Baja (Sedlock, 1988a, 1996, 1999). Kinematic data from the Puerto Nuevo region were interpreted to indicate subhorizontal east-west to northwest-southeast extension (Sedlock, 1999). Additional kinematic data are presented below.

Faults That Place Kv and KJe atop Younger Units

Phase 4 normal faults place strata of the Valle Group, Eugenia Formation, or both atop ophiolitic rocks at a continuous but locally sinuous contact that roughly encircles the uplifted core of the study area (Figs. 2 and 8). Re-entrants and promontories are common at map scale, and abrupt changes in attitude, including sharp right-angle turns and complex undulations, are common at outcrop scale. Klippe of the Valle Group above the Triassic ophiolite (F-6, K-9, L-9) probably are remnants of a structural lid across the entire range that has been breached by erosion (Fig. 13A)

Phase 4 faulting produced a variety of effects in the hanging-wall strata and in the ophiolitic rocks of the footwall. The contact is marked by black gouge that ranges in thickness from 5 cm to 40 cm and by local contact-parallel calcite veins from which other veins cut up into the hanging wall. Strata of the Valle Group locally are bleached or blackwalled (probably by magnesium

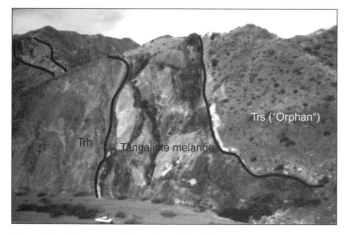

Figure 12. View west of southern belt of Tangaliote mélange along margin of Tangaliote fault zone, west wall of Arroyo San Cristóbal. Light-colored blocks in mélange are diabase and gabbro; mélange continues on hill at upper left. Core of fault zone consists of strata of unit Trs that form structural base of "the Orphan," a large tectonic inclusion within the fault zone.

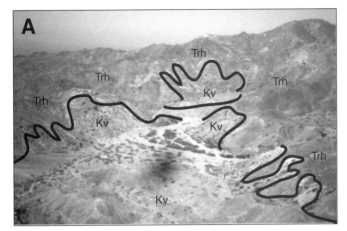

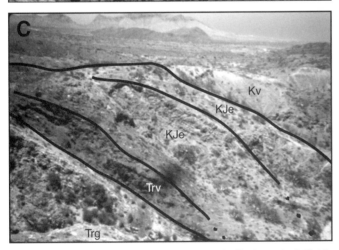

Figure 13. Phase 4 normal faulting. A: Klippe of Valle Group (unit Kv) in Arroyo San Cristóbal that overlie ultramafic rocks of the ophiolite (unit Trh). Undulatory contact results from eroded dip slopes of fault surface (with overlying Valle strata) that dip into this structural bowl on south (left), west, and north sides. B: Normal fault within Valle Group places bedded-sandstones (Kv ss) atop mudstone (Kv mst). Fault is synthetic to, and merges downward into (see foreground) major normal fault separating Valle Group from underlying ophiolitic gabbro (unit Trg). C: Nested structural horses within a Phase 4 fault zone between Valle Group strata (unit Kv) in hanging wall and ophiolitic gabbro (unit Trg) in footwall. Horses include basalt and diabase of Trv (structurally lowest) and two distinct slices of lower member of Eugenia Formation (unit KJe).

metasomatism; Moore, 1983) up to 20 m from the contact. Anastomosing fractures and normal drag folding are locally developed within the Valle Group along these faults. However, these and other fault-related structures generally extend no more than 50 m into the hanging wall strata. Footwall rocks typically show a moderately spaced foliation along the contact that becomes unrecognizable 5–10 m from the contact.

East of Arroyo Puerto Nuevo, the Phase 4 fault that places the Valle Group above ophiolitic ultramafic rocks contains abundant striae and mullions that indicate horizontal east-northeast–west-southwest extension (Fig. 14, diagrams B and C). Farther east, data from a highly undulatory zone of klippe and re-entrants indicate horizontal east-southeast–west-northwest extension (Fig. 14, diagram A).

Northwest of Arroyo Puerto Nuevo, fault-striae data indicate horizontal east-southeast–west-northwest extension in the western edge of the mapped area (Fig. 14, diagram F). Fault-striae data beneath and north of the klippe of Valle at F-6 (Fig. 14, diagrams D, E) did not produce well-defined kinematic axes (low eigenvalues).

Faults within Kv

Phase 4 faults strongly disrupt the original lateral continuity of strata and facies within the Valle Group (Fig. 13B). Most normal faults within the Valle are sharp and fairly planar, though a few show gouge zones up to 20 cm thick. Slip ranges from ≤1 cm to >100 m. Fault-striae data collected between Arroyo Puerto Nuevo and Arroyo San Cristóbal (Fig. 14, diagram G) indicate horizontal east-northeast–west-southwest extension, matching the result from the underlying Kv/Trh contact (Fig. 14, diagrams B and C).

Horses Containing KJe, Trv, Trs

Many Phase 4 faults with hanging-wall Valle or Eugenia and footwall Trh or Trg contain structural horses of units Trv, Trs, and KJe. Most of these horses are far too small to show at the scale of Figure 2, rarely exceeding 100 m in length and 20 m in structural thickness. A mappable horse along the Valle/Trg contact at I-8 consists of nested horses of Trv and KJe in their original stratigraphic order (Fig. 13C).

All exposures of unit Trs are horses within Phase 4 fault zones in the eastern half of the map area (L,M-1; P-1; R-3; R-4; K,L,M-5; O,P-8). The klippe of unit Trs at K,L-5 lies as little as 100 m from a Phase 1 fault that places unit Trg over unit Trh, and faults at the base of the Trs klippe form a high angle to the Phase 1 fault. These relations suggest that the klippe and horses of unit Trs were produced by normal faults of Phase 4 rather than Phase 1. The Trs klippe at K,L,M-5 are probably the last remnants of a depression in an undulatory Phase 4 fault (akin to those at F-6, K-9, L-9) from which the entire Valle hanging wall has been eroded away.

Pre-Valle Extension

As noted above, the stratigraphic section of "the Orphan" within the Tangaliote fault zone is thinned by numerous normal faults. A 50-m-thick fault zone that contains slices of hanging-wall and footwall rocks places the Eugenia Formation atop unit

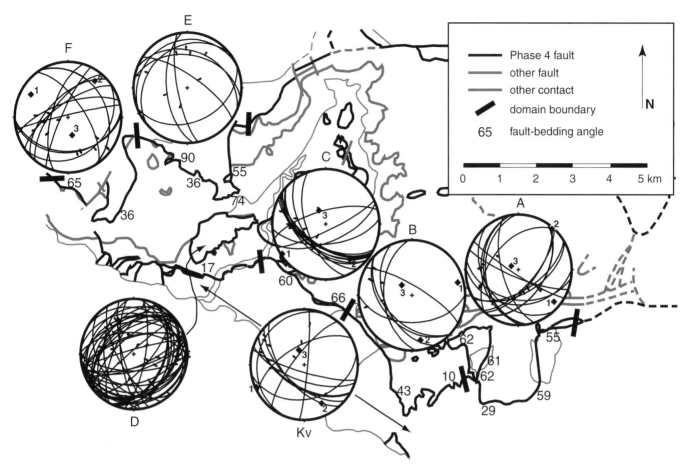

Figure 14. Structural data from Phase 4 faults. Solid numbers along major faults refer to angle between fault surface and adjacent homoclinal strata in hanging wall. Plot labeled Kv shows results from within Valle Group between Arroyo Puerto Nuevo and Arroyo San Cristobal (lateral coverage shown by arrowed lines). All other plots show results from major faults that place Valle Group or Eugenia Formation against underlying ophiolitic units. Short black line segments bound domains. Arrows show slip direction of hanging wall. 1, 2, 3: extension, intermediate, and shortening axes of unweighted moment tensor.

Trs of the ophiolite (Fig. 2, L-8; also see Fig. 6). This fault zone cannot be a Phase 1 structure because it affects the Upper Jurassic–Lower Cretaceous Eugenia Formation. I interpret this feature as an early Phase 4 fault and conclude that the onset of Phase 4 extension occurred prior to cessation of Phase 3 strike-slip faulting on the Tangaliote fault zone (see Discussion section).

Puerto Nuevo Mélange

As noted in an earlier section, the faults bounding the exotic-block mélange were interpreted as margins of a zone of thrusting by Moore (1986) and thus could have formed during Phase 2 deformation. However, structural data suggest that the dip-slip component of slip on the mélange-bounding faults was normal rather than thrust. The serpentinite-matrix mélange is cut by pervasive minor faults; where slip was sufficiently small, separations can be recognized by restoring offset resistant blocks of serpentinized ultramafic rock. Dip separations on these faults are almost exclusively normal, implying a component of roughly horizontal extension. The mélange-bounding faults dip gently

(≤30°) in these areas. Thus, if the minor faults and the mélange-bounding faults formed together, the sense of dip slip on the latter most likely was normal rather than thrust.

I infer that the faults that bound the exotic-block mélange are Phase 4 normal faults, supporting my earlier (Sedlock, 1988a) speculative correlation of the Puerto Nuevo mélange with mélanges in the crustal-scale normal fault zones elsewhere in western Baja California. Structural discontinuities in the lateral extent of the exotic-block mélange (D-6,7, I-6,7, I,J-3) thus represent the juxtaposition of the overlying and underlying serpentinite-matrix mélange by younger faults that cut out the previously assembled exotic-block mélange.

DISCUSSION

Early Extension

The Sierra de San Andres ophiolite probably formed in a suprasubduction-zone environment such as a forearc, intra-arc, or backarc, judging from the Late Triassic volcanogenic strata that overlie it; geochemical data seem to show an arc (i.e., supra-subduction-zone) component (Moore, 1985), but similar arc signatures also have been recognized in modern mid-oceanic-ridge basalt (MORB) (Sturm et al., 2000) and in ophiolites of probable mid-oceanic-ridge origin (Moores et al., 2000). After its Late Triassic formation, but prior to Late Jurassic to earliest Cretaceous shortening, the ophiolite was subjected to extension that produced Phase 1 normal faults. Extension may have occurred either at the Triassic spreading center that generated the Sierra de San Andres ophiolite (Fig. 15A) or during subsequent Jurassic plate-margin tectonism of unclear type or cause.

Marine studies of modern spreading centers have documented tectonic denudation, detachment faulting, and large extensional strains (e.g., Karson, 1998; Macdonald, 1998). Phase 1 structures are somewhat similar to those described from modern spreading centers with an intermediate rate of spreading and a moderate magma supply. Modern and fossil spreading centers with slow spreading rates, low magma budgets, or both are cut by widespread detachment faults that typically show greater levels of tectonic thinning than in the study area; modern and fossil spreading centers with fast spreading rates, high magma budgets, or both are cut by fewer, comparatively high-angle faults (Dilek et al., 1998; Karson, 1998; Macdonald, 1998). However, it is unclear whether such comparisons are appropriate given the probable suprasubduction-zone origin of the Sierra de San Andres ophiolite. More data are needed from modern suprasubduction-zone environments to resolve this issue.

Metamorphic assemblages, the weak foliation in the gabbros, and the absence of foliation in the basalts and diabase of unit Trv, imply that faulting probably occurred at intermediate to low temperatures and involved negligible plastic strain. Sparse orientation data suggest that Phase 1 faults formed with initial dips of ~30°–50°.

Shortening of the Ophiolite

If the Sierra de San Andres ophiolite formed in a suprasubduction-zone environment outboard of the North American plate, then Phase 2 thrusting and folding may have been caused by accretion of the ophiolite to North America (Fig. 15B). Such accretion has been inferred from sedimentologic studies showing that continentally derived clasts supplemented or replaced volcanogenic clasts by 150 Ma in the Vizcaíno peninsula (Hickey, 1984) and by the early Late Jurassic on Isla Cedros (Boles and Landis, 1984).

Alternatively, Phase 2 thrusting and folding may reflect shortening of the ophiolite in a suprasubduction-zone environment within the North American plate. In this scenario, the ophiolite retains its plate parentage and is not accreted, so Phase 2 strain and the coeval provenance change may reflect intraplate shortening of unclear origin.

Evidence for Phase 2 deformation is not widespread or penetrative, and the paucity of Phase 2 structures prevents determination of the direction and magnitude of shortening. This deformation phase is much more subtly expressed than the regional "Nevadan" event envisioned by Rangin (1978) and Rangin et al. (1983) but is largely rejected by subsequent workers.

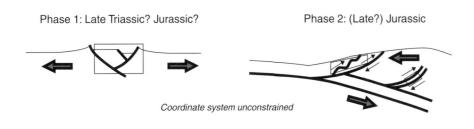

Figure 15. Schematic cartoon showing inferred tectonic setting of deformation phases that affected study area. Boxed areas denote schematic position of study area at different times. Heavy lines are faults, with slip sense shown by small arrows. Large arrows show relative plate motions. Phase 1 extension probably occurred at Late Triassic spreading center or in unspecified Jurassic suprasubduction-zone environment. Phase 2 shortening produced thrusting and folding of ophiolite and formation of exotic-block mélange. Deformation may have occurred within North American plate rather than at a subduction zone as shown (see text). Phase 3 sinistral slip and early Phase 4 extension probably occurred within continental forearc subjected to transpression. Phase 4 syn-subduction extension, with concomitant exhumation of underplated blueschist (Sedlock 1996, 1999), continued into Late Cretaceous and Paleogene.

Early Cretaceous Strike Slip

The geometry of the Tangaliote fault zone strongly suggests that it was dominated by strike slip, and cross-cutting relations indicate Early Cretaceous slip. The Tangaliote must be younger than Phase 2 thrusts, which it cuts, and the Tithonian to Neocomian lower member of the Eugenia Formation, which it incorporates as a tectonic inclusion. The Tangaliote fault zone must be older than the Valle Group, which occupies the hanging walls of Phase 4 faults that truncate the Tangaliote fault zone in lower Arroyo Los Ajos and at the eastern edge of the study area.

I suggest that the Tangaliote fault zone accommodated sinistral slip based on the regional distribution of the Eugenia Formation. The Eugenia crops out in several areas northwest of the map area but does not crop out south of the Tangaliote fault zone. I speculate that the Eugenia fragment in the Tangaliote fault zone has been left-laterally displaced 15–20 km from the vicinity of Puerto Escondido, where it overlies Triassic sedimentary and volcanic rocks (Moore, 1983) similar to those that it overlies in the Tangaliote fault zone.

Sinistral slip is also consistent with the current distribution of outcrops of Late Triassic sedimentary and volcanic rocks of unit Trs. As noted in the preceding paragraph, rocks similar to those of unit Trs in the Tangaliote fault zone crop out 15–20 km to the west-northwest (i.e., north of the westward projection of the Tangaliote fault zone) near Puerto Escondido (Fig. 1B). Moore (1985) mapped similar rocks ~20 km east of the study area, slightly south of the eastward projection of the Tangaliote fault zone (solid circle in Fig. 1B). These western and eastern outcrops have significantly greater stratigraphic thickness and slightly different stratigraphy than unit Trs within the Tangaliote fault zone, possibly due to tectonic thinning along Phase 1 faults in the latter.

The magnitude estimates and inferred sense of slip on the Tangaliote fault zone are based strictly on current outcrop distribution and would need to be reconsidered if Eugenia or Trs strata underlie Kv or were eroded away in this region. However, the Early Cretaceous timing of such sinistral slip would be consistent with the interpretation of left-oblique convergence between western North America and oceanic plates in the eastern Pacific until ca. 100 Ma (Engebretson et al., 1985, 1992) (Fig. 15C).

Moore (1983, 1985) mapped the Tangaliote fault zone of this paper as part of the Sierra Placeres mélange, a tectonic unit that separates the Vizcaíno Norte and Vizcaíno Sur terranes (Fig. 1B). Moore interpreted the Sierra Placeres mélange as a major fault zone with uncertain slip sense and magnitude. If the Tangaliote and Sierra Placeres are parts of the same structural unit, I speculate that both accommodated Early Cretaceous sinistral slip.

Phase 4 Extension

Timing

Relations in the map area indicate only that Phase 4 extension was active after Cenomanian deposition of the Valle Group, but analysis of the Valle on nearby Isla Cedros indicates that extension began by the early Cenomanian. Smith and Busby (1993) mapped early Cenomanian breccias that formed in a half-graben caused by a syndepositional normal fault. They did not recognize syndepositional extension in Albian strata of the Valle on Cedros.

A strong argument can be made that Phase 4 extension in the study area began in the Early Cretaceous. Within the Tangaliote fault zone (Phase 3), a Phase 4 normal fault places the Tithonian-Neocomian lower member of the Eugenia Formation against underlying Triassic sedimentary rocks. The normal fault is cut by the Tangaliote fault zone, yet the Tangaliote fault zone is truncated by a younger Phase 4 normal fault at its western termination (Figs. 2 and 6). The simplest interpretation of these observations is that Phase 3 and Phase 4 both were active in the Early Cretaceous, but Phase 3 ceased during the (late?) Early Cretaceous and was cut by Late Cretaceous faults of Phase 4 (Fig. 15, C and D).

An Early Cretaceous onset of Phase 4 is consistent with tectonic interpretations of the region. Phase 4 extension probably formed due to prolonged exhumation of blueschists in an active subduction zone (Sedlock, 1999). Key factors that drove exhumation were (1) episodic tectonic underplating of blueschist-facies materials, and (2) the subduction of oceanic lithosphere that was sufficiently old and cold to continuously refrigerate the North American forearc, preserving those blueschist assemblages. These factors probably applied not only to the Late Cretaceous and Paleogene (Sedlock, 1999), but also to the Early Cretaceous, as suggested by Early Cretaceous plate ages and motions (Engebretson et al., 1985, 1992).

General Features

Phase 4 normal faulting produced not only large-slip normal faults that place hanging-wall strata of the Valle Group and Eugenia Formation against footwall ophiolitic units, but also normal faults and vein systems that cut hanging-wall strata throughout the study area. Field and structural data indicate that these structures are almost certainly coeval and cogenetic. First, hanging-wall faults sole downward into and merge with the large-slip faults at the base of the hanging-wall Valle or Eugenia. Second, kinematics of the two fault populations are very similar (Fig. 14, compare diagrams B and C with diagram G). Throughout the study area, fault-striae data from Phase 4 extension show roughly east-west extension, implying that the deformation was approximately coaxial.

Fault-bedding Relations

To further characterize Phase 4 extension, I investigated the angular relations between hanging-wall Valle and Eugenia strata and the large-slip faults that separate them from footwall ophiolitic rocks. Attitudes of hanging-wall strata were chosen from homoclinal outcrops that were not affected by fault-related tilting, rotation, or disruption. Sites were rejected if appropriate outcrops lay more than 300 m from contact exposures; factoring in fault dip, this distance corresponds to 100–200 m from the contact in three dimensions. Sites were also rejected in areas where undularity of the fault surface prevented me from assigning an average orientation with confidence.

Of the 17 sites that successfully met these criteria, nine yielded angles between 55° and 65°, and 13 yielded angles between 35° and 74° (Fig. 14). The other four sites are at or near bends (not small-scale undulations) in the fault surface.

The concentration of fault-bedding angles near 60° suggests that, assuming an Andersonian geometry (e.g., Twiss and Moores, 1992), much of the Valle and Eugenia were roughly horizontal before major slip on the bounding faults. Anderson's criteria probably apply because faulting must have occurred within a few kilometers of the surface, given the negligible metamorphism of the Valle and Eugenia. Deviation from an initial 60° dip may have been produced in at least two ways: (1) over some areas, the fault(s) formed more steeply or shallowly due to variations in stress field, fault geometry, or rheology; (2) Valle and Eugenia strata were tilted before major slip on the bounding faults, presumably by older Phase 4 faulting.

Fault Undulations

Several arguments support the interpretation of undulations in the major Phase 4 faults as intrinsic features that were not caused by later folding, arching, or buckling. (1) The undulations are not symmetrical, cylindrical, or regularly distributed, and thus should not be considered corrugations produced by fault-parallel shortening. (2) The only tectonic folds in hanging-wall strata are drag folds along normal faults. (3) In several locations, a sharply undulatory fault clearly truncates homoclinal bedding in hanging-wall strata that spans the undulation with no change in orientation.

Significance of Bedding Orientation in Mesozoic Strata

The orientation of bedding in the Eugenia Formation and Valle Group provides useful tests of various aspects of the structural history of the study area. Attitudes from the Valle were divided into seven domains (Fig. 7) based on geographic location and internal consistency. The southeastern and southern domains (A and B) dip gently to moderately south to east; the northwestern domains (F, G, and JKe) dip gently to moderately north to west. These differences probably reflect Cenozoic folding of the study area along a north-northeast–trending axis that ultimately produced the structural window along Arroyo Puerto Nuevo.

Domains C and D, near the mouth of Arroyo Puerto Nuevo, dip gently to moderately south and west, i.e., roughly intermediate between the southeast and northwest dips of the domains to their southeast and northwest, respectively. I attribute these intermediate dips to the position of these domains along the Cenozoic antiformal axis. The greater variability of orientations in Domain C may be due to minor deformation of the hanging wall above a jog in the Phase 4 normal fault that crops out along the southern margin of the range. Although the jog is mostly buried beneath alluvium of the arroyo, it crops out in several places along the base of the east wall of the arroyo (F, G-7).

Domain E, the klippe of Valle strata just west of lower Arroyo Puerto Nuevo, contains subgroups that dip southeastward and northwestward, i.e., parallel to strata in the southeastern and northwestern domains, respectively. This klippe probably represents a depression in an undulatory Phase 4 fault that was originally continuous with other Phase 4 faults now exposed to the southeast, southwest, and northwest.

Cenozoic Uplift

The study area comprises a structural dome, with a core of older ophiolitic units mantled by the Cretaceous Valle Group. This dome forms part of a continuous belt of ranges in the western Vizcaíno peninsula that exposes Mesozoic rocks. The uplift of all these ranges occurred after deposition of the Valle, which is as young as Campanian in the northern Vizcaíno and as young as Eocene in the southern Vizcaíno (Kimbrough et al., 2001).

It is tempting to ascribe uplift to compression at restraining bends or stepovers in strike-slip faults that formed due to Miocene and younger transform motion, but the distribution of dextral faults along western Baja makes such a cause unlikely. Northwest-striking faults in the southern Baja borderland are interpreted as Cenozoic dextral-slip faults (Normark et al., 1987; Fenby and Gastil, 1991); most of these faults lie west of the Vizcaíno peninsula and Isla Cedros (Fig. 1A), though minor splays may cut the Vizcaíno near Puerto Escondido and Bahía Tortugas (Robinson, 1979; Moore, 1983). North of the Vizcaíno, offshore dextral faults lie within Bahía Vizcaíno (Fig. 1A), and their southeastward projections would bound the study area on the east. If these fault zones are coeval, they define a right step whose releasing geometry is highly unlikely to have produced surface uplift of the Vizcaíno region.

The Vizcaíno may have been uplifted due to extension, normal faulting, and rotation, possibly associated with development of the transtensional boundary in the Miocene. Such faults are present along the west coast of Baja on nearby Isla Cedros (Sedlock, 1988c) and on Isla Santa Margarita at 24.5°N (Sedlock, unpubl. mapping; Fletcher et al., 2000).

Alternatively, the uplifts may have formed during crustal shortening, presumably prior to the Miocene development of the transform boundary. Regionally, the Valle Group is gently folded about northwest-southeast axes, and sparse Miocene strata are tilted even less (Robinson, 1979). These observations indicate Cenozoic folding persisted until at least 15 Ma (J. Helenes, 2000, personal commun.).

CONCLUSIONS

New field and structural data from superb exposures of Mesozoic oceanic rocks in the northern Vizcaíno peninsula help unravel the complex tectonic history of this region.

1. The Sierra de San Andres ophiolite underwent Phase 1 extension after its Late Triassic formation but prior to Late Jurassic to earliest Cretaceous shortening. Extension may have occurred at or near the host spreading-center shortly after ophiolite formation or during Jurassic plate-margin tectonism of unclear type or cause. Detachment faults are present, but they do

not appear to have produced extreme thinning as seen at many modern spreading centers.

2. Spreading-center extensional features are cut by folded thrusts that developed during Jurassic to earliest Cretaceous Phase 2 shortening. If the ophiolite formed in a suprasubduction-zone environment outboard of the North American plate, then Phase 2 thrusting and folding may have been caused by accretion of the ophiolite to North America. Alternatively, Phase 2 thrusting and folding may reflect shortening of the ophiolite in a suprasubduction-zone environment within the North American plate. Thrusting and folding are not widely or penetratively developed, suggesting that total shortening was minor. This contrasts greatly with the penetrative ductile strains that accompanied roughly synchronous accretion of an ophiolite on Isla Santa Margarita at 24.5°N (Sedlock, 1993).

3. After accretion, the ophiolite formed part of the structural lid atop the subduction complex in the North American forearc. Left-oblique convergence of North America with plate(s) in the eastern Pacific (e.g., Farallon) may have spawned a left-lateral strike-slip fault zone (Phase 3) that cut the ophiolitic rocks in the forearc during the Early Cretaceous.

4. Exhumation of deep-level coherent blueschists triggered synsubduction extension (Phase 4) that began in the Early Cretaceous, continued through the Late Cretaceous, and ceased in the Paleogene. Extension affected the ophiolitic rocks as well as Late Jurassic to Cenomanian strata that were deposited in different forearc environments.

ACKNOWLEDGMENTS

The early work of Gordon Gastil and his colleagues laid the groundwork for all later studies of the geology of the Vizcaíno. Tom Moore provided essential contributions for establishing the geologic framework of the study area and for understanding its geologic history. Yildirim Dilek and Tom Moore provided thoughtful, thorough reviews that greatly improved the manuscript. I thank the students and teaching assistants in San Jose State University's summer field classes of 1995, 1998, and 2001 for their contributions, particularly John Baldwin, John Siskowic, and Ante Mlinarevic. Finally, un mil gracias to Marcos Lucero Arce and his family for their hospitality and friendship.

REFERENCES CITED

Baldwin, S.L., and Harrison, T.M., 1989, Geochronology of blueschists from west-central Baja California and the timing of uplift of subduction complexes: Journal of Geology, v. 97, p. 149–163.

Baldwin, S.L., and Harrison, T.M., 1992, The P-T-t history of blocks in serpentinite-matrix mélange, west-central Baja California: Geological Society of America Bulletin, v. 104, p. 18–31.

Barnes, D.A., 1982, Basin analysis of volcanic arc derived, Jura-Cretaceous sedimentary rocks, Vizcaíno peninsula, Baja California Sur, México (Ph.D. dissertation): Santa Barbara, University of California, 240 p.

Blake, M.C., Jr., Jayko, A.S., Moore, T.E., Chavez, V., Saleeby, J.B., and Seel, K., 1984, Tectonostratigraphic terranes of Magdalena Island, Baja California Sur, *in* Frizzell, V.A., Jr., ed., Geology of the Baja California

peninsula: Pacific Section, Society of Economic Paleontologists and Mineralogists, v. 39, p. 183–191.

Boles, J.A., and Landis, C.A., 1984, Jurassic sedimentary mélange and associated facies, Baja California, México: Geological Society of America Bulletin, v. 95, p. 513–521.

Bonini, J.A., 1994, ^{40}Ar/^{39}Ar geochronology of accreted terranes from southwestern Baja California Sur [M.S. thesis]: Tucson, University of Arizona, 37 p.

Bonini, J.A., and Baldwin, S.L., 1994, Mesozoic metamorphic and middle to late Tertiary magmatic events on Magdalena and Santa Margarita Islands, Baja California Sur, Mexico: Implications for the tectonic evolution of the Baja California continental borderland: Geological Society of America Bulletin, v. 110, p. 1094–1104.

Coleman, R.G., 1977, Ophiolites: Ancient oceanic lithosphere?: New York, Springer-Verlag, 229 p.

Coney, P.J., and Campa, M.F., 1987, Lithotectonic terrane map of Mexico (west of the 91st meridian): U.S. Geological Survey Miscellaneous Field Studies Map MF-1874-D, scale 1:2,500,000, 1 sheet.

Couch, R.W., Ness, G.E., Sanchez-Zamora, O., Calderón-Riveroll, G., Doguin, P., Plawman, T., Coperude, S., Huehn, B., and Gumma, W., 1991, Gravity anomalies and crustal structure of the Gulf and peninsular province of the Californias, in Dauphin, J.P., and Simoneit, B.R.T., eds., The Gulf and peninsular province of the Californias: American Association of Petroleum Geologists Memoir 47, p. 25–45.

Dilek, Y., Moores, E.M., and Furnes, H., 1998, Structure of modern oceanic crust and ophiolites and implications for faulting and magmatism at oceanic spreading centers, *in* Buck, W.R., Delaney, P.T., Karson, J.A., and Lagabrielle, Y., eds., Faulting and magmatism at mid-ocean ridges: American Geophysical Union Geophysical Monograph 106, p. 219–265.

Engebretson, D.C., Cox, A., and Gordon, R.G., 1985, Relative motions between oceanic and continental plates in the Pacific Basin: Boulder, Colorado, Geological Society of America Special Paper 206, 64 p.

Engebretson, D.C., Kelley, K.P., Cashman, H.J., and Richards M.A., 1992, 180 million years of subduction: GSA Today, v. 2, no. 5, p. 93–100.

Fenby, S. S., and Gastil, R.G., 1991, Geologic-tectonic map of the Gulf of California and surrounding areas, *in* Dauphin, J.P., and Simoneit, B.R.T., eds., The Gulf and peninsular province of the Californias: American Association of Petroleum Geologists Memoir 47, p. 79–83.

Fletcher, J.M., Eakins, B.A., Sedlock, R.L., Mendoza-Borrunda, R., Walter, R.C., Edwards, R.L., and Dixon, T.H., 2000, Quaternary and Neogene slip history of the Baja-Pacific plate margin: Bahía Magdalena and the southwestern borderland of Baja California [abs.]: Eos (Transaction, American Geophysical Union), v. 81, p. F1232.

Gastil, R.G., Phillips, R.P., and Allison, E.C., 1975, Reconnaissance geology of the state of Baja California: Boulder, Colorado, Geological Society of America Memoir 140, 170 p.

Hickey, J.J., 1984, Stratigraphy and composition of a Jura-Cretaceous volcanic arc apron, Punta Eugenia, B.C., Sur, México, *in* Frizzell, Jr., V.A., ed., Geology of the Baja California peninsula: Los Angeles, California, Pacific Section, Society of Economic Paleontologists and Mineralogists, v. 39, p. 149–160.

Karson, J.A., 1998, Internal structure of oceanic lithosphere: A perspective from tectonic windows, *in* Buck, W.R., Delaney, P.T., Karson, J.A., and Lagabrielle, Y., eds., Faulting and magmatism at mid-ocean ridges: American Geophysical Union Geophysical Monograph 106, p. 177–218.

Kimbrough, D.L., 1982, Structure, petrology, and geochronology of Mesozoic paleooceanic terranes on Cedros Island and the Vizcaíno peninsula, Baja California Sur, México [Ph.D. dissertation]: Santa Barbara, University of California, 395 p.

Kimbrough, D.L., 1985, Tectonostratigraphic terranes of the Vizcaíno peninsula and Cedros and San Benito Islands, Baja California, México, *in* Howell, D.G., ed., Tectonostratigraphic terranes of the Circum-Pacific region: Circum-Pacific Council for Energy and Mineral Resources, Earth Science Series, no. 1, p. 285–298.

Kimbrough, D.L., Smith, D.P., Mahoney, J.B., Moore, T.E., Grove, M., Gastil, R.G., Ortega-Rivera, A., and Fanning, C.M., 2001, Forearc-basin sedimentary response to rapid Late Cretaceous batholith emplacement in the Peninsular Ranges of southern and Baja California: Geology, v. 29, p. 491–494.

Macdonald, K.C., 1998, Linkages between faulting, volcanism, hydrothermal activity and segmentation on fast spreading centers, *in* Buck, W.R., Delaney, P.T., Karson, J.A., and Lagabrielle, Y., eds., Faulting and magmatism at mid-ocean ridges: American Geophysical Union Geophysical Monograph 106, p. 27–58.

Moore, T.E., 1983, Geology, petrology, and tectonic significance of the Mesozoic paleooceanic terranes of the Vizcaíno peninsula, Baja California Sur, México [Ph.D. dissertation]: Stanford, California, Stanford University, 376 p.

Moore, T.E., 1985, Stratigraphy and tectonic significance of the Mesozoic tectonostratigraphic terranes of the Vizcaíno peninsula, Baja California Sur, México, *in* Howell, D.G., ed., Tectonostratigraphic terranes of the Circum-Pacific region: Circum-Pacific Council for Energy and Mineral Resources, Earth Science Series, no. 1, p. 315–329.

Moore, T.E., 1986, Petrology and tectonic implications of the blueschist-bearing Puerto Nuevo mélange complex, Vizcaíno peninsula, Baja California Sur, México, *in* Evans, B.W., and Brown, E.H., eds., Blueschists and eclogites: Geological Society of America Memoir 164, p. 43–58.

Moores, E.M., Kellogg, L.H., and Dilek, Y., 2000, Tethyan ophiolites, mantle convection, and tectonic "historical contingency": A resolution of the "ophiolite conundrum," *in* Dilek, Y., Moores, E.M., Elthon, D., and Nicolas, A., eds., Ophiolites and oceanic crust: New insights from field studies and the Ocean Drilling Program: Boulder, Colorado, Geological Society of America Special Paper 349, p. 3–12.

Normark, W.R., Spencer, J.E., and Ingle, Jr., J.C., 1987, Geology and Neogene history of the Pacific continental margin of Baja California Sur, México, *in* Scholl, D.W., Grantz, A., and Vedder, J.G., eds., Geology and resource potential of the continental margin of western North America and adjacent ocean basins—Beaufort Sea to Baja California: Circum-Pacific Council for Energy and Mineral Resources Earth Science Series, p. 449–472.

Rangin, C., 1978, Speculative model of Mesozoic geodynamics, central Baja California to northeastern Sonora (México), *in* Howell, D.G., and McDougall, K.A., eds., Mesozoic paleogeography of the western United States: Los Angeles, California, Pacific Section, Society of Economic Paleontologists and Mineralogists, p. 85–106.

Rangin, C., Girard, D., and Maury, R., 1983, Geodynamic significance of Late Triassic to Early Cretaceous volcanic sequences of Vizcaíno peninsula and Cedros Island, Baja California, Mexico: Geology, v. 11, p. 552–556.

Robinson, J.W., 1979, Structure and stratigraphy of the northern Vizcaíno peninsula with a note on the Miocene reconstruction of the peninsula, *in* Abbott, P.L., and Gastil, R.G., eds., Baja California Geology: San Diego, Department of Geological Sciences, San Diego State University, p. 77–82.

Sedlock, R.L., 1988a, Tectonic setting of blueschist and island-arc terranes of west-central Baja California, México: Geology, v. 16, p. 623–626.

Sedlock, R.L., 1988b, Metamorphic petrology of a high-pressure, low-temperature subduction complex in west-central Baja California, México: Journal of Metamorphic Geology, v. 5, p. 205–233.

Sedlock, R.L., 1988c, Lithology, petrology, structure, and tectonics of blueschists and associated rocks in west-central Baja California, México [Ph.D. dissertation]: Stanford, California, Stanford University, 223 p.

Sedlock, R.L., 1993, Mesozoic geology and tectonics of blueschist and associated oceanic terranes in the Cedros-Vizcaíno-San Benito and Magdalena-Santa Margarita regions, Baja California, México, *in* Dunne, G.C., and McDougall, K.A., eds., Mesozoic paleogeography of the western United States-II: Los Angeles, California, Pacific Section, Society of Economic Paleontologists and Mineralogists Book 71, p. 113–125.

Sedlock, R.L., 1996, Syn-subduction forearc extension and blueschist exhumation in Baja California, México, *in* Bebout, G.E., Scholl, D.W., Kirby, S.H., and Platt, J.P., eds., Dynamics of subduction: American Geophysical Union Monograph 96, p. 155–162.

Sedlock, R.L., 1999, Evaluation of exhumation mechanisms for coherent blueschists in western Baja California, México, *in* Ring, U., Brandon, M.T., Lister, G.S., and Willett, S.D., eds., Exhumation processes: Normal faulting, ductile flow, and erosion: Geological Society of London Special Publication 154, p. 29–54.

Sedlock, R.L., and Isozaki, Y., 1990, Lithology and biostratigraphy of Franciscan-like chert and associated rocks, west-central Baja California, México: Geological Society of America Bulletin, v. 102, p. 852–864.

Sedlock, R.L., Ortega-Gutiérrez, F., and Speed, R.C., 1993, Tectonostratigraphic terranes and tectonic evolution of México: Boulder, Colorado, Geological Society of America Special Paper 278, 153 p.

Silver, L.T., and Chappell, B.W., 1988, The Peninsular Ranges Batholith: An insight into the evolution of the Cordilleran batholiths of southwestern North America: Transactions of the Royal Society of Edinburgh: Earth Sciences, v. 79, p. 105–121.

Smith, D.P., and Busby, C.J., 1993, Mid-Cretaceous crustal extension recorded in deep-marine half-graben fill, Cedros Island, Mexico: Geological Society of America Bulletin, v. 105, p. 547–562.

Sturm, M.E., Klein, E.M., Karsten, J.L., and Karson, J.A., 2000, Evidence for subduction-related contamination of the mantle beneath the southern Chile Ridge: Implications for ambiguous ophiolite compositions, *in* Dilek, Y., Moores, E.M., Elthon, D., and Nicolas, A., eds., Ophiolites and oceanic crust: New insights from field studies and the Ocean Drilling Program: Boulder, Colorado, Geological Society of America Special Paper 349, p. 13–20.

Twiss, R.J., and Moores, E.M., 1992, Structural geology: New York, W.H. Freeman and Company, 532 p.

MANUSCRIPT ACCEPTED BY THE SOCIETY JUNE 2, 2003

Geological Society of America
Special Paper 374
2003

Mesozoic tectonic evolution of the Peninsular Ranges of southern and Baja California

Paul H. Wetmore
Department of Earth Sciences, University of Southern California, Los Angeles, California 90831-0740, USA

Charles Herzig
Earth Sciences Department, El Camino College, Torrance, California 90506, USA

Helge Alsleben
Michelle Sutherland
Department of Earth Sciences, University of Southern California, Los Angeles, California 90831-0740, USA

Keegan L. Schmidt
Lewis and Clark College, Lewiston, Idaho

Paul W. Schultz
Scott R. Paterson
Department of Earth Sciences, University of Southern California, Los Angeles, California 90831-0740, USA

ABSTRACT

The Mesozoic evolution of the Peninsular Ranges of southern California, USA, and Baja California, México, remains a controversial aspect of Cordilleran tectonics with multiple, often mutually exclusive, models potentially viable. A fundamental reason for the lack of agreement between the proposed tectonic models is that they are based on one dimensional, arc perpendicular observations of the batholith from widely separated locations on opposite sides of the ancestral Agua Blanca fault, an active strike-slip fault with an earlier Mesozoic history. North of the ancestral Agua Blanca fault, the Late Triassic through Jurassic was characterized by deep to moderately deep marine sedimentation of continentally derived turbidite sequences of the Bedford Canyon Complex. These strata were deformed within an accretionary prism setting and were subsequently uplifted and beveled by subaerial erosion. During the Early Cretaceous the continental margin arc associated with the earlier-formed accretionary prism migrated westward and developed within and on the Bedford Canyon Complex.

South of the ancestral Agua Blanca fault Jurassic strata are only preserved locally in the central zone. During the Early Cretaceous this part of the arc subsided below sea level and became the site of turbidite sedimentation before being uplifted and dominated by the deposition of submarine sediment, succeeded by subaerial volcanics derived from the continental margin arc present in the central and eastern zones. Outboard, the Alisitos arc, developed through and on oceanic crust, began to impinge upon the continental margin in the Early Cretaceous (ca. 115 and 108 Ma). During accretion of the Alisitos arc across the Main Mártir thrust and ancestral Agua Blanca fault, the Late Triassic–Jurassic accretionary prism (correlative to the Bedford Canyon Complex) was structurally removed from between the arc and the continent by forcible subduction. If

Wetmore, P.H., Herzig, C., Alsleben, H., Sutherland, M., Schmidt, K.L., Schultz, P.W., and Paterson, S.R., 2003, Mesozoic tectonic evolution of the Peninsular Ranges of southern and Baja California, *in* Johnson, S.E., Paterson, S.R., Fletcher, J.M., Girty, G.H., Kimbrough, D.L., and Martín-Barajas, A., eds., Tectonic evolution of northwestern México and the southwestern USA: Boulder, Colorado, Geological Society of America Special Paper 374, p. 93–116. For permission to copy, contact editing@geosociety.org. © 2003 Geological Society of America.

this model is correct, it implies that the Late Cretaceous uplift of the central zone of the Peninsular Ranges batholith, both north and south of the ancestral Agua Blanca fault, was not driven by accretion-related deformation at the trench.

Keywords: Peninsular Ranges batholith, Santiago Peak arc, Alisitos arc, Agua Blanca fault, Mesozoic, tectonics.

INTRODUCTION

The Mesozoic tectonic evolution of the Peninsular Ranges province of southern California, USA, and Baja California, México, remains a poorly constrained component of North American Cordilleran geology. Although a variety of tectonic models have been proposed, they differ in their most fundamental aspects, such as whether or not arc-continent collision occurred (e.g., Todd et al., 1988; Thomson and Girty, 1994). Differences between these models, at least in part, result from models being based on observations made in locations separated by faults that are interpreted to have been active during the Mesozoic. The presence of these faults calls into question the validity of extrapolating the findings of local, one-dimensional studies to the entire Peninsular Ranges and beyond (e.g., Dickinson and Lawton, 2001). To address some of the long-standing geologic problems associated with the Peninsular Ranges we present a compilation of multiple data sets from several widely distributed parts of the central and western Peninsular Ranges to identify along-strike variations in the character of this region and to provide better constraint to the Mesozoic tectonic evolution.

This study is an expansion of results reported in Wetmore et al. (2002). There, evidence was presented to support the conclusion that during the Early Cretaceous the western zone of the Peninsular Ranges (defined below) evolved as two distinct tectonic blocks, a continental margin arc to the north and an island arc to the south. The two arcs were ultimately joined due to the accretion of the southern island arc near the end of the Early Cretaceous. In this paper the details of the Mesozoic depositional, structural, and paleogeographical evolution of the Peninsular Ranges are discussed to fully describe, evaluate, and justify the earlier proposed model.

GEOLOGIC BACKGROUND

The geology of the Peninsular Ranges is intrinsically tied to the Peninsular Ranges batholith, which forms the core of this province. The Peninsular Ranges batholith is the southernmost segment of a chain of North American Mesozoic batholiths that extend from Alaska to the southern tip of Baja California. It is exposed from the Transverse Ranges in southern California to as far as the 28th parallel. Recent studies have also correlated the intrusives of the Los Cabos block in southernmost Baja California Sur with those of the Peninsular Ranges batholith to the north (Kimbrough et al., 2002). To the east, the Peninsular Ranges batholith is bounded by the San Andreas–Gulf of California transform-rift system. To the west, the batholith is bounded by the Continental Borderlands, a collage of Mesozoic rocks variably

formed and deformed within trench, forearc, and arc tectonic settings (e.g., Sedlock et al., 1993). Paleogeographic relationships between the Continental Borderlands and the Peninsular Ranges are highly speculative due to the Mesozoic and Cenozoic history of a series of strike-slip faults within the Borderlands (e.g., Busby et al., 1998), some of which may coincide with active structures (e.g., Legg et al., 1991).

The Peninsular Ranges batholith is a world-class example of a laterally zoned batholith; it has a mafic western (outboard) zone and a felsic eastern zone. Several data sets document the existence of east-west transitions between eastern, central, and western batholith-parallel zones (Fig. 1). These data include rare earth elemental (REE) abundances (Gromet and Silver, 1987); oxygen isotopic signatures (Taylor and Silver, 1978); and Sr initial ratios and ε_{Nd} determinations from plutonic rocks (DePaolo, 1981). In addition, the Fe-Ti oxide mineralogy of the batholith exhibits an east-west transition where plutons of the western zone contain magnetite and ilmenite while those of the eastern zone contain the latter mineral phase (Gastil et al., 1990).

The above plutons intrude four major lithostratigraphic belts that parallel the long axis of the batholith (Gastil, 1993). These are, from east to west, Late Precambrian to Permian miogeoclinal strata, Ordovician to Permian (Early Triassic?) slope basin deposits, (Late?) Triassic to Cretaceous "back-arc" sedimentary rocks, and Jurassic(?) to Cretaceous arc volcanics. The Peninsular Ranges has thus been subdivided into three zones (eastern, central, and western), the trends of which parallel the batholith (Fig. 1). Typically, the eastern zone includes miogeoclinal and slope basin deposits, the central zone "back-arc" sedimentary rocks, locally overlying slope basin strata, and the western zone volcanic arc rocks (Gastil, 1993).

These across-strike variations of the host rock stratigraphy and batholith have been interpreted to reflect a change in basement composition (e.g., DePaolo, 1981; Silver and Chappell, 1988), where the primitive western zone, with its island arc signature, is underlain by oceanic lithosphere, and the eastern zone is underlain by older lithosphere of continental composition. These observations and interpretations have provided the foundations for a series of tectonic models that seek to explain the juxtaposition of these disparate lithospheric types.

Tectonic Models for the Mesozoic Evolution of the Peninsular Ranges Batholith

The models most often proposed for the Mesozoic tectonic evolution of the Peninsular Ranges batholith may be distilled down to three end members: (1) a single, eastward-migrating arc developed across a pre-Triassic join between oceanic and

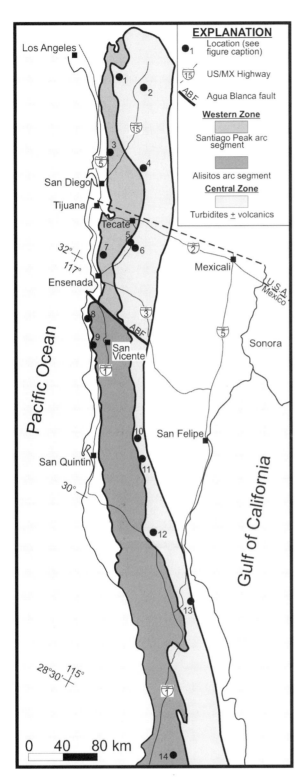

Figure 1. Map of southern California and Baja California Norte showing western and central zones of Peninsular Ranges batholith and localities discussed in text (modified from Gastil, 1993). Locations: 1—Santa Ana Mountains; 2—Winchester area; 3—central San Diego County; 4—eastern San Diego County; 5—Rancho San Marcos; 6—Rancho Vallecitos; 7—Cañon La Mision; 8—Punta China; 9—Erendira; 10—northern Sierra San Pedro Mártir; 11—southern Sierra San Pedro Mártir; 12—El Marmol; 13—Sierra Calamajue; 14—El Arco.

continental lithospheres (Walawender et al., 1991; Thomson and Girty, 1994); (2) an exotic island arc accreted to the North American margin across a non-terminal suture (Johnson et al., 1999a; Dickinson and Lawton, 2001); and (3) the reaccretion of a rifted and fringing arc to the North American margin (Gastil et al., 1981; Busby et al., 1998).

Models requiring the western zone of the Peninsular Ranges batholith to have been initially rifted from the continental margin and subsequently reaccreted are generally based on regional observations of stratigraphy and structure. However, the specific geometric requirements a rift-reaccretion model implies are inconsistent with existing data sets. For example, a rift-reaccretion model requires that a basin form between the arc and the continental margin, implying that the ages of strata should oldest within the arc and the continent and youngest in the basin. However, as discussed below, the ages of the stratigraphy of the Peninsular Ranges decrease continuously toward the west.

The observations that led to the single migrating arc (Thomson and Girty, 1994) and exotic arc (Johnson et al., 1999a) models were derived from studies of the Peninsular Ranges batholith in southern California and the Sierra San Pedro Mártir, México (Fig. 1), respectively. These two areas are separated by the Agua Blanca fault, an active dextral strike-slip fault south of Ensenada, México (e.g., Allen et al., 1960; Rockwell et al., 1989; Suarez-Vidal et al., 1991). Gastil et al. (1981), on the basis of observed differences in the apparent age and environment of deposition of the volcanics on either side of the Agua Blanca fault, identified it as an inherited structure with a Mesozoic origin as a transform fault. Gastil et al. (1981), thus, divided the western zone into northern and southern arc segments called the Santiago Peak and Alisitos, respectively. In their model, the fault is interpreted as having accommodated the diachronous accretion of the two arc segments to the western margin of North American.

Since the Gastil et al. (1981) study, our understanding of the geology of the Peninsular Ranges has improved substantially through numerous studies of the regional geology. However, few have followed up on the apparent discontinuity within the western and central zone of the batholith across the Agua Blanca fault. Here we focus on the along-strike variation in the Peninsular Ranges north and south of the Agua Blanca fault with particular attention paid to the temporal evolution of sedimentation and deformation within the central zone and the structural relationships of the central zone strata to western zone volcanics. We will use these observations to constrain the tectonic evolution of the Peninsular Ranges batholith during the Late Triassic through the Early Cretaceous and will reevaluate recently proposed models (e.g., Busby et al., 1998; Dickinson and Lawton, 2001) for the Peninsular Ranges batholith evolution.

PENINSULAR RANGES NORTH OF THE AGUA BLANCA FAULT

The pre-batholithic stratigraphy of the central and western zones of the Peninsular Ranges north of the Agua Blanca fault range in age from Late Triassic through Early Cretaceous. They can be subdivided into pre-Cretaceous and Cretaceous groups on the basis of general lithology, deformational history, and depositional setting. Those of the older group largely represent turbidite sequences (e.g., Germinario, 1993) subsequently deformed within an accretionary prism adjacent to the North American continental margin (e.g., Criscione et al., 1978). The Early Cretaceous sequences are generally the volcanic products of the western arc developed on and through the older accretionary prism.

Late Triassic through Jurassic: Turbidite Sedimentation and Deformation

Late Triassic through Jurassic sedimentary strata have been described from several locations in both southern and Baja California north of the Agua Blanca fault and have commonly been given local formational names. In the north, from west to east, these are the Bedford Canyon Formation, French Valley Formation, Julian Schist, and, further south, the Vallecitos Formation. In this group we also include the volcaniclastic-rich turbidite sequences of central San Diego County (Fife et al., 1967; Balch et al., 1984) west of exposures of the Julian Schist for reasons discussed below. Collectively, these strata have long been interpreted to be a correlative group of deep to moderately deep submarine fan deposits (Gastil, 1993).

Bedford Canyon Formation

The Bedford Canyon Formation, exposed in the Santa Ana Mountains of Orange and Riverside Counties of southern California, is the most well-studied of the middle Mesozoic turbidite sequences. The formation is dominated by alternating lithic and feldspar-rich sandstones (litharenites to lithic arkoses) and shales, with lesser amounts of limestone, conglomerate, chert, pebbly mudstones, and tuffaceous sequences (Moscoso, 1967; Moran, 1976). Additionally, isolated exposures of serpentinite occur along fault contacts within the formation (Moran, 1976; Criscione et al., 1978; Herzig, 1991). Deformation of the Bedford Canyon Formation is characterized by disrupted bedding, well-formed bedding-parallel foliation, and a second axial planar

foliation associated with abundant tight to isoclinal folds, and ubiquitous small faults, typically subparallel to bedding and with some ramp-flat geometries.

The age of the Bedford Canyon Formation has been difficult to resolve because the formation is, in general, poorly fossiliferous with the exception of several allochthonous (olistostromal?) limestone blocks that contain Bajocian to Callovian (176.5 to 159.4 Ma) fossils (Silberling et al., 1961; Imlay, 1963, 1964; Moscoso, 1967). Moscoso (1967) noted that the ages of these fossils become older towards the east (Fig. 2). The age of deposition for the Bedford Canyon Formation, therefore, may be constrained as being between Bajocian (176.5 Ma, Gradstein et al., 1994) and the age of the overlying basal unit of the Santiago Peak Volcanics (127 ± 2 Ma; Herzig, 1991). Isotopic studies of the Bedford Canyon Formation are similar with a 175.8 ± 3.2 Ma Rb/Sr whole-rock isochron age (Criscione et al., 1978). The lower intercept age of 210 ± 49 Ma derived from a mixed detrital zircon population (Bushee et al., 1963; Gastil and Girty, 1993) is also consistent with a Middle Jurassic age of deposition for the Bedford Canyon Formation.

French Valley Formation

The French Valley Formation crops out east of the Elsinore fault near Winchester, California (Fig. 1). It is dominated by immature sandstones and shales with lesser amounts of conglomerate and chert as well as horizons composed of olistostromes (Schwartz, 1960), all indicative of a medium to deep submarine fan depositional setting. The French Valley Formation has been isoclinally folded and pervasively cleaved.

The age of the French Valley Formation may be Late Triassic, as indicated by bivalves (Lamb, 1970). Detrital zircons yield a U/Pb lower intercept age of 285 ± 130 Ma (Gastil and Girty, 1993). However, a Rb/Sr whole-rock isochron age of 151 ± 11 Ma (Davis and Gastil, 1993) appears far too young and may reflect a subsequent metamorphic event rather than the age of deposition.

Julian Schist

The Julian Schist of eastern San Diego County (Fig. 1) is composed predominantly of sandstones and shales with minor

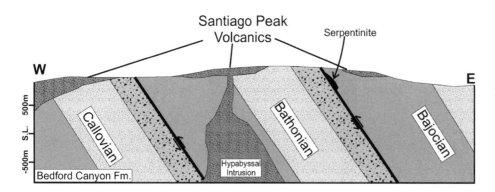

Figure 2. Schematic cross section through the Santa Ana Mountains based on our own field observations and age constraints from Moscoso (1967).

amounts of carbonate and other lithologies (Germinario, 1993). The depositional environment, therefore, is inferred to have been within the distal and medial portions of a deep to moderately deep submarine fan. Unlike many of the other correlative formations the Julian Schist has been metamorphosed to amphibolite grade and strongly deformed by the emplacement of multiple intrusives and episodes of deformation associated with the Cuyamaca Laguna Mountain Shear Zone (Thomson and Girty, 1994).

The age of the Julian Schist is poorly constrained. Only one fossil has ever been reported from these strata, the imprint of an ammonoid that was interpreted to be Triassic (Hudson, 1922). Unfortunately, the sample was subsequently lost. U/Pb analyses of detrital zircons collected from the Julian Schist are strongly discordant with a poorly constrained lower intercept at ca. 260 Ma (Gastil and Girty, 1993). However, this lower intercept is at least consistent with the age of the Harper Creek gneiss, which had a tonalite to granite protolith (Leeson, 1989) and intruded the Julian Schist at 156 ± 16 Ma (Girty et al., 1993).

Contact relationships with younger stratigraphy, such as the Santiago Peak Volcanics, have not been described for the Julian Schist. Thus, correlation of tectonic events observed elsewhere (e.g., Santa Ana Mountains) cannot be unambiguously established in the eastern part of San Diego County.

Vallecitos Formation

Further south, the Vallecitos Formation, described from the Rancho Vallecitos and Rancho San Marcos areas in northern Baja California (Fig. 1; Reed, 1993), is characterized by lithologies and internal structures similar to those of other turbidite sequences further north. Reed (1993) interprets recrystallized sandstones and shales within the formation to have been deposited in the distal portions of a submarine fan. The presence of pebbly mudstones and large (5 km²) olistostromal blocks of Ordovician miogeoclinal strata (Lothringer, 1993; Gehrels et al., 2002) further indicate proximity to a slope of the continental margin. U/Pb analysis of detrital zircons from these Ordovician strata by Gehrels et al. (2002) indicates a North American source, providing a definitive tie between the Mesozoic turbidite sequences and the continent.

Deformation and metamorphism of the Vallecitos Formation is variable and largely dependent upon proximity to the multiple large intrusive bodies present near Rancho Vallecitos (Reed, 1993; Sutherland and Wetmore, unpubl. mapping). Away from intrusives, recrystallization, cleavage, bedding-parallel faulting, and folding are perceived to be only slightly more intense than conditions observed for the Bedford Canyon Formation in the Santa Ana Mountains.

The age of the Vallecitos Formation, like that of more northerly formations, is not well defined. Gastil and Girty (1993) report the lower intercept of a mixing line formed from U/Pb analyses of detrital zircons collected from the formation to be 369 ± 59 Ma. A Rb/Sr whole-rock isochron age of 206 ± 12 Ma is suggested by Davis and Gastil (1993), indicating a possible Late Triassic age of deposition.

Tres Hermanos–Santa Clara Area

The descriptions of the southernmost exposures of turbidite sequences north of the Agua Blanca fault are from the Tres Hermanos–Santa Clara area (Fig. 1; Chadwick, 1987). Sandstone- and shale-dominated strata of this area have been intruded by several plutonic bodies that have imparted a strong metamorphic overprint as well as a significant component of emplacement-related deformation. Chadwick (1987) reports the presence of earlier-formed northwest-trending structures that include tight and isoclinal folds and associated foliation within the turbidite sequences.

Isotopic ages for the Tres Hermanos–Santa Clara strata are similar to those described for the Vallecitos Formation. U/Pb analyses of detrital zircons yield a lower intercept of 302 ± 61 Ma and a Rb/Sr isochron age of 167 ± 9 Ma (Chadwick, 1987), suggesting a slightly younger Early Jurassic age of deposition. The ages of intrusive bodies within these strata are all Early Cretaceous (132 ± 1.25 Ma or younger; Chadwick, 1987), approximately the same age as the Santiago Peak Volcanics, and thus they do not constrain the age of these pre-Cretaceous strata well.

Central San Diego County Volcaniclastics

Exposed in central San Diego County (Fig. 1) are a series of volcaniclastics, deposited in a submarine environment (named the Santiago Peak volcaniclastics by Balch et al. (1984)), that have long been correlated with the Santiago Peak Volcanics (Fife et al., 1967; Balch et al., 1984). Based on depositional environment, structural relationship to younger volcanic sequences, and age of these strata, we believe that they are better correlated with the pre-Cretaceous continentally derived turbidite sequences (e.g., the Bedford Canyon Formation) for reasons discussed below.

These volcaniclastics are epiclastically reworked breccias composed of andesites, dacites, and latites of similar composition to the Santiago Peak Volcanics. However, the deposits are exposed in sections containing turbidites, sandstones, and shales that are interpreted to have been deposited in medial to distal submarine fans (Balch et al., 1984) similar to those of the Bedford Canyon Formation. The Santiago Peak Volcanics, sensu stricto, on the other hand, are interpreted to have been deposited in a subaerial environment.

The volcaniclastic-rich sandstones and shales in central San Diego Country are characterized by a penetrative bedding parallel foliation, tight and locally overturned(?) folds, and brittle faults, all of which are truncated along the structural top of the section by an erosional surface. Overlying these volcaniclastic-rich turbidite sequences across an angular unconformity are volcanics that dip moderately and do not possess the deformational features of the underlying volcaniclastic and turbidite sequences. This relationship is remarkably similar to that observed between the Bedford Canyon Formation and the Santiago Peak Volcanics in the Santa Ana Mountains (described below).

The age of the strata in central San Diego was established by Fife et al. (1967) as Tithonian based upon the presence of the fossil *Buchia Piochii*. A recent U/Pb zircon age of ca. 152 Ma from a volcanic flow (dike?) within the package of volcaniclastic-rich turbidites is consistent with the fossil age (Anderson, 1991). Thus, given that the basal unit of the Santiago Peak Volcanics in the type section of the sequence yields an age of 127 ± 2 Ma (Herzig, 1991), correlation between the strata of central San Diego County and the Santiago Peak Volcanics is unfounded and a more appropriate correlation is made with the Bedford Canyon Formation.

The above descriptions of the turbidite sequences of the central zone north of the Agua Blanca fault reveal consistent features throughout this region. These include similar lithologies and inferred depositional environment, style and magnitude of deformation, presence of olistoliths, detrital zircon populations that indicate a source that included Precambrian to Triassic exposures, and a general depositional age that ranges between Late Triassic and Jurassic. The identification of volcaniclastic layers within these strata is common only within those formations that are clearly identified as being deposited in the Jurassic (e.g., Bedford Canyon Formation and the volcaniclastic-rich turbidite sequences of central San Diego County). Additionally, a general east-to-west younging of the strata is indicated for the southern California sequences. That is, the French Valley Formation, which contains Early Triassic fossils, is east of the Bedford Canyon Formation, which is Jurassic and which exhibits an intraformational westward younging (Bajocian on the east to Callovian on the west; Fig. 2). Similarly, the Triassic Julian Schist in eastern San Diego County is east of the volcaniclastic-rich turbidites of central San Diego County, which yield Tithonian fossils. Because of the similarities between all of these formations, we propose that they should be incorporated under a single group named the Bedford Canyon Complex.

Early Cretaceous: Development of the Santiago Peak Arc

The Early Cretaceous arc volcanic strata of the western zone north of the Agua Blanca fault are dominated by the Santiago Peak Volcanics. Rocks of similar stratigraphic position, such as the Estelle Mountain Volcanics, exposed east of the San Andreas fault east of the Santa Ana Mountains, named the Temescal Wash quartz latite porphyry by Larsen (1948), possess almost identical ages, contact relationships, petrologies, geochemistries, and degree of deformation to those described for the Santiago Peak Volcanics (Herzig, 1991). Hence, it seems practical to include all such units with the Santiago Peak Volcanics.

The most extensive studies of the Santiago Peak Volcanics have been completed in the Santa Ana Mountains (Larsen, 1948; Peterson, 1968; Gorzolla, 1988; Herzig, 1991) and central and northern San Diego County (Hanna, 1926; Adams, 1979; Tanaka et al., 1984; Anderson, 1991; Reed, 1992; Carrasco et al., 1993). South of the international border, where exposures are substantially better than to the north, studies have been completed in Cañon La Mision (Meeth, 1993) and the Ensenada area (Schroeder, 1967). Together these studies provide a relatively coherent picture of the volcanic products of the Early Cretaceous arc that existed along the western side of the Peninsular Ranges batholith north of the Agua Blanca fault.

Lithology and Petrochemistry of the Santiago Peak Volcanics

The Santiago Peak Volcanics are composed of flows, volcaniclastic breccias, welded tuffs, hypabyssal intrusions, and relatively rare epiclastic deposits (Larsen, 1948; Schroeder, 1967; Adams, 1979; Gorzolla, 1988; Herzig, 1991; Reed, 1992; Carrasco et al., 1993; Meeth, 1993). The volcanics are inferred to be subaerially deposited based on the abundance of accretionary lapilli and preserved paleosols, as well as the apparent absence of pillow lavas, thick and laterally extensive epiclastic deposits, and other marine deposits.

A relatively large body of geochemical data exists for northern exposures of the Santiago Peak Volcanics (e.g., Hawkins, 1970; Tanaka et al., 1984; Gorzolla, 1988; Herzig, 1991; Reed, 1992; Meeth, 1993), and, when combined with that for intrusive bodies from the same arc segment (e.g., Taylor and Silver, 1978; DePaolo, 1981; Gromet and Silver, 1987; Silver and Chappell, 1988; Carollo and Walawender, 1993), a coherent picture of the generation and evolution of magmas from the region may be drawn. Many of the early geochemical investigations focused on the intrusive suites and variations in the REE and isotopic compositions across the batholith (e.g., Taylor and Silver; 1978; DePaolo, 1981; Gromet and Silver, 1987). These studies identify a relatively primitive western zone underlain by oceanic lithosphere juxtaposed with a relatively evolved central/eastern zone underlain by transitional to continental lithosphere.

Major and trace element data, as well as isotopic data from the Santiago Peak Volcanics, indicate that they, like the plutons that intrude them (e.g., DePaolo, 1981), were derived from a depleted, oceanic mantle source (Herzig, 1991). However, some significant modification of the magmas is indicated by observed low Ni concentrations and slightly more evolved Nd and Sr isotopic values for the rhyolites. Such observations are suggestive of fractional crystallization of olivine and clinopyroxene, which was likely promoted by the hydrous character of the magmas (e.g., Nicholls and Ringwood, 1973), as well as the assimilation of some minor amounts of radiogenic crustal material. The most likely assimilant, at least near the level of extrusion, would have been continentally derived turbidite sequences, such as the Bedford Canyon Complex, which have an isotopic signature consistent with a continental provenance (Criscione et al., 1978). Assimilation, however, is not perceived to be significantly more than ~10% for even the most felsic units because of the relatively consistent concentrations of the incompatible elements (Herzig, 1991; Herzig and Wetmore, unpubl. data).

One hundred and seventeen whole rock geochemical analyses of the Santiago Peak Volcanics define a wide range of lava types from basalts to rhyolites (Fig. 3). Major elemental determinations of lava type indicate that the Santiago Peak arc was most likely a continental margin arc rather than an island arc (e.g., Todd et al., 1988). For example, although the mean lava type is andesite, basalts are uncommon and rhyolites comprise ~25% of the samples analyzed. This is in direct contrast with the petrologic characterization of active island arc systems as being largely composed of basalts and basaltic andesites (Marsh, 1979). Fur-

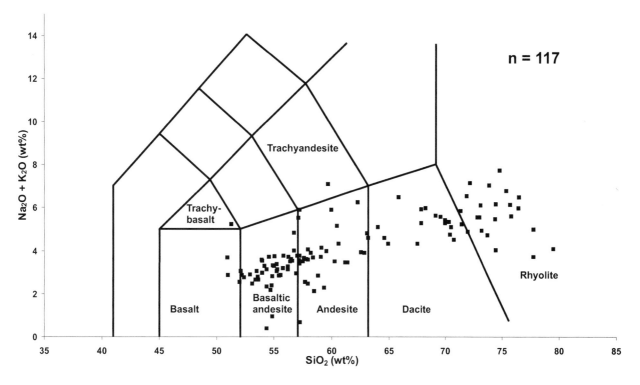

Figure 3. Chemical classification plot of lava types from Santiago Peak Volcanic. Diagram of Le Maitre et al. (1989). Data compiled from Tanaka et al. (1984), Gorzolla (1988), and Herzig (1991).

thermore, rhyolites comprise the basal unit to the Santiago Peak Volcanics in many studied localities (e.g., Santa Ana Mountains; Herzig, 1991). This precludes the possibility that the Santiago Peak arc was long-lived and evolved into a felsic island arc.

Contact Relations between the Santiago Peak Volcanics and the Bedford Canyon Complex

Contact relations between the Santiago Peak Volcanics and the underlying Bedford Canyon Formation in the Santa Ana Mountains are often not well exposed. This fact has led to some erroneous interpretations that have been perpetuated in the literature through the years. Initial descriptions by Larsen (1948) as well as later descriptions by Schoellhamer et al. (1981) indicated a depositional contact between the two stratigraphic units. However, several subsequent studies concluded that the two were juxtaposed across a low-angle fault (e.g., Peterson, 1968; Criscione et al., 1978). Many of the early tectonic models for the Mesozoic evolution of the Peninsular Ranges batholith were thus based on an accretionary juxtaposition of the Santiago Peak arc to the North American continent (e.g., Gastil et al., 1981; Todd et al., 1988). Such interpretations have persisted and can be found the most recent models (e.g., Dickinson and Lawton, 2001).

We have reexamined several key exposures of the contact between the Santiago Peak Volcanics and the Bedford Canyon Complex in the Santa Ana Mountains (Herzig, 1991; M. Sutherland, 2002, personal commun.) in central San Diego County

(Wetmore and Herzig, unpubl. mapping), and the Vallecitos area in Baja California (Sutherland and Wetmore, unpubl. mapping). In each area, several observations strongly suggest that this contact is an angular unconformity. At each exposure the moderately well-formed cleavage and folds, characteristic of the Bedford Canyon Complex strata, were oriented at high angles to, and truncated at, the contact (Fig. 4, A–C). In the Santa Ana Mountains and Vallecitos area, an uncleaved conglomerate composed of chert and/or sandstone pebbles in a muddy matrix is locally preserved along the contact. Also in the Santa Ana Mountains, the basal flows of the Santiago Peak Volcanics contain accidental sandstone and greywacke clasts from the underlying Bedford Canyon Formation. Additionally, Herzig (1991) reports that feeder dikes within the Bedford Canyon Formation may be traced directly into flows of the overlying Santiago Peak Volcanics (Fig. 4D).

Geochronology of the Santiago Peak Volcanics

Recent U/Pb geochronology studies of the Santiago Peak Volcanics by D.L. Kimbrough (San Diego State University) and his students (Anderson, 1991; Meeth, 1993; Carrasco et al., 1995) yield mildly discordant ages that range from 128 to 116 Ma. For example, a welded tuff that unconformably overlies the Bedford Canyon Formation in the Santa Ana Mountains and is inferred to be the basal unit of the Santiago Peak Volcanics yielded a U/Pb zircon age of 127 ± 2 Ma (Anderson, 1991;

Figure 4. A: Contact between Bedford Canyon Formation and Santiago Peak Volcanics (top) in Santa Ana Mountains. B: Basal conglomerate exposed along contact between Bedford Canyon Formation and Santiago Peak Volcanics in Santa Ana Mountains. C: Contact between Vallecitos Formation (left) and Santiago Peak Volcanics (right) with basal conglomerate preserved along contact. Steep dip of contact results from rotation in structural aureole of nearby pluton. D: Hypabyssal intrusion of Santiago Peak Volcanics feeding overlying flows.

Herzig, 1991). The 116 Ma age, derived from a sample 200 m below the mapped top of the Santiago Peak Volcanics section in the Cañon La Mision area (Fig. 1; Meeth, 1993), is assumed to be a minimum age for the end of Santiago Peak arc magmatism. Overlying the Santiago Peak Volcanics are course clastic forearc strata of Late Cretaceous age (e.g., Rosario Formation). Thus, the true termination of Santiago Peak volcanism is at some time between 116 Ma and the Late Cretaceous.

Observed discordance in the U/Pb zircon analyses of the Santiago Peak Volcanics and associated plutonics provide further evidence that the Santiago Peak arc was a continental margin arc and not an island arc. Anderson (1991) suggests that the observed discordances resulted from both minor lead

loss and some inheritance of radiogenic lead. The possibility of inheritance is also supported by the presence of a small fraction of discolored zircons that did not appear to be consistent with the majority of clear, euhedral zircons. This observation suggests that these discolored crystals are xenocrysts derived from minor amounts of incorporation of the country rock through which the plutons were emplaced (e.g., Bedford Canyon Complex). This is consistent with the observation that sandstone and greywacke xenoliths of apparent Bedford Canyon Formation are entrained within the basal flows and volcaniclastic units of the Santiago Peak (Herzig, 1991). These observed discordances and inferred inheritance occur in study areas for the entire length of the Santiago Peak arc segment including the

San Diego area (Anderson, 1991), and the Cañon La Mision area south of the international border (Meeth, 1993).

Deformation of the Santiago Peak Volcanics

The deformational history of the Santiago Peak Volcanics is one of the most poorly constrained aspects of this part of the Peninsular Ranges. Our mapping in the Santa Ana Mountains and central San Diego County has resulted in the identification of high-angle brittle faults, and gentle to moderately steep tilting and open folding of the volcanics. However, the observed cleavage and ductile strain that characterize the underlying Bedford Canyon Complex are not observed within the Santiago Peak Volcanics. In fact, fabric ellipsoids determined from lithic-rich samples from the Santiago Peak Volcanics do not significantly deviate from those measured in undeformed volcanics (Sutherland et al., 2002).

Mapping of the Santiago Peak Volcanics near Rancho Vallecitos (Sutherland et al., 2002) and north of the active Agua Blanca fault (Figs. 1 and 5B) has resulted in the identification of pronounced tilting and ductile strain, including a pervasive and well-developed cleavage, within the structural aureoles of plutons in both these areas. However, away from these intrusive bodies the minor deformation observed affecting the volcanics is similar to that characterizing exposures north of the international boarder.

In the area of the Agua Blanca fault the Santiago Peak Volcanics are cut discordantly by both the northwest-trending active strike slip structure and an older dip slip fault. The structural trends defined by the axes of open folds, minor west-vergent faults, and the average strike of bedding are truncated at the southern extent of the Santiago Peak Volcanics by what has been mapped as the ancestral Agua Blanca fault (Fig. 5) (Wetmore et al., 2002). This steeply northeast-dipping shear zone parallels the active Agua Blanca fault but is located ~2 km to the south. While some deflection (drag) of these regional trends is observed with proximity to the active structure, no deflection has been observed associated with the older ancestral Agua Blanca fault. Furthermore, no increases in finite ductile strain or in metamorphic grade are observed in proximity to the latter structure.

The above description of the Santiago Peak Volcanics and associated intrusives indicates that the Santiago Peak arc north of the Agua Blanca fault developed atop the Late Triassic through Jurassic Bedford Canyon Complex. Initiation of Early Cretaceous arc magmatism began after the earlier-formed strata were deformed, uplifted, and erosionally beveled as indicated by the truncation of fabrics and structures within the Bedford Canyon Complex at the contact with the overlying volcanics. Evidence supporting a depositional contact between the volcanics and the Bedford Canyon Complex strata includes a basal conglomerate present along the contact, xenoliths and xenocrysts of Bedford Canyon Complex derivation in the Santiago Peak Volcanics, a pronounced break in deformation across the contact without any indication of shear, and hypabyssal intrusions that cross-cut the turbidites and the contact to feed the overlying volcanics.

PENINSULAR RANGES SOUTH OF THE AGUA BLANCA FAULT

South of the Agua Blanca fault there have been considerably fewer geologic investigations of the western and central zones of the Peninsular Ranges. However, those that have been completed describe a Late Triassic through Early Cretaceous history that is markedly different from that to the north of the Agua Blanca fault. The most salient differences south of the ancestral Agua Blanca fault are the lack a Late Triassic through Jurassic accretionary prism (cf., Bedford Canyon Complex) to the south, and that the western and central zones are juxtaposed across a well-defined, east-dipping ductile shear zone (Main Mártir thrust). Below we describe the central and western zones in this region (Fig. 1).

Central Zone

Sierra San Pedro Mártir

The central zone in the Sierra San Pedro Mártir area (Fig. 1) has been the focus of two recent structural studies: Johnson et al. (1999a) in the northern part and Schmidt (2000) in the southern. In each area, the thermal affect of the numerous plutonic bodies has metamorphosed most of the preserved country rock screens to at least amphibolite grade. As such, the age and depositional environments of these prebatholithic strata cannot be constrained unequivocally. However, along the westernmost exposures of the central zone in both the northern and southern Sierra San Pedro Mártir, the strata are comprised of calc-silicates, metavolcanics, and quartzo-feldspathic metapelites.

Plutons that intrude these strata range in age from ca.134 Ma to ca. 97 Ma (Johnson et al., 1999a; Schmidt, 2000) with two major pulses, one between 134 and ca. 128 Ma and the other associated with the Late Cretaceous La Posta event between 100 and 94 Ma (e.g., Walawender et al., 1990). Thus, the age of the host stratigraphy must be greater than ca.134 Ma. Correlation of these strata with the Paleozoic through Early Triassic continental slope basin deposits exposed elsewhere in the central zone south of the Agua Blanca fault is possible, but the presence of a volcanic component is inconsistent with lithologies described for the older strata. Therefore, we suggest that the strata exposed in the westernmost exposures of the Sierra San Pedro Mártir were most likely deposited during the Middle to Late Jurassic, when arc magmatism is known to have been active in eastern parts of the central zone (Schmidt, 2000) and further east in mainland México (e.g., Damon et al., 1983).

Two deformational events are recorded within the prebatholithic strata of the central zone in the Sierra San Pedro Mártir (Schmidt, 2000). The oldest event predates the ca. 134 Ma plutons as host rocks in the middle part of the central zone possess a foliation which is not preserved in the intrusives of the earlier pulse of magmatism. In the western part of the central zone, this earlier fabric is strongly overprinted by mylonitic fabrics associated with the Main Mártir thrust (discussed below), the west-directed shear zone that juxtaposes the central and western

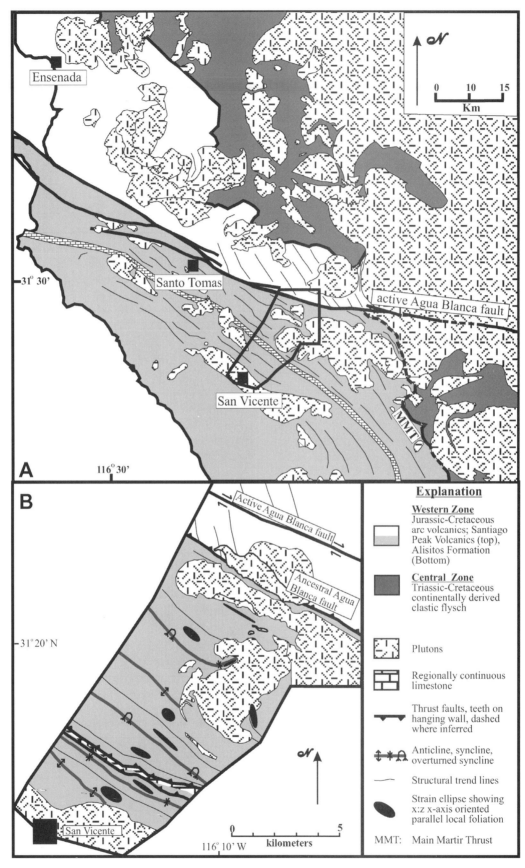

Figure 5. A: Geologic map of Agua Blanca fault based on Gastil et al. (1975); B: Map of northern Alisitos arc northeast of San Vicente (mapping by Wetmore).

zones (Johnson et al., 1999a). Deformation associated with this structure may have initiated as early as ca. 132 Ma, as indicated by igneous sheets and high-temperature subsolidus fabrics in plutons of this age along the western margin of the central zone (Schmidt, 2000).

El Marmol Area

The El Marmol area is located ~80 km south-southeast of the southern Sierra San Pedro Mártir (Fig. 1). Paleozoic to Early Triassic continental slope basin deposits that overlapped older North American miogeoclinal assemblages (Buch and Delattre, 1993; Campbell and Crocker, 1993; Gastil, 1993) are exposed in this part of the central zone. The slope basin stratigraphy is comprised of thin-bedded argillite, sandstones, and cherts with interbedded calcareous quartzarenite and pebble conglomerates with clast compositions of chert and quartzite. These strata are interpreted to have been deposited by sediment gravity flows with intervening intervals of quiescent pelagic sedimentation (Buch and Delattre, 1993).

Paleozoic to Early Triassic strata are overlain with angular discordance by the Early Cretaceous Olvidada Formation in the El Marmol area (Fig. 6; Phillips, 1993). Phillips (1993) describes lower, middle, and upper members of this formation. The lower member is composed of boulder-pebble conglomerate and sand-stone, some containing volcanogenic detritus and minor lime-stones clasts. Phillips (1993) interprets this member to represent shallow marine deposition. The gradationally overlying middle member is composed of rhythmically bedded cherts, sandstones, and shales and is interpreted to represent deep-water slope basin to abyssal plain deposition. These marine strata are unconform-ably overlain by the upper member of the formation, which con-sists of cobble conglomerate containing clasts that appear to be derived from the middle member of the formation, and sandstones and shales. The section is capped by vesicular andesites that are interpreted to have been deposited in a subaerial environment.

Deformation of the Paleozoic through Early Cretaceous strata of the El Marmol area includes an overall east-tilting of the entire section as well as two generations of folding (Buch and Delattre, 1993). The first generation of folding appears to affect only the Paleozoic to Early Triassic strata and is characterized by tight to isoclinal folds with northwest-striking, northeast-dipping axial planes. The second generation of folding affects both pre-Cretaceous and Early Cretaceous strata but not Late Cretaceous dikes and sills. Similar to the earlier-formed folds, the second generation are tight to isoclinal but with axial surfaces that strike more westerly than those of the former generation. Tertiary and Quaternary volcanic and sedimentary deposits obscure contact relationships between the strata of the El Marmol area and that of the western zone volcanics.

Sierra Calamajue

The Sierra Calamajue is located ~80 km south-southeast of the El Marmol area. Mapping in this area was completed by Griffith and Hoobs (1993) and is being remapped as part of a regional transect by H. Alsleben (2002, personal commun.). The stratigraphy of the Sierra Calamajue is dominated by metavolca-nics and volcaniclastics with subordinate amounts of carbonate, phyllite, chert, and limestone pebble to cobble conglomerate. According to Griffith and Hoobs (1993), the strata in the Sierra Calamajue range from Mississippian through the Early Creta-ceous. However, many of the U/Pb age determinations for the volcanic stratigraphy in this part of the central zone are presently being reevaluated by D.L. Kimbrough. An early observation from this work indicates that the analyzed units from this area are that all of the volcanics are Early Cretaceous and contain some component of inherited Precambrian zircons.

Perhaps similar to the El Marmol section, the westernmost exposures in the Sierra Calamajue are comprised of Paleozoic(?) deep marine strata, represented here by the Cañon Calamajue unit (Griffith and Hoobs, 1993). Lower parts of the Cañon Calamajue unit are composed of cherts and phyllites. Near the top of the unit is a limestone cobble conglomerate with a volcaniclastic matrix. The blocks of this conglomerate yield Chesterian age conodonts of North American affinity. However, the allochthonous nature of these blocks suggests that this is a lower age limit and not neces-sarily the age of deposition for the Cañon Calamajue unit.

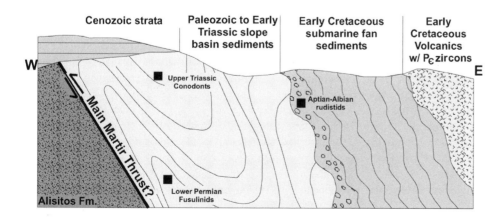

Figure 6. Schematic cross section through El Marmol area based on map-ping and descriptions from Buch and Delattre (1993) and Phillips (1993).

Northeast of the Cañon Calamajue unit are a series of meta-volcanic units with interbedded limestones juxtaposed with phyllite-dominated units, with lesser volcanics, across northeast-dipping, southwest-vergent thrust faults (Griffith and Hoobs, 1993). Deformation of these units reaches a maximum intensity within Cañon Calamajue, an observation that led Griffith and Hoobs (1993) to suggest that this zone was the suture between North America and the Alisitos arc of the western zone. However, due to the presence of strata with North American provenance and volcanics containing Precambrian zircons west of the faults, alternative interpretations are possible (see Discussion section).

In summary, the along-strike lithologic character of the central zone south of the ancestral Agua Blanca fault does exhibit several differences from place to place, but as a whole it appears to have experienced a broadly similar evolution throughout along its length. For example, while the ratio of volcanics to marine sediments is not the same for any of the two areas, the observation of marine deposition of clastic sediments succeeded by the subaqueous deposition of volcanics is common to all three. This indicates the presence of a basin or basins along the southwestern margin during the latest Jurassic through much of the Early Cretaceous. However, missing from each of these areas are the Late Triassic through Jurassic turbidite sequences that dominate the central zone of the Peninsular Ranges north of the ancestral Agua Blanca fault.

Western Zone

Early Cretaceous arc strata of the western zone of the Peninsular Ranges batholith south of the Agua Blanca fault are included in the Alisitos Formation. Most early studies of the Alisitos Formation focused on stratigraphy, paleontology, and depositional environment (e.g., Allison, 1955, 1974; Silver et al., 1963; Suarez-Vidal, 1986) with many of these studies confined to northernmost exposures. Recent studies have expanded the understanding of the stratigraphy to more southern areas (e.g., Beggs, 1984; Fackler-Adams, 1997) and have begun to focus on the structural/tectonic and magmatic evolution of this part of the Peninsular Ranges (e.g., Goetz, 1989; Johnson et al., 1999a, 1999b; Tate and Johnson, 2000; Tate et al., 1999; Schmidt, 2000; Wetmore et al., 2002).

Alisitos Formation

The Alisitos Formation is composed of reworked or epiclastic volcaniclastics, volcanogenic argillites and sandstones, several primary volcanic flows and breccias, and a regionally extensive prominent limestone/marble member that can be traced continuously from Punta China to the northern Sierra San Pedro Mártir (Silver et al., 1963; Figs. 1 and 5). Subaqueous deposition dominated during the emplacement of the Alisitos Formation based on the observed volcaniclastic lithologies and the abundant fossils preserved within them and the presence of several basaltic lava flows exhibiting pillow structures (Leedom, 1967; Reed, 1967; Allison, 1974; Beggs, 1984; Suarez-Vidal, 1986, 1993; Fackler-Adams, 1997). Subaerial deposition occurs locally near inferred volcanic edifices (Fackler-Adams, 1997).

Petrologic classifications of the volcanics of the Alisitos Formation lack the support of the large geochemical data set that exists for the Santiago Peak Volcanics. However, descriptions from hand samples and thin sections suggest that the two share a similar range in composition (e.g., Fackler-Adams, 1997). However, based on the few published stratigraphic columns (e.g., Leedom, 1967, Allison, 1974), combined with our own mapping near San Vicente (Wetmore, unpubl. mapping; Fig. 1) and in the western part of the Sierra San Pedro Mártir (Schmidt, 2000), we believe that basalts, basaltic andesites, and andesites overwhelmingly dominate and that more siliceous volcanics (e.g., rhyolites) are uncommon.

The presence of moderate- to deep-water clastic and biochemical sediments in the Alisitos Formation is reported by Suarez-Vidal (1986, 1993) and Johnson et al. (1999a). Suarez-Vidal (1993) suggested that the package of clastic sediments that he mapped in an area south of the Agua Blanca fault and near El Arco (Fig. 1) represented a regionally continuous depositional package that comprised the northern and eastern exposures of the Alisitos Formation. In the northern Sierra San Pedro Mártir, Johnson et al. (1999a) mapped a north-northwest–trending belt of equivalent rocks bounded on the west and east by west-vergent thrust faults. Suarez-Vidal (1993) further suggested that the presence of such rocks indicated deposition in a "tectonically quiet" setting, such as a backarc environment.

Paleontological investigations of the Alisitos Formation have consistently yielded Early Cretaceous fauna (e.g., Allison, 1955, 1974; Silver et al., 1963). Some early confusion may have existed concerning the exact age (Aptian-Albian) of some of the fossils, but ultimately an Albian age was determined by Allison (1974). Subsequently, a small number of U/Pb zircon ages have been reported for both of the volcanics as well as for some of the plutons that intrude the Alisitos Formation. Carrasco et al. (1995) and Johnson et al. (2003) report ages from the volcanics of 116 ± 2 and 115 ± 1.1 Ma. Johnson et al. (1999a) further report ages from plutons of the western part of the Sierra San Pedro Mártir area that range from 116.2 ± 0.9 to 102.5 ± 1.6 Ma. None of these U/Pb zircon studies have resulted in the observation of any component of Precambrian inheritance.

The most complete geochemical data sets for the Alisitos arc segment are from the Zarza Intrusive Complex and the San José tonalite of the northern Sierra San Pedro Mártir area reported in Tate et al. (1999) and Johnson et al. (2003; Fig. 1). Similar to data from the Santiago Peak Volcanics, these intrusive bodies yield major and trace element and isotopic signatures that are consistent with melt derivation from within depleted oceanic lithosphere. However, unlike the Santiago Peak Volcanics, interpreted contamination is consistent with assimilation of metabasite (Tate et al., 1999) rather than more silicic continentally derived clastic sedimentary sequences. Combined with the observed lack of any inherited component to the zircons from either intrusive or extrusive igneous rocks, these observations strongly suggest that the basement of the Alisitos arc segment does not contain continentally derived materials (Johnson et al., 1999a; Wetmore et al., 2002).

Structures Attending the Boundaries of the Alisitos Arc

Contact relationships between the Alisitos Formation and the continentally derived clastic sedimentary sequences of the central zone have been described in both the northern and southern Sierra San Pedro Mártir (Goetz, 1989; Johnson et al., 1999a; Schmidt, 2000). In each of these areas the two lithostratigraphic belts are juxtaposed across a large, east-over-west ductile shear zone known as the Main Mártir thrust. A similar structural juxtaposition exists between the Alisitos Formation and the Santiago Peak Volcanics to the north across a newly identified southwest-vergent reverse fault (Wetmore et al., 2002). To the south of the Sierra San Pedro Mártir, the presence of structures clearly juxtaposing the Alisitos Formation with central zone, deep-water sediments and successive volcanics have not, as yet, been clearly identified.

Studies of deformation within the Alisitos Formation have traditionally focused on structures developed along the eastern margin of this part of the western zone (Goetz, 1989; Johnson et al., 1999a; Schmidt, 2000). Strong deformation characterizes each of these areas where west-vergent ductile shear zones place the continentally derived clastic sedimentary sequences over the Alisitos Formation (Figs. 1 and 5). Our reconnaissance work across the Alisitos indicates that the intensity of deformation increases from shallowly west-dipping strata without observed internal fabrics in western exposures to openly folded strata with horizontal axes, moderate- to well-formed cleavages, and intermediate strain intensities (<~40% shortening in the z-direction), and finally to isoclinally folded strata with inclined axes, strongly developed foliations, and large strain intensities (>~60% shortening) adjacent to the Main Mártir thrust. The overall across strike width of this fold and thrust belt is ~25 km (Johnson et al., 1999a; Wetmore et al., 2002). The Main Mártir thrust also corresponds to the transition between intrusives to the west that exhibit no observed inherited older zircons and yield primitive isotopic signatures from those to the east that do possess Precambrian zircons and evolved isotopic signatures (Johnson et al., 1999a). Johnson et al. (1999a) constrain the timing of the main pulse of deformation within this fold-and-thrust belt and across the Main Mártir thrust to be between ca. 115 and 108 Ma.

In the northern part of the arc segment, the fold-and-thrust belt that includes the Main Mártir thrust is deflected into sub-parallelism with the trace of the ancestral Agua Blanca fault (Wetmore et al., 2002). This deflection involves as much as 50° of strike rotation in a counterclockwise sense (i.e., from ~N15°W to N65°W; Fig. 5). Our recent mapping in the area south of the fault (Fig. 5) has resulted in the identification of a pronounced break in strain intensity across a northeast-dipping ancestral Agua Blanca fault. Overall deformation increases dramatically with proximity to the ancestral Agua Blanca fault with folds becoming isoclinal and strain intensities becoming immeasurably large. Shear sense determined from lineation-parallel sections within the underlying Alisitos Formation suggests a strong component of northeast side-up motion across the fault, which is also consistent with the southwest vergence of all folds developed in this region. However, additional kinematic information was obtained from

sections perpendicular to the lineation suggesting an additional component of sinistral shear opposite current motion across the nearby active brittle fault.

In summary, the western zone of the Peninsular Ranges south of the ancestral Agua Blanca fault is composed of plutons that intrude the Albian Alisitos Formation, which is characterized by subaqueous volcanic deposits (dominantly basalts and andesites), epiclastically reworked volcanic sediments, and subordinate amounts of carbonate. Geochemical and geochronological studies of the volcanics and plutonics suggest derivation from a depleted mantle source without contamination from continental crust or continentally derived deposits. The northern and eastern boundaries of the Alisitos arc are characterized by broad (>20 km), southwest-vergent fold-and-thrust belts with the Main Mártir thrust and the ancestral Agua Blanca fault juxtaposing the Alisitos arc with the central zone and Santiago Peak Volcanics, respectively.

DISCUSSION

The above descriptions of the western and central zones of the Peninsular Ranges batholith clearly illustrate the dramatic differences north and south of the ancestral Agua Blanca fault. These differences include the following (Table 1): (1) the presence or absence of Late Triassic through Jurassic continentally derived turbidite sequences (north) and/or Early Cretaceous submarine sedimentary strata (south), (2) the environment of deposition of Early Cretaceous western zone volcanics and contact relations between these volcanics and the continentally derived strata of the central zone (depositional, north; fault, south), (3) the presence of xenocrystic Precambrian zircons in plutons and volcanic flows of the western zone (present, north; absent, south), (4) the frequency of lava types of western zone volcanics (abundant rhyolites, north; abundant basalts, south), (5) the general distribution and intensity of deformation within the western zone (minor to moderate, north; 20-km-wide fold-and-thrust belt, south), and (6) the character of deformation associated with the ancestral Agua Blanca fault (truncation, north; deflection into subparallelism, south). We believe that these differences unambiguously indicate that the ancestral Agua Blanca fault is the along-strike continuation of the Main Mártir thrust, as together they served as a nonterminal suture accommodating the juxtaposition of the Alisitos arc segment to the Santiago Peak arc segment and North America. Additionally, the ancestral Agua Blanca fault must have accommodated the tectonic removal of the pre-Cretaceous accretionary prism, which is represented by the Bedford Canyon Complex to the north, prior to Alisitos arc accretion.

An Alternative Tectonic Model

During the Late Triassic through at least the Jurassic, the southwestern margin of North America north of the ancestral Agua Blanca fault was the site of a considerable amount of turbidite sedimentation. These deposits, which contain Precambrian zircons and olistostromal blocks of miogeoclinal quartzite,

TABLE 1. VARIATIONS IN THE PENINSULAR RANGES BATHOLITH NORTH AND SOUTH OF THE AGUA BLANCA FAULT

	North of Agua Blanca fault	South of Agua Blanca fault
Late Triassic–Jurassic clastic turbidites	Yes	No
Early Cretaceous clastic sediments	No	Yes
Depositional environment of Cretaceous arc volcanics	Subaerial	Submarine
Contact relations between Early Cretaceous volcanics and continentally derived sediments of the central zone	Depositional unconformity	Large ductile southwest-vergent shear zone
Inherited zircons	Observed within both volcanics and plutonics	Not observed in either volcanics or plutonics
Lava types	Andesites most abundant, but with ~40% comprised of dacites and rhyolites	Dominated by basalts, basaltic andesites, and andesites, dacites and rhyolites rare
Regional deformation	Weak to moderate, somewhat elevated in southeastern exposures	Intense within fold-and-thrust belt along eastern and northern limits to the Alisitos arc
Deformation associated with Agua Blanca fault	Truncation of regional structures without increases in strain or metamorphism	Regional change in structural trend, high strain intensities and amphibolite? Grade metamorphism

clearly exhibit a North American provenance. The presence of an arc to the east (Damon et al., 1983; Saleeby et al., 1992) active contemporaneously with turbidite sedimentation, combined with observed serpentinite blocks and olistostromes included within the section, suggest that the deformation exhibited by these sequences likely resulted during incorporation into an accretionary prism (Fig. 7, A and B [line A–A′]). Additionally, the westward younging of east-dipping Bedford Canyon Complex strata is consistent with the imbrication of coherent (albeit internally deformed) packages of stratified sediments forming duplexes in an accretionary wedge associated with a west-facing arc (e.g., Lash, 1985; Sample and Fisher, 1986; Sample and Moore, 1987).

Following deposition of the Bedford Canyon Complex and deformation within an accretionary prism setting, the entire section appears to have been uplifted, subaerially exposed, and erosionally beveled (Fig. 7, A and B [line C–C′]). This is consistent with the interpretation that the Santiago Peak Volcanics were deposited in a subaerial environment. A specific uplift event is not necessary to expose these strata to subaerial erosion. Rather, the strata of the Bedford Canyon Complex could have simply moved toward higher elevations as a function of time and the addition of material into the wedge, both scraped off at the toe, as the Bedford Canyon Complex appears to have been, and underplated in a manner similar to the process proposed for the uplift and exhumation of the Cascadia accretionary wedge in Washington state (Brandon et al., 1998).

In addition to uplift of strata within the accretionary prism, a further result of long-term accretion of material to the southwestern margin of North American may have been the apparent westward migration of magmatism that impinged upon the Late Triassic to Jurassic accretionary prism in the Early Cretaceous

(Fig. 7A). However, other options could include a steepening of the subducting slab or the initiation of a new subduction zone outboard of an older one.

Regardless of the mechanism that caused the apparent migration, by the Early Cretaceous the Santiago Peak arc was being built on and through the southwestern margin of the North American continent. This interpretation is supported by observed depositional contacts between the volcanics and the continentally derived stratigraphy, xenocrysts of Precambrian zircon in both the volcanics and plutonics, xenoliths of sandstone and greywacke in the volcanic flows, and the presence of intrusions that cut the Bedford Canyon Complex and can be traced directly into the overlying flows of the Santiago Peak Volcanics. Furthermore, while the overall chemistry of the intrusive and extrusive magmas of the Santiago Peak arc are clearly derived from a depleted mantle source, the volcanics have been altered by moderate amounts of assimilation of silicic material and fluid-enhanced fractional crystallization, such that the overall distribution of lava types is strongly skewed away from the typical island arc assemblage and toward more silicic dacites and rhyolites.

Deformation within the Santiago Peak Volcanics is somewhat enigmatic in that the regional deformation is typically very minor (upright, open folds and minor offset (< 50 m), high-angle brittle faults). While part of this deformation may have been Early Cretaceous in age, some proportion of it must be Late Cretaceous or younger, given the steeply west-dipping paraconformity(?) between the Santiago Peak Volcanics and the overlying Late Cretaceous clastic strata of the western Santa Ana Mountains.

South of the ancestral Agua Blanca fault, the fact that Late Triassic through Jurassic sedimentary strata have been preserved only locally (i.e., Sierra San Pedro Mártir) suggests that this part of the Peninsular Ranges was largely emergent during this time

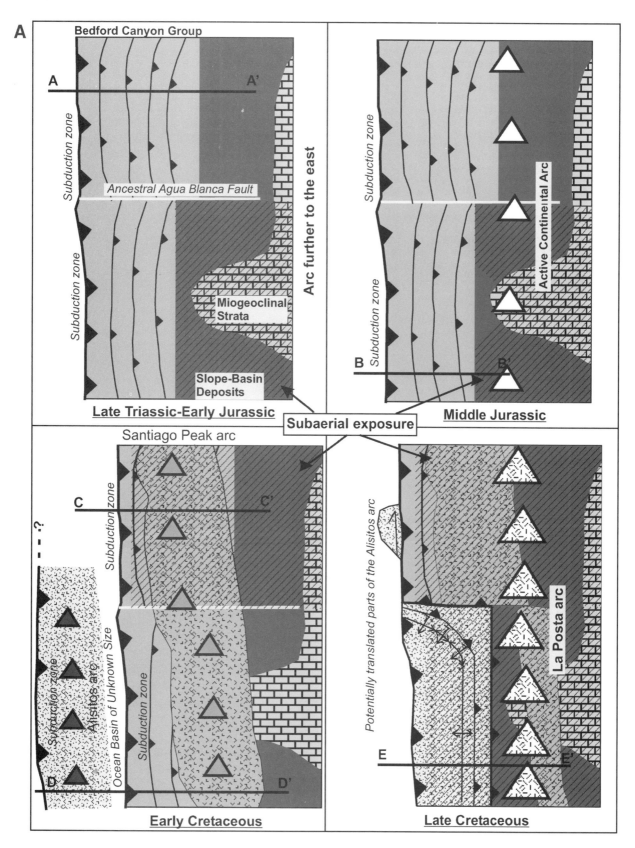

Figure 7 (on this and following page). A: Schematic depiction of Late Triassic through Early Cretaceous tectonic evolution of Peninsular Ranges batholith. B: Cross sections (lines A–A′ and C–C′) through Peninsular Ranges batholith north of ancestral Agua Blanca fault. C: Cross sections (lines B–B′, D–D′, and E–E′) through Peninsular Ranges batholith south of ancestral Agua Blanca fault.

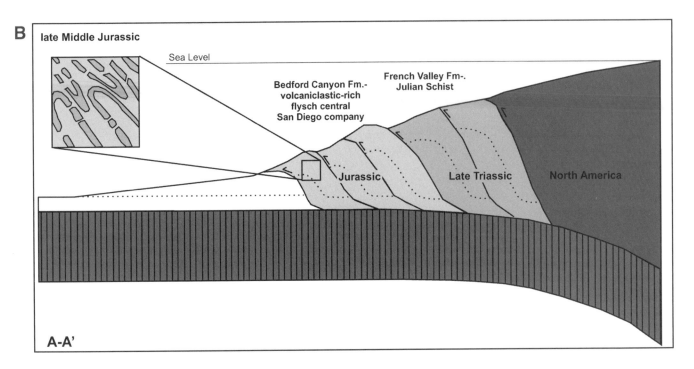

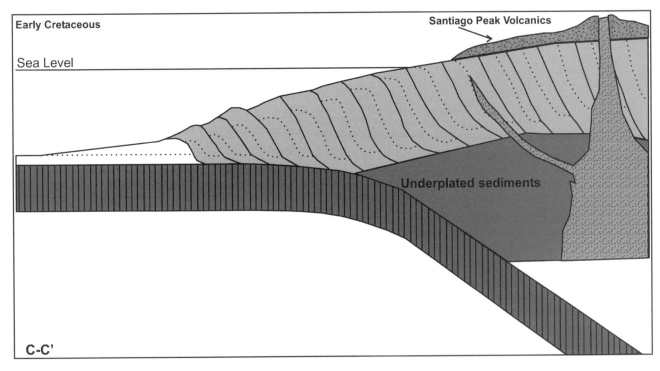

Figure 7. (*continued*)

(Fig. 7, A and C [line B–B′]). However, given that a continental arc did exist in Sonora, México (Ramon et al., 1983) during this time, the trench and accretionary prism represented by the Bedford Canyon Complex to the north of the ancestral Agua Blanca fault should have existed west of the present location of the central zone south of the fault. This is also consistent with the presence of earliest Cretaceous plutonics in the hanging wall of the Main Mártir thrust (Johnson et al., 1999a).

During the Early Cretaceous, the central zone of the Peninsular Ranges batholith south of the Agua Blanca fault subsided

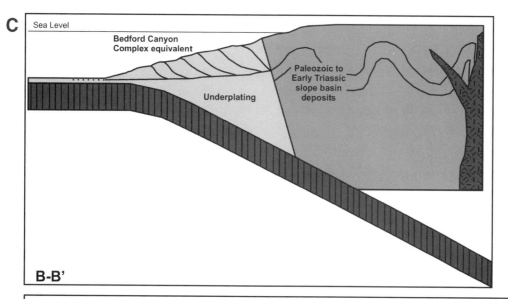

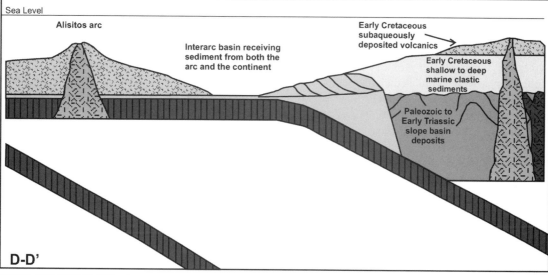

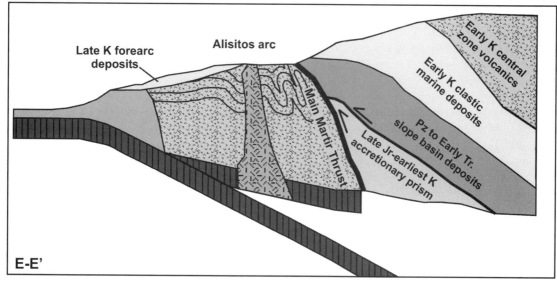

Figure 7. (*continued*)

below sea level and began receiving clastic detritus from the east (Fig. 7, A and C [line D–D′]). This appears to have been a relatively short-lived condition, as later in the Early Cretaceous this basin was filled or uplifted as clastic sedimentation gave way to the deposition of proximally derived volcanics that grade upward from submarine into subaerial deposits. The transition between clastic and volcanic deposits was contemporaneous with the deformation of the clastic strata prior to the deposition of the volcanics. This suggests that termination of the latest Jurassic–Early Cretaceous central zone basin may have, in part, resulted from tectonic closure. Like the Santiago Peak Volcanics, the volcanics of the central zone bear the signature of contamination by continentally derived materials implying derivation from sources within the central zone or from further east.

The volcanics/volcaniclastics of the Alisitos arc south of the Agua Blanca fault were deposited in a submarine environment and are believed to be everywhere in fault juxtaposition with continentally derived strata, including the Santiago Peak Volcanics (Fig. 7A). In fact, volcanic strata of the Alisitos arc exhibit no indication that rocks of continental derivation exist in the basement of this arc, such as the presence of xenocrystic Precambrian zircons in volcanics or plutonics, or evolved magmas, which can be shown to have otherwise assimilated clastic detrital material (e.g., Bedford Canyon Complex). Such observations led Johnson et al. (1999a) to argue that the Alisitos arc originated as an island arc developed on oceanic crust not previously associated with North America. This interpretation is also consistent with the prominence of basalts and andesites as the dominant petrologic types for volcanics of the Alisitos Formation.

The deformation within the Alisitos arc is much more widespread and, along its eastern margin, of greater intensity than that observed within the Santiago Peak arc to the north. Deformation associated with the southwest-vergent fold-and-thrust belt that parallels the Main Mártir thrust, the structure juxtaposing the Alisitos arc with North America–derived strata, suggests a causal link between the two. Goetz (1989), Johnson et al. (1999a), and Schmidt (2000) all concluded that this belt of deformation reflects the accretion of the western arc to the continental margin in the Early Cretaceous (Fig. 7, A and C [line E–E′]). In all of these models, the accretion occurs across an east-dipping subduction zone along the western continental margin. This is consistent with the southwest vergence of all identified structures associated with the Main Mártir thrust (Johnson et al., 1999a; Wetmore et al., 2002).

The fold-and-thrust belt along the ancestral Agua Blanca fault represents a unique feature within the Peninsular Ranges and is best explained by tectonic juxtaposition of the two western zone arc segments. The counterclockwise rotation of structures within the Alisitos arc into subparallelism with the ancestral Agua Blanca fault in the northern part of the arc, along with observed kinematics in this area, suggest that the northern Alisitos was strongly affected by sinistral transpression, opposite to displacements along the active fault. The continuation of southwest-vergent thrusting from the Main Mártir thrust to the ancestral Agua

Blanca fault suggests that the latter represents the northward continuation of the former. Therefore, the ancestral Agua Blanca fault is interpreted to be the suture zone that juxtaposes the two arc segments of the western zone (e.g., Wetmore et al., 2002).

Additional Tectonic Considerations

The above tectonic history brings several additional tectonic problems into focus. Deformations observed in eastern San Diego County associated with the Cuyamaca Laguna Mountain Shear Zone and those of the faults in Cañon Sierra Calamajue have been interpreted as sutures juxtaposing western and central zones. However, these interpretations are not reconcilable with the above models; thus, further discussion is necessary. Additionally, the above tectonic model implies that components of the central zone observed north of the ancestral Agua Blanca fault must have been tectonically removed to the south of the fault and that the Late Cretaceous exhumation observed in the central zone along the length of the Peninsular Ranges cannot everywhere be attributed to accretion-related deformation.

Deformation in the Sierra Calamajue was interpreted by Griffith and Hoobs (1993) as resulting from the accretion of the Alisitos arc to the North American margin. However, as noted above, volcanics containing inherited Precambrian zircons and limestone conglomerates yielding North American affinity fossils are observed west of the faults mapped in Cañon Calamajue. These observations are inconsistent with those from the Alisitos Formation in other areas of the Peninsular Ranges. Two alternative explanations for this apparent inconsistency are: (1) these faults do not form the suture, rather the suture likely exists further to the west of the Sierra Calamajue; and (2) continentally derived sediments and volcanics dominated the Early Cretaceous basin that existed between the continent and the Alisitos arc such that these strata were deposited across the intervening trench and onto the Alisitos island arc. The existence of the suture at a more westerly position than the Cañon Calamajue is certainly possible given the overall lack of mapping in this area combined with the large areas to the west of the Sierra Calamajue covered by Tertiary and Quaternary strata. The latter alternative is likewise possible given that the Alisitos arc was largely submerged throughout its depositional history and as the arc got progressively close to North America, the basin between the two continually shrank. However, more detailed study of this part of the Peninsular Ranges is required before these or other potential alternatives can be distinguished.

The Cuyamaca Laguna Mountain Shear Zone is a northwest-striking, east-dipping ductile shear zone exposed in eastern San Diego County (Fig. 1). This structure, which is approximately coincident (±10 km) with many of the major chemical, petrological, and mineralogical transitions between western and central zones (e.g., Silver et al., 1979), has traditionally been identified as an Early Cretaceous suture between North America and the Santiago Peak arc (e.g., Todd et al., 1988). However, a recent study of the structural evolution of the Cuyamaca Laguna Mountain Shear Zone by Thomson and Girty (1994) reinterpreted Early Cretaceous deformation exhibited by the structure as having resulted from strain concentration upon a preexisting lithospheric boundary between oceanic and continental crusts. This interpretation is largely based on the observation that the continentally derived strata of the Julian Schist are present on both sides of the Cuyamaca Laguna Mountain Shear Zone. We believe that the designation of this structure as an intra-arc shear zone by Thomson and Girty (1994) and not an arc-continent suture is consistent with the data present in this paper.

The juxtaposition of the Alisitos arc with the central zone implies that a substantial portion of the western margin of North America in this region must have been removed. This follows from the interpretation that the Alisitos arc, unlike the Santiago Peak arc, did not develop across the former accretionary prism, but rather was built on oceanic crust. Wetmore et al. (2002) proposed two hypotheses to explain the missing accretionary prism: strike-slip translation and subduction. They cite the observed steep lineation with northeast side-up sense of shear along the Main Mártir thrust and lack of kinematic indicators suggesting a component of lateral translation in this area to argue that the missing terrane was most likely subducted beneath the central zone. They further suggest a means of testing this hypothesis through the geochemical study of the Late Cretaceous intrusive bodies present in the central zone of the Peninsular Ranges batholith north and south of the Agua Blanca fault. In this instance, if the accretionary prism south of the Agua Blanca fault was subducted, it should be recorded as an identifiable chemical signature within these magmas (e.g., Ducea, 2001) and be absent from magmas north of the fault.

During the Late Cretaceous, the entire length of the central zone of the Peninsular Ranges batholith experienced an enormous amount of exhumation and denudation (Lovera et al., 1999; Schmidt, 2000; Kimbrough et al., 2001). This uplift and associated deformation has commonly been attributed to the accretion or reaccretion of the western zone of the batholith (e.g., Todd et al., 1988). However, if the above model is correct, then accretion only affected the central zone adjacent to the Alisitos arc and no such mechanism can be called upon to drive the Late Cretaceous event in the central zone adjacent to the Santiago Peak arc. Additionally, the timing of this exhumation event, between 100 and 85 Ma (Schmidt, 2000), is as much as 15 m.y. after the accretion of the Alisitos arc to the continental margin, indicating that even this accretion event was unlikely to have been fully responsible for the observed Late Cretaceous uplift and exhumation in the central zone adjacent to the Alisitos arc.

In the absence of terrane accretion to drive deformation and exhumation within the central zone of the Peninsular Ranges batholith, two alternative models have been proposed: increased coupling between the subducting and overriding plates at the trench (Schmidt et al., 2002) and the thermal-mechanical effects associated with the emplacement of the volumetrically large La Posta suite of plutons (Kimbrough et al., 2001). While the temporal overlap between magmatism, uplift, and exhumation are

enticing, the true mechanism(s) for magmatism to drive the latter two processes are simply too vague at present to view the coincidence of these events as one of cause and effect. Additionally, the temporal overlap between the La Posta event and uplift is only partial in that uplift both predates and postdates the magmatic event by several million years each. Conversely, the relative plate motion vectors (Engebretson et al., 1985) indicate a high angle of convergence between North America and subducted oceanic crust from 100 to 85 Ma at relatively high velocities, which is consistent with an increased coupling model. However, additional constraints are required to further resolve the mechanism(s) that drove this Late Cretaceous event.

Alternative Models

Tectonic models discussed in the introduction require some dramatically different processes and tectonic geometries to explain the present configuration of the Peninsular Ranges. These include the rifted-fringing arc model (Gastil et al., 1981; Busby et al., 1998) and the accretion of an exotic, east-facing island arc (Dickinson and Lawton, 2001). The rifted-fringing arc model proposed by Gastil et al. (1981) and Busby et al. (1998) is based on observations made from areas both north and south of the ancestral Agua Blanca fault as well as from the Continental Borderlands terrane.

Rifted-fringing arc models are commonly based on the observation of large amounts of "flysch-like" strata preserved within the central portions of batholiths (e.g., Gastil et al., 1981). Gastil et al. (1981) and Gastil (1993) correlated the turbidite sequences of the central zone for the entire length of the batholith and identified them all as "Triassic-Cretaceous back-arc clastics." The Gastil et al. (1981) model proposes that the entire western zone of the batholith was rifted from the continental margin in the Triassic, allowing for the deposition of these strata. Several objections can be raised against this interpretation. First, if a fragment of the continent were rifted in the Triassic, then the ages of the strata should be older in the western and eastern zones and younger in the central zone. This is not consistent with the age distributions of central zone strata, either north or south of the ancestral Agua Blanca fault. Second, if rifting and the development of a fringing arc occurred in the Triassic, volcanics and plutonics of this age should be present in the western zone. While some Jurassic epiclastically reworked volcaniclastics have been observed in central San Diego County, their volume is far too small to support their interpretation as a Triassic-Jurassic arc. Triassic magmatism may be preserved in eastern San Diego County (e.g., Thomson, 1994), but no evidence supports the existence of an arc of this age in the western zone north or south of the ancestral Agua Blanca fault.

The rifted-fringing arc model proposed by Busby et al. (1998) is based on observations of the western zone in the area south of the ancestral Agua Blanca fault (e.g., Fackler-Adams and Busby, 1998) and stratigraphy preserved in the Continental Borderlands terrane. Busby et al. (1998) partition their model into three phases that correspond to time periods of 220–130 Ma (Phase one), 140–100 Ma (Phase two), and 100–50 Ma (Phase three). In their model, phase one represents the rifting of a fragment of the continental margin to form the extensional fringing arc of phase two, which ultimately becomes reaccreted to the continental margin during phase three. In general, the same objections that were raised against the Gastil et al. (1981) rifted-fringing arc model apply here. However, there are several additional complications from the Busby et al. (1998) model, which are described below.

The identification of an early Mesozoic accretionary prism with continental ties (e.g., Boles and Landis, 1984; Sedlock and Isozaki, 1990) within the Continental Borderlands terrane implies that this terrane must have been adjacent to a continent in the Mesozoic. However, strike-slip translation of this terrane during the Late Cretaceous (e.g., Busby et al., 1998) makes correlations between the Jurassic and Early Cretaceous strata of the Continental Borderlands terrane and the Early Cretaceous arc strata of the western zone of the Peninsular Ranges batholith highly suspect. Presently, no data exist to indicate that the Continental Borderlands were adjacent to the Peninsular Ranges batholith prior to the Late Cretaceous (e.g., Klinger et al., 2000).

Detailed mapping of structures and stratigraphic sequences within the Alisitos arc southeast of San Quintin by Fackler-Adams and Busby (1998) form the basis for events in phases one and two of the Busby et al. (1998) model. It is important to note here that strata and intrusive bodies older than ca. 116 Ma have not been identified within the Alisitos arc. Therefore, none of phase one and only the last 16 m.y. of phase two are potentially represented in this part of the Peninsular Ranges. This 16 m.y. coincides with the time during which we propose that the Alisitos arc is characterized by northeast-southwest–directed shortening associated with its accretion to the continental margin. Fackler-Adams and Busby (1998), however, argue for a rifting arc model based on the presence of northeast-trending dikes and small-offset (<100 m) normal faults, and ~200 m of basalts capping the section. Their interpretation of these observations seems somewhat overstated, given that all extensional features are oriented perpendicular to the extension directions suggested by Busby et al. (1998) and that the magnitudes of fault offset and dike-induced dilation are all very small. Furthermore, the relatively small volume of basaltic lavas termed "rift-related" could just as easily be ocean island basalts. No reported chemical data support the designation of these lavas as rift-related. Similarly, north of the ancestral Agua Blanca fault, the identification of extensional structures and volcanic petrologies have not been reported, even though a greater percentage of the time discussed in the Busby et al. (1998) model is preserved in the rock record.

Finally, if the western zone of the Peninsular Ranges were to have developed as a reaccreted rifted-fringing arc, at least some proportion of the basement of this fringing arc must have be composed of continentally derived materials. It therefore follows that volcanics and plutonics that were emplaced through and deposited on these continental materials should show some contamination, such as that observed in the Santiago Peak Volcanics and associated plutonic units. However, this is not the case for the Alisitos arc. We conclude that models identifying any portion of the western zone of the Peninsular Ranges batholith as having evolved as a rifted-fringing arc are untenable.

Dickinson and Lawton (2001) recently proposed a tectonic model wherein the western zone of the Peninsular Ranges batholith, as part of the larger Guerrero superterrane, originated as an east-facing island arc, exotic to North America. This proposal arises from supposition that the only proposed alternative models are one of a rifted-fringing arc (e.g., Busby et al., 1998), combined with the observation of east-directed thrusting along the eastern margin of the Guerrero superterrane in mainland México (Tardy et al., 1994). Dickinson and Lawton (2001) are also influenced by the tradition of poorly constrained age ranges for magmatism within the western zone of the batholith (i.e., that it was a Jurassic-Cretaceous arc).

As discussed in detail above, models supporting the accretion of the western zone of the Peninsular Ranges batholith (either as an east- or west-facing arc) do not conform with the relationships that exist between the Santiago Peak Volcanics and the continentally derived Bedford Canyon Complex. South of the ancestral Agua Blanca fault the Alisitos arc and adjacent central zone define a broad, west-vergent fold-and-thrust belt. Early Cretaceous east-vergent structures are known locally from the eastern zone in the southern Sierra San Pedro Mártir area (Schmidt, 2000) but have not been reported from most areas of the Peninsular Ranges. If the Alisitos arc had been an east-facing arc, and a contemporaneous west-facing continental arc existed in the present location of the central zone, then the accretionary prisms associated with both of these arcs have been removed. Given the lack of east-directed structures to support this interpretation, the additional tectonic complexity of having to remove an additional accretionary prism seems unnecessary. Finally, given the descriptions here of the dramatic differences in the tectonic evolution of the Peninsular Ranges across the ancestral Agua Blanca fault, there is no reason to think that similar transitions cannot exist between the Peninsular Ranges and its on-strike continuation in mainland México.

CONCLUSIONS

The Mesozoic tectonic evolution of the southwestern margin of North America has been one of the more poorly resolved aspects of Cordilleran geology. A variety of competing models have been proposed, most of which disagree on even the most fundamental aspects (e.g., origin of the western zone of the Peninsular Ranges). We suggest that, in part, the differences between these models are the result of tectonically significant variations in the along-strike character of the continental margin, and they are most pronounced across the Agua Blanca fault of northern Baja California, México. Variations across this fault can be observed in the geology of the central and western zones of the Peninsular Ranges strata as old as the Late Triassic. These variations include: (1) the presence or absence of Late Triassic through Jurassic continentally derived turbidite sequences (north) and/or

Early Cretaceous submarine sedimentary strata (south), (2) the environment of deposition of Early Cretaceous western zone volcanics, and contact relations between these volcanics and the continentally derived strata of the central zone (depositional, north; fault, south), (3) presence of xenocrystic Precambrian zircons in plutons and volcanic flows of the western zone (present, north; absent, south), (4) the frequency of lava types of western zone volcanics (abundant rhyolites, north; abundant basalts, south), (5) the general distribution and intensity of deformation within the western zone (minor to moderate, north; 20-km-wide fold-and-thrust belt, south), and (6) the character of deformation associated with the ancestral Agua Blanca fault (truncation, north; deflection into subparallelism, south).

We propose a model that specifically incorporates these along-strike variations. During Late Triassic through Jurassic, the central zone north of the ancestral Agua Blanca fault was the site of sedimentation of continentally derived turbidite sequences that were ultimately incorporated and deformed within an accretionary prism setting. We propose that all of the Late Triassic through Jurassic strata of similar lithology present within the central zone north of the Agua Blanca fault should be included within a single stratigraphic group, here termed the Bedford Canyon Complex. By the Early Cretaceous, the Bedford Canyon Complex was uplifted and exposed to subaerial erosion prior to the unconformable deposition of the Santiago Peak Volcanics. South of the ancestral Agua Blanca fault, the central zone was generally not submerged as it was to the north; thus, deposits of Late Triassic through Jurassic age are only preserved locally in this part of the batholith. The outboard trench and accretionary prism (i.e. Bedford Canyon Complex correlative strata) presumably existed west of the present-day central zone. During the latest Jurassic through Early Cretaceous, the central zone south of the ancestral Agua Blanca fault became submerged below sea level and filled with shallow to deep water sediments overlain by subaqueously deposited volcanics that grade upward into subaerial volcanics. Contemporaneous with Early Cretaceous basin sedimentation in the central zone, the Alisitos arc initiated on oceanic crust not previously associated with the continental margin. The Alisitos arc impinged upon the continental trench between 115 and 108 Ma and during accretion, forcing the subduction of the associated accretionary prism. While some uplift and exhumation of the central zone adjacent to the Alisitos arc may be attributed to the accretion of that arc, the majority of this central zone event was most likely caused by increased coupling between the continent and the subducting slab.

ACKNOWLEDGMENTS

This study was supported by grants from the Geological Society of America, Sigma Xi, National Science Foundation (EAR-9614682), and the Department of Earth Sciences, University of Southern California. Many thanks are due to Scott Johnson, David Bottjer, David Kimbrough, Francisco Suarez-Vidal, Erwin Melis, Robert Trzebski, Chris Kopf, John Fletcher, and Tom Rockwell for numerous insightful discussions. We also wish to thank Pedro Morenco, Marcos Marin, and Melisa Paramo for field assistance and Brendan McNulty and Gary Girty for very helpful reviews of the manuscript.

REFERENCES CITED

Adams, M.A., 1979, Stratigraphy and petrology of the Santiago Peak Volcanics east of Rancho Santa Fe, California [M.S. thesis]: San Diego State University, 123 p.
Allen, C., Silver, L., and Stehil, F., 1960, Agua Blanca fault—A major transverse structure of northern Baja California, Mexico: Geological Society of America Bulletin, v. 71, p. 457–482.
Allison, E.C., 1955, Middle Cretaceous Gastropoda from Punta China, Baja California, Mexico: Journal of Paleontology, v. 29, p. 400–432.
Allison, E.C., 1974, The type Alisitos Formation (Cretaceous, Aptian-Albian) of Baja California and its bivalve fauna, in Gastil, G., and Lillegraven, J., eds., Geology of peninsular California: Los Angeles, California, American Association of Petroleum Geologists Pacific Section, p. 20–59.
Anderson, C.L., 1991, Zircon uranium-lead isotopic ages of the Santiago Peak Volcanics and spatially related plutons of the Peninsular Ranges batholith, southern California [M.S. thesis]: San Diego, San Diego State University, 111 p.
Balch, D.C., Bartling, S.H., and Abbott, P.L., 1984, Volcaniclastic strata of the Upper Jurassic Santiago Peak Volcanics, San Diego, California, in Crouch, J.K., and Bachman, S.B., eds., Tectonics and sedimentation along the California margin: Los Angeles, Society of Economic Paleontologist and Mineralogists Pacific Section, p. 157–170.
Beggs, J.M., 1984, Volcaniclastic rocks of the Alisitos Group, Baja California, Mexico, in Frizzell, V., Jr., ed., Geology of the Baja California peninsula: Los Angeles, Society of Economic Paleontologists and Mineralogists Pacific Section, p. 43–52.
Boles, J.R., and Landis, C.A., 1984, Jurassic sedimentary mélange and associated facies, Baja California, Mexico: Geological Society of America Bulletin, v. 95, p. 513–521.
Brandon, M.T., Roden-Tice, M.K., and Garver, J.I., 1998, Late Cenozoic exhumation of the Cascadia accretionary wedge in the Olympic Mountains, northwest Washington State: Geological Society of America Bulletin, v. 110, p. 985–1009.
Buch, I.P., and Delattre, M.P., 1993, Permian and Lower Triassic stratigraphy along the 30th parallel, eastern Baja California Norte, Mexico, in Gastil, R.G., and Miller, R.H., eds., The pre-batholithic stratigraphy of peninsular California, p. 77–90.
Busby, C., Smith, D., Morris, W., and Fackler-Adams, B., 1998, Evolutionary model for convergent margins facing large ocean basins: Mesozoic Baja California, Mexico: Geology, v. 26, p. 227–230.
Bushee, J., Holden, J., Geyer, B., and Gastil, R.G., 1963, Lead-alpha dates for some basement rocks of southwestern California: Geological Society of America Bulletin, v. 74, p. 803–806.
Campbell, M., and Crocker, J., 1993, Geology west of the Canal de Las Ballenas, Baja California, Mexico, in Gastil, R.G., and Miller, R.H., eds., The pre-batholithic stratigraphy of peninsular California, p. 61–76.
Carrasco, A.P., Kimbrough, D.L., Herzig, C.T., and Meeth, G.L., 1993, Discovery of accretionary lapilli in the Santiago Peak Volcanics of southern and Baja California, in Abbott, P.L., Sangines, E.M., and Rendina, M.A., eds., Geologic investigations in Baja California: Santa Ana, California, South Coast Geological Society, p. 145–150.
Carrasco, A.P., Kimbrough, D.L., and Herzig, C.T., 1995, Cretaceous arc-volcanic strata of the western Peninsular Ranges: Comparison of the Santiago Peak Volcanics and Alisitos Group: Abstracts of Peninsular Geological Society International Meeting on Geology of the Baja California Peninsula, v. III, p. 19.
Carollo, G.F., and Walawender, M.J., 1993, Geochemistry and petrography of leucogranite plutons from the western zone of the Peninsular Ranges batholith, San Diego, California, in Abbott, P.L., Sangines, E.M., and Rendina, M.A., eds., Geologic investigations in Baja California: Santa Ana, California, South Coast Geological Society, p. 151–162.
Chadwick, B., 1987, The geology, petrography, geochemistry, and geochronology of the Tres Hermanos–Santa Clara region, Baja California, Mexico [M.S. Thesis]: San Diego State University, 601 p.

Criscione, J.L., Davis, T.E., and Ehlig, P., 1978, The age of sedimentation/ diagenesis for the Bedford Canyon Formation and Santa Monica Formation in southern California—A Rb/Sr evaluation, *in* Howell, D.G., and McDougall, K.A., eds., Mesozoic paleogeography of the western United States: Los Angeles, Society of Economic Paleontologists and Mineralogists Pacific Section, p. 385–396.

Damon, P.E., Shafiqullah, M., and Clark, K.F., 1983, Geochronology of the porphyry copper deposits and related mineralization of Mexico: Canadian Journal of Earth Sciences, v. 20, p. 1052–1071.

Davis, T.E., and Gastil, R.G., 1993, Whole-rock Rb/Sr isochrons of fine-grained clastic rocks in peninsular California, *in* Gastil, R.G., and Miller, R.H., eds., The prebatholithic stratigraphy of peninsular California: Boulder, Colorado, Geological Society of America Special Paper 279, p. 157–158.

DePaolo, D.J., 1981, A neodymium and strontium isotopic study of the Mesozoic calc-alkaline granitic batholiths of the Sierra Nevada and Peninsular Ranges, California: Journal of Geophysical Research, v. 86, p. 10,470–10,488.

Dickinson, W.R., and Lawton, T.F., 2001, Carboniferous to Cretaceous assembly and fragmentation of Mexico: Geological Society of America Bulletin, v. 113, p. 1142–1160.

Ducea, M., 2001, The California arc: Thick granitic batholiths, eclogite residues, lithospheric-scale thrusting, and magmatic flare-ups: GSA Today, v. 11, no. 11, p. 4–10.

Engebretson, D.C., Cox, A., and Gordon R.G., 1985, Relative motions between oceanic and continental plates in the pacific basin: Boulder, Colorado, Geological Society of America Special Paper 206, p. 1–59.

Fackler-Adams, B., 1997, Volcanic and sedimentary facies, processes, and tectonics of intra-arc basins: Jurassic continental arc of California and Cretaceous oceanic arc of Baja California [Ph.D. Dissertation]: University of California at Santa Barbara, 248 p.

Fackler-Adams, B.N., and Busby, C.J., 1998, Structural and stratigraphic evolution of extensional oceanic arcs: Geology, v. 26, p. 735–738.

Fife, D.L., Minch, J.A., and Crampton, P.J., 1967, Late Jurassic age of the Santiago Peak Volcanics, California: Geological Society of America Bulletin, v. 78, p. 229–304.

Gastil, R.G., 1993, Prebatholithic history of Peninsular California, *in* Gastil, R.G., and Miller, R.H., eds., The prebatholithic stratigraphy of peninsular California: Boulder, Colorado, Geological Society of America Special Paper 279, p. 145–156.

Gastil, R.G., and Girty, M.S., 1993, A reconnaissance U-Pb study of detrital zircon in sandstones of peninsular California and adjacent areas, *in* Gastil, R.G., and Miller, R.H., eds., The prebatholithic stratigraphy of peninsular California: Boulder, Colorado, Geological Society of America Special Paper 279, p. 135–144.

Gastil, R.G., Phillips, R., and Allison, E., 1975, Reconnaissance geology of the State of Baja California: Boulder, Colorado, Geological Society of America Memoir 140, p. 170.

Gastil, R.G., Morgan, G.J., and Krummenacher, D., 1981, The tectonic history of peninsular California and adjacent Mexico, *in* Ernst, W.G., ed., The geotectonic development of California: Englewood Cliffs, New Jersey, Prentice-Hall, p. 284–306.

Gastil, R.G., Diamond, C.K., Walawender, M.J., Marshal, M., Boyles, C., and Chadwick, B., 1990, The problem of the magnetite/ilmenite boundary in southern and Baja California, *in* Anderson, J.L., ed., The nature and origin of Cordilleran magmatism: Boulder, Colorado, Geological Society of America Memoir 174, p. 19–32.

Gehrels, G.E., Stewart, J.H., and Ketner, K.B., 2002, Cordilleran-margin quartzites in Baja California—Implications for tectonic transport: Earth and Planetary Science Letters, v. 199, p. 201–210.

Germinario, M., 1993, The early Mesozoic Julian Schist, Julian, California, *in* Gastil, R.G., and Miller, R.H., eds., The prebatholithic stratigraphy of peninsular California: Boulder, Colorado, Geological Society of America Special Paper 279, p. 107–118.

Girty, G.H., Thomson, C.N., Girty, M.S., Miller, J., and Bracchi, K., 1993, The Cuyamaca-Laguna Mountains shear zone, Late Jurassic plutonic rocks and Early Cretaceous extension, Peninsular Ranges, southern California, *in* Abbott, P.L., Sangines, E.M., and Rendina, M.A., eds., Geologic investigations in Baja California, Mexico: Southern California Geological Society annual field trip guide, no. 21, p. 173–181.

Goetz, C.W., 1989, Geology of the Rancho El Rosarito area, Baja California, Mexico: Evidence for latest Albian compression along a terrane boundary in the Peninsular Ranges [M.S., thesis]: San Diego, San Diego State University, 134 p.

Gorzolla, Y., 1988, Geochemistry and petrography of the Santiago Peak Volcanics, Santa Margarita and Santa Ana Mountains, southern California [M.S. thesis]: San Diego, San Diego State University, 145 p.

Gradstein, F.M., Agterberg, F.P., Ogg, J.G., Hardenbol, J., van Veen, P., Thierry, J., and Huang, Z., 1994, A Mesozoic time scale: Journal of Geophysical Research, v. 99, p. 24,051–24,074.

Griffith, R., and Hoobs, J., 1993, Geology of the southern Sierra Calamajue, Baja California Norte, Mexico, *in* Gastil, R.G., and Miller, R.H., eds., The prebatholithic stratigraphy of peninsular California: Boulder, Colorado, Geological Society of America Special Paper 279, p. 43–60.

Gromet, L.P., and Silver, L.T., 1987, REE variations across the Peninsular Ranges batholith: Implications for batholithic petrogenesis and crustal growth in magmatic arcs: Journal of Petrology, v. 28, p. 75–125.

Hanna, M.A., 1926, Geology of the La Jolla quadrangle, California: University of California Department of Geological Sciences Bulletin, v. 16, p. 187–246.

Hawkins, J.W., 1970, Metamorphosed Late Jurassic andesites and dacites of the Tijuana-Tecate area, California, Pacific slope geology of northern Baja California and adjacent Alta California: San Diego, California, Pacific section, American Association of Petroleum Geologists/Society of Economic Paleontologists and Mineralogists/ Society of Exploration Geophysicists, p. 25–29.

Herzig, C.T., 1991, Petrogenetic and tectonic development of the Santiago Peak Volcanics, northern Santa Ana Mountains, California [Ph.D. dissertation]: Riverside, University of California, 376 p.

Hudson, F.S., 1922, Geology of the Cuyamaca region of California: California University Department of Geological Science Bulletin, v. 13, p. 175–253.

Imlay, R.W., 1963, Jurassic fossils from southern California: Journal of Paleontology, v. 38, p. 97–107.

Imlay, R.W., 1964, Middle Jurassic and Upper Jurassic fossils from southern California: Journal of Paleontology, v. 38, p. 505–509.

Johnson, S.E., Tate, M.C., and Fanning, C.M., 1999a, New geologic mapping and SHRIMP U-Pb data in the Peninsular Ranges batholith, Baja California, Mexico: Evidence of a suture?: Geology, v. 27, p. 743–746.

Johnson, S.E., Paterson, S.R., and Tate, C.M., 1999b, Structure and emplacement history of a multiple-center, cone-sheet-bearing ring complex: The Zarza intrusive complex, Baja California, Mexico: Geological Society of America Bulletin, v. 111, p. 607–619.

Johnson, S.E., Fletcher, J.M., Fanning, C.M., Paterson, S.R., Vernon, R.H., and Tate, M.C., 2003, Structure and emplacement of the San Jose tonalite pluton, Peninsular Ranges batholith, Baja California, Mexico: Journal of Structural Geology (in press).

Kimbrough, D.L., Smith, D.P., Mahoney, J.B., Moore, T.E., Grove, M., Gastil, R.G., Ortega-Rivera, A., and Fanning, C.M., 2001, Forearc-basin sedimentary response to rapid Late Cretaceous batholith emplacement in the Peninsular Ranges of southern and Baja California: Geology, v. 29, p. 491–494.

Kimbrough, D.L., Gastil, R.G., Garrow, P.K., Grove, M., Aranda-Gomez, J., and Perez-Venzor, J.A., 2002, A potential correlation of plutonic suites from the Los Cabos Block and Peninsular Ranges batholith, *in* VI International Meeting on Geology of the Baja California Peninsula, La Paz, Baja California Sur, Mexico, p. 9.

Klinger, M., Mahoney, J.B., and Kimbrough, D.L., 2000, Whole rock geochemistry of conglomerate clasts from Valle Group Cretaceous forearc basin strata, Baja California: Geological Society of America Abstracts with Programs, v. 32, no. 6, p. 22.

Lamb, T.N., 1970, Fossiliferous Triassic metasedimentary rocks near Sun City, Riverside County, California: Geological Society of America Abstracts with Programs, v. 2, no. 2, p. 110–111.

Larsen, E.S., 1948, Batholith and associated rocks of Corona, Elsinore, and San Luis Rey quadrangles, southern California: New York, Geological Society of America Memoir 29, p. 182.

Lash, G.G., 1985, Accretion-related deformation of an ancient (early Paleozoic) trench-fill deposit, central Appalachian orogen: Geological Society of America Bulletin, v. 96, p. 1167–1178.

Leedom, S., 1967, A stratigraphic study of the Alisitos Formation near San Vicente, Baja California, Mexico [Senior thesis]: San Diego, California, San Diego State University, 21 p.

Leeson, R.T., 1989, Fabric analysis of the Cuyamaca-Laguna Mountains shear zone [M.S. thesis]: San Diego, San Diego State University, 136 p.

Legg, M.R., Wong, O.V., and Suarez-Vidal, F., 1991, Geologic structure and tectonics of the inner continental borderland of northern Baja California, *in* Dauphin, J.P.E., and Simoneit, B.R.T., eds., The Gulf and peninsular province of the Californias: Tulsa, Oklahoma, American Association of Petroleum Geologists Memoir 47, p. 145–177.

Le Maitre, R.W., Bateman, P., Dudek, A., Keller, J., Lemeyre, J., Le Bas, M.J., Sabine, P.A., Schmid, R., Sorensen, H., Streckeisen, A., Wooley, A.R., and Zenttin, B., 1989, Igneous rocks: A classification of igneous rocks and glossary of terms: Cambridge, United Kingdom, Cambridge University Press, 236 p.

Lothringer, C.J., 1993, Allochthonous Ordovician strata of Rancho San Marcos, Baja California Norte, Mexico, *in* Gastil, R.G., and Miller, R.H., eds., The prebatholithic stratigraphy of peninsular California: Boulder, Colorado, Geological Society of America Special Paper 279, p. 11–22.

Lovera, O.M., Grove, M., Kimbrough, D.L., and Abbott, P.L., 1999, A method for evaluating basement exhumation histories from closure age distributions of detrital minerals: Journal of Geophysical Research, v. 104, p. 29,419–29,438.

Marsh, B.D., 1979, Island-arc volcanism: American Scientist, v. 67, p. 161–172.

Meeth, G., 1993, Stratigraphy and petrology of the Santiago Peak Volcanics east of La Mision, Baja California [B.S. thesis]: San Diego, San Diego State University, 16 p.

Moran, A.I., 1976, Allochthonous carbonate debris in Mesozoic flysch deposits in Santa Ana Mountains, California: American Association of Petroleum Geologists Bulletin, v. 60, p. 2038–1043.

Moscoso, B.A., 1967, A thick section of "flysch" in the Santa Ana Mountains, southern California [M.S. thesis]: San Diego, San Diego State University, 106 p.

Nicholls, I.A., and Ringwood, A.E., 1973, Effect of water on olivine stability in tholeiites and the production of silica saturated magmas in the island arc environment: Journal of Geology, v. 81, p. 285–300.

Peterson, G.L., 1968, Structure of the late Mesozoic pre-batholithic rocks north of San Diego, California, *in* Abstracts for 1967: Boulder, Colorado, Geological Society of America Special Paper 115, p. 347.

Phillips, J.R., 1993, Stratigraphy and structural setting of the mid-Cretaceous Olvidada Formation, Baja California Norte, Mexico, *in* Gastil, R.G., and Miller, R.H., eds., The prebatholithic stratigraphy of peninsular California: Boulder, Colorado, Geological Society of America Special Paper 279, p. 97–106.

Reed, R.G., 1967, Stratigraphy and structure of the Alisitos Formation near El Rosario, Baja California, Mexico [M.S. thesis]: San Diego State University, 118 p.

Reed, B.C., 1992, Petrology and eruptive setting of the Santiago Peak Volcanics, in the Mission Gorge–San Diego area, San Diego, California [M.S. thesis]: San Diego, San Diego State University, 108 p.

Reed, J., 1993, Rancho Vallecitos Formation, Baja California Norte, Mexico, *in* Gastil, R.G., and Miller, R.H., eds., The prebatholithic stratigraphy of peninsular California: Boulder, Colorado, Geological Society of America Special Paper 279, p. 119–134.

Rockwell, T.K., Muhs, D.R., Kennedy, G.L., Hatch, M.E., Wilson, S.H., and Klinger, R.E., 1989, Uranium-series ages, faunal correlations and tectonic deformation of marine terraces within the Agua Blanca Fault zone at Punta Banda, northern Baja California, Mexico, *in* Abbott, P.L., ed., Geologic studies in Baja California: Los Angeles, Society for Sedimentary Geology (SEPM) Pacific Section, p. 1–16.

Saleeby, J.B., Busby-Spera, C., Oldow, J.S., Dunne, G.C., Wright, J.E., Cowan, D.S., Walker, N.W., and Allmendinger, R.W., 1992, Early Mesozoic tectonic evolution of the western U.S. Cordillera, *in* Burchfiel, B.C., Lipman, P.W., and Zoback, M.L., eds., The Cordilleran Orogen: Conterminous U.S.: Boulder, Colorado, Geological Society of America, Geology of North America, v. G3, p. 107–168.

Sample, J.C., and Fisher, D.M., 1986, Duplex accretion and underplating in an ancient accretionary complex, Kodiak Islands, Alaska: Geology, v. 14, p. 160–163.

Sample, J.C., and Moore, J.C., 1987, Structural style and kinematics of an underplated slate belt, Kodiak and adjacent islands, Alaska: Geological Society of America Bulletin, v. 99, p. 7–20.

Schmidt, K.L., 2000, Investigations of arc processes: Relationships among deformation magmatism, mountain building, and the role of crustal anisotropy in the evolution of the Peninsular Ranges batholith, Baja California [Ph.D. dissertation]: Los Angeles, University of Southern California, 324 p.

Schmidt, K.L., Wetmore, P.H., Paterson, S.R., and Johnson, S.E., 2002, Controls on orogenesis along an ocean-continental margin transition in the Jura-Cretaceous Peninsular Ranges batholith, *in* Barth, A., ed., Contributions to crustal evolution of the southwestern United States: Boulder, Colorado, Geological Society of America Special Paper 365, p. 49–71.

Schoellhamer, J.E., Vedder, J.G., Yerkes, R.F., and Kinney, D.M., 1981, Geology of the northern Santa Ana Mountains: U.S. Geological Survey, 109 p.

Schroeder, J.E., 1967, Geology of a portion of the Ensenada Quadrangle Baja California, Mexico [M.S. thesis]: San Diego, San Diego State University, 65 p.

Schwartz, H.P., 1960, Geology of the Winchester-Hemet area, Riverside County, California [Ph.D. dissertation]: Pasadena, California Institute of Technology, 317 p.

Sedlock, R.L., and Isozaki, Y., 1990, Lithology and biostratigraphy of Franciscan-like chert and associated rocks in west-central Baja California, Mexico: Geological Society of America Bulletin, v. 102, p. 852–864.

Sedlock, R.L., Ortega-Gutierrez, F., and Speed, R.C., 1993, Tectonostratigraphic terranes and tectonic evolution of Mexico: Boulder, Colorado, Geological Society of America Special Paper 278, 153 p.

Silberling, N.J., Schoellhamer, J.E., Gray, C.H., and Imlay, R.W., 1961, Upper Jurassic fossil from the Bedford Canyon Formation, southern California: America Association of Petroleum Geologists Bulletin, v. 4745, p. 1746–1765.

Silver, L.T., and Chappell, B.W., 1988, The Peninsular Ranges batholith: An insight into the evolution of the Cordilleran batholiths of southwestern North America: Transactions of the Royal Society of Edinburgh, v. 79, p. 105–121.

Silver, L.T., Stehli, G.G., and Allen, C.R., 1963, Lower Cretaceous pre-batholithic rocks of northern Baja California, Mexico: Bulletin of the American Association of Petroleum Geologists, v. 47, p. 2054–2059.

Silver, L.T., Taylor, H.P.J., and Chappell, B.W., 1979, Some petrological, geochemical, and geochronological observations of the Peninsular Ranges batholith near the international border of the U.S.A., and Mexico, *in* Abbott, P.L., and Todd, V.R., eds., Mesozoic crystalline rocks: Peninsular Ranges batholith and pegmatites, Point Sal Ophiolite: San Diego, California, San Diego State University, p. 83–110.

Suarez-Vidal, F., 1986, Alisitos Formation calcareous facies: Early Cretaceous episode of tectonic calm: American Association of Petroleum Geologists Bulletin, v. 70, p. 480.

Suarez-Vidal, F., 1993, The Aptian-Albian on the west coast of the state of Baja California, a mixture of marine environments, *in* Abbott, P.L., Sangines, E.M., and Rendina, M.A., eds., Geologic investigations in Baja California: Santa Ana, California, South Coast Geological Society, p. 125–138.

Suarez-Vidal, F., Armijo, R., Morgan, G., Bodin, P., and Gastil, R.G., 1991, Framework of recent and active faulting in northern Baja California, *in* Dauphin, J.P.E., and Simoneit, B.R.T., eds., The gulf and peninsular province of the Californias: Tulsa, Oklahoma, p. 285–300.

Sutherland, M., Wetmore, P.H., Herzig, C., and Paterson, S.R., 2002, The Early Cretaceous Santiago Peak arc: A continental margin arc built on the North American Triassic-Jurassic accretionary prism of southern and Baja California: Geological Society of America Abstracts with Programs, v. 34, no. 6, p. 43.

Tanaka, H., Smith, T.E., and Huang, C.H., 1984, The Santiago Peak volcanic rocks of the Peninsular Ranges batholith, southern California; volcanic rocks associated with coeval gabbros: Bulletin Volcanologique, v. 47, p. 153–171.

Tardy, M., Lapierre, H., Freydier, C., Coulon, C., Gill, J.B., Mercier de Lepinay, B., Beck, C., Martinez, J., Talavera, O., Ortiz, E., Stein, G., Bourdier, J.L., and Yta, M., 1994, The Guerrero suspect terrane (western Mexico) and coeval arc terranes (the Greater Antilles and the Western Cordillera of Colombia): A late Mesozoic intra-oceanic arc accreted to cratonal America during the Cretaceous: Tectonophysics, v. 230, p. 49–73.

Tate, M.C., and Johnson, S.E., 2000, Subvolcanic and deep-crustal tonalite genesis beneath the Mexican Peninsular Ranges: Journal of Geology, v. 108, p. 720–728.

Tate, M.C., Norman, M.D., Johnson, S.E., Fanning, C.M., and Anderson, J.L., 1999, Generation of tonalite and trondhjemite by subvolcanic fractionation and partial melting in the Zarza Intrusive Complex, western Peninsular Ranges batholith, northwestern Mexico: Journal of Petrology, v. 40, p. 983–1010.

Taylor, H.P, Jr., and Silver, L.T., 1978, Oxygen isotope relationships in plutonic igneous rocks of the Peninsular Ranges batholith, southern and Baja California, *in* Zartman, R.E., ed., Short papers of the fourth international conferences on geochronology, cosmochronology, isotope geology: U.S. Geological Survey Open-File Report 78-701, p. 423–426.

Thomson, C.N., 1994, Tectonic implications of Jurassic magmatism and Early Cretaceous mylonitic deformation, Cuyamaca-Laguna Mountains shear zone, southern California [M.S. thesis]: San Diego, San Diego State University, 85 p.

Thomson, C.N., and Girty, G.H., 1994, Early Cretaceous intra-arc ductile strain in Triassic-Jurassic and Cretaceous continental margin arc rocks, Peninsular Ranges, California: Tectonics, v. 13, p. 1108–1119.

Todd, V.R., and Shaw, S.E., 1985, S-type granitoids and an I-S line in the Peninsular Ranges batholith, southern California: Geology, v. 13, p. 231–233.

Todd, V.R., Erskine, B.G., and Morton, D.M., 1988, Metamorphic and tectonic evolution of the northern Peninsular Ranges batholith, *in* Ernst, W.G., ed., Metamorphism and crustal evolution of the western United States: Rubey Volume VII: Englewood Cliffs, New Jersey, Prentice-Hall, p. 894–937.

Walawender, M.J., Gastil, R.G., Clinkenbeard, W.V., McCormick, W.V., Eastman, B.G., Wernicke, M.S., Wardlaw, M.S., Gunn, S.H., and Smith, B.M., 1990, Origin and evolution of the zoned La Posta–type plutons, eastern Peninsular Ranges batholith, southern and Baja California, *in* Anderson, J.L., ed., The nature and origin of Cordilleran magmatism: Boulder, Colorado, Geological Society of America Memoir 174, p. 1–18.

Walawender, M.J., Girty, G.H., Lombardi, M.R., Kimbrough, D., Girty, M.S., and Anderson, C., 1991, A synthesis of recent work in the Peninsular Ranges batholith, *in* Walawender, M.J., and Hanan, B.B., eds., Geological excursions in southern California and Mexico: San Diego, California, San Diego State University, p. 297–318.

Wetmore, P.H., Schmidt, K.L., Paterson, S.R., and Herzig, C., 2002, Tectonic implications for the along-strike variation of the Peninsular Ranges batholith, southern and Baja California: Geology, v. 30, p. 247–250.

MANUSCRIPT ACCEPTED BY THE SOCIETY JUNE 2, 2003

Geological Society of America
Special Paper 374
2003

A model for the North America Cordillera in the Early Cretaceous: Tectonic escape related to arc collision of the Guerrero terrane and a change in North America plate motion

Paul J. Umhoefer

Department of Geology, Northern Arizona University, Flagstaff, Arizona 86011, USA

ABSTRACT

Many terranes of western North America were accreted to the continent from Middle Jurassic to Early Cretaceous time. The location of many of these accretionary events is poorly known because the amount of syn- and post-accretion translation is widely debated. Thus, the paleogeography of the Cordillera before the Cenozoic is poorly known. Here I present a selective summary of a few key interpretations and data sets that I believe indicate that the Early Cretaceous of the North American Cordillera had a major arc-continent collision in the south and northward tectonic escape in the north. I adopt four conclusions from other workers for the Early Cretaceous: (i) the Guerrero terrane was an oceanic arc that collided with México ca. 120 Ma; (ii) the Baja British Columbia (Baja B.C.) block had a moderate amount of northward translation (~1600–1800 km) in the Late Cretaceous–early Tertiary that was preceded by ~800 km of southward translation (sinistral faulting) in the Early Cretaceous; (iii) the sinistral faulting in Baja B.C. occurred at the same time as a dextral fault system of 200–400 km offset in eastern California, Nevada, and Idaho; and (iv) the large volume of magmatism in the Sierra Nevada in the Late Cretaceous (100–85 Ma) was mainly due to lithospheric-scale underthrusting of North America under the Sierra Nevada and not arc processes. These conclusions lead to a speculative model. In the model, the earliest Cretaceous (145–125 Ma) was dominated by sinistral oblique convergence. There was a major change in the tectonics of the Cordillera at 125–120 Ma that may have been driven by an arc-continent collision of the Guerrero terrane in México and a change to more westerly absolute motion of the North America plate. These events resulted in major tectonic escape of the central and northern Cordillera to the north away from the arc collision in México from 125 to 105 Ma. The collision and escape were accompanied by renewed eastward thrusting in the Sevier–Rocky Mountain thrust belt. This model is similar in many ways to the modern tectonics of the eastern Mediterranean to Caucasus region. The 105–85 Ma interval had major convergence and the beginning of northward translation of the Baja B.C. block. This interval had two belts of magmatism, one from subduction and the other the result of the underthrusting of North American crust to the west.

Keywords: Cordillera, Early Cretaceous, oblique convergence, paleogeography, western North America, tectonics.

Umhoefer, P.J., 2003, A model for the North America Cordillera in the Early Cretaceous: Tectonic escape related to arc collision of the Guerrero terrane and a change in North America plate motion, *in* Johnson, S.E., Paterson, S.R., Fletcher, J.M., Girty, G.H., Kimbrough, D.L., and Martín-Barajas, A., eds., Tectonic evolution of northwestern México and the southwestern USA: Boulder, Colorado, Geological Society of America Special Paper 374, p. 117–134. For permission to copy, contact editing@geosociety.org. © 2003 Geological Society of America.

INTRODUCTION

During Mesozoic time, the Cordillera of western North America had many orogenies that resulted in a complex and poorly understood mountain belt. The main process for continental growth was terrane accretion, which is the addition of fragments from the oceans (volcanic island arcs, ocean plateaux, ocean floor, seamounts) to the continents along convergent and oblique convergent plate boundaries (Coney et al., 1980; Schermer et al., 1984). The timing of the accretion of many terranes to North America is known, but the location and details of the accretionary events is widely debated. The margin of western North America is thought to have been one or more subduction zones continuously from the Late Triassic to the Late Cretaceous (e.g., Burchfiel et al., 1992). Much of the western Cordillera was added to the North American continent in the Mesozoic, especially from Middle Jurassic through Early Cretaceous time (Fig. 1) (Coney et al., 1980; Oldow et al., 1989; Burchfiel et al., 1992). Little of the Cordillera was accreted after the early Late Cretaceous, or ca. 90 Ma.

Despite agreement on the dominance of a convergent margin and the accretion of many terranes in western North America, there is an astonishing lack of agreement on the basic paleogeography before the Cenozoic. One example of this disagreement is the various paleogeographies for the Late Jurassic to earliest Cretaceous (cf. Oldow et al., 1989; Pavlis, 1989; Saleeby and Busby-Spera, 1992; Burchfiel et al., 1992, Cowan et al., 1997; Moores, 1998). The source of the difficulty in making paleogeographic maps that are widely accepted is the lack of understanding of the amount and relative role of translation and convergence, the number of subduction zones along the margin at one time, and the polarity of subduction zones.

The component of translation during the Mesozoic appears to be driven primarily by oblique convergence (Engebretson et al., 1985; Oldow et al., 1989). The motion of North America away from Eurasia and Africa since 180 Ma is a first-order component of the motion of plates along the western side of the continent. The motion was to the northwest, then more westerly in the Early Cretaceous, then finally to the west-southwest in the Late Cretaceous to present. This early northwestward motion path suggests that southward translation (and sinistral faulting) should be expected from the Middle Jurassic to earliest Cretaceous, while northward translation (dextral faulting) should dominate from Late Cretaceous to present. The fault offset and paleomagnetic data agree with the latter inference, as they suggest that translation during and after the Late Cretaceous was northward for the U.S. and Canadian Cordillera (e.g., Engebretson et al., 1985; Irving et al., 1985; Umhoefer, 1987; Irving et al., 1996; Cowan et al., 1997; Stamatakos et al., 2001). The amount of northward translation from the middle Cretaceous to the early Tertiary (Baja B.C.), however, is widely debated as being from 1000 to 4000 km (e.g., Irving et al., 1996; Cowan et al., 1997; Butler et al., 2001; Stamatakos et al., 2001; Enkin et al., 2002). Whatever the amount of translation, it is clearly substantial based on cumulative fault offsets across Baja B.C. of at least 1000 km, which are a mini-

mum estimate of translation (Umhoefer, 2000). Many researchers argue for a sinistral oblique-convergent margin along the western United States in the Middle to Late Jurassic (e.g., Avé Lallemant and Oldow, 1988; Wyld and Wright, 1988; Saleeby and Busby-Spera, 1992; Moores, 1998; Schermer et al., 2001). In contrast, for the Early Cretaceous there is good evidence for both major dextral *and* sinistral fault systems in the United States and Canada (e.g., Monger et al., 1994; Hurlow, 1993; Wyld and Wright, 2001).

As Moores (1998) noted, despite plate tectonics and the terrane concept being fully engrained in the literature, there are few models for the accretion of terranes in western North America that account for a complex evolution, even though much data suggests a complex evolution. In this paper, I explore a speculative model for the Early Cretaceous of western North America with an emphasis on a novel interpretation for the enigmatic 125–105 Ma time interval. This is not a comprehensive review, such as those from previous *Decade of North American Geology* volumes (Oldow et al., 1989; Burchfiel et al., 1992) or Sedlock (this volume). But instead I select four conclusions on very different tectonic events that occurred at the same time. I then explore how these events can be tied together into a coherent model. One purpose of this paper is to stimulate discussion of the relation of data and models that have been proposed from disparate parts of the Cordillera and thus are seemingly not related. My main hypothesis is that an arc-continent collision in México from 125 to 110 Ma, and the change to a more westward motion of North America at 125–120 Ma, drove the northward tectonic escape of a large wedge of the central and northern Cordillera.

RATIONALE AND METHODS OF RECONSTRUCTIONS

My approach here is to present three paleogeographic maps for the Early Cretaceous based on four main interpretations. In the Discussion section I briefly explore alternatives to these interpretations and the affect they may have on the reconstructions presented here. Because I am using these selective interpretations, I must ignore other interpretations (any reconstruction for the Early Cretaceous would have this condition), and thus these are highly interpretive paleogeographic maps.

These four conclusions from other workers are used in this analysis. (i) The Guerrero terrane was an oceanic arc that collided with the southern Cordillera about 120 Ma (Tardy et al., 1994; Dickinson and Lawton, 2001), though the polarity of the arc on the Guerrero terrane is debated (compare Johnson et al., 1999; Wetmore et al., 2002). The Guerrero terrane north of the Agua Blanca fault was built on the edge of western North America (e.g., Schmidt et al., 2002). (ii) The Baja British Columbia (Baja B.C.) block had a moderate amount of northward translation (~1600–1800 km) in the Late Cretaceous–early Tertiary (Stamatakos et al., 2001; Kodama and Ward, 2001; Enkin et al., 2002) that was preceded by ~800 km of southward translation (sinistral faulting) in the Early Cretaceous (Monger et al., 1994). (iii) The sinistral faulting in Baja B.C. occurred at the same time, and at the same latitude, as dextral faulting of 200–400 km in the backarc region (Wyld and Wright,

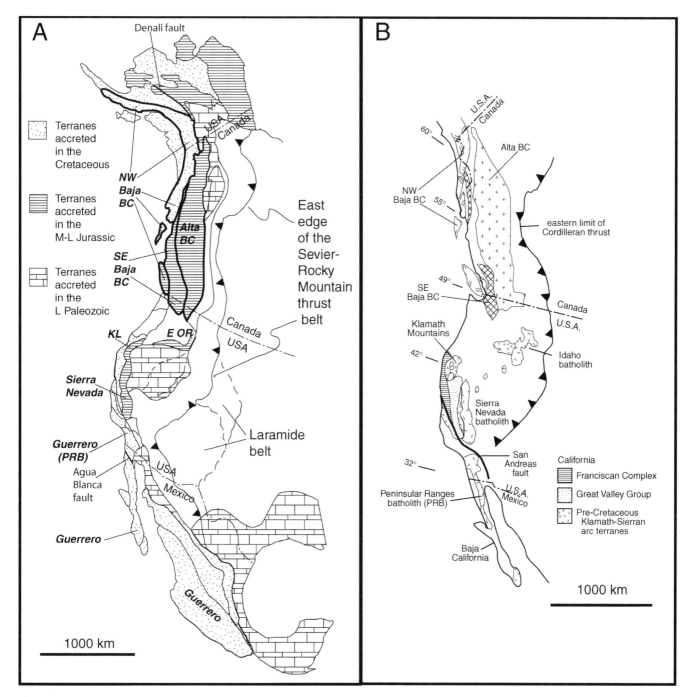

Figure 1. A: Terranes of western North America coded by when they are accreted. Terranes and crustal blocks that are labeled are those names used in this paper. See Table 1 for correlation of these names to other commonly used names. Terranes in white were accreted in Cenozoic. KL—Klamath Mountains; E OR—eastern Oregon. B: Major terranes and crustal blocks that are key elements of paleogeographic maps (Figs. 6, 7, and 8) are shown in patterns similar to those in maps in Figures 6, 7, and 8 and labeled with names used in text.

2001; Umhoefer et al., 2002). (iv) The large volume of magmatism in the Sierra Nevada in the Late Cretaceous (100–85 Ma) was mainly due to lithospheric-scale underthrusting of North America under the Sierra Nevada (Ducea, 2001).

The main consequence of these conclusions is that there was a large fault-bounded wedge that opened to the north from a point

in southern California. The wedge had major strike-slip faults of opposing sense of offset on either side. This geometry is what has been called tectonic escape in Turkey (Şengör et al., 1985) and is proposed here for the Early Cretaceous of the Cordillera. A second consequence of the four conclusions is that the southern point of the wedge is close to the northern edge of the proposed collision of the oceanic arc that comprises the Guerrero terrane of México. This spatial relation suggests a link between the collision and the proposed tectonic escape and further similarity to the Neogene of the eastern Mediterranean region (see Pavlis, 1989, for an earlier idea of tectonic escape farther north in the Cordillera).

The elements used for the three paleogeographic maps are explained in more detail below. My main interest here is to present a new hypothesis for the 125–105 Ma period, when it appears the collision and tectonic escape occurred. Because of the dependence of one tectonic configuration on the previous one, I also include a reconstruction for the 145–125 Ma time period. I include the reconstruction for the 105–85 Ma period so I can briefly explore the apparent conflict with the moderate-translation Baja B.C. model used here and the classic California geology for the middle Cretaceous. The time boundaries for the reconstructions are all at times when major changes occurred in the Cordillera as discussed further below. I have attempted to minimize the names of terranes and other geologic blocks (see Table 1).

GENERAL TECTONIC SETTING

To properly explore the 145–85 Ma time interval of the three reconstructions, it is necessary to review the whole Middle Jurassic to Late Cretaceous period when many of the terranes that are key parts of the reconstructions were accreted. Here is a brief overview of major events of this time period in the Cordillera of North America. Some type of convergent margin was along the western edge of North America during most of this interval (Burchfiel et al., 1992), but the amount and sense of obliquity is uncertain. In addition, even though there was nearly continuous magmatism in the Sierra Nevada through the Middle Jurassic to early Cretaceous, the flux

from 145 to 105 Ma was considerably below rates typical for island arcs (Ducea, 2001). The southern Baja B.C. block has a similar pattern of magmatism, but the interval of decreased magmatism was from 145 to 120 Ma (Friedman and Armstrong, 1995).

The North America plate was moving northwestward from 180 Ma to ca. 80 Ma because of the opening of the central Atlantic (Engebretson et al., 1985). The absolute motion was especially fast from 180 to 145 Ma, then slowed and became more northwestward. There was a change to more westward motion at 125–120 Ma (Scotese, 2002). This history of motion favors sinistral (or southward) motion of plates and terranes along the western side of North America in the Middle Jurassic to Early Cretaceous (Avé Lallemant and Oldow, 1988), but the true relative motion depends on the plates in the Pacific basin. Although there is not a large area of Jurassic to early Cretaceous crust preserved in the Pacific Ocean, Engebretson et al. (1985) analyzed the relative motion of plates based on hot spot tracks. They showed that the Farallon plate likely moved with a southward component relative to North America from 180 to 135 Ma in the northern Cordillera (from the Klamath Mountains northward) and from 180 to 119 Ma in the southern Cordillera (Engebretson et al., 1985). All along the North American margin there is a change at 119 Ma to northeastward relative motion of the Farallon plate relative to North America that persisted through the Cretaceous (Engebretson et al., 1985). Northward motion favors moderate dextral-oblique convergence. This analysis depends critically on the shape of the margin of western North America, which is difficult to reconstruct.

The Middle Jurassic was a time of accretion of many terranes to western North America (Fig. 1). Middle Jurassic accretion has been demonstrated for the Intermontane terrane of central British Columbia (Monger et al., 1982; Ricketts et al., 1992), and the terranes of

TABLE 1. CORRELATION OF TERRANES AND CRUSTAL BLOCKS USED IN MAPS
IN FIGURES 6, 7, AND 8 WITH TERRANE NAMES IN THE LITERATURE

Crustal block or terrane used here	Other names used in literature for all or part of the crustal block in column one
Northwest Baja B.C.	Insular terrane; Wrangellia terrane; Peninsular-Wrangellia-Alexander terrane
Southeast Baja B.C.	Coast belt; Cascade–southern Coast belt orogen; Bridge River, Cadwallader, Methow, Shuksan, Harrison Lake terranes
Alta B.C.	Intermontane terrane (Stikine, Cache Creek, Quesnellia terranes)
Northeastern Oregon	Blue Mountains terrane; Wallowa, Baker, Izee terranes
Klamath Mountains	Rattlesnake Creek terrane; Triassic-Paleozoic composite terrane; Eastern Klamath terrane
Sierra Nevada	Foothills, Calaveras terranes; many others
Peninsular Ranges batholith	Guerrero terrane; Cortes and Santa Ana terranes; Santiago Peak and Alisitos arcs
Forearc basins	Great Valley Group; Santa Ana terrane
Accretionary complex	Franciscan, Vizcaíno, Magdalena complexes
Guerrero terrane	Guerrero, Cortes, Juarez, Mixteca, Santa Ana, Xolapa terranes

Note: Terrane names after Coney et al. (1980), Monger et al. (1982), Pavlis (1989), Silberling et al. (1992), and Cowan (1994).

western British Columbia were accreted in either Middle Jurassic or Early Cretaceous time (Monger et al., 1982; Thorkelson and Smith, 1989; Monger et al., 1994). Many terranes in the Klamath Mountains and Sierra Nevada were also accreted in the Middle Jurassic (Burchfiel et al., 1992). In Alaska, some of the same terranes were accreted in the middle Cretaceous (Cole et al., 1999).

After the widespread Middle Jurassic accretionary event, many workers have suggested that sinistral oblique convergence dominated western North America in the Middle to Late Jurassic and into the Early Cretaceous (e.g., Anderson and Schmidt, 1983; Oldow et al., 1989; Saleeby and Busby-Spera, 1992). Plate motions (Engebretson et al., 1985) and local evidence, though fragmentary, support this model. Based on regional considerations of fault kinematics, there is a sinistral component in the Jurassic Pine Nut fault zone in Nevada (Avé Lallement and Oldow, 1988). Based on the orientation of dikes in the ophiolite and the surrounding geology in the Klamath Mountains, the formation of the Josephine ophiolite in the Late Jurassic was likely in a back-arc basin within an overall sinistral oblique convergent setting (Harper and Wright, 1984; Saleeby and Busby-Spera, 1992). The complex occurrence of both extensional and contractional structures in the Mojave desert has led some workers to the conclusion that sinistral oblique convergence is the best model for that region (Saleeby and Busby-Spera, 1992; Schermer et al., 2001).

Paleomagnetic evidence in British Columbia shows that the Wrangellia and Stikine terranes (part of Baja B.C. and Alta B.C., respectively) (Fig. 1 and Table 1) were near the latitude of the Pacific northwest relative to North America in Late Triassic to Early Jurassic time (Irving and Wynne, 1990). Paleontologic data from ammonites in the Early Jurassic strata of these same terranes suggest they were at latitudes of Nevada to Washington (Tipper, 1981), which agrees with the paleomagnetic data. Despite the amount of translation being controversial, all data point to these terranes lying at least 1000–1500 km south of their present latitude relative to North America in the late Early Cretaceous (ca. 100 Ma) (see summary below). These paleomagnetic and paleontologic data sets seem to require major southward (sinistral) translation between Early Jurassic and Early Cretaceous time, followed by northward translation from Late Cretaceous to early Tertiary.

Like the Late Jurassic, many locations along the Cordillera in Early Cretaceous time have local evidence for either sinistral tectonics or mixed contractional and extensional structures that were interpreted to have formed in an oblique margin. This evidence in the northern Cordillera includes opening of the Canada basin and related sinistral faults in Alaska (Avé Lallement and Oldow, 1988; Oldow et al., 1989). There is evidence based on paleogeographic relations for major sinistral faulting through the arc in the Insular and Intermontane terranes of British Columbia (Monger et al., 1994). Local structural data on major strike-slip faults (Lawrence, 1978; Miller, 1988; Hurlow, 1993) support the sinistral-slip model of Monger et al. (1994). In the western United States, evidence for sinistral oblique convergence includes local sinistral transpressive faults in northeastern Oregon (Avé Lallement and Oldow, 1988). Most of these areas with evidence that

supports sinistral oblique convergence have been overprinted by later structures, and alternative explanations are also possible. In México, in the Middle to Late Jurassic, the Guerrero arc south of the Agua Blanca transform fault was migrating to the east toward North America and closing an ocean basin (Dickinson and Lawton, 2001). From ~125 to 115 Ma, this arc collided with North America (Gastil et al., 1981; Johnson et al., 1999; Schmidt et al., 2002; Wetmore et al., 2002).

What emerges from this overview is that there is evidence from many types of data that the 145–105 Ma period was likely one of sinistral oblique convergence in the Cordillera from the western United States to the north, while an arc collision was progressing in the south. This points to a major triple junction or boundary in the convergent margin at approximately the latitude of the México–California border. Most workers, however, have downplayed or not considered the sinistral oblique aspect of the convergent margin and consider western North America to be an Andean-style margin through this interval. My contention is that strike-slip faulting was an important component of the Cordillera in the Early Cretaceous and many aspects of the 145–105 Ma period are not typical of convergent margins.

SELECTED TECTONIC EVENTS FROM NORTH TO SOUTH

The paleogeographic maps presented below cover western North America from Alaska to southern México and Central America. A thorough review of the Early Cretaceous geology of this entire region is beyond this paper. Instead, in this section I discuss the main features and events from north to south that are key to the reconstructions.

Alaska

Much of central Alaska north of the Denali fault was accreted to North America by or during Early Cretaceous time (Fig. 1) (Coney et al., 1980). The large Wrangellia composite terrane (also known as Peninsula-Wrangellia-Alexander terrane and northwestern Baja B.C. in southern Alaska, see Table 1 and Fig. 1) was accreted by closing of the Kahiltna-Nutzotin ocean in early Late Cretaceous time (Jones et al., 1982; Nokleberg et al., 1994). The youngest marine rocks are Cenomanian age (ca. 95–90 Ma) and indicate the time of the closure of the Kahiltna-Nutzotin ocean (Jones et al., 1982). The final accretion of the northern Wrangellia composite terrane was during deposition (ca. 80–70 Ma) and deformation (70–60 Ma) of the largely nonmarine lower Cantwell basin (Cole et al., 1999). Closing of the Kahiltna-Nutzotin ocean was accompanied by widespread large-magnitude extension across central to northern Alaska in an arc setting (Miller and Hudson, 1991; Pavlis et al., 1993). Pavlis (1989) presented a model of the Wrangellia composite terrane approaching central Alaska and closing the Kahiltna-Nutzotin ocean in a scissor-like fashion from south in British Columbia to north in Alaska. The trench along the interior of Alaska and northeastern Canada was retreating in this model and causing the wide extensional belt much like the eastern Mediterranean to Caucasus of the late Cenozoic. Rubin et al. (1995) presented a model in which trench retreat along Alaska also played a role in the regional extension. They envisioned trench retreat being related to the opening of the Canada basin in the Arctic Ocean and to old oceanic crust of the subducting Farallon plate.

British Columbia

The large Intermontane terrane (Alta B.C. on Fig. 1 and SS + CC + QN on Fig. 2) was accreted to North America in the early Middle Jurassic (Ricketts et al., 1992). A Middle Jurassic to Early Cretaceous magmatic belt (arc?) lies on the southern Intermontane terrane in south-central British Columbia. Another magmatic belt lies along the entire Insular terrane in western British Columbia and into Alaska, where it is the Wrangellia composite terrane (WR + AX + PE on Fig. 2A) (Monger et al., 1994). This pattern appears to form a doubling up of magmatic belts in southwestern British Columbia. The magmatic belt in the Intermontane terrane is tied to North America by the Middle Jurassic, while the magmatic belt in the Insular terrane has variable interpretations for accretion from Middle Jurassic to early Cretaceous (e.g., Monger et al., 1982; Thorkelson and Smith, 1989; van der Heyden, 1992). Monger et al. (1994) presented a model where the two magmatic belts of British Columbia formed one long arc that was then cut obliquely in the late Early Cretaceous by a major sinistral fault of about 800 km offset. The large sinistral offset caused the doubling of the arc in southwestern British Columbia (Fig. 2B). Sinistral faulting is found in numerous locations in southern Baja B.C., but no through-going fault has been found. The Pasayten fault on the eastern boundary of southern Baja B.C. does show good evidence for sinistral motion from 120 to 105 Ma and is a long fault (Greig et al., 1992; Hurlow, 1993), but there are no reliable estimates of the amount of sinistral offset. Magmatism in the Insular terrane was widespread in the Middle to Late Jurassic and mid-Cretaceous but much less voluminous in the Neocomian (ca. 145 to 120 Ma) (Armstrong, 1988; Friedman and Armstrong, 1995). Based on stratigraphic changes in the Tyaughton basin and the first indication of major input of sediments from the west in Hauterivian age strata, Umhoefer et al. (2002) suggested that the sinistral faulting caused substantial doubling of the arc by Hauterivian time (ca. 130 Ma) (Fig. 2C), much earlier than Monger et al. (1994) suggested. Reasonable rates of strike-slip faulting in oblique convergent margins also seem to demand that a fault system with 800 km of offset formed in a few tens of millions of years (for example, at 20 km/m.y. the fault would have a duration of 40 m.y.).

Central California to W Idaho

The Sierra Nevada had major pulses of magmatism at 160–150 Ma and 105–85 Ma with relatively little magmatism in the intervening 150–105 Ma period, and there was particularly little from 145 to 120 Ma (Ducea, 2001). Little magmatism or defor-

mation was occurring in the Klamath Mountains in the Early Cretaceous as well (Hacker and Ernst, 1993). Isotopic evidence from granites and xenoliths, and structural data, all suggest that the Sierra Nevada arc was thrust eastward over North American crust beginning at ca.125 Ma (Figs. 3 and 4) (Ducea, 2001). Ducea further suggested that the major pulse of magmatism that occurred later at 105–85 Ma was caused by the earlier underthrusting of the arc by the volatile-rich continental crust (Fig. 4).

The major dextral strike-slip Mojave–Snow Lake fault cuts through the Sierra Nevada (Lahren et al., 1990) and continues north as the western Nevada shear zone of Wyld and Wright (2001) (Fig. 3). It may continue to the north in the Salmon River suture zone of western Idaho. The Salmon River shear zone involved the accretion of the Wallowa terrane to North America in either a contractional (Strayer et al., 1989; Selverstone et al., 1992; Manduca et al., 1993) or transpressional setting (Lund and Snee, 1988; McClelland et al., 2000). The regional dextral fault zone has 200–400 km of offset (Lahren et al., 1990; Wyld and Wright, 2001; Lewis and Girty, 2001). The timing of this fault system is not well constrained but is broadly known to be from 150 to 110 Ma. The western Nevada shear zone had dextral shear between ca. 140 and 108 Ma (Wyld and Wright, 2001), while the Salmon River shear zone was active from ca. 130 to 115 Ma (McClelland et al., 2000). A dextral transpressional shear zone cut the crest of the Sierra Nevada in middle Cretaceous time (ca. 105–85 Ma) and has up to ~40 km of offset (Busby-Spera and Saleeby, 1990; Tikoff and St. Blanquat, 1997).

Interior Western United States and Canadian Rocky Mountains

A major pulse of thrusting and foreland basin subsidence occurred in Nevada and the Utah-Idaho trough in the Middle to Late Jurassic (Fig. 3) (e.g., Bjerrum and Dorsey, 1995; Wyld et al., 2001). A similar Middle Jurassic to earliest Cretaceous succession is present in the Alberta basin, where it has been correlated to the accretion of the Intermontane terrane (Cant and Stockmal, 1989). After deposition of the Upper Jurassic to earliest Cretaceous Morrison and Kootenay formations, there are major unconformities in the foreland basin in Utah with episodic sedimentation from ca. 145 to 120 Ma (Currie, 1998) and from ca. 135 to 120 Ma in the Alberta basin in Canada (Cant and Stockmal, 1989). Many ideas have been proposed for why the Sevier foreland basin has little stratigraphic record and many major unconformities in earliest Cretaceous time, including tectonic quiescence (Heller and Paola, 1989), thermal doming before thrusting (Heller and Paola, 1989), subduction-related dynamic uplift (Lawton, 1994), and migration of a forebulge (Currie, 1998). Another major succession of foreland basin subsidence began in the early Aptian (ca. 120 Ma) in both the Sevier and Alberta basins (Heller et al., 1986; Cant and Stockmal, 1989). There was widespread thrusting and basin subsidence after middle Albian time (105–100 Ma) (Gillespie and Heller, 1995; DeCelles and Currie, 1996).

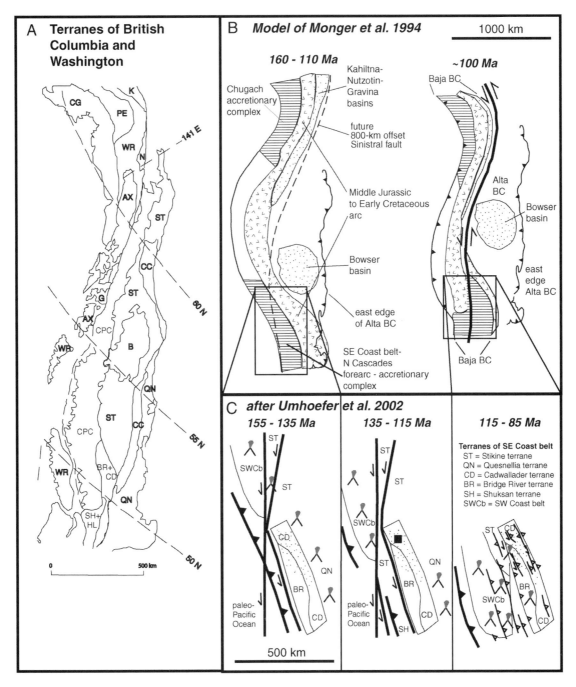

Figure 2. A: Map of terranes of northern Cordillera. Terrane names and other features: Ax—Alexander; B—Bowser basin; BR + CD—Bridge River + Cadwallader; CC—Cache Creek; CG—Chugach; CPC—Coast Plutonic Complex; G—Gravina basin; N—Nutzotin basin; K—Kahiltna basin; PE—Peninsular; QN—Quesnellia; SH+HL—Shuksan + Harrison Lake; ST—Stikine; WR—Wrangellia. B: Model of Monger et al. (1994) with a sinistral fault system with ~800 km of offset postulated to have cut obliquely across Coast and Insular belts of British Columbia and Alaska (Baja B.C.) in late Early Cretaceous. In this model, one arc system lied along continent's western margin. Northern part of arc lied on Baja B.C. block while southern end lied on Alta B.C. block. C: Model of Umhoefer et al. (2002) for same sinistral fault system in southwestern British Columbia and northwestern Washington but occurring over a longer time interval from ca. 155 to 115 Ma. Longer period of faulting is based on two lines of evidence discussed in text. Southwestern Coast belt (SWCb) and Cadwallader (CD on maps), Bridge River (BR), Shuksan (SH), and a small part of Stikine (ST) terranes form southern end of Baja B.C. block. Stikine terrane in north and arc rocks east of Cadwallader terrane on Quesnellia terrane (QN) are part of Alta B.C. block.

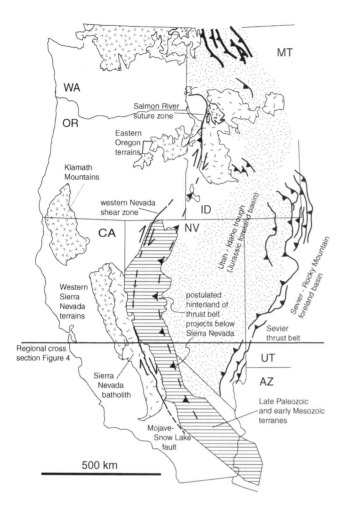

Figure 3. Map of some major tectonic belts of western United States showing some primary features used in paleogeographic maps presented here. Position of Mojave–Snow Lake fault, western Nevada shear zone, and Salmon River suture zone, all of which have been suggested to be parts of an Early Cretaceous dextral strike-slip fault system with 200–400 km of offset, are from Wyld and Wright (2000). Dashed thrust fault east of Sierra Nevada is major hinterland belt along which Sierra Nevada is postulated to have thrust east over North American crust (Ducea, 2001). Position of this hinterland thrust system is speculative.

Southern California to Northwestern México

The Peninsular Ranges batholith is a major Early Cretaceous to early Late Cretaceous magmatic arc exposed in southern California and northern Baja California, México (Fig. 1). There is also a belt of Late Jurassic plutons in northern Baja California (Schmidt et al., 2002). Gastil et al. (1981) proposed that a fringing oceanic arc lay west of a cratonal arc in the Late Jurassic to Early Cretaceous. They further suggested that the fringing arc was accreted to North America in the late Early Cretaceous. Johnson et al. (1999) refined this model with new geochronology and structural studies in northern Baja California to suggest that the suture between the arcs formed at ca. 115–108 Ma. Wetmore et al. (2002) and Schmidt et al. (2002) showed how there was a major change in the Peninsular Ranges batholith across the Agua Blanca fault near Ensenada, México. North of the Agua Blanca fault, the Santiago Peak volcanic arc contains evidence that it was formed on oceanic basement that was previously accreted to North America. The evidence includes zircons with Precambrian inheritance, subaerial volcanics, and a depositional unconformity between the arc and the basin on the west that indicates this was an intact forearc basin. South of the Agua Blanca fault, the Alisitos arc was oceanic based on no zircon inheritance, no sedimentary rocks that are continentally derived, dominantly submarine sedimentary and volcanic rocks, and juxtaposition of the arc against rocks of definite North America origin across a major thrust fault (Johnson et al., 1999; Wetmore et al., 2002). Wetmore et al. (2002) and Schmidt et al. (2002) suggested that the Alisitos was a west-facing arc that collided with North America as an ocean basin closed and that the Agua Blanca fault was a Cretaceous sinistral fault that separated the Santiago Peak and Alisitos arcs.

The Guerrero terrane of México encompasses the Peninsular Ranges batholith and a discontinuous belt of magmatic arc rocks from Baja California to southwestern México near Acapulco (Tardy et al., 1994). The magmatic rocks are Late Jurassic to Early Cretaceous in age and are correlated with similar arc rocks in the Greater Antilles and northern South America (Fig. 5) (Tardy et al., 1994; Dickinson and Lawton,

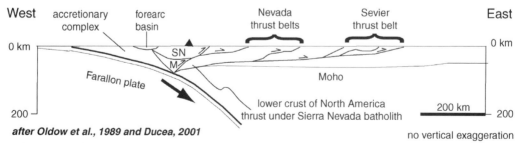

Figure 4. Possible regional cross-section from Sierra Nevada arc system to Sevier and Laramide foreland at 120–115 Ma. Sierra Nevada portion is modified from Ducea (2001), while eastern portion is modified from Oldow et al. (1989). SN—Sierra Nevada magmatic belt; M—accreted mantle material below crustal rocks of Sierra Nevada as proposed by Ducea (2001).

2001). The petrology, geochemistry, and isotopic signature of these rocks indicate that they were formed in an intraoceanic arc and rest on ophiolites. The upper part of the terrane includes Aptian (ca. 120–112 Ma) volcaniclastic strata that are overlain by Albian-Cenomanian (112–90 Ma) carbonate platforms similar to facies in eastern México (Dickinson and Lawton, 2001). The stratigraphic evidence suggests that the Guerrero terrane accreted to North America at ca. 120 Ma (Fig. 5). In a thorough review of the evidence, Dickinson and Lawton (2001) favored the Guerrero arc being an east-facing arc that approached North America and closed the oceanic Mezcalera plate (following Tardy et al.,

1994, and others) (Fig. 5, A and B). At the same time in the Early Cretaceous, the east side of the Mezcalera plate was subducting to the east under México with the trench rolling back. The Guerrero arc collided with México at ca. 120 Ma and the Farallon plate then started to subduct beneath the new western margin of México to form the middle Cretaceous arc that is superimposed on the Guerrero arc (Dickinson and Lawton, 2001). This model is similar to the conclusions of recent workers (Johnson et al., 1999; Schmidt et al., 2002; Wetmore et al., 2002) for the Peninsular Ranges batholith, except that these latter authors favor a west-facing Guerrero arc.

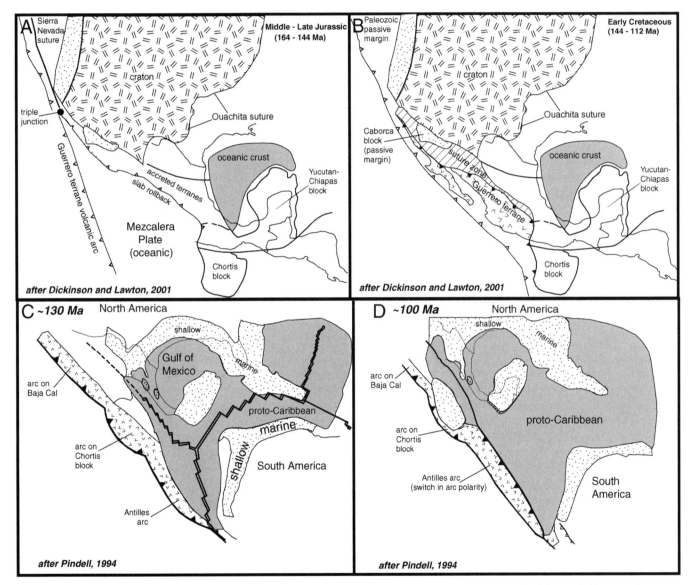

Figure 5. Reconstructions of southern Cordillera from México to Central America. A: Reconstruction modified from Dickinson and Lawton (2001) for Middle to Late Jurassic. B: Reconstruction modified from Dickinson and Lawton (2001) for Early Cretaceous. Suture zone from Guerrero arc-continent collision started to form at 120 Ma. C and D: Reconstructions from Pindell (1994) for Early Cretaceous showing interpretation that western arc (Guerrero terrane in A) was first west-facing and then between ca. 120 and 105 Ma it switched polarity and became east-facing along its southern part. East-facing arc evolved to become Greater Antilles arc as Caribbean plate formed and evolved.

Greater Antillean Arc

The Greater Antillean arc is currently found as the roots of many of the Caribbean islands on the north and south sides of the Caribbean plate (Burke, 1988; Pindell, 1993, 1994). It began as a magmatic arc in Early Cretaceous time and is postulated to have been west-facing and the southern end of the Peninsular Ranges batholith and Guerrero arc on the west side of the future Caribbean (Fig. 5C). There was a major period of thrusting in the Greater Antillean arc that led to a reversal of arc polarity by the late Albian (105–100 Ma) (Fig. 5D) (Draper et al., 1996; Snoke and Noble, 2001). Subsequently, the east-facing Antillean arc migrated to the east to form the Caribbean ocean.

PROPOSED RECONSTRUCTIONS

The time interval of most interest here is from 125 to 105 Ma, though I include a paleogeographic map for the intervals before and after for continuity. There appears to have been a major change in many parts of the Cordillera at ca. 125–120 Ma and a large increase in magmatism at ca. 105 Ma, and thus these are natural time boundaries for the reconstructions. To construct these maps, I started with the hypothesis A reconstruction of Cowan et al. (1997) for 90 Ma. Cowan et al. (1997) used latitudes relative to North America from Van Fossen and Kent (1992) and restored the following elements. They removed 275 km of extension east of the Sierra Nevada in the Basin and Range; removed 300 km of dextral slip on the San Andreas fault; the Salinian block was moved to the southern end of the Sierra Nevada by removing 500 km of dextral faulting; and the Gulf of California was closed. Changes to the Cowan et al. (1997) map needed for the present reconstructions are discussed below.

105–85 Ma

To make the reconstructions for the Early Cretaceous, one has to make a decision about the amount of northward translation of the eastern and western parts of British Columbia relative to North America (Baja B.C. hypothesis). I chose to use 1650 km of northward translation for northwestern Baja B.C. (southern and southeastern Alaska) from the latest paleomagnetic investigation of the McColl Ridge Formation in southern Alaska (1650 ± 890 km from Stamatakos et al., 2001). Because the amount of strike-slip offset on the Denali fault in southern Alaska is uncertain (Cole et al., 1999) and strike-slip on the Coast shear zone is controversial (McClelland et al., 2000), it is difficult to extrapolate the result of Stamatakos et al. (2001) to southeastern Baja B.C.

On the Figure 6 map, I used 200 km of offset on the Denali fault and ~1500 km of northward translation for southern Baja B.C. However, note that the recent paleomagnetic result (1800 ± 500 km) from the Methow block near the U.S.-Canadian border (Enkin et al., 2002) is compatible with the result from southern Alaska. This ~1500–1800 km estimate of translation is conservative as the range of suggested translation is from 1000 to 4000 km (Wynne et al., 1995; Ague and Brandon, 1996; Butler et al., 2001; Kodama and Ward, 2001; Stamatakos et al., 2001; Enkin et al., 2002). However, there are two lines of evidence for this moderate translation. A recent combined paleomagnetic and paleogeographic study of fossil rudists suggested <1500 km of translation for Baja B.C. (Kodama and Ward, 2001).

There was an extensive study of the Spences Bridge volcanic rocks (Irving et al., 1995) that lie immediately east of southern Baja B.C. across the Yalakom and related faults. The Spences Bridge rocks give a translation of 1100 ± 600 km for eastern British Columbia in the Late Cretaceous to early Tertiary. I used this 1100 km of translation for the eastern British Columbia block called Alta B.C. (Figs. 1 and 6). When the 250 km of dextral offset on the Yalakom and Fraser fault systems (Umhoefer and Schiarizza, 1996) is added to the 1100 ± 600 km, it gives 1350 ± 600 km of translation for southern Baja B.C., which broadly agrees with the ~1500 km estimate from the other studies.

The reconstruction for 105–85 Ma (Fig. 6) has subduction along all of western North America and a related volcanic arc on Baja B.C., the Peninsular Ranges batholith, and its southern extension in México. There was also magmatism on Alta B.C. and in the Sierra Nevada to western Idaho belts. Southern Baja B.C. likely accreted to North America in the Middle Jurassic (Monger et al., 1994), while northern Baja B.C. in Alaska accreted in the middle Cretaceous (during this reconstruction) as the Kahiltna-Nutzotin ocean closed (Jones et al., 1982). The northeast Washington–central British Columbia block (Alta B.C.) lay along the northern Sierra Nevada, Klamath Mountains, Blue Mountains, and the stable parts of easternmost British Columbia. This part of the reconstruction is taken directly from Cowan et al. (1997). In the present reconstruction, Baja B.C. lies along the southern Sierra Nevada and the translated Alta B.C. block.

The Kula plate formed from the northern part of the Farallon plate in this interval (Engebretson et al., 1985), and I speculate that the triple junction lay at the southern end of Baja B.C. (Umhoefer, 1987; Umhoefer et al., 1989). The Kula–North America boundary was highly oblique convergent (Engebretson et al., 1985). The Kula–Farallon–North America triple junction would have moved northward rapidly as dextral strike-slip faults formed in the arc and backarc (Umhoefer et al., 1989). I speculate that these strike-slip faults were the major Baja B.C. fault (Cowan et al., 1997) and the modest dextral faults along the crest of the Sierra Nevada to western Idaho. Because of this *northward* translation of Baja B.C. following just a few million years after the *southward* translation of Baja B.C. (in reconstructions discussed below), southern Baja B.C. lay along the California margin for a relatively short period of time in the middle Cretaceous.

In general, this was a time of a large volume of magmatism in the Sierra Nevada (Ducea, 2001) and the Baja B.C. block (Friedman and Armstrong, 1995). The area of magmatism was wider in the central and northern Cordillera (United States and Canada) than normal arcs and formed two belts. One magmatic belt was from the Sierra Nevada to Idaho and north to Alta B.C., and the other belt was on Baja B.C. These

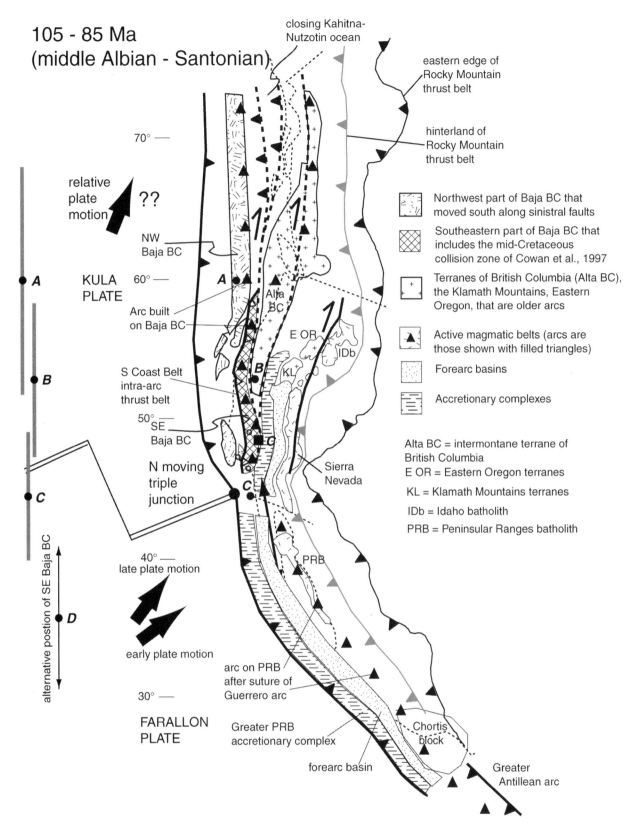

105 - 85 Ma
(middle Albian - Santonian)

closing Kahitna-
Nutzotin ocean

eastern edge of
Rocky Mountain
thrust belt

hinterland of
Rocky Mountain
thrust belt

relative
plate
motion ??

70°

NW
Baja BC

KULA
PLATE

60°

A

Arc built
on Baja BC

Alta
BC

E OR

IDb

S Coast Belt
intra-arc
thrust belt

B

KL

50°

SE
Baja BC

C

N moving
triple
junction

Sierra
Nevada

40°
late plate motion

early plate motion

30°

FARALLON
PLATE

arc on PRB
after suture of
Guerrero arc

Greater PRB
accretionary complex

forearc basin

PRB

Chortis
block

Greater
Antillean arc

alternative postion of SE Baja BC

Northwest part of Baja BC that
moved south along sinistral faults

Southeastern part of Baja BC that
includes the mid-Cretaceous
collision zone of Cowan et al., 1997

Terranes of British Columbia (Alta BC),
the Klamath Mountains, Eastern
Oregon, that are older arcs

Active magmatic belts (arcs are
those shown with filled triangles)

Forearc basins

Accretionary complexes

Alta BC = intermontane terrane of
British Columbia
E OR = Eastern Oregon terranes
KL = Klamath Mountains terranes
IDb = Idaho batholith
PRB = Peninsular Ranges batholith

Figure 6. Paleogeographic map for 105–85 Ma time interval. Creation of this map and assumptions used are discussed in text. Main features are long convergent margin with triple junction near U.S.-México border. Kula plate formed out of northern Farallon plate late in this time interval. Northernmost Cordillera had small ocean basin closing across former convergent margin. Baja B.C. block is divided in two parts based on sinistral offset on a proposed strike-slip fault system (Monger et al., 1994) that was active in previous time intervals (Figs. 7 and 8). Position of three paleomagnetic studies used here is marked by black dots and bold A for southern Alaska site (Stamatakos et al., 2001), B for Spences Bridge site on southern end of Alta B.C. (Irving et al., 1995), and C for Methow block sites on southeast Baja B.C. eastern boundary (Enkin et al., 2002). Note that dot for Methow block study (C) is at its paleolatitude and not where it lies on Baja B.C., which is shown as a black square. Latitude and analytical errors (gray bars) of three paleomagnetic studies are shown on left side. An alternative paleomagnetic interpretation based on study of Wynne et al. (1995) is shown as point D and its error is shown as black arrow on left side. Baja B.C. block included an arc, forearc basin, and accretionary complex. Other parts of Cordillera are shown as one of four main elements: accretionary complex, forearc basin, magmatic arc, or retroarc thrust belt. Large arrows on Kula and Farallon plates are motion of plates relative to North America after Engebretson et al. (1985).

two belts of magmatism do not seem compatible with a simple convergent margin.

Ducea (2001) provided an explanation for the large flux of magmatism in the Sierra Nevada that is compatible with the two belts of magmatism in my reconstruction. The large volume of magmatism in the Sierra Nevada was explained as being related to major thrusting of western North America under the Sierra Nevada (Ducea, 2001) in the hinterland of the Sevier thrust belt (Fig. 3). There was a 15–20 m.y. lag between the initiation of the thrusting under the Sierra Nevada at ca. 120 Ma and the large increase in magmatism beginning at 105 Ma. The thrusting of continental crust under the Sierra Nevada is thought to have caused a large amount of crustal melting due to a much higher content of volatiles in the overthickened crust (Ducea, 2001). This mechanism for large volumes of magmatism in the Sierra Nevada may also explain magmatism in western Idaho and the Alta B.C. block as all of these areas were immediately west of the Sevier–Rocky Mountain thrust belt (Fig. 3).

A new magmatic arc formed along the western margin of México where the Guerrero arc had previously accreted. The southern part of this arc switched polarity at the beginning of this time interval and is now east-facing and will become the Greater Antillean arc of the late Cretaceous to Neogene Caribbean.

125–105 Ma

The reconstruction is for the early part of this period, when the Guerrero arc had just accreted and the wedge of tectonic escape was forming (Fig. 7). This reconstruction presents an interpretation of the Cordillera in which apparently disparate and unrelated features are united in one model. The main features of the reconstruction are the arc-continent collision of the formerly oceanic Guerrero arc with México (Tardy et al., 1994; Dickinson and Lawton, 2001) and the tectonic escape of the central and northern Cordillera between two major strike-slip fault systems. Supporting this tectonic escape is the observation that three major tectonic belts terminated in southern California and northern Baja California. The major *dextral* fault system on the east apparently ends in the Mojave desert of southern California (Fig. 3). The southern end of Baja B.C. and therefore the major *sinistral* fault is also at the latitude of southern California according to recent estimates of the translation of Baja B.C. summarized above. The Baja B.C. block is not in its southernmost position in figure 7 because this shows a reconstruction for the early part of the 125–105 Ma time interval. At the same time that the southern ends of the two major strike-slip faults were in southern California, the northern end of the greater Guerrero arc (Peninsular Ranges batholith)–continent suture zone lay along the Agua Blanca fault also close to the latitude of southern California.

The southern end of the tectonically escaping block included part of the southeastern Coast Belt of British Columbia (southern Baja B.C.), and the Franciscan, Great Valley Group basin, and Sierra Nevada magmatic belts of California (Fig. 7). North of those blocks are the western Intermontane terrane of British Columbia on the west (Alta B.C.) and the Klamath Mountains, Blue Mountains, and eastern Intermontane terrane on the east. The block was escaping along a major sinistral fault system on its west that cut obliquely through Baja B.C. I assign 400 km of the total 800 km of offset on this sinistral fault system (Monger et al., 1994) during this period. The 800 km of sinistral slip is spread over 40 m.y. (145–105 Ma) to give a reasonable rate of intra-arc faulting and to honor the data from the Tyaughton basin (Umhoefer et al., 2002). I also put 300–400 km of sinistral offset on the Pasayten fault that lay between southern Baja B.C. and the Klamaths and southern Alta B.C. This offset on the Pasayten fault is totally speculative as there are no known offset markers, but there is good evidence for sinistral slip. On the east side of the escaping block, a major dextral fault system runs through the Sierra Nevada and up through northwestern Nevada and along the western side of Idaho with 200–400 km of offset. Thus, the Pasayten and eastern dextral faults may have had a similar amount of offset but the opposite sense of motion.

The northern Cordillera had a major extensional belt along the east side of the escaping block (now in Yukon and east-central Alaska) and a convergent margin with slab rollback along the west side. The entire margin of western North America had a convergent margin with the southern part being a continental arc and the northern part being an oceanic arc. The Kahiltna-Nutzotin ocean was closing east of the northern arc. Magmatism started to be more rigorous in the Baja B.C. block at 115–110 Ma, perhaps because of the more orthogonal motion of the converging Farallon plate (Fig. 7).

The southern Cordillera had a major episode of east-vergent thrusting in México during and after suturing of the Guerrero terrane to México. This is the southern part of the Cordillera-long Sevier–Rocky Mountain thrust belt, which had a major increase in activity starting in the Aptian, or early in this period, at the same time as the arc-continent collision.

145–125 Ma

This reconstruction (Fig. 8) combines the sinistral oblique convergence in the northern Cordillera (Avé Lallement and Oldow, 1988; Monger et al., 1994) with the paleogeography proposed by Dickinson and Lawton (2001) for México. In the north, the first 400 km of the 800 km of sinistral faulting suggested by Monger et al. (1994) is inferred to have started at the beginning of the Cretaceous. This early initiation of sinistral faulting through Baja B.C. is supported by the first firm evidence of a major source of sediments from the west in the Tyaughton basin in Hauterivian strata (ca. 130 Ma) (Umhoefer et al., 2002). The Tyaughton basin lies on the southeastern Baja B.C. block and would have sat immediately east of the proposed sinistral fault system of Monger et al. (1994). To move the block out to the west of the Tyaughton basin, the sinistral faulting must have begun well before 130 Ma. The inferred Farallon plate motion relative to North America suggests a component of southward oblique convergence along the northern Cordillera early in this

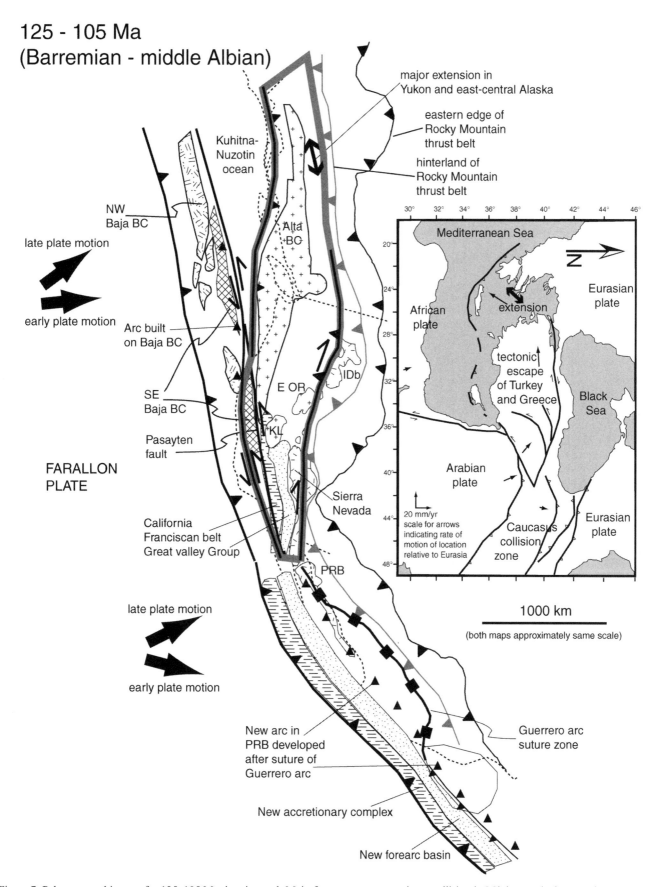

125 - 105 Ma
(Barremian - middle Albian)

major extension in
Yukon and east-central Alaska

eastern edge of
Rocky Mountain
thrust belt

hinterland of
Rocky Mountain
thrust belt

Kuhitna-
Nuzotin
ocean

NW
Baja BC

late plate motion

early plate motion

Arc built
on Baja BC

SE
Baja BC

Pasayten
fault

FARALLON
PLATE

California
Franciscan belt
Great valley Group

late plate motion

early plate motion

New arc in
PRB developed
after suture of
Guerrero arc

New accretionary complex

New forearc basin

Alta
BC

E OR

IDb

KL

Sierra
Nevada

PRB

Guerrero arc
suture zone

Inset

Mediterranean Sea

N

extension

African
plate

Eurasian
plate

tectonic
escape
of Turkey
and Greece

Black
Sea

Arabian
plate

20 mm/yr
scale for arrows
indicating rate of
motion of location
relative to Eurasia

Caucasus
collision
zone

Eurasian
plate

30° 32° 34° 36° 38° 40° 42° 44° 46°

20°
24°
28°
32°
36°
40°
44°
48°

1000 km

(both maps approximately same scale)

Figure 7. Paleogeographic map for 125–105 Ma time interval. Main features are arc-continent collision in México producing a major suture zone and tectonic escape of central and northern Cordillera away from collision zone (zone of escape is bounded by wide gray line). Text explains how this map was created from previous map (Fig. 6). Inset is map of eastern Mediterranean to Caucasus region turned so that north is at left; simplified after Şengör et al. (1985) and Dewey et al. (1986). Both maps are approximately same scale. Note similar basic patterns of collision and tectonic escape between western North America at ca. 120 Ma and present eastern Mediterranean region. The two regions differ in that one is an arc-continent collision and other is a continent-continent collision, which is addressed in text. Symbols and abbreviations are same as in Figure 6.

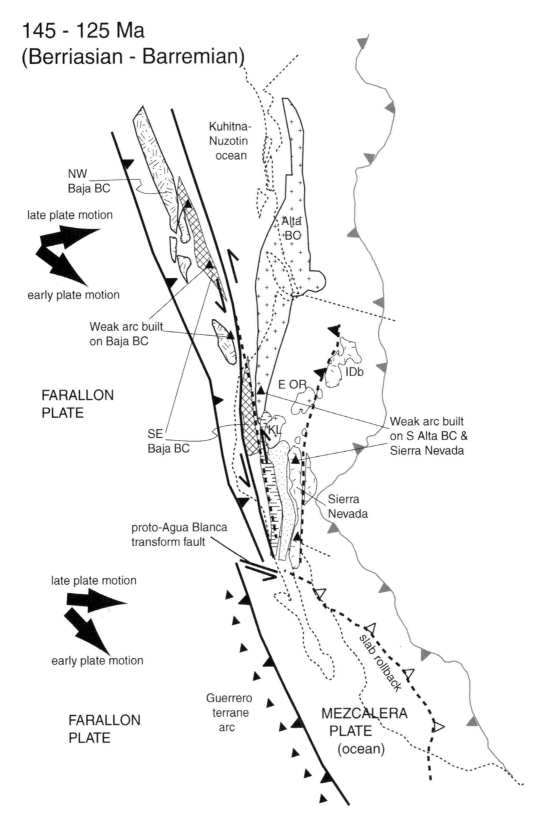

145 - 125 Ma
(Berriasian - Barremian)

Kuhitna-
Nuzotin
ocean

NW
Baja BC

late plate motion

early plate motion

Alta
BC

Weak arc built
on Baja BC

IDb

FARALLON
PLATE

E OR

SE
Baja BC

KL

Weak arc built
on S Alta BC &
Sierra Nevada

proto-Agua Blanca
transform fault

Sierra
Nevada

late plate motion

slab rollback

early plate motion

FARALLON
PLATE

Guerrero
terrane
arc

MEZCALERA
PLATE
(ocean)

Figure 8. Paleogeographic map for 145–125 Ma time interval. Main features are a sinistral oblique convergent margin in northern Cordillera and a doubly subducting plate (Mezcalera plate, after Dickinson and
Lawton, 2001) disappearing in southern Cordillera. Text explains how this map was created from previous
map (Fig. 7). Mezcalera plate was nearly subducted by end of this time period. There was little magmatism in
arc and only minor thrust faulting in backarc region during this phase of oblique convergence. Symbols and
abbreviations are same as in Figure 6.

period, which is compatible with sinistral faulting. There was much reduced magmatism within the northern Cordillera during this time, and the Sevier–Rocky Mountain thrust belt may have experienced reduced activity, as is suggested by major unconformities and thin conglomerates.

In the southern Cordillera, the Mezcalera plate was subducting east beneath México with slab rollback and little magmatism. The Mezcalera plate was decreasing in size as it was also subducting to the west beneath the oceanic Guerrero arc. The Farallon plate was moving southward relative to North America, but its motion relative to the Mezcalera plate is unknown.

DISCUSSION

Alternatives to the Reconstructions

There are clearly many alternatives to speculative reconstructions like those presented here. Here I briefly discuss selected alternative interpretations and how they would affect the reconstructions.

For the 105–85 Ma map, one alternative is that the Baja B.C. block may have had less translation (Butler et al., 2001) so that it lay mainly north of the Sierra Nevada. The two magmatic belts overlap along a shorter distance in this model. This interpretation also negates the tectonic escape model for the 125–105 Ma period, but it still has two large strike-slip fault systems with opposing sense of slip at the same latitude. Using the interpretations with larger translation of Baja B.C. (3000–4000 km, e.g., Wynne et al., 1995; Cowan et al., 1997) puts much of Baja B.C. along the western side of México. This then begs the question of why there were two magmatic belts at that latitude, though the Ducea (2001) model may apply here as well. In addition, the agreement of the position of the southern end of Baja B.C. in the Triassic–early Jurassic based on paleomagnetic data and the 800 km of subsequent southward motion works well in my model, but in the larger translation of Baja B.C., ~2000 km of southward motion is missing.

As summarized above, many workers suggest that the Guerrero arc was west-facing and not east-facing as Dickinson and Lawton (2001) suggested and I used. Changing the polarity of this arc is not critical for my model. A west-facing Guerrero arc would cause lower rates of closing of the Mezcalera ocean. A west-facing oceanic arc would cause the arc to collide with the accretionary complex along North America, while in the model discussed here two accretionary complexes collide initially.

An Eastern Mediterranean Analog

The 125–105 Ma reconstruction of western North America with an arc-continent collision in the south and tectonic escape in the central and northern Cordillera has many aspects similar to the modern eastern Mediterranean to Caucasus region (Fig. 7). This analog is an expansion and variation of a model proposed by Pavlis (1989) for the northern Cordillera. In this comparison, the

Arabian-Eurasia collision zone in the Caucasus region and Iran (Şengör et al., 1985; Dewey et al., 1986) is similar to the Guerrero arc-continent collision. The tectonic escape of Turkey to the west, away from the collision zone, is similar to the northward escape proposed here for the central to northern Cordillera. The two settings also are similar in that the major strike-slip faults of opposite sense of slip converge toward the outer edge of the collision zone, and the tectonic escape is lateral relative to the collision zone. In the reconstruction for western North America, it is clear that the *northern* end of the Guerrero arc-continent collision was in southern California and that the *southern* end of the major Mojave–Snow Lake–Western Nevada–Salmon River dextral fault was likely in the Mojave region of southern California. Placing the southern end of the major sinistral fault system also in southern California is dependent on the "moderate" reconstruction of Baja B.C. used here. Using reconstructions of Baja B.C. much farther south or north would not be compatible with the tectonic escape model. The major extension in Greece today is similar to the major extension across interior Alaska and northwestern Canada as Pavlis (1989) noted. The free facing convergent plate boundary south of Crete and Greece facilitates escape as the trench is retreating (Dewey, 1988). In the northern Cordillera, Pavlis (1989) and Rubin et al. (1995) suggested that the subduction zone along interior Alaska (and the northern part of Alta B.C. on Fig. 7) was likely to have been retreating in the Early Cretaceous.

There are also important differences between the present model and the modern eastern Mediterranean. In detail, the northern Cordillera is different than the eastern Mediterranean because there was a second convergent plate boundary along the west side of the Baja B.C. (Wrangellia composite terrane) and the Kuhitna-Nutzotin ocean was closing between the two convergent boundaries (Fig. 7). Another critical difference between the Early Cretaceous western North America and the modern eastern Mediterranean to Caucasus is that the colliding block, the Arabian plate, is much larger than the Guerrero oceanic arc. This begs the question of whether there were other major forces driving the escape in western North America. The free-facing plate boundary in Alaska with possible trench retreat would have promoted tectonic escape of the central Cordillera toward the north. In addition, a major change in the absolute direction of motion of the North America plate favored tectonic escape and related events. Note that in the Cordillera at ca. 120 Ma, the northward tectonic escape was accompanied by the start of major east-vergent thrusting across the backarc region. This included the major thrusting of the Sierra Nevada arc eastward over North American crust and the renewal of major thrusting and foreland basin subsidence in the Sevier–Rocky Mountain thrust belt discussed above.

These events at ca. 125–120 Ma were likely driven by a profound change in the motion of North America as well as the coincident collision of the Guerrero arc in the southern Cordillera. From ca. 180–125 Ma, the North America plate was moving northwestward away from Africa as the central Atlantic opened (Morgan, 1983; Engebretson et al., 1985; Klitgord and Schouten, 1986). At ca. 125–120 Ma, the North America plate changed to a more westerly absolute motion (Scotese, 2002). That change in plate motion may have been the main cause of the increased thrusting in the North American Cordillera and promoted the tectonic escape. The initial opening of the equatorial Atlantic at ~120 Ma (Jones et al., 1995) may have triggered the change in the motion of the North America plate.

Finally, we can ask, why did the Guerrero terrane and southern Cordillera not start to move north in the 105–85 Ma interval, as did the central and northern Cordillera? I suggest that this is because the southern boundary of the Kula plate formed at the previous boundary at the northern end of the colliding Guerrero arc. In addition, the central and northern Cordillera were cut by numerous large strike-slip faults (Figs. 7 and 8) that were readily reactivated as dextral-slip faults along which northward motion occurred.

ACKNOWLEDGMENTS

Over the years, there have been way too many colleagues who discussed Mesozoic paleogeography and the problem of Baja B.C. to mention them all here, so a few key people who sparked this paper will have to do. The organizers of the Penrose conference on terrane translation, Brian Mahoney, Julie Maxson, and Basil Tikoff, are thanked for a stimulating conference that rekindled my interest in that general problem and Baja B.C. specifically. The coincidence of my reading at essentially the same time in 2001 the important papers by Dickinson and Lawton, Wyld and Wright, and Ducea jelled the ideas for this paper and I thank them for three excellent papers. The recent important paper on the paleomagnetism of the Methow block by Randy Enkin and others also encouraged me. I thank Darrel Cowen for many fruitful discussions on Mesozoic paleogeography over the years and for providing the figures from his seminal 1997 paper. Darrel Cowen and Sandra Wyld are thanked for thoughtful reviews, and Scott Paterson was patient and an excellent editor. Finally, I thank Eldridge Moores for being an inspiring pioneer with papers on the Cordillera in which he dares to face the probable complexity of the mountain belt head on, and I thank Gordon Gastil for many discussions of the geology of northwestern México, both in the Mesozoic and the Cenozoic.

REFERENCES CITED

Ague, J.J., and Brandon, M.T., 1996, Regional tilt of the Mt. Stuart batholith, Washington, determined using aluminum-in-hornblende barometry: Implications for northward translation of Baja British Columbia: Geological Society of America Bulletin, v. 108, p. 471–488.

Anderson, T.H., and Schmidt, V.A., 1983, The evolution of Middle America and the Gulf of Mexico-Caribbean Sea region during Mesozoic time: Geological Society of America Bulletin, v. 94, p. 941–966.

Armstrong, R.L., 1988, Mesozoic and early Cenozoic magmatic evolution of the Canadian Cordillera: Boulder, Colorado, Geological Society of America Special Paper 218, p. 55–91.

Avé Lallemant, H.G., and Oldow, J.S., 1988, Early Mesozoic southward migration of Cordilleran transpressional terranes: Tectonics, v. 7, p. 1057–1088.

Bjerrum, C.J., and Dorsey, R.J., 1995, Tectonic controls on deposition of Middle Jurassic strata in a retroarc foreland basin, Utah-Idaho trough, western interior, United States: Tectonics, v. 14, p. 962–978.

Burchfiel, B.C., Cowan, D.S., and Davis, G.A., 1992, Tectonic overview of the Cordillera orogen in the western United States, in Burchfiel, B.C.,

Lipman, P.W., and Zoback, M.L., eds., The Cordilleran Orogen: Conterminous U.S.: Boulder, Colorado, Geological Society of America, Geology of North America, v. G-3, p. 407–479.

Burke, K., 1988, Tectonic evolution of the Caribbean: Annual Reviews of Earth and Planetary Science, v. 16, p. 201–230.

Busby-Spera, C.J., and Saleeby, J.B., 1990, Intra-arc strike-slip fault exposed at batholithic levels in the southern Sierra Nevada, California: Geology, v. 18, p. 255–259.

Butler, R.F., Gehrels, G.E., and Kodama, K.P., 2001, A moderate translation alternative to the Baja British Columbia hypothesis: GSA Today, v. 11, no. 6, p. 4–10.

Cant, D.J., and Stockmal, G.S., 1989, The Alberta foreland basin: Relationship between stratigraphy and Cordilleran terrane-accretion events: Canadian Journal of Earth Sciences, v. 26, p. 1964–1975.

Cole, R.B., Ridgway, K.D., Layer, P.W., Drake, J., 1999, Kinematics of basin development during the transition from terrane accretion to strike-slip tectonics, Late Cretaceous–early Tertiary Cantwell Formation, south central Alaska: Tectonics, v.18, p. 1224–1244.

Coney, P.J., Jones, D.L., and Monger, J.W.H., 1980, Cordilleran suspect terranes: Nature, v. 188, p. 329–333.

Cowan, D.S., 1994, Alternative hypotheses for the mid-Cretaceous paleogeography of the western Cordillera: GSA Today, v. 4, p. 181–186.

Cowan, D.S., Brandon, M.T., and Garver, J.I., 1997. Geological tests of hypotheses for large coastwise displacements: a critique illustrated by the Baja British Columbia controversy: American Journal of Science, v. 297, p. 117–173.

Currie, B.S., 1998, Upper Jurassic–Lower Cretaceous Morrison and Cedar Mountain formations, NE Utah–NW Colorado: Relationships between nonmarine deposition and early Cordilleran foreland-basin development: Journal of Sedimentary Research, v. 68, p. 632–652.

DeCelles, P.G., and Currie, B.S., 1996, Long-term sediment accumulation in the Middle Jurassic–early Eocene Cordilleran retroarc foreland-basin system: Geology, v. 24, p. 591–594.

Dewey, J.F., 1988, Extensional collapse of orogens: Tectonics, v. 7, p.1123–1139.

Dewey, J.F., Hempton, M.R., Kidd, W.S.F., Saroglu, F., and Şengör, A.M.C., 1986, Shortening of continental lithosphere: The neotectonics of Eastern Anatolia—a young collision zone, in Coward, M.P., and Ries, A.C., eds., Collision tectonics: Geological Society of London Special Publication 19, p. 3–36.

Dickinson, W.R., and Lawton, T.F., 2001, Carboniferous to Cretaceous assembly and fragmentation of Mexico: Geological Society of America Bulletin, v. 113, p. 1142–1160.

Draper, G., Gutiérrez, G., and Lewis, J.F., 1996, Thrust emplacement of the Hispaniola peridotite belt: Orogenic expression of the mid-Cretaceous Caribbean arc polarity reversal?: Geology, v. 24, p. 1143–1146.

Ducea, M., 2001, The California arc: Thick granite batholiths, eclogitic residues, lithospheric-scale thrusting, and magmatic flare-ups: GSA Today, v. 11, no. 11, p. 4–10.

Engebretson, D.G., Cox, A., and Gordon, R.G., 1985, Relative motions between oceanic and continental plates in the Pacific basin: Boulder, Colorado, Geological Society of America Special Paper 206, 59 p.

Enkin, R.J., Mahoney, J.B., Baker, J., Liessling, M., and Haugerud, R.A., 2002, Syntectonic remagnetization in the southern Methow block: Resolving large displacements in the southern Canadian Cordillera: Tectonics, v. 21, p. 18-1–18-18.

Friedman, R.M., and Armstrong, R.L., 1995, Jurassic and Cretaceous geochronology of the southern Coast Belt, British Columbia, 49°-51°N, in Miller, D.M., and Busby, C., eds., Jurassic magmatism and tectonics of the North American Cordillera: Boulder, Colorado, Geological Society of America Special Paper 299, p. 95–139.

Gastil, R.G., Morgan, G.J., and Krummenacher, D., 1981, The tectonic history of peninsular California and adjacent Mexico, in Ernst, W.G., ed., The geotectonic development of California (Rubey volume I): Englewood Cliffs, New Jersey, Prentice Hall, p. 284–306.

Gillespie, J.M., and Heller, P.L., 1995, Beginning of foreland subsidence in the Columbian-Sevier belts, southern Canada and northwest Montana: Geology, v. 23, p. 723–726.

Greig, C.J., Armstrong, R.L., Harakal, J.E., Runkle, D., and van der Heyden, P., 1992, Geochronology of the Eagle plutonic complex and the Coquihalla area, southwestern British Columbia: Canadian Journal of Earth Sciences, v. 29, p. 812–829.

Hacker, B.R., and Ernst, W.G., 1993, Jurassic orogeny in the Klamath Mountains; a geochronological analysis, in Dunne, G.C., and McDougall, K.A., eds., Mesozoic paleogeography of the Western United States; II: Los Angeles, California, SEPM (Society for Sedimentary Geology) Pacific Section, v. 71, p. 37–59.

Harper, G.D., and Wright, J.E., 1984, Middle to Late Jurassic tectonic evolution of the Klamath Mountains, California-Oregon: Tectonics, v. 3, p. 759–772.

Heller, P.L., Bowdler, S.S., Chambers, H.P., Coogan, J.C., Hagen, E.S., Shuster, M.W., Winslow, N.S., and Lawton, T.F., 1986, Time of initial thrusting in the Sevier orogenic belt, Idaho-Wyoming and Utah: Geology, v. 14, p. 388–391.

Heller, P.L., and Paola, C., 1989, The paradox of Lower Cretaceous gravels and the initiation of thrusting in the Sevier orogenic belt, United States Western Interior: Geological Society of America Bulletin, v. 101, p. 864–875.

Hurlow, H.A., 1993, Mid-Cretaceous strike-slip and contractional fault zones in the western Intermontane terrane, Washington, and their relation to the North Cascades–southwestern Coast Belt orogen: Tectonics, v. 12, p. 1240–1257.

Irving, E., and Wynne, P.J., 1990, Paleomagnetic evidence bearing on the evolution of the Canadian Cordillera: Philosophical Transactions of the Royal Society of London, v. A331, p. 487–509.

Irving, E., Woodsworth, G.J., Wynne, P.J., and Morrison, A., 1985, Paleomagnetic evidence for displacement from the south of the Coast Plutonic Complex, British Columbia: Canadian Journal of Earth Sciences, v. 22, p. 584–598.

Irving, E., Thorkelson, D.J., Wheadon, P.M., and Enkin, R.J., 1995, Paleomagnetism of the Spences Bridge Group and northward displacement of the Intermontane Belt, British Columbia; a second look: Journal of Geophysical Research, v. 100, p. 6057–6071.

Irving, E., Wynne, P.J., Thorkelson, D.J., and Schiarizza, P., 1996, Large (1000 to 4000 km) northward movements of tectonic domains in the northern Cordillera, 83 to 45 Ma: Journal of Geophysical Research, v. 101, p. 17901–17916.

Johnson, S.E., Tate, M.C., and Fanning, C.M., 1999, New geologic mapping and SHRIMP U-Pb zircon data in the Peninsular Ranges batholith, Baja California, Mexico: Evidence for a suture?: Geology, v. 27, p. 743–746.

Jones, D.L., Silberling, N.J., Gilbert, W., and Coney, P.J., 1982, Character, distribution, and tectonic significance of accretionary terranes in the central Alaska Range: Journal of Geophysical Research, v. 87, p. 3709–3717.

Jones, E.J.W., Cande, S.C., and Spathopoulos, F., 1995, Evolution of a major oceanographic pathway: The equatorial Atlantic, *in* Scrutton, R.A., Stoker, M.S., Shimmield, G.B., and Tudhope, A.W., eds., The tectonics, sedimentation and palaeoceanography of the North Atlantic region: Geological Society [London] Special Publication 90, p. 199–213.

Klitgord, K.D., and Schouten, H., 1986, Plate kinematics of the central Atlantic, *in* Vogt, P.R., and Tocholke, B.E., eds., The western North Atlantic region: Boulder, Colorado, Geological Society of America, Geology of North America, v. M, p. 351–378.

Kodama, K.P., and Ward, P.D., 2001, Compaction-corrected paleomagnetic paleolatitudes for Late Cretaceous rudists along the Cretaceous California margin: Evidence for less than 1500 km of post-Late Cretaceous offset for Baja British Columbia: Geological Society of America Bulletin, v. 113, p. 1171–1178.

Lahren, M.M., Schweickert, R.A., Mattinson, J.M., and Walker, J.D., 1990, Evidence of uppermost Proterozoic to Lower Cambrian miogeoclinal rocks and the Mojave-Snow Lake fault: Snow Lake pendant, central Sierra Nevada, California: Tectonics, v. 9, p. 1585–1608.

Lawrence, R.D., 1978, Tectonic significance of petrofabric studies along the Chewack-Pasayten fault, north-central Washington: Geological Society of America Bulletin, v. 89, p. 731–743.

Lawton, T.F., 1994, Tectonic setting of Mesozoic sedimentary basins, Rocky Mountain region, United States, *in* Caputo, M.V., Peterson, J.A., and Franczyk, K.J., eds., Mesozoic systems of the Rocky Mountain region, USA: Denver, Colorado, SEPM (Society for Sedimentary Geology) Rocky Mountain Section, p. 1–25.

Lewis, J.G., and Girty, G.H., 2001, Tectonic implications of a petrographic and geochemical characterization of the Lower to Middle Jurassic Sailor Canyon Formation, northern Sierra Nevada, California: Geology, v. 29, p. 627–630.

Lund, K., and Snee, L.W., 1988, Metamorphism, structural development, and age of the continental-island arc juncture in west-central Idaho, *in* Ernst, W.G., ed., Metamorphism and crustal evolution, western conterminous United States, Rubey Volume VII: Englewood Cliffs, New Jersey, Prentice Hall, p. 296–331.

Manduca, C.A., Kuntz, M.A., and Silver, L.T., 1993, Emplacement and deformation history of the western margin of the Idaho batholith near McCall, Idaho:

Influence of a major terrane boundary: Geological Society of America Bulletin, v. 105, p. 749–765.

McClelland, W.C., Tikoff, B., and Manduca, C.A., 2000, Two-phase evolution of accretionary margins: Examples from the North America Cordillera: Tectonophysics, v. 326, p. 37–55.

Miller, M.G., 1988, Possible pre-Cenozoic left-lateral slip on the Yalakom fault, southwestern British Columbia: Geology, v. 16, p. 584–587.

Miller, E.L., and Hudson, T.L., 1991, Mid-Cretaceous extensional fragmentation of a Jurassic–Early Cretaceous compressional orogen: Tectonics, v. 10, p. 781–796.

Monger, J.H.W., Price, R.A., and Tempelman-Kluit, D.J., 1982, Tectonic accretion and the origin of the two major metamorphic and plutonic welts in the Canadian Cordillera: Geology, v. 10, p. 70–75.

Monger, J.W.H., van der Heyden, P., Journeay, J.M., Evenchick, C.A., and Mahoney, J.B., 1994, Jurassic-Cretaceous basins along the Canadian Coast Belt; their bearing on pre-Mid-Cretaceous sinistral displacements: Geology, v. 22, p. 175–178.

Moores, E.M., 1998 Ophiolites, the Sierra Nevada, "Cordillera," and orogeny along the Pacific and Caribbean margins of North and South America: International Geology Review, v. 40, p. 40–54.

Morgan, W.J., 1983, Hotspot tracks and the early rifting of the Atlantic: Tectonophysics, v. 94, p. 123–139.

Nokleberg, W.J., Plafker, G., Wilson, F.H., 1994, Geology of south-central Alaska, *in* Plafker, G., and Berg, H.C., eds., The geology of Alaska: Boulder, Colorado, Geological Society of America, Geology of North America, v. G-1, p. 311–366.

Oldow, J.S., Bally, A.W., Avé Lallemant, H.G., and Leeman, W.P., 1989, Phanerozoic evolution of the North American Cordillera, United States and Canada, *in* Bally, A.W., and Palmer, A.R., eds., Geology of North America—An overview: Boulder, Colorado, Geological Society of America, Geology of North America, v. A, p. 139–232.

Pavlis, T.L., 1989, Middle Cretaceous orogenesis in the northern Cordillera: A Mediterranean analog of collision-related extensional tectonics: Geology, v. 17, p. 947–950.

Pavlis, T.L., Plafker, G., Sisson, V.B., Foster, H.L., and Nokleberg, W.J., 1993, Mid-Cretaceous extensional tectonics of the Yukon-Tanana terrane, Trans-Alaska crustal transect (TACT), east-central Alaska: Tectonics, v. 12, p. 103–122.

Pindell, J.L., 1993, Regional synopsis of Gulf of Mexico and Caribbean evolution, *in* Pindell, J.L., and Perkins, B.F., eds., 13th Annual Research Conference Proceedings, Mesozoic and Early Cenozoic development of the Gulf of Mexico and Caribbean region: SEPM (Society for Sedimentary Geology) Gulf Coast Section, p. 251–274.

Pindell, J.L., 1994, Evolution of the Gulf of Mexico and the Caribbean, *in* Caribbean Geology: An Introduction: Kingston, Jamaica, University of West Indies Publishers Association, p. 13–39.

Ricketts, B.D., Evenchick, C.A., Anderson, R.G., and Murphy, D.C., 1992, Bowser basin, northern British Columbia: Constraints on the timing of initial subsidence and Stikinia– North America terrane interactions: Geology, v. 20, p. 1119–1122.

Rubin, C.M., Miller, E.L., and Toro, J., 1995, Deformation of the northern circum-Pacific margin: Variations in tectonic style and plate-tectonic implications: Geology, v. 23, p. 897–900.

Saleeby, J.B., and Busby-Spera, C., 1992, Early Mesozoic tectonic evolution of the western U.S., Cordillera, *in* Burchfiel, B.C., Lipman, P.W., and Zoback, M.L., eds., The Cordilleran Orogen: Conterminous U.S.: Boulder, Colorado, Geological Society of America, Geology of North America, v. G-3, p. 107–168.

Schermer, E.R., Howell, D.G., and Jones, D.L., 1984, The origin of allochthonouos terranes: Perspectives on the growth and shaping of continents: Annual Review of Earth and Planetary Science, v. 12, p. 107–131.

Schermer, E.R., Stephens, K.A., and Walker, J.D., 2001, Paleogeographic and tectonic implications of the geology of the Tiefort Mountains, northern Mojave Desert, California: Geological Society of America Bulletin, v. 113, p. 920–938.

Schmidt, K.L., Wetmore, P.H., Paterson, S.R., and Johnson, S.E., 2002, Controls on orogenesis along an ocean-continent margin transition in the Jura-Cretaceous Peninsular Ranges batholith: Boulder, Colorado, Geological Society of America Special Paper 365, p. 49–71.

Scotese, C.R., 2002, Plate Tectonic animation, Jurassic to Quaternary: http://www.scotese.com, (PALEOMAP Web site) (January 2002).

Selverstone, J., Wernicke, B., and Aliberti, E., 1992, Intracontinental subduction and hinged uplift along the Salmon River suture zone in west central Idaho: Tectonics, v. 11, p. 355–372.

Şengör, A.M.C., Gorur, N., and Saroglu, F., 1985, Strike-slip faulting and related basin formation in zones of tectonic escape: Turkey as a case study, *in* Biddle, K.T., and Christie-Blick, N., eds., Strike-slip deformation, basin formation, and sedimentation: SEPM (Society for Sedimentary Geology) Special Publication 37, p. 227–264.

Silberling, N.J., Jones, D.L., Monger, J.W.H., and Coney, P.J., 1992, Lithotectonic terrane map of the North America Cordillera: U.S. Geological Survey Miscellaneous Investigation Series, Map I-2176, scale 1:500,000, 2 sheets.

Snoke, A.W., and Noble, P.J., 2001, Ammonite-radiolarian assemblage, Tobago Volcanic Group, Tobago, West Indies—Implications for the evolution of the Great Arc of the Caribbean: Geological Society of America Bulletin, v. 113, p. 256–264.

Stamatakos, J.A., Trop, J.M., and Ridgway, K.D., 2001, Late Cretaceous paleogeography of Wrangellia: Paleomagnetism of the MacColl Ridge Formation, southern Alaska, revisited: Geology, v. 29, p. 947–950.

Strayer, L.M.IV, Hyndman, D.W., Sears, J.W., and Myers, P.E., 1989, Direction and shear sense during suturing of the Seven Devils—Wallowa terrane against North America in western Idaho: Geology, v. 17, p. 1025–1028.

Tardy, M., Lapierre, H., Freydier, C., Coulon, C., Gill, J.-B., Mercier de Lepinay, B., Beck, C., Martinez, J., Talavera, O., Ortiz, E., Stein, G., Bourdier, J.-L., and Yta, M., 1994, The Guerrero suspect terrane (western Mexico) and coeval arc terranes (the Greater Antilles and the Western Cordillera of Colombia): A late Mesozoic intra-oceanic arc accreted to cratonal America during the Cretaceous: Tectonophysics, v. 320, p. 49–73.

Thorkelson, D.J., and Smith, A.D., 1989, Arc and intraplate volcanism in the Spences Bridge Group: Implications for Cretaceous tectonics in the Canadian Cordillera: Geology, v. 17, p. 1093–1096.

Tikoff, B., and de St. Blanquat, M., 1997, Transpressional shearing and strike-slip partitioning in the Late Cretaceous Sierra Nevada magmatic arc, California: Tectonics, v. 16, p. 442–459.

Tipper, H.W., 1981, Offset of an upper Pliensbachean geographic zonation in the North America Cordillera by transcurrent movement: Canadian Journal of Earth Sciences, v. 18, p. 1788–1792.

Umhoefer, P.J., 1987, Northward translation of "Baja British Columbia" along the Late Cretaceous to Paleocene margin of western North America: Tectonics, v. 6, p. 377–394.

Umhoefer, P.J., 2000, Where are the missing faults in translated terranes?: Tectonophysics, v. 326, p. 23–35.

Umhoefer, P.J., and Schiarizza, P., 1996, Southeastern part of the Late Cretaceous to early Tertiary dextral strike-slip Yalakom fault system, southwestern British Columbia and implications for regional tectonics: Geological Society of America Bulletin, v. 108, p. 768–785.

Umhoefer, P.J., Dragovich, J., Cary, J., and Engebretson, D.C., 1989, Refinements of the "Baja British Columbia" plate-tectonic model for northward translation along the margin of western North America, *in* Hillhouse, J.W., ed., Deep structure and past kinematics of accreted terranes: American Geophysical Union Geophysical Monograph/International Union of Geodesy and Geophysics, v. 5, p. 101–111.

Umhoefer, P.J., Schiarizza, P., and Robinson, M., 2002, Relay Mountain Group, Tyaughton-Methow Basin, southwest British Columbia: A major Middle Jurassic to Early Cretaceous terrane-overlap assemblage: Canadian Journal of Earth Sciences, v. 39, p. 1143–1167.

van der Heyden, P., 1992, A Middle Jurassic to early Tertiary Andean-Sierran subduction model for the Coast Belt of British Columbia: Tectonics, v. 11, p. 82–97.

Van Fossen, M.C., and Kent, D.V., 1992, Paleomagnetism of the Front Range (Colorado) Morrison Formation and an alternative model of Late Jurassic North American apparent polar wander: Geology, v. 20, p. 223–226.

Wetmore, P.H., Schmidt, K.L., Paterson, S.R., and Herzig, C., 2002, Tectonic implications for the along-strike variation of the Peninsular Ranges batholith, southern and Baja California: Geology, v. 30, p. 247–250.

Wyld, S.J., and Wright, J.E., 1988, The Devils Elbow ophiolite remnant and overlying Galice Formation: New constraints on the Middle to Late Jurassic evolution of the Klamath Mountains, California: Geological Society of America Bulletin, v. 100, p. 29–44.

Wyld, S.J., and Wright, J.E., 2001, New evidence for Cretaceous strike-slip faulting in the United States Cordillera and implications for terrane-displacement, deformation patterns, and plutonism: American Journal of Science, v. 301, p. 150–181.

Wyld, S.J., Rogers, J.W., and Wright, J.E., 2001, Structural evolution within the Luning-Fencemaker Nevada; progression from back-arc basin closure to intra-arc shortening fold-thrust belt: Journal of Structural Geology, v. 23, p.1971–1995.

Wynne, P.J., Irving, E., Maxson, J.A., and Kleinspehn, K.L., 1995, Paleomagnetism of the Upper Cretaceous strata of Mount Tatlow: Evidence for 3000 km of northward displacement of the eastern Coast Belt, British Columbia: Journal of Geophysical Research, v. 100, p. 6073–6091.

MANUSCRIPT ACCEPTED BY THE SOCIETY JUNE 2, 2003

Geological Society of America
Special Paper 374
2003

Paleomagnetism and geobarometry of the La Posta pluton, California

D.T.A. Symons*
Department of Earth Sciences, University of Windsor, Windsor, Ontario N9B 3P4, Canada

M.J. Walawender
Department of Geological Sciences, San Diego State University, San Diego, California, 92182, USA

T.E. Smith
S.E. Molnar*
M.J. Harris
W.H. Blackburn
Department of Earth Sciences, University of Windsor, Windsor, Ontario N9B 3P4, Canada

ABSTRACT

The 94 Ma La Posta granitoid pluton is intruded into the eastern zone of the Peninsular Ranges batholith in southern and Baja California. Paleomagnetic analyses of 297 specimens from 25 sites along an ~45 km stretch of Interstate I-8 across the pluton isolated a characteristic remanence direction of declination = 340.4°, inclination = 60.4°, α_{95} = 3.6°, N = 20 that resides mostly in single to pseudosingle domain titanomagnetite and magnetite. After correcting for the Neogene opening of the Gulf of California, the pluton's position indicates nonsignificant translation (1° ± 5° southward) and rotation (4° ± 7° counterclockwise) relative to the North American cratonic reference pole. Thus, the pluton and its eastern zone host rocks have not been tilted since emplacement, nor were they translated or rotated during the Late Cretaceous or Paleogene.

New amphibole (36) and plagioclase (32) analyses, plus recalculated previously published analyses, yield 14 emplacement pressure estimates averaging 4.7 ± 0.4, 2.5 ± 0.2, and 3.6 ± 0.2 kbar for the eastern and western sides of the La Posta pluton and for the adjacent tonalites in the western zone of the Peninsular Ranges batholith, respectively, using the Al-in-hornblende geobarometer with plagioclase geothermometer corrections. These pressure differences are interpreted to reflect uplift differences of ~8 km on the Cuyamaca–Laguna Mountain Shear Zone that lies west of the pluton and of ~3 km on a postulated fault cutting north-south through the center of the pluton.

Utilizing a pole position calculated from published data for 43 sites in Early Cretaceous plutonic rocks in the western Peninsular Ranges batholith zone of southern California gives significant estimates for translation of 11° ± 4° northward and rotation of 20° ± 6° clockwise between 124 and 94 Ma. This shows that the eastern and western zones have very different terrane histories and that the intervening suture, approximately marked by the Magnetite-Ilmenite line that also marks major differences in many other geological, geochemical, and geophysical properties, is a major tectonic boundary. Evidently the western zone originated as a volcanic arc ~2700 ± 1000 km southwest of Baja California, was carried northeast on the Farallon plate, impacted against the craton in northern México in a dextral transpressive mode as

*E-mail, Symons: dsymons@uwindsor.ca. Present address, Molnar: School of Earth and Ocean Sciences, University of Victoria, Victoria, British Columbia V8W 3P6, Canada.

Symons, D.T.A., Walawender, M.J., Smith, T.E., Molnar, S.E., Harris, M.J., and Blackburn, W.H., 2003, Paleomagnetism and geobarometry of the La Posta pluton, California, *in* Johnson, S.E., Paterson, S.R., Fletcher, J.M., Girty, G.H., Kimbrough, D.L., and Martín-Barajas, A., eds., Tectonic evolution of northwestern México and the southwestern USA: Boulder, Colorado, Geological Society of America Special Paper 374, p. 135–155. For permission to copy, contact editing@geosociety.org. © 2003 Geological Society of America.

the plate subducted under the craton, and was accreted to the craton until the Neo-gene opening of the Gulf of California.

Keywords: paleomagnetism, geobarometry, Cretaceous, Peninsular Ranges batholith, geotectonics.

INTRODUCTION

Gastil (1975) was first to describe fundamental geologic differences between the eastern and western sides of the Peninsular Ranges batholith (Fig. 1). Since then many such basic differences have been noted (Table 1), and Gastil et al. (1990) have pointed out that the Magnetite-Ilmenite line running down the middle of the Peninsular Ranges batholith closely marks a change in virtually every measurable property of the eastern and western zones. These include size, deformation style, geochemical and isotopic character, emplacement ages, and oxidation state of the granitoid plutons as well as country rock lithologies and their metamorphic grades.

Important to this study is the fact that the Magnetite-Ilmenite line runs along the western contact of the La Posta pluton and the line must be related to a major tectonic event to result in such contrasting terranes. For example, Walawender et al. (1990) attributed the contrast to the emplacement of the Peninsular Ranges batholith at varying distances above an eastward-dipping middle Cretaceous subduction zone. If true, then the two terranes should yield similar paleomagnetic characteristic remanent magnetization (ChRM) directions for co-eval rock units. Also, one would reasonably expect to be able to correlate some presubduction geologic features across the Magnetite-Ilmenite line, but this has not been observed to date. In their initial paleomagnetic study on rocks of the western zone of the Peninsular Ranges batholith, Teissere and Beck (1973) reported that the remanences of these rocks indicated that they had undergone 1210 km of northward displacement and 26° of clockwise rotation, and they attributed these motions to transform faulting between the Pacific and North American plates. In this scenario, the two terranes should give similar ChRM declinations because of a common rotation but differing inclinations across the Magnetite-Ilmenite line because of coast-parallel translation. To date, however, no geologic correlation has been made between either terrane and the rocks about 1210 km to the southeast on the Mexican coastline. A third alternative is that the two terranes originated in differing geologic environments on different plates and were assembled together upon subduction (e.g., Gastil et al., 1981; Todd and Shaw, 1985; Johnson et al., 1999). In this case, the two terranes would be expected to record ChRM directions that differ in both declination and inclination, as will be shown, and to have distinctly different geologies on either side of the Magnetite-Ilmenite line, as Table 1 records.

Some two dozen paleomagnetic studies have been done since 1973 on Mesozoic and Paleogene rocks to the west of the San Andreas fault in southern and Baja California (see summary of Dickinson and Butler, 1998). Most paleomagnetists have favored northward translations of >500 km plus >10° of clockwise rotation for this region, arguing that it was split off the North American plate and carried northward along the craton's margin (e.g., Beck,

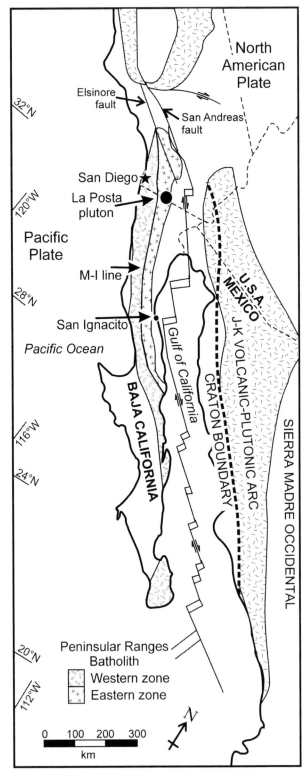

Figure 1. Location map of study area.

TABLE 1. SOME IMPORTANT CONTRASTING GEOLOGIC PROPERTIES
OF THE EASTERN VERSUS WESTERN ZONES OF THE PENINSULAR RANGE BATHOLITH

Property	Eastern Zone	Magnetite-Ilmenite line	Western Zone	Reference
Plutons				
size	larger, 400–1400 km^2		smaller, <100 km^2	1
granitoid type	I & S type		I type	2,4,7
titanite	primary		secondary	4
muscovite	present and primary		none	4
other lithology	no gabbros		gabbros present	2
metamorphism	unmetamorphosed		greenschist–lower amphibolite	1
ages, U-Pb zircon	90–100 Ma		105–120 Ma	3,4
K-Ar hornblende	95–80 Ma		115–95 Ma	8
K-Ar biotite	90–75 Ma		110–90 Ma	8
Sr abundance	>500 ppm		<300 ppm (except cumulates)	3,4,9
Sr$_i$ isotope ratio (whole rock)	≥ 0.705		≤0.705	3,4
δO^{18} values (whole rock)	≥ 9.0 per mil		≤8.5 per mil	3,4
^{206}Pb/^{204}Pb	19.0–19.5		18.5–18.9	13
REE patterns	steep, no Eu anomaly		flat HREE, (-) Eu anomaly	9,10
primary oxides	ilmenite		magnetite and ilmenite	3,4
deformation/foliation	weak, synplutonic		locally intense, penetrative	7
tectonism	late to post-tectonic		syntectonic assemblage	7
Bouger gravity	0 to ~–60 mgal		0 to ~+60 mgal	1,11
model density	~2.7 gm/cm^3		~2.8 gm/cm^3	6,12
magnetic susceptibily	0.91, 1.15, 1.35		24, 51, 114	5,6,15,16
	(Q$_1$,M,Q$_3$ x 10^{-4} SI)		(Q$_1$,M,Q$_3$ x 10^{-4} SI)	
aeromagnetic anomalies	≤20γ		20–2000 γ	16
Host rocks	biotite schist, marble, quartzite, amphibolite		Santiago Peak volcanics, micaceous schist, quartzite	1,7,9
Host rock metamorphism	mid amphibolite to upper migmatitic		greenschist to lower amphibolite	7,9
Depth to M-discontinuity	15–25 km		35–45 km	14
Magnetite-Ilmenite line, step discontinuity		in major elements of plutons		9
		in trace elements of plutons		10
		~-80 W/m^2 in heat flow trend		2

Note: 1—Gastil (1975); 2—Todd and Shaw (1985); 3—Silver et al. (1979); 4—Walawender et al. (1990); 5—Diamond (1985); 6—Gastil et al. (1986); 7—Todd and Shaw (1979); 8—Krummenacher et al. (1975); 9—Walawender and Girty (1991); 10—Gromet and Silver (1979); 11—Oliver (1980); 12—Baird and Meisch (1984); 13—Patterson et al.,(1956), Silver and Chappell (1988); 14—Ichinose et al. (1996); 15—Knaack (1985); 16—Gastil et al. (1990).

1991). This hypothesis has been challenged on geologic (e.g., Gastil, 1991) and paleomagnetic grounds (e.g., Butler et al., 1991; Dickinson and Butler, 1998). The latter authors argue that simple closure of the Gulf of California plus correction of the paleomagnetic data for sedimentary compaction, plutonic tilt, etc., are sufficient to account for both the geologic offsets and paleomagnetic results. It should be noted, however, that all of the paleomagnetic data, which come from a few hundred sites, have come from the western zone of the Peninsular Ranges batholith and its associated strata, except perhaps for four sites near San Ignacito in Baja California (Hagstrum et al., 1985), whereas the geologic evidence for minimal displacement has come mostly from the eastern zone. Thus, it is a reasonable hypothesis that substantial dextral transcurrent motion may have occurred along the Magnetite-Ilmenite line of the Peninsular Ranges batholith that displaced the western zone northward by some hundreds of kilometers relative to the eastern zone and that the eastern zone has only been offset minimally

because of the opening of the Gulf of California. This study is the first to report a substantial body of paleomagnetic data from a unit in the eastern zone, and the results support the notion that the Magnetite-Ilmenite line is a suture between two distinct terranes with a substantial apparent transcurrent dextral offset.

GEOLOGY OF THE LA POSTA PLUTON

The La Posta pluton is part of a belt of large, inwardly zoned granitic plutons that extends along the eastern side of the Peninsular Ranges batholith from Riverside County in California southward at least 300 km into Baja California (Gastil et al., 1990; Walawender et al., 1990) (Fig. 1). Although zircon U-Pb emplacement ages for these units generally vary between 90 and 100 Ma, several older bodies exhibit La Posta–like features but reverse zoning. The small (~100 km^2) El Topo pluton, 10 km south of the La Posta pluton, is reversely zoned from a garnet-muscovite tonalite margin to a hornblende-biotite tonalite core and has a zircon age of 107 Ma (Wernicke, 1987). The El Pinal pluton (Duffield, 1968) is about 10 km west of the El Topo pluton and west of the Magnetite-Ilmenite line. It has no published zircon ages but must be older than its 95 Ma K-Ar age (Krummenacher et al., 1975).

Straddling the international border, the La Posta pluton (Fig. 2) has an estimated exposure area of 1400 km² and exhibits structural and textural features indicative of inward crystallization of a static melt chamber (Walawender et al., 1990). Concentric lithologic zones range from tonalite margins to a monzogranite core with internal contacts that are typically gradational over distances of several tens of meters. The outer two facies, the Hornblende-Biotite and Large Biotite facies (Fig. 2), are characterized by sharply euhedral crystals of amphibole, biotite, and titanite, whereas the Small Biotite facies contains large oikocrystic alkali feldspar grains up to 5 cm in length. Both features suggest that the unit has undergone little in the way of subsolidus deformation. The core of the pluton, the Muscovite-Biotite facies, unlike the three outer facies, has an isotopic signature that includes a component enriched in radiogenic Sr and ¹⁸O. Foliation is present only within the first 2 km of the exposed margins around the pluton where it is defined by a subparallel alignment of centimeter-sized amphibole crystals and sparse, highly elongate mafic inclusions up to several meters in length.

The western half of the La Posta pluton intrudes into a series of older tonalitic rocks and highly deformed Jurassic orthogneisses. The contact is steep and parallel to the magmatic foliation developed in the westernmost Hornblende-Biotite facies (Fig. 2).

The eastern margin of the pluton is covered by younger sedimentary materials related to the development of the Salton Sea Trough. Foliation directions in the easternmost facies are shallow and east dipping. The two halves are separated by an elongate metasedimentary screen, the Julian Schist, which is terminated to the north by the Elsinore fault but extends approximately 50 km south of the international border. Within the La Posta pluton, the metamorphic rocks exhibit an eastward-directed shift from a flysch-like sequence to one dominated by migmatitic biotite gneiss, marble, and quartzite. Walawender et al. (1991) suggested that the eastern side of this screen may be Ordovician or older, whereas the western side is more likely Triassic in age. In addition, Krummenacher et al. (1975) pointed to the abrupt decrease in biotite cooling ages from the western to the eastern side of the La Posta pluton. Thus, the screen appears to hold a large structural feature of yet undetermined significance. Small monzogranite plutons within the screen have emplacement ages that are a few million years younger than the La Posta pluton (Walawender et al., 1990) and indicate that deformation in the screen continued beyond emplacement of the larger enclosing pluton.

PALEOMAGNETISM

Sampling

Twenty-five sites were sampled from across the La Posta pluton to represent it both petrographically and spatially (Fig. 2).

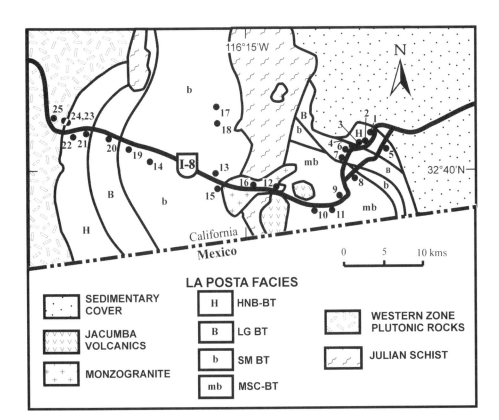

Figure 2. Geology of La Posta pluton along Interstate I-8 transect with site locations. La Posta facies are Hornblende-Biotite (HNB–BT), Large Biotite (LG BT), Small Biotite (SM BT), and Muscovite-Biotite (MSC-BT).

Despite efforts to find unweathered rock elsewhere, sufficiently "fresh" rock was found only in road cuts along Interstate I-8, old Highway 8 that runs parallel to it, and the McCain Valley Road at sites 17 and 18. Of the 25 sites, 22 sampled the main phases of the La Posta pluton (Table 2), site 23 sampled the pluton's strongly foliated border facies with older igneous rocks of the western zone of the Baja batholith, and sites 24 and 25 were from the western zone granitoids next to the contact and about 2 km west of the contact with the eastern zone, respectively.

At each site, six or more cores were drilled and oriented in situ using a sun compass. Most cores yielded two standard 2.54 cm diameter "right-cylindrical" paleomagnetic specimens, creating a total collection of 297 specimens.

Equipment

The specimens were stored in a magnetically shielded room with an ambient field of ~0.2% of the Earth's magnetic field intensity for about six months prior to measuring to allow their viscous remanent magnetization (VRM) components to decay, and all subsequent measurements were done without removing the specimens from the room.

The paleomagnetic measurements were done using an automated Canadian Thin Films DRM-420 cryogenic magnetometer, a Sapphire Instruments SI-4 alternating field (AF) demagnetizer, a Sapphire Instruments SI-6 pulse magnetizer, and a Magnetic Measurements MMTD-80 thermal (TH) demagnetizer.

Natural Remanent Magnetization

The natural remanent magnetization (NRM) of each specimen was first measured. Excluding the specimens from the western zone granitoids, i.e., sites 24 and 25, the remaining 272 specimens from the La Posta pluton had a weak median NRM intensity of 2.10×10^{-4} A/m with lower and upper quartiles of 1.02×10^{-4} A/m and 3.58×10^{-4} A/m, respectively. Two pilot specimens with typical NRM directions and intensities were selected to represent each site in step demagnetization tests.

AF Step Demagnetization

One pilot specimen per site was AF demagnetized in 14 steps up to 140 mT. Although some of them gave fairly well-defined ChRM vectors (Fig. 3, A and B), the majority either failed to define

TABLE 2. SITE MEAN REMANENCE DIRECTIONS AND SITE LOCATIONS

Site	Site location		Specimen numbers				Mean ChRM direction				Rock type
	lat (N)	long (W)	s.d.	e	r	n	D (°)	I (°)	α_{95} (°)	k	
1	32°42.5'	116°04.0'	7	3	2	5	195.1	−62.1	7.5	105	Hb-Bt
2	32°42.0'	116°04.5'	14	13	1	14	333.0	69.3	5.4	55	Hb-Bt
3	32°42.0'	116°05.0'	13	13	0	13	4.5	67.6	4.2	99	Hb-Bt
4	32°42.0'	116°05.0'	11	10	0	10	323.9	58.4	5.2	88	Lg Bt
5	32°42.5'	116°03.5'	8	8	0	8	61.7	65.7	8.5	43	Hb-Bt
6	32°41.5'	116°05.5'	9	9	0	9	353.8	58.6	8.0	42	Lg Bt
7	32°41.0'	116°05.5'	9	5	4	9	330.5	58.1	8.7	36	Lg Bt
8	32°40.5'	116°06.0'	10	7	2	9	318.9	64.8	9.2	32	Ms-Bt
9	32°39.0'	116°06.0'	8	6	2	8	345.8	56.7	8.0	50	Ms-Bt
10	32°38.0'	116°09.0'	9	1	7	8	333.5	63.2	5.5	103	Ms-Bt
11	32°38.0'	116°07.5'	8	2	4	6	350.7	61.5	9.5	51	Ms-Bt
12	32°39.0'	116°12.0'	9	5	4	9	340.1	56.1	10.2	26	Monz
13	32°40.0'	116°14.5'	8	1	3	4	324.1	58.9	13.8	45	Sm Bt
14	32°41.0'	116°20.0'	8	4	2	6	347.3	65.2	17.0	18	Sm Bt
15	32°39.5'	116°14.0'	8	3	3	6	350.2	55.8	8.9	58	Sm Bt
16	32°39.5'	116°12.5'	8	2	5	7	329.6	59.7	11.6	28	Monz
17	32°45.0'	116°16.5'	9	4	1	5	339.6	44.1	14.5	29	Sm Bt
18	32°43.5'	116°16.0'	9	3	5	8	338.1	57.7	7.3	58	Sm Bt
19	32°43.0'	116°23.0'	7	3	3	6	319.1	57.2	4.7	207	Sm Bt
20	32°43.5'	116°24.5'	8	3	4	7	353.0	59.7	7.9	59	Lg Bt
21	32°43.5'	116°26.0'	10	3	3	6	344.2	57.5	18.1	50	Hb-Bt
22	32°43.5'	116°26.5'	9	7	1	8	1.1	54.2	9.7	34	Hb-Bt
23	32°43.5'	116°28.5'	8	8	0	8	3.3	54.9	9.5	35	Bord
24	32°43.5'	116°28.5'	9	8	1	9	15.0	53.9	6.5	63	W.Z.
25	32°44.0'	116°28.5'	9	9	0	9	358.6	57.1	7.8	45	W.Z.

Note: Site location is the standard latitude (lat) in degrees and minutes north (N) and longitude (long) in degrees and minutes west (W). Specimen numbers are the number step demagnetized (s.d.), the number giving endpoints (e), the number found by remagnetized circles (r) and the number used to calculate the site mean direction (n). The mean ChRM direction is defined by its declination (D), inclination (I) and radius of 95% cone of confidence (α_{95}) all in degrees and the precision parameter (k) after Fisher (1953). The rock type abbreviations are hornblende biotite granite (Hb-Bt), large biotite granite (Lg Bt), muscovite-biotite granite (Ms-Bt), small biotite granite (Sm Bt), contact phase (Bord) and western zone granitoids (W.Z.).

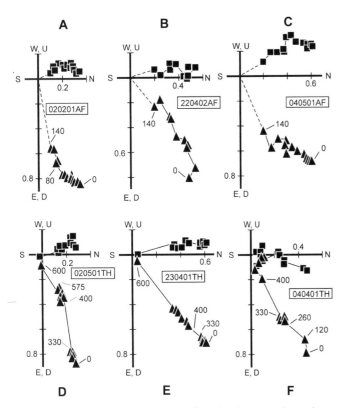

Figure 3. Step demagnetization vector plots showing example specimens. A–C: Alternating field (AF). D–F: Thermal (TH) demagnetization. Vectors in horizontal plane (north, N; east, E; south, S; west, W) are denoted by squares, and vectors in the vertical plane (N—north; D—down; S—south; U—up) by triangles. Axial values are expressed as ratio of specimen's natural remanent magnetization intensity. In A–C some step demagnetization field intensities are given in mT, and in D–F some are given in °C.

a vector or gave a vector that did not project to the origin of a vector component plot (Fig. 3C) so that its ChRM direction was in doubt.

The ChRM directions were determined using a combination of the visual inspection of stereonet and vector component plots (Zijderveld, 1967), component analysis using least-squares fitting (Kirschvink, 1980), and remagnetization circles (Halls, 1978).

TH Step Demagnetization

The second pilot specimen for each site was TH demagnetized in 13 steps up to 600 °C. Most of these specimens gave readily interpretable ChRM vectors either directly (Fig. 3, D and E) or by remagnetization circles (Fig. 3F).

The majority of the La Posta pluton's specimens gave distributed intensity decay spectra on TH demagnetization that are indicative of the unblocking of titanomagnetite. Of these, most had well-defined intensity drops between 500° and 600 °C that indicate the presence of some relatively pure fine-grained magnetite. A few specimens lost nearly all of their remanence intensity at close to 575°, indicating that magnetite was the sole significant magnetic mineral. The remaining few specimens lost more than 70% of their

remanence intensity by 120 °C, which, coupled with the evident yellow-brown weathering color, indicates that goethite was present, probably from the alteration of pyrite. Finally, many of the specimens retained a small percentage of residual remanence intensity after the 600 °C step, indicating the presence of hematite. In some cases, the hematite was clearly generated by the oxidation of magnetite in the oven starting at ~550 °C. In other cases, the likely cause was progressive conversion of goethite and/or pyrite beginning at ~350 °C. It is possible that some hematite was primary and formed by exsolution of titanomagnetite to ilmenite and hematite, because the isolated ChRM is collinear with the titanomagnetite-magnetite ChRM. Unfortunately, it was not possible to investigate the nature of the hematite component in detail by TH demagnetization above 600 °C because most of the specimens were already severely disintegrating by this temperature.

Recognizing that the La Posta pluton had given some Al-in-hornblende geobarometric values that indicated uplift from depths of as much as 15 km (Clinkenbeard and Walawender, 1989), it was evident that the best hope for isolating its primary remanence would be to use TH step demagnetization in the 500–600 °C range on the remaining specimens. This assumes a geothermal gradient of ~30 °C/km and utilizes the time-blocking curves for magnetite of Pullaiah et al. (1975; see also Dunlop, 1990). A consequence of this decision was that glued specimens that had been broken during collecting but would have been suitable for AF step demagnetization had to be abandoned, thereby reducing the representation of many sites to an average of nine specimens (Table 2). Fortunately, ~88% of the remaining 225 specimens gave usable ChRM directions.

Saturation Remanence

Twelve specimens, 10 from the La Posta pluton and two from the western zone tonalites, representing the various lithologies along the I-8 transect were subjected to saturation isothermal remanent magnetization (SIRM) testing. They were pulse magnetized in a direct field in 13 steps to 1900 mT and then AF demagnetized in six steps to 160 mT.

The SIRM acquisition curves confirm the presence of both titanomagnetite-magnetite and hematite. Some specimens (e.g., Fig. 4A, specimens 1, 10, 16, 22, and 23) display rapid intensity acquisition, reaching near saturation by ~350 mT, which is typical of magnetite. Others show the presence of hematite as a remanence carrier in minor (Fig. 4A, specimens 2, 9, and 18) to moderate (Fig. 4A, specimens 6, 12) amounts. Similarly, upon AF step demagnetization of the SIRM, the initial rapid to moderate decrease of the SIRM intensity indicates that the titanomagnetite-magnetite ranges from single domain to pseudosingle domain in nature, and the flat tails of the curves above ~100 mT, crossing the single domain boundary for magnetite, indicate the variable percentage of hematite.

Site Mean Directions

The specimens' ChRM directions, derived almost entirely from TH step demagnetization in the 500–600 °C range, were com-

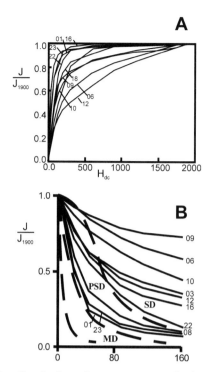

Figure 4. Saturation isothermal remanent magnetization intensity. A: Acquisition in a direct current field (H_{dc}). B: Demagnetization in an alternating field (H_{af}). The intensity (J) is expressed as ratio of saturation intensity in a 1900 mT field. Type curves (dashed lines) locate boundaries between single (SD), pseudosingle (PSD), and multidomain (MD) magnetite. Specimen curves (solid lines) are identified by site number.

bined to get the site mean directions using Fisher (1953) statistical methods (Table 2). Most of the specimens' ChRMs are very weak magnetizations that were isolated in the low 10^{-5} A/m range. All of the sites give a reasonably coherent cluster of ChRM vectors. The radii of their cones of 95% confidence (α_{95}) range between 4.2° and 18.1° with a median value of 8.5°, and their precision parameters (k) range between 18 and 207 with a median value of 50.

Examination of the distribution of the site means (Fig. 5) shows that the direction from site 5 is clearly anomalous, and so it is not considered further in the statistical analysis. The cause of the discordance is uncertain, but the site is in a canyon and adjacent to a very steep mountain slope so that it is likely that a huge slump block has been sampled. The site 1 mean direction has a reverse polarity but it is antipodal to the main cluster of site mean directions and so it is reversed into the normal polarity population. Excluding the two sites from the western zone granitoid host rocks that could carry a remanence component that predates intrusion of the La Posta pluton (Fig. 5, triangles), the remaining 22 sites (N) of the pluton give a unit mean direction of D = 342.7°, I = 60.0°, α_{95} = 3.6° and k = 75.9 (declination, inclination, radius of 95% confidence and precision parameter of Fisher [1953]).

Examination of the La Posta site mean ChRM directions shows that the directions from the foliated sites 22 and 23 beside the western contact are deviated clockwise from the main cluster of directions and similar to those of sites 24 and 25 in the western zone of the Peninsular Ranges batholith. As will be discussed, the cause of the clockwise deviation is believed to be distributed strain from post-emplacement tectonism. Accordingly, the site 22 and site 23 mean directions have also been excluded from the La Posta pluton mean, resulting in a unit mean direction of D = 340.4°, I = 60.4°, N = 20, α_{95} = 3.6°, k = 82.6 (Table 3). Note that the exclusion of the two sites shifts the unit mean direction by only 2.5° of arc. The unit mean direction from the 20 accepted sites for the La Posta pluton gives a pole position of 72.2°N, 171.7°W (d_p = 4.2°, d_m = 5.5°) (latitude, longitude, semi-axes of the oval of 95% confidence along and perpendicular to the sampling site-pole great-circle, respectively) (Fig. 6).

GEOTHERMOBAROMETRY

Sampling and Analytic Methods

The data used in this study are taken from an earlier study (Clinkenbeard and Walawender, 1989), and they are supplemented

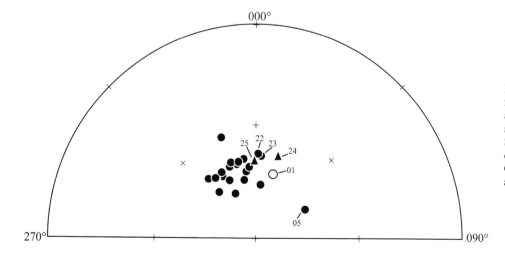

Figure 5. Site mean characteristic remanent magnetization directions plotted on an equal-area stereogram segment, with some site means identified by number. Solid (shaded) symbols indicate normal down (reversed antipodal) directions. Circles denote sites in La Posta pluton and triangles in western zone tonalites.

TABLE 3. PALEOMAGNETIC RESULTS FROM THE WESTERN ZONE
OF THE PENINSULAR RANGES BATHOLITH IN SOUTHERN CALIFORNIA

Entry, Unit	N	D (°)	I (°)	α₉₅ (°)	k	References
A La Posta pluton	20	340.4	60.4	3.6	82.6	This study
B La Posta sites 22 and 23	2	2.2	54.6	n.a.	>999	This study
C Western zone granitoids—sites 24 and 25	2	7.2	55.8	n.a.	137.0	This study
D San Marcos gabbro	11	4.5	49.0	6.9	44.4	Teissere and Beck (1973)
E Western zone granitoids	7	1.0	49.5	8.7	49.7	Teissere and Beck (1973)
F Entries D + E	18	3.3	49.4	5.0	48.3	Teissere and Beck (1973)
G San Marcos gabbro	14	9.2	47.8	3.4	136.7	Hagstrum et al. (1985)
H Santiago Peak volcanics	9	354.0	58.4	5.7	67.1	Yule and Herzig (1994)
I Entries C + F + G + H	43	3.9	51.2	2.7	66.4	

Note: N—number of sites; n.a.—not applicable. Other abbreviations as in Table 2.

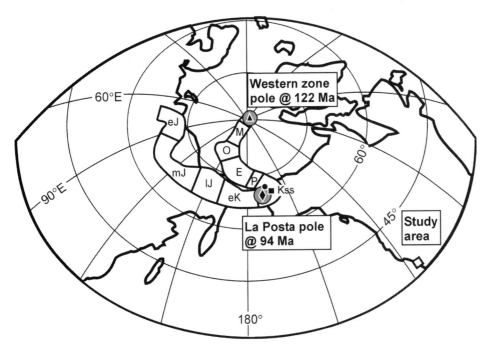

Figure 6. Apparent polar wander path for North American craton since Early Jurassic (eJ) from Besse and Courtillot (1991). Reference poles for 94 Ma and 122 Ma are shown as solid circle and solid square, respectively. The poles, with cones of 95% confidence, are shown for La Posta pluton in eastern zone of Peninsular Ranges batholith (solid diamond) and for plutonic rocks in western zone (solid triangle) after corrections for Neogene opening of Gulf of California.

by 32 new plagioclase analyses and 36 new amphibole analyses (Tables 4–7). The new analyses represent samples taken from each of the major facies of the La Posta pluton that were previously analyzed. The analytical methods used to determine the old data are more completely described in Clinkenbeard and Walawender (1989). The chemical analyses were made on minerals in polished thin sections using the three-channel CAMECA MBX electron microprobe at Scripps Institute of Oceanography. Operating conditions used an accelerating potential of 15 kV, sample current of 15 nA on brass, and a spot size ranging from 2 to 10 μm. Standards used for calibration were well-characterized silicates and oxides. Data reduction was performed on-line using the program COR-REX, which is part of the MBXCOR package. Here the chemical analyses were made for hornblende-plagioclase pairs on a JEOL JXA-8600 Superprobe at the University of Western Ontario using a probe current of 10 nA, accelerating voltage of 15 kV, a beam focus of 2 μm for amphibole and 5 μm for plagioclase, and a 20 s accumulation time. The calibration standards for amphibole were: Si—Cr-augite, USNM 164905; Al—anorthite; Cr—chromite, USNM 117075; Fe—fayalite; Mn—rhodonite; Mg—olivine, USNM 111312; Ca—Wakefield diopside; Na—Amelia albite; K—orthoclase; F—LiF; Cl—Tugtupite; O—by difference. The calibration standards for plagioclase were: Si—Amelia albite; Al—anorthite; Fe—fayalite; Ca—anorthite; Na—Amelia albite; K—orthoclase; and O—by difference. A fuller description of the methods can be found in Symons et al. (2000).

The results from the new analyses show good agreement with those derived from the older data, in terms of amphibole

TABLE 4. AVERAGE RIM COMPOSITIONS OF PLAGIOCLASE IN THE UNITS OF THE LA POSTA PLUTON AND THE ADJACENT WESTERN ZONE TONALITES

	Western zone tonalites				West-side La Posta pluton				Large biotite facies		Small biotite facies		East side La Posta pluton		
---	---	---	---	---	---	---	---	---	west side	east side					
Site:	25 (LP2)	25 (LP23)	25 (LP39)	25	21 (LP22)	21	22 (LP8)	22	20 (LP4)	4 (LP35)	14 (LP5)	1 (LP12)	2 (12)	3 (LP20)	5 (LP19)
n:	3	3	3	6	3	9	3	5	3	3	3	3	(12)	3	3
SiO_2	57.29	55.60	57.21	59.10	59.80	60.15	59.02	61.35	59.27	58.91	57.91	58.01	59.65	59.07	57.88
Al_2O_3	26.78	27.96	27.01	25.94	25.04	24.66	25.25	24.32	25.10	26.01	26.36	26.17	25.09	25.88	26.26
Fe_2O_3	0.15	0.14	0.15	0.12	0.09	0.13	0.10	0.11	0.08	0.04	0.08	0.05	0.11	0.07	0.12
CaO	8.10	9.63	7.93	7.58	6.42	6.27	6.30	5.62	6.57	6.78	7.42	7.42	6.63	7.00	7.67
Na_2O	7.02	6.53	7.38	7.15	8.13	7.67	8.10	8.22	7.81	7.87	7.44	7.47	7.49	7.91	7.38
K_2O	0.24	0.18	0.12	0.13	0.27	0.14	0.21	0.19	0.27	0.20	0.19	0.20	0.21	0.21	0.19
Total	99.59	100.03	99.81	100.00	99.75	99.03	98.99	99.80	99.09	99.81	99.40	99.32	99.18	100.14	99.50
Si	10.31	10.02	10.28	10.55	10.70	10.80	10.64	10.91	10.67	10.54	10.42	10.45	10.71	10.54	10.42
Al	5.66	5.94	5.72	5.46	5.28	5.22	5.37	5.10	5.33	5.49	5.59	5.56	5.31	5.45	5.57
Fe^{3+}	0.02	0.02	0.02	0.02	0.01	0.02	0.01	0.01	0.01	0.01	0.01	0.01	0.02	0.01	0.02
Tot Z	16.02	15.98	16.02	16.02	15.99	16.03	16.02	16.02	16.00	16.03	16.03	16.01	16.03	16.00	16.00
Ca	1.56	1.86	1.53	1.45	1.23	1.21	1.22	1.07	1.27	1.30	1.43	1.43	1.28	1.34	1.48
Na	2.45	2.28	2.57	2.46	2.82	2.67	2.83	2.83	2.73	2.73	2.60	2.61	2.61	2.74	2.57
K	0.06	0.04	0.03	0.03	0.06	0.03	0.05	0.04	0.06	0.05	0.04	0.05	0.05	0.05	0.04
Tot A	4.07	4.18	4.13	3.95	4.11	3.91	4.10	3.95	4.05	4.07	4.07	4.09	3.93	4.12	4.10
An	38.41	44.46	37.03	36.75	29.97	30.87	29.72	27.14	31.24	31.87	35.13	35.03	32.44	32.46	36.10
Ab	60.24	54.57	62.31	62.47	68.55	68.29	69.10	71.79	67.25	66.99	63.80	63.86	66.34	66.40	62.83
Or	1.36	0.97	0.66	0.78	1.48	0.84	1.18	1.08	1.51	1.14	1.07	1.11	1.22	1.14	1.07

Note: Site numbers are from this study (Fig. 2); notations in parentheses are from Clinkenbeard and Walawender (1989). Structural formulae are calculated on the basis of 32 oxygens.

TABLE 5. THE AVERAGE RIM COMPOSITIONS OF AMPHIBOLES IN THE UNITS OF THE LA POSTA PLUTON AND THE ADJACENT WESTERN ZONE TONALITES

	Western zone tonalites				West side La Posta pluton				Large biotite facies		Hornblende facies (east side)			
									West side	East side				
Site:	25 (LP2)	25 (LP23)	25 (LP39)	25	21 (LP22)	21	22 (LP8)	22	20 (LP4)	4 (LP35)	1 (LP12)	2	3 (LP20)	5 (LP19)
n:	3	3	3	6	3	9	3	9	3	3	3	12	3	3
SiO_2	42.93	44.82	46.42	45.46	47.30	47.48	47.62	47.32	47.76	45.54	43.81	43.82	44.25	44.46
TiO_2	1.23	1.12	1.29	0.87	1.06	1.09	1.01	0.93	1.01	1.11	1.22	0.86	1.02	1.08
Al_2O_3	10.17	9.64	8.81	9.03	7.13	6.09	6.74	6.45	6.46	8.85	10.24	10.37	10.68	10.20
Fe_2O_3	3.78	3.55	4.29	3.16	1.72	0.96	1.65	1.98	1.71	1.29	2.37	1.66	1.56	1.49
Cr_2O_3	0.02	0.02	0.04	0.06	0.05	0.03	0.00	0.01	0.02	0.05	0.02	0.02	0.02	0.01
FeO	15.38	13.99	12.57	13.90	16.95	17.13	16.86	17.02	17.15	16.85	16.67	17.94	17.13	16.86
MnO	0.43	0.41	0.39	0.40	0.64	0.43	0.60	0.58	0.59	0.55	0.47	0.42	0.45	0.44
MgO	9.60	10.76	12.01	10.98	10.22	10.21	10.52	9.95	10.48	9.73	9.19	8.33	9.06	9.24
NiO	0.02	0.01	0.02	0.00	0.05	0.00	0.02	0.00	0.02	0.04	0.00	0.00	0.02	0.03
CaO	12.18	12.09	11.47	11.80	11.73	11.77	11.91	11.84	12.02	11.88	12.15	11.96	12.00	11.78
Na_2O	1.18	1.06	1.16	1.10	1.02	0.98	0.98	0.96	0.93	1.04	1.16	1.07	1.15	1.15
K_2O	1.10	0.71	0.41	0.52	0.60	0.49	0.51	0.62	0.57	0.86	0.97	1.06	1.01	0.89
P_2O_5	0.11	0.10	0.11	0.00	0.10	0.00	0.10	0.00	0.09	0.10	0.11	0.00	0.10	0.10
H_2O	1.97	2.00	2.03	1.99	2.01	1.97	2.01	1.98	2.01	1.99	1.98	1.96	1.99	1.98
F	0.00	0.00	0.00	0.00	0.00	0.00	0.00	0.00	0.00	0.00	0.00	0.00	0.00	0.00
Cl	0.00	0.00	0.00	0.03	0.00	0.03	0.00	0.02	0.00	0.00	0.00	0.04	0.00	0.00
Total	100.10	100.28	101.03	99.28	100.57	98.65	100.53	99.66	100.82	99.88	100.36	99.50	100.44	99.71
Si	6.481	6.657	6.774	6.780	7.034	7.173	7.075	7.101	7.088	6.836	6.587	6.653	6.631	6.693
[4]Al	1.519	1.343	1.226	1.220	0.966	0.827	0.925	0.899	0.912	1.164	1.413	1.347	1.369	1.307
Tot T	8.000	8.000	8.000	8.000	8.000	8.000	8.000	8.000	8.000	8.000	8.000	8.000	8.000	8.000
[6]Al	0.290	0.344	0.289	0.368	0.284	0.257	0.255	0.242	0.218	0.402	0.401	0.510	0.518	0.503
Ti	0.140	0.125	0.142	0.097	0.119	0.123	0.113	0.105	0.113	0.125	0.138	0.0098	0.115	0.122
Cr	0.002	0.002	0.005	0.007	0.006	0.003	0.000	0.001	0.002	0.006	0.002	0.000	0.002	0.001
Fe^{3+}	0.429	0.396	0.471	0.355	0.192	0.109	0.185	0.223	0.191	0.146	0.268	0.190	0.176	0.169
Fe^{2+}	1.942	1.738	1.478	1.728	2.108	2.164	2.094	2.136	2.129	2.115	2.096	2.278	2.147	2.123
Mn	0.033	0.010	0.000	0.000	0.020	0.045	0.021	0.059	0.026	0.024	0.035	0.034	0.016	0.004
Mg	2.161	2.383	2.613	2.442	2.266	2.299	2.330	2.225	2.319	2.177	2.060	1.885	2.024	2.074
NiO	0.002	0.001	0.002	0.000	0.006	0.000	0.002	0.000	0.002	0.005	0.000	0.000	0.002	0.004
Ca	0.000	0.000	0.000	0.000	0.000	0.000	0.000	0.007	0.000	0.000	0.000	0.000	0.000	0.000
Tot C	5.000	5.000	5.000	5.000	5.000	5.000	5.000	5.000	5.000	5.000	5.000	5.000	5.000	5.000
Mn	0.022	0.042	0.048	0.048	0.061	0.011	0.055	0.015	0.048	0.046	0.025	0.020	0.041	0.052
Fe^{2+}	0.000	0.000	0.056	0.005	0.000	0.000	0.000	0.000	0.000	0.000	0.000	0.000	0.000	0.000
Ca	1.958	1.913	1.782	1.886	1.859	1.905	1.885	1.897	1.902	1.900	1.946	1.944	1.916	1.889
Na	0.020	0.045	0.114	0.061	0.081	0.085	0.060	0.088	0.050	0.054	0.029	0.036	0.042	0.059
Tot B	2.000	2.000	2.000	2.000	2.000	2.000	2.000	2.000	2.000	2.000	2.000	2.000	2.000	2.000
Na	0.325	0.260	0.215	0.258	0.213	0.202	0.233	0.192	0.218	0.249	0.309	0.279	0.292	0.277
K	0.212	0.135	0.076	0.099	0.114	0.094	0.097	0.119	0.108	0.165	0.186	0.206	0.193	0.171
Tot A	0.537	0.395	0.291	0.356	0.327	0.296	0.319	0.310	0.325	0.414	0.495	0.485	0.485	0.448
OH	2.000	2.000	2.000	1.993	2.000	1.990	2.000	1.994	2.000	2.000	2.000	1.989	2.000	2.000
F	0.000	0.000	0.000	0.000	0.000	0.001	0.000	0.001	0.000	0.000	0.000	0.000	0.000	0.000
Cl	0.000	0.000	0.000	0.007	0.000	0.008	0.000	0.005	0.000	0.000	0.000	0.010	0.000	0.000
Mg#	0.53	0.58	0.63	0.58	0.52	0.52	0.53	0.51	0.520	0.510	0.500	0.450	0.490	0.490
Fe^{3+}#	0.18	0.19	0.24	0.17	0.08	0.05	0.08	0.09	0.080	0.060	0.110	0.080	0.080	0.070

Note: Site numbers are from this study (Fig. 2); notations in parentheses are from Clinkenbeard and Walawender (1989). Structural formulae are calculated on the basis of 22 oxygens following the procedures of Schumacher (1997). Fe_2O_3, FeO and H_2O calculated from stoichiometry.

TABLE 6. THE AVERAGE RIM COMPOSITIONS OF BIOTITES IN THE UNITS OF THE LA POSTA PLUTON AND THE ADJACENT WESTERN ZONE TONALITES

	Western zone tonalites				Hornblende facies West side pluton		Large biotite facies		Small biotite facies	Hornblende facies East side			Muscovite-biotite facies East side			
							West side	East side	West side							
	25 (LP1)	25 (LP2)	25 (LP23)	25 (LP39)	21 (LP22)	22 (LP8)	20 (LP4)	4 (LP35)	(LP5)	1 (LP12)	3 (LP20)	5 (LP19)	8 (LP11)	8 (LP14)	9 (LP18)	8 (LP32)
	4	4	5	5	5	4	3	5	5	4	4	7	3	4	5	4
SiO_2	35.06	35.33	34.74	35.86	34.85	35.78	35.61	35.33	35.80	35.86	35.66	34.74	34.28	33.55	33.81	34.32
TiO_2	3.00	2.82	3.09	3.26	3.64	3.59	3.70	2.82	4.18	3.26	3.31	3.09	3.15	3.39	3.50	3.33
Al_2O_3	16.56	16.49	16.10	13.30	14.88	15.52	15.23	16.49	14.74	16.30	16.59	16.10	18.66	17.40	18.72	18.19
Fe_2O_3	4.39	4.63	4.30	4.13	4.83	4.85	5.02	4.68	5.03	4.66	4.51	4.55	5.58	5.50	5.36	5.45
Cr_2O_3	0.02	0.02	0.03	0.04	0.02	0.02	0.04	0.02	0.02	0.04	0.02	0.03	0.02	0.01	0.04	0.01
FeO	15.79	16.67	15.46	14.87	17.39	17.47	18.07	16.85	18.10	16.77	16.22	16.37	20.06	19.81	19.28	19.60
MnO	0.33	0.32	0.26	0.28	0.35	0.33	0.31	0.32	0.40	0.28	0.27	0.26	0.35	0.34	0.41	0.27
MgO	8.77	9.54	9.60	9.72	8.97	9.50	9.20	9.54	8.88	9.72	9.67	9.60	5.72	5.98	5.65	5.94
NiO	0.03	0.01	0.02	0.03	0.03	0.01	0.02	0.01	0.01	0.03	0.01	0.02	0.01	0.02	0.02	0.03
CaO	0.27	0.01	0.01	0.02	0.28	0.01	0.02	0.01	0.03	0.00	0.01	0.15	0.01	0.01	0.04	0.01
Na_2O	0.13	0.09	0.15	0.19	0.09	0.10	0.09	0.09	0.14	0.28	0.09	0.07	0.07	0.08	0.07	0.08
K_2O	9.13	9.10	9.12	8.82	9.04	9.07	9.55	8.94	9.36	9.06	9.11	8.93	9.69	8.91	8.93	9.04
P_2O_5	0.01	0.01	0.01	0.01	0.00	0.01	0.09	0.01	0.10	0.01	0.02	0.01	0.11	0.01	0.01	0.02
H_2O	3.82	3.87	3.80	3.86	3.80	3.90	3.90	3.87	3.89	3.92	3.90	3.82	3.89	3.78	3.84	3.85
Total	97.31	98.91	96.69	97.39	98.17	100.17	100.85	98.99	100.69	100.19	99.39	97.74	101.60	98.80	99.67	100.14
Si	5.506	5.476	5.485	5.570	5.495	5.503	5.477	5.473	5.514	5.485	5.478	5.450	5.288	5.322	5.283	5.343
^{iv}Al	2.494	2.524	2.515	2.430	2.505	2.497	2.523	2.527	2.486	2.515	2.522	2.550	2.712	2.678	2.717	2.657
Tot Z	8.000	8.000	8.000	8.000	8.000	8.000	8.000	8.000	8.000	8.000	8.000	8.000	8.000	8.000	8.000	8.000
^{vi}Al	0.571	0.489	0.481	0.554	0.260	0.316	0.238	0.483	0.190	0.423	0.481	0.427	0.681	0.575	0.730	0.680
Ti	0.354	0.329	0.367	0.381	0.432	0.415	0.428	0.329	0.484	0.375	0.382	0.365	0.366	0.404	0.411	0.390
Cr	0.002	0.002	0.004	0.005	0.002	0.002	0.005	0.002	0.002	0.005	0.002	0.004	0.002	0.001	0.005	0.001
Fe^{3+}	0.519	0.540	0.511	0.483	0.573	0.562	0.581	0.546	0.583	0.536	0.521	0.537	0.647	0.657	0.630	0.638
Fe^{2+}	2.075	2.161	2.042	1.932	2.293	2.247	2.324	2.184	2.332	2.145	2.084	2.148	2.589	2.628	2.519	2.552
Mn	0.044	0.042	0.035	0.037	0.047	0.043	0.040	0.042	0.052	0.036	0.035	0.035	0.046	0.046	0.054	0.036
Ni	0.004	0.001	0.003	0.004	0.004	0.001	0.002	0.001	0.001	0.004	0.001	0.003	0.001	0.003	0.003	0.004
Mg	2.053	2.204	2.260	2.251	2.108	2.178	2.110	2.203	2.039	2.216	2.214	2.245	1.315	1.414	1.316	1.379
Tot Y	5.622	5.769	5.702	5.646	5.719	5.765	5.729	5.790	5.684	5.740	5.723	5.762	5.647	5.728	5.669	5.680
Ca	0.043	0.000	0.000	0.001	0.047	0.000	0.000	0.000	0.000	0.000	0.000	0.023	0.000	0.000	0.004	0.000
Na	0.040	0.027	0.046	0.057	0.028	0.030	0.027	0.027	0.042	0.083	0.027	0.021	0.021	0.025	0.021	0.024
K	1.829	1.799	1.837	1.748	1.818	1.779	1.874	1.767	1.839	1.768	1.785	1.787	1.907	1.803	1.780	1.795
Tot A	1.912	1.826	1.883	1.806	1.893	1.809	1.901	1.794	1.881	1.851	1.812	1.831	1.928	1.828	1.806	1.819
OH	4.000	4.000	4.000	4.000	4.000	4.000	4.000	4.000	4.000	4.000	4.000	4.000	4.000	4.000	4.000	4.000
X_{Mg}	0.500	0.500	0.530	0.540	0.480	0.490	0.480	0.500	0.470	0.510	0.520	0.510	0.340	0.350	0.340	0.350
$Fe^{3+}\#$	0.200	0.200	0.200	0.200	0.200	0.200	0.200	0.200	0.200	0.200	0.200	0.200	0.200	0.200	0.200	0.200

Note: Site numbers are from this study (Fig. 2); notations in parentheses are from Clinkenbeard and Walawender (1989). Structural formulae are calculated on the basis of 20 oxygens using a fixed $Fe^{3+}/(Fe^{3+}+Fe^{2+})$ of 0.20.

TABLE 7. SITE AVERAGED COMPOSITIONS OF CO-EXISTING AMPHIBOLE AND PLAGIOCLASE AND BIOTITE AND
GEOTHERMOBAROMETRIC DATA, LA POSTA PLUTON AND ADJACENT WESTERN ZONE TONALITES

Site	Plagioclase (Ab%)	Amphibole (avg Al^{TOT})	Amphibole (avg Si)	P (kbar) (Schmidt)	T (°C) B&H	P (kbar) (A&S)	Biotite (avg ^{vi}Ti)	T (°C) (biotite)
Western zone tonalites								
25 (LP1)	n.d.	1.463	6.673	3.95	n.d.	n.d.	0.354	509
25 (LP2)	60.2	1.809	6.481	4.89	784	3.64	0.329	472
25 (LP23)	54.6	1.687	6.657	3.98	774	3.37	0.367	530
25 (LP39)	62.3	1.515	6.774	3.50	735	3.40	0.381	553
25	62.5	1.590	6.780	4.56	728	3.85	n.d.	n.d.
Hornblende facies (west)								
21 (LP22)	68.6	1.250	7.034	2.94	680	2.90	0.432	647
21	68.3	1.090	7.173	2.18	661	2.26	n.d.	n.d.
22 (LP8)	69.1	1.180	7.075	2.61	675	2.61	0.415	614
22	71.8	1.140	7.101	2.42	666	2.48	n.d.	n.d.
Large biotite facies (west)								
20 (LP4)	67.3	1.130	7.088	2.37	679	2.34	0.428	639
Small biotite facies (west)								
14 (LP4)	63.8	n.d.	n.d.	n.d.	n.d.	n.d.	0.484	760
Hornblende facies (east)								
1 (LP12)	63.9	1.814	6.587	5.62	749	4.45	0.375	543
2	66.3	1.860	6.653	5.84	718	5.22	n.d.	n.d.
3 (LP20)	66.4	1.887	6.631	5.97	727	5.19	0.382	554
5 (LP19)	62.8	1.810	6.693	5.61	730	4.79	0.365	526
Large biotite facies (east)								
4 (LP35)	67.0	1.566	6.836	4.44	706	4.09	0.329	506
Muscovite-biotite facies (east)								
8 (LP11)	n.d.	n.d.	n.d.	n.d.	n.d.	n.d.	0.366	528
8 (LP14)	n.d.	n.d.	n.d.	n.d.	n.d.	n.d.	0.404	593
9 (LP18)	n.d.	n.d.	n.d.	n.d.	n.d.	n.d.	0.411	606
8 (LP32)	n.d.	n.d.	n.d.	n.d.	n.d.	n.d.	0.390	568

Note: Site numbers are from this study (Fig. 2); notations in parentheses are from Clinkenbeard and Walawender (1989).
n.d.—not determined.

nomenclature and the temperatures and pressures derived from geothermobarometric calculations, and thus it was not considered necessary to reanalyze all of the samples. However, the structural formulae of all of the minerals were recalculated (Tables 4–6), paying special attention to charge balance in the assignment of cations to structural sites. There are several different ways to calculate the structural formulae for amphiboles, and the procedure followed here is that of Schumacher (1997). This method was chosen because it was specially written for use with the newly recommended amphibole classification (Leake et al., 1997) and is also suitable for use with the hornblende-plagioclase geothermometer used herein (Blundy and Holland, 1990, see below). For a review of the chemistry of amphiboles and biotites and the normalization techniques used in calculating their structural formulae, see Samson et al. (1999). In addition, new equations, unavailable at the time of the previous study (Blundy and Holland, 1990; Schmidt, 1992;

Anderson and Smith, 1995), have been used to calculate equilibration temperatures and pressures from coexisting plagioclase and amphibole and equilibration temperatures from biotite data (see Harris et al., 1999).

Amphibole Geochemistry

The amphiboles are classified according to the criteria of Leake et al. (1997) and occupy three separate fields on the calcic-amphibole diagram (Fig. 7A). Those of the western zone tonalites have higher Mg/(Mg + Fe^{2+}) ratios than those of the La Posta pluton, and are all magnesio-hornblendes with the exception of one tschermakite. Amphiboles from the west side of the La Posta pluton are all magnesio-hornblendes and have higher Si values than those from the east side of the pluton. One of the east side amphiboles is a magnesio-hornblende whereas the remainder plot in the ferro-hornblende field.

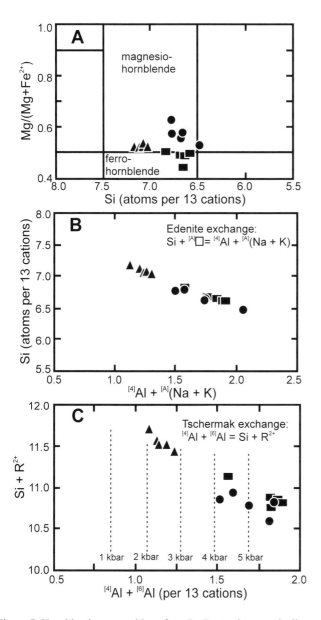

Figure 7. Hornblende compositions from La Posta pluton and adjacent western zone tonalites. A: Simplified calcic-amphibole classification plot after Leake et al. (1997); B–C: Amphibole exchange reactions.

Plots testing the temperature sensitive edenite exchange reaction, albite + tremolite = 4 quartz + edenite (Si + [A]□ = [4]Al + [A](Na + K), where [A]□ stands for an A-site vacancy), and the pressure sensitive tschermakite-forming reaction (2 quartz + 2 anorthite + biotite = orthoclase + tschermakite ([4]Al + [6]Al = Si + R²⁺) are shown in Figures 7B and C. On both diagrams the amphiboles outline three separate linear fields that are indicative of equilibration over three separate ranges of temperature and pressure. On the edenite diagram (Fig. 7B), the data from the east side and those from the west side of the La Posta pluton define separate fields which, when taken together, define a very good

linear trend. The west side data all lie at higher Si and lower [4]Al +[A](Na + K) values than those of the east side and are indicative of lower temperatures of equilibration. The western zone tonalite data lie on a line below that of the La Posta east side data and are indicative of the same general range of temperature of equilibration for that suite also. Similarly, there are three separate fields on the tschermakite exchange diagram, and the data from the east and west sides of the La Posta pluton taken together define a good linear trend that lies above the field defined by the western zone tonalites. The east side data all lie at lower Si + R²⁺ and higher [4]Al + [6]Al values than the west side data and suggest that the La Posta west side rocks equilibrated at pressures in the range of 2–3 kbar whereas the east side rocks and the western zone tonalites equilibrated at pressures on the order of 4–6 kbar (Schmidt, 1992). Thus, these diagrams suggest that the La Posta east side rocks crystallized at higher pressure and temperatures than those of the west side of the pluton. Further, the Peninsular Ranges batholith western zone plutons crystallized at temperatures and pressures that are similar to those of the east side rocks.

Crystallization Temperatures and Pressures

The Al-in-hornblende geobarometer and the amphibole-plagioclase geothermometer were used to estimate the pressures and temperatures of equilibration for the La Posta pluton and the adjacent western zone tonalites. The procedures followed here, and the theory behind them, have been detailed elsewhere (Blundy and Holland, 1990; Schmidt, 1992; Anderson and Smith, 1995; Harris et al., 1996, 1997). The information is supplemented using the Ti-in-biotite geothermometer developed by Blackburn (see Harris et al., 1999, for discussion). This geothermometer is not well constrained and the absolute temperatures calculated are not considered to be very accurate. However, the relative differences in temperatures yielded for the different suites are considered to be significant. The results of the geothermobarometric calculations are given in Table 7 and confirm that the western zone tonalites, the west side of the La Posta pluton, and the east side of the La Posta pluton each have different thermobarometric histories.

Preliminary pressures derived from the Schmidt (1992) geobarometer range from 3.50 to 4.89 kbar in the western zone tonalites, from 2.18 to 2.61 kbar on the west side of La Posta pluton, and from 4.44 to 5.97 kbar on the east side of the La Posta pluton. The temperatures derived from the Blundy and Holland (1990) amphibole-plagioclase geothermometer range from 728° to 784 °C in the western zone tonalites, from 661° to 680 °C on the west side of La Posta pluton, and from 706° to 749 °C on the east side of the La Posta pluton. These temperatures are used to correct the preliminary pressures using the Anderson and Smith (1995) geobarometer. The final pressures thus determined a range from 3.37 to 3.85 kbar in the western zone tonalites, from 2.26 to 2.90 kbar on the west side of La Posta pluton, and from 4.09 to 5.22 kbar on the east side of the La Posta pluton. There is little or no overlap in the ranges of temperature and pressure, determined from the hornblende and plagioclase compositions, for each of the three rock suites. Thus, the western zone tonalites equilibrated at both higher pressures and higher temperatures than the rocks of the adjacent west side of the La Posta pluton and at lower pressures but similar temperatures when compared to the east side of the La Posta pluton. The rocks of the west side of the La Posta pluton equilibrated at much lower pressures and temperatures than those on the east side of the pluton. The Ti-in-biotite geothermometer yields temperature ranges of 472–553 °C for the western zone tonalites, 614–760 °C for the west side of the La Posta pluton, and 506–606 °C for the east side of the La Posta pluton. The majority of the biotite temperatures are sub-solidus and their exact meaning is unclear. However, they do confirm that the thermal history of each of the three suites studied differs and may help explain differences in apparent K-Ar ages determined for biotites (Krummenacher et al., 1975) and U-Pb zircon ages from each side of the La Posta pluton. The rocks from both sides of the La Posta pluton yield U-Pb zircon ages of 94 Ma, three K-Ar biotite ages on the west side average 89 Ma, and the average of three ages from the east side is 75 Ma. The average hornblende and biotite equilibration temperatures of the hornblende-bearing facies on the west side of the pluton are 672° and 633 °C, respectively, and on the east they are 726° and 532 °C. Thus, the relatively large difference in zircon and biotite ages (ca. 20 Ma) on the east side of the La Posta pluton, as compared to the difference on the west side (ca. 5 Ma), may be explained as resulting from their having very different thermal histories, i.e., a much longer cooling interval prior to biotite equilibration on the east side (ca. 200 °C) as compared to the west side (ca. 40 °C).

The geobarometric data indicate that the Peninsular Ranges batholith western zone tonalites crystallized at higher pressure than the rocks of the adjacent west side of the La Posta pluton, indicating the presence of a structural discontinuity between them. The higher pressures of equilibration on the east side of the La Posta pluton as compared to the west side may be explained in two ways. Either the pluton has been tilted since emplacement or there is a structural discontinuity somewhere between the sites at which the pressure determinations were made. These problems are discussed later using the paleomagnetic data in conjunction with the geothermobarometric data.

DISCUSSION

Cratonic Reference Poles

The choice of reference poles for the North American craton at 94 Ma can be done in one of two ways. One approach, which Dickinson and Butler (1998) have taken, is to subjectively select a few "key" poles from the North American database that give an average age that is close to the desired age. The other more objective approach, used here, is to interpolate the reference pole from an apparent polar wander path (APWP) for the North American craton that has been constructed using all available paleomagnetic data that meet the minimum threshold requirements. The most robust path is the Besse and Courtillot (1991) APWP, which gives a characteristic reference pole for 94 Ma of 72.6°N,

166.8°W (A$_{95}$ = 4.3°) and for 122 Ma of 71.0°N, 162.0°W (A$_{95}$ = 4.3°) (Fig. 6). Both poles are <2° of arc from the reference pole used by Dickinson and Butler (1998).

La Posta Pluton Tectonics

Prior to calculating the pole positions for tectonic analysis, it helps to first correct for the Neogene opening of the Gulf of California as the Peninsular Ranges batholith moved northwest away from the Mexican coastline (Atwater, 1970; Abbott and Smith, 1989; Lonsdale, 1991). Following the analysis of Dickinson (1996; see also Dickinson and Butler, 1998), this results in a translation correction of 295 km southeastward of the Peninsular Ranges batholith and a counterclockwise rotation correction of 3.0° ± 0.6° for the study area. Upon correction, the ChRM direction for the La Posta pluton in the eastern zone yields a pole position of 69°N, 168°W (A$_{95}$ = 5°) (latitude, longitude, [radius of cone of 95% confidence]). Given the uncertainties in the tectonic corrections, it is reasonable to round these values off to the nearest degree. Furthermore, when the corrected La Posta pole is compared to the 94 Ma reference pole for the North American craton, the amount of relative post-emplacement translation and rotation can be calculated for the time between 94 Ma and the onset of opening of the Gulf of California in the Neogene. For the La Posta pluton, this calculation yields 1° ± 5° (110 ± 550 km) southward coast-parallel translation, or away from the reference pole, and 4° ± 7° of counterclockwise rotation.

Two conclusions are immediately evident. First, neither the translation nor the rotation estimate for the La Posta pluton in the eastern zone has statistical significance at the 95% confidence level. Thus, the results support the geologic evidence that the eastern zone of the Peninsular Ranges batholith was emplaced very close to its "fit" position and has been displaced only by the Neogene opening of the Gulf of California with respect to the North American craton. This supports the model of Butler et al. (1991) and Dickinson and Butler (1998), who have argued that Baja and southern California to the west of the San Andreas fault have not been translated significantly northward if Neogene opening of the Gulf of California is excluded. Second, the estimates also indicate that the La Posta pluton has not been subjected to post-emplacement tilting and, therefore, that differences in the depth estimates from Al-in-hornblende geobarometry must be accommodated by vertical fault motions and not by tilting. This result conflicts with the model of Butler et al. (1991) and Dickinson and Butler (1998) that explains the paleomagnetic results from plutonic rocks in the western zone of the Peninsular Ranges batholith by a northeast-side-up, ~21° tilt of the Peninsular Ranges batholith, which is further discussed later in this paper.

Given that the La Posta pluton is not tilted, then the geobarometric difference of 2.2 ± 0.3 kbar, or ~8 km, between its eastern side determination of 4.7 ± 0.4 kbar (N = 5) and western side of 2.5 ± 0.2 kbar (N = 5) needs to be explained. Although the absolute error on the eastern and western side determinations is likely ~ ±0.6 kbar, the variation in Al content as a function of pressure is essentially linear so that the difference between two determinations is likely accurate to ~0.2 kbar. We speculate that there is a major high-angle normal(?) fault—termed here for convenience the "mid La Posta Fault"—within the metasedimentary screen that cuts through the center of the pluton (Fig. 2). This proposed structure drops the western side of the La Posta pluton down relative to the eastern side. It may also be responsible for juxtaposing the different metasedimentary protoliths on either side of the roof pendant and for the abrupt decrease in K-Ar cooling ages east of the roof pendant. When the mean remanence directions are averaged for the La Posta sites to the east (sites 1–4, 6–11) and west (sites 12–23) of the proposed fault, they give mean directions of D = 342.8°, I = 63.1°, N = 10, α_{95} = 5.6°, k = 76.3, and D = 342.7°, I = 57.4°, N = 11, α_{95} = 4.8°, and k = 82.9, respectively, that are not statistically significantly different at 95% confidence using the test of McFadden and Lowes (1981). This result implies that the fault motion is linear (e.g., simple uplift) without a significant rotational component. If the directions from sites 12 and 16, which are in a plutonic phase that could postdate faulting (Fig. 2), are excluded, the test gives the same insignificant difference between the eastern and western halves of the pluton.

Also requiring explanation is the 1.1 ± 0.2 kbar difference from geothermobarometry between the western zone granitoids at 3.6 ± 0.2 kbar (N = 4) and the western side of the La Posta pluton at 2.5 ± 0.2 kbar. The obvious candidate fault for accommodating the difference is the north-northwest–trending Cuyamaca–Laguna Mountain Shear Zone that terminates at the western edge of the La Posta pluton. This steeply dipping structure records at least two distinct periods of movement (Thomson and Girty, 1994). An older (118–115 Ma) compressional event is reflected in a pervasive northwest-striking, steep northeast-dipping mylonitic fabric. The early phases of the La Posta pluton were emplaced into this shear zone and crosscut it in places. The younger deformation event records northeast-southwest extension between ca. 105 and 93 Ma that, in concert with the proposed mid-La Posta fault, has dropped the western half of the pluton relative to both its eastern side and the western Peninsular Ranges batholith. In this scenario, the difference of ~4 km between the two sides of the Cuyamaca–Laguna Mountain Shear Zone is deemed to be transtensional in character (Fig. 8).

Paleomagnetism of Plutonic Rocks in the Western Zone

In their study, Teissere and Beck (1973) analyzed 202 specimens paleomagnetically from 30 sites in the Peninsular Ranges batholith of southern California between the Mexican border and San Bernardino. Of the 30 sites, 11 from 17 sites in gabbros and 7 from 13 sites in granitoid rocks gave good paleomagnetic data, and all 18 accepted sites are located in the western zone of the Peninsular Ranges batholith. All five of their sites in the eastern zone, including site 6 in the La Posta pluton, failed to give an acceptable result. The 11 San Marcos gabbro sites and 7 granitoid sites give statistically similar ChRM directions at the 95% confidence level (Table 3, entries D and E) and give a combined mean direction of D = 3.3°, I = 49.5°, α_{95} = 5.0°. Using a mean cratonic reference pole of 69°N, 175°E for the Cretaceous, corresponding to a D = 337°, I = 61° for their study area, led Teissere and Beck (1973) to calculate a northward translation of 11.5° (1210 km) and a clockwise rotation of 26° for the Peninsular Ranges batholith.

Erskine and Marshall (1980) reported in an abstract that they tested samples paleomagnetically from 30 sites throughout the same general area of the northern Peninsular Ranges batholith in southern California. Of these, 11 sites from western zone plutons gave paleomagnetic results that suggested an 11° northward translation and a 40° clockwise rotation relative to stable North America. An unspecified number of their sites were in eastern zone plutons but apparently did not yield useful paleomagnetic data.

Thereafter, Hagstrum et al. (1985), as part of a much larger study of various rock units throughout the length of Baja California and southern California, reported results from 99 specimens from 14 sites (their sites PM1–PM14) to the north of San Diego in the San Marcos gabbro of the western zone of the Peninsular Ranges batholith. Their sites correspond in part to some of the sites sampled by Teissere and Beck (1973). All 14 sites gave accepted results with a mean of D = 9.2°, I = 47.8°, α_{95} = 3.4° (Table 3, entry G). They compared their San Marcos gabbro pole to a Cretaceous cratonic reference pole at 67.3°N, 173.7°W and calculated the northward translation to be 13.9° ± 4.0° and clockwise rotation to be 35.4° ± 5.0°.

The fourth relevant paleomagnetic study was by Yule and Herzig (1994) on the gently dipping Santiago Peak volcanic rocks

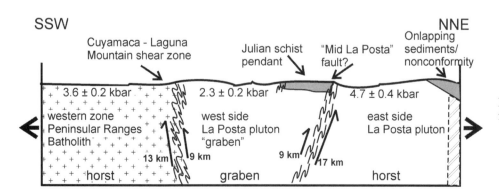

Figure 8. Sketch cross-section proposed for La Posta pluton (not to scale).

near San Diego. They reported that 9 of 15 flow units from four locations gave a tilt-corrected ChRM direction of D = 354.0°, I = 58.4°, α_{95} = 5.7°. They also report results from interbedded sediments; however, compaction shallowing of their inclination may well have given an erroneous result for them, and hence they are omitted (Dickinson and Butler, 1998; Kodama and Ward, 2001).

The western zone tonalite intrusives have yielded U-Pb zircon ages ranging from 120 to 105 Ma (Silver et al., 1979; Walawender et al., 1990), and so their sites are deemed to have a magnetization age of 113 Ma on average. There are no radiometric ages available for the San Marcos gabbros. Based on mineral assemblages and chemistry, the western zone tonalites and San Marcos gabbros of the Peninsular Ranges batholith represent two distinct igneous events with the latter probably older (Walawender and Smith, 1980). The gabbros are typical volcanic arc gabbros and represent subvolcanic magma chambers from which the Santiago Peak volcanic rocks erupted (Smith et al., 1983). This interpretation has been supported by trace element studies of the volcanics (Tanaka et al., 1984) and by chemical analysis showing that the phenocrysts in the volcanics are very similar to the cumulate minerals in the gabbros (Herzig, 1991). Although there is paleontologic evidence from sediments interbedded with the volcanic rocks of a Tithonian age (Fife et al., 1967; Jones and Miller, 1982), the most concordant of the U-Pb zircon dates indicate that the Santiago Peak volcanic rocks, and therefore the San Marcos gabbros, are 124 ± 4 m.y. old (Yule and Herzig, 1994; Herzig and Kimbrough, 1998; Dickinson and Butler, 1998).

The four paleomagnetic studies of plutonic rocks discussed above give quite similar results for the western zone of the Peninsular Ranges batholith in southern California. Thus, the data from the 18 accepted sites of Teissere and Beck (1973), the 14 accepted sites of Hagstrum et al. (1985), the nine accepted flow sites of Yule and Herzig (1994), and the two sites (24 and 25) from this study have been combined to represent the western zone of the Peninsular Ranges batholith in southern California for comparison with the La Posta pluton results. Weighting by site, their combined mean direction is D = 3.9°, I = 51.2°, N = 43, α_{95} = 2.7°, k = 66.1 (Table 3, entry H), and their combined mean age is 122 ± 4 Ma. This direction results in a pole position at 86.5°N, 8.2°W (d_p = 2.5°, d_m = 3.7°) (Fig. 6).

Western Zone Tectonics

When the pole for the western zone plutonic rocks is corrected to close the Gulf of California, it moves to 89°N, 60°W (A_{95} = 3°). Although the western zone plutonic pole matches the present Earth's rotational pole and long-term axial dipole, there is no reason to believe that it records a modern remagnetization because: (1) it incorporates data from a wide variety of lithologies, both intrusive and extrusive, and both mafic and felsic; (2) the sites are distributed over a large area in a variety of topographic settings; and (3) the pole does not match the present Earth's magnetic pole. When compared to the 122 Ma reference pole for the North American craton, the western zone is shown

to have been translated northward (poleward) by 11° ± 4° (1210 ± 440 km) and rotated clockwise by 20° ± 6°. Statistically, these translation and rotation values are clearly significantly different at >>95% confidence from the values for the North American craton and those obtained from the La Posta pluton representing the eastern zone of the Peninsular Ranges batholith.

As suggested by Butler et al. (1991) and Dickinson and Butler (1998), one possible explanation for the discordance of the western zone motion estimates compared to those of the craton and eastern zone would be a northeastern-side-up tilt of the western zone by ~21°. We disagree, noting that: (1) the Santiago Peaks volcanic rocks are nearly flat-lying and co-magmatic with the San Marcos gabbros and older than the tonalites (Tanaka et al., 1984); (2) the <124 Ma sedimentary strata in the western zone of the Peninsular Ranges batholith have variable dips but show no mapped bias favoring the tilt model; (3) the western and eastern zones of the Peninsular Ranges batholith have similar geothermobarometry trends (Ague and Brandon, 1992), and the La Posta paleomagnetic results show that the latter has not been tilted; and (4) the Ague and Brandon (1992) interpretation of their geobarometry data is credible and supports a translation and tilt interpretation of the paleomagnetic data as they noted.

The alternative explanation is that the western zone of the Peninsular Ranges batholith has been translated northward by 12° ± 5° (11° ± 4°) and rotated clockwise by 24° ± 7° (20° ± 6°) relative to the eastern zone (craton). Two basic possible models merit consideration to explain these relative motions. One is that the western zone was formed as part of the margin of the North American craton some 1210 ± 440 km south-southeast of its location when the Gulf of California started to open. This moves Baja California south-southeast by about the length of the peninsula so that southern California would be at about the latitude of México City.

The other possibility is that the western zone was formed offshore of the craton and then moved northeastward to accrete with the eastern zone. The time span available for relative transport of the western zone is fairly tightly constrained. At the older end, the constraint is the 124-Ma age of the Santiago Peaks volcanic rocks and San Marcos gabbros. At the younger end, the constraint is a swarm of rare-element pegmatitic dykes that stitch the eastern and western zones together in the southern California study area (M.J. Walawender, 2002, personal commun.). This swarm of veins gives [40]Ar-[39]Ar ages ranging from 100 to 93 Ma (Smee and Foord, 1991) and is thought to have been derived from partial melting of micaceous sediments using heat from the La Posta plutonic intrusion at ca. 93 Ma. Using the 30 Ma time span from 124 Ma to 94 Ma, the minimum translation rate is about 4.4 ± 1.8 cm/a. This is a minimum or paleolatitudinal rate because it does not include any paleolongitudinal component. Also, if the paleontologists are correct and the Santiago Peak volcanics and San Marcos gabbros prove to be ~22 m.y. older than 124 Ma, the minimum rate would drop to ~2.5 ± 1.1 cm/a.

There are good reasons to believe that the coast-parallel model for western zone translation is not realistic. Many investigators over the past quarter century have looked without success for a plausible geologic match for Baja California against the southwestern Mexican coastline. Conversely, the San Marcos gabbros and Santiago Peak volcanic rocks are typical volcanic arc rocks, as previously noted, and the tonalites could have been accreted to the arc as it moved toward California. Second, analysis of seafloor magnetic patterns have yielded a seafloor spreading direction that is about perpendicular to the Mexican coastline during the Early Cretaceous (Engebretson et al., 1985), whereas a northwesterly component of motion is required for the coast-parallel model to be valid. Translation from the southwest offshore, on the other hand, would give a close match of the seafloor spreading rate and direction of Engebretson et al. (1985) to the paleolatitudinal and velocity values required by the paleomagnetic data (Fig. 9). For example, following Engebretson et al. (1985), if North America is held fixed and the "Farallon Plate south of the Mendocino" model is used, then the Farallon plate moves toward North America on an azimuth of N43°E at 11.3 cm/a from ca. 124 Ma to 94 Ma. For Baja and southern California to translate sufficiently to move 11° ± 4° poleward (i.e., to the north pole at the time), the western zone arc would have to start ~2700 ± 1200 km offshore to the southwest and be carried northeast on the Farallon plate at a rate of 9 ± 4 cm/a. This latter rate is a reasonable fit within error to the 11.3 cm/a plate rate, and it implies that the western zone arc need only to have ridden on the Farallon plate for about 24 of the available 30 Ma; thus, the western zone could have docked with the eastern zone at ca. 100 Ma. It should be noted that either the coast-parallel or

the ocean arc model could easily explain the 20° ± 6° clockwise rotation given by the paleomagnetic data.

Ague and Brandon (1992) fitted a trend surface to 35 Al-in-hornblende geobarometric values from across the northern end of the Peninsular Ranges batholith in southern California to estimate tilt in the batholith. Their results, when projected into our study area, give reasonably similar answers. Their calculated depths of erosion appear to be ~15% greater for equivalent places, such as the eastern margin of the La Posta pluton, probably because they assumed a different rock density and did not use a temperature correction. For example, their data show ~7 km of west-side-up difference between the eastern and western zones of the Peninsular Ranges batholith to the north of our study area on the Cuyamaca–Laguna Mountain Shear Zone compared to the ~4 km difference that we found beside the La Posta pluton on the Cuyamaca–Laguna Mountain Shear Zone. Their survey, however, did not locate the ~8 km displacement recorded on the "mid La Posta" fault. From their data, Ague and Brandon (1992) suggested that the eastern and western zones of the Peninsular Ranges batholith record northeast-side-up tilts of 14° ± 3° and 18° ± 5°, respectively, about an axis trending at 341° ± 8° or about parallel to the trend of the batholith. Applying their tilt corrections to the paleomagnetic data of Teissere and Beck (1973) and Hagstrum et al. (1985), they obtained a pre-Neogene estimate of 700 ± 450 km for northward translation and a nonsignificant counterclockwise rotation estimate of 1° ± 9°. Acceptance of the Ague and Brandon (1992) model does not affect the results of this study in a substantive way. It simply requires that:

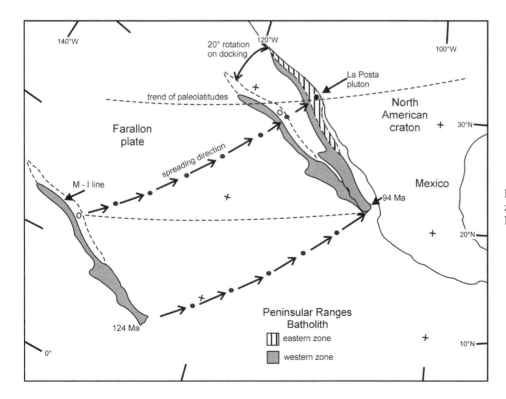

Figure 9. Proposed accretion of western zone of Peninsular Ranges batholith to North American craton, about to scale.

(a) the translation distance of the volcanic arc on the Farallon plate be reduced from ~2700 ± 1200 km to ~1800 ± 800 km so that the poleward displacement is reduced from ~1100 km to 700 km between 124 Ma and 94 Ma; and (b) the inbound arc is subjected to a northeast side-up tilt of ~16° ± 5° on accretion to the craton rather than a clockwise rotation of ~20° ± 6° before the La Posta pluton is emplaced. The fundamental conclusion that the Magnetite-Ilmenite line marks a major terrane boundary remains unchanged. A positive feature of the Ague and Brandon (1992) model is that it indicates no post-emplacement rotation of the terrane, giving a counterclockwise value of 1° ± 9° that agrees with the La Posta value of counterclockwise of 4° ± 7°.

Two recent paleomagnetic studies from western zone plutonics in México merit consideration. Böhnel and Delgado-Argote (2000) reported a pole position of 75.5°N, 189.1°W, α_{95} = 3.7° from 16 sites in the massive ca. 95 Ma San Telmo gabbroic to granodioritic batholith, about 250 km south-southeast of San Diego. Based on vertical jointing that is deemed to be primary and dips of <20° in roof pendants, they considered that the batholith is essentially untilted. Using a reference pole at 74.1°N, 192.5°E, about 1.6° from the pole used in this study, they determined an apparent counterclockwise rotation angle of 1° ± 7° and a northward translation of 3° ± 6° after closing the Gulf of California. Noting that both values are nonsignificant, they favored a hypothesis that the western zone was accreted to North America upon intrusion in the Late Cretaceous. Thus, the San Telmo result agrees with the La Posta pluton result and indicates accretion to North America of the western and eastern zones of the Peninsular Ranges batholith by ca. 95 Ma.

In a second paleomagnetic study, Böhnel et al. (2002) have reported a combined pole position of 86.6°N, 112°W, α_{95} = 4.8° from 17 sites in the 120 ± 1 Ma San Marcos basaltic to rhyolitic dike swarm and from 15 sites in the truncating El Testerazo hornblende-biotite pluton about 100 km southeast of San Diego. The dikes strike ~N30°W and dip ~75°NE suggesting ~15° of northeast-side-up tilt if vertical emplacement is assumed. Comparison with a 122 Ma reference pole at 72.3°N, 174.4°W, α_{95} = 3.3°, based on averaging poles from three selected North American units, the apparent clockwise rotation is a significant 15° ± 6°, and the poleward translation is a nonsignificant 2° ± 5° after closing the Gulf of California. The alternative and favored interpretation is that the terrane underwent ~11° of northeast-side-up tilt. We note that the San Marcos–El Testerazo pole is not statistically significantly different from the composite pole calculated for the western zone of the Peninsular Ranges batholith in this study. Thus, if the dip of the dikes is a primary rather than secondary tilting feature so that a tilt event is not required and the reference pole of this study is used, then the San Marcos–El Testerazo result also agrees with the Farallon Plate transport model hypothesis of this study.

An obvious concern with cessation of motion for the western zone of the Peninsular Ranges batholith by ca. 94 Ma is the fact that some 20 studies have reported paleomagnetic results for about 35 Upper Cretaceous and Paleogene sedimentary units—

mainly turbidites—that onlap, or are adjacent to, the western zone of the Peninsular Ranges batholith. When Neogene motion on the San Andreas fault system is removed, these units show apparent paleolatitudinal transport of 10° northward on average, as the analysis of Dickinson and Butler (1998) summarizes. These authors argue that these sedimentary units were subjected to pervasive compaction-shallowing of their remanence inclination to explain the 10° difference and cite experimental results, data comparisons from sequences with contrasting lithologies, etc., to support their argument. We agree, as Kodama and Davi (1995) and Tan and Kodama (1998) have found for some of the units in question, that compaction-shallowing of the inclination is a significant contributor to the 10° difference. In essence, the La Posta results and our interpretation of the paleomagnetic data from Early Cretaceous igneous rock units in southern California support their claim that the Peninsular Ranges batholith was static with respect to the North American craton during Late Cretaceous and Paleocene time. Another mechanism of inclination shallowing is also probable. Noting that sedimentary rocks are usually much more porous and permeable than plutonic rocks, it is likely that post-depositional subsurface fluid flows and/or modern weathering have generated significant stable magnetization or remagnetization components in the sedimentary rocks by both reduction and oxidation reactions depending on the local environment in the rocks. Such components, because of the route of the North American APWP relative to southern California, will result in apparent inclination shallowing (i.e., northward translation) and clockwise rotation of the ChRM in the sedimentary units if such components are undetected and are not entirely removed by the demagnetization and analysis procedures.

The difference in eastern and western zone paleomagnetic results, both in terms of paleolatitude and timing, indicates that the Magnetite-Ilmenite line marks the locus of significant dextral transcurrent motion and that it defines a major terrane boundary. Also, the clockwise rotation of ~25° and the inclination shallowing of ~6° of the ChRM vectors at sites 22 and 23 in the foliated margin of the La Posta pluton bring them about into alignment with the values at sites 24 and 25 that are just across the contact in the western zone tonalites. The rotation is entirely consistent with drag-folding deformation in response to dextral shear, probably a few tens of kilometers at most, along the pluton's contact shortly after its intrusion. Also, the shallowing is entirely consistent with the Cuyamaca–Laguna Mountain Shear Zone, accommodating about 4 km more uplift on its west side than east side. Together, the rotation and shallowing imply a dextral transpressive shear zone with a minor component of thrust fault motion following emplacement of the La Posta pluton.

Finally, it is obvious that more paleomagnetic studies are needed on the Peninsular Ranges batholith rocks. In particular, the La Posta result provides the only paleomagnetic data available to represent the entire eastern zone of the batholith. There are several other La Posta–type plutons along the eastern zone that are obvious targets for further geobarometric and paleomagnetic research. Also in the western zone, the tonalite intrusions and Santiago Peak volcanic rocks are, in addition to being candidates for definitive age dating studies, certainly candidates for more thorough paleomagnetic analysis.

CONCLUSIONS

The 94 Ma La Posta pluton—and presumably other La Posta–type plutons in the Peninsular Ranges batholith—has proven suitable for paleomagnetic study. However, there are caveats. First, they have a weak stable ChRM in the low 10^{-5} A/m range, requiring a good cryogenic magnetometer to measure. Second, they have substantial viscous remanence components so that several months of storage and measurement in a good magnetic field-free space is desirable prior to, and during, measurement. Third, these plutons were emplaced at depths of 10 ± 5 km so that their ChRM is best isolated by thermal steps between ~500° and 600 °C. And fourth, extensive modern hematitic weathering is present in outcrops so that "fresh" rock sites such as road cuts must be available to sample. The ChRM is carried by titanomagnetite and magnetite primarily that is shown to be single or pseudosingle domain in nature by saturation isothermal remanence studies.

Aluminum-in-hornblende geobarometry with temperature corrections determined using the amphibole-plagioclase geothermometer were successful in giving useful results for the outer Hornblende-Biotite and Large Biotite facies of the La Posta pluton in the eastern zone of the Peninsular Ranges batholith and from the tonalites in the adjacent western zone of the Peninsular Ranges batholith. The calculated intrusion pressures show a geographic distribution with the eastern and western sides of the La Posta pluton and the tonalites in the western zone of the Peninsular Ranges batholith giving average values of 4.7 ± 0.4 kbar, 2.5 ± 0.2 kbar and 3.6 ± 0.2 kbar, respectively.

The ChRM direction for the La Posta pluton gives a pole position within $1° \pm 5°$ of the 94 Ma reference pole for the North American craton both before and after tectonic correction to close the Gulf of California, moving Baja and southern California back to their pre-Neogene position. This indicates that the La Posta pluton was emplaced into the eastern zone of the Peninsular Ranges batholith when it was part of the craton. Similarly, the pluton's pole position gives a nonsignificant $4° \pm 7°$ of counterclockwise rotation to further affirm its cratonic origin.

These nonsignificant translation and rotation values indicate also that the La Posta pluton has not been tilted since emplacement and, therefore, the eastern zone of the Peninsular Ranges batholith has not been tilted subsequently as has been suggested to explain paleomagnetic results from plutonic rocks in the adjacent western zone of the Peninsular Ranges batholith. Thus it is proposed that the differences in geobarometric depth determinations are explained by: (a) an inferred high-angle normal (?) fault located within the metasedimentary screen running through the La Posta pluton that has about 8 km of differential vertical motion between the eastern and western sides of the pluton; and (b) an eastward-dipping normal fault—probably the Cuyamaca-

Laguna Mountain Shear Zone—located west of the La Posta pluton with about 4 km of differential vertical motion. Clockwise rotation of ~25° of the ChRM vectors for the two sides in the pluton's western margin indicate several to tens of kilometers of dextral transcurrent displacement on the Cuyamaca–Laguna Mountain Shear Zone also.

In contrast to the eastern zone of the Peninsular Ranges batholith, previously published paleomagnetic data from Early Cretaceous plutonic rocks in the western zone of southern California record $11° \pm 4°$ (1210 ± 440 km) of northward translation and $20° \pm 6°$ of clockwise rotation. Radiometric, geochemical and geologic evidence constrains the translation to occur between ca. 124 Ma and 94 Ma, giving a minimum translation rate of 4.0 ± 1.5 cm/a. Noting the petrologic and geochemical evidence suggesting that the western zone of the Peninsular Ranges batholith originated as a volcanic arc, we propose that it was formed about 2700 km offshore to the southwest of North America. Starting at ca. 124 Ma, the arc was translated northeastward on the Farallon plate, accreting to North America in a dextral transpressive regime by ca. 94 Ma. Thus, the western zone of the Peninsular Ranges batholith is a terrane distinct from the eastern zone and explains the observed differences between the eastern and western zones in virtually every geological, geochemical, and geophysical property that has been measured to date. Therefore, the dividing line between the zones, closely approximated by the Magnetite-Ilmenite line that runs down the center of the Peninsular Ranges batholith for most of its length, is a major terrane boundary.

ACKNOWLEDGMENTS

The authors gratefully thank Shelie Ann Cascadden who did much of the specimen preparation, Mryl Beck and Ken Kodama, who provided helpful reviews that significantly improved the manuscript, and the Natural Sciences and Engineering Research Council of Canada, who funded this research through a grant to the lead author.

REFERENCES CITED

Abbott, P.L., and Smith, T.E., 1989, Sonora, Mexico, source of the Eocene Poway Conglomerate of southern California: Geology, v. 17, p. 329–332.

Ague, J.J., and Brandon, M.T., 1992, Tilt and northward offset of Cordilleran batholiths resolved using igneous barometry: Nature, v. 360, p. 146–149.

Anderson, J.L., and Smith, D.R., 1995, The effects of temperature and fO_2 on the Al-in-hornblende barometer: American Mineralogist, v. 80, p. 549–559.

Atwater, T., 1970, Implications of plate tectonics for the Cenozoic tectonic evolution of western North America: Geological Society of America Bulletin, v. 81, p. 3513–3536.

Baird, A.K., and Meisch, A.T., 1984, Batholithic rocks of southern California: A model for the petrochemical nature of their source materials: U.S. Geological Survey Professional Paper 1284, 42 p.

Beck Jr., M.E., 1991, Case for northward transport of Baja and coastal southern California: Paleomagnetic data, analysis, and alternatives: Geology, v. 19, p. 506–509.

Besse, J., and Courtillot, V., 1991, Revised and synthetic apparent polar wander paths of the African, Eurasian, North American and Indian plates, and true polar wander since 200 Ma: Journal of Geophysical Research, v. 96, p. 4029–4050.

Blundy, J.D., and Holland, T.J.B., 1990, Calcic amphibole equilibria and a new amphibole-plagioclase geothermometer: Contributions to Mineralogy and Petrology, v. 104, p. 208–224.

Böhnel, H., and Delgado-Argote, L., 2000, Paleomagnetic data from northern Baja California (Mexico): Results from the Cretaceous San Telmo batholith, in Delgado-Granados, H., Aguirre-Diaz, G., and Stock, J.M., eds., Cenozoic tectonics and volcanism of Mexico: Boulder, Colorado, Geological Society of America Special Paper 334, p. 157–165.

Böhnel, H., Delgado-Argote, L.A., and Kimbrough, D.L., 2002, Discordant paleomagnetic data for middle-Cretaceous intrusive rocks from northern Baja California: Latitude displacement, tilt, or vertical axis rotation?: Tectonics, v. 21, no. 5, p. 13-1–12-12.

Butler, F.R., Dickinson, W.R., and Gehrels, G.E., 1991, Paleomagnetism of coastal California and Baja California: Alternatives to large-scale northward transport: Tectonics, v. 10, p. 561–576.

Clickenbeard, J.P., and Walawender, M.J., 1989, Mineralogy of the La Posta Pluton: Implications for the origin of zoned plutons in the eastern Peninsular Ranges batholith, southern and Baja California: American Mineralogist, v. 74, p. 1258–1269.

Diamond, J.L., 1985, The magnetite/ilmenite line in the Peninsular Ranges batholith, Baja California, Mexico [B.Sc. thesis]: San Diego, San Diego State University, 54 p.

Dickinson, W.R., 1996, Kinematics of transrotational tectonism in the California Transverse Ranges and its contribution to cumulative slip along the San Andreas transform fault system: Boulder, Colorado, Geological Society of America Special Paper 305, 46 p.

Dickinson, W.R., and Butler, R.F., 1998, Coastal and Baja California paleomagnetism reconsidered: Geological Society of America Bulletin, v. 110, p. 1268–1280.

Dunlop, D.J., 1990, Developments in rock magnetism: Report of Progress in Physics, v. 53, p. 707–792.

Duffield, W.A., 1968, The geology and structure of El Pinal tonalite, Baja California, Mexico: Geological Society of America Bulletin, v. 79, p. 1351–1374.

Engebretson, D.C., Cox, A., and Gordon, R., 1985, Relative motions between oceanic and continental plates in the Pacific Basin: Boulder, Colorado, Geological Society of America Special Paper 206, 59 p.

Erskine, B.G., and Marshall, M., 1980, Magnetite- and ilmenite-series sub-belts in the northern Peninsular Ranges batholith, southern California: Their association with magnetic stability and implications on source regions: Eos (Transactions, American Geophysical Union), v. 62, p. 849.

Fife, D.L, Minch, J.A., and Crampton, P.J., 1967, Late Jurassic age of the Santiago Peak Volcanics, California: Geological Society of America Bulletin, v. 78, p. 229–304.

Fisher, R.A., 1953, Dispersion on a sphere: Royal Astronomical Society of London Proceedings, v. A217, p. 295–305.

Gastil, R.G., 1975, Plutonic zones in the Peninsular Ranges of southern California and northern Baja California: Geology, v. 3, p. 361–363.

Gastil, G., 1991, Is there a Oaxaca-California megashear? Conflict between paleomagnetic data and other elements of geology: Geology, v. 19, p. 502–505.

Gastil, R.G., Morgan, G.J., and Krummenacher, D., 1981, The tectonic history of peninsular California and adjacent Mexico, in Ernst, W.G., ed., The geotectonic development of California: Englewood Cliffs, New Jersey, Prentice-Hall, p. 294–306.

Gastil, R.G., Diamond, J., and Knaack, C., 1986, The magnetite-ilmenite line in peninsular California: Geological Society of America Abstracts with Programs, v. 18, p. 109.

Gastil, G., Diamond, J., Knaack, C., Walawender, M., Marshall, M., Boyles, C., Chadwick, B., and Erskine, B., 1990, The problem of the magnetite/ilmenite boundary in southern and Baja California, in Anderson, J.L., ed., The nature and origin of Cordilleran magmatism: Boulder, Colorado, Geological Society of America Memoir 174, p. 19–32.

Gromet, L.P., and Silver, L.T., 1979, Profile of rare earth elements characteristics across the Peninsular Ranges batholith near the international border, southern California, U.S.A., and Baja California, Mexico, in Abbott, P.L., and Todd, V.R., eds., Mesozoic crystalline rocks: Peninsular Ranges batholith and pegmatites, Point Sal Ophiolite: San Diego, San Diego State University, p. 133–141.

Hagstrum, J.T., McWilliams, M., Howell, D.G., and Grommé, S., 1985, Mesozoic paleomagnetism and northward translation of the Baja California peninsula: Geological Society of America Bulletin, v. 96, p. 1077–1090.

Halls, H.C., 1978, The use of converging remagnetization circles in paleomagnetism: Physics of the Earth and Planetary Interiors, v. 16, p. 1–11.

Harris, M.J., Symons, D.T.A., Blackburn, W.H., and Hart, C.J.R., 1996, Paleomagnetic study of the Mount McIntyre pluton, Whitehorse map area (105D), southern Yukon Territory: Yukon Exploration and Geology 1996: Exploration and Geological Services Division, Yukon, Indian and Northern Affairs Canada, Whitehorse, p. 122–130.

Harris, M.J., Symons, D.T.A., Blackburn, W.H., and Hart, C.J.R., 1997, Paleomagnetic and geobarometric study of the mid-Cretaceous Whitehorse Pluton Yukon Territory: Canadian Journal of Earth Sciences, v. 34, p. 1379–1391.

Harris, M.J., Symons, D.T.A., Hart, C.J.R., and Blackburn, W.H., 1999, Jurassic plate motions of the Stikine Terrane, southern Yukon: A paleomagnetic and geothermometric study of the Teslin Crossing Pluton (105E/7), in Roots, C.F., and Edmond, D.S., eds., Yukon Exploration and Geology 1998: Exploration and Geological Services Division, Yukon, Indian and Northern Affairs Canada, Whitehorse, p. 155–166.

Herzig, C.T., 1991, Petrogenetic and tectonic development of the Santiago Peak Volcanics, northern Santa Ana Mountains, California [Ph.D. thesis]: Riverside, University of California, 376 p.

Herzig, C.T., and Kimbrough, D.K., 1998, Provenance of silicic volcanic clasts in upper Cretaceous strata, northern Ana Mountains, California, in Behl, R.J., ed., Late Cretaceous denudation history of the Peninsular Ranges as recorded in upper Cretaceous-Paleocene sedimentary rocks, northern Santa Ana Mountains: Long Beach, Department of Geological Sciences, California State University, p. 23–29.

Ichinose, G., Day, S.M., Magistrale, H., Prush, T., Vernon, F.L., and Edelman, A., 1996, Crustal thickness variations beneath the Peninsular Ranges, southern California: Geophysical Research Letters, v. 23, p. 3095–3098.

Johnson, S.E., Tate, M.C., and Fanning, C.M., 1999, New mapping and SHRIMP U-Pb zircon data in the Peninsular Ranges batholith, Baja California, Mexico: Evidence for a suture?: Geology, v. 27, p. 743–746.

Jones, D.A., and Miller, R.H., 1982, Jurassic fossils from the Santiago Peak Volcanics, San Diego County, California, in Abbott, P.L., ed., Geologic studies in San Diego: San Diego Association of Geologists Publication, p. 93–103.

Kirschvink, J.L., 1980, The least-squares line and plane and the analysis of paleomagnetic data: Royal Astronomical Society Geophysical Journal, v. 62, p. 699–618.

Kodama, K.P., and Davi, J.M., 1995, A compaction correction for the paleomagnetism of the Cretaceous Pigeon Point Formation of California: Tectonics, v. 14, p. 1153–1164.

Kodama, K.P., and Ward, P.D., 2001, Compaction-corrected paleomagnetic paleolatitudes for Late Cretaceous rudists along the Cretaceous California margin: Evidence for less than 1500 km of post-Late Cretaceous offset for Baja British Columbia: Geological Society of America Bulletin, v. 113, p. 1171–1178.

Knaack, C.M., 1985, Magnetite/ilmenite division within Peninsular Ranges, southern California [B.S. thesis]: San Diego, San Diego State University, 54 p.

Krummenacher, D., Gastil, R.G., Bushee, J., and Doupont, J., 1975, Potassium-argon apparent ages in the Peninsular Ranges batholith, southern California and northeastern Mexico: Geological Society of America Bulletin, v. 86, p. 760–768.

Leake, B.E., and 21 others, 1997, Nomenclature of amphiboles: Report of the sub-committee on amphiboles of the International Mineralogical Association, Commission on New Minerals and Mineral Names: Canadian Mineralogist, v. 35, p. 219–246.

Lonsdale, P., 1991, Structural patterns of the Pacific floor offshore of peninsular California, in Dauphin, J.P., and Simoneit, B.R.T., eds., The gulf and peninsular provinces of the Californias: Tulsa, Oklahoma, American Association of Petroleum Geologists Memoir 47, p. 87–125.

McFadden, P.L., and Lowes, F.J., 1981, The discrimination of mean directions drawn from Fisher distributions: Geophysical Journal of the Royal Astronomical Society, v. 67, p. 19–33.

Oliver, H.W., 1980, Interpretation of the gravity map of California and the continental region: California Division of Mines and Geology Bulletin v. 205, 52 p.

Patterson, C., Silver, L., and McKinney, C., 1956, Lead isotopes and magmatic differentiation: Report of the International Geological Congress XX, Session 11b, México City: México D.F., p. 221–222.

Pullaiah, G., Irving, E., Buchan, K.L., and Dunlop, D.J., 1975, Magnetization changes caused by burial uplift: Earth and Planetary Science Letters, v. 28, p. 133–143.

Samson, I.M., Blackburn, W.H., and Gagnon, J.E., 1999, Paragenesis and composition of amphibole and biotite in the MacLellan gold deposit, Lynn

Lake greenstone belt, Manitoba, Canada: Canadian Mineralogist, v. 37, p. 1405–1421.

Schmidt, M.W., 1992, Amphibole composition in tonalite as a function of pressure: An experimental calibration of the Al-in-hornblende barometer: Contributions to Mineralogy and Petrology, v. 110, p. 304–310.

Schumacher, J.C., 1997, The estimation of the proportion of ferric iron in electron-microprobe analysis of amphiboles: Canadian Mineralogist, v. 35, p. 238–246.

Silver, L.T., and Chappell, B.W., 1988, The Peninsular Ranges batholith; An insight into the evolution of the Cordilleran batholiths of southwestern North America: Transactions of the Royal Society of Edinburgh, Earth Sciences, v. 79, p. 105–121.

Silver, L.T., Taylor, H.P., Jr., and Chappell, B., 1979, Some petrological, geochemical, and geochronological observations of the Peninsular Ranges batholith near the international border of the U.S.A., and Mexico, *in* Abbott, P.L., and Todd, V.R., eds., Mesozoic crystalline rocks: Peninsular Ranges batholith and pegmatites, Point Sal Ophiolite: San Diego, San Diego State University, p. 83–110.

Smee, I.W., and Foord, E.E., 1991, ⁴⁰Ar/³⁹Ar thermochronology of granitic pegmatites and host rocks, San Diego County, California: Geological Society of America Abstracts with Programs, v. 23, no. 5, p. A479.

Smith, T.E., Huang, C.H., Walawender, M.J., Cheung, P., and Wheeler, C., 1983, The gabbroic rocks of the Peninsular Ranges Batholith, southern California: Cumulate rocks associated with calc-alkalic basalts and andesites: Journal of Volcanology and Geothermal Research, v. 18, p. 249–278.

Symons, D.T.A., Williams, P.R., McCausland, P.J.A., Harris, M.J., Hart, C.J.R., and Blackburn, W.H., 2000, Paleomagnetism and geobarometry of the Big Creek batholith suggests that the Yukon-Tanana terrane has been a parautochthon since Early Jurassic: Tectonophysics, v. 326, p. 57–72.

Tan, X., and Kodama, K.P., 1998, Compaction-corrected inclinations from southern California marine sedimentary rocks indicate no paleolatitudinal offset for the Peninsular Ranges terrane: Journal of Geophysical Research, v. 103, p. 27,164–27,192.

Tanaka, H., Smith, T.E., and Huang, C.H., 1984, The Santiago Peak Volcanic rocks of the Peninsular Ranges batholith, southern California, volcanic rocks associated with co-eval gabbros: Bulletin Volcanolgique, v. 47, p. 153–171.

Teissere, R.F., and Beck, M.E., Jr., 1973, Divergent Cretaceous paleomagnetic pole positions for the southern California batholith, U.S.A.: Earth and Planetary Science Letters, v. 18, p. 296–300.

Thomson, C.N., and Girty, G.H., 1994, Earth Cretaceous intra-arc ductile strain in Triassic-Jurassic and Cretaceous continental margin arc rocks, Peninsular Ranges, California: Tectonics, v. 13, p. 1108–1119.

Todd, V.R., and Shaw, S.E., 1979, Structural, metamorphic and intrusive framework of the Peninsular Ranges batholith in southern San Diego County, California, *in* Abbott, P.L., and Todd, V.R., eds., Mesozoic crystalline rocks: Peninsular Ranges batholith and pegmatites, Point Sal Ophiolite: San Diego, San Diego State University, p. 177–231.

Todd, V.R., and Shaw, S.E., 1985, S-type granitoids and an I-S line in the Peninsular Ranges batholith, southern California: Geology, v. 13, p. 231–233.

Walawender, M.J., and Smith, T.E., 1980, Geochemical and petrologic evolution of the basic plutons of the Peninsular Ranges batholith, southern California: Journal of Geology, v. 88, p. 233–242.

Walawender, M.J., Gastil, R.G., Clinkenbeard, J.P., McCormick, W.V., Eastman, B.G., Wernicke, R.S., Wardlaw, M.S., Gunn, S.H., and Smith, B.M., 1990, Origin and evolution of the zoned La Posta-type plutons, eastern Peninsular Ranges batholith, southern and Baja California, *in* Anderson, J.L., ed., The nature and origin of Cordilleran magmatism: Boulder, Colorado, Geological Society of America Memoir 174, p. 1–18.

Walawender, M.J., Girty, G.H., Lombardi, M.R., Kimbrough, D., Girty, M.S., and Anderson, C., 1991, A synthesis of recent work in the Peninsular Ranges batholith, *in* Walawender, M.J., and Hana, B.B., eds., Geological excursions in southern California and Mexico: San Diego, Department of Geological Sciences, San Diego State University, p. 297–318.

Wernicke, R.S., 1987, Origin and emplacement of the El Topo pluton, northern Baja California [M.S. thesis]: San Diego, San Diego State University, 205 p.

Yule, J.D., and Herzig, C.T., 1994, Paleomagnetic evidence for no large-scale northward translation of the southern California Peninsular Ranges: Geological Society of America Abstracts with Program, v. 26, no. 7, p. A-461.

Zijderveld, J.D.A., 1967, A.C. demagnetization of rocks: Analysis of results, *in* Collinson, D.W., Creer, K.M., and Runcorn, S.K., eds., Methods in paleomagnetism: Amsterdam, Elsevier, p. 254–286.

MANUSCRIPT ACCEPTED BY THE SOCIETY JUNE 2, 2003

Geological Society of America
Special Paper 374
2003

Jurassic peraluminous gneissic granites in the axial zone of the Peninsular Ranges, southern California

S.E. Shaw
GEMOC (National Key Centre for the Geochemical Evolution and Metallogeny of Continents), Department of Earth and Planetary Sciences, Macquarie University, NSW 2109, Australia

V.R. Todd
Palomar College, San Marcos, California 92069, USA

M. Grove
Department of Earth and Space Sciences, University of California, Los Angeles, California 90095, USA

ABSTRACT

The Peninsular Ranges batholith of southern California and Baja California, México, is well recognized as a prime example of an I-type Cretaceous batholith. Often overlooked, however, is a volumetrically significant amount of pre-Cretaceous gneissic granite in the axial zone of the batholith. New U-Pb zircon age data confirm that the metaluminous and peraluminous plutonic bodies were emplaced during the middle Jurassic. Also reported in this paper is a Jurassic U-Pb age for a metaluminous (I-type) tonalite-quartz diorite pluton that is spatially related to the peraluminous suites. This result suggests that other unrecognized Jurassic I-type plutons may also be present in the batholith.

Within San Diego County, Todd and Shaw (1985) recognized and mapped two suites of strongly deformed gneissic granites and migmatites. One is peraluminous (Harper Creek suite) while the other is transitional between metaluminous and peraluminous (Cuyamaca Reservoir suite). These rocks bear a striking resemblance to deformed and, in places, migmatitic, peraluminous (S-type) examples from the Lachlan fold belt and New England batholith of eastern Australia. The gneissic granite suites are known to extend north along the axial zone of the Peninsular Ranges batholith and cover an area at least 45 km wide by 150 km long. To the south, rocks of similar type are known to extend into Baja California, México, for at least 300 km. Chemical and isotopic studies of these Jurassic suites confirm that they meet the criteria necessary to define them as S-type and transitional I- to S-type, respectively. However, unlike the majority of Lachlan fold belt S-type granites that are high level and often associated with their volcanic equivalents, the Peninsular Ranges batholith suites were emplaced at much deeper levels, possibly as much as 11–16 km. The Harper Creek suite, of S-type gneissic granodiorite-tonalite plutons and associated Stephenson Peak migmatitic schist and gneiss facies, is strongly peraluminous and contains biotite, cordierite, sillimanite, abundant graphite, and ilmenite. It has elevated $\delta^{18}O$ up to +20 per mil, initial $^{87}Sr/^{86}Sr$ ratios (Sr$_i$) to 0.713, a high aluminum saturation index, and Na$_2$O/K$_2$O ratios that overlap those of the Lachlan fold belt S-type granites. The Cuyamaca Reservoir suite contains gneissic granodiorite-tonalite plutons, transitional between metaluminous and moderately peraluminous (I- to S-type), containing reduced biotite, subaluminous amphibole, orthopyroxene, titanite

Shaw, S.E., Todd, V.R., and Grove, M., 2003, Jurassic peraluminous gneissic granites in the axial zone of the Peninsular Ranges, southern California, *in* Johnson, S.E., Paterson, S.R., Fletcher, J.M., Girty, G.H., Kimbrough, D.L., and Martín-Barajas, A., eds., Tectonic evolution of northwestern México and the southwestern USA: Boulder, Colorado, Geological Society of America Special Paper 374, p. 157–183. For permission to copy, contact editing@geosociety.org. © 2003 Geological Society of America.

and ilmenite. It has values of $\delta^{18}O$ and Sr_i greater than the Cretaceous I-type granites but less than the Harper Creek suite. Leucosome melt phase accumulation from the Julian Schist diatexites to produce a restite-rich magma is considered the most likely origin for the Harper Creek suite. For the Cuyamaca Reservoir suite, possible source components include young mantle-derived magma, metaigneous and metasedimentary rocks formed in an arc environment, and evolved basinal fill metasedimentary rocks. The evolved metasedimentary component may be of Julian type.

The Harper Creek and Cuyamaca Reservoir suites comprise deformed, steep-walled, north-northwest–trending bodies up to 20 km long and with length-to-width ratios of 4:1. Textures range from strongly foliated to gneissic or mylonitic. Internal foliation that strikes parallel to the long dimension of the bodies and dips steeply to the east is defined by alignment of relict magmatic feldspar and quartz grains and recrystallized aggregates of quartz and biotite. The concordance of magmatic and subsolidus foliations in the Jurassic plutons and the continuity of these structures with regionally developed metamorphic fabrics in their wallrocks indicate that magmatic foliation was overprinted by high-temperature, post-magmatic solid-state foliation. Foliation in the Harper Creek and Cuyamaca Reservoir suites is concordant with the axial planes of outcrop-scale isoclinal folds, and map patterns suggest that the plutons underwent regional-scale isoclinal folding. Their fabric probably formed during multiple episodes of synintrusive deformation that began at least by the Late Jurassic and culminated by the middle Cretaceous.

Keywords: granite, Jurassic, migmatite, Peninsular Ranges, peraluminous, restite, S-type.

INTRODUCTION

The Peninsular Ranges batholith of southern California, U.S.A., and Baja California, México, is traditionally viewed as having evolved wholly within the Cretaceous (Silver and Chappell, 1988). Although the great volume of the Cretaceous plutonic rocks has obscured the earlier Mesozoic history of the region, the recognition of gneissic granites of Jurassic age (Thomson and Girty, 1994) in the southern California segment of the batholith extends the history of igneous activity significantly and increases the similarity of the Peninsular Ranges batholith to the Sierra Nevada batholith and other arc terranes along the western margin of North America.

Palinspastic restoration of Neogene dextral displacement on the San Andreas fault system places the Peninsular Ranges batholith about 300 km south of its present position, opposite the Sonoran Desert region (Gastil, 1993; Fackler-Adams et al., 1997). Whereas an impressive body of data describes Jurassic volcanic and plutonic rocks of the Mojave-Sonoran arc (Busby-Spera et al., 1990; Staude and Barton, 2001), relatively little is known about contemporaneous magmatic activity to the west, oceanward of the continental margin. However, Jurassic igneous activity and metamorphism has been reported in the Sierra San Pedro Martir area, Baja California (Schmidt and Paterson, 2002). Data presented in this paper show that a substantial part of the central Peninsular Ranges batholith, extending from at least 33° 45′ N in southern California and possibly as far south as 28° N in Baja California, is composed of a belt of Jurassic gneissic granites (Todd et al., l99la; Thomson and Girty, 1994) (Fig. 1).

Broad regions of gneissic granitoids (Stonewall Peak granodiorite) and "mixed rocks" that were believed to be of potential Jurassic age were originally mapped by Everhart (1951) and Merriam (1958) within the axial zone of the Peninsular Ranges batholith near 33°N. Todd and Shaw (1979, 1985) studied these heterogeneous rocks in more detail and were able to recognize two compositionally distinct suites. Following the original definition of I-type and S-type granites of Chappell and White (1974) and White and Chappell (1977) for rocks in southeastern Australia, and noting the similarity, Todd and Shaw (1985) described an S-type Harper Creek suite and a transitional I-type to S-type Cuyamaca Reservoir suite. Although the application of these terms to plutons in the Peninsular Ranges batholith was questioned by White et al. (1986), the gneissic granites more than meet the criteria of Chappell and White (1974, 1992) and White and Chappell (1988) for I- and S-type granites from southeastern Australia. The distinctive criteria used to define S-types in the Lachlan fold belt (and which also characterize the Jurassic peraluminous gneissic granites of the Peninsular Ranges batholith) include: molecular $Al_2O_3/(Na_2O+K_2O+CaO) >1.1$, equivalent to CIPW normative corundum >1%; relatively high SiO_2 values (>63%); less regular inter-element variations than I-types; reduced nature of the granites as reflected by ilmenite rather than magnetite and reduced biotite with red-brown pleochroism; cordierite present in less deformed granites; and elevated $Sr_i > 0.708$ and $\delta^{18}O > +10$ per mil.

The word "granite" as used in this paper is a general term to describe coarse-grained rocks of igneous origin consisting mainly of quartz and feldspars. The term "gneissic granite" is used to

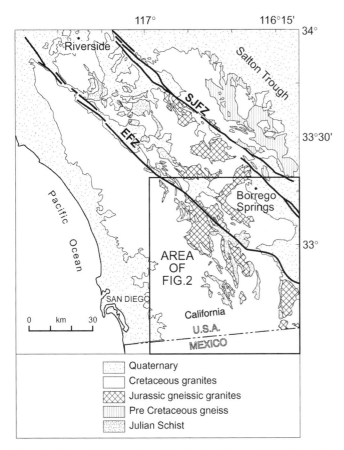

Figure 1. Map of the northern Peninsular Ranges batholith (after Jachens et al., 1986; Todd, 1978, 1994a, 1994b; Todd et al., 1988). Neogene San Andreas fault system: SJFZ—San Jacinto fault zone; EFZ—Elsinore fault zone. Jurassic gneissic granites are in part mylonitic and include migmatite and gneiss facies. Pre-Cretaceous gneiss includes orthogneiss (Harper Creek–type?) and anatexites, mylonites, and Paleozoic metasedimentary rocks.

describe metamorphosed "granite," although compositionally, the rocks referred to in this paper are mainly granodiorite and tonalite. The term "restite" is used to describe residuum or solid material in equilibrium with the melt that remains after a melting event.

The Jurassic Peninsular Ranges batholith gneissic granite belt is essentially parallel to, but lies ~150 km west of, the intracontinental Jurassic-Triassic Mojave-Sonoran arc. An estimate of the outcrop area of the Peninsular Ranges batholith Jurassic gneissic granites from Figure 1 is ~1155 km². While this represents a small proportion of the more than 1200-km-long, mainly Cretaceous batholith, the area is comparable with Australian examples of mixed I-S-type batholiths in the Lachlan fold belt: Berridale (1670 km²), Murrumbidgee (1470 km²), and Kosciusko (4000 km²). The Cooma S-type gneiss complex in the Lachlan fold belt is, however, minor at 14 km² (areas from Chappell et al., 1991).

This paper discusses in more detail the field relationships, mineralogy, U-Pb zircon geochronology, and geochemistry of the Jurassic Harper Creek and Cuyamaca Reservoir suites.

Descriptions of the metaluminous tonalite–quartz diorite plutons of the East Mesa suite that are also partly Jurassic in age (this paper) are given in Todd et al. (this volume).

GEOLOGICAL SETTING

The north-northwest–trending Peninsular Ranges batholith exhibits a strong transverse asymmetry (DePaolo, 1981; Gastil, 1983; Taylor, 1986; Gromet and Silver, 1987; Silver and Chappell, 1988; Todd et al., 1988; Walawender et al., 1990). This asymmetry is considered to reflect the change in granite source regions from oceanic lithosphere on the west to continental lithosphere on the east. The resultant west-to-east lithologic variations within plutons and prebatholithic wallrocks, which were noted in early studies, have been confirmed by subsequent geochemical, isotopic, and geophysical studies (Taylor, 1986; Shaw et al., 1986; Silver and Chappell, 1988; Walawender et al., 1990; Jachens et al., 1986, 1991; Wooden et al., 1997; Premo et al., 1998). Todd and Shaw (1985, their fig. 2) summarized significant boundaries and gradients that roughly divide the batholith into western and eastern zones. Prebatholithic rocks in the western zone of the Peninsular Ranges batholith are chiefly Mesozoic volcanic and sedimentary rocks, whereas those in the eastern zone consist of sedimentary and lesser volcanic rocks of Mesozoic, Paleozoic, and Late Proterozoic age (Gastil, 1993). The batholith in the western zone comprises relatively magnetic, dense plutonic and volcanic rocks and in the eastern zone, less dense, virtually nonmagnetic plutons and metasedimentary wallrocks. Jachens et al. (1986) modeled the steep gravity-magnetic gradient that separates the two zones as the expression of a tectonic boundary dipping 45–60° to the east and extending to depths of at least 10–12 km. In San Diego County, this geophysical boundary coincides with the western limit of Jurassic granites and is shown in Figure 2 as the boundary PSGR (pseudogravity gradient). Jachens (1992, personal commun.) defines the pseudogravity gradient as the "mathematical transformation of magnetic field data into pseudogravity anomalies, effectively converting the magnetic field to the gravity field that would be produced if all magnetic material were replaced by proportionally dense material." The $\delta^{18}O$ step of Taylor and Silver (1978) corresponds closely with the PSGR boundary (Fig. 2). In a depth subdivision of the Peninsular Ranges batholith (Gastil, 1975), the boundary of intermediate Zone B with deepest Zone C lies approximately 5 km to the west of the PSGR boundary. In the threefold subdivision of the Peninsular Ranges batholith based on rare earth element (REE) variation (Gromet and Silver, 1987), the boundary between western and central regions lies ~5 km to the east of the PSGR boundary (Fig. 2).

On the basis of these discontinuities, Todd and Shaw (1985) suggested they are indicative of a fundamental change in the Mesozoic crust, possibly a suspect terrane boundary formed by the accretion of a western volcanic arc and the continental margin to the east. Schmidt and Paterson (2002) describe the boundary as an example of crustal transition between juxtaposed oceanic and continental floored arcs, with Jurassic-Cretaceous deforma-

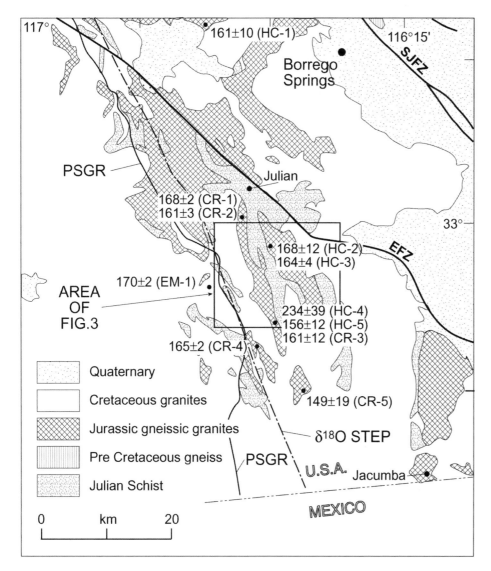

Figure 2. Enlargement of southern part of Figure 1 shows approximate sample locations for dated Jurassic plutons (see section on age of Jurassic granites). Locations have been combined in cases where samples were collected within a distance of approximately 1–2 km. HC—Harper Creek suite: HC-1, this paper; HC-2, Todd et al. (1991b, Rb/Sr whole-rock age); HC-3, this paper; HC-4 and HC-5, Thomson and Girty (1994). CR—Cuyamaca Reservoir suite: CR-1 and CR-2, this paper; CR-3, Thomson and Girty (1994); CR-4, this paper; CR-5, Thomson and Girty (1994). EM—East Mesa suite: EM-1, this paper. Unless specified, all are zircon ^{206}Pb/^{238}U ages. Also shown are SJFZ (San Jacinto fault zone) and EFZ (Elsinore fault zone) (as in Fig. 1), δ^{18}O step of Taylor and Silver (1978) and gravity-magnetic gradient of Jachens et al. (1986, 1991) shown here as pseudogravity gradient boundary (PSGR). Gravity and magnetic gradients coincide closely with western edge of terrain of Jurassic granites and Julian Schist and I-S line of Todd and Shaw (1985).

tion evident for at least 800 km along the axis of the Peninsular Ranges batholith.

The oldest plutonic rocks recognized in the Peninsular Ranges batholith are highly deformed Middle and Late Jurassic metaluminous to peraluminous granodiorites and tonalites that are exposed in the central zone of the Peninsular Ranges batholith (Todd et al., 1991a, 1991b; Thomson and Girty, 1994). Plutonic rocks of possible Triassic age (Girty et al., 1994) have also been reported (see section on age below). By the Late Jurassic, these granites and their Mesozoic and Paleozoic wallrocks formed a belt at least 45 km wide along the southwestern margin of North America (Figs. 1 and 2). About 40 km west of the belt of Middle and Late Jurassic plutons are scattered exposures of Late Jurassic to Early Cretaceous volcaniclastic-volcanic island arc-type rocks (Balch et al., 1984; Anderson, 1991; Wetmore et al., 2001). The Jurassic plutons were subsequently intruded on the west by volu-

minous Early Cretaceous I-type plutons (ca. 120–105 Ma, Silver and Chappell, 1988; Premo et al., 1998). Contacts between the Cretaceous and Jurassic plutons are broadly concordant and in detail suggest mylonitization and localized incipient melting of the Jurassic granites during Cretaceous intrusion. Between ca. 105 and 89 Ma, with a major pulse centered on 95 Ma (Wala-wender et al., 1990; Kimbrough et al., 2001), large hornblende-bearing trondhjemite to garnetiferous two-mica monzogranite plutons (La Posta–type) intruded the eastern part of the Jurassic belt. Unlike the Early Cretaceous intrusions, the middle to Late Cretaceous plutons are largely discordant to the belt, crosscutting it in the southern part of the study area (Fig. 3) and fragmenting or obliterating large parts of the belt on its eastern side.

The Peninsular Ranges block, deeply eroded and pene-planed by Eocene time (Krummenacher et al., 1975), underwent renewed uplift and apparent westward tilting as a result of the

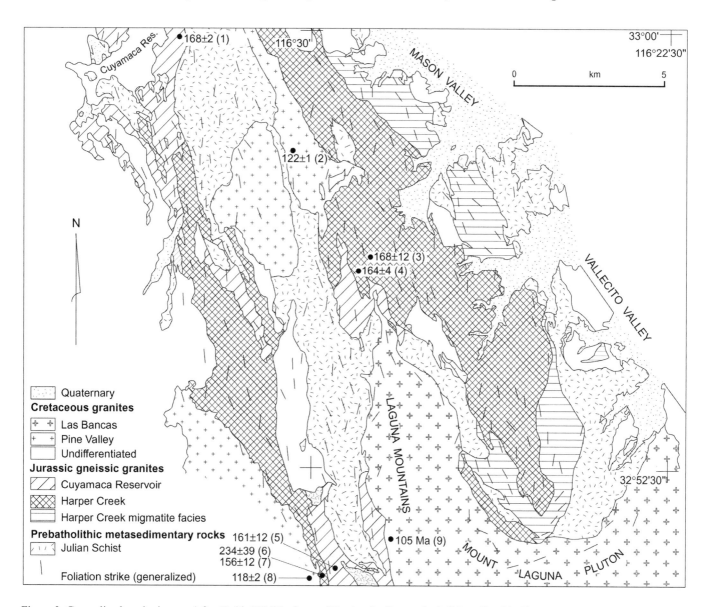

Figure 3. Generalized geologic map (after Todd, 1994b) of part of Peninsular Ranges batholith outlined in Figure 2. Trend lines indicate strike orientation of steep (60–90°), mainly eastward-dipping foliation in Jurassic gneissic granites, Cretaceous plutons, and metamorphic screens. Entire area is within Cretaceous Cuyamaca–Laguna Mountains shear zone of Todd et al. (1988). References for numbered age locations are: (1) this paper; (2) D.L. Kimbrough, 1994, personal commun.; (3) Todd et al. (1991b); 4) this paper; (5), (6), and (7) Thomson and Girty (1994); (8) D.L. Kimbrough, 1994, personal commun.; (9) L.T. Silver, 1979, personal commun.

Neogene development of the Gulf of California San Andreas rift-transform system. Progressively deeper levels of erosion across the batholith have produced a depth profile from volcanic levels on the west to mid-crustal depths in the central to eastern Peninsular Ranges batholith (see Grove et al., this volume).

PREBATHOLITHIC FRAMEWORK

The Jurassic granites were emplaced into predominantly metasedimentary country rocks, which are now represented as a large number of steeply dipping, northwest-trending screens (Figs. 1 and 2). The granites appear to have intruded as a north-trending belt across two or more northwest-trending prebatholithic wallrock terrains. Flyschlike metasedimentary wallrocks in the western part of the Jurassic intrusive belt are assigned to the early Mesozoic Julian Schist of Hudson (l922), whereas wallrocks in the central part of the intrusive belt, although lithologically similar to the Julian Schist, differ in that they contain ubiquitous thin marble interbeds as well as a greater proportion of quartz-rich metasandstone (Engel and Schultejann, 1984; Gastil and Miller, 1984; Todd et al., 1987, 1988). Todd et al. (1987) informally named these rocks the Jacumba sequence and tentatively assigned them a Mesozoic and/or Late Paleozoic age.

Julian Schist

The Julian Schist comprises a diverse assemblage of fine-grained, quartzo-feldspathic micaceous metasedimentary rocks, grading from semi-pelitic to feldspathic metaquartzite (metapelites to metapsammites). The predominant schists and quartzites are interlayered with smaller amounts of amphibolite and mafic schist, calc-silicate quartzite and gneiss, metaconglomerate, and rare marble and talc schist (Berggreen and Walawender, 1977; Todd and Shaw, 1979; Germinario, 1993; Grove, 1987). The overall fine grain size, relict thin bedding, and well-preserved turbidite structures in the larger western screens of the Julian Schist indicate that the sediments were deposited on submarine fans or fan complexes (Detterman, 1984; Grove, 1987; Germinario, 1993; Reed, 1993; Gastil, 1993) in a forearc or interarc basin or basins (sandstone-shale belt of Gastil and Miller, 1984).

Amphibolite and mafic schist, which compose ~5% of the Julian Schist, are interlayered with and share the metamorphic foliation of the schist and quartzite facies of the unit. The sizes and geometries of these mafic bodies, as well as their interfingering contacts with the enclosing sediments, suggest penecontemporaneous extrusion of mafic volcanic rocks during marine deposition (Schwarcz, 1969; Grove, l987; Todd, l994a). In addition to mafic intrusives, some metaconglomerates contain clasts (silicic tuff and a variety of intermediate to mafic lithic clasts), indicating that parts of a Late Triassic–Early Jurassic arc were undergoing erosion during deposition.

Fragmentary fossil data for the Julian Schist and possible correlative units of the sandstone-shale belt in southern California and Baja California, México, suggest a Late Triassic–Jurassic depositional age (Gastil and Miller, 1984; Todd et al., l988; Gastil, l993; Germinario, l993). Whole-rock Rb-Sr isochrons of the Julian Schist and related units also suggest a Jurassic and Triassic depositional age for the protoliths (Hill, 1984; Gastil et al., l988; Davis and Gastil, 1993). Therefore, because the Julian Schist was intruded by Middle and Late Jurassic plutons (this paper; Todd et al., 1991a, 1991b; Thomson and Girty, 1994), its minimum age is constrained.

Metamorphism of the Julian Schist

Between the Middle Jurassic and the Late Cretaceous, the Julian Schist underwent multiphase synkinematic metamorphism. In general, peak-grade metamorphic conditions ranged from lower amphibolite facies in the western part of the Julian Schist belt to upper amphibolite facies in the eastern part (Todd et al., 1988). The metamorphic assemblage in pelitic and semi-pelitic schists in the lower amphibolite facies is quartz-biotite-muscovite-plagioclase-K-feldspar-andalusite $\pm$ staurolite $\pm$ sillimanite $\pm$ cordierite $\pm$ garnet. In the upper amphibolite facies, pelitic rocks contain quartz-biotite-muscovite-plagioclase-K-feldspar-

sillimanite $\pm$ cordierite $\pm$ garnet. Based upon thermobarometry of metapelitic rocks of the Julian Schist across the central and eastern zones of the batholith, Grove (1994) recognized the existence of a steep, north-northwest–trending gradient, across which metamorphic pressures varied from ~3 kbar on the west to ~4.5 kbar on the east. The implied increase in structural depth, from ~7 to 10 km in the western zone to ~11 to 16 km in the eastern zone, was interpreted by Grove (1994) to mark the trace of a Late Cretaceous ductile, west-directed thrust or reverse fault.

Many of the larger screens have antiformal or synformal structures, the limbs of which host steeply plunging minor folds (Schwarcz, 1969; Berggreen and Walawender, 1977; Grove, 1987). Relict bedding and sedimentary structures are preserved within the larger screens in the western part of the Julian Schist belt, but in the eastern part, migmatization and polyphase folding have obliterated primary structures (Grove, 1987; Lampe, 1988; Todd, 1994a).

Jacumba Sequence

Sillimanite-grade, flysch-like metasedimentary wallrocks in the central part of the Jurassic intrusive belt (Jacumba Mountains, Fig. 2) consist of interbedded pelitic and semi-pelitic schists, metaquartzite, thin-bedded calcitic and dolomitic marble, calc-silicate rocks, amphibolite and mafic schist, and metachert. The metamorphic assemblage in pelitic rocks is the same as the upper amphibolite facies Julian Schist. Slight lithologic differences between these rocks and the Julian Schist suggest that the two units differ in lithofacies and/or depositional age (Engel and Schultejann, l984; Gastil and Miller, l984; Todd et al., l988; Walawender et al., 1991; Gastil, l993).

The Jacumba sequence and possible equivalent rocks in northeastern San Diego County are generally similar in lithology to middle Paleozoic to Triassic clastic rocks of slope-basin facies and interbedded basalt flows in Baja California (Leier-Engelhardt, 1993). For this reason, they are tentatively assigned a Mesozoic and/or late Paleozoic age.

Paleozoic Miogeoclinal Rocks

Several large wallrock screens occur in the northern and eastern Peninsular Ranges batholith, north of Borrego Springs and east of the San Jacinto Fault Zone (SJFZ; vertical ruling in Fig. 1). These screens are described as consisting of high-grade, pre-Cretaceous metamorphic and metaplutonic rocks, Cretaceous plutonic rocks, and Paleozoic metasedimentary rocks of miogeoclinal and slope-basinal affinities (Sharp, 1967; Theodore and Sharp, 1975; Miller and Dockum, 1983; Hill, 1984; Erskine, 1986; Wagner, 1996). Common features of the pre-Cretaceous screens include upper amphibolite facies, quartz- and carbonate-rich metasedimentary rocks and anatexites, orthogneiss, and minor amphibolite. As the orthogneiss component of some of these screens is lithologically similar to the Stephenson Peak migmatite and gneiss facies of the Harper Creek suite, we speculate that Jurassic metamorphism and plutonism may have extended eastward into Paleozoic miogeoclinal crust.

JURASSIC GRANITES

Gneissic granites form a belt at least 45 km wide and 150 km long within the axial zone of the Peninsular Ranges batholith, and similar rocks are reported as far south as 28° N in Baja California (R.G. Gastil, 1991, personal commun.). In San Diego County, they occupy an area of ~1155 km^2, or roughly 13% of the total outcrop area of prebatholithic-screen rocks and granite. The western margin of the Jurassic intrusive belt approximates to the δ^{18}O step and the pseudogravity gradient (PSGR) boundary (Fig. 2), with most plutons in the western part of the intrusive belt located within an Early Cretaceous north-northwest–trending ductile shear zone in the Cuyamaca and Laguna Mountains, named the Cuyamaca–Laguna Mountains shear zone (Todd et al., 1988; Thomson and Girty, 1994). The area of Figure 3 is entirely within the Cuyamaca–Laguna Mountains shear zone, the shear zone being approximately 15 km wide at this location. The eastern part of the Jurassic intrusive belt extends at least to Borrego Springs (Fig. 1) and the Borrego Valley.

The Jurassic granites can generally be distinguished in outcrop from adjacent Cretaceous granites by their reddish-, orange-, and yellow-brown weathered surfaces (resulting from the oxidation of biotite and trace iron sulfides), abundant quartz and mica, aluminous minerals, and strong deformational fabric. The granites have been mapped as two intrusive suites, each of which comprises a number of separate bodies. The suites are: (1) the strongly peraluminous Harper Creek gneissic granite, which includes the Stephenson Peak facies of associated migmatitic schist and gneiss; and (2) the metaluminous to moderately peraluminous Cuyamaca Reservoir gneissic granite. The Cuyamaca Reservoir suite and more obvious magmatic facies of the Harper Creek suite consist chiefly of fine- to coarse-grained, medium- to dark-gray gneissic granodiorite and tonalite, but in addition, the Harper Creek suite includes mixtures of orthogneiss with paragneiss, feldspathic metapsammitic and metapelitic migmatites (diatexites), and discrete enclaves of refractory metasedimentary rocks and amphibolite. The Jurassic suites are ilmenite-series granites according to the classification of Ishihara (1977).

The intrusive relationships between the two suites change gradually from west to east across the Jurassic belt in a manner reminiscent of Gastil's (1975) depth zonation. To the west, where contacts between the two suites are sharp, the marginal Cuyamaca Reservoir rocks display higher level features and tend to be relatively coarse-grained and leucocratic (near-pegmatitic) or may be notably fine-grained against the Harper Creek gneissic granite, suggesting that, at least in these localities, Cuyamaca Reservoir magma intruded near-solidus Harper Creek plutons. Further east, contacts between the Harper Creek and Cuyamaca Reservoir plutons are more diffuse and marked by zones of gradation and interlayering that typically measure from a few meters to several tens of meters across strike. In the central part of the

Jurassic belt, intrusive relationships appear to be at a deeper level and gradational zones may be as broad as 1 km, with Harper Creek plutons containing scarce lenticular bodies of Cuyamaca Reservoir gneissic granite up to 2 km long and oriented parallel to gneissic foliation. From west to east across the intrusive belt, gradational changes in the Cuyamaca Reservoir suite include mafic to felsic compositions, amphibole rather than biotite as the dominant mafic mineral, and an increase in mica-rich clots and enclaves of Julian-type metasediments.

Harper Creek Suite

The Harper Creek gneissic granites consist of biotite granodiorite and tonalite with lesser monzogranite. Where magmatic textures are least altered by subsolidus deformation and recrystallization, the mineral assemblage is plagioclase, quartz, biotite, and (in granodiorite and monzogranite) K-feldspar. Accessories include variable amounts of graphite, muscovite, tourmaline, sillimanite (mostly fibrous habit), cordierite, garnet, and andalusite.

The suite is characterized by abundant, uniformly distributed metasedimentary enclaves in a wide range of sizes. Rocks of the suite have a gneissic texture that is marked by thin (0.5 mm to several millimeters thick), evenly spaced lenticular biotite-rich clots oriented parallel to the foliation. In outcrop, reaction rims between larger, relatively intact pelitic schist enclaves and surrounding orthogneiss suggest that the biotitic clots are remnants of partly melted and/or assimilated metapelitic rocks. Mylonitic gneiss and schist of the Harper Creek suite contain thin (less than 1 cm thick) quartz stringers, boudins, and blebs that are oriented parallel to foliation. Modal analyses show that some mylonitic Harper Creek rocks plot in the field of quartz-rich granitoids on the International Union of Geological Sciences modal diagram of Streckeisen (1973). In addition to the abundant metasedimentary enclaves, the Harper Creek gneissic granite contains large tabular enclaves, or rafts, of calc-silicate–bearing metaquartzite and amphibolite, ranging in size from 10 cm to several meters long, aligned parallel to the foliation (Fig. 4A). Both enclaves and rafts are similar in lithology and metamorphic grade to the most refractory rocks in the adjacent metasedimentary screens. The fact that their distribution appears to be unrelated to distance from pluton walls suggests that the majority of these enclaves represent restite (possible "resistate" in the sense of Vernon et al., 2001) from the partial melting of supracrustal rocks in the source region rather than material stoped from walls or roof during intrusion. Amphibolite inclusions that grade laterally over short distances to bodies of fine-grained metagabbro are similar to layers of totally reconstituted amphibolite/gabbro intercalated within the Julian Schist. The majority of these inclusions are inferred to have been inherited from similar refractory mafic rocks in the Julian Schist.

Eastern plutons of the Harper Creek suite are partly mantled by migmatitic and gneissic rocks that Todd (1978) referred to as "the migmatitic schist and gneiss of Stephenson Peak." Now included as a facies of the Harper Creek suite, these rocks comprise orthogneiss, mylonitic gneiss, paragneiss, and pelitic and psam-

Figure 4. A: Horizontal outcrop surface of Harper Creek suite showing steeply dipping northwest-southeast foliation. Gneissic foliation is defined by flattened to ovoid enclaves of calc-silicate quartzite, amphibolite, and biotite schist that are similar in lithology to adjacent Julian Schist. B: Outcrop photograph of folded migmatite gneiss showing development of patchy granite leucosome and large single grains of white magmatic plagioclase. These rocks, The Stephenson Peak migmatitic schist and gneiss, are included as facies of Harper Creek suite and commonly occur as mantle around its plutons. Granite-migmatite relationships are similar to high-grade migmatitic zone surrounding the western part of Cooma granodiorite gneiss, Lachlan fold belt, eastern Australia (Vernon et al., 2001).

mitic schist with abundant granitic leucosome. From west to east, the bodies of Stephenson Peak facies vary from thin (<100 m) discontinuous rinds, to more sizable (1.5–2 km) envelopes surrounding Harper Creek plutons, to broad regions of diatexites and orthogneiss. Contacts between metamorphic enclaves and layers and Harper Creek orthogneiss are diffuse, wispy, or gradational. The migmatitic schist and gneiss of the Stephenson Peak facies is characterized by irregular patches of granite leucosome (Fig. 4B), discrete tabular layers of calc-silicate–bearing metaquartzite, amphibolite, mafic schist, and marble that range in thickness from a few centimeters to tens of meters. The facies forms the dominant part of the eastern Jurassic belt and appears to be composed of ultrametamorphosed metasedimentary and metavolcanic pre-batholithic rocks similar to those of the Julian Schist and Jacumba sequence. The association of Harper Creek gneissic granite and migmatite in the Peninsular Ranges batholith closely parallels that of the Cooma granodiorite gneiss and migmatite in the Lachlan fold belt of eastern Australia (Vernon et al., 2001). They proposed that partial melting of prebatholithic feldspathic metapsammitic rocks was the major process in the formation of the Cooma granodiorite and that enclaves of metapsammitic residuum were restite. The formation of metapelitic (as distinct from metapsammitic) migmatite was considered by Vernon et al. (2001) to have formed earlier in the pressure-temperature (P-T) melting path, with partial melting to form leucosome limited by the amount of water released through breakdown reactions involving muscovite in the metapelite. Vernon et al. (2001) would view the earlier-formed metapelitic migmatite enclaves as "resistate" rather than restite. Similarly, outcrop-sized enclaves of metapelitic migmatite in eastern Harper Creek plutons suggest a resistate rather than restite association if the Harper Creek magma formed dominantly by partial melting of feldspar-rich metapsammitic rocks.

To the northeast of the study area in the Peninsular Ranges batholith (Fig. 3), Engel and Schultejann (1984, p. 662–663) suggested that pre-Cretaceous metamorphic wallrocks were "…feldspathized and granitized and in many places converted to gneisses almost indistinguishable from sheared and refoliated margins of [Cretaceous] plutons." These authors attributed the origin of the gneisses, which are described as containing "…relict marble beds, patches of laminated feldspathic quartzite, and other relict sedimentary features…," to amphibolite-grade metamporphism and metasomatism of wallrocks during Cretaceous intrusion. We suggest that such gneisses are S-type orthogneisses derived in large part from the anatexis of metasedimentary rocks, the character and age of which varied from west to east across the Jurassic continental margin.

Cuyamaca Reservoir Suite

Gneissic granite plutons of the Cuyamaca Reservoir suite are composed mainly of biotite- and hypersthene-biotite granodiorite and tonalite. The least deformed and recrystallized rocks contain plagioclase, subequal quartz and biotite ± orthopyroxene ± hornblende ± K-feldspar as a relict magmatic assemblage. Less common magmatic mafic assemblages in the granodiorite and tonalite are biotite–subaluminous amphibole (actinolite) and hypersthene-actinolite-biotite. Many samples fall between granodiorite and tonalite. The suite contains fine-grained mafic to intermediate enclaves (microgranitoid enclaves of Vernon, 1983) and, less commonly, enclaves of the Julian Schist. Many enclaves are dioritic in composition, have a mineral assemblage similar to that of the host rock, and have relict igneous textures. Lenticular biotite-rich aggregates ~1–2 cm long are aligned parallel to a strong mineral foliation in the Cuyamaca Reservoir plutons. Next to contacts with the larger metasedimentary screens, narrow marginal zones consist of mylonitic gneiss, in which metasedimentary enclaves (mainly calc-silicate quartzite) are concentrated near the contacts.

The Cuyamaca Reservoir suite is similar in many ways to the Hillgrove suite of the New England batholith, eastern Australia (Flood and Shaw, 1977; Shaw and Flood, 1981). Plutons of the Hillgrove suite have intruded into trench-complex metasedimentary rocks, are biotite-rich, and contain actinolitic amphibole and ilmenite. Several plutons have well-developed regional metamorphic aureoles with migmatites and gneisses developed at the highest grades. The Hillgrove suite has elevated Sr_i and $\delta^{18}O$ relative to the main I-type suites of the New England batholith but lower than those in the Cuyamaca Reservoir suite.

U-Pb Geochronology

The suspicion that some deformed granites in the Cuyamaca and Laguna Mountains were Jurassic was initially based on their regional metamorphic fabric (e.g., Everhart, 1951). This was confirmed by conventional U-Pb ion exchange mass spectrometry on zircon separates (Girty et al., 1994; Thomson and Girty, 1994; G.H. Girty, 1998, personal commun.). Two samples of the Harper Creek suite yielded minimum crystallization ages of 156 ± 12 Ma and 234 ± 39 Ma, and a sample of the Cuyamaca Reservoir suite yielded an age of 161 ± 12 Ma. Girty et al. (1994) reported zircon from two additional Harper Creek samples as having Late Jurassic minimum ages. Thomson and Girty (1994), on the basis of the 234 Ma age, suggested that Triassic and Jurassic rocks are interleaved in the Scove Canyon segment of the Cuyamaca–Laguna Mountains shear zone. However, large uncertainties in age determinations due to an inherited Proterozoic (ca. 1.5 Ga) zircon component limit the significance of age correlations. Todd et al. (1991a) reported a Rb-Sr isochron age for five whole-rock samples of 168 ± 12 Ma from a separate body of Harper Creek. A summary of available age data is plotted in Figures 2 and 3.

In Situ Single Zircon Ages—LAM-ICPMS

U-Pb age data are presented (Table 1, Fig. 5, A and B) for 26 single zircon grains from a sample of Harper Creek gneissic granite (WS-91-1). The analyses were conducted at Macquarie University on mounted and polished zircon grains using a laser ablation microprobe attached to a quadrapole induced-coupled-plasma mass spectrometer (LAM-ICPMS) following the procedure of Jackson et al. (1992). A Tera and Wasserburg (1972) isochron diagram (Fig. 5B) shows that, as pointed out by Thomson and Girty (1994), the Harper Creek gneissic granite contains a large inherited component of Paleozoic and Proterozoic zircons, many exhibiting contamination with common Pb. As the source of the common Pb is probably from the surface of the epoxy mount rather than the zircon itself (see discussion below), $^{207}Pb/^{206}Pb$ ages, although included in Table 1, are not considered meaningful. However, the six zircons that are grouped about a Jurassic age give, when pooled, a $^{207}Pb/^{206}Pb$ versus $^{238}U/^{206}Pb$ isochron age of 160.7 ± 9.5 Ma at 2σ standard errors (Fig. 5B).

In Situ Single Zircon Ages—Ion Microprobe

Zircons were extracted from seven samples of potential Jurassic age (Cuyamaca Reservoir samples CP-153, CP-178, WD-81A; Harper Creek sample 5-84-SS-2; East Mesa I-type tonalite samples TS-23, CP-155, 6-139-17) using conventional crushing, density, and magnetic methods. Epoxy mounts were prepared for hand-selected grains that were chosen on the basis of euhedral form and absence of inclusions and/or optically visible cores of older zircon. Isotopic U-Pb analysis was performed using the University of California at Los Angeles Cameca ims 1270 ion microprobe. AS-3 standard zircon (Paces and Miller, 1993) was employed to determine Pb/U relative sensitivity (see Appendix 3 of Grove et al., this volume, for additional analysis details). Unless otherwise stated, all U-Pb ages quoted are $^{206}Pb/^{238}U$ ages with $\pm 1\sigma$ standard error. Calculation of radiogenic Pb/U ratios was performed using ^{204}Pb as a proxy for common Pb. The vast majority of common Pb affecting our analyses is contributed from the periphery of the ion pit rather than the sample itself (Dalrymple et al., 1999). Direct measurement of this component indicates that it is very similar in composition to anthropogenic Pb from the Los Angeles basin (see Sañudo-Wilhelmy and Flegal, 1994).

Zircon U-Pb age results (Table 2) are illustrated in $^{206}Pb/^{238}U$ versus $^{207}Pb/^{235}U$ concordia plots in Figure 6. As indicated, Middle Jurassic U-Pb zircon ages were obtained from three samples of the Cuyamaca Reservoir suite (CP-178, CP-153, and WD-81A), one sample of the Harper Creek suite (5-84-SS-2), and one sample of the East Mesa suite. The two additional samples from the East Mesa suite (CP-155 and 6-139-17) contained zircon that yielded Early Cretaceous U-Pb ages between 115 and 120 Ma and have not been included in Figure 6. The East Mesa suite consists of pyroxene-biotite-hornblende quartz diorite and tonalite plutons restricted to the Cuyamaca–Laguna Mountains shear zone. Because some East Mesa plutons interfinger with and/or grade to adjacent plutons of the Cuyamaca Reservoir suite, and others have petrographic and textural characteristics that indicate a Cretaceous age, Todd et al. (this volume) designate the suite as Jurassic and Cretaceous.

The typically low (50–200 ppm) U-contents of zircons from all of the samples that we have studied result in relatively poorly defined $^{207}Pb/^{235}U$ ages after correction for common Pb have been applied. Hence, while all results statistically overlap concordia, the $^{207}Pb/^{235}U$ age data we obtained are too imprecise to allow us to conclude with a high degree of confidence that the U-Pb analyses are unaffected by inheritance and/or Pb loss. In spite of this, only one sample (WD-81A) yielded clear evidence for such phenomena (Fig. 6C). For example, the rim of one zircon from WD-81A yielded an early Cretaceous $^{206}Pb/^{238}U$ age (118 ± 3 Ma), while two additional highly discordant (and much older) results from other grains indicated incorporation of inherited Pb. Given the high spatial resolving power of the ion microprobe, we feel that it is likely that we were generally able to avoid the problematic behavior that was prominent in previous isotope dilution studies of these rocks (Thomson and Girty, 1994; Murray and Girty, 1996) and

TABLE 1. ZIRCON Pb/U SINGLE GRAIN ANALYSES, HARPER CREEK, SAMPLE WS-91-1

	Isotope ratios				Age estimates (Ma)			
Grain	$^{206}Pb/^{238}U$	± 2σ	$^{207}Pb/^{206}Pb$	± 2σ	$^{206}Pb/^{238}U$	± 2σ	$^{207}Pb/^{206}Pb$	± 2σ
24	0.02391	0.00106	0.06769	0.00850	152.3	6.7	858.9	250.4
2	0.02496	0.00072	0.08143	0.00710	158.9	4.5	1231.9	166.3
34	0.02533	0.00080	0.04898	0.00622	161.3	5.0	147	285.2
15	0.02538	0.00084	0.07407	0.00750	161.6	5.0	1043.5	197.7
8	0.02562	0.00140	0.07471	0.01914	163.1	8.9	1060.7	476.3
18	0.02585	0.00094	0.07421	0.00682	164.5	6.0	1047.1	180.2
32	0.03017	0.00170	0.05125	0.01598	191.6	10.7	252.1	647.6
11	0.04359	0.00092	0.05778	0.00274	275.1	5.7	521.2	102.4
26	0.04393	0.00134	0.05581	0.00528	277.2	8.2	444.4	204.2
12	0.04514	0.00130	0.06860	0.00444	284.6	8.0	886.8	131.1
6	0.04793	0.00178	0.09596	0.00818	301.8	11.0	1547	156.1
20	0.05016	0.00096	0.08176	0.00314	315.5	5.9	1239.7	73.9
16	0.05055	0.00150	0.07677	0.00538	317.9	9.3	1115.1	136.8
23	0.05316	0.00128	0.08693	0.00432	333.9	7.9	1359	94.0
4	0.05478	0.00134	0.09440	0.00448	343.8	8.2	1516.1	88.2
21	0.05648	0.00154	0.10040	0.00658	354.2	9.4	1631.5	119.4
22	0.05696	0.00122	0.05955	0.00292	357.1	7.5	587.1	104.5
14	0.06080	0.00132	0.07379	0.00338	380.5	8.1	1035.7	91.0
29	0.07263	0.00250	0.09176	0.00710	452.0	15.0	1462.3	143.8
3	0.07479	0.00192	0.08872	0.00428	465.0	11.5	1398	90.8
13	0.09337	0.00208	0.09805	0.00394	575.4	12.2	1587.4	74.0
30	0.11549	0.00354	0.06685	0.00436	704.6	20.5	833	133.2
7	0.11947	0.00358	0.11013	0.00626	727.5	20.7	1801.6	101.8
28	0.15504	0.00492	0.07923	0.00488	929.1	27.5	1178	119.3
25	0.18826	0.00580	0.08266	0.00540	1111.9	31.5	1261.2	125.0
31	0.25992	0.00714	0.09923	0.00418	1489.4	36.5	1609.7	77.4

Note: LAM-ICPMS analyses, Macquarie University. Ablation diameter, 30 μ. Wavelength, 266 nm. Method after Jackson et al. (1992).

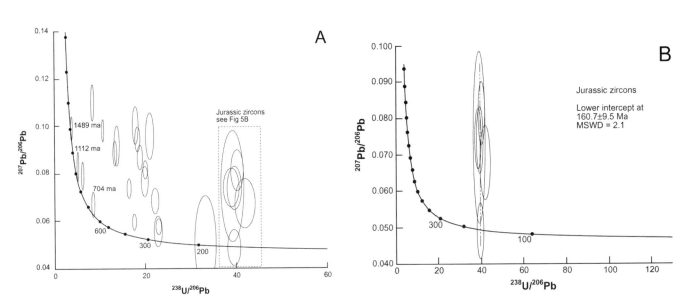

Figure 5. A: Tera and Wasserburg (1972) isochron diagram of 26 single zircon analyses for Harper Creek sample WS-91-1, dated using procedure of Jackson et al. (1992). Ages based on $^{238}U/^{206}Pb$ show spread from Jurassic to Proterozoic. Discordant $^{207}Pb/^{206}Pb$ ages in some zircons indicate significant common lead in direction of elongation of error ellipses and are probably due to surface contamination of sample. B: Isochron diagram of six Jurassic zircons outlined in A, giving calculated least-squares age of 160.7 ± 9.5 Ma 2σ standard errors. MSWD is mean square of weighted deviates.

TABLE 2. IONPROBE* U-Pb AGE RESULTS FROM ZIRCON

Sample	Suite	$^{206}Pb/^{238}U^{\#}$ ± 1σ (Ma)	$^{207}Pb/^{235}U^{\#}$ ± 1σ (Ma)	MSWD	Number**
CP-153	Cuyamaca Reservoir	168 ± 2	169 ± 18	3.5	11/11
CP-178	Cuyamaca Reservoir	161 ± 3	154 ± 25	3.4	10/10
WD-81A	Cuyamaca Reservoir	165 ± 2	164 ± 11	4.9	11/14##
5-84-SS-2	Harper Creek	164 ± 4	171 ± 44	3.8	10/10
TS-23	Tonalite of East Mesa	170 ± 2	157 ± 29	3.5	10/10

*UCLA's Cameca ims1270. See Appendix 3 in Grove et al. (this volume, Chapter 14) for additional details on analytical methods.
#Weighted mean age. Standard errors (± 1σ) shown have been scaled by the square root of the MSWD.
**Number of spot analyses used to calculate mean age/total number of spot analyses measured.
##Excluded measurements for WD-81A include rim analysis of 118 Ma, and two highly discordant analyses (206 Ma and 564 Ma).

provisionally conclude that the majority of the analyses we have obtained are concordant and represent the age of intrusion.

Weighted mean U-Pb ages are presented in Table 2 and represented diagrammatically by filled ellipses in Figure 6. In nearly all instances, the mean square of weighted deviates (MSWD) values are significantly in excess of unity. We believe that the most likely explanation for this is instrumental instability that is not adequately taken into account in the error propagation. For example, our ability to reproduce $^{206}Pb/^{238}U$ ratios from AS-3 standard zircon during the analysis sessions in question was about ±2 % (1σ). Hence, to ensure that the stated uncertainties adequately reflect instrumental instability, we have scaled the calculated standard errors by the square root of the MSWD in all cases where the MSWD exceeded unity.

Complete data tables for each sample investigated have been included in the GSA Data Repository[1] and on the CD-ROM that accompanies this volume.

Structural Characteristics

The Jurassic granites form steep-walled, north-northwest–elongate bodies as much as 5 km wide and 20 km long. The rock textures in these bodies range from strongly foliated to mylonitic gneiss. Foliation strikes parallel to the sides of the bodies; dips are steep (60–90°) to the east and defined by alignment of relict magmatic feldspar and quartz grains and by recrystallized aggregates of quartz and biotite. A penetrative lineation plunges steeply to the east in the plane of the foliation. Modification of igneous grains in the Jurassic granites is variable: foliated granites contain aligned relict subhedral feldspar and biotite phenocrysts and interstitial quartz, whereas in mylonitic rocks, the larger feldspar

grains are porphyroclasts. Most quartz and biotite form fine-grained recrystallized aggregates, and both K-feldspar and biotite are extensively altered to muscovite. The apparent concordance of magmatic and subsolidus foliations in the Jurassic plutons and the continuity of plutonic foliation and lineation with regionally developed metamorphic fabrics in their wallrocks indicate that magmatic foliation was overprinted by a high-temperature, post-magmatic solid-state foliation (Paterson et al., 1989).

The foliation in the Harper Creek and Cuyamaca Reservoir suites is axial planar to scattered outcrop-scale isoclinal folds, and map patterns of several bodies suggest that the plutons were folded isoclinally about near-vertical axes on a regional scale prior to Cretaceous intrusion (Todd, 1994b). The fabric of these suites probably formed during multiple episodes of synintrusive deformation that began at least as early as the Late Jurassic (Todd et al., 1994) and culminated by the middle Cretaceous (Thomson and Girty, 1994). Apparently, deformation was under way in the Triassic and Jurassic prior to Jurassic intrusion, because both the Harper Creek and Cuyamaca Reservoir plutons contain, as xenoliths, the detached hinges of isoclinal folds in refractory Julian Schist rock-types.

The fabric of the Jurassic plutons was not fully developed until the Cretaceous period. Foliation and lineation in Early Cretaceous plutons within the Cuyamaca–Laguna Mountains shear zone have orientations similar to those in the adjacent Jurassic plutons, and in some outcrops, foliation is continuous between Jurassic and Cretaceous plutons, crossing contacts at a high angle (Fig. 3). Studies of two bodies of the Harper Creek gneissic granite located within the Cuyamaca–Laguna Mountains shear zone indicate that mylonitic fabrics developed during two periods of Cretaceous deformation (Girty et al., 1993; Thomson and Girty, 1994).

Petrographic Characteristics

Microstructures of the Harper Creek (including the Stephenson Peak facies) and Cuyamaca Reservoir suites indicate high-temperature strain and recrystallization of magmatic grains during late-(?) and post-crystallization subsolidus deformation. In thin section, quartz and feldspar are generally similar between

[1]GSA Data Repository item 2003173, chemical analyses of 62 Jurassic plutonic rocks and 12 eastern-zone prebatholithic rocks (Julian Schist, orthoamphibolites), Peninsular Ranges batholith, San Diego County, is available on request from Documents Secretary, GSA, P.O. Box 9140, Boulder, CO 80301-9140, USA, at editing@geosociety.org, www.geosociety.org/pubs/ft2003.htm, or on the CD-ROM accompanying this volume

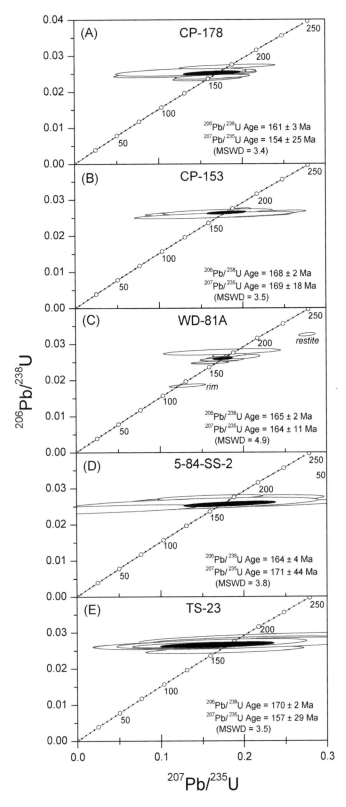

Figure 6. Concordia diagrams $^{206}Pb/^{238}U$ versus $^{207}Pb/^{235}U$ for zircons from five Jurassic plutons. Cuyamaca Reservoir suite CP-178 (A), CP-153 (B), and WD-81A (C). Harper Creek suite 5-84-SS-2 (D). East Mesa suite TS-23 (E). Age errors in Ma ± 1σ standard error.

the two suites, although mafic and accessory mineral contents are different (discussed below).

The plagioclase composition of the Harper Creek suite varies from oligoclase to andesine, whereas that of the Cuyamaca Reservoir suite is typically sodic to calcic andesine. Cores of normally zoned grains in both suites are as calcic as sodic labradorite, and rims are as sodic as oligoclase. Scarce relict grains showing magmatic zoning are more common in the Cuyamaca Reservoir suite (Fig. 7A) than in the Harper Creek suite (Fig. 7B), but typically zoning has been obliterated by wedge-shaped deformation twins, kinking, fracturing, and incipient subgrain formation.

Quartz in the Harper Creek and Cuyamaca Reservoir suites varies in size and degree of recrystallization from strain-shadowed relict magmatic grains ~10 mm long to granoblastic grains less than 1 mm across (Fig. 7B). In rocks that contain the best-preserved magmatic quartz, the mineral commonly is interstitial to plagioclase. In both suites, 3–5 mm lenticular multi-grain quartz aggregates and, in the most deformed rocks, ribbons from 0.5 to 1 cm long are common.

Potassium feldspar content in the granodiorite and tonalite of both suites ranges from almost zero to 12%. The mineral is present as 5- to 15-mm-long lenticular grains, as smaller 0.5–3 mm irregular interstitial grains, and as fine-grained granoblastic aggregates with quartz, plagioclase, and biotite. Although most K-feldspar is recrystallized, the larger grains within the granodiorite of both suites show relict magmatic textures such as subhedral shape, small euhedral inclusions of early-crystallized plagioclase, and Carlsbad twins. Potassium feldspar-plagioclase grain contacts are marked by extensive replacement of plagioclase by K-feldspar as well as by nucleation of myrmekite on plagioclase grains (replacement of adjacent K-feldspar, Vernon, 1991).

Biotite in both suites has a "foxy" red-brown color characteristic of reduced biotite in S-type granites (Hine et al., 1978; Shaw and Flood, 1981; Clemens and Wall, 1988). The mineral occurs as 1- to 5-mm-long, relatively equant, subhedral and anhedral grains (relict magmatic grains) and as 2- to 5-mm-long decussate aggregates interleaved with graphite, ilmenite, and minor chlorite (recrystallized biotite).

Muscovite is restricted to the Harper Creek suite, varies in abundance from zero to 13%, and is present mainly as decussate aggregates with biotite and also as sheaf-like aggregates or ragged poikiloblasts that overprint feldspar grains and quartzofeldspathic matrix. The mineral is most abundant in mylonitic Harper Creek rocks. Nearly equant, 1–4 mm anhedral or subhedral muscovite grains with ragged terminations (possibly relict phenocrysts) are rare. Textural relations suggest that most muscovite of the Harper Creek suite is secondary.

Prominent accessory minerals of the Harper Creek suite are graphite, apatite, zircon, tourmaline, allanite, and ilmenite. Graphite, in some cases, makes up more than 2% of the rock. Zircon in Harper Creek rocks occurs as two populations: (1) small colorless euhedral magmatic grains; and (2) pale yellowish-tan, subhedral, broken or rounded inherited grains. Irregular to subhedral partial overgrowths of colorless zircon were observed on some colored

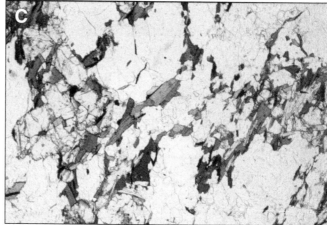

Figure 7. A: Metaluminous sample of Cuyamaca Reservoir gneissic granite, SY-17. Least deformed examples show zoned subrectangular plagioclase with bent deformation twinning. Quartz subgrains, mottled biotite, and hornblende (against plagioclase) display metamorphic grain boundaries. Crossed polars; base of photo 4.4 mm. B: Photomicrograph of strongly foliated Harper Creek gneissic granite, J-91-4. Larger grains of quartz are elongate in plane of foliation and show undulose extinction. Mica folia and smaller grains of quartz wrap around brittle porphyroclast of zoned plagioclase. Deformation in S-type relative to I-type granites has been attributed to higher contents of weak and/or ductile minerals such as quartz and mica (Vernon and Flood, 1988). Crossed polars; base of photo 4.4 mm. C: Harper Creek gneissic granite, D-94-12 with subrectangular grains of cordierite. Cordierite is elongate in foliation direction and is partially surrounded by mica folia. Plane polarized light; base of photo is 4.4 mm.

zircon cores. Tourmaline that is pleochroic in yellow and yellow-green hues is a minor but ubiquitous mineral in Harper Creek rocks. It occurs mainly as 1–2 mm anhedral grains within micaceous folia and, more rarely, as euhedral or subhedral grains.

One or more of the aluminosilicate minerals sillimanite, cordierite, garnet, and andalusite are found in variable amounts in Harper Creek rocks. Fibrous sillimanite and, less commonly, small prismatic grains of sillimanite, are associated with muscovite along biotite cleavage planes and in relict plagioclase and K-feldspar phenocrysts. Harper Creek rocks in the Cuyamaca and Laguna Mountains contain pseudomorphous aggregates of muscovite, chlorite, and fibrous sillimanite that partly or wholly replace cordierite. Other rocks contain relatively abundant, essentially unaltered cordierite, mostly as small anhedral grains associated with biotite, but, in a few rocks, phenocryst-sized rectangular grains are interpreted as magmatic (Fig. 7C). Scarce garnet porphyroblasts occur within aggregates of biotite, muscovite, and quartz in some thin sections.

The least altered Cuyamaca Reservoir rocks contain the relict magmatic assemblage plagioclase-orthopyroxene-quartz-biotite ± K-feldspar ± hornblende. Less common magmatic mafic assemblages in granodiorite and tonalite are biotite-subalumi-

nous amphibole (actinolite) and hypersthene-actinolite-biotite. Hypersthene, in places partly bordered by clinopyroxene, is relatively abundant as subhedral or skeletal magmatic grains and is commonly replaced by biotite and/or actinolite. Actinolite also occurs as well-formed prismatic grains in both granodiorite and tonalite, with or without pyroxene. Accessory minerals include titanite, apatite, ilmenite, zircon, and allanite. Titanite, which may form as much as 1% of the rock, occurs as: (1) subhedral grains up to 3 mm across associated with, or included within, biotite (primary titanite); or (2) narrow rims on ilmenite, tiny seed-like grains along biotite cleavage planes, and in decussate aggregates of mafic minerals (secondary titanite). Clear, colorless zircons are euhedral to subhedral, with a few zoned grains containing rounded (inherited?) tan cores.

Geochemical Variations

Whole-Rock Chemistry

Analyses of eight typical granites from the Harper Creek and Cuyamaca Reservoir suites for major oxides and trace elements are listed in Table 3. Sixty-two analyses from these two suites and 20 from the Julian Schist and interbedded rocks are included in

TABLE 3. COMPOSITION OF SELECTED JURASSIC GNEISSIC GRANITES

	Harper Creek suite				Cuyamaca Reservoir suite				Cooma*
	MP-35	10-79-E	WS-91-1	WS-94-4	D-29	CP-178	ML-5	SY-12	C1
SiO_2	66.07	68.27	71.10	74.47	61.75	65.49	68.49	73.13	72.00
TiO_2	0.54	0.59	0.58	0.48	1.06	1.12	0.88	0.65	0.54
Al_2O_3	15.70	14.65	13.96	12.61	15.29	14.25	14.03	12.93	13.72
Fe_2O_3	0.71	1.04	0.87	0.35	0.78	0.45	0.43	0.57	0.59
FeO	2.74	2.97	2.42	1.77	4.48	4.06	4.25	1.94	3.48
MnO	0.10	0.05	0.06	0.03	0.12	0.08	0.06	0.04	0.06
MgO	1.85	1.72	1.66	1.25	3.95	3.29	1.47	0.78	1.76
CaO	3.22	3.26	1.71	2.63	5.89	4.47	3.07	2.14	0.95
Na_2O	2.75	2.34	1.54	2.70	2.89	2.63	2.54	2.44	1.49
K_2O	4.03	2.96	3.37	1.60	2.71	2.79	3.39	3.96	3.73
P_2O_5	0.30	0.12	0.06	0.14	0.16	0.19	0.16	0.08	0.13
H_2O+	1.26	1.10	1.47	0.76	1.12	1.14	0.47	0.55	n.d.
H_2O-	0.10	0.09	0.21	0.10	0.07	0.11	0.09	0.05	n.d.
CO_2	0.20	0.74	2.71	1.74	0.13	0.31	0.03	0.18	n.d.
Carbon	n.d.	n.d.	0.74	0.47					

Trace elements (parts per million)

	MP-35	10-79-E	WS-91-1	WS-94-4	D-29	CP-178	ML-5	SY-12	C1
Ba	1360	1261	1353	715	959	867	942	1697	765
Cr	39	63	77	47	171	150	31	25	56
Cu	18	41	20	16	19	23	36	<1	n.d.
Ga	19	17	14	15	24	17	19	12	n.d.
Nb	n.d.	n.d.	14	12	12	13	n.d.	9	n.d.
Ni	18	25	7	6	16	16	5	8	24
Pb	22	25	18	11	14	12	18	16	35
Rb	143	98	126	71	101	112	139	135	153
Sr	224	223	182	221	174	156	133	106	127
Th	<2	12	8	8	16	10	11	14	22
U	4	4	3	<3	<3	<3	4	5	4
V	52	108	122	68	126	71	59	39	39
Y	35	22	23	21	50	38	48	40	23
Zn	100	108	82	78	89	75	106	41	n.d.
Zr	81	179	155	176	136	192	228	214	201
C (CIPW)#	0.96	1.67	4.67	1.66			0.60	0.74	5.51
ASI (Mol)**	1.065	1.129	1.503	1.151	0.831	0.921	1.045	1.061	1.670

*Cooma gneiss, Lachlan fold belt, eastern Australia (taken from White and Chappell, 1988).
#C (CIPW) is percent normative corundum.
**ASI (Mol) is the aluminum saturation index as $Al_2O_3/(Na_2O+K_2O+CaO)$.
n.d. = not determined.
Sample locations as latitude and longitude:

Sample	Lat. W	Long. N	Sample	Lat. W	Long. N
MP-35	116°29′03″	32°56′19″	D-29	116°33′13″	32°48′03″
10-79-E	116°31′26″	32°48′52″	CP-178	116°33′58″	33°00′02″
WS-91-1	116°38′49″	33°18′26″	ML-5	116°28′32″	32°51′15″
WS-94-4	116°35′58″	32°22′44″	SY-12	116°40′18″	33°01′31″

the GSA Data Repository and on the CD-ROM accompanying this volume (see footnote 1). Unlike the Early Cretaceous I-type granites of the Peninsular Ranges batholith, the Jurassic granites have the characteristics of Australian S-type granites (Hine et al., 1978; White and Chappell, 1988; Shaw and Flood, 1981). In particular, the Jurassic granites have higher values of K_2O, TiO_2, MgO, Ba, Rb, Cr, Ni, Zn, and Nb, and lower values of CaO and Na_2O (Todd and Shaw, 1995). In common with S-type granites of eastern Australia, the Harper Creek rocks have a limited SiO_2 range from 65% to 75% and compositionally overlap for most major oxides, but there are some differences. The Harper Creek suite and the Lachlan fold belt S-type granites, including a Cooma gneiss sample, are plotted on a Na_2O versus K_2O diagram in Figure 8. The Lachlan fold belt analyses, apart from a few scattered points, have Na_2O values equal to or less than K_2O, a consequence of Na_2O being differentially removed from sediments by seawater (Hine et al., 1978). Overall, the trend of the Lachlan fold belt S-types is toward increasing Na_2O with increasing K_2O, and in general the Harper Creek analyses fall within the field of the Lachlan fold belt S-types (Fig. 8). However, at lower K_2O values, the data field overlaps that of the Julian Schist, indicating a significant restite (migmatite) component of the Harper Creek magma at those values.

The Harper Creek gneissic granites are strongly peraluminous while those of the Cuyamaca Reservoir vary between meta-luminous and peraluminous compositions (Table 3). The range in composition of these rocks is shown on an aluminum saturation index (ASI) versus FeO (total) diagram in Figure 9. For comparison, a sample of Cooma gneiss analyzed by White and Chappell (1988) is included in Table 3 and Figure 9. The horizontal line at an ASI of 1 (Fig. 9) separates metaluminous and peraluminous compositions. Two arrowhead lines define a field that constrains the Julian Schist and the Harper Creek granites. The field converges toward the minimum-melt composition of the simple granite system, based on an ASI value of 1 and an FeO value close to 1.5%. The extreme scatter of Harper Creek points, ranging from Julian metasedimentary compositions toward the granite minimum, is strongly suggestive of a considerable proportion of restite being retained during ascent and emplacement.

Also plotted on Figure 9 are samples of the metaluminous to peraluminous Cuyamaca Reservoir suite. In general, the trend is toward the granite minimum, but some are significantly peraluminous and not in the direction that would be expected from crystal fractionation of an initial metaluminous Cuyamaca Reservoir magma alone, suggesting at least two source components were involved. It has already been noted that the composition of the Cuyamaca Reservoir suite varies geographically. To the west, the plutons are more mafic and contain amphibole as a major phase, while to the east, biotite becomes significant,

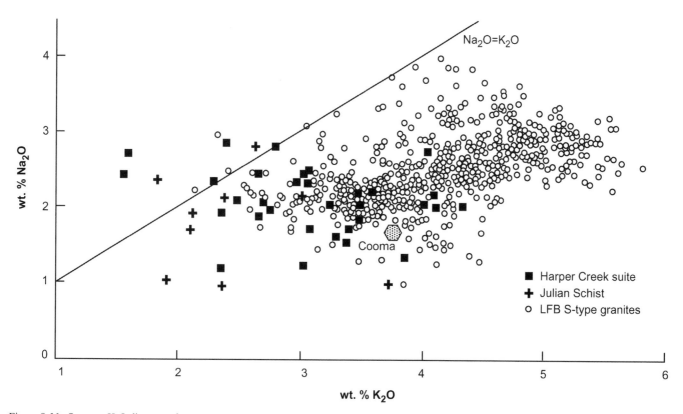

Figure 8. Na_2O versus K_2O diagram of Harper Creek suite and Julian Schist. Lachlan fold belt (LFB) S-type granites of eastern Australia, including Cooma granodiorite gneiss, are added for comparison. Samples of Harper Creek with $Na_2O > K_2O$ probably reflect high restite content of Julian Schist.

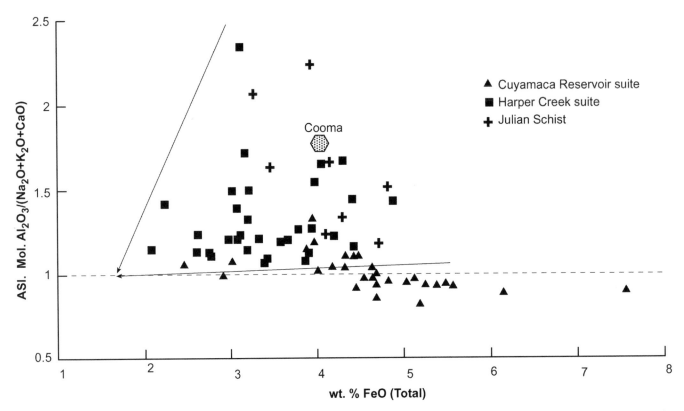

Figure 9. ASI versus FeO (total) diagram of Harper Creek S-type suite, Cuyamaca Reservoir transitional I-type to S-type suite, and Julian Schist. Lachlan fold belt Cooma granodiorite gneiss is added for comparison. Harper Creek samples are bounded by two arrowed lines and converge toward minimum melt composition of simple granite system at an ASI value of 1.0. Harper Creek samples lie between Julian Schist compositions and granite minimum. Cuyamaca Reservoir samples are metaluminous at mafic compositions but at lower FeO values are weakly to moderately peraluminous, reflecting a compositional change from west to east across the Jurassic intrusive belt.

the plutons are more felsic, and enclaves of Julian Schist are more common.

Geochemical variations between the Harper Creek and Cuyamaca Reservoir granites (Fig. 10) show considerable overlap on Harker diagrams for TiO_2, P_2O_5, CaO, Ba, Rb, Y, Zr, and Sr. Neither suite is characterized by regular chemical trends, apart from CaO and Ba for the Cuyamaca Reservoir. The scatter of data points for the Harper Creek samples, particularly at high SiO_2 contents, would suggest that crystal fractionation and restite removal at minimum melt compositions were ineffective.

$^{87}Sr/^{86}Sr$ *and* $\delta^{18}O$

Published Sr_i data for the Cretaceous granites demonstrate a general uniformity parallel to the axis of the batholith and a progressive increase from about 0.704 in the western margin of the batholith to 0.709 in the eastern zone (Silver and Chappell, 1988). Isopleths of $\delta^{18}O$ for the Cretaceous granites show a similar parallelism to the axis of the batholith, rising from +6 to +8.5 per mil in the western zone to +9 to +12 per mil in the eastern zone (Taylor and Silver, 1978). The higher eastern values indicate that these magmas were derived from, or had interacted

with, rocks formed at near-surface environments (O'Neil and Chappell, 1977; O'Neil et al., 1977; Taylor and Silver, 1978; Shaw and Flood, 1981).

The Jurassic gneissic granites are significantly different from the Cretaceous I-type granites with respect to Sr_i and $\delta^{18}O$ values. Table 4 lists whole-rock $\delta^{18}O$ and Sr_i values for the Harper Creek and Cuyamaca Reservoir suites. The age used for the Sr_i calculation was taken as 161 Ma. Initial $^{87}Sr/^{86}Sr$ ratios and $\delta^{18}O$ values of the Harper Creek suite vary from 0.7100 to 0.7129 and +14.8 to +19.8 per mil respectively, whereas those of the more mafic Cuyamaca Reservoir suite vary from 0.7064 to 0.7082 and +11.8 to +13.8 per mil. Differences between the Jurassic and Cretaceous plutonic rocks are clearly shown on a plot of Sr_i versus $\delta^{18}O$ (Fig. 11) where all Cretaceous samples have Sr_i values <0.705 and $\delta^{18}O$ values ≤+9.0 per mil. The Jurassic gneissic granites form a roughly linear array with a positive slope in the high Sr_i-high $\delta^{18}O$ region of the diagram, with a small but distinct gap between the Harper Creek and Cuyamaca Reservoir samples.

The $\delta^{18}O$ and Sr_i values of the Jurassic gneissic granites, particularly those of the Harper Creek suite, are higher than those reported for the Lachlan fold belt and New England batholith

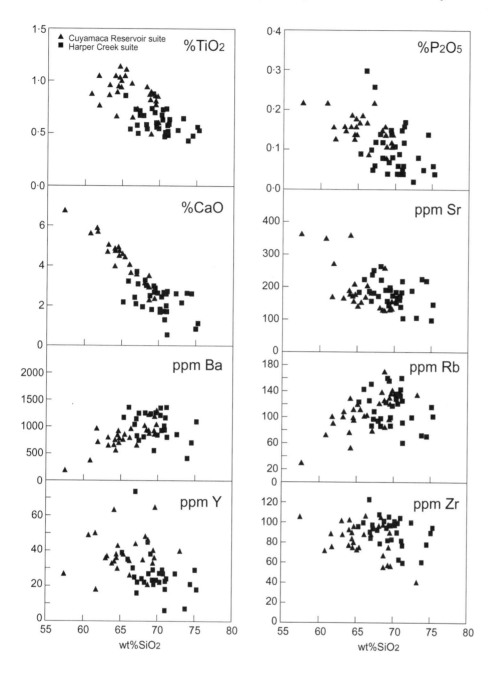

Figure 10. Selected Harker diagrams of Harper Creek and Cuyamaca Reservoir suites. Neither suite displays regular inter-element variation patterns and there is considerable compositional overlap between the suites.

S-type granites (O'Neil and Chappell, 1977; O'Neil et al., 1977), and therefore it is unlikely that these rocks were derived solely from a mantle or young oceanic crustal source (O'Neil et al., 1977; Shaw and Flood, 1981). While the overall chemical and isotopic behavior of the Harper Creek can well be explained by partial melting of a single component, the Julian Schist and related metasediments, the Cuyamaca Reservoir suite from Figure 11 requires an origin from possibly three source components: (1) young mantle-derived magma; (2) relatively young metaigneous and metavolcanic mafic rocks formed in an arc environment; and (3) a metasedimentary component equivalent

to or less isotopically evolved than the Julian Schist. The isotopic gap between the mantle-derived Cretaceous (and presumably the Jurassic) I-type magmas and the Cuyamaca Reservoir suite (Fig. 11) would not argue strongly for a significant contribution from a young mantle-derived component.

Rare Earth Compositions
REE data for 17 rock samples of Julian Schist, Harper Creek suite, and Cuyamaca Reservoir suite are presented in Table 5. The values are plotted as standard chondrite-normalized (C1, Sun and McDonough, 1989) patterns in Figure 12.

TABLE 4. STRONTIUM AND OXYGEN ISOTOPIC ANALYSES

Sample number	Rb (ppm)	Sr (ppm)	^{87}Rb/^{86}Sr measured	^{87}Sr/^{86}Sr present day	^{87}Sr/^{86}Sr initial	δ^{18}O* rock
Cuyamaca Reservoir suite[#]						
CP-153	114	150	2.200	0.71143	0.70640	11.8
CP-178	113	160	2.054	0.71202	0.70732	12.1
D-29	111	167	1.934	0.71153	0.70710	11.8
J-38	139	154	2.605	0.71401	0.70805	13.8
TS-55	81.2	180	1.303	0.71083	0.70785	12.4
TS56	103	168	1.772	0.71224	0.70818	12.4
10-79-H	79	176	1.300	0.70970	0.70672	12.2
Harper Creek suite[#]						
J-40	133.7	189.9	2.041	0.71613	0.71146	15.9
Jhc-1	65.8	50.2	3.796	0.72159	0.71290	19.8
Jhc-2	159.3	164.2	2.813	0.71740	0.71096	17.9
10-79-E	99	216	1.319	0.71372	0.71070	14.8
10-79-G	90	211	1.238	0.71332	0.71049	16.2
10-79-M	119	185	1.862	0.71509	0.71083	16.7
MP-35	143.1	227.4	1.824	0.71415	0.70998	17.7
Julian Schist**						
10-79-I	76	320	0.686	0.71308	0.71103	16.7
10-79-J	121	258	1.358	0.71342	0.70936	17.2
10-79-L1	145	144	2.922	0.71874	0.71001	17.7

Note:Oxygen analyses, USGS Menlo Park. Analyst J. R. O'Neil. Rb and Sr analyses by XRF, Macquarie University. Variance ^{87}Rb/^{86}Sr=1.0%. ^{87}Sr/^{86}Sr analyses, Centre for Isotope Studies, Sydney. Variance = 0.5%.
*Relative to SMOW as per mil.
[#]Initial ^{87}Sr/^{86}Sr calculated at 161 Ma.
**Initial ^{87}Sr/^{86}Sr calculated at 210 Ma.
Sample locations as latitude and longitude:

Sample	Lat. N	Long. W	Sample	Lat. N	Long. W
CP-153	33°00′22″	116°3′20″	10-79-L1	33°00′13″	116°37′17″
CP-178	33°00′02″	116°33′58″	J-40	33°05′33″	116°31′26″
D-29	32°48′03″	116°33′13″	Jhc-1	32°56′02″	116°26′49″
J-38	33°05′26″	116°34′48″	Jhc-2	32°56′02″	116°26′49″
TS-55	33°00′10″	116°42′29″	10-79-E	32°48′52″	116°31′26″
TS-56	32°57′22″	116°38′50″	10-79-G	32°46′46″	116°38′29″
10-79-H	32°46′41″	116°38′31″	10-79-M	32°04′14″	116°32′36″
10-79-I	32°53′35″	116°26′49″	MP-35	32°56′19″	116°29′03″
10-79-J	33°03′47″	116°34′06″			

Gromet and Silver (1987), in a regional study of the Cretaceous gabbros and granites across the batholith, showed a systematic variation of REE patterns from west to east. Their western region, which corresponds to the western zone as used in this paper, has as its boundary to the east regional discontinuities such as the PSGR gradient and δ^{18}O gap, age step, gabbro line, and other major isopleths (Todd and Shaw, 1985, their Fig. 2). Despite the widespread distribution of Jurassic foliated granites in Gromet and Silver's (1987) central and eastern regions, there

was no acknowledgment of the existence of pre-Cretaceous granites within the Peninsular Ranges batholith.

The Julian Schist samples consisting of quartz-feldspar-mica semi-pelitic schist are characterized by REE abundances (Fig. 12A) between 10 and 100 times chondrite. The light rare earth elements (LREE) are enriched relative to heavy rare earth elements (HREE), slopes from middle rare earth elements (MREE) to HREE are fairly flat, and there are minor negative Eu anomalies. The patterns show a remarkable similarity to the

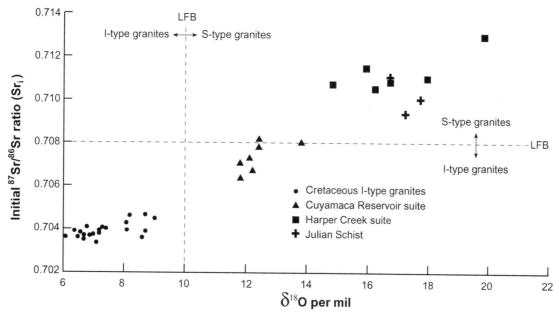

Figure 11. Initial $^{87}Sr/^{86}Sr$ (Sr_i) versus $\delta^{18}O$ diagram. Samples of Harper Creek suite plot in S-type field of Lachlan fold belt granites, eastern Australia (Chappell and White, 1992). Samples of Cuyamaca Reservoir suite are transitional between I-type and S-type fields and plot between isotopically depleted Cretaceous granites from eastern zone of Peninsular Ranges batholith and isotopically evolved Julian Schist metasediments.

TABLE 5. RARE EARTH ELEMENT ANALYSES OF JURASSIC GNEISSIC GRANITES

Sample number	La	Ce	Nd	Sm	Eu	Tb	Yb	Lu
Cuyamaca Reservoir suite								
10-79-H	11.70	24.30	12.70	4.72	0.90	1.04	2.82	0.36
CP-153	25.50	51.50	24.80	6.31	0.88	0.96	2.92	0.43
CP-178	26.49	52.99	24.61	6.27	0.96	0.94	3.06	0.45
D-29	26.52	54.78	27.12	7.39	1.05	1.16	4.23	0.59
J-38	14.96	35.63	23.28	7.20	0.95	1.35	6.38	0.90
ML-30	26.60	54.90	27.00	7.69	1.00	1.34	5.40	0.75
TS-55	22.84	44.63	21.27	5.18	1.10	0.79	2.98	0.43
Harper Creek suite								
10-79-E	34.90	64.30	28.80	6.13	1.16	0.70	2.07	0.28
10-79-G	24.91	48.42	21.22	5.09	1.38	0.66	1.98	0.28
10-79-M	36.67	69.52	31.73	6.58	1.09	0.75	2.37	0.35
J-40	40.80	78.30	32.40	7.16	1.05	0.80	2.79	0.45
MP-35	10.20	19.40	10.00	3.38	1.19	0.79	3.18	0.45
Julian Schist								
5-84-10	39.00	71.60	30.90	6.83	1.34	0.93	3.59	0.53
5-84-SS-1	29.00	56.60	23.50	5.36	1.22	0.74	3.03	0.41
10-79-I	22.14	37.32	17.22	4.11	0.89	0.58	2.48	0.37
10-79-J	38.46	74.06	33.43	7.51	1.50	1.08	3.74	0.56
10-79-L1	32.25	59.90	28.31	5.63	0.98	0.72	2.80	0.43

Note: REE, as parts per million, by neutron activation, USGS, Reston.
Sample locations not mentioned in Tables 3 and 4 as latitude and longitude:

Sample	Lat. N	Long. W
ML-30	32°44'36"	116°27'08"
5-84 10	33°03'47"	116°34'06"
5-84-SS-1	32°58'47"	116°31'28"

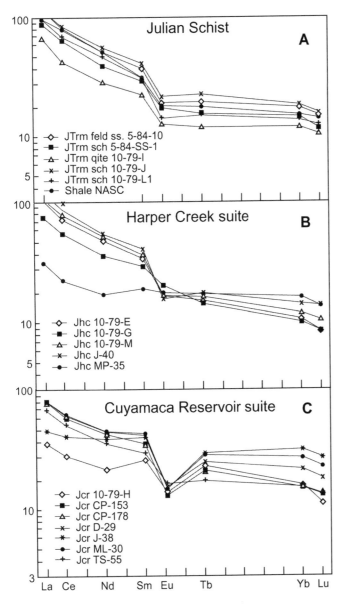

plutons containing elongate migmatitic and other lithologically recognizable enclaves of Julian Schist in varying stages of disaggregation to migmatized schist and gneiss. The Harper Creek suite REE patterns are similar to the pattern of the Bundarra S-type suite of the New England batholith, eastern Australia, for which an origin by partial melting of trench-complex metasedimentary rocks has been proposed (Shaw and Flood, 1981).

The Cuyamaca Reservoir suite (Fig. 12C) has flatter REE patterns and moderately negative Eu anomalies relative to the Harper Creek suite. In terms of REE patterns, mineralogy (reduced biotite and subaluminous amphibole), elevated Sr_i, and transitional I- to S-type character, the Cuyamaca Reservoir relates closely to the Hillgrove suite of the New England batholith (Shaw and Flood, 1981). The Cuyamaca Reservoir REE patterns also show similarities to the 50-Ma granodiorites from the eastern Gulf of Alaska (fig. 14 *in* Barker et al., 1992), which is considered to have formed from the partial melting of young trench-fill sediments in a forearc environment. The Alaskan granodiorites also have ASI values greater than 1, Sr_i values between 0.705 and 0.707, and FeO/Fe_2O_3 ratios greater than 3:1. Values greater than 3:1 indicate reducing conditions during crystallization, probably with ilmenite as an oxide phase (Todd et al., this volume).

PETROGENESIS

Plate-Tectonic Setting of Jurassic Magmatism

The new U-Pb zircon crystallization ages for the Harper Creek and Cuyamaca Reservoir suites presented in this paper cluster in the Middle Jurassic and include a Jurassic age for an I-type pluton in the Peninsular Ranges batholith (East Mesa pluton). Plate-tectonic models for the early Mesozoic (Triassic–Early Jurassic) Cordilleran convergent margin characterize the forearc region as a zone of extension or transtension within which a system of ophiolite-floored basins and fringing volcanic arcs developed (Busby-Spera, 1988; Saleeby and Busby-Spera, 1992; Busby and Saleeby, 1993; Busby et al., 1998). Available data for the Peninsular Ranges indicate that Late Triassic to Early and Middle(?) Jurassic deposition took place in fault-bounded forearc and/or interarc basins formed during crustal extension and rifting marginal to North America (Germinario, 1993; Gastil, 1993; Busby and Saleeby, 1993, Busby et al., 1998). The depositional basin of the Julian Schist is inferred to have been floored by oceanic or ophiolitic crust (Todd et al., 1988; Saleeby and Busby-Spera, 1992; Grove, 1993; Gastil, 1993; Thomson and Girty, 1994). Thomson and Girty (1994, p. 1114) speculated that the "....plutons of the Cuyamaca–Laguna Mountains shear zone were emplaced into the sedimentary fill of a composite forearc basin which was probably floored by highly extended continental crust on the east and oceanic crust on the west." Titanium enrichment in metabasalt layers within the Julian Schist suggests that these were distal volcanic products of fringing intraoceanic arc. Miogeoclinal metasedimentary rocks of predominantly Paleozoic age in the eastern part of the Peninsular Ranges are

Figure 12. Various REE patterns chondrite-normalized. A: Julian Schist (quartzo-feldspathic sandstones and semi-pelites). North American shale composite (of Haskin et al., 1966) is added for comparison. B: Harper Creek suite, similar to (A) but with more pronounced negative Eu anomaly. C: Cuyamaca Reservoir suite, transitional between metaluminous and peraluminous compositions. Patterns similar to Cretaceous quartz gabbro of Gromet and Silver (1987) but with distinctly higher $\delta^{18}O$ and Sr_i values.

North American shale composite (of Haskin et al., 1966) and to a group of Australian shales and lithic sandstones from Proterozoic to Triassic in age (Nance and Taylor, 1976).

The Harper Creek suite, apart from one sample, has REE patterns (Fig. 12B) that significantly overlap the field of the Julian Schist and differ only in having slightly more pronounced negative Eu anomalies. This close relationship is to be expected, as the Harper Creek suite shows a complete gradation from igneous considered to have overlain fragments of the North American craton (Gastil, 1993). Schmidt and Paterson (2002) suggest the Cuyamaca–Laguna Mountains shear zone is a segment of a Jurassic-Cretaceous deformation zone that extends along the length of the Peninsular Ranges batholith and which marks a lithospheric discontinuity between an oceanic floored arc to the west and a continental floored arc to the east.

The major source of the Julian Schist detritus apparently lay to the east, where late Proterozoic sialic basement and overlying Precambrian-derived epiclastic rocks were undergoing rapid uplift and unroofing as the result of early Mesozoic sinistral transform faulting along the truncated Paleozoic miogeocline (Saleeby and Busby-Spera, 1992; Busby and Saleeby, 1993; Germinario, 1993). Mafic flows and tuffs were extruded onto submarine fans and mafic feeder dikes and/or sills intruded these sediments. Gastil et al. (1981) and Gastil (1993) reported arc-volcanic strata of Triassic and Jurassic age in Baja California. Multigrain U-Pb analysis of detrital zircons in the Julian Schist and related units in southern California, however, are interpreted as mixtures of late Paleozoic and early Mesozoic grains, together with a blend of Precambrian zircons averaging ca. 1540 Ma (Gastil et al., 1988; Gastil and Girty, 1993).

Along the California Cordilleran margin, continental arc magmatism is interpreted to have persisted at least into the Middle Jurassic, migrating westward during Early and Middle Jurassic time across earlier-accreted Triassic–Early Jurassic forearc terranes and subduction complexes (Saleeby and Busby-Spera, 1992). A profound tectonic change occurred in the late Middle to early Late Jurassic with the development of spreading ocean basins floored by ophiolite west of North America. Fringing island arcs grew along the outer edge of the ophiolite basins in a tectonic regime dominated by extension and dextral transform faulting parallel to the continental margin. By the Late Jurassic, ophiolite development had reached its final stages, the fringing arcs were being accreted to North America, and transform faulting along the continental margin had shifted to sinistral motion (Saleeby and Busby-Spera, 1992; Busby and Saleeby, 1993).

Models for the origin of the Harper Creek and Cuyamaca Reservoir suites depend upon the timing of pre-Cretaceous magmatic activity in the Peninsular Ranges. As discussed above, plutons of the Harper Creek and Cuyamaca Reservoir suites yield Middle and Late Jurassic isotopic ages. The seemingly anomalous location of the Peninsular Ranges Jurassic magmatic "arc," 100–200 km west of the Jurassic Mojave-Sonoran continental-margin arc, apparently reflects its plate-tectonic environment in the upper plate of the Mojave–Sonoran subduction zone. In the late Middle Jurassic to Late Jurassic, the Mojave-Sonoran arc was the site of regional extension or transtension coeval with batholithic emplacement (Tosdal et al., 1989; Busby-Spera et al., 1990; Saleeby and Busby-Spera, 1992; Schermer and Busby, 1994). During this period, subduction was sinistral-oblique and sinistral strike-slip faults cut across the arc. In southeastern California and southwestern Arizona, regional metamorphism was contemporaneous with crustal extension and rifting and with Middle and Late Jurassic granitic intrusion. Tosdal et al. (1989) speculated

that an oceanic rift- and transform-fault system existed west of the continental-margin arc in Middle to Late Jurassic time. In this region, fault-bounded basins floored by ophiolite were the site of volcanogenic, oceanic, and terrigenous deposition and of mafic igneous activity while fragments of continental crust and/ or rifted island arc segments were transported southward along sinistral transform faults. In southeastern California and southwestern Arizona, regional metamorphism was contemporaneous with crustal extension and rifting and with coeval Middle and Late Jurassic granitic intrusion (Tosdal et al., 1989).

Mixing of Source Rock Components

Variations within the Cretaceous I-type plutonic rocks of the Peninsular Ranges batholith have been explained as the result of mixing of two or more source components (DePaolo, 1981; Hill et al., 1986; Shaw et al., 1986; Gromet and Silver, 1987). Notably, DePaolo (1981), through Sm-Nd and Rb-Sr isotopic studies, proposed a model for the batholith that involved the interaction of depleted mantle magmas with a crustally derived component. The crustal component was considered to have been derived mostly from the mantle in the Mesozoic, but the remaining material was of Precambrian continental derivation. The models of DePaolo (1981) and Gromet and Silver (1987) do not address the origin of the volumetrically significant Jurassic metaluminous and peraluminous granites within the batholith.

In southeastern Australia, Chappell and White (1974, 1992) proposed that magmas of the Lachlan fold belt were derived by partial melting of either an igneous or a sedimentary source component to produce an I-type (metaluminous) or an S-type (peraluminous) partial melt and a residuum (restite). They suggested that most of the variation within an intrusive suite was a result of variable degrees of separation of melt and restite prior to crystallization, with crystal fractionation important only after all restite was removed. In the case of some granites transitional between metaluminous and peraluminous compositions, a single physically mixed-source of infracrustal and supracrustal rocks has been suggested (Shaw and Flood, 1981; Barker et al., 1992). Shaw and Flood (1981) argue that the Uralla plutonic suite of the New England batholith, although metaluminous, has some S-type affinities, such as higher Sr_i than the main I-type suites of the batholith, $\delta^{18}O$ only marginally less than +10 per mil, ilmenite present in some plutons, and high normative hypersthene to diopside ratios. They envisage the source rocks as a physical mixture possibly along interfaces of trench-complex metasedimentary and metaigneous material. Barker et al. (1992) describe three granodiorite plutons intrusive into an accretionary prism in the eastern Gulf of Alaska. The composition and isotopic overlap of the granodiorite and flyschoid sediments, together with modelling and reference to melting experiments of greywacke indicate the granodiorite was derived from the metasediments by high degrees of melting. Basaltic magmas are seen by Barker et al. (1992) as a heat source rather than a melt component mixing with other crustally derived magmas.

As an alternative to the single component model of Chappell and White (1974), two- and three- source components for granites of the Lachlan fold belt have been proposed (Gray, 1984, 1995; Collins, 1996, 1998). Collins (1998) argued that the observed range of both I- and S-type magmas could be generated by a sequence of melting and mixing events involving three source components: (1) mantle generated basalt; (2) melts of older, lower crustal metaigneous rocks; and (3) melts of younger, middle crustal sedimentary rocks. According to Collins (1998), the S-type magmas of the Lachlan fold belt were generated in the middle crust by mixing and hybridization of partly melted Ordovician sedimentary rocks (diatexites) with I-type tonalite magmas derived by melting and mixing of mantle basalts and Cambrian greenstones; in effect, S-type magmas are interpreted as highly contaminated I-type magmas.

Origin of the Harper Creek Suite

Similarities in mineralogy, whole-rock geochemistry and isotopic ratios between the Harper Creek metagranite and metasedimentary rocks of the Julian Schist indicate that the Harper Creek magmas originated by partial melting of the abundant quartzo-feldspathic and semi-pelitic rocks of the Julian Schist and perhaps other, similar flysch units such as the Jacumba sequence. Supporting evidence for such an origin includes field relations, in which Harper Creek plutons are: (1) mantled by anatectic migmatites and gneisses of the Stephenson Peak facies; (2) interlayered and interfolded with migmatites and high-grade metasedimentary rocks; and (3) contain uniformly distributed Julian Schist–type refractory enclaves (restite).

The intermediate chemical compositions and moderately high Sr_i and $\delta^{18}O$ values of the Harper Creek suite reflect the heterogeneous character of the metasedimentary source (Julian Schist and related units). Protoliths of these metasedimentary rocks include Late Proterozoic and Paleozoic crustal sources as well as immature Mesozoic arc materials (Gastil et al., 1988; Gastil and Girty, 1993); the arc materials are, in part, only slightly older than Harper Creek magmatic activity.

According to the paleotectonic reconstruction discussed above, Middle to Late Jurassic Harper Creek magmatism occurred in a setting dominated by extension and transform faulting parallel to the continental margin, which was located well to the east. Peraluminous Harper Creek magmas may have formed at mid-crustal depths in the predominantly metasedimentary early Mesozoic forearc wedge during subduction and rifting, continentward (east) of the site of Cuyamaca Reservoir melting and mixing. Unlike Cuyamaca Reservoir plutons, Harper Creek plutons do not contain metaigneous enclaves or metaluminous phases, which indicates that I-type basaltic/tonalitic magmas did not rise and mingle with this segment of the wedge. Because there are no Harper Creek Sr_i and $\delta^{18}O$ values that differ significantly from the Julian Schist, the contribution of an I-type lower crustal component must have been minor. The relative lack of pegmatites in Harper Creek plutons and the fact that muscovite is

largely of subsolidus origin suggest that the magmas were relatively anhydrous and thus able to rise to considerable distances above the site of melting. Accordingly, we consider that the discrete Harper Creek plutons on the western side of the Jurassic belt resulted from crystallization of S-type magma that rose into the upper crust.

Stephenson Peak Facies

We suggest that the Stephenson Peak facies of the Harper Creek suite in the eastern Peninsular Ranges batholith represents the metasedimentary partial melts and restite (diatexites) that remained at or near the mid-crustal source region after more homogeneous bodies of Harper Creek magma separated from restite and rose into the upper crust. In the source region, up to 50% melt plus restite phases may have co-existed in near-equilibrium (White and Chappell, 1977). Low degrees of separation of melt and restite in the Stephenson Peak facies suggest that this material did not rise in the crust more than a few kilometers from the site of melting. The Cooma granodiorite gneiss of the Lachlan fold belt may be an analogous case: Collins (1998) suggested that Cooma-type melts were hydrous and thus incapable of moving far from their middle crustal source region. He further suggested that large migmatitic complexes such as those of Cooma-type in the Lachlan fold belt (and of Stephenson Peak–type in the Peninsular Ranges batholith?) underlie S-type granitoid batholiths to be exposed only after deformation and uplift.

Origin of the Cuyamaca Reservoir Suite

The westernmost Cuyamaca Reservoir plutons have relatively mafic compositions that in thin section suggest disequilibrium among the minerals hornblende, prismatic actinolite, hypersthene, clinopyroxene, and biotite. Preliminary pressure estimates based on hornblende geobarometry from a western pluton, sample SY-91-1, are 3.5 kb (J.M. Hammarstrom, 2002, personal commun.), implying a minimum depth of crystallization of 9–11 km. In the field, Cuyamaca Reservoir plutons contain metaigneous enclaves and interfinger/grade to bodies of I-type hornblende tonalite-quartz diorite (of East Mesa type?), suggestive of magma mingling and/or mixing. Concurrent with a west-to-east change from mafic to felsic compositions, there is a tendency for: (1) inclusions of Julian Schist and mica-rich clots to be more abundant in eastern Cuyamaca Reservoir plutons; (2) a change from amphibole to biotite as the main mafic phase; and (3) contacts between Cuyamaca Reservoir and Harper Creek plutons to become broader and more diffuse. Compositionally, Cuyamaca Reservoir major- and trace-element variation diagrams (Fig. 10) do not follow the systematic trends of fractionation or simple two-component mixing models. The aluminum saturation index varies from metaluminous to moderately peraluminous compositions (Fig. 9) in a manner that cannot be explained by crystal fractionation of an originally homogeneous magma. On a Sr_i versus $\delta^{18}O$ diagram (Fig. 11), the data fall between the more primitive Cretaceous I-type granite suites and the more evolved Harper Creek suite.

The above characteristics suggest an origin for the Cuyamaca Reservoir magmas from a possible three-component source rather than from a single component of arc-type metaigneous and metasedimentary rocks. Three possible source components are: (1) young mantle-derived magma; (2) a mixture of arc-type metaigneous and metasedimentary rocks; and (3) a mid-crustal, predominantly metasedimentary source consisting of Julian Schist and related units.

During Middle to Late Jurassic time, basaltic melts probably underplated and invaded Early Mesozoic-accreted arc-type terranes that comprised the oceanic crust west of the Mojave-Sonoran continental-margin arc and above its subduction zone (Tosdal et al., 1989, and references cited therein). In this offshore zone of rifting and extension, young mantle-derived magma may have been associated with a fringing island arc or may have risen through fracture zones and/or thinned oceanic crust, acting both as a source of heat necessary for partial melting and possibly as a component, mixing with the generated crustal melts. Probable Middle and Late Jurassic fringing island arc-type rocks are reported from several localities in the Continental Borderland of southern California (Gastil et al., 1981; Sorenson, 1988) and metamorphosed Late Jurassic arc rocks underlie Early Cretaceous volcanic rocks unconformably in coastal San Diego County (Balch et al., 1984; Anderson, 1991; C.T. Herzig, 1994, personal commun.). On the Vizcaíno Peninsula, two Late Jurassic source terranes, one exposing I-type hornblende tonalite and a second exposing high-level peraluminous granitic rocks with abundant red-brown biotite (Harper Creek–type?) were contributing detritus to the Eugenia and Perforada Formations by latest Jurassic-earliest Cretaceous time (Kimbrough et al., 1987). These formations overlap an early Mesozoic oceanic arc assemblage.

The older arc-type metaigneous and metasedimentary rocks that formed the oceanic crust outboard of the Mojave-Sonoran arc probably included accreted Triassic and Early Jurassic arc-ophiolites, subduction complexes, and forearc/interarc supracrustal sediments. Partial melting in lower crustal I-type granulitic-amphibolitic rocks, with or without magma mixing of the young mantle-derived magma, could have produced rocks such as the East Mesa suite, the bodies of hornblende tonalite that occur within western Cuyamaca Reservoir plutons, and perhaps other, as yet unrecognized, Jurassic I-type plutons. The Agua Caliente tourmaline-bearing biotite tonalite of Baja California (Schmidt and Paterson, 2002) with an age of 164.3 ± 2.3 Ma could be a further example.

A possible third source-component for the Cuyamaca Reservoir magmas may have been mid-crustal, predominantly metasedimentary assemblages such as the Julian Schist and related units, which, as noted above, are composed of mixtures of Precambrian-derived detritus with early Mesozoic arc material. Mixing of mantle-derived melts and/or lower crustal I-type melts with melts of such isotopically evolved material could produce the Cuyamaca Reservoir magmas with their transitional geochemical and isotopic characteristics.

CONCLUSIONS

Systematic variations within the Jurassic plutonic belt reflect an increasing depth profile from west to east across the Peninsular Ranges batholith (due largely to late Cenozoic tectonism), as well as apparent west-to-east changes in the source regions of the Jurassic magmas. Plutons of the relatively mafic Cuyamaca Reservoir suite are located in the western part of the belt, closer to an inferred western, oceanic lithospheric source. Sharp contacts and little or no evidence of in-situ mixing between those plutons and plutons of the Harper Creek suite in this area suggest that plutons of both suites are relatively high level. Deeper crustal levels and higher proportions of continentally derived source material are suggested in the central part of the belt by more peraluminous compositions of Cuyamaca Reservoir plutons and by broad gradational contacts between them and Harper Creek plutons, which in turn display a greater degree of mixing. In the eastern part of the belt, broad regions appear to be underlain by pre-Cretaceous peraluminous orthogneisses and metasedimentary anatectic migmatites (diatexites). In San Diego County, these rocks comprise the Stephenson Peak facies of the Harper Creek suite. In the Sierra San Pedro Martir area, Baja California, Schmidt and Paterson (2002) have described similar deformation structures as those that occur in the Cuyamaca–Laguna Mountains shear zone. Migmatitic orthogneisses are present in both areas, an extended history of Jurassic-Cretaceous magmatism and deformation is similar in both areas, and significant changes in recorded metamorphic pressures occur across the Cuyamaca–Laguna Mountains shear zone and Sierra San Pedro Martir deformation zones. To account for the strong lateral changes in rheology in the Sierra San Pedro Martir deformation zone, Schmidt and Paterson (2002) reason that extensive heating by sheeted tonalites along the lithospheric discontinuity was sufficient to focus deformation. Although Jurassic tonalite magmas from the East Mesa suite (Todd et al., this volume) could provide some heat in the Cuyamaca–Laguna Mountains shear zone and to the east, it is more probable that the Cuyamaca Reservoir suite and possible volcanic precursors may have been a significant source of heat for the high-grade metamorphism, partial melting of the Julian Schist, and magma mixing within and to the east of the Cuyamaca–Laguna Mountains shear zone.

The geochemical and isotopic characteristics of the Jurassic granites of the Peninsular Ranges batholith suggest that aspects of both mixing and restite models may be required to explain the observed variations within the Harper Creek and Cuyamaca Reservoir suites. In the broadest sense, geochemical variations within both suites can be explained by mixing of recycled Precambrian crustal materials with Mesozoic arc rocks. However, at the present level of erosion, separation of refractory metasedimentary and metavolcanic restite from peraluminous melts during partial fusion of early Mesozoic forearc strata in the middle crust was apparently a dominant process for the Harper Creek magmas. For the Cuyamaca Reservoir suite, partial melting of forearc strata with input of younger basalt/tonalite magma originating in the mantle/lower crust may have been a primary process in its formation. The Middle Jurassic age of an I-type pluton (East Mesa suite) that is spatially associated with a Cuyamaca Reservoir pluton suggests partial melting of early Mesozoic lower crustal metaigneous rocks with input of young mantle basalt as indicated by the presence of synplutonic basaltic dikes and mafic enclaves. Widespread mixing of magmas among these three suites at the present level of erosion is considered minimal, although complex contacts suggest that some interaction occurred.

SPECULATIONS

The above petrogenetic model for the origin of the Peninsular Ranges batholith Jurassic granites relies on offshore Jurassic plate-margin reconstructions for the southwestern Cordillera that are not well documented; therefore, the model is speculative. However, the data available for the Jurassic suites support a west-to-east transition from oceanic-ophiolitic source to a transitional source or one with a significant contribution of North American cratonic detritus. More speculative is the suggested vertical transition from lower crustal to mid-crustal sites of melting. As stated, our model for the Middle to Late Jurassic images the complex earlier Mesozoic history of subduction and arc magmatism, repeated seafloor spreading, and margin-parallel tectonic transport and accretion, although the degree of transpression relative to orthogonal convergence has been questioned (Schmidt and Paterson, 2002). We envision the Peninsular Ranges batholith Jurassic intrusive belt as forming in a setting offshore of the Mojave-Sonoran arc during a period of seafloor spreading, ophiolite development, and growth of transient fringing oceanic arcs. The production of Jurassic magmas ranging from I-type through S-type reflects the complexity of the oceanic-ophiolitic or transitional crust inherited from similar early Mesozoic plate-tectonic events. The presence of craton-derived Mesozoic metasedimentary strata as a source component is not unique to the Peninsular Ranges batholith, although the recognition of S-type granitic plutons apparently is. Based upon different chemical and isotopic compositions of plutons in the Sierra Nevada batholith, Kistler (1990) recognized and named two different lithospheric source regions separated by a tectonic boundary: a western Panthalassan type and an eastern North American type. Isotopic ratios of plutons that intruded Panthalassan lithosphere indicate a greater sedimentary component in the source than that for plutons that intruded North American lithosphere. Plutons that intruded Panthalassan lithosphere were described as "strongly contaminated" by Ague and Brimhall (1988). Indeed, such plutons might be expected to occur wherever subduction/magmatic arc activity results in the recycling of material from ancient cratons.

Jachens et al. (1991) speculated that the western edge of the Jurassic intrusive belt and the coincident gravity-magnetic boundary represent either: (1) a suture between continental crust of North America and an inferred Jurassic magmatic arc built on oceanic crust; or (2) a younger fault that reactivated the suture. We suggest that this structure originated as a Late Jurassic-Early

Cretaceous transpressive or orthogonal convergent (Schmidt and Paterson, 2002) deformation zone located outboard of the craton at what was to become the new continental margin. In San Diego County, this deformation zone separated an eastern terrane of mid- to upper-crustal S-type, transitional I- to S-type, and I-type plutons and their high-grade metamorphic wallrocks from the shallow, volcanic part of a Late Jurassic island arc on the west (east-over-west sense of motion). By latest Jurassic-earliest Cretaceous time, both I-type and mixed I-S-type "arcs" were accreted to North America, were emergent, and actively undergoing erosion (Kimbrough et al., 1987). Cretaceous plutons stitched across the boundary, welding the Jurassic arc to North America. The prominent geophysical, geochemical, and isotopic asymmetries of the Cretaceous batholith are a direct reflection of the eastward migration of the Cretaceous magmatic arc across this fundamental boundary (Todd et al., this volume). Moreover, since deformation occurs more readily in quartz- and mica-rich granites (Vernon and Flood, 1988), we believe that the Jurassic belt localized Early Cretaceous contractional strain, leading to the development of the Cuyamaca–Laguna Mountains shear zone.

ACKNOWLEDGMENTS

This paper benefited from many discussions with colleagues at the U.S. Geological Survey, Menlo Park; University of California, Riverside; and Macquarie University. We thank R.H. Flood and R.H. Vernon for comments on an earlier draft of the paper, GSA reviewers K. Howard and K. Schmidt for their valuable appraisals, and S.R. Paterson for his comments. Analyses and facilities at Macquarie University were supported by a Macquarie University Research Grant. The University of California at Los Angeles Keck Center for Geochemistry is supported by a grant from the National Science Foundation's Instrumentation and Facilities Program.

REFERENCES CITED

Ague, J.J., and Brimhall, G.H., 1988, Magmatic arc asymmetry and distribution of anomalous plutonic belts in the batholiths of California: Effects of assimilation, crustal thickness, and depth of crystallization: Geological Society of America Bulletin, v. 100, p. 912–927.
Anderson, C.L., 1991, Zircon uranium-lead isotope ages of the Santiago Peak volcanics and spatially related plutons of the Peninsular Ranges batholith, southern California [M.S. thesis]: San Diego, San Diego State University, 111 p.
Balch, D.C., Bartling, S.H., and Abbott, P.L., 1984, Volcaniclastic strata of the Upper Jurassic Santiago Peak Volcanics, San Diego, California, in Crouch, J.K., and Bachman, S.B., eds., Tectonics and sedimentation along the California margin: Los Angeles, California, Society of Economic Paleontologists and Mineralogists Pacific Section, v. 38, p. 157–170.
Barker, F., Farmer, G.L., Ayuso, R.A., Plafker, G., and Lull, J.S., 1992, The 50 Ma granodiorite of the eastern Gulf of Alaska: Melting in an accretionary prism in the forearc: Journal of Geophysical Research, v. 97, p. 6757–6778.
Berggreen, R.G., and Walawender, M.J., 1977, Petrography and metamorphism of the Morena Reservoir roof pendant, southern California: California Division of Mines and Geology Special Report 129, p. 61–66.
Busby, C., and Saleeby, J.B., 1993, Paleogeographic and tectonic setting of axial and western metamorphic framework rocks of the southern Sierra Nevada, California, in Dunn, G., and McDougall, K., eds., Mesozoic

paleogeography of the western United States—II: Los Angeles, California Society of Economic Paleontologists and Mineralogists Pacific Section, Book 71, p. 197–226.

Busby, C., Smith, D., Morris, W., and Fackler-Adams, B., 1998, Evolutionary model for convergent margins facing large ocean basins: Mesozoic Baja California, Mexico: Geology, v. 26, p. 227–230.

Busby-Spera, C.J., 1988, Speculative tectonic model for the early Mesozoic arc of the southwest Cordilleran United States: Geology, v. 16, p. 1121–1125.

Busby-Spera, C.J., Mattinson, J.M., Riggs, N.R., and Schermer, E.R., 1990, The Triassic-Jurassic magmatic arc in the Mojave-Sonoran Deserts and the Sierran-Klamath region; similarities and differences in paleogeographic evolution, *in* Harwood, D.S., and Miller, M.M., eds., Paleozoic and early Mesozoic paleogeographic relations; Sierra Nevada, Klamath Mountains, and related terranes: Boulder, Colorado, Geological Society of America Special Paper 255, p. 325–337.

Chappell, B.W., and White, A.J.R., 1974, Two contrasting granite types: Pacific Geology, v. 8, p. 173–174.

Chappell, B.W., and White, A.J.R., 1992, I- and S-type granites in the Lachlan Fold Belt: Transactions of the Royal Society of Edinburgh, Earth Sciences, v. 83, p. 1–26.

Chappell, B.W., White, A.J.R., and Williams, I.S., 1991, A transverse section through granites of the Lachlan Fold Belt: Second Hutton Symposium Excursion Guide, Canberra: Bureau of Mineral Resources, Geology and Geophysics, Record 22, 125 p.

Clemens, J.D., and Wall, J., 1988, Controls on the mineralogy of S-type volcanic and plutonic rocks: Lithos, v. 21, p. 53–66.

Collins, W.J., 1996, S- and I-type granitoids of the eastern Lachlan fold belt: Products of three-component mixing: Transactions of the Royal Society of Edinburgh, Earth Sciences, v. 88, p. 171–179.

Collins, W.J., 1998, Evaluation of petrogenetic models for Lachlan Fold Belt granitoids: Implications for crustal architecture and tectonic models: Australian Journal of Earth Sciences, v. 45, p. 483–500.

Dalrymple, G.B., Grove, M., Lovera, O.M., Harrison, T.M., Hulen, J.B., and Lanphere, M.A., 1999, Age and thermal history of The Geysers plutonic complex (felsite unit), Geysers geothermal field, California; a $^{40}Ar/^{39}Ar$ and U-Pb study: Earth and Planetary Sciences Letters, v. 30, p. 285–298.

Davis, T.E., and Gastil, R.G., 1993, Whole-rock Rb/Sr isochrons of fine-grained clastic rocks (Appendix), *in* Gastil, R.G., and Miller, R.H., eds., The prebatholithic stratigraphy of peninsular California: Boulder, Colorado, Geological Society of America Special Paper 279, p. 157–158.

DePaolo, D.J., 1981, A neodymium and strontium isotopic study of the Mesozoic calc-alkaline granitic batholiths of the Sierra Nevada and Peninsular Ranges, California: Journal of Geophysical Research, v. 86, p. 10470–10488.

Detterman, M.E., 1984, Geology of the Metal Mountain district, In-ko-pah Mountains, San Diego County California [M.S. thesis]: San Diego, San Diego State University, 216 p.

Engel, A.E.J., and Schultejann, A., 1984, Late Mesozoic and Cenozoic tectonic history of south-central California: Tectonics, v. 3, p. 659–675.

Erskine, B.G., 1986, Mylonitic deformation and associated low-angle faulting in the Santa Rosa mylonitic zone, southern California [Ph.D. dissertation]: Berkeley, University of California, 247 p.

Everhart, D.L., 1951, Geology of the Cuyamaca Peak quadrangle, San Diego County, California: California Division of Mines Bulletin 159, p. 51–115.

Fackler-Adams, B.N., Busby, C.J., and Mattinson, J.M., 1997, Jurassic magmatism and sedimentation in the Palen Mountains, southeastern California: Implications for regional tectonic controls on the Mesozoic continental arc: Geological Society of America Bulletin, v. 109, p. 1464–1484.

Flood, RH., and Shaw, S.E., 1977, Two "S-type" granite suites with low initial $^{87}Sr/^{86}Sr$ ratios from the New England batholith, Australia: Contributions to Mineralogy and Petrology, v. 61, p. 163–173.

Gastil, R.G., 1975, Plutonic zones in the Peninsular Ranges of southern California and northern Baja California: Geology, v. 3, p. 361–363.

Gastil, R.G., 1983, Mesozoic and Cenozoic granitic rocks of southern California and western Mexico: Boulder, Colorado, Geological Society of America Memoir 159, p. 265–275.

Gastil, R.G., 1993, Prebatholithic history of peninsular California, *in* Gastil, R.G., and Miller, R.H., eds., The prebatholithic stratigraphy of peninsular California: Boulder, Colorado, Geological Society of America Special Paper 279, p. 145–156.

Gastil, R.G., and Girty, M.S., 1993, A reconnaissance U-Pb study of detrital zircon in sandstones of peninsular California and adjacent areas, *in* Gastil, R.G., and Miller, R.H., eds., The prebatholithic stratigraphy of peninsular

California: Boulder, Colorado, Geological Society of America Special Paper 279, p. 135–144.

Gastil, R.G., and Miller, R.H., 1984, Prebatholithic paleogeography of peninsular California and adjacent Mexico, *in* Frizzell, A., Jr., ed., Geology of Baja California peninsula: Los Angeles, California, Society of Economic Paleontologists and Mineralogists Pacific Section, v. 39, p. 9–16.

Gastil, R.G., Morgan, G.J., and Krummenacher, D., 1981, The tectonic history of peninsular California and adjacent Mexico, *in* Ernst, W.G., ed., The geotectonic development of California, Rubey Volume I: Englewood Cliffs, New Jersey, Prentice-Hall, p. 284–306.

Gastil, R.G., Girty, G., Wardlaw, M., and Davis, T., 1988, Correlation of Triassic-Jurassic sandstone in peninsular California: Geological Society of America Abstracts with Programs, v. 20, no. 3, p. 162.

Germinario, M., 1993, The early Mesozoic Julian Schist, Julian, California, *in* Gastil, R.G., and Miller, R.H., eds., The prebatholithic stratigraphy of peninsular California: Boulder, Colorado, Geological Society of America Special Paper 279, p. 107–118.

Girty, G.H., Thomson, C.N., Girty, M.S., Miller, J., and Bracchi, K., 1993, The Cuyamaca–Laguna Mountains shear zone, Late Jurassic plutonic rocks and Early Cretaceous extension, Peninsular Ranges, southern California, *in* Abbott, L., Sangines, E.M., and Rendina, M.A., eds., Geological investigations in Baja California, Mexico: South Coast Geological Society Annual Field Trip Guidebook 21, p. 173–181.

Girty, M.S., Thomson, C., Girty, G.H., Bracchi, K.A., and Miller, J., 1994, U-Pb zircon geochronology, CLMSZ (Cuyamaca–Laguna Mountains shear zone), Peninsular Ranges, southern California: Geological Society of America Abstracts with Programs, v. 26, no. 2, p. 54.

Gray, C.M., 1984, An isotopic mixing model for the origin of granitic rocks in southeastern Australia: Earth and Planetary Science Letters, v. 70, p. 47–60.

Gray, C.M., 1995, Discussion of 'Lachlan and New England: Fold belts of contrasting magmatic and tectonic development' by B.W. Chappell: Journal and Proceedings of the Royal Society of New South Wales, v. 128, p. 29–32.

Gromet, L.P., and Silver, L.T., 1987, REE variations across the Peninsular Ranges batholith: Implications for batholithic petrogenesis and crustal growth in magmatic arcs: Journal of Petrology, v. 28, p. 75–125.

Grove, M., 1987, Metamorphism and deformation of prebatholithic rocks in the Box Canyon area, eastern Peninsular Ranges, San Diego County, California [M.S. thesis]: Los Angeles, University of California, 174 p.

Grove, M., 1993, Thermal histories of southern California basement terranes [Ph.D. dissertation]: Los Angeles, University of California, 419 p.

Grove, M., 1994, Contrasting denudation histories within the east-central Peninsular Ranges batholith (33°N), *in* McGill, S.F., and Ross, T.M., eds., Geological investigations of an active margin: Geological Society of America Cordilleran Section Guidebook, p. 235–240.

Haskin, L.A., Frey, F.A., Schmitt, R.A., and Smith, R.H., 1966, Meteoritic, solar and terrestrial rare-earth distributions, *in* Ahrens, L.H., ed., Physics and chemistry of the earth: New York, Pergamon Press, v. 7, p. 169–321.

Hill, R.I., 1984, Petrology and petrogenesis of batholithic rocks, San Jacinto Mountains, southern California [Ph.D. dissertation]: Pasadena, California Institute of Technology, 660 p.

Hill, R.I., Silver, L.T., and Taylor, H.P., Jr., 1986, Coupled Sr-O isotope variations as an indicator of source heterogeneity for the northern Peninsular Ranges batholith: Contributions to Mineralogy and Petrology, v. 92, p. 351–361.

Hine, R., Williams, I.S., Chappell, B.W., and White, A.J.R., 1978, Contrasts between I- and S-type granitoids of the Kosciusko Batholith: Journal of the Geological Society of Australia, v. 25, p. 219–234.

Hudson, F.S., 1922, Geology of the Cuyamaca region of California, with special reference to the origin of nickeliferous pyrrhotite: University of California Department of Geological Sciences Bulletin 13, p. 175–252.

Ishihara, S., 1977, The magnetite-series and ilmenite-series granitic rocks: Journal of Mineralogy and Geology, v. 27, p. 175–252.

Jachens, R.C., Simpson, R.W., Griscom, A., and Mariano, J., 1986, Plutonic belts in southern California defined by gravity and magnetic anomalies: Geological Society of America Abstracts with Programs, v. 18, no. 2, p. 120.

Jachens, R.C., Todd, V.R., Morton, D.M., and Griscom, A., 1991, Constraints on the structural evolution of the Peninsular Ranges batholith, California, from a new aeromagnetic map: Geological Society of America Abstracts with Programs, v. 23, no. 2, p. 38.

Jackson S.E., Longerich, H.P., Dunning, G.R., and Fryer, B.J., 1992, The application of laser ablation microprobe-inductively coupled plasma-mass spectrometry (LAM-ICP MS) to in-situ trace element determinations in minerals: Canadian Mineralogist, v. 30, p. 1049–1064.

Kimbrough, D.L., Hickey, J.J., and Tosdal, R.M., 1987, U-Pb ages of granitoid clasts in upper Mesozoic arc-derived strata of the Vizcaino peninsula, Baja California, Mexico: Geology, v. 15, p. 26–29.

Kimbrough, D.L., Smith, D.P., Mahoney, J.B., Moore, T.E., Grove, M., Gastil, R.G., and Ortega-Rivera, A., 2001, Forearc-basin sedimentary response to rapid Late Cretaceous batholith emplacement in the Peninsular Ranges of southern and Baja California: Geology, v. 29, p. 491 494.

Kistler, R.W., 1990, Two different lithosphere types in the Sierra Nevada, California, in Anderson, J.L., ed., The nature and origin of Cordilleran magmatism: Boulder, Colorado, Geological Society of America Memoir 174, p. 271–281.

Krummenacher, D., Gastil, R.G., Bushee, J., and Doupont, J., 1975, K-Ar apparent ages, Peninsular Ranges batholith, southern California: Geological Society of America Bulletin, v. 86, p. 760–768.

Lampe, C.M., 1988, Geology of the Granite Mountain area: Implications of the extent and style of deformation along the southeast portion of the Elsinore fault [M.S. thesis]: San Diego, San Diego State University, 150 p.

Leier-Engelhardt, P., 1993, Middle Paleozoic strata of the Sierra Las Pintas, northeastern Baja California Norte, Mexico, in Gastil, R.G., and Miller, R.H., eds., The prebatholithic stratigraphy of peninsular California: Boulder, Colorado, Geological Society of America Special Paper 279, p. 23–41.

Merriam, R.H., 1958, Geology of the Santa Ysabel quadrangle, San Diego County, California: California Division of Mines Bulletin, v. 177, p. 7–20.

Miller, R.H., and Dockum, M.S., 1983, Ordovician conodonts from metamorphosed carbonates of the Salton Trough, California: Geology, v. 11, p. 410–412.

Murray, G.T., and Girty, G.H., 1996, Pre-Jurassic deformation and metamorphism within the Julian Schist, Peninsular Ranges, Southern California: Geological Society of America Abstracts with Programs, v. 28, no. 5, p. 95.

Nance, W.B., and Taylor, S.R., 1976, Rare earth element patterns and crustal evolution – I. Australian post-Archean sedimentary rocks: Geochimica et Cosmochimica Acta, v. 40, p. 1539–1551.

O'Neil, J.R., and Chappell, B., 1977, Oxygen and hydrogen isotope relations in the Berridale batholith: Journal of the Geological Society of London, v. 133, p. 559–571.

O'Neil, J.R., Shaw. S.E., and Flood, R.H., 1977, Oxygen and hydrogen isotope compositions as indicators of granite genesis in the New England batholith, Australia: Contributions to Mineralogy and Petrology, v. 62, p. 313–328.

Paces, J.B., and Miller, J.D., 1993, Precise U-Pb ages of Duluth Complex and related mafic intrusions, northeastern Minnesota; geochronological insights to physical, petrogenetic, paleomagnetic, and tectonomagnetic processes associated with the 1.1 Ga Midcontinent Rift System: Journal of Geophysical Research, v. 98, p.13997–14013.

Paterson, S.R., Vernon, R.H., and Tobisch, O.T., 1989, A review of criteria for the identification of magmatic and tectonic foliations in granitoids: Journal of Structural Geology, v. 11, p. 349–363.

Premo, W.R., Morton, D.M., Snee, L.W., Naeser, N.D., and Fanning, C.M., 1998, Isotopic ages, cooling histories, and magmatic origins for Mesozoic tonalitic plutons from the N. Peninsular Ranges batholith, S. California: Geological Society of America Abstracts with Programs, v. 30, no. 5, p. 59.

Reed, J., 1993, Rancho Vallecitos Formation, Baja California Norte, Mexico, in Gastil, R.G., and Miller, R.H., eds., The prebatholithic stratigraphy of peninsular California: Boulder, Colorado, Geological Society of America Special Paper 279, p. 119–134.

Saleeby, J.B., Busby-Spera, C., and 6 contributors, 1992, Early Mesozoic tectonic evolution of the U.S., Cordilleran orogen, in Burchfield, B.L., Zoback, M.L., and Lipman, P., eds., The Cordilleran Orogen: Conterminous U.S.: Boulder, Colorado, Geological Society of America, The Geology of North America, v. G-3, p. 107–168.

Sañudo-Wilhelmy, S.A., and Flegal, A.R., 1994, Temporal variations in lead concentrations and isotopic composition in the southern California bight: Geochimica et Cosmochimica Acta, v. 58, p. 3315–3320.

Schermer, E.R., and Busby, C., 1994, Jurassic magmatism in the central Mojave Desert: Implications for arc paleogeography and preservation of continental volcanic sequences: Geological Society of America Bulletin, v. 106, p. 767–790.

Schmidt, K.L., and Paterson, S.R., 2002, A doubly vergent fan structure in the Peninsular Ranges batholith: Transpression or local complex flow around a continental margin buttress?: Tectonics, v. 21, 19 p.

Schwarcz, H.P., 1969, Pre-Cretaceous sedimentation and metamorphism in the Winchester area, northern Peninsular Ranges, California: Boulder, Colorado, Geological Society of America Special Paper 100, p. 61.

Sharp, R.V., 1967, San Jacinto fault zone in the Peninsular Ranges of southern California: Geological Society of America Bulletin, v. 78, p. 705–730.

Shaw, S.E., and Flood, R.H., 1981, The New England batholith, eastern Australia: Geochemical variations in time and space: Journal of Geophysical Research, v. 86, p. 10,530–10,544.

Shaw, S.E., Cooper, J.A., O'Neil, J.R., Todd, V.R., and Wooden, J.L., 1986, Strontium, oxygen, and lead isotope variations across a segment of the Peninsular Ranges batholith, San Diego County, California: Geological Society of America Abstracts with Programs, v. 18, no. 2, p. 183.

Silver, L.T., and Chappell, B.W., 1988, The Peninsular Ranges batholith: An insight into the evolution of the Cordilleran batholiths of southwestern North America: Transactions of the Royal Society of Edinburgh: Earth Sciences, v. 79, p. 105–121.

Sorenson, S., 1988, Tectonometamorphic significance of the basement rocks of the Los Angeles basin and the inner California Continental Borderland, in Ernst, W.G., ed., Metamorphism and crustal evolution of the western United States, Rubey Volume VII: Englewood Cliffs, New Jersey, Prentice Hall, Inc., p. 998–1022.

Staude, J.M.G., and Barton, M.D., 2001, Jurassic to Holocene tectonics, magmatism, and metallogeny of northwestern Mexico: Geological Society of America Bulletin, v. 113, p. 1357–1374.

Streckeisen, A.L., 1973, Plutonic rocks, classification and nomenclature recommended by the I.U.G.S., Subcommission on the Systematics of Igneous Rocks: Geotimes, v. 18, no. 10, p. 26–30.

Sun, S.-S., and McDonough, W.F., 1989, Chemical and isotopic systematics of oceanic basalts: Implications for mantle composition and processes, in Saunders, A.D., and Norry, M.J., eds., Magmatism in the ocean basins: Geological Society of London Special Publication 42, p. 313–345.

Taylor, H.P., Jr., and Silver, L.T., 1978, Oxygen isotope relationships in plutonic igneous rocks of the Peninsular Ranges batholith, southern and Baja California, in Zartman, R.E., ed., Short papers of the Fourth International Conference on geochronology, cosmochronology and isotope geology: U.S. Geological Survey Open-File Report 78-701, p. 423–426.

Taylor, H.P., 1986, Igneous rocks: ii. Isotopic case studies of circum-Pacific magmatism, in Valley, J.W., Taylor, H.P., Jr., and O'Neil, J.R., eds., Stable isotopes in high-temperature geological processes: Washington, D.C., Mineralogical Society of America, Reviews in Minerology, v. 16, p. 273–317.

Tera, F., and Wasserburg, G.J., 1972, U-Th-Pb systematics in three Apollo basalts and the problem of initial Pb in lunar rocks: Earth and Planetary Science Letters, v. 14, p. 281–304.

Theodore, T.G., and Sharp, R.V., 1975, Geologic map of the Clark Lake quadrangle, San Diego County, California: U.S. Geological Survey Miscellaneous Field Study Map MF-644, scale 1:24,000.

Thomson, C.N., and Girty, G.H., 1994, Early Cretaceous intra-arc ductile strain in Triassic-Jurassic and Cretaceous continental margin arc rocks, Peninsular Ranges, California: Tectonics v. 13, p. 1108–1119.

Todd, V.R., 1978, Geologic map of the Monument Peak quadrangle, San Diego County, California: U.S. Geological Survey Open-File Report 78-697, 47 p.

Todd, V.R., 1994a, Geologic map of the Julian 7.5' quadrangle, San Diego County, California: Unpublished U.S. Geological Survey Open-File Report 94-16, Director's Approval 1-3-94, 36 p.

Todd, V.R., 1994b, Preliminary geologic map of the El Cajon 30' x 60' quadrangle, San Diego and Imperial Counties, California: Unpublished U.S. Geological Survey Open-File Report 94-18, Director's Approval 1-3-94, 45 p.

Todd, V.R., and Shaw, S.E., 1979, Structural, metamorphic and intrusive framework of the Peninsular Ranges batholith in southern San Diego County, California, in Abbott, L., and Todd, V.R., eds., Mesozoic crystalline rocks: Department of Geological Sciences, San Diego State University, Guidebook for Geological Society of America Annual Meeting, San Diego, California, November 1979, p. 177–231.

Todd, V.R., and Shaw, S.E., 1985, S-type granitoids and an I-S line in the Peninsular Ranges batholith, southern California: Geology, v. 13, p. 231–233.

Todd, V.R., and Shaw, S.E., 1995, Petrology and preliminary geochemistry of plutonic rocks in the Peninsular Ranges batholith, southern San Diego County, California: Unpublished U.S. Geological Survey Open-File Report 95-524, Director's Approval 6-21-95, 132 p.

Todd, V.R., Kilburn, J.E., Detra, D.E., Griscom, A., Kruse, F.A., and McHugh, E.L., 1987, Mineral resources of the Jacumba (In-ko-pah) Wilderness Study Area, Imperial County, California: U.S. Geological Survey Bulletin 1711-D, p. 18.

Todd, V.R., Erskine, B.G., and Morton, D.M., 1988, Metamorphic and tectonic evolution of the northern Peninsular Ranges batholith, *in* Ernst, W.G., ed., Metamorphism and crustal evolution of the western United States, Rubey Volume VII: Englewood Cliffs, New Jersey, Prentice-Hall, p. 894–937.

Todd, V.R., Girty, G.H., Shaw, S.E., and Jachens, R.C., 1991a, Geochemical, geochronologic, and structural characteristics of Jurassic plutonic rocks, Peninsular Ranges, California: Geological Society of America Abstracts with Programs, v. 23, no. 5, A249.

Todd, V.R., Shaw, S.E., Girty, G.H., and Jachens, R.C., 1991b, A probable Jurassic plutonic arc of continental affinity in the Peninsular Ranges batholith, southern California: Tectonic implications: Geological Society of America Abstracts with Programs, v. 23, no. 2, p. 104.

Todd, V.R., Kimbrough, D.L., and Herzig, C.T., 1994, The Peninsular Ranges batholith from western volcanic arc to eastern mid-crustal intrusive and metamorphic rocks, San Diego County, California, *in* McGill, S.F., and Ross, T.M., eds., Geological investigations of an active margin: Geological Society of America Cordilleran Section Guidebook, San Bernardino, California, p. 227–235.

Tosdal, R.M., Baxel, G.B., and Wright, J.E., 1989, Jurassic geology of the Sonoran Desert region, southern Arizona, southeastern California, and northernmost Sonora: Construction of a continental-margin magmatic arc, *in* Jenney, J.P., and Reynolds, S.J., eds., Geologic evolution of Arizona: Arizona Geological Society Digest 17, p. 397–434.

Vernon, R.H., 1983, Restite, xenoliths and microgranitoid enclaves in granites: Royal Society of New South Wales Journal and Proceedings, v. 116, p. 77–103.

Vernon, R.H., 1991, Questions about myrmekite in deformed rocks: Journal of Structural Geology, v. 13, p. 979–985.

Vernon, R.H., and Flood, R.H., 1988, Contrasting deformation of S- and I-type granitoids in the Lachlan Fold Belt, eastern Australia: Tectonophysics, v. 147, p. 127–143.

Vernon, R.H., Richards, S.W., and Collins, W.J., 2001, Migmatite-granite relationships: Origin of the Cooma Granodiorite magma, Lachlan Fold Belt, Australia: Physics and Chemistry of the Earth (A), v. 26, no. 4–5, p. 267–271.

Wagner, D.L., 1996, Geologic map of the Tubb Canyon 7.5' quadrangle, San Diego County, California: California Division of Mines and Geology Open-File Report 96-06.

Walawender, M.J., Gastil, R.G., Clinkenbeard, J.P., McCormick, W.V., Eastman, B.G., Wernicke, R.S., Wardlaw, M.S., and Gunn, S.H., 1990, Origin and evolution of the zoned La Posta-type plutons, eastern Peninsular Ranges batholith, southern and Baja California, *in* Anderson, J.L., ed., The nature and origin of Cordilleran magmatism: Boulder, Colorado, Geological Society of America Memoir 174, p. 1–18.

Walawender, M.J., Girty, G.H., Lombardi, M.R., Kimbrough, D., Girty, M.S., and Anderson, C, 1991, A synthesis of recent work in the Peninsular Ranges batholith, *in* Walawender, M.J., and Hanan, B.B., eds., Geological excursions in southern California and Mexico: Geological Society of America Guidebook for the 1991 Annual Meeting, San Diego, California, p. 297–318.

Wetmore, P.H., Schmidt, K.L., Paterson, S.R., and Herzig, C., 2001, Tectonic implications for the along-strike variation of the Peninsular Ranges Batholith, Southern and Baja California: Geology, v. 30, p. 247–250.

White, A.J.R., and Chappell, B.W., 1977, Ultrametamorphism and granitoid genesis: Tectonophysics, v. 43, p. 7–22.

White, A.J.R., and Chappell, B.W., 1988, Some supracrustal (S-type) granites of the Lachlan Fold Belt: Transactions of the Royal Society of Edinburgh: Earth Sciences, v. 79, p. 169–181.

White, A.J.R., Clemens, J.D., Holloway, J.R., Silver, L.T., Chappell, B.W., and Wall, V.J., 1986, S-type granites and their probable absence in southwestern North America: Geology, v. 14, p. 115–118.

Wooden, J.L., Kistler, R.W., and Morton, D.M., 1997, Rb-Sr WR ages and Sr, O, and Pb isotopic systematics of plutons in the northern Peninsular Ranges batholith, southern California: Geological Society of America Abstracts with Programs, v. 29, no. 5, p. 74.

MANUSCRIPT ACCEPTED BY THE SOCIETY JUNE 2, 2003

Geological Society of America
Special Paper 374
2003

Cretaceous plutons of the Peninsular Ranges batholith, San Diego and westernmost Imperial Counties, California: Intrusion across a Late Jurassic continental margin

Victoria R. Todd
Earth Sciences Department, Palomar College, 1140 West Mission Road, San Marcos, California 92069, USA

Stirling E. Shaw
GEMOC, Department of Earth and Planetary Sciences, Macquarie University, New South Wales 2109, Australia

Jane M. Hammarstrom
U.S. Geological Survey, 954 National Center, Reston, Virginia 20192, USA

ABSTRACT

The Peninsular Ranges batholith of southern California and Baja California was emplaced across the lithospheric boundary between North America and Pacific plates in the Jurassic and Cretaceous. In San Diego County, the locus of Cretaceous plutonism migrated eastward from oceanic lithosphere across the continental margin into early Mesozoic continental lithosphere. Uplift and westward tilting of the Peninsular Ranges block associated with Late Cenozoic rifting and transform faulting resulted in an erosional depth profile from volcanic levels in the west to mid-crustal depths in the east.

Our study of Cretaceous plutons in a west-to-east transect across southern San Diego County shows that granitic plutons have distinctive geophysical, geobarometric, mineralogical, geochemical, and isotopic characteristics that vary systematically with geographic position within the batholith. On the basis of these characteristics, granitic plutons are grouped into 10 plutonic suites, each comprising numerous plutons. In the west-central part of the study area, five granitic suites and a suite of gabbroic rocks comprise a series of large, concentrically zoned plutonic complexes, possibly the roots of ring complexes. Mingling contacts between granitic and gabbroic plutons within these complexes indicate that mafic and silicic magmas were produced simultaneously during Cretaceous intrusion.

Published and unpublished isotopic age studies indicate that plutons in the western zone are primarily of Early and middle Cretaceous age, whereas those in the eastern zone are middle to Late Cretaceous in age. Our delineation of western and eastern plutonic zones on the basis of the geographic distribution and compositions of plutonic suites corresponds closely with previously noted west-to-east batholithic asymmetries. Primary among these is a steep aeromagnetic- and gravity-anomaly gradient that coincides closely with the westernmost limit of Jurassic granites and their early Mesozoic wallrocks. This gradient is interpreted as the manifestation of an east-dipping fault of crustal dimensions that formed in the latest Jurassic-earliest Cretaceous (Shaw et al., this volume, Chapter 7). The distribution of Cretaceous plutonic suites, together with compositional changes in suites that intrude both western and eastern zones of the batholith, are considered to reflect the location of the fault-bounded Late Jurassic continental margin.

Todd, V.R., Shaw, S.E., and Hammarstrom, J.M., 2003, Cretaceous plutons of the Peninsular Ranges batholith, San Diego and westernmost Imperial Counties, California: Intrusion across a Late Jurassic continental margin, *in* Johnson, S.E., Paterson, S.R., Fletcher, J.M., Girty, G.H., Kimbrough, D.L., and Martín-Barajas, A., eds., Tectonic evolution of northwestern México and the southwestern USA: Boulder, Colorado, Geological Society of America Special Paper 374, p. 185–235. For permission to copy, contact editing@geosociety.org. © 2003 Geological Society of America.

Geochemical studies of the Cretaceous granitic suites characterize them as low-K_2O, low-Rb, calcic granites with aluminum saturation indices (ASI, i.e., molar $Al_2O_3/[CaO+Na_2O+K_2O]$ values) between 0.67 and 1.20, reflecting a significant mantle contribution in the source of the magmas. Systematic variations in the geographic distribution and composition of suites indicate that partial melting of chemically inhomogeneous, p_{H_2O}-variable metaigneous rocks in the lower crust played a significant role in their origin. Constant rare earth element patterns and low Sr_i values of the suites across the gravity-magnetic boundary suggest that source rocks of basaltic composition were present in the lower crust and/or as underplates from earlier subduction episodes beneath both western and eastern zones of the batholith. Compositional differences among five tonalite suites reflect (1) increasing crustal thickness from the Early to Late Cretaceous, causing a phase transition of the zone of melting toward eclogitic assemblages, and/or (2) increasing depths of melting as the magmatic arc migrated eastward. Four leucogranite suites apparently represent fractionates of mafic melts, and/or partial melts of lithologically distinctive lower crustal rocks. The interaction of magmas with western and eastern lithosphere during ascent and emplacement was responsible for slight eastward increases in K_2O, Rb, initial $^{87}Sr/^{86}Sr$ ratios, $\delta^{18}O$ values, and radiogenic lead isotopes in suites that are present in both western and eastern zones, as well as for east-west differences in opaque oxide mineralogy and magnetic properties.

The data of our study imply that a single Cretaceous magmatic arc migrated eastward across a pre-existing Late Jurassic-earliest Cretaceous lithospheric boundary. The distribution of known Cretaceous pluton ages does not require an age break, but rather suggests continuous eastward migration of a single magmatic arc. Published and unpublished Early Cretaceous pluton ages for the oldest Cretaceous granitic suite that stitched across the lithospheric boundary in this region constrain the minimum age of its formation as early Early Cretaceous.

Keywords: Cordilleran batholith, petrogenesis, Cretaceous tectonics, southwestern California, Peninsular Ranges batholith, granites.

INTRODUCTION

The Peninsular Ranges batholith is a continental-margin batholith composed of Jurassic metagranitic rocks of mainly continental affinity (Shaw et al., this volume, Chapter 7, and references therein) and voluminous Cretaceous I-type (i.e., chiefly derived from igneous source materials) granitic and gabbroic plutons. Northwest-southeast–elongate remnants of highly deformed Jurassic plutons including those whose source region contained a significant sedimentary or metasedimentary component (S-type granitoids of Chappell and White, 1992; Todd et al., 1991) are present in a central belt that extends southward from ~33°45′N latitude in southern California into Baja California (Schmidt and Paterson, 2002).

During the period from ca. 140 to 80 Ma (Silver and Chappell, 1988; Anderson, 1991; Premo et al., 1998), an east-dipping subduction zone and magmatic arc existed at, or near, the southwestern margin of North America. A decrease in the dip of the subducted slab in the Late Cretaceous and early Tertiary (Dickinson, 1981; Gastil et al., 1981) has been invoked to explain the eastward migration of the Cretaceous arc into the North American craton. In San Diego County, the locus of Cretaceous magmatism migrated from oceanic lithosphere eastward across the continental margin into young, early Mesozoic continental lithosphere.

The oldest Cretaceous plutons in the county were comagmatic with subaerial island-arc volcanic rocks (the Santiago Peak Volcanics), which are preserved in coastal San Diego County (Herzig and Kimbrough, 1991; Anderson, 1991) (Fig. 1). Scattered remnants of deformed Late Jurassic island-arc metavolcanic and volcaniclastic rocks lie nonconformably beneath these Early Cretaceous volcanic strata (Balch et al., 1984; Anderson, 1991; C.T. Herzig, 1994, personal commun.).

In the central Peninsular Ranges batholith, Early Cretaceous plutons intruded crust composed of a Jurassic-Triassic forearc sequence plus the Middle and Late Jurassic granitic plutons that were in part derived from this sequence (Shaw et al., this volume, Chapter 7). On the east side of the batholith, middle to Late Cretaceous plutons intruded Mesozoic and Paleozoic metasedimentary rocks; the latter are considered to be tectonically displaced remnants of the Paleozoic miogeocline of western North America (Gastil, 1993).

Late Cenozoic rifting and transform faulting resulted in broad uplift and westward tilting of the Peninsular Ranges block in southern California (Gastil et al., 1975; Ague and Brandon, 1992), and erosion has exposed a substantial crustal section of the batholith from volcanic levels in coastal San Diego County to mid-crustal depths (at least 11–16 km, Grove, 1989) in the desert ranges of eastern San Diego and westernmost Imperial Counties.

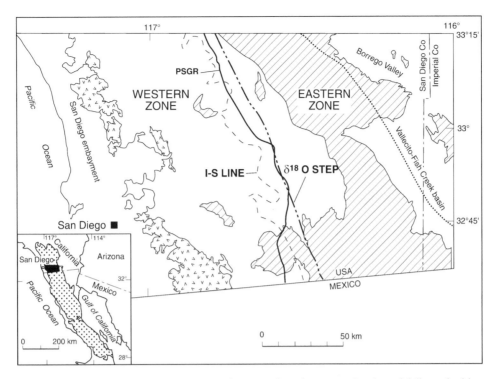

Figure 1. Schematic map showing western and eastern plutonic zones and major axial discontinuities of Cretaceous Peninsular Ranges batholith in study area; prebatholithic screens and Jurassic metagranites not shown. Inset shows Peninsular Ranges batholith in southern and Baja California, study area in black. Western zone (unpatterned) includes Early to middle Cretaceous granitic and gabbroic plutons; eastern zone (lined pattern) is composed of middle to Late Cretaceous granitic plutons and minor gabbro. Dashed line marks western limit of Jurassic metagranites (I-S line), which extend at least to longitude of Borrego Valley; dot-dash line is δ^{18}O-step of Silver et al. (1979); heavy solid line is approximate location of steep magnetic-gravity gradient mathematically transformed into a pseudogravity gradient (PSGR)(Jachens, 1992); denser, more magnetic batholithic rocks lie to west of gradient and less dense, virtually non-magnetic rocks to east. V pattern = Santiago Peak Volcanics; San Diego embayment includes Eocene marine and nonmarine sedimentary rocks and younger Cenozoic deposits (Kennedy, 1975); Vallecito–Fish Creek basin = Neogene sedimentary and volcanic rocks and younger deposits (Kerr and Kidwell, 1991). Prebatholithic rocks west of I-S line are mainly metavolcanic; those between I-S line and dotted line are metamorphosed early Mesozoic flysch; Paleozoic metasedimentary rocks lie east of dotted line.

This paper describes field, petrographic, and geochemical characteristics of the Cretaceous plutons in a west-to-east transect across the Peninsular Ranges batholith between ~33° and 32°37′30″N latitudes (Fig. 1). Gabbroic plutons, while not the subject of this paper, invariably show evidence of mingling with adjacent granitic plutons and are therefore discussed briefly. The final section of the paper discusses the implications of our study for the tectonic evolution of the Cretaceous magmatic arc in southwestern California.

EAST-WEST GRADIENTS ACROSS NORTHERN PENINSULAR RANGES BATHOLITH

Most workers have divided the Cretaceous batholith into "western" and "eastern" zones (Fig. 1) on the basis of lithologic differences between plutons on the west side of the batholith (mafic granitic rocks and gabbro predominant) and those on the

eastern side (trondhjemites, leucogranodiorites, and two-mica granites) (Gastil, 1983; Todd et al., 1988, and references therein). Prebatholithic wallrocks also differ, from volcanic rocks on the west, to flysch-like metasedimentary rocks in the central part, to predominantly miogeoclinal metasedimentary rocks on the east. Geochemical, isotopic, and geophysical studies by many workers confirm the systematic west-east asymmetry of the Peninsular Ranges batholith. This asymmetry has been explained as (1) the product of two Cretaceous magmatic arcs (Silver and Chappell, 1988; Johnson et al., 1999a); (2) due chiefly to a change in plate-tectonic motions in the middle Cretaceous (Walawender et al., 1990); and (3) largely the consequence of a single Cretaceous arc migrating eastward across a fundamental pre-Cretaceous lithospheric boundary (Thomson and Girty, 1994; Shaw et al., this volume, Chapter 7).

Figure 1 shows significant axis-parallel geophysical, lithologic, and isotopic discontinuities that are considered to divide the

Peninsular Ranges batholith into western and eastern zones. Notably, a steep north-northwest–trending gradient in aeromagnetic anomalies separates a western region of highly magnetic plutons/ metavolcanic rocks from an essentially nonmagnetic eastern region (Jachens et al., 1986, 1991; Jachens, 1992). Almost coincident with, but displaced 3–4 km east, of the magnetic gradient is a steep gravity gradient (R.C. Jachens, 1990, personal commun.), which separates relatively dense, mafic batholithic rocks on the west from a region of less dense, more silicic Late Cretaceous plutons, Jurassic metagranites, and metasedimentary wallrocks on the east. The gravity-magnetic gradient coincides closely with the "I-S line" of Todd and Shaw (1985), defined as the westernmost limit of Jurassic S- and I-S–type metagranite plutons and their Jurassic-Triassic metasedimentary wallrocks. New U-Pb zircon dating has led to the first recognition of a Jurassic I-type pluton within the belt of Jurassic granites in the study area ($^{206}Pb/^{238}U$ age of 170 ± 2 Ma, Shaw et al., this volume, Chapter 7). The S-type granitoids described in Shaw et al. (this volume, Chapter 7) include Jurassic metagranitic rocks having a substantial sedimentary source component and meeting all of the criteria cited by Chappell and White (1992) for S-type granites in the Lachlan Fold Belt of eastern Australia. The designation "S-type" does not include weakly peraluminous, Early Cretaceous silicic plutons of the western zone of the Peninsular Ranges batholith or peraluminous two-mica granites associated with middle to Late Cretaceous plutons in the eastern zone.

Cross-sections showing the west-to-east magnetic anomaly reveal that, in many cases, the I-S line and magnetic-anomaly break coincide exactly (R.C. Jachens, 1991, personal commun.). In Figure 1, the combined gravity-magnetic gradient is portrayed as a "pseudogravity" gradient (PSGR) created by the mathematical transformation of the magnetic-anomaly field data into the gravity field that would be produced if all magnetic material were replaced by proportionately dense material (Jachens, 1992). Jachens et al. (1986, 1991) modelled the aeromagnetic gradient as a planar boundary dipping 45–60°E to depths of at least 10–12 km. Shaw et al. (this volume, Chapter 7) suggest that this boundary is a latest Jurassic-earliest Cretaceous margin-parallel transpressive(?) fault that bounded the offshore region of the Late Jurassic-Early Cretaceous southwestern North American continental margin. Late Jurassic granitic clasts of S-type affinity are present in latest Jurassic-earliest Cretaceous flysch deposited across the Vizcaino peninsula in the Baja California borderland (Kimbrough et al., 1987), thus documenting the emergence of parts of the Late Jurassic-Early Cretaceous continental margin during this period.

Many workers have noted other east-west asymmetries in the Cretaceous batholith, e.g., (1) the $\delta^{18}O$ "step" of Silver et al. (1979), which divides the batholith into a western zone of low, "normal" igneous $\delta^{18}O$ values (+6 to +8.5) and an eastern zone of higher values (+9 to +12) (Fig. 1), reflecting prior access of magmas to surface or near-surface materials at some time in their history; (2) the "gabbro line" of Todd et al. (1988), which separates a western region of large and abundant gabbro plutons from an eastern one in which gabbro is scarce or absent; and (3) the change from predominantly mafic western-zone plutons to more

silicic, eastern-zone plutons of La Posta type (Walawender et al., 1990). We consider that the axial trend of most of these discontinuities reflects the eastward migration of the Cretaceous magmatic arc across the Late Jurassic–Early Cretaceous continental margin. Magmas that interacted with oceanic crust west of this margin crystallized magnetite ± ilmenite, whereas those that intruded reduced (low Fe^{3+}/Fe^{2+}) crust east of the margin crystallized ilmenite and are thus largely nonmagnetic.

In the study area, exceptions are mafic I-type plutons east of the I-S line whose large size apparently protected their weakly to moderately magnetic cores from interaction with surrounding reduced crust (e.g., the Mount Laguna pluton and the Cuyamaca Mountains gabbroic complex [Jachens, 1992]).

BATHOLITHIC STRUCTURE

Batholithic structure, as expressed by the orientation of plutons, wallrock screens, and plutonic and metamorphic foliation varies in a west-to-east direction across the study area (Todd and Shaw, 1979; Todd, 2004). We observe that many Cretaceous plutons in the west-central part of the study area display the following characteristics that are commonly ascribed to syntectonic plutons: (1) concordance of pluton shapes with regional batholithic structure; (2) parallelism of magmatic and subsolidus foliations and their concordance with foliation in adjacent metamorphic wallrocks (the latter present as steeply dipping tabular screens within and between plutons); (3) isoclinally folded dikes whose axial planes are parallel to magmatic foliation; and (4) chiefly dynamothermal rather than static thermal aureoles in wallrocks. These and other characteristics indicate overlap of magmatic flow with regional-scale ductile deformation (Pitcher and Berger, 1972; Hutton, 1981; Soula, 1982; Paterson et al., 1989; Paterson and Vernon, 1995).

Evidence of magmatic flow is present in virtually all plutons as shown by the preferred orientation of euhedral to subhedral feldspar, hornblende, and/or pyroxene crystals (Fig. 2A and 2B) and by the parallel alignment of variably flattened mafic and intermediate enclaves (equivalent to the microgranitoid enclaves of Vernon (1983), and referred to herein as "mafic enclaves"). Flattening and elongation of enclaves is greatest in areas where magmatic flow was most intense, in the extreme case resulting in thin, laterally continuous mafic schlieren. Commonly, mineral foliation passes across contacts between mafic enclave and granitic host without deviation, as though the viscosity of the enclaves was similar to that of the enclosing magma. This geometry is generally considered to indicate that enclaves were not completely crystallized when magmatic flow occurred (Vernon, 1983). Magmatic foliation is overprinted to varying degrees by near-solidus and/or subsolidus dynamothermal recrystallization of quartz and biotite grains, and, to a lesser extent, of feldspar and hornblende (Fig. 2C and 2D). This has resulted in variable modification of feldspar and the formation of lenticular quartz and mafic aggregates that are aligned parallel to magmatic foliation.

Cretaceous plutons in the western zone of the study area display a variable structural pattern (Fig. 3) (Todd, 2004): west-

Figure 2. Outcrops and photomicrographs of Japatul Valley suite showing magmatic and subsolidus textures. A: Near-vertical magmatic foliation in biotite-hornblende tonalite with approximately blocky mafic enclaves. B: Photomicrograph (crossed Nicols, long dimension ~8 cm) of tonalite in (A) shows subhedral plagioclase, hornblende, and biotite; interstitial quartz grains with at least one planar face against plagioclase, which has oscillatory zoning and minor deformation twinning; magmatic sphene and epidote. Magmatic foliation in this view is subvertical. C: Steeply inclined magmatic-subsolidus foliation in hornblende-biotite granodiorite whose mafic enclaves are strongly flattened and elongated in plane of foliation. D: Photomicrograph of granodiorite in (C) shows substantial recrystallization/grain-size reduction of quartz, bitotite, and K-feldspar to fine-grained submosaic matrix containing seedlike sphene. Blocky plagioclase porphyroclasts contain very fine-grained (solid-state) mafic inclusions, sericite, and epidote; grains are fractured and margins intergrown with matrix. Foliation in this view is from lower right to middle left, but orientation of some plagioclase phenocrysts suggests presence of a secondary foliation.

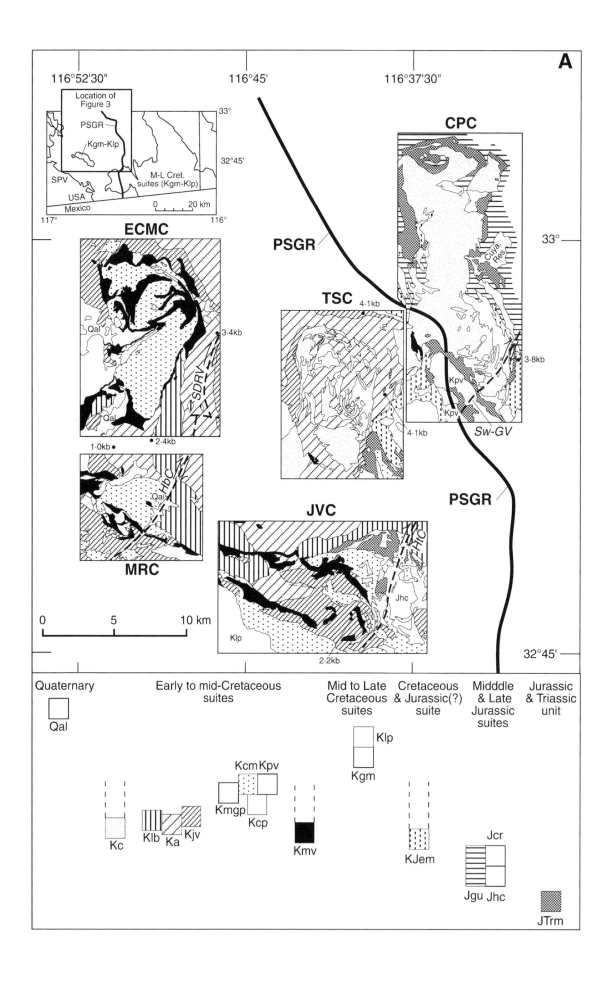

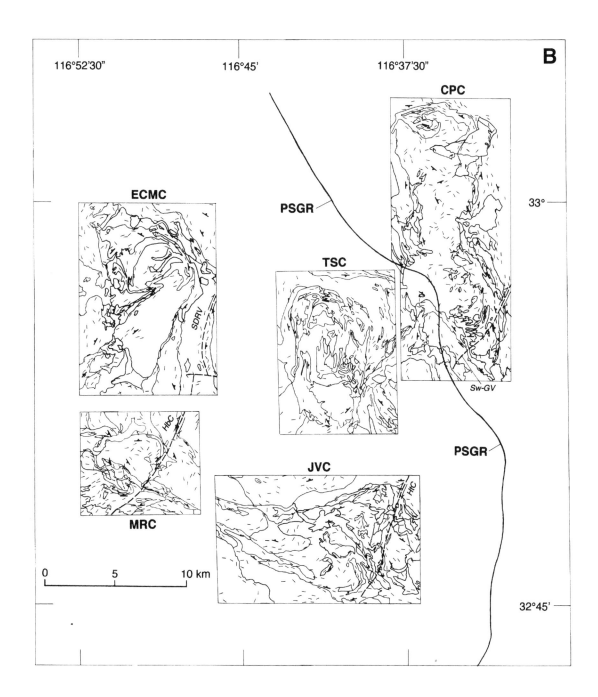

Figure 3 on this and previous page. Five Early Cretaceous zoned plutonic complexes in west-central Peninsular Ranges batholith. A: ECMC—El Cajon Mountain complex; MRC—McClain Ranch complex; JVC—Japatul Valley complex; TSC—Tule Springs complex; CPC—Cuyamaca Peak complex. Light solid lines are lithologic contacts; heavy dashed lines are ductile-to-brittle shear zones, arrows indicate sense of lateral displacement: SDRV—San Diego River Valley; HbC—Harbison Canyon; HtC—Horsethief Canyon; Sw-GV—Sweetwater River–upper Green Valley. Heavy solid line is pseudogravity gradient (PSGR). Solid dots are emplacement pressures from hornblende geobarometry (± 1.5 kb). Legend shows patterning and field-based age relations of suites: Kc—Cuyamaca Gabbro; Klb—Las Bancas; Ka—Alpine; Kjv—Japatul Valley; Kmgp—Mother Grundy Peak; Kcp—Chiquito Peak; Kcm—Corte Madera; Kpv—Pine Valley; Kmv—metavolcanic rocks; Kgm—Granite Mountain; Klp—La Posta; KJem—East Mesa; Jcr—Cuyamaca Reservoir; Jhc—Harper Creek; Jgu—undivided Jurassic granitic rocks; JTrm—Julian Schist. On maps, Chiquito Peak suite is unpatterned with no unit symbol shown. B: Foliation trends in plutonic and metamorphic rocks (light dashes) and generalized steep (70–90°) dips of foliation and contacts (heavy strike-and-dip symbols) within plutonic complexes of (A).

northwest to northerly trending sheetlike tonalite plutons and wallrock screens with steeply dipping contacts were intruded by a series of gabbroic and granitic "diapirs" whose shapes are typically elongate parallel to their hosts. Within both granitic and gabbroic plutons of the western zone (1) a weak to strong magmatic foliation is generally concordant with pluton margins but is locally discordant to both internal and external pluton contacts; (2) a steeply plunging mineral and/or mafic enclave lineation may be present within the plane of foliation; (3) recrystallization and subsolidus foliation development overprint magmatic foliation, especially near pluton margins; and (4) two, or even three, superimposed mineral foliations may be present.

The youngest batholithic structures in the west-central zone are ductile-to-brittle shear zones that coincide with lineaments visible on high-altitude photographs. These shear zones, which range in length from ~6 to 20 km and are marked by mylonitic rocks and/or fault breccia, offset batholithic contacts from 0.5 to >1 km (Todd, 2004; R.R. Rector, 1993, personal commun.; Rector, 1993). We interpret them as late-batholithic ductile shear zones that continued to be active as zones of brittle faulting after intrusion had ceased. Shown on Figure 3 are four approximately northeast-striking shear zones with right-lateral displacement, two of which are spatially associated with plutons of the Corte Madera suite (see below).

Cretaceous plutons within the central part of the study area were emplaced into the Early Cretaceous (ca. 118 to ca. 114 Ma) Cuyamaca–Laguna Mountains shear zone (Todd et al., 1988; Thomson and Girty, 1994). In Figure 1, the Cuyamaca–Laguna Mountains shear zone coincides approximately with the belt of Jurassic and Cretaceous plutons located between the I-S line on the west and the northwest-trending main western boundary of middle to Late Cretaceous Granite Mountain and La Posta plutons on the east. Thomson and Girty (1994) and their colleagues demonstrated that Jurassic plutons and two adjacent Cretaceous plutons have mylonitic fabrics that indicate alternating intra-arc contraction and extension. We observe that most Cretaceous plutons within the Cuyamaca–Laguna Mountains shear zone exhibit moderate to strong, locally protomylonitic foliation that is parallel to the north-northeast strike of the shear zone. Associated with this foliation is a steeply plunging lineation marked by aligned mineral grains and aggregates and elongate mafic enclaves. We interpret the fabric of the Cretaceous plutons as resulting from syn- and post-intrusive ductile flow within the Cuyamaca–Laguna Mountains shear zone.

In the eastern zone of the Peninsular Ranges batholith, plutons of the Granite Mountain suite are marked by moderate to strong, magmatic ± near-solidus or subsolidus foliation near pluton margins to weak magmatic foliation in the interiors. However, ~6 km north of the study area (Granite Mountain–Blair Valley area), Granite Mountain plutons consist of protomylonite to mylonitic gneiss developed during an episode of mid-Cretaceous, regional-scale synintrusive deformation (Grove, 1987). While foliation in the interior part of the large La Posta pluton in the eastern zone of the study area is weak or moderate to absent,

the marginal zone of the northern part of the pluton underwent significant syn- and/or post-intrusive strain and recrystallization over distances up to >5 km from its walls (Todd, 1977a).

Tectonite fabrics within magmatic arcs may be attributed to localized deformation as rising magma bodies inflate and older plutons and volcanic roof rocks are compressed and displaced downward by return flow (e.g., Gastil (1979) in the Peninsular Ranges batholith; Tobisch et al. (1986) in the Sierra Nevada batholith). In a wide-ranging study of syntectonic diapirs, Paterson and Vernon (1995) explained subsolidus deformation as the result of syn- to post-intrusive regional deformation. Evidence of both local and regional-scale processes is present in the study area: the concordance of foliation with plutonic contacts indicates that mineral grains and mafic enclaves were oriented by flow during ascent and emplacement of magma, but, in numerous examples, magmatic foliation transects contacts between pluton-and-pluton, pluton-and-dike, or pluton-and-wallrock screen (Fig. 4). Dikes that cut across magmatic foliation in the host pluton and are themselves foliated and/or folded concordantly with foliation in the host are ubiquitous in this part of the batholith. Observing that magmatic foliation in "concentrically expanded plutons" (or syntectonic nested diapirs) ignores or transects internal contacts and compositional gradations within plutons, Paterson and Vernon (1995) suggest that magmatic foliation forms at a relatively late stage when magma contains less than ~30% melt. Our general observations imply that magmatic foliation in the Peninsular Ranges batholith developed in magmas with high crystal:melt ratios, in some places passing imperceptibly into subsolidus foliation by solid-state recrystallization.

EMPLACEMENT DEPTHS OF PLUTONS

The Peninsular Ranges batholith displays a general west-to-east increase in the emplacement depths of plutons and the metamorphic grade of country rocks (Silver et al., 1979; Gastil, 1983; Todd et al., 1988; Ague and Brimhall, 1988; Grove, 1987, 1989; Rothstein and Manning, 1994). This eastward depth increase is considered to reflect the effects of Neogene regional uplift and westward tilting of the Peninsular Ranges block that accompanied rifting of the Gulf of California (Gastil et al., 1975; Silver et al., 1979; Kerr and Kidwell, 1991). The application of empirical Al-in-hornblende barometry to granitic plutons in San Diego County supports the eastward increase in emplacement depths indicated by the above and many other studies.

Previous pressure estimates based on Al-in-hornblende barometry for I-type granitic plutons in the study area displayed west-to-east pressure zonation from 0.2 to 0.4 GPa (1 kbar = 0.1 Gpa) in the westernmost zone, to 0.4–0.9 GPa in the west-central zone, 0.2–0.4 GPa in the east-central zone, and 0.4–0.7 GPa in the easternmost zone (Hammarstrom, 1992; Shaw et al., 1994). This pattern of zonation is similar to that reported by Smith et al. (1991) in the northernmost part of the Peninsular Ranges batholith where a ridge of high crystallization pressures in the axial zone of the batholith is flanked to the west and east by lower

Figure 4. Typical relations between foliation and internal plutonic contacts. A: Large, fine-grained synplutonic mafic dike cuts across magmatic foliation in Chiquito Peak granodiorite, western zone; mafic dike is back-veined by host magma that itself contains inclusions of dike, some partly assimilated to mafic streaks; hammer parallel to steep foliation in host and dike. B: Remnants of a mafic dike that originally cut across gneissic, heterogeneously banded East Mesa quartz diorite at low angle, eastern zone (CLMSZ); granitic back-veining material (not obviously derived from host pluton in this outcrop) has "taken over," with fragments of original mafic dike present only as inclusions; pencil parallel to secondary foliation and to healed offsets in dike wall. C: Closeup of highly deformed mafic dike in strongly foliated Chiquito Peak monzogranite, eastern zone (CLMSZ); axial-plane foliation of Chiquito Peak veinlets in dike and mineral foliation in dike and host are approximately parallel to knife.

pressures. Clinkenbeard and Walawender (1989) found an eastward increase in pressure estimates across the La Posta pluton in southeastern San Diego County, from 0.19 GPa on the west side to 0.52 GPa on the east. Emplacement pressure estimates for tonalite of the Granite Mountain suite along the western margin of the La Posta pluton averaged 0.45 GPa.

We revisited previous hornblende data for the study area in light of a new formulation of the barometer by Anderson and Smith (1995) based on the experimental work of Schmidt (1992) and Johnson and Rutherford (1989). The new barometer incorporates a temperature term and provides pressure estimates that more closely approximate pressures for wallrocks based on well-constrained metamorphic mineral barometers. Additional samples were analyzed to fill in gaps in west-to-east coverage within individual Cretaceous granitic suites and to check earlier anomalously high estimates of pressure for some samples. We documented mineral assemblages, analyzed mineral compositions by electron microprobe, and recalculated previous data for a total of 49 samples (Table 1). For 39 samples that have appropriate mineral assemblages and hornblende compositions, we estimated pressures of emplacement based on Al-in-hornblende barometry using both empirically and experimentally based formulations. For about half of the samples, we had sufficient data on hornblende-plagioclase pairs to estimate temperatures using the Holland and Blundy (1994) method. Tables of sample locations and mineral assemblages, pressure and temperature estimates, and microprobe data for hornblendes are included as supplemental data for this paper.[1]

Monzogranite samples from leucogranite suites (the Chiquito Peak, Corte Madera, and Pine Valley suites; see Table 2 for suites) all contain iron-rich hornblendes with Fe/(Fe+Mg) >0.64; these amphibole compositions exceed the recommended range for barometry and yield elevated pressure estimates that are geologically unrealistic, even when temperature is considered. For tonalites and granodiorites of the Alpine, Japatul Valley, Granite Mountain, and La Posta suites, amphibole compositions and mineral assemblages are appropriate for applying the barometer. Pressures estimated using Schmidt's (1992) equation are typically 0.05 GPa (0.5 kbar) higher than pressures estimated from empirical formulations. This observation suggests that, in general, hornblende in these granitoids crystallized at temperatures somewhat above the water-saturated tonalite solidus used in the Schmidt experiments (700–655 °C over a pressure range of 0.25–1.3 GPa). From hornblende-plagioclase thermometry, temperature estimates for the Peninsular Ranges batholith samples range from ~650 to 800 °C. Pressures computed from the Anderson and Smith (1995) equation

(incorporating these temperature estimates) are within ± 0.15 GPa (1.5 kbar) of estimates from empirical barometers.

Figure 5 shows the results of the empirical-barometer pressure estimates for the 39 suitable samples from the study area in terms of pressure groups. At all latitudes, the westernmost samples fall into the low-pressure (<0.3 GPa) group. No low-pressure samples are observed east of the trace of the Elsinore fault in the eastern part of the batholith. Intermediate pressures (0.3–0.5 GPa) dominate the central part of the batholith. High pressures (>0.5 GPa) in the central part of the batholith represent the easternmost samples of the Alpine, Japatul Valley, and Granite Mountain suites. Slightly more mafic lithologies (quartz diorite) or proximity to ductile shear zones (gneissic fabric) may explain the elevated pressure estimates for these samples. Hornblendes in rocks proximal to shear zones tend to be strongly zoned with blue-green, Al-rich rims; these hornblendes may have reequilibrated due to combined effects of a protracted thermal history and post-magmatic deformation and recrystallization.

For the samples where temperature corrections were applied, 80% of the pressure estimates fall in the same pressure group indicated by the empirical barometer. Two samples plot in the next lower pressure group and two plot in the next higher pressure group as a result of the correction. The disparity in pressure estimates for the eastern (P > 0.5 GPa) and western (P < 0.3 GPa) margins of the La Posta pluton noted by Clinkenbeard and Walawender (1989) is apparent in our data set, even with temperature corrections. Many of our samples are from the same localities. We recalculated their hornblende and plagioclase data with the Anderson and Smith (1995) equation and determined that the pressure difference is not simply a function of temperature or iron-rich compositions. Differences in K/Ar biotite and hornblende cooling ages east and west of a screen of metamorphic rocks in the central part of the La Posta pluton suggest the possibility of a structural discontinuity (Krummenacher et al., 1975). Walawender et al. (1990) showed that U-Pb zircon ages from the western (95 ± 1 Ma) and eastern (93 ± 1 Ma) hornblende-bearing facies of the La Posta pluton do not rule out the possibility of such a fault. They concluded that movement along such a fault, if present, was dominantly vertical and exposed different levels of a single pluton.

CRETACEOUS PLUTONIC SUITES

Introduction

In our original subdivision of the batholith (Todd and Shaw, 1979), we used the term "plutonic unit" to define and map groups of granitic plutons that display similar mineralogical, textural, and petrographic variations. Subsequent chemical, mineralogical, and isotopic studies confirm our belief that these plutonic units represent a relatively small number of granitic magma types, each of which comprises many plutons. The term plutonic unit is equivalent to "suite" as used by Chappell (1978), Shaw and Flood (1981), and White et al. (2001) in the Lachlan Fold Belt

[1]GSA Data Repository item 2003174, chemical analyses of 183 Cretaceous plutonic rocks and 10 western-zone prebatholithic rocks, and complete data tables for hornblende geobarometry, Peninsular Ranges batholith, San Diego County, is available on request from Documents Secretary, GSA, P.O. Box 9140, Boulder, CO 80301-9140, USA, editing@geosociety.org, at www.geosociety.org/pubs/ft2003.htm, or on the CD-ROM accompanying this volume

TABLE 1. HORNBLENDE GEOBAROMETRY

Plutonic suite/pluton	Lithology	Sample No.	Latitude N	Longitude W	Fe/(Fe+Mg)	$Fe^3/(Fe^2+Fe^3)$	Al^{Total}_{all}	Al^{Total}_{rims}	$P_{all-Schmidt}$	$P_{rims-Schmidt}$	P_{all-HZ}	$P_{rims-HZ}$	T_{ed-ri} P=0	T_{ed-ri} P=5	P_{AS}-T0	P_{AS}-T5	Plotted	Comment
Chiquito Peak Monzogranite																		
	granodiorite	86CA013	32° 54' 31"	116° 34' 32"	0.77	0.16	2.38	2.54	N.A.	N.A.	N.A.	N.A.	N.D.	N.D.	N.A.	N.A.	no	Too Fe-rich
Cuyamaca Reservoir Granodiorite																		
	granodiorite	SY91-1	33° 04' 26"	116° 43' 06"	0.54	0.25	1.48	1.47	4.0	4.0	3.5	3.5	N.D.	N.D.	N.A.	N.A.	yes	
	tonalite	86CA046	33° 22' 29"	116° 37' 16"	0.43	0.25	1.20	1.24	N.A.	N.A.	N.A.	N.A.	N.D.	N.A.	N.A.	N.A.	N.A.	N of map area
	granodiorite	86CA007	32° 49' 22"	116° 33' 19"	0.36	0.17	0.65	0.78	N.A.	N.A.	N.A.	N.A.	N.D.	N.A.	N.A.	N.A.	N.A.	Altered
Tonalite of Alpine																		
	tonalite	A-98	32° 52' 14"	116° 51' 05"	0.42	0.51	1.32	1.35	3.3	3.4	2.7	2.8	802	815	1.3	1.0	yes	
	tonalite	A-99	32° 52' 33"	116° 49' 11"	0.43	0.35	1.48	1.48	4.0	4.0	3.5	3.5	770	778	2.6	2.4	yes	
	tonalite	A-1	32° 50' 19"	116° 46' 49"	0.52	0.27	1.50	N.D.	4.1	N.A.	3.6	N.A.	N.D.	N.A.	N.A.	N.A.	yes	
	tonalite	86CA036	32° 50' 23"	116° 46' 01"	0.48	0.27	1.50	1.51	4.1	4.2	3.6	3.7	N.D.	N.A.	N.A.	N.A.	yes	
	tonalite	4-125-2	32° 57' 03"	116° 45' 33"	0.46	0.50	1.46	1.47	3.9	4.0	3.4	3.4	N.D.	N.A.	N.A.	N.A.	yes	
	tonalite	VM-75	32° 50' 06"	116° 44' 59"	0.47	0.45	1.37	N.D.	3.5	N.A.	3.0	N.A.	N.D.	N.A.	N.A.	N.A.	yes	
	tonalite	TS-37	32° 54' 16"	116° 38' 42"	0.54	0.23	1.59	N.D.	4.6	N.A.	4.1	N.A.	N.D.	N.A.	N.A.	N.A.	yes	
	quartz diorite	TS-65	32° 57' 03"	116° 38' 17"	0.51	0.38	2.04	2.51	6.7	8.9	6.3	8.7	N.D.	N.A.	N.A.	N.A.	yes	
Corte Madera Monzogranite																		
	granodiorite	86CA005	32° 45' 14"	116° 46' 26"	0.64	0.14	1.52	1.50	N.A.	N.A.	N.A.	N.A.	N.D.	N.A.	N.A.	N.A.	no	Too Fe-rich
	granodiorite	D-94	32° 45' 44"	116° 36' 47"	0.67	0.21	1.77	N.D.	N.A.	N.A.	N.A.	N.A.	N.D.	N.A.	N.A.	N.A.	no	Too Fe-rich
	granodiorite	D-94*	32° 45' 44"	116° 36' 47"	0.69	0.44	2.05	N.D.	N.A.	N.A.	N.A.	N.A.	N.D.	N.A.	N.A.	N.A.	no	Too Fe-rich
	monzogranite	6-21-6	32° 50' 29"	116° 36' 33"	0.84	0.26	1.65	1.81	N.A.	N.A.	N.A.	N.A.	N.D.	N.A.	N.A.	N.A.	no	Too Fe-rich
Tonalite of Granite Mountain																		
	granodiorite	86CA001	32° 48' 12"	116° 49' 10"	0.46	0.37	1.17	1.14	2.5	2.4	1.9	1.8	N.D.	N.A.	N.A.	N.A.	yes	
	tonalite	A-9	32° 45' 47"	116° 47' 16"	0.41	0.31	1.21	N.D.	2.8	N.A.	2.2	N.A.	N.D.	N.A.	N.A.	N.A.	yes	
	tonalite	86CA006	32° 45' 23"	116° 46' 13"	0.41	0.60	1.26	1.26	3.0	3.0	2.4	2.4	755	764	2.0	1.8	yes	
	tonalite gneiss	SY92-1	33° 06' 26"	116° 39' 24"	0.52	0.39	1.94	1.96	6.2	6.3	5.9	5.9	N.D.	N.A.	N.A.	N.A.	yes	High P due to shear zone?

(continued)

TABLE 1. HORNBLENDE GEOBAROMETRY (continued)

Plutonic suite/pluton	Lithology	Sample No.	Latitude N	Longitude W	Fe/(Fe+Mg)	$Fe^3/(Fe^2+Fe^3)$	Al^{Total}_{all}	Al^{Total}_{rms}	$P_{all-Schmidt}$	$P_{rims-Schmidt}$	P_{all-HZ}	$P_{rims-HZ}$	$T_{ed-ri}\ P=0$	$T_{ed-ri}\ P=5$	P_{AS}-T0	P_{AS}-T5	Plotted	Comment
Tonalite of Granite Mountain, continued																		
	tonalite	J-32A	33° 07' 35"	116° 36' 01"	0.49	0.39	1.50	1.42	4.1	3.7	3.6	3.2	699	724	3.5	3.2	yes	
	diorite	J-32B	33° 07' 35"	116° 36' 01"	0.60	0.51	1.97	1.97	N.A.	N.A.	N.A.	N.A.	N.D.	N.A.	N.A.	N.A.	no	Too mafic
	tonalite	MR-12	32° 42' 29"	116° 30' 47"	0.49	0.48	1.68	1.73	5.0	5.2	4.5	4.8	642	686	5.5	5.1	yes	
	tonalite	86CA023	32° 36' 44"	116° 30' 03"	0.50	0.49	1.75	1.78	5.3	5.5	4.9	5.0	N.D.	N.D.	N.A.	N.A.	yes	
	tonalite	ML-42	32° 43' 47"	116° 28' 41"	0.47	0.34	1.59	1.65	4.5	4.8	4.1	4.4	709	728	4.4	4.1	yes	
	tonalite	CPO-4	32° 36' 44"	116° 28' 30"	0.50	0.36	1.56	1.66	4.4	4.9	3.9	4.4	719	747	4.3	3.8	yes	
	granodiorite	SS-83-2	32° 36' 28"	116° 28' 25"	0.57	0.30	1.64	1.64	4.8	N.A.	4.3	4.3	735	746	4.0	3.8	yes	
	tonalite	EV-3	33° 02' 20"	116° 24' 32"	0.47	0.20	1.45	N.D.	3.9	N.A.	3.4	N.A.	N.D.	N.D.	N.A.	N.A.	yes	
	tonalite	ML-17	32° 46' 44"	116° 24' 03"	0.45	0.43	1.55	1.49	4.4	4.1	3.9	3.6	684	720	4.0	3.5	yes	
	tonalite	86CA026	32° 57' 13"	116° 18' 10"	0.52	0.19	1.95	2.23	6.3	7.6	5.9	7.3	N.D.	N.D.	N.A.	N.A.	yes	
Hot Springs Mountain pluton																		
	granodiorite	86CA045	33° 16' 24"	116° 37' 19"	0.51	0.28	1.67	1.71	5.0	5.1	4.5	4.7	719	740	4.5	4.2	yes	
	tonalite	86CA043	33° 16' 11"	116° 36' 58"	0.50	0.19	1.63	1.64	4.8	4.8	4.3	4.3	705	726	4.4	4.1	yes	
East Mesa Quartz Diorite																		
	tonalite	86CA015	32° 54' 44"	116° 33' 06"	0.52	0.21	1.37	1.41	3.5	3.7	3.0	3.2	638	663	4.0	3.8	yes	
	quartz diorite	CP-117A	32° 54' 30"	116° 30' 34"	0.48	0.31	1.11	1.49	2.3	4.1	1.7	3.6	N.D.	N.D.	N.A.	N.A.	yes	
Japatul Valley Tonalite																		
	tonalite	86CA003	32° 48' 22"	116° 48' 23"	0.46	0.32	1.26	1.49	3.0	4.1	2.4	3.5	N.D.	N.D.	N.A.	N.A.	yes	
	tonalite	86CA004	32° 46' 14"	116° 41' 52"	0.50	0.22	1.45	1.57	3.9	4.5	3.3	4.0	N.D.	N.D.	N.A.	N.A.	yes	
	tonalite	VM-39	32° 46' 19"	116° 40' 16"	0.45	0.49	1.50	1.50	4.1	N.A.	3.6	3.6	774	793	2.6	2.2	yes	
	tonalite/ granodiorite	5-1-1	32° 45' 25"	116° 39' 40"	0.60	0.25	1.80	N.D.	5.5	N.A.	5.1	N.A.	N.D.	N.D.	N.A.	N.A.		SS data
	tonalite/ granodiorite	5-1-1*	32° 45' 25"	116° 39' 40"	0.58	0.45	1.76	1.78	5.4	5.5	4.9	5.0	665	697	5.6	5.2	yes	JH data— these agree within expected error

(continued)

TABLE 1. HORNBLENDE GEOBAROMETRY (continued)

Plutonic suite/pluton	Lithology	Sample No.	Latitude N	Longitude W	Fe/(Fe+Mg)	$Fe^3/(Fe^2+Fe^3)$	Al^{Total}_{all}	Al^{Total}_{rims}	$P_{all-Schmidt}$	$P_{rims-Schmidt}$	P_{all-HZ}	$P_{rims-HZ}$	T_{ed-ri} P=0	T_{ed-ri} P=5	P_{AS}-T0	P_{AS}-T5	Plotted	Comment
Japatul Valley Tonalite, continued																		
	tonalite	TS-17	32° 57' 20"	116° 39' 32"	0.50	0.31	1.54	1.54	4.3	N.A.	3.8	3.8	672	695	4.3	4.1	yes	
	tonalite	WD-100	32° 45' 45"	116° 36' 46"	0.59	0.30	1.84	N.D.	5.7	N.A.	5.3	N.A.	N.D.	N.D.	N.A.	N.A.		SS data
	tonalite	WD-100*	32° 45' 45"	116° 36' 46"	0.57	0.49	1.88	1.96	5.9	6.3	5.5	5.9	N.D.	N.D.	N.A.	N.A.	yes	JH data—these agree within expected error
Las Bancas Tonalite																		
	tonalite	POW-2-2	32° 58' 01"	117° 00' 45"	0.46	0.50	1.26	1.27	3.0	3.0	2.4	2.4	756	765	2.0	1.8	yes	
	tonalite	86CA039	33° 04' 07"	116° 43' 13"	0.37	0.24	1.10	1.17	2.2	2.6	1.6	2.0	653	672	2.7	2.6	yes	
Tonalite of La Posta																		
	granodiorite	CPO-3	32° 37' 35"	116° 28' 25"	0.52	0.93	1.39	1.41	3.6	3.7	3.1	3.2	N.D.	N.D.	N.A.	N.A.	yes	
	granodiorite/ tonalite	86CA009	32° 43' 23"	116° 26' 35"	0.51	0.19	1.14	1.15	2.4	2.5	1.8	1.9	N.D.	N.D.	N.A.	N.A.	yes	
	granodiorite	CC-10-1	32° 43' 16"	116° 23' 50"	0.50	0.40	1.18	1.09	2.6	2.2	2.0	1.6	N.D.	N.D.	N.A.	N.A.	yes	
	tonalite	86CA031	32° 41' 53"	116° 03' 41"	0.56	0.19	1.97	2.03	6.4	6.7	6.0	6.3	672	701	6.7	6.3	yes	
	tonalite	W-6-6	32° 42' 22"	116° 03' 09"	0.55	0.28	1.97	1.97	6.4	6.4	6.0	6.0	694	710	6.1	5.9	yes	
Monzogranite of Pine Valley																		
	granodiorite	CP-132	32° 57' 28"	116° 30' 26"	0.73	0.19	2.06	2.13	N.A.	N.A.	N.A.	N.A.	N.D.	N.D.	N.A.	N.A.	no	Too Fe-rich
	granodiorite	WD-80A	32° 50' 10"	116° 32' 09"	0.72	0.26	1.98	1.88	N.A.	N.A.	N.A.	N.A.	N.D.	N.D.	N.A.	N.A.	no	Too Fe-rich

Note: N.D.—no data; N.A.—not applicable
P_{all-S}—Pressures calculated using all hornblende analyses with Schmidt (1992) equation.
P_{rims-S}—Pressures calculated using only rim hornblende analyses with Schmidt (1992) equation.
P_{all-HZ}—Pressures calculated using all hornblende analyses with Hammarstrom and Zen (1986) equation.
$P_{rims-HZ}$—Pressures calculated using only rim hornblende analyses with Hammarstrom and Zen (1986) equation.
T_{ed-ri} P=0—T at 0 kbar from Holland and Blundy (1994) hornblende-plagioclase thermometer.
T_{ed-ri} P=5—T at 5 kbar from Holland and Blundy (1994) hornblende-plagioclase thermometer.
P_{AS}-T0—Pressures calculated using Anderson and Smith (1995) with above T.
P_{AS}-T5—Pressures calculated using Anderson and Smith (1995) with above T.
JH, SS data—Jane Hammarstrom, Stirling Shaw, analysts.

*Duplicate sample numbers.

TABLE 2. CRETACEOUS PLUTONIC SUITES

Plutonic suite	Early to mid-Cretaceous granitic suites			
	Las Bancas	East Mesa	Alpine	Japatul Valley
Geographic location*; (no. of plutons; see Fig. 6)	Western and eastern zones (5)	Eastern zone; restricted to CLMSZ (5)	Western zone (6)	West-central zone (as very small bodies, present up to 10 km east of gravity-magnetic gradient) (8)
Lithology (Streckeisen, 1973)	Hypersthene-biotite tonalite, quartz diorite, granodiorite, diorite, quartz norite	Pyroxene-biotite-hornblende quartz diorite, tonalite, diorite, granodiorite; minor monzodiorite and gabbro	Biotite-hornblende tonalite and quartz diorite; rare borderline tonalite-granodiorite	Biotite-hornblende tonalite, hornblende-biotite tonalite, borderline tonalite-granodiorite, and granodiorite
Essential minerals	Plagioclase (An$_{60}$-An$_{40}$), quartz, biotite, hypersthene ± hornblende ± augite ± K-feldspar	Plagioclase (An$_{60}$-An$_{40}$), hornblende + subaluminous amphibole, quartz, biotite ± opx ± cpx ± K-feldspar; reddish-brown (reduced) biotite	Plagioclase (An$_{60}$-An$_{30}$), quartz, hornblende, biotite ± opx ± cpx ± K-feldspar	Plagioclase (An$_{50}$-An$_{30}$), quartz, hornblende, biotite ± opx ± cpx ± K-feldspar
Accessory minerals	Apatite, zircon, magnetite ± ilmenite ± pyrite, sphene, allanite, local subhedral epidote; secondary epidote	Apatite, zircon, ilmenite, pyrite, sphene, allanite; secondary clinozoisite	Apatite, zircon, magnetite ± ilmenite, pyrite, molybdenite, sphene, allanite, local subhedral epidote; secondary clinozoisite	Apatite, zircon, magnetite ± ilmenite, sphene, allanite, local subhedral epidote; secondary clinozoisite
Foliation	Weak to moderate magmatic, minor subsolidus recrystallization; local protomylonite	Strong magmatic, widespread subsolidus recrystallization; locally gneissic, protomylonitic	Moderate to strong magmatic, minor but widespread subsolidus recrystallization	Moderate to strong magmatic, local substantial subsolidus recrystallization
Distinguishing field characteristics	Fresh rock dark-gray to black (C.I. 22-32), homogeneous, medium-grained, hypidiomorphic-granular with biotite oikocrysts <2.5 cm; xenoliths, mafic enclaves absent; minor cumulate rock containing 5- to 10-cm, ovoid basaltic inclusions	Medium- to dark-gray or black, fine- to medium-grained, hypidiomorphic to porphyritic; C.I. 23-52; narrow sheeted plutons and dikes; complex internal contacts suggest repeated intrusion of magma into CLMSZ; mafic enclaves ± metasedimentary xenoliths	Medium- to dark-gray, coarse-grained, hypidiomorphic-granular, heterogeneous (mafic mineral aggregates and biotite oikocrysts <3 cm); C.I. 20-37; large sheetlike plutons, abundant mafic enclaves including enclaves of coeval gabbro; xenocrystic quartz	Light- to medium-gray, medium- to coarse-grained, hypidiomorphic-granular, subhedral hornblende, plagioclase; C.I. 12-32; abundant mafic enclaves, scarce granitic enclaves; spatially associated with metavolcanic screens to form zones of contact breccia and migmatite up to 4 km wide
Alkali-lime index	63	N.D.	66	65
A.S.I.	0.82-1.00	0.67-1.00	0.84-0.95	0.84-1.00
Initial $^{87}Sr/^{86}Sr$ (no. of samples)	0.70373-0.70461 (5)	0.70446 (1)	0.70336-0.70369 (3)	0.70351-0.70391 (3)
$\delta^{18}O$#	7.0-8.2	9.0	6.5-7.1	6.4-6.7
U-Pb zircon age (see Fig. 6)	109 ± 2 Ma (Anderson, 1991); ca.104 Ma, ca.119 Ma (L.T. Silver, oral commun., 1979)	170 ± 2 Ma; 118 ± 1 Ma, 117 ± 1 Ma (Shaw et al., this volume)	108 ± 2 Ma (Anderson, 1991); $^{40}Ar/^{39}Ar$ minimum age of emplacement ca. 105 Ma (Snee et al., 1994) (Fig. 6)	No absolute age determinations; $^{40}Ar/^{39}Ar$ minimum emplacement ages are ca. 112 Ma and ca.104 Ma (Snee et al., 1994) (Fig. 6)

(continued)

TABLE 2. CRETACEOUS PLUTONIC SUITES (continued)

Early to mid-Cretaceous granitic suites (continued)

Plutonic suite	Chiquito Peak	Corte Madera	Mother Grundy Peak	Pine Valley
		Western zone (8)	Western zone (2)	Eastern zone (4)
Geographic location (no. of plutons*; see Fig. 6)	West-central zone (present up to10 km east of gravity-magnetic gradient) (9)			
Lithology (Streckeisen, 1973)	Hornblende-biotite monzogranite, granodiorite, leucomonzo-granite, minor tonalite; abundant leucocratic dikes	Biotite leucomonzogranite, leucogranodiorite, syenogranite, abundant leucocratic dikes	Hornblende-biotite leucomonzogranite, leucogranodiorite, borderline granodiorite-tonalite	Biotite leucomonzogranite, leucogranodiorite, minor tonalite, abundant leucocratic dikes
Essential minerals	Plagioclase (An$_{40}$-An$_{20}$), quartz, K-feldspar, biotite ± hornblende	Quartz, plagioclase (An$_{30}$-An$_{10}$), K-feldspar, biotite ± small relict hornblende grains	Plagioclase, quartz, K-feldspar, biotite, slender subhedral hornblende phenocrysts <1 cm	Quartz, plagioclase (An$_{40}$-An$_{20}$), K-feldspar, biotite ± hornblende (relict grains within biotite)
Accessory minerals	Apatite, zircon, magnetite ± ilmenite, sphene, allanite, epidote (locally subhedral)	Apatite, zircon, magnetite ± ilmenite, molybdenite, sphene, allanite	Apatite, zircon, ilmenite, allanite, secondary sphene, epidote	Apatite, zircon, sphene, allanite, garnet, muscovite, schorl
Foliation	Strong magmatic, widespread subsolidus recrystallization; local porphyroclastic gneiss	Weak to strong magmatic; locally, marginal zones and dikes are protomylonitic	Strong magmatic; moderate to strong subsolidus recrystallization	Strong magmatic; variable subsolidus recrystallization, locally gneissic or protomylonitic
Distinguishing field characteristics	Light-colored (grayish hue), medium-grained, hypidiomorphic-granular; aligned decussate mafic aggregates 2 to 5 mm long (C.I. 2-16); mafic enclaves in granodiorite, tonalite	Light-colored (peach hue), medium-to coarse-grained, hypidiomorphic-granular; widely spaced (0.5-2 cm) aligned decussate mafic aggregates, quartz and K-feldspar lenticles <3 cm long; C.I. 1-11; rare mafic enclaves; local gradation to silicic metavolcanic screens (westernmost plutons)	Light-colored, medium- to very coarse-grained, hypidiomorphic-granular, homogeneous; quartz lenticles <4 cm; next to gabbro plutons, margins contain synplutonic basaltic dikes and enclaves	Light-colored, medium- to coarse-grained (quartz, K-feldspar phenocrysts <3 cm long); hypidiomorphic-granular; foliation marked by decussate 2-cm-long mafic aggregates and 4-cm-long quartz lenticles; C.I. 4-10; local segregation of quartz-feldspar and mafic minerals produces gneissic texture; tonalite carries subhedral hornblende, mafic enclaves
Alkali-lime index	65	66	N.D.	65
A.S.I.	0.94-1.18	1.00-1.20	N.D.	1.00-1.17
Initial ^{87}Sr/^{86}Sr (no. of samples)	0.70388, 0.70395 (2)	0.70375, 0.70407 (2)	N.D.	0.70358, 0.70447 (2)
δ^{18}O$^{\#}$	8.1, 8.7	7.2, 7.3	N.D.	8.6, 9.0
U-Pb zircon age (see Fig. 6)	ca.117 Ma, ca. 113 Ma (L.T. Silver, oral commun., 1979)	111 ± 2 Ma (D.L. Kimbrough, oral commun., 1994)	No absolute age determinations; Early Cretaceous?	118 ± 2 Ma, 122 ± 1 Ma (D.L. Kimbrough, oral commun., 1994)

(continued)

TABLE 2. CRETACEOUS PLUTONIC SUITES (continued)

Plutonic suite	Mid- to Late-Cretaceous granitic suites		Mafic plutonic rocks	
	Granite Mountain	La Posta	Cuyamaca Gabbro	Mafic and intermediate dikes
Geographic location (no. of plutons*; see Fig. 6)	Western and eastern zones (plutons larger and more numerous in eastern zone) (7)	Western and eastern zones; two small plutons in west-central zone and vast La Posta pluton in eastern zone (4)	Large plutons, dike swarms in west-central zone; scarce small bodies, dikes in eastern zone (21)	Dikes ubiquitous throughout study area, but largest and most numerous in CLMSZ; dikes intrude all granitic suites, gabbroic plutons, and pre-existing mafic dikes
Lithology (Streckeisen, 1973)	Hornblende-biotite tonalite, granodiorite, quartz diorite; eastern-zone plutons zoned inward from tonalite to biotite granodiorite with voluminous leucocratic phases(Todd et al., 1987; Lampe, 1988)	Hornblende-biotite trondhjemite and granodiorite in west-central zone; biotite trondhjemite, granodiorite, garnetiferous two-mica leucogranodiorite and leucomonzogranite plus abundant leucocratic dikes in eastern zone	Interior of large plutons: hornblende-bearing troctolite; anorthositic gabbro ± amphibole ± opx ± olivine; amphibole-olivine gabbronorite; minor hornblende diorite, leucodiorite ± pyroxene ± biotite. Smaller plutons, marginal zones of large plutons: hornblende gabbro ± opx ± cpx ± biotite	Quartz diorite, tonalite, gabbro, diorite, rare granodiorite
Essential minerals	Plagioclase (An_{55}–An_{35}), quartz, biotite, hornblende ± opx ± cpx ± K-feldspar	Plagioclase (An_{40}–An_{20}), quartz, biotite ± hornblende ± K-feldspar	Olivine- and pyroxene-bearing gabbro: plagioclase An_{90}–An_{80}; hornblende gabbro: plagioclase An_{70}–An_{60}	Hornblende, plagioclase (An_{80}–An_{40}), biotite ± cpx ± sphene ± quartz
Accessory minerals	Apatite, zircon, ilmenite, pyrite, allanite, sphene, garnet; secondary clinozoisite	Apatite, zircon, ilmenite, allanite, euhedral yellow-to-brown sphene; rare subhedral epidote; minor retrograde alteration of biotite to chlorite, epidote/clinozoisite, sericite	Magnetite, ilmenite, pyrite, zircon; green spinel in most mafic compositions; secondary epidote/clinozoisite, sphene, chlorite, rutile	Zircon, ilmenite, pyrite, sphene, allanite; secondary epidote/clinozoisite rims on allanite
Foliation	Weak to strong magmatic; variable subsolidus recrystallization; local protomylonite to mylonitic gneiss in eastern zone	Weak to moderate magmatic to ~massive; northern part of La Posta pluton has marginal zone of moderate subsolidus recrystallization up to 5 km wide; La Posta dikes deformed with wallrocks	Moderate to strong magmatic; variable, locally substantial, subsolidus recrystallization; foliation of aligned, partly recrystallized plagioclase and mafic minerals may be subparallel to or discordant to primary cumulate and compositional layering	Igneous quenched textures overprinted variably by subsolidus recrystallization; in CLMSZ, many dikes isoclinally folded, disrupted
Distinguishing field characteristics	Light- to medium-gray, medium- to coarse-grained, locally subporphyritic (2-cm-long hornblende, plagioclase phenocrysts); C.I. 17-27; scarce 2-cm K-feldspar oikocrysts; sparse to common mafic enclaves in tonalite; scarce rhythmically layered (felsic, mafic layers) cumulate rafts	Light-colored (white), medium- to coarse-grained, idiomorphic, homogeneous; bipyramidal 1-cm quartz phenocrysts, euhedral biotite "barrels" <1.5 cm long, unevenly distributed 2- to 5-cm K-feldspar oikocrysts, hornblende in scattered acicular phenocrysts; C.I. 6-15; rare mafic enclaves, layered cumulate rafts	Large plutons: variable compositions, hypidiomorphic or allotriomorphic, gradational internal contacts, cumulate texture; zonation inward from fine-grained hornblende gabbro to more mafic, coarser grained interior compositions; all phases carry hornblende in poikilitic grains (<5 cm) and hornblende aggregates that produce banding	Fine- to medium-grained, locally porphyritic (plagioclase phenocrysts); most dikes are synplutonic, i.e., back-veined by granitic host pluton; some dikes grade to East Mesa or gabbro plutons
Alkali-lime index	65	65	N.D.	N.D.
A.S.I.	0.85–1.07	0.94–1.20	N.D.	N.D.
Initial $^{87}Sr/^{86}Sr$ (no. of samples)	0.70370–0.70402 (5)	0.70383–0.70563 (6)	0.70362–0.70408 (5)	N.D.
$\delta^{18}O$#	6.6-7.8	7.2-11.0	6.1-6.9	N.D.
U-Pb zircon age (see Fig. 6)	101 ± 2 Ma (D.L. Kimbrough, oral commun., 1994); 98 ± 3 Ma (L.T. Silver, oral commun., 1979)	104 ± 2 Ma, ca.102 Ma (D.L. Kimbrough, oral commun., 1994); 94 ± 2 Ma (La Posta pluton, Walawender et al., 1990)	107 ± 2 Ma (D.L. Kimbrough, oral commun., 1994); ca. 99 Ma (M. Taylor, oral commun., 1990)	No absolute age determinations

Note: Essential minerals in order of modal abundance; accessory minerals are magmatic unless noted otherwise; N.D.—not determined. CLMSZ—Cuyamaca–Laguna Mountains shear zone.
*Most plutonic suites include a number of bodies too small to depict in Figure 6.
#Rb-Sr-O isotopes determined for same group of samples.

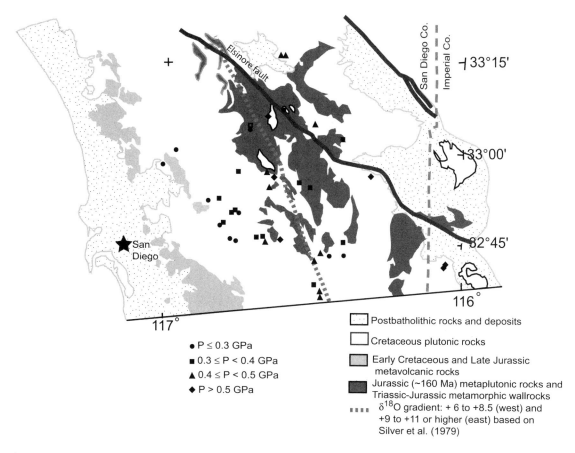

Figure 5. Regional trends in pressures of emplacement of Cretaceous (solid symbols) and Jurassic (open symbols) plutons in Peninsular Ranges batholith, San Diego County. Pressures are based on Al-in-hornblende geobarometry. Samples represent plutonic suites (number of samples): Las Bancas suite (2), East Mesa suite (2), Alpine suite (8), Japatul Valley suite (6), Granite Mountain suite (13), La Posta suite (5), Hot Springs Mountain pluton (2), and Jurassic Cuyamaca Reservoir suite (1).

and New England orogen, Australia. As discussed by White et al. (2001), a similar usage has been adopted by workers in other Cordilleran batholiths, for example, the "unit" of Pitcher in the Coastal batholith of Peru and the "intrusive suite" of Bateman in the Sierra Nevada batholith. Larsen (1948) formally named units, or plutonic rock types, that form multiple plutons over a large area of the northern Peninsular Ranges batholith. In this paper, we adopt the terminology of White et al. (2001) by which a suite is designated as a group of granitic rocks with common textural, mineralogical, and compositional characteristics. The authors note that plutons within a suite need not be of the same age, although their ages are likely to be similar.

Table 2 describes major characteristics of the Cretaceous plutonic suites as determined in our study, and Figure 6 shows their geographic distribution together with published and unpublished pluton ages. In Table 2, suites are listed in an approximate older-to-younger intrusive age order, which corresponds generally to the west-to-east zonation of this part of the batholith, and they are also listed in an order that emphasizes possible petroge-

netic relationships. Detailed descriptions of the original plutonic units and newly defined suites can be found in Todd (2004). Absolute ages given in Table 2 and Figure 6 were determined at the Baylor Brooks Institute of Isotope Geology, San Diego State University, by G.H. Girty, M.S. Girty, and D.L. Kimbrough using U-Pb zircon and monazite methods; and by Marty Grove at the University of California, Los Angeles, by ion microprobe analysis. Preliminary U-Pb zircon ages for other plutons in the study area were reported orally to Todd by L.T. Silver (1979) and are depicted in Fig. 3 *in* Silver et al. (1979).

Distribution of Plutonic Suites

Plutons of the Las Bancas suite occur from the westernmost part of the study area to the eastern desert ranges but are largest (e.g., >100 km² for each of two plutons north of the study area; Rogers, 1965; Todd, reconnaissance mapping) and most numerous in the west-central Peninsular Ranges batholith (Fig. 6B). East Mesa plutons are present in a 10–12-km zone in the central part of the study

A. GEOGRAPHIC LOCATIONS MENTIONED IN TEXT

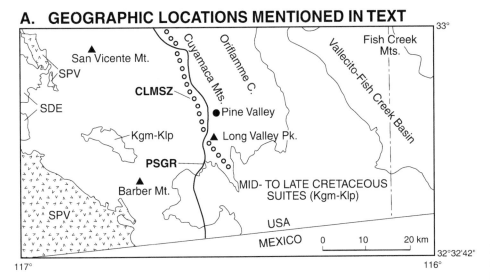

B. LAS BANCAS AND EAST MESA SUITES

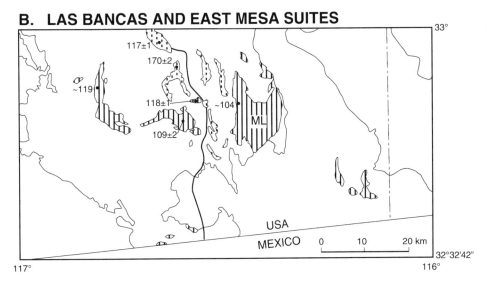

C. ALPINE AND JAPATUL VALLEY SUITES

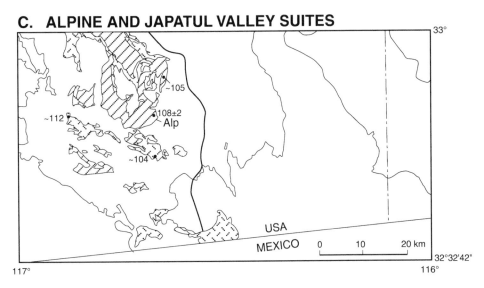

Figure 6. Distribution of plutonic suites and dated plutons in study area. Pluton ages are indicated by solid circles and listed in Table 2. A: Geographic locations mentioned in text; SPV = Santiago Peak Volcanics; SDE = San Diego embayment; heavy solid line is pseudogravity gradient (PSGR); line of circles marks approximate western boundary of Cuyamaca-Laguna Mountains shear zone (CLMSZ); middle to Late Cretaceous suites: Granite Mountain (Kgm), La Posta (Klp). B: Las Bancas (vertical lines) and East Mesa (vertical dashes) suites; ML—Mount Laguna pluton. C: Alpine (inclined lines) and Japatul Valley (dashes) suites; Alp—pluton underlying town of Alpine. D: Leucogranite suites: wide horizontal lines— Chiquito Peak; stipple—Corte Madera; dark gray—Mother Grundy Peak; close horizontal lines—Pine Valley; OC—Oriflamme Canyon pluton, RV—Rattlesnake Valley pluton. E: Granite Mountain (dark stipple) and La Posta (light stipple) suites; LR—Loveland Reservoir pluton, MR—Morena Reservoir pluton. F: Cuyamaca Gabbro (black); PM—Poser Mountain pluton.

D. LEUCOGRANITE SUITES

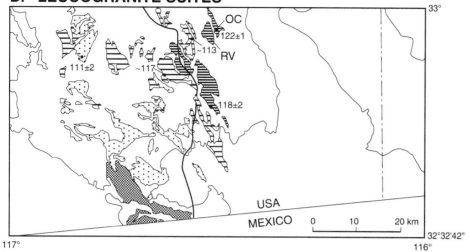

E. GRANITE MOUNTAIN AND LA POSTA SUITES

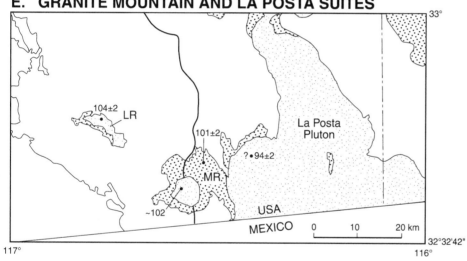

Figure 6. (*continued*)

F. CUYAMACA GABBRO

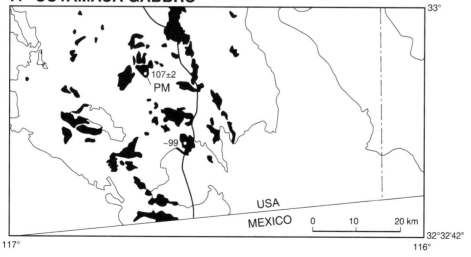

area (Fig. 6B), a zone that coincides with the Cuyamaca–Laguna Mountains shear zone. The Alpine suite is restricted to the western zone of the batholith, where it forms large, sheetlike plutons that are bordered by plutons of the Las Bancas and/or the Japatul Valley suites (Fig. 6C). Mapped relationships suggest that tonalite plutons of the Las Bancas, Alpine, and Japatul Valley suites in the western zone may have been emplaced as subvertical sheets prior to intrusion of more silicic granitic plutons and gabbroic plutons. Plutons of the Japatul Valley suite are largely restricted to the western zone of the Peninsular Ranges batholith, with the exception of small bodies (not shown in Fig. 6C) that occur as much as 10 km east of the Alpine suite. These bodies grade to, and are intruded by, plutons of the Chiquito Peak suite.

Of four leucogranite suites in the study area (the Chiquito Peak, Corte Madera, Pine Valley, and Mother Grundy Peak suites), the geographic distribution of the Chiquito Peak suite is similar to that of the Japatul Valley suite (Fig. 6D), whereas the Corte Madera suite is restricted to the western zone, where it appears to form the youngest intrusions within several zoned plutonic complexes (Fig. 3A, and see discussion below). The Mother Grundy Peak suite forms at least two plutons in the southwestern part of the study area, to the south of the Corte Madera suite, whereas the Pine Valley suite is present only in the eastern zone of the Peninsular Ranges batholith (Fig. 6D). All four leucogranite suites are associated with voluminous dikes of leucogranite, pegmatite, and aplite.

Plutons of the Granite Mountain and La Posta suites are regionally extensive and varied in composition within the eastern Peninsular Ranges batholith in southern California and Baja California (Walawender et al., 1990; Gastil et al., 1991). We find that the Granite Mountain suite comprises plutons in both western and eastern zones of the study area but is most voluminous in the eastern zone (Fig. 6E; see also the type Granite Mountain pluton northeast of the study area, Grove, 1987). In the western zone, Granite Mountain plutons form narrow, discontinuous envelopes around plutons of the La Posta suite. In the central-eastern zone, large Granite Mountain plutons are zoned from biotite-hornblende and hornblende-biotite tonalite to biotite granodiorite and are associated with abundant leucocratic dikes (Fish Creek Mountains pluton, Todd et al. (1987); Granite Mountain pluton, Grove (1987), Lampe (1988); Morena Reservoir pluton, Rector (1994)). The La Posta suite consists of a single vast pluton and at least two smaller plutons in the eastern zone (Kimzey, 1982; Clinkenbeard, 1987; Walawender et al., 1990; Todd, 2004). Like the Granite Mountain suite, with which it is spatially associated, the La Posta also forms small plutons in the west-central zone of the batholith (Rector, 1993; Todd, 2004) (Fig. 6E).

Gabbroic rocks (assigned to the Cuyamaca Gabbro of Everhart, 1951) compose (1) large plutons and dikes in a west-central zone of the batholith that extends ~8 km east of the gravity-magnetic gradient; and (2) scarce, small bodies and dikes in the eastern zone (Fig. 6F). Three gabbro plutons in the west-central zone are composed of intersheeted, mingling gabbro and granitic rocks (e.g., Tule Springs complex and Cuyamaca Peak complex,

Fig. 3A). Some of the larger gabbro plutons in this zone are continuous with smaller bodies and dikes composed of fine- to medium-grained hornblende gabbro.

All granite and gabbro plutons in the study area are cut by fine-grained, dark-colored mafic and intermediate dikes ("mafic dikes") of uncertain age and affinity. Older dikes are cut by younger dikes, indicating at least two generations of dike emplacement. Some of the dikes, which are largest and most numerous in the Cuyamaca–Laguna Mountains shear zone, are continuous with adjacent gabbro plutons whereas others appear to grade to plutons of the East Mesa suite.

Evidence of Coeval Mafic-Silicic Intrusion

Mingling and mixing relations between mafic and silicic magmas have been recognized in plutonic environments (Todd and Shaw, 1979; Vernon, 1984, 1990; Frost and Mahood, 1987; Barbarin, 1988; Foster and Hyndman, 1990). In the case of mingling, mafic and silicic magmas retain their identities and interact to form heterogeneous mafic/silicic rocks, whereas in mixing, they combine to form approximately homogeneous hybrid magma. In the Peninsular Ranges batholith, contacts evidencing magma mingling are common between gabbroic and granitic plutons, the latter ranging in composition from quartz diorite to leucomonzogranite. Where the volumetric ratio of gabbroic to granitic magma is high, small volumes of hybrid rocks may be present as well (e.g., the Tule Springs and Cuyamaca Peak complexes, see below).

Further evidence of coeval mafic and granitic intrusion are the mafic dikes, many of which are back-veined by fine-grained aplitic material, which in some cases is observably continuous with the host granitic pluton (synplutonic dikes) (Fig. 4). Aplitic veinlets are folded into concordance with foliation in the granitic host and, where back-veining is extensive, the mafic dike may be present only as fragments of dark-colored dike rock within aplite (Fig. 4B). Locally, dismemberment of mafic dikes has given rise to swarms of mafic enclaves in the host granitic pluton. Synplutonic mafic dikes are common in the East Mesa, Japatul Valley, and in all leucogranite suites.

Field observations suggest that at least some of the mafic enclaves in tonalite and granodiorite plutons were incorporated into the host magma from an adjacent gabbroic magma chamber and/or from disrupted mafic dikes. The igneous compositions and textures of the enclaves indicate an origin as globules of mafic or intermediate magma quenched in a lower-temperature silicic granitic magma (Vernon, 1990, 1991). In granitic plutons that contain abundant synplutonic mafic dikes, disruption of dikes at deeper and earlier stages of intrusion may have yielded fragments that were incorporated as enclaves in the host granite.

Zoned Plutonic Complexes and Ring-Dike Complexes

The west-central Peninsular Ranges batholith exposes a number of large, concentrically sheeted plutonic complexes composed of granitic rocks and gabbro (Todd, 2004). Complexes such as the Tule Springs and Cuyamaca Peak complexes contain more gabbro than granitic rocks, with evidence of mingling and minor mixing of mafic and silicic magmas. Other plutonic complexes are dominated by granitic suites but contain small volumes of fine- to medium-grained, plagioclase–porphyritic hornblende gabbro. Contacts between individual plutons within the complexes range from sharp to gradational, and plutons of all granitic suites show normal and oscillatory zoning in plagioclase, hornblende reaction rims on pyroxene and biotite rims on hornblende, and interstitial grains of quartz and K-feldspar, i.e., evidence of progressive crystallization from a melt. The relative ages of the plutonic suites based upon field relations are summarized schematically in the legend of Figure 3A. Five zoned plutonic complexes are described below.

El Cajon Mountain Complex

The north-northeast–elongate El Cajon Mountain complex (Fig. 3A) (Todd, 1983) consists of leucomonzogranite and leucogranodiorite of the Corte Madera suite intersheeted with amphibolite-facies metavolcanic rocks (silicic and intermediate tuff, tuff-breccia, and flows, with minor basalt) and small volumes of hornblende gabbro. Lithologic contacts and both magmatic and metamorphic foliations within the complex dip steeply (~70–90°) and variably both inward and outward, but the predominant dips are outward, away from the axis of the complex (Fig. 3B).

The Corte Madera pluton that forms the core of the El Cajon Mountain complex intrudes tightly folded metavolcanic rocks and an approximately symmetrical sequence of north-striking, subvertical sheeted tonalite plutons belonging to the Las Bancas, Alpine, and Japatul Valley suites. These plutons include a central sheet of Las Bancas tonalite, which gives way to the east and west to Alpine tonalite, then to Japatul Valley tonalite, the latter flanked and intruded by monzogranite plutons of the Chiquito Peak suite (Fig. 3A). Contacts between the Las Bancas and Alpine sheets vary from sharp to gradational or interfingering, and chilled Las Bancas dikes intrude the eastern Alpine pluton concordantly with host foliation. Plutons of the Alpine suite are in relatively sharp contact with, and also grade to, plutons of the Japatul Valley suite. Locally, these plutons are separated by thin discontinuous screens of metavolcanic rocks and/or gabbro. Contacts between Japatul Valley and Chiquito Peak plutons suggest that (1) plutons of the two suites are partly gradational and (2) Chiquito Peak dikes intrude Japatul Valley plutons.

The southern end of the El Cajon Mountain complex intrudes and transects Las Bancas and Alpine tonalite sheets whereas the northern end consists of a series of large, fringing leucocratic Corte Madera dikes that intrude metavolcanic rocks and tonalite of the Japatul Valley suite (Fig. 3A). Plutons and dikes of the Corte Madera suite in fact intrude all other plutonic suites in the western zone. The El Cajon Mountain complex is bordered on the east by the ~10-km-long, northeast-striking San Diego River Valley lineament (offset in a right-lateral sense by a short west-northwest–trending fault, Fig. 3), across which batholithic contacts show ~0.5 km right-lateral displacement.

Notable features of the El Cajon Mountain complex are (1) 0.5-km-wide zones of breccia/migmatite between metavolcanic screens and the Corte Madera pluton and its fringing dikes (Fig. 7A), and (2) transitional hypabyssal-volcanic rocks that occur near the margins of the Corte Madera pluton as well as adjacent to silicic metavolcanic screens within the pluton. In the area of San Vicente Mountain (Fig. 6A), the pluton consists largely of microgranite containing evenly distributed, patchy areas of flow-banded rhyolite. Contacts between gabbro and metabasalt also suggest a plutonic-to-volcanic transition: the larger bodies of gabbro in Figure 3A grade along strike into metabasalt that is clearly part of the metavolcanic screens.

The strike of magmatic foliation within the central Corte Madera pluton and its fringing dikes; in metavolcanic screens within and outside of the complex; and in the surrounding tonalite-granodiorite plutons is grossly concordant with the concentric structure of the El Cajon Mountain complex (Fig. 3B). However, north-trending foliation in the Alpine and Las Bancas plutons south of the central Corte Madera pluton is overprinted locally by a secondary foliation that appears to wrap around the southern end of the complex. A U-Pb zircon date of 111 ± 2 Ma was obtained at the southern end of the central Corte Madera pluton (Table 2). West of the El Cajon Mountain complex, metavolcanic screens that, prior to intrusion, were probably continuous with screens in the El Cajon Mountain complex (and in the McClain Ranch complex to the south) yield isotopic ages of (1) 122 and 126 ± 3 Ma and (2) 128–119 Ma (Anderson, 1991).

McClain Ranch Complex

The McClain Ranch complex is located about 3.5 km south of the El Cajon Mountain complex and was emplaced into the same sequence of subvertically sheeted tonalite plutons (Fig. 3A) (Todd, 1980). The McClain Ranch complex consists of a central pluton of leucomonzogranite and syenogranite of the Corte Madera suite that on the west-southwest side intrudes a large metavolcanic screen, gabbro, and tonalite of the Japatul Valley suite and on the northeast side appears to have "ballooned" into the sheeted tonalites. The tonalite plutons on the north side of the complex were apparently truncated by the central Corte Madera pluton. For example, from west to east, a Japatul Valley–Alpine contact ends against the diapir, the Alpine pluton is cut off, and foliation trends in the Las Bancas and Alpine plutons north and northeast of the McClain Ranch complex appear to be deflected to the east around the Corte Madera pluton (Fig. 3B).

The McClain Ranch complex is bordered on the east by the Harbison Canyon lineament (Fig. 3), which on high-altitude photographs is the southern extension of the San Diego River Valley lineament. On the southeastern side of the Harbison Canyon lineament, the Las Bancas pluton widens by ~1 km whereas the Corte Madera pluton is present only as a narrow "tail." This part of the lineament appears to be a shear zone with right-lateral-oblique(?) motion, as suggested by (1) a possible 0.5-km right-lateral jog in the Las Bancas–Alpine contact (northeastern part of the McClain Ranch complex map); (2) the apparent right-bending of tonalite

Figure 7. Mingling reactions between granitic magma, metavolcanic wall-rocks, and coeval gabbroic magma in zoned plutonic complexes. A: El Cajon Mountain complex: contacts between metavolcanic screens and granodiorite of central Corte Madera pluton are marked by contact breccia/migmatite containing a variety of approximately intact to assimilated migmatitic metavolcanic inclusions, small-scale folds; hammer parallel to near-vertical foliation. B: McClain Ranch complex: Japatul Valley tonalite magma and derivative aplite-pegmatite mingling with pillows of fine-grained hornblende gabbro; contacts range from sharp to irregular-diffuse, and one pillow is transected by a shear healed by aplitic material; hammer parallel to steep mineral foliation. C: Tule Springs complex: heterogeneous gneissic Japatul Valley tonalite with variably assimilated volcanic inclusions; to left of pencil (which is parallel to steep foliation) is a partly melted silicic inclusion; below pencil point and to right, amphibolitic fragments; mafic volcanics also reacted with magma to form dark, fine-grained ovoid to streaky amphibolitic bodies. D: Cuyamaca Peak complex: margins of concentric Chiquito Peak monzogranite dike in gabbro pluton is embayed by hornblende gabbro whose grain size varies from chilled next to dike, to fine-grained and hornblende-porphyritic away from contact with dike, to fine- and medium-grained (on left). Knife for scale.

foliations in the vicinity of the Harbison Canyon lineament; and (3) the juxtaposition across it of the central Corte Madera pluton on the northwest against the Corte Madera dike on the southeast. South of the McClain Ranch complex, batholithic contacts cross the Harbison Canyon shear zone with no apparent offset.

Metavolcanic screens on the western and southwestern flanks of the McClain Ranch complex consist of amphibolite (metabasalt); silicic and intermediate metatuff, tuff-breccia, and flows; and minor calc-silicate rocks. These rocks interfinger with bodies of hornblende gabbro, which grade along strike to amphibolite of the screens. As in the El Cajon Mountain complex, the metavolcanic rocks that immediately border the central Corte Madera pluton consist of meta-rhyolite and dacite tuff that may grade to a marginal microgranite phase of the pluton. On the southwestern flank of the McClain Ranch complex, Japatul Valley tonalite interfingers with metavolcanic rocks and mingles with gabbro (Fig. 7B); all of these rocks are cut by concordant leucocratic Corte Madera dikes. Contacts and foliation within the McClain Ranch complex dip steeply and variably both inward and outward. To our knowledge, none of the plutons of the McClain Ranch complex has been dated.

Tule Springs Complex

Gabbro forms a significant part of the north-elongate, ellipsoidal Tule Springs complex (Fig. 3A); the largest gabbro body is the 107 ± 2 Ma Poser Mountain pluton at the southern end of the complex (Fig. 6F) (Todd, 1982). In the northern part of the complex, smaller bodies of fine- to medium-grained hornblende gabbro are surrounded by Alpine tonalite in contacts that indicate mingling of tonalite and gabbro magmas. The overall composition of tonalite within the Tule Springs complex is more mafic than average tonalite of the Alpine suite, probably due to contamination by the abundant gabbroic dikes and inclusions. In contrast with Alpine plutons outside the Tule Springs complex, tonalite of the complex carries as much as 50% gabbro bodies, many of which appear to be disrupted gabbroic dikes. About 5–6 km to the northwest, in the same tonalite pluton, scattered zones as broad as several hundred meters consist of closely packed, "blob-like" bodies of fine-grained gabbro that are thinly veined and partly surrounded by tonalite; these zones may extend for at least a kilometer along the strike of foliation (Todd, 1983). We interpret such zones as representing the breakup of small "tongues" of gabbroic magma surrounded by, and mingling with, coeval tonalite magma.

The largest bodies of fine-grained hornblende gabbro in the northern part of the Tule Springs complex contain veins and small cuspate patches of tonalite that, where best developed, appear to be remnants of tonalite magma that was interstitial to closely packed gabbroic magma globules. Concentric dikes and irregular bodies of the Chiquito Peak suite, some trondhjemitic, intrude and interfinger with gabbro; in many places, subequal layers of trondhjemite and gabbro form gneissic bands >2 m long. Magmatic foliation within components of the Tule Springs complex is approximately concordant with the overall shape of the complex and with internal lithologic contacts. The dip of foliation and

contacts appears to have been controlled by the eastward regional dip of plutons and screens in the Cuyamaca–Laguna Mountains shear zone. Evidence supporting such an interpretation includes foliation and contacts within the Tule Springs complex that dip steeply (80–90°) eastward on the western side of the complex and outward on the northern and eastern sides (Fig. 3B).

Japatul Valley Complex

The Japatul Valley complex (Fig. 3A) (Todd, 1978) has the shape of a flattened "U" whose long dimension is oriented approximately northwest. The core of the complex is composed of quartz norite of the Las Bancas suite, surrounded discontinuously by tonalite of the Alpine suite, which grades outward to tonalite of the Japatul Valley suite. The central Japatul Valley pluton grades inward from pyroxene-biotite-hornblende tonalite adjacent to the contact with the Alpine pluton to hornblende-biotite tonalite and granodiorite. The Japatul Valley complex is bounded on the east by the northeast-striking, ~8-km-long Horsethief Canyon shear zone, across which batholithic contacts show ~0.5 km right-lateral bending and local offset.

The Japatul Valley pluton intrudes amphibolite-facies metavolcanic screens (chiefly andesitic and dacitic-rhyolitic compositions) and is intruded by ringlike Corte Madera dikes that are grossly concordant with the overall shape of the complex. On the northeast margin of the Japatul Valley complex, one of these dikes appears to have "ballooned" to form an approximately triangular (in plan view) Corte Madera leucomonzogranite pluton, which contains numerous small bodies and dikes of fine-grained hornblende gabbro. At outcrop scale, these gabbroic bodies invariably are back-veined by the Corte Madera host. Foliation within plutonic rocks and metamorphic screens of the Japatul Valley complex are roughly concordant with lithologic contacts and the overall shape of the complex (Fig. 3B). There is an overall outward dip of contacts and foliation in the complex within the regional northward dip of contacts in this part of the batholith.

The Japatul Valley complex is spatially associated with screens of metavolcanic rocks with which it forms complex zones of contact breccia and migmatite as much as 3–4 km wide. Contamination-assimilation reactions between tonalite magma and stoped metavolcanic inclusions as well as partial melting of silicic volcanic rocks have produced a great variety of compositions and textures in these zones, from intact metavolcanic blocks with concordant dikes of gneissic tonalite-granodiorite to extensively modified tonalite (including pegmatite and aplite) containing barely recognizable volcanic remnants (Fig. 7C). Other contact-breccia zones in tonalite of the Japatul Valley complex consist of (1) large fine-grained mafic blocks, some of which show mingling textures with tonalite; and (2) marginal zones containing angular enclaves of fine-grained tonalite that may represent early-emplaced Japatul Valley magma quenched against the walls or roof and subsequently disrupted and incorporated by flow of magma (solid quench fragments, Vernon, 1983).

No absolute ages have been determined for plutons or metavolcanic screens of the Japatul Valley complex, but concordant

^{40}Ar/^{39}Ar ages averaging ~104 Ma were determined for horn-blende from a sample of Japatul Valley tonalite collected from the complex (Table 2). Based upon field relations, we speculate that all of the plutons involved in the complex are relatively close in age, in the following intrusive order: dikes of the Las Bancas suite are common in the Alpine tonalite north of the Japatul Valley complex; the Alpine and Japatul Valley tonalites are in part gradational; and the Corte Madera suite forms dikes in all tonalitic plutons of the complex. The volcanic component of the Japatul Valley complex may have originated by sinking of superjacent volcanic rocks.

Cuyamaca Peak Complex

The Cuyamaca Peak complex consists of voluminous gabbro intersheeted with lesser monzogranite of the Chiquito Peak suite (Fig. 3A) (Todd, 1977b). The complex consists of a north-northwest–elongate, continuous chain of centrally located gabbro "plutons" in the Cuyamaca Mountains (Fig. 6F), the largest of which underlie (from north to south) North, Middle, and Cuyamaca peaks. Although it is not connected to this chain at the present level of erosion, a gabbro pluton to the south-southeast may also be part of the Cuyamaca Peak complex. Gabbro and monzogranite sheets of the Middle and North Peak plutons are also intersheeted with the Late Jurassic Cuyamaca Reservoir suite in contacts suggestive of remobilization of tightly folded Jurassic granitic and Julian Schist wallrocks during Early Cretaceous intrusion (V.R. Todd, unpublished mapping). The orientation of external and internal contacts and foliation within the Cuyamaca Peak complex is variable; in part, it reflects a strong eastward regional dip, but gabbro foliation may tend to dip steeply inward toward the north-northwest axis of the complex (Fig. 3B).

On its western margin, the contact of the Cuyamaca Peak complex with tonalite of the Alpine suite is marked by a broad (up to 1.5 km) zone of hybrid gabbroic and granitic rocks, plus scarce metavolcanic and metasedimentary inclusions. This zone appears to represent a zone of mingling and mixing between coeval mafic and tonalitic magmas. Although not all of the following rocks are present in any given traverse westward from the Cuyamaca Peak gabbro pluton, the following sequence is typical: (1) round bodies of fine- to medium-grained, plagioclase-porphyritic hornblende gabbro plus scarce orbicular gabbro in a biotite-hornblende diorite or quartz-bearing gabbro matrix (the latter comprising ~25% of the rock); grading to (2) subequal, fine-grained gabbro and tonalite in contacts suggesting that tonalite magma intruded and surrounded gabbroic melt globules; to (3) predominantly tonalitic rocks in which large gabbro inclusions and dikes make up ~50% of the rock; to (4) tonalite with discrete mafic enclaves whose abundance decreases away from the contact with the gabbro pluton.

On its eastern side, the Cuyamaca Peak complex is intersheeted with monzogranite of the Chiquito Peak suite (Fig. 3A) in exposures that attest to magma mingling and minor mixing over short (<10 m) distances at the present level of erosion. Concentric monzogranite dikes in gabbro have contacts that display (1) pillowing of fine-grained gabbro into monzogranite (Fig. 7D);

(2) monzogranite embaying and surrounding gabbro pillows; (3) chilling of gabbro against monzogranite; and (4) local chilling of monzogranite against gabbro. Small volumes of hybrid diorite-tonalite are present along some gabbro-monzogranite contacts.

Two additional components of the Cuyamaca Peak complex deserve mention. The complex is intruded on the southeast by two plutons of the Pine Valley suite and on both the southeastern and southwestern flanks by plutons of the East Mesa suite (Fig. 3A). To the south of the Cuyamaca Peak complex, diffuse contacts between Pine Valley and Chiquito Peak plutons suggest that the Chiquito Peak may not have have been completely solidified or else remained at near-magmatic temperatures, at the time of Pine Valley intrusion. Contacts between bodies of the Chiquito Peak and East Mesa suites in the Cuyamaca Peak complex suggest overlapping intrusion and mingling, as is the case elsewhere in the Cuyamaca–Laguna Mountains shear zone (e.g., the Long Valley Peak area, Fig. 6A). However, the close spatial association with, and apparent synplutonic contacts between, some East Mesa plutons and Jurassic metagranite plutons (Shaw et al., this volume, Chapter 7) suggested to us that the East Mesa magma type spanned the Jurassic and Cretaceous. This speculation was confirmed by new U-Pb zircon ^{206}Pb/^{238}U ion microprobe ages of 117 ± 1 Ma and 118 ± 1 Ma for two East Mesa plutons in the study area, and an age of 170 ± 2 Ma for a third (Table 2). In addition, Schmidt and Paterson (2002) report U-Pb zircon ages of ~164 Ma for two bodies of biotite orthogneiss in the southern Sierra San Pedro Martir.

Interpretation of Zoned Plutonic Complexes

We suggest that the zoned plutonic complexes in the west-central Peninsular Ranges batholith represent the roots of ring complexes that underlay volcanic centers, possibly collapse calderas as described by Merriam (1946), Gastil (1990b), and Johnson et al. (1999b). In each complex, gabbroic and granitic magmas mingle, indicating that magmas of both compositions were produced throughout Cretaceous magmatism. The truncation of plutonic contacts and deflection of foliation in tonalite "sheets" adjacent to relatively late Corte Madera diapirs (e.g., McClain Ranch complex, above) plus local overprinting of tonalite foliation by a secondary foliation parallel to the margins of the Corte Madera pluton (e.g., El Cajon Mountain complex), suggest that the sheeted tonalites were in a near-solidus condition at the time of Corte Madera intrusion. The north-northeast direction of "ballooning" of the Corte Madera pluton in the Japatul Valley complex and right-lateral sense of shear along the adjacent Horsethief Canyon shear zone (Fig. 3) are similar to movement directions in the McClain Ranch complex to the west. This suggests a genetic relationship between Corte Madera intrusion and localized right-lateral ductile shear in the west-central Peninsular Ranges batholith. The apparent overlap of magmatic near-solidus or hot subsolidus structures in some of the complexes implies that the intrusive ages of the plutons that compose them may differ by only a few million years.

Due to the northwest strike and westward tilting of the Peninsular Ranges batholith, the depth of emplacement represented by the zoned plutonic complexes should increase from southwest to northeast (Fig. 5, Table 1). From field relations, the El Cajon Mountain complex and McClain Ranch complex are the shallowest of the five complexes, and lithologic transitions between volcanic and plutonic rocks, plus local mingling between plutons and metavolcanic screens suggest that these are ring complexes that underlay bimodal volcanic centers. Two samples of Alpine tonalite collected between the El Cajon Mountain complex and McClain Ranch complex (Fig. 3A) give hornblende pressures of ~1.0 kb and ~2.4 kb (all hornblende pressures are ± 1.5 kb), whereas a sample of Alpine tonalite from the eastern side of the El Cajon Mountain complex yields ~3.4 kb. Farther to the east, the Japatul Valley complex apparently represents a deeper level of the Early Cretaceous crust: metavolcanic screens are as abundant in the Japatul Valley complex as in the two western complexes, but contacts between the Japatul Valley tonalite and metavolcanic rocks suggest that major processes in the emplacement of the Japatul Valley complex included (1) stoping, partial melting, and incorporation of downward-transported volcanic rocks, perhaps following or accompanying (2) fracturing of volcanic cover rocks by early-stage intrusion and upward doming as described by Johnson et al. (1999b). A hornblende pressure of ~2.2 kb was obtained from Japatul Valley tonalite in the southern part of the Japatul Valley complex.

From field relations, the Tule Springs complex represents a deeper level of intrusion: metavolcanic inclusions are much smaller and far less abundant than in the western complexes. Hornblende from Japatul Valley tonalite in the northernmost part of the Tule Springs complex gives a pressure of ~4.1 kb, while Alpine tonalite from the eastern side also yields ~4.1 kb. The Cuyamaca Peak complex apparently represents the deepest of the five complexes; few remnants of Cretaceous volcanic roof or wallrocks are present within it. A hornblende pressure of ~3.8 kb was obtained for an East Mesa sample from the southeastern part of the Cuyamaca Peak complex. In the Tule Springs and Cuyamaca Peak complexes, mingling and minor mixing of gabbroic and granitic magmas apparently were major processes.

The concentrically zoned plutonic complexes described above are similar in both external and internal features to the "concentrically expanded plutons" described by Paterson and Vernon (1995) and Paterson et al. (1996). In numerous studies of zoned "ballooned" plutons, these authors documented 30% or less expansion of country rocks during emplacement, with regional deformation occurring both during and after emplacement. They propose a model of a syntectonic, nested diapir in which vertical movement of country rocks and magma is the dominant mechanism for material transfer and further propose that such compositionally zoned plutons result from multiple pulses of separately evolved parental magmas utilizing the same magma plumbing system. The ubiquitous presence in the Peninsular Ranges batholith complexes of steeply oriented contacts, foliation, and lineation in both plutons and wallrock screens

would seem to suggest that material transfer during emplacement was mainly vertical at the present level of erosion (i.e., magma moved upward and wallrocks and older plutonic rocks moved downward), as proposed in the model of Paterson and Vernon (1995). Variations in lithology, the nature of zoning, and mechanisms of emplacement/material transfer among these complexes are intriguing, but until each complex is dated and studied in detail, interpretation of these variations remains speculative.

GEOCHEMISTRY OF GRANITIC SUITES

Two-hundred-nine granitic rocks were analyzed for major-element oxides and selected trace elements (Fig. 8; Table 3). Complete geochemical tables are included as supplemental material with the volume (see footnote 1 or the CD-ROM accompanying this volume). Of these analyses, major oxides for 176 rocks were analyzed by X-ray fluorescence (XRF) method at Macquarie University and those for 33 samples by rapid rock method at the U.S. Geological Survey; trace elements for all samples were determined at Macquarie. Although all plutonic suites are represented, only a small number of samples were collected from any one pluton; therefore, the results do not address the geochemical variability of individual plutons. Plotted on Harker diagrams, the data show scatter that may reflect variations among plutons of a given suite, but overall, patterns are generally consistent with field and mineralogical variations among the suites.

All of the Cretaceous granites are characterized as low-K_2O, low-Rb granites (Silver et al., 1979; Silver and Chappell, 1988; S.E. Shaw, unpublished data). As a group, they have high CaO:total alkalies ratios with Peacock indices ranging from 63 to 66 (calcic granites) and aluminum saturation indices (ASI; Zen, 1986) in the metaluminous to weakly peraluminous range (Table 2). Of the Early to middle Cretaceous granitic suites, the Las Bancas suite includes the most mafic plutons of the transect (SiO_2 ~51–66%, basaltic andesite to dacite) (Fig. 8), overlapping the most mafic rocks of the combined Alpine and Japatul Valley suites (SiO_2 ~54–71%, andesite to dacite). Four Early to middle Cretaceous granitic suites are characterized by high SiO_2: the Chiquito Peak suite, ~65–77%; the combined Corte Madera and Mother Grundy Peak suites, ~71–78%; and the Pine Valley suite, ~68–75% SiO_2. Corte Madera and Mother Grundy Peak samples are combined on geochemical diagrams because of a possible genetic relation between the two suites (see below). Of the two middle to Late Cretaceous, mainly eastern-zone granitic suites, the Granite Mountain suite is more mafic (SiO_2 ~51–69%) than the La Posta (SiO_2 ~64–77%) and, in general, western plutons of both suites are more mafic than eastern plutons.

Harker Diagrams

For the most part, major and trace element abundances of the Cretaceous granitic rocks plot in similar trends/fields on variation diagrams (Fig. 8). Exceptions are K_2O, Rb, Ba, TiO_2, P_2O_5, Cr, Sr, Y, and Zr, which may show differences between suites or

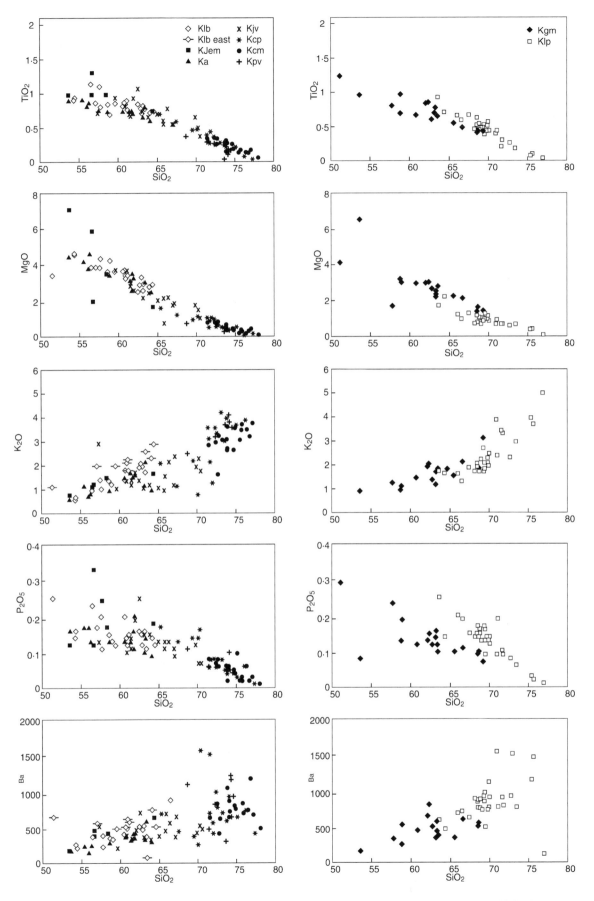

Figure 8. Geochemical variations of selected oxides/trace elements in Cretaceous granitic suites. Early to middle Cretaceous suites: Klb—Las Bancas; KJem—East Mesa; Ka—Alpine; Kjv—Japatul Valley; Kcp—Chiquito Peak; Kcm—Corte Madera plus samples of Mother Grundy Peak; Kpv—Pine Valley. Klb east—Las Bancas samples collected east of I-S line. Middle to Late Cretaceous suites: Kgm—Granite Mountain, Klp—La Posta.

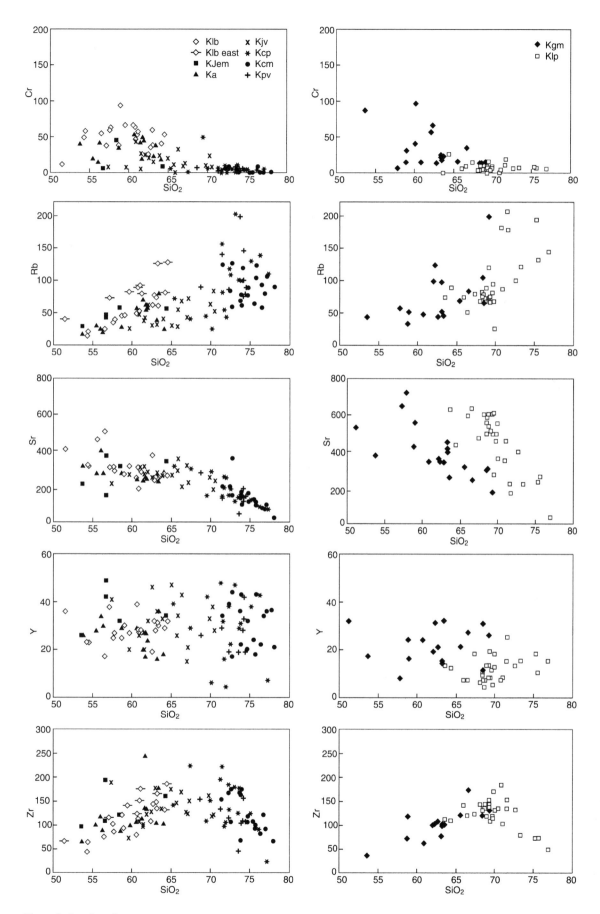

Figure 8. *(continued)*

TABLE 3. MAJOR OXIDES AND TRACE ELEMENTS OF SELECTED CRETACEOUS PLUTONIC ROCKS, SAN DIEGO COUNTY, CALIFORNIA

Sample no.:	VM89	ML12	ECM1	ML26	A2	ML13	A61	R19	SP923	A931	CP117A	TS23A	TS76
Latitude:	32 49 02	32 47 32	32 52 57	32 51 15	32 49 13	32 48 43	32 49 31	33 04 53	33 04 52	32 48 02	32 54 30	32 54 53	32 58 10
Longitude:	116 44 12	116 23 35	116 48 43	116 26 10	116 46 34	116 24 12	116 47 34	116 47 03	116 54 12	116 44 53	116 30 34	116 36 52	116 40 13
					Las Bancas suite							East Mesa suite	
SiO_2	65.59	63.37	62.83	61.01	60.83	59.58	58.91	56.48	54.39	54.30	58.42	56.62	53.64
TiO_2	0.71	0.77	0.78	0.89	0.84	0.85	0.69	1.14	0.94	0.90	0.98	0.98	0.97
Al_2O_3	15.06	15.18	15.29	15.57	15.97	16.00	16.23	18.24	18.04	18.46	16.37	16.32	16.13
Fe_2O_3	1.87	1.13	2.00	1.89	1.86	1.84	2.28	3.79	2.49	2.88	0.85	0.77	1.15
FeO	3.08	4.26	3.90	4.26	4.58	4.58	4.73	3.52	6.09	6.01	6.96	5.37	5.55
MnO	0.10	0.12	0.11	0.14	0.14	0.13	0.22	0.15	0.14	0.16	0.15	0.12	0.10
MgO	1.72	2.98	2.87	3.23	3.42	3.64	4.27	3.89	4.55	4.63	3.45	5.84	7.06
CaO	4.48	5.42	5.81	6.25	6.29	6.68	7.21	7.58	8.73	8.76	7.14	8.93	10.24
Na_2O	3.51	2.94	3.22	2.89	3.18	2.91	3.28	3.72	2.73	3.00	3.33	2.49	2.84
K_2O	2.94	2.57	2.06	2.25	1.80	1.99	1.20	0.93	0.72	0.58	1.50	1.10	0.77
P_2O_5	0.13	0.11	0.12	0.12	0.14	0.12	0.13	0.23	0.16	0.14	0.17	0.12	0.12
H_2O+	0.96	0.88	0.76	0.80	0.90	1.11	0.89	0.55	0.54	0.53	0.47	0.52	1.42
H_2O-	0.06	0.05	0.03	0.03	0.07	0.07	0.06	0.08	0.06	0.01	0.05	0.07	0.14
CO_2	0.04	0.06	0.05	0.05	0.04	0.11	0.04	0.09	0.05	0.01	0.02	0.02	0.12
Total	100.25	99.84	99.83	99.38	100.06	99.61	100.14	100.39	99.63	100.37	99.86	99.27	100.25
Ba	510	68	461	608	472	478	330	358	201	257	403	374	170
Cr	8	205	35	53	46	66	93	55	58	49	45	389	638
Cu	47	56	28	16	21	22	35	7	18	23	34	12	7
Ga	17	15	17	20	14	20	19	19	16	19	19	17	16
Nb	N.D.*	6	N.D.	N.D.	N.D.	N.D.	N.D.	6	4	2	N.D.	N.D.	4
Ni	8	32	15	12	12	19	30	16	18	23	23	11	12
Pb	13	13	8	5	4	10	8	7	29	8	11	9	4
Rb	99	126	77	91	52	83	46	24	21	14	58	46	29
Sr	215	246	231	292	186	309	265	504	307	312	310	148	210
Th	11	14	7	5	4	7	5	3	8	<2	4	4	2
U	5	6	3	3	4	3	<3	<3	3	<3	3	3	<3
V	110	266	130	147	136	149	151	167	208	209	202	155	285
Y	41	30	32	28	27	27	30	17	23	23	32	42	26
Zn	75	57	61	69	57	80	70	91	87	75	99	86	88
Zr	189	166	143	151	113	141	93	76	65	45	121	109	97

(continued)

TABLE 3. MAJOR OXIDES AND TRACE ELEMENTS OF SELECTED CRETACEOUS PLUTONIC ROCKS, SAN DIEGO COUNTY, CALIFORNIA (continued)

Sample no.:	41252	TS37	VM75	VM19	A945	A46	TS65	EC1	EC4	VM511	TS17	VM922	POT1
Latitude:	32 57 03	32 54 16	32 50 06	32 50 06	32 51 41	32 46 13	32 57 03	32 48 41	32 46 43	32 45 25	32 57 20	32 45 41	32 36 02
Longitude:	116 45 33	116 38 42	116 44 59	116 43 05	116 50 41	116 49 43	116 38 17	116 52 32	116 59 50	116 39 40	116 39 32	116 40 25	116 35 27
	Alpine suite							Japatul Valley suite					
SiO_2	64.06	63.17	61.72	61.40	58.63	56.01	55.41	70.50	67.12	65.82	65.02	62.53	59.65
TiO_2	0.60	0.65	0.75	0.74	0.74	0.81	0.91	0.41	0.65	0.78	0.60	0.74	0.81
Al_2O_3	16.27	16.38	16.11	16.03	17.16	18.16	17.87	14.65	14.62	14.62	16.25	16.44	16.84
Fe_2O_3	1.52	1.85	1.03	1.50	2.19	3.57	2.39	1.32	1.47	1.51	1.81	2.42	2.29
FeO	3.56	4.00	5.43	4.47	4.95	4.72	5.60	2.13	2.95	4.33	3.23	3.62	4.48
MnO	0.09	0.11	0.12	0.12	0.13	0.16	0.16	0.07	0.17	0.10	0.09	0.12	0.13
MgO	2.45	2.84	3.52	3.09	3.43	3.75	4.17	0.87	1.72	2.17	2.13	2.15	3.65
CaO	5.70	6.44	7.00	6.15	7.38	8.11	7.66	3.37	4.29	5.08	5.75	5.71	7.20
Na_2O	3.47	3.07	2.82	3.11	3.08	3.30	3.25	4.20	3.73	3.12	3.51	3.80	2.98
K_2O	0.98	1.18	1.51	1.68	0.98	0.73	1.15	1.95	1.95	2.16	0.97	1.07	1.36
P_2O_5	0.09	0.10	0.11	0.13	0.13	0.17	0.17	0.08	0.12	0.13	0.11	0.15	0.10
H_2O+	1.20	0.39	0.27	1.06	0.79	1.14	1.06	0.84	1.07	0.34	0.48	0.72	1.10
H_2O-	0.13	0.03	0.25	0.09	0.05	0.08	0.08	0.10	0.11	0.18	0.12	0.07	0.07
CO_2	0.05	0.02	0.06	0.10	0.11	0.04	0.05	0.09	0.10	0.01	0.01	0.05	0.00
Total	100.17	100.23	100.70	99.67	99.75	100.75	99.93	100.58	100.07	100.35	100.08	99.59	100.66
Ba	279	322	344	316	285	146	234	377	425	563	449	360	385
Cr	19	24	49	43	35	16	20	6	10	33	11	15	42
Cu	19	17	3	34	15	14	19	4	12	18	56	6	13
Ga	17	17	17	16	16	20	18	15	16	16	17	19	16
Nb	2	N.D.	N.D.	4	4	N.D.	5	N.D.	N.D.	N.D.	N.D.	4	2
Ni	5	12	3	18	13	7	11	4	6	13	32	5	9
Pb	6	7	8	14	9	10	9	6	4	10	4	9	8
Rb	25	39	55	71	29	25	31	61	64	68	29	32	47
Sr	278	264	249	229	275	396	272	192	247	196	286	273	305
Th	2	4	2	5	5	4	2	3	6	5	<2	3	2
U	<3	<3	<3	<3	<3	<3	<3	3	<3	<3	<3	3	<3
V	102	125	136	145	168	178	191	44	87	123	95	112	180
Y	18	16	27	20	29	34	28	33	33	43	26	36	26
Zn	61	69	75	67	70	91	89	46	57	72	63	69	71
Zr	104	105	136	115	89	119	101	157	161	169	146	176	109

(continued)

TABLE 3. MAJOR OXIDES AND TRACE ELEMENTS OF SELECTED CRETACEOUS PLUTONIC ROCKS, SAN DIEGO COUNTY, CALIFORNIA (continued)

Sample no.:	SS831	SPL3	D9413	VM68	CP925	TS92A	D6216	A944	VM935	ECM2	BL9	D94	D58	WD80A	CP175A
Lat:	32 55 12	33 00 13	32 51 51	32 50 16	32 55 59	32 54 59	32 50 29	32 50 40	32 45 35	32 53 32	32 37 55	32 45 44	32 49 02	32 50 10	32 57 28
Long:	116 52 52	116 57 25	116 36 30	116 39 35	116 34 29	116 39 23	116 36 33	116 49 44	116 41 58	116 49 26	116 41 15	116 36 47	116 31 01	116 32 09	116 30 26
			Chiquito Peak suite				Corte Madera–Mother Grundy Peak suites							Pine Valley suite	
SiO_2	75.05	73.61	72.46	72.32	71.55	70.07	76.74	75.72	75.50	74.69	73.74	72.40	74.48	72.35	68.70
TiO_2	0.22	0.25	0.26	0.36	0.30	0.48	0.13	0.17	0.14	0.26	0.28	0.35	0.17	0.27	0.37
Al_2O_3	13.34	13.64	14.34	13.78	14.49	15.49	12.68	13.07	13.11	13.08	13.59	14.39	13.63	14.37	16.30
Fe_2O_3	0.71	0.86	0.75	1.08	0.80	1.35	0.22	0.50	0.72	0.75	0.86	1.07	0.32	0.39	0.67
FeO	1.08	1.36	1.69	1.72	2.04	2.06	1.23	1.06	0.61	1.28	1.60	2.04	1.49	2.30	2.80
MnO	0.05	0.04	0.06	0.06	0.05	0.04	0.04	0.03	0.01	0.04	0.03	0.06	0.05	0.06	0.06
MgO	0.50	0.42	0.50	0.91	0.47	1.08	0.27	0.28	0.14	0.40	0.39	0.73	0.32	0.64	0.73
CaO	1.67	1.94	2.22	2.63	2.47	4.62	1.02	1.46	1.39	2.02	1.89	2.79	1.76	2.35	3.62
Na_2O	3.81	3.77	3.78	3.02	3.55	3.84	4.16	3.61	3.70	4.02	3.85	3.54	3.64	3.48	3.51
K_2O	3.68	3.04	3.33	3.61	3.15	0.80	3.22	3.47	3.06	2.64	3.06	2.95	3.59	3.20	2.53
P_2O_5	0.03	0.05	0.06	0.08	0.06	0.14	0.02	0.02	0.02	0.05	0.06	0.07	0.04	0.08	0.11
H_2O+	0.24	0.33	0.43	0.33	0.50	0.43	0.23	0.35	0.35	0.65	0.40	0.42	0.29	0.31	0.72
H_2O-	0.10	0.03	0.12	0.04	0.06	0.09	0.11	0.08	0.08	0.03	0.04	0.12	0.05	0.03	0.16
CO_2	0.04	0.03	0.12	0.02	0.06	0.07	0.10	0.12	0.05	0.05	0.14	0.01	0.05	0.15	0.05
Total	100.52	99.37	100.12	99.96	99.55	100.56	100.17	99.94	98.88	99.96	99.93	100.94	99.88	99.98	100.33
Ba	646	591	785	990	1532	254	1185	831	2662	788	1049	824	942	816	1110
Cr	5	3	<3	9	6	7	<3	<3	<3	5	4	5	5	<3	7
Cu	<1	<1	<1	13	<1	10	12	<1	<1	2	<1	56	22	40	<1
Ga	11	14	13	15	15	14	15	13	12	16	13	16	15	16	19
Nb	7	5	6	N.D.	7	2	N.D.	5	3	N.D.	6	N.D.	N.D.	N.D.	8
Ni	<2	3	5	7	3	3	6	2	<2	3	2	32	19	10	3
Pb	13	6	8	10	8	6	7	7	5	4	4	8	14	13	8
Rb	130	102	109	119	81	25	80	104	64	76	68	84	90	86	90
Sr	123	134	188	150	192	281	70	107	122	159	131	197	143	191	273
Th	15	13	15	7	7	<2	6	10	9	6	5	7	9	9	3
U	4	3	<3	4	<3	3	4	<3	3	<3	<3	4	5	4	<3
V	10	18	20	36	29	77	6	12	4	17	22	28	12	27	37
Y	38	29	22	38	32	6	34	18	24	20	36	39	19	19	26
Zn	23	28	31	37	44	41	23	13	12	17	32	39	32	50	67
Zr	124	125	131	184	222	151	121	93	107	106	167	167	109	115	154

(continued)

TABLE 3. MAJOR OXIDES AND TRACE ELEMENTS OF SELECTED CRETACEOUS PLUTONIC ROCKS, SAN DIEGO COUNTY, CALIFORNIA (continued)

Sample no.:	MR2	SY16	SS833	ML42	EV931	A47	J32A	A3	W3	W4	A48	CC1	CC10	CC11	R17
Lat:	32 39 40	33 04 44	32 37 16	32 43 47	33 00 44	32 48 09	33 07 35	32 46 52	32 40 28	32 42 45	32 48 12	32 43 07	32 43 16	32 43 34	33 02 27
Long:	116 33 13	116 37 53	116 36 46	116 28 41	116 23 27	116 49 13	116 36 01	116 47 43	116 16 52	116 22 50	116 49 03	116 15 28	116 23 50	116 26 02	116 49 27
			Granite Mountain suite								La Posta suite				
SiO_2	68.60	66.60	65.47	63.19	62.10	60.79	57.77	73.38	71.02	69.82	69.44	68.74	68.50	67.43	64.35
TiO_2	0.44	0.49	0.55	0.70	0.85	0.67	0.81	0.18	0.44	0.52	0.38	0.54	0.63	0.67	0.71
Al_2O_3	15.40	15.97	16.14	16.65	16.79	17.10	21.39	14.11	15.37	15.65	15.50	16.00	16.00	15.91	16.46
Fe_2O_3	1.22	1.11	1.42	1.76	1.00	2.23	0.60	0.62	0.87	0.18	1.17	0.34	0.65	0.67	1.60
FeO	2.34	3.13	3.08	3.05	4.79	3.90	3.53	1.08	1.37	2.42	1.88	2.29	2.22	2.61	3.05
MnO	0.06	0.09	0.09	0.08	0.09	0.14	0.09	0.04	0.04	0.05	0.07	0.06	0.04	0.06	0.05
MgO	1.59	2.09	2.24	2.16	3.00	2.94	1.69	0.63	0.89	1.07	1.25	1.02	1.11	1.26	2.21
CaO	3.97	4.60	5.24	5.72	6.20	6.47	7.58	2.25	3.37	3.67	3.78	3.88	4.07	4.37	5.06
Na_2O	3.56	3.32	3.94	3.85	2.66	3.44	4.05	3.74	3.85	3.74	3.58	4.13	3.85	3.85	3.65
K_2O	1.86	2.11	1.57	1.16	1.90	1.45	1.24	2.95	2.36	2.26	2.05	1.72	2.02	1.86	1.63
P_2O_5	0.10	0.11	0.10	0.16	0.13	0.12	0.23	0.06	0.19	0.13	0.09	0.16	0.17	0.15	0.14
H_2O+	0.62	0.73	0.58	0.98	0.96	1.18	0.98	0.94	0.77	0.46	0.53	0.72	0.63	0.75	0.76
H_2O-	0.10	0.06	0.08	0.03	0.06	0.07	0.04	0.08	0.09	0.03	0.11	0.14	0.06	0.15	0.07
CO_2	0.01	0.11	0.09	0.06	0.14	0.04	0.06	0.04	0.03	0.02	0.05	0.06	0.21	0.11	0.03
Total	99.87	100.52	100.59	99.55	100.67	100.54	100.06	100.10	100.66	100.02	99.88	99.80	100.16	99.85	99.77
Ba	544	604	359	356	652	448	335	768	772	763	497	755	759	625	462
Cr	14	36	17	21	57	15	7	7	4	7	16	6	10	15	26
Cu	N.D.	9	13	6	11	24	8	2	36	56	16	5	<1	<1	3
Ga	N.D.	14	15	20	17	21	25	18	20	20	17	22	20	18	16
Nb	N.D.	7	3	N.D.	7	N.D.	N.D.	N.D.	N.D.	N.D.	5	6	7	9	4
Ni	N.D.	13	8	7	10	8	5	4	5	32	5	6	3	5	8
Pb	N.D.	6	10	7	11	<3	6	15	12	15	13	7	10	8	8
Rb	45	57	47	35	68	33	39	82	59	64	45	46	51	54	60
Sr	305	245	317	454	360	347	727	220	461	497	273	540	500	475	436
Th	N.D.	4	9	4	6	<2	4	5	3	8	3	5	5	6	5
U	N.D.	<3	<3	<3	<3	<3	4	<3	3	<3	<3	<3	<3	3	<3
V	70	80	83	87	107	114	47	16	19	33	50	20	46	51	98
Y	31	27	21	14	19	24	8	15	8	5	19	4	9	18	12
Zn	49	45	65	74	85	81	73	24	66	75	54	89	73	75	75
Zr	384	173	121	77	100	63	430	78	101	114	107	134	123	123	109

(continued)

TABLE 3. MAJOR OXIDES AND TRACE ELEMENTS OF SELECTED CRETACEOUS PLUTONIC ROCKS, SAN DIEGO COUNTY, CALIFORNIA (continued)

Sample no.:	Cuyamaca Gabbro								Mafic and intermediate dikes				Mafic enclaves		
	A28	TS43	A74	VM40	ML36	TS57	D59	ECM11	VM76A	W5	D52	D103B	A5X#	EC4X	VM75X
Lat:	32 49 40	32 55 34	32 49 10	32 45 02	32 46 09	32 55 33	32 49 26	32 55 45	32 51 29	32 43 21	32 50 29	32 49 10	32 43 17	32 46 43	32 50 06
Long:	116 50 07	116 37 53	116 50 14	116 40 19	116 29 14	116 39 34	116 30 14	116 51 05	116 40 59	116 28 28	116 36 33	116 32 34	116 52 53	116 59 50	116 44 59
SiO_2	57.63	52.26	51.60	49.57	47.82	47.60	44.07	43.49	63.40	59.00	56.32	50.33	56.58	55.92	53.77
TiO_2	0.74	0.96	0.30	0.57	0.53	1.12	1.35	1.80	0.75	0.93	1.03	0.95	0.75	0.75	0.80
Al_2O_3	17.21	18.63	16.01	19.50	17.99	17.74	19.58	15.97	16.67	18.78	17.95	17.52	16.04	15.88	17.39
Fe_2O_3	2.69	3.12	1.39	1.28	0.81	2.61	4.19	7.28	2.28	2.15	2.13	2.20	2.03	1.93	4.22
FeO	4.81	6.51	5.09	6.73	8.92	8.27	7.98	9.24	3.42	3.64	5.34	7.20	5.51	5.97	4.89
MnO	0.16	0.16	0.15	0.14	0.19	0.19	0.19	0.17	0.09	0.09	0.14	0.18	0.26	0.25	0.17
MgO	4.14	5.38	7.62	7.06	10.41	6.06	6.89	5.90	2.17	2.57	3.38	6.19	4.30	4.17	5.27
CaO	7.94	9.90	13.10	12.62	9.72	10.92	12.64	11.78	5.39	6.88	8.38	9.90	8.12	7.59	9.44
Na_2O	3.15	2.63	1.21	1.24	1.19	2.81	1.39	1.31	3.45	3.76	3.13	2.85	4.23	4.85	3.17
K_2O	1.00	0.34	0.35	0.17	0.11	0.40	0.22	0.26	1.60	1.27	1.11	1.24	1.34	1.08	1.15
P_2O_5	0.11	0.14	0.02	0.09	0.01	0.12	0.13	0.06	0.14	0.21	0.16	0.10	0.10	0.08	0.10
H_2O+	0.92	0.14	2.34	0.23	1.63	1.70	0.36	1.91	1.14	0.36	1.33	1.82	1.19	0.83	0.18
H_2O-	0.08	0.08	0.24	0.18	0.13	0.11	0.10	0.23	0.10	0.09	0.16	0.11	0.09	0.10	0.06
CO_2	0.05	0.09	0.09	0.31	0.48	0.17	0.01	0.42	0.06	0.04	0.05	0.07	0.04	0.11	0.04
Total	100.63	100.34	99.51	99.69	99.94	99.82	99.10	99.82	100.66	99.77	100.61	100.66	100.58	99.51	100.65
Ba	24	149	47	80	67	67	81	44	441	246	321	204	408	192	221
Cr	61	80	380	231	48	74	107	<3	7	19	40	83	52	46	69
Cu	41	14	20	14	14	62	41	126	18	5	34	15	15	22	47
Ga	17	17	13	18	13	18	21	18	19	23	19	18	17	18	19
Nb	N.D.	N.D.	2	N.D.	3	3	N.D.	2	N.D.	N.D.	N.D.	N.D.	N.D.	N.D.	N.D.
Ni	24	14	34	7	28	15	N.D.	14	17	4	15	17	11	11	8
Pb	6	8	12	7	8	11	10	11	10	13	7	<3	4	10	10
Rb	30	11	12	5	<1	4	3	5	48	27	54	35	35	26	37
Sr	275	356	250	361	267	281	374	353	322	718	272	312	209	209	291
Th	4	4	<2	3	4	4	2	<2	2	3	7	<2	3	2	2
U	<3	<3	<3	<3	<3	<3	<3	<3	3	<3	4	<3	3	<3	<3
V	166	242	170	117	175	414	436	1001	112	127	197	281	182	195	190
Y	29	18	11	11	7	22	23	19	16	19	34	22	80	52	31
Zn	73	103	65	73	96	105	107	109	82	90	91	80	98	117	98
Zr	74	32	21	23	13	60	28	37	78	126	50	44	75	49	43

Note: Analytical methods and limits of detection for trace elements are given in an appendix in the GSA Data Repository (see footnote 1) and on the CD-ROM accompanying this volume. Latitude and longitude in degrees, minutes, seconds.

*N.D.—not determined.

#X suffix = mafic enclaves in Japatul Valley samples A5 and EC4, and in Alpine sample VM75.

may differ within a single suite according to east-west location within the batholith. Additionally, suites that form plutons both west and east of the I-S line (Las Bancas, Granite Mountain, and La Posta, Fig. 6) display west-to-east geochemical differences in FeO/Fe_2O_3 ratios.

A Harker diagram for K_2O of the granitic suites (Fig. 8) consists of a series of overlapping positive linear fields. The Las Bancas and East Mesa samples overlap with tonalites of the Alpine and Japatul Valley suites in the low-SiO_2 range, but eastern Las Bancas samples (most from the Mount Laguna pluton, Fig. 6B), are enriched in K_2O relative to western Las Bancas, East Mesa, Alpine, and Japatul Valley samples. The abundance of K_2O in the leucogranite suites (Chiquito Peak, combined Corte Madera–Mother Grundy Peak, and Pine Valley) increases rapidly with increasing SiO_2, and the Chiquito Peak and Pine Valley samples show slight separation from the Corte Madera samples toward higher K_2O values. The Granite Mountain and La Posta suites form a curvilinear trend in which the Granite Mountain samples overlap the Alpine-Japatul Valley field and the La Posta samples overlap the Japatul Valley and leucogranite fields.

The variation of Ba in the Cretaceous granites (Fig. 8) is similar to that of K_2O, except for greater overlap among the Las Bancas–East Mesa and Alpine-Japatul Valley suites. Compared to western Las Bancas samples and samples of the East Mesa, Alpine, and Japatul Valley suites, the eastern Las Bancas samples are slightly enriched in Ba. The Granite Mountain samples are similar to Early to middle Cretaceous tonalites, whereas the La Posta samples are somewhat Ba-enriched compared to Japatul Valley and leucogranite samples having the same SiO_2 contents. In Figure 8, Rb variation defines a series of irregular, positive trends for all suites: the eastern Las Bancas samples are notably enriched in Rb relative to other tonalites, and the leucogranites define a steep array in which >50% of Corte Madera samples plot in a lower-Rb field than the Chiquito Peak and Pine Valley samples.

The Cretaceous suites as a group define smooth, negative linear/curvilinear trends on Harker diagrams for TiO_2, MgO, and P_2O_5 (Fig. 8), as might be expected since all of these elements tend to be concentrated in minerals that formed early (Mg and Ti in mafic minerals; P in apatite). As would be predicted by the lower mafic mineral content of the middle to Late Cretaceous suites, the Granite Mountain suite is lower in MgO than most Early Cretaceous tonalites while the La Posta overlaps the leucogranite field. However, the La Posta suite is distinctly enriched in P_2O_5 and slightly enriched in TiO_2 relative to other suites. Higher P in the La Posta may be reflected by the greater abundance of apatite in eastern-zone plutons as noted by Silver and Chappell (1988).

Although there is some overlap among samples of the Early to middle Cretaceous tonalite suites in Cr contents (Fig. 8), the Las Bancas suite tends to greater Cr enrichment relative to the Alpine suite, as reflected by more abundant pyroxene in Las Bancas plutons; both suites are enriched in Cr relative to the Japatul Valley suite. All but four (eastern) Granite Mountain samples have Cr contents similar to those of the Alpine-Japatul

Valley suites, whereas the La Posta suite overlaps the most silicic samples of the Japatul Valley and leucogranite suites.

Harker plots for Sr show that the La Posta suite is notably enriched in Sr relative to the Early to middle Cretaceous suites, whereas the Granite Mountain suite tends to occupy a field intermediate between the two. Lower Sr values in western samples of the La Posta suite are similar to those of the western Early to middle Cretaceous suites. On the Y-SiO_2 diagram, the Early to middle Cretaceous suites form a diffuse, approximately flat array, whereas the La Posta samples occupy a distinctly lower-Y field and the Granite Mountain samples are generally intermediate between the two. Like the pattern for Y, Pb variation (not shown in Fig. 8) of the Cretaceous suites (Pb ~4–15 ppm) shows no distinct trends with the exception of the La Posta suite, whose samples define a steep, positive linear trend with Pb values rising to 21 ppm.

Zr trends on Harker diagrams (Fig. 8) for both Cretaceous age groups have the convex-upward shape that reflects early concentration of Zr in the melt, followed by Zr saturation and then depletion as zircon crystallizes. Inflection points for both groups lie between 65–70% SiO_2, but that of the combined Early to middle Cretaceous suites is ~65% whereas that of the Granite Mountain and La Posta suites is closer to 70%. Eastern samples of the Las Bancas suite and those of the East Mesa suite are enriched in Zr compared to western Las Bancas, Alpine, and Japatul Valley samples.

West-to-East Chemical Variations

Variation in Fe contents among Cretaceous granitic suites is demonstrated in a plot of Fe_2O_3 versus FeO (Fig. 9), which suggests a significant change in the oxidation state of granitic magmas from west to east across the gravity-magnetic gradient and I-S line described in an earlier section. The magnetite-ilmenite boundary of Gastil (1990a) divides the Peninsular Ranges batholith into a western region of plutons containing magnetite and an eastern region whose plutons are ilmenite-bearing. In the study area, these two regions are shown by Gastil (1990a) as separated by a narrow zone in which both magnetite and ilmenite may be present. Gastil's magnetite-ilmenite boundary overlaps with other north-northwest–trending discontinuities in the Peninsular Ranges batholith and is probably the mineralogical expression of Jachens' (1992) magnetic anomaly gradient.

An FeO/Fe_2O_3 ratio of 3:1 is chosen as the boundary between oxidized and reduced fields because it most closely defines the field of S-type granites in the New England batholith, Australia. That value lies approximately midway between the average FeO/Fe_2O_3 of the magnetite-bearing Moonbi I-type suite and the ilmenite-bearing Uralla I-type suite of the New England batholith (Shaw and Flood, 1981). In that batholith, I-type granites with magnetite ± ilmenite as opaque oxide phases span both oxidized and reduced fields, whereas S-type granites are typically reduced and ilmenite bearing. The Alpine, Japatul Valley, and Corte Madera–Mother Grundy Peak suites, which form plutons in the western zone, typically have ferrous:ferric ratios >3:1 and plot mainly in the oxidized field of the diagram. The

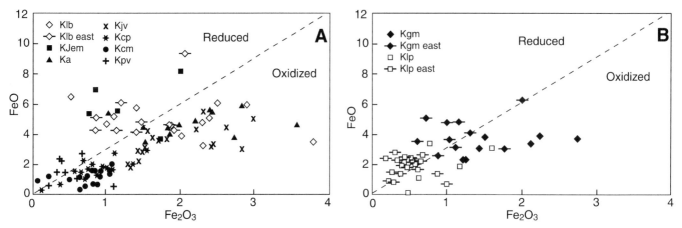

Figure 9. Variation of FeO with Fe_2O_3 in Cretaceous granitic suites. A: Early to middle Cretaceous suites. B: middle to Late Cretaceous suites. Symbols for suites are same as those in Figure 8, except in (B), Kgm east and Klp east are samples from plutons located east of I-S line. Boundary between oxidized (magnetite- ± ilmenite-bearing) and reduced (ilmenite-bearing) fields in diagram is taken as $3:1$ FeO:Fe_2O_3. That value lies approximately midway between average FeO:Fe_2O_3 of magnetite-bearing Moonbi I-type suite (4.03) and ilmenite-bearing Uralla I-type suite (2.18) of New England batholith (Shaw and Flood, 1981).

East Mesa and Pine Valley suites, which are present only in the eastern zone, plot for the most part in the reduced portion of the diagram. Samples from suites that are present in both western and eastern zones (Las Bancas, Granite Mountain, and La Posta) are both oxidized and reduced; except for late differentiates and dikes, samples that crop out in the western zone are for the most part oxidized while those from the eastern zone are reduced. An exception is the Chiquito Peak suite, all of whose samples fall within, or close to, the oxidized field, even though four samples are from plutons located east of the I-S line.

Rare Earth Element Patterns

Rare Earth Element (REE) abundances for 64 powdered rock samples were determined at the U.S. Geological Survey, Reston, using neutron activation methods (Table 4). The values are plotted as standard chondrite normalized patterns (C1, Sun and McDonough, 1989) in Figure 10. In a regional study of the Peninsular Ranges batholith, Gromet and Silver (1987) showed a systematic variation of REE patterns across the batholith, which they divided into western, central, and eastern regions. Their western region coincides approximately with the western zone of this paper, its boundary to the east marked by regional discontinuities such as the $\delta^{18}O$ step.

REE patterns of four gabbros in the study area, two of which lie in the eastern zone (CP-26, D-59), are relatively flat with light rare earth element (LREE) enrichment 5–20 times chondrite (Fig. 10A), similar to the pyroxene gabbros of Gromet and Silver (1987). Europium has a slight to moderate positive anomaly consistent with plagioclase enrichment by cumulate processes.

The Las Bancas suite comprises plutons in the western and eastern zones of the Peninsular Ranges batholith, whereas the East

Mesa is present only in the east. Regardless of their geographic distribution and the variable mafic mineralogy of the suites, the REE patterns of eight samples (Fig. 10B) conform to a remarkably tight field. The overall pattern is 40–70 times enriched in LREE, with the enrichment decreasing to ~20 in the middle rare earth elements (MREE) and heavy rare earth elements (HREE), producing a concave-upward pattern. The removal of hornblende, sphene, and, to a lesser extent, clinopyroxene as cumulate phases would preferentially reduce MREE relative to LREE (Campbell and Gorton, 1980). There is a marked negative Eu anomaly for each sample, suggesting cumulate removal of plagioclase. The REE patterns most closely resemble Gromet and Silver's (1987) western tonalite and quartz diorite.

Seven samples from the Alpine and Japatul Valley suites (Fig. 10C) have a lower overall slope than the Las Bancas–East Mesa pattern, are flatter in MREE and LREE, have no (or only minor) negative Eu anomalies, and have lower LREE enrichments of 15–25. Variable patterns, particularly in MREE and HREE, suggest variable fractionation behavior such as unmixing of hornblende and plagioclase. Alternatively, magma mingling and mixing with gabbro (Fig. 7B) could well explain part of the REE diversity. Gabbro would impart lower REE abundances, flatter slopes, and minor or no Eu anomalies. It is interesting to note that the two specimens with the lowest REE abundances (TS-37 and 4–125–2) have those field characteristics. There do not appear to be any REE patterns equivalent to the Alpine-Japatul Valley rocks in the subdivision of Gromet and Silver (1987).

The leucogranite suites are present in both western and eastern zones of the batholith. The mainly western Chiquito Peak and Corte Madera suites are spatially associated with one another and with the more mafic Las Bancas, Alpine, and Japatul Valley suites in zoned complexes, whereas the Pine Valley suite is

TABLE 4. RARE EARTH ELEMENT ABUNDANCES OF SELECTED CRETACEOUS PLUTONIC ROCKS, SAN DIEGO COUNTY, CALIFORNIA

Plutonic suite:	Las Bancas					East Mesa			Alpine				Japatul Valley		
Sample no.:	A61	D28A	ECM1	ML12	ML13	VM89	CP117A	TS23A	41252	A46	TS37	VM75	EC1	VM3	TS17
Latitude:	32°49'31"	32°49'22"	32°52'57"	32°47'32"	32°48'43"	32°49'02"	32°54'30"	32°54'53"	32°57'03"	32°46'13"	32°54'16"	32°50'06"	32°48'41"	32°46'03"	32°57'20"
Longitude:	116°47'34"	116°35'43"	116°48'43"	116°23'35"	116°24'12"	116°44'12"	116°30'34"	116°36'52"	116°45'33"	116°49'43"	116°38'42"	116°44'59"	116°52'32"	116°41'38"	116°39'32"
La*	12.48	14.40	15.22	20.70	16.18	18.60	16.94	16.90	10.20	8.47	9.87	10.99	9.51	8.58	9.88
Ce	26.34	32.50	32.36	43.60	34.26	38.70	34.52	34.80	18.60	20.80	18.41	22.97	19.33	20.18	20.40
Nd	14.22	19.60	16.77	21.30	18.16	19.50	18.59	19.90	9.71	14.70	8.83	11.56	11.03	13.26	11.70
Sm	3.99	5.41	4.53	5.08	4.41	5.20	4.72	5.46	2.49	4.66	2.27	3.47	3.40	3.92	3.23
Eu	0.96	1.00	0.91	0.89	0.96	0.93	1.00	0.97	0.71	1.07	0.75	0.83	0.83	1.00	0.87
Tb	0.66	0.91	0.70	0.75	0.66	0.87	0.75	0.98	0.46	0.82	0.40	0.61	0.63	0.68	0.55
Yb	2.69	3.34	2.90	2.77	2.22	3.36	2.86	3.60	1.65	2.46	1.56	2.50	3.01	2.96	2.15
Lu	0.40	0.51	0.43	0.36	0.34	0.51	0.40	0.52	0.25	0.39	0.25	0.36	0.46	0.42	0.32
Abundance with respect to chondrite															
La	37.82	43.64	46.12	62.73	49.03	56.36	51.33	51.21	30.91	25.67	29.91	33.30	28.82	26.00	29.94
Ce	29.93	36.93	36.77	49.55	38.93	43.98	39.23	39.55	21.14	23.64	20.92	26.10	21.97	22.93	23.18
Nd	23.70	32.67	27.95	35.50	30.27	32.50	30.98	33.17	16.18	24.50	14.72	19.27	18.38	22.10	19.50
Sm	22.04	29.89	25.03	28.07	24.36	28.73	26.08	30.17	13.76	25.75	12.54	19.17	18.78	21.66	17.85
Eu	13.91	14.46	13.19	12.93	13.91	13.52	14.49	13.99	10.33	15.51	10.87	12.03	12.03	14.49	12.67
Gd (calculated)	14.16	20.00	14.97	16.28	14.63	18.93	16.29	21.29	10.15	18.43	8.65	13.07	13.08	14.40	11.79
Tb	14.04	19.45	14.89	15.87	14.04	18.57	15.96	20.74	9.83	17.40	8.51	12.98	13.40	14.47	11.62
Yb	13.45	16.70	14.50	13.85	11.10	16.80	14.30	18.00	8.25	12.30	7.80	12.50	15.05	14.80	10.75
Lu	11.76	15.06	12.65	10.59	10.00	14.88	11.76	15.24	7.29	11.41	7.35	10.59	13.53	12.35	9.32

(continued)

TABLE 4. RARE EARTH ELEMENT ABUNDANCES OF SELECTED CRETACEOUS PLUTONIC ROCKS, SAN DIEGO COUNTY, CALIFORNIA (continued)

Plutonic suite:	Chiquito Peak			Corte Madera				Pine Valley			Cuyamaca Gabbro			
Sample no.:	CP101	VM5	VM69	D6216	ECM2	VM77	WD49	CP25	D58	WD80A	CP26	D59	VM40	A72
Latitude:	32°57'46"	32°50'29"	32°50'16"	32°50'29"	32°53'32"	32°47'25"	32°45'52"	32°55'48"	32°49'02"	32°50'10"	32°55'48"	32°49'26"	32°45'02"	32°47'33"
Longitude:	116°34'10"	116°39'13"	116°39'35"	116°36'33"	116°49'26"	116°39'38"	116°34'44"	116°36'15"	116°31'01"	116°32'09"	116°36'15"	116°30'14"	116°40'19"	116°51'23"
La	22.00	28.33	27.30	19.20	14.91	25.10	20.13	24.42	27.00	25.00	1.33	4.32	3.72	8.13
Ce	42.00	56.46	54.10	34.70	26.13	47.30	36.34	44.28	48.10	44.10	2.96	11.70	7.99	15.99
Nd	17.30	23.06	22.80	15.80	9.81	18.70	15.62	19.09	17.70	16.20	1.77	7.38	5.41	7.25
Sm	4.51	5.87	5.96	3.84	2.56	4.14	3.42	4.32	3.54	3.56	0.61	2.92	1.56	1.63
Eu	0.44	0.62	0.76	0.78	0.59	0.20	0.36	0.50	0.62	0.75	0.36	0.99	0.65	0.59
Tb	0.98	0.97	0.96	0.65	0.41	0.58	0.52	0.68	0.45	0.47	0.11	0.53	0.29	0.29
Yb	4.32	4.22	3.95	3.25	2.15	2.05	2.07	2.87	1.77	1.61	0.46	1.97	1.12	1.28
Lu	0.63	0.64	0.58	0.48	0.33	0.28	0.30	0.43	0.24	0.25	0.05	0.28	0.16	0.21
Abundance with respect to chondrite														
La	66.67	85.85	82.73	58.18	45.18	76.06	61.00	74.00	81.82	75.76	4.03	13.09	11.27	24.64
Ce	47.73	64.16	61.48	39.43	29.69	53.75	41.30	50.32	54.66	50.11	3.36	13.30	9.08	18.17
Nd	28.83	38.43	38.00	26.33	16.35	31.17	26.03	31.82	29.50	27.00	2.95	12.30	9.02	12.08
Sm	24.92	32.43	32.93	21.22	14.14	22.87	18.90	23.87	19.56	19.67	3.37	16.13	8.62	9.01
Eu	6.33	8.99	11.07	11.25	8.55	2.84	5.22	7.25	8.93	10.93	5.22	14.29	9.38	8.55
Gd (calculated)	20.65	20.55	20.43	13.32	8.32	12.68	11.21	14.49	9.82	10.31	2.35	11.49	6.31	6.12
Tb	20.81	20.64	20.32	13.81	8.72	12.28	11.06	14.47	9.66	9.94	2.34	11.21	6.19	6.17
Yb	21.60	21.10	19.75	16.25	10.75	10.25	10.35	14.35	8.85	8.05	2.30	9.85	5.60	6.40
Lu	18.41	18.82	16.94	14.21	9.71	8.15	8.82	12.65	7.15	7.29	1.47	8.09	4.74	6.18

(continued)

TABLE 4. RARE EARTH ELEMENT ABUNDANCES OF SELECTED CRETACEOUS PLUTONIC ROCKS, SAN DIEGO COUNTY, CALIFORNIA (continued)

Plutonic suite:	Granite Mountain								La Posta					
Sample no.:	A47	A71	A9	EV3	ML42	SS832	SS833	A3	ACS49	W1	W2	W3	W4	WSP16
Latitude:	32°48'09"	32°47'25"	32°45'48"	33°02'20"	32°43'47"	32°36'28"	32°37'16"	32°46'52"	32°54'37"	32°38'02"	32°39'37"	32°40'28"	32°42'45"	32°52'08"
Longitude:	116°49'13"	116°50'30"	116°47'16"	116°24'32"	116°28'41"	116°28'25"	116°36'46"	116°47'43"	116°19'10"	116°08'47"	116°05'38"	116°16'52"	116°22'50"	116°13'50"
La	10.60	7.01	8.53	40.80	8.55	97.20	12.56	15.18	24.60	24.60	29.06	17.49	23.40	21.05
Ce	25.30	17.00	18.82	76.20	19.78	164.00	25.84	28.62	47.10	47.10	54.92	34.66	42.60	39.77
Nd	15.80	13.00	11.40	29.60	11.70	45.20	12.77	12.29	21.00	18.70	23.43	15.46	15.30	17.68
Sm	4.17	3.72	3.64	6.43	3.41	6.72	3.48	2.53	4.33	4.04	4.39	2.90	2.33	3.56
Eu	0.89	0.94	0.89	1.10	0.91	0.87	0.79	0.61	1.01	1.06	1.04	0.78	0.85	1.07
Tb	0.59	0.55	0.60	0.86	0.44	0.69	0.51	0.32	0.35	0.39	0.38	0.24	0.17	0.32
Yb	2.01	1.73	1.89	2.21	1.30	2.61	1.86	1.32	0.71	0.99	0.75	0.42	0.60	0.57
Lu	0.30	0.25	0.26	0.31	0.19	0.38	0.29	0.20	0.10	0.14	0.09	0.08	0.08	0.09
Abundance with respect to chondrite														
La	32.12	21.24	25.85	123.64	25.91	294.55	38.06	46.00	74.55	74.55	88.06	53.00	70.91	63.79
Ce	28.75	19.32	21.39	86.59	22.48	186.36	29.36	32.52	53.52	53.52	62.41	39.39	48.41	45.19
Nd	26.33	21.67	19.00	49.33	19.50	75.33	21.28	20.48	35.00	31.17	39.05	25.77	25.50	29.47
Sm	23.04	20.55	20.11	35.52	18.84	37.13	19.23	13.98	23.92	22.32	24.25	16.02	12.87	19.67
Eu	12.91	13.58	12.90	15.94	13.19	12.57	11.45	8.84	14.64	15.36	15.07	11.30	12.33	15.51
Gd (calculated)	13.03	12.39	13.43	19.82	9.93	15.11	11.16	6.85	8.30	8.99	8.95	5.71	3.82	7.60
Tb	12.53	11.77	12.77	18.36	9.36	14.77	10.85	6.81	7.51	8.32	8.09	5.11	3.68	6.81
Yb	10.05	8.65	9.45	11.05	6.50	13.05	9.30	6.60	3.55	4.95	3.75	2.10	3.01	2.85
Lu	8.71	7.35	7.65	9.03	5.59	11.09	8.53	5.88	2.81	3.97	2.65	2.35	2.26	2.65

Notes: Rare earth element in ppm determined by I.N.A.A. method by G.A. Wandless, U.S. Geological Survey, Reston, Virginia.
*X-axis position on figure 11: La, 1.00; Ce, 2.00; Nd, 4.00; Sm, 6.00; Eu, 7.00; Gd (calculated), 8.00; Tb, 9.00; Yb, 14.00; Lu, 15.00.

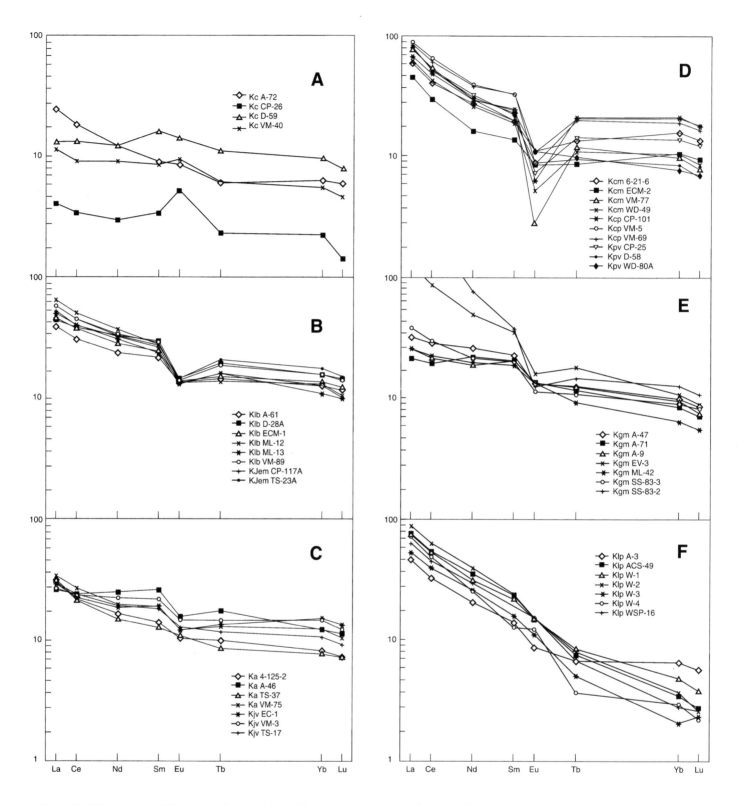

Figure 10. REE patterns of Cretaceous plutonic suites of Peninsular Ranges batholith normalized to chondrite. A: Various gabbros (Kc), including pyroxene and hornblende gabbros. B: Las Bancas (Klb)–East Mesa (KJem) (inferred pyroxene-plagioclase source). C: Alpine (Ka)-Japatul Valley (Kjv) (hornblende-plagioclase source). D: leucogranite suites (Corte Madera, Kcm; Chiquito Peak, Kcp; Pine Valley, Kpv) (removal of plagioclase by fractionation). E: Granite Mountain (Kgm) (transitional hornblende-plagioclase to garnet-bearing source). F: La Posta (Klp) (garnet-bearing source, plagioclase-poor or plagioclase-absent).

entirely eastern. Their REE patterns (Fig. 10D) show significant differences in absolute REE abundance, though all are enriched in LREE, 40–90 times chondrite. The middle to heavy REE slopes are lower than in either the Las Bancas–East Mesa or Alpine-Japatul Valley patterns, although, in common with the Las Bancas–East Mesa, the overall pattern is concave-upward. All samples have a negative Eu anomaly and, in several, the anomaly is very large. The closest equivalents of these leucogranite suites are the leucogranodiorite and low-K granodiorite of Gromet and Silver (1987). These authors point out that late leucocratic melt fractionates could contain most of the REE in the system, particularly the minerals allanite, zircon, apatite, and sphene. Therefore, if the early crystallizing phases such as plagioclase and biotite, which act as REE dilutants, are removed, there will be an overall increase in REE abundance without necessarily changing other aspects of the pattern.

The Granite Mountain and La Posta suites intrude both western and eastern zones of the batholith. Compared with the Early Cretaceous suites, both are higher in Sr and the La Posta is slightly enriched in Ba; Sr appears to increase from west to east in the La Posta. The westernmost Granite Mountain REE patterns (A-47, A-71, A-9) are most like the patterns of the Alpine suite, but further to the east the LREE abundance increases from a modest 20 times chondrite to over 300 (Fig. 10E). Other changes include a noticeable increase in a negative Eu anomaly and an overall steeper slope. The La Posta suite mirrors the change of increasing LREE abundance from west to east but differs from the Granite Mountain in having much steeper slopes and more depleted HREE (Fig. 10F). These patterns correspond to the eastern tonalites, low-K granodiorite, and granodiorite and monzogranite of Gromet and Silver (1987).

To have such steep slopes and significant LREE to HREE enrichment in some Granite Mountain and in all La Posta samples, garnet must be a stable restite phase at the zone of magma generation. Although not directly affecting the LREE, garnet removal will cause the melt to increase in LREE relative to HREE. High Ca and Sr in the melt would further suggest that plagioclase was not stable, indicating the melting of source rocks in the region encompassing amphibolite-granulite-eclogite mineralogies. Gromet and Silver (1987) consider a garnet-bearing, plagioclase-poor eclogitic assemblage as an appropriate source rock for the eastern region, i.e., for the eastern Granite Mountain and La Posta suites. We believe that these suites in the western region formed from similar source-rock compositions but at depths less than in the eastern zone, where plagioclase was more stable and garnet less stable.

Isotopic Data

Rb, Sr, and O isotopic data were determined for 35 samples from the 10 Cretaceous granitic suites described in this paper and from two prebatholithic rocks (Table 5). In addition, K-feldspar and/or plagioclase separates from 20 plutons, both Cretaceous and Jurassic, and from two prebatholithic rocks, were analyzed for common Pb (Table 6).

Rb-Sr-O

Published initial $^{87}Sr/^{86}Sr$ ratios (Sr_i) of Cretaceous plutons of the Peninsular Ranges batholith in general display a uniform distribution parallel to its axis and a progressive eastward increase from ratios <0.704 on the west to those as high as 0.708 at the eastern margin of the batholith (Silver and Chappell, 1988; J.L. Wooden, personal commun., 1995). Isopleths of $^{18}O/^{16}O$ ($\delta^{18}O$) of Cretaceous plutons exhibit a similar parallelism to the axis of the batholith, ranging from +6 to +8.5 per mil on the west to +9 to +12 on the east (Taylor and Silver, 1978). The Jurassic plutons of the study area have calculated Sr_i and $\delta^{18}O$ values that are substantially higher than those reported for most Cretaceous plutons in the Peninsular Ranges batholith (Shaw et al., this volume, Chapter 7).

Figure 11 is a plot of Sr_i versus $\delta^{18}O$ for 35 plutonic rocks representing all of the Cretaceous suites of this report plus two western metavolcanic rocks. The samples define a roughly linear positive trend with all but four samples having Sr_i values <0.705 and $\delta^{18}O$ values <9.5. The four plutonic samples with Sr_i between 0.705 and 0.706 and $\delta^{18}O$ between 10.0 and 11.0 are from the regionally extensive, Late Cretaceous La Posta pluton in the eastern part of the study area. In addition to the west-to-east geochemical variations described above for some Cretaceous suites, the pattern of Sr and O isotopic ratios also reflects the east-west geographic position of the samples. Those samples having the lowest Sr_i and $\delta^{18}O$ values ($Sr_i \sim< 0.7042$ and $\delta^{18}O < 7.5$) include the Alpine, Corte Madera, Cuyamaca Gabbro, western Las Bancas, three of four Japatul Valley, and two western metavolcanic samples. Two gabbro samples having $Sr_i > 0.704$ are from the eastern zone. An intermediate group of samples with relatively low Sr_i ($\sim<0.7047$) but $\delta^{18}O > 7.5$ were all collected from plutons located in the vicinity of, or east of, the I-S line: the single East Mesa sample, two eastern Las Bancas samples, the easternmost Japatul Valley sample, two eastern Chiquito Peak samples, and two Pine Valley samples. Of four Granite Mountain samples, the two with the lowest Sr_i and $\delta^{18}O$ values are from the westernmost Granite Mountain pluton; likewise, the lowest La Posta Sr and O isotopic ratios are for two samples from the Loveland Reservoir pluton in the western zone (Fig. 6E).

Pb isotopes

Common lead isotopic compositions of potassium feldspar and/or plagioclase from 15 Cretaceous plutons and five Jurassic plutons were analyzed and the data were plotted on a $^{207}Pb/^{204}Pb$ versus $^{206}Pb/^{204}Pb$ diagram, along with data from a single western metavolcanic rock and a metasandstone from the Julian Schist (Table 6, Fig. 12). Plotted together, the Peninsular Ranges batholith samples define a roughly linear, positive array in which 207/204 ranges from ~15.60 to 15.68 and 206/204 from ~18.65 to 18.95. Whereas the five Jurassic plutonic samples are among the most radiogenic, there is overlap between them and the most radiogenic Cretaceous plutons, the latter including an East Mesa sample, a sample from the easternmost Las Bancas–type pluton, and three eastern La Posta samples (Fig. 12). The less radiogenic

TABLE 5. Rb-Sr AND OXYGEN ISOTOPIC DATA FOR CRETACEOUS PLUTONS AND METAVOLCANIC ROCKS, SAN DIEGO COUNTY, CALIFORNIA

Sample no.	Latitude (N)	Longitude (W)	Plutonic suite	Lithology	Rb (ppm)	Sr (ppm)	Rb/Sr	^{87}Rb/^{86}Sr	^{87}Sr/^{86}Sr	Initial ^{87}Sr/^{86}Sr*	δ^{18}O (rock)	δ^{18}O (quartz)
A2	32°49'13"	116°46'34"	Las Bancas	tonalite	64.6	239.1	0.27	0.78225	0.70530	0.70402	7.4	
A61	32°49'31"	116°47'34"	"	quartz diorite	45.3	262.9	0.17	0.49885	0.70454	0.70373	7.0	
ECM1	32°52'57"	116°48'43"	"	tonalite	74.7	227.9	0.33	0.94902	0.70546	0.70391	7.2	
ML12	32°47'32"	116°23'35"	"	granodiorite	122.2	243.3	0.50	1.45443	0.70698	0.70461	8.2	
ML13	32°48'43"	116°24'12"	"	tonalite	78.5	302.9	0.26	0.75566	0.70551	0.70428	8.1	
CP117	32°54'30"	116°30'34"	East Mesa	quartz diorite	63.7	297.2	0.21	0.62057	0.70547	0.70446	9.0	10.6
A46	32°46'13"	116°49'43"	Alpine	tonalite	21.8	391.5	0.06	0.16120	0.70395	0.70369	6.7	
VM75	32°50'06"	116°44'59"	"	tonalite	54.2	248.6	0.22	0.63118	0.70439	0.70336	7.1	
TS37	32°54'16"	116°38'42"	"	tonalite	36.0	262.8	0.14	0.39657	0.70425	0.70360	6.5	
VM3	32°46'03"	116°41'38"	Japatul Valley	tonalite	39.3	259.6	0.15	0.43827	0.70433	0.70361	6.5	
EC1	32°48'41"	116°52'32"	"	granodiorite	59.4	188.4	0.32	0.91282	0.70500	0.70351	6.7	
VM39	32°46'19"	116°40'16"	"	tonalite	27.0	255.6	0.11	0.30581	0.70441	0.70391	6.4	
CP49	32°53'56"	116°33'22"	"	tonalite	54.0	328.0	0.16	0.48100	0.70543	0.70463	8.7	
VM5	32°50'29"	116°39'13"	Chiquito Peak	monzogranite	197.0	117.2	1.68	4.86977	0.71184	0.70388	8.7	9.9
VM69	32°50'16"	116°39'35"	"	granodiorite	149.3	137.9	1.08	3.13579	0.70907	0.70395	8.1	
ECM2	32°53'32"	116°49'26"	Corte Madera	granodiorite	73.7	158.1	0.47	1.43931	0.70642	0.70407	7.3	
WD49	32°45'52"	116°34'44"	"	monzogranite	91.2	75.4	1.21	3.50342	0.70948	0.70375	7.2	
CP25	32°55'50"	116°36'17"	Pine Valley	monzogranite	136.9	126.8	1.08	3.12720	0.70958	0.70447	9.0	9.9
D58	32°49'02"	116°31'01"	"	monzogranite	86.7	122.4	0.71	2.05116	0.70693	0.70358	8.6	
A9	32°45'47"	116°47'16"	Granite Mountain	tonalite	20.7	417.9	0.05	0.14339	0.70391	0.70370	6.6	
A47	32°48'09"	116°49'13"	"	tonalite	30.0	342.4	0.09	0.25365	0.70420	0.70382	7.0	
ML42	32°43'47"	116°28'41"	"	tonalite	33.2	456.9	0.07	0.21036	0.70422	0.70391	7.8	
SS833	32°37'16"	116°36'46"	"	tonalite	47.3	313.7	0.15	0.43647	0.70467	0.70402	7.3	
SS832	32°36'28"	116°28'25"	La Posta	granodiorite	139.1	189.8	0.73	2.12293	0.70743	0.70397	N.D.#	
A3	32°46'52"	116°47'43"	"	granodiorite	76.9	213.5	0.36	1.04285	0.70539	0.70383	7.6	
A48	32°48'12"	116°49'03"	"	tonalite	43.9	264.6	0.17	0.47987	0.70466	0.70394	7.2	
W2	32°39'37"	116°05'38"	"	tonalite	44.1	543.0	0.08	0.23516	0.70598	0.70563	10.4	
W3	32°40'28"	116°16'52"	"	granodiorite	56.3	452.7	0.12	0.36009	0.70578	0.70524	10.9	12.8
W4	32°42'45"	116°22'50"	"	granodiorite	60.9	491.4	0.12	0.35884	0.70581	0.70527	10.7	
WSP16	32°52'08"	116°13'50"	"	tonalite	34.9	646.5	0.05	0.15630	0.70564	0.70541	11.0	12.7
A28	32°49'40"	116°50'07"	Cuyamaca Gabbro	hornblende gabbro	26.9	270.1	0.10	0.28832	0.70409	0.70362	6.1	
CP26	32°55'50"	116°36'17"	"	gabbronorite	3.0	446.3	0.01	0.01946	0.70410	0.70407	6.8	
D59	32°49'26"	116°30'14"	"	hornblende gabbro	3.3	371.9	0.01	0.02569	0.70387	0.70383	6.6	
VM40	32°45'02"	116°40'19"	"	gabbronorite	6.5	357.4	0.02	0.05281	0.70417	0.70408	6.8	
A72	32°47'33"	116°51'23"	"	hornblende gabbro	27.7	292.9	0.09	0.27378	0.70413	0.70368	6.9	
TS45	32°55'02"	116°39'02"	Metavolcanic	amphibolite	1.8	143.0	0.01	0.03610	0.70392	0.70386	5.3	
1079C1	32°49'35"	116°51'44"	"	amphibolite	1.5	138.0	0.01	0.03168	0.70387	0.70382	6.1	

Note: Rubidium-strontium isotopic analyses performed by J.A. Cooper at the University of Adelaide, Adelaide, Australia. Oxygen isotope determinations were made by J.R. O'Neil at U.S. Geological Survey, Menlo Park, California.

*Based upon age of 115 Ma except for Granite Mountain and La Posta suites, which are based on 105 Ma.

#N.D.—not determined.

TABLE 6. FELDSPAR-Pb ISOTOPIC DATA FOR PLUTONS AND PREBATHOLITHIC ROCKS,
SAN DIEGO COUNTY, CALIFORNIA

Sample No.	Latitude (N)	Longitude (W)	Plutonic suite	Lithology	Mineral analyzed	206/204	207/204	208/204
				Jurassic plutons				
J38	33°05'26"	116°34'48"	Cuyamaca Reservoir	granodiorite	K-feldspar	18.846	15.665	38.665
ML5	32°51'15"	116°28'32"	"	granodiorite	K-feldspar	18.875	15.673	38.787
TS55	33°00'10"	116°42'29"	"	tonalite	Plagioclase	18.882	15.638	38.689
J40	33°05'33"	116°35'41"	Harper Creek	granodiorite	K-feldspar	18.824	15.653	38.738
MP35	32°56'19"	116°29'03"	"	granodiorite	K-feldspar	18.796	15.655	38.647
				Cretaceous plutons				
A2*	32°49'13"	116°46'34"	Las Bancas	tonalite	K-feldspar	18.737	15.637	38.525
					Plagioclase	18.785	15.644	38.565
ML12	32°47'32"	116°23'35"	"	granodiorite	K-feldspar	18.791	15.635	38.585
15M65A	32°40'27"	116°11'32"	"	quartz norite	K-feldspar	18.875	15.667	38.760
					Plagioclase	18.865	15.658	38.704
CP117	32°54'30"	116°30'34"	East Mesa	quartz diorite	Plagioclase	18.828	15.667	38.697
EC1	32°48'41"	116°52'32"	Japatul Valley	granodiorite	K-feldspar	18.724	15.621	38.465
SS831	32°55'12"	116°52'52"	Chiquito Peak	monzogranite	K-feldspar	18.746	15.624	38.501
VM69	32°5016"	116°39'35"	"	granodiorite	K-feldspar	18.763	15.640	38.573
D58	32°49'02"	116°31'01"	Pine Valley	monzogranite	K-feldspar	18.709	15.618	38.506
SY3	33°05'56"	116°39'47"	"	monzogranite	K-feldspar	18.792	15.632	38.590
SS833	32°37'16"	116°36'46"	Granite Mountain	tonalite	K-feldspar	18.725	15.626	38.503
SS832	32°36'28"	116°28'25"	"	granodiorite	K-feldspar	18.755	15.615	38.500
A3	32°46'52"	116°47'43"	La Posta	granodiorite	K-feldspar	18.688	15.601	38.410
W4	32°42'45"	116°22'50"	"	granodiorite	K-feldspar	18.867	15.659	38.695
ACS54	32°53'50"	116°15'31"	"	granodiorite	K-feldspar	18.886	15.666	38.726
					Plagioclase	18.875	15.648	38.671
5848	32°45'53"	116°10'00"	"	leucomonzogranite	K-feldspar	18.911	15.649	38.694
					Plagioclase	18.955	15.661	38.770
				Prebatholithic rocks				
1079B	32°51'16"	116°51'05"	Metavolcanic	dacitic tuff-breccia	K-feldspar	18.770	15.632	38.545
58410	33°03'47"	116°34'06"	Julian Schist	quartzo-feldspathic semischist	K-feldspar	18.749	15.650	38.702

Note: Lead isotopic analyses by J.L. Wooden, U.S. Geological Survey, Menlo Park, California.
*Both K-feldspar and plagioclase analyzed for samples A2,15M65A, ACS54, and 5848.

Cretaceous group contains samples of the Las Bancas, Japatul Valley, Chiquito Peak, Pine Valley, and Granite Mountain suites, along with the single metavolcanic rock; the least radiogenic sample was collected from the La Posta pluton at Loveland Reservoir in the western zone.

Discussion of Isotopic Data

Initial $^{87}Sr/^{86}Sr$ ratios and $\delta^{18}O$ values for Cretaceous suites in southern San Diego County show a general correspondence with the geographic position of plutons west or east of the western margin of the Jurassic–Triassic plutonic/metamorphic terrain (I-S line), which we infer to mark the trace of a major latest Jurassic–earliest Cretaceous fault. In general, the trend of all Cretaceous suites is toward high $\delta^{18}O$ but not toward high Sr_i (<0.706). A number of the western-zone, Early to middle Cretaceous plutons can be characterized as having low Sr_i (<0.7038) and low $\delta^{18}O$ (<7.0), which indicates a primitive mantle source. However, an equal number have Sr_i > 0.7038 and $\delta^{18}O$ > 7.0,

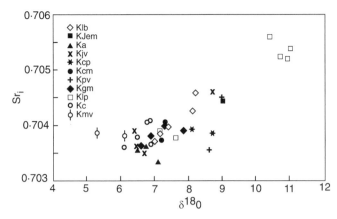

Figure 11. Plot of Sr$_i$ versus $\delta^{18}O$ for Cretaceous plutonic suites, Peninsular Ranges batholith. Symbols for suites same as Figure 8 except for Kc (Cuyamaca Gabbro) and Kmv (Cretaceous? metavolcanic rock).

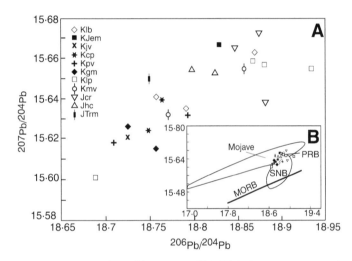

Figure 12. Plot of $^{207}Pb/^{204}Pb$ versus $^{206}Pb/^{204}Pb$ for Cretaceous and Jurassic plutonic suites, Peninsular Ranges batholith. Symbols same as Figure 11 except for Jcr (Cuyamaca Reservoir suite), Jhc (Harper Creek suite), and JTrm (Julian Schist).

suggesting limited mixing of mantle material with Mesozoic supracrustal material (cf. De Paolo, 1981). Plutons that intruded into, or east of, the I-S line apparently underwent a greater degree of mixing with more evolved crust.

Feldspar common Pb isotopic ratios of granitic suites indicate little difference between western- and eastern-zone Cretaceous plutons; with the exception of A-3 (from the westernmost La Posta pluton), all Cretaceous samples are somewhat radiogenic ($^{206}Pb/^{204}Pb > 18.7$, J.L. Wooden, 1985, personal commun.). This suggests derivation from older igneous rocks whose source may have been mantle melts that mixed with continental lead, possibly during early Mesozoic subduction/accretion of continentally derived sediments. Only the La Posta and, possibly, the Las Bancas Pb samples show a correlation with east-west position in the batholith.

PETROGENESIS OF CRETACEOUS GRANITIC SUITES

The western margin of North America was the site of subduction and continental-arc magmatism for much of the Mesozoic (Lipman, 1992). In southeastern California, a Triassic-Jurassic arc was constructed on the truncated southwestern margin of the North American craton, across cratonal and Paleozoic miogeoclinal settings (Tosdal et al., 1989; Barth et al., 1990; Busby-Spera et al., 1990). Shaw et al. (this volume, Chapter 7) describe Middle and Late Jurassic magmatism that occurred oceanward of the Jurassic Mojave–Sonoran continental-margin arc in the site of the present-day Peninsular Ranges batholith. Subduction and arc magmatism resumed in the Early Cretaceous in the Peninsular Ranges region and continued into the Late Cretaceous as the new arc migrated eastward across the Late Jurassic–Early Cretaceous continental margin (Silver et al., 1979; Gastil et al., 1981; Todd et al., 1988; Walawender et al., 1990).

Geochemical and isotopic data indicate a significant mantle component in the source region of Cordilleran magmatic arcs, and current models for granite genesis agree that magmas are derived by partial melting of oceanic and/or continental subarc lower crustal rocks and/or underplated basalt and gabbro. The common occurrence of magma mingling and synplutonic mafic dikes in granite batholiths suggests that large bodies of basaltic to andesitic magma may underlie silicic magma chambers, providing heat for melting and the potential for mixing (Vernon, 1983). Ascending magmas may be modified by assimilation of middle to upper crustal materials, whereas fractional crystallization during magma ascent and cooling also contributes to the variety of granite compositions (Chappell and White, 1992; Collins, 1998).

Petrogenetic models for the Peninsular Ranges batholith based upon mixing of two or more source components (DePaolo, 1981; Hill et al., 1986; Shaw et al., 1986; Silver and Chappell, 1988; Walawender et al., 1991) differ as to the relative contributions of mantle and crustal components. Our study indicates that by the onset of Cretaceous magmatism, the crust in the site of the present-day Peninsular Ranges of southern California was heterogeneous: the western crustal region was mainly composed of young, mantle-derived igneous rocks, whereas the eastern region was predominantly made up of early Mesozoic igneous and sedimentary rocks, with admixtures of Late Proterozoic cratonal material. On the west side, earlier accreted fringing arc-ophiolite systems, volcaniclastic aprons, and subduction complexes (Gastil et al., 1981; Busby-Spera, 1988; Saleeby and Busby-Spera, 1992; Busby and Saleeby, 1993) constituted basement for a Late Jurassic island arc. On the east side, Middle and Late Jurassic S-type, transitional I-S–type, and I-type plutons intruded, and were in part derived from, deposits of an early Mesozoic forearc basin(s) that had formed in an arc-ophiolite setting. These fore-arc deposits included Late Proterozoic–Paleozoic detritus derived from sialic basement highlands to the east plus intercalated volcanic and shallow-plutonic fringing-arc rocks. Thus, with the exception of the easternmost Peninsular Ranges batholith, which may be underlain in part by North American craton

(Gastil, 1993; Grove, 1993), lower crustal source regions for eastern Cretaceous magmas were primarily young and igneous. Although lithologic differences between the western and eastern crustal regions apparently had significant effects on some mineralogical and geochemical characteristics of the Cretaceous granitic magmas, the many shared chemical and isotopic characteristics of these magmas must reflect similar lower crustal source rocks on both sides of the Late Jurassic continental margin.

Las Bancas and East Mesa Suites

The Las Bancas suite forms large, homogeneous, restite-free plutons in the western zone of the Peninsular Ranges batholith where it is spatially associated with the Alpine and Japatul Valley suites. The Las Bancas suite includes the most mafic granitic plutons of the study area, has orthopyroxene > hornblende, scarce cumulate layering, and small volumes of pyroxene granodiorite. Las Bancas plutons are present in both western and eastern zones of the batholith, which suggests partial melting of a widespread and uniform preexisting basalt-gabbro underplate. The available ages of Las Bancas plutons (Fig. 6B) suggest that progressively younger magmas intruded from west to east during Cretaceous magmatism.

With the exception of a small area ~8 km south-southeast of the town of Pine Valley, plutons of the Las Bancas and East Mesa suites are mutually exclusive. Fine grain size and porphyritic textures, along with the sheet- and dike-like aspects of many East Mesa bodies, suggest that some East Mesa magma was emplaced into the Cuyamaca–Laguna Mountains shear zone during a period of intra-arc extension. Complex internal contacts within most plutons further suggest that (1) movements in the shear zone created pressure shadows that attracted partial melt, (2) melt partly crystallized, and (3) renewed movements repeated these processes. The interiors of two larger East Mesa plutons consist of biotite-hypersthene-hornblende tonalite that is finer-grained than, but otherwise lithologically similar to, tonalite of the Las Bancas suite. Early Cretaceous ages for two East Mesa plutons and REE patterns of two samples that fit closely with those of the Las Bancas suite (Fig.10B) indicate that some East Mesa plutons are the result of intrusion of Las Bancas magma into the Cuyamaca–Laguna Mountains shear zone. The fact that one East Mesa pluton has yielded a Middle Jurassic age (Shaw et al., this volume, Chapter 7) indicates that subarc source materials of Las Bancas-type were also present during Jurassic magmatism.

Alpine and Japatul Valley Suites

As discussed above, plutons of the Las Bancas suite are spatially associated with, and chemically similar to, the most mafic rocks of the Alpine and Japatul Valley suites. However, the Alpine and Japatul Valley suites are restricted to the western zone of the batholith, with the exception of several small Japatul Valley bodies that occur as much as 10 km to the east of the eastern limit of Alpine plutons. Hornblende is the dominant mafic mineral in plutons of the Alpine and Japatul Valley suites, and plutons of

both suites are heterogeneous due to abundant mafic enclaves and scattered restitic(?) inclusions of granitic- and volcanic-rock fragments and quartz xenocrysts. The Japatul Valley suite generally comprises the outermost tonalite pluton, or sheet, of western zoned plutonic complexes and has extensive stoping and assimilation-contamination reactions with adjacent metavolcanic screens.

The restriction of the Alpine and Japatul Valley suites to the western zone indicates derivation of the Alpine parental magma by partial melting of amphibolitic lower crustal rocks in the western lithospheric region. Gradational contacts and systematic geochemical variations between Alpine and Japatul Valley plutons suggest that the Japatul Valley magma was derived by fractional crystallization of parental magma of Alpine type. Abundant evidence of mingling between Alpine magma and the basaltic magma that gave rise to the Cuyamaca Gabbro, together with variable Alpine-Japatul Valley REE patterns (Fig. 10C), imply that interactions between tonalitic and basaltic magmas at greater depths may have significantly altered the Alpine and Japatul Valley magmas. The close spatial and temporal association of these tonalite suites with the Las Bancas suite further suggests that their magmas formed from similar souce materials but at different crustal levels, temperatures, or p_{H_2O} of melting: i.e., the Alpine and Japatul Valley magmas formed at lower temperatures where amphibole was a stable phase, and the Las Bancas at higher temperatures where pyroxene was stable.

Leucogranite Suites

The Chiquito Peak, Corte Madera, Mother Grundy Peak, and Pine Valley suites are characterized by high SiO_2, low mafic mineral content with biotite > hornblende, and, in comparison to western-zone tonalites, they are moderately enriched in alkalies (K, Ba, Rb). All four suites intrude tonalitic plutons of the study area and the Corte Madera and Pine Valley suites intrude the Chiquito Peak. The leucogranite suites are weakly peraluminous with few samples having ASI values >1.1 (Table 2). Their peraluminous character is considered to result from fractionation of parental I-type (metaluminous) magmas; therefore, the leucogranites are not S-type granites. Major- and trace-element variations of the suites are similar, with the possible exception of the combined Corte Madera–Mother Grundy Peak suites (see below), whose samples form a tighter grouping than those of the other leucogranites and which, in the high-SiO_2 range (73–77%), may have slightly lower K_2O and Rb. Three leucogranite suites have similar enriched REE abundances with large negative Eu anomalies, patterns that are characteristic of fractionation processes (Fig. 10D). Since all three approach minimum-melt compositions and yet have different geographic distributions, they cannot be related by fractional crystallization of a single parental magma. However, all four suites have given rise to abundant silicic differentiates, indicating that fractional crystallization was a major process.

The similar geographic distributions of Chiquito Peak and Japatul Valley plutons and local gradational contacts between them suggest a genetic relationship by crystal fractionation of

Alpine-type parental magma. Based upon the 45–60° eastward dip of the inferred latest Jurassic-earliest Cretaceous fault represented by the gravity-magnetic gradient and I-S line, the location of Chiquito Peak and Japatul Valley plutons as much as 10 km east of these boundaries further suggests that western oceanic crust extended for some distance to the east beneath eastern continental crust. Thus, eastern Chiquito Peak plutons rose through continental crust, which might account for the greater compositional range and geochemical heterogeneity of the suite.

We have included the few chemical analyses of Mother Grundy Peak rocks with those of the Corte Madera suite even though the age relationship between the two is unclear. Both suites were emplaced west of the gravity-magnetic boundary, and although they are similar in their field-determined age relations to western-zone tonalitic and gabbroic plutons, the two suites are almost mutually exclusive in outcrop area (Fig. 6D). The exception is the Barber Mountain area (Fig. 6A), where Corte Madera(?) leucogranite dikes intrude a Mother Grundy Peak pluton. The chief difference between the two suites is that the Mother Grundy Peak contains well-crystallized hornblende, whereas in the Corte Madera, hornblende is absent or occurs only as scarce relict grains.

The Corte Madera–Mother Grundy Peak suites are more siliceous (71–78% SiO_2) than the Chiquito Peak or Pine Valley suites, which suggests a less mafic, more homogeneous source and/or a different evolutionary path. Carollo and Walawender (1993) suggest that high SiO_2 in the leucogranites in general may have been due to low degrees of partial melting or hydrothermally introduced silica in the source. Lower K_2O and Rb of the Corte Madera–Mother Grundy Peak suites suggest that the source was already depleted in large ion lithophile elements (LILE) prior to the melting event that gave rise to the parental magmas, possibly by an earlier partial melting event.

The Pine Valley plutons in the eastern zone are predominantly composed of leucomonzogranite and leucogranodiorite ± muscovite ± garnet. However, at least two plutons (the Rattlesnake Valley and Oriflamme Canyon plutons, Fig. 6D, Cameron, 1980; Bracchi et al., 1993) have compositions that range from tonalite through granodiorite and monzogranite to leucogranite. This compositional range, plus the suite's low Sr_i values (Table 5), suggest that the more silicic Pine Valley plutons that we sampled for this study may have originated by fractional crystallization of a basaltic parental magma, e.g., one derived by partial melting of lower crustal amphibolites generated in eastern-zone crust during a pre-Cretaceous melting event.

La Posta and Granite Mountain Suites

Plutons of the Granite Mountain and La Posta suites are present in both western and eastern zones, although both form much smaller plutons in the western zone (Fig. 6E). Similarities such as idiomorphic texture, K-feldspar oikocrysts, and euhedral sphene suggest a close petrogenetic relationship between the two suites (Table 2). Overall, the Granite Mountain is more mafic than the La Posta, and both suites are more mafic in the western zone.

Although many geochemical abundances of the Granite Mountain and La Posta suites overlap those of the other granitic suites, the La Posta suite has significantly higher Sr and lower Y, and the Granite Mountain tends to be intermediate between the La Posta and western granitic suites (Fig. 8). Absolute age determinations suggest an eastward decrease in emplacement age of the La Posta suite by as much as ca. 10 Ma (Fig. 6E).

Field, petrographic, and geochemical similarities between the Granite Mountain and La Posta suites argue for a unique origin for both parental magmas. The La Posta, especially the vast La Posta pluton itself, is extremely homogeneous; this, plus its distinctive chemistry, requires a voluminous and relatively homogeneous basaltic source (see also Walawender et al., 1990). Possible source rocks present beneath both western and eastern zones include lower crustal amphibolites and underplated basalt and gabbro. Rare-earth element patterns suggest that garnet was a restite phase after melting and removal of the La Posta magma from the source (Gromet and Silver, 1987; Fig. 10F). If so, this implies lower crustal melting in the garnet stability field where plagioclase is not stable (i.e., eclogite depths >60 km). At such depths, source rocks would have lost LILE and most volatiles; the magmas would therefore be high-temperature and would not retain restite phases. Garnet retains HREE and the HREE-like element Y, whereas the melting of almost all plagioclase will produce magma with high Ca and Sr. Figure 8 shows a general increase in Y with increasing SiO_2 of ~20–25 ppm for the Granite Mountain suite and a slightly smaller increase for the La Posta suite. No such increase is evident for the Early to middle Cretaceous suites, with the possible exception of the Las Bancas–East Mesa samples. The minerals hornblende, sphene, allanite, and epidote, which have distribution coefficients >1 for Y, should contain most of the element. However, these minerals are also present, even relatively abundant, in most Cretaceous granitic suites. Therefore, it seems likely that low Y in the Granite Mountain and La Posta suites is related to greater depths of melting.

The more mafic character and lower Sr_i and $\delta^{18}O$ of western La Posta plutons probably reflects shallower depths of melting and the inability of mafic restite to separate from the melt during ascent and emplacement. Eastern-zone La Posta magmas with little or no restite, elevated Sr_i and $\delta^{18}O$, and enriched LREE and depleted HREE patterns crystallized from melt produced in a hotter(?), deeper zone with more complete separation of restite. The highest La Posta Sr_i value in our study is 0.7056 for an eastern sample, but values as high as 0.7074 are known from a La Posta-type pluton in Baja California (Walawender et al., 1990). Elevated Sr_i of the La Posta would seem to indicate either considerable aging of igneous materials in the eastern crustal region prior to melt generation and/or a contribution from sialic crust.

Both the Granite Mountain and La Posta suites indicate a period of deeper melting during the middle to Late Cretaceous. In the Peninsular Ranges batholith, crustal thickening has been proposed as a mechanism for bringing lower crustal garnet-bearing amphibolites to eclogite depths: e.g., in a west-to-east plutonic transect at ~31°N latitude in Baja California, Johnson et al. (1999a) proposed crustal thickening due to collision of the Alisitos island arc with the continental margin between 115 and 110 Ma. North of the international border, Shaw et al. (this volume, Chapter 7) suggest that accretion of a Late Jurassic island arc to an eastern terrain of Middle and Late Jurassic granitic rocks occurred in the latest Jurassic and/or early Early Cretaceous. Evidence for a Late Jurassic island arc west of the continental margin consists of scattered exposures of deformed, lower greenschist-grade, Late Jurassic island-arc volcaniclastic and volcanic rocks that unconformably underlie the non-accretionary Early Cretaceous Santiago Peak Volcanics (Herzig and Kimbrough, 1991; C.T. Herzig, 1994, personal commun.).

Additional causes of Cretaceous crustal thickening in the northern Peninsular Ranges batholith might be (1) the addition of great volumes of Early Cretaceous plutons to the crust (Gastil et al., 1992); (2) in the eastern zone, Late Cretaceous west-directed thrusting and tectonic thickening (Grove, 1994; George and Dokka, 1994); and (3) changes in plate-convergence directions and/or rates between North American and Pacific plates, and/or shallowing of the subducted slab (Walawender et al., 1990; Livaccari, 1991; Thomson and Girty, 1994). George and Dokka (1994) interpret fission-track ages in the San Jacinto Mountains of the eastern Peninsular Ranges batholith as recording Late Cretaceous synintrusive mylonitization between ca. 99 and 92 Ma. According to the authors, west-directed thrusting beginning as early as 99 Ma produced an unstable crustal welt; thickening was followed closely by detachment faulting and extensional collapse, also associated with intrusion, at ca. 94–93 Ma. To the south, in San Diego County, Grove (1994) used $^{40}Ar/^{39}Ar$ isotopic studies to identify two episodes of Late Cretaceous uplift and exhumation following final emplacement of the batholith at ca. 100–90 Ma. He suggested that rapid cooling between 88–85 Ma and 76–72 Ma reflects episodes of west-directed thrusting along faults such as the Chariot Canyon fault. Because this fault coincides with the protomylonitic western margin of the tonalite pluton at Granite Mountain, we suggest that west-directed thrusting and crustal thickening may have begun as early as 98 ± 3 Ma, the presumed age of the Granite Mountain pluton (L.T. Silver, 1979, personal commun.).

Whatever the cause, or causes, of crustal thickening, partial melting of a lower crustal source at eclogite-facies depths is considered to have given rise to the La Posta magma, whereas the Granite Mountain depth of melting was transitional between amphibolite-granulite and eclogite-facies conditions.

Conclusions

The systematic REE patterns and low Sr_i values of the Cretaceous granitic suites across the gravity-magnetic boundary do not indicate a fundamental west-to-east change in the source of the magmas. We suggest that source rocks of basaltic composition were present in the lower crust and/or as underplates from earlier subduction episodes beneath both western and eastern zones of the Peninsular Ranges batholith. Differences among the three major tonalite types—Las Bancas–East Mesa, Alpine–Japatul

Valley, and Granite Mountain–La Posta—may simply reflect the production of different restite and melt phases due to increasing depths of partial melting as Cretaceous intrusion proceeded eastward and the crust thickened.

The REE patterns of the granite suites are interpreted as the result of the modification of the relatively flat REE patterns of the basaltic source material during partial melting, ascent, and emplacement. Because of the regional nature of these patterns, we would agree with Gromet and Silver (1987) that they represent changes in equilibrium between residual mineral assemblages and granite magma. Our studies suggest that three principal REE patterns account for the changes in mineralogy of the residual assemblages: (1) Alpine-Japatul Valley–type patterns, slightly LREE-enriched, nearly flat MREE to HREE, and minor to no negative Eu anomalies indicating hornblende and variable plagioclase as residual phases under relatively high p_{H_2O}; (2) Las Bancas–East Mesa–type patterns, enriched LREE relative to MREE, low MREE to HREE slopes, and moderate negative Eu anomalies indicating pyroxene and significant plagioclase extraction under lower p_{H_2O}; and (3) Granite Mountain–La Posta–type patterns, transitional between patterns of the Alpine–Japatul Valley and Las Bancas–East Mesa suites and those with high MREE to LREE slopes, depleted HREE, and little or no Eu anomalies indicating garnet extraction and plagioclase breakdown.

Variations in temperature and volatile content (e.g., water from the subducting slab or from hydrated minerals) determine the degree of partial melting and the identity of restite minerals in equilibrium with the melt, thus providing an opportunity for melting at a range of crustal depths. Moderate dehydration melting of basalt in the amphibole stability field produces granodiorite to tonalite compositions (Beard and Lofgren, 1989) with amphibole as both a residual phase and a mineral in equilibrium with the melt. Higher degrees of partial melting, or melting at lower crustal depths, result in the breakdown of amphibole and the presence of pyroxene and plagioclase as melt and restite phases (Bryant et al., 1997). Granulite residual assemblages at higher pressures/temperatures give way to garnet and eventually to plagioclase breakdown, significantly enriching the magma in Ca and Sr.

Very low Sr_i ratios of gabbro plutons (Table 5) require a primitive mantle source. Smith et al. (1983) and Walawender et al. (1991) propose that basaltic melt was derived in the mantle wedge above the subduction zone under hydrous conditions and subsequently underwent fractionation to ultramafic cumulates and mafic compositions. In the Early Cretaceous, voluminous Peninsular Ranges batholith magmas (Las Bancas and Alpine) were generated from lower crustal materials and/or basaltic underplates by dehydration reactions involving hornblende and biotite in the amphibolite-granulite fields. Fractionation of these magmas led to the production of cumulates (Las Bancas) and intermediate to silicic magmas (Las Bancas, Alpine, Japatul Valley, and Chiquito Peak suites). Broadly, the leucogranite magmas represent fractionates of the above mafic melts, possibly modified by interaction with altered lower and/or middle crustal igneous/sedimentary rocks. By the middle to Late Cretaceous, melting was occurring in the granulite-eclogite fields and producing the Granite Mountain and La Posta magmas.

Although the Cretaceous magmatic arc tapped a similar basaltic source as it migrated across the Late Jurassic-earliest Cretaceous continental margin, some mineralogical, geophysical, and chemical characteristics of the Cretaceous granitic suites suggest that interactions of magmas with western oceanic crust and/or eastern, young continental crust played a secondary role in their petrogenesis. These characteristics include west-to-east variations in opaque oxide mineralogy and magnetic properties plus the relatively abrupt change from oxidized western to reduced eastern plutons, as well as geochemical and isotopic variations in plutons that intruded across the continental margin. The latter include elevated P, Pb, Sr_i, and $\delta^{18}O$ in eastern samples of the La Posta suite; higher K_2O, Rb, Zr, Sr_i, and $\delta^{18}O$ in eastern samples of the Las Bancas suite; and the relatively radiogenic Pb ratios of eastern samples of the Las Bancas and La Posta suites, and of the East Mesa suite. Such variations probably reflect interaction of the parental magmas with western and eastern crust during ascent and emplacement and possibly also reflect greater initial thickness of the crust in the eastern region (Grove, 1993).

The relatively steep $^{207}Pb/^{204}Pb$ versus $^{206}Pb/^{204}Pb$ trend of the Peninsular Ranges batholith samples differs from those of other Cordilleran batholiths in the region, with the exception of the Sierra Nevada batholith (Fig. 12) (J.L. Wooden, 1995, personal commun.). Figure 12 suggests that the granitic rocks of San Diego County are similar to, but slightly more radiogenic than, those of the Sierra Nevada batholith. Broadly, the San Diego County trend can be viewed as a mixing line between two end-members, primitive mantle and Precambrian, probably Proterozoic upper crust, but the $^{208}Pb/^{204}Pb$ versus $^{206}Pb/^{204}Pb$ trend is not toward Mojave Proterozoic crust. J.L. Wooden (1985, personal commun.) suggested Proterozoic crust of Arizona as a possible source component, which is consistent with ~300 km northwest displacement of the Peninsular Ranges batholith along the San Andreas fault system (Gastil et al., 1981). The Rb-Sr-O data are more complex and require at least three components. Mixing curves of Sr_i and $\delta^{18}O$ for Jurassic and Cretaceous plutons in the study area (Shaw et al., 1986) suggest a common, primitive mantle-derived source plus at least two crustal components: a $\delta^{18}O$- and Sr_i-enriched quartzo-feldspathic to pelitic source (early Mesozoic sediments containing a component of Precambrian upper crust) and a $\delta^{18}O$-enriched metabasaltic (ophiolitic) source (base of the early Mesozoic crust?).

The widespread occurrence of synplutonic mafic and intermediate dikes in the study area suggests the existence of a relatively mafic (basaltic to andesitic) magma chamber, or chambers, below the present level of erosion. Such magma chambers may also have been the source for the mafic enclaves, which, in preliminary study, appear to be geochemically similar to the mafic dikes. The mafic enclaves may represent fragments of mafic dikes that were injected into granitic magma chambers at an early stage of intrusion and disrupted during ascent and emplacement (Frost and Mahood, 1987).

TECTONIC IMPLICATIONS AND SPECULATIONS

The distribution of Cretaceous plutonic suites and fabrics across the San Diego County segment of the Peninsular Ranges batholith has implications for the tectonic evolution of the batholith. Plutons as old as 137(?) Ma (Anderson, 1991) to as young as 93 ± 1 Ma (Walawender et al., 1991) were emplaced from west to east across a latest Jurasssic-earliest Cretaceous discontinuity, or "suture," which we interpret as the fault-bounded Late Jurassic continental margin. Plutons of the Las Bancas suite that stitched across the suture apparently range in age from ca. 119 Ma (pluton at El Capitan Reservoir, L.T. Silver, 1979, personal commun.) on the west, through ca. 111 Ma (a body adjacent to the Corte Madera gabbro pluton, L.T. Silver, 1979, personal commun.) and 109 Ma in the vicinity of the suture, to ca. 104 Ma on the east (Table 2; Figs. 6B and 13). Plutons of the La Posta suite decrease in age from ca. 104 Ma west of the suture, through ca. 102 and 101 Ma near the suture, to ca. 94 Ma on the east side. These age ranges constrain accretion of the Late Jurassic island arc to the belt of Middle and Late Jurassic plutons to the early Early Cretaceous.

Regional-scale isoclinal folds involving the Julian Schist and the Jurassic metagranites, whose cores apparently served as loci for Cretaceous intrusion (Shaw et al., this volume, Chapter 7; Todd, 2004), indicate an episode of Late Jurassic–earliest Cretaceous shortening prior to development of the Early Cretaceous Cuyamaca–Laguna Mountains shear zone. We speculate that this episode of shortening represents accretion of the Late Jurassic fringing arc and closing of a backarc basin (Busby et al., 1998). Prebatholithic screens that contain mixtures of volcanogenic and continentally derived metasedimentary rocks (felsic tuff-breccia and amphibolite grading to fine-grained gabbro; calc-silicate rocks, biotite schist, and metaquartzite) form a discontinuous, ~1-km-wide zone between screens of bona fide Julian Schist and metamorphosed Santiago Peak Volcanics (Fig. 13C). These rocks may be remnants of backarc basin sediments and young ocean floor caught up between the two Jurassic arcs.

Thomson and Girty (1994) determined that Jurassic and Cretaceous plutons in the Cuyamaca–Laguna Mountains shear zone underwent an Early Cretaceous (ca. 118–114 Ma) episode of northeast-southwest shortening, followed on the western side of the shear zone by extension from ca. 105 to 94 Ma (middle Cretaceous). They speculated that shortening resumed in the Late Cretaceous (ca. 99–94 Ma) as evidenced by development of east-over-west thrusts in the eastern Peninsular Ranges mylonite zone (Erskine, 1986; George and Dokka, 1994) and of the west-directed reverse-sense Chariot Canyon fault in central-eastern San Diego County (Grove, 1994). This sequence of intra-arc deformations coincides approximately with changes in subduction dynamics from nearly orthogonal convergence in the Early Cretaceous (Glazner, 1991), to oblique-dextral subduction (Engebretson et al., 1985) and possible shallowing of the subducting slab in the middle Cretaceous (Walawender et al., 1990).

We agree with Thomson and Girty (1994) that the Cuyamaca–Laguna Mountains shear zone formed when Cretaceous intrusion

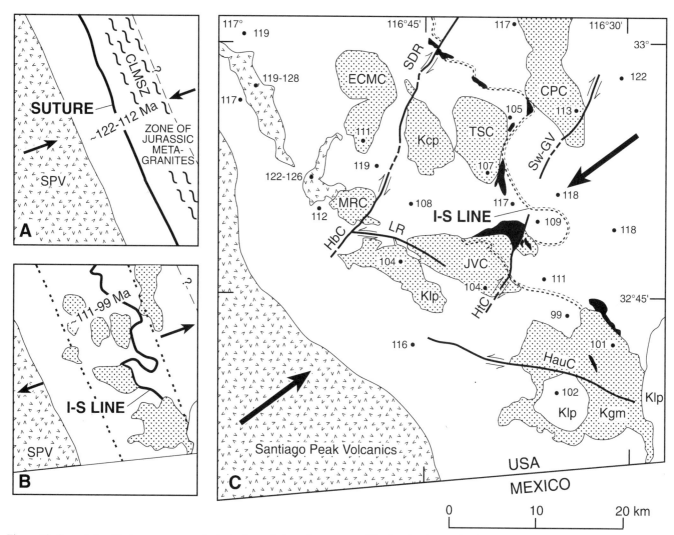

Figure 13. Speculative Cretaceous magmatic-tectonic evolution of west-central Peninsular Ranges batholith. A: Plutonic arc between ca. 122 and 112 Ma, arrows indicate intra-arc shortening. V pattern, Early Cretaceous Santiago Peak Volcanics (SPV), representing volcanic cover of western plutons + hypabyssal plutons (latter not shown); CLMSZ—Cuyamaca–Laguna Mountains shear zone; latest Jurassic–Early Cretaceous suture shown by heavy solid line; queried dashed line represents eastern margin of Cretaceous arc (not to scale). B: Plutonic arc between ca. 111 and ca. 99 Ma; stippled areas between double-dot lines suggest plutonic material added during this period by diapiric intrusion; arrows indicate inferred direction of intra-arc extension. I-S line (western limit of Jurassic metagranites) is trace of intruded, deformed suture. C: West-central Peninsular Ranges batholith in middle to Late Cretaceous. Stippled plutons are diapiric plutonic complexes: ECMC—El Cajon Mountain complex; MRC—McClain Ranch complex; JVC—Japatul Valley complex; TSC—Tule Springs complex; CPC—Cuyamaca Peak complex; Kcp—pluton of Chiquito Peak suite; Kgm—pluton of Granite Mountain suite; Klp—plutons of La Posta suite. Heavy solid lines (dashed where inferred) are ductile-to-brittle shear zones: SDR—San Diego River Valley; HbC—Harbison Canyon; HtC—Horsethief Canyon; Sw-GV—Sweetwater River–upper Green Valley; LR—Loveland Reservoir; HauC—Hauser Canyon; arrows indicate sense of lateral displacement. I-S line indicated by double-dash line; black bodies in vicinity of I-S line are screens of metamorphosed mixed volcanogenic/continental sedimentary rocks. Solid circles = locations of U-Pb isotopic ages from L.T. Silver (1979, oral commun.); Matthew Taylor (1990, oral commun.); Anderson (1991); Bracchi et al. (1993); D.L. Kimbrough (1994, oral commun.); Shaw et al. (this volume, Chapter 7). Solid arrows indicate inferred direction of intra-arc shortening.

localized strain in thermally weakened crust in the vicinity of the suture between oceanic and young continental lithosphere. We further suggest that the quartz- and biotite-rich S- and I-S-type Jurassic granitic plutons were especially susceptible to deformation (Vernon and Flood, 1988). Given the eastward dip of the suture, the Cuyamaca–Laguna Mountains shear zone apparently developed within the relatively thin uppermost wedge of the

Late Jurassic continental margin. Cretaceous plutons that were emplaced east of the suture may have originated as dikes that expanded through the addition of magma but retained strong north-northwest flattening fabrics.

We speculate that the contrast in structural trends between western and eastern zones in this part of the Peninsular Ranges batholith (Todd and Shaw, 1979; Todd, 2004) may reflect the above sequence of Cretaceous intra-arc deformational episodes. Cartooned in Figure 13A are first, intrusion of Early Cretaceous (ca. 122–112 Ma) plutons during the episode of shortening represented by the Cuyamaca–Laguna Mountains shear zone on the east and possibly by a belt of strongly foliated granitic plutons in the San Vicente Reservoir 7.5-minute quadrangle on the west (R.G. Gastil, 1986, personal commun.). This episode was followed by intrusion of diapiric plutonic complexes during a period of extension from ca. 111 to 99 Ma (Fig. 13B), during which time the width of the west-central part of the batholith increased significantly and the suture was deformed. Swarms of mafic and intermediate dikes may have been emplaced into the Cuyamaca–Laguna Mountains shear zone during this period of extension. Figure 13C shows the present-day configuration.

The youngest batholithic structures in the study area are a series of in part synintrusive, ductile-to-brittle shear zones that we speculate may be associated with a middle to Late Cretaceous change to oblique subduction and the resumption of shortening. Two north-northeast- to northeast–trending, right-lateral shear zones are spatially associated with Corte Madera diapirs, while two northwest- to west-northwest–trending left-lateral shear zones are associated with Granite Mountain–La Posta intrusion (Rector, 1993; Todd, 2004) (Fig. 13C). The northeast shear zones caused small (~0.5 km) right-lateral bends and/or offsets of plutonic contacts, and the asymmetric diapiric shapes of the associated Corte Madera bodies suggest that leucogranite magma was intruded preferentially into these zones. In the case of the northwest to west-northwest shear zones, contacts are locally folded and/or offset left-laterally by as much as 1–2 km. The geometric and age relationships between the northeast- and northwest–trending shear zones is unclear; however, if they represent conjugate shear zones, then the maximum principal stress was oriented at a high angle to the continental margin, indicating shortening (Fig. 13C). Alternatively, northeast-directed, right-lateral shear associated with Corte Madera intrusion may have preceded northwest-directed, left-lateral shear that accompanied Granite Mountain–La Posta intrusion. Sinistral-sense shear zones may have formed in response to development of a major tranverse structure in the magmatic arc related to voluminous La Posta intrusion. In this regard, it may be noteworthy that the eastern margin of the Santiago Peak Volcanics shows a northwest bend that is aligned with a west-northwest protrusion in the western margin of the La Posta–Granite Mountain plutonic belt and with the west-northwest Hauser Canyon lineament (Fig. 13C).

The nature and age of the boundary between western and eastern zones of the Peninsular Ranges batholith has been the subject of study for more than 2 decades (Silver et al., 1979; Gastil et al., 1981). Walawender et al. (1991) argued for distinct western and eastern static arcs, with each arc evolving in relation to prevailing Cretaceous subduction rates and geometries. Gastil (1993) suggested that the apparent Cretaceous "double-arc" in the Peninsular Ranges was the result of the opening of a Jurassic intra-arc sea that closed ca. 100 Ma. Based upon mapping of

prebatholithic terrains in Baja California, Griffiths and Hoobs (1993) described the boundary as defined by "structural interleaving" of Upper Jurassic volcanic/volcaniclastic rocks (containing uncommon quartz metasandstone), Paleozoic sedimentary rocks, and rocks of the Lower Cretaceous Alisitos Group that occurred between 103 ± 4 and ca. 100 Ma. In northern Baja California, Johnson et al. (1999a) interpreted a 2-km-wide gradient in lithology, metamorphic grade, and deformational style plus a 10 Ma plutonic age break to mark a suture between western and eastern arcs; this suture is considered to have formed ca. 110 Ma. In the southern Sierra San Pedro Martir, Schmidt and Paterson (2002) describe the transition from western to eastern zones as marked by a narrow zone of long-lived multiphase Jurassic and Cretaceous deformation that occurred during the period from >132 Ma (the age of a plutonic complex that intrudes the transition zone) to ca. 85 Ma.

We propose that a number of the mineralogical and geophysical characteristics of the Cretaceous suites described in this paper reflect intrusion across a pre-existing tectonic boundary between oceanic and continental lithosphere. Similarly, Thomson and Girty (1994) proposed that a poorly understood lithospheric boundary in the axial zone of the Peninsular Ranges batholith separates oceanic crust on the west from transitional or continental crust on the east. Their model differs from ours in that they consider the Triassic Julian Schist to be an overlap assemblage, which would require juxtaposition of the two lithospheric plates to have occurred prior to the Triassic. We consider that the Triassic–Jurassic Julian Schist is restricted to the eastern lithospheric region and that two unlike Jurassic "terrains" are juxtaposed across the boundary. Overlap assemblages include (1) Late Jurassic–Early Cretaceous sedimentary assemblages in the borderland of northern Baja California (Kimbrough et al., 1987) and possibly (2) a discontinuous strip of mixed metavolcanic and metasedimentary rocks lying between the Julian Schist and screens of probable metamorphosed Santiago Peak Volcanics in the western zone (Fig. 13C).

All of the above tectonic models account for the striking west-to-east asymmetry of the Cretaceous batholith; differences in the inferred age of the suture may reflect its time-transgressive nature, which results from somewhat different tectonic settings from north to south along the arc. In the study area, an original latest Jurassic–earliest Cretaceous margin-parallel transform(?) fault separating oceanic and young continental lithosphere (represented by the gravity-magnetic gradient and I-S line) may have been reactivated in the Early Cretaceous as the Cuyamaca–Laguna Mountains shear zone, and again in the middle to Late Cretaceous as a zone of extension "healed" by intrusion of diapiric plutonic complexes. Detailed mapping of plutons and prebatholithic rocks and dating of Peninsular Ranges plutons are required to resolve this problem.

We conclude that, in southern California, a single Cretaceous arc migrated from west to east across a preexisting Late Jurassic–earliest Cretaceous lithospheric boundary. The distribution of available Cretaceous pluton ages in San Diego County does not require an age break, but rather suggests approximately continuous eastward intrusion in response to changing subduction dynamics/crustal thickness. The presence of the Cretaceous Las Bancas, Granite Mountain, and La Posta suites in both western and eastern zones further indicates that the source of their magmas was independent of the crustal boundary that runs the length of the Peninsular Ranges batholith.

ACKNOWLEDGMENTS

This study benefitted from many discussions and field trips with colleagues from the U.S. Geological Survey, San Diego State University, and Macquarie University. Aaron Yoshinobu, Calvin Miller, and Gary Girty offered helpful reviews of the manuscript.

REFERENCES CITED

Ague, J.J., and Brandon, M.T., 1992, Tilt and northward offset of Cordilleran batholiths resolved using igneous barometry: Nature, v. 360, p. 146–149.

Ague, J.J., and Brimhall, G.H., 1988, Magmatic arc asymmetry and distribution of anomalous plutonic belts in the batholiths of California: Effects of assimilation, crustal thickness, and depth of crystallization: Geological Society of America Bulletin, v. 100, p. 912–927.

Anderson, C.L., 1991, Zircon uranium-lead isotopic ages of the Santiago Peak volcanics and spatially related plutons of the Peninsular Ranges batholith, southern California [M.S. thesis]: San Diego, San Diego State University, 111 p.

Anderson, J.L., and Smith, D.R., 1995, The effects of temperature and fO_2 on the Al-in-hornblende barometer: American Mineralogist, v. 80, p. 549–559.

Balch, D.C., Bartling, S.H., and Abbott, P.L., 1984, Volcaniclastic strata of the Upper Jurassic Santiago Peak Volcanics, San Diego, California, *in* Crouch, J.K., and Bachman, S.B., eds., Tectonics and sedimentation along the California margin: Los Angeles, California, Society of Economic Paleontologists and Mineralogists Pacific Section, v. 38, p. 157–170.

Barbarin, B., 1988, Field evidence for successive mixing and mingling between the Piolard Diorite and the Saint-Julien-la-Vetre Monzogranite, (Nord-Forez, Massif Central, France): Canadian Journal of Earth Sciences, v. 25, p. 49–59.

Barth, A.P., Tosdal, R.M., and Wooden, J.L., 1990, A petrologic comparison of Triassic plutonism in the San Gabriel and Mule Mountains, southern California: Journal of Geophysical Research, v. 95, p. 20075–20096.

Beard, J.S., and Lofgren, G.E., 1989, Effect of water on the composition of partial melts of greenstone and amphibolite: Science, v. 244, p. 195–197.

Bracchi, K.A., Girty, G.H., and Girty, M.S., 1993, CLMSZ, Garnet Mountain area, southern California: A collisonally generated contractional shear zone: Geological Society of America Abstracts with Programs, v. 25, no. 5, p. 13.

Bryant, C.J., Arculus, R.J., and Chappell, B.W., 1997, Clarence River Supersuite: 250-Ma Cordilleran tonalitic I-type intrusions in eastern Australia: Journal of Petrology, v. 38, p. 975–1001.

Busby-Spera, C.J., 1988, Evolution of a Middle Jurassic back-arc basin, Cedros Island, Baja California: Evidence from a marine volcaniclastic apron: Geological Society of America Bulletin, v. 100, p. 218–233.

Busby, C., and Saleeby, J.B., 1993, Paleogeographic and tectonic setting of axial and western metamorphic framework rocks of the southern Sierra Nevada, California, *in* Dunne, G., and McDougall, K., eds., Mesozoic paleogeography of the western United States: Los Angeles, California, Society of Economic Paleontologists and Mineralogists Pacific Section, Book 71, p. 197–226.

Busby, C., Smith, D., Morris, W., and Fackler-Adams, B., 1998, Evolutionary model for convergent margins facing large ocean basins: Mesozoic Baja California, Mexico: Geology, v. 26, p. 227–230.

Busby-Spera, C.J., Mattinson, J.M., Riggs, N.R., and Schermer, E.R., 1990, The Triassic-Jurassic magmatic arc in the Mojave-Sonoran Deserts and the Sierran-Klamath region: Similarities and differences in paleogeographic evolution, *in* Harwood, D.S., and Miller, M.M., eds., Paleozoic and early Mesozoic paleogeographic relations, Sierra Nevada, Klamath Mountains, and related terranes: Boulder, Colorado, Geological Society of America Special Paper 255, p. 325–337.

Cameron, J.L., 1980, The Lucky Five pluton in the Southern California batholith: A history of emplacement and solidification under stress [M.S. thesis]: Los Angeles, University of California, 145 p.

Campbell, I.H., and Gorton, M.P., 1980, Accessory phases and the generation of LREE-enriched basalts—a test for disequilibrium melting: Contributions to Mineralogy and Petrology, v. 72, p. 157–163.

Carollo, G.F., and Walawender, M.J., 1993, Geochemistry of leucogranite plutons from the western zone of the Peninsular Ranges batholith, San Diego, California: Geological Society of America Abstracts with Programs, v. 25, no. 5, p. 18.

Chappell, B.W., 1978, Granitoids of the Moonbi District, New England batholith, eastern Australia: Journal of the Geological Society of Australia, v. 25, p. 267–283.

Chappell, B.W., and White, A.J.R., 1992, I- and S-type granites in the Lachlan Fold Belt: Transactions of the Royal Society of Edinburgh: Earth Sciences, v. 83, p. 1–26.

Clinkenbeard, J.P., 1987, The mineralogy, geochemistry, and geochronology of the La Posta pluton, San Diego and Imperial Counties, California [M.S. thesis]: San Diego, San Diego State University, 215 p.

Clinkenbeard, J.P., and Walawender, M.J., 1989, Mineralogy of the La Posta pluton: Implications for the origin of zoned plutons in the eastern Peninsular Ranges batholith, southern and Baja California: American Mineralogist, v. 74, p. 1258–1269.

Collins, W.J., 1998, Evaluation of petrogenetic models for Lachlan Fold Belt granitoids: Implications for crustal architecture and tectonic models: Australian Journal of Earth Sciences, v. 45, p. 483–500.

DePaolo, D.J., 1981, A neodymium and strontium isotopic study of the Mesozic calc-alkaline granitic batholiths of the Sierra Nevada and Peninsular Ranges, California: Journal of Geophysical Research, v. 86, p. 10470–10488.

Dickinson, W.R., 1981, Plate tectonic evolution of the southern Cordillera, *in* Dickinson, W.R., and Payne, W.D., eds., Relations of tectonics to ore deposits in the southern Cordillera: Arizona Geological Society Digest, v. 14, p. 113–135.

Engebretson, D.C., Cox, A., and Gordon, R.G., 1985, Relative motions between continental plates in the Pacific basin: Boulder, Colorado, Geological Society of America Special Paper 206, 59 p.

Erskine, B.G., 1986, Mylonitic deformation and associated low-angle faulting in the Santa Rosa mylonite zone, southern California [Ph.D. dissertation]: Berkeley, University of California, 247 p.

Everhart, D.L., 1951, Geology of the Cuyamaca Peak quadrangle, San Diego County, California: California Division of Mines Bulletin 159, p. 51–115.

Foster, D.A., and Hyndman, D.W., 1990, Magma mixing and mingling between synplutonic mafic dikes and granite in the Idaho-Bitterroot batholith, *in* Anderson, J.L., ed., The nature and origin of Cordilleran magmatism: Boulder, Colorado, Geological Society of America Memoir 174, p. 347–358.

Frost, T.P., and Mahood, G.A., 1987, Field, chemical, and physical constraints on mafic-felsic magma interaction in the Lamarck Granodiorite, Sierra Nevada, California: Geological Society of America Bulletin, v. 99, p. 272–291.

Gastil, R.G., 1979, A conceptual hypothesis for the relation of differing tectonic terranes to plutonic emplacement: Geology, v. 7, p. 542–544.

Gastil, G., 1983, Mesozoic and Cenozoic granitic rocks of southern California and western Mexico: Boulder, Colorado, Geological Society of America Memoir 159, p. 265–275.

Gastil, G., 1990a, The boundary between the magnetite-series and ilmenite-series granitic rocks in Peninsular California: Tokyo, University Museum, the University of Tokyo, Nature and Culture 2, p. 91–100.

Gastil, G., 1990b, Zoned plutons of the Peninsular Ranges in southern and Baja California: University Museum, the University of Tokyo, Nature and Culture 2, p. 77–90.

Gastil, R.G., 1993, Prebatholithic history of peninsular California, *in* Gastil, R.G., and Miller, R.H., eds., The prebatholithic stratigraphy of peninsular California: Boulder, Colorado, Geological Society of America Special Paper 279, p. 145–156.

Gastil, R.G., Phillips, R.P., and Allison, E.C., 1975, Reconnaissance geology of the state of Baja California: Boulder, Colorado, Geological Society of America Memoir 140, 170 p.

Gastil, G., Morgan, G.J., and Krummenacher, D., 1981, The tectonic history of peninsular California and adjacent Mexico, *in* Ernst, W.G., ed., The geotectonic development of California: Rubey Volume I: Englewood Cliffs, New Jersey, Prentice-Hall, p. 284–306.

Gastil, G., Kimbrough, J., Tainosho, Y., Shimizu, M., and Gunn, S., 1991, Plutons of the eastern Peninsular Ranges, southern California, U.S.A., and Baja California, Mexico, *in* Walawender, M.J., and Hanan, B.B., eds., Geological excursions in southern California and Mexico: Geological Society of America Annual Meeting Field Trip Guidebook: San Diego, San Diego State University, Department of Geological Sciences, p. 319–331.

Gastil, G., Wracher, M., Strand, G., Kear, L.L., Ely, D., Chapman, D., and Anderson, C., 1992, The tectonic history of the southwestern United States and Sonora, Mexico, during the past 100 m.y.: Tectonics, v. 11, p. 990–997.

George, P.G., and Dokka, R.K., 1994, Major Late Cretaceous cooling events in the eastern Peninsular Ranges, California, and their implications for Cordilleran tectonics: Geological Society of America Bulletin, v. 106, p. 903–914.

Glazner, A.F., 1991, Plutonism, oblique subduction, and continental growth: An example from the Mesozoic of California: Geology, v. 19, p. 784–786.

Griffiths, R., and Hoobs, J., 1993, Geology of the southern Sierra Calamajue, Baja California Norte, Mexico, *in* Gastil, R.G., and Miller, R.H., eds., The prebatholithic stratigraphy of peninsular California: Boulder, Colorado, Geological Society of America Special Paper 279, p. 43–60.

Gromet, L.P., and Silver, L.T., 1987, REE variations across the Peninsular Ranges batholith: Implications for batholithic petrogenesis and crustal growth in magmatic arcs: Journal of Petrology, v. 28, p. 75–125.

Grove, M., 1987, Metamorphism and deformation of prebatholithic rocks in the Box Canyon area, eastern Peninsular Ranges, San Diego County, California [M.S. thesis]: Los Angeles, University of California, 174 p.

Grove, M., 1989, Nature of the metamorphic discontinuity across the Chariot Canyon fault, east-central Peninsular Ranges batholith, California: Geological Society of America Abstracts with Programs, v. 21, no. 5, p. 87.

Grove, M., 1993, Thermal histories of southern California basement terranes [Ph.D. dissertation]: Los Angeles, University of California, 419 p.

Grove, M., 1994, Contrasting denudation histories within the east-central Peninsular Ranges batholith (33°N), *in* McGill, S.F., and Ross, T.M., eds., Geological investigations of an active margin: Geological Society of America Cordilleran Section Meeting Field Trip Guidebook: Redlands, California, San Bernardino County Museum Association, p. 235–240.

Hammarstrom, J.M., 1992, Mineral chemistry of Cretaceous plutons: Hornblende geobarometry in southern California and southeastern Alaska: Geological Society of America Abstracts with Programs, v. 24, no. 5, p. 30.

Herzig, C., and Kimbrough, D.L., 1991, Early Cretaceous zircon ages prove a non-accretionary origin for the Santiago Peak Volcanics, northern Santa Ana Mountains, California: Geological Society of America Abstracts with Programs, v. 23, no. 2, p. 35.

Hill, R.I., Silver, L.T., and Taylor, Jr., H.P., 1986, Coupled Sr-O isotope variations as an indicator of source heterogeneity for the northern Peninsular Ranges batholith: Contributions to Mineralogy and Petrology, v. 92, p. 351–361.

Holland, T., and Blundy, J., 1994, Non-ideal interactions in calcic amphiboles and their bearing on amphibole-plagioclase thermometry: Contributions to Mineralogy and Petrology, v. 116, p. 433–447.

Hutton, D., 1981, The Main Donegal granite: Lateral wedging in a synmagmatic shear zone [abs.], *in* Coward, M.P., ed., Diapirism and gravity tectonics: Report of a tectonic studies group: Journal of Structural Geology, v. 3, p. 93.

Jachens, R.C., 1992, Aeromagnetic map of the El Cajon 1:100,000 scale quadrangle, California: U.S. Geological Survey Open-File Report 92-548, 4 p.

Jachens, R.C., Simpson, R.W., Griscom, A., and Mariano, J., 1986, Plutonic belts in southern California defined by gravity and magnetic anomalies: Geological Society of America Abstracts with Programs, v. 18, no. 2, p. 120.

Jachens, R.C., Todd, V.R., Morton, D.M., and Griscom, A., 1991, Constraints on the structural evolution of the Peninsular Ranges batholith, California, from a new aeromagnetic map: Geological Society of America Abstracts with Programs, v. 23, no. 2, p. 38.

Johnson, M.C., and Rutherford, M.J., 1989, Experimental calibration of the aluminum-in-hornblende geobarometer with application to the Long Valley caldera (California) volcanic rocks: Geology, v. 17, p. 837–841.

Johnson, S.E., Tate, M.C., and Fanning, C.M., 1999a, New geological mapping and SHRIMP U-Pb zircon data in the Peninsular Ranges batholith, Baja California, Mexico: Evidence for a suture?: Geology, v. 27, p. 743–746.

Johnson, S.E., Paterson, S.R., and Tate, M.C., 1999b, Structure and emplacement history of a multiple-center, cone-sheet-bearing ring complex: The Zarza Intrusive Complex, Baja California, Mexico: Geological Society of America Bulletin, v. 111, p. 607–619.

Kennedy, M.P., 1975, Geology of the San Diego Metropolitan area, California, Section A: California Division of Mines and Geology Bulletin 200, p. 9–39.

Kerr, D.R., and Kidwell, S.M., 1991, Late Cenozoic sedimentation and tectonics, western Salton Trough, California, *in* Walawender, M.J., and Hanan, B.B., eds., Geological excursions in southern California and Mexico: Geological

Society of America Annual Meeting Field Trip Guidebook: San Diego, San Diego State University, Department of Geological Sciences, p. 397–416.

Kimbrough, D.L., Hickey, J.J., and Tosdal, R.M., 1987, U-Pb ages of granitoid clasts in upper Mesozoic arc-derived strata of the Vizcaíno Peninsula, Baja California, Mexico: Geology, v. 15, p. 26–29.

Kimzey, J.A., 1982, Petrology and geochemistry of the La Posta granodiorite [M.S. thesis]: San Diego, San Diego State University, 81 p.

Krummenacher, D., Gastil, R.G., Bushee, J., and Doupont, J., 1975, K-Ar apparent ages, Peninsular Ranges batholith, southern California and Baja California: Geological Society of America Bulletin, v. 86, p. 760–768.

Lampe, C.M., 1988, Geology of the Granite Mountain area: Implications of the extent and style of deformation along the southeast portion of the Elsinore fault [M.S. thesis]: San Diego, San Diego State University, 150 p.

Larsen, E.S., Jr., 1948, Batholith and associated rocks of Corona, Elsinore, and San Luis Rey quadrangles, southern California: New York, Geological Society of America Memoir 29, 182 p.

Lipman, P.W., 1992, Magmatism in the Cordilleran United States, *in* Burchfiel, B.L., Lipman, P.W., and Zoback, M.L., eds., The Cordilleran Orogen: Conterminous U.S.: Boulder, Colorado, Geological Society of America, The Geology of North America, v. G-3, p. 481–514.

Livaccari, R.G., 1991, Role of crustal thickening and extensional collapse in the tectonic evolution of the Sevier-Laramide orogeny, western United States: Geology, v. 19, p. 1104–1107.

Merriam, R.H., 1946, Igneous and metamorphic rocks of the southwestern part of the Ramona quadrangle, San Diego County, California: Geological Society of America Bulletin, v. 57, p. 223–260.

Paterson, S.R., and Vernon, R.H., 1995, Bursting the bubble of ballooning plutons: A return to nested diapirs emplaced by multiple processes: Geological Society of America Bulletin, v. 107, p. 1356–1380.

Paterson, S.R., Vernon, R.H., and Tobisch, O.T., 1989, A review of criteria for the identification of magmatic and tectonic foliations in granitoids: Journal of Structural Geology, v. 11, p. 349–363.

Paterson, S.R., Fowler, Jr., T.K., and Miller, R.B., 1996, Pluton emplacement in arcs: A crustal-scale exchange process: Transactions of the Royal Society of Edinburgh: Earth Sciences, v. 87, p. 115–123.

Pitcher, W.S., and Berger, A.R., 1972, The geology of Donegal: A study of granite emplacement and unroofing: New York, John Wiley, 435 p.

Premo, W.R., Morton, D.M., Snee, L.W., Naeser, N.D., and Fanning, C.M., 1998, Isotopic ages, cooling histories, and magmatic origins for Mesozoic tonalitic plutons from the N. Peninsular Ranges batholith, S. California: Geological Society of America Abstracts with Programs, v. 30, no. 5, p. 59.

Rector, R.R., 1993, The interior nature and dynamics of the Long Potrero pluton, San Diego County, California: Geological Society of America Abstracts with Programs, v. 25, no. 6, p. A-303.

Rector, R.R., 1994, Petrology and emplacement of the Morena Reservoir pluton, San Diego County, California [M.S. thesis]: San Diego, San Diego State University, 260 p.

Rogers, T.H., 1965, Geologic map of California, Santa Ana sheet: California Division of Mines and Geology, scale 1:250,000, 1 sheet.

Rothstein, D.A., and Manning, C.E., 1994, Metamorphic P-T conditions in the eastern Peninsular Ranges batholith, Baja California, Mexico—Implications for geothermal gradients in magmatic arcs: Geological Society of America Abstracts with Programs, v. 26, no. 7, p. A42–A43.

Saleeby, J.B., and Busby-Spera, C.J, 1992, Early Mesozoic tectonic evolution of the western U.S., Cordillera, *in* Burchfiel, B.C., Lipman, P.W, and Zoback, M.L., eds., The Cordilleran Orogen: Conterminous U.S.: Boulder, Colorado, Geological Society of America, Geology of North America, v. G-3, p. 107–168.

Schmidt, M.W., 1992, Amphibole composition in tonalite as a function of pressure: An experimental calibration of the Al-in-hornblende barometer: Contributions to Mineralogy and Petrology, v. 110, p. 304–310.

Schmidt, K.L., and Paterson, S.R., 2002, A doubly vergent fan structure in the Peninsular Ranges batholith: Transpression or local complex flow around a continental margin buttress?: Tectonics, v. 21, 19 p.

Shaw, S.E., and Flood, R.H., 1981, The New England batholith, eastern Australia: Geochemical variations in time and space: Journal of Geophysical Research, v. 86, p. 10530–10544.

Shaw, S.E., Cooper, J.A., O'Neil, J.R., Todd, V.R., and Wooden, J.L., 1986, Strontium, oxygen, and lead isotope variations across a segment of the Peninsular Ranges batholith, San Diego County, California: Geological Society of America Abstracts with Programs, v. 18, no. 2, p. 183.

Shaw, S.E., Todd, V.R., and Hammarstrom, J.M., 1994, Preliminary mafic mineral chemistry of Jurassic and Early Cretaceous plutons of the Peninsular

Ranges batholith, California: Geological Society of America Abstracts with Programs, v. 26, no. 2, p. 91.

Silver, L.T., and Chappell, B.W., 1988, The Peninsular Ranges batholith: An insight into the evolution of the Cordilleran batholiths of southwestern North America: Transactions of the Royal Society of Edinburgh: Earth Sciences, v. 79, p. 105–121.

Silver, L.T., Taylor, H.P., Jr., and Chappell, B., 1979, Some petrological, geochemical and geochronological observations of the Peninsular Ranges batholith near the International Border of the U.S.A., and Mexico, *in* Abbott, P.L., and Todd, V.R., eds., Mesozoic crystalline rocks: Geological Society of America Annual Meeting Field Trip Guidebook: San Diego, San Diego State University, Department of Geological Sciences, p. 83–110.

Smith, T.E., Huang, C.H., Walawender, M.J., Cheung, P., and Wheeler, C., 1983, The gabbroic rocks of the Peninsular Ranges batholith, southern California: Cumulate rocks associated with calc-alkaline basalts and andesites: Journal of Volcanology and Geothermal Research, v. 18, p. 249–278.

Smith, D.K., Morton, D.M., and Miller, F.K., 1991, Hornblende geobarometry and biotite K-Ar ages from the northern part of the Peninsular Ranges batholith, southern California: Geological Society of America Abstracts with Programs, v. 23, no. 5, p. A273.

Snee, L.W., Naeser, C.W., Naeser, N.D., Todd, V.R., and Morton, D.M., 1994, Preliminary ⁴⁰Ar/³⁹Ar and fission-track cooling ages of plutonic rocks across the Peninsular Ranges batholith, southern California: Geological Society of America Abstracts with Programs, v. 26. no. 2, p. 94.

Soula, J.C., 1982, Characteristics and mode of emplacement of gneiss domes and plutonic domes in central-eastern Pyrenees: Journal of Structural Geology, v. 4, p. 313–342.

Streckeisen, A.L., 1973, Plutonic rocks, classification and nomenclature recommended by the I.U.G.S. Subcommission on the Systematics of Igneous Rocks: Geotimes, v. 18, no. 10, p. 26–30.

Sun, S.-S., and McDonough, W.F., 1989, Chemical and isotopic systematics of oceanic basalts: Implications for mantle composition and processes, *in* Saunders, A.D., and Norry, M.J., eds., Magmatism in the ocean basins: Geological Society of London Special Publication 42, p. 313–345.

Taylor, H.P., and Silver, L.T., 1978, Oxygen isotope relationships in plutonic igneous rocks of the Peninsular Ranges batholith, southern and Baja California, *in* Zartman, R.E., ed., Short papers of the Fourth International Conference on geochronology, cosmochronology and isotope geology: U.S. Geological Survey Open-file Report 78-701, p. 423–426.

Thomson, C.N., and Girty, G.H., 1994, Early Cretaceous intra-arc ductile strain in Triassic-Jurassic and Cretaceous continental margin arc rocks, Peninsular Ranges, California: Tectonics, v. 13, p. 1108–1119.

Tobisch, O.T., Saleeby, J.B., and Fiske, R.S., 1986, Structural history of continental volcanic arc rocks in the eastern Sierra Nevada, California: A case for extensional tectonics: Tectonics, v. 5, p. 65–94.

Todd, V.R., 1977a, Geologic map of the Agua Caliente Springs 7.5′ quadrangle, San Diego County, California: U.S. Geological Survey Open-File Report 77-742, 20 p., scale 1:24,000.

Todd, V.R., 1977b, Geologic map of the Cuyamaca Peak 7.5′ quadrangle, San Diego County, California: U.S. Geological Survey Open-File Report 77-405, 13 p., scale 1:24,000.

Todd, V.R., 1978, Geologic map of the Viejas Mountain 7.5′ quadrangle, San Diego County, California: U.S. Geological Survey Open-File Report 78-113, 30 p., scale 1:24,000.

Todd, V.R., 1980, Geologic map of the Alpine 7.5′ quadrangle, San Diego County, California: U.S. Geological Survey Open-File Report 80-82, 42 p., scale 1:24,000.

Todd, V.R., 1982, Geologic map of the Tule Springs 7.5′ quadrangle, San Diego County, California: U.S. Geological Survey Open-File Report 82-221, 23 p., scale 1:24,000.

Todd, V.R., 1983, Geologic map of the El Cajon Mountain 7.5′ quadrangle, San Diego County, California: U.S. Geological Survey Open-File Report 83-781, 20 p., scale 1:24,000.

Todd, V.R., 2004, Preliminary geologic map of the El Cajon 1:100,000 scale quadrangle, San Diego and Imperial Counties, California: Unpublished U.S. Geological Survey Open-File Report 94-18, Director's Approval 1-3-94, 45 p. (in press).

Todd, V.R., and Shaw, S.E., 1979, Structural, metamorphic and intrusive framework of the Peninsular Ranges batholith in southern San Diego County, California, *in* Abbott, P.L., and Todd, V.R., eds., Mesozoic Crystalline Rocks: Geological Society of America Annual Meeting Field Trip Guidebook: San Diego, California, San Diego State University, Department of Geological Sciences, p. 177–231.

Todd, V.R., and Shaw, S.E., 1985, S-type granitoids and an I-S line in the Peninsular Ranges batholith, southern California: Geology, v. 13, p. 231–233.

Todd, V.R., Detra, D.E., Kilburn, J.E., Griscom, A., Kruse, F.A., and Campbell, H.W., 1987, Mineral resources of the Fish Creek Mountains Wilderness Study Area, Imperial County, California: U.S. Geological Survey Bulletin 17711-C, 14 p.

Todd, V.R., Erskine, B.G., and Morton, D.M., 1988, Metamorphic and tectonic evolution of the northern Peninsular Ranges batholith, *in* Ernst, W.G., ed., Metamorphism and crustal evolution of the western United States: Rubey Volume VII: Englewood Cliffs, New Jersey, Prentice-Hall, p. 894–937.

Tosdal, R.M., Haxel, G.B., and Wright, J.E., 1989, Jurassic geology of the Sonoran Desert region, southern Arizona, southeastern California, and northernmost Sonora: Construction of a continental-margin magmatic arc, *in* Jenney, J.P., and Reynolds, S.J., eds., Geologic evolution of Arizona: Arizona Geological Society Digest 17, p. 397–434.

Vernon, R.H., 1983, Restite, xenoliths and microgranitoid enclaves in granites: Royal Society of New South Wales Journal and Proceedings, v. 116, p. 77–103.

Vernon, R.H., 1984, Microgranitoid enclaves: Globules of hybrid magma quenched in a plutonic environment: Nature, v. 304, p. 438–439.

Vernon, R.H., 1990, Crystallization and hybridism in microgranitoid enclave magmas: Microstructural evidence: Journal of Geophysical Research, v. 95, p. 17849–17859.

Vernon, R.H., 1991, Interpretation of microstructures of microgranitoid enclaves, *in* Didier, J., and Barbarin, B., eds., Enclaves and granite petrology: New York, Elsevier, p. 277–291.

Vernon, R.H., and Flood, R.H., 1988, Contrasting deformation of S- and I-type granitoids in the Lachlan Fold Belt, eastern Australia: Tectonophysics, v. 147, p. 127–143.

Walawender, M.J., Gastil, R.G., Clinkenbeard, J.P., McCormick, W.V., Eastman, B.G., Wernicke, R.S., Wardlaw, M.S., and Gunn, S.H., 1990, Origin and evolution of the zoned La Posta-type plutons, eastern Peninsular Ranges batholith, southern and Baja California: *in* Anderson, J.L., ed., The nature and origin of Cordilleran magmatism: Boulder, Colorado, Geological Society of America Memoir 174, p. 1–18.

Walawender, M.J., Girty, G.H., Lombardi, M.R., Kimbrough, D., Girty, M.S., and Anderson, C., 1991, A synthesis of recent work in the Peninsular Ranges batholith, *in* Walawender, M.J., and Hanan, B.B., eds., Geological excursions in southern California and Mexico: Geological Society of America Annual Meeting Field Trip Guidebook: San Diego, San Diego State University, Department of Geological Sciences, p. 297–318.

White, A.J.R., Allen, C.M., Beams, S.D., Carr, P.F., Champion, D.C., Chappell, B.W., Wyborn, D., and Wyborn, L.A.I., 2001, Granite suites and supersuites of eastern Australia: Australian Journal of Earth Sciences, v. 48, p. 515–530.

Zen, E-an, 1986, Aluminum enrichment in silicate melts by fractional crystallization: Some mineralogic and petrographic constraints: Journal of Petrology, v. 27, p. 1095–1117.

MANUSCRIPT ACCEPTED BY THE SOCIETY JUNE 2, 2003

Geological Society of America
Special Paper 374
2003

Geology and geochronology of granitic batholithic complex, Sinaloa, México: Implications for Cordilleran magmatism and tectonics

Christopher D. Henry
Nevada Bureau of Mines and Geology, University of Nevada, Reno, Nevada 89557, USA

Fred W. McDowell
Department of Geological Sciences, The University of Texas at Austin, Austin, Texas 78712, USA

Leon T. Silver
Division of Geological and Planetary Sciences, California Institute of Technology, Pasadena, California 91125, USA

ABSTRACT

Most of southern Sinaloa is underlain by a large, composite batholith, a continuation of the better-known Cordilleran batholiths of California and Baja California. Field relations and extensive K-Ar and U-Pb dating within a 120-km-wide and 120-km-deep transect show that the Sinaloa batholith formed in several stages. Early layered gabbros have hornblende K-Ar ages of 139 and 134 Ma, although whether these record emplacement age, cooling from metamorphism, or excess Ar is unresolved. A group of relatively mafic tonalites and granodiorites were emplaced before or during an episode of deformation and are restricted to within 50 km of the coast. These plutons, referred to here as syntectonic, show dynamic recrystallization textures that suggest deformation between 300° and 450 °C. A U-Pb zircon date on a probable syntectonic intrusion is 101 Ma. Hornblende K-Ar ages on definite syntectonic intrusions range between 98 and 90 Ma; these may record cooling soon after emplacement or following regional metamorphism.

Numerous posttectonic intrusions crop out from within ~20 km of the coast, where they intrude syntectonic rocks, to the eastern edge of the Sierra Madre Occidental, where they are covered by middle Cenozoic ash-flow tuffs. Posttectonic rocks are dominantly more leucocratic granodiorites and granites. Three samples were analyzed by both U-Pb and K-Ar methods. Their biotite and hornblende ages are concordant at 64, 46, and 19 Ma and agree within analytical uncertainties with U-Pb zircon ages of 66.8, 47.8, and 20 Ma. These data and field relations demonstrate that posttectonic intrusions were emplaced at shallow depths and cooled rapidly. Therefore, concordant K-Ar age pairs and hornblende ages in discordant samples approximate the time of emplacement. Discordance of biotite-hornblende age pairs is largely if not entirely a result of reheating by younger plutons. The combined age data indicate that posttectonic intrusions were emplaced nearly continuously between 90 and 45 Ma. One intrusion is 20 Ma. Based on outcrop area, volumes of intrusions were relatively constant through time.

The combined geochronological data indicate that posttectonic magmatism shifted eastward between 1 and 1.5 km/Ma. Whether syntectonic magmatism also

---

*E-mails: chenry@unr.edu; mcdowell@mail.utexas.edu; lsilver@gps.caltech.edu.

Henry, C.D., McDowell, F.W., and Silver, L.T., 2003, Geology and geochronology of granitic batholithic complex, Sinaloa, México: Implications for Cordilleran magmatism and tectonics, *in* Johnson, S.E., Paterson, S.R., Fletcher, J.M., Girty, G.H., Kimbrough, D.L., and Martín-Barajas, A., eds., Tectonic evolution of northwestern México and the southwestern USA: Boulder, Colorado, Geological Society of America Special Paper 374, p. 237–273. For permission to copy, contact editing@geosociety.org. © 2003 Geological Society of America.

migrated is uncertain. Ages of middle and late Tertiary volcanic rocks indicate that magmatism shifted rapidly (10–15 km/Ma) westward from the Sierra Madre Occidental after ca. 30 Ma.

The Sinaloa batholith is borderline calc-alkalic to calcic. SiO_2 contents of analyzed rocks range from 47 to 74%; the lower limit excluding two gabbros is 54%. Syntectonic rocks are more mafic on average than posttectonic rocks. SiO_2 contents of seven out of nine analyzed syntectonic rocks range narrowly between 59 and 62%, with one each at 65 and 67%. The posttectonic rocks show a wider range from 54 to 74% SiO_2, but only border phases and small intrusions have SiO_2 less than ~63%. Combined with their distribution, these data indicate that intrusions become more silicic eastward. The fact that the 20 Ma intrusion is relatively mafic (61% SiO_2) and lies near the coast with syntectonic rocks indicates that composition is related to location rather than to age.

The Sinaloa batholith shows both marked similarities and differences from batholiths of the Peninsular Ranges, Sonora, Cabo San Lucas (Baja California Sur), and Jalisco. The greatest similarities are in types of intrusions, a common sequence from early gabbro through syntectonic to posttectonic rocks, and general eastward migration of magmatism. However, the end of deformation recorded by syntectonic rocks may be different in each area. Sinaloa rocks show a similar wide range of compositions as rocks of the Peninsular Ranges and Sonora but are more potassic than the calcic Peninsular Ranges. Rare earth element patterns are most like those of the eastern part of the Peninsular Ranges and central part of Sonora, both areas that are underlain by Proterozoic crust or crust with a substantial Proterozoic detrital component. However, southern Sinaloa lies within the Guerrero terrane, which is interpreted to be underlain by accreted Mesozoic crust. The greatest differences are in distance and rate of eastward migration. Published data show that magmatism migrated eastward at ~10 km/Ma across the Peninsular Ranges and Sonora and from Jalisco southeast along the southwestern México coast. The area of slower eastward migration roughly correlates with the location of the Guerrero terrane and of possibly accreted oceanic crust that is no older than Jurassic.

Keywords: batholith, geochronology, tectonics, Cordillera, México, Sinaloa.

INTRODUCTION

The chain of granitic batholiths along the North American Cordillera (Fig. 1) may be its most prominent feature and the most permanent record of plate convergence and related magmatism. Batholiths of the Sierra Nevada and Peninsular Ranges of California and northern Baja California have been extensively studied (e.g., Evernden and Kistler, 1970; Gastil, 1975; Gastil et al., 1975; Krummenacher et al., 1975; Silver et al., 1979; Walawender and Smith, 1980; Baird and Miesch, 1984; Silver and Chappell, 1988; Todd et al., 1988; Walawender et al., 1990, 1991; Bateman, 1992; Ortega-Rivera et al., 1997; Johnson et al., 1999a, 1999b; Tate et al., 1999; Kimbrough et al., 2001; Ortega-Rivera, this volume, Chapter 11), but little has been done on the batholiths of mainland México (Anderson and Silver, 1974; Damon et al., 1983a, b; McDowell et al., 2001). Indeed, as with many aspects of Mexican geology that are critical to a full understanding of the geology and evolution of North America, there seems to be little recognition even of their existence.

The Sinaloa batholith is the southward continuation of the Peninsular Ranges batholith and related granitic rocks in Sonora

(Fig. 1). Batholithic rocks of the Cabo San Lucas block at the southern end of the Baja California peninsula and the Jalisco area of southwestern México are a further southward continuation (Gastil et al., 1978; Frizzell, 1984; Frizzell et al., 1984; Aranda-Gómez and Perez-Venzor, 1989; Moran-Zenteno et al., 2000; Schaaf et al., 2000; Kimbrough et al., 2002).

Our geologic map of a 10,000 km² area in southern Sinaloa forms the basis for an understanding of granitic magmatism in this part of the Cordillera (Fig. 2; Plate 1 [on the CD-ROM accompanying this volume]; Henry and Fredrikson, 1987). Although semitropical weathering is a problem in Sinaloa, cover by younger rocks is much less than in Sonora. The map area is underlain dominantly by batholithic rocks and extends from the Pacific coast to the Sierra Madre Occidental, where the granitic rocks are covered by largely Oligocene and Miocene ash-flow tuffs. This area constitutes a 120-km-wide and 120-km-deep transect across the batholith and is the only area with a relatively comprehensive map between southern Sonora and Nayarit (Gastil and Krummenacher, 1977; Gastil et al., 1978; Gans, 1997). To supplement data from this area, we also sampled granitic rocks in a reconnaissance transect in northern Sinaloa (Fig. 1). Based

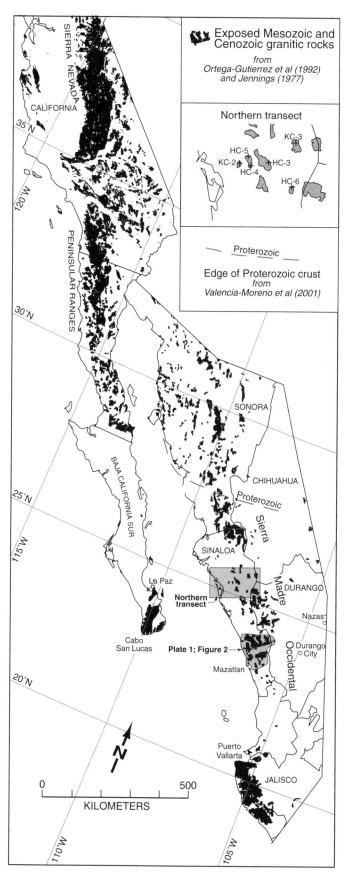

Figure 1. Mesozoic and Cenozoic granitic rocks of southern California and western México (from Jennings, 1977, and Ortega-Gutiérrez et al., 1992). Cretaceous to Tertiary granitic rocks crop out discontinuously from Sierra Nevada through Peninsular Ranges of California and Baja California, to Cabo San Lucas block of Baja California Sur, and through Sonora and Sinaloa to Jalisco. Granitic rocks continue southeast along southern México coast (Schaaf et al., 1995; Moran-Zenteno et al., 2000). They once formed a continuous belt that has been disrupted by displacement along San Andreas and other major faults and by seafloor spreading in Gulf of California. Shaded areas denote area of Figure 2 and northern Sinaloa transect. Northern transect is blown up to show sample locations.

upon this mapping we have acquired an extensive suite of K-Ar and U-Pb data that are critical to understanding the character and development of the Sinaloa batholith.

Locations mentioned here variably cite them relative to Mazatlan, for reference to the simplified geologic map of Figure 2, or more precisely in reference to Plate 1, the more detailed geologic map of Henry and Fredrikson (1987).

GEOCHRONOLOGY AND COMPARISON OF K-AR AND U-PB DATA

Fifty-seven samples representing a large but unknown number of separate granitic intrusions and seven samples of minor intrusions or volcanic rocks were dated by the K-Ar method; both biotite and hornblende were analyzed in 34 of these (Table 1, Appendix 1). Fifty-one of the granitic samples and all of the minor rocks are from the map area in southern Sinaloa, with a concentration of data from a well-constrained, east-northeast transect along the Rio Piaxtla (Plate 1, Fig. 2). The other six dated samples are from the northern reconnaissance transect (Fig. 1). Four samples that spanned most of the range of K-Ar ages and include both syn- and posttectonic intrusions were further dated by U-Pb on zircon (Table 2, Appendix 1). K-Ar ages range from ca. 145 Ma to 19 Ma, but, as explained below, the oldest apparent age probably reflects excess Ar. U-Pb ages on the four samples, ca. 101, 67, 48, and 20 Ma, are critical for establishing the overall age range and significance of K-Ar ages.

If biotite and hornblende ages from a single sample are indistinguishable within analytical uncertainty, we consider them to be concordant. By this criterion, twenty-one biotite-hornblende pairs in Table 1 are concordant. Samples with younger ages are more commonly concordant than are samples with older ages. Geologic inferences drawn from concordant results depend largely upon the age of the rock in question. A small absolute age difference for a relatively young (e.g., 50 Ma) rock implies rapid cooling from ~500 °C through ~300 °C (the estimated blocking temperatures for hornblende and biotite; Harrison, 1981; Harrison et al., 1985). In most cases, such rapid cooling would immediately follow emplacement. A granitic intrusion emplaced at great depth might stay at elevated temperature for

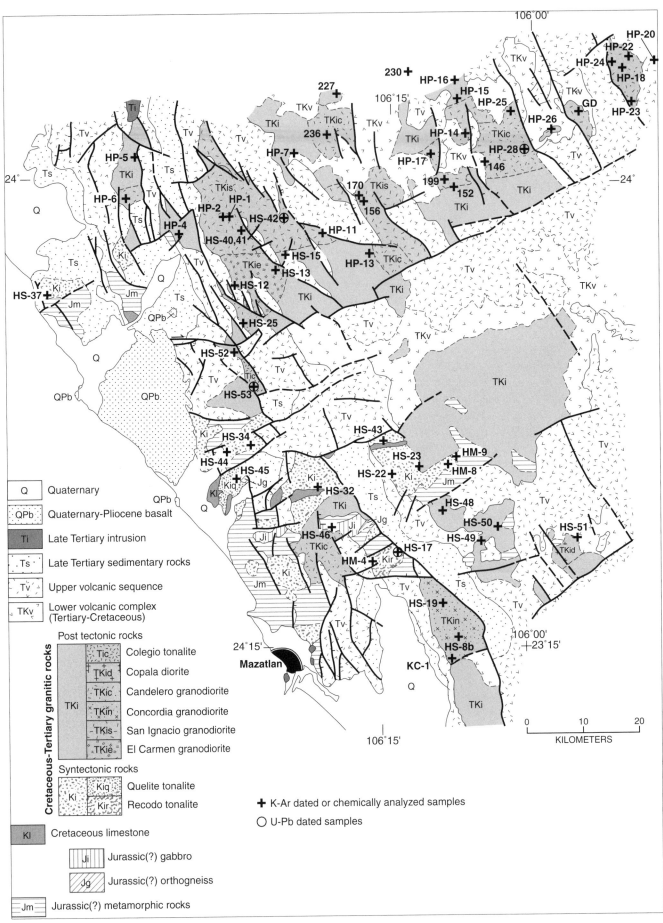

Figure 2. Simplified geologic map of southern Sinaloa showing locations of isotopically dated or chemically analyzed samples (from Henry and Fredrikson, 1987; see Plate 1 for detailed map).

TABLE 1. K-Ar DATA

Sample	Rock type	Mineral[1]	% K	$^{40}Ar^*$ (x10^{-5} scc/gm)	% $^{40}ArRad$	Age[2] (Ma)	±1σ
Gabbro (Jm)							
HS-46	Gabbro	H	0.158	0.0801	61.8	**133.8**	3.0
			0.146	0.0832	57.2		
			0.150				
HS-48	Gabbro	H	0.427	0.2392	73.0	**138.6**	3:1
			0.423	0.2365	70.2		
Syntectonic Intrusions (Ki)							
HS-17	Recodo tonalite	B	6.50	1.329	74.4	**52.7**	0.6
			6.44	1.359	90.4		
		H #1	0.350	0.1885	72.3	**125.0**	10.1
			0.362	0.1682	55.1		
			0.352				
		H #2	0.347	0.1904	65.8	**130.5**	2.9
			0.365				
			0.374				
			0.360				
		P	0.195	0.1116	55.6	**145.4**	3.3
			0.185				
			0.189				
HM-4	Recodo tonalite	B	6.75	2.649	92.7	**98.3**	1.1
			6.68	2.621	92.6		
		H	0.720	0.2794	85.1	**96.3**	2.2
			0.733				
			0.729				
HS-43	Amphibolite	H	0.745	0.2837	76.4	**94.5**	1.1
			0.756				
			0.756				
HS-22	Foliated granodiorite	B	7.06	2.564	94.5	**91.3**	1.0
			7.02	2.557	89.0		
		H	0.552	0.2177	89.4	**97.9**	2.2
			0.565				
			0.553				
HM-8	Quartz diorite	B	6.82	1.853	84.5	**69.0**	0.8
			6.74				
		H	0.664	0.2540	86.9	**95.6**	2.1
			0.668				
HS-45	Quelite tonalite	B	7.41	2.369	94.3	**80.7**	0.9
			7.36				
		H	0.855	0.3225	92.1	**94.3**	2.1
			0.859				
HS-44	Granodiorite	B	6.92	2.563	90.4	**92.8**	1.0
			6.94	2.564	80.5		
HS-34	Foliated granodiorite	B	6.97	2.511	95.2	**89.8**	1.0
			7.07				
			7.01				
		H	0.781	0.2538	83.5	**86.5**	1.9
			0.774	0.2818	84.7		
KC-2	Gneissic granodiorite	B	6.65	2.377	87.1	**89.7**	1.0
			6.66				
		H	0.525	0.1926	69.9	**93.4**	2.1
			0.510				
HC-4A	Foliated quartz diorite	B	6.82	2.499	90.5	**92.1**	1.0
			6.80				
		H	0.555	0.1984	80.8	**90.4**	2.0
			0.547				
HS-37	Foliated quartz diorite	B	7.10	2.415	90.9	**85.7**	1.0
			7.06				
		H	1.130	0.4030	88.7	**89.1**	2.0
			1.140				

1—B=biotite; H=hornblende; P=pyroxene; A=adularia (with a lot of quartz).
2—Ages calculated with decay constants of Steiger and Jager (1977); $^{40}K = 0.01167$ atom percent, $\lambda_\beta = 4.962 \times 10^{-10}$ yr^{-1}; $\lambda_\epsilon = 0.581 \times 10^{-10}$ yr^{-1}.

(continued)

TABLE 1. K-Ar DATA *(continued)*

Sample	Rock type	Mineral[1]	% K	$^{40}Ar^*$ (x10^{-5} scc/gm)	% $^{40}ArRad$	Age[2] (Ma)	±1σ
Syntectonic Intrusions (Ki) *(continued)*							
HS-32	Hypersthene granodiorite	B	6.80 6.69	1.8780	86.8	**69.4**	0.8
		P	0.130 0.121 0.135	0.0439 0.0440	13.3 6.2	**85.8**	1.9
Posttectonic Intrusions (Tki)							
HC-5	Granodiorite	B	6.56 6.59	2.321	88.9	**88.6**	1.0
		H	0.659 0.660	0.2352	83.6	**89.5**	2.0
HS-12	El Carmen granodiorite	B	6.61 6.54	1.946 1.964	88.2 91.6	**75.0**	0.8
		H	0.696 0.687	0.1953 0.2048	65.5 88.9	**73.0**	1.6
HS-13	El Carmen granodiorite	B	6.42 6.47	1.604	82.7	**62.9**	0.7
		H	0.650 0.648	0.1922	55.6	**74.7**	1.7
HS-25	El Carmen granodiorite	B	1.98 1.96	0.5085	77.0	**65.2**	0.7
		H	0.372 0.377	0.1090	74.9	**73.4**	1.6
HS-50	Quartz diorite	B	5.60 5.61	1.493	8.0	**67.2**	0.8
		H	0.502 0.502	0.1405	65.4	**70.6**	1.6
HC-6	Monzogranite	B	4.88 4.92	1.277	85.0	**65.9**	0.8
		H	0.515 0.508	0.1365	65.9	**67.4**	1.5
HS-15	Granodiorite	B	5.62 5.60	1.411 1.415	83.7 92.2	**63.7**	0.7
		H	0.506 0.507	0.1320	79.1	**65.9**	1.4
152	Quartz diorite	B	3.66 3.53 3.52	0.7215	77.5	**51.3**	0.6
		H	0.437 0.434	0.1157	73.2	**67.1**	1.5
HP-4	Quartz diorite	B	6.77 6.76	1.724 1.722	81.4 91.3	**64.4**	0.7
		H	0.447 0.452	0.1138 0.1162	67.7 80.9	**64.7**	1.4
HS-42	San Ignacio granodiorite main San Ignacio pluton	B	6.92 6.93	1.735	91.6	**63.4**	0.7
		H	0.520 0.518	0.1316	52.4	**64.1**	1.4
HS-40	San Ignacio granodiorite main San Ignacio pluton	B	7.32 7.26	1.853	91.9	**64.3**	0.7
HS-41	San Ignacio granodiorite main San Ignacio pluton	B	6.94 7.02	1.764	93.2	**63.9**	0.7
HP-2	San Ignacio granodiorite main San Ignacio pluton	B	5.71 5.75	1.169	81.1	**51.7**	0.6
		H	0.701 0.685 0.708	0.1757	76.6	**63.7**	1.4

1—B=biotite; H=hornblende; P=pyroxene; A=adularia (with a lot of quartz).
2—Ages calculated with decay constants of Steiger and Jager (1977); ^{40}K = 0.01167 atom percent, λ_β = 4.962 x 10^{-10} yr^{-1}; λ_e = 0.581 x 10^{-10} yr^{-1}.

(continued)

TABLE 1. K-Ar DATA *(continued)*

Sample	Rock type	Mineral[1]	% K	^{40}Ar* (x10^{-5} scc/gm)	% ^{40}ArRad	Age[2] (Ma)	±1σ
Posttectonic Intrusions (Tki) *(continued)*							
HP-11	San Ignacio granodiorite	B	5.78	1.388	83.2	**61.0**	0.7
	main San Ignacio pluton		5.71	1.382	89.5		
		H	0.552	0.1377	64.5	**63.7**	1.4
			0.545	0.1385	77.7		
156	San Ignacio granodiorite	B	6.60	1.672	83.0	**64.6**	0.7
			6.46				
			6.57				
		H	0.610	0.1531	61.4	**63.2**	1.4
			0.615				
170	San Ignacio granodiorite	B	6.20	1.588	86.5	**64.5**	0.7
			6.26				
146	San Ignacio granodiorite	B	6.54	1.207	80.4	**47.0**	0.5
			6.51				
		H	0.564	1.116	63.4	**49.6**	1.1
			0.586				
			0.563				
			0.56				
HP-26	San Ignacio granodiorite	B	6.42	1.245	79.6	**49.3**	0.6
			6.41				
HS-23	Granodiorite	B	5.88	1.446	84.6	**62.1**	0.7
			5.91				
HS-51	Granodiorite	B	4.09	0.9365	80.2	**57.9**	0.7
			4.11				
HM-9	Monzogranite	B	4.68	1.007	83.5	**55.0**	0.6
			4.61				
KC-3	Monzogranite	B	6.86	1.384	84.1	**52.6**	0.6
			6.66				
KC-1	Granodiorite	B	1.74	0.3960	63.3	**57.3**	0.6
			1.76				
HS-8b	Concordia granodiorite	B	4.89	1.045	79.9	**53.8**	0.6
			4.91	1.033	86.2		
		H	0.620	0.1329	56.8	**54.2**	1.2
			0.613	0.1307	82.0		
HP-16	Granodiorite	B	5.80	1.232	85.2	**54.5**	0.6
			5.79	1.261	87.5		
		H	0.515	0.1131	56.4	**55.6**	1.2
			0.516				
230	Granodiorite	B	6.23	1.342	89.4	**53.8**	0.6
			6.29	1.316	91.4		
		H	0.673	0.1443	60.8	**53.6**	1.2
			0.671	0.1398	62.0		
HP-7	Diorite	H	0.238	0.0494	44.4	**52.2**	1.2
			0.242				
HP-5	Granodiorite	B	3.75	0.7798	75.0	**52.4**	0.6
			3.80				
HP-15	Granodiorite	B	7.27	1.363	86.3	**47.7**	0.5
			7.25				
HP-22	Granodiorite	B	6.87	1.226	87.1	**45.5**	0.5
			6.88	1.234	89.6		
		H	0.526	0.0936	52.0	**46.2**	1.0
			0.524	0.0974	54.2		
HP-6	Candelero granodiorite	B	5.77	1.907	89.3	**81.8**	0.9
			5.83	1.864	90.3		
		H	0.634	0.2078	76.6	**82.6**	1.8
			0.633	0.2083	81.3		

1—B=biotite; H=hornblende; P=pyroxene; A=adularia (with a lot of quartz).
2—Ages calculated with decay constants of Steiger and Jager (1977); ^{40}K = 0.01167 atom percent, $\lambda_\beta = 4.962 \times 10^{-10}$ yr^{-1}; $\lambda_\epsilon = 0.581 \times 10^{-10}$ yr^{-1}.

(continued)

TABLE 1. K-Ar DATA *(continued)*

Sample	Rock type	Mineral[1]	% K	^{40}Ar* (x10^{-5} scc/gm)	% ^{40}ArRad	Age[2] (Ma)	±1σ
Posttectonic Intrusions (Tki) *(continued)*							
HP-13	Candelero granodiorite	B	6.45 6.47	1.564	73.3	**61.3**	0.7
		H	0.722 0.705 0.740	0.1770	69.3	**62.0**	1.4
236	Candelero granodiorite	B	2.29 2.33	0.5099 0.4436	66.3 75.1	**52.4**	5.1
		H	0.338 0.339	0.0698 0.0668	46.9 58.8	**51.2**	1.6
HC-3	Candelero granodiorite	B	4.87 4.88	0.9829	80.5	**51.2**	0.6
		H	0.336 0.328 0.339	0.0654	51.9	**49.7**	1.1
HP-14	Candelero granodiorite main Candelero pluton	B	4.20 4.08	0.7605 0.7727	75.7 83.4	**47.0**	0.5
		H	0.480 0.475	0.0876	39.2	**46.6**	1.0
HP-25	Candelero granodiorite main Candelero pluton	B	6.26 6.15	1.134	76.8	**46.4**	0.5
HP-28	Candelero granodiorite main Candelero pluton	B	6.18 6.14	1.115	91.4	**46.0**	0.5
		H	0.352 0.348	0.0613 0.0615	53.2 38.5	**44.6**	1.0
HP-18	Candelero granodiorite	B	4.59 4.58	0.7919	66.7	**43.9**	0.5
HP-23	Candelero granodiorite	B	6.01 5.93	0.7842	70.6	**33.5**	0.4
HP-20	Candelero granodiorite	B	3.92 3.96	0.4770	66.7	**30.9**	0.3
199	Quartz diorite	B	6.53 6.52	1.267	83.4	**49.3**	0.6
GD	Granodiorite	B	2.24 2.27	0.3883	58.4	**43.8**	0.5
Hbl por	Porphyritic granodiorite	H	0.415 0.410	0.0771	60.3	**47.5**	1.0
227	Granodiorite	B	4.58 4.56	0.5210	59.0	**29.1**	0.3
HS-53	Colegio tonalite	B	7.54 7.45	0.5467	65.4	**18.7**	0.2
		H #1	0.609 0.651 0.629 0.639	0.0468 0.0494	39.8 54.3	**19.5**	0.4
		H #2	0.525 0.530	0.0401 0.0381	53.4 43.0	**19.0**	0.4
Upper volcanic rocks							
HP-24	quartz diorite dike	B	6.76 6.80	0.8473 0.8400	76.4 78.0	**31.8**	0.4
HP-10	andesite dike	H	0.490 0.488	0.0565	58.1	**29.5**	0.7
174	rhyolite dike	B	4.65 4.71	0.4357	38.7	**23.8**	0.3
HM-1	dacite lava	B	5.20 5.13	0.4501	66.8	**22.3**	0.2
Miscellaneous							
Adularia	Vein	A	0.597 0.593	0.0947 0.0965	46.9 50.1	**40.9**	0.4

1—B=biotite; H=hornblende; P=pyroxene; A=adularia (with a lot of quartz).
2—Ages calculated with decay constants of Steiger and Jager (1977); ^{40}K = 0.01167 atom percent, λ_β = 4.962 x 10^{-10} yr^{-1}; λ_ϵ = 0.581 x 10^{-10} yr^{-1}.

TABLE 2. U-Pb DATA

	Zircon weight (mg)	Concentration (ppm)		Observed isotope ratios		Atom ratios[1]			Ages (Ma)[2]		
		U	Pb[1]	$^{206}Pb/^{204}Pb$	$^{207}Pb/^{204}Pb$	$^{206}Pb/^{238}U$	$^{207}Pb/^{235}U$	$^{207}Pb/^{206}Pb$	$^{206}Pb/^{238}U$	$^{207}Pb/^{235}U$	$^{207}Pb/^{206}Pb$
HS-53 Colegio tonalite (best age estimate = 20.0 ± 0.4 Ma)											
Fraction 1											
conc. 1	306.2	206.7	0.899	55.61	17.66	0.00362	0.02889	0.05404	23.3 ± 1.0	26.9 ± 3.0	373 ± 260
conc. 2		203.0	0.777	92.50	19.28	0.00319	0.02172	0.04938	20.5 ± 0.4	21.8 ± 2.5	166 ± 140
Fraction 2											
conc.	594.1	253.7	0.943	96.25	19.29	0.00311	0.02013	0.04705	20.0 ± 0.4	20.2 ± 1.3	52 ± 190
HP-28 Candelero granodiorite (best age estimate = 47.8 ± 1.0 Ma)											
Fraction 1											
conc.	308.6	983.4	7.873	614.09	43.64	0.00748	0.04828	0.04691	48.0 ± 1.0	47.9 ±1.0	46 ± 30
comp.				719.69	48.90			0.04750			
Fraction 2											
conc.	439.1	802.0	6.312	364.88	32.14	0.00740	0.04858	0.04759	47.6 ± 1.0	48.2 ± 1.2	79 ± 40
HS-42 San Ignacio granodiorite (best age estimate = 66.8 ± 1.3 Ma)											
Fraction 1											
conc.	384.4	371.9	4.043	118.68	20.24	0.01035	0.06575	0.04550	66.4 ± 1.3	64.7 ± 3.2	<0 ± >85
Fraction 2											
conc.	187.9	514.7	5.641	149.79	21.86	0.01046	0.06855	0.04770	67.1 ± 1.3	67.3 ± 2.7	84 ± 85
comp.				169.20	22.51			0.04531			
HS-17 Recodo tonalite (best age estimate = 101.2 ± 2.0 Ma)											
Fraction 1											
conc.	270.5	297.2	4.623	191.70	24.19	0.01583	0.10585	0.04901	101.2 ± 2.0	102.2 ± 3.3	148 ± 65
comp.				206.18	24.88			0.04951			
Fraction 2											
conc.	439.1	335.6	5.380	327.70	30.45	0.01581	0.10478	0.04810	101.1 ± 2.0	101.2 ± 2.4	104 ± 40

1—radiogenic; corrected for $^{206}Pb/^{204}Pb$ = 18.60 and $^{207}Pb/^{204}Pb$ = 15.60 for both blank and common Pb.
2—calculated with $^{238}U/^{235}U$ = 137.88; $\lambda238$ = 1.55125 x 10^{-10}, $\lambda235$ = 9.8485 x 10^{-10}
conc.—concentration, aliquot spiked with ^{208}Pb and ^{235}U before fusion; comp. = composition, unspiked.

long time, but except in unusual circumstances, it would not then cool rapidly enough to generate concordant ages. For older rocks, e.g., ≥100 Ma, the larger absolute age difference of between 3 and 4 Ma allowed by concordance of our criteria does not allow such a simple explanation. A concordant 100 Ma rock could have cooled relatively slowly following emplacement. Further interpretation must recognize the geologic setting.

The best estimate of the U-Pb age for samples HP-28, HS-42, and HS-17, where both fractions are concordant, is taken to be the average of the two $^{206}Pb/^{238}U$ ages (Table 2). The best estimate of the U-Pb age for sample HS-53, where only fraction 2 is concordant, is taken to be that fractions $^{206}Pb/^{238}U$ age. We choose the $^{206}Pb/^{238}U$ ages because most of the ^{207}Pb in all samples is from the reagents used in analysis, especially the borate flux. Therefore, uncertainties in the $^{207}Pb/^{235}U$ and $^{207}Pb/^{206}Pb$ ages are large. Nevertheless, most $^{206}Pb/^{238}U$ and $^{207}Pb/^{235}U$ ages are indistinguishable within the analytical uncertainties, and only the $^{207}Pb/^{206}Pb$ ages of sample HS-53 differ significantly from the $^{206}Pb/^{238}U$ and $^{207}Pb/^{235}U$ ages. An obvious implication of these data is that the rocks contain little if any inherited zircon.

The U-Pb zircon ages for three posttectonic samples are all slightly greater than but mostly within analytical uncertainties of concordant biotite-hornblende pairs: HS-42 zircon 66.8 ± 1.3 Ma, biotite 63.4 ± 0.7 Ma, hornblende 64.1 ± 1.4 Ma; HP-28 zircon 47.8 ± 1.0 Ma, biotite 46.0 ± 0.5 Ma, hornblende. 44.6 ± 1.0 Ma; HS-53 zircon 20.0 ± 0.04 Ma, biotite 18.7 ± 0.2 Ma, hornblende 19.0 ± 0.4 and 19.5 ± 0.4 Ma (Fig. 3, Tables 1 and 2). Results from the fourth sample, HS-17, which is pre-or syntectonic, are complicated. The U-Pb age is well defined at 101.2 ±

2.0 Ma. The K-Ar biotite age is much younger, at 52.7 ± 0.6 Ma. Apparent ages of three different separates of amphibole-pyroxene mixtures are impossibly old at 125.0 ± 10.1 to 145.4 ± 3.3 Ma. In thin section, hornblende encloses and is intimately intergrown with clinopyroxene and a colorless amphibole, probably tremolite. It was not possible to get pure separates of any phase, but the roughly inverse relation between K and Ar contents (Table 1) suggests that pyroxene contains excess Ar. The high, ~0.19% K content of the pyroxene concentrate indicates significant contamination with another phase, either or both of the amphiboles. The U-Pb and K-Ar results for sample HS-17 can be further illuminated by comparison with K-Ar ages of sample HM-4, which we interpret to be the same intrusion. Sample HM-4, collected just 5.5 km to the west of HS-17, is petrographically and compositionally similar to sample HS-17 (Table 3). The biotite (98.3 ± 1.1 Ma) and hornblende (96.3 ± 2.2 Ma) ages from sample HM-4 are, as with the posttectonic samples, slightly younger than and partly overlap with the U-Pb age.

The general agreement between the K-Ar and U-Pb ages for at least three and possibly all four samples implies that the sampled intrusions cooled rapidly following emplacement. This interpretation is strongest for the posttectonic intrusions, for which 18 of 25 sample pairs are concordant (Fig. 3, Table 1). It is consistent with field observations that infer shallow emplacement for the posttectonic intrusions. They generally intrude and commonly dome possibly contemporaneous volcanic rocks (Lower volcanic complex of Fig. 2 and Plate 1). Contacts are sharp, discordant, and commonly brecciated. The intrusions are locally porphyritic and associated with numerous epithermal ore deposits (Smith et

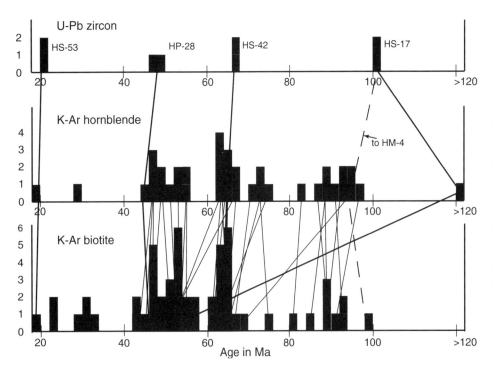

Figure 3. Histogram of U-Pb zircon, K-Ar hornblende, and K-Ar biotite ages. Tie lines connect dates on separates from same samples. Heavy lines connect four samples for which U-Pb dating was employed. Dashed line connects U-Pb age of sample HS-17 to K-Ar ages of sample HM-4, which is from same Recodo tonalite. K-Ar ages are slightly lower but mostly within analytical uncertainties of U-Pb ages. Our interpretation is that concordant biotite and hornblende K-Ar ages, those that agree within analytical uncertainty, generally record rapid cooling immediately following emplacement.

TABLE 3. CHEMICAL ANALYSES OF GRANITIC ROCKS, SINALOA BATHOLITH

	Gabbro		Syntectonic rocks								
Unit:			Quelite tonalite					Recodo tonalite			
Sample:	HS-46	HS-48	HS-37	HS-34	HS-44	HS-45	HS-22	HM-8	HS-17	HM-4	HS-32
Map unit:	Ji	Ji	Ki	Ki	Ki	Kiq	Ki	Ki	Kir	Kir	Ki
SiO_2	47.28	49.34	61.32	64.97	67.19	58.57	60.99	59.25	58.67	60.48	59.86
TiO_2	1.84	2.34	0.87	0.69	0.58	0.79	0.94	0.96	0.91	0.79	0.94
Al_2O_3	23.92	16.08	16.22	15.77	15.48	17.42	16.67	16.85	16.38	15.34	16.41
Fe_2O_3	1.63	12.16	6.65	1.78	2.16	4.77	2.10	0.29	1.90	6.96	1.97
FeO*	6.13			2.71	1.84	2.19	3.97	7.08	5.43		4.17
MnO	0.12	0.20	0.11	0.07	0.04	0.12	0.10	0.12	0.13	0.09	0.10
MgO	2.15	5.98	2.39	2.41	1.94	3.44	2.98	3.64	4.39	3.46	3.81
CaO	12.66	10.19	4.88	4.48	3.69	6.47	5.53	6.03	7.09	5.74	6.03
Na_2O	3.22	1.87	3.56	4.24	4.17	4.07	3.92	3.64	3.25	3.26	4.00
K_2O	0.19	0.71	2.99	2.67	2.76	1.97	2.59	1.90	1.63	2.19	2.46
P_2O_5	0.85	0.08	0.58	0.20	0.15	0.20	0.21	0.25	0.21	0.18	0.24
H_2O+											
H_2O-											
LOI	0.58	1.17	0.41	0.85	1.69	0.93	0.86	0.59	1.69	1.62	0.60
Total*	100.58	100.13	99.97	100.85	101.69	100.93	100.86	100.59	101.69	100.13	100.60
Sc xrf	20	33	17	10	9	20	16	18	25	20	18
V	207	403	119	109	58	175	142	191	177	184	156
Ni	25	<20	44	31	14	27	13	110	28	<20	146
Cu	164	74	19	35	244	60	51	56	35	30	43
Zn	104	98	83	74	36	90	90	96	98	79	80
Ga	24	20	20	20	12	19	19	20	18	19	19
As	5	<5	<5	<5	6	<5	11	8	7	<5	6
Rb	4	21	105	55	57	62	69	56	50	67	48
Sr	705	347	327	647	360	551	520	541	467	405	651
Sr xrf	705	346	336	608	546	529	493	534	439	400	623
Y	25	23.9	28.5	10	5	18	19	19	19	20.4	15
Y xrf	26	23.0	28.0	11	7	20	19	19	18	20.0	16
Zr	65	94	225	157	81	158	284	139	112	155	194
Zr xrf	68	125	236	168	130	139	257	144	115	157	201
Nb	17.6	15.2	7.4	7.3	4.2	7.0	12.5	9.1	7.2	5.3	9.8
Mo	9.0	<2	<2	5.6	13.5	7.3	4.9	9.2	5.7	<2	5.7
Sb	0.9	0.4	-0.2	0.5	0.3	0.3	1.2	3.1	0.8	0.6	0.5
Cs	2.2	3.8	5.0	1.8	3.7	3.1	3.0	2.2	2.4	4.8	1.2
Ba	127	382	991	1070	557	541	863	717	550	712	1083
Ba xrf	134	368	994	1072	882	535	871	731	533	678	1068
La	15.5	16.1	29.4	27.1	12.9	22.2	27.6	21.4	19.6	23.1	26.5
Ce	38.9	35.3	58.3	50.9	23.7	43.3	52.9	45.3	39.3	46.1	51.2
Pr	5.62	4.49	6.96	5.72	2.59	4.97	6.12	5.55	4.59	5.33	5.88
Nd	27.9	18.7	27.5	21.2	9.0	20.1	22.8	22.7	18.4	20.7	22.7
Sm	6.8	4.33	5.84	4.0	1.5	4.2	4.6	4.9	3.9	4.24	4.4
Eu	3.04	1.34	1.22	1.08	0.56	1.19	1.20	1.29	1.20	1.03	1.28
Gd	7.1	4.06	5.20	2.8	1.3	3.8	3.9	4.4	4.0	3.51	3.7
Tb	1.0	0.75	0.90	0.4	0.2	0.6	0.7	0.7	0.6	0.61	0.5
Dy	4.9	4.36	5.03	1.9	0.8	3.1	3.2	3.2	3.2	3.54	2.6
Ho	1.1	0.86	0.98	0.4	0.2	0.7	0.8	0.8	0.8	0.71	0.6
Er	2.5	2.46	2.85	1.0	0.4	1.9	2.1	1.9	1.9	2.02	1.6
Tm	0.31	0.363	0.408	0.14	0.07	0.29	0.30	0.29	0.31	0.315	0.22
Yb	1.60	2.24	2.51	0.90	0.30	1.90	1.80	1.7	1.80	1.88	1.40
Lu	0.24	0.335	0.379	0.13	0.07	0.27	0.31	0.26	0.28	0.282	0.23
Hf	1.7	3.0	6.0	4.2	2.3	4.5	7.5	3.8	3.2	4.2	4.9
Ta	1.67	0.09	<0.01	1.13	0.74	0.91	1.44	0.64	0.91	<0.01	1.00
Tl	<0.1	0.14	0.53	0.2	0.2	0.2	0.3	0.3	0.2	0.37	0.2
Pb	4	5	20	7	2	9	14	10	7	18	13
Th	0.50	3.78	9.05	5.77	2.92	4.94	7.52	3.19	5.09	8.26	2.90
U	0.35	0.74	2.45	1.95	1.14	1.62	1.66	1.14	1.32	1.42	1.03
Eree	116.5	95.7	147.5	117.7	53.6	108.5	128.3	114.4	99.9	113.4	122.8
La_N/yb_N	6.71	4.97	8.10	20.85	29.77	8.09	10.62	8.72	7.54	8.47	13.11
Eu/Eu*	1.32	0.95	0.66	0.93	1.19	0.89	0.84	0.83	0.91	0.79	0.94
Sri											

Note: Analyses by Actlabs by ICP-optical (major oxides) and mass spectrometry (trace elements) and XRF (as noted).

(continued)

TABLE 3. CHEMICAL ANALYSES OF GRANITIC ROCKS, SINALOA BATHOLITH (continued)

					Posttectonic rocks						
Unit:	San Ignacio granodiorite								El Carmen granodiorite	Concordia granodiorite	
Sample:	HP-1	HS-40	HS-41	HS-42	HP-11	156	146	HP-26	HS-25	HS-8B	HS-19
Map unit:	TKis	TKis	TKis	TKis	TKis	TKis	TKis	TKis	TKie	TKicn	TKicn?
SiO_2	62.84	66.92	54.94	66.41	66.44	64.32	63.30	65.14	66.73	65.13	58.10
TiO_2	0.78	0.57	0.96	0.64	0.61	0.68	0.76	0.67	0.52	0.60	0.77
Al_2O_3	16.13	15.96	21.28	14.92	15.17	15.06	15.91	15.97	15.65	16.22	17.49
Fe_2O_3	1.21	1.98	3.67	4.54	0.66	5.96	5.47	1.50	1.11	0.67	2.62
FeO*	4.96	2.56	2.83		4.01			3.21	3.29	3.66	4.85
MnO	0.11	0.06	0.08	0.07	0.08	0.10	0.08	0.08	0.09	0.08	0.14
MgO	2.53	1.93	2.58	1.85	1.95	2.16	2.44	2.04	2.02	2.29	3.94
CaO	4.92	4.73	7.15	3.81	3.95	4.42	5.01	4.30	4.40	4.73	7.01
Na_2O	3.02	3.45	4.19	3.08	3.20	3.21	3.39	3.58	3.49	3.57	3.24
K_2O	3.34	1.67	1.99	3.80	3.79	3.48	2.84	3.36	2.55	2.91	1.64
P_2O_5	0.15	0.16	0.35	0.12	0.13	0.13	0.17	0.15	0.15	0.14	0.19
H_2O+											
H_2O-											
LOI	0.39	1.05	1.08	0.33	0.45	0.78	0.92	0.74	1.56	1.29	1.59
Total*	100.39	101.05	101.08	99.58	100.45	100.31	100.28	100.74	101.56	101.29	101.59
Sc xrf	19	5	6	13	12	14	13	11	12	16	22
V	181	82	79	94	82	128	114	106	95	103	168
Ni	26	44	13	<20	16	<20	<20	21	5	25	23
Cu	50	<5	53	25	75	40	29	39	24	13	8
Zn	66	77	97	-30	57	72	74	83	62	110	104
Ga	17	18	21	16	12	17	20	19	17	18	19
As	9	<5	44	<5	5	<5	<5	9	<5	7	<5
Rb	139	49	54	172	140	149	147	190	81	154	48
Sr	330	650	884	273	211	270	385	381	403	443	511
Sr xrf	316	638	940	269	262	269	379	364	389	430	479
Y	24	6	8	23.9	18	24.0	22.6	23	13	22	16
Y xrf	24	6	9	23.0	23	23.0	22.0	23	13	22	16
Zr	251	143	320	227	141	203	202	213	105	187	91
Zr xrf	255	154	362	232	181	213	198	223	104	206	90
Nb	10.5	4.6	6.3	6.7	6.7	15.3	9.0	11.1	6.5	9.2	4.9
Mo	7.4	6.0	4.4	<2	5.4	7	2	14.2	7.6	9.2	5.2
Sb	2.4	<0.1	1.8	0.4	6.1	0.6	0.3	1.7	0.3	1.2	0.6
Cs	5.8	2.2	2.9	7.5	5.2	9.0	7.2	9.2	3.8	4.2	2.5
Ba	827	926	1075	816	573	862	708	732	877	824	641
Ba xrf	837	937	1192	795	761	836	675	747	887	827	631
La	24.3	12.1	18.2	20.7	26.9	25.8	27.1	39.7	24.2	21.5	14.2
Ce	49.5	21.3	32.1	45.3	51.8	53.6	57.2	77.6	43.0	47.4	29.7
Pr	5.86	2.28	3.53	5.52	5.62	6.21	6.67	8.51	4.68	5.83	3.58
Nd	22.3	8.7	12.7	21.5	20.0	23.8	25.2	30.4	16.1	23.7	14.9
Sm	4.9	1.6	2.3	4.70	3.9	4.94	5.14	5.9	3.0	5.0	3.3
Eu	1.15	0.79	1.08	0.864	0.74	0.944	1.09	1.11	0.79	0.98	1.00
Gd	4.8	1.2	1.9	4.03	3.5	4.23	4.21	5.0	2.6	4.1	3.2
Tb	0.8	0.2	0.3	0.71	0.6	0.74	0.71	0.8	0.4	0.8	0.5
Dy	4.1	1.0	1.3	4.06	2.9	4.17	3.90	4.0	2.0	3.5	2.8
Ho	1.0	0.2	0.3	0.80	0.7	0.80	0.73	0.9	0.5	0.9	0.7
Er	2.5	0.6	0.8	2.32	1.8	2.32	2.07	2.3	1.3	2.3	1.8
Tm	0.39	0.10	0.14	0.356	0.27	0.356	0.303	0.35	0.20	0.35	0.28
Yb	2.4	0.70	0.90	2.17	1.7	2.19	1.93	2.2	1.30	2.30	1.70
Lu	0.35	0.11	0.17	0.330	0.25	0.326	0.283	0.31	0.20	0.33	0.28
Hf	7.0	4.0	7.1	6.4	4.0	5.7	5.3	6.1	3.0	5.4	2.6
Ta	0.90	1.09	0.70	0.91	0.71	11.4	2.62	1.28	1.44	2.03	0.73
Tl	0.4	0.3	0.3	0.69	1.8	0.82	0.79	0.9	0.4	0.5	0.3
Pb	10	10	8	12	205	31	20	19	11	9	9
Th	16.08	2.94	3.79	19.4	17.88	24.3	24.8	30.28	8.25	18.32	2.88
U	4.83	1.51	1.87	4.62	4.37	6.22	6.71	9.28	1.82	6.51	0.87
Σree	124.4	50.9	75.7	113.3	120.7	130.4	136.6	179.1	100.3	119.0	77.9
La_N/yb_N	7.01	11.97	14.00	6.60	10.96	8.17	9.73	12.49	12.89	6.47	5.78
Eu/Eu*	0.71	1.66	1.52	0.59	0.60	0.61	0.69	0.60	0.84	0.64	0.92
Sri			0.7030			0.7062			0.7058		

Note: Analyses by Actlabs by ICP-optical (major oxides) and mass spectrometry (trace elements) and XRF (as noted).

(continued)

TABLE 3. CHEMICAL ANALYSES OF GRANITIC ROCKS, SINALOA BATHOLITH *(continued)*

					Posttectonic rocks						
Unit:				Candelero granodiorite							
Sample:	HP-6	HP-13	236	HP-25	HP-28	HP-23	HC-3	HS-49	HS-50	HS-15	HS-23
Map unit:	TKica	TKica	Tkica	TKica	TKica	TKica	Tkica	TKi	TKi	Tki	Tki
SiO$_2$	63.86	71.70	70.48	65.22	68.90	67.47	66.84	63.63	62.96	61.86	67.05
TiO$_2$	0.65	0.36	0.31	0.71	0.50	0.62	0.49	0.69	0.60	0.82	0.54
Al$_2$O$_3$	16.30	14.08	14.29	16.76	15.39	15.64	16.35	16.16	16.96	16.60	15.56
Fe$_2$O$_3$	1.02	0.11	3.05	0.94	0.71	3.85	1.10	2.73	2.29	1.98	1.25
FeO*	4.48	2.92		3.29	2.43		2.83	2.87	3.26	4.13	2.88
MnO	0.10	0.06	0.06	0.07	0.04	0.06	0.07	0.09	0.10	0.10	0.08
MgO	2.41	0.88	1.11	1.67	1.23	1.16	1.62	2.58	2.52	2.80	1.99
CaO	4.86	2.49	2.94	4.59	3.33	3.73	4.14	4.87	5.55	5.42	4.18
Na$_2$O	3.73	3.20	3.23	4.21	3.80	4.14	4.29	3.36	3.67	3.47	3.49
K$_2$O	2.40	4.11	3.95	2.32	3.54	2.74	2.11	2.85	1.90	2.63	2.83
P$_2$O$_5$	0.18	0.10	0.08	0.21	0.13	0.20	0.16	0.16	0.19	0.18	0.14
H$_2$O+											
H$_2$O−											
LOI	0.58	0.29	0.76	0.44	0.35	0.65	0.54	1.49	1.41	1.40	1.05
Total*	100.58	100.29	100.25	100.44	100.35	100.27	100.54	101.49	101.41	101.40	101.05
Sc xrf	11	5	7	6	6	6	8	18	16	18	13
V	133	61	60	94	65	73	82	127	114	136	88
Ni	21	21	<20	18	18	<20	16	18	8	37	17
Cu	17	15	<10	9	9	16	17	28	13	45	27
Zn	78	46	34	77	51	66	47	90	90	798	62
Ga	18	17	14	20	18	22	19	18	18	18	16
As	<5	<5	<5	<5	<5	<5	<5	7	<5	8	47
Rb	74	165	149	79	135	86	50	120	57	119	108
Sr	488	226	236	545	367	508	514	380	523	390	354
Sr xrf	464	217	239	541	360	507	505	369	514	376	341
Y	15	13	10.8	11	11	12.2	9	22	15	21	16
Y xrf	15	13	12.0	12	12	12.0	10	22	16	22	16
Zr	149	125	109	148	129	159	102	165	128	165	126
Zr xrf	145	118	99	150	140	159	101	171	128	181	142
Nb	8.4	7.2	6.8	7.3	5.5	9.6	7.2	8.8	6.8	10.1	6.9
Mo	8.4	20.7	2	7.4	7.7	2	8.2	10.8	7.3	7.5	7.4
Sb	1.7	1.4	0.2	1.1	1.0	<0.2	3.3	0.6	0.4	8.8	2.1
Cs	2.5	4.6	4.5	2.7	3.3	1.5	2.2	6.6	2.3	6.6	4.3
Ba	952	672	729	720	713	926	807	778	954	683	904
Ba xrf	976	679	711	760	745	884	823	796	957	701	927
La	27.1	30.3	13.0	22.7	22.1	28.6	21.3	38.2	21.3	23.0	18.9
Ce	50.6	59.1	25.6	46.0	42.2	58.1	39.4	74.8	41.4	47.7	36.9
Pr	5.56	6.31	2.82	5.38	4.79	6.72	4.27	8.17	4.89	5.62	4.31
Nd	20.4	22.2	10.1	21.0	17.7	25.2	15.3	29.6	18.1	21.6	16.2
Sm	3.9	4.0	1.84	4.2	3.4	4.60	2.8	5.4	3.6	4.4	3.4
Eu	1.07	0.81	0.533	1.16	0.86	1.22	0.84	0.99	1.08	1.11	0.89
Gd	3.4	3.2	1.64	3.4	2.9	3.36	2.4	4.7	3.0	4.4	3.1
Tb	0.5	0.5	0.28	0.5	0.4	0.47	0.3	0.7	0.5	0.7	0.5
Dy	2.5	2.2	1.64	2.1	2.0	2.29	1.6	3.7	2.6	3.7	2.6
Ho	0.6	0.5	0.34	0.4	0.4	0.40	0.4	0.9	0.6	0.9	0.6
Er	1.5	1.2	1.02	1.0	1.0	1.07	0.9	2.3	1.7	2.2	1.6
Tm	0.24	0.19	0.169	0.14	0.15	0.151	0.14	0.34	0.25	0.35	0.26
Yb	1.6	1.2	1.15	0.9	0.90	0.95	1.0	2.20	1.70	2.20	1.60
Lu	0.23	0.17	0.195	0.13	0.15	0.140	0.14	0.32	0.26	0.34	0.25
Hf	4.2	3.7	3.2	4.2	3.9	4.3	2.9	4.9	3.5	4.6	3.6
Ta	0.87	0.75	2.86	0.60	0.65	5.49	0.89	1.67	0.92	1.66	2.18
Tl	0.4	0.9	0.74	0.3	0.6	0.45	0.2	0.6	0.3	0.6	0.5
Pb	13	14	22	8	10	24	8	14	12	86	11
Th	8.73	36.15	38.0	10.53	25.87	15.2	6.21	12.98	4.98	16.05	11.13
U	2.89	8.61	12.4	3.20	6.31	3.11	2.20	4.71	1.55	4.02	3.89
Eree	119.2	131.9	60.4	109.0	99.0	133.3	90.8	172.3	101.0	118.2	91.1
La$_N$/yb$_N$	11.73	17.48	7.87	17.46	17.00	20.94	14.75	12.02	8.67	7.24	8.18
Eu/Eu*	0.87	0.67	0.91	0.90	0.81	0.90	0.96	0.58	0.97	0.76	0.82
Sri		0.7050				0.7044					

Note: Analyses by Actlabs by ICP-optical (major oxides) and mass spectrometry (trace elements) and XRF (as noted).

(continued)

TABLE 3. CHEMICAL ANALYSES OF GRANITIC ROCKS, SINALOA BATHOLITH *(continued)*

				Posttectonic rocks					Colegio tonalite
Unit:									
Sample:	HM-9	HP-16	HP-5	HP-7	199	HS-52	HP-17	HP-22	HS-53
Map unit:	Tki	Tki	Tki	TKi	Tki	TKi	TKi	TKi	Tic
SiO_2	74.16	65.96	64.91	53.81	57.39	58.68	67.22	63.25	60.88
TiO_2	0.12	0.63	0.67	0.73	0.94	0.97	0.59	0.80	0.81
Al_2O_3	13.70	14.92	15.51	21.41	16.92	16.47	14.91	15.73	16.72
Fe_2O_3	<0.01	5.08	5.11	1.58	7.84	3.76	0.31	5.51	6.24
FeO*	2.54			4.51		3.10	4.45		
MnO	0.06	0.08	0.08	0.09	0.12	0.11	0.08	0.09	0.12
MgO	0.37	2.12	2.12	3.66	3.72	4.06	1.78	2.70	2.77
CaO	1.10	4.34	4.63	9.33	6.85	6.51	3.85	5.17	5.52
Na_2O	3.84	3.12	3.56	3.05	3.29	3.54	3.10	3.25	4.13
K_2O	4.04	3.32	2.93	1.72	1.61	2.54	3.59	2.92	2.06
P_2O_5	0.07	0.13	0.14	0.11	0.21	0.25	0.12	0.18	0.22
H_2O+									
H_2O-									
LOI	0.45	0.44	0.60	1.35	0.95	1.53	0.50	0.63	0.41
Total*	100.45	100.13	100.26	101.35	99.83	101.53	100.50	100.21	99.87
Sc xrf	4	13	13	18	18	22	11	15	14
V	44	120	133	184	175	171	143	113	128
Ni	29	<20	<20	34	<20	33	25	<20	<20
Cu	11	12	25	181	20	59	15	33	11
Zn	44	60	33	79	-30	121	55	80	79
Ga	18	18	18	20	15	19	17	19	19
As	<5	<5	9	13	<5	10	7	<5	<5
Rb	140	132	110	98	64	78	149	141	80
Sr	162	284	362	560	428	564	282	373	451
Sr xrf	157	278	357	536	441	541	265	377	453
Y	21	26.4	21.6	13	23.3	21	24	24.0	18.5
Y xrf	19	26.0	21.0	13	22.0	21	23	25.0	18.0
Zr	61	149	173	90	157	220	183	200	111
Zr xrf	64	154	167	95	162	204	160	213	115
Nb	7.6	5.9	6.0	5.4	6.7	9.7	8.5	6.4	6.1
Mo	8.9	4	3	6.1	2	6.6	11.6	4	<2
Sb	2.1	0.5	0.8	4.3	-0.2	1.0	1.6	0.4	0.2
Cs	3.5	4.3	5.2	14.1	3.1	1.6	5.9	5.2	3.2
Ba	716	748	857	358	633	677	873	710	634
Ba xrf	714	727	813	358	628	678	875	691	615
La	18.0	31.0	27.6	14.9	22.8	25.6	24.4	24.8	22.3
Ce	34.7	61.7	56.8	29.0	47.1	53.6	48.8	54.3	43.8
Pr	3.86	6.91	6.60	3.25	5.63	6.51	5.66	6.68	5.09
Nd	13.4	25.4	24.7	12.2	22.4	25.5	21.6	26.3	19.9
Sm	3.1	5.11	4.90	2.7	4.76	5.1	4.6	5.46	3.86
Eu	0.47	0.999	0.955	0.89	1.29	1.29	1.05	1.09	1.13
Gd	3.1	4.36	4.06	2.6	4.12	4.9	4.5	4.55	3.26
Tb	0.6	0.76	0.67	0.4	0.72	0.7	0.7	0.77	0.56
Dy	3.3	4.39	3.71	2.2	4.06	3.6	3.7	4.25	3.15
Ho	0.8	0.87	0.72	0.5	0.79	0.9	0.9	0.79	0.64
Er	2.1	2.46	2.05	1.2	2.18	2.3	2.4	2.27	1.89
Tm	0.34	0.375	0.312	0.21	0.316	0.34	0.36	0.328	0.283
Yb	2.1	2.39	1.92	1.2	1.98	2.10	2.4	2.05	1.84
Lu	0.31	0.355	0.289	0.18	0.284	0.33	0.34	0.308	0.279
Hf	2.2	4.6	4.6	2.7	3.6	6.1	5.2	5.5	3.5
Ta	1.08	<0.01	<0.01	0.45	<0.01	1.10	1.15	<0.01	<0.01
Tl	0.8	0.76	0.38	1.1	0.16	0.3	0.7	0.75	0.51
Pb	21	24	12	12	<5	9	11	21	24
Th	7.85	18.2	23.3	5.57	8.88	9.41	12.98	21.9	8.69
U	4.18	5.11	5.16	1.65	2.21	2.82	4.84	6.46	2.51
Eree	86.2	147.2	135.3	71.4	118.5	132.8	121.4	133.9	107.9
La_N/yb_N	5.93	8.99	9.95	8.60	7.99	8.44	7.04	8.38	8.42
Eu/Eu*	0.46	0.63	0.63	1.01	0.87	0.77	0.69	0.65	0.94
Sri									

Note: Analyses by Actlabs by ICP-optical (major oxides) and mass spectrometry (trace elements) and XRF (as noted).

al., 1982; Henry and Fredrikson, 1987; Clarke and Titley, 1988; Staude and Barton, 2001). These data indicate that posttectonic intrusions were emplaced at depths less than 3 km.

Interpretation of K-Ar results for the syn- or pretectonic intrusions is not so straightforward. Only three of nine syntectonic pairs are concordant, and the larger absolute age difference allowed by concordance for them does not preclude slow cooling. Syntectonic granites intrude Jurassic(?) metasedimentary rocks. A few amphibolite outcrops of uncertain protolith are the only possible volcanic equivalent. The syntectonic rocks were probably emplaced at greater depths than were posttectonic rocks, although how deep is unknown. Moreover, the syntectonic rocks show common evidence of deformation at moderate temperatures during or following emplacement.

The small absolute differences between U-Pb and K-Ar ages are nearly constant in percentage, but they may still reflect geologic environments. The 20 Ma Colegio granodiorite (HS-53) probably was emplaced into the coolest environment, as no other dated plutons of that age have been recognized. The 48 Ma Candelero and 67 Ma San Ignacio granodiorites were emplaced during a period of intense and repeated batholithic intrusion, and thus presumably into a much hotter environment; their larger absolute age differences could reflect this. Finally, the 101 Ma Recodo tonalite was also emplaced during repeated granitic intrusion and possibly at greater depth and has the largest absolute difference between K-Ar and U-Pb ages.

A further observation is that the biotite age in some discordant samples is indistinguishable from the concordant age of a younger, adjacent intrusion, whereas the hornblende age appears unaffected. This is best illustrated by three samples of the El Carmen granodiorite in the north-central part of the map area (Fig. 2). Sample HS-12, which was collected at least 6 km from the nearest younger intrusion, has concordant biotite (75.0 ± 0.8 Ma) and hornblende (73.0 ± 1.6 Ma) ages. Biotite in two other samples is significantly younger. Biotite in sample HS-13 gives 62.9 ± 0.7 Ma, which is indistinguishable from numerous ages for the nearby San Ignacio granodiorite. Hornblende in HS-13 gives 74.7 ± 1.7 Ma, indistinguishable from the concordant age of HS-12, which suggests it lost no measurable Ar despite reheating that completely reset biotite. Similarly, biotite in sample HS-25 gives 65.2 ± 0.7 Ma, whereas hornblende is 73.4 ± 1.6 Ma. No other dated rock lies near sample HS-25, so it is not possible to determine whether the biotite age has been completely reset. Nevertheless, the hornblende has again been unaffected.

The 52.7 Ma biotite age of the 101 Ma sample HS-17 is also indistinguishable from the concordant ages of an adjacent pluton. Sample HS-8B of the Concordia granodiorite gives 53.8 ± 0.6 (biotite) and 54.2 ± 1.2 Ma (hornblende). Although sample HS-8B was collected ~17 km to the southeast, we traced the Concordia pluton to within 6 km of sample HS-17. Notably, neither biotite nor hornblende in sample HM-4, which is farther from the Concordia intrusion, appears affected.

Finally, sample 146 was collected from within what appears to be a 1-km-diameter inclusion of San Ignacio granodiorite in

Candelero granodiorite and no more than 100 m from the contact. The biotite age of 146 (47.0 ± 0.5 Ma) has been reset to the concordant age of the Candelero intrusion. Assuming the San Ignacio body was emplaced at ca. 64 Ma, the hornblende age of 49.6 ± 1.1 Ma suggests it has been only partly reset.

A reasonable interpretation of these data is that biotites were completely reset by intrusion of the younger plutons, but that hornblendes were generally unaffected. Discordance appears to be solely the result of reheating by younger plutons, not of slow cooling of any origin. Therefore, hornblende ages, even in discordant samples, indicate the time of emplacement. As with our interpretation of concordant data, this interpretation is most straightforward for the posttectonic intrusions.

An alternative interpretation, that the concordant ages represent rapid uplift long after emplacement, seems impossible. Samples with concordant ages range from 98 to 20 Ma. This alternative would require a checkerboard of small uplifts, on the order of 10^2 km^2, throughout that time while adjacent areas remained at depth. This situation is in marked contrast to that in the Peninsular Ranges batholith, where regional uplift long after intrusion has exposed deep levels of the eastern part of the batholith and generated a regional pattern of generally discordant K-Ar dates that are variably younger than U-Pb zircon ages (Gastil, 1975; Krummenacher et al., 1975; Silver et al., 1979; Silver and Chappell, 1988; Snee et al., 1994; Ortega-Rivera et al., 1997; Ortega-Rivera, this volume, Chapter 11).

PRE-BATHOLITHIC ROCKS

Pre-batholithic rocks include quartz diorite orthogneiss and metasedimentary and metavolcanic(?) rocks. Orthogneiss crops out in two small areas in the south-central part of the map area, where it is in contact with both metamorphosed sandstone and early gabbro of the batholith (Fig. 2, Plate 1). Gneiss in the eastern area 26 km northeast of Mazatlan contains quartz, plagioclase, biotite, hornblende, and minor orthoclase. Metamorphism has produced aligned aggregates of biotite and hornblende and mosaics of strained quartz. Foliation strikes N 40° E and dips steeply northwest, concordant with bedding and schistosity in the metasedimentary rocks (Plate 1). Strongly foliated granitic gneiss is exposed ~30 km north of Mazatlan west of Highway 15. The gneiss is cut by many tabular amphibolite bodies, possibly metamorphosed diabase dikes. Foliation and axes of recumbent folds in the gneiss and amphibolite strike irregularly northwest. This early granitic magmatism is distinguished from the main batholith by the strong metamorphic overprint and a possible Jurassic age. Similar metamorphosed granitic rocks are recognized in northern Sinaloa (Mullan, 1978). Anderson et al. (1972) and Damon et al. (1984) identified a Jurassic (190–160 Ma) magmatic arc that extends the length of México; gneiss in Sinaloa may be part of this arc.

Two sequences of metasedimentary rocks crop out in southern Sinaloa. The most extensive consists of phyllitic sandstone, quartzite, and quartz-biotite-muscovite schist (Jm). Protolith was finely laminated to massive muddy sandstone and siltstone. Detrital grains are mostly quartz with lesser chert and plagioclase. Marble and minor quartz-, garnet-, and epidote-bearing calc-silicate rock (Kl) form an east-trending belt 30 km north of Mazatlan (Fig. 2, Plate 1). The metamorphosed clastic rocks and marble generally do not occur together. Nevertheless, bedding and schistosity in both are parallel and strike east to east-northeast, suggesting they have undergone the same deformation. Amphibolite was found in scattered locations, some adjacent to marble, but its protolith is uncertain. The amphibolite could be in part metamorphosed mafic to intermediate volcanic rock, which in turn could be volcanic equivalents of some early batholithic rocks.

The metamorphosed clastic rocks are probably Jurassic, and the marble is almost certainly Aptian. Similar metamorphic rocks of the El Fuerte Group in northern Sinaloa are Jurassic or older (Mullan, 1978), and slightly metamorphosed graywackes and argillites, the oldest known rocks in Nayarit, are thought to be Late Jurassic (Gastil et al., 1978). Holguin (1978) and Bonneau (1970) demonstrated that metamorphosed carbonate rocks throughout Sinaloa are early Aptian (ca. 113–110 Ma), which precedes emplacement of the main batholith. Bonneau also found andesite, conglomerate, limestone, and dolomite underlying the marble in central Sinaloa. Amphibolite in southern Sinaloa could be equivalent to the andesite found by Bonneau.

THE SINALOA BATHOLITH

Intrusive rocks of the Sinaloa batholith crop out over much of the region west of the Sierra Madre Occidental (Figs. 1 and 2, Plate 1). In our Sinaloa map area, more than a third of outcrop is granitic; the rest of the outcrop is mostly younger cover. Batholiths are exposed in all the deeply incised canyons of the Sierra Madre, so they are probably extensive beneath it. Therefore, batholithic rocks probably underlie most of the region.

Granodiorite is the most abundant rock, but compositions of individual intrusions range from gabbro to granite (Figs. 4 and 5, Appendix 2). Plagioclase, quartz, alkali feldspar, biotite, hornblende, and clinopyroxene are major phases. Sphene is a common accessory, and large euhedral grains are particularly prominent in some posttectonic rocks. Other accessory minerals, present in most samples, are zircon, apatite, and epidote.

The Sinaloa batholith developed in three stages. Early gabbros may have been emplaced ca. 135 Ma. Foliated pre- or syntectonic rocks were emplaced before ca. 90 Ma, apparently while the region was being deformed and are mostly tonalites. Posttectonic rocks were emplaced between ca. 90 and 45 Ma, with one intrusion at 20 Ma, and after deformation had ceased; they are predominantly granodiorite.

Early Gabbro

Gabbro crops out as two tabular, east-northeast–striking bodies that intrude the clastic metasedimentary rocks (Fig. 2, Plate 1). A western body, ~7 km by 12 km, crops out in roadcuts

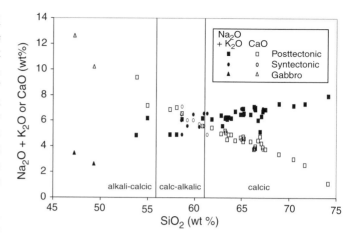

Figure 4. Plot of weight % $Na_2O + K_2O$ and CaO versus SiO_2 for chemical analyses of Sinaloa batholith, which is borderline calc-alkalic or calcic.

along Highway 15, 20 km north of Mazatlan. An eastern body, ~7 km by 4 km, is well exposed in roadcuts along the paved road to La Noria, 20 km northeast of Mazatlan and ~8 km south of La Noria. A small, unmapped quartz-bearing gabbro (HS-48) crops out within granitic rocks ~10 km east of the eastern body. Cumulus layering in the two main bodies generally strikes east-northeast and dips northwest parallel to contacts, bedding, and schistosity in the enclosing metasedimentary rocks. Marble appears to be in depositional contact on the western gabbro in a quarry off Highway 15.

Plagioclase-clinopyroxene ± hornblende gabbro is the dominant rock type in both bodies, but pyroxenite and anorthosite are also present. The rocks are variably altered or metamorphosed. Plagioclase is commonly bent and shows minor sericitic alteration. Pyroxene locally contains cores of a colorless, fibrous amphibole, probably tremolite, and is rimmed by hornblende. Mafic minerals are locally replaced by actinolite.

Hornblendes from the eastern intrusion and from the unmapped quartz-bearing gabbro give similar K-Ar ages of 133.8 ± 3.0 (HS-46) and 138.6 ± 3.1 (HS-48). These ages could indicate the time of emplacement, excess Ar, or cooling following metamorphism. If marble is early Aptian and depositionally on gabbro, then the gabbro is at least ca. 110 Ma. Given the resistance of hornblende to Ar loss and the improbability that metamorphism reached 500 °C, the ages are not likely to reflect cooling following metamorphism. Because the amphiboles have very different K contents (~0.15 and 0.42) but similar ages, excess Ar would have had to have been incorporated proportional to their K contents. Excess Ar has been interpreted for similar gabbros of the Peninsular Ranges, which give hornblende K-Ar ages as old as 146 Ma (Krummenacher et al., 1975), because no U-Pb ages are that old (Silver et al., 1979). Similar metamorphosed mafic intrusions are present in northern Sinaloa but are not dated (Mullan, 1978; Ortega-Gutiérrez et al., 1979).

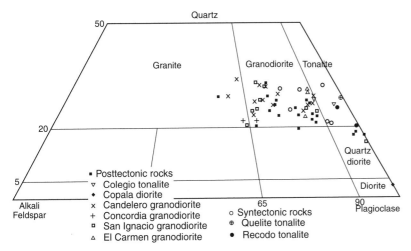

Figure 5. Quartz-alkali feldspar-plagioclase modes of granitic rocks of Sinaloa batholith. Syntectonic rocks are relatively mafic tonalites and granodiorites containing 15–28% mafic minerals. Posttectonic rocks are mostly more leucocratic granodiorites and lesser granite containing 5–20% mafic minerals; minor posttectonic quartz diorites are border phases or small intrusions. The 20 Ma Colegio tonalite, which contains 24% mafic minerals, is compositionally like the pre–90 Ma syntectonic rocks and crops out in same location near coast.

Syntectonic Intrusions

A group of tonalites to mafic granodiorites commonly showing evidence of deformation during or following emplacement are at least 90 Ma based on both K-Ar and U-Pb dates. These pre- or syntectonic intrusions are distinctly more mafic than later, posttectonic intrusions (up to 28 modal percent mafic minerals; Fig. 5) and contain lesser potassium feldspar, which is mostly microcline. For simplicity, we refer to these rocks as syntectonic, although whether they intruded during or before deformation is unknown.

All known syntectonic rocks crop out within 50 km of the coast, both in the map area and in northern Sinaloa (Figs. 1 and 2, Plate 1); they are the dominant intrusive type within ~25 km. Individual plutons are probably no more than 8 km in diameter, as individual rock types cannot be traced more than that distance. Two distinctive plutons have been mapped separately. A strongly foliated, mafic, coarse-grained tonalite (Quelite tonalite; Kiq) crops out ~35 km north of Mazatlan. A fine-grained, non-foliated tonalite ~25 km northeast of Mazatlan (Recodo tonalite; Kir; dated at 101 Ma by U-Pb) contains 25–28% mafic minerals and minor microcline. Quartz occurs as poikilitic grains enclosing plagioclase, biotite, and hornblende. Hornblende surrounds cores of augite and tremolite(?). Foliation in the Quelite tonalite (Kiq) strikes east-northeast, is vertical, and is approximately parallel to foliation in enclosing marble. The orientation of foliation in other syntectonic rocks is unknown because the foliation is generally faint and because most outcrops in the coastal plain consist of large residual boulders that are not strictly in place.

Most syntectonic rocks are weakly foliated and show varying degrees of dynamic recrystallization. The most common texture is recrystallization or subgrain development of quartz, which is strongly to weakly developed in different samples. In rocks showing the greatest quartz recrystallization, a few plagioclase grains are cut by through-going fractures and filled with quartz. A few plagioclase grains in the same rock are bent, but we see little or no evidence of subgrain development in plagioclase. In

less deformed rocks, interstitial microcline shows undulatory extinction, suggesting incipient recrystallization. Myrmekite is common at microcline-plagioclase boundaries; although locally present in posttectonic rocks, this texture is far more common in syntectonic rocks and may be an indicator of deformation (Vernon, 1991). Dynamic recrystallization textures are most intense in samples HS-34, HS-44, HS-45, and HS-37 near the coast, less intense in samples HS-22 and HM-8 to the east, and absent in the Recodo tonalite samples farther south (HS-17 and HM-4). The petrographic characteristics suggest deformation occurred above 300 °C, the temperature required for dynamic recrystallization of quartz, but probably below 450 °C, the minimum temperature of plagioclase recrystallization (Tullis, 1983).

Timing of emplacement for these rocks is anchored by the U-Pb age of Recodo tonalite (HS-17; 101.2 ± 2.0 Ma) and concordant K-Ar dates on sample HM-4 (biotite: 98.3 ± 1.1 Ma; hornblende 96.3 ± 2.2) from the same pluton. However, the time of emplacement of other syntectonic rocks and of the end of deformation are not fully understood. Only two other samples are concordant (HC-4A, biotite 92.1 ± 1.0, hornblende, 90.4 ± 2.0; HC-5, biotite 88.6 ± 1.0, hornblende 89.5 ± 2.0 Ma), and both of these are from the northern Sinaloa transect. All other samples are discordant, with hornblende ages slightly to 27 Ma greater than biotite ages (Table 1); hornblende from sample HS-34 is 3.3 Ma less than the biotite age of 89.8 Ma. Notably, neither of the two samples of Recodo tonalite show evidence of deformation. Sample HC-4A and all other dated syntectonic samples show hand specimen and petrographic evidence of deformation, and their hornblende ages range from 97.9 ± 2.2 Ma (HS-22) to 86.5 ± 1.9 Ma (HS-34).

These observations imply that either (1) all the deformed intrusions are older than 101 Ma and neither concordant nor hornblende ages of the syntectonic rocks record time of emplacement, or (2) the Recodo tonalite was emplaced before or during deformation but remained undeformed. The latter is possible, because deformation could have been concentrated in narrow zones with intervening areas showing less or no deformation. The syntectonic rocks can be no older than ca. 110 Ma, the approximate age of Aptian marble that they intrude.

For all deformed intrusions to be older than 101 Ma, they would have to have been maintained at temperatures greater than ~500 °C until the maximum apparent hornblende age of 97.9 ± 2.2 Ma (HS-22), i.e., any regional cooling must have been ≤98 Ma. The ca. 98 Ma concordant ages of sample HM-4 require the same conditions. However, the irregular pattern of hornblende (as well as biotite) ages demonstrates they are not all recording the same regional cooling. The younger hornblende ages could reflect variable reheating by later plutons, yet the data of sample 146 suggest that such reheating is only able to reset hornblende if the younger pluton is closer than 100 m. Therefore, it seems more likely that they do indicate time of emplacement. These uncertainties leave open the question as to when deformation recorded by the syntectonic plutons occurred or ended. Additional U-Pb as well as $^{40}Ar/^{39}Ar$ dating is needed.

Posttectonic Intrusions

Most intrusions lack foliation or dynamic recrystallization textures and were emplaced after any observable deformation ended. Their ages in the main map area range from ca. 82 to 45 Ma, with one intrusion at 20 Ma. Where observed, contacts indicate that posttectonic rocks cut syntectonic rocks. These posttectonic intrusions range from diorite to monzogranite but are predominantly granodiorite and distinctly less mafic than the syntectonic rocks. Potassium feldspar, which is orthoclase in all samples, and quartz are more abundant than in the syntectonic rocks (Fig. 5), and mafic minerals make up no more than 15 modal percent. Posttectonic rocks are widely distributed throughout the map area from within ~12 km of the coast, where they intrude syntectonic rocks and metasedimentary rocks, to the Sierra Madre Occidental, where they crop out in the deep canyons of the Rio Piaxtla and Rio Presidio (Plate 1) and intrude the lower volcanic complex. Posttectonic intrusions also crop out along the transect in northern Sinaloa, where they are as old as 89 Ma.

Posttectonic intrusions are generally larger than syntectonic intrusions and reach dimensions of 15 km north-south and up to 25 km east-west. However, these latter dimensions are parallel to extension that affected southern Sinaloa and a large region surrounding the Gulf of California (Henry, 1989; Stock and Hodges, 1989; Henry and Aranda-Gómez, 2000). Their true, unextended east-west dimensions are probably closer to 15 km also.

Six posttectonic rock types have been mapped separately, but only two are sufficiently distinctive and widespread to be correlated beyond individual plutons. The Candelero granodiorite (TKic) is named for exposures in the Arroyo de Candelero in the northeastern part of the map area (HP-14; Fig. 2). It forms several plutons across the northern part as well as one cutting syntectonic rocks 20 km north of Mazatlan (Fig. 2, Plate 1). The Candelero granodiorite is medium grained and distinguished by 5–10% large euhedral biotite and hornblende. Euhedral sphene

to 2 mm is common. Field relations indicate that the Candelero is the youngest major intrusive phase. Similar rocks crop out ~20 km north of the northwest corner of Figure 2 and 200 km to the northwest along our northern transect (HC-3; Fig. 1). Furthermore, the Candelero granodiorite is petrographically similar to the La Posta type of the Peninsular Ranges batholith and Los Cabos block (Walawender et al., 1990; Kimbrough et al., 2001; Kimbrough et al., 2002) and to the Half Dome Granodiorite of the central Sierra Nevada (Bateman, 1992).

The San Ignacio granodiorite (TKis) is named for exposures along the Rio Piaxtla at San Ignacio in the north-central part of the map area (HS-42; Fig. 2, Plate 1). The San Ignacio granodiorite is fine- to medium-grained equigranular and contains 7–15% anhedral biotite, hornblende, and clinopyroxene. Similar rocks crop out throughout the map area but cannot definitely be correlated. The modes of both San Ignacio and Candelero rock types vary considerably and overlap (Fig. 5).

Numerous other granitic rocks of small size or less distinctive character could not be assigned to either of these major posttectonic types. Four were correlated locally. Mafic quartz diorite (Colegio quartz diorite; Tic) petrographically and compositionally similar to some of the syntectonic types but lacking foliation crops out ~50 km north of Mazatlan. Fine-grained granodiorite and quartz monzonite (Concordia granodiorite; TKin) containing 1 cm poikilitic orthoclase underlie a 100 km² area ~30 km east-northeast of Mazatlan. Coarse-grained, K-feldspar-poor granodiorite (El Carmen granodiorite; TKie) underlies an area of ~100 km² south of the San Ignacio pluton. A small (15 km²), fine-grained diorite pluton (Copala diorite; TKid) crops out around Copala along the Mazatlan-Durango highway, is commonly hydrothermally altered, and hosts some of the silver ore there.

Posttectonic intrusions are well dated by both U-Pb and K-Ar (Tables 1 and 2). The range of ages are again anchored by U-Pb dates of 67 Ma (HS-42; San Ignacio granodiorite), 48 Ma (HP-28; Candelero granodiorite), and 20 Ma (HS-53; Colegio quartz diorite). Most intrusions are concordant, which we interpret to date the time of emplacement. The oldest posttectonic rocks are sample HP-6 (biotite, 81.8 ± 0.9 Ma; hornblende, 82.6 ± 1.8 Ma) in the main map area and sample HC-5 (biotite, 88.6 ± 1.0 Ma; hornblende, 89.5 ± 2.0 Ma) from the northern transect. The next youngest intrusion is El Carmen granodiorite, with samples HS-12, HS-13, and HS-25 indicating an emplacement age ca. 75 Ma. Ages on other samples range nearly continuously to 46 Ma with a clear geographic pattern. Oldest ages are near the coast, and youngest ages are farthest inland.

In addition to the U-Pb and K-Ar dates on sample HS-42, concordant dates or hornblende dates on the type pluton of the San Ignacio granodiorite are all ca. 64 Ma (HP-2, HS-40, HS-41, HP-11; Table 1). Samples 170 and 156 to the east are also ca. 64 Ma and may be from a continuation of the main pluton. San Ignacio granodiorite intrusions farther east have more equivocal ages. Biotite from sample 146 was reset to the 47 Ma age of the adjoining Candelero granodiorite, and hornblende was at least partly reset to 50 Ma. Sample HP-26 has only a biotite date of

49 Ma, which may indicate resetting also. Emplacement ages of these eastern intrusions are not established.

The K-Ar and U-Pb data indicate that the Candelero granodiorite was emplaced at several different times. The pluton of the type locality is ca. 48 Ma (HP-28, HP-14, HP-25). The western-most example is concordant at ca. 82 Ma (HP-6), two different plutons in the north-central part of the map area are concordant at 62 Ma (HP-13) and 52 Ma (236), and the easternmost example has only a biotite date of 44 Ma (HP-18). A Candelero pluton along the northern transect is concordant at ca. 51 Ma. As is recognized in other Cordilleran batholiths, petrographically similar intrusions can be emplaced at different times.

Several biotite ages on major intrusions are still younger at 29 Ma (227), 31 Ma (HP-20), and 34 Ma (HP-23). The latter two samples are part of a Candelero-type intrusion in the north-easternmost part of the map area. The true age of this intrusion is at least 44 Ma, the biotite age of HP-18, and the young ages probably indicate reheating. Later magmatism in that area is indicated by a biotite age of 32 Ma on a large diorite dike (HP-24). The young biotite age of sample 227 in the north-central part of the map area may also indicate reheating rather than time of emplacement. Later magmatism in that area is indicated by a hornblende age of 30 Ma on an andesite dike (HP-10).

Similar posttectonic intrusions are recognized along the northern Sinaloa transect. In addition to the 89 Ma intrusion (HC-5) and the Candelero type intrusion with a concordant age of ca. 51 Ma, sample HC-6 gives a concordant age of ca. 67 Ma. The northeasternmost sample along that transect has only a 53-Ma biotite date (KC-3).

Chemical Characteristics and Spatial Variation

Granitic rocks of the Sinaloa batholith show compositional variations similar to those of other Cordilleran batholiths (Table 3, Figs. 4 and 6). Sinaloa granitic rocks are calcalkaline but border on calcic. Including two gabbros, SiO_2 contents of analyzed rocks range from 47 to 74%; excluding gabbro, the lower limit is 54%. Most major oxides vary smoothly with SiO_2, with incompatible elements such as TiO_2 and CaO decreasing and K_2O increasing with SiO_2. Trace elements show considerably more scatter, but Rb, Ba, La, Th, and U clearly increase with increasing SiO_2; V, Sr, and Sc decrease. Some, including Y and Nb, show no apparent correlation with SiO_2.

Both gabbros contain less than 50% SiO_2 and have very low K_2O and Rb concentrations (Table 3). Although sample HS-48 contains modal quartz, it is chemically equivalent to the gabbros, not the quartz gabbros, of the Peninsular Ranges batholith (Silver and Chappell, 1988). Rare earth element patterns are relatively flat (Fig. 7). Sample HS-46 shows a positive Eu anomaly that suggests it is a plagioclase cumulate, which is consistent with its high CaO and Al_2O_3 concentrations. Gabbros commonly plot off trends of the other granitic rocks, which is most noticeable for TiO_2 and Nb. Compared to the gabbros of the Peninsular Ranges, the Sinaloa gabbros have similar patterns but much higher total rare earth elements

(REE) (Gromet and Silver, 1987) and higher concentrations of many incompatible trace elements (Walawender and Smith, 1980).

Syntectonic intrusions are dominated by mafic compositions. SiO_2 contents of seven out of nine analyzed syntectonic rocks range narrowly between 59 and 62% with one each at 65 and 67%. They are moderately enriched in light rare earth elements (LREE) and have small negative Eu anomalies (Fig. 7). Sample HS-44 is unusual in having very low total REE and a small positive Eu anomaly.

Posttectonic rocks have a wider compositional range, from 54 to 74% SiO_2, than do syntectonic rocks (Table 3, Figs. 4 and 6). However, with a few exceptions, rocks containing less than ~63% SiO_2 are minor border phases (HS-40, HS-41) or small intrusions (HP-7, 199). Other major oxides and trace elements reflect these differences. Posttectonic rocks have lower TiO_2, Al_2O_3, CaO, total Fe, MnO, MgO, P_2O_5, and Sr and higher K_2O, Rb, Th, and U than do syntectonic rocks. Some trace elements show little difference, including Zr, Nb, and, despite general correlation with SiO_2, Ba and most REE.

The Candelero and San Ignacio granodiorites are compositionally as well as petrographically distinct. SiO_2 concentrations in Candelero samples range from ~64 to 72% and in San Ignacio samples from 62 to 67% (excluding sample HS-41, which is a plagioclase concentrate). They also have distinct REE patterns with similar LREE abundances, but Candelero samples have lower heavy rare earth element (HREE) (and Y) abundances and smaller Eu anomalies (Fig. 7).

Samples HS-40 and HS-41 are border phases of the San Ignacio pluton. The composition of HS-41, including high Al_2O_3, CaO, and Sr, indicates it is a plagioclase concentrate, which is consistent with outcrop and thin section characteristics. Sample HS-40 also has high Sr, but other major oxides are not unusual. Both samples have unusual REE patterns with low total REE and prominent positive Eu anomalies consistent with plagioclase accumulation.

The granitic rocks show distinct spatial chemical patterns (Fig. 8). With much scatter, SiO_2 increases eastward, which reflects the distribution of more mafic syntectonic and more silicic posttectonic rocks. Many major oxides and trace elements show similar patterns reflecting their variation with SiO_2 but also with considerable scatter. The Colegio tonalite is a notable exception of a relatively mafic posttectonic intrusion. It is compositionally like the syntectonic rocks and crops out near the coast with them. This indicates that the compositional variation is a function of location, not of age.

To look for spatial patterns that are not simply related to variations in SiO_2, we used a subset of the granitic rocks containing between 63 and 67% SiO_2 (Fig. 9). For this subset, SiO_2 shows at most a slightly negative correlation with location. Nevertheless, TiO_2, Rb, Zr, Pb, Th, U, and REEs show considerable scatter but increase eastward; Ba and Sr decrease. Other elements show no discernible pattern. Location apparently plays a role in magma composition, and a variety of factors are possible, but further interpretation is beyond the scope of this report.

Sr isotope compositions determined on six samples range from 0.7030 to 0.7062 (Table 3) (Note: analysis of sample HP-28

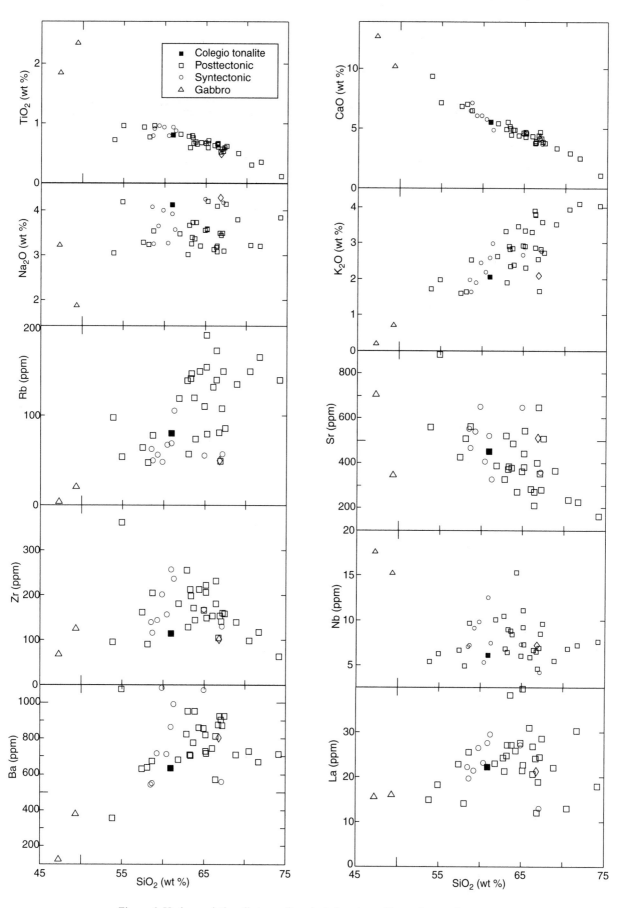

Figure 6. Harker variation diagrams for selected major oxides and trace elements.

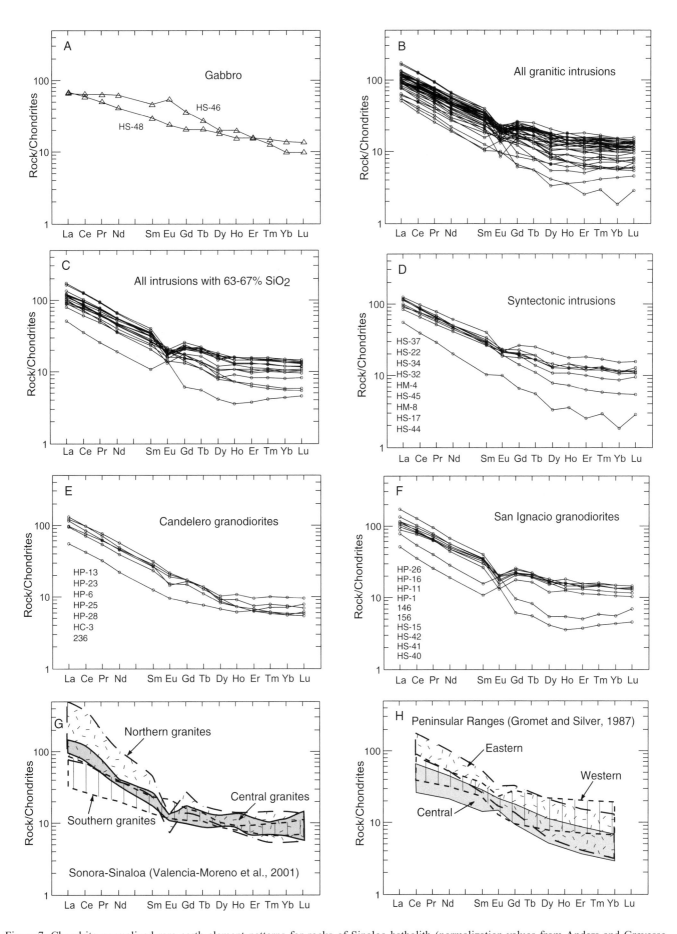

Figure 7. Chondrite normalized rare earth element patterns for rocks of Sinaloa batholith (normalization values from Anders and Grevesse, 1989). A: Gabbros. B: All granitic rocks of batholith. C: All granitic rocks containing between 63 and 67% SiO₂. D: Syntectonic intrusions. E: Candelero granodiorites. F: San Ignacio granodiorites. In D, E, and F, sample numbers are listed in decreasing order of La abundance. G: REE patterns of batholithic rocks of Sonora and Sinaloa from Valencia-Moreno et al. (2001). H: REE patterns of rocks of Peninsular Ranges batholith from Gromet and Silver (1987).

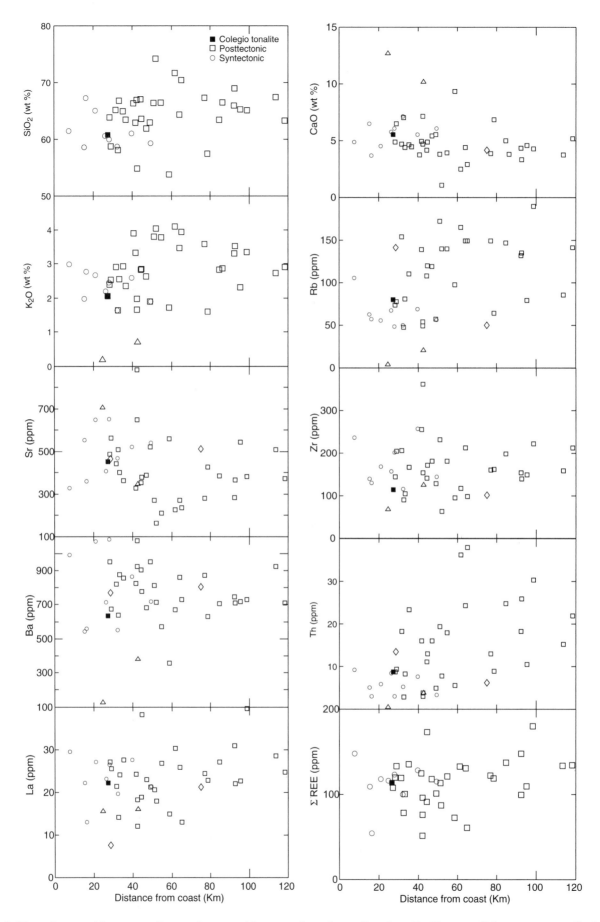

Figure 8. Plots of composition versus distance from an arbitrary, north-northwest line along Pacific coast. SiO₂ content generally increases eastward but with considerable scatter. Much of gradient reflects lack of more mafic plutons to east, such as syntectonic intrusions. Similarly, major oxides (CaO and K₂O) and trace elements that vary with SiO₂ (Rb, Sr, Zr, Ba, and Th) also show compositional gradients but also with considerable scatter. The 20 Ma Colegio tonalite is compositionally like syntectonic rocks and crops out near coast with them. This indicates that compositional variation is a function of location, not age.

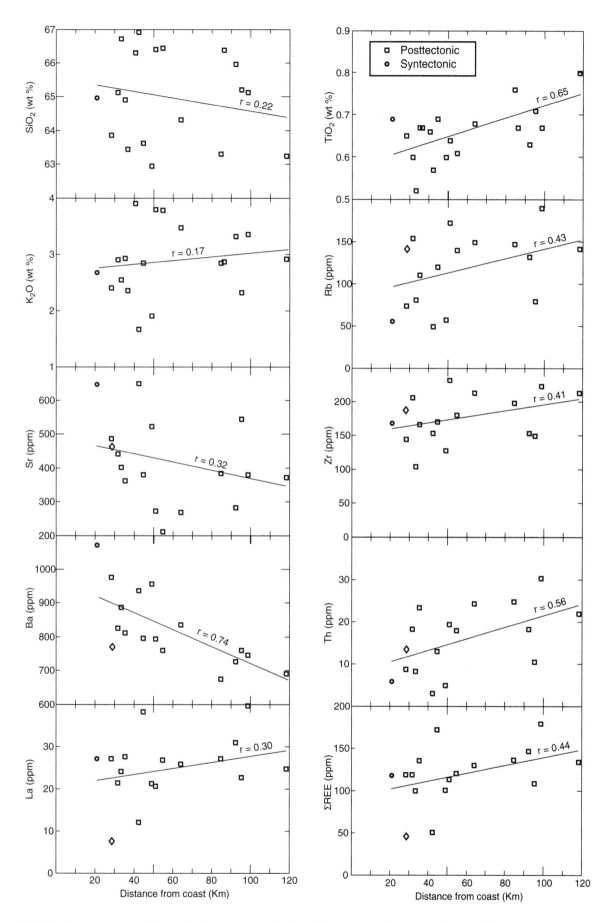

Figure 9. Variation in composition with location for samples containing between 63 and 67% SiO$_2$. Location is based on same arbitrary, north-northwest line used in Figure 8. SiO$_2$ exhibits at most slightly negative correlation with location. With considerable scatter, some major oxides and trace elements increase with distance from coast, including TiO$_2$, Rb, Zr, Th, U, La, and total REE. Ba and Sr decrease with distance. (Regression lines and correlation coefficients using IgPet99 developed by Michael Carr).

in 1973 gave an initial ratio of 0.7163. Reanalysis in 2002 gave 0.7044. We interpret the reanalysis to be correct and the old analysis to have been contaminated. An initial ratio of 0.7044 is consistent with the other five samples and the following published data). Damon et al. (1983a) report initial ratios on granitic rocks of 0.7042 from La Azulita, which lies ~6 km northeast of sample HS-23, and of 0.7052 from Cosala, which lies ~40 km north of sample HP-5 (Fig. 2, Plate 1). Valencia-Moreno et al. (2001) report an initial ratio of 0.70256 on their sample Mal-74, which is from the same location as sample KC-1. All ratios fall within the range of southern granites of Valencia-Moreno et al., which lie within the Guerrero terrane underlain by Mesozoic volcanic and sedimentary rocks and apparently lacking a Precambrian basement.

Lower Volcanic Complex (TKv)

The informally named lower volcanic complex (McDowell and Keizer, 1977) crops out chiefly in the eastern parts of the map area (upper Rio Piaxtla and Rio Presidio valleys of Plate 1), where it is generally coextensive with the posttectonic plutons. The lower volcanic complex is composed of andesitic to rhyolitic lavas, rhyolitic ash-flow and air-fall tuffs, volcaniclastic sedimentary rocks, and minor hypabyssal intrusions. They are almost everywhere propylitically altered and are the hosts of many of the major ore deposits of the area and the Sierra Madre Occidental (Wisser, 1965; Smith et al., 1982; Clarke and Titley, 1988).

The best-studied sections are around the major silver-gold mines of Tayoltita in the northeastern part of the map area and Panuco in the southeastern part (locations GD and HS-51 of Fig. 2, Plate 1). At Tayoltita, ~1200 m of rhyolitic tuffs and possible lavas are overlain by ~800 m of andesite lavas (Nemeth, 1976). Rhyolites contain phenocrysts of quartz, sanidine, biotite, and plagioclase. Feldspars are commonly altered to sericite and biotite to chlorite. One of the tuffs contains clasts up to 1 m in diameter, indicating a local source. Andesites and possibly equivalent hypabyssal intrusions contain plagioclase and clinopyroxene phenocrysts in fine-grained groundmasses of plagioclase laths, clinopyroxene, and opaque minerals. At Panuco, rhyolitic tuffs and lava flows predominate over andesitic lavas (T. Albinson and R.L. Leisure, personal commun., 1982).

Field relations and a few K-Ar ages suggest that lower volcanic rocks are broadly contemporaneous with the posttectonic rocks and are, in part, their extrusive equivalents. The lower volcanic rocks are generally coextensive with the posttectonic rocks. Lower volcanic rocks directly overlie metasedimentary rocks near Panuco (HS-51) and north of San Ignacio (HS-42) and have not undergone regional metamorphism or folding. Most contacts between lower volcanic rocks and granitic rocks are intrusive. However, lower volcanic rocks accumulated upon the eroded surface of a granodiorite pluton (HS-51 with a minimum age of 51 Ma) near Panuco, and Albinson (personal commun., 1982) found granitic boulders in agglomerates within the lower volcanic complex near Panuco. Granitic inclusions occur in tuffs of the lower volcanic complex in the northern part of the map area. Rhyolitic dikes, petrographically

similar to rocks of the lower volcanic complex, cut granitic rocks in several areas near San Ignacio.

Because of their alteration, lower volcanic rocks have not been directly dated anywhere in the map area. On the northeast side of the Sierra Madre Occidental, an andesitic lava 40 km southeast of Durango City is dated at 53 Ma (McDowell and Keizer, 1977), and a sequence of rhyolitic tuffs and andesitic lavas near Nazas 130 km north of Durango are 51–40 Ma (Aguirre-Díaz and McDowell, 1991). Recent U-Pb dating of equivalent rocks in eastern Sonora has shown that they are as much as 20 Ma older than the plutons of the local batholith (McDowell et al., 2001). Hence, the Lower Volcanic Complex may be an important but undated part of the magmatic record in southern Sinaloa.

Oligocene-Miocene Magmatism

Both U-Pb and K-Ar ages confirm the young, 20 Ma age of the Colegio quartz diorite. Although 25 Ma younger than any other dated large intrusion, it is, nevertheless, a large, coarse-grained intrusion indistinguishable from other posttectonic intrusions without the isotopic ages. It demonstrates major batholithic activity at 20 Ma.

Although the Colegio quartz diorite is the only identified large young pluton, several other minor intrusions, as well as lavas and tuffs, are also Oligocene or Miocene (Henry and Fredrikson, 1987; Aranda-Gómez et al., 1997). In addition to the dikes mentioned above, these include a dacite lava in the southernmost part of the map area (22 Ma; HM-1), a rhyolite dike in the northern part (24 Ma; 174), an ash-flow tuff in the north-central part (28 Ma; 3–7-10), and another tuff in the central part (17 Ma; F-8–1). Except for the two tuffs, for which source calderas are not known, these indicate local magmatism at those times.

These young ages indicate that magmatism in Sinaloa was coeval with the upper volcanic sequence of the Sierra Madre Occidental and the Miocene volcanic arc of Baja California. All known sources of ash-flow tuffs of the upper volcanic sequence are in Durango to the east (McDowell and Keizer, 1977; Swanson et al., 1978; McDowell and Clabaugh, 1979; Swanson and McDowell, 1984; Nieto-Samaniego et al., 1999). However, our reconnaissance mapping in Sinaloa could have missed major sources. Furthermore, many other small intrusions, lavas, and ash-flow tuffs are undated but undoubtedly late Tertiary (e.g., Tif of Plate 1). The Miocene volcanic arc, so well known around the rest of the Gulf of California (Gastil et al., 1979; Hausback, 1984; Sawlan, 1991; Mora-Alvarez and McDowell, 2000), may be well represented in Sinaloa. Basaltic volcanism at 11 and 2–3 Ma records the change to extension around the Gulf of California (Aranda-Gómez et al., 1997; Henry and Aranda-Gómez, 2000).

DISCUSSION

Overall Timing and Continuity of Magmatism

The combined U-Pb and K-Ar ages document magmatism in southern Sinaloa from at least 101 to 20 Ma. Gabbro intrusion may have begun as early as 138 Ma, but interpretation of the two gabbro hornblende ages is uncertain. Intrusion of mafic tonalites began at least by 101 Ma. Many of the tonalites show dynamic crystallization textures that indicate they were emplaced before or during a major deformational episode. Based upon the estimated maximum of 450 °C, this deformation would likely have reset biotite K-Ar ages but probably not hornblende K-Ar ages. Uncertainty in the interpretation of the K-Ar ages of these early mafic rocks allows either that they were emplaced between ca. 101 and 90 Ma, or that many if not all of the syntectonic intrusions were emplaced before 101 Ma. Some must be younger than ca. 110 Ma, the probable age of Aptian limestone that they intrude. Posttectonic intrusions began to be emplaced as early as 82 Ma in southern Sinaloa and 89 Ma in northern Sinaloa. No intrusions are dated between 89 and 82 Ma, between 82 and 75 Ma, or between 62 and 56 Ma, but batholith emplacement was apparently prolific between 75 and 45 Ma (Fig. 3). Another gap for granitic bodies falls between 45 and 20 Ma, although several minor intrusions fall within that time.

Whether these gaps indicate a real hiatus in magmatism or simply incomplete exposure or sampling and to what extent the volumes of magmatism have varied through time are uncertain. The mapped area is large but still small relative to the distribution of batholiths along the west coast of México (Fig. 1). Based on the distribution of ages, intrusions between 89 and 82 and 82 and 75 Ma should reside near the Pacific coast (Figs. 2 and 10A, Plate 1). Much of that area is covered by younger rocks, especially along our Rio Piaxtla transect. Additional sampling of granitic rocks in areas where access has improved since 1972 would be useful. Except for the possible age gaps, outcrop areas and therefore probable volumes of granitic rocks seem similar between 100 and 45 Ma. Certainly, no time dominates the record of batholithic emplacement in southern Sinaloa.

The age pattern suggests that intrusions younger than 45 Ma might be present beneath tuffs of the upper volcanic sequence, which form a nearly complete cover in the Sierra Madre Occidental. The ash-flow tuffs of the Sierra Madre erupted from numerous calderas (Swanson and McDowell, 1984) that are almost certainly underlain by major plutons (Lipman, 1984). Aguirre-Díaz and McDowell (1991), who found the 51–40 Ma rhyolites and andesites near Nazas (Fig. 1), identified many other areas of Eocene magmatism around the Sierra Madre and suggested that an Eocene volcanic field was as extensive as the mostly Oligocene Sierra Madre (McDowell and Clabaugh, 1979).

Eastward Shift of Magmatism

The Sinaloa batholith shows a distinct geographic age pattern (Fig. 10, A and B). Oldest ages are near the coast and youngest ages are farthest east. Whether the syntectonic rocks are part of an eastward trend is uncertain. The first uncertainty is whether the K-Ar ages of the syntectonic rocks indicate emplacement times. Even if they do, the ages do not correlate with location

(Fig. 10A). Both the age uncertainty and the narrow, 50-km width of their distribution limit interpretations about migration.

The age pattern for the posttectonic rocks is more obvious. Ages decrease progressively eastward from between 80 and 90 Ma, within 20–30 km of the Pacific Ocean, to ca. 45 Ma, 120 km from the coast at the western edge of the Sierra Madre Occidental. The six samples from our reconnaissance transect in northern Sinaloa show the same pattern. Critical information that is missing is timing for the volcanic rocks of the Lower Volcanic Complex. Though they are generally presumed to be coeval with the batholiths, they have not been dated in Sinaloa and most other areas due to their general alteration. One andesitic lava, possibly equivalent to the Lower Volcanic Complex in Durango has a K-Ar age of 53 Ma (McDowell and Keizer, 1977).

Another uncertainty is whether rocks of the upper volcanic sequence of the Sierra Madre Occidental represent a continuation of the same eastward shift of magmatism. Near Durango City, these rocks are 32–28 Ma (McDowell and Keizer, 1977; Swanson et al., 1978; Aranda-Gómez et al., 1997), which is consistent with a continued eastward migration of magmatism.

Upper Cretaceous and Eocene igneous rocks near Nazas (Fig. 1) constitute the only known exceptions to the regional age pattern. A small (<1 km^2) diorite intrusion is dated at 88 Ma (hornblende K-Ar), two ash-flow tuffs (with unknown sources) are 51 and 43 Ma, and andesite lavas are 49 and 45 Ma (Aguirre-Díaz and McDowell, 1991). The pluton and andesites are as old as or older than the youngest plutons dated in southern Sinaloa and demonstrate that older igneous activity did occur locally along the eastern margin of the Sierra Madre Occidental.

The age data also indicate a westward return, but that is based on many fewer ages and with significant data gaps and uncertainties in source locations (Fig. 10B). Within the Sierra Madre, the youngest (23 Ma) volcanic rocks are in its western part, just east of the area of Figure 2 (McDowell and Keizer, 1977). Ages for the 20 Ma Colegio quartz diorite (HS-53) and the 22 Ma rhyodacite lava (HM-1) may reflect a continuation of this westward trend.

Rates of migration can be calculated from the age patterns (Fig. 10), but distances need to be corrected for Miocene and later extension (Henry, 1989; Henry and Aranda-Gómez, 2000). Extension affected only the region west and east of the relatively unextended block of the Sierra Madre Occidental. Fortunately, extension paralleled the direction of igneous migration. Using only the ages of posttectonic rocks and correcting for as much as 50% east-northeast extension, the eastward sweep progressed at a rate of ~1.5 km/Ma. If syntectonic rocks or at least the 101-Ma Recodo tonalite are included, the rate decreases to as little as 1 km/Ma. The apparent westward shift was much more rapid at ~12–15 km/Ma.

Comparison with Batholiths of Western México

The Sinaloa batholith shares marked similarities and some differences with the Peninsular Ranges batholith of southern and Baja

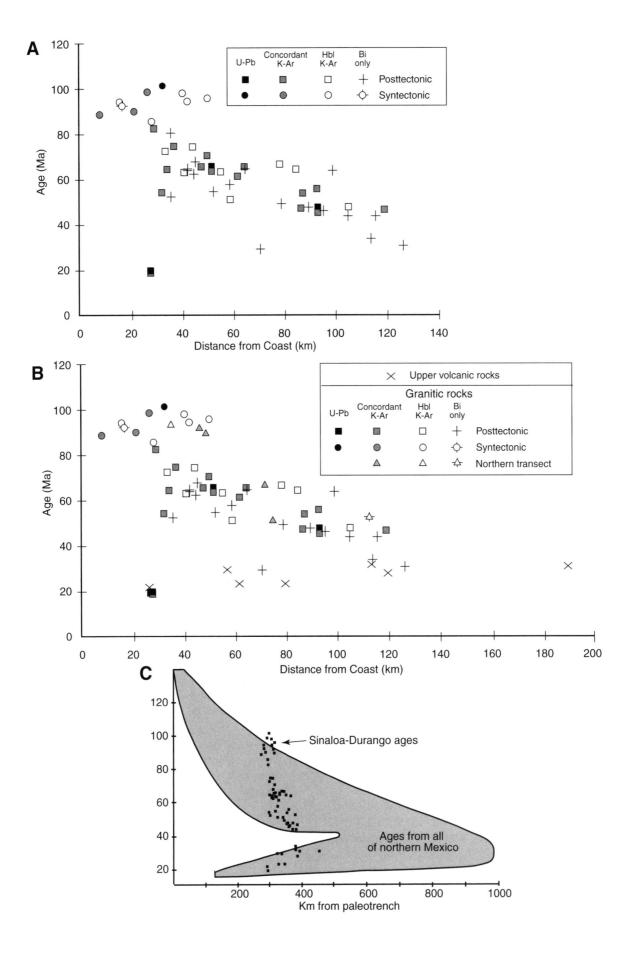

Figure 10. Plots of age of igneous rocks versus distance from same north-northwest line as Figure 8. A: Only granitic rocks from main map area (Fig. 2). Posttectonic rocks show eastward migration of intrusion between ca. 82 and 45 Ma. Allowing for ~50% east-northeast extension across area, calculated rate of eastward migration is ~1.5 km/Ma. The 45 Ma rocks crop out in deep canyons of Sierra Madre Occidental; any granitic rocks, if present farther east, are covered by tuffs of Sierra Madre. Although oldest, syntectonic rocks are closest to coast, individual syntectonic rocks show no apparent relation between K-Ar age and location. B: Age plot with addition of granitic rocks from northern Sinaloa transect and upper volcanic rocks from southern Sinaloa and Durango (Durango data from McDowell and Keizer, 1977). Granitic rocks of northern transect show same pattern as do rocks of southern Sinaloa. Ages and distribution of upper volcanic rocks suggest westward return of magmatism at ~12–15 km/Ma. C: Age plot of Figure 10A placed on plot of Damon et al. (1981), who compiled ages including data of this study from a 1200-km-long, arc-parallel swath from México-U.S. border to Trans-Mexican volcanic belt onto single east-northeast transect. Magmatism migrated much farther inland in northern México and southwestern United States (Coney and Reynolds, 1977) than it did in southern Sinaloa.

California, batholithic rocks of Sonora, the batholith of the Cabo San Lucas block at the tip of Baja California, and batholithic rocks that crop out from Jalisco eastward to Oaxaca in southern México (Fig. 1, Table 4). The strongest similarities are in the types of intrusions, sequence from early gabbro through commonly mafic pre- or syntectonic intrusions to more felsic posttectonic intrusions, and general eastward migration of magmatism. The major differences are in the range of ages, probable depth of emplacement and concordant versus discordant ages, and the short distance and much slower rate of eastward migration in Sinaloa compared to that for either the Peninsular Ranges or Jalisco-Oaxaca. Sinaloa rocks share both compositional similarities and differences with rocks of the Peninsular Ranges and Sonora.

Early, layered gabbros that range from peridotite to anorthosite and hornblende gabbro and which formed through cumulate processes are prominent in all four areas (Gastil, 1975; Gastil et al., 1979; Todd and Shaw, 1979; Walawender and Smith, 1980; Hausback, 1984; Aranda-Gómez and Perez-Venzor, 1989; Schaaf et al., 1997). Hornblende K-Ar ages of gabbros in the Peninsular Ranges range from 129 to 146 Ma (Krummenacher et al., 1975) but have been interpreted to indicate excess Ar (Silver et al., 1979). The only U-Pb date on gabbro in the Peninsular Ranges is 111 ± 1 Ma, and this dated gabbro is not the same as the gabbros that have K-Ar ages (Silver et al., 1979). A single hornblende K-Ar date of 115 Ma is reported in the Cabo San Lucas block (Hausback, 1984). Gabbro is abundant in western Jalisco near the Pacific Ocean, but the only dated rock is a gabbro pegmatite farther east in Nayarit (hornblende, 97.6 ± 2.6 Ma; Gastil et al., 1979; Schaaf et al., 1995). Gabbros of Sinaloa appear to be similar to those of the Peninsular Ranges batholith in major oxides but have substantially higher REE and incompatible trace element concentrations (Walawender and Smith, 1980; Gromet and

Silver, 1987; Walawender et al., 1991). A significant difference is that gabbro in Sinaloa crops out with granitic rocks that were contemporaneous with the younger (105–80 Ma), eastern part of the Peninsular Ranges batholith, where gabbro is minor.

Pre- or syntectonic intrusions are concentrated in the western part of the Peninsular Ranges (Silver et al., 1979; Todd and Shaw, 1979; Todd et al., 1988). They are also present in the western part of the Cabo San Lucas block (Aranda-Gómez and Perez-Venzor, 1989), but their distribution is otherwise unknown. No data are available about the presence of syntectonic rocks in the Jalisco-Oaxaca transect. Syntectonic rocks are dominantly low-K tonalites to granodiorites rocks in Sinaloa, the Peninsular Ranges, and Cabo San Lucas but include more felsic rocks in the Peninsular Ranges and Cabo San Lucas (Silver et al., 1979; Todd and Shaw, 1979; Aranda-Gómez and Perez-Venzor, 1989).

All areas except Jalisco, for which no data are available, and probably Sonora underwent intense deformation, but the end of this deformation may be different in each area. Deformation ended ca. 105 Ma in the Peninsular Ranges, essentially coincident with the beginning of eastward migration of magmatism (Silver et al., 1979; Todd and Shaw, 1979; Todd et al., 1988; Johnson et al., 1999b). An intensely deformed zone that forms the boundary between older, deformed granitic rocks on the west and younger, undeformed rocks on the east may be a suture between an outlying island arc and North America (Todd et al., 1988; Johnson et al., 1999b). The end of deformation is not as well dated in Sinaloa or in the Cabo San Lucas block. Depending on interpretations of K-Ar dates on syntectonic rocks in Sinaloa, the change may have occurred before 101 Ma but definitely had occurred by 89 Ma. Whether the beginning of eastward migration of magmatism accompanied this change, as it did in the Peninsular Ranges, is unclear because the eastward migration was so slow. The change probably occurred ca. 98 Ma in the Cabo San Lucas block (Frizzell et al., 1984; Aranda-Gómez and Perez-Venzor, 1989).

Posttectonic intrusions were emplaced in all areas and were accompanied by an eastward migration, although whether an eastward migration also occurred in the Cabo San Lucas block is unknown. Post- and syntectonic intrusions overlap spatially more in Sinaloa than in the Peninsular Ranges, where posttectonic rocks are almost exclusively east of the belt of syntectonic intrusions. Posttectonic intrusions are heavily dominated by more potassic and felsic compositions, mostly >62% SiO_2 in Sinaloa, the Peninsular Ranges (Silver et al., 1979; Silver and Chappell, 1988; Baird and Miesch, 1984), and Sonora (Roldan-Quintana, 1991; McDowell et al., 2001; Valencia-Moreno et al., 2001; Housh and McDowell, unpublished data) and high-K granodiorite and granite in the Cabo San Lucas block (Aranda-Gómez and Perez-Venzor, 1989; Kimbrough et al., 2002). A difference is that tonalite dominates in the Peninsular Ranges whereas granodiorite dominates in Sinaloa and Sonora. Also, the Peninsular Ranges are calcic whereas the Sinaloa batholith is borderline calc-alkalic (Fig. 4; Table 4). Granodiorite and monzongranite dominate in the Cabo San Lucas block (Frizzell, 1984; Aranda-Gómez and Perez-Venzor, 1989; Schaaf et al., 1997; Kimbrough et al., 2002).

TABLE 4. BATHOLITH COMPLEXES OF WESTERN MEXICO

Batholith complex	Sinaloa	Peninsular Ranges—Sonora		Cabo San Lucas	Jalisco—SW Mexico
		Peninsular Ranges	Sonora		
General composition	Borderline calc-alkalic/calcic	Calcic	Calc-alkalic		Calc-alkalic
Dominant rock type or range	Granodiorite	Tonalite	Granodiorite to quartz monzonite	Tonalite to monzogranite	Tonalite to granite
Rock type progression	Early gabbro, syntectonic mafic quartz diorite and tonalite, posttectonic granodiorite and monzogranite	Early gabbro, syntectonic tonalite to granodiorite, posttectonic leucocratic tonalite and granodiorite	Unknown	Early gabbro, syntectonic tonalite and trondhjemite, posttectonic granodiorite and monzogranite	?
Granitic duration	101–45 Ma, 20 Ma	140–80 Ma	90–50 Ma	ca.100–70 Ma	ca.103–25 Ma
Gabbro type	Layered gabbro, pyroxenite, anorthosite	Layered hornblende gabbro, norite, and peridotite	Gabbro not present	Layered hornblende gabbro, pyroxenite, anorthosite	Gabbro present but not described
K-Ar ages	134, 139 Ma	129, 133, 146 Ma		115 Ma; 129 ± 15 Ma (Rb/Sr)	98 Ma
Distribution	Mostly in west	Very dominantly in west		Dominantly in west?	Abundant in west near Puerto Vallarta, but also reported farther east at Zihuatanejo
Syntectonic rocks	Mafic quartz diorite and granodiorite	Mafic quartz diorite and granodiorite accompanied by leucocratic granodiorite	Not present?	Tonalite and trondhjemite	Not definitely present
Posttectonic rocks	Granodiorite and granite	Tonalite to low-K granodiorite		Tonalite to granite	Quartz diorite to granite
S-type (Two-mica granites) present?	No	Yes, mostly on east side	Minor two-mica granite	Yes	Yes in Jalisco (Puerto Vallarta batholith) at west end
Syn/post overlap?	Yes, a lot	Very little?	?	Yes	?
Syn/Post change	>90 Ma	105 Ma	?	98 Ma	?
Eastward migration	Yes after 90 Ma, before?	Yes after 105 Ma	Yes	Probably	Yes
Distance	Batholiths: 120 km; adding Oligocene volcanic: 200 km	~100 km width of Peninsular Ranges	Plutons: 250 km @28°N - minus extension; total for northern Mexico ~1000 km		~600 km perpendicular to paleotrench
Rate	1–1.5 km/Ma	10 km/Ma	~6 km/Ma	?	~10 km/Ma
Westward return?	Yes, at ~12–15 km/Ma		Yes, rate very high (>50 km/Ma)	Yes, Miocene volcanic rocks abundant north of La Paz	No
Basement	Jurassic? metasedimentary	West: Mesozoic metasedimentary; east: Paleozoic and Mesozoic metasedimentary	North=Precambrian; central and south=Paleozoic eugeoclinal	Jurassic? metasedimentary, but granitic Sr and Nd data indicate Precambrian	Highly variable. West=Jurassic metasedimentary, but granitic Sr and Nd data indicate Precambrian; east=Precambrian
U-Pb/K-Ar age difference	Posttectonic=no; syntectonic=?	West: no; east: yes	West: unknown; east: probably		West: probably; east: no
Age gaps	89–82, 82–75, 62–56 Ma	No	No	?	No?
Sr initial	9 samples: .7026 to .7062	.703 on west to .708 on east	.706–.709; one value .713	average: .706; gabbros: .7035–.7039; granitic: .7048–.707; two-mica granites: .712–.716	Jalisco: .7038–.712; Oaxaca: .7037–.7054

Note: Data from this study and Anderson and Silver, 1974; Aranda-Gómez and Perez-Venzor, 1989; Damon et al., 1981; Ferrari et al., 2000; Frizzell, 1984; Frizzell et al., 1984; Gastil, 1975; Gastil et al., 1975, 1978, 1979; Hausback, 1984; Housh and McDowell, 1999, unpublished; Kimbrough et al., 2001; Martiny et al., 2000; McDowell and Mauger, 1994; McDowell et al., 2001; Moran-Zenteno et al., 2000; Nieto-Samaniego et al., 1999; Ortega-Rivera et al., 1997, this volume; Roldan-Quintana, J., 1991; Schaaf et al., 1995, 1997, 2000; Sedlock et al., 1993; Silver et al., 1979; Silver and Chappell, 1988; Stein et al., 1994; Tardy et al., 1994; Todd et al., 1988; Todd and Shaw, 1979; Todd et al., 1988; Valencia-Moreno et al., 2001; Walawender et al., 1990, 1991; Zimmermann et al., 1988

Muscovite-bearing rocks with high $^{87}Sr/^{86}Sr$ are present as minor components everywhere except Sinaloa.

The concordance of K-Ar ages and U-Pb ages in southern Sinaloa is a significant difference from the Peninsular Ranges, where K-Ar ages are generally discordant and the absolute age difference compared to U-Pb zircon ages increases eastward (Gastil, 1975; Krummenacher et al., 1975; Silver et al., 1979; Silver and Chappell, 1988; Snee et al., 1994; Ortega-Rivera et al., 1997; Ortega-Rivera, this volume, Chapter 11). The age pattern in Sinaloa indicates shallow emplacement and rapid cooling, which is consistent with the geological evidence of shallow emplacement (coexistence of plutons with coeval(?) volcanic rocks and epithermal ore deposits). The regional discordance in the Peninsular Ranges reflects regional uplift that exposed deep levels of the eastern part of the batholith long after intrusion.

Similarities and differences between Sinaloa, the Peninsular Ranges, and Cabo San Lucas extend to individual intrusive types. The Candelero granodiorite, La Posta suite, and Las Cruces granite are prominent, petrographically similar posttectonic intrusive types in Sinaloa, the Peninsular Ranges, and Cabo San Lucas respectively (Walawender et al., 1990; Kimbrough et al., 2001, 2002). They overlap compositionally and share petrographic characteristics of euhedral biotite and hornblende plus large, euhedral, honey-colored sphene (Walawender et al., 1990). REE patterns are indistinguishable with similar steep slopes and negligible Eu anomalies (Gromet and Silver, 1987). Marked differences include the lack of muscovite-bearing cores in Candelero intrusions, which are common in La Posta intrusions. Also, La Posta intrusions were emplaced over a narrow age range of 99–94 Ma in both the Peninsular Ranges and Cabo San Lucas (Kimbrough et al., 2001; Kimbrough et al., 2002), whereas Candelero rocks are younger and emplaced between 82 and 48 Ma.

Sinaloa rocks show compositional similarities and differences with batholithic rocks of the Peninsular Ranges and Sonora. The wide compositional range is a notable characteristic of the Sinaloa batholith (Figs. 4 and 6). Sinaloa rocks span most of the SiO_2 range seen in the Peninsular Ranges but are distinctly more potassic (Silver and Chappell, 1988). Similarly, they span most of the range of K_2O-SiO_2 relations seen for a suite of granitic rocks that extends from northern Sonora into Sinaloa as far south as the area of this study (Valencia-Moreno et al., 2001).

Comparison of the REE characteristics of Sinaloa rocks to those of the Peninsular Ranges and Sonora is particularly interesting (Fig. 7). Gromet and Silver (1987) found that REE patterns varied systematically west to east across the Peninsular Ranges batholith with abrupt changes between the western, central, and eastern belts. Similarly, Valencia-Moreno et al. (2001) found systematic changes south to north from southern Sinaloa to northern Sonora. In both cases, the REE patterns paralleled other compositional and isotopic trends and correlated with the nature of the crust. In the Peninsular Ranges, crust changes from Mesozoic turbidites or volcanic equivalents of the batholith on the west to Mesozoic and Paleozoic metasedimentary rocks containing Precambrian detritus on the east (Silver and Chappell,

1988). The northwestern Mexican transect is inferred to pass from poorly known Mesozoic rocks in Sinaloa, as in this area, northward to Proterozoic crust covered by Paleozoic eugeoclinal rocks in southern Sonora, and to exposed Proterozoic crust in central and northern Sonora (Housh and McDowell, 1999; Valencia-Moreno et al., 2001).

REE characteristics of Sinaloa rocks are most like those of the eastern belt of the Peninsular Ranges and central granites of Sonora (Fig. 7). However, Sinaloa rocks span much of the range of the western and central belts of the Peninsular Ranges and of the southern and northern granites of northwestern México. Compared to the Peninsular Ranges, Sinaloa rocks do not show the low HREE abundances of rocks of the central belt or the very flat overall patterns of the western belt. Compared to the Sonora-Sinaloa rocks, the Sinaloa rocks overlap with all but the most LREE-enriched of the northern granites and the most LREE-depleted of the southern granites. This is despite the fact that Sinaloa lies within the defined southern group of Valencia-Moreno et al. (2001) and that one of their samples comes from this field area (Mal-74 is from the same location as KC-1; Fig. 2, Plate 1). Indeed, the REE pattern of Mal-74 is distinctly unlike any of our samples. If these REE patterns are in any way related to the nature of the crust, they suggest that southern Sinaloa is underlain by either Precambrian crust or crust that has a significant Precambrian detrital component. However, Sr initial ratios are low and similar to other ratios reported by Valencia-Moreno et al. (2001) for their southern granites. Our U-Pb data show little or no evidence of old zircons in the four dated samples. The low Sr ratios and lack of old zircons support the interpretation that old crust is not present in southern Sinaloa or in the area of southern granites. Little is known about the nature of crust beneath central México east of Sinaloa (Sedlock et al., 1993), but Nd model ages of lower crustal xenoliths suggest that most of the area is underlain by Phanerozoic, probably Paleozoic, crust and lacks Precambrian basement (Ruiz et al., 1988).

Batholithic magmatism continued far later in Sinaloa, to 45 Ma, than in the Peninsular Ranges (80 Ma; Silver et al., 1979; Silver and Chappell, 1988; Ortega-Rivera, this volume, Chapter 11) or the Cabo San Lucas block (70–80 Ma; Frizzell, 1984; Frizzell et al., 1984; Aranda-Gómez and Perez-Venzor, 1989; Kimbrough et al., 2002). However, these differences disappear when one considers likely eastern limits to the overall distribution of magmatism. Magmatism of Sinaloa continued to ca. 30 Ma if the upper volcanic sequence of the Sierra Madre Occidental of Durango is included (McDowell and Keizer, 1977; Swanson et al., 1978; Aranda-Gómez et al., 1997). Similarly, plutonism continued eastward of the Peninsular Ranges to ca. 60 Ma in eastern Sonora (Anderson and Silver, 1974; Damon et al., 1981; McDowell et al., 2001). Caldera-forming volcanism with underlying plutons continued to ca. 30 Ma throughout the Sierra Madre Occidental and as far eastward as Chihuahua and Trans–Pecos Texas, which is the easternmost magmatism at that latitude (Henry et al., 1991; McDowell and Mauger, 1994). Tertiary rhyolitic magmatism in the southern Sierra Madre Occidental, opposite the southern end of the Cabo San Lucas block, and in Jalisco (Fig. 1), appears to roughly parallel that in Sinaloa and Durango (Nieto-Samaniego et al., 1999; Ferrari et al., 2000). Numerous rhyolite domes were emplaced ca. 30 Ma in the Mesa Central east of the Sierra Madre. Volcanism then shifted westward with a major pulse ca. 24 Ma in the Sierra Madre (Nieto-Samaniego et al., 1999). Magmatism appears to have migrated only eastward, to as young as ca. 25 Ma, farther south along the Jalisco–Oaxaca trend (Schaaf et al., 1995; Martiny et al., 2000; Moran-Zenteno et al., 2000). However, the western and eastern halves of this trend had markedly different Late Cenozoic histories. Eastward displacement of the Chortis block truncated the southern México continental margin of the eastern half and removed older parts of the magmatic arc (Schaaf et al., 1995; Moran-Zenteno et al., 2000). The presence of ca. 100–70 Ma intrusions in Jalisco indicate that the western half was not truncated (Gastil et al., 1978; Zimmermann et al., 1988; Schaaf et al., 1995).

Regardless of whether this inland activity is included, differences still remain in the distances that magmatism migrated eastward and the rates at which it did so. The Sierra Madre Occidental is ~400 km east of a projected paleotrench (Fig. 10C) and was 300–350 km before "proto-Gulf" extension (Stock and Hodges, 1989; Henry and Aranda-Gómez, 2000). Likewise, Trans–Pecos Texas is ~1000 km east of the paleotrench and originally was 700–900 km, allowing for various estimates of extension across the Gulf of California, Sonora, and Chihuahua (Stock and Hodges, 1989; Gans, 1987; Henry and Aranda-Gómez, 2000). Magmatism migrated ~600 km northeastward perpendicular to the inferred position of the late Mesozoic–early Cenozoic paleotrench in the Jalisco–Oaxaca trend, but truncation of the continental margin again complicates interpretation (Schaaf et al., 1995; Moran-Zenteno et al., 2000).

The difference in migration rates is also striking. Silver and Chappell (1988) estimated that magmatism migrated at ~10 km/Ma across the Peninsular Ranges. A similar rate can be calculated farther east for migration across the eastern segment in Sonora, Chihuahua, and Texas. We calculate a similar 10 km/Ma for the Jalisco–Oaxaca trend, which is different than the still higher rates calculated in part along strike of the subduction zone by Schaaf et al. (1995). These rates are far more rapid than the 1–1.5 km/Ma rate across Sinaloa and the central Sierra Madre Occidental.

These variations in distance and rate of migration along the magmatic arc were not considered in earlier interpretations about patterns of arc magmatism in México and the southwestern United States. An eastward progression of magmatism was first well established for the southwestern United States and northwestern México (Anderson and Silver, 1974; Coney and Reynolds, 1977; Silver et al., 1979; Damon et al., 1981; Silver and Chappell, 1988; Ortega-Rivera, this volume, Chapter 11). For example, Damon et al. (1981) extrapolated age information from a 1200-km-long, arc-parallel swath from the U.S. border to the Trans-Mexican volcanic belt onto a single east-northeast reference line (Fig. 10C). Their included arc length is greater than the length of their cross-arc traverse. By combining data from near

the México-U.S. border, where magmatism migrated far inland, and from the Sinaloa-Durango area, where magmatism did not, the compilation of Damon et al. (1981) gives the impression that magmatism migrated far inland throughout México. Their compilation includes our data from southern Sinaloa, yet it yields markedly different results than we derive for our data alone.

We do not know the location of the change from the long and rapid migration of magmatism in northern and southern México to the much slower and shorter migration in Sinaloa, or whether the change is gradational or abrupt. The zone of shorter migration approximately coincides with the Guerrero or Tahue terrane, which is composed of possibly accreted rocks that are no older than Jurassic (Campa and Coney, 1983; Coney and Campa, 1987; Sedlock et al., 1993; Tardy et al., 1994). Our data and published work (e.g., Ortega-Rivera, this volume, Chapter 11) require that the northern change be north of north-central Sinaloa and south of central Sonora, a location that approximately coincides with the southern edge of North American Proterozoic basement (Fig. 1; Valencia-Moreno et al., 2001). The location between Sinaloa and Jalisco-Oaxaca is even less certain. However, the basement in Jalisco is Jurassic(?) metasedimentary rock (Gastil et al., 1978), similar to basement in Sinaloa, whereas Precambrian metamorphic rocks are encountered in the middle and eastern parts of the Jalisco–Oaxaca trend (Centeno-Garcia et al., 1993; Schaaf et al., 1995). A very speculative possibility is that the character of the overriding plate influenced the pattern of subduction. Rapid subduction has been suggested as the cause of the eastward migration of magmatism (Hamilton, 1988), but rapid subduction occurred along the whole length of northwestern México, so that cannot be the only factor (Engebretson et al., 1985; Stock and Molnar, 1988).

Arc Migration and Plate Tectonics

A detailed analysis of the relationship between arc migration and regional plate tectonic history is beyond the scope of this paper. Many authors (e.g., Hamilton, 1988) have suggested that the position of a magmatic arc relative to its convergent margin (the arc-trench gap) is a function of the dip angle of the subducting plate, which in turn is related to the orthogonal rate of plate convergence. Analysis of changes in the arc position generally assumes a uniform dip angle for the entire length of the slab and a similar depth range for the first stage of magma production. These conditions predict a regular migration of the arc through time and an increase in the width of the arc away from the trench. However, the actual magmatic record for the Cordillera of North America is anything but uniform. The abrupt drop-off of Late Cretaceous–early Tertiary igneous activity eastward from the Sierra Nevada batholith, and the Eocene-Oligocene ignimbrite flare-ups of the western United States—and especially the Sierra Madre Occidental—are two examples of the episodic nature of magma production in time and space. Hence, as indicators of subduction history, compilations of age versus location are incomplete indicators without attempts to assess variations in the rate of magma production through time. Moreover, most existing compilations for the Late Cretaceous–

early Tertiary magmatic history of western North America have failed to include data for the "coeval" volcanic rocks, which can provide a critical part of a region's magmatic history (McDowell et al., 2001). Indeed, the lower volcanic complex remains undated in Sinaloa. Finally, the order of magnitude contrast between the migration rate calculated for the magmatic arc of Baja-Sonora and that of Sinaloa requires significant contrasts in subduction pattern along the Farallon–North America plate margin.

Ties across the Gulf of California

Both granitic and volcanic rocks of Sinaloa are similar to those of the east coast of Baja California Sur, which lay adjacent to Sinaloa before transform opening of the Gulf of California. Based on reconstructions by Stock and Hodges (1989), Mazatlan lay approximately opposite La Paz, and our northern transect lay between Loreto and Bahia Concepcion (Fig. 11), rare locations where granitic rocks are exposed in Baja California Sur north of La Paz. As discussed above, the granitic rocks of the Cabo San Lucas block are similar to those of Sinaloa (Frizzell et al., 1984; Hausback, 1984; Aranda-Gómez and Perez-Venzor, 1989; Kimbrough et al., 2002). A slightly lineated, mafic quartz diorite dated at 87.4 ± 2.0 Ma (K-Ar, biotite; McLean, 1988) crops out just northwest of Loreto. The rock type and age are indistinguishable from those of the syntectonic rocks of Sinaloa. Host rocks for the quartz diorite are greenschist facies volcaniclastic sandstone and conglomerate, probably similar to rocks described by Bonneau (1970) in our northern transect but only somewhat like the metasedimentary rocks of southern Sinaloa. McFall (1968) reports "granodiorite to quartz monzonite" with a 80.4 ± 2.8 Ma biotite age at Bahia Concepcion.

Miocene volcanic and intrusive rocks are also similar along the east coast of Baja California Sur and in Sinaloa. In an extensive study of volcanism in and north of La Paz, Hausback (1984) reports volcanic rocks ranging from 28 to 12 Ma. Rocks from ca. 24 to 12 Ma include rhyolitic ash-flow tuffs and andesite to rhyodacite lavas erupted from sources in Baja California. Rhyolitic tuffs with ages between 28 and 24 Ma appear to have erupted from sources east of Baja California, presumably in either Sinaloa or Durango. McLean (1988) reports 22 Ma ages on two ash-flow tuffs and a 19 Ma hornblende andesite intrusion near Loreto, and McFall (1968) reports a 29 Ma ash-flow tuff and a 20.5 Ma tonalite stock near Bahia Concepcion.

Although these similarities support matching southern Sinaloa across the Gulf of California to the area between La Paz and Loreto, it must be noted that similar rocks occur all along the length of the former Cordilleran arc. More distinct rock types need to be correlated (Gastil et al., 1981; Oskin et al., 2001), because the general character of rocks is not diagnostic. Ash-flow tuffs of the upper volcanic sequence are probably the most promising for correlation. They are known to crop out on both sides of the Gulf, individual tuffs would have been sufficiently widespread to extend across the now separated terranes, and highly precise methods such as ^{40}Ar/^{39}Ar dating are available for correlation.

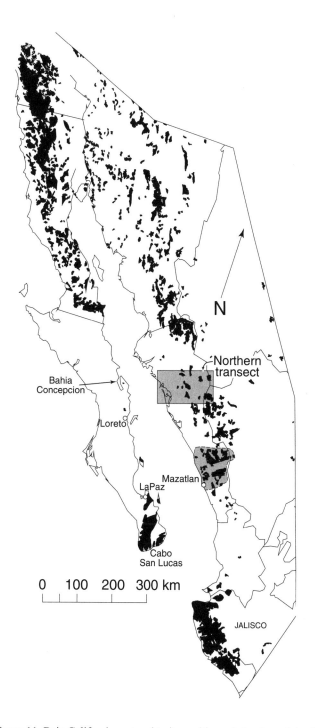

Figure 11. Baja California restored to its position relative to mainland México before transform faulting that began ca. 6 Ma but after Miocene, east-northeast extension that began ca. 12 Ma (adapted from Stock and Hodges, 1989). Mazatlan lay opposite La Paz and north of Cabo San Lucas block. Our northern transect lay between Loreto and Bahia Concepcion. Miocene extension affected area between eastern edge of Baja California and western edge of Sierra Madre Occidental and also east of Sierra Madre (Henry and Aranda-Gómez, 2000). Before this extension, Cabo San Lucas batholith may have lain more closely between batholiths of southern Sinaloa and Jalisco (Schaaf et al., 1997, 2000; Moran-Zenteno et al., 2000).

ACKNOWLEDGMENTS

We thank Goran Fredrikson, Steve Clabaugh, Michel Bonneau, and Ken Clark for discussions about the geology of Sinaloa and Jim Faulds for discussions about the character and significance of dynamic recrystallization textures. Geologists and staff of Minas de San Luis, now Luismin, provided hospitality, a place to stay, and a thorough knowledge of the geology around Tayoltita, in the northeastern part of the area. Paul Guenther provided instruction and guidance during K-Ar dating; Thomas Anderson, Jaime Alvarez, and Gerri Silver did the same for U-Pb work. The opportunity for this study arose through a project of the Instituto de Geologia, Universidad Nacional Autónoma de México, funded by the state of Sinaloa. The project provided expenses, vehicles, and other supplies during 1971 and 1972. Funds were also provided by a Penrose Research Grant from the Geological Society of America and by National Science Foundation Grant GA-16080 to Steve Clabaugh. We also thank Stan Keith for providing many of the chemical analyses reported here and Mark Barton and Joaquin Ruiz for providing the data to construct the map of granitic rocks in western México. Finally, we thank Jim Mattinson and David Kimbrough for helpful, constructive reviews.

APPENDIX 1: DATING PROCEDURES

Both K-Ar and U-Pb dating were done in the early 1970s using procedures and instruments of that time.

K-Ar Procedures

Samples for K-Ar dating were crushed and ground and a 60–80 mesh fraction was used for mineral separation. Biotite and hornblende were separated using standard magnetic and heavy liquid techniques. K analyses were done with a Baird-Atomic model KY-2 flame photometer using a Li internal standard and Na buffering. Approximately 100 mg aliquots were dissolved with HF, HCL, and HNO_3, dried and redissolved twice with HCL to get rid of fluoride ions, and then taken up with HCl and deionized H_2O. Final solutions were made to have K concentrations of less than 8 ppm.

Ar analysis was done in an on-line extraction system where the furnace, purification system, ^{38}Ar spike and calibration system, and mass spectrometer are part of one vacuum system. Approximately 100 mg of biotite and 500 mg of hornblende were spiked and then melted in a Mo crucible at ~1700 °C using a RF induction heater. The released gas was purified, passed through a series of zeolite traps and Ti getters. Isotopic analysis was done on a CEC 21–615 cycloidal focusing residual gas analyzer operated in the static mode.

Pooled standard deviations for K content were 0.61% and 1.14% and for Ar content 0.95% and 1.97% for biotite and hornblende, respectively. These values combine to give an estimated analytical precision of 1.13% and 2.28% for biotite and hornblende ages, which have been applied here.

U-Pb Procedures

Crushed and ground samples were passed across a Wilfley table and further concentrated with a hand magnet and heavy liquids. The sample was sieved, and the various size fractions were run through a magnetic separator. The least magnetic material was washed with hot HNO_3 and rerun through the magnetic separator. Resulting zircon fractions have been demonstrated to show a correlation between U content, magnetic susceptibility, and size, with the most magnetic and smallest zircons having the highest U content (Silver, 1963).

Separate aliquots of ~200–400 mg of each zircon size-magnetic susceptibility fraction were fused with a borate flux and analyzed for U and Pb concentration and Pb isotopic composition. The concentration aliquot was spiked with ^{208}Pb and ^{235}U before fusion. Pb purification for each aliquot was done by double dithizone extraction. Purified Pb was picked up in silica gel and loaded on a single Re filament. U purification was done by anion exchange. U was picked up in HNO_3 and loaded on a single Ta filament. Mass spectrometry was performed on a 60° sector field, single-focusing instrument with a 12-in (30.5 cm) radius of curvature.

Observed Pb isotope ratios were corrected for both blank and natural common Pb using $^{206}Pb/^{204}Pb = 18.60$ and $^{207}Pb/^{204}Pb = 15.60$. Borate flux was the major source of blank Pb and constituted most of the ^{207}Pb in almost all analyses.

REFERENCES CITED

Aguirre-Díaz, G.J., and McDowell, F.W., 1991, The volcanic section at Nazas, Durango, México, and the possibility of widespread Eocene volcanism within the Sierra Madre Occidental: Journal of Geophysical Research, v. 96, p. 13,373-13,388.

Anders, E., and Grevesse, N., 1989, Abundances of the elements: Meteoritic and solar: Geochimica et Cosmochimica Acta, v. 53, p. 197–214.

Anderson, T.H., and Silver, L.T., 1974, Late Cretaceous plutonism in Sonora, México, and its relationship to Circum-Pacific magmatism: Geological Society of America Abstracts with Programs, v. 6, p. 484.

Aranda-Gómez, J.J, Henry, C.D., Luhr, J.F., and McDowell, F.W., 1997, Cenozoic volcanism and tectonics in NW México—A transect across the Sierra Madre Occidental volcanic field and observations on extension related magmatism in the southern Basin and Range and Gulf of California tectonic provinces, in Aguirre-Díaz, G.J., Aranda-Gómez, J.J., Carrasco-Nuñez, G., and Ferrari, L., eds., Magmatism and tectonics in the central and northwestern México—A selection of the 1997 IAVCEI General Assembly excursions: Universidad Nacional Autónoma México, Instituto de Geologia, México, D.F., Excursión 11, 41–84.

Aranda-Gómez, J.J., and Perez-Venzor, J.A., 1989, Estratigrafia del complejo cristalino de la region de Todos Santos, Estado de Baja California Sur: Revista de Universidad Nacional Autónoma de México, Instituto de Geologia, v. 8, p. 149–170.

Baird, A.K., and Miesch, A.T., 1984, Batholithic rocks of Southern California—A model for the petrochemical nature of their source materials: U.S. Geological Survey Professional Paper 1284, 42 p.

Bateman, P.C., 1992, Plutonism in the central part of the Sierra Nevada Batholith, California: U.S. Geological Survey Professional Paper 1483, 186 p.

Bonneau, M., 1970, Una nueva area Cretacica fosilifera en el estado de Sinaloa: Boletin Sociedad Geologica Mexicana, v. 32, p. 159-167.

Campa, M.F., and Coney, P.J., 1983, Tectono-stratigraphic terranes and mineral resource distributions in México: Canadian Journal of Earth Sciences, v. 20, p. 1040–1051.

Cibula, D.A., 1975, The geology and ore deposits of the Cosala mining district, Cosala municipality, Sinaloa, México: [M.S. thesis]: University of Iowa, 145 p.

APPENDIX 2. SAMPLE DESCRIPTIONS AND LOCATIONS

Sample	Rock name (granitic type)	Location and type of outcrop	Mineralogy
HM-1	Rhyodacite flow (upper volcanic sequence)	Roadcut on Highway 40 8 km northeast of Villa Union	30% plagioclase, 5% biotite and 2% quartz phenocrysts in groundmass of quartz, alkali feldspar, and plagioclase
HM-4	Mafic tonalite (R e c o d o)	Outcrop in stream bed 7 km southwest of Rocodo	53% plagioclase, 16% quartz, 6% microcline, 14% biotite, 11% hornblende
HM-8	Mafic tonalite	Erosional remnant 7 km east of San Marcos	52% plagioclases 21% quartz, 4% microcline, 14% biotite, 13% hornblende
HM-9	Monzogranite	Outcrop in stream bed 8 km east of San Marcos	33% plagioclase, 19% quartz, 41% orthoclase, 6% biotite
HP-2	Granodiorite (San Ignacio)	Erosional remnant 7 km east of Ixpalino	39% plagioclase, 27% quartz, 17% orthoclase, 9% biotite, 7% hornblende
HP-4	Quartz diorite	Outcrop in stream bed in Ixpalino	61% plagioclase, 13% quartz, 10% biotite, 11% hornblende, 3% pyroxene, 3% opaques
HP-5	Granodiorite	Erosional remnant 8.5 km north of Elota	51% plagioclase, 21% quartz, 16% orthoclase, 7% biotite, 5% hornblende
HP-6	Granodiorite (Candelero)	Roadcut on Highway 15, 2 km west of Elota	55% plagioclase, 20% quartz, 9% orthoclase, 8% biotite, 8% hornblende
HP-7	Quartz diorite	Erosional remnant 10 km north of San Ignacio	71% plagioclase, 7% quartz, 22% hornblende and clinopyroxene
HP-10	Andesite dike (upper volcanic sequence)	Outcrop in stream bed 12 km north of San Ignacio	7% hornblende, 4% altered plagioclase phenocrysts in groundmass of plagioclase, hornblende and opaques
HP-11	Granodiorite (San Ignacio)	Erosional remnant 6 km east of San Ignacio	43% plagioclase, 24% quartz, 19% orthoclase, 9% biotite, 5% hornblende
HP-13	Monzogranite (Candelero)	Roadcut 6 km east of San Juan	37% plagioclase, 34% orthoclase, 27% quartz, 4% biotite, 2% hornblende
HP-14	Granodiorite (Candelero)	Outcrop in stream bed near Candelero	47% plagioclase, 22% quartz, 18% orthoclase, 8% biotite, 3% hornblende
HP-15	Granodiorite	Outcrop in stream bed 4 km south of Pueblo Viejo	47% plagioclase, 24% quartz, 15% microcline, 11% biotite, 4% hornblende
HP-16	Granodiorite	Stream bed 4 km northeast of Pueblo VieJo	41% plagioclase, 22% quartz, 17% orthoclase, 11% biotite, 9% hornblende
HP-18	Granodiorite (Candelero)	Outcrop in Rio Piaxtla 10 km northeast of Tayoltita	46% plagioclase, 26% quartz, 19% orthoclase, 8% biotite, 1% hornblende
HP-20	Granodiorite (Candelero)	Outcrop in stream bed 18 km northeast of Tayoltita	46% plagioclase, 25% quartz, 14% orthoclase, 8% biotite, 4% hornblende
HP-22	Granodiorite	Outcrop in stream bed 12 km northeast of Tayoltia	51% plagioclase, 19% quartz, 11% orthoclase, 10% biotite, 7% hornblende
HP-23	Granodiorite (Candelero)	Outcrop in Arroyo Chicaral 9 km east of Tayoltita	59% plagioclase, 22% quartz, 14% orthoclase, 5% biotite
HP-24	Quartz diorite dike (upper volcanic sequence)	Outcrop in Arroyo San Geronimo 9 km northeast of Tayoltita	67% plagioclase, 15% biotite, 10% hornblende, 8% quartz
HP-25	Granodiorite (Candelero)	Outcrop in A r r o y o 8 km west of Tayoltita	56% plagioclase, 17% quartz, 14% orthoclase, 9% biotite, 2% hornblende
HP-26	Granodiorite (San Ignacio)	Outcrop in Rio Piaxtla 4 km southwest of Tayoltita	40% plagioclase, 25% quartz, 20% orthoclase, 10% biotite, 5% hornblende
HP-28	Granodiorite (Candelero)	Outcrop in Rio Piaxtla 8 km southwest of Tayoltita	46% plagioclase, 25% quartz, 22% orthoclase, 6% biotite, 1% hornblende

(continued)

Sample	Rock name (granitic type)	Location and type of outcrop	Mineralogy
HS-8B	Porphyritic granodiorite (Concordia)	Roadcut on Highway 40, 5 km west of Concordia	43% Plagioclase, 20% quartz, 21% orthoclase as l cm porphyroblasts, 6% biotite, 10% hornblende
HS-12	Granodiorite (El Carmen)	Roadcut on San Ignacio Road	50% plagioclase, 25% quartz, 10% orthoclase, 8% biotite, 7% hornblende (with pyroxene cores)
HS-13	Granodiorite (El Carmen)	Roadcut on San Ignacio Road	45% plagioclase, 25% quartz, 15% orthoclase) 10% biotite, 5% hornblende
HS-15	Granodiorite	Roadcut on San Ignacio Road	50% plagioclase, 25% quartz, 15% orthoclase, 7% biotite, 3% hornblende (with pyroxene cores)
HS-17	Mafic tonalite (Recodo)	Outcrop in stream bed 2 km southwest of Recodo	60% plagioclase, 14% quartz, 1% Kfeldspar, 10% biotite, 15% hornblende (with pyroxene cores)
HS-22	Foliated tonalite	Erosional remnant 2 km southwest of San Marcos	54% plagioclase, 16% quartz, 13% biotite, 5% microcline, 12% hornblende (with Py.roxene cores)
HS-23	Granodiorite	Outcrop in stream bed near Rio Presidio, 5 km east of San Marcos	40% plagioclase, 30% quartzi 20% orthoclase, 6% biotite, 4% hornblende
HS-25	Porphyritic granodiorite (El Carmen)	Roadcut on dirt road 8 km east of Coyotitan	45% plagioclase (some as porphyroblasts up to 2 cm long), 35% quartz, 10% orthoclase, 7% biotite, 3% hornblende
HS-32	Hypersthene granodiorite	Erosional remnant 6 km west of La Noria	55% plagioclase, 15% hypersthene-, 12% biotite, 10% quartz, 8% orthoclase
HS-34	Foliated granodiorite	Erosional remnant 4.km northwest of Quelite	48% plagioclase, 25% quartz, 6% biotite, 8% microcline, 13% hornblende
HS-37	Foliated granodiorite	Outcrop in jungle 10 km south of La Cruz	40% plagioclase, 23% quartz, 18% biotite, 8% hornblende (with rare cores of pyroxene), 11% microcline
HS-40	Tonalite (San Ignacio border phase)	Outcrop in stream bed near San Javier	50% plagioclase, 40% quartz, 9% biotite, 1% hornblende
HS-4l	Quartz diorite (San Ignacio border phase)	Outcrop in stream bed near San Javier (same as HS-40)	70% plagioclase, 20% biotite, 10% quartz
HS-42	Granodiorite (San Ignacio)	Outcrop in bed of Rio Piaxtla across from San Ignacio	43% plagioclase, 27% quartz, 15% orthoclase, 9% biotite, 6% hornblende (cores of pyroxene)
HS-43	Amphibolite (possible metadiorite?)	Outcrop in stream bed 5 km north of San Marcos	50% plagioclase, 40% hornblende, 5% quartz, 1% biotite
HS-44	Tonalite	Erosional remnant 1 km west of Highway 15 near Quelite	55% plagioclase, 27% quartz, 15% biotite, 3% microcline
HS-45	Gneissic tonalite (Quelite tonalite)	Quarry along road to Marmol	55% plagioclase, 20% quartz, 12% biotite, 11% hornblende, 2% microcline
HS-46	Hornblende gabbro	Roadcut 6 km southwest of La Noria	70% plagioclase, 15% hornblende, 15% fibrous alteration
HS-48	Hornblende gabbro	Outcrop in stream bed 3 km north of Cofradia	60% plagioclase 30% hornblende, 7% alteration, 3% quartz
HS-49	Tonalite	Roadcut 7 km northeast of El Verde	66% plagioclase, 25% quartz, 5% biotite, 3% orthoclase, 1% hornblende and clinopyroxene
HS-50	Tonalite	Outcrop in stream bed 1 km south of Naranjos	50% plagioclase, 30% quartz, 10% biotite, 5% hornblende, 2% orthoclase
HS-51	Granodiorite	Outcrop in stream bed at Panuco	40% quartz, 35% plagioclase, 10% orthoclase, 10% biotite
HS-53	Tonalite (Colegio tonalite)	Erosional remnant 9 km north of Quelite	60% plagioclase, 14% quartz, 2% orthoclase, 11% biotite, 13% hornblende
146	Granodiorite (San Ignacio)	Outcrop in Rio Piaxtla 1.5 km north of Los Brasiles	48% plagioclase, 20% quartz, 11% orthoclase, 13% biotite, 8% hornblende

(continued)

APPENDIX 2. SAMPLE DESCRIPTIONS AND LOCATIONS *(continued)*

Sample	Rock name (granitic type)	Location and type of outcrop	Mineralogy
152	Granodiorite	Outcrop in Rio Piaxtla 7 km southwest of Los Brasiles	50% plagioclase, 20% quartz, 8% orthoclase, 10% biotite, 11% hornblende
156	Granodiorite (San Ignacio)	Outcrop in Rio Piaxtla 11 km east of San Ignacio	42% plagioclase, 21% quartz, 20% orthoclase, 9% biotite, 7% hornblende
170	Monzogranite (San Ignacio)	Outcrop in Rio Piaxtla 11 km east-northeast of San Ignacio	44% plagioclase, 18% quartz, 24% orthoclase, 6% biotite, 7% hornblende
174	Rhyolite dike	Outcrop in stream bed 3 km southwest of Ajoya	2% biotite and 1% plagioclase phenocrysts in groundmass of plagioclase, quartz, K feldspar and biotite
199	Tonalite	Outcrop in Rio Piaxtla 8 km southwest of Los Brasiles	69% plagioclase, 15% quartz, 8% biotite, 5% hornblende and clinopyroxene, 3% orthoclase
227	Granodiorite	Outcrop in Rio Verde 6.5 km north of Ajoya	55% plagioclase, 12% quartz, 14% orthoclase, 16% biotite, 3% hornblende
230	Granodiorite	Outcrop in Arroyo Verano near Verano north of map area	49% plagioclase, 16% quartz, 13% orthoclase, 15% biotite, 8% hornblende
236	Monzogranite (Candelero)	Quarry in Ajoya	37% plagioclase, 26% quartz, 26% orthoclase, 5% biotite, 6% hornblende
GD	Granodiorite	Mine workings at Tayoltita	56% plagioclase, 20% quartz, 20% orthoclase, 4% biotite
Hbl	Porphyritic granodiorite	Mine workings at Tayoltita	1% hornblende phenocrysts in highly altered groundmass of plagioclase, opaques, and quartz
Adul	Adularia-bearing vein	Mine Workings at Tayoltita	quartz, adularia, and calcite filling vein in andesite
HC-3	Granodiorite (Candelero)	Erosional remnant 15 km northwest of Badiraguato, northern Sinaloa	55% plagioclase, 24% quartz, 7% orthoclase, 10% biotite, 4% hornblende
HC-4	Mafic foliated tonalite	Outcrop in stream bed 20 km north of Pericos, northern Sinaloa	50% plagioclase, 20% biotite, 15% quartz, 10% hornblende, 5% microcline
HC-5	Granodiorite	Erosional remnant 10 km north of HC-4, northern Sinaloa	50% plagioclase, 25% quartz, 10% orthoclase, 10% biotite, 5% hornblende
HC-6	Monzogranite	Outcrop in stream bed 10 km east of Presa Adolfo L. Mateos, northern Sinaloa	35% plagioclase, 25% quartz, 25% orthoclase, 8% biotite, 7% hornblende
KC-1	Granodiorite	13 km east of Villa Union near Malpica	quartz, plagioclase, orthoclase, biotite, hornblende
KC-2	Gneissic Granodiorite	5 km southwest of Tabora, northern Sinaloa	49% plagioclase, 21% quartz, 13% microcline, 7% biotite, 8% hornblende
KC-3	Monzogranite	2 km east of Tameapa, northern Sinaloa	30% plagioclase, 20% quartz, 40% orthoclase, 10% biotite and altered hornblende

Clarke, M., and Titley, S.R., 1988, Hydrothermal evolution in the formation of silver-gold veins in the Tayoltita Mine, San Dimas District: Economic Geology, v. 83, p.1830–1840.

Coney, P.J., and Campa, M.F., 1987, Lithotectonic terrane map of Mexico (west of the 91st meridian): U.S. Geological Survey Miscellaneous Field Studies Map MF-1874-D, scale 1:2,500,000, 1 sheet.

Coney, P.J., and Reynolds, S.J., 1977, Cordilleran Benioff zones: Nature, v. 270, p. 403- 406.

Damon, P.E., Shafiqullah, M., and Clark, K.F., 1981, Age trends of igneous activity in relation to metallogenesis in the southern Cordillera: Arizona Geological Society Digest, v. 14, p. 137-154.

Damon, P.E., Shafiqullah, M., and Clark, K.F., 1983a, Geochronology of the porphyry copper deposits and related mineralization of México: Canadian Journal of Earth Science, v. 20, p. 1052–1071.

Damon, P.E., Shafiqullah, M., Roldan, J., and Cocheme, J.J., 1983b, El batolito Laramide (90–40 Ma) de Sonora: Asociacion de Ingenieros de Minas, Metalurgistas y Geologos de México, A.C., Memoria, XV Convencion Nacional, p. 63–95.

Damon, P.E., Shafiqullah, M., and Roldan-Quintana, J., 1984, The Cordilleran Jurassic arc from Chiapas (southern Mexico) to Arizona: Geological Society of America Abstracts with Programs, v. 16, no. 6, p. 482.

Engebretson, A.C., Cox, A., and Gordon, R.G., 1985, Relative motions between oceanic and continental plates in the Pacific Basin: Boulder, Colorado, Geological Society of America Special Paper 206, 59 p.

Evernden, J.F., and Kistler, R.W., 1970, Chronology of emplacement of Mesozoic batholithic complexes in California and western Nevada: U.S. Geological Survey Professional Paper 623, 42 p.

Ferrari, L., Pasquare, G., Venegas-Salgado, S., and Romero-Rios, F., 2000, Geology of the western Mexican volcanic belt and adjacent Sierra Madre

Occidental and Jalisco Block: Boulder, Colorado, Geological Society of America Special Paper 334, p. 65–83.

Frizzell, V.A, Jr., 1984, The geology of the Baja California peninsula; An introduction, in Frizzell, V.A., Jr., ed., Geology of the Baja California Peninsula: Los Angeles, Society of Economic Paleontologists and Mineralogists Pacific Section, v. 39, p. 1–7.

Frizzell, V.A, Jr., Fox, L.K., Moser, F.C., and Ort, K.M., 1984, Late Cretaceous granitoids, Cabo San Lucas Block, Baja California Sur, México: Eos (Transactions, American Geophysical Union), v. 65, p.1151

Gans, P.B., 1997, Large-magnitude Oligo-Miocene extension in southern Sonora: implications for the tectonic evolution of northwest México: Tectonics, v. 16, p. 388–408.

Gastil, R.G., 1975, Plutonic zones in the Peninsular Ranges of southern California and Baja California: Geology, v. 3, p. 361–363.

Gastil, R.G., and Krummenacher, D., 1977, Reconnaissance geology of coastal Sonora between Puerto Lobos and Bahia Kino: Geological Society of America Bulletin, v. 88, p. 189-198.

Gastil, R.G., Phillips, R.P., and Allison, E.C., 1975, Reconnaissance geology of the state of Baja California: Boulder, Colorado, Geological Society of America Memoir 140, 170 p.

Gastil, R.G., Krummenacher, D., and Jensky, W.E., 1978, Reconnaissance geologic map of the west-central part of Nayarit, México: Geological Society of America, Map and Chart Series Map MC-24, scale 1:200,000, 1 sheet.

Gastil, R.G., Krummenacher, D., and Minch, J., 1979, The record of Cenozoic volcanism around the Gulf of California: Geological Society of America Bulletin, v. 90, p. 839-857.

Gastil, R.G., Morgan, G.J., and Krummenacher, D., 1981, The tectonic history of peninsular California and adjacent México, in Ernst, W.G., ed., The geotectonic development of California, Rubey Volume I: Englewood Cliffs, New Jersey, Prentice-Hall, p. 284–305.

Gromet, L.P., and Silver, L.T., 1987, REE variations across the Peninsular Ranges batholith: Implications for batholithic petrogenesis and crustal growth in magmatic arcs: Journal of Petrology, v. 28, p. 75–125.

Hamilton, W., 1988, Tectonic setting and variations with depth of some Cretaceous and Cenozoic structural and magmatic systems of the western United States, in Ernst, W.G., ed., Metamorphism and crustal evolution of the western United States, Rubey Volume VII: Englewood Cliffs, New Jersey, Prentice-Hall, p. 1–40.

Harrison, T.M., 1981, Diffusion of ^{40}Ar in hornblende: Contributions to Mineralogy and Petrology, v. 78, p. 324-331.

Harrison, T.M., Duncan, I., and McDougall, I., 1985, Diffusion of ^{40}Ar in biotite; temperature, pressure and compositional effects: Geochimica et Cosmochimica Acta, v. 49, p. 2461–2468.

Hausback, B.P., 1984, Cenozoic volcanic and tectonic evolution of Baja California Sur, México, in Frizzell, V.A., Jr., ed., Geology of the Baja California peninsula: Los Angeles, Society of Economic Paleontologists and Mineralogists Pacific Section, v. 39, p. 219-236.

Henry, C.D., 1989, Late Cenozoic Basin and Range structure in western México adjacent to the Gulf of California: Geological Society of America Bulletin, v. 101, p. 1147–1156.

Henry, C.D., and Aranda-Gómez, J.J., 2000, Plate interactions control middle-late Miocene, proto-Gulf and Basin and Range extension in the southern Basin and Range: Tectonophysics, v. 318, p. 1–26.

Henry, C.D., and Fredrikson, G., 1987, Geology of southern Sinaloa adjacent to the Gulf of California: Geological Society of America, Map and Chart Series, Map MCH063, 1 sheet.

Henry, C.D., Price, J.G., and James, E.W., 1991, Mid-Cenozoic stress evolution and magmatism in the southern Cordillera, Texas and México: Transition from continental arc to intraplate extension: Journal of Geophysical Research, v. 96, p. 13545–13560.

Holguin, Q.N., 1978, Estudio estratigrafico del Cretacico inferior en el norte de Sinaloa, México: Revista del Instituto Mexicano del Petroleo, January 1978, p. 6–13.

Housh, T., and McDowell, F.W., 1999, Delineation of basement provinces in northwestern México through isotopic studies of late-Cretaceous to mid-Tertiary igneous rocks [abs.]: Eos (Transactions, American Geophysical Union), v. 80, p. F987.

Jennings, C.W., compiler, 1977, Geologic map of California: California Division of Mines and Geology, Geologic Data Map No. 2, scale: 1:750,000, 1 sheet].

Johnson, S.E., Paterson, S.R., and Tate, M.C., 1999a, Structure and emplacement history of a multiple-center, cone-sheet-bearing ring complex: The Zarza intrusive complex, Baja California, México: Geological Society of America Bulletin, v. 111, p. 607–619.

Johnson, S.E., Tate, M.C., and Fanning, C.M., 1999b, New geologic mapping and SHRIMP U-Pb zircon data in the Peninsular Ranges Batholith, Baja California, México: Evidence for a suture?: Geology, v. 27, p. 743–746.

Kimbrough, D.L., Smith, D.P., Mahoney, J.B., Moore, T.E., Grove, M., Gastil, R.G., Ortega-Rivera, A., Fanning, C.M., 2001, Forearc-basin sedimentary response to rapid Late Cretaceous batholith emplacement in the Peninsular Ranges of Southern and Baja California: Geology, v. 29, p. 491–494.

Kimbrough, D.L., Gastil, R.G., Garrow, P.K., Grove, M., Aranda-Gómez, J.J., Perez-Venzor, J.A., and Fletcher, J., 2002, A potential correlation of plutonic suites from the Los Cabos block and Peninsular Ranges batholith: VI International Meeting on Geology of the Baja California Peninsula, La Paz, Baja California Sur: Universidad Autónoma de Baja California, p. 9.

Lipman, P.W., 1984, The roots of ash flow calderas in western North America: Windows into the tops of granitic batholiths: Journal of Geophysical Research, v. 89, p. 8801–8841.

Krummenacher, D., Gastil, R.G., Bushee, J., and Doupont, J., 1975, K-Ar apparent ages, Peninsular Ranges batholith, southern California and Baja California: Geological Society of America Bulletin, v. 86, p. 760–768.

Martiny, B., Martinez Serrano, R.G., Moran Zenteno, D.J., Macias Romo, C., and Ayuso, R.A., 2000, Stratigraphy, geochemistry and tectonic significance of the Oligocene magmatic rocks of western Oaxaca, southern México: Tectonophysics, v. 318, p. 71–98.

McDowell, F.W., and Clabaugh, S.E., 1979, Ignimbrites of the Sierra Madre Occidental and their relation to the tectonic history of western México: Boulder, Colorado, Geological Society of America Special Paper 180, p. 113–124.

McDowell, F.W., and Keizer, R.P., 1977, Timing of mid-Tertiary volcanism in the Sierra Madre Occidental between Durango City and Mazatlan, México: Geological Society of America Bulletin, v. 88, p. 1479-1486.

McDowell, F.W., and Mauger, R.L., 1994, K-Ar and U-Pb zircon chronology of Late Cretaceous and Tertiary magmatism in central Chihuahua state, México: Geological Society of America Bulletin, v. 106, p. 118–132.

McDowell, F.W., Roldan-Quintana, J., and Connelly, J.N., 2001, Duration of Late Cretaceous–early Tertiary magmatism in east-central Sonora, México: Geological Society of America Bulletin, v. 113, p. 521–531.

McFall, C.C., 1968, Reconnaissance geology of the Concepcion Bay area, Baja California, México: Stanford, Stanford University, Geological Sciences, v. 10, no. 5, 25 p.

McLean, H., 1988, Reconnaissance geologic map of the Loreto and part of the San Javier quadrangles, Baja California Sur, México: U.S. Geological Survey Miscellaneous Field Studies Map MF-2000, scale 1:50,000, 1 sheet.

Mora-Alvarez, G., and McDowell, F.W., 2000, Miocene volcanism during late subduction and early rifting in the Sierra Santa Ursula of western Sonora, México, in Delgado-Granados, H., Aguirre-Díaz, G., and Stock, J.M., eds., Cenozoic tectonics and volcanism of México: Boulder, Colorado, Geological Society of America Special Paper 334, p. 123–141.

Moran-Zenteno, D.J., Martiny, B., Tolson, G., Solis-Pichardo, G., Alba-Aldave, L., Hernandez-Bernal, M.D.S., Macias-Romo, C., Martinez-Serrano, R.G., Schaaf, P., and Silva-Romo, G., 2000, Geocronologia y caracteristicas geoquimicas de las rocas magmaticas terciarias de la Sierra Madre del Sur: Boletin de la Sociedad Geologica Mexicana, v. 53, p. 27–58.

Mullan, H.S., 1978, Evolution of part of the Nevadan orogen in northwestern México: Geological Society of America Bulletin, v. 89, p. 1175-1188.

Nemeth, K.E., 1976, Petrography of the Lower Volcanic group, Tayoltita–San Dimas district, Durango, México [M.A. thesis]: Austin, The University of Texas, 141 p.

Nieto-Samaniego, A.F., Ferrari, L., Alaniz-Alvarez, S.A., Labarthe-Hernández, G., and Rosas-Elguera, R., 1999, Variation of Cenozoic extension and volcanism across the southern Sierra Madre Occidental volcanic province, México: Geological Society of America Bulletin, v. 111, p. 347–363.

Ortega-Gutiérrez, F., Prieto-Velez, R., Zuniga, Y., and Flores, S., 1979, Una secuencia volcano-plutonica-sedimentaria Cretacica en el norte de Sinaloa; un complejo ofiolitico?: Revista de Universidad Nacional Autónoma de México, Instituto de Geologia,, v. 3, p.1–8.

Ortega-Gutiérrez, F., Mitre-Salazar, L.M., Roldan-Quintana, J., Aranda-Gómez, J., Moran-Zenteno, D., Alaniz-Alvarez, S., and Nieto-Samaniego, A., 1992, Carta geologica de la Republica Mexicana: Instituto de Geologia, Universidad Nacional Autónoma de México, scale 1:2,000,000, 1 sheet.

Ortega-Rivera, A., Farrar, E., Hanes, J.A., Archibald, D.A., Gastil, R.G., Kimbrough, D.L., Zentilli, M., Lopez-Martinez, M., Feraud, G., and Ruffet, G., 1997, Chronological constraints on the thermal and tilting history of the Sierra San Pedro Martir pluton, Baja California, México, from U/Pb, ^{40}Ar/^{39}Ar, and fission-track geochronology: Geological Society of America Bulletin, v. 109, p. 728–745.

Oskin, M., Stock, J., and Martín-Barajas, A., 2001, Rapid localization of Pacific-North American plate motion in the Gulf of California: Geology, v. 29, p. 459–462.

Roldan-Quintana, J., 1991, Geology and chemical composition of the Jaralito and Aconchi batholiths in east-central Sonora, *in* Perez-Segura, E., and Jacques-Ayala, C., eds., Studies of Sonoran geology: Boulder, Colorado, Geological Society of America Special Paper 254, p. 69–80.

Ruiz, J., Patchett, P.J., and Ortega-Gutiérrez, F., 1988, Proterozoic and Phanerozoic basement terranes of México from Nd isotopic studies: Geological Society of America Bulletin, v. 100, p. 274–281.

Sawlan, M.G., 1991, Magmatic evolution of the Gulf of California rift, *in* Dauphin, J.P., and Simoneit, B.A., eds., The Gulf and Peninsular Province of the Californias: American Association of Petroleum Geologists Memoir, v. 47, p. 301–369.

Schaaf, P., Moran-Zenteno, D., Hernandez-Bernal, M.d.S., Solis-Pichardo, G., Tolson, G., and Kohler, H., 1995, Paleogene continental margin truncation in southwestern México: Geochronological evidence: Tectonics, v. 14, p. 1339–1350.

Schaaf, P., Boehnel, H., and Perez-Venzor, J.A., 1997, Isotopic data on Los Cabos and Jalisco block granitoids: Paleogeographic implications [abs.]: Eos (Transactions, American Geophysical Union), v. 78, p. F844.

Schaaf, P., Boehnel, H., and Perez-Venzor, J.A., 2000, Pre-Miocene palaeogeography of the Los Cabos block, Baja California Sur: Geochronological and palaeomagnetic constraints: Tectonophysics, v. 318, p. 53–69.

Sedlock, R.L., Ortega-Gutiérrez, F., and Speed, R.C., 1993, Tectonostratigraphic terranes and tectonic evolution of México: Boulder, Colorado, Geological Society of America Special Paper 278, 153 p.

Sillitoe, R.H., 1976, A reconnaissance of the Mexican porphyry copper belt: Institute of Mining and Metallurgy Transactions, v. 85, p. B170-B189.

Silver, L.T., 1963, The relation between radioactivity and discordance in zircons: National Academy of Science, National Research Council, Nuclear Geophysics Publication 1075, p. 34–42.

Silver, L.T., and Chappell, B.W., 1988, The Peninsular Ranges batholith: An insight into the evolution of the Cordilleran batholiths of southwestern North America: Transactions of the Royal Society of Edinburgh: Earth Sciences, v. 79, p. 105–121.

Silver, L.T., Taylor, H.P., Jr., and Chappell, B., 1979, Some petrological, geochemical and geochronological observations of the Peninsular Ranges batholith near the International Border of the U.S.A., and México, *in* Abbott, P.L., and Todd, V.R., eds., Mesozoic crystalline rocks—Peninsular Ranges batholith and pegmatites, Point Sal ophiolite, *in* Geological Society of America Annual Meeting Guidebook: San Diego, San Diego State University, p. 83–110.

Smith, D.M., Albinson, T., and Sawkins, F.J., 1982, Geologic and fluid inclusion studies of the Tayoltita silver-gold vein deposit, Durango, México: Economic Geology, v. 77, p. 1120-1145.

Snee, L.W., Naeser, C.W., Naeser, N.D., Todd, V.R., and Morton, D.M., 1994, Preliminary ^{40}Ar/^{39}Ar and fission-track cooling ages of plutonic rocks across the Peninsular Ranges batholith, southern California: Geological Society of America Abstracts with Programs, v. 26, no. 2, p. A94.

Staude, J.-M.G., and Barton, M.D., 2001, Jurassic to Holocene tectonics, magmatism, and metallogeny of northwestern México: Geological Society of America Bulletin, v. 113, p. 1357–1374.

Steiger, R.H., and Jäger, E., 1977, Subcommission on geochronology: Convention on the use of decay constants in geo- and cosmochronology: Earth and Planetary Science Letters, v. 36, p. 359–362.

Stein, G., LaPierre, H., Monod, O., Zimmermann, J.L., and Vidal, R., 1994, Petrology of some Mexican Mesozoic-Cenozoic plutons: Sources and tectonic environments: Journal of South American Earth Sciences, v. 7, p. 1–7.

Stock, J.M., and Hodges, K.V., 1989, Pre-Pleistocene extension around the Gulf of California and the transfer of Baja California to the Pacific Plate: Tectonics, v. 8, p. 99–115.

Stock, J., and Molnar, P., 1988, Uncertainties and implications of the Late Cretaceous and Tertiary positions of North America relative to the Farallon, Kula, and Pacific plates: Tectonics, v. 7, p. 1339-1384.

Swanson, E.R., and McDowell, F.W., 1984, Calderas of the Sierra Madre Occidental volcanic field, western México: Journal of Geophysical Research, v. 89, p. 8787-8799.

Swanson, E.R., Keizer, R.P., Lyons, J.I., and Clabaugh, S.E., 1978, Tertiary volcanism and caldera development near Durango City, Sierra Madre Occidental, México: Geological Society of America Bulletin, v. 89, p. 1000–1012.

Tardy, M., Lapierre, H., Freydier, C., Coulon, C., Gill, J.B., Mercier de Lepinay, B., Beck, C., Martinez, R.J., Talavera, M.O., Ortiz, H.E., Stein, G., Bourdier, J.L., and Yta, M., 1994, The Guerrero suspect terrane (western México) and coeval arc terranes (the Greater Antilles and the western Cordillera of Colombia): A late Mesozoic intra-oceanic arc accreted to cratonal America during the Cretaceous: Tectonophysics, v. 230, p. 49–73.

Tate, M.C., Norman, M.D., Johnson, S.E., Fanning, C.M., and Anderson, J.L., 1999, Generation of tonalite and trondhjemite by subvolcanic fractionation and partial melting in the Zarza Intrusive Complex, western Peninsular Ranges batholith, northwestern México: Journal of Petrology, v. 40, p. 983–110.

Todd, V.R., and Shaw, S.E., 1979, Structural, metamorphic and intrusive framework of the Peninsular Ranges batholith in southern San Diego County, California, *in* Abbott, P.L., and Todd, V.R., eds., Mesozoic crystalline rocks—Peninsular Ranges batholith and pegmatites, Point Sal ophiolite: Geological Society of America Annual Meeting Guidebook: San Diego, San Diego State University, p. 177–232.

Todd, V.R., Erskine, B.G., and Morton, D.M., 1988, Metamorphic and tectonic evolution of the northern Peninsular Ranges batholith, southern California, *in* Ernst, W.G., ed., Metamorphism and crustal evolution of the western United States, Rubey Volume VII: Englewood Cliffs, New Jersey, Prentice-Hall, p. 894–937.

Tullis, J., 1983, Deformation of feldspars, *in* Ribbe, P.H., ed., Feldspar mineralogy: Mineralogical Society of America Reviews in Mineralogy, v. 2, p. 297–323.

Valencia-Moreno, M., Ruiz, J., Barton, M.D., Patchett, P.J., Zurcher, L., Hodkinson, D.G., and Roldan-Quintana, J., 2001, A chemical and isotopic study of the Laramide granitic belt of northwestern México: Identification of the southern edge of the North American Precambrian basement: Geological Society of America Bulletin, v. 113, p. 1409–1422.

Vernon, R.H., 1991, Questions about myrmekite in deformed rocks: Journal of Structural Geology, v. 13, p. 979–985.

Walawender, M.J., and Smith, T.E., 1980, Geochemical and petrologic evolution of the basic plutons of the Peninsular Ranges batholith, southern California: Journal of Geology, v. 88, p. 233–242.

Walawender, M.J., Gastil, R.G., Clinkenbeard, J.P., McCormick, W.V., Eastman, B.G., Wardlaw, R.S., Gunn, S.H., and Smith, B.M., 1990, Origin and evolution of the zoned La Posta-type plutons, eastern Peninsular Ranges batholith, southern and Baja California, *in* Anderson, J.L., ed., The nature and origin of Cordilleran magmatism: Boulder, Colorado, Geological Society of America Memoir 174, p. 1–18.

Walawender, M.J., Girty, G.H., Lombardi, M.R., Kimbrough, D., Girty, M.S., and Anderson, C., 1991, A synthesis of recent work in the Peninsular Ranges batholith, *in* Walawender, M.J., and Hanan, B.B., eds., Geological excursions in southern California and México: San Diego, San Diego State University, p. 297–317.

Wisser, E., 1965, The epithermal precious metal province of northwest México: Nevada Bureau of Mines Report 13, Part C, p. 63–92.

Zimmermann, J.L., Stussi, J.M., Gonzalez-Partida, E., and Arnold, M., 1988, K-Ar evidence for age and compositional zoning in the Puerto Vallarta–Rio Santiago batholith (Jalisco, México): Journal of South American Earth Sciences, v. 1, p. 267–274.

MANUSCRIPT ACCEPTED BY THE SOCIETY JUNE 2, 2003

Geological Society of America
Special Paper 374
2003

Paired plutonic belts in convergent margins and the development of high Sr/Y magmatism: Peninsular Ranges batholith of Baja-California and Median batholith of New Zealand

Andrew J. Tulloch
Institute of Geological and Nuclear Sciences, Private Bag 1930, Dunedin, New Zealand

David L. Kimbrough
Department of Geological Sciences, San Diego State University, San Diego, California 92182, USA

ABSTRACT

Cretaceous plutons of the eastern Peninsular Ranges batholith and the Separation Point suite of New Zealand represent major fluxes of relatively high Na, Sr, and low Y heavy rare earth element (HREE) magmas. They have similarities to Archean trondhjemite-tonalite-granodiorite (TTG) granitoids and Cenozoic adakites, but their genesis in Phanerozoic subduction zone settings is controversial. The well-documented margin-normal asymmetry of the Peninsular Ranges batholith is similar to that observed in the Median batholith of New Zealand. In both areas, similar-sized belts of high Na, Al, Sr, and low Y (here termed HiSY, after high Sr/Y) diorite-tonalite-granodiorite plutons developed continentalward of, and 10–15 m.y. after, parallel belts of low Sr/Y (LoSY) gabbro-diorite-granite plutons representing 30–40 m.y. of convergent margin magmatism. In the Peninsular Ranges batholith, the HiSY La Posta suite (≈ 99–92 Ma) lies inboard of a western belt of LoSY plutons (≈130–104 Ma) over the ≈ 800 km length of the batholith. In New Zealand, plutons of the HiSY Separation Point suite (126–105 Ma) mostly lie inboard of the LoSY Median suite (mostly 170–128 Ma). Chemical and isotopic links between HiSY and LoSY belts indicate genetic relationships between the paired belts within each area. Comparative features from both margins support a model that involves underthrusting of the outboard LoSY arc base during shallowing subduction to a deeper, more continentalward position. The mafic arc base is then partially melted under high-pressure conditions, resulting in plagioclase-poor or absent garnet-bearing residual mineral assemblages that produce high Sr/Y partial melts. The La Posta plutons appear to represent mixtures of HiSY magmas and Paleozoic metasedimentary crust.

Keywords: granite, granodiorite, tonalite, diorite, batholith, adakite, TTG, Cretaceous, New Zealand, California, Baja California, subduction.

INTRODUCTION

Plutons that form major components of the Peninsular Ranges batholith and the Median batholith of New Zealand represent major fluxes of unusually high Sr/Y continental margin magmas. Recognition of such granitoids in these areas and elsewhere has led to their comparison with Archean trondhjemite-tonalite-granodiorite (TTG) plutons and adakitic volcanic rocks, both of which are generally considered to have been derived by partial melting of eclogitized oceanic slab (e.g., Drummond and Defant, 1990; Smithies, 2000). Key questions regarding the genesis of Na, Al, and Sr-rich rocks in Phanerozoic convergent margins are: what are the source materials

Tulloch, A.J., and Kimbrough, D.L., 2003, Paired plutonic belts in convergent margins and the development of high Sr/Y magmatism: Peninsular Ranges batholith of Baja-California and Median batholith of New Zealand, *in* Johnson, S.E., Paterson, S.R., Fletcher, J.M., Girty, G.H., Kimbrough, D.L., and Martín-Barajas, A., eds., Tectonic evolution of northwestern México and the southwestern USA: Boulder, Colorado, Geological Society of America Special Paper 374, p. 275–295.

(oceanic slab versus underplated crust and/or older continental crust), and what settings and events triggered the rapid development of such magmas in subduction settings? Placing constraints on the genesis of high Na, Al, Sr plutons rocks is important because the increasing volumes of such rocks now being recognized means that they represent a potentially very significant and under-emphasized crustal growth mechanism. They have also been implicated in ore-forming processes (Kay and Mpodozis, 2001).

Margin-normal magmatic arc asymmetry is especially well documented from the Peninsular Ranges batholith of Baja California and California (hereafter simplified to Baja-California). In the northern part of the batholith this asymmetry is well illustrated by systematic variations in rock type, initial $^{87}Sr/^{86}Sr$ ratios, REE, and $\delta^{18}O$ (e.g., Gastil, 1975; Taylor and Silver, 1978; Silver et al., 1979; Gromet and Silver, 1987; Ague and Brimhall, 1988). A similar asymmetry in which high Sr, Na, Al plutons form the continentalward sides of the batholiths also exists in New Zealand (Tulloch and Kimbrough, 1995). The high Sr, Na, Al belts of plutons comprise the La Posta suite of the eastern Peninsular Ranges batholith in Baja-California (Walawender et al., 1990) and the Separation Point suite in New Zealand (Tulloch, 1983; Tulloch, 1988; Tulloch and Rabone, 1993; Muir et al., 1995, 1998). The origin of the La Posta magmas remains controversial. Both Hill et al. (1986) and Gromet and Silver (1987) favored a hydrothermally altered ocean crust source, whereas Walawender et al. (1990) concluded that continental crust was an important component. Tate et al. (1999) reported a comprehensive and detailed study of the mostly mafic Zarza Intrusive Complex of the western Peninsular Ranges batholith. They suggested that relatively shallow generation of small bodies of tonalite and trondhjemite at Zarza might also apply to such rocks elsewhere in the batholith. However, none of the Zarza rocks have the high Sr/Y compositions typical of the eastern Peninsular Ranges batholith tonalites.

Broad similarities in the character and settings of the two major Cretaceous belts of high Sr, Na, Al rocks in Baja-California and western New Zealand suggest a common process is involved in their generation. In particular, the high Sr, Na, Al pluton belts are preceded by, and lie continentalward of, more "typical" convergent margin plutonic arcs and associated volcanic rocks. In this paper we compare and contrast the major features of these belts in Baja-California and New Zealand and suggest a provisional model for the genesis of high Sr, Na, Al magmatism in continental margin settings.

Plutonic Rock Nomenclature

High Na, Al, Sr, Sr/Y (>40) and low Y are characteristic of adakites and Archean TTG granitoids (e.g., Drummond and Defant, 1990; Smithies and Champion, 2000), and these terms have both also been applied to some extent to Phanerozoic plutonic rocks. We consider that the term adakite is inappropriate for describing the plutonic rocks discussed here because it refers to volcanic rocks, to rocks that have a reasonably clear association with slab melting, and because adakites are mafic-intermediate

rocks with relatively high Mg-numbers. We also have reservations about using the term TTG to describe rocks suites and associations in Baja-California and New Zealand because it was defined from, and carries an association with, Archean rocks, for which many models (e.g., Drummond and Defant, 1990; Smithies, 2000) argue for a slab-melting origin. Furthermore, the term TTG does not include dioritic rocks that are a major part of the compositional spectrum in the especially deeply exhumed crustal section in New Zealand. Other writers (e.g., Petford and Atherton, 1996) simply refer to such high Na, Al, Sr, low Y rocks as "sodic." Unfortunately, the term sodic does not convey many other significant features, such as rock type, trace element chemistry, and plutonic nature. Nor does it immediately indicate distinction from sodic undersaturated alkaline rocks. The term basalt-andesite-rhyolite-dacite (BARD) used by Petford and Atherton (1996) to describe the non-high Na, Al, Sr etc. rocks is not favored for use in this context because it emphasizes volcanic rocks. Rock suites in New Zealand and Baja-California (and elsewhere, e.g., Cordillera Blanca versus Coastal batholith) do tend to define compositional trends and ranges, but rock types are not sufficiently unique to use for practical purposes.

A nongenetic terminology. We suggest a terminology based on the single most definitive parameter, Sr/Y: HiSY for the high Sr/Y, Na, Al, Sr, low Y rocks (Sr/Y > 40 after Drummond and Defant; typical range of rock types includes diorite, tonalite, and [leuco-] granodiorite; and LoSY for the complementary low Sr/Y, etc., rocks (typical rock types include gabbro, diorite, and granite).

As will be shown below, it is not appropriate to distinguish the LoSY rocks as calc alkaline or calcic etc., implying that the Na, Al, Sr rocks are not calc alkaline or calcic. The paired HiSY/LoSY magmatic belts are described below in terms of inboard (toward the continental interior) and outboard (toward the Pacific Ocean) belts, respectively. We use suite as a group of plutonic rocks of similar age, mineralogy, and chemistry, and batholith simply as a >100 km² contiguous area of plutonic rock, regardless of age (Tulloch, 1988).

COMPARISON OF BATHOLITHIC BELTS IN BAJA-CALIFORNIA AND NEW ZEALAND

Spatial Distribution of Paired Belts

In both New Zealand and Baja-California, adjacent plutonic belts parallel the Cretaceous continental margins and represent the coarse scale structure of margin-normal asymmetry. Fine-scale structure of this asymmetry is best displayed in the exceptionally well-exposed Peninsular Ranges batholith. Both inboard belts are predominantly comprised of relatively high-Na, Al, Sr/Y HiSY plutons. Paired belts in both New Zealand and Baja-California occur over similar lengths parallel to their respective continental margins, have similar apparent widths, and both have grossly similar proportions of HiSY:LoSY plutons (≈ 50/50).

SiO₂ histograms (Fig. 1) giving an indication of rock types show that both areas contain the full range of gabbro-granite com-

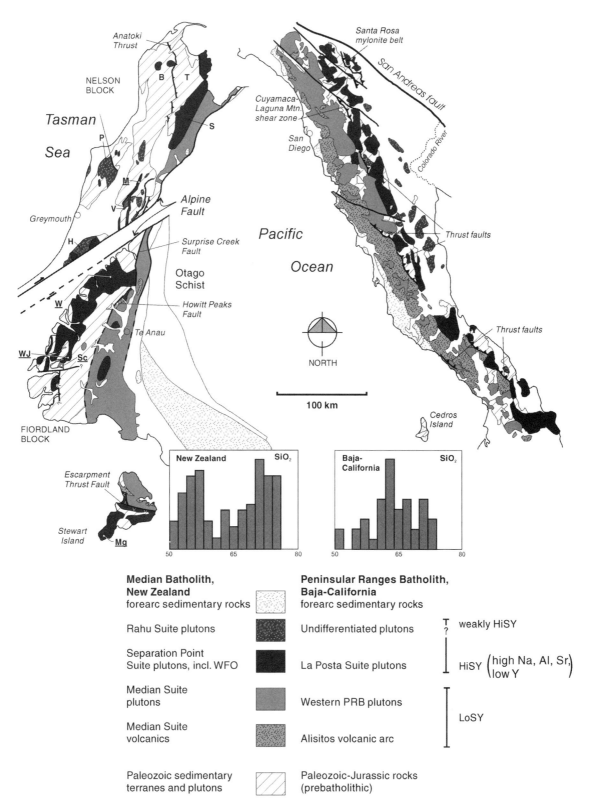

Figure 1. Maps showing similarly asymmetric distribution of HiSY and LoSY plutonic belts in Baja-California and New Zealand. Note that our simplified use of the term Baja-California includes both California (U.S.) and Baja California (México). New Zealand Pacific margin segment comprises Nelson and Fiordland blocks restored to pre-Alpine fault configuration and is compiled largely from Bradshaw, 1990; Tulloch, 1983, 1988; Muir et al., 1995, 1998; Tulloch, 2001; Allibone and Tulloch, 1997, unpubl. data). B—Buller Paleozoic metasedimentary terrane, T—Takaka Paleozoic metasedimentary terrane, P—Paparoa Range, V—Victoria Range, H—Hohonu Range and associated lower plates of metamorphic core complexes. Underlined M, W, WJ, SC and Mg are dated samples: Macey Granodiorite, Worsley Monzodiorite, Wet Jacket diorite, Supper Cove Diorite, and Magog leucogranodiorite, respectively. Western Fiordland Orthogneiss is Western Fiordland Orthogneiss of Separation Point suite. Possible forearc is Murihiku Super Group/terrane. Baja-California map is from Kimbrough et al. (2001). SiO$_2$ histograms are from same data set as for Figure 3, but with addition of all Western Fiordland Orthogneiss samples reported in Bradshaw (1985), to more appropriately represent large Western Fiordland Orthogneiss body. Y-axis is number of analyses (maximum value = 9). PRB—Peninsular Ranges batholith; WFO—Western Fiordland Orthogneiss.

positions (see also Table 1). However, the plots show a striking contrast in the distribution of SiO_2 content between New Zealand and Baja-California (both belts included). Whereas the Baja-California distribution is unimodal and dominated by values typical of tonalites and granodiorites (60–70% SiO_2), the New Zealand distribution is distinctly bimodal, with peaks at values typical of gabbro/monzodiorite (52–58%) and leucogranodiorite-granite (70–75%). Within each area both belts contain a similar range of rock types, although gabbroic rocks are more abundant in the outboard belts and granodiorite is more abundant in the inboard belts. Tate and Johnson (2000) noted that the narrow compositional range observed within the Peninsular Ranges batholith has hindered interpretations of these rocks.

New Zealand

Paired belts forming the Median batholith extended along the ancient Gondwana margin some 500 km from Stewart Island through the South Island into Nelson. Prior to Late Cenozoic offset along the Alpine fault, Nelson and Fiordland blocks formed an essentially contiguous segment of the Gondwana margin. Offshore oil well intersections extend the on-shore belt to the north (Mortimer et al., 1997) and to the southeast for a total of at least 900 km. Onshore, extensive rainforest cover often obscures field relationships, and it was not until recently that the extent and continuity of the composite marginal batholith was recognized (Median batholith; Mortimer et al., 1999b). Although the bulk of the New Zealand Mesozoic magmatism is included within the Median batholith (Mortimer et al., 1999b), isolated smaller batholiths and plutons occur west of the batholith in the Nelson block (Tulloch, 1983; Muir et al., 1997).

The inboard belt is dominated by low-K, high-Na (HiSY) rocks of the Separation Point suite (Tulloch, 1988; Tulloch and Rabone, 1993; Muir et al., 1995). In the mostly shallow Nelson block this suite comprises tonalite-granodiorite-granite (Tulloch and Rabone, 1993), whereas in the deeper level Fiordland block it is dominated by diorite-monzodiorite of the Western Fiordland Orthogneiss (Bradshaw, 1990). The continentalward side of the inboard belt is dominated by the weakly HiSY Rahu suite (Tulloch, 1983, 1988; Waight et al., 1997). The outboard belt is dominated by dioritic rocks of the Median suite, although large gabbroic plutons and smaller granite bodies are common (Kimbrough et al., 1994; Mortimer et al., 1999b; Rattenbury et al., 1998; Blattner and Graham, 2000).

Although minor bodies of LoSY rocks occur in the HiSY belt and vice versa (e.g., Mortimer et al., 1999b; Wandres et al., 1998), the belts are mostly sharply bounded against each other, often apparently faulted. In Fiordland, it is unclear how much of that faulting (e.g., Surprise Creek fault, Howitt Peaks fault; Kimbrough et al., 1994; Bradshaw, 1990) is Cenozoic and related to the Alpine fault and current Australia–Pacific plate boundary (e.g., Mortimer et al., 1999b), although the overall Mesozoic margin asymmetry is intact. On Stewart Island, however, more distant from the overprinting effects of the Alpine fault, a major outboard-verging thrust fault (Escarpment fault) currently separates the two belts (Allibone and Tulloch, 1997) and likely formed during 125–105 Ma HiSY magmatism.

TABLE 1. SUMMARY OF CHARACTERISTICS OF PAIRED HiSY AND LoSY PLUTONIC BELTS IN NEW ZEALAND AND BAJA-CALIFORNIA

	Setting	Suite/ belt	Rock type	Age (Ma)	Emplacement pressure	Assoc. volcs?	Assoc. basic rock	Peacock (MALI)	At 65% SiO_2: Na_2O/K_2O	Sr ~	Average Sr/Y	Typical Sm/Yb[†]	Sri	^{18}O	Residual minerals
LoSY	Outboard NZ	Median	gabbro, diorite, granite	170–127	mostly <3 kb	Y	Yes	ca-ac	1.25	400	23	2.1	0.7033–0.7038	4.4–7.6	pyrox, plag
	Outboard Baja Calif.	W-PRB	gabbro-tonalite-monzogranite	125–102	mostly <3 kb	Y	Yes	c	2 (1.8)	220	11	1.6	0.7028–0.7050	5.5–8.5	pyrox, plag
HiSY	Inboard NZ	Separation Point Suite including WFO	2 pyx diorite-monzodiorite; hnb, bio granodiorite (± musc, gnt grdiorite)	127–105*	2–7 kb	N	Yes, with WFO	ca-ac	2	1050	178	5.2	0.7038–0.7045	5.8–6.9	amph, gnt (± plag)
	Inboard Baja Calif	E-PRB (La Posta)	tonalite-granodiorite	100–92	mostly >3 kb	N	Rare	c	1.7 (2.5)	650	60	5.2	0.7058–0.7084	9–11.5	amph, gnt (± plag)
LoSY– HiSY	Far inboard NZ	Rahu	(hnb) biotite granodiorite-granite	120–105	0–6 kb	(Y)	Rare	ca	1.25	550	27	3.71	0.7058–0.7094	9.4–10.0	amph, gnt (± plag)

Note: minor older HiSY plutons occur in outboard belt. MALI—modified alkali-lime index of Frost et al. (2001). W-PRB—western Peninsular Ranges batholith. E-PRB—eastern Peninsular Ranges batholith. WFO—Western Fiordland Orthogneiss. NZ—New Zealand.
[†]Sm/Yb values from samples plotted in Figure 6.

Plutons of the inboard belt intrude extensive Paleozoic sedimentary, metasedimentary, and plutonic rocks (e.g., Cooper and Tulloch, 1992), but with one possible exception (Beresford et al., 1996): no pre-(Mesozoic) batholith sedimentary sequences have been recognized in the Median suite-dominated eastern/outboard belt, although Carboniferous granite and diorite plutons are a widespread minor component (Kimbrough et al., 1994).

Much of the country rock in the Nelson block is greenschist metamorphic facies, although isolated windows of higher-grade rocks mainly representing the lower plates of metamorphic core complexes do occur (Tulloch and Kimbrough, 1989; Spell et al., 2000). By contrast, much of Fiordland represents the lower plate of a metamorphic core complex (Gibson et al., 1988), and many rocks record pressures, though transient, in the range 10–16 kb (Clarke et al., 2000; Daczko et al., 2001). The Fiordland block represents a deeper level along-strike equivalent of the Nelson block. High Vp (7.4 km s^{-1} at 15 km depth) suggests Western Fiordland Orthogneiss-like rocks extend to ≈40 km (Eberhart-Phillips and Reyners, 2001). High-grade rocks in both blocks were exhumed during middle Cretaceous extension, but the Fiordland block was further exhumed during Miocene–Holocene uplift on the Alpine fault/present plate boundary. Thus, currently exposed rocks in western New Zealand represent crustal levels spanning some 50 vertical km of Early Cretaceous magmatism.

Within the eastern margin of the Median batholith, Triassic to Early Cretaceous plutons intrude the Permian oceanic arc Brook Street Terrane (Mortimer et al., 1999a). Further to the east, Brook Street Terrane is in fault contact with several terranes of New Zealand Eastern Province, which accreted in the Jurassic. To the west, exposure of plutonic rocks is restricted to the narrow coastal strip, but the Late Cretaceous rifted margin against oceanic crust of the Tasman Sea is not far offshore. (Tulloch et al., 1991). Formation of this rifted margin was the culmination of some 20 m.y. of continental extension that overlapped with final episodes of plutonism (Tulloch and Kimbrough, 1989).

Baja-California

The ≈800-km-long Peninsular Ranges batholith comprises longitudinally distinct western and eastern zones that are distinguished on the basis of differences in age, petrology, style, and depth of emplacement, prebatholithic wall rock, geophysical parameters, and exhumation history (Gastil, 1975; Silver et al., 1979; Gromet and Silver, 1987; Silver and Chappell, 1988; Lovera et al., 1999). Western intrusions are older (≈ 130–105 Ma), range from gabbro to monzogranite, have relatively primitive island-arc geochemical affinities, and are variably overprinted by subsolidus ductile deformation. These plutons occur as relatively small sheet and diapirs that were shallowly emplaced into the Early Cretaceous supracrustal volcanic/sedimentary sequence.

The most striking feature of the younger eastern zone intrusions is the large volume of relatively homogeneous hornblende and/or biotite tonalite and low-K granodiorite. Walawender et al. (1990) coined the term "La Posta–type plutons" for these intrusions, and we refer to them here as the La Posta suite. The La Posta plutons occur as a series (12 or more) of large, internally zoned intrusive centers with outcrop areas that range up to at least 1400 km^2 (Gastil et al., 1990; Silver and Chappell, 1988; Fig. 1). Weakly peraluminous muscovite-bearing granodiorite core-zones (locally expressed as cross-cutting dikes) contain inherited zircon and have sharp intrusive contacts.

The western and eastern zones of the Peninsular Ranges batholith are locally juxtaposed by a series of synbatholithic west-vergent Early Cretaceous ductile thrust faults that separate rocks with contrasting metamorphic grade and deformational history (Johnson et al., 1999; Fig. 1). La Posta–type plutons postdate this Early Cretaceous episode of contractional deformation. They are intruded across the fan-like structure in the east-central Peninsular Ranges batholith and stitch thrust faults that separate western and eastern zone rocks.

Age Relationships

New high-precision U-Pb age determinations on selected New Zealand plutons (Fig. 1) are presented in Figure 2 and Table 2. Worsley Monzodiorite (OU49122) is typical of much of the granulite-facies Western Fiordland Orthogneiss plutons of the Separation Point suite. The age of 124 ± 1 Ma reported here is a refinement on one of the samples previously reported as 120–130 Ma age by Mattinson et al. (1986). The youngest-known phase of the Western Fiordland Orthogneiss is Wet Jacket diorite (OU49113). The weighted mean $^{206}Pb/^{238}U$ age of 116.6 ± 1.2 Ma is based on the data reported in Mattinson et al. (1986) and a new concordant age reported in Table 1. The Macey Granodiorite (OU49210) at 126 ± 1 Ma is the oldest dated pluton in the Separation Point suite. Magog leucogranodiorite (OU49155) is the youngest Separation Point suite rock dated, with an age of 105 ± 1 Ma. Like several Separation Point suite granites and most Rahu suite plutons, the Magog sample contains a component of inherited zircon, which here is suggestive of a Late Proterozoic mean age. Supper Cove Diorite (OU49114) is a medium-K rock of the Median suite, which forms an isolated pluton well west of the bulk of this suite. We report an age of 128 ± 1 Ma, which is the youngest age reported for this suite.

Age comparisons of HiSY (mostly inboard) and LoSY (mostly outboard) plutonic belts from Baja-California and New Zealand are shown in Figure 3. All ages used are U-Pb zircon (± monazite) with uncertainties <5 Ma. The main points are:

1. In both cases most HiSY rocks are younger than associated LoSY plutons, although some (older) HiSY rocks do occur in the older outboard belts of both Baja-California and New Zealand.

2. The total episode of continuous magmatism is slightly longer in New Zealand (≈65 m.y.) than in Baja-California (≈50 m.y.). Assuming similar plutonic areas in each region, and thus volumes, this indicates a somewhat greater magma flux in Baja-California.

3. In both regions, outboard LoSY magmatism appears to form a symmetrical peak and had nearly ceased when a burst of HiSY magmatism rapidly appeared, then gradually waned.

TABLE 2. U-PB ISOTOPIC AGE DATA FOR ZIRCON AND MONAZITE FROM FOUR NEW ZEALAND PLUTONS

Sample fractions	Weight (µg) (a)	U (ppm)	Pb (ppm)	Pb(c) (pg) (b)	$\frac{206Pb}{204Pb}$ (c)	$\frac{208Pb}{206Pb}$ (d)	$\frac{206Pb}{238U}$ (e)	err (2s%)	$\frac{207Pb}{235U}$ (e)	err (2s%)	$\frac{207Pb}{206Pb}$ (e)	err (2s%)	$\frac{206Pb}{238U}$	$\frac{207Pb}{235U}$	$\frac{207Pb}{206Pb}$	corr. coef.
OU49122 (Western Fiordland Orthogneiss, Separation Point Suite)																
Z2	35.2	124	3.0	13.3	423.2	0.272	0.019366	(0.30)	0.1293	(0.37)	0.04844	(0.20)	123.6	123.5	120.7	0.84
Z4	1.7	123	2.7	2.1	145.3	0.266	0.019558	(1.54)	0.1343	(3.73)	0.04979	(3.22)	124.9	127.9	185.0	0.52
Z5	1.6	243	5.2	3.9	142.3	0.241	0.019322	(1.56)	0.1296	(2.07)	0.04864	(1.27)	123.4	123.7	130.7	0.79
OU49114 (Supper Cove dioritic orthogneiss, Median Suite)																
Z1	6.0	205	4.4	1.6	1022.2	0.188	0.020021	(0.26)	0.1339	(0.52)	0.04851	(0.44)	127.8	127.6	124.1	0.55
Z2	4.9	121	2.5	1.0	814.0	0.170	0.020014	(0.27)	0.1340	(0.59)	0.04855	(0.50)	127.7	127.7	126.2	0.54
Zr cg*	3200	125	2.6		3488.3		0.019548	(0.27)	0.1311	(0.36)	0.04863	(2.33)	124.8	125.1	130.1	0.76
OU49210 (Macey Granodiorite, Separation Point Suite)																
Z1 (1)	16.0	694	15.7	41.4	355.3	0.081	0.019792	(0.13)	0.1325	(0.34)	0.04855	(0.30)	126.3	126.3	126.2	0.48
Z2 (1)	4.3	2175	43.4	9.1	1318.4	0.092	0.019765	(0.16)	0.1324	(0.21)	0.04857	(0.13)	126.2	126.2	127.2	0.79
Zr fg*	6400	1748	34.3		7692		0.019771	(0.32)	0.1327	(0.34)	0.04868	(0.10)	126.2	126.5	132.6	0.96
Zr cg*	10100	1561	30.3		13643		0.019768	(0.31)	0.1327	(0.33)	0.04868	(0.09)	126.2	126.5	132.4	0.96
Zr< 44*	3200	965	18.4		61310		0.011967	(0.27)	0.1320	(0.28)	0.04866	(0.05)	125.6	125.9	131.3	0.98
OU49155 (Magog Granodiorite, Separation Point Suite)																
>150um*	6500	511	7.5		7369		0.017048	(0.26)	0.1143	(0.30)	0.04864	(0.06)	108.9	109.8	130.0	0.90
<150um*	9500	560	8.7		8928		0.017921	(0.26)	0.1223	(0.46)	0.04952	(0.05)	114.4	117.1	172.0	0.90
fgr*	10500	911	13.6		968		0.017269	(0.26)	0.1166	(0.46)	0.04899	(0.40)	110.2	111.9	147.0	0.95
mz fg* mz† 2x	1200	2547	182.9		1023		0.016513	(0.38)	0.109117	(0.52)	0.04793	(0.33)	105.6	105.2	95.6	0.77
150um	11.0	2919	311.3	15	2244	0.156	0.016363	(1.00)	0.018867	(1.44)	0.04825	(1.00)	104.6	104.9	111.7	0.72
OU49113 (Wet Jacket diorite, Western Fiordland Orthogneiss, Separation Point Suite)																
zr cg*	3800	128	2.6		4198		0.018201	(0.28)	0.121394	(0.32)	0.04837	(0.15)	116.3	116.3	117.5	0.88

Note: (a) Sample weights are estimated by using a video monitor and are known to within 40%; (b) Total common-Pb in analyses; (c) Measured ratio corrected for spike and fractionation only; (d) Radiogenic Pb; (e) Corrected for fractionation, spike, blank, and initial common Pb. Mass fractionation correction of 0.15%/amu ± 0.04%/amu (atomic mass unit) was applied to single-collector Daly analyses and 0.12%/amu ± 0.04% for dynamic Faraday-Daly analyses. Total procedural blank <0.6 pg for Pb and <0.1 pg for U. Blank isotopic composition: $\frac{206Pb}{204Pb}$ = 19.10 ± 0.1, $\frac{207Pb}{204Pb}$ =15.71 ± 0.1, $\frac{208Pb}{204Pb}$ = 38.65 ± 0.1. Corr. coef.—correlation coefficient. Age calculations are based on the decay constants of Steiger and Jäger (1977). Common-Pb corrections were calculated by using the model of Stacey and Kramers (1975) and the interpreted age of the sample. Analyses undertaken at Dept of Earth and Planetary Sciences, MIT, except:
*Analyzed at San Diego State University, see Kimbrough et al. (1994).
†Analyzed by N. Walker; GNS Dunedin, and Brown University.

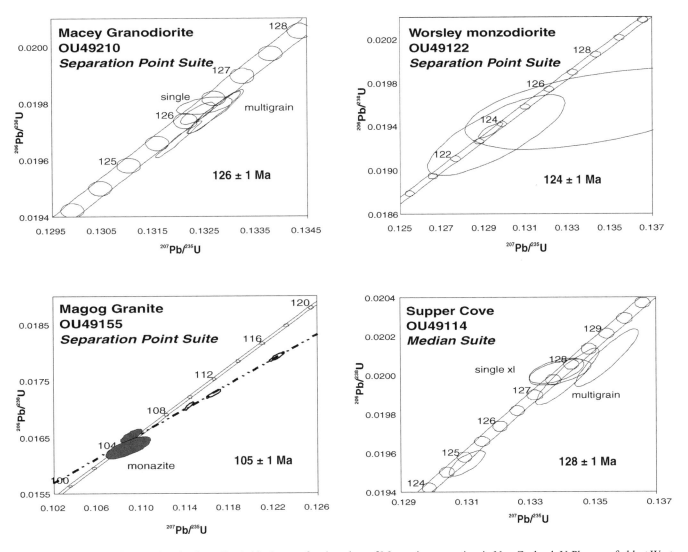

Figure 2. U-Pb concordia plots showing bounding/critical ages of main pulses of Mesozoic magmatism in New Zealand: U-Pb ages of oldest Western Fiordland Orthogneiss (Worsley Monzodiorite), oldest Separation Point suite (Macey Granodiorite), youngest Separation Point (Magog granodiorite), and youngest Median suite (Supper Cove Diorite). Isoplot-calculated uncertainties (Ludwig, 1999) are rounded up to 1 m.y. (see text).

4. The peak of inboard HiSY magmatism in both regions occurs within 5 m.y. of virtual cessation of LoSY magmatism and 15–20 m.y. after the peak of outboard HiSY magmatism.

5. There is no strong evidence for margin-parallel age variations in Baja-California or New Zealand. Although Muir et al. (1998) postulated a southward age-increase in New Zealand HiSY plutons, the oldest (Macey; 126 Ma) and youngest (Magog; 105 Ma) plutons reported here are from the north and south, respectively (Figs. 1 and 3). In Baja-California, the data of Kimbrough et al. (2001) suggest a slight decrease southwards, but this may be influenced by hornblende Ar/Ar dates in the south.

6. Margin-normal age variations are, however, well established in both areas. In New Zealand this can be extended to earlier Permian and Triassic magmatism on the oceanward side of the Median batholith, through to Rahu suite magmatism at

≈110 Ma. Within the LoSY Median suite, the oldest plutons (Triassic) are restricted to the east (Kimbrough et al., 1994; Mortimer et al., 1999b), whereas the youngest rocks (<135 Ma) are restricted to isolated plutons west of the Median batholith that intrude Paleozoic sequences (Crow Granite; Muir et al., 1997; Supper Cove orthogneiss, Fig. 2). Even younger (100–80 Ma; Tulloch et al., 1994; Waight et al., 1997) post-orogenic alkaline magmatism related to New Zealand-Australia-Antarctic breakup occurs at the western extreme of the New Zealand exposures.

Geochemical Comparisons

Chemical analyses of representative rocks are presented in Table 3. Plots that serve to illustrate differences between the

A.J. Tulloch and D.L. Kimbrough

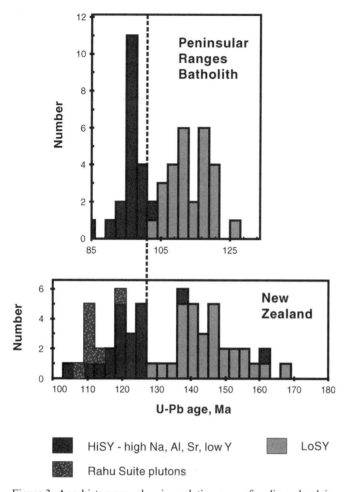

Figure 3. Age histograms showing relative ages of sodic and calcic, calc-alkaline plutonic belts. In both areas HiSY rocks follow major bursts/flare-ups of subduction-related LoSY magmatism by ca. 20 Ma. Histograms are aligned at time of initiation of major HiSY flux; ca. 126 Ma in New Zealand, ca. 100 Ma in Baja-California. Rahu suite plutons in New Zealand have variable HiSY-LoSY and dominate the younger end of range in New Zealand. New Zealand age sources include representative U-Pb ages from Ireland and Gibson (1998), Kimbrough et al. (1998), Mattinson et al. (1986), Mortimer et al. (1999), Muir et al. (1994, 1997, 1998), Waight et al. (1997), Table 2, and unpublished data of Tulloch and Kimbrough (OU49109, OU44211, OU49213, OU49208, OU49113, OU49212, OU49203, P57107). Peninsular Ranges batholith age sources are U-Pb data from Silver et al. (1979), Tate et al. (1999) and Kimbrough et al. (2001).

HiSY and LoSY suites are presented in Figures 4, 5, and 6. Data are restricted to those samples or plutons that have been U-Pb dated. The Peninsular Ranges batholith data set is consistent with the averages reported by Silver and Chappell (1988) and Wala-wender and Smith (1980). Some gabbroic rocks have been omitted because of suspicions that they may be extreme cumulates (e.g., Windfall Creek gabbro: Mattinson et al., 1986; an outboard pluton with location/age/lithology characteristic of LoSY belt, but it has high Sr/Y).

Magma series terminology: Alkali-lime values of HiSY/LoSY plutons

Silver et al. (1979) and Silver and Chappell (1988) have emphasized the particularly calcic nature of both outboard and inboard belts of the Peninsular Ranges batholith, and this feature is also evident on the modified alkali-lime index (MALI) plot (Fig. 4A) of Frost et al. (2001). The calcic nature of the Peninsular Ranges batholith as a whole is due to both relatively high Ca, and relatively low Na and K, as compared with New Zealand plutons. In New Zealand Separation Point, suite rocks that dominate the inboard belt were classified as alkali-calcic by Tulloch and Rabone (1993), although considerable scatter about the regression lines suggests overlap into the calc-alkali field. Rocks of the New Zealand outboard belt, dominated by the Median suite, have been loosely regarded as calc alkaline (e.g., Kimbrough et al., 1994), but on the MALI plot they (and the inboard plutons) extend over both calc alkaline and alkali-calcic fields. Casual use of calc alkaline as a bag term for continental margin or "Cordilleran" magmatism should be discouraged.

Within each area (Baja-California and New Zealand) the inboard HiSY rocks share the same MALI field as the outboard LoSY rocks. Thus:

1. the HiSY rocks are not associated with a particular magma series, as defined by alkali-lime indices, and it is therefore inappropriate to distinguish the associated LoSY rocks in terms of these series (e.g., Frost et al., 2001; Wareham et al., 1997; Tulloch and Kimbrough, 1995), or to implicate a particular magma series in the development of HiSY magmas.

2. Specific alkali-lime indices appear to be characteristic of each individual margin. Similar values for HiSY and LoSY rocks within each area are consistent with the genetic association of the inboard and outboard belts within each area.

Isotopic Variations

Sri-εNd (McCulloch et al., 1987; Pickett and Wasserburg, 1989; Muir et al., 1995, 1998; Waight et al., 1998; A.J. Tulloch, D.L. Kimbrough, and N.W. Walker, unpubl. data) and $\delta^{18}O$ (Blattner and Williams, 1991; P. Blattner and A.J. Tulloch, unpubl. data) ranges for outboard Median suite, inboard Separation Point suite, and far-inboard Rahu suite are summarized in Table 1. Sri and ^{18}O values in both areas increase from outboard to inboard, and the well-known age/compositional step (at $^{18}O \approx 8.5$) in the Peninsular Ranges batholith margin–normal trend (e.g., Silver et al., 1979) separates the inboard HiSY belt (La Posta) from the outboard LoSY belt (western Peninsular Ranges batholith). However, in New Zealand the $\delta^{18}O$ and Sri step is apparently further inboard, within the HiSY belt. Some values less than 5‰ $\delta^{18}O$ may be due to alteration or contamination.

Major and Trace Element Comparisons

In both Baja-California and New Zealand, the HiSY rocks that dominate the inboard belts (La Posta and Separation Point suites, respectively) are distinguished by relatively high Na, Al, and Sr, and low Y (Fig. 4). For Na, Sr, and Sr/Y, New Zealand

Legend: ■ HiSY - high Na, Al, Sr, low Y ▨ LoSY ▨ Rahu Suite plutons

TABLE 3. XRF ANALYSES OF REPRESENTATIVE ROCKS FROM NEW ZEALAND AND BAJA-CALIFORNIA BATHOLITHS, INCLUDING THE ROCKS FOR WHICH U-PB AGES ARE REPORTED IN FIGURE 2

	New Zealand								Baja California					
	Far Inboard Rahu Suite		Inboard (HiSY) Separation Pt. Suite			Outboard (LoSY) Median Suite			Outboard (LoSY) W PRB			Inboard (HiSY) E PRB–La Posta Suite		
	R6168	P52246	OU49155	P46102	OU49122	OU49127	OU49114	OU49140	ML9025*	ML9028*	GD37†	BCA	CVP-2	LTS218
SiO_2	74.25	69.73	73.03	71.81	58.64	53.60	56.56	67.83	57.54	62.04	72.00	63.62	66.14	69.44
TiO_2	0.15	0.29	0.11	0.12	1.04	0.91	1.07	0.52	0.91	0.80	0.37	0.49	0.75	0.49
Al_2O_3	14.67	15.62	15.43	15.86	17.40	15.89	17.09	15.42	16.69	16.15	14.40	15.33	16.42	14.93
Fe_2O_3	1.03	2.92	1.18	1.17	5.69	11.37	8.38	3.42	8.25	5.98	3.09	2.85	4.27	2.63
MnO	0.05	0.11	0.02	0.03	0.08	0.19	0.13	0.10	0.13	0.10	0.06	0.04	0.07	0.03
MgO	0.28	0.91	0.11	0.19	3.17	5.58	3.09	0.88	4.32	2.70	0.80	1.52	1.56	0.85
CaO	1.58	3.06	1.61	1.57	5.10	8.16	6.13	2.17	7.73	5.81	2.72	3.53	4.39	3.32
Na_2O	4.09	3.54	5.03	5.49	4.79	3.05	3.81	4.27	3.12	3.54	3.64	4.45	3.64	3.38
K_2O	3.98	3.37	2.77	3.20	3.02	1.14	1.96	4.00	1.37	1.63	3.54	2.33	1.89	3.25
P_2O_5	0.05	0.16	0.03	0.03	0.38	0.18	0.35	0.18	0.14	0.14	0.08	0.14	0.18	0.11
LOI	0.44	0.57	0.55	0.48	0.17	0.10	0.48	0.86	n	n	n	0.05	0.59	0.63
Total	100.57	100.28	99.87	99.95	99.48	100.17	99.05	99.65	100.20	98.89	100.70	94.35	99.90	99.06
Ba	1122	1092	1124	1097	1072	287	546	860	263	470	654	716	975	1640
Ce	52	55	45	20	63	32	55	54	26	26	37	n	n	48
Cr	13	1	1	4	63	37	14	1	63	37	5	59	17	14
Cu	3	3	2	<1	12	132	52	44	n	n	3	4	10	1
Ga	21	19	16	15	22	19	20	17	17	13	15	21	21	19
La	19	29	21	14	34	13	30	25	12	13	18	n	n	23
Nb	12	16	3	5	8	1	6	10	6	13	6	10	10	8
Ni	3	2	1	<1	39	19	17	3	18	16	9	12	10	4
Pb	29	24	27	24	21	18	3	24	7	9	9	17	11	15
Rb	160	125	45	80	59	42	53	126	58	63	146	71	73	89
Sc	1	4	1	<1	11	34	14	5	32	19	13	n	n	4
Sr	482	705	778	741	1236	589	685	301	269	250	146	705	454	513
Th	2	14	9	3	6	4	4	17	7	6	14	n	n	11
U	1	5	1	<1	1	1	3	3	2	2	3	n	n	2
V	14	31	11	7	114	273	151	45	191	117	47	60	62	35
Y	12	24	3	7	13	27	30	30	27	20	31	11	10	8
Zn	72	55	30	33	89	95	92	52	78	71	39	67	103	68
Zr	111	154	113	86	211	75	149	244	126	159	196	137	169	146
Sr/Y	40	29	259	106	95	22	23	10	10	13	5	64	45	64

Note: n—not determined; P and R—GNS National Petrology Reference Collection; OU—Otago University Geology Department collection.
*Lombardi (1992).
†Corollo (1993).

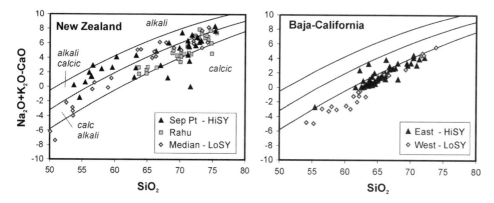

Figure 4. Modified alkali-lime index (MALI; Frost et al., 2001) plots showing that HiSY rocks have (a) developed within different magma series and (b) have same MALI magma series compositions as LoSY outboard plutonic belts. Data set as for Figure 5.

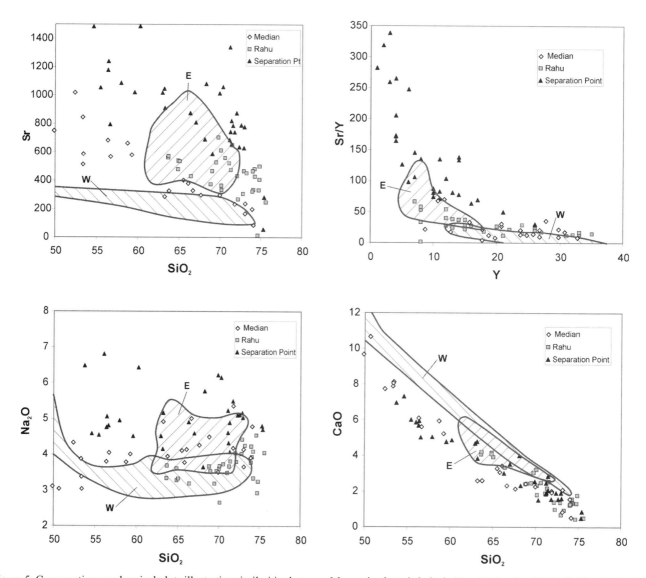

Figure 5. Comparative geochemical plots illustrating similarities between Mesozoic plutonic belts in New Zealand and Baja-California. E and W are fields for Eastern and Western Peninsular Ranges batholith, respectively. New Zealand plutons have higher Sr/Y, Sr, and Na values in both belts. New Zealand data sources include representative analyses from Graham and White (1991), Kimbrough and Tulloch (1989), McCulloch et al. (1987), Muir et al. (1997, 1998), Waight et al. (1998), Table 3, and unpublished analyses of Tulloch (including all relevant samples from Kimbrough et al., 1994, and analyses of the dated samples listed in Figure 3 caption). Peninsular Ranges batholith data sources are Corollo (1993) and Lombardi (1992) for western Peninsular Ranges batholith and Hill et al. (1988), Kimbrough (unpublished) for eastern Peninsular Ranges batholith.

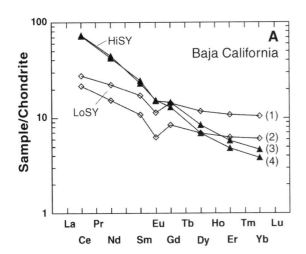

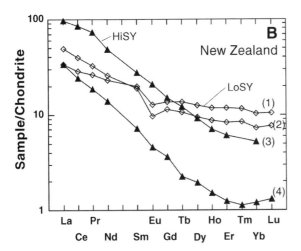

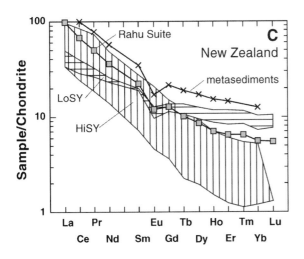

Figure 6. Chondrite-normalized REE patterns for HiSY and LoSY suites in Baja-California and New Zealand. Weak-absent Eu anomalies and HREE-depletion in the HiSY suites indicate paucity of plagioclase and presence of garnet in source rock residuum. Rahu suite pattern (Fig. 6C) reflects mixing of HiSY Separation Point suite and metasedimentary rocks; these subdued HiSY characteristics are similar to HiSY suites of Peninsular Ranges batholith. Baja-California data are summarized from Gromet and Silver (1987). FM90A and RPK2.26 are from McCulloch et al. (1987). Other data are reported in Table 4, analyzed by ICP-MS at Western Washington State University. Sample details, with (Sm/Dy): A: averages from Gromet and Silver, (1987) (1) low-K granodiorite western Peninsular Ranges batholith (1.67); (2) tonalite western Peninsular Ranges batholith (1.5); (3) low-K granodiorite eastern Peninsular Ranges batholith (4.9); (4) tonalite, eastern Peninsular Ranges batholith (5.6). B: (1). OU49147 quartz monzonite, Median Suite (2.4); (2) OU49127 gabbronorite, Median Suite (1.8); (3) FM90A diorite, Separation Point Suite (4.9); (4) OU49138 granodiorite, Separation Point Suite (2.2). C: (1) metasediment RPK2.26 (2.6); (2) Rahu Suite is average of P46072 and P46002L (3.7). LoSY, HiSY from B, above.

rocks show a more clear distinction between HiSY and LoSY rocks and attain significantly higher values for these variables (Table 1). Thus, Sr/Y maximum in La Posta suite is ≈160, whereas in New Zealand the bulk of the trend Separation Point suite reaches 350, with some rocks reaching ≈ 800. Sr/Y values for the outboard suite in New Zealand (Median suite) are also higher than those for the western Peninsular Ranges batholith, thus the entire New Zealand margin appears more HiSY. Most New Zealand HiSY rocks of the Separation Point suite fit the Drummond and Defant (1990) TTG/adakite definition of Sr/Y > 40, whereas ≈1/3 of La Posta rocks have Sr/Y < 40. In our data set, Peninsular Ranges batholith values of <40 are mostly from the San Jacinto pluton (Hill et al., 1988). HiSY rocks in the La Posta suite tend to occur as zones or pulses within composite or zoned plutons that also contain rocks with Sr/Y < 40. Within the large San Jacinto and Sierra San Pedro Martir plutons, Sr/Y ranges from 20 to 138 and 33–205, respectively (Hill et al., 1988; Kimbrough et al., 1998).

The distinction between HiSY and LoSY suites is emphasized in chondrite-normalized REE patterns (Fig. 6). HiSY plutons exhibit LREE enrichment and HREE depletion compared to the LoSY plutons. The HREE depletion reflects the presence of garnet in the source region residue upon partial melting. Absent (New Zealand) to weak negative (Baja-California) Eu-anomalies in the HiSY suites reflect plagioclase-free and plagioclase-poor compositions of the source rock residue, respectively.

The plutons of the La Posta suite are similar to the far-inboard Rahu suite of New Zealand, in terms of the presence of zoned plutons, within-pluton range of Sr/Y, which often spans the LO-HiSY boundary (Sr/Y = 40), and Sri. Absence of Eu anomalies in the core of the large Sierra San Pedro Martir pluton in the eastern Peninsular Ranges batholith indicates that Sr/Y variation cannot be due to shallow level fractionation of plagio-

TABLE 4. REE ANALYSES OF TYPICAL NEW ZEALAND PLUTONS

Sample	Ce	La	Pr	Nd	Sm	Eu	Gd	Tb	Dy	Ho	Er	Tm	Yb	Lu
OU49127	27.21	12.31	3.61	16.44	4.6	1.1	4.13	0.79	4.77	1	2.91	0.41	2.56	0.4
OU49147	37.88	18.09	4.45	18.37	4.35	0.84	3.49	0.61	3.61	0.74	2.06	0.3	1.81	0.29
OU49138	23.26	12.42	2.56	9.78	1.64	0.4	1.11	0.13	0.74	0.13	0.31	0.04	0.3	0.05
FM90A	80.9	35.7	10	34.6	6.4	1.8	4.6	0.7	3.5	0.6	1.5		1.3	
Typical Rahu	64.81	36.11	6.79	25.33	5.16	1.03	3.86	0.59	3.25	0.6	1.62	0.23	1.39	0.21

Note: Analyses by ICP-MS at Western Washington State University, except for fm90a (McCulloch et al., 1987). See Figure 6 caption for sample details.

clase. Elevated Sr_i and $\delta^{18}O$ and the presence of inherited zircon indicate that the Rahu suite contains a variable, but significant, component of Paleozoic metasedimentary crust (Tulloch, 1983; Waight et al., 1998). Correlation of increasing Sr_i with decreasing Sr/Y in the Buckland pluton (Fig. 7; Graham and White, 1990) of the Paparoa Range is consistent with mixing of Separation Point suite and Ordovician Greenland Group to produce Rahu suite magmas, although compositional control by differing source H_2O (Waight et al., 1998) and depth of melting may also be factors in variable Sr/Y. A chondrite-normalized REE pattern for the Rahu suite (Fig. 6C) is also consistent with HiSY (Separation Point suite)-metasediment mixing. The Rahu REE pattern is similar to those for the La Posta suite of Baja-California, displaying relatively subdued HiSY characteristics. The Rahu pattern is not consistent with mixing of metasedimentary rock and LoSY plutons. Similar supracrustal material has also been postulated as a source component for eastern Peninsular Ranges batholith La Posta magmas (Walawender et al., 1990). A strongly HiSY Separation Point suite-equivalent has not yet been recognized in the Peninsular Ranges batholith.

Variations in degree of "HiSY" signature have previously been ascribed to depth of melting, source rock chemical and mineralogical composition, degree of melting, H_2O content, and assimilation of crustal material (e.g., Drummond and Defant, 1990; Rushmer, 1991; Petford and Atherton, 1996; Waight et al., 1998).

TECTONIC SETTING

New Zealand

The paired plutonic belts in both New Zealand and Baja-California were initiated during subduction, which was directly responsible for formation of at least the outboard belts. The location of the Jurassic–Early Cretaceous trench is not clear (for differing opinions, see Mortimer et al., 1999b; Muir et al., 1995). An accretionary prism may be represented by the Torlesse greywackes and Otago Schist in New Zealand, perhaps analogous to the Franciscan Complex in California, but it is also possible that the relevant trench was located further west. The widespread presence of an Albian unconformity suggests that subduction had ceased by ≈100 Ma. Bradshaw (1993) suggested the change

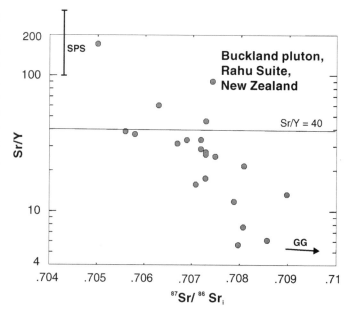

Figure 7. Sr/Y vs. $^{87}Sr/^{86}Sr_i$ plot for Buckland and associated plutons of Rahu suite (Paparoa Range) showing possible mixing between more strongly HiSY rocks of Separation Point suite and Paleozoic Greenland Group greywacke (GG; 0.720, 2). Separation Point restricted to Y<10. Data from Graham and White (1990).

from convergence to extension coincided with ridge subduction at ≈105–100 Ma. By the cessation of inboard belt HiSY magmatism, the setting was strongly extensional and metamorphic core complexes had developed (Tulloch and Kimbrough, 1989); ultimately New Zealand rifted from Australia and Antarctica by ca. 80 Ma (Tulloch et al. 1994).

Crustal Thickening

We interpret that widespread crustal thickening preceded the major pulse of HiSY plutonism by 10–20 m.y., as evidenced by events both within the arc and adjacent terranes:

1. Subduction-related magmatism of the outboard LoSY belt peaked at ≈140 Ma (Kimbrough et al., 1994), indicating rapid addition of material to the continental lithosphere at this time.

2. Absence of complete isotopic resetting of Paleozoic horn-blende (Tulloch and Challis, 2000) in currently exposed lower plates of metamorphic core complexes in the inboard belt indicates only a transient burial of these rocks to mid-crustal levels prior to rapid uplift. This lack of thermal equilibrium suggests burial occurred <10–20 m.y. prior to the start of uplift ca. 110 Ma (Spell et al., 2000). The Ar/Ar thermal history of the Barrytown Pluton (Spell et al., 2000) also suggests ca. 140 Ma uplift in upper plate rocks of the continental margin to the west of the batholith. To the east, Ar/Ar data suggest crustal thickening in the Otago Schist from ca. 150 to 120 Ma (Gray and Foster, 2002; Little et al., 1999).

3. Paleozoic orthogneiss country-rock to the Western Fiord-land Orthogneiss in northern Fiordland experienced granulite-facies metamorphism (e.g., Clarke et al., 2000; Brown, 1996) and strong contractional deformation (Daczko et al., 2001) in the Early Cretaceous. Apparent recrystallization of zircon and major resetting of U-Pb in the Paleozoic Milford Orthogneiss suggests some of this metamorphism may be related to burial at the zircon reset age of 134 ± 2 Ma (Tulloch et al., 2000).

A second, more widely recognized high-pressure event at ca. 120 Ma (Bradshaw, 1990) immediately followed mid-crustal emplacement of at least the northern Western Fiordland Orthogneiss (124 Ma), forming granulite facies orthogneisses and ca. 120 Ma metamorphic rims on zircon (Tulloch et al., 2000). The timing of this event and its short duration, relatively low-strain nature has led to magma-loading models to explain high-pressure metamorphism in Fiordland (Brown, 1996). This event may represent a continuation of the earlier crustal thicken-ing episode postulated above.

Inboard/Outboard Plutonic Belt Boundary

The original boundary between the inboard and outboard belts in the South Island is to some extent reactivated and obscured by Late Cenozoic deformation, but on Stewart Island the Escarpment fault (Allibone and Tulloch, 1997) represents a Pacific Ocean-verg-ing thrust fault, perhaps analogous to the west-verging structures in the Peninsular Ranges batholith (Johnson et al., 1999; Schmidt et al., 1999). However, whereas the Peninsular Ranges batholith features apparently predate the La Posta suite, the Escarpment fault (or its associated Gutter high strain zone) was formed dur-ing HiSY magmatism. Thus, early HiSY plutons (ca. 120 Ma) are deformed in the hanging wall of this shear zone, but correlatives of the youngest pluton (105 ± 1 Ma; Fig. 2) cut the shear zone.

Baja-California

The Peninsular Ranges batholith was emplaced during a period of unusually voluminous subduction-related magmatism that corresponds closely in age to the Cretaceous normal super-chron (118–83 Ma). The convergent margin plate setting for the Cretaceous Peninsular Ranges batholith is well established. The batholith, with associated forearc basin sediments (Nilsen and Abbott, 1981; Busby-Spera and Boles, 1986), and subduction complex high pressure/temperature (P/T) metamorphic rocks on Catalina and Cedros Islands (Platt, 1975; Grove and Bebout, 1995; Baldwin and Harrison, 1989), provide a relatively intact cross-section view across an Andean-type convergent margin. Well-established isotopic and paleontologic ages demonstrate that arc magmatism and batholith emplacement was coeval with forearc basin sedimentation and high P/T metamorphism of sub-duction complex rocks (Kimbrough et al., 2001).

Engebretson et al. (1985) presented a model for displacement history between western North America and adjacent oceanic plates for the past 180 m.y. based on construction of global plate circuits. During the period of intense circum-Pacific plutonism (ca. 120–80 Ma), the rate of Farallon–North America conver-gence was moderately high (50–100 km/m.y.), although major flare-ups in magmatism, as represented by the La Posta suite (98–93 Ma), and large volume intrusions in the Sierra Nevada batholith (Coleman and Glazner, 1998) cannot be correlated with variations in convergence rates (Ducea, 2001). However, very rapid Farallon–North America plate convergence between 75 and 40 Ma (the Laramide orogeny), which reached a peak of ≈150 km/m.y. was not accompanied by extensive plutonism.

The western and eastern zones (Kimbrough et al., 2001) of the Peninsular Ranges batholith are locally juxtaposed by a series of intrabatholith west-vergent Early Cretaceous ductile thrust faults that separate rocks with contrasting metamorphic grade and deformational history (Johnson et al., 1999; Fig. 1). In the Sierra San Pedro Martir region, these west-vergent contractional structures form the western flank of a ≈ 20-km-wide, doubly ver-gent fan structure (Schmidt et al., 1999). This fan-like structure appears to be a through-going element in the Baja California segment of the Peninsular Ranges batholith. Contemporaneous crustal shortening is recorded to the west of the thrust belt by shallow-level kilometer-scale upright to west-vergent folds and related axial-plane faults in supracrustal volcanic and sedimentary rocks (Fig. 1; Gastil et al., 1975). Synbatholithic crustal shorten-ing occurred from ca. 115 to 108 Ma in the Sierra San Pedro Martir region (Johnson et al., 1999) and from 118 to 105 Ma north of the border (Todd et al., 1988; Thomson and Girty, 1994). La Posta–type plutons postdate this Early Cretaceous episode of contractional deformation. They are intruded across the fan-like structure in the east-central Peninsular Ranges batholith and stitch thrust faults that separate western and eastern zone rocks. The locus of La Posta–type intrusive centers corresponds closely with the most deeply exhumed part of the Peninsular Ranges batholith. Middle- to upper amphibolite facies mineral assem-blages within characteristically migmatitic wall-rock indicate 5–20 km emplacement depths for La Posta–type plutons north of lat. 30° N (Todd et al., 1988; Grove, 1993; Rothstein, 1997). South of lat. 30° N, emplacement depths for La Posta–type intru-sions are somewhat shallower (5–15 km), with upper greenschist facies and lower amphibolite facies assemblages more prevalent (Gastil et al., 1975; Rothstein, 1997).

Available thermochronologic data indicate that La Posta–type plutons were significantly exhumed (>10 km) immediately fol-lowing emplacement (Lovera et al., 1999; Schmidt et al., 1999).

Final denudation of the eastern Peninsular Ranges batholith occurred in the Late Cretaceous/Early Tertiary, perhaps in response to Laramide flat subduction beneath southwestern North America (Goodwin and Renne, 1991; Grove, 1993; Schmidt et al., 1999; Lovera et al., 1999; Axen et al., 2000). Amphibolite facies rocks from Catalina and Cedros Islands yield cooling ages of ca. 115–110 Ma, which are anomalously young compared to the middle Jurassic ages of similar high-grade rocks from the Franciscan Complex farther north (e.g., Baldwin and Harrison, 1989). High-grade assemblages such as this have been interpreted to reflect initiation of a new subduction zone (Platt, 1975). Grove and Bebout (1995) proposed that the Catalina rocks record slab abandonment and initiation of a new subduction zone in the borderland region. In their model, thermal equilibration and eventual melting of the abandoned slab produced the La Posta suite magmas.

DISCUSSION

Similarities between New Zealand and Baja-California

1. Both areas developed paired belts of similar dimensions parallel to their respective margins in which the inboard belt is dominated by HiSY plutons (Separation Point and La Posta suites in New Zealand and Baja-California, respectively). This repeated pairing strongly suggests that the inboard belt is genetically related to the outboard LoSY belt (western Peninsular Ranges batholith, Median suite in New Zealand). The inboard belt axis is located ~40 km from the outboard belt axis in New Zealand and Baja-California (and Peru; Petford and Atherton, 1996).

2. Within each area (New Zealand and Baja-California), alkali-lime indices (MALI) are similar for both suites, also suggesting a genetic linkage between suites, albeit possibly indirect. Similar relative differences in Na, Sr enrichment in inboard versus outboard, in both areas, also suggests an inboard/outboard genetic relationship.

3. In both areas, HiSY rocks rapidly developed some 15–20 m.y. after the peak of 30–40 m.y. of LoSY subduction-related magmatism. The scale of this time lag is similar to the time required to underthrust, heat, and partially melt large volumes of crustal rocks.

4. Minor HiSY magmatism occurred periodically within the older, outboard LoSY belts, but no LoSY magmatism occurred anywhere once the major bursts of HiSY magmatism began.

5. In both areas, the boundaries between the two belts are apparently marked by contractional faults that were active at some stage during HiSY magmatism.

6. Both inboard belts experienced rapid exhumation post emplacement (e.g., Tulloch and Kimbrough, 1989; Tulloch and Palmer, 1990; Spell et al., 2000; Mattinson et al., 1986; Gibson et al., 1988), consistent with rebounding of overthickened crust. Development of core complexes (Tulloch and Kimbrough, 1989) in the far-inboard region in New Zealand suggests maximum crustal thickening occurred here. McNulty and Farber (2002) have recently identified detachment faults associated with the

exhumation of the La Posta suite, perhaps analogous to the detachments associated with the Rahu suite.

Differences between Median and Peninsular Ranges Batholiths

1. Significantly deeper crustal levels are exposed in New Zealand. The Median batholith of New Zealand is cut by the Alpine fault with the eastern half uplifted by a combination of Cretaceous extension and Neogene transpression to reveal 9–12 kb granulite facies rocks in Fiordland, and thus providing a unique view into deep crustal processes operating in arcs.

2. A greater compositional range is evident in New Zealand HiSY plutons from dioritic rocks of the Western Fiordland Orthogneiss (predominant in Fiordland) to relatively abundant leucogranodiorite (Fig.1).

3. HiSY characteristics are more strongly developed in New Zealand (average Sr/Y≈ 180) than Baja-California (average Sr/Y≈ 60). HiSY plutons (La Posta suite of the Peninsular Ranges batholith) are equivalent to the weakly HiSY, far-inboard Rahu suite in New Zealand, which is interpreted here as Separation Point suite plus continental crust. A Separation Point suite-type strongly HiSY endmember equivalent is not currently exposed in Baja-California, nor are Western Fiordland Orthogneiss-type mafic rocks.

4. Alkali-lime (MALI) values are very different in New Zealand (scattered alkali-calcic to calc-alkali) and Baja-California (well defined linear calcic to calc-alkali), which indicates that these HiSY rocks are not associated with a particular magma series.

5. The shorter interval over which the La Posta suite was emplaced suggests a significantly higher magma flux rate in Baja-California. The enormous outcrop area (15,000 km^2) and the narrow emplacement interval for the La Posta HiSY suite indicate unusually high magma-production rates (75–100 km^3/km/m.y.) for the Peninsular Ranges batholith from ≈98–93 Ma (Silver and Chappell, 1988; Kimbrough et al., 1998).

6. Both New Zealand belts are associated with lesser relative volumes of volcanic rocks than are observed in Baja-California. These proportions may be consistent with the greater contraction of the lithosphere (Kay and Mpodozis, 2001) in New Zealand, suggested by higher Sr/Y and thus lithosphere thickness.

HISY ROCKS FROM OTHER AREAS

Other continental margin batholiths include both HiSY and LoSY rocks. In the Ross orogen (Transantarctic Mountains; Allibone et al., 1993) the high Na, Al, Sr, and low Y DV1b suite postdates the DV1a suite by ≈20 m.y. Cox et al., 2000). In Peru, the Cordillera Blanca batholith (e.g., Petford and Atherton, 1996) follows 60 m.y. of voluminous plutonism. Minor HiSY magmatism also occurs within the eastern (inboard) side of Coastal batholith itself (Tiabaya suite; Pitcher et al., 1985), being ≈10 m.y. younger than the associated LoSY plutons. In North Victoria Land, Antarctica, Borg et al. (1987) determined subduction polarity on the basis of regional trace element trends of the Admiralty intrusives.

However, the relative disposition of the paired belts described here would suggest that increasing Sr, Na, Al and decreasing Y might also suggest a trend away from the subduction zone. In Western Palmer Land of the Antarctic Peninsula, Wareham et al. (1997) report rocks apparently spanning 230–90 Ma, of which one-third are sodic and have Sr/Y values ranging from 40 to 60. No age relationships are apparent, and there is no clear spatial distribution, although HiSY rocks are possibly relatively more abundant inboard. Within each of these areas, both HiSY and associated LoSY rocks have internally similar MALI values (Fig. 7). In northwestern North America, the Idaho batholith includes paired sodic and main series rocks, perhaps analogous to HiSY and LoSY rocks (Hyndman, 1983).

MODELS FOR HISY MAGMATISM

Widespread common features of continental margin HiSY magmatism described above from New Zealand, Baja-California, and to some extent from Antarctica, suggest a common process of formation. HiSY rocks have geochemical signatures (high Na, Al, Sr, low K, Y, etc.) and isotopic compositions indicative of an origin by partial melting of plagioclase-poor to absent, hornblende, garnet-bearing (i.e., at depths > ≈ 40 km) rocks of basaltic composition (e.g., Drummond and Defant, 1990; Rushmer, 1991). There are two main potential sources for the basaltic material.

Partial Melting of Oceanic Crust

Comparison with TTG and adakites (Drummond and Defant, 1990) suggests possible derivation of HiSY granitoids from slab melting. Possible tectonic scenarios include: melting of young, actively subducting crust <5–10 Ma (Peacock et al., 1994); melting of an abandoned slab (e.g., Grove and Bebout, 1995); and melting of mid-oceanic ridge basalt within a subducted ridge (e.g., Parada et al., 1999).

Problems with these models are considerable for the rocks described here: the thermal structure of subduction zones predicts the location of such rocks outboard of the LoSY magmatism (Peacock et al., 1994; Petford and Atherton, 1996); volumes of oceanic crust are probably insufficient to produce the large volumes and flux rates of observed HiSY plutons by the small to moderate degrees of partial melting indicated. Magmas produced from a subducted ridge would likely be produced at variable times along the margin depending on the arrival times of the ridge along the margin. However, there is no evidence for along-margin variation within the two 800-km-long margin segments studied here.

Partial Melting of Subduction Zone-Derived Basaltic Crustal Underplate

HiSY magmas could be generated from subduction-generated deep mafic crust within several tectonic environments: (a) a mafic underplate grows downward by magmatic accretion into the garnet stability field (Petford and Atherton, 1996); (b) subduction underthrusting carries the older outboard belt into the garnet stability field (Muir et al., 1995; Tulloch and Kimbrough, 1995); and (c) vertical thickening of an underplated margin by plate convergence and compression.

Features in favor of this group of models include: the HiSY belts described above (at least from New Zealand, Baja-California, and Peru) are always accompanied by significant volumes of outboard LoSY magmatism; chemical linkages (Na, Sr, MALI) exist between the inboard and outboard suites in each area; and the time lag of 10–20 Ma between LoSY outboard belt and HiSY inboard magmatism is consistent with the likely time required to reach melting temperatures. This model is supported by the εNd-Sri data of McCulloch et al. (1987), who argued that the large Western Fiordland Orthogneiss (Separation Point suite) intrusion was not slab-derived and that the source rocks probably had a significant, but limited, crustal prehistory prior to Early Cretaceous intrusion. Decoupling of εNd and Sri evolution is consistent with derivation from a protolith with similar Rb/Sr and Sm/Nd contents to the slightly older (LoSY) Darran leucogabbros of the Median suite. The highest Sr/Y and Sm/Dy values may denote regions of previously thickest mafic crust. Dehydration melting of such crust to form HiSY magmas likely released fluids that may have been instrumental in forming significant ore deposits in the Andes (Kay and Mpodozis, 2001). In New Zealand, Mo-mineralization associated with belt of Separation Point suite leucogranite stocks (Tulloch and Rabone, 1993) may have formed in this manner. Crustal-scale Pb-isotope homogenization (Kimbrough and Tulloch, 1995) may be related to fluid flux on such a scale. We conclude that this group of models based on partial melting of a mafic underplate is more consistent with the observations from both areas than with a slab-melt origin.

Similar, sodic, high Sr/Y plutons have been described from the Peruvian Andes, where HiSY rocks of the Cordillera Blanca batholith lie inboard of the Coastal batholith. The relatively recent formation of this batholith provides a clearer picture of its tectonic setting and associated crustal structure. Petford and Atherton (1996) describe a model of magmagenesis in which earlier mantle wedge-derived magmas underplate the margin and the crust grows downwards, extending further into the garnet stability field, producing higher Sr/Y rocks with time.

However, the Petford and Atherton model fails to explain the inboard location of the HiSY belt relative to the LoSY belt of plutons, a distinctive feature of the New Zealand and Baja-California convergent margins, among others. In these and the Peruvian examples, the separation of the belt axes is ~40 km and normal to the margin. There is minor overlap of young HiSY into older LoSY arc in New Zealand, clear separation in Baja-California, and in Peru the relatively minor Cordillera Blanca batholith is well separated from the (significantly older) Coastal batholith, which itself includes some HiSY rocks on its eastern margin (see Fig. 8). Magmatic underplating is more likely to form a symmetrical root (Petford and Atherton, 1996), which would place the HiSY source directly under the LoSY pluton belt. Such a

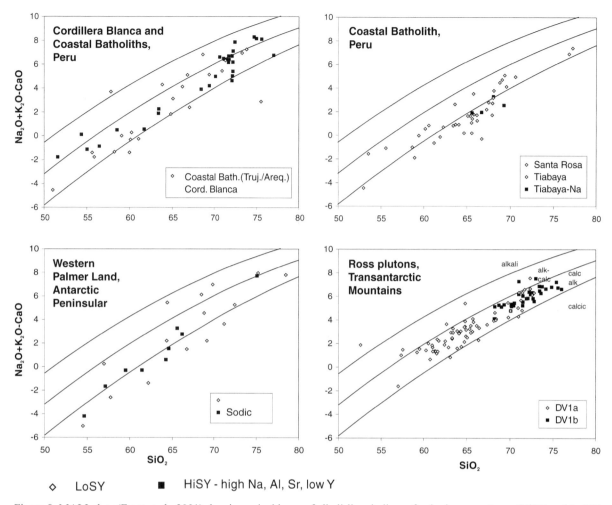

Figure 8. MALI plots (Frost et al., 2001) showing coincidence of alkali-lime indices of paired magma types (HiSY and LoSY) in other areas: Peru (Petford and Atherton 1996; Pitcher et al., 1985); Western Palmer Land (Wareham et al., 1997); and Ross Orogen, Transantarctic Mountains (Allibone et al., 1993).

model would predict voluminous intrusion of older LoSY plutons by younger HiSY plutons. Absence of younger LoSY plutons in such a model might be ascribed to freezing as they tried to intrude the partially melted underplate, but older plutons should form much of the upper crust prior to HiSY magmatism.

A Model

The HiSY magmas are most likely derived from a source region comprised of the base of the LoSY arc and thrust inboard and downward into the garnet stability field. Subduction and/or underthrusting of the LoSY arc and tectonically underplating inboard crust is the most likely mechanism to achieve this (Fig. 9), as previously suggested by Muir et al. (1995) and Tulloch and Kimbrough (1995) for New Zealand. In Peru, the generation of such magmas has been linked to flat or shallowing subduction (Kay et al., 1991). In Baja-California, subduction did shallow after generation of the older outboard belt of the Peninsular Ranges batholith; a corollary might be that the subduction

angle flattened at the end of LoSY magmatism in New Zealand. Gutscher et al. (2000) showed that a common reason for flat slab subduction is the arrival of oceanic plateaus. The large Hikurangi Plateau (Mortimer and Parkinson, 1996) may have collided with continental New Zealand at around this time. However, the effectively synchronous development of LoSY magmas along ≈800 km of margin would require the unlikely scenario of synchronous collision along the margin.

Shallowing subduction angle not only allows increased coupling (e.g., Schmidt et al., 1999) and underthrusting of the outboard LoSY arc, but also explains (a) cessation of further LoSY magmatism by shutting off access of slab-derived fluids to the mantle wedge, and (b) a single voluminous burst of HiSY magmatism, because the source is finite and nonreplenishing.

Crustal Assimilation in Inboard Plutons

The "HiSY" characteristics of the Separation Point suite in New Zealand are significantly stronger than those of the La Posta

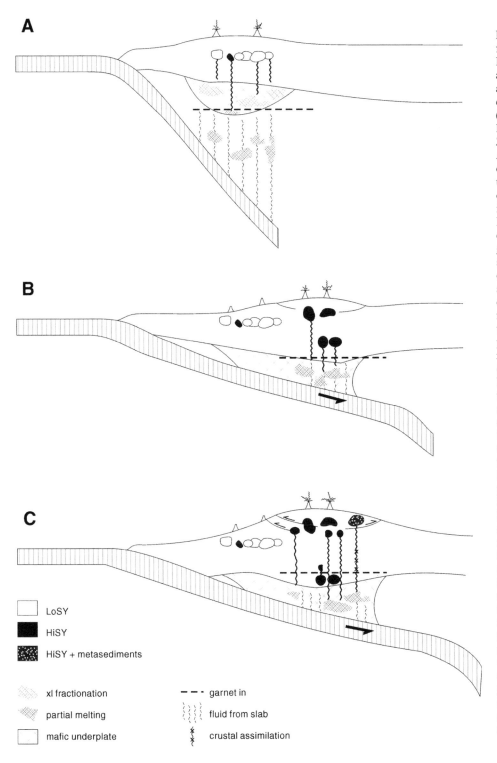

Figure 9. Schematic provisional model for development of paired inboard HiSY–outboard LoSY magmatic belts along convergent margins. Considerable vertical exaggeration. The model extends that of Petford and Atherton (1996) to explain predominant inboard location of HiSY pluton belts in New Zealand, Baja-California, and Peru. A: Long-lived, steady subduction generates thick, mantle-derived basaltic underplate from which LoSY gabbro-diorite-granite plutons fractionate and intrude upper crust of outer margin. Periodically, during rapid magma generation from mantle wedge, underplate grows briefly downwards into garnet stability field generating minor volumes of HiSY magmas. B: Shallowing of subduction zone, possibly as a result of subduction of an oceanic plateau or ridge, leads to underthrusting of mafic underplate to depths that produce amphibole and garnet-bearing assemblages ± plagioclase. Early dioritic HiSY rocks (Western Fiordland Orthogneiss) in New Zealand (and Peru) formed by partial melting of underplate at relatively shallow depths and moderate degrees of partial melting. Contractional deformation thickens continental crust. C: Continued underthrusting and crustal thickening leads to a) burial and granulite facies metamorphism of early HiSY plutons in New Zealand (Western Fiordland Orthogneiss) and generation of leucogranites from small melt fractions at greater depths of partial melting of underplate, indicated by trace-element indications of garnet-rich, plagioclase-absent compositions. Some plutons (Rahu, NZ; La Posta, Baja-California) assimilate metasedimentary rocks. Older HiSY plutons are cut by thrust faults; younger plutons intrude same fault systems. Subsequent extensional collapse of overthickened and thermally weakened margin leads to decompression melting and emplacement of final plutons, rapid post-emplacement uplift of HiSY plutons, and (in NZ at least) extensive development of metamorphic core complexes in inboard region.

suite or most other Phanerozoic batholiths worldwide. In many respects, the La Posta suite is more similar to the Rahu suite of far-inboard New Zealand than the coeval Separation Point suite. Significantly, the Rahu suite is entirely restricted (Tulloch 1983, 1988) to the westernmost, more continental Paleozoic terrane (Buller), strongly suggesting that the reduced Sr, increased K, Sri, etc. of Rahu are due to contamination of Separation Point magmas with Paleozoic continental crust (e.g., Waight et al., 1998). A similar model has been proposed by Miller et al. (1997) for trondhjemitic to granodioritic plutonism in the southern Appalachians.

This supports a similar argument for the La Posta suite: higher Sri and $\delta^{18}O$ and lower Sr/Y values are found in plutons that occur above over old crust in the northern Peninsular Ranges batholith (San Jacinto), compared with southern plutons (Sierra San Pedro Martir), which occur over younger crust. The depleted HREE and high Sr, etc. of the La Posta suite could be used as an argument against crustal contamination. However, compared to the Separation Point suite, these depletions/enrichments are modest, and it is possible that a Separation Point–type end member exists at depth in the (less deeply exhumed) Peninsular Ranges batholith.

CONCLUSIONS

Similarities between the Peninsular Ranges batholith of Baja-California and the Median batholith of New Zealand include: both batholiths comprise paired belts with a relatively high-Na, Al, Sr and low-Y belt (HiSY) inboard, contrasted with an (LoSY) outboard belt type. The HiSY magmas were emplaced in a major 10–20 m.y. pulse that formed the endstage of more than 30–40 m.y. of LoSY magmatism. The belts locally overlap geographically but are generally separated by ocean-verging contractional faults. Both inboard belts experienced at least 10–30 km post emplacement uplift, possibly reflecting the axis of greatest crustal overthickening of the respective margins. Significant differences include: New Zealand batholiths exhibit a greater compositional range in both belts; from gabbro and diorite through tonalite and granodiorite to granite. HiSY characteristics are significantly stronger in New Zealand (Separation Point suite) than in Baja-California (La Posta suite). In New Zealand, subduction apparently ceased with ridge subduction immediately following emplacement of the inboard suite. However, in Baja-California, subduction not only continued, but rates increased (albeit shallow and largely amagmatic), following emplacement of the La Posta suite.

The above features, collectively, including chemical linkages between HiSY and LoSY belts, favor a two-stage model in which mafic mantle-derived magmas first underplate the margin during outboard magmatism. Secondly, this thickened crust is underthrust to a deep inboard location by subduction and partially melted, leaving plagioclase-poor to absent and garnet-bearing residua. Continentalward translation of the underplate to a position under the inboard belt better explains the consistently inboard location of the HiSY magmas, than the symmetrical, downward-accreting model of Petford and Atherton (1996).

The La Posta suite of Baja-California has the more subdued HiSY characteristics of the far-inboard Rahu suite in New Zealand. Sri, eNd, $\delta^{18}O$ and REE compositions, and the presence of inherited Precambrian zircons, are all consistent with the derivation of the Rahu suite from mixtures of Separation Point suite magmas and Paleozoic metasedimentary crust endmembers. By analogy, we suggest that the La Posta suite is likewise a mixture of a primary HiSY magma (similar to the Separation Point suite of New Zealand, but not currently exposed in Baja-California) and continental metasedimentary rocks.

Despite similarities with adakites and Archean TTG, formation of the inboard magmas by dehydration melting of oceanic slab is ruled out by the inboard location (Drummond and Defant, 1990), the >5 Ma age of the slab in Baja-California, chemical and isotopic compositions, and the inability of thin oceanic crust to rapidly supply partial melts in sufficient volume.

The occurrence of similar paired rock suites in the Ross orogen of the Transantarctic Mountains, the Antarctic peninsular, the Peruvian batholiths, and elsewhere indicate that crust-derived HiSY and mantle wedge-derived LoSY plutons represent two major subgroups of magmas in convergent margin batholiths.

ACKNOWLEDGMENTS

We thank Nick Mortimer for several reviews that greatly improved this manuscript and for frequent discussions. Bob Weibe, Bill Collins, and Gordon Gastil and many others also provided helpful discussions. Keith Klepeis and Calvin Miller provided constructive and helpful journal reviews. Belinda Smith-Lyttle, John Simes, Neville Orr, Ben Morrison (Institute of Geological and Nuclear Sciences), Beverley Elliott, and Joan Kimbrough (San Diego State University) provided sterling technical support. Peter Blattner (Institute of Geological and Nuclear Sciences) provided oxygen isotope data and discussion. Jahan Ramezani (Massachusetts Institute of Technology) provided high-precision U-Pb data. Funding for Tulloch was provided from the New Zealand public good science fund managed by the Foundation for Research, Science, and Technology. We greatly appreciate the opportunity to honor the geological work of Gordon Gastil.

REFERENCES CITED

Ague, J.J., and Brimhall, G.H., 1988, Magmatic arc asymmetry and distribution of anomalous plutonic belts in the batholiths of California: Effects of assimilation, crustal thickness, and depth of crystallization: Geological Society of America Bulletin, v. 100, p. 912–927.

Allibone, A.H., and Tulloch, A.J., 1997, Metasedimentary, granitoid and gabbroic rocks from central Stewart Island, New Zealand: New Zealand Journal of Geology and Geophysics, v. 40, p. 53–68.

Allibone, A.H., Cox, S.C., and Smillie, R.W., 1993, Granitoids of the Dry Valleys region, southern Victoria Land, Antarctica: Geochemistry and evolution along the Early Paleozoic Antarctic craton margin: New Zealand Journal of Geology and Geophysics, v. 36, p. 299–316.

Axen, G.J., Grove, M., Stockli, D., Lovera, O.M., Rothstein, D.A., Fletcher, J.M., Farley, K., and Abbott, P.L., 2000, Thermal evolution of Monte Blanco Dome; low-angle normal faulting during Gulf of California rifting and late Eocene denudation of the eastern Peninsular Ranges: Tectonics, v. 19, p. 197–212.

Baldwin, S.L., and Harrison, T.M., 1989, Geochronology of blueschists from west-central Baja California and the timing of uplift in subduction complexes: Journal of Geology, v. 97, p. 149–163.

Beresford, S.W., Bradshaw, J.D., Weaver, S.D., and Muir, R.J., 1996, Echinus Granite and Pepin Group of Pepin Island, northeast Nelson, New Zealand. Drumduan Terrane basement or exotic fragment in the Median tectonic zone?: New Zealand Journal of Geology and Geophysics, v. 39, p. 265–270.

Blattner, P., and Graham, I.G., 2000, New Zealand's Darran Complex and Mackay intrusives – Rb/Sr whole-rock isochrons in the Median tectonic zone: American Journal of Science, v. 300, p. 603–629.

Blattner, P., and Williams, J.G., 1991, The Largs high-latitude oxygen isotope anomaly (New Zealand) and climatic controls of oxygen isotopes in magma: Earth and Planetary Science Letters, v. 103, p. 270–284.

Borg, S.G., Stump, E., Chappell, B.W., McCulloch, M.T., Wyborn, D., Armstrong, R.C., and Holloway, J.R., 1987, Granitoids of northern Victoria Land, Antarctica: Implications of chemical and isotopic variations to regional crustal structure and tectonics: American Journal of Science, v. 287, p. 127–165.

Bradshaw, J.D., 1993, A review of the Median tectonic zone: Terrane boundaries and terrane amalgamation near the Median tectonic line: New Zealand Journal of Geology and Geophysics, v. 36, p. 117–126.

Bradshaw, J.Y., 1989, Origin and metamorphic history of an Early Cretaceous polybaric granulite terrain, Fiordland, southwest New Zealand: Contributions to Mineralogy and Petrology, v. 103, p. 346–360.

Bradshaw, J.Y., 1990, Geology of crystalline rocks of northern Fiordland: Details of the granulite facies Western Fiordland Orthogneiss and associated rocks: New Zealand Journal of Geology and Geophysics, v. 33, p. 465–484.

Brown, E.H., 1996, High-pressure metamorphism caused by magma loading in Fiordland, New Zealand: Journal of Metamorphic Geology, v. 14, p. 441–452.

Busby-Spera, C.J., and Boles, J.R., 1986, Sedimentation and subsidence styles in a Cretaceous forearc basin, southern Vizcaino peninsula, Baja California Sur, (Mexico), in Abbott, P.L., ed., Cretaceous stratigraphy, western North America: Los Angeles, Society for Sedimentary Geology (SEPM) Pacific Section, v. 46, p. 79–90.

Clarke, G.L., Klepeis, K.A., and Daczko, N.R., 2000, Cretaceous high-P granulites at Milford Sound, New Zealand: Metamorphic history and emplacement in a convergent margin setting: Journal of Metamorphic Geology, v. 18, p. 359–374.

Coleman, D.S., and Glazner, A.F., 1998, The Sierra Crest Magmatic Event: Rapid formation of juvenile crust during the Late Cretaceous in California, in Ernst, W.G., and Nelson, C.A., eds., Integrated Earth and environmental evolution of the southwestern United States: Columbia, Maryland, Bellwether Publishing for the Geological Society of America, p. 253–272.

Cooper, R.A., and Tulloch, A.J., 1992, Early Paleozoic terranes in New Zealand and their relationship to the Lachlan Fold Belt: Tectonophysics, v. 214, p. 129–144.

Cox, S.C., Parkinson, D.L., Allibone, A.H., and Cooper, A.F., 2000, Isotopic character of Cambro-Ordovician plutonism, southern Victoria Land, Antarctica: New Zealand Journal of Geology and Geophysics, v. 43, p. 501–520.

Daczko, N.R., Klepeis, K.A., and Clarke, G.L., 2001, Evidence of early Cretaceous collisional-style orogenesis in northern Fiordland, New Zealand and its effects on the evolution of the lower crust: Journal of Structural Geology, v. 23, p. 693–713.

Drummond, M.S., and Defant, M.J., 1990, A model for trondhjemite-tonalite-dacite genesis and crustal growth via slab melting: Archean to modern comparisons: Journal of Geophysical Research, v. 95, p. 21,503–21,521.

Ducea, M., 2001, The California arc: Thick granitic batholiths, eclogite residues, lithospheric-scale thrusting, and magmatic flare-ups: GSA Today, v. 11, no. 11, p. 4–10.

Eberhart-Phillips, D., and Reyners, M., 2001, A complex, young subduction zone imaged by three-dimensional seismic velocity, Fiordland, New Zealand: Geophysical Journal International, v. 146, p. 731–746.

Engebretson, D.C., Cox, A., and Gordon, R.G., 1985, Relative motions between oceanic and continental plates in the Pacific Basin: Geological Society of America Special Paper 206, 59 p.

Frost, B.R., Barnes, C.G., Collins, W.J., Arculus, R.J., Ellis, D.J., and Frost, C.D., 2001, A geochemical classification for granitic rocks: Journal of Petrology, v. 42, p. 2033–2048.

Gastil, R.G., 1975, Plutonic zones in the Peninsular Ranges of southern California and northern Baja California: Geology, v. 3, p. 361–363.

Gastil, G., Diamond, J., Knapp, C., Walawender, M., Marshall, M., Boyles-Reaber, C., and Erskine, B., 1990, The problem of the magnetite-ilmenite boundary in Southern and Baja California: Boulder, Colorado, Geological Society of America Memoir 174, p. 19–32.

Gibson, G.M., McDougall, I., and Ireland, T.R., 1988, Age constraints on metamorphism and the development of a metamorphic core complex in Fiordland, southern New Zealand: Geology, v. 16, p. 405–408.

Goodwin, L.B., and Rennie, P.R., 1991, Effects of progressive mylonitization on Ar retention in biotites from the Santa Rosa mylonite zone, California, and thermochronologic implications: Contributions to Mineralogy and Petrology, v. 108, p. 283–297.

Graham, I.J., and White, P.J., 1990, Rb-Sr dating of Rahu suite granitoids from the Paparoa Range, North Westland, New Zealand: New Zealand Journal of Geology and Geophysics, v. 33, p. 11–22.

Gray, D.R., and Foster, D.A., 2002, New ^{40}Ar/^{39}Ar data from Otago Schist: Record of Gondwana deformation and breakup, in Gondwana 11 Correlations and Connections, Programme and Abstracts: Christchurch, New Zealand, University of Canterbury.

Gromet, L.P., and Silver, L.T., 1987, REE variations across the Peninsular Ranges batholith: Implications for batholith petrogenesis and crustal growth in magmatic arcs: Journal of Petrology, v. 28, p. 75–125.

Grove, M., 1993, Thermal histories of southern California basement terranes [Ph.D. thesis]: University of California, Los Angeles, 307 p.

Grove, M., and Bebout, G.E., 1995, Jurassic and Cretaceous tectonic evolution of coastal southern California: Insights from the Catalina Schist: Tectonics, v. 14, p. 1290–1308.

Gutscher, M., Spakman, W., and Bijwaard, H., 2000, Geodynamics of flat subduction: Seismicity and tomographic constraints from the Andean margin: Tectonics, v. 19, p. 814–833.

Hill, R.I., Chappell, B.W., and Silver, L.T., 1988, San Jacinto intrusive complex 2. Geochemistry: Journal of Geophysical Research, v. 93, p. 10,349–10,372.

Hill, R.I., Silver, L.T., and Taylor, H.P.J., 1986, Coupled Sr-O isotope variations as an indicator of source heterogeneity for the northern Peninsular Ranges batholith: Contributions to Mineralogy and Petrology, v. 92, p. 351–361.

Hyndman, D.W., 1983, The Idaho batholith and associated plutons, Idaho and western Montana: Boulder, Colorado, Geological Society of America Memoir 159, p. 213–240.

Ireland, T.R., and Gibson, G.M., 1998, SHRIMP monazite and zircon geochronology of high-grade metamorphism in New Zealand: Journal of Metamorphic Geology, v. 16, p. 149–167.

Johnson, S.E., Tate, M.C., and Fanning, C.M., 1999, New geologic mapping and SHRIMP U-Pb zircon data in the Peninsular Ranges batholith, Baja California, Mexico: Evidence for a suture?: Geology, v. 27, p. 743–746.

Kay, S.M., and Mpodozis, C., 2001, Central Andean ore deposits linked to evolving shallow subduction systems and thickening crust: GSA Today, v. 11, no. 3, p. 4–9.

Kay, S.M., Mpodozis, C., Ramos, V.A., and Munizaga, F., 1991, Magma source variations for mid-late Tertiary magmatic rocks associated with a shallowing subduction zone and a thickening crust in the central Andes (28 to 33° S): Geological Society of America Special Paper 265, p. 113–137.

Kimbrough, D.L., and Tulloch, A.J., 1989, An early Cretaceous age of orthogneiss from the Charleston Metamorphic Group, New Zealand: Earth and Planetary Science Letters, v. 95, p. 130–140.

Kimbrough, D.L., and Tulloch, A.J., 1995, Uniform Pb-isotopic composition of chemically diverse Cretaceous batholithic suites in western New Zealand: A reflection of regional scale Pb isotopic homogenization processes: Eos (Transactions, American Geophysical Union), v. 76, p. 304.

Kimbrough, D.L., Tulloch, A.J., Coombs, D.S., Landis, C.A., Johnston, M.R., and Mattinson, J.M., 1994, Uranium-lead zircon ages from the Median tectonic zone, New Zealand: New Zealand Journal of Geology and Geophysics, v. 37, p. 393–419.

Kimbrough, D.L., Magistrale, H., and Gastil, R.G., 1998, Enormous Late Cretaceous magma flux associated with TTG batholith emplacement; eastern Peninsular Ranges batholith of southern and Baja California: Geological Society of America Abstracts with Programs, v. 30, no. 7, p. 258.

Kimbrough, D.L., Smith, D.P., Mahoney, J.B., Moore, T.E., Grove, M., Gastil, R.G., and Ortega-Rivera, A., 2001, Forearc-basin sedimentary response to rapid Late Cretaceous batholith emplacement in the Peninsular Ranges of southern and Baja California: Geology, v. 29, p. 491–494.

Little, T.A., Mortimer, N., and McWilliams, M., 1999, An episodic cooling model for the Otago-Marlborough Schist, New Zealand, based on ^{40}Ar/^{39}Ar ages: New Zealand Journal of Geology and Geophysics, v. 42, p. 305–326.

Lovera, O.M., Grove, M., Kimbrough, D.L., and Abbott, P.L., 1999, A method for evaluating basement exhumation histories from closure age distributions of detrital minerals: Journal of Geophysical Research, v. 104, p. 29,419–29,438.

Ludwig, K.L., 1999, ISOPLOT/Ex version 2.06: Berkeley, California, Berkeley Geochronology Centre Special Publication 1a, 54 p.

Mattinson, J.M., Kimbrough, D.L., and Bradshaw, J.Y., 1986, Western Fiordland Orthogneiss: Early Cretaceous arc magmatism and granulite facies metamorphism, New Zealand: Contributions to Mineralogy and Petrology, v. 92, p. 383–392.

McCulloch, M.T., Bradshaw, J.Y., and Taylor, S.R., 1987, Sm-Nd and Rb-Sr isotopic and geochemical systematics in Phanerozoic granulites from Fiordland, southwest New Zealand: Contributions to Mineralogy and Petrology, v. 97, p. 183–195.

McNulty, B., and Farber, D., 2002, Active detachment faulting above the Peruvian flat slab: Geology, v. 30, p. 567–570.

Miller, C.F., Fullagar, P.D., Sando, T.W., Kish, S.A., Soloman, G.C., Russell, G.S., and Wood-Steltenpohl, L.F., 1997, Low-potassium, trondhjemitic to granodioritic plutonism in the eastern Blue Ridge, southwestern North Carolina-northeastern Georgia: Boulder, Colorado, Geological Society of America Memoir 191, p. 235–254.

Mortimer, N., and Parkinson, D., 1996, Hikurangi Plateau: A Cretaceous large igneous province in the southwest Pacific Ocean: Journal of Geophysical Research, v. 10, p. 687–696.

Mortimer, N., Tulloch, A.J., and Ireland, T.R., 1997, Basement geology of Taranaki and Wanganui Basins, New Zealand: New Zealand Journal of Geology and Geophysics, v. 40, p. 223–236.

Mortimer, N., Gans, P.B., Calvert, A., and Walker, N.W., 1999a, Geology and thermochronometry of the east edge of the Median batholith (Median tectonic zone): A new perspective on Permian to Cretaceous crustal growth of New Zealand: The Island Arc, v. 8, p. 404–425.

Mortimer, N., Tulloch, A.J., Spark, R.N., Walker, N.W., Ladley, E., Allibone, A.H., and Kimbrough, D.L., 1999b, Overview of the Median batholith, New Zealand: A new interpretation of the geology of the Median tectonic zone and adjacent rocks: Journal of African Earth Sciences, v. 29, p. 257–268.

Muir, R.J., Weaver, S.D., Bradshaw, J.D., Eby, G.N., and Evans, J.A., 1995, The Cretaceous Separation Point batholith, New Zealand. Granitoid magmas formed by melting of mafic lithosphere: Journal of the Geological Society, v. 152, p. 689–701.

Muir, R.J., Ireland, T.R., Weaver, S.D., Bradshaw, J.D., Waight, T.E., Jongens, R., and Eby, G.N., 1997, SHRIMP U-Pb geochronology of Cretaceous magmatism in northwest Nelson-Westland, South Island, New Zealand: New Zealand Journal of Geology and Geophysics, v. 40, p. 453–463.

Muir, R.J., Ireland, T.R., Weaver, S.D., Bradshaw, J.D., Evans, J.A., Eby, G.N., and Shelley, D., 1998, Geochronology and geochemistry of a Mesozoic magmatic arc system, Fiordland, New Zealand: Journal of the Geological Society of London, v. 155, p. 1037–1053.

Nilsen, T.H., and Abbott, P.L., 1981, Paleogeography and sedimentology of Upper Cretaceous turbidites, San Diego, California: American Association of Petroleum Geologists Bulletin, v. 65, p. 1256–1284.

Parada, M.A., Nystrom, J.O., and Levi, B., 1999, Multiple sources for the Coastal batholith of central Chile (31–34S): Geochemical and Sr-Nd isotopic evidence and tectonic implications: Lithos, v. 46, p. 505–521.

Peacock, S.M., Rushmer, T., and Thompson, A.B., 1994, Partial melting of subducting oceanic crust: Earth and Planetary Science Letters, v. 121, p. 227–244.

Platt, J.P., 1975, Metamorphic and deformational processes in the Franciscan Complex, California; some insights from the Catalina Schist terrane: Geological Society of America Bulletin, v. 86, p. 1337–1347.

Petford, N., and Atherton, M., 1996, Na-rich partial melts from newly underplated basaltic crust; The Cordillera Blanca Batholith, Peru: Journal of Petrology, v. 37, p. 1491–1521.

Pickett, D.A., and Wasserburg, G.J., 1989, Neodymium and strontium isotopic characteristics of New Zealand granitoid and related rocks: Contributions to Mineralogy and Petrology, v. 103, p. 131–142.

Pitcher, W.S., Atherton, M.P., Cobbing, E.J., and Beckinsale, R.D., 1985, Magmatism at a plate edge: The Peruvian Andes: New York, Halsted Press, John Wiley & Sons, 328 p.

Rattenbury, M.S., Cooper, R.A., and Johnston, M.R., 1998, Geology of the Nelson area: Institute of Geological and Nuclear Sciences Geological Map 9: Lower Hutt, New Zealand, Institute of Geological & Nuclear Sciences, scale: 1:250 000.

Rothstein, D.A., 1997, Metamorphism and denudation of the eastern Peninsular Ranges batholith, Baja California Norte, Mexico [Ph.D. thesis]: Los Angeles, University of California, 445 p.

Rushmer, T., 1991, Partial melting of two amphibolites: Contrasting experimental results under fluid-absent conditions: Contributions to Mineralogy Petrology, v. 107, p. 41–59.

Schmidt, K.L., Paterson, S.R., Blythe, A.E., and Kopf, C.F., 1999, Linked but temporally distinct episodes of tectonism, magmatism, and exhumation in the Cretaceous Peninsular Ranges: Geological Society of America, Abstracts with Programs, v. 31, no. 7, p. 482.

Silver, L.T., and Chappell, B.W., 1988, The Peninsular Ranges batholith: An insight into the evolution of the Cordilleran batholiths of southwestern North America: Transactions of the Royal Society of Edinburgh, v. 79, p. 105–121.

Silver, L.T., Taylor, H.P., and Chappell, B., 1979, Some petrological, geochemical and geochronological observations of the Peninsular Ranges batholith near the international border of the U.S.A., and Mexico, *in* Patrick, L.A., and Todd, V.R., eds., Mesozoic crystalline rocks: Peninsular Ranges batholith and pegmatites Point Sal Ophiolite: Manuscripts & road logs Geological Society of America Annual Meeting: San Diego, California, Department of Geological Sciences, San Diego State University.

Smithies, R.H., 2000, The Archaean tonalite-trondhjemite-granodiorite (TTG) series is not an analogue of Cenozoic adakite: Earth and Planetary Science Letters, v. 182, p. 115–125.

Smithies, R.H., and Champion, D.C., 2000, The Archaean High-Mg Diorite Suite: Links to Tonalite-Trondhjemite-Granodiorite magmatism and implications for Early Archean crustal growth: Journal of Petrology, v. 41, no. 12, p. 1653–1671.

Spell, T., McDougall, I., and Tulloch, A.J., 2000, Thermochronologic constraints on the breakup of the Pacific Gondwana margin: The Paparoa metamorphic core complex, South Island, New Zealand: Tectonics, v. 19, p. 433–451.

Stacey, J.S., and Kramers, J.D., 1975, Approximation of terrestrial lead evolution by a two-stage model: Earth and Planetary Science Letters, v. 26, p. 207–221.

Steiger, R.H., and Jäger, E., 1977, Subcommission on geochronology: Convention on the use of decay constants in geo- and cosmochronology: Earth and Planetary Science Letters, v. 36, p. 359–362.

Tate, M.C., and Johnson, S.E., 2000, Subvolcanic and deep-crustal tonalite genesis beneath the Mexican Peninsular Ranges: Journal of Geology, v. 108, p. 721–728.

Tate, M.C., Norman, M.D., Johnson, S.E., Anderson, J.L., and Fanning, C.M., 1999, Generation of tonalite and trondhjemite by subvolcanic fractionation and partial melting in the Zarza Intrusive Complex, western Peninsular Ranges batholith, northwestern Mexico: Journal of Petrology, v. 40, p. 983–1010.

Taylor, H.P., and Silver, L.T., 1978, Oxygen isotope relationships in plutonic igneous rocks of the Peninsular Ranges batholith, southern and Baja California, *in* Zartman, R.E., ed., Short papers of the fourth international conference, geochronology, cosmochronology, isotope geology: U.S. Geological Survey Open-File Report 78-0701, p. 423–426.

Thomson, C.N., and Girty, G.H., 1994, Early Cretaceous intra-arc ductile strain in Triassic-Jurassic and Cretaceous continental margin arc rocks, Peninsular Ranges, California: Tectonics, v. 13, p. 1108–1119.

Todd, V.R., Erskine, B.G., and Morton, D.M., 1988, Metamorphic and tectonic evolution of the northern Peninsular Ranges batholith, southern California, *in* Ernst, W.G., ed., Metamorphism and crustal evolution of the western United States: Englewood Cliffs, New Jersey, Prentice Hall, p. 894–937.

Tulloch, A.J., 1983, Granitoid rocks of New Zealand—A brief review: Boulder, Colorado, Geological Society of America Memoir 159, p. 5–20.

Tulloch, A.J., 1988, Batholiths, plutons and suites: nomenclature for granitoid rocks of Westland-Nelson, New Zealand: New Zealand Journal of Geology and Geophysics, v. 31, p. 505–509.

Tulloch, A.J., and Challis, G.A., 2000, Reconnaissance hornblende-Al geobarometry of plutonic rocks from western New Zealand: New Zealand Journal of Geology and Geophysics, v. 43, p. 555–567

Tulloch, A.J., and Kimbrough, D.L., 1989, The Paparoa Metamorphic Core Complex, Westland-Nelson, New Zealand: Cretaceous extension associated with fragmentation of the Pacific margin of Gondwana: Tectonics, v. 8, p. 1217–1234.

Tulloch, A.J., and Kimbrough, D.L., 1995, Mesozoic plutonism in western New Zealand—A 50 Ma transition from subduction to continental rifting on the Gondwana margin, Brown, M., and Piccoli, P.M., eds., The origin of granites and related rocks: Third Hutton Symposium Abstracts, U.S. Geological Survey Circular, v. 1129, p. 153–154.

Tulloch, A.J., and Palmer, K., 1990, Tectonic implications of granite cobbles from the mid-Cretaceous Pororari Group, SE Nelson, New Zealand: New Zealand Journal of Geology and Geophysics, v. 33, p. 5–18.

Tulloch, A.J., and Rabone, S.D.C., 1993, Mo-bearing granodiorite porphyry plutons of the Early Cretaceous Separation Point suite west Nelson, New Zealand: New Zealand Journal of Geology and Geophysics, v. 36, p. 401–408.

Tulloch, A.J., Kimbrough, D.L., and Wood, R.A., 1991, Carboniferous granite basement dredged from a site on the Challenger Plateau, Tasman Sea: New Zealand Journal of Geology and Geophysics, v. 34, p. 121–126.

Tulloch, A.J., Kimbrough, D.L., and Waight, T., 1994, The French Creek granite, north Westland, New Zealand—Late Cretaceous A-type plutonism on the Tasman passive margin, *in* Evolution of the Tasman Sea Basin: Rotterdam, Balkema, p. 65–66.

Tulloch, A.J., Ireland, T.R., Walker, N.W., and Kimbrough, D.L., 2000, U-Pb zircon ages from the Milford Orthogneiss, northern Fiordland: Paleozoic igneous emplacement and Early Cretaceous metamorphism: Lower Hutt, New Zealand, Institute of Geological and Nuclear Sciences, 2000/6.

Waight, T.E., Weaver, S.D., Ireland, T.R., Maas, R., Muir, R.J., and Shelley, D., 1997, Field characteristics, petrography, and geochronology of the Hohonu batholith and the adjacent Granite Hill complex, North Westland, New Zealand: New Zealand Journal of Geology and Geophysics, v. 40, p. 1–17.

Waight, T.E., Weaver, S.D., Muir, R.J., Maas, R., and Eby, G.N., 1998, The Hohonu batholith of North Westland, New Zealand: Granitoid compositions controlled by source H_2O contents and generated during tectonic transition: Contributions to Mineral Petrology, v. 130, p. 225–239.

Walawender, M.J., and Smith, T.E., 1980, Geochemical and petrologic evolution of the basic plutons of the Peninsular Ranges batholith, southern California: Journal of Geology, v. 88, p. 233–242.

Walawender, M.J., Gastil, R.G., Clinkenbeard, J.P., McCormick, W.V., Eastman, B.G., Wernicke, R.S., Wardlaw, M.S., and Smith, B.M., 1990, Origin and evolution of the zoned La Posta-type plutons, eastern Peninsular Ranges batholith, southern and Baja California: Boulder, Colorado, Geological Society of America Memoir 174, p. 1–18.

Wandres, A.M., Weaver, S.D., Shelley, D., and Bradshaw, J.D., 1998, Change from calc-alkaline to adakitic magmatism recorded in the Early Cretaceous Darran Complex, Fiordland, New Zealand: New Zealand Journal of Geology and Geophysics, v. 41, p. 1–14.

Wareham, C.D., Millar, I.L., and Vaughan, A.P.M., 1997, The generation of sodic granite magmas, western Palmer land, Antarctic peninsular: Contributions to Mineralogy and Petrology, v. 128, p. 81–96.

MANUSCRIPT ACCEPTED BY THE SOCIETY JUNE 2, 2003

Geological Society of America
Special Paper 374
2003

Geochronological constraints on the tectonic history of the Peninsular Ranges batholith of Alta and Baja California: Tectonic implications for western México

Amabel Ortega-Rivera*

Department of Geological Sciences, Queen's University, Kingston, Ontario K7L 3N6, Canada

ABSTRACT

A compilation of existing age data and new age determinations (U/Pb zircon, $^{40}Ar/^{39}Ar$ step-heating, K/Ar, Rb/Sr, and apatite fission-track dates) for the Peninsular Ranges batholith of Alta and Baja California provides geochronological constraints on the tectonic history of the Peninsular Ranges batholith and tectonic implications for western México.

The plutons of the Peninsular Ranges batholith of Alta and Baja California between the 28°N and 34°N parallels were emplaced from west to east between ca. 140 to 80 Ma (U/Pb zircon dates) and display $^{40}Ar/^{39}Ar$ hornblende and biotite plateau dates that range from 118 to 83 Ma and 116 to 80 Ma, respectively, and biotite K/Ar dates as young as 65 Ma. Rapid cooling is indicated by the small differences in U/Pb zircon and cogenetic $^{40}Ar/^{39}Ar$ dates. Mineral pairs having mainly concordant dates yield $^{40}Ar/^{39}Ar$ cooling ages that also systematically decrease from southwest to northeast. Also, monotonic eastward younging of ages across various plutons was found as was previously reported for the Sierra San Pedro Mártir pluton. This systematic regional and local eastward younging may be explained by the systematic tilting of pluton-sized crustal blocks containing individual plutons.

The northeastward decrease in ages across the Peninsular Ranges batholith and across some of its plutons is therefore attributed to regional eastward migration of granitic intrusion foci combined with rapid differential exhumation histories and superimposed east-side-up tilting of crustal-sized blocks containing some of the plutons and beautifully exemplifies the short time lapse between batholith intrusion and unroofing.

Additionally, apatite fission-track dates from samples collected across the Peninsular Ranges, from Ensenada to San Felipe, Baja California, establish that there is also an eastward younging of ages from 104 Ma to 51 Ma from west to east and also across plutons. The fission-track dates indicate that from west to east, present exposures of much of the western Peninsular Ranges batholith have been within «3 km of the Earth's surface (assuming a normal geothermal gradient) since Early Cretaceous time, and that by early Tertiary time, much of the eastern Peninsular Ranges had cooled to temperatures of <110 °C, indicating rapid exhumation histories and superimposed east-side-up tilting shown by younger ages at higher elevations within plutons. The batholith was sufficiently uplifted during the Late Cretaceous such that 10–15 km of material was eroded off the eastern part of the batholith by early Paleocene time.

Furthermore, a regional compilation of the available age data for western México provides geochronological constraints for the testing of controversial paleomagnetic and geological models for the tectonic evolution of the Mesozoic Peninsular Ranges of

---

*Present address: CGEO-UNAM, Campus-Juriquilla, Domicilio Conocido, Km 15 carretera Qro.-S.L.P. Centro de Geociencias, Juriquilla, Querétaro, Qro., México C.P. 76230; amabel@geociencias.unam.mx.

Ortega-Rivera, A., 2003, Geochronological constraints on the tectonic history of the Peninsular Ranges batholith of Alta and Baja California: Tectonic implications for western México, *in* Johnson, S.E., Paterson, S.R., Fletcher, J.M., Girty, G.H., Kimbrough, D.L., and Martín-Barajas, A., eds., Tectonic evolution of northwestern México and the southwestern USA: Boulder, Colorado, Geological Society of America Special Paper 374, p. 297–335. For permission to copy, contact editing@geosociety.org. © 2003 Geological Society of America.

Alta and Baja California. The controversy involves different perceptions of the geologic development of western México and southernmost California as derived from paleomagnetic studies and regional geology. Whereas the geological data and plate tectonic reconstructions seem to indicate a northward motion between 300 and 500 km with respect to the rest of North America since the Cretaceous, paleomagnetic data suggest much greater movement involving at least 2500 km of northward translation for the Baja peninsula during the same period.

The proposed block tilting about an axis subparallel to the trend of the batholith can largely account for the observed discordant paleomagnetic inclinations without appealing to large-scale northward transport. This further supports the geological evidence and corroborates that the Peninsular Ranges province has been part of the North America craton since at least Early Cretaceous time, has undergone only limited northward tectonic displacement, and, that by simply closing the Gulf of California to its pre-Miocene opening, as indicated by the geological evidence, the peninsula is restored to its original position.

Interpretation of the regional age patterns suggests that a long, linear subduction zone analogous to the modern setting of the South American Andes formed the basic framework of the western margin of México; that subduction took place along a uniform north-northwest–south-southeast–oriented paleotrench paralleling the Pacific coast of México during the Mesozoic to at least Eocene time; and that the apparent eastward migration of the locus of magmatism, along México's western margin, is due to changes of dip and velocity of the subducting slab.

Keywords: Peninsular Ranges batholith, Alta and Baja California, geochronology, U/Pb, $^{40}Ar/^{39}Ar$, K/Ar, Rb/Sr, and fission-track dates, tectonics, tilting, paleomagnetism, large-scale northward translation.

INTRODUCTION

During Cretaceous time the western Cordillera of North America was flanked by an essentially continuous Andean-type, west-facing convergent margin (e.g., Hamilton, 1969, 1988a, 1988b). This subduction margin has undergone substantial modification as a result of crustal shortening in association with the Sevier and Laramide orogenies, subsequent stretching and large-scale strike-slip faulting from Late Cretaceous through Cenozoic time, and Cenozoic magmatism and sedimentation (e.g., Ziegler et al., 1985). The best-preserved element of the Cretaceous subduction margin is the arc itself. Erosion has stripped off most of the volcanic cover, leaving a series of batholiths as a clue that marks its general trend. The Peninsular Ranges batholith of southern California and Baja California (Fig. 1) has played a central role in the debate over Cretaceous paleogeography of the western Cordillera. Thus, it is important to better understand its tectonothermal history.

Geologic Background

The rocks of the Peninsular Ranges province have been divided into prebatholithic, batholithic, and postbatholithic rocks.

Prebatholithic Rocks

The prebatholithic rocks range in age from Paleozoic to middle Cretaceous (Albian). They flank the batholithic spine of the province on both the west and the east (Fig. 2). Lithologies

include hornblende and mica schist, phyllite, slate, quartzofeldspathic schists and gneiss, weakly metamorphosed volcanic rocks of various types, bedded cherts, quartzites, and crystalline limestones. These rocks are commonly found as screens and pendants between and within batholithic plutons. The foliation of these rocks dips steeply eastward and generally strikes parallel to the north-northwest outcrop trend of the batholith (Duffield, 1968). Many workers have reported systematic prebatholithic variations normal to the long axis of the batholith (i.e., compositional, structural, geochemical, isotopic, and geochronological variations; Fig. 2) between the western and more eastern portions of it (e.g., Early and Silver, 1973; Silver et al., 1979; Krummenacher et al., 1975; DePaolo, 1981; Gastil, 1981, 1983; Todd and Shaw, 1985). Silver et al. (1979) and Gastil (1983) divided the prebatholithic rocks into two main subprovinces, the western and eastern Peninsular Ranges (Fig. 2). Todd and Shaw (1979) delimit a broad, "prebatholithic boundary" denoting the transition from volcaniclastic to flysch-type country rocks and their increasing metamorphic grade from west to east. They suggest that this transition is the result of greater uplift and depth of erosion of the eastern parts of the batholith.

The western Peninsular Ranges consist of relatively unmetamorphosed sedimentary, volcanic, and volcaniclastic strata (locally attaining greenschist facies) ranging in age from Late Jurassic to Early Cretaceous. These rocks are known as the Santiago Peak Volcanics (Larsen, 1948) north of the Agua Blanca fault and as the Alisitos Formation (Santillán and Barrera, 1930; Allison, 1974) south of the Agua Blanca fault. This continuous

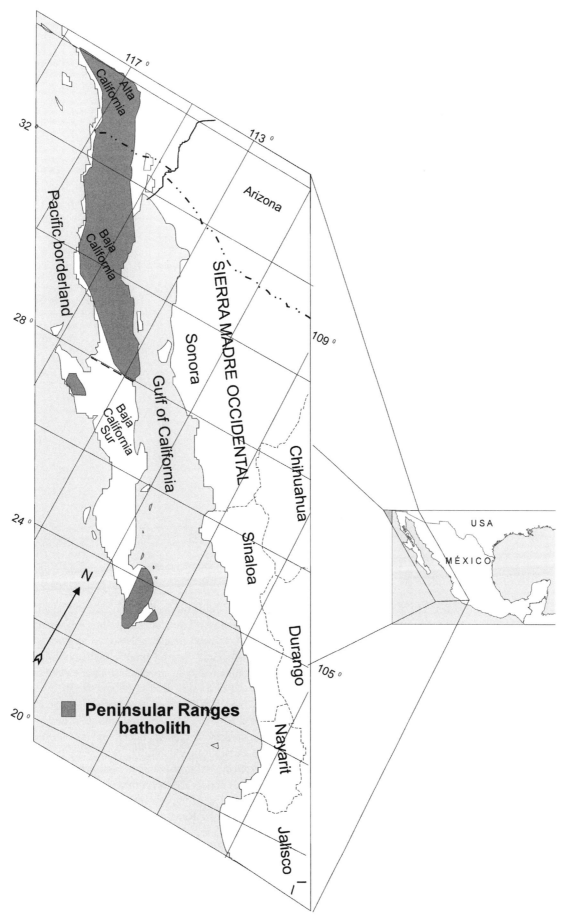

Figure 1. Index map showing Peninsular Ranges batholith of southern California and Baja California and part of western México (modified after Gastil, 1983).

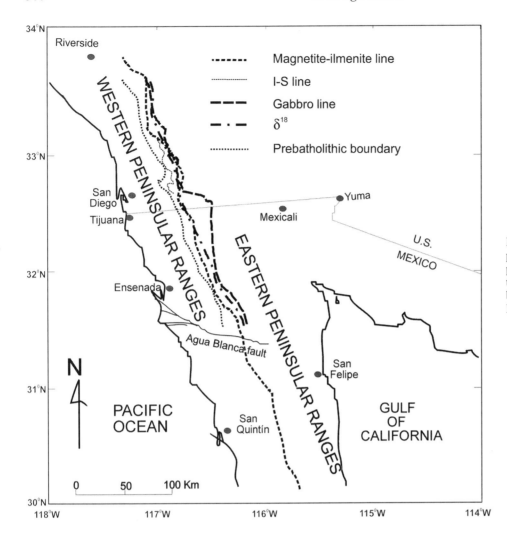

Figure 2. Superposed prebatholithic, compositional, geochronological, and isotopic discontinuities (see text for explanation) that define approximate boundary between eastern and western Peninsular Ranges (modified after Todd et al., 1988).

belt of volcanic and volcaniclastic rocks extends from the Santa Ana Mountains of southern California to the Vizcaíno Desert of Baja California Sur. Only rocks of Late Jurassic age (Tithonian, based on fossils from a few exposed windows of marine volcanogenic sedimentary strata through the volcanic cover) have been recognized in the Santiago Peak Volcanics north of the Agua Blanca fault. On the other hand, south of the Agua Blanca fault the Alisitos Formation is considered to be Aptian-Albian in age (also based on fossils, Allison, 1955).

In contrast to the mainly sedimentary and volcanic lithofacies exposed in the western Peninsular Ranges, metasedimentary rocks dominate the eastern Peninsular Ranges. In the central part of the eastern Peninsular Ranges, the country rocks consist mainly of metamorphosed Triassic and Jurassic marine clastic strata (flysch-type) whose protoliths include greywacke, arkosic sandstone, and argillite. The easternmost ranges of the peninsula, however, contain more abundant metamorphosed quartz-rich sandstones and rare carbonates and associated sedimentary rocks of Early Ordovician to middle Cretaceous age (Gastil et al., 1975; Gastil, 1993). The grade of regional metamorphism increases eastward, reaching a maximum near the central-eastern zone boundary, coinciding with the Gulf escarpment, where the rocks have been metamorphosed to amphibolite facies. East of the main Gulf escarpment, the metamorphic grade decreases, eventually reaching greenschist facies near San Felipe on the Gulf of California (Todd and Shaw, 1979).

Therefore, the western basement of the Peninsular Ranges batholith or the western prebatholithic rocks is inferred to be from an oceanic crust in nature, and the basement for the eastern Peninsular Ranges batholith or eastern prebatholithic rocks is inferred to be transitional to continental crust. Although some tectonic models based on local studies have been proposed for the evolution of the prebatholithic terranes and development of the Peninsular Ranges batholith, these can be compiled in two major models: (a) a single inboard-propagating arc developed across a pre-Triassic join between oceanic and continental lithospheres (Walawender et al., 1991; Thomson and Girty, 1994), or (b) an exotic island arc accreted to the continent between 115 and 108 Ma (Johnson et al., 1999a). But, the timing of the juxtaposition of these two disparate lithospheric types and the processes responsible for it remain controversial and unresolved issues in the Peninsular Ranges geology (Wetmore et al., 2002).

The Peninsular Ranges Batholith

Bounded by the Pacific Borderland to the west and the Gulf of California to the east (Fig. 1), the Peninsular Ranges batholith of Alta and Baja California is a northwest-trending and nearly continuously exposed series of granitoid rocks having an average width of ~100 km. The batholith is well exposed over a distance of more than 800 km, from the 34th parallel near Riverside, southern California, southward to the 28th parallel in Baja California. South of the 28th parallel, the batholith is thought to continue beneath the Cenozoic cover for ~900 km to the tip of the Baja peninsula (Gastil et al., 1975), where it again outcrops. Before the Late Miocene opening of the Gulf of California, the rocks of the batholith were contiguous with the plutonic rocks of Sonora and Sinaloa, México (e.g., Gastil et al., 1975, 1991; Silver, 1996; Henry et al., this volume). The Sinaloa batholith is the southward continuation of the Peninsular Ranges batholith, related granitic rocks in Sonora, and batholithic rocks of the Cabo San Lucas block at the southern end of the Baja California peninsula, and the Jalisco area of southwestern Mexico are a further southward continuation (Gastil et al., 1978; Frizzell, 1984; Frizzell et al.,1984; Aranda-Gómez and Pérez-Venzor, 1989; Morán-Zenteno et al., 2000; Schaaf et al., 2000; Henry et al., this volume).

Gastil (1975) also divided the batholith into western and eastern zones (Fig. 2), based on east-west variations in the type of host rock, age, style of emplacement, and pluton composition. These zones are moreover distinguished by their distinct chemical and isotopic properties that parallel the long axis of the batholith and which also delimit the boundary between the eastern and western Peninsular Ranges in Alta and northern Baja California (Fig. 2). Such boundaries (Figs. 2, 3, and 4) include: a magnetite-ilmenite line (Ishihara, 1977; Gastil, 1990a, 1990b; Gastil, 1986) and an isotopic age and ‰ $\delta^{18}O$ step (Taylor and Silver, 1978; Silver et al., 1979; DePaolo, 1981; Silver and Chappell, 1988). The isotopic composition of the western Peninsular Ranges batholith suggests a primitive island arc–type source, while for the eastern Peninsular Ranges batholith the isotopic compositions suggest older, deeper reservoirs with eclogitic mineral assemblages (Silver and Chappell, 1988; Silver, 1992).

The rocks of the western Peninsular Ranges batholith (Figs. 2–4) consist of Jurassic-Cretaceous (ca. 140–105 Ma) plutons intruding cogenetic (125–118 Ma) volcanic rocks (Silver et al., 1979; Silver and Chappell, 1988; Todd et al., 1988; Silver, 1992; Todd et al., 1994; Kimbrough et al., 2001). The western Peninsular Ranges batholith (Figs. 2–4) consists of gabbro to tonalite plutons that intrude lower grade greenschist-grade metamorphose volcanic arc and flysch assemblages (Silver et al., 1979; Silver and Chappell, 1988; Todd et al., 1988; Silver, 1992; Todd et al., 1994; Walawender et al., 1991). Is also known as the "gabbro belt" (Gastil, 1975; Silver et al., 1979), and it is characterized by typically small (<15 km in diameter), magnetite-bearing plutons and the presence of gabbroic rocks. The plutons of the western Peninsular Ranges batholith range in composition from peridotite to granite (but predominantly comprise gabbros, tonalities, and lesser amounts of granodiorite) and are generally irregularly shaped and reversely zoned (Gastil, 1975; Silver et al., 1979; Gastil et al., 1991; Todd et al., 1994; Tate et al., 1999).

The plutons comprising the eastern Peninsular Ranges batholith (Figs. 2–4) are younger, from 105 to ca. 80 Ma (Silver and Chappell, 1988; although the youngest intrusions (<95 Ma) appear to be limited to small (1–2 km scale) two-mica granites and associated pegmatites, and not volumetrically abundant). They are also predominantly tonalitic-to-granodioritic in composition. These plutons have invaded Phanerozoic metasedimentary, metavolcanic, and metaplutonic rocks (flysch, slope-basin, and miogeoclinal assemblages) as young as Cretaceous and together comprise the "La Posta" granite rock province (Fig. 4) of the eastern Peninsular Ranges batholith (Clinkenbeard, 1987; Duffield, 1968; Gastil et al., 1975, 1991; Gunn, 1985; Hill, 1984; Kimzey, 1982; McCormick, 1986; Thomson and Girty, 1994; Walawender et al., 1990, 1991). The "La Posta–type" plutons are large, comprising composite bodies measuring hundreds of square kilometers in surface exposure. Individual plutons display well-developed compositional zonation with roughly concentric outer margins and gradational inner margins, and regional metamorphism and deformation occurred concurrently with intrusion of the Peninsular Ranges batholith (Todd et al., 1994; Gastil, 1990a; Gastil et al., 1991).

Postbatholitic Rocks

To the west of the Peninsular Ranges province, Late Cretaceous to Eocene marine clastic strata rest nonconformably on prebatholithic and batholith rocks and probably were derived from erosion of the Alisitos–Santiago Peak Formations and Peninsular Ranges batholith (Gastil et al., 1975). Cretaceous strata exposed south of the Agua Blanca fault contain abundant *Coralliochama Orcutti*, a warm water rudist coral that indicates an Aptian-Albian age to the Alisitos Formation. The Cenozoic record in this region, in general, is characterized by: (a) the accumulation of a great thickness of continental sediments that crop out in numerous localities; (b) the development of marine sedimentary deposits, particularly on the western edge of the peninsula; and (c) widespread volcanic activity that partly covered the batholith. During the Paleocene and Eocene, sedimentary rocks accumulated in nearshore and deltaic environments (Gastil et al., 1975). These are covered by Miocene and Pliocene basalts, tuffs, and fluvial and shallow marine clastic sedimentary rocks in the west. However, to the east, relatively undeformed and locally thick accumulations of postbatholithic sedimentary and volcanic rocks unconformably overlie the prebatholithic and batholithic rocks (Duffield, 1968).

Previous Geochronology

Silver et al. (1969, 1979) and Silver and Chappell (1988) reported U/Pb zircon dates for the Peninsular Ranges in southern

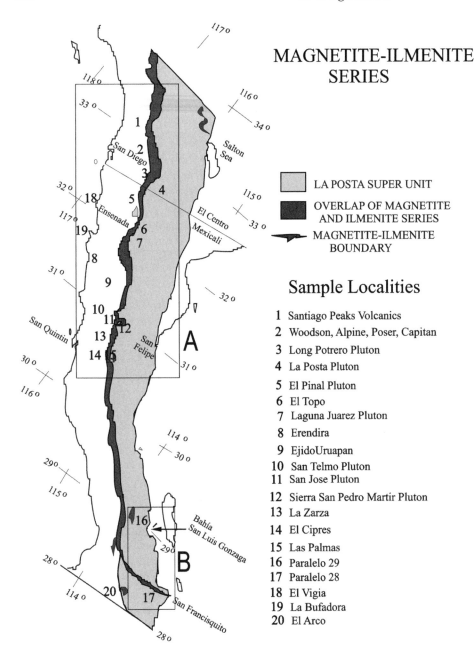

MAGNETITE-ILMENITE
SERIES

LA POSTA SUPER UNIT

OVERLAP OF MAGNETITE
AND ILMENITE SERIES

MAGNETITE-ILMENITE
BOUNDARY

Sample Localities

1 Santiago Peaks Volcanics
2 Woodson, Alpine, Poser, Capitan
3 Long Potrero Pluton
4 La Posta Pluton
5 El Pinal Pluton
6 El Topo
7 Laguna Juarez Pluton
8 Erendira
9 EjidoUruapan
10 San Telmo Pluton
11 San Jose Pluton
12 Sierra San Pedro Martir Pluton
13 La Zarza
14 El Cipres
15 Las Palmas
16 Paralelo 29
17 Paralelo 28
18 El Vigia
19 La Bufadora
20 El Arco

Figure 3. Magnetite-ilmenite limit and La Posta Super Unit (modified after Gastil, 1990a, 1990b). Location of two regional study areas A, B along Peninsular Ranges batholith of Alta and Baja California (and main sample localities for this study).

California, east of San Diego, that established that the plutons were emplaced over a 60 m.y. time interval, from 140 Ma to 80 Ma. In the area lying between the Agua Blanca fault in Baja California (31°N) and »50 km north of the international border in southern California (Fig. 3), Krummenacher et al. (1975) conducted the first extensive K/Ar study of the northern Peninsular Ranges batholith in Alta and Baja California. Based on 159 K/Ar dates, they constructed chrontours—age contours—that showed a systematic decrease of hornblende dates (from 125 to 80 Ma) and biotite dates (from 115 to 65 Ma) toward the northeast, with hornblende dates being typically 5–10 Ma older than coexisting biotite dates from the same rock.

In a comparison of their U/Pb zircon dates with K/Ar dates from Krummenacher et al. (1975), Silver et al. (1979) and Silver and Chappell (1988) confirmed that for the eastern Peninsular Ranges batholith there is an eastward younging trend of dates across the batholith as shown by the K/Ar dates (Fig. 5). Those authors also suggested that the pattern of U/Pb zircon ages for the westernmost plutons (ranging from 140 to 105 Ma) exhibit no preferred distribution and concluded, therefore, that these plutons were formed in a static magmatic arc. On the other hand, they argued that the U/Pb zircon ages for the eastern plutons become progressively younger toward the east (from 105 to 80 Ma) and interpreted this belt of eastern

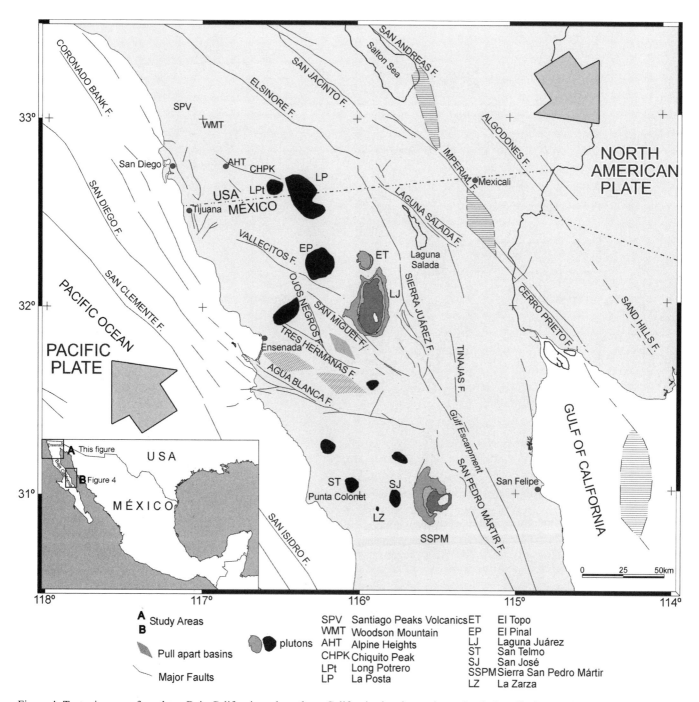

Figure 4. Tectonic map of northern Baja California and southern California showing major active faults affecting northern Peninsular Ranges batholith and major plutons. Arrows—direction of relative motion between Baja California and North America since the Pliocene (modified after Ortega-Rivera, 1988, and Suárez-Vidal et al., 1991).

plutons to have resulted as the manifestation of an eastward-migrating arc (Fig. 5). However, Walawender et al. (1990) have presented additional U/Pb zircon age data for the eastern Peninsular Ranges batholith and argue that their data do not support the eastward migration. Thus, they suggested that two distinct periods of static-arc magmatism occurred across the Peninsular

Ranges batholith, with the younger magmatic event spanning a tighter range of 98–89 Ma.

More recently, Ortega-Rivera et al. (1994, 1997) presented a detailed ^{40}Ar/^{39}Ar step-heating study, combined with U/Pb, K/Ar, fission-track, and Al-in-hornblende geobarometry data of the Sierra San Pedro Mártir pluton in the eastern Peninsular Ranges

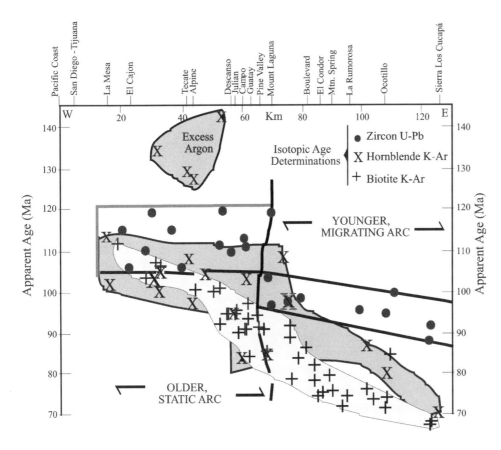

Figure 5. Comparison of U/Pb zircon ages and K/Ar ages in a transverse profile across northern Peninsular Ranges batholith east of San Diego, California. Shaded area represents pattern for biotite cooling ages; dashed show hornblende cooling ages, and dotted lines enveloping U/Pb dots represent static magmatic arc to west and migrating arc to east, defined by U/Pb ages. See text for details (from Silver and Chappell, 1988).

(SSPM, Fig. 3). They demonstrated that an eastward younging of K/Ar dates is also observed south of the Agua Blanca fault, and concluded that »15° east-side-up tilting about a north-south horizontal axis subparallel to the trend of the batholith best explains the eastward younging age pattern within the crustal block containing the pluton.

Tectonic History

It has been suggested that during the Late Cretaceous, terranes were transported northward with respect to North America as a consequence of oblique plate convergence, and therefore, that much of the western Cordillera of North America is composed of "suspect terranes" (e.g., Coney, 1981; Dickinson, 1981a, 1981b; Engebretson et al., 1984, 1985; Beck, 1991; Sedlock et al., 1993). These "suspect terranes" have been studied in detail in western Canada and the western United States. However, few data exist to constrain tectonic models for western México (Fig. 1); consequently, contrasting views have arisen. For example, Campa and Coney (1983) proposed that most of western México was accreted to the North American continent during the late Mesozoic or early Cenozoic. On the basis of paleomagnetic data obtained from Alta and Baja California, Champion et al. (1984) further suggested that, during the Oligocene, the Baja peninsula and adjoining regions formerly lying to the west

moved northward relative to the North American craton, arriving at their pre–Gulf of California position no later than earliest Miocene time. Similarly, other researchers studying this active margin have interpreted their paleomagnetic data obtained from coastal California and western Baja California to indicate that the "suspect terranes" were tectonically transported in post–middle Cretaceous time as much as 2500 km northward with respect to interior North America (e.g., Teissere and Beck, 1973; Hagstrum et al., 1985; Ziegler et al., 1985; Morris et al., 1986; Howell et al., 1987; Hagstrum and Filmer, 1990; Lund and Bottjer, 1991; Beck, 1991, 1992a, 1992b; Hagstrum and Sedlock, 1998).

In contrast, several lines of geologic evidence (e.g., Gastil and Miller, 1984; Silver and Chappell, 1988; and Silver, 1996) suggest that, prior to the well-documented late Miocene opening of the Gulf of California, Baja California had been attached to the North American craton for at least 500 Ma. Total displacements along the San Andreas fault system during Neogene time has been estimated at ~300 km (Clarke and Nielsen, 1973). The geologic evidence for only small displacement of Baja California (of the order of 200–400 km) has not been particularly compelling for some workers, but it has prompted other workers to challenge the reliability of the paleomagnetic observations in Baja California, creating a heated controversy over that issue.

Furthermore, according to Silver (1996), palinspastic restorations of the Gulf of California show that the Peninsular Ranges is not an accreted terrane, but rather that it uniquely fits with the rocks of the Sonoran batholith exposed along the eastern Gulf coast. Moreover, geological and tectonic interpretations of data obtained along the Gulf of California indicate that the Peninsular Ranges batholith was contiguous with the Sonoran and Sinaloan batholith (mainland México) during Cretaceous time and prior to the late Cenozoic opening of the Gulf of California, and that no more than the »300 km of northward motion associated with that opening is required for a good fit (e.g., Gastil, 1983; Gastil and Miller, 1984; Silver, 1996; Ortega-Rivera, 1997; Henry et al., this volume).

Moreover, as stated by Schaaf et al. (2000), paleogeographic reconstructions for the Baja California peninsula have yielded very controversial models in papers published over the past 20 years. Most hypothesis place the peninsula at lower paleolatitudes—in front of Central America or southern Mexico—with subsequent 10–20° northward translation(s) between late Cretaceous and Miocene. Many models include clockwise rotations from 25° to 45°. They also acknowledged that it is still not clear whether the Baja California peninsula acted as a uniform unit during that translation or if it is composed of lithologically distinguishable blocks with different ages and origins; and, if the later scenario is more likely, what is the provenance and time of accretion of the crustal mosaic that constitutes the present day peninsula of Baja California? This question has not been completely answered yet.

Tilting versus Translation?

Butler et al. (1991) made the intriguing observation that the paleomagnetic inclinations from the Peninsular Ranges batholith, which require ~900 of northward transport, can be reconciled by westward tilting of the batholith about a horizontal axis. Therefore, if we make the correction for westward tilting of the batholith to the paleomagnetic data, these restored paleomagnetic directions would require a minimum northward translation for the restoration of the Baja Peninsula, consistent with the Gastil (1993) geologic restoration. But on the other hand, there are several lines of evidence however, that tend to negate the westward tilting model of the entire batholith.

1. The tilt model does not explain paleomagnetic results from sedimentary rocks on the peninsula that provide an unambiguous paleohorizontal and yield similar anomalously shallow inclination data in agreement with results from plutons. A careful paleomagnetic study of the Cretaceous Valle Group strata on Cedros Island rules out remagnetization or burial compaction as mechanisms to explain shallow magnetic inclinations (Smith and Busby, 1993).

2. Grove (1994) presents field, petrologic, and Ar isotopic data from the east-central Peninsular Ranges batholith at 33°N that indicate that post-intrusive, Late Cretaceous, west-directed thrusting caused uplift of the deeper rocks exposed along the eastern side of the batholith. Segmentation of the batholith by thrust faults may better explain uplift rather than regional tilting of a single crustal block.

3. Although tilting of Cordilleran batholiths has been recognized elsewhere (i.e., Ague and Brandon, 1992), Beck (1992a, 1992b) argues on the basis of thermal modeling that large-scale tilting cannot explain the shallow inclination data from the batholith. These arguments mean that we must continue to consider large terrane motion as an explanation for the shallow inclination data.

Additionally, work in the Peninsular Ranges batholith (Figs. 2–5) has documented beautifully developed transverse asymmetries in geochronologic, mineralogic, geochemical, and isotopic properties (Gastil, 1975; Gastil et al., 1975; Krummenacher et al., 1975; Silver and Chappell, 1988; Silver et al., 1979; Todd and Shaw, 1979; Gromet and Silver, 1987; Ortega-Rivera, 1997, 2000a, 2000b, 2000c; Ortega-Rivera et al., 1997; Silver and Chappell, 1988; Silver et al., 2000; Kimbrough et al., 2001). A similar asymmetric trend to that previously observed regionally has been shown to occur locally within at least one of the major plutons of the Peninsular Ranges batholith (Gastil et al., 1994; Ortega-Rivera et al., 1994 a, 1994b, 1997). A detailed geochronologic study of the Sierra San Pedro Mártir pluton in the eastern northern Peninsular Ranges batholith (Fig. 3) that involved $^{40}Ar/^{39}Ar$ step-heating, U/Pb, conventional K/Ar and fission-track methods, combined with Al-in-hornblende geobarometry was presented by Ortega-Rivera et al. (1994a, 1994b, 1997). They showed that a monotonic eastward younging of ages (U/Pb zircon ages, hornblende, muscovite and biotite $^{40}Ar/^{39}Ar$ plateau ages, and apatite fission-track ages) occurs *within the pluton*, a trend similar to that previously observed regionally to the north (Krummenacher et al., 1975; Silver and Chappell, 1988; Ortega-Rivera, 1997, 2000a, 2000b, 2000c, 2003). Using a model that assumes progressive inside cooling of the pluton from its margin, the authors (Ortega-Rivera et al., 1994a, 1994b, 1997; and Ortega-Rivera, 1997, 2000a, 2000b, 2000c, 2003) concluded that ~15° of east-side-up tilting of crustal blocks of pluton-scale about a north-south horizontal axis, subparallel to the strike of the batholith, may best explain the eastward-younging age pattern within the crustal block containing the Sierra San Pedro Mártir pluton.

In addition, it has been suggested that tilting of plutons about subhorizontal axes and shallowing of paleomagnetic inclinations during compaction of marine sedimentary rocks, rather than large-scale latitudinal movement, may explain paleomagnetic inclination anomalies such as those observed for the Peninsular Ranges batholith (Irving and Archibald, 1990; Butler et al., 1989, 1991; Constanzo-Alvarez and Dunlop, 1988; Marquis and Irving, 1990; Ague and Brandon, 1992; Ortega-Rivera et al., 1997; Ortega-Rivera, 1997, 1998a, 1998b, 2000a, 2000b, 2000c, 2003; Dickinson and Butler, 1998). Ortega-Rivera et al. (1997) thus suggested that restoration of local east-side-up ~15–20° tilting of individual plutons within the Peninsular Ranges batholith can be used to calculate the paleo-horizon at the time of pluton crystallization.

Until Ortega-Rivera et al. (1997), paleo-horizons for plutons of the Peninsular Ranges had not been determined by combining geochronologic and geochemical methods. Butler et al. (1991) attributed the observed pattern of eastward younging of plutons and increasing denudation of the Peninsular Ranges batholith to uniform (~15–20°) northeast-side-up tilting of the *entire batholith* about an axis with an azimuth ~330–340°. Although such a tilt of the entire batholith could explain aspects of the regional pattern of eastward younging ages found across the Peninsular Ranges batholith, this interpretation would lead to enormous uplift in the eastern Peninsular Ranges (~35–55 km), for which there is no evidence. In contrast, the postulated systematic *tilt of small crustal blocks of pluton scale* (Ortega-Rivera et al., 1997) about an axis subparallel to the trend of the batholith can largely account for some of the observed discordant paleomagnetic inclinations without appealing to *large-scale northward transport or tilt of the entire batholith.*

If monotonic eastward younging of ages across other plutons similar to that shown for the Sierra San Pedro Mártir pluton by Ortega-Rivera et al. (1997) can be found across the Peninsular Ranges batholith, and if systematic tilting of crustal blocks (of pluton-scale size) containing individual plutons can be shown to have occurred at a regional scale across the Peninsular Ranges batholith, this can be used to calculate the paleo-horizon at the time of pluton crystallization. An additional implication of this model is that both small-scale northward transport (~300 km) and local crustal size tilting may explain some of the anomalous paleomagnetic inclinations reported for the Peninsular Ranges batholith. On the other hand, if the plutons do not reveal the age pattern observed at Sierra San Pedro Mártir pluton but instead display uniform cooling ages only, these should later be tested for paleomagnetic direction. If they show slightly discordant paleomagnetic directions, then the peninsula has not been significantly displaced. As a result, to distinguish between the different competing plate tectonic models, the tectonothermal history of the Peninsular Ranges batholith and inferences derived from it have become a key question that needs to be answered. This can be tested by looking for tectonothermal history of the Peninsular Ranges batholith and for systematic tilting of other plutons across the Peninsular Ranges batholith.

Building on these results, this paper presents a geochronologic study of the Mesozoic Peninsular Ranges batholith and western México to: provide a geochronological framework for the evolution of the Peninsular Ranges batholith, extend the geochronological coverage to the region south of the Agua Blanca Fault (Figs. 3 and 4), assess the significance of the apparent eastward younging (Krummenacher et al., 1975; Silver et al., 1979) of the foci of granitoid emplacement of the Peninsular Ranges batholith, and provide additional constraints on the merits of the competing models suggested for the tectonic evolution of the Peninsular Ranges province and western México, by a compilation of new and existing age data (U/Pb zircon, $^{40}Ar/^{39}Ar$ step-heating, K/Ar, Rb/Sr, and apatite fission-track) from the Peninsular Ranges batholith and western México.

SAMPLING AND ANALYTICAL METHODS

One-hundred-nine fresh plutonic and volcanic rock samples were collected from two study areas (Figs. 3 and 4 [insets A and B]) for a detailed $^{40}Ar/^{39}Ar$ step-heating study along the Peninsular Ranges batholith of Alta and Baja California, from the 34th parallel in southern California north of the international border to the 28th parallel in Baja California. Forty-seven of the samples were obtained from the western Peninsular Ranges batholith and the Santiago Peak Volcanics Formation north of the international border (Figs. 3 and 4) and 55 samples were obtained from between the 30th and 32nd parallels across both the western and eastern Peninsular Ranges batholith, and from the Alisitos Formation (Figs. 3 and 4). Further to the south, three samples were collected from the eastern Peninsular Ranges batholith (Figs. 3 and 4) near San Luis Gonzaga north of the 29th parallel (G.C.-29 pluton), and four samples were taken from a region east of El Arco (Figs. 3 and 4) and north of the 28th parallel (G.C.-28 pluton). Fourteen cogenetic samples were selected for fission-track analyses. Sample localities are given in the Appendices[1].

$^{40}Ar/^{39}Ar$ Method

In total, five pyroxene, 144 hornblende, 162 biotite, 19 muscovite, 46 plagioclase, and four K-feldspar mineral separates were obtained from 109 rock samples. These mineral separates were extracted from crushed rock samples using a Frantz magnetic separator, heavy organic liquids, and hand-picking under a binocular microscope. The mineral separates, along with four whole-rock grain separates, replicates and flux-monitors (LP-6), and samples of known age (i.e., intralaboratory and interlaboratory monitors, respectively) were individually packaged in disc-shaped aluminum-foil pouches and stacked in aluminum irradiation cans (11.5 cm long and 2.0 cm in diameter). The positions of samples and monitors within the can were carefully measured.

The irradiation cans were exposed to fast neutrons in position 5C of the McMaster University Nuclear Reactor (Hamilton, Ontario) for 14.5 hours. Typically, 10 or more monitors and two to four sample replicates (used for inter-can comparison) for each can were used to determine the neutron-flux. J-values for individual samples were determined by a second-order polynomial interpolation. Seventeen separate irradiations were required to accommodate all of the samples. After irradiation, the mineral separates, replicates, and monitors were loaded into niobium crucibles. These crucibles were heated within a Lindberg furnace in a pure-silica tube (GE214) connected to a bakeable, ultra-high vacuum stainless-steel argon-extraction system operated online to a substantially modified A.E.I. MS-10 mass-spectrometer run in the static mode. For the isochron correlation analyses, step-heating blank runs were measured. The ^{40}Ar blank volume (less than 2% for hornblendes and 1% for biotites) varied between 0.4×10^{-10} (lowest temperature steps) and 0.9×10^{-10} cm³ STP (highest temperature steps). The ^{37}Ar and ^{39}Ar blanks were very

[1]GSA Data Repository Item 2003175, Appendices 1–4, is available on request from Documents Secretary, GSA, P.O. Box 9140, Boulder, CO 80301-9140, USA, editing@geosociety.org, at www.geosociety.org/pubs/ft2003.htm, or on the CD-ROM accompanying this volume.

small compared with the signals measured in the analysis of the samples. The ^{36}Ar blanks were at or below the limit of detection of the MS-10 mass spectrometer.

Measured mass-spectrometric ratios were extrapolated to zero time, normalized to the ^{40}Ar/^{36}Ar atmospheric ratio, and corrected for neutron-induced ^{40}Ar from potassium and ^{39}Ar and ^{36}Ar from calcium. Dates and errors were calculated using formulae given by Dalrymple et al. (1981) and the constants recommended by Steiger and Jäger (1977). All errors shown represent the analytical precision at 2σ and include the analytical uncertainties of the monitor analyses (J-uncertainties) but assume that the error for the age of the monitor is zero. A conservative estimate of 0.5% in the error of the J-value should be added for comparison with samples using a different monitor. The Queen's University dates are referenced to LP-6 biotite standard at 128.5 Ma (Roddick, 1983b). Mineral replicates (see Table 1 and Appendices 1 and 2 [see footnote one]) irradiated within the same can and in different cans were analyzed to monitor reproducibility. For information on size, blanks, and errors in J for the mineral and whole-rock separates, see Appendix 1 (see footnote 1).

U/Pb Zircon Dates

Available U/Pb zircon ages from the literature from samples across the northern Peninsular Ranges batholith of Alta and Baja California were compiled (Silver et al., 1979; Silver and Chappell, 1988; Walawender et al., 1990; Johnson et al., 1999a, 1999b; Kimbrough et al., 2001 [and references therein]; Schmidt, 2002) to choose between contrasting published interpretations of U/Pb zircon results and to constrain emplacement ages.

Fission-Track Method

Fourteen apatite mineral separates were obtained using standard heavy liquid and magnetic mineral separation techniques at Queen's University. The fission-track analyses were done at the Fission-track Research Laboratory of Dalhousie University. Mineral separation, grain mounting, polishing, etching, irradiation, and counting were all done by standard techniques using the external detector method described in Ravenhurst and Donelick (1992) and Ravenhurst et al. (1994).

RESULTS

Integrated and plateau dates for the new ^{40}Ar/^{39}Ar step-heating analyses on mineral separate and whole rocks are presented in Table 1. Samples are arranged by plutons and from north to south, and within plutons from west to east (this is important for the discussions); since there is no major difference between plateau and correlation dates, these last dates are not included. The complete data set is given in the Appendices 1 and 2 (see footnote 1). Representative age spectra from the Peninsular Ranges batholith are illustrated in Figure 6. The ^{40}Ar/^{39}Ar plateau ages were combined with U/Pb, K/Ar, ^{40}Ar/^{39}Ar, Rb/Sr, and fission

track dates available from the literature and are plotted in Plate 1 (on the CD-ROM accompanying this volume). K/Ar dates were recalculated by the constants recommended by Steiger and Jäger (1977), when necessary (see Appendices 3 and 4, see footnote 1).

In Table 2, new ^{40}Ar/^{39}Ar plateau dates (Table 1) from cogenetic hornblende and biotite mineral separates (where available) are listed with their cogenetic U/Pb zircon dates (Kimbrough et al., 2001) for samples collected across the northern Peninsular Ranges batholith of southern California, east of San Diego and from the literature. Apatite fission-track analyses from Ortega-Rivera (1997) were combined with the apatite fission-track analyses from samples collected from the Sierra San Pedro Mártir pluton (Ortega-Rivera et al., 1994a, 1994b, 1997) and their cogenetic ^{40}Ar/^{39}Ar hornblende and biotite data. Dates are summarized in Table 3. ^{40}Ar/^{39}Ar and apatite fission-track data are shown with respect to their sample localities in Figures 7 and 8. (Samples for U/Pb and fission track are also arranged from north to south and from west to east.)

DISCUSSIONS

U/Pb Zircon Dates

U/Pb zircon dates (Table 2) from cogenetic ^{40}Ar/^{39}Ar hornblende and biotite dates taken across the northern Peninsular Ranges batholith of Alta and Baja California (Ortega-Rivera, 1997; Kimbrough et al., 2001) are considered to represent emplacement ages, and almost all but two are younger than their cogenetic hornblende ^{40}Ar/^{39}Ar dates (due to possible excess radiogenic argon [^{40}Ar$_{rad}$]; see ^{40}Ar/^{39}Ar dates section). The rest of them are almost concordant, or normal discordant, defining a slightly younger age range than their cogenetic U/Pb zircons and indicating rapid cooling immediately after emplacement, from west to east. U/Pb zircon dates from the western intrusions range from ca. 140 to ca. 105 Ma; in contrast, the eastern plutons were emplaced between 105 to 80 Ma, but most of these intrusions were emplaced in a surprisingly brief interval (99–92 Ma) during which ~47% of the total surface exposure of the plutonic rocks of the entire batholith were composed. These last rocks belong to the La Posta suite. The Santiago Peak Volcanics range in age from 130 to 116 Ma based on U/Pb zircon dates, while the plutons for the western Peninsular Ranges batholith range between 120 and 103 Ma. These data suggest that Early Cretaceous arc volcanism preceded the inception of voluminous intrusive activity in the western part of the batholith by ca. 10 Ma. However, there is an overlap in ages. Meanwhile, the Alisitos Volcanics has an U/Pb zircon age ca. 127 Ma (Krummenacher et al., 1975).

These U/Pb zircon dates were combined with previous U/Pb zircon data from Silver et al. (1979), Silver and Chappell (1988), Walawender et al. (1990), Johnson et al. (1999a, 1999b), Tate et al. (1999), Schmidt (2002), Kimbrough et al. (2001 and references therein), and Böhnel et al. (2002) in an attempt to distinguish between the contrasting interpretations of

TABLE 1. ^{40}AR/^{39}AR INTEGRATED AND PLATEAU DATES FOR SAMPLES COLLECTED ACROSS THE PENINSULAR RANGES BATHOLITH OF ALTA AND BAJA CALIFORNIA, FROM NORTH TO SOUTH (SEE FIGURE 4 AND PLATE 1)

WESTERN PENINSULAR RANGES BATHOLITH									
Sample no.	Sample name	Mineral	Laboratory run	GSA DR pages	Integrated Date (Ma)	2 σ error ±	Plateau date (Ma)	2σ error ±	Vol. ^{39}Ar (%)
Santa Anas Area									
101	BS102	hb	AOR-849	1	209.5	97.0	121.5	5.2	91.0
102	CS14	hb	AOR-875	2	124.9	3.9	124.9	3.9	100.0
103	CS31/35	hb	AOR-850	4	151.9	4.7	154.4	3.8	90.9
104	CS119	hb	AOR-867	3	118.0	7.4	118.0	7.4	100.0
Santiago Peak Volcanics									
105	SP770	hb	AOR-855	5	121.2	4.7	121.2	4.7	100.0
106	SP18A	hb	AOR-859	6	122.9	5.0	122.9	5.0	100.0
107	SP215	bt	AOR-1024	7	118.5	0.8	119.6	0.8	94.9
108	RSF	hb	AOR-908	8	132.0	13.6	132.0	13.6	100.0
Riverside									
110	RED MOUNTAIN	hb	AOR-516	9	116.5	14.8	114.9	8.7	97.2
110	RED MOUNTAIN	hb	AOR-741	10	141.1	54.7	114.5	9.1	97.3
110	RED MOUNTAIN	hb	AOR-1522	11	114.3	3.7	114.3	3.7	100.0
111	RICE CANYON	hb	AOR-522	12	155.6	105.9	118.6	72.3	68.8
112	MERRIAM MT	hb	AOR-515	13	112.1	72.9	112.1	72.9	100.0
109	PBV-1	hb	AOR-858	14	193.8	6.5	193.8	6.5	100.0
139	THOMAS-MT	hb	AOR-518	15	106.0	3.8	106.0	3.8	100.0
139	THOMAS-MT	hb	AOR-690	16	112.2	12.9	106.9	5.3	94.2
139	THOMAS-MT	hb	AOR-1505	17	106.3	8.8	106.3	8.8	100.0
140	NC-2	pg	AOR-249	18	106.1	3.2	85.6	0.9	86.7
Woodson Mountain pluton									
113	WM-1	hb	AOR-889	19	109.6	1.3	111.2	1.3	72.6
113	WM-1	bt	AOR-1040	20	107.0	0.6	107.4	0.6	97.4
114	WOODSON(W)	hb	AOR-1090	21	111.5	2.8	111.5	2.8	100.0
114	WOODSON(W)	bt	AOR-1083	22	107.1	1.2	107.5	1.2	97.5
114	WOODSON(W)	bt	AOR-1303	23	107.4	8.6	107.3	1.5	90.6
115	RAMONA	*bt	AOR-3	24	98.2	1.1	98.7	1.1	91.5
115	RAMONA	bt	AOR-233	25	100.1	0.3	100.2	0.3	94.2
115	RAMONA	bt	AOR-1242	26	99.8	0.9	100.1	0.7	97.3
116	HC-1	hb	AOR-914	27	101.2	2.9	101.2	2.9	100.0
116	HC-1	bt	AOR-1195	28	99.6	0.9	99.7	0.9	97.1
La Mesa									
117	MGQ	hb	AOR-874	29	109.2	1.3	109.2	1.3	100.0
117	MGQ	bt	AOR-925	30	103.1	0.7	104.8	0.7	93.2
El Cajón									
118	WILLOW-ROAD	bt	AOR-1025	31	102.5	0.9	105.9	1.0	85.4
119	WILD-CAT	hb	AOR-866	32	107.1	2.0	107.1	2.0	100.0
119	WILD-CAT	bt	AOR-1005	33	102.9	0.7	104.8	0.8	89.4
120	BVR-1A	wr	AOR-995	34	94.3	4.3	99.1	4.4	83.4
121	EL-CAPITAN	hb	AOR-904	35	106.9	1.6	106.9	1.6	100.0
121	EL-CAPITAN	bt	AOR-1207	36	104.2	1.2	104.7	1.2	92.8
Alpine Pluton									
122	LOVELAND-RES	hb	AOR-895	37	100.9	1.7	100.9	1.7	100.0
122	LOVELAND-RES	bt	AOR-1185	38	98.6	1.9	98.2	1.9	99.6
122	LOVELAND-RES	bt	AOR-1492	39	98.5	1.5	98.5	1.5	100.0
123	ALPINE	hb	AOR-408	40	114.3	2.2	110.7	2.8	67.8
123	ALPINE	bt	AOR-699	41	105.1	0.4	105.3	0.4	92.7
123	ALPINE	pg	AOR-662	42	107.5	2.0	103.8	1.8	98.5
124	ALPINE-HT-5	bt	AOR-1062	43	105.1	1.2	105.6	1.2	95.4
125	VICTORIA	hb	AOR-115	44	106.0	1.3	106.0	1.3	100.0
125	VICTORIA	bt	AOR-117	45	102.6	1.0	103.3	1.0	96.1
126	LBT-1	hb	AOR-909	46	107.2	2.6	107.2	2.6	100.0
126	LBT-1	bt	AOR-1203	47	104.0	1.2	104.3	1.2	96.8

(continued)

| | | | | | WESTERN PENINSULAR RANGES BATHOLITH | | | | |
Sample no.	Sample name	Mineral	Laboratory run	GSA DR pages	Integrated Date (Ma)	2 σ error ±	Plateau date (Ma)	2σ error ±	Vol. ^{39}Ar (%)
Poser									
127	POSER DYKE	hb	AOR-514	48	108.4	16.0	105.9	5.9	59.2
127	POSER DYKE	hb	AOR-673	49	127.7	7.1	105.2	3.0	84.1
127	POSER DYKE	hb	AOR-523	50	157.5	11.9	138.5	3.1	96.4
128	POSER-HB-PEG.	hb	AOR-684	51	140.0	21.5	112.5	13.1	97.2
128	POSER-HB-PEG.	hb	AOR-1497	52	439.1	388.2	61.4	22.1	87.8
128	POSER-HB-PEG.	hb	AOR-1511	53	87.7	37.8	118.7	4.3	95.3
128	POSER-HB-PEG.	pg	AOR-1496	54	177.0	108.7	146.4	106.3	55.3
128	POSER-HB-PEG.	pg	AOR-1516	55	93.4	153.4	131.9	16.6	48.9
129	POSER PX BRING	px	AOR-1507	56	165.6	20.1	165.6	20.1	100.0
129	POSER PX BRING	hb	AOR-510	57	150.5	39.4	126.5	12.1	73.3
129	POSER PX BRING	hb	AOR-704	58	123.5	157.7	118.6	14.7	87.6
129	POSER PX BRING	pg	AOR-1514	59	132.1	20.2	132.1	20.2	100.0
Lake Morena									
130	CHIQUITO-PEAK	hb	AOR-881	60	106.9	1.0	107.3	1.0	91.0
130	CHIQUITO-PEAK	bt	AOR-880	61	103.7	9.4	100.0	2.1	94.5
131	LBT-2	hb	AOR-899	62	108.4	1.1	108.4	1.1	100.0
131	LBT-2	bt	AOR-1199	63	107.5	1.7	107.5	1.7	100.0
132	LOS PINOS SUMMIT	hb	AOR-517	64	146.6	57.7	113.6	9.4	98.0
132	LOS PINOS SUMMIT	hb	AOR-740	65	380.2	200.7	113.4	14.4	93.5
132	LOS PINOS SUMMIT	hb	AOR-1499	66	129.9	53.6	129.9	53.6	100.0
133	CORTE-MADERA	hb	AOR-886	67	95.5	1.0	95.5	1.0	100.0
133	CORTE-MADERA	bt	AOR-971	68	94.8	0.7	95.1	0.7	93.6
134	LK-MORENA	hb	AOR-869	69	96.5	2.6	96.5	2.6	100.0
Descanso									
135	CHERRY-F.C.	hb	AOR-509	70	662.7	793.9	216.0	44.9	57.7
135	CHERRY-F.C.	hb	AOR-682	71	550.1	101.1	301.5	73.2	80.9
136	PINE-VALLEY	bt	AOR-981	72	90.1	0.8	93.0	0.9	74.1
137	LPG-1a	hb	AOR-1506	73	121.9	57.5	105.9	31.4	98.3
137	LPG-1a	pg	AOR-1515	74	28.6	25.7	28.6	25.7	100.0
138	LPG-1b	px	AOR-1521	75	105.7	41.9	105.7	41.9	100.0
138	LPG-1b	pg	AOR-1520	76	129.1	356.2	129.1	356.2	100.0
Long Potrero pluton									
141	LP-R60B	hb	AOR-976	77	113.7	29.5	104.3	37.7	74.1
141	LP-R60B	bt	AOR-1016	78	101.1	0.8	101.1	0.8	100.0
142	LONG-POTRERO-O	hb	AOR-890	79	101.6	1.6	101.6	1.6	100.0
142	LONG-POTRERO-O	bt	AOR-1000	80	99.7	0.8	99.9	0.8	99.5
143	LP-R27A	bt	AOR-1191	81	97.6	0.8	98.3	0.8	82.6
144	LP-R15	hb	AOR-903	82	107.7	1.8	102.4	1.8	84.8
144	LP-R15	bt	AOR-1010	83	99.3	1.0	101.1	1.0	78.0
145	LP-R837	hb	AOR-915	84	99.8	5.6	100.1	5.8	96.9
145	LP-R837	bt	AOR-1019	85	96.3	1.0	98.1	0.9	88.8
146	LP-R236	bt	AOR-961	86	94.9	1.0	95.6	1.1	81.6
147	LP-R497B	hb	AOR-891	87	99.1	1.7	99.1	1.7	100.0
148	LP-R591	hb	AOR-910	88	100.8	2.7	110.8	2.7	100.0
148	LP-R591	bt	AOR-966	89	93.8	0.9	94.1	0.9	99.3
149	LONG-POTRERO-I	bt	AOR-920	90	90.5	0.6	92.8	0.7	71.0

(continued)

TABLE 1. ^{40}AR/^{39}AR INTEGRATED AND PLATEAU DATES FOR SAMPLES COLLECTED
ACROSS THE PENINSULAR RANGES BATHOLITH OF ALTA AND BAJA CALIFORNIA,
FROM NORTH TO SOUTH (SEE FIGURE 4 AND PLATE 1) *(continued)*

| | | | | | WESTERN PENINSULAR RANGES BATHOLITH | | | | |
Sample no.	Sample name	Mineral	Laboratory run	GSA DR pages	Integrated Date (Ma)	2 σ error ±	Plateau date (Ma)	2σ error ±	Vol. ^{39}Ar (%)
El Pinal Pluton									
100	PINAL	hb	AOR-130	91	100.9	1.6	100.9	1.5	95.4
100	PINAL	hb	AOR-275	92	99.6	1.5	99.6	1.5	100.0
100	PINAL	bt	AOR-144	93	96.3	1.1	96.7	1.2	95.6
100	PINAL	bt	AOR-272	94	96.3	1.2	96.9	1.1	98.3
San Antonio									
99	SAN ANTONIO	bt	AOR-622	95	79.2	0.6	96.5	1.2	13.6
99	SAN ANTONIO	ks	AOR-678	96	102.6	0.7	102.9	0.7	94.8
Alisitos Volcanics									
97	LA BUFADORA	wr	AOR-590	97	94.6	1.3	96.7	1.3	89.4
97	LA BUFADORA	wr	AOR-583	98	95.9	1.3	97.0	1.3	97.3
97	LA BUFADORA	pg	AOR-595	99	88.9	2.7	87.2	4.7	46.4
97	LA BUFADORA	pg	AOR-600	100	80.6	3.2	86.7	3.3	72.4
98	EL VIGIA	wr	AOR-540	101	105.0	0.9	106.0	0.9	96.6
98	EL VIGIA	pg	AOR-533	102	103.3	1.0	104.0	1.0	97.4
98	EL VIGIA	pg	AOR-535	103	103.1	2.8	104.6	3.4	78.3
Los Encinos and Valle Pedregoso plutons									
93	LE-1	hb	AOR-1038	104	107.3	5.1	107.3	5.1	100.0
93	LE-1	bt	AOR-1075	105	100.9	1.1	101.5	1.1	82.6
93	LE-1	pg	AOR-1079	106	41.2	28.8	74.3	15.4	90.5
94	LE-2	hb	AOR-1088	107	101.8	3.5	101.8	3.5	100.0
94	LE-2	bt	AOR-1073	108	97.6	1.1	100.7	1.1	83.2
95	LE-1-7-88-3	*bt	AOR-5	109	99.2	1.6	100.3	1.0	70.3
95	LE-1-7-88-3	bt	AOR-230	110	97.3	7.8	99.0	1.6	64.5
95	LE-1-7-88-3	bt	AOR-237	111	97.2	0.5	99.1	0.5	81.9
96	VP-8	hb	AOR-1051	112	101.8	2.5	101.8	2.5	100.0
96	VP-8	bt	AOR-1058	113	98.1	1.2	100.3	1.2	87.0
Eréndira Area									
83	E-32	hb	AOR-383	114	92.3	2.4	98.6	3.4	56.0
84	E-31	hb	AOR-392	115	111.5	3.2	111.5	3.2	100.0
84	E-31	pg	AOR-396	116	113.1	8.2	111.6	3.3	80.3
84	E-31	pg	AOR-403	117	111.0	3.4	111.0	3.4	100.0
85	E-29	hb	AOR-400	118	106.5	1.1	105.6	1.2	86.4
85	E-29	pg	AOR-404	119	106.4	1.4	106.4	1.4	100.0
85	E-29	pg	AOR-1288	120	113.4	23.0	105.0	7.2	89.4
86	E-M-26	hb	AOR-379	121	99.9	35.7	103.3	13.1	93.5
86	E-M-26	pg	AOR-386	122	102.0	3.6	100.1	4.1	86.7
86	E-M-26	pg	AOR-399	123	102.2	3.3	102.2	3.3	100.0
87	E-26	hb	AOR-382	124	102.0	10.4	105.8	9.8	81.9
87	E-26	pg	AOR-387	125	78.5	2.6	85.7	4.1	57.1
87	E-26	pg	AOR-1307	126	104.6	4.2	101.7	4.4	87.7
Ejido Uruapan Area									
88	EUBC-21	*hb	AOR-15	127	99.9	3.1	100.2	3.0	99.4
88	EUBC-21	hb	AOR-1047	128	110.8	4.2	110.8	4.2	100.0
88	EUBC-21	*bt	AOR-21	129	104.5	1.3	105.2	1.1	98.0
88	EUBC-21	bt	AOR-1285	130	111.1	2.0	111.1	2.0	100.0
89	EUBC-9	*bt	AOR-2	131	103.1	1.0	103.8	0.8	87.8
89	EUBC-9	bt	AOR-1300	132	110.3	1.5	110.3	1.5	100.0
90	EUBC-1	hb	AOR-1045	133	113.4	2.4	113.4	2.4	100.0
90	EUBC-1	bt	AOR-1292	134	115.3	2.9	112.3	2.4	67.5
90	EUBC-1	pg	AOR-1314	135	111.9	2.9	113.1	3.9	67.0
90	EUBC-1	pg	AOR-1316	136	114.4	3.8	112.5	1.1	91.4
91	EUBC-2	*hb	AOR-18	137	94.0	5.4	99.7	4.5	90.1
91	EUBC-2	hb	AOR-1048	138	101.0	3.2	101.0	3.2	100.0
92	EUBC-22	*hb	AOR-24	139	102.7	6.1	100.0	4.6	97.7
92	EUBC-22	hb	AOR-1052	140	102.2	2.5	102.2	2.5	100.0

(continued)

TABLE 1. ⁴⁰AR/³⁹AR INTEGRATED AND PLATEAU DATES FOR SAMPLES COLLECTED
ACROSS THE PENINSULAR RANGES BATHOLITH OF ALTA AND BAJA CALIFORNIA,
FROM NORTH TO SOUTH (SEE FIGURE 4 AND PLATE 1) *(continued)*

Sample no.	Sample name	Mineral	Laboratory run	GSA DR pages	Integrated Date (Ma)	2 σ error ±	Plateau date (Ma)	2σ error ±	Vol. ³⁹Ar (%)
WESTERN PENINSULAR RANGES BATHOLITH									
San José Pluton									
81	T.23.SJ	hb	AOR-916	141	103.7	5.3	103.7	5.3	100.0
81	T.23.SJ	bt	AOR-1069	142	98.1	1.8	100.4	1.8	92.4
81	T.23.SJ	pg	AOR-988	143	69.8	16.0	72.2	4.8	65.2
81	T.23.SJ	pg	AOR-989	144	66.8	18.1	93.2	18.0	69.9
82	SJ-2	hb	AOR-930	145	101.3	2.5	101.3	2.5	100.0
82	SJ-2	bt	AOR-937	146	99.0	0.9	100.9	0.9	94.7
82	SJ-2	pg	AOR-846	148	132.3	17.3	84.4	6.0	84.1
82	SJ-2	pg	AOR-1020	147	117.7	33.1	94.5	4.7	88.2
Las Palmas									
77	LP-1-90	px	AOR-1091	149	112.3	2.6	115.8	2.3	81.8
77	LP-1-90	hb	AOR-1032	150	95.0	9.7	101.2	7.4	87.4
77	LP-1-90	pg	AOR-1235	151	130.9	7.7	95.6	4.6	79.1
78	LP-2-90	bt	AOR-255	152	113.9	1.1	114.5	1.2	39.9
78	LP-2-90	pg	AOR-1255	153	108.2	2.6	108.2	2.6	100.0
78	LP-2-90	pg	AOR-1256	154	108.7	2.0	108.3	2.1	90.7
79	LP-3-90	hb	AOR-1041	155	113.4	3.6	112.7	3.6	98.4
79	LP-3-90	bt	AOR-1251	156	101.0	7.0	116.6	7.2	53.5
79	LP-3-90	pg	AOR-1266	157	103.6	7.0	103.6	7.0	100.0
79	LP-3-90	pg	AOR-1276	158	111.2	1.8	111.2	1.8	100.0
80	LP-4-90	bt	AOR-260	159	113.5	1.1	114.7	1.1	71.1
80	LP-4-90	ks	AOR-539	160	106.0	1.5	104.3	1.4	98.8
El Ciprés									
72	EL-C-1-90	hb	AOR-1089	161	112.5	2.5	117.5	2.5	82.2
72	EL-C-1-90	pg	AOR-1271	162	106.0	2.2	106.0	2.2	100.0
72	EL-C-1-90	pg	AOR-1277	163	106.0	3.8	106.0	3.8	100.0
73	EL-C-1.2-90	hb	AOR-1042	164	128.2	8.8	119.0	6.1	63.7
73	EL-C-1.2-90	pg	AOR-1267	165	107.8	18.6	108.3	2.3	89.9
73	EL-C-1.2-90	pg	AOR-1272	166	109.1	1.8	110.4	1.9	85.1
74	EL-C-2-90	hb	AOR-1033	167	102.4	15.0	109.7	12.3	92.9
74	EL-C-2-90	pg	AOR-1247	169	136.1	13.4	115.5	4.1	77.2
74	EL-C-2-90	pg	AOR-1259	168	113.0	3.7	115.9	3.8	89.5
75	EL-C-3-90	pg	AOR-398	171	110.0	1.1	106.8	1.0	80.2
75	EL-C-3-90	pg	AOR-1296	170	111.3	2.0	109.2	1.2	83.4
76	EL-C-4-90	bt	AOR-246	172	113.1	1.2	115.7	1.3	70.1
76	EL-C-4-90	pg	AOR-245	173	105.7	1.0	106.0	1.0	92.1
76	EL-C-4-90	pg	AOR-1278	174	113.4	3.1	110.7	2.8	88.7
76	EL-C-4-90	pg	AOR-251	175	107.2	23.9	104.7	10.3	79.3
76	EL-C-4-90	pg	AOR-1282	176	107.9	2.1	107.5	1.5	90.3
76	EL-C-4-90	ks	AOR-252	177	110.8	1.3	109.9	1.3	90.4

(continued)

the U/Pb zircon results. The plutons of the Peninsular Ranges batholith were emplaced over a 60 m.y. interval, from 140 Ma to 80 Ma. Although the U/Pb zircon data are sparse, a west-east profile for the U/Pb zircon dates (Fig. 9) and a corresponding crude U/Pb chrontour (i.e., age contour) map were constructed (Fig. 10).

Fault offsets across the Agua Blanca fault and the other major faults for this regional study were ignored in Figure 10, as possible offsets of the chrontours by the faults are relatively small compared with the data density, although at a local scale

major displacement on the Agua Blanca fault should be considered (Gastil, 1975; Ortega-Rivera, 1988; Wetmore et al., 2002). Figures 9 and 10 suggest that the U/Pb zircon dates obtained for the northern Peninsular Ranges batholith young steadily from west to east, lacking the age "step" at the boundary of the western and eastern Peninsular Ranges batholith, as originally inferred by Silver and Chappell (1979, 1988). Even though two or three older plutons ca. 134 Ma have been reported (found around the east-west boundary of the Peninsular Ranges batholith in southern California and Baja California, or where the

EASTERN PENINSULAR RANGES BATHOLITH

Sample no.	Sample name	Mineral	Laboratory run	GSA DR pages	Integrated Date (Ma)	2 σ error ±	Plateau date (Ma)	2σ error ±	Vol. ^{39}Ar (%)
El Topo pluton									
56	EL-TOPO-CONT	bt	AOR-121	178	84.5	1.2	86.7	1.3	83.1
57	EL-TOPO-5	bt	AOR-146	179	84.4	1.1	86.2	1.1	85.4
58	EL-TOPO-H-88	hb	AOR-266	180	89.1	1.0	89.4	1.0	98.9
58	EL-TOPO-H-88	hb	AOR-1053	181	91.2	2.0	90.9	1.4	95.5
58	EL-TOPO-H-88	bt	AOR-264	182	82.5	0.7	83.3	0.7	86.9
58	EL-TOPO-H-88	bt	AOR-1162	183	85.7	1.3	85.7	1.3	100.0
59	EL-TOPO-7-30-1	bt	AOR-286	184	82.0	0.6	82.6	0.6	94.6
60	EL-TOPO-7-30-2	bt	AOR-285	185	80.6	1.3	81.0	1.3	98.6
Laguna Juárez pluton									
61	LJ-7-31-2	*hb	AOR-9	186	91.3	3.0	91.5	2.7	98.3
61	LJ-7-31-2	hb	AOR-26	187	93.5	1.2	93.5	1.2	100.0
61	LJ-7-31-2	hb	AOR-120	188	93.8	1.5	93.8	1.5	100.0
61	LJ-7-31-2	hb	AOR-239	189	93.1	16.4	93.3	0.9	99.5
61	LJ-7-31-2	hb	AOR-493	190	93.7	1.7	93.7	1.7	*100.0
61	LJ-7-31-2	hb	AOR-639	191	93.5	1.4	93.5	1.4	100.0
61	LJ-7-31-2	hb	AOR-1037	192	92.3	1.0	92.8	0.9	91.4
61	LJ-7-31-2	hb	AOR-1086	193	92.5	0.9	92.5	2.3	100.0
61	LJ-7-31-2	hb	AOR-1179	194	93.9	1.1	93.9	1.1	100.0
61	LJ-7-31-2	*hb	AOR-494	195	93.6	1.6	93.6	1.6	*100.0
61	LJ-7-31-2	*hb	AOR-495	196	94.7	1.3	94.7	1.3	*100.0
61	LJ-7-31-2	bt	AOR-407	197	88.1	1.0	88.6	1.0	89.7
61	LJ-7-31-2	bt	AOR-240	198	88.5	0.4	88.9	0.4	96.8
61	LJ-7-31-2	bt	AOR-114	199	89.6	1.0	89.9	1.0	99.5
61	LJ-7-31-2	bt	AOR-668	200	88.3	0.9	88.8	0.9	95.5
61	LJ-7-31-2	bt	AOR-1262	201	89.9	1.4	89.9	1.4	100.0
62	LJ-7-31-4	bt	AOR-172	202	85.9	1.1	87.9	1.1	88.4
63	LJ-M2	ms	AOR-257	203	87.6	0.8	86.7	0.8	98.0
63	LJ-M2	bt	AOR-262	204	83.8	0.7	86.1	0.8	78.9
64	LJ-7-31-9	ms	AOR-730	205	89.3	1.1	89.1	1.1	94.5
64	LJ-7-31-9	bt	AOR-310	206	76.5	1.6	80.3	1.7	90.0
EN-91 (W-E) Regional Transect									
40	EN-1-91	hb	AOR-161	207	106.0	1.7	106.0	1.2	100.0
40	EN-1-91	bt	AOR-28	208	104.1	1.1	106.1	1.2	89.2
40	EN-1-91	bt	AOR-279	209	104.8	0.6	106.2	0.6	62.2
41	EN-2-91	hb	AOR-64	210	105.8	3.6	105.8	3.6	100.0
41	EN-2-91	hb	AOR-153	211	97.7	1.7	105.4	1.8	86.0
41	EN-2-91	bt	AOR-303	212	106.7	1.6	109.0	2.3	46.2
42	EN-3-91	px	AOR-890	213	103.0	3.4	103.0	3.4	100.0
42	EN-3-91	hb	AOR-150	214	103.2	2.7	103.2	2.7	100.0
42	EN-3-91	hb	AOR-657	215	147.5	55.1	103.0	2.1	95.0
42	EN-3-91	bt	AOR-94	216	99.3	0.7	100.0	0.7	94.2
42	EN-3-91	bt	AOR-672	217	100.9	0.8	102.4	0.8	92.4
42	EN-3-91	pg	AOR-647	218	105.6	6.5	94.7	5.4	97.4
42	EN-3-91	pg	AOR-1237	219	88.3	10.0	96.1	2.6	91.2
43	EN-4-91	hb	AOR-410	220	103.8	2.3	103.7	1.8	97.9
43	EN-4-91	pg	AOR-1236	221	65.1	18.2	65.1	18.2	100.0
44	EN-5-91	bt	AOR-718	222	99.5	1.1	101.7	1.2	90.2
45	EN-6-91	px	AOR-652	223	105.7	7.2	103.0	3.2	85.7
45	EN-6-91	hb	AOR-651	224	120.4	16.8	101.2	4.5	98.9
46	EN-7-91-SJ	hb	AOR-148	225	100.6	1.3	100.1	1.3	99.8
46	EN-7-91-SJ	bt	AOR-174	226	96.2	1.2	97.0	1.2	99.3
47	EN-8-91-SJ	hb	AOR-159	227	98.9	1.5	97.9	1.4	99.1
47	EN-8-91-SJ	bt	AOR-152	228	94.9	1.1	95.3	1.1	94.6
48	EN-9-91-SJ	hb	AOR-156	229	98.9	1.4	99.4	1.4	87.9
48	EN-9-91-SJ	bt	AOR-157	230	95.2	1.2	96.4	1.2	70.1
49	EN-10-91-SJ	hb	AOR-167	231	96.6	1.2	97.6	1.3	85.3
49	EN-10-91-SJ	hb	AOR-1046	232	98.2	1.9	98.2	1.9	100.0
49	EN-10-91-SJ	bt	AOR-223	233	93.3	0.3	94.9	0.3	85.0
49	EN-10-91-SJ	bt	AOR-1101	234	93.3	1.6	94.5	1.7	95.8

(continued)

EASTERN PENINSULAR RANGES BATHOLITH

Sample no.	Sample name	Mineral	Laboratory run	GSA DR pages	Integrated Date (Ma)	2 σ error ±	Plateau date (Ma)	2σ error ±	Vol. ^{39}Ar (%)
EN-91 (W-E) Regional Transect *(continued)*									
50	EN-11-91-PSM	ms	AOR-61	235	82.5	0.5	82.6	0.5	99.4
50	EN-11-91-PSM	ms	AOR-242	236	82.2	0.4	82.3	0.4	99.6
50	EN-11-91-PSM	ms	AOR-729	237	83.6	1.1	83.3	1.0	96.1
50	EN-11-91-PSM	bt	AOR-294	238	85.1	1.2	88.4	2.3	34.2
50	EN-11-91-PSM	bt	AOR-663	239	79.2	1.2	82.5	1.8	47.9
51	EN-12-91-PSM	hb	AOR-29	240	114.0	5.9	96.6	3.2	70.4
51	EN-12-91-PSM	hb	AOR-241	241	87.5	1.2	89.1	1.0	83.3
51	EN-12-91-PSM	hb	AOR-405	242	90.3	1.6	90.3	1.6	100.0
51	EN-12-91-PSM	bt	AOR-1231	243	85.0	1.6	87.8	2.1	68.9
52	EN-13-91-PSM	hb	AOR-87	244	88.5	0.9	88.8	0.9	88.2
52	EN-13-91-PSM	hb	AOR-1076	245	88.4	3.8	88.4	3.8	100.0
52	EN-13-91-PSM	bt	AOR-154	246	83.7	0.7	84.3	0.7	96.7
52	EN-13-91-PSM	bt	AOR-1227	247	84.1	0.4	84.7	0.4	84.4
53	EN-14-91-SF	hb	AOR-166	248	84.4	1.4	84.4	1.4	100.0
53	EN-14-91-SF	bt	AOR-58	249	82.4	0.5	82.9	0.5	97.7
53	EN-14-91-SF	bt	AOR-281	250	81.6	0.5	82.1	0.5	98.0
54	EN-15-91-SF	hb	AOR-409	251	88.8	1.6	88.8	1.6	100.0
54	EN-15-91-SF	bt	AOR-33	252	83.4	0.9	84.4	0.8	95.9
55	EL FARO	hb	AOR-617	253	79.8	1.7	82.8	1.8	80.7
55	EL FARO	bt	AOR-612	254	79.7	0.4	80.4	0.4	82.7
55	EL FARO	ks	AOR-606	255	73.9	0.4	73.8	0.4	99.4
G.C.-29 pluton									
65	CATAVIÑA	ms	AOR-62	256	88.9	0.6	89.4	0.6	98.1
65	CATAVIÑA	bt	AOR-55	257	88.3	0.5	89.8	0.5	79.5
66	G.C.-7-29-88-5	hb	AOR-11	258	92.9	1.7	93.2	1.6	93.5
66	G.C.-7-29-88-5	hb	AOR-194	259	93.4	0.8	93.4	0.8	100.0
66	G.C.-7-29-88-5	hb	AOR-1036	260	97.5	15.3	93.4	1.2	92.4
66	G.C.-7-29-88-5	bt	AOR-132	261	93.2	1.2	93.5	1.2	98.4
66	G.C.-7-29-88-5	bt	AOR-1241	262	94.3	0.9	94.9	1.0	87.7
67	G.C.-7-29-88-4	hb	AOR-306	263	92.4	1.5	92.4	1.5	100.0
67	G.C.-7-29-88-4	bt	AOR-719	264	91.9	1.0	92.3	1.0	79.2
G.C.-28 pluton									
68	G.C.-7-28-8-#9	*hb	AOR-14	265	91.6	2.0	92.4	1.6	93.1
68	G.C.-7-28-8-#9	hb	AOR-186	266	91.9	2.9	92.6	3.0	94.6
68	G.C.-7-28-8-#9	bt	AOR-84	267	87.9	0.6	89.1	0.6	92.4
69	G.C.-7-28-88-5	hb	AOR-113	268	92.5	1.5	92.5	1.5	100.0
69	G.C.-7-28-88-5	bt	AOR-205	269	90.0	0.8	90.0	0.8	98.5
69	G.C.-7-28-88-5	bt	AOR-224	270	89.7	0.8	89.9	0.8	98.6
69	G.C.-7-28-88-5	bt	AOR-225	271	89.8	0.9	90.0	0.9	96.7
69	G.C.-7-28-88-5	bt	AOR-1182	272	90.2	1.0	90.4	1.0	99.4
69	G.C.-7-28-88-5	bt	AOR-1215	273	90.8	0.8	90.8	0.8	100.0
70	G.C.-7-28-88-3	bt	AOR-48	274	90.7	0.7	91.3	0.7	96.2
70	G.C.-7-28-88-3	bt	AOR-1223	275	90.2	1.4	91.2	1.6	90.3
71	G.C.7-28-88-4	hb	AOR-391	276	91.5	2.5	91.5	2.5	100.0
71	G.C.7-28-88-4	bt	AOR-1211	277	90.0	1.1	90.0	1.1	96.0

Note: Plate 1 is on the CD-ROM accompanying this volume. Volume ^{39}Ar (%) refers to the total volume of ^{39}Ar used in the calculation of each respective plateau date. The integrated ages are shown for comparison with the K-Ar dates available from the literature (see text). Since the plateau dates are almost perfect and correspond to the correlation-isochrone ages, these are not shown. For sample locality, spectra, isochrones and complete data set, see Appendices 1 and 2 in the GSA Data Repository (GSA DR) item #2003175 (see footnote 1). The laboratory run is given to distinguish, between sample runs from the same site (sample number or name), different mineral, and replicates. There is no isochrone plot or correlation date for minerals marked with an asterisk (*); hb—hornblende, bt—biotite, ms—muscovite, g—plagioclase, wr—whole rock, ks—K-feldspar.

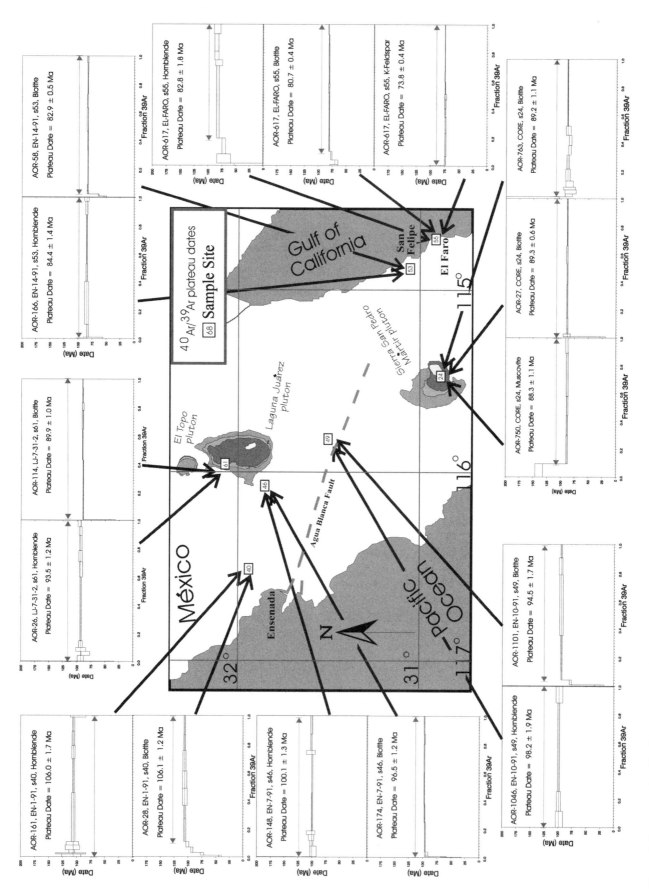

Figure 6. Representative $^{40}Ar/^{39}Ar$ age spectra for Peninsular Ranges batholith. (Number inside squares represents sample number; see Appendices [see footnote 1] for entire data set).

TABLE 2. COMPILATION OF NEW $^{40}AR/^{39}AR$ HORNBLENDE AND BIOTITE PLATEAU DATES WITH COGENETIC AVAILABLE U/PB ZIRCON DATES FROM SAMPLES ACROSS THE PENINSULAR RANGES BATHOLITH AND RELATED VOLCANICS OF SOUTHERN CALIFORNIA, EAST OF SAN DIEGO

Sample locality	Sample no.	Sample name	Hb $^{40}Ar/^{39}Ar$ plateau date (Ma)	Bt $^{40}Ar/^{39}Ar$ plateau date (Ma)	Zircon U-Pb date (Ma)
Santa Anas	101	BS	121.5 ± 5.2	N/A	124
			N/A	N/A	123
Volcanics	102	CS	124.9 ± 3.9	N/A	127
	104		118.0 ± 7.4	N/A	N/A
Santiago Peak Volcanics	105	SP	121.2 ± 4.7	N/A	120
	106		122.9 ± 5.0	N/A	130
	107		N/A	119.6 ± 0.8	N/A
	108		123.8 ± 6.5	N/A	N/A
RSF	109	RSF	132.0 ± 13.6	N/A	124
Woodson	113	WM-1	111.2 ± 1.3	107.4 ± 0.6	118
	114		111.5 ± 2.8	107.5 ± 1.2	N/A
			N/A	107.3 ± 1.5	N/A
	116	HC-1	N/A	99.7 ± 0.9	101
La Mesa	117	MGQ	109.2 ± 1.3	104.8 ± 0.7	106
			N/A	N/A	119
El Cajon	119	WILD CAT	N/A	104.8 ± 0.8	112
	121	EL CAPITAN	106.7 ± 1.6	104.7 ± 1.2	111
Alpine	122	LOVELAND RESERVOIR	100.9 ± 1.7	98.2 ± 1.9	104
			N/A	98.5 ± 1.5	N/A
	126	LBT-1	107.2 ± 2.6	104.3 ± 1.2	108
Poser	128	POSER MOUNTAIN PPX	124.6 ± 10.7	N/A	107
			124.9 ± 11.1	N/A	106
	129	POSER MOUNTAIN PPH	105.9 ± 5.8	N/A	108
			105.2 ± 3.0	N/A	107
Lake-Morena	131	LBT-2	108.4 ± 1.1	107.5 ± 1.7	109
	134	LAKE-MORENA	97.5 ± 2.6	N/A	102
Descanso	136	PINE VALLEY	93.0 ± 0.88	N/A	118
			N/A	N/A	105
	139	THOMAS MOUNTAIN	95.2 ± 4.5	N/A	101
			106.0 ± 4.0	N/A	N/A
Long Potrero	142	LONG POTRERO OUTER	101.0 ± 1.6	99.9 ± .0.8	104

Note: After Ortega-Rivera (1997) and Kimbrough et al. (2001).

western and eastern zones of the Peninsular Ranges batholith are locally juxtaposed), they are deformed and belong to the prebatholithic rocks or to the very early stages of the Peninsular Ranges batholith (Todd et al., 1988; Thomson and Girty, 1994; Johnson et al., 1999; Schmidt, 2000). Until more of these older plutons are found spread across the Peninsular Ranges batholith, and their origin constrained, for the purpose of this study they will be considered as part of the host rocks, and not as part of the Peninsular Ranges batholith as suggested by, for example, Johnson et al. (1999a, 1999b) and Schmidt, (2000).

To summarize, emplacement ages for the Peninsular Ranges batholith show a distinct geographic age pattern: oldest ages are near the Pacific coast and youngest ages are toward the Gulf. This implies a steady eastward migration of the foci of granitic intrusion and associated volcanic arc during the Mesozoic.

$^{40}Ar/^{39}Ar$ Dates

For a better understanding of the $^{40}Ar/^{39}Ar$ dates we should remember that ideally a mineral date determined by this method

TABLE 3. COMPILATION OF ^{40}AR/^{39}AR PLATEAU DATES FOR HORNBLENDE AND BIOTITE DATES
WITH AVAILABLE APATITE FISSION-TRACK DATES

Sample no.	Sample name	Elevation (m)	hb ^{40}Ar/^{39}Ar plateau date (Ma)	bt ^{40}Ar/^{39}Ar plateau date (Ma)	Fission-track apatite date (Ma)	Track length (μm)
40	EN-1-91	500	106.0 ± 1.7	106.2 ± 0.6	95 ± 6.9	14.0 ± 1.1
			106.1 ± 1.2	N/A	N/A	N/A
42	EN-3-91	850	103.2 ± 2.7	100.0 ± 0.7	90 ± 5.2	N/A
			103.0 ± 2.1	102.4 ± 0.8	N/A	N/A
43	EN-4-91	855	103.7 ± 1.8	N/A	83 ± 13.0	14.1 ± 1.2
46	EN-7-91	1000	100.1 ± 1.3	97.0 ± 1.2	57 ± 2.5	14.0 ± 1.1
47	EN-8-91	1200	97.9 ± 1.4	95.3 ± 1.1	68 ± 3.3	13.9 ± 1.1
48	EN-9-91	1100	99.4 ± 1.4	96.4 ± 1.2	73 ± 3.1	14.0 ± 1.3
49	EN-10-91	1200	97.6 ± 1.3	93.3 ± 0.3	58 ± 2.5	14.3 ± 1.1
			98.2 ± 1.9	93.3 ± 1.6	N/A	N/A
51	EN-12-91	850	97.5 ± 3.0	87.8 ± 2.1	51 ± 3.0	13.8 ± 1.1
			89.1 ± 1.0	N/A	N/A	N/A
			90.3 ± 1.6	N/A	N/A	N/A
52	EN-13-91	600	88.8 ± 0.9	84.3 ± 0.7	55 ± 3.6	13.9 ± 1.2
			88.4 ± 3.8	84.7 ± 0.4	N/A	N/A
2	CONT-1.1	1500	99.6 ± 1.4	94.1 ± 1.1	59 ± 5.0	13.8 ± 1.2
4	WC-(2M)	1975	94.7 ± 0.8	93.0 ± 0.7	72 ± 3.7	13.6 ± 1.0
			94.6 ± 0.6	93.5 ± 1.0	N/A	N/A
			94.6 ± 2.7	94.7 ± 1.2	N/A	N/A
			94.4 ± 1.2	93.7 ± 0.8	N/A	N/A
				93.2 ± 2.0	N/A	N/A
10	E.8-20.6	2450	94.0 ± 2.5	90.7 ± 0.5	62 ± 5.6	14.0 ± 1.1
			92.4 ± 0.7	89.3 ± 0.2	N/A	N/A
24	CORE	2050	N/A	88.3 ± 0.6	57 ± 7.4	10.1 ± 1.3
				89.2 ± 0.6	N/A	N/A
				88.5 ± 0.8	N/A	N/A
61	LJ-7-31-2	1800	91.5 ± 2.7	88.1 ± 1.0	70 ± 6.1	14.0 ± 1.3
			93.5 ± 1.2	88.5 ± 0.4	N/A	N/A
			93.8 ± 1.5	89.6 ± 1.0	N/A	N/A
			93.3 ± 0.9	88.3 ± 0.9	N/A	N/A
			93.7 ± 1.7	89.9 ± 1.4	N/A	N/A
			93.5 ± 1.4	N/A	N/A	N/A
			92.8 ± 0.9	N/A	N/A	N/A
			92.5 ± 2.3	N/A	N/A	N/A
			93.9 ± 1.1	N/A	N/A	N/A
			93.6 ± 1.6	N/A	N/A	N/A
			94.7 ± 1.3	N/A	N/A	N/A
66	G.C.-7-29-88-5	475	93.2 ± 1.6	93.5 ± 1.2	78 ± 8.0	14.0 ± 1.1
			93.4 ± 0.8	94.9 ± 1.0	N/A	N/A
			93.4 ± 1.2	N/A	N/A	N/A
69	G.C.-7-28-88-5	625	92.5 ± 1.5	90.0 ± 0.8	65 ± 4.6	12.7 ± 1.3
			N/A	89.9 ± 0.8	N/A	N/A
			N/A	90.4 ± 1.0	N/A	N/A
			N/A	90.8 ± 0.8	N/A	N/A
90	EUBC-1	300	113.4 ± 2.4	112.3 ± 2.4	104 ± 14.0	13.4 ± 1.3
96	VP-8	920	101.8 ± 2.5	100.3 ± 1.2	95 ± 6.9	14.0 ± 1.1

Note: Samples from across the western and eastern Peninsular Ranges batholith of Baja California, México (after Ortega-Rivera, 1997; 2000; Ortega-Rivera et al., 1994, 1997). Sample localities given in Appendix 2, GSA Data Repository item 2003175 (see footnote 1).

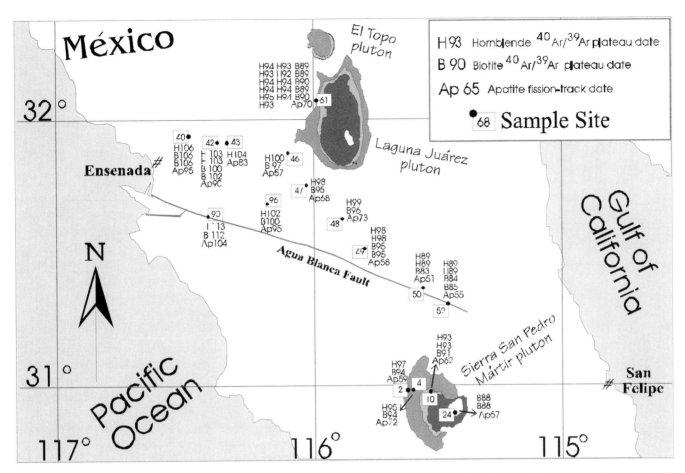

Figure 7. Apatite fission-track and $^{40}Ar/^{39}Ar$ plateau dates for cogenetic hornblende and biotite from Laguna Juárez pluton, Sierra San Pedro Mártir pluton, and from a regional west-east transect from Ensenada to Gulf Escarpment (Fig. 2).

marks the time when that mineral became closed to diffusion of argon or when it cooled below their closure temperature for ^{40}Ar diffusion. Closure of a particular mineral to diffusion is controlled chiefly by temperature, to a small extent by cooling rate (Dodson, 1973), and possibly by chemical composition or strain.

In general, by stepwise-heating a mineral sample, the argon is released for each increment, and an age-spectrum of a mineral is obtained. An age-spectrum, consequently, is a record of the distribution of argon within the mineral. Later, an age is calculated from the argon released for each increment. If the mineral has a relative simple thermal history, the argon will be evenly distributed, and the age of the majority of the released argon will be the same within error across the spectrum, and a "plateau" is defined. But if the mineral has had a complex thermal history, a "disturbed spectrum" is commonly exhibited. Disturbed spectra typically result because $^{40}Ar_{rad}$ has been lost or gained with respect to the amount that corresponds to the real age and different shaped spectra result. Therefore, in general, if a sample has gained $^{40}Ar_{rad}$, this will produce an "excess-argon spectra" that is characterized by anomalously old first steps, and in some cases also the last steps will be old, producing a U-shape or saddle-

shape pattern. Alternatively, a sample that has lost $^{40}Ar_{rad}$ will produce an "argon-loss spectra" characterized by the increase in the age of increasing temperature steps.

Because Ar-closure temperatures are well known, $^{40}Ar/^{39}Ar$ step-heating age-spectrum dating has valuable applications not only for cooling ages, but also for evaluating thermal histories of a region. Each mineral that is appropriated for argon dating has a characteristic closure temperature that is known with a precision of ~20 °C. The closure temperature is higher for minerals that cooled rapidly, and vice-versa. Commonly accepted closure temperatures range from rapid cooling (1000 °C/Ma) to slow cooling (5 °C/Ma) and are between 580 °C and 480 °C for hornblende (Harrison, 1981), 325–270 °C for muscovite (Purdy and Jäger, 1976), and 300–260 °C for biotite (Harrison and McDougall, 1985). For simplicity, intermediate closure temperatures for intermediate cooling rates of 100–500 °C/Ma are assumed in this study: for hornblende, ~500 °C (Harrison, 1981; Harrison et al., 1985); muscovite, ~350 °C (Purdy and Jäger, 1976); biotite, ~280 °C (Harrison et al., 1985); and plagioclase, ~220 °C (Harrison and Clarke, 1979).

Most of the new $^{40}Ar/^{39}Ar$ age-spectra reported in this study are either completely flat or climb over the first ~20% of ^{39}Ar

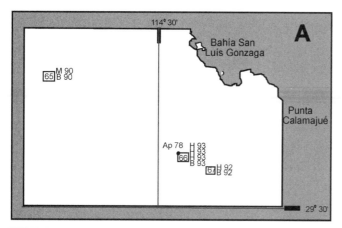

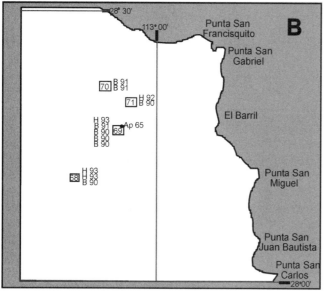

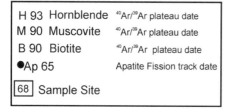

Figure 8. Apatite fission-track and ^{40}Ar/^{39}Ar plateau dates for cogenetic hornblende, biotite, and muscovite for samples from southeastern Peninsular Ranges batholith, Baja California, near the Gulf of California; data are shown plotted in two local maps (i.e., sample locality map A located around 29°30″N lat–114°30″W long, and sample locality map B located near 28°N lat–133°W long [see points 16, corresponding to map A, and 17, corresponding to map B] in the B inset of the regional locality map in Figure 4.

released to a plateau for the remaining ~80% of the spectrum. Within analytical error (2σ), the ^{40}Ar/^{39}Ar plateau dates typically agree with the integrated and isotope correlation dates for the same analysis, suggesting that these samples are relatively undisturbed. As a result, plateau dates are used. Since the Pen-

insular Ranges batholith dates present almost concordant dates or "normal" discordant dates with their cogenetic minerals, the plateau dates (Table 1) are interpreted as the times at which these minerals cooled below their closure temperature for ^{40}Ar diffusion. If cogenetic hornblende and/or muscovite and/or biotite dates from a single sample are indistinguishable within analytical uncertainty, they are considered to be concordant. And since the hornblende ^{40}Ar/^{39}Ar dates are mainly indistinguishable within analytical uncertainties with their cogenetic U/Pb zircon dates across the Peninsular Ranges batholith accordingly, they are considered to be concordant. Thus, where we do not have U/Pb zircon ages, we may use them to represent emplacement ages if they are not disturbed or reset.

In the western Peninsular Ranges batholith, pyroxene, hornblende, biotite, and/or plagioclase dates were obtained from rocks that range in composition from gabbro to tonalite. In the eastern Peninsular Ranges batholith, hornblende, biotite, plagioclase, and muscovites were dated from predominantly tonalites and, less commonly, granodiorites. Four pyroxene-hornblende and five biotite-muscovite mineral pairs were available, the plateau dates are generally concordant, and, moreover, 18 of 20 of the dates determined from cogenetic mineral pairs are statistically indistinguishable. Therefore, these dates are interpreted to record rapid cooling through their respective closure temperature for ^{40}Ar immediately after emplacement.

In general, for the Peninsular Ranges batholith, cogenetic minerals from a given pluton show minor "normal" discordance, with the ^{40}Ar/^{39}Ar hornblende plateau dates being ca. 1–5 Ma younger than the U/Pb zircon age, and ca. 1–6 Ma older than corresponding ^{40}Ar/^{39}Ar biotite plateau dates. Such discordance is consistent with an interpretation of rapid cooling of the plutons. Only two samples (#109 and #128 [Table 2]) have reverse discordance, where the hornblende ^{40}Ar/^{39}Ar plateau dates are older than the U/Pb zircon dates (see Appendices; see footnote 1). This is attributed to excess argon. The geologic interpretation suggested from undisturbed spectra and "normal discordance" or concordant results depends largely upon the age of the rocks in question. Younger samples present more commonly concordant dates than older rock samples. As a general rule, this concordance represents rapid cooling. Such rapid cooling would immediately follow emplacement. When a granitic intrusion is emplaced at great depth, it might remain at elevated temperature for long time, and greatly discordant dates will be shown. But if these rocks present undisturbed spectra and lightly normal discordant or concordant dates, such "normal" discordance will be consistent with an interpretation of a relatively rapid cooling of the plutons, as shown for the Peninsular Ranges batholith.

U/Pb zircon dates (Table 2) have been combined with their cogenetic hornblende and biotite ^{40}Ar/^{39}Ar plateau dates (Table 1) from this work, and ^{40}Ar/^{39}Ar plateau dates reported by Ortega-Rivera et al. (1994a, 1994b, 1997), to produce a crude west-east age profile (Fig. 11). Notice that all of the dates in general are progressively younger from west to east and present normal discordance or are concordant. Much of the scatter

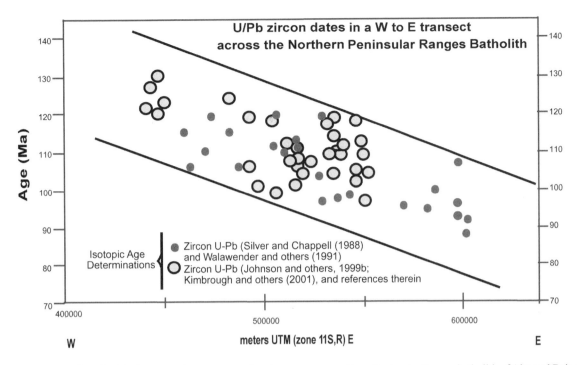

Figure 9. West-to-east profile of U/Pb zircon emplacement ages for plutons across northern Peninsular Ranges batholith of Alta and Baja California. Figure was drawn by plotting available U/Pb zircon data presented in Table 2 versus sample localities from appendices. Data from a distance of >130 km were projected onto the section. Locations of some samples (Silver and Chappell, 1988; Walawender et al., 1991) are approximate, as precise locations have not been published. Heavy lines underline new interpretation of eastward migration of foci of granitoid emplacements.

in ages at a given position can be attributed to projection of the data a considerable distance onto the profile and from U/Pb dates from prebatholithic rocks. Cogenetic samples show rapid cooling patterns. Also, these dates have firmly established an Early Cretaceous age (130~116 Ma) for the SPV and conflicts with the Late Jurassic (Tithonian) age assignment of the SPV based on fossils from isolated exposures of marine volcanogenic sedimentary strata in western San Diego County. Meanwhile, the Alisitos (SPV?) Volcanics near Ensenada (north of the Agua Blanca fault) and la Bufadora (on the Agua Blanca fault) are younger than the SPV and present $^{40}Ar/^{39}Ar$ whole rock dates that range between 107 and 97 Ma; the Alisitos was also considered to be of Aptian-Albian age based on fossils.

At a regional scale, hornblende $^{40}Ar/^{39}Ar$ plateau ages across the Peninsular Ranges batholith young from 118 Ma in the west to 83 Ma in the east, and similarly, biotite $^{40}Ar/^{39}Ar$ plateau dates young from 116 Ma in the west to 80 Ma in the east. Plagioclase $^{40}Ar/^{39}Ar$ plateau dates display a comparable eastward younging of dates (115–70 Ma); when combined with the K/Ar dates from the literature, the biotites are as young as 65 Ma. Some of the plagioclase mineral separates are sericitized (particularly for samples from the western batholith) and the ages may be effectively sericite dates. The plagioclase ages are generally younger than their cogenetic pyroxene, hornblende, and/or biotite, and most display almost perfect plateau ages. K-feldspars show excellent plateau age spectra, and their $^{40}Ar/^{39}Ar$ plateau dates

range from 109 Ma to 74 Ma. Representative age spectra are shown in Figure 6.

Age profiles such as those in Figure 11 are useful for showing the overall pattern of ages but tend to mask the details. Since the $^{40}Ar/^{39}Ar$ ages for the Peninsular Ranges batholith are mainly plateau ages, it can be assumed that they have not been disturbed, and these can be compared and combined with the K/Ar ages reported for the Peninsular Ranges batholith. According to these, age chrontours for hornblende and biotite from the Peninsular Ranges batholith (Figs. 12 and 13, respectively) have been drawn using the 279 new $^{40}Ar/^{39}Ar$ plateau dates reported here, 105 $^{40}Ar/^{39}Ar$ plateau dates reported by Ortega-Rivera et al. (1994, 1997), 10 K/Ar dates reported by Evernden and Kistler (1970), 26 K/Ar dates reported by Armstrong and Suppe (1973), 160 K/Ar dates reported by Krummenacher et al. (1975), 19 $^{40}Ar/^{39}Ar$ biotite dates reported by Schmidt (2000), 7 $^{40}Ar/^{39}Ar$ dates reported by Delgado-Argote et al. (1995), and 25 $^{40}Ar/^{39}Ar$ total fusion biotite dates reported by Grove (1994). Localities for the last two sets of data were approximated because the precise locations have not been published. K-Ar dates previous to 1977 were recalculated with the new constants. K-Ar and plateau $^{40}Ar/^{39}Ar$ cooling ages can be combined for the chrontours since the age spectra (and cogenetic U/Pb and F-T dates) show that they mainly have not been disturbed or reheated after emplacement and denudation. Also, K-Ar and $^{40}Ar/^{39}Ar$ cooling ages for biotites from the country rocks around some of the plutons are essentially concordant. All dates are plotted on Plate 1.

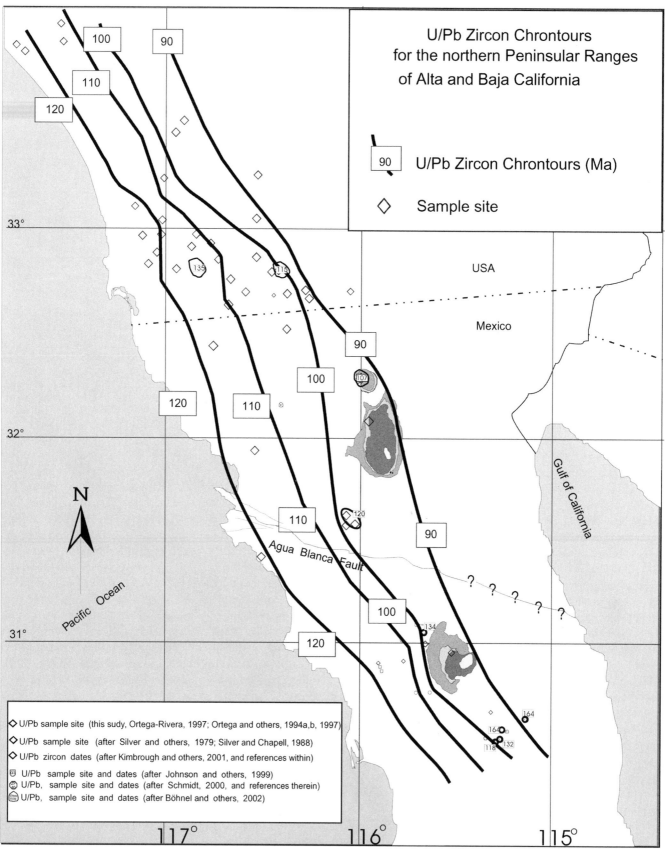

Figure 10. U/Pb zircon chrontours for northern Peninsular Ranges batholith of Alta and Baja California (U/Pb zircon data after Silver et al., 1979; Silver and Chappell, 1988; Walawender et al.,1990; and Kimbrough et al., 2001); compiled ages are shown in Plate 1 (on the CD-ROM accompanying this volume).

U/Pb and ^{40}Ar/ ^{39}Ar plateau dates

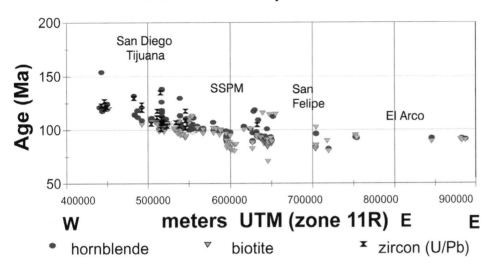

Figure 11. West-to-east profile of all zircon U/Pb dates and hornblende and biotite ^{40}Ar/^{39}Ar plateau dates from samples collected across Peninsular Ranges batholith of Alta and Baja California. Note that even though all dates from Tables 1 and 2 and from Ortega-Rivera et al. (1997) are plotted from west to east, without discriminating latitudinal positions, along Baja peninsula between the 34th and 28th parallels north latitude. Dates are progressively younger from west to east. SSPM—Sierra San Pedro Mártir pluton.

As described above for the zircon chrontours, possible offset of the chrontours by faults was ignored.

Timing of uplift and denudation of the Peninsular Ranges batholith is constrained by biotite K/Ar and ^{40}Ar/^{39}Ar apparent ages that show a progressive eastward younging of ages from ca. 116 to 65 Ma from west to east across the batholith, such that the youngest ages correspond to the deepest levels of erosion (2kb in the west to 9kb in the east [Hammarstrom, 1992]).

The chrontours (Figs. 12 and 13) reveal a monotonic younging of hornblende and biotite dates from southwest to northeast, with hornblende and biotite K/Ar-^{40}Ar/^{39}Ar chrontours paralleling the U/Pb zircon chrontours (Fig. 10). The pattern of the chrontours is similar to that originally presented by Krummenacher et al. (1975), but the present work shows refined chrontours and that they can be extended south of the Agua Blanca fault; also, there is almost no discordance between hornblende and biotite pairs, and they are basically concordant. Furthermore, the almost ubiquitous occurrence of plateaus in the ^{40}Ar/^{39}Ar age spectra, when coupled with the chrontour patterns, is best explained as result of a sequential west to east emplacement of the granitoid intrusions followed by sequential rapid cooling of both the intrusive rocks and the surrounding metamorphic rocks. This was one of the five plausible interpretations for the K/Ar apparent-age gradient put forward by Krummenacher et al. (1975).

West to east age profiles have been constructed for hornblende and biotite ^{40}Ar/^{39}Ar plateau dates (Figs. 14 and 15) from: (1) examples from local west-to-east transects across some of the major plutons of the Peninsular Ranges batholith; and (2) a *regional* detailed transect from the Pacific to the Gulf (west-east (EN-91) transect from Ensenada to San Felipe; Fig. 7).

Figure 14C shows a detailed west-to-east regional profile of the dates across the Baja Peninsula (EN-91 transect; Fig. 7), which clearly shows the monotonic nature of eastward younging across the region. At a more local scale, and across various

individual plutons (Fig. 14, A–C, and Fig. 15, A–C), it can be seen that the pattern of eastward younging and rapid cooling of dates is also present within the individual plutons of the Peninsular Ranges batholith. It should be pointed out that the symbols are greater than the 2σ errors given for each date. Also, the rapid cooling history is shown for each pluton by the overlap of ages shown for each cogenetic sample (i.e., concordant) across each pluton, the same patterns as the one described for the Sierra San Pedro Mártir pluton by Ortega-Rivera et al. (1994, 1997). They made a detailed study of the pattern of local eastward younging across the Sierra San Pedro Mártir pluton of the eastern Peninsular Ranges batholith (Figs. 7 and 15A). Therein, they concluded rapid uplift and that »15° east-side-up tilting of the crustal block containing the Sierra San Pedro Mártir pluton, about a horizontal axis subparallel to the trend of the batholith, best explains the observed eastward younging of the ages.

This local pattern of *intrapluton* eastward younging of the ^{40}Ar/^{39}Ar plateau dates (illustrated by the west-east profiles across some of the sampled plutons across the Peninsular Ranges batholith in Figs. 14 and 15, as in other plutons dated in this study) is similar to that observed for the Sierra San Pedro Mártir pluton (Fig. 15A). Taking this into consideration, the local pattern of eastward younging of dates within these plutons has been interpreted, therefore, as to have resulted from rapid uplift and denudation, and localized east-side-up tilting of the crustal size blocks containing the individual plutons. This interpretation is also consistent with previous estimates of »15° tilt for the eastern Peninsular Ranges (Ague and Brandon, 1992) and rapid uplift and denudation rates found by Ortega-Rivera (1997), Lovera et al. (1999), and Kimbrough et al. (2002). More recently (during the review of this paper), based on paleomagnetic data interpretations, Böhnel et al. (2002) concluded that as suggested by Ortega-Rivera (1997) and Ortega-Rivera et al. (1997), tilt of a small crustal block of pluton-scale, not regional tilt, may explain

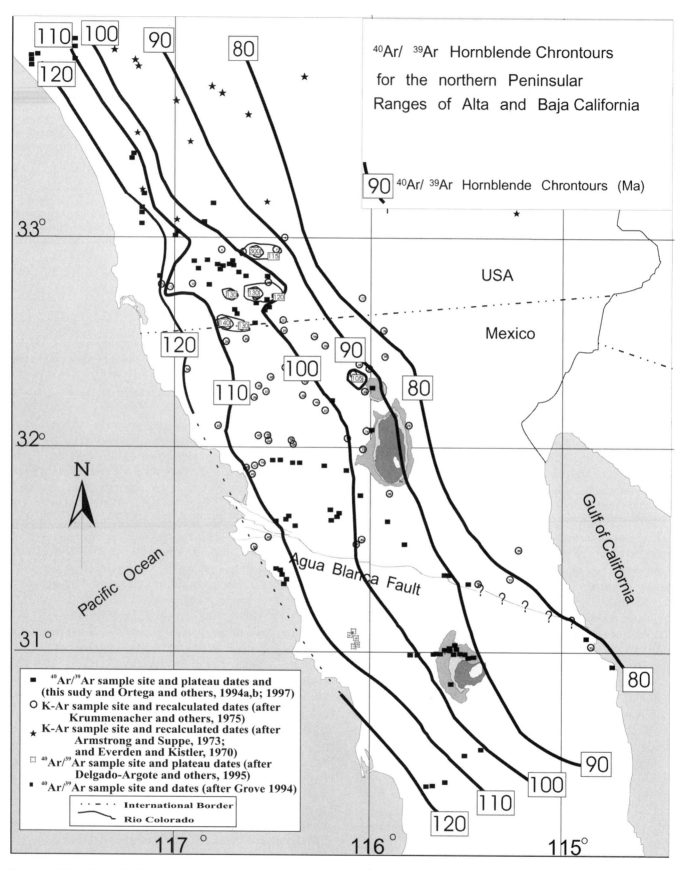

Figure 12. Hornblende ^{40}Ar/^{39}Ar plateau and K/Ar chrontours for northern Peninsular Ranges batholith; compiled ages are shown in Plate 1 (on the CD-ROM accompanying this volume).

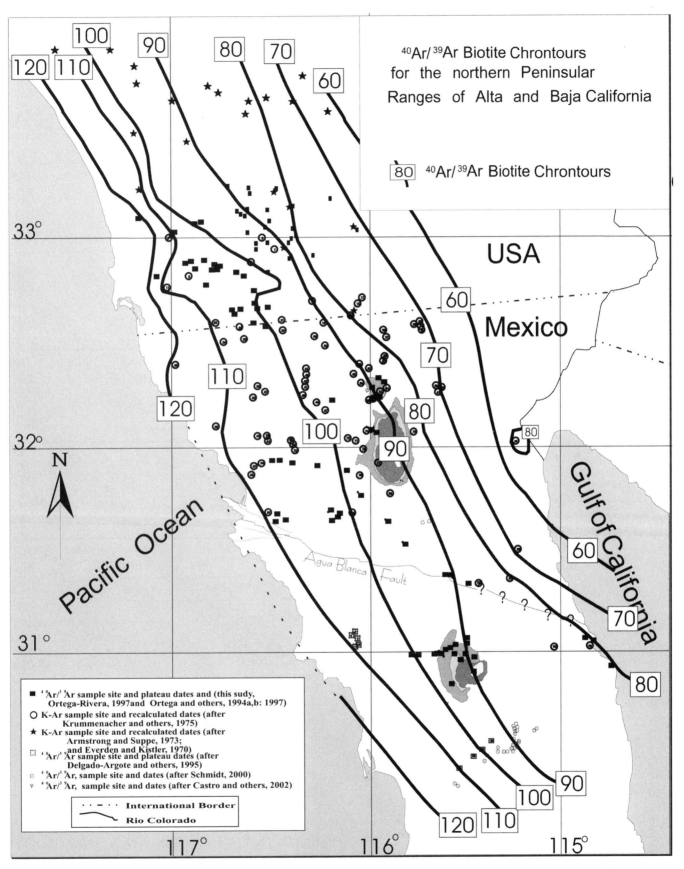

Figure 13. Biotite ^{40}Ar/^{39}Ar plateau and K/Ar chrontours for northern Peninsular Ranges batholith; compiled ages are shown in Plate 1 (on the CD-ROM accompanying this volume).

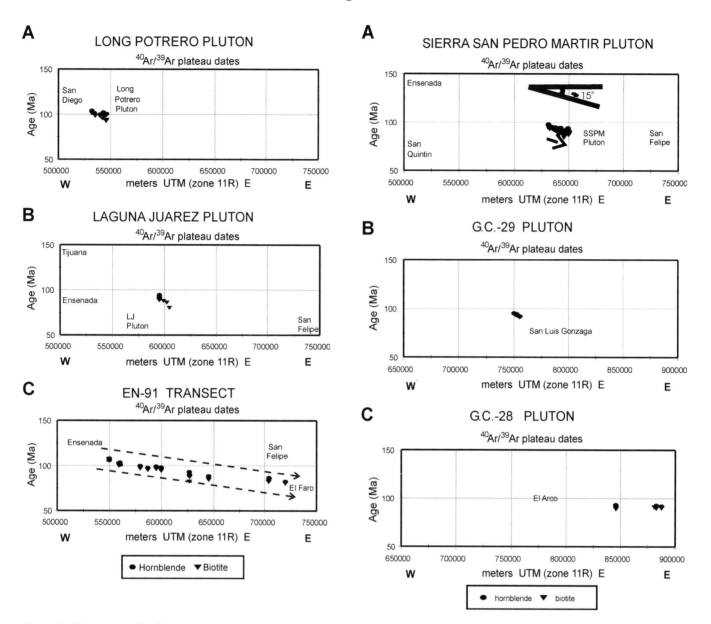

Figure 14. West-to-east ⁴⁰Ar/³⁹Ar plateau age profiles from plutons of Peninsular Ranges batholith. A: Long Potrero pluton from western Peninsular Ranges east of San Diego (Figs. 2 and 4); B: Laguna Juárez pluton from eastern Peninsular Ranges, west of Gulf Escarpment (Figs. 2, 4, 6, and 7); and C: West-to-east transect across batholith from Ensenada to El Faro (dashed lines represent pattern of eastward younging of cooling ages).

Figure 15. West-to-east ⁴⁰Ar/³⁹Ar plateau age profiles from plutons of Peninsular Ranges batholith. A: Sierra San Pedro Mártir pluton (Figs. 2, 4, 6, and 7) (data from Ortega-Rivera et al., 1997); B: G.C.-29 area near Gulf of California (Figs. 4 and 8); and C: pluton from G.C.-28 area east of El Arco (Figs. 4 and 8).

some of the discordant paleomagnetic data observed for the El Testerazo pluton and the San Marcos dike swarm in northern Baja California.

Fission-Track Dates and Rb/Sr Dates

Although sparse, apatite fission-track dates (ranging between 104 and 51 Ma, Table 3, Figs. 7 and 8) show a west-southwest to east-northeast decrease in age perpendicular to the trend of the batholith, comparable to the trend of U/Pb zircon and ⁴⁰Ar/³⁹Ar ages. Track length data show that the apatites have experienced moderate annealing (Table 3). The fission-track dates are interpreted as apatite closure ages (i.e., the time of cooling through ~110 °C; Naeser, 1979). If a normal geothermal gradient (30°/km) is assumed, present surface exposures of the batholith must have been within «3 km of Earth's surface since Late Albian–Early Campanian (104–83 Ma) time (Palmer, 1983) in the western Peninsular Ranges, and since early Campanian–early Eocene (78–51 Ma) time, much of the eastern Peninsular Ranges had cooled to temperatures of <110 °C. This indicates that the Peninsular Ranges batholith plutons were significantly exhumed immediately following emplacement in the Late Cretaceous–early Tertiary. This is shown in Table 3 and Figures 7 and 8 if we compare the $^{40}Ar/^{39}Ar$ and fission track dates plotted by locality from west to east. We should also point out that the fission track dates within plutons get younger toward the east, paralleling their cogenetic $^{40}Ar/^{39}Ar$ dates, and that the younger ages are also at higher elevations within the plutons. This pattern of eastward younging could also be explained by tilting (Ortega-Rivera et al., 1997).

Therefore, by comparing the obtained ages for cogenetic biotites and apatites, it can be seen that the measured fission-track dates support the hypothesis that these rocks cooled through the apatite closure temperature a minimum of 5–30 Ma after the closure to Ar of their cogenetic biotite. However, the track-length data for some samples show that there has been moderate annealing and that cooling of the plutonic rocks was therefore protracted. As can be seen in Figure 7 and Table 3, cogenetic $^{40}Ar/^{39}Ar$ and fission track ages display patterns of very rapid cooling and therefore rapid uplift and exhumation histories. Eastward younging of the fission-track dates suggests diachronous cooling through ± 110 °C. If the late diachronous cooling implied by the eastward younging of the fission-track ages across the region is real, such diachroneity could also be explained as a consequence of rapid uplift and denudation and by the mechanisms previously proposed as some minor tilting taking place after the biotites close to Ar-loss or post-biotite closure ages.

Rb/Sr dates are sparse and show the same eastward younging as the other dating methods (Plate 1).

TECTONIC IMPLICATIONS

Peninsular Ranges Batholith

The plutons of the Peninsular Ranges batholith of Alta and Baja California between the 28°N and 34°N parallels were emplaced from west to east between ca. 140 to 90 Ma and few small plutons up to 80 Ma (U/Pb zircon dates) and display $^{40}Ar/^{39}Ar$ hornblende and biotite plateau cooling dates that range from 118 to 83 Ma and 116 to 80 Ma, respectively, and biotite K-Ar dates as young as 65 Ma. Mineral pairs having mainly concordant dates yield $^{40}Ar/^{39}Ar$ cooling ages that systematically decrease also from SW to NE. Rapid cooling is indicated by the small differences in U/Pb zircon and cogenetic $^{40}Ar/^{39}Ar$ dates. Thus, the Peninsular Ranges batholith intrudes from west to east a series of northwest-trending prebatholithic lithostraigraphic assemblages. In general, from west to east, these lithostratigraphic assemblages include a Triassic–Cretaceous continental borderland assemblage; a Jurassic–Cretaceous volcanic arc assemblage; a Triassic(?)–middle Cretaceous clastic assemblage; and volcanic flysch; an Ordovician–Permian slope-basin clastic; and Upper Proterozoic–Permian miogeoclinal carbonate-siliciclastic (Gastil, 1993). The prebatholithic structures get older toward the east.

The northeastward-decreasing pattern of U/Pb zircon and nearly concordant $^{40}Ar/^{39}Ar$ hornblende ages across the Peninsular Ranges batholith is in accord with a model implying a regional scale eastward migration of the locus of magmatism. Relatively rapid cooling of the Mesozoic Peninsular Ranges batholith is indicated by the small differences in U/Pb zircon dates and $^{40}Ar/^{39}Ar$ hornblende plateau dates (Table 2), suggesting that uplift and erosion of the batholith occurred shortly after intrusion of the individual plutons (140 to ca. 80 Ma). U/Pb and $^{40}Ar/^{39}Ar$ data for the eastern Peninsular Ranges batholith also indicate that the bulk of these intrusions were emplaced in a remarkably short interval between 92 and 98 Ma, following cessation of west-directed compression that, according to Johnson et al. (1999), dominated the arc system until ca. 105 Ma. Preliminary hornblende geobarometry suggests that depths of emplacement were ~6 km in the west to ~27 km in the east (Hammarstrom, 1992), suggesting that the emplacement depths also increase eastward across the batholith (Silver et al., 1979; Gastil, 1983; Ague and Brimhall, 1988a, 1988b). The La Posta–type plutons, therefore, correspond closely to the most deeply exhumed part of the Peninsular Ranges batholith (Kimbrough et al., 2001). Moreover, the fission-track dates (Figs. 7 and 8, Table 3) indicate that present exposures of the western and eastern Peninsular Ranges batholith had cooled to temperatures of <110 °C and therefore were within «3 km of Earth's surface, between 104 and 83 Ma and between 78 and 51 Ma, respectively. As a result, it can be said that the thermochronology data, then, indicate that the Peninsular Ranges batholith plutons were significantly exhumed immediately following emplacement in the Late Cretaceous–early Tertiary. These dates are compatible with the time of the initial onset of Laramide tectonism, and are consistent with uplift due to low-angle subduction and with later tectonic erosion caused by subsequent underplating of lower crustal rocks (Coney, 1972; Engebretson et al. 1985). Low-angle subduction beneath southwestern North America in late Cretaceous through mid-Tertiary time is supported by spatial, temporal and chemical patterns of igneous rocks in California, Arizona and New Mexico (Helmstaedt and Doig, 1975; Keith, 1978; and Dumitru, 1990). As a result, it can be said that the thermochronology data, then, indicates that the PRB plutons were significantly exhumed immediately following emplacement in the Late Cretaceous–early Tertiary. The eastward younging on a local scale within plutons (see Table 1 and Figs. 14 and 15) is consistent with superimposed local northeast-side-up tilting of the crustal-scale blocks containing the plutons that form the Peninsular Ranges batholith, perhaps in response to Laramide flat subduction beneath North America as it has been suggested (Gastil, 1983; Goodwin and Renne, 1991; Grove, 1993; Ortega-Rivera et al., 1997; Lovera et al., 1999; Axen et al., 2000).

Burchfield et al. (1992) have shown that the juxtaposition of oceanic and continental lithosphere may have occurred during a late Paleozoic–Early Triassic event that truncated western North

American continental lithosphere before the emplacement of the Peninsular Ranges batholith. Other studies of the Peninsular Ranges province have also demonstrated that the two compositionally distinctive western and eastern terranes more likely reflect different source regions (e.g., Silver and Chapell, 1988). The boundary separating these compositional crustal components was defined by geophysical, geochemical, and petrological discontinuities (e.g., Early and Silver, 1973; Silver et al., 1979; Krummenacher et al., 1975; DePaolo, 1981; Gastil, 1983; Todd and Shaw, 1985; Silver and Chapell, 1988). The more mafic and older western section formed in oceanic lithosphere, and the more silicic and younger section developed in continental lithosphere (Gastil, 1979; Silver et al., 1979). Based on their interpretation that the flysch assemblage overlaps the batholith discontinuity in southern California and is Triassic or older as determined from a Triassic age obtained from a pluton that intrudes part of the flysch assemblage, Thomson and Girty (1994) proposed that the transition is a pre-Triassic crustal boundary that possibly corresponds to the ancient North American rifted margin. This compositional boundary, as suggested by Thomson and Girty (1994) and Schmidt (2000), has provided a mechanical weakness in the crust along which the Mesozoic intra-arc strain was concentrated. Also, Magistrale and Sanders (1995) proposed that Quaternary fault development also has been localized at this discontinuity.

The origin between the western and eastern basement in the Peninsular Ranges batholith and the deformation belt within it has been controversial for some time. Other workers (Gastil et al., 1981; Rangin, 1978; Todd et al., 1988; Griffith and Hoobs, 1993; and Busby et al., 1998) suggested that the transition was initiated as far back as Jurassic time with formation of a back-arc basin and development of a fringing western arc that was sutured back to the continent following collapse of the basin in the middle Cretaceous. Alternatively, in a local study of the northern Sierra San Pedro Mártir pluton, Johnson et al. (1999b), suggested that the western zone of the Peninsular Ranges batholith represents an exotic island arc terrane based on the paucity of continental components in both volcanics and plutons. They proposed suturing of this terrane on the Main Mártir thrust between 115 and 108 Ma, although they are uncertain as to the extent of this suture.

Until the geologists settle on which of the models is the correct one, the geochronology tells us that the plutons were emplaced into a series of northwest-trending prebatholithic lithostraigraphic assemblages that get younger toward the west. Also, the overprinting of the K/Ar systematics for the metamorphic rocks across the province by the intruding eastward migration of plutons of the Peninsular Ranges batholith can be interpreted as that the prebatholithic rocks were in place before intrusions as suggested by Burchfield et al. (1992), i.e., the juxtaposition of oceanic and continental lithosphere may have occurred during a late Paleozoic–Early Triassic time, before the emplacement of the Peninsular Ranges batholith, and therefore the metamorphic rocks are overprinted in concordance with the regional eastward younging of intrusion of the plutons and with rapid uplift and tilting. The next step to further corroborate this is to find the faults

or shear zones between the tilted blocks necessary to accommodate such motion; these structures are highly undocumented, and some may be masked by the ductile deformation aureoles around some of the plutons and some by the country rock metamorphic fabric parallel to the grain of the batholith. Paleomagnetic data across the west-east transect across the plutons is also needed to further corroborate tilting and deformation.

Continentward migration of the foci of plutonic magmatism has long been recognized at other "Andean-type" margins. Isotopic ages young continentward in Japan (Nozawa, 1983), the Sierra Nevada (Bateman, 1983; Dumitru, 1990), the Central Andes of northern Chile (Farrar et al., 1970; Clark et al., 1976), and the Coastal batholith of British Columbia (Roddick, 1983a). The migration of magmatism has been related to progressively downdip changes and effects of more rapid convergence in the site of initiation of magmatism with time at a long-lived convergent margin (Farrar et al., 1970; Clark and Zentilli, 1972; Zentilli, 1974; Clark et al., 1976), to changes in the dip of a single subduction zone (Clark et al., 1982) as a result of the increasing convergence velocity of the oceanic (e.g., Farallon) plate relative to the overriding continental (e.g., North American) plate (Coney and Reynolds, 1977). This change caused a shallowing of the subduction angle beneath the North American plate and a displacement of the locus of magmatism to the east (e.g., Clark et al., 1982). According to Gastil (1983), the flattening of the subducting slab may have resulted also, at least in part, from dilation resulting from sialic accretion. In addition, it is thought that changes in plate motions were responsible for the development of the classic Laramide deformation of the Cordilleran foreland (Coney, 1972; George and Dokka, 1994). These changes included an increase in convergence rates, an increase in the absolute velocity of the North American plate with respect to the subducting plate, and a change in the direction of convergence between plates, and shallowing of the subduction angle (Engebretson et al., 1985). Consequently, the locus of arc magmatism began to migrate far to the east into the interior of the North American plate as revealed by the thermochronology and is best shown by the chrontour biotite map. If variation of the angle of subduction is the correct explanation for the Laramide orogeny, the Late Cretaceous shallowing of the subduction shifted the Peninsular Ranges batholith from a magmatic arc setting into a forearc setting of low heat flows and gradients, the result of underflow of a relatively cold subducting plate, similar to the setting described by Dumitru (1990) for the Sierra Nevada. In consequence, a similar situation could have existed at this time all along the Cordilleran batholiths. This tectonic regime remained for the Peninsular Ranges batholith until at least the late Cenozoic, when the plate margin for most of Alta California and Baja California shifted over to the current San Andreas–Gulf of California plate regime.

Tilting versus Translation?

Recent studies have emphasized that much of the western coast of North America consists of a collage of geologically unrelated fragments, some of which appear far-traveled

(Coney et al., 1980; Jones et al., 1982; Johnson et al., 1999a; Dickinson and Lawton, 2001). These rocks have been found to record shallow paleomagnetic inclinations that may be indicative of magnetization at low paleolatitudes. Paleomagnetic data for coastal California and western Baja California have been interpreted to suggest that as much as 2500 km of post–middle Cretaceous northward tectonic transport of peninsular California has occurred with respect to interior North America (Teissere and Beck, 1973; Champion et al., 1984; Hagstrum et al., 1985; Ziegler et al., 1985; Morris et al., 1986; Howell et al., 1987; Lund and Bottjer, 1991; Beck, 1991). This model is consistent with a Cretaceous paleolatitude for the Peninsular Ranges terrane of 15–20° less than that of the adjacent (prior to the Gulf opening) North American stable craton. It is important to point out that this hypothesis rests exclusively on paleomagnetic results from rocks collected mainly in westernmost California and Baja California (see figure 9 *in* Lund and Bottjer, 1991). No paleomagnetic data are available, however, for the eastern Peninsular Ranges batholith. The paleomagnetic data have been interpreted by Lund and Bottjer (1991) to indicate that the Peninsular Ranges province was part of the North American craton before the Miocene, but that it was located ~15° south of its present position with respect to the craton, perhaps along the margin of southern México. Those authors also indicated that the peninsular terrane moved northward to its current position with respect to North America at some time between 40 and 20 Ma.

In contrast, several lines of evidence suggest that, prior to the well-documented Late Miocene opening of the Gulf of California, Baja California had been attached to the North American craton for at least 500 Ma. For example, in a multidisciplinary study of the early Eocene marine and terrestrial biostratigraphy and paleomagnetism of central Baja California, Flynn et al. (1989) provided constraints for the early Cenozoic plate-tectonic motion of the peninsula. Their results indicate that in its pre–Gulf of California position (prior to the middle early Eocene), Baja California was sutured to the northern Mexican mainland. Furthermore, Abbot and Smith (1978) point out the occurrence of the distinctive Eocene Poway Group conglomerate ties terranes of the California continental borderland to the Peninsular Ranges, and more recently Abbot and Smith (1989) have demonstrated that the occurrence of this distinctive alluvial fan conglomerate in San Diego ties the California continental borderland to a source rock in western Sonora, México. Also, features such as the Mesozoic magmatic arc appear to extend from peninsular California into Sinaloa and Sonora (Fig. 1). For example, Silver and Chappell (1988) and more recently Silver (1996) have suggested that chemical and isotopic zonations within the Cretaceous batholith can be correlated across the gulf without apparent large-scale separation, and perhaps most importantly, according to Gastil (1991, 1993), no faults have been found that would permit large-scale northward transport. Moreover, Gastil (1991, 1993) suggested that facies boundaries from Ordovician to Cretaceous strata within the peninsula cross the Gulf of California, indicating that peninsular California had been adjacent to this part of the

North American craton at least for 500 Ma prior to the Cenozoic opening of the Gulf of California and that peninsular California was not an oceanic platform or rise that docked against the edge of the continent at the end of the Cretaceous.

Preliminary paleomagnetic results from the westernmost Peninsular Ranges batholith reported by Yule and Herzig (1994) show that the paleomagnetic inclinations obtained from a lava flow within the Santiago Peak Volcanics agree with expected pre-Neogene (Cretaceous?) paleolatitudes determined for cratonic North America and with geologic reconstructions that emphasize that the southern Californian Peninsular Ranges batholith underwent no large-scale northward transport.

The aforementioned studies, therefore, indicate that peninsular California, prior to the Late Tertiary opening of the Gulf of California, had been attached to the North American craton for at least 500 Ma, that there has been no significant pre– and post–early Eocene northward transport of the Baja peninsula, and that the latitudinal separation is only the 2° created by the opening of the Gulf of California (Hagstrum et al., 1985, 1987; Gastil, 1991). Indeed, the data discussed herein suggest a close association of peninsular California with mainland México since at least the Early-Late Cretaceous. However, no one has yet satisfactorily reconciled the contradictory geologic and paleomagnetic data for the Peninsular Ranges province. Until fault zones capable of accommodating the translations proposed from the paleomagnetic inclination data are identified, it is perhaps more sensible to consider the paleomagnetic data, rather than the rocks, as suspect.

It has been suggested that the tilting of batholiths or plutons about subhorizontal axes and the shallowing of paleomagnetic inclinations during compaction in marine sedimentary rocks may explain the observed paleomagnetic inclination anomalies rather than large-scale latitudinal movement (e.g., Irving and Archibald, 1990; Butler et al., 1989, 1991; Constanzo-Alvarez and Dunlop, 1988; Marquis and Irving, 1990; Ague and Brandon, 1992). Furthermore, for a correct interpretation of the paleomagnetic data, the paleo-horizontal at the time of pluton crystallization must be known.

Until now, data concerning the paleo-horizontals for plutons of the Peninsular Ranges have not been determined (e.g., Gastil and Miller, 1984; Gastil et al., 1991; Gastil, 1991; Butler et al., 1991). Although Butler et al. (1991) attributed the observed pattern of eastward younging and denudation of the Peninsular Ranges batholith to uniform (~15–20°), northeast-side-up tilting of the entire batholith about an axis with an azimuth ~330–340°, and although such a tilt of the entire batholith could explain aspects of the eastward younging of $^{40}Ar/^{39}Ar$ ages, this interpretation is not correct, since it would lead to enormous uplift (»45 km) in the eastern Peninsular Ranges, for which is there no evidence. These results suggest that in the Peninsular Ranges batholith, tilting cannot be ruled out.

On the other hand, tectonic tilting of small scale blocks, containing individual plutons is inferred by the eastward younging of the ages (U/Pb, $^{40}Ar/^{39}Ar$, K/Ar and fission-track data) within plutons, as shown before by Ortega-Rivera et al., (1997), and Ortega-Rivera, 2003. These results, therefore, suggest that in the Peninsular Ranges batholith, tilting cannot be ruled out. If this is true, the paleo-horizontals at the time of pluton crystallization can be calculated by correcting for the local east-side-up crustal size block tilting (~15° to 20°) of individual plutons as is indicated by the eastward younging of the ages. Then, minor tilting of individual plutons rather than large-scale latitudinal movement is the explanation for the majority of the discordant paleomagnetic inclinations observed for the Peninsular Ranges batholith.

Keppie and Dostal (2001), in their evaluation of the Baja B.C. controversy, conclude that the weight of evidence in favor of geological piercing points across the Gulf of California appears to outweigh the paleomagnetic evidence for 1000–2500 km of apparent northward displacement of Baja California in the Cretaceous and early Tertiary, which may be explained by ~15° to 20° tilting of fault blocks recorded by $^{40}Ar/^{39}Ar$ cooling ages as suggested by Ortega-Rivera (1997, 2003). Therefore, Baja California has not moved more than was required to open the Gulf of California.

Western México

The Peninsular Ranges batholith shows both marked similarities to and differences from the batholiths of Sonora, Sinaloa, Cabo San Lucas (Baja California Sur), and Jalisco. Henry et al. (this volume), have shown that the greatest similarities of these batholiths are in types of intrusions, a common sequence from early gabbro through syntectonic to posttectonic rocks and general eastward migration of magmatism. However, they have also shown that the end of deformation recorded by syntectonic rocks may be different in each area. The calcic Peninsular Ranges batholith rocks show a similar wide range of compositions as rocks of Sonora, Sinaloa, and Jalisco, but are more calcic than the potassic sinaloan batholith. The REE patterns for the Sinaloa batholith are most like those of the eastern part of the Peninsular Ranges and central part of Sonora, both areas that are underlain by Proterozoic crust or crust with a substantial Proterozoic detrital component. However, southern Sinaloa lies within the Guerrero terrane, which is interpreted to be underlain by accreted Mesozoic crust. The greatest differences are in distance and rate of eastward migration (Henry et al., this volume).

As stated above, the northeastward-decreasing pattern of U/Pb zircon and nearly concordant $^{40}Ar/^{39}Ar$ hornblende biotite and fission track ages across the Peninsular Ranges batholith indicating relatively rapid cooling of the Mesozoic Peninsular Ranges batholith are in accord with a model implying a regional-scale, eastward migration of the locus of magmatism throughout western Mexico. Published age data also show that magmatism migrated eastward at ~10 km/Ma in the Peninsular Ranges and Sonora and from Jalisco southeast along the southwestern Mexico coast. The area of slower eastward migration roughly correlates with the location of the Guerrero terrane and of possibly accreted oceanic crust that is no older than Jurassic (Henry et al., this volume).

Based on the compiled and new geochronological data, this paper provides sufficient coverage in Baja California to test whether the eastward younging of the foci of granitic intrusion continues in formerly contiguous (pre-Gulf) Sonora (Figs. 16 and 17). Since biotite dates are the main mineral dated in the literature, these were used to construct a regional biotite chrontour map for western México (the Baja Peninsula and the mainland) by combining available $^{40}Ar/^{39}Ar$ plateau biotite dates and K-Ar dates from the literature (Evernden and Kistler, 1970; López-Ramos, 1978; Gastil and Krummenacher, 1977; Gastil et al., 1978; Damon et al., 1983; Henry and Fredrikson, 1987; Wallace and Carmichel, 1989; Righter et al., 1995; Schaaf et al., 1995; Ortega-Rivera et al., 1997 (and references therein); and Kimbrough et al., 2001 (and references therein), with the age data from the Peninsular Ranges presented above (Fig. 16A). A reconstructed map to the pre–Gulf of California opening (with the Los Cabos batholith near Puerto Vallarta) was used to draw the new chrontours (Fig. 16B, Table 1, and Appendices 3 and 4; see footnote 1).

The regional biotite chrontours (Fig. 16) indicate that, by moving the Baja Peninsula to its position prior to the opening of the Gulf in Late Miocene time, the chrontours match well with those of mainland México and indicate that the eastward migration of intrusion foci, documented for Baja California, continued in the mainland until at least 40 Ma. Geologic and geochronologic data outweigh the paleomagnetic evidence for 1000–2500 km of apparent northward displacement of Baja California during Cretaceous-Tertiary times, which may be explained by the ~15–20° tilting of fault bocks recorded by the $^{40}Ar/^{39}Ar$ cooling ages. It is considered that it would be most unlikely that such a good fit would result from chance; therefore, it is concluded that systematic tilting combined with small-scale northward movement, rather than large northward translation of "suspect terranes," better explains the paleomagnetic data.

Two age profiles across the Peninsular Ranges batholith and western mainland México (top profile, Fig. 17) and for southwestern México (bottom profile, Fig. 17) show that, starting during the Mesozoic along the Pacific coast and extending at least to Eocene time in mainland México, the locus of magmatism migrated from west to east across this part of western North America. This indicates that a long, linear subduction zone (as suggested for southwestern México by Schaaf et al., 1995), analogous to the modern setting of the South American Andes, formed the basic framework of western México and southwestern México during this time.

Summarizing, the northeastward decrease in U/Pb, $^{40}Ar/^{39}Ar$, K/Ar, Rb/Sr, and fission-track dates (Plate 1, Figs. 6, 7, 8, and 11) across the Peninsular Ranges is a regional age pattern and also a local age pattern (Figs. 14 and 15). This consistency cannot be attributed only to the eastward younging of the foci of granitoid emplacement. As indicated by the eastward younging of chrontours and the age patterns presented for individual plutons in this paper for the Peninsular Ranges, the eastward migration of the foci of magmatic emplacement in combination with local tilting of crustal sized blocks containing these plutonic rocks about an axis

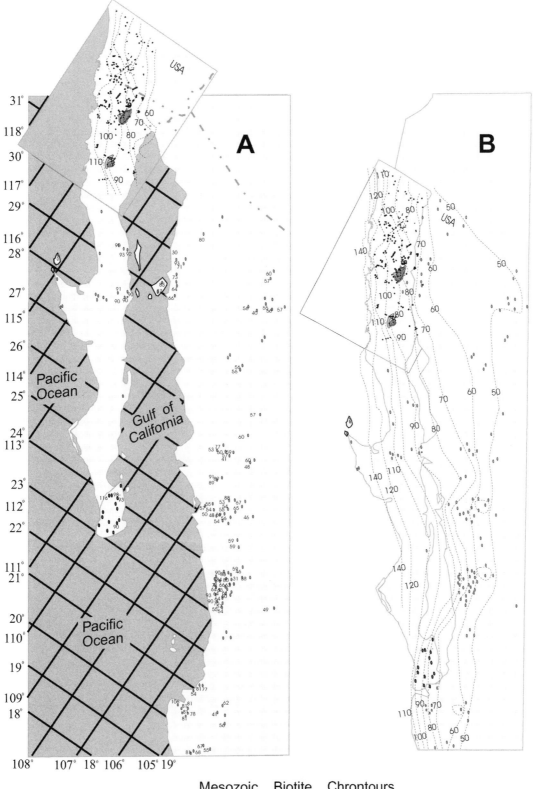

Mesozoic Biotite Chrontours
Peninsular Ranges batholith and western Mexico

^{40}Ar/^{39}Ar and K/Ar Biotite Chrontours (Ma)

Sample sites ^{95} date

Figure 16. Biotite ^{40}Ar/^{39}Ar plateau and K/Ar chrontours for reconstructed Mesozoic peninsular Ranges batholith and western mainland México. Rectangle indicates area of Figure 13.

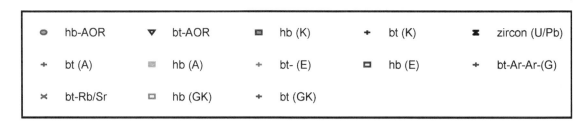

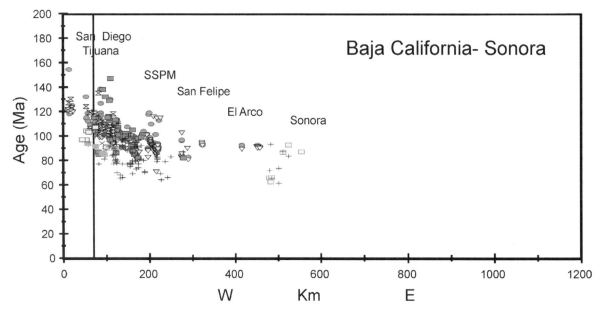

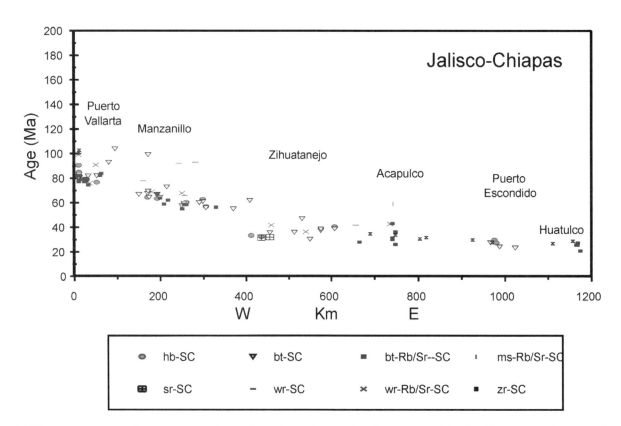

Figure 17. West-to-east transects illustrating eastward younging of locus of magmatism for western mainland México. Top profile extends from Baja California to Sonora (base line is 117°W Meridian). AOR—^{40}Ar/^{39}Ar plateau dates, this study; K—recalculated K/Ar dates after Krummenacher et al. (1975); A—recalculated K/Ar dates after Armstrong and Suppe (1973); E—recalculated K/Ar dates after Evernden and Kistler (1970); GK—recalculated K/Ar dates after Gastil and Krummenacher (1977), Gastil et al. (1978); G—^{40}Ar/^{39}Ar plateau dates from Grove (1994); bt/Rb/Sr after DePaolo (1981); Kimbrough et al. (2001) and references within; Schmidt (1992). Bottom profile is from Sinaloa to Chiapas (after Schaaf et al., 1995), SC—Schaaf et al., 1995, and references therein). Legend: hb—hornblende, bt—biotite, sr—sericite, wr—whole rock, Rb/Sr—^{87}Rb/^{86}Sr analyses.

subparallel to the trend of the batholith explains the pattern of east-ward younging and can largely account for the observed discordant paleomagnetic directions without large-scale northward transport. As illustrated by the regional biotite chrontours (Fig. 16B), the simple closing of the Gulf of California restores the peninsula to its original position with respect to the rest of North America, a position that it had throughout the Phanerozoic. The history proposed above contrasts with models requiring major Cenozoic northward translation based on paleomagnetic inclinations.

CONCLUSIONS

By combining the new geochronological data of the present study and age data available from the literature with geological and paleomagnetic observations also from the literature, the following conclusions can be made:

1. The eastward younging of U/Pb zircon dates for the northern Peninsular Ranges batholith confirms an eastward migration of the focus of magmatism as a result of a migrating magmatic arc during the Mesozoic.

2. The $^{40}Ar/^{39}Ar$ plateau dates for cogenetic hornblende and/or biotite and/or muscovite are concordant or show a normal sequence of cooling ages. This small difference in dates, from the same samples, suggests rapid cooling. However, across the Peninsular Ranges batholith, hornblende plateau dates decrease from 118 Ma in the west to 83 Ma in the east, and biotite plateau dates decrease from 116 Ma in the west to 80 Ma in the east, paralleling the decreasing U/Pb zircon dates.

The almost ubiquitous occurrence of plateaus in the $^{40}Ar/^{39}Ar$ age spectra, when coupled with the chrontour patterns, is best explained as result of a sequential west-to-east emplacement of the granitoid intrusions, followed by sequential rapid cooling and uplift and denudation of both the intrusive rocks and the surrounding metamorphic rocks. This was one of three interpretations for the K/Ar apparent-age gradient put forward by Krummenacher et al. (1975) that is corroborated by the new $^{40}Ar/^{39}Ar$ data, and this conclusion can be extended south of the Agua Blanca fault.

3. Apatite fission-track dates (ranging from 104 to 51 Ma) show that there is a comparable southwest to northeast decrease in cooling ages across the batholith. The dates indicate (assuming a normal geothermal gradient) that present exposures of the western and eastern Peninsular Ranges batholith had cooled to temperatures of <110 °C and therefore were within «3 km of Earth's surface by between 104 and 83 Ma and between 78 and 51 Ma, respectively.

The northeastward decrease in ages across the zone is therefore attributed to regional eastward migration of granitic intrusion foci combined with differential and rapid exhumation histories and superimposed east-side-up tilting of crustal-scale sized blocks containing the plutons and exemplifies the brief lapse between batholith intrusion and unroofing

4. Detailed studies of several plutons reveal that the $^{40}Ar/^{39}Ar$ and fission-track dates also show a systematic decrease in age toward the east within the individual plutons that cannot be

related to the overall eastward migration of magmatism. These local age gradients are attributed to minor (~15°) east-side-up tilting of the crustal-scale blocks containing the plutons. This interpretation contrasts with that of Butler et al. (1991), who suggest regional-scale tilting of the entire Peninsular Ranges.

5. Local block tilting about an axis subparallel to the trend of the batholith can largely account for the observed discordant paleomagnetic inclinations (as suggested by Butler et al., 1991) without large-scale northward transport. This suggests that the Peninsular Ranges province has been part of the adjacent North America craton since the Early Cretaceous and has undergone only limited northward displacement (~300 km) consistent with the opening of the Gulf of California.

6. Restoration of Alta and Baja California to its pre-drift position with respect to the rest of México (with the Los Cabos batholith near Puerto Vallarta) confirms that the new $^{40}Ar/^{39}Ar$ and K/Ar biotite chrontours cross into the western part of mainland México from Sonora to Jalisco without disruption.

7. The amalgamation of new and published age data reveals that the systematic eastward younging can be extended from southern California as far south as Jalisco, central México toward the east, and Chiapas toward the southeast. This suggests that starting during the Mesozoic along the Pacific coast and extending at least to Eocene time in mainland México, the locus of magmatism migrated from west to east across this part of western North America. This indicates that a long and linear (north-northwest–south-southeast) subduction zone, analogous to the modern setting of the South American Andes, formed the basic framework of western México during this time.

8. The weight of evidence in favor of correlating geological piercing points across the Gulf of California, and the continuation of chrontours into the mainland, appear to outweigh paleomagnetic evidence for 1000–2500 km of apparent northward displacement of Baja California in the Cretaceous and early Tertiary, which may be explained by ~15–20° tilting of fault blocks recorded by $^{40}Ar/^{39}Ar$ cooling ages.

ACKNOWLEDGMENTS

This study could not have been carried out without a research scholarship provided to AOR by the Mexican Science Foundation (CONACyT) and Research CONACyT grant #33100-T, research and teaching assistantships from the Department of Geological Sciences at Queen's University, and by grants to E. Farrar and J.A. Hanes from the Natural Sciences and Engineering Research Council (NSERC). The Centro de Investigación Científica y Educación Superior de Ensenada, San Diego State University, family, and friends provided logistical assistance, without which the field research would have been impossible. Travel grants from the School of Graduate Studies at Queen's University and the Department of Geology at San Diego State University are gratefully acknowledged. We also thank Ed Farrar, John Hanes, Doug Archibald, Gordon Gastil, Dave Kimbrough, Marcos Zentilli, Vicky Todd, Paul Wetmore, Marty Grove, Chris Henry, and Fred McDowell for suggestions and critical reviews on the structure and content of this paper.

REFERENCES CITED

Abbott, P.L., and Smith, T.E., 1978, Trace-element comparison of clasts in Eocene conglomerates, southwestern California and northwestern México: Journal of Geology, v. 86, p. 753–762.

Abbott, P.L., and Smith, T.E., 1989, Sonora, México, source for the Eocene Poway Conglomerate of southern California: Geology, v. 17, p. 329–332.

Ague, J.J., and Brandon, M.T., 1992, Tilt and northward offset of Cordilleran batholiths resolved using igneous barometry: Nature, v. 360, p. 146–149.

Ague, J.J., and Brimhall, G.H., 1988a, Magmatic arc asymmetry and distribution of anomalous plutonic belts in the batholiths of California: Effects of assimilation, crustal thickness, and depth of crystallization: Geological Society of America Bulletin, v. 100, p. 912–927.

Ague, J.J., and Brimhall, G.H., 1988b, Regional variations in bulk chemistry, mineralogy, and the compositions of mafic and accessory minerals in the batholiths of California: Geological Society of America Bulletin, v. 100, p. 891–911.

Allison, E.C., 1955, Middle Cretaceous gastropoda from Punta China, Baja California, Mexico: Journal of Paleontology, v. 29, p. 400–432.

Allison, E.C., 1974, The type Alisitos Formation (Cretaceous, Aptian-Albian) of Baja California and its bivalve fauna, in Geology of peninsular California: Pacific Sections AAPG, SEPM, and SEG, Guidebook, p. 20–59.

Aranda-Gómez, J.J., Pérez-Venzor, J.A., 1989, Estratigrafía del complejo cristalino de la región de Todos Santos, Estado de Baja California Sur: Revista de Universidad Nacional Autónoma de Mexico, Instituto de Geologia, v. 8, p.149–170.

Armstrong, R.L., and Suppe, J., 1973, Potassium-argon geochronometry of Mesozoic igneous rocks in Nevada, Utah, and southern California: Geological Society of America Bulletin, v. 84, p. 1375–1392.

Axen, G.J., Grove, M., Stockli, D., Lovera, O.M., Rothstein, D.A., Fletcher, J.M., Farley, K., and Abbot, P.L., 2000, Thermal evolution of Monte Blanco Dome: Low-angle normal faulting during Gulf of California rifting and late Eocene denudation of the eastern Peninsular Ranges: Tectonics, v. 19, p. 197–212.

Bateman, B.C., 1983, A summary of critical relations in the central part of the Sierra Nevada batholith, California, U.S.A, in Roddick, J.A., ed., Circum-Pacific plutonic terranes: Geological Society of America Memoir 159, p. 241–254.

Beck, M.E., 1991, Case for northward transport of Baja and coastal southern California: paleomagnetic data, analysis, and alternatives: Geology, v. 19, p. 506–509.

Beck, M.E., 1992a, Tectonic significance of paleomagnetic results for the western conterminous United States, in Burchfiel, B.C., Lipman, P.W., and Zoback, M.L., eds., The Cordilleran orogen: Conterminous U.S.: Boulder, Colorado, Geological Society of America, Geology of North America, v. G-3, p. 683–697.

Beck, M.E., 1992b, Some thermal and paleomagnetic consequences of tilting a batholith: Tectonics, v. 11, p. 197–302.

Böhnel, H., Delgado-Argote, L., and Kimbrough, D.K., 2002, Discordant paleomagnetic data for middle-Cretaceous, intrusive, intrusive rocks from northern Baja California: Latitude displacement, tilt, or vertical axis rotation?: Tectonics, v. 21, no. 5, p. 1049.

Burchfiel, B.C., Cowan, D.S., and Davis, G.A., 1992, Tectonic overview of the Cordilleran orogen in the western United States, in Burchfiel, B.C., Lipman, P.W., and Zoback, M.L., eds., The Cordilleran orogen, conterminous U.S.: Boulder Colorado, Geological Society of America, Geology of North America, v. G-3, p. 407–482.

Busby, C., Smith, D., Morris, W., and Fackler-Adam, B., 1998, Evolutionary model for convergent margins facing large ocean basins: Mesozoic Baja California Mexico: Geology, v. 26, p. 227–230.

Butler, R.F., Gehrels, G.E., McClelland, W. C., May, S.R., and Klepacki, D., 1989, Discordant paleomagnetic poles from the Canadian Coast Plutonic Complex: Regional tilt rather than large-scale displacement?: Geology, v. 17, p. 691–694.

Butler, R.F., Dickinson, W.R., and Gehrels, G.E., 1991, Paleomagnetism of coastal California and Baja California: Alternatives to large-scale northward transport: Tectonics, v. 10, p. 561–576.

Campa, M.F., and Coney, P.J., 1983, Tectonostratigraphic terranes and mineral resource distributions in Mexico: Canadian Journal Earth Sciences, v. 26, p. 1040–1051.

Chadwick, B., 1987, The geology, petrography, geochemistry, and geochronology of the Tres Hermanas–Santa Clara region, Baja California, Mexico, [M.Sc. thesis]: San Diego, San Diego State University, 208 p.

Champion, D.E., Howell, D.G., and Gromme, C.S., 1984, Paleomagnetic and geological data indicating 2500 km of northward displacement for the Salinian and related terranes, California: Journal of Geophysical Research, v. 89, p. 7736–7752.

Clark, A.H., Farrar, E., Caelles, J.C., Haynes, S.J., Lortie, R.B., McBride, S.L., Quirt, G.S., Robertson, R.C.R., and Zentilli, M., 1976, Longitudinal variations in the metallogenetic evolution of the central Andes: A progress report, in Strong, D.F., ed., Metallogeny and plate tectonics: Geological Association of Canada Special Paper 14, p. 23–58.

Clark, K.F., Foster, C.T., and Damon, P.E., 1982, Cenozoic mineral deposits and subduction-related magmatic arcs in Mexico: Geological Society of America Bulletin, v. 93, p. 533–544.

Clarke, S.H., Jr.; Nilsen, and Tor H., 1973, Displacement of Eocene strata and implications for the history of offset along the San Andreas Fault, central and northern California, in Conference on tectonic problems of the San Andreas Fault system, Proceedings: Stanford University Publications, Geological Sciences 13, p. 358–367

Clinkenbeard, J.P., 1987, The mineralogy, geochemistry, and geochronology of the La Posta pluton, San Diego and Imperial Counties, California [M.Sc. thesis]: San Diego, San Diego State University, 148 p.

Coney, P.J., 1972, Cordilleran tectonics and North American plate motion: American Journal of Science, v. 275-A, p. 136–148.

Coney, P.J., 1981, Plate tectonic evolution of the southern Cordillera, in Dickinson, W.R., and Payne, W.D., eds., Relations of tectonics to ore deposits in the southern Cordillera: Tucson, Arizona, Arizona Geological Society Digest, Volume XIV, p.113–136.

Coney, P.J., and Reynolds, S.J., 1977, Cordilleran Benioff zones: Nature, v. 270, p. 403–406.

Coney, P.J., Jones, D.L., and Monger, J.W.H., 1980, Cordilleran suspect terranes: Nature, v. 288, p. 329–333.

Constanzo-Alvarez, V., and Dunlop, D.J., 1988, Paleomagnetic evidence for post 2.55 Ga tectonic tilting and 1.1 Ga reactivation in the southern Kapuskasing zone, Ontario, Canada: Journal of Geophysical Research, v. 93, p. 9126–9136.

Dalrymple, G.B., Alexander Jr., E.C., Lanphere, M.A., and Kraker, G.P., 1981, Irradiation of samples for $^{40}Ar/^{39}Ar$ dating using the Geological Survey TRIGA Reactor: U.S. Geological Survey Professional Paper 1176, 55 p.

Damon, P.E., Shafiquillah, M., and Clark, K.F., 1983, Geochronology of the porphyry copper deposits and related mineralization of Mexico: Canadian Journal Earth Sciences, v. 20, p. 1052–1071.

DePaolo, D.J., 1981, A neodymium and strontium isotopic study of the Mesozoic calc-alkaline granitic batholiths of the Sierra Nevada and Peninsular Ranges, California: Journal of Geophysical Research, v. 86, p. 10470–10488.

Delgado-Argote, L.A., López-Martínez, M., Pérez-Flores, M.A., and Fernández-Tomé, R., 1995, Emplacement of the nucleus of the San Telmo Pluton, Baja California, from geochronologic, fracture and magnetic data, in Jacques-Ayala, C., González-León, C.M., and Roldán-Quintana, J. eds., Studies on the Mesozoic of Sonora and adjacent areas: Geological Society of America Special Paper 301, p. 191–204.

Dickinson, W.R., 1981a, Plate tectonic evolution of the southern Cordillera, in Dickinson, W.R., and Payne, W.D., eds., Relations of tectonics to ore deposits in the southern Cordillera: Tucson, Arizona, Arizona Geological Society Digest, Volume XIV, p. 113–136.

Dickinson, W.R., 1981b, Plate tectonics and the continental margin of California, in Ernst, W.G., ed., The geotectonic development of California. Volume I: Englewood Cliffs, New Jersey, Prentice Hall, p. 29–50.

Dickinson and Lawton, 2001, Carboniferous to Cretaceous assembly and fragmentation of Mexico: Geological Society of America Bulletin, v. 113, p. 1142–1160.

Dodson, M.H., 1973, Closure temperature in cooling geochronogical and petrological systems: Contributions for Mineralogy and Petrology, v. 4, p. 259–274.

Duffield, W.A., 1968, The petrology and structure of the El Pinal tonalite, Baja California, México: Geological Society of America Bulletin, v. 79, p. 1351–1374.

Dumitru, T.A., 1990, Subnormal Cenozoic geothermal gradients in the extinct Sierra Nevada magmatic arc: Consequences of Laramide and post-

Laramide shallow-angle subduction: Journal of Geophysical Research, v. 95, p. 4925–2941.

Early, T.O., and Silver, L.T., 1970, Rb-Sr isotopic systematics in the Peninsular Ranges of southern and Baja California: Eos (Transactions, American Geophysical Union), v. 54, p. 494.

Engebretson, D.C., Cox, A., and Thompson, G.A., 1984, Correlation of plate motions with continental tectonics: Laramide to Basin-Range: Tectonics, v. 3, p. 115–119.

Engebretson, D.C., Cox, A., and Gordon, R.G., 1985, Relative motions between oceanic and continental plates in the Pacific Basin: Boulder, Colorado, Geological Society of America Special Paper 206, 59 p.

Evernden, J.K., and Kistler, R.W., 1970, Chronology of emplacement of Mesozoic batholithic complexes in California and western Nevada: U.S. Geological Survey Professional Paper 623, 42 p.

Farrar, E., Clark, A.H., Haynes, S.J., Quirt, G.S., and Zentilli, M., 1970, K-Ar evidence for the post-Paleozoic migration of granitic intrusion foci in the Andes of northern Chile: Earth and Planetary Science Letters, v. 10, p. 60–66.

Flynn, J.J., Cipolletti, R.M., and Novacek, M.J., 1989, Chronology of early Eocene marine and terrestrial strata, Baja California, Mexico: Geological Society of America Bulletin, v. 101, p. 1182–1196.

Frizzell, V.A., Jr., 1984, The geology of the Baja California peninsula; An introduction, in Frizzell, V.A., Jr., ed., Geology of the Baja California peninsula: Los Angeles, Society for Sedimentary Geology (SEPM) Pacific Section, v. 39, p. 1–7.

Frizzell, V.A., Jr., Fox, L.K., Moser, F.C., and Ort, K.M.,1984, Late Cretaceous granitoids, Cabo San Lucas Block, Baja California Sur, Mexico: Eos (Transactions, American Geophysical Union), v. 65, p. 1151.

Gastil, R.G., 1975, Plutonic zones in the Peninsular Ranges of southern California and northern Baja California: Geology, v. 3, p. 361–363.

Gastil, R.G., 1981, The tectonic history of peninsular California and adjacent Mexico, in Ernst, W.G., ed., The geotectonic development of California: Rubey Symposium, Vol. I: Englewood Cliffs, New Jersey, Prentice Hall, p. 284–305.

Gastil, R.G., 1983, Mesozoic and Cenozoic granitic rocks of southern California and western Mexico, in Roddick, J.A., ed., Circum-Pacific plutonic terranes: Boulder, Colorado, Geological Society of America Memoir 159, p. 265–275.

Gastil, R.G., 1985, Terranes of peninsular California and adjacent Sonora, in Howell, D.G., ed., Tectonostratigraphic terranes of the circum-Pacific region: Circum-Pacific Council for Energy and Mineral Resources Earth Science Series 1, p. 273–283.

Gastil, R.G., 1990a, Zoned plutons of the Peninsular Ranges in southern and Baja California, in Shimizu, M., and Gastil, G., eds., Recent advances in concepts concerning zoned plutons in Japan and southern and Baja California: The University Museum, The University of Tokyo, Nature and Culture 2, p. 77–90.

Gastil, R.G., 1990b, The boundary between the magnetite-series and ilmenite-series granitic rocks in peninsular California, in Shimizu, M., and Gastil, G., eds., Recent advances in concepts concerning zoned plutons in Japan and southern and Baja California: The University Museum, The University of Tokyo, Nature and Culture 2, p. 91–100.

Gastil, R.G., 1991, Is there a Oaxaca-California megashear? Conflict between paleomagnetic data and other elements of geology: Geology, v. 19, p. 502–505.

Gastil, R.G., 1993, Prebatholithic history of peninsular California: Boulder, Colorado, Geological Society of America Special Paper 279, p. 145–156.

Gastil, R.G., and Krummenacher, D., 1977, Reconnaissance geology of coastal Sonora between Punta Lobos and Bahía Kino: Geological Society of America Bulletin, v. 88, p. 189–198.

Gastil, R.G., and Miller, R.H., 1984, Prebatholithic paleogeography of peninsular California and adjacent Mexico, in Frizzel, V.A., Jr. ed., Geology of the Baja California peninsula: Los Angeles, Society for Sedimentary Geology (SEPM) Pacific Section, v. 39, p. 9–16.

Gastil, R.G., Phillips, R.P., and Allison, E.C., 1975, Reconnaissance geology of the state of Baja California and reconnaissance geologic map of the state of Baja California: Boulder, Colorado, Geological Society of America Memoir 140, 170 p.

Gastil, R.G., Krummenacher, D., and Jensky II, W.A., 1978, Reconnaissance geology of west-central Nayarit, Mexico: Geological Society of America Map Chart Series MC-24, scale 1:200,000, 1 sheet.

Gastil, R.G., Morgan, G.J., and Krummenacher, D., 1981, The tectonic history of peninsular California and adjacent Mexico, in Ernst, W.G., ed., The geotectonic development of California: Rubey Symposium, Vol. I: Englewood Cliffs, New Jersey, Prentice Hall, p. 284–306.

Gastil, R.G., Diamond, J., and Knaack, C. 1986, The magnetite ilmenite-line in peninsular California: Geological Society of America Abstracts with Programs, v. 18, p. 109.

Gastil, R.G., Diamond, J., Knaack, C., Walawender, M., Marshall, M., Boyles, C., and Chadwick, B., 1990, The problem of the magnetite/ilmenite boundary in southern and Baja California, in Anderson , J.L., ed., The nature and origin of Cordilleran magmatism: Boulder, Colorado, Geological Society of America Memoir 174, p. 19–32.

Gastil, R.G., Tanasi, Y., Chammies, M., and Gunn, S., 1991, Plutons of the eastern Peninsular Ranges, Southern California, U.S.A., and Baja California, México, in Walawender, M.J., and Hainan, B., eds., Geological excursions in southern California and México: Geological Society of America, 1991 Annual Meeting, Guidebook, p. 124–145.

Gastil, R.G., Ortega Rivera, A., and López, M., 1994, History of the Sierra San Pedro Mártir pluton crystallization: Geological Society of America Abstracts with Programs, v. 26, no. 2, p. 54.

George, P.G., and Dokka, R.K., 1994, Major Late Cretaceous cooling events in the eastern Peninsular Ranges, California, and their implications for Cordilleran tectonics: Geological Society of America Bulletin, v. 106, p. 903–914.

Goodwin, L.B., and Renne, P.R., 1991, Effects of progressive mylonitization on Ar retention in biotites from the Santa Rosa mylonite zone. California, and thermochronologic implications: Contributions to Mineralogy and Petrology, v. 8, p. 283–297.

Griffith, R., and Hobbs, J., 1993, Geology of the southern Sierra Calamajue, Baja California Norte, Mexico, in Gastil, R.G., and Miller, R.D., eds., The prebatholithic stratigraphy of peninsular California: Boulder, Colorado, Geological Society of America Special Paper 279, p. 43–60.

Gromet, L.P., and Silver, L.T., 1987, REE variations across the Peninsular Ranges batholith: Implications for batholithic petrogenesis and crustal growth in magmatic arcs: Journal of Petrology, v. 28, p. 77–125.

Grove, M., 1994, Contrasting denudation histories within the east-central Peninsular Ranges batholith (33°N): Geological Society of America, Cordilleran Section, Guidebook: San Bernardino, California, p. 235–240.

Gunn, S.H., 1985, Geology, petrology, and geochemistry of the Laguna Juárez pluton, Baja California, México: [M.Sc. thesis]: San Diego, San Diego State University, 166 p.

Hagstrum, J.T., and Filmer, P.E., 1990, Paleomagnetic and tectonic constraints on the Late Cretaceous to early northward translation of the Baja California peninsula: Geofisica Internacional, v. 29, p. 175–184.

Hagstrum, J.T., and Sedlock, R.L., 1998, Remagnetization of Cretaceous forearc strata on Santa Margarita and Magdalena islands, Baja California Sur; implications for northward transport along the California margin: Tectonics, v. 17, no. 6 (199812), p. 872–882.

Hagstrum, J.T., McWilliams, M., Howell, D.G., and Gromme, S., 1985, Mesozoic paleomagnetism and northward translation of the Baja California peninsula: Geological Society of America Bulletin, v. 96, p. 1077–1090.

Hagstrum, J.T., Sawlan, M.G., Hausback, B.P., Smith, J.G. , and Gromme, C.S., 1987, Miocene paleomagnetism and tectonic setting of the Baja California peninsula, Mexico: Journal of Geophysical Research, v. 92, p. 2627–2639.

Hamilton, W., 1969, Mesozoic California and the underflow of Pacific mantle, Geological Society of America Bulletin, Vol. 80, p. 2409–2430.

Hamilton, W., 1988a, Plate tectonics and island arcs: Geological Society of America Bulletin, v. 100, p. 1503–1527.

Hamilton, W., 1988b, Tectonic setting and variations with depth of some Cretaceous and Cenozoic structural and magmatic systems of the western United States, in Ernst, W.G., ed., Metamorphism and crustal evolution of the western United States, Rubey Symposium, Vol. VII: Englewood Cliffs, New Jersey, Prentice Hall, p. 894–937.

Hammarstrom, J.M., 1992, Mineral chemistry of Cretaceous plutons; hornblende geobarometry in Southern California and southern Alaska: Geological Society of America Abstracts with Programs, v. 24, no. 5, p. 30.

Harrison, T.M., 1981, Diffusion of ^{40}Ar in hornblendes: Contributions to Mineralogy and Petrology, v. 78, p. 324–331.

Harrison, T.M., and Clarke, G.K.C., 1979, A model of the thermal effects of igneous intrusion and uplift as applied to Quottoon Pluton, British Columbia: Canadian Journal of Earth Sciences, v. 16, p. 411–420.

Harrison, T.M., Duncan, I., and McDougall, I., 1985, Diffusion of ^{40}Ar in biotite: Temperature, pressure and compositional effects: Geochimica et Cosmochimica Acta, v. 49, p. 2461–2468.

Helmstaedt, H., and Doig, R., 1975, Eclogite nodules from kimberlite pipes of the Colorado Plateau-samples of subducted Franciscan-type oceanic lithosphere, *in* Ahrens, L.H., Dawson, J.B., Duncan, A.R., and Erlank, A.J. eds., Physics and chemistry of the Earth: New York, Pergamon Press, p. 95–111.

Henry, C.D., and Fredrikson, G., 1987, Geology of part of southern Sinaloa, México adjacent to the Gulf of California: Geological Society of America Map and Chart Series MCH063, scale 1:250,000, 14 sheets.

Hill, R.I., 1984, Petrology and petrogenesis of batholithic rocks, San Jacinto Mountains, southern California [Ph.D. thesis]: Pasadena, California, California Institute of Technology, 234 p.

Howell, D.G., Champion, D.E., and Vedder, J.G, 1987, Terrane accretion, crustal kinematics, and basin evolution, southern California, *in* Ingersoll, R.A., and Ernst, W.G., eds., Rubey Symposium, Vol. VI: Englewood Cliffs, New Jersey, Prentice Hall, p. 242–258.

Irving, E., and Archibald, D.A., 1990, Bathozonal tilt corrections to paleomagnetic data from mid-Cretaceous plutonic rocks: Examples from the Omenica Belt, British Columbia: Journal of Geophysical Research, v. 95, p. 4579–4585.

Ishihara, S., 1977, The magnetite-series and ilmenite-series granitic rocks: Mining Geology, v. 27, p. 293–305.

Johnson, S.E., Paterson, S.R., and Tate, M.C., 1999a, Structure and emplacement history of a multiple-center, cone-sheet-bearing ring complex: The Zarza Intrusive Complex, Baja California, Mexico: Geological Society of America Bulletin, v. 111, p. 607–619.

Johnson, S.E., Tate, M.C., and Fanning, C.M., 1999b, New geologic mapping and SHRIMP U-Pb zircon data in the Peninsular Ranges batholith, Baja California, Mexico: Evidence for a suture?: Geology, v. 27, p. 643–746.

Jones, D.L., Silberling, N.L., Gilbert, W., and Coney, P., 1982, Character distribution and tectonic significance of accretionary terranes in the central Alaska range: Journal of Geophysical Research, v. 87, p. 3709–3717.

Keith, S.B., 1978, Paleosubduction geometries inferred from Cretaceous and Tertiary magmatic patterns in southwestern North America: Geology, v. 6, p. 516–521.

Keppie, J.D., and Dostal, J., 2001, Evaluation of the Baja controversy using paleomagnetic and faunal data, plume magmatism, and piercing points: Tectonophysics, v. 339, p. 427–442.

Kimbrough, D.L., Smith, D.P., Mahoney, J.B., Moore, T.E., Grove, M., Gastil, R.G., Ortega- Rivera, A., and Fanning, C.M., 2001, Forearc-basin sedimentary response to rapid late Cretaceous batholith emplacement in the Peninsular ranges of southern and Baja California: Geology, v. 29, p. 491–493.

Kimzey, J.A., 1982, Petrography and geochemistry of the La Posta granodiorite [M.Sc. thesis]: San Diego, California, Department of Geological Sciences, San Diego State University, 150 p.

Krummenacher, D., Gastil, R.G., Bushee, J., and Doupont, J., 1975, K-Ar apparent ages, Peninsular Ranges batholith, southern California and Baja California: Geological Society of America Bulletin, v. 86, p. 760–768.

Larsen, E.S., 1948, Batholith and associated rocks of Corona, Elsinore, and San Luis Rey quadrangles, southern California: Boulder, Colorado, Geological Society of America, Memoir 29, 182 p.

López-Ramos, E., 1978, Geología de México, Tomo III: México City, Universidad Nacional Autónoma de México, 453 p.

Lovera, O.M., Grove, M., Kimbrough, D.L., and Abbot, P.L., 1999, A method for evaluating basement exhumation histories from closure age distribution of detrital minerals: Journal of Geophysical Research, v. 104, p. 29419–29438.

Lund, S.P., and Bottjer, D.J., 1991, Paleomagnetic evidence for microplate tectonic development of southern California and Baja California, *in* Dauphin, J.P., and Simoneit, B.R.T., eds., The gulf and peninsular province of the Californias: American Association of Petroleum Geologists Memoir 47, p. 231–248.

Magistrale, H., and Sanders, C., 1995, P wave image of the Peninsular Ranges batholith, southern California: Geophysical Research Letters, v. 22, p. 2549–2552.

Marquis, G., and Irving, E., 1990, Observing tilts in midcrustal rocks by paleomagnetism: Examples from British Columbia: Tectonics, v. 9, p. 925–934.

McCormick, W.V., 1986, The geology, mineralogy, and geochronology of the Sierra San Pedro Mártir pluton, Baja California, México [M.Sc. thesis]: San Diego, California, San Diego State University, 123 p.

Moran-Zenteno, D.J., Martiny, B., Tolson, G., Solis-Pichardo, G., Alba-Aldave, L., Hernandez- Bernal, M.D.S., Macias-Romo, C., Martínez-Serrano, R.G., Schaaf, P., and Silva-Romo, G., 2000, Geocronología y características geoquímicas de las rocas magmáticas terciarias de la Sierra Madre del Sur: Boletín de la Sociedad Gemológica Mexicana, v. 53, p. 27–58.

Morris, L.K., Lund, S.P., and Bottjer, D.J., 1986, Paleolatitude drift history of displaced terranes in southern and Baja California: Nature, v. 321, p. 844–847.

Naeser, C.W., 1979, Fission-track dating and geologic annealing of fission tracks, *in* Jäger, E., and Hunziker, J.C., eds., Lectures in isotope geology: New York, Springer-Verlag, p. 154–169.

Nozawa, T., 1983, Felsic plutonism in Japan, *in* Roddick, J.A., ed., Circum-Pacific plutonic terranes: Boulder, Colorado, Geological Society of America Memoir 159, p. 105–122.

Ortega-Rivera, A., 2003, Eastward younging of ages across the Peninsular Ranges Batholith and Plutons: Geochronological constraints for the tectonic history of the Peninsular Ranges Batholith and tectonic implications for Western México: Geological Society of America Abstracts with Programs, v. 35, no. 4, p. 73.

Ortega-Rivera, M.-A., 1988, Neotectónica de un sector de la falla de Agua Blanca, Valle de Agua Blanca (Rancho la Cocina–Rancho Agua Blanca), Baja California México [M.Sc. thesis]: Ensenada, Baja California, México, Departamento de Sismología, División de Ciencias de la Tierra, Centro de Investigación Científica y Educación Superior de Ensenada, 146 p.

Ortega-Rivera, M.-A., 1997, Geochronological constraints on the thermal and tilting history of the Peninsular Ranges batholith of Alta California and Baja California: Tectonic implications for southwestern North America (México) [Ph.D. thesis]: Kingston, Ontario, Canada, Department of Geological Sciences, Queen's University, 582 p.

Ortega-Rivera, M.-A., 1998a, New geochronological framework for the evolution of the Mesozoic Peninsular Ranges batholith of Alta and Baja California, *in* Alaniz Alvarez, S., Ferrari, L., Nieto Samaniego, A.F., and Ortega-Rivera, M.A., eds., Primera Reunión Nacional de Ciencias de la Tierra, Libro de Resúmenes, p. 108.

Ortega Rivera, M.-A., 1998b, Geochronological constrains for testing controversial paleomagnetic and geological models on the tectonic evolution of the Peninsular Ranges Province [abs]: Reunión Anual de la Unión Geofísica Mexicana, Puerto Vallarta, México, Abstracts with Programs, v. 18, p. 346.

Ortega Rivera, M.-A., 2000a, Geochronological constraints for testing controversial paleomagnetic and geological models on the tectonic evolution of the Peninsular Ranges province, *in* Proceedings, 31st International Geological Congress: Rio de Janeiro, Brazil, Geological Survey of Brazil (CD-ROM).

Ortega Rivera, M.-A., 2000b, Geochronological constraints on the tectonic history of the Peninsular Ranges batholith of Alta and Baja California: Tectonic implications for Western México, *in* Education and Planning Workshop: Rupturing of the continental lithosphere in the Gulf of California and the Salton Trough [abs.]: MARGINS, Puerto Vallarta Abstracts Volume 1, p. 82–84 (CD-ROM).

Ortega Rivera, M.-A., 2000c, Restricciones geocronológicas para la historia tectónica del batolito de las sierras peninsulares de Alta y Baja California: Implicaciones tectónicas para el oeste de México (Geochronological constraints on the tectonic history the peninsular ranges batholith of Alta and Baja California: tectonic implications for western México), *in* Reunión Internacional sobre la Geología de la Península de Baja California, 5ª. Memorias (International Meeting on Geology of the Baja California Peninsula) [abs.]: Loreto, Baja California Sur, Universidad Autónoma de Baja California, Universidad Nacional Autónoma de México, Sociedad Geológica Peninsular, p. 79–80.

Ortega-Rivera, M.-A., Farrar, E., Hanes, J.A., Archibald, D.A., Gastil, R.G., López-Martínez, M., and Féraud, G., 1994a, Cooling history of the Sierra San Pedro Mártir pluton, Baja California, México, from $^{40}Ar/^{39}Ar$ geochronology: Geological Society of America, Abstracts with Programs, v. 26, no. 2, p. A78.

Ortega-Rivera, M.-A., Hanes, J.A., Archibald, D.A., Farrar, E., and Gastil, R.G., 1994b, $^{40}Ar/^{39}Ar$ results for La Posta-Type plutons, Baja California: Geological Society of America Abstracts with Programs, v. 26, no. 7, p. A-196.

Ortega-Rivera, M.-A., Farrar, E., Hanes, J.A., Archibald, D.A., Gastil, R.G., Kimbrough, D., López-Martínez, M., Féraud, G., and Zentilli, M., 1997, Chronological constraints on the thermal and tilting history of the Sierra San Pedro Mártir Pluton, Baja California, México, from U/Pb, $^{40}Ar/^{39}Ar$, and fission track geochronology: Geological Society of America Bulletin, v. 109, no. 6, p. 728–745.

Palmer, A.R., 1983, The decade of North American geology 1983, Geologic Time Scale: Geology, v. 11, p. 503–504.

Purdy, J.W., and Jäger, E., 1976, K-Ar ages of rock-forming minerals from the central Alps: The Institute of geology and mineralogy: University of Padua, Italy, Memoir 30, 31 p.

Rangin, C., 1978, Speculative model of Mesozoic geodynamics, central Baja California to northeastern Sonora (Mexico), *in* Howell, D.G., and McDougall, K.A., eds., Mesozoic paleogeography of the western United States, *in* Proceedings, Pacific Coast Paleogeography Symposium 2: Los Angeles, Pacific Section, Society of Economic Paleontologists and Mineralogists, p. 65–106.

Ravenhurst, C.E., and Donelick, R.A., 1992, Fission track thermochronology, *in* Zentilli, M., and Reynolds, P.H., eds., Short Course on Low temperature thermochronology: Halifax, Nova Scotia, Mineralogical Association of Canada, Short Course, v. 20, p. 21–42.

Ravenhurst, C.E., Willet, S.D., Donelick, R.A., and Beaumont, C., 1994, Apatite fission track thermochronometry from central Alberta: Implications for the thermal history of the western Canada sedimentary basin: Journal of Geophysical Research, v. 99, p. 20023–20041.

Righter, K., Carmichel, I.S.E., Becker, T.A., and Renne, R.P., 1995, Pliocene to Quaternary volcanism and tectonism at the intersection of the Mexican Volcanic Belt and the Gulf of California: Geological Society of America Bulletin, v. 107, p. 612–626.

Roddick, J.A., ed., 1983a, Circum-Pacific plutonic terranes: An overview, *in* Roddick, J.A., ed., Circum-Pacific plutonic terranes: Boulder, Colorado, Geological Society of America Memoir 159, p. 1–3.

Roddick, J.C., 1983b, High precision intercalibration of $^{40}Ar/^{39}Ar$ standards: Geochimica et Cosmochimica Acta, v. 47, p. 887–898.

Santillán, M., and T. Barrera, 1930, Las posibilidades petrolíferas en la Costa Occidental de Baja California, entre los paralelos 30° y 32° de latitud norte: México (City) Universidad Nacional, Instituto de Geología Anales, v. 5, p. 1–37.

Schaaf, P., Morán-Zenteno, D., Hernandez-Bernal, M., Solis-Pichardo, G., Tolson, G., and Köhler, H., 1995, Paleogene continental margin truncation in southwestern Mexico: Geochronological evidence: Tectonics, v. 14, p. 1339–1350.

Schaaf, P., Böhnel, H., and Pérez-Venzor, J.A., 2000, Pre-Miocene palaeogeography of the Los Cabos block, Baja California Sur: Geochronological and paleomagnetic constraints: Tectonophysics, v. 318, p. 53–69.

Schmidt, M.W., 1992, Amphibole composition in tonalite as a function of pressure: An experimental calibration of the Al-in-hornblende barometer: Contributions to Mineralogy and Petrology, v. 110, p. 304–310.

Schmidt, K.L., 2002, Investigations of arc processes: Relationships among deformation magmatism, mountain building, and the role of crustal anisotropy in the evolution of the Peninsular Ranges batholith, Baja California [PhD. thesis]: Los Angeles, University of Southern California, 310 p.

Sedlock, R.L., Ortega-Gutierrez, F., and Speed, R.C., 1993, Tectonostratigraphic terranes and the tectonic evolution of Mexico: Boulder, Colorado, Geological Society of America Special Paper 278, 153 p.

Silver, L.T., 1992, Tracking magmatic arc evolution across contrasting lithosphere setting in southwestern North America, *in* Brown, P.E., and Chappell, B.W., eds., 2nd Hutton Symposium, The origin of granites and related rocks: Boulder, Colorado, Geological Society of America Special Paper 272, 499 p.

Silver, L.T., 1996, Petrogenetic implications of petrographic and geochemical zonations in the integrated Peninsular Ranges and Sonoran batholiths: Geological Society of America Abstracts with Programs, v. 28, no. 5, p. 112.

Silver, L.T., and Chappell, B., 1988, The Peninsular Ranges batholith: An insight into the Cordilleran batholiths of southwestern North America: Transactions of the Royal Society of Edinburgh, Earth Sciences, v. 79, p. 105–121.

Silver, L.T., Stehli, F.G., and Allen, C.R., 1963, Lower Cretaceous pre-batholithic rocks of northern Baja California, Mexico: American Association of Petroleum Geologists Bulletin, v. 47, p. 2054–2059.

Silver, L.T., Allen, C.R., and Stehli, F.G., 1969, Geological and geochronological observations on a portion of the Peninsular Ranges batholith of northwestern Baja California, México: Boulder, Colorado, Geological Society of America Special Paper 121, p. 279–280.

Silver, L.T., Taylor, H.P., and Chappell, B., 1979, Some petrological, geochemical, and geochronological observations of the Peninsular Ranges batholith near the international border of the U.S.A. and Mexico, *in* Abbott, P.L., and Todd, V.R., eds., Mesozoic crystalline rocks: Geological Society of America Annual Meeting Guidebook, p. 83–110.

Silver, L.T., Chappell, B.W., and Anderson, T., 2000, An integrated view of the Cretaceous Circum-Gulf batholiths of northern Baja California an Sonora, Mexico, *in* Calmus, T, and Pérez Segura, E., eds., Cuarta Reunión sobre la Geología del Noroeste de México y áreas adyacentes, Libro de resúmenes: Hermosillo, Universidad Nacional Autónoma de México, Instituto de Geología, Estación Regional del Noroeste; Universidad de Sonora, Publicaciones Ocasionales, 2, p. 123.

Smith, D.P., and Busby, C.J., 1993, Shallow magnetic inclinations in the Cretaceous Valle Group, Baja California: Remagnetization, compaction, or terrane translation?: Tectonics, v. 12, no. 5, p. 1258–1266.

Steiger, R.H., and Jäger, E., 1977, Subcommission on geochronology: Convention on the use of decay constants in geo- and cosmo-chronology: Earth and Planetary Science Letters, v. 36, p. 359–362.

Suárez-Vidal, F., Armijo, R., Morgan, G., Bodin, P., Gastil, R.G., 1991, Framework of Recent and active faulting in northern Baja California Serie/ Fuente: The Gulf and Peninsular Province of the Californias, *in* Dauphin, J.P., and Simoneit, B.R.T., eds., The Gulf and Peninsular Province of the Californias, Tulsa, Oklahoma, USA, American Association of Petroleum Geologists Memoir 47, p. 285–300.

Tate, M.C., Norman, M.D., Johnson, S.E., Fanning, C.M., and Anderson, J.L, 1999, Generation of tonalite and trondhjemite by subvolcanic fractionation and partial melting in the Zarza Intrusive Complex, western Peninsular Ranges batholith, northwestern Mexico: Journal of Petrology, v. 10, p. 983–1010.

Taylor, H.P., and Silver, L.T., 1978, Oxygen isotope relationships in plutonic igneous rocks of the Peninsular Ranges batholith, southern and Baja California: U.S. Geological Survey Open-File Report 78-701, p. 423–426.

Teissere, R.F., and Beck, M.E., Jr., 1973, Divergent Cretaceous paleomagnetic pole position for the southern California batholith, U.S.A.: Earth and Planetary Science Letters, v. 18, p. 296–300.

Thomson, C.N., and Girty, G.H., 1994, Early Cretaceous intra-arc ductile strain in Triassic-Jurassic rocks and Cretaceous continental margin arc rocks, Peninsular Ranges, California: Tectonics, v. 13, p. 1108–1119.

Todd, V.R., and Shaw, S.E., 1979, Structural, metamorphic and intrusive framework of the Peninsular Ranges batholith in San Diego County, California, *in* Abbott, P.L., and Todd, V.R., eds., Mesozoic crystalline rocks: Geological Society of America Annual Meeting Guidebook, p. 177–231.

Todd, V.R., and Shaw, S.E., 1985, S-type granitoids and an I-S line in the Peninsular Ranges batholith, southern California: Geology, v. 13, p. 231–233.

Todd, V.R., Erskine, B.G., and Morton, D.M., 1988, Metamorphic and tectonic evolution of the northern Peninsular Ranges batholith, southern California, *in* Ernst, W.G., ed., Metamorphism and crustal evolution of the western United States, Rubey Symposium, Vol. VII: Prentice Hall, p. 894–937.

Todd, V.R., Kimbrough, D.L., and Herzig, C.T., 1994, The Peninsular Ranges batholith in San Diego County, California, from volcanic arc to mid-crustal intrusive and metamorphic rocks: Geological Society of America, Cordilleran Section, Guidebook: San Bernardino California, p. 227–235.

Walawender, M.J., Gastil, R.G., Clinkenbeard, J.P., McCormick, W.V., Eastman, B.G., Wardlaw, R.S., Gunn, S.H., and Smith, B.M., 1990, Origin and evolution of the zoned La Posta-type plutons, eastern Peninsular Ranges Batholith, southern and Baja California, *in* Anderson, J.L., ed., The nature and origin of Cordilleran magmatism: Boulder, Colorado, Geological Society of America Memoir 174, p. 1–18.

Walawender, M.J., Girty, G.H, Lombardi, M.R., Kimbrough, D., Girty, M.S., and Anderson, C., 1991, A synthesis of recent work in the Peninsular Ranges batholith, *in* Walawender, M.J., and Hanan, B. eds., Geological excursions in southern California and México: Geological Society of America Annual Meeting Guidebook, p. 297–312.

Wallace, P., and Carmichel, I.S.E., 1989, Minette lavas and associated leucitites from the western front of the Mexican Volcanic Belt: Petrology, chemistry and origin: Contributions to Mineralogy and Petrology, v. 103, p. 470–492.

Wetmore, P.H., Schmidt, K.L., Patterson, S.R., and Herzig, C., 2002, Tectonic implications for the along-strike variation of the Peninsular Ranges batholith, southern and Baja California: Geology, v. 30, p. 247–250.

Yule, J.D., and Herzig, C., 1994, Paleomagnetic evidence for no large-scale northward translation of the southern California Peninsular Ranges: Geological Society of America Abstracts with Programs, v. 26, no. 7, p. A-461.

Zentilli, M., 1974, Geological evolution and metallogenetic relationships in the Andes of northern Chile between 26° and 29° South: [Ph.D. thesis]: Kingston, Ontario, Canada, Queen's University, 446 p.

Ziegler, A.M., Rowley, D.B., Lottes, A.L., Sahagian, D.L., Hulver, M.L., and Gierlowski, T.C., 1985, Paleogeographic interpretation: With an example from the mid-Cretaceous: Annual Review of Earth and Planetary Sciences, v. 13, p. 385–425.

MANUSCRIPT ACCEPTED BY THE SOCIETY JUNE 2, 2003

Geological Society of America
Special Paper 374
2003

Geothermal gradients in continental magmatic arcs: Constraints from the eastern Peninsular Ranges batholith, Baja California, México

David A. Rothstein*
Craig E. Manning
Department of Earth and Space Sciences, University of California, Los Angeles, California 90095-1567, USA

ABSTRACT

In continental arcs the extension of geothermal gradients derived from shallow crustal levels to depth predicts widespread melting at pressures that are inconsistent with seismic studies. Numerical models of low-pressure metamorphism in continental arcs suggest that these extrapolations are problematic because magmatic advection is the dominant mechanism of heat transport in these terranes rather than conduction from the base of the lithosphere. Metamorphic thermobarometry data from the middle crust of the eastern Peninsular Ranges batholith in Baja California, México, provide a useful field test of these models. Graphite-bearing pelitic and semi-pelitic schists record peak metamorphic temperatures of 475–720 °C at pressures of 3–6 kbar. These data bridge a gap between shallow and deep crustal levels of continental magmatic arcs in the southwestern United States and Baja California, México. Recognition of the transient, isobaric heating that accompanies contact metamorphism allows the definition of a gradient in minimum wall-rock temperatures of ~22 °C/km from 10 to 25 km with thermobarometric data from the eastern Peninsular Ranges batholith and other continental arcs that have relatively simple thermal histories. This gradient defines a maximum background geotherm that reconciles the results of geophysical and numerical models with wall-rock thermobarometry and is consistent with the formation of granulites in the lower crust of sub-arc regions and numerical models of the thermal effects of nested plutons. Recognition of the proposed geotherm may lower estimates of the depth to the seismic Moho and increase strain and unroofing rates inferred from structural and thermochronologic studies, respectively.

Keywords: geothermal gradients, continental arc, contact metamorphism.

INTRODUCTION

The thermal structure of the crust plays a central role in the geophysical, geologic, geochemical, and isotopic evolution of continental magmatic arcs. Because rheology is sensitive to temperature, the distribution of heat in continental arcs affects seismic velocities and strain partitioning (Birch, 1961; Christensen, 1982; Christensen and Fountain, 1975; Furlong and Fountain, 1986; Paterson and Tobisch, 1992). The distribution of thermal energy also influences crustal melting (e.g., Wyllie, 1979). Cooling histories derived from mineral thermochronometers are also sensitive to assumptions about temperature changes during cooling (e.g., McDougall and Harrison, 1988).

Understanding the thermal evolution of continental arcs requires quantitative observational constraints on changes in

* Present address: DRP Consulting, Inc., 2825 Wilderness Place, Suite 1000, Boulder, Colorado 80301, USA.

Rothstein, D.A., and Manning, C.E., 2003, Geothermal gradients in continental magmatic arcs: Constraints from the eastern Peninsular Ranges batholith, Baja California, México, *in* Johnson, S.E., Paterson, S.R., Fletcher, J.M., Girty, G.H., Kimbrough, D.L., and Martín-Barajas, A., eds., Tectonic evolution of northwestern México and the southwestern USA: Boulder, Colorado, Geological Society of America Special Paper 374, p. 337–354. For permission to copy, contact editing@geosociety.org. © 2003 Geological Society of America.

temperature with depth across a broad range of crustal levels. Hereafter we refer to these depth-dependent variations in temperature as geothermal gradients or geotherms without implying specific conditions regarding basal heat fluxes, thermal conductivity, or surface heat flow. In continental arcs geothermal gradients are relatively well understood to depths of 10–15 km from surface heat flow data and metamorphic petrology (Table 1), whereas geotherms in the middle and deep crust are poorly known. The absence of well-studied examples of deeper crustal levels of magmatic arcs requires extending geothermal gradients inferred from shallow depths to higher pressure. However, seismic velocity and other geophysical studies indicate that the inferred geotherms do not extrapolate sensibly to the deeper crust (e.g., Giese, 1994). Accounting for the temperature and pressure dependence of thermal conductivity, surface heat-flow values of 90 mW/m^2 measured in the Andes extrapolate to >1400 °C at 35 km (Arndt et al., 1997), which exceeds the basalt liquidus (e.g., Yoder and Tilley, 1962; Thompson, 1972). If melt is present in the central Andean crust, it constitutes less than 20% of the volume (Schilling and Partzsch, 2001). Heating to solidus temperatures at >25 km in the Cascades would yield higher inland heat flow than is currently observed (Morgan, 1984).

These observations require that geothermal gradients change in the middle crust of active continental arcs. In this paper, we exploit exposures of the middle crust in the eastern Peninsular Ranges batholith in Baja California, México, (Rothstein, 1997) to constrain the nature of this change and the range of possible background geothermal gradients that attend the middle crust during arc magmatism. The paper begins by reviewing observations that constrain geothermal gradients in continental arcs. We then describe a conceptual framework for using wall-rock thermobarometry to constrain the background geothermal gradient in continental magmatic arcs. A description of the geologic setting and metamorphism of the eastern Peninsular Ranges batholith in Baja California México follows. We combine these results with pressure-temperature (P-T) data from other relatively simple

continental arcs to bridge the gap in data between the upper and lower crust of these terranes. We then use two-dimensional numerical models to investigate the validity of the proposed mid-crustal geothermal gradient and conclude by discussing the geotherm's implications for seismic velocity, structural, and thermochronologic data from continental arc terranes.

OBSERVATIONS FROM CONTINENTAL ARCS

Geothermal gradients have been derived from surface heat flow for two well-studied, active continental arcs, the Andean and Cascade ranges. Heat flow in these systems is low in the outer arc region (~40 mW/m^2), high in the active arc and backarc (>80 mW/m^2), and low toward the stable craton (~40 mW/m^2; Uyeda and Watanabee, 1982; Giese, 1994; Henry and Pollack, 1988; Springer and Förster, 1998). Modeled geothermal gradients for the active portions of these arcs range from 30 to 35 °C/km in the upper 35 km (Morgan, 1984; Giese, 1994) to ~50 °C/km in the upper 12 km (Henry and Pollack, 1988). Surface heat-flow in other active arcs can range to higher values (e.g., Gill, 1981; Furukawa and Uyeda, 1989), which requires even higher geothermal gradients.

Studies of shallow crustal levels exhumed from ancient magmatic arcs show that variations in peak metamorphic temperatures at low pressures broadly agree with geothermal gradients inferred from surface heat flow. The metamorphic belts associated with those terranes are typically exhumed from ≤15 km, where P-T-t paths imply rapid, nearly isobaric metamorphism with metamorphic field gradients of 35–150 °C/km (e.g., Barton et al., 1988; De Yoreo et al., 1991). Gradients significantly exceeding the 35–50 °C/km suggested by surface heat flow can plausibly be explained by local, transient heating events during contact metamorphism, which produce peak metamorphic temperatures that do not reflect a background crustal geotherm (Barton et al., 1988)

Estimates of geothermal gradients from both heat-flow and metamorphic data are subject to significant uncertainty. For

TABLE 1. GEOTHERMAL GRADIENTS IN MAGMATIC ARCS

Data/method	Geothermal gradient	Source
Cascade Arc heat flow Transition zone half widths	45°C/km (0–15 km)	Blackwell et al. (1982) Blackwell et al. (1990)
Andean and Cascade heat flow Transition zone half widths	25–30°C/km (0–25 km); <20°C/km below 25 km	Morgan (1984)
Basaltic underplating/numerical models	~30°C/km (0–25 km) ~17°C/km (25–40 km)	Wells (1980)
Regional granitic sill emplacement/numerical models	~35°C/km (0–25 km) ~10°C/km (25–50 km)	Wells (1980) Lux et al. (1985)
Multiple diapirism/numerical models	~45°C/km (0–10 km) 15–18°C/km (10–30 km)	Barton and Hanson (1989)
Metamorphic field gradients	~50°C/km (0–15 km)	Turner (1981)
Granulite thermobarometry	30–35°C/km (0–25 km)	Bohlen (1991)
Melt-residue xenolith thermobarometry	~20°C/km (0–40 km)	Miller et al. (1992)

example, hydrothermal effects (e.g., Blackwell et al., 1982, 1990; Ingebritsen et al., 1989, 1992) and parameter uncertainty and geometric assumptions (Furlong et al., 1991) complicate the interpretation of heat-flow data. Similarly, estimating geothermal gradients from metamorphic P-T-t paths is complicated by the fact that metamorphism is a response to a perturbed geotherm, by uncertainties in regional gradients in the age of peak metamorphic temperatures, and by post-metamorphic shortening or extension (e.g., England and Richardson, 1977; Miller et al., 1992a). But even if these considerations can be addressed adequately, a more important uncertainty arises from the fact that much of the heat in continental arcs may be advected by magma. The thermal histories that result from such heating are very different from those in classical regional metamorphic terranes, where regionally extensive heat sources in the lower crust and upper mantle drive metamorphism (e.g., England and Thompson, 1984).

In magmatic arcs experiencing little tectonically driven disruption of the crust, magmatic advection is the predominant heat transport mechanism that influences temperatures in the upper and middle crust (e.g., Barton and Hanson, 1989; De Yoreo et al., 1991). Numerical models suggest that geothermal gradients in continental arcs are high in the upper crust but decrease significantly below the level at which magmas pond. For example, simple, one-dimensional models of the thermal effects of granitic sills in the middle crust predict high geothermal gradients (~50 °C/km) in the upper 20–25 km of the crust that decrease rapidly to ~15 °C/km between 25–40 km (e.g., Lux et al., 1985; Wells, 1980; Rothstein and Hoisch, 1994). Two-dimensional models accounting for the emplacement of multiple diapiric plutons predict that thermal gradients in the upper 10 km of the crust are high (~35 °C/km) but decrease to ~20 °C/km between 10 and 35 km (Barton and Hanson, 1989; Hanson, 1995). These results indicate that interpretations of geothermal gradients from metamorphic thermobarometers require a conceptual framework that recognizes magmatic advection as the dominant heat transport process in the middle and upper crust of continental arcs.

CONCEPTUAL FRAMEWORK

The numerical models described above make specific predictions about metamorphism in magmatic arcs. Magmas pond and crystallize in the middle and upper crust, liberating heat that drives contact metamorphism. Numerical and analytical models of heat transport, metamorphic mineral assemblages and textures, and mineral thermochronometers consistently show that the high-temperature history of wall rocks is short-lived, on the order of 1–3 m.y. depending on the size of the individual pluton (e.g., Lovering, 1935; Jaeger, 1964; Hanson and Gast, 1967; Krumenacher et al., 1975; Harrison and Clarke, 1979; Joesten and Fisher, 1988; Mahon et al., 1988; Miller et al., 1988; Lovera et al., 1999). The intruded crust experiences localized, contact metamorphism during short-lived events that, when integrated over the history of the construction of the arc, give the appearance of a regionally metamorphosed terrane that was simultane-

ously at moderate to high temperature (e.g., Barton and Hanson, 1989). However, there is a "background" geothermal gradient that characterizes much of the crust throughout the duration of arc magmatism. At any given time, only a fraction of the crust is at temperatures elevated above this background geotherm, and the timing of these departures varies throughout the terrane. Because the background geotherm is not significantly elevated on a regional scale, the metamorphic field gradient does not reflect the geothermal gradient. If the metamorphic record is inconsistent with these predictions, then alternative models must be considered, such as elevated heat flux from the mantle (e.g., Miyashiro, 1973; Wickham and Oxburgh, 1985, 1987; Bodorokos et al., 2002).

Figure 1 presents a schematic illustration of this conceptual framework. In stable shield lithosphere, there is little lateral variation in temperature within the crust (Condition I, Fig. 2A). The emplacement of plutons in the crust creates thermal perturbations in the wall rocks. The magnitude of the perturbations varies depending on proximity to the pluton. Heating is greatest in the roof pendants above plutons and at sidewall contacts (Condition IV and III, respectively, Fig. 2A) and diminishes with distance from pluton margins (Condition II, Fig. 2A). Because wall rocks adjacent to the tops and sides of a pluton will record increasing peak metamorphic temperatures as the pluton is approached, the minimum wall-rock temperature will represent the smallest excursion from ambient wall-rock temperature during arc magmatism. The locus of these P-T points traces a maximum background geotherm (MBG) that represents the highest possible ambient temperatures at a given depth (Fig. 1B). The actual ambient temperature distribution (Condition II) lies anywhere between the upper limit defined by the MBG and the lower limit defined by a geothermal gradient that was present prior to arc magmatism. We conservatively assume this lower limit to be 18 °C/km, which is broadly characteristic of stable continental lithosphere before the inception of arc magmatism.

Syn- to post-magmatic tectonism will, to varying degrees, drive departures from the idealized thermal structure shown in Figure 1. Cases where terranes have complex polymetamorphic histories, such as the Idaho batholith, United States (Wiswall and Hyndman, 1987; Hyndman and Foster, 1989), the Coast Plutonic Complex of British Columbia, Canada, (e.g., Rusmore and Woodsworth, 1994; Crawford et al., 1987) and deep-seated rocks from the Andean batholith (e.g., Grissom et al., 1991; Kohn et al., 1995; Lucassen and Franz, 1996) were excluded for this reason. Our goal is to focus on continental arc terranes with metamorphic histories that were relatively unaffected by syn- to post-batholithic deformation, such as the eastern Peninsular Ranges batholith.

EASTERN PENINSULAR RANGES BATHOLITH

Geologic Setting

Our study focused on the petrology of wall rocks from the eastern Peninsular Ranges batholith in Baja California Norte,

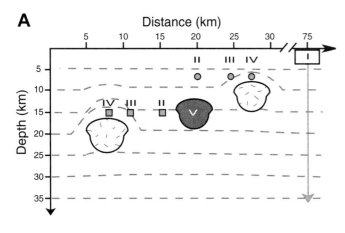

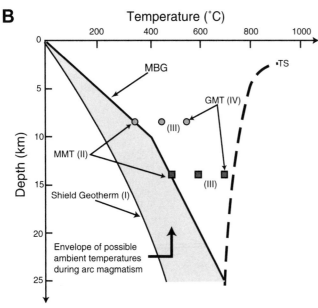

Figure 1. Conceptual model of temperature gradients in continental magmatic arcs. A: Summary of five different temperature conditions that can result during arc magmatism; diagram represents distribution of temperatures at a single time. (I) Continental crust before magmatism; note break in horizontal distance axis. (II) Ambient temperature distribution in between plutons. (III) Temperatures adjacent to side of diapiric pluton. (IV) Temperatures above diapiric plutons. (V) Older pluton that has cooled completely and no longer affects temperatures of adjacent wall rocks. B: Schematic diagram of P-T paths in continental arcs, modified after Lister and Baldwin (1993). Rocks are subjected to thermal effects of emplacement of two plutons at different structural levels. Labels in roman numerals are similar situation to that described for (A). Pluton crystallization isobarically heats wall rocks at two different crustal levels. Greatest metamorphic temperatures (GMT) are recorded by rocks close to igneous contacts (IV), whereas minimum metamorphic temperatures (MMT) are recorded distal to igneous contacts (II). Locus of minimum metamorphic temperatures defines maximum background geotherm (MBG, see text). Actual background temperatures extant during arc magmatism (condition (II) in 2A) are indicated by gray shaded area lying between MBG and shield geotherm (a).

México. The eastern Peninsular Ranges batholith is a fragment of a continental arc that was accreted to the western edge of the North American plate in Jurassic–Cretaceous time. The batholith extends for 1100 km from southern California, USA, to Baja California Sur, México (Fig. 2). Belts of longitudinal consistency and transverse asymmetry occur in prebatholithic lithologies, and in the major element geochemistry, rare earth element signatures, isotopic composition, geochronological signature of plutonic rocks, and grade of wall-rock metamorphism (Gastil et al., 1975; Taylor, 1986; Silver and Chappell, 1988, and references therein; Todd et al., 1988). The transverse asymmetry may result from the juxtaposition of a western island arc terrane against the eastern Peninsular Ranges batholith continental arc at 115–108 Ma (Johnson et al., 1999). The batholithic rocks are predominantly tonalitic to granitic in composition (e.g., Walawender et al., 1990). Available U-Pb zircon measurements and field mapping indicate that >90% of the batholith preserved west of the San Andreas transform was emplaced between 120 and 90 Ma (Silver and Chappell, 1988); the remainder of the batholith consists of earlier Jurassic intrusions (Todd et al., 1991).

Structural, geochronologic, and thermochronometric considerations are important for understanding the metamorphic history of the eastern Peninsular Ranges batholith. Most of the plutons in the eastern Peninsular Ranges batholith were emplaced between ca. 100 and 90 Ma. Mesozoic mylonitic deformation in the Jurassic Cuyamaca–Laguna Mountains Shear Zone and Late Cretaceous Eastern Peninsular Ranges Mylonite Zone involved west-directed thrusting with minor extensional overprinting (e.g., Sharp, 1967; Simpson, 1984; Erskine and Wenk, 1985; Thomson and Girty, 1994). Thermochronologic data from these shear zones indicate that deformation post-dated cooling from temperatures >450 °C ca. 10–30 m.y. after plutonism, such that high-temperature history of the wall rocks was relatively unaffected by deformation (Goodwin and Renne, 1991; Grove, 1993, 1994).

In the Baja California, México, portion of the batholith, thermochronologic data suggest the high temperature portion of the thermal history batholith was relatively unaffected by post-magmatic deformation. Contractional deformation proceeded intermittently from the Late Devonian–Mississippian Antler orogeny to the emplacement of Cretaceous plutons (Gastil and Miller, 1993, and references therein). Biotite K-Ar and $^{40}Ar/^{39}Ar$ apparent ages of 85–80 Ma in the eastern batholith and K-feldspar cooling histories from the northeastern batholith indicate slow cooling after magmatism (Krumenacher et al., 1975; Ortega-Rivera et al., 1997; Rothstein, 1997). A well-developed Eocene erosional surface on the granitic rocks over much of the crest of the Peninsular Ranges, combined with regionally consistent 60–70 Ma apatite fission track results from the central and eastern batholith and stratigraphic constraints suggest ~1 mm/yr unroofing rates throughout much of the batholith between the end of magmatism in Late Cretaceous and Eocene time (Minch, 1979; Dokka, 1984; George and Dokka, 1994). $^{40}Ar/^{39}Ar$ age distributions of detrital K-feldspars in forearc basin strata derived from the northern Peninsular Ranges batholith record closure temperatures from 105 to 75 Ma, which overlap closure ages of basement rocks in

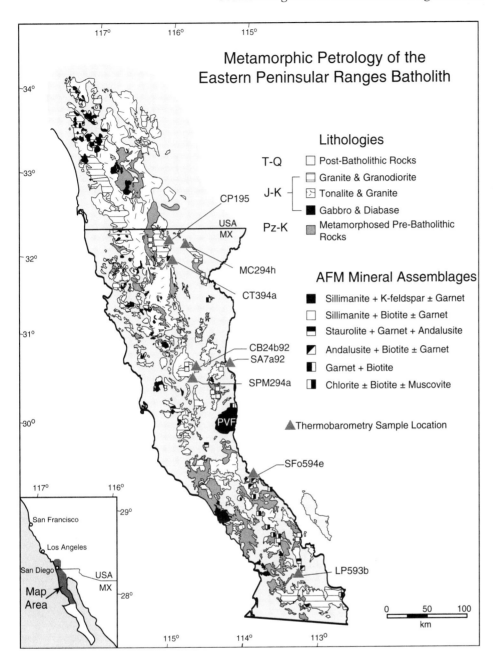

Figure 2. Simplified geologic map of Peninsular Ranges batholith showing observed mineral assemblages and thermobarometry sample locations in eastern Peninsular Ranges batholith. Geology generalized after regional maps from Krumenacher et al. (1975) and Gastil and Miller (1993). PVF—Puertocitos volcanic field.

the eastern Peninsular Ranges batholith (e.g., Lovera et al., 1999). Neogene extensional and strike-slip faulting largely generated the Basin and Range topography of the eastern batholith (Stock and Hodges, 1989) but did not affect the cooling histories of the exposed mid-crustal metamorphic rocks (Rothstein et al., 1995).

Regional Metamorphism in the Eastern Peninsular Ranges Batholith

Both miogeoclinal and slope-basin deposits contain graphite-bearing pelitic and semi-pelitic protoliths (Fig. 2). The predominance of wall rocks with semi-pelitic bulk compositions limits the number of lithologies suitable for thermobarometry. We identified eight metapelite localities for thermobarometry studies. Table 2 summarizes the mineralogy of these rocks. Important index minerals include chlorite in the greenschist facies, garnet ± andalusite ± staurolite in the lower amphibolite facies, and garnet ± sillimanite ± cordierite ± K-feldspar in the upper amphibolite facies. Graphite and ilmenite are ubiquitous accessory phases; apatite, zircon, monazite, and tourmaline are also common. Migmatitic textures are common in the higher-grade rocks. Rothstein (1997) discussed the distribution and composition of mineral assemblages throughout the eastern Peninsular Ranges batholith in Baja California Norte in detail.

TABLE 2. MINERAL ASSEMBLAGES FOR THERMOBAROMETRY SAMPLES

Sample number	Qtz	And	Sil	Grt	Crd	St	Bt	Chl	Ms	Pl	Kfs	Gr	Ilm	Ap	Zrn
MC294h	x		x	5			10–15			30	x	tr	tr	tr	
CP195	x		x	8			20	tr		20	x	tr	tr	tr	tr
CT394a	x		10–15	5–10	<5		30–40			20	x	tr			tr
SA7a92	x		x	3			3	x	5–10			tr	tr	tr	
CB24b92	x		<5	10–15	5–10		50–60	x				tr	tr	tr	tr
SPM294a	x		x	15			40–50	tr		10–20	x	tr	tr		
SFo594e	x	5–10		<5			10–15	x	15–20	5–10		tr	tr	tr	tr
LP593b	x		<3	10			5	x	15	10–15		tr		tr	

Note: Where given, numbers refer to estimated percentage of mode. Mineral abbreviations after Kretz (1983); tr—trace. Sample locations are posted on Figure 2.

Observed textures indicate varying degrees of dynamic recrystallization. Rock matrices are dominated by lepidoblastic muscovite + biotite folia interlayered with granoblastic quartz + plagioclase bands. Garnet, staurolite, cordierite, andalusite, and crystalline, and fibrolitic sillimanite form pre-, syn-, and postkinematic porphyroblasts in amphibolite facies rocks. Textural evidence does not indicate that there was significant thermal overprinting after peak metamorphism, which is consistent with the thermochronologic database from the region.

Mineral compositions were determined by electron microprobe analysis. Compositions are given in Tables 3–6. Garnets are compositionally homogenous with the exception of steep compositional gradients at the outer 1–3 microns of some grains. Analyses from these regions were excluded from the data used in the thermobarometry calculations. Chlorite is a minor retrograde phase in some rocks, partially replacing biotite, garnet, and staurolite. Rocks with evidence of strong retrogression were excluded from the thermobarometry study.

Thermobarometry

The P-T calculations used TWEEQU software (Berman, 1991). Table 7 summarizes the equilibria used to calculate temperatures and pressures. The temperature calculations employ the Fe-Mg exchange reaction between garnet and biotite. The anorthite breakdown reaction allows the calculation of pressure for six plagioclase-bearing assemblages. Equilibria among garnet, biotite, muscovite, sillimanite, and quartz allow the calculation of pressure for one sample (SA7a92a). Equilibrium between cordierite, garnet, sillimanite, and quartz allows the calculation of pressure for one cordierite-bearing sample (CB24b92). The calculations employ activity models of Berman (1990) for garnet, McMullin et al. (1991) for biotite, and Fuhrman and Lindsley (1988) for feldspar model for plagioclase. An ideal solution model is used for hydrous cordierite.

Sources of errors in metamorphic barometry include the accuracy of the experimentally located end-member reaction, analytical imprecision, uncertainties in microprobe standards and α-factors, thermometer calibration errors, variations in activity

models, and compositional heterogeneities (e.g., Kohn and Spear, 1991). Of these, uncertainties in compositional heterogeneities are the most difficult to treat with a global statistical approach. In this study, compositional heterogeneities were considered by performing multiple P-T calculations that accounted for the range of compositions determined for the relevant mineral phases. Rothstein (1997) discussed the details of these calculations.

Figure 3 plots the results of the P-T calculations. Temperatures range from 475 to 720°C at pressures of 3–6 kbars; most of these results fall between the classical Barrovian and Buchan metamorphic field series. The reported P-T estimates and their associated uncertainties are the average and standard deviation (1σ), respectively, of each set of calculations.

Comparison to P-T Data from Other Continental Arcs

Available P-T data from other continental arc settings include heat flow values from active continental arcs and thermobarometric estimates from other studies. Table 8 lists these data and their sources; they were selected from continental arcs where available structural and thermochronologic constraints indicated relatively simple thermal histories dominated by magmatic heating rather than tectonic thickening.

The previously described heat flow data from the Andean and Cascade volcanic arcs indicate a wide range of thermal gradients in the upper crust. Metamorphic thermobarometry data from the middle crust of continental arcs include data from the Sierra Nevada batholith, United States; the Ryoke metamorphic belt, Japan; and the Peninsular Ranges batholith from southern California, United States, and Baja California, México. In the Mesozoic Sierra Nevada batholith, P-T estimates are from thermobarometry or phase relations in contact metamorphic rocks that formed pluton roof pendants (Loomis, 1966; Morgan, 1975; Ferry, 1989; Hanson et al., 1993; Davis and Ferry, 1993). These data indicate high temperatures (~450 °C) at low (2–3 kbar) pressures, consistent with numerical results that predict strong heating directly over crystallizing intrusions. Thermobarometry from pelitic rocks in the Tehachapi Mountains in the southern Sierra Nevada batholith indicate P-T conditions of 620–770 °C at 5.3–6.0 kbar (Dixon et

TABLE 3. AVERAGE GARNET COMPOSITIONS

(avg. of)	MC294h (10)	CP195 (23)	CT594 (15)	SA7a92 (32)	CB24b92 (32)	SPM294a (24)	SFo594e (16)	LP593b (10)
SiO_2	38.58 (0.22)	38.19 (0.21)	38.41 (0.49)	38.92 (0.43)	38.04 ((0.23)	38.19 (0.22)	37.29 (0.43)	38.15 (0.17)
Al_2O_3	21.94 (0.13)	21.84 (0.11)	21.88 (0.30)	21.65 (0.25)	21.68 (0.13)	21.53 (0.18)	21.14 (0.17)	21.71 (0.11)
TiO_2	0.02 (0.02)	0.02 (0.02)	0.02 (0.02)	0.02 (0.02)	0.04 (0.04)	0.04 (0.03)	0.22(0.45)	0.01 (0.01)
Cr_2O_3	0.06 (0.04)	0.05 (0.02)	0.01 (0.02)	0.03 (0.03)	0.04 (0.03)	0.07 (0.04)	0.03 (0.03)	0.04 (0.03)
FeO	32.06 (0.45)	33.94 (0.29)	31.85 (1.66)	28.02 (0.77)	36.76 (0.66)	31.36 (0.30)	25.67 (2.14)	30.28 (0.35)
MnO	1.56 0.53)	2.39 (0.31)	4.63 (0.19)	8.66 (1.70)	0.77 (0.08)	5.12 (0.32)	13.27 (2.42)	8.20 (0.16)
MgO	5.26 (0.81)	3.64 (0.24)	3.73 (0.13)	4.21 (0.85)	3.38 (0.30)	2.98 (0.22)	1.73 (0.26)	2.53 (0.05)
CaO	1.28 (0.06)	1.40 (0.05)	0.87 (0.11)	1.07 (0.19)	0.91 (0.08)	1.97 (0.14)	1.11 (0.22)	0.98 (0.16)
Na_2O								
K_2O								
Total	100.76 (0.38)	101.46 (0.40)	102.03 (0.63)	102.59 (0.90)	101.61 (0.60)	101.26 (0.46)	100.44 (0.41)	101.89 (0.46)
Sum O	12	12	12	12	12	12	12	12
Si	3.02	3.01	3.02	3.03	3.01	3.02	3.01	3.02
Al	2.02	2.03	2.03	1.98	2.02	2.01	2.02	2.02
Ti	0.00	0.00	0.00	0.00	0.00	0.00	0.01	0.00
Cr	0.00	0.00	0.00	0.00	0.00	0.00	0.00	0.00
Fe	2.10	2.23	2.09	1.82	2.43	2.07	1.73	2.00
Mn	0.10	0.16	0.31	0.57	0.05	0.34	0.91	0.55
Mg	0.61	0.43	0.44	0.49	0.40	0.35	0.21	0.30
Ca	0.11	0.12	0.07	0.09	0.08	0.17	0.10	0.08
Na	0.00	0.00	0.00	0.00	0.00	0.00	0.00	0.00
K	0.00	0.00	0.00	0.00	0.00	0.00	0.00	0.00
Total	7.97	7.98	7.96	7.98	7.98	7.97	7.98	7.97
Alm	0.72 (0.02)	0.76 (0.00)	0.72 (0.01)	0.61 (0.01)	0.82 (0.01)	0.71 (0.00)	0.59 (0.05)	0.68 (0.01)
Pyr	0.21 (0.03)	0.15 (0.01)	0.15 (0.00)	0.16 (0.03)	0.13 (0.01)	0.12 (0.01)	0.07 (0.01)	0.10 (0.00)
Sps	0.04 (0.01)	0.05 (0.01)	0.11 (0.00)	0.19 (0.04)	0.02 (0.00)	0.12 (0.01)	0.31 (0.06)	0.19 (0.00)
Grs	0.04 (0.00)	0.04 (0.01)	0.03 (0.00)	0.03 (0.01)	0.03 (0.00)	0.06 (0.00)	0.03 (0.01)	0.03 (0.00)

Note: All microprobe analyses were performed at the University of California at Los Angeles with a Cameca Camebax electron microprobe. Operating conditions were 15 kV accelerating potential, a 15 nA beam current, and 20 second counting times. A focused (~1 μm) was used for all minerals except plagioclase, for which the beam was defocused to ~5 μm. A ZAF correction algorithm was used with well-characterized natural and synthetic standards. Mineral abbreviations are after Kretz (1983). All oxide data in Table 3 to Table 6 are reported as averages of analyses. Numbers in parentheses after averages give 1σ standard deviation. Calculated mole fractions of mineral components are reported as averages with 1σ standard deviation following in parentheses. Number in parentheses below sample number at top of each column indicates number of analyses averaged.

al., 1994): these are some of the deepest exposures of the Sierra Nevada magmatic arc. P-T conditions in the Japanese Ryoke metamorphic belt are from pelitic thermobarometers (Nakajima, 1994; Okudaira, 1996) and constrain temperatures in the middle crust (10–25 km) of the Cretaceous Eurasian continental arc (e.g., Miyashiro, 1972). The P-T data presented in this study, coupled with other P-T estimates in the Peninsular Ranges batholith from north of the international border (Hill, 1984; Grove, 1986) constrain temperatures from 10 to 25 km.

Granulite and xenolith thermobarometry data were not used to constrain the MBG for several reasons. These include uncertainties regarding geothermal gradients, particularly during

the Archean when many granulites formed (e.g., Nutman and Collerson, 1991), the tectonic environment that generated granulite facies metamorphism (e.g., Bohlen, 1991; Brown, 1993), and the environment of xenolith equilibration (e.g., Miller et al., 1992). P-T data are available from granulites and xenoliths from the Mesozoic continental arc exposed in the central and eastern Mojave Desert (e.g., Henry and Dokka, 1992; Miller et al., 1992; and Hanchar et al., 1994). The data were included in Table 8 to illustrate the range of temperature conditions recorded by rocks exhumed from the lower crust of a Mesozoic arc in southwestern North America but were not considered in defining the MBG described below.

TABLE 4. REPRESENTATIVE BIOTITE COMPOSITIONS

(avg. of)	MC294h (4)	CP195 (10)	CT394 (12)	SA7a92 (4)	CB24b92 (11)	SPM294a (14)	SFo594e (10)	LP593b (5)
SiO_2	35.76 (0.11)	35.36 (0.28)	35.80 (0.12)	36.30 (1.14)	34.83 (0.20)	35.80 (0.32)	35.08 (0.57)	35.12 (0.23)
Al_2O_3	17.71 (0.03)	19.57 (0.24)	19.15 (0.24)	18.70 (0.50)	20.00 (0.44)	19.67 (0.26)	19.93 (0.28)	21.14 (0.14)
TiO_2	4.56 (0.01)	3.14 (0.16)	3.61 (0.19)	2.74 (0.19)	2.99 (0.48)	2.36 (0.15)	1.45 (0.20)	1.63 (0.13)
Cr_2O_3	0.03 (0.02)	0.05 (0.03)	0.06 (0.01)	0.05 (0.02)	0.07 (0.04)	0.16 (0.05)	0.05 (0.03)	0.01 (0.02)
FeO	13.63 (0.09)	19.37 (0.54)	18.72 (0.37)	15.57 (0.67)	20.55 (0.62)	18.13 (0.24)	18.38 (0.26)	19.42 (0.51)
MnO	0.04 (0.24)	0.09 (0.04)	0.15 (0.05)	0.45 (0.02)	0.03 (0.03)	0.10 (0.0.)	0.29 (0.07)	0.24 (0.02
MgO	11.89 (0.11)	8.02 (0.26)	8.98 (0.15)	11.12 (0.30)	7.66 (0.44)	8.94 (0.10)	9.54 (0.53)	8.38 (0.20)
CaO	0.00 (0.06)	0.02 (0.01)	0.01 (0.02)	0.02 (0.03)	0.01 (0.01)	0.02 (0.02)	0.05 (0.06)	0.03 (0.02)
Na_2O	0.49 (0.16)	0.16 (0.08)	0.11 (0.04)	0.12 (0.03)	0.33 (0.03)	0.19 (0.02)	0.03 (0.02)	0.05 (0.02)
K_2O	9.01 (0.02)	9.82 (0.16)	9.58 (0.11)	9.80 (0.41)	8.20 (0.22)	8.79 (0.13)	8.73 (0.38)	9.52 (0.21))
Total	93.13 (0.01)	95.61 (0.16)	96.17 (0.48)	94.88 (3.02)	94.66 (0.34)	94.17 (0.66)	93.51 (1.21)	95.53 (0.48)
Sum O	11	11	11	11	11	11	11	11
Si	2.74	2.71	2.69	2.73	2.66	2.74	2.71	2.68
Al	1.60	1.77	1.70	1.66	1.80	1.78	1.81	1.90
Ti	0.26	0.18	0.20	0.15	0.17	0.14	0.08	0.09
Cr	0.00	0.00	0.00	0.00	0.00	0.01	0.00	0.00
Fe	0.87	1.24	1.18	0.98	1.32	1.16	1.19	1.24
Mn	0.00	0.01	0.01	0.03	0.00	0.01	0.02	0.02
Mg	1.36	0.92	1.01	1.25	0.87	1.02	1.10	0.95
Ca	0.00	0.00	0.00	0.00	0.00	0.00	0.00	0.00
Na	0.07	0.02	0.02	0.02	0.05	0.03	0.01	0.01
K	0.88	0.96	0.92	0.94	0.80	0.86	0.86	0.93
Total	7.80 (0.01)	7.81 (0.01)	7.72 (0.01)	7.76 (0.01)	7.68 (0.02)	7.74 (0.02)	7.78 (0.02)	7.83 (0.01)
Al^{iv}	1.26 (0.01)	1.29 (0.01)	1.31 (0.01)	1.27 (0.00)	1.34 (0.01)	1.26 (0.01)	1.29 (0.02)	1.32 (0.01)
Al^{vi}	0.34 (0.01)	0.48 (0.01)	0.39 (0.02)	0.39 (0.02)	0.47 (0.04)	0.52 (0.02)	0.52 (0.02)	0.59 (0.01)
Sum Oct	2.85 (0.01)	2.82 (0.01)	2.79 (0.01)	2.80 (0.01)	2.83 (0.01)	2.85 (0.01)	2.91 (0.03)	2.89 (0.01)
Sum A	0.96 (0.01)	0.99 (0.02)	0.94 (0.01)	0.96 (0.01)	0.85 (0.02)	0.89 (0.01)	0.87 (0.03)	0.94 (0.02)
Fe/Fe+Mg	0.39 (0.01)	0.57 (0.01)	0.54 (0.01)	0.44 (0.00)	0.60 (0.02)	0.53 (0.00)	0.52 (0.02)	0.57 (0.01)
Xmg	0.48 (0.00)	0.32 (0.01)	0.36 (0.01)	0.45 (0.00)	0.31 (0.02)	0.36 (0.00)	0.38 (0.02)	0.33 (0.01)
Xfe	0.31 (0.00)	0.44 (0.01)	0.42 (0.01)	0.35 (0.01)	0.46 (0.01)	0.41 (0.00)	0.41 (0.01)	0.43 (0.01)
Xal	0.12 (0.00)	0.17 (0.00)	0.14 (0.01)	0.14 (0.01)	0.17 (0.01)	0.18 (0.01)	0.18 (0.01)	0.20 (0.00)
Xti	0.09 (0.00)	0.06 (0.00)	0.07 (0.00)	0.06 (0.00)	0.06 (0.01)	0.05 (0.00)	0.03 (0.00)	0.03 (0.00)

THE MAXIMUM BACKGROUND GEOTHERM (MBG)

Figure 4 illustrates the continental arc P-T database and the MBG inferred from it. The MBG is shown as a bold line that traces the locus of minimum contact metamorphic temperature estimates between depths of ~10–25 km. In detail, the MBG traces a geothermal gradient of 40 °C/km from 2 to 10 km, which was based on the heat flow data from active arcs discussed above. Where constrained by metamorphic P-T data from 10 to 25 km, the MBG is ~22 °C/km. This gradient is significantly lower than earlier inferences based on metamorphic P-T data (see Table 1). While the overall geometry of the MBG inferred from petrologic data is similar to the "75% minimum metamorphic gradient" modeled by Barton and Hanson (1989), it is somewhat higher than the 15–18 °C/km minimum meta-

morphic gradients between 10 and 20 km depths predicted by Barton and Hanson (1989). This is expected because the MBG is inferred from petrologic data and is consistent with the MBG placing an upper limit on ambient temperature conditions during arc magmatism. The scatter of data that record wall-rock temperatures above the MBG is also consistent with the isobaric heating model for low-pressure metamorphism discussed by Barton and Hanson (1989). Numerical models are used in the following section to investigate further the validity of the proposed MBG.

THERMAL MODELING

The difference in temperatures that lie between a shield geotherm and the MBG implies that ambient temperatures in wall

TABLE 5. REPRESENTATIVE PLAGIOCLASE COMPOSITIONS

(avg. of)	MC294h (6)	CP195 (12)	CT394a (10)	SPM294a (7)	SFo594e (11)	LP593b (5)
SiO_2	60.68 (0.58)	58.54 (0.59)	62.43 (0.56)	57.04 (0.75)	63.97 (0.65)	61.92 (0.27)
Al_2O_3	24.81 (0.36)	25.40 (0.41)	24.19 (0.31)	26.89 (0.43)	22.05 (0.70)	24.05 (0.16)
TiO_2	0.01 (0.01)	0.02 (0.02)	0.01 (0.01)	0.03 (0.03)	0.02 (0.012)	0.02 (0.01)
Cr_2O_3	0.01 (0.01)	0.02 (0.02)	0.02 (0.02)	0.02 (0.03)	0.02 (0.02)	0.01 (0.03)
FeO	0.10 (0.06)	0.05 (0.04)	0.02 (0.03)	0.13 (0.03)	0.17 (0.06)	0.03 (0.02)
MnO	0.01 (0.01)	0.02 (0.01)	0.03 (0.02)	0.02 (0.02)	0.03 (0.04)	0.03 (0.02)
MgO	0.00 (0.00)	0.00 (0.00)	0.00 (0.00)	0.00 (0.00)	0.00 (0.02)	0.00 (0.00)
CaO	6.18 (0.65)	7.61 (0.40)	5.18 (0.39)	9.13 (0.31)	3.29 (0.27)	5.16 (0.05)
Na_2O	8.18 (0.32)	7.41 (0.23)	8.80 (0.24)	6.50 (0.29)	9.75 (0.29)	8.84 (0.04)
K_2O	0.19 (0.17)	0.17 (0.03)	0.18 (0.05)	0.06 (0.02)	0.07 (0.02)	0.27 (0.06)
Total	100.17 (0.13)	99.23 (0.55)	100.85 (0.27)	99.81 (0.47)	99.37 (0.65)	100.33 (0.23)
Sum O	8	8	8	8	8	8
Si	2.70	2.64	2.75	2.56	2.84	2.74
Al	1.30	1.35	1.25	1.42	1.15	1.25
Ti						
Cr						
Fe						
Mn						
Mg						
Ca	0.29	0.37	0.24	0.44	0.16	0.24
Na	0.70	0.65	0.75	0.56	0.84	0.75
K	0.01	0.01	0.01	0.00	0.00	0.01
Total	5.01 (0.01)	5.02 (0.01)	5.01 (0.01)	5.01 (0.01)	5.00 (0.03)	5.02 (0.00)
Xan	0.29 (0.03)	0.36 (0.02)	0.24 (0.02)	0.44 (0.02)	0.16 (0.01)	0.24 (0.00)
Xab	0.70 (0.02)	0.63 (0.02)	0.75 (0.02)	0.56 (0.02)	0.84 (0.01)	0.75 (0.00)
Xor	0.01 (0.01)	0.01 (0.00)	0.01 (0.00)	0.00 (0.00)	0.00 (0.00)	0.01 (0.00)

Note: Mineral abbreviations after Kretz (1983).

rocks of the upper and middle crust lie within a narrow envelope during continental arc magmatism. To see if the MBG inferred from petrologic data could develop in the upper and middle crust of a continental arc, we ran two-dimensional heat transfer models that simulate the heating of wall rocks between plutons in the middle crust (5–25 km) and tracked the development of gradients in predicted minimum metamorphic temperatures (PMMT). We also used the models to further investigate whether the MBG represents a reasonable upper boundary for ambient temperature distributions in the middle crust of continental arcs.

Parameterization

Details regarding the numerical approach and parameterization of the simulations are described in the Appendix and Table A1. Sixteen different simulations were run with varying parameters (Table A2). The simulations employ two different initial temperature distributions: linear geothermal gradient of 18°C, which represents a stable shield geotherm, and the MBG described in this study. The use of the MBG as an initial temperature distribution is to test whether the resulting PMMT grossly

exceed the temperatures determined via wall-rock thermobarometry and predict widespread crustal melting at unrealistically shallow levels of the crust. The use of a stable shield geotherm is to investigate whether rapid and abundant plutonism will yield PMMT gradients that are reasonably close to the MBG inferred from petrologic data. The models also investigate the effects of variations in the volume and rate of plutonism by simulating wall-rock temperatures between two 20-km-wide plutons. Varying the distance between the two plutons simulates the effects of 50% and 75% pluton volume, which is broadly consistent with mapping from Baja California, México (e.g., Gastil et al., 1975).

The models use two rates of pluton emplacement: simultaneous intrusion and a 5 m.y. difference in the timing of emplacement (hereafter referred to as the simultaneous and differential emplacement models, respectively). Because the characteristic time constant ($t_{char} = L^2/4\kappa$; see Appendix for abbreviations) for a 20-km-wide pluton is <1 m.y., a 5 m.y. difference allows nearly complete cooling after the emplacement of the first pluton. The simultaneous emplacement models have run times of 5 m.y.; the differential emplacement models have 10 m.y. run times. Such time scales are broadly consistent with the geochronology and thermochronology

TABLE 6. REPRESENTATIVE MUSCOVITE AND
CORIERITE ANALYSES

(avg. of)	SA7a92 Muscovite (8)		CB24b92 Cordierite (14)
SiO$_2$	46.54 (0.55)	SiO$_2$	48.39 (0.24)
Al$_2$O$_3$	35.13 (0.29)	Al$_2$O$_3$	32.46 (0.24)
TiO$_2$	0.12 (0.15)	TiO$_2$	0.01 (0.02)
Cr$_2$O$_3$	0.01 (0.01)	Cr$_2$O$_3$	0.01 (0.02)
FeO	1.10 (0.13)	FeO	9.45 (0.21)
MnO	0.05 (0.08)	MnO	0.07 (0.02)
MgO	1.07 (0.31)	MgO	7.23 (0.09)
CaO	0.02 (0.02)	CaO	0.01 (0.01)
Na$_2$O	0.71 (0.19)	Na$_2$O	0.40 (0.03)
K$_2$O	10.13 (0.37)	K$_2$O	0.00 (0.01)
Total	95.18 (0.87)	Total	98.05 (0.52)
Sum O	11	Sum O	18
Si	3.10	Si	5.02
Al	2.76	Al	3.97
Ti	0.01	Ti	0.00
Cr	0.00	Cr	0.00
Fe	0.06	Fe	0.82
Mn	0.00	Mn	0.01
Mg	0.11	Mg	1.12
Ca	0.00	Ca	0.00
Na	0.09	Na	0.08
K	0.89	K	0.00
Total	7.01 (0.01)	Total	11.03 (0.01)
Al^{iv}	0.90 (0.02)	Xfe	0.42 (0.01)
Al^{vi}	1.85 (0.02)	Xmg	0.58 (0.01)
Sum A	0.98 (0.01)		
Xal	0.91 (0.01)		
Xk	0.90 (0.03)		
Xna	0.09 (0.03)		

of the eastern Peninsular Ranges batholith and track the high-temperature portion of the wall rocks' thermal history.

Results

Table A2 summarizes the results from the modeling. Figure 5 shows isotherms from three simulations that varied the volume and rate of plutonism. The region of interest in the simulations lies between the plutons because this most closely simulates wall rocks between nested plutons. Monitoring gradients in the predicted minimum metamorphic temperatures at different depths allows comparison with the MBG recorded by wall-rock thermobarometry as described above. Figure 5 shows the geometry of gradients in the PMMT. In cases of simultaneous emplacement of 50% and 75% plutons, gradients in PMMT are vertical and located midway between the plutons (Figs. 5A and 5B). For models that simulate differential timing of emplacement, the PMMT gradients are subvertical and displaced slightly

toward the second pluton emplaced during the simulation (Fig. 5C). Because the simulations in this study emplace plutons between depths of 5–25 km, the results of the modeling are not relevant for crustal levels deeper than 25 km.

The sensitivity of the simulation results varies with different parameters. The assumption of a uniform magmatic emplacement temperature versus pressure-dependent temperatures dictated by the tonalite solidus produces a negligible effect on the PMMT. Increasing the volume of plutons from 50% to 75% has a strong effect and increases the predicted temperatures at a given structural level by ~50–75 °C (Fig. 6A). Changing the rate of plutonism from simultaneous to differential emplacement increases temperatures by a similar amount (Fig. 6B). The sensitivity of the simulations to the volume and rate of plutonism does not change significantly with the higher initial temperature distribution dictated by the MBG, which suggests it represents a reasonable initial temperature distribution.

The geometry of the PMMT gradients is sensitive to the initial temperature distribution. Models that employ the linear 18°C/km geothermal gradient predict minimum metamorphic temperature gradients that are ~10–20 °C/km lower than models that use the MBG as an initial temperature distribution. The linear geotherm models start at lower temperatures, so the PMMT gradients from this initial condition show greater curvature than the models that use the MBG as an initial condition (Fig. 6A). The mid-crustal (10–25 km) PMMT gradients from the MBG models (20–22 °C/km) are higher than the 15 °C/km temperature gradients from 10 to 30 km predicted by Barton and Hanson (1989), who used a background thermal gradient of ~22 °C/km.

Numerical simulations that employ an initial geothermal gradient of 18 °C/km yield PMMT gradients that nearly coincide with the MBG from 10 to 20 km depths when a high volume of plutons are emplaced rapidly (Fig. 6C). Because numerical simulations that use a stable shield geotherm as an initial condition yield PMMT gradients that are close to the petrologic MBG only when magmatism is most voluminous and rapid, the results suggest the MBG is a reasonable constraint on the maximum background temperatures during arc magmatism. Numerical simulations that employ the MBG as the initial temperature distribution yield PMMT that fall below the wet tonalite solidus from 0 to 20 km and above the solidus at greater depths (Fig. 6A). This result also supports the validity of the MBG, as the use of this temperature distribution as an initial condition does not predict widespread partial melting of the upper crust during the rapid emplacement of a high volume of plutons. The result is also broadly consistent with the observation of migmatites in mid-crustal wall rocks (>20 km) in much of the eastern Peninsular Ranges batholith in Baja California, México (e.g., Rothstein, 1997). Taken together, the results of the numerical modeling suggest that the MBG inferred from petrologic data effectively brackets the upper end of maximum ambient wall-rock temperatures in the middle crust of an active arc. In the following section, we discuss the implications of this geotherm for seismic velocity, structural, and thermochronologic studies in these terranes.

TABLE 7. THERMOBAROMETRY RESULTS

Sample	Locality*	Latitude Longitude	Reaction[†]	P (kbar) (±1σ)	T (°C)[§] (±1σ)
MC294h	Sierra Cucapah (100)	N32°32' W115°38.7'	GASP	4.0 (0.65)	575 (8)
CP195	Cantu Palms (250)	N32°21.22' W115°49.74'	GASP	4.6 (0.47)	720 (25)
CT394a	Cañon Tajo (300)	N32°15.7' W115°49'	GASP	4.7 (0.50)	676 (10)
SA7a92	Sierra Abandonada (350)	N31°3.2' W114°52.8'	SGAM	5.0 (0.04)	614 (7)
CB24b92	Cañade El Baroso (50)	N30°48.8' W115°13.7'	CASBGQ	5.8 (0.07)	668 (37)
SPM294a	Agua Caliente (75)	N30°37.56' W115°16.1'	GASP	4.1 (0.18)	610 (6)
SFo594e	Sierra San Francisquito (500)	N29°37.46' W114°16.1'	GASP	3.2 (0.54)	476 (18)
LP593b	Los Paredones (500)	N28°39.1' W113°23.9'	GASP	3.1 (0.26)	590 (18)

*Number beneath locality denotes distance, in meters, of sample from pluton–wall rock contact.
[†]Reactions are as follows (mineral abbreviations after Kretz, 1983):
  GASP: 3 pl = grs + 2 sil + qtz
  SGAM: alm + ms = ann + sil + qtz
          pyr + ms + ann + 2 sil + qtz
  CASBGQ: 2 pyr + 5 qtz + 4 sil = 3 crd
           2 ann + 3 crd = 4 sil + 5 qtz + 2 phl + 2 alm
[§]All temperatures were calculated from the garnet-biotite Fe-Mg exchange reaction: alm + phl = pyr + ann

DISCUSSION

Relatively low (~22 °C/km) geothermal gradients in the middle crust of continental arcs are an important consideration in modeling tectonism from seismic velocity. Because seismic velocities decrease with increasing temperature (Griffin and O'Reilly, 1987; Weaver and Tarney, 1984; Stüwe and Sandiford, 1994), the background geotherm is important in the interpretation of the broad structure and composition of the crust in continental arcs. The inferred position of the Mohorovicic (Moho) discontinuity will depend on the geothermal gradient at the time such velocities are measured. Cooler geothermal gradients in continental arcs will also yield estimates of more intermediate compositions for the lower crust from seismic data (Griffin and O'Reilly, 1987) that are consistent with petrologic models and the composition of granulite terranes.

The geothermal gradient described in this study is also relevant to the mechanical evolution of continental arcs. Low geothermal gradients accelerate cooling and diminish the length scales of thermal "softening" or weakening around crystallizing plutons in the shallow and middle crust (Lister and Baldwin, 1993). This accentuates the pulses of multiple, short-lived deformation events in heavily intruded regions that are inferred from numerical investigations and combined structural, geochronologic, and thermochronologic studies (e.g., Sandiford et al., 1991; Paterson and Tobisch, 1992; Stüwe et al., 1993; Lister and Baldwin, 1993;

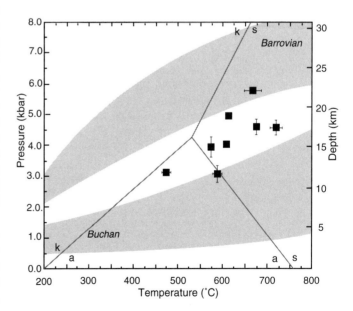

Figure 3. P-T diagram of thermobarometry estimates from eastern Peninsular Ranges batholith presented in this study. Error bars of 1σ are shown where estimates are not smaller than size of symbol. Aluminum silicate stability fields taken from Bohlen et al. (1991). Buchan and Barrovian P-T fields after Turner (1981).

TABLE 8. P-T CONSTRAINTS FROM CONTINENTAL MAGMATIC ARCS

Location	Data	P-T constraints	Source
Andes Mountains Bolivia	prospecting boreholes	25–57 °C/km	Giese (1994)
Bolivia and Peru	mineral and petroleum exploration holes	20–35 °C/km	Henry and Pollack (1988)
Peninsular Ranges Mount San Jacinto	pelitic/calc-silicate phase relations/thermobarometry	620–800 °C, 3.2–3.4 kb	Hill (1984)
Santa Rosa Mountains	pelitic and calc-silicate phase relations	650–800 °C, 2.0–4 .0 kb	Erskine (1986)
Box Canyon	pelitic phase relations and thermobarometry	650–700 °C, 4.0–5.0 kb	Grove (1993; 1994)
Sierra Nevada Tehachapi Mountains	pelitic phase relations and thermobarometry	680–770 °C, 5.3–6.0 kb	Dixon et al. (1994)
Lake Isabella	pelitic phase relations and thermobarometry	620–720 °C, 5.5–6.0 kb	Dixon et al. (1994)
Ritter Range	calc-silicate phase relations	>450–500 °C, 1.8–2.3 kb	Hanson et al. (1993)
Hope Valley	calc-silicate phase relations and thermobarometry	440–540 °C, 1.8–2.3 kb	Ferry (1989)
Twin Lakes	calc-silicate phase relations and thermobarometry	600–650 °C, 2.5–3.5 kb	Davis and Ferry (1993)
Mount Morrison	calc-silicate phase relations	500–600 °C, 2.5–3.0 kb	Morgan (1975)
Mount Tallac	metavolcanic phase relations	625–650 °C,1.8–2.0 kb	Loomis (1966)
Cascade Mountains Washington	water wells and mineral and geothermal exploration holes	45 °C/km	Blackwell et al. (1990)
Northern Oregon	water wells and geothermal exploration holes	65 °C/km	Blackwell et al. (1982)
Japan Ryoke Belt	pelitic phase relations and thermobarometry	650–700 °C, 3.5–4.0 kb	Nakajima (1994)
Ryoke Belt	pelitic phase relations and thermobarometry	460–590 °C, 2.5–3.5 kb 630–690 °C, 3.0–5.0 kb 730–770 °C, 5.5–6.5 kb	Okudaira (1996)
Granulites scattered locations	thermobarometry	750–850 °C, 6.5–8.5 kb	Bohlen (1991)
Mojave Desert Eastern Mojave Xenoliths Waterman Complex	phase relationships and thermobarometry	750 °C, 11 kb 750–800 °C, 10–12 kb	Hanchar et al. (1994) Henry and Dokka (1992)

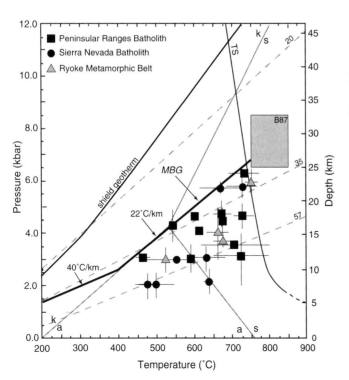

Figure 4. Summary P-T diagram showing available P-T estimates from Table 8 used to constrain maximum background geotherm (MBG). Aluminosilicate stability fields from Bohlen et al. (1991), stable shield geotherm from Morgan (1984), and vapor-saturated tonalite solidus (TS) from Johannes (1984) are shown for reference. Large gray box labeled *B87* encloses mean granulite P-T conditions from Bohlen (1987); other symbols are explained in diagram. Dashed gray lines labeled 20, 35, and 57 delineate linear geothermal gradients of 20 °C/km, 35 °C/km, and 57 °C/km extrapolated from heat flow data in active arcs (see Table 2). For clarity, thermobarometry symbols denote mean P-T estimates summarized in Table 8. Error bars are taken from reported studies when available or are arbitrarily set at ±50 °C and 0.5 kbar. Symbols with no error bars have reported errors that are smaller than size of the symbol. Black dashed line labeled MBG represents maximum background geotherm inferred from P-T data. Where constrained by P-T data from 10 to 30 km, locus of minimum temperatures in array defines a gradient of ~22 °C/km. Assumption of a 40 °C/km gradient from 0 to 10 km is consistent with available P-T constraints and studies summarized in Table 1.

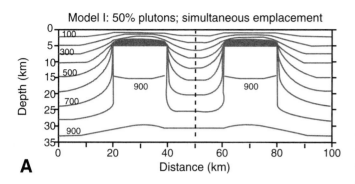

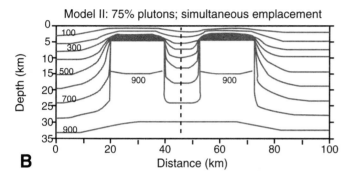

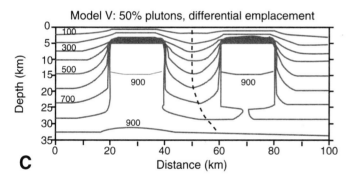

Figure 5. Examples of thermal modeling grids. Grid measures 100 km across and 40 km deep. Plutons are 20 km wide and 20 km tall and are emplaced instantaneously between depths of 5 and 25 km. Solid gray lines labeled 100–900 in 200 °C increments are isotherms showing distribution of maximum temperatures recorded in each simulation. Dashed black lines trace locus of minimum temperatures recorded from surface to 35 km depths and represent gradients in predicted minimum metamorphic temperatures (PMMT). A: Maximum temperature distribution for model I with 50% pluton volume and simultaneous rate of emplacement; PMMT is vertical and located midway between plutons. B: Maximum temperature distribution for model II with 75% pluton volume and simultaneous rate of emplacement. PMMT is vertical and located midway between plutons, which are more closely spaced to simulate 75% pluton volume. C: Maximum temperature distribution for model V with 50% pluton volume and 5 m.y. time lag between emplacement of plutons. PMMT is sub-vertical and displaced toward second pluton. See appendix for details regarding parameterization of each model (indicated by roman numeral).

Karlstrom et al., 1993). Because this structural setting will consist of domains of high strain rates localized around cooling plutons, their spatial and temporal distribution will be discontinuous. However, over the duration of arc emplacement, these domains may coalesce to form broad belts of deformation that appear continuous in the geologic record (e.g., Patterson and Tobisch, 1992).

The recognition of low geothermal gradients in batholiths also affects modeling denudation histories from thermochronologic data. Assuming high mid-crustal geothermal gradients in arc terranes will lead to underestimating the rate of unroofing recorded by various mineral thermochronometers. Low temperature (~450 °C) $^{40}Ar/^{39}Ar$ mica and K-feldspar and fission-track apparent ages from rocks in arc terranes such as the Peninsular Ranges batholith are typically on the order of 5–10 m.y. younger than pluton emplacement ages recorded by zircon U-Pb systematics. Because thermal anomalies from mid-crustal magma sources will likely decay over 5–10 m.y. toward low (~20 °C/km) pre-magmatic temperatures, it may be inappropriate to use geothermal gradients inferred from short-lived peak temperatures recorded in contact metamorphic mineral assemblages to interpret exhumation processes that operate over longer time scales.

CONCLUSION

Metamorphic data from the eastern Peninsular Ranges batholith in Baja California, México, provide constraints on the distribution of wall-rock temperatures in a continental magmatic arc characterized by a relatively simple thermal history. Combining these data with P-T data from similar terranes results in the recognition of a maximum background geotherm characterized by a significant change from ~40 °C/km in the upper crust (0–10 km) to ~22 °C/km in the middle crust (10–25 °C/km). This mid-crustal geotherm is significantly lower than the 35–50 °C/km gradients previously inferred from low-pressure metamorphic P-T data and is somewhat higher than 15–18 °C/km gradients in the middle crust inferred from previous modeling studies. Recognition of this geothermal gradient may lower estimates of the depth to the seismic Moho and increase strain and unroofing rates inferred from structural and thermochronologic studies, respectively.

APPENDIX

The numerical approach uses the explicit finite-difference method to approximate the analytical solution to the heat flow equation in two dimensions:

$$\frac{dT}{dt} = \kappa \left(\frac{\partial^2 T}{\partial x^2} + \frac{\partial^2 T}{\partial z^2} + \frac{A}{K} \right)$$

where κ is the thermal diffusivity ($m^2 s^{-1}$), T is temperature (°C), t is time (s), x is the horizontal Cartesian coordinate (m), z is the depth and increases downward (m), A is the radiogenic heat production ($mW m^{-3}$), and k is the thermal conductivity ($mW m^{-1} °C^{-1}$). The numerical code for these simulations was written

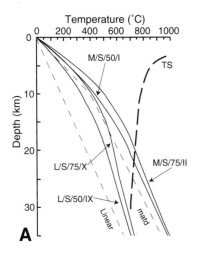
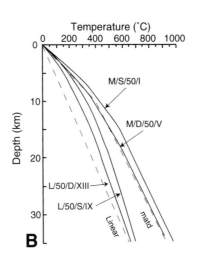
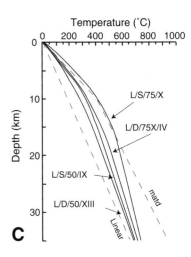

Figure 6. Comparison of gradients in predicted minimum metamorphic temperatures (PMMT). Abbreviations refer to following initial conditions for thermal models: L—18 °C/km initial temperature distribution; M— maximum background geotherm (MBG) initial temperature distribution; D—differential emplacement (5 m.y. between plutons); S—Simultaneous emplacement; 50—50% pluton volume; 75—75% pluton volume. Roman numerals correspond to simulations described in Table A2. A: Comparison of simultaneous intrusion models that illustrate sensitivity of PMMT to pluton volume. TS indicates wet tonalite solidus (Johannes, 1984). B: Comparison of 50% pluton volume models that illustrate sensitivity of PMMT to simultaneous versus differential pluton emplacement. C: Comparison of various simulations using 18 °C/km linear ambient geotherm. Most rapid and highest volume of plutonism yields a gradient in PMMT that is broadly similar to MBG described in this study. See text for further discussion.

by Hanson and Barton (1989) but was modified to allow implementation of different initial temperature distributions. Although implicit methods exhibit greater numerical stability (Carnahan et al., 1969; Harrison and Clarke, 1979), the explicit finite-difference method facilitates the adjustment of thermal parameters after each time step to better simulate the thermal effects of the latent heat of crystallization, which require consideration to accurately model temperatures near intrusive contacts (Jaeger, 1964).

The numerical code was compared with an implicit finite-difference code written by O.M. Lovera to test the stability and convergence of the explicit finite-difference method and the accuracy of the results. For the simple case of wall rock heating from the crystallization of a single pluton, the numerical approaches are indistinguishable from each other. The Hanson and Barton code facilitates the simulation of varying combinations of pluton volume and timing of intrusions. The code uses a nine-point approximation of the second derivative of temperature in the heat flow equation to improve numerical stability and convergence (Hanson and Barton, 1989).

The finite difference grid consists of a two-dimensional array of equally spaced points. The spacing between points was 1 km, such that the grid measured 100 km wide and 40 km deep (Fig. 6). Each simulation involves the instantaneous emplacement of two 20-km-wide plutons between depths of 5 and 25 km in each simulation. Instantaneous emplacement yields lower temperatures than for cases of forceful emplacement, a preheated ascent path, or if extension facilitates intrusion.

The models use one of two temperature distributions for the tonalitic plutons. One set of models assumes uniform emplacement temperatures of 900 °C throughout the 20-km-thick magma chamber. The other set of models uses experimental constraints on the wet tonalite solidus (Johannes, 1984) to dictate depth-dependent variations in emplacement temperatures. Averaging the supersolidus temperatures in magma chambers after each time step simulates the thermal effects of convection; this approximation maximizes heat flow from the intrusions and tends to overestimate temperatures near the pluton contact. While temperatures are above the solidus, halving the thermal diffusivity inside the magma chamber simulates the thermal effects of the latent heat of crystallization (see Table A1). The thermal effects of this approximation diminish rapidly from the pluton contact and are negligible at distances greater than one-quarter of the width of the pluton. The effect of the latent heat of crystallization is calculated over a 100 °C temperature interval between emplacement and solidus temperatures.

The thermal effects of metamorphic and melting reactions are ignored because it is unclear that uniformly adjusting the wall-rock thermal diffusivity over a particular temperature interval appropriately simulates the thermal effects of the discontinuous and continuous metamorphic reactions that occur within a crust that has a heterogeneous bulk composition. Because this study is most concerned with gradients in temperature rather than absolute values, this simplification had a negligible effect on our interpretation of the model results.

The models also ignore alternative heat sources for low-pressure metamorphism. Because the Peninsular Ranges batholith lacks large-scale regional metasomatic zones, the models do not consider the regional effects of heat transport by migrating

TABLE A1. PARAMETERIZATION OF THERMAL MODELS

Parameter/Symbol	Definition/Value
MBG	Maximum ambient temperature distribution 40 °C/km (0–10 km); 22 °C/km (10–35 km)
Linear Geotherm	Linear geothermal gradient. 18 °C/km
K	Thermal conductivity
κ_m	Thermal diffusivity, supersolidus. 5.0×10^{-7} cm^2 s^{-1}
κ_r	Thermal diffusivity, solidified pluton and country rocks. 1.0×10^{-6} cm^2 s^{-1}
T_s	Solidus temperature. 1000 °C, 5–10 km; 920 °C, 10–15 km; 860 °C, 15–20 km; 820 °C, 20–25 km
T_o	Surface temperature. 0 °C
T	Initial temperature at depth z.
Δt	Time step increment. 10,000 years.
ΔX	Spacing between nodes in x-direction. 1000 m (100 km wide grid)
ΔZ	Spacing between nodes in z-direction. 1000 m (40 km deep grid)
A	Internal heat production at depth z.

TABLE A2. THERMAL MODELING INITIAL CONDITIONS AND RESULTS

Model	Initial Temperature	Pluton Temperature	Pluton Volume	Timing	PMMT
I	MBG	Tonalite Solidus	50%	Simultaneous	44 °C/km 0–10 km 22 °C/km 10–35 km
II	MBG	Tonalite Solidus	75%	Simultaneous	50 °C/km 0–10 km 20 °C/km 10–35 km
III	MBG	Uniform (900 °C)	50%	Simultaneous	44 °C/km 0–10 km 22 °C/km 10–35 km
IV	MBG	Uniform (900 °C)	75%	Simultaneous	50 °C/km 0–10 km 20 °C/km 10–35 km
V	MBG	Tonalite Solidus	50%	Differential	40 °C/km 0–10 km 22 °C/km 10–35 km
VI	MBG	Tonalite Solidus	75%	Differential	41 °C/km 0–10 km 22 °C/km 10–35 km
VII	MBG	Uniform (900 °C)	50%	Differential	40 °C/km 0–10 km 22 °C/km 10–35 km
VIII	MBG	Uniform (900 °C)	75%	Differential	41 °C/km 0–10 km 22 °C/km 10–35 km
IX	Linear	Tonalite Solidus	50%	Simultaneous	33 °C/km 0–10 km 16 °C/km 10–35 km
X	Linear	Tonalite Solidus	75%	Simultaneous	41 °C/km 0–10 km 13 °C/km 10–35 km
XI	Linear	Uniform (900 °C)	50%	Simultaneous	33 °C/km 0–10 km 16 °C/km 10–35 km
XII	Linear	Uniform (900 °C)	75%	Simultaneous	41 °C/km 0–10 km 13 °C/km 10–35 km
XIII	Linear	Tonalite Solidus	50%	Differential	27 °C/km 0–10 km 16 °C/km 10–35 km
XIV	Linear	Tonalite Solidus	75%	Differential	32 °C/km 0–10 km 15 °C/km 10–35 km
XV	Linear	Uniform (900 °C)	50%	Differential	27 °C/km 0–10 km 16 °C/km 10–35 km
XVI	Linear	Uniform (900 °C)	75%	Differential	31 °C/km 0–10 km 15 °C/km 10–35 km

Note: Linear geotherm is 18 °C/km; the MBG is 40 °C from 0–10 km and 22 °C/km from 10–30 km as defined in the text. Simultaneous and differential timing of pluton emplacement are explained in the text. PMMT—predicted minimum metamorphic temperature; MBG—Maximum background geotherm.

low-density metamorphic fluids (e.g., Hoisch, 1987; Hanson, 1995). The models did not investigate the consequences of magmatic underplating, which may be important in the generation of granulites at depths >35 km in batholiths. Because the geologic, petrologic, geochronologic, and thermochronologic frameworks of the batholith suggest that Cretaceous plutonism dominated the thermal evolution of the Peninsular Ranges batholith, the thermal models focused on this aspect of the metamorphic evolution of the batholith.

Numerous crustal models and combinations of the physical and thermal properties of rocks allow the calculation of the ~18 °C/km geotherm used in the linear geotherm models in this study (e.g., Turcotte and Schubert, 1982). An example of parameters that yield an ~18°C/km geotherm for a one-layer, 35-km-thick crust with a homogenous distribution of radioactive elements includes internal heat production (A) = 0.7 mW/m^3; basal heat flux (Q*) = 25 W/m^2; thermal conductivity (k) = 2.5 W m^{-1} °C^{-1}.

ACKNOWLEDGMENTS

This work benefited from our association with M. Grove, O. Lovera, R. Jones, and R. Alkaly at the University of California at Los Angeles and through discussions with G. Gastil, V. Todd, D. Kimbrough, J. Lee, and J. Stock on various aspects of field geology in the eastern batholith and from reviews by J. Fletcher and D. Whitney. This research was in part supported by National Science Foundation grant EAR 94-05999 (C.M.) and grants from Sigma Xi and the California chapter of the American Mineralogical Society (D.R.).

REFERENCES CITED

Arndt, J., Bartel, T., Scheuber, E., and Schuilling, F., 1997, Thermal and rheological properties of granodioritic rocks from the central Andes, north Chile: Tectonophysics, v. 271, p. 75–88.

Barton, M.D., and Hanson, R.B., 1989, Magmatism and the development of low-pressure metamorphic belts: implications from the western United States and thermal modeling: Geological Society of America Bulletin, v. 101, p. 1051–1065.

Barton, M.D., Battles, D.A., Bebout, G.A., Capo, R.C., Christensen, J.N., Davis, S.R., Hanson, R.B., Michelsen, C.J., and Trim, H.E., 1988, Mesozoic contact metamorphism in the western United States, in Ernst, W.G., ed., Metamorphism and crustal evolution of the western conterminous United States, Rubey Volume VII: Englewood Cliffs, N.J., Prentice-Hall, p. 110–178.

Berman, R.G., 1990, Mixing properties of Ca-Mg-Fe-Mn garnets: American Mineralogist, v. 75, p. 328–344.

Berman, R.G., 1991, Thermobarometry using multi-equilibrium calculations: A new technique, with petrological applications: Canadian Mineralogist, v. 29, p. 833–855.

Birch, F., 1961, The velocity of compressional waves in rocks to 10 kbar: Journal of Geophysical Research, v. 66, p. 2199–2224.

Blackwell, D.D., Bowen, R.G., Hull, D.A., Riccio, J., and Steele, J.L., 1982, Heat flow, arc volcanism, and subduction in northern Oregon: Journal of Geophysical Research, v. 87, p. 8735–8754.

Blackwell, D.D., Steele, J.L., Kelley, S., and Korosec, M.A., 1990, Heat flow in the state of Washington and thermal conditions in the Cascade Range: Journal of Geophysical Research, v. 95, p. 19,495–19,516.

Bodorokos, S., Sandiford, M., Oliver, N.H.S., and Cawood, P.A., 2002, High-T, low-P metamorphism in the paleoproterozoic Halls Creek orogen, northern Australia: The middle crustal response to a mantle-related transient thermal pulse: Journal of Metamorphic Geology, v. 20, p. 217–237.

Bohlen, S.R., 1991, On the formation of granulites: Journal of Metamorphic Geology, v. 9, p. 223–229.

Bohlen, S.R., Montana, A., and Kerrick, D.M., 1991, Precise determinations of the equilibria kyanite = sillimanite and kyanite = andalusite and a revised triple point for Al$_2$SiO$_5$ polymorphs: American Mineralogist, v. 76, p. 677–680.

Brown, M., 1993, P-T-t evolution of orogenic belts and the causes of regional metamorphism: Journal of the Geological Society of London, v. 150, p. 227–241.

Carnahan, B., Luther, H.A., and Wilkes, J.O., 1969, Applied numerical methods: New York, John Wiley & Sons, 603 p.

Christensen, N.I., 1982, Seismic velocities, in Carmichael, R.S., ed., Handbook of physical properties of rocks, V. 2: Boca Raton, Florida, CRC Press, p. 1–228.

Christensen, N.I., and Fountain, D.M., 1975, Constitution of the lower continental crust based on experimental studies of seismic velocities in granulite: Geological Society of America Bulletin, v. 86, p. 227–236.

Crawford, M.L., Hollister, L.S., and Woodsworth, G.J., 1987, Crustal deformation and regional metamorphism across a terrane boundary, Coast Plutonic Complex, British Columbia: Tectonics, v. 6, p. 343–361.

Davis, S.R., and Ferry, J.M., 1993, Fluid infiltration during contact metamorphism of interbedded marble and calc-silicate hornfels, Twin Lakes area, central Sierra Nevada, California: Journal of Metamorphic Geology, v. 11, p. 71–88.

De Yoreo, J.J., Lux, D.R., and Guidotti, C.V., 1989, The role of crustal anatexis and magma migration in the thermal evolution of regions of thickened continental crust, in Daly, J.S., Cliff, R.A., and Yardley, B.W.D., ed., Evolution of metamorphic belts: Geological Society [London] Special Publication 42, p. 187–202.

De Yoreo, J.J., Lux, D.R., and Guidotti, C.V., 1991, Thermal modeling in low-pressure/high-temperature metamorphic belts: Tectonophysics, v. 188, p. 209–238.

Dixon, E.T., Essene, E.J., and Halliday, A.N., 1994, Critical tests of hornblende barometry, Lake Isabella to Tehachapi area, southern Sierra Nevada, California [abs.]: Eos (Transactions, American Geophysical Union), v. 74, p. 744.

Dokka, R.K., 1984, Fission-track geochronologic evidence for Late Cretaceous mylonitization and Early Paleocene uplift of the northeastern Peninsular Ranges, California: Geophysical Research Letters, v. 11, p. 46–49.

England, P.C., and Richardson, S.W., 1977, The influence of erosion upon the mineral facies of rocks from different metamorphic environments: Journal of the Geological Society of London, v. 134, p. 201–213.

England, P.C., and Thompson, A.B., 1984, Pressure-temperature-time paths of regional metamorphism I. Heat transfer during the evolution of regions of thickened continental crust: Journal of Petrology, v. 25, p. 894–928.

Erskine, B.G., 1986, Mylonitic deformation and associated low-angle faulting of the northern Peninsular Ranges batholith, southern California [Ph.D. dissertation thesis]: Berkeley, University of California.

Erskine, B.G., and Wenk, H.R., 1985, Evidence for Late Cretaceous crustal thinning in the Santa Rosa mylonite zone, southern California: Geology, v. 13, p. 274–277.

Ferry, J.M., 1989, Contact metamorphism of roof pendants at Hope Valley, Alpine County, California, USA: Contributions to Mineralogy and Petrology, v. 101, p. 402–417.

Fuhrman, M.L., and Lindsley, D.H., 1988, Ternary-feldspar modeling and thermometry: American Mineralogist, v. 73, p. 201–215.

Furlong, K.P., and Fountain, D.M., 1986, Continental crustal underplating: Thermal considerations and seismic-petrologic consequences: Journal of Geophysical Research, v. 91, p. 8285–8294.

Furlong, K.P., Hanson, R.B., and Bowers, J.R., 1991, Modeling thermal regimes, in Kerrick, D.A., ed., Contact metamorphism: Reviews in Mineralogy, v. 26, p. 437–505.

Furukawa, Y., and Uyeda, S., 1989, Thermal state under Tohoku arc with consideration of crustal heat generation: Tectonophysics, v. 164, p. 175–187.

Gastil, R.G., and Miller, R.H., 1993, The prebatholithic stratigraphy of peninsular California: Boulder, Colorado, Geological Society of America Special Paper 279, 163 p.

Gastil, R.G., Philips, R.P., and Allison, E.C., 1975, Reconnaissance geology of the state of Baja California: Boulder, Colorado, Geological Society of America Memoir 140, 170 p.

George, P.G., and Dokka, R.K., 1994, Major Late Cretaceous cooling events in the eastern Peninsular Ranges, California, and their implications for Cordilleran tectonics: Geological Society of America Bulletin, v. 106, p. 903–914.

Giese, P., 1994, Geothermal structure of the central Andean crust-implications for heat transport and rheology, in Reutter, K.J., Scheuber, E., and Wigger,

P.J., eds., Tectonics of the southern central Andes; structure and evolution of an active continental margin: Berlin, Springer-Verlag, p. 69–76.

Gill, J., 1981, Orogenic andesites and plate tectonics: Berlin, Springer, 390 p.

Goodwin, L.B., and Renne, P.R., 1991, Effects of progressive mylonitization on Ar retention in biotites from the Santa Rosa mylonite zone, California, and thermochronologic implications: Contributions to Mineralogy and Petrology, v. 108, p. 283–297.

Griffin, W.L., and O'Reilly, S.Y., 1987, Is the continental Moho the crust-mantle boundary?: Geology, v. 15, p. 241–244.

Grissom, G.C., DeBari, S.M., Page, S.P., Page, R.F.N., Villar, L.M., Coleman, R.G., and de Ramirez, M.V., 1991, The deep crust of an early Paleozoic arc; the Sierra de Riambalá, northwestern Argentina, in Harmon, R.S., and Rapela, C.W., eds., Andean magmatism and its tectonic setting: Boulder, Colorado, Geological Society of America Special Paper 265, p. 189–200.

Grove, M., 1986, Metamorphism of the Box Canyon area, Peninsular Ranges batholith, San Diego County, California [M.S. thesis]: Los Angeles, University of California, 174 p.

Grove, M., 1993, Thermal histories of southern California basement terranes [Ph.D. thesis]: Los Angeles, University of California, 419 p.

Grove, M., 1994, Contrasting denudation histories within the east-central Peninsular Ranges batholith (33°N): Geological Society of America Cordilleran Section Field Trip Guidebook, p. 235–240.

Hanchar, J.M., Miller, C.F., Wooden, J.L., Bennett, V.C., and Staude, J.M.G., 1994, Evidence from xenoliths for a dynamic lower crust, eastern Mojave Desert, California: Journal of Petrology, v. 35, p. 1377–1415.

Hanson, B.N., and Gast, P.W., 1967, Kinetic studies in contact metamorphic zones: Geochimica et Cosmochimica Acta, v. 31, p. 1119–1153.

Hanson, R.B., 1995, The hydrodynamics of contact metamorphism: Geological Society of America Bulletin, v. 107, p. 595–611.

Hanson, R.B., and Barton, M.D., 1989, Thermal development of low-pressure metamorphic belts: Results from two-dimensional numerical models: Journal of Geophysical Research, v. 94, p. 10,363–10,377.

Hanson, R.B., Sorensen, S.S., Barton, M.D., and Fiske, R.S., 1993, Long-term evolution of fluid-rock interactions in magmatic arcs: Evidence from the Ritter Range pendant, Sierra Nevada, California, and numerical modeling: Journal of Petrology, v. 34, p. 23–62.

Harrison, T.M., and Clarke, G.K.C., 1979, A model of the thermal effects of igneous intrusion and uplift as applied to Quottoon pluton, British Columbia: Canadian Journal of Earth Sciences, v. 16, p. 411–420.

Henry, D.J., and Dokka, R.K., 1992, Metamorphic evolution of exhumed middle to lower crustal rocks in the Mojave Extensional Belt, southern California: Journal of Metamorphic Geology, v. 10, p. 347–364.

Henry, S.G., and Pollack, H.N., 1988, Terrestrial heat flow above the Andean subduction zone in Bolivia and Peru: Journal of Geophysical Research, v. 93, p. 15,153–15,162.

Hill, R.I., 1984, Petrology and petrogenesis of batholithic rocks, San Jacinto Mountains, southern California [Ph.D.] dissertation: Pasadena, California, California Institute of Technology, 660 p.

Hoisch, T.D., 1987, Heat transport by fluids during Late Cretaceous regional metamorphism in the Big Maria mountains, southeastern California: Geological Society of America Bulletin, v. 98, p. 549–553.

Hyndman, D.W., and Foster, D.A., 1989, Plutonism at deep crustal levels: The Idaho batholith, Montana and Idaho, in Hyndman, D.W., Rutland, C., and Hardyman, R.F., eds., Cordilleran volcanism, plutonism, and magma generation at various crustal levels, Montana and Idaho: Washington, D.C., American Geophysical Union, p. 3–15.

Ingebritsen, S.E., Sherrod, D.R., and Mariner, R.H., 1989, Heat flow and hydrothermal circulation in the Cascade Range, north-central Oregon: Science, v. 243, p. 1458–1462.

Ingebritsen, S.E., Sherrod, D.R., and Mariner, R.H., 1992, Rates and patterns of groundwater flow in the Cascade range volcanic arc, and the effect of subsurface temperatures: Journal of Geophysical Research, v. 97, p. 4599–4627.

Jaeger, J.C., 1964, Thermal effects of intrusions: Reviews in Geophysics, v. 2, p. 433–466.

Joesten, R., and Fisher, G., 1988, Kinetics of diffusion-controlled mineral growth in the Christmas Mountains (Texas) contact aureole: Geological Society of America Bulletin, v. 100, p. 714–732.

Johannes, W., 1984, Beginning of melting in the granite system Qz-Or-Ab-An-H_2O: Contributions to Mineralogy and Petrology, v. 86, p. 264–273.

Johnson, S.E., Tate, M.C., and Fanning, C.M., 1999, New geologic mapping and SHRIMP U-Pb zircon data in the Peninsular Ranges batholith, Baja California, México: Evidence for a suture?: Geology, v. 27, p. 743–746.

Karlstrom, K.E., Miller, C.F., Kingsbury, J.A., and Wooden, J.L., 1993, Pluton emplacement along an active ductile thrust zone, Piute Mountains, southeastern California: Interactions between deformational and solidification processes: Geological Society of America Bulletin, v. 105, p. 213–230.

Kohn, M.J., and Spear, F.S., 1991, Error propagation for barometers: 2. Application to rocks: American Mineralogist, v. 76, p. 138–147.

Kohn, M.J., Spear, F.J., Harrison, T.M., and Dalziel, I.W.D., 1995, $^{40}Ar/^{39}Ar$ geochronology and P-T-t paths from the Cordillera Darwin metamorphic complex, Tierra del Fuego, Chile: Journal of Metamorphic Geology, v. 13, p. 251–270.

Kretz, R., 1983, Symbols for rock-forming minerals: American Mineralogist, v. 68, p. 277–279.

Krumenacher, D., Gastil, R.G., Bushee, J., and Doupont, J., 1975, K-Ar apparent ages, Peninsular Ranges batholith, southern California and Baja California: Geological Society of America Bulletin, v. 86, p. 760–768.

Lister, G.S., and Baldwin, S.L., 1993, Plutonism and the origin of metamorphic core complexes: Geology, v. 21, p. 607–610.

Loomis, A.A., 1966, Contact metamorphic reactions and processes in the Mt. Tallac roof remnant, Sierra Nevada, California: Journal of Petrology, v. 7, p. 221–245.

Lovera, O.M., Grove, M., Kimbrough, D.L., and Abbott, P.L., 1999, A method for evaluating basement exhumation histories from closure age distributions of detrital minerals: Journal of Geophysical Research, v. 104, no. B2, p. 29,419–29,438.

Lovering, T.S., 1935, Theory of heat conduction applied to geological problems: Geological Society of America Bulletin, v. 48, p. 69–94.

Lucassen, F., and Franz, G., 1996, Magmatic arc metamorphism: Petrology and temperature history of metabolic rocks in the coastal Cordillera of northern Chile: Journal of Metamorphic Geology, v. 14, p. 249–265.

Lux, D.R., De Yoreo, J.J., Guidotti, C.V., and Decker, E.R., 1985, Role of plutonism in low-pressure metamorphic belt formation: Nature, v. 323, p. 794–797.

Mahon, K.I., Harrison, T.M., and Drew, D.A., 1988, Ascent of a granitoid diapir in a temperature-varying medium: Journal of Geophysical Research, v. 93, p. 1174–1188.

McDougall, I., and Harrison, T.M., 1988, Geochronology and Thermochronology by the $^{40}Ar/^{39}Ar$ Method: Oxford, Oxford University Press, 212 p.

McMullin, D.W.A., Berman, R.G., and Greenwood, H.J., 1991, Calibration of the SGAM thermobarometer for pelitic rocks using data from phase-equilibrium experiments and natural assemblages: Canadian Mineralogist, v. 29, p. 889–908.

Miller, C.F., Watson, E.B., and Harrison, T.M., 1988, Perspectives on the source, segregation, and transport of granitoid magmas: Transactions of the Royal Society of Edinburgh, Earth Sciences, v. 79, p. 135–156.

Miller, C.F., Hanchar, J.M., Wooden, J.L., Bennet, V.C., Harrison, T.M., Wark, D.A., and Foster, D.A., 1992, Source region of a granite batholith: Evidence from lower crustal xenoliths and inherited accessory minerals: Transactions of the Royal Society of Edinburgh, Earth Sciences, v. 83, p. 49–62.

Minch, J.A., 1979, The Late Mesozoic–Early Tertiary framework of continental sedimentation, northern Peninsular Ranges, Baja California, México, in Abbott, P.L., ed., Eocene depositional systems, San Diego, California: Los Angeles, Society of Economic Paleontologists and Mineralogists Pacific Section, p. 43–67.

Miyashiro, A., 1972, Metamorphism and related magmatism in plate tectonics: American Journal of Science, v. 272, p. 629–656.

Miyashiro, A., 1973, Metamorphism and metamorphic belts: New York, Wiley, 492 p.

Morgan, B.A., 1975, Mineralogy and origin of skarns in the Mount Morrison pendant, Sierra Nevada, California: American Journal of Science, v. 275, p. 119–142.

Morgan, P., 1984, The thermal structure and thermal evolution of the continental lithosphere, in Pollack, H.N., ed., Structure and evolution of the continental lithosphere: Physics and Chemistry of the Earth, v. 15, p. 107–193.

Nakajima, T., 1994, The Ryoke plutonometamorphic belt: Crustal section of the Cretaceous Eurasian continental margin: Lithos, v. 33, p. 51–66.

Norton, D.L., 1984, Theory of hydrothermal systems: Annual Reviews of Earth and Planetary Sciences, v. 12, p. 155–177.

Nutman, A.P., and Collerson, K.D., 1991, Very early Archean crustal-accretion complexes preserved in the North Atlantic craton: Geology, v. 19, p. 791–794.

Okudaira, T., 1996, Temperature-time path for the low-pressure Ryoke metamorphism, Japan, based on chemical zoning in garnet: Journal of Metamorphic Geology, v. 14, p. 427–440.

Ortega-Rivera, M.A., Farrar, E., Hanes, J.A., Archibald, D.A., Gastil, R.G., Kimbrough, D.L., Zentilli, M., Lopez Martinez, M., Feraud, G., and Ruffet, G., 1997, Chronological constraints on the thermal and tilting history of the Sierra San Pedro Martir pluton, Baja California, México, from U/Pb, ^{40}Ar/^{39}Ar, and fission-track geochronology: Geological Society of America Bulletin, v. 109, p. 728–745.

Paterson, S.R., and Tobisch, O.T., 1992, Rates of processes in magmatic arcs: Implications for the timing and nature of pluton emplacement and wall rock deformation: Journal of Structural Geology, v. 14, p. 291–300.

Rothstein, D.A., 1997, Metamorphism and denudation of the eastern Peninsular Ranges batholith, Baja California, México [Ph.D. dissertation]: Los Angeles, University of California, 445 p.

Rothstein, D.A., and Hoisch, T.D., 1994, Multiple intrusions and low-pressure metamorphism in the central Old Woman Mountains, south-eastern California: Constraints from thermal modelling: Journal of Metamorphic Geology, v. 12, p. 723–734.

Rothstein, D.A., Manning, C.E., and Grove, M., 1994, Denudation patterns in the eastern Peninsular Ranges batholith, Baja California, México [abs.]: Eos (Transactions, American Geophysical Union), v. 75, p. 229.

Rothstein, D.A., Grove, M., and Manning, C.E., 1995, Role of the main gulf escarpment in the denudation of the east-central Peninsular Ranges batholith, Baja California, México: Insights from high resolution ^{40}Ar/^{39}Ar thermochronology [abs.]: Eos (Transactions, American Geophysical Union), v. 76, p. 639.

Rusmore, M.E., and Woodsworth, G.J., 1994, Evolution of the eastern Waddington thrust belt and its relation to the mid-Cretaceous Coast Mountains arc, western British Columbia: Tectonics, v. 13, p. 1052–1067.

Sandiford, M., Martin, N., Zhou, S., and Graser, G., 1991, Mechanical consequences of granite emplacement during high-T, low-P metamorphism and the origin of "anticlockwise" PT paths: Earth and Planetary Science Letters, v. 107, p. 164–172.

Schilling, F.R., and Parzsch, G.M., 2001, Quantifying partial melt fraction in the crust beneath the central Andes and the Tibetan plateau: Physics and Chemistry of the Earth, Part A, v. 26, p. 239–246.

Sharp, R.V., 1967, San Jacinto fault zone in the Peninsular Ranges of southern California: Geological Society of America Bulletin, v. 78, p. 706–729.

Silver, L.T., and Chappell, B.W., 1988, The Peninsular Ranges batholith: An insight into the evolution of the Cordilleran batholiths of southwestern North America: Transactions of the Royal Society of Edinburgh; Earth Sciences, v. 79, p. 105–121.

Simpson, C., 1984, Borrego Springs–Santa Rosa mylonite zone: A Late Cretaceous west-directed thrust in southern California: Geology, v. 12, p. 8–11.

Springer, M., and Förster, A., 1998, Heat-flow density across the central Andean subduction zone: Tectonophysics, v. 291, p. 123–139.

Stock, J.M., and Hodges, K.V., 1989, Pre-Pliocene extension around the Gulf of California and the transfer of Baja California to the Pacific Plate: Tectonics, v. 8, p. 99–115.

Stüwe, K., and Sandiford, M., 1994, Contribution of deviatoric stresses to metamorphic P-T paths: An example appropriate to low-P, high-T metamorphism: Journal of Metamorphic Geology, v. 12, p. 445–454.

Stüwe, K., Sandiford, M., and Powell, R., 1993, Episodic metamorphism and deformation in low-pressure, high-temperature terranes: Geology, v. 21, p. 829–832.

Taylor, H.P., 1986, Igneous rocks: II. Isotopic case studies of Circum-Pacific magmatism, *in* Valley, J.W., Taylor, H.P., and O'Neil, J.R., eds., Stable isotopes in high temperature geological processes: Reviews in Mineralogy, v. 16, p. 273–318.

Thomson, C.N., and Girty, G.H., 1994, Early Cretaceous intra-arc ductile strain in Triassic-Jurassic and Cretaceous continental margin arc rocks, Peninsular Ranges, California: Tectonics, v. 13, p. 1108–1119.

Thompson, R.N., 1972, Melting behavior of two Snake River lavas at pressures up to 35 kb: Carnegie Institution of Washington Yearbook, v. 71, p. 406–410.

Todd, V.R., Erskine, B.G., and Morton, D.M., 1988, Metamorphic and tectonic evolution of the northern Peninsular Ranges batholith, southern California, *in* Ernst, W.G., ed., Metamorphism and crustal evolution of the western United States (Rubey Volume VII): Englewood Cliffs, N.J., Prentice-Hall, p. 894–937.

Todd, V.R., Girty, G.H., Shaw, W.E., and Jachens, R.C., 1991, Geochemical, geochronologic, and structural characteristics of Jurassic plutonic rocks, Peninsular Ranges, California [abs]: Geological Society of America Abstracts with Programs, v. 23, p. A249.

Turcotte, D.L., and Schubert, G., 1982, Geodynamics: Applications of continuum physics to geological problems: New York, John Wiley and Sons, 450 p.

Turner, F.J., 1981, Metamorphic petrology: Mineralogical, field, and tectonic aspects: New York, McGraw-Hill, 524 p.

Uyeda, W., and Watanabee, T., 1982, Terrestrial heat flow in western South America: Tectonophysics, v. 83, p. 63–70.

Walawender, M.J., Gastil, R.G., Clinkenbeard, J.P., McCormick, W.V., Eastman, B.G., Wernicke, F.S., Wardlaw, M.S., Gunn, S.H., and Smith, B.M., 1990, Origin and evolution of the zoned La Posta-type plutons, eastern Peninsular Ranges batholith, southern and Baja California, *in* Anderson, J.L., ed., The nature and origin of Cordilleran magmatism: Boulder, Colorado, Geological Society of America Memoir 174, p. 1–18.

Weaver, B.L., and Tarney, J., 1984, Empirical approach to estimating the composition of the continental crust: Nature, v. 310, p. 575–577.

Wells, P.R.A., 1980, Thermal models for the magmatic accretion and subsequent metamorphism of continental crust: Earth and Planetary Science Letters, v. 46, p. 253–265.

Wickham, S.R., and Oxburgh, E.R., 1985, Continental rifts as a setting for regional metamorphism: Nature, v. 318, p. 330–333.

Wickham, S.R., and Oxburgh, E.R., 1987, Low-pressure regional metamorphism in the Pyrenees and its implications for the thermal evolution of rifted continental crust: Philosophical Transactions of the Royal Society of London, v. 321A, p. 219–242.

Wiswall, C.G., and Hyndman, D.W., 1987, Emplacement of the main plutons of the Bitterroot lobe of the Idaho batholith, *in* Vallier, T.L., and Brooks, H.C., eds., Geology of the Blue Mountains Region of Oregon, Idaho, and Washington: The Idaho batholith and its border zone: U.S. Geological Survey Professional Paper 1436, p. 59–72.

Wyllie, P.J., 1979, Petrogenesis and the physics of the earth, *in* Yoder, H.S., ed., The evolution of the igneous rocks: Fiftieth anniversary perspectives: Princeton, Princeton University Press, p. 483–520.

Yoder, H.S., Jr., and Tilley, C.E., 1962, Origin of basalt magmas: An experimental study of natural and synthetic rock systems: Journal of Petrology, v. 3, p. 342–532.

MANUSCRIPT ACCEPTED BY THE SOCIETY JUNE 2, 2003

Geological Society of America
Special Paper 374
2003

Late Cretaceous cooling of the east-central Peninsular Ranges batholith (33°N): Relationship to La Posta pluton emplacement, Laramide shallow subduction, and forearc sedimentation

Marty Grove
Oscar Lovera
Department of Earth and Space Sciences, University of California, 595 Charles Young Drive E,
Los Angeles, California 90095-1567, USA

Mark Harrison
Department of Earth and Space Sciences, University of California, 595 Charles Young Drive E, Los Angeles, California
90095-1567, USA, and Research School of Earth Sciences, The Australian National University, Canberra, ACT 0200, Australia

ABSTRACT

Biotite and K-feldspar ^{40}Ar/^{39}Ar systems from the east-central Peninsular Ranges batholith near 33°N were affected by two distinct phases of Late Cretaceous rapid cooling. The 85-Ma biotite K-Ar isochron separates comparatively shallow rocks in the southwest that record earlier cooling (91–86 Ma) from deeper rocks in the northeast that record later cooling (78–68 Ma). Samples close to 85 Ma isochron record both episodes of rapid cooling as well as slower cooling between 86 and 78 Ma. Although the 85 Ma isochron also coincides with a steep (1–2 m.y./km) K-Ar age gradient, only localized faulting has been detected along it. We attribute 91–86 Ma cooling to denudation related to emplacement of the voluminous suite of 96 ± 3 Ma La Posta plutons. In contrast, we link cooling after 78 Ma to Laramide shallow subduction beneath the Peninsular Ranges batholith. Our numerical simulations indicate that the latter cooling phase can be explained by either pure erosional denudation or by subduction refrigeration beginning at 80 Ma. In the latter case, erosional denudation occurs during steady-state shallow subduction. While final erosion depths predicted by the two models differ significantly (as much as 20 km for pure erosional denudation but only 11 km for subduction refrigeration followed by erosion), both are within the range indicated by independent thermobarometry of the eastern batholith. Based upon the similarity of independently determined denudation histories from Peninsular Ranges batholith basement rocks and forearc sediments that onlap the northern Peninsular Ranges batholith, we conclude that erosional denudation was probably the most important process between 78 and 68 Ma. We speculate that removal of lower crust and lithospheric mantle beneath the east-central Peninsular Ranges batholith during Laramide shallow subduction triggered erosional denudation and localized thrusting until the density balance between the crust and mantle was restored in latest Cretaceous–early Tertiary time.

Keywords: Peninsular Ranges, denudation, Laramide, La Posta plutons, thermochronology.

*Marty@oro.ess.ucla.edu

Grove, M., Lovera, O., and Harrison, M., 2003, Late Cretaceous cooling of the east-central Peninsular Ranges batholith (33°N): Relationship to La Posta pluton emplacement, Laramide shallow subduction, and forearc sedimentation, *in* Johnson, S.E., Paterson, S.R., Fletcher, J.M., Girty, G.H., Kimbrough, D.L., and Martín-Barajas, A., eds., Tectonic evolution of northwestern México and the southwestern USA: Boulder, Colorado, Geological Society of America Special Paper 374, p. 355–379. For permission to copy, contact editing@geosociety.org. © 2003 Geological Society of America.

INTRODUCTION

A fundamental and incompletely understood aspect of crustal evolution is the response of continental arcs to the thermal and gravitational anomalies generated by batholith emplacement. The first-order expectation is that exhumation, erosion, faulting, and related processes should be most active during and immediately following intrusion when temperatures are high and rocks are weak (e.g., Tobisch et al., 1995). As thermal and density differences decay, the crust should strengthen and resist further mechanical readjustment unless triggered by changes in tectonic regime (e.g., Chase and Wallace, 1986).

Episodes of enormous magmatic influx are a hallmark of the medial Cretaceous Cordilleran magmatic arcs distributed along western North America (e.g., Bateman and Chappell, 1979; Barton et al., 1988; Coleman and Glazner, 1998; Ducea, 2001). The best examples of this phenomena are the distinctive large-volume, Late Cretaceous tonalite-trondhjemite-granodiorite (TTG) plutons that dominate the eastern Peninsular Ranges batholith of southern and Baja California (Gastil et al., 1975; Gastil, 1983; Taylor, 1986; Gromet and Silver, 1987; Todd et al., 1988; Silver and Chappell, 1988; Hill, 1988; Clinkenbeard and Walawender, 1989; Walawender et al., 1990; Gastil et al., 1991; Fig. 1). Walawender et al. (1990) coined the term "La Posta–type plutons" for these intrusions, and we follow Kimbrough et al. (1998) in referring to them as the La Posta TTG suite. The La Posta TTG suite is the defining characteristic of the Peninsular Ranges batholith. It occurs as a series of large, internally zoned intrusive centers that are distributed throughout the entire 800 km length of the batholith (Fig. 1; Gastil et al., 1975; Kimbrough et al., 2001). Available geologic mapping, geochronology, and geophysical data indicate that the suite represents a magmatic influx of ~75–100 km^3/km of batholith strike length/m.y. from ca. 98 to 93 Ma and that roughly 50% of the total exposed area of the batholith was emplaced during this surprisingly brief interval (Silver and Chappell, 1988; Kimbrough et al., 1998; Kimbrough et al., 2001).

It is reasonable to expect that injection of such an enormous quantity of magma into the crust over a short period would have had a profound impact upon the evolution of the Peninsular Ranges batholith. At its more shallowly denuded southern end, this appears to have been the case. Extensive exhumation and forearc sedimentation directly overlapped intrusion of the La Posta TTG and little has happened since (Kimbrough et al., 2001). In the more heavily denuded northern Peninsular Ranges batholith, however, the correspondence between La Posta TTG intrusion and exhumation is less clear. The ambiguity stems from the fact that large tracts of the La Posta TTG–dominated eastern batholith apparently required 15–20 Ma from the time of their emplacement to cool below the ~350 °C temperature (McDougall and Harrison, 1999) required for Ar diffusion in biotite (Armstrong and Suppe, 1973; Krummenacher et al., 1975; Goodwin and Renne, 1991; George and Dokka, 1994; Grove, 1994; Naeser et al., 1996; Ortega-Rivera et al., 1997; Ortega-Rivera, 2003). In fact, it has been proposed that significant post-

batholithic, Late Cretaceous cooling of the northern Peninsular Ranges batholith occurred primarily in two discrete phases: 1) closely following emplacement of the La Posta TTG during the Cenomanian–Turonian, and (2) 15–20 m.y. later during the Late Campanian–Maastrictian (Grove, 1994; Lovera et al., 1999).

The Late Campanian–Maastrictian timing of the second phase of cooling suggests a causal relationship with Laramide shallow subduction (ca. 80–45 Ma; e.g., Coney and Reynolds, 1977; Dickinson and Snyder, 1978; Bird, 1988). The probability that Laramide shallow subduction occurred beneath the Peninsular Ranges is indicated by the Late Cretaceous migration of the magmatic arc into the formerly adjacent regions of Sonora (e.g., Gastil and Krummenacher, 1977; Silver et al., 1996; Staude and Barton, 2001; McDowell et al., 2001; Ortega-Rivera, 2003). Grove et al. (2003b) and Barth et al. (2003) have demonstrated that the Rand Schist and schists of Sierra de Salinas and Portal Ridge were underplated beneath the southern Sierra Nevada batholith and equivalent arc rocks of Salinia (southern Coast Ranges) and the western Mojave region between ca. 90 and 75 Ma. Projecting this relationship to the south, Grove et al. (2003b) speculated that schist of equivalent age was also accreted beneath the Peninsular Ranges batholith. Hence, subduction-induced cooling (i.e., subduction refrigeration) of the overriding crust may have occurred beneath the Peninsular Ranges in the same manner as has been proposed for the Sierra Nevada batholith (Dumitru, 1990; Dumitru et al., 1991).

Subduction refrigeration is not the only expected consequence of Laramide shallow subduction. The most widely accepted model for the underplating of the Pelona and related schists involves tectonic removal of the mantle lithosphere and basal crust (Burchfiel and Davis, 1981; Crowell, 1981; Hamilton, 1988; Bird, 1988). There is evidence that a similar process could have occurred beneath the eastern Peninsular Ranges batholith. During emplacement of the La Posta TTG, the thickness of the crust in the eastern batholith is likely to have been well in excess of 40 km (Gromet and Silver, 1987). Present-day crustal thickness within the eastern Peninsular Ranges batholith is considerably thinner (~30 km) with no Airy root evident (Lewis et al., 2001). Removal of the crustal root during Laramide shallow subduction would have destabilized the overlying crust and triggered denudation to restore the density balance between the crust and the upper mantle.

In this study, we have performed detailed thermochronology based upon ^{40}Ar/^{39}Ar analysis of K-feldspar (Lovera et al., 1989, 1993, 1997, 2002) and other phases to better understand the timing and magnitude of cooling in an area (east-central Peninsular Ranges batholith at 33°N) that is dominated by the immense (>1500 km^2) La Posta pluton sensu stricto (e.g., Miller, 1935) and related TTG intrusives (Fig. 2). Our analysis clearly documents the two phases of rapid cooling described above. We further conclude that Cretaceous, east-side up, semi-ductile to brittle fault zones may locally have played an important role in producing the sharp age gradients observed across the 85 Ma biotite K-Ar isochron. Using our thermal history results, we numerically evaluate

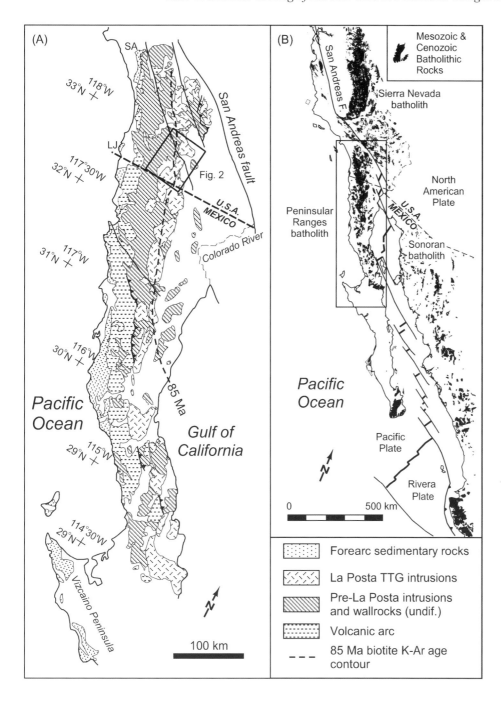

Figure 1. A: Geologic sketch map of Peninsular Ranges batholith. Distribution of La Posta tonalite-trondjemhite-granodiorite (TTG) suite is after Kimbrough et al. (2001). Location of the 85 Ma biotite K-Ar age isochron based upon Krummenacher et al. (1975), Ortega-Rivera (2003), and this study. Location of Figure 2 (east-central batholith at 33°N) is indicated by the open box. SA—Santa Ana Mountains and LJ—La Jolla are localities from the Peninsular Ranges batholith forearc that are discussed in text. B: Geologic setting of Peninsular Ranges batholith. Distribution of Mesozoic and Cenozoic granitic rocks of southwestern North America adapted from Jennings (1977) and Ortega-Gutierrez et al. (1992). Northern Peninsular Ranges batholith (outlined by rectangular area) was adjacent to Sonoran batholith of mainland México prior to late Tertiary opening of Gulf of California and development of San Andreas transform system.

crustal scale models for the Late Cretaceous (erosion denudation and Laramide subduction refrigeration). We conclude that while both processes can account for important aspects of our thermal history results, independent data from the forearc favors erosional denudation as the controlling mechanism.

BACKGROUND

The Peninsular Ranges batholith is characterized by numerous strike parallel, lithologic, and compositional belts and has traditionally been subdivided into western and eastern zones on the basis of pluton composition, size, style of emplacement, age, and isotopic considerations (Fig. 2; Gastil et al., 1975; DePaolo, 1981; Baird and Miesch, 1984; Taylor, 1986; Hill et al., 1986; Jachens et al., 1986; Gromet and Silver, 1987; Silver and Chappell, 1988; Todd et al., 1988; Ague and Brimhall, 1988; Gastil et al., 1990, 1991; Walawender et al., 1990).

Prebatholithic host rocks of the studied area are predominately early Mesozoic(?) "sandstone-shale" slope margin lithologies (Gastil, 1993) that are referred to locally as Julian Schist

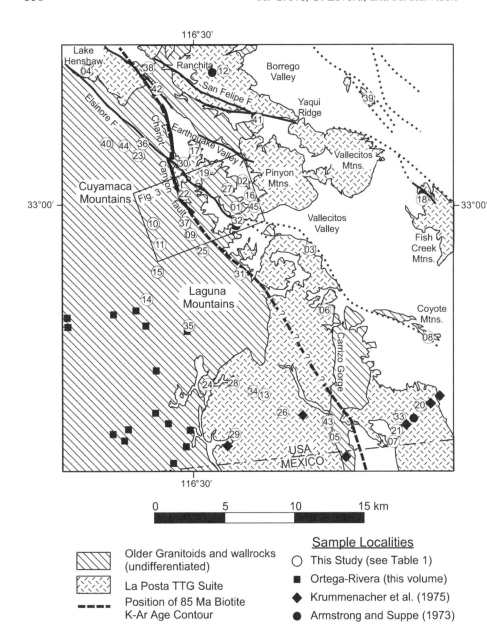

Figure 2. Geologic sketch map of study area in east-central Peninsular Ranges batholith near 33°N. Distribution of La Posta tonalite-trondhjemite-granodiorite (TTG) intrusions from Strand (1962), Todd (1977a, 1978, 1979), Todd and Shaw (1979), Todd et al. (1988), personal communication with Dave Kimbrough, and field observations. Position of 85-Ma biotite isochron is same as in Figure 1. Numbers within circles correspond to sample localities in Table 1. Lake Henshaw–Ranchita, Chariot Canyon–Granite Mountain, Carrizo Gorge, Yaqui Ridge, and Coyote Mountains are localities discussed in text. Location of Figure 3 is outlined by rectangle.

(Fig. 3). Quartz and carbonate-rich wallrocks with miogeoclinal affinities and radiogenic $^{87}Sr/^{86}Sr$ (0.706–0.708) in plutons from the northeastern part of Figure 2 (Yaqui Ridge, Coyote Mountains) suggest that this portion of the study area is probably underlain by cratonal basement (DePaolo, 1981; Silver and Chappel, 1988; Gastil, 1993).

Most intrusions in the western portion of Figure 2 are ca. 125–100 Ma "I-type" quartz monzonite, granodiorite, tonalite, and gabbro bodies with generally primitive isotopic signatures (Everhart, 1951; Merriam, 1958; DePaolo, 1981; Todd and Shaw, 1985; Taylor, 1986; Silver and Chappell, 1988; Todd and Shaw, 2003). Also abundant are Middle Jurassic, compositionally heterogeneous and strongly deformed "S-type" granitoids (referred to as Cuyamaca granodiorite and Harper Creek gneiss

in Fig. 3) that are intimately intermingled with Julian Schist over broad regions (Everhart, 1951; Merriam, 1958; Todd and Shaw, 1985; Todd et al., 1988; Thomson and Girty, 1994; Shaw et al., 2003). While Cretaceous western zone plutons are generally only weakly deformed throughout the western batholith, they are locally strongly attenuated within the Cuyamaca–Laguna Mountains shear zone (Fig. 3; Todd and Shaw, 1979; Todd et al., 1988; Thomson and Girty, 1994; Todd et al., 2003).

Intrusion of the La Posta TTG suite postdated shearing along the Cuyamaca–Laguna Mountains shear zone (Todd et al., 2003). The rocks underlie much of the eastern portion of Figure 2. Generally massive to strongly foliated, hornblende-biotite ± clinopyroxene tonalite of Granite Mountain is intruded by the massive and generally undeformed La Posta pluton (Todd,

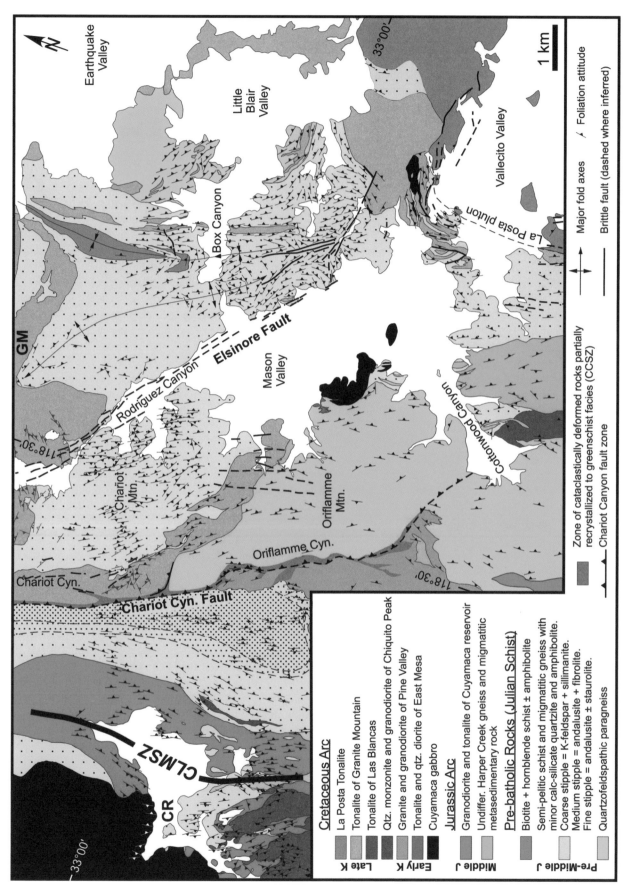

Cretaceous Arc

Late K	La Posta Tonalite
	Tonalite of Granite Mountain
	Tonalite of Las Blancas
Early K	Qtz. monzonite and granodiorite of Chiquito Peak
	Granite and granodiorite of Pine Valley
	Tonalite and qtz. diorite of East Mesa
	Cuyamaca gabbro

Jurassic Arc

Middle J	Granodiorite and tonalite of Cuyamaca reservoir
	Undiffer. Harper Creek gneiss and migmatitic metasedimentary rock

Pre-batholic Rocks (Julian Schist)

Middle J	Biotite + hornblende schist ± amphibolite
	Semi-pelitic schist and migmatitic gneiss with minor calc-silicate quartzite and amphibolite.
Pre-Middle J	Coarse stipple = K-feldspar + sillimanite.
	Medium stipple = andalusite + fibrolite.
	Fine stipple = andalusite ± staurolite.
	Quartzofeldspathic paragneiss

Zone of cataclastically deformed rocks partially recrystallized to greenschist facies (CCSZ)

Chariot Canyon fault zone

⤢ Major fold axes

⤙ Foliation attitude

—— Brittle fault (dashed where inferred)

Figure 3. Geologic map of Cuyamaca Reservoir (CR)–Granite Mountain (GM) region. Map relations from Phillips (1964), Todd (1977a, 1977b, 1978, 1979), Grove (1987), Lampe (1988), Germinario (1994), and Grove (1994). Cuyamaca–Laguna mountain shear zone (CLMSZ) and Chariot Canyon fault zone are discussed in text.

1977a, Todd et al., 1988; see Fig. 3). Ubiquitous, fine- grained biotite granodiorite and swarms of subparallel tabular garnetiferous, two-mica pegmatites are broadly coeval with the La Posta suite rocks (Gastil et al., 1991). Locally, tight- to isoclinal, shallow-plunging, north- to northwest-trending folds deform both the tonalite of Granite Mountain and the granite pegmatites that crosscut it (Fig. 3; Phillips, 1964; Grove, 1987). Foliation surfaces expressed in both intrusive units are transposed parallel to the intrusive contact with the essentially undeformed and areally extensive La Posta pluton.

Basement Cooling Age Patterns and Relationships to Intrusion and Forearc Sedimentation

Post–93 Ma cooling within the east-central Peninsular Ranges batholith near 33°N occurred after volumetrically significant intrusion had ceased within the region (Silver et al., 1988; Todd et al., 1988). Late Cretaceous plutonism (to ca. 65 Ma; Anderson and Silver, 1974; Gastil and Krummenacher, 1977; Silver and Chappell, 1988) took place within formerly adjacent rocks (e.g., Silver et al., 1996) across the Gulf of California in mainland Sonora, México (Fig. 1). However, the thermal effects related to these bodies would have reached only a few kilometers at most (e.g., Hanson and Barton, 1989). West of the Salton Trough, a few zircon U/Pb and mineral Rb/Sr isochron ages determined for pegmatites and muscovite granites are younger than 93 Ma (e.g., Parrish, 1990). Because they are volumetrically minor, these late-stage intrusions are unlikely to have induced significant ^{40}Ar loss from K-Ar thermochronometers through the eastern Peninsular Ranges batholith. Moreover, although larger granitoids yielding U-Pb zircon ages as young as 75 Ma do crop out within the Sierra Cucapá (Silver and Chappell, 1988; Grove et al., 2003a) and other ranges within the Salton Trough (Premo et al., 1998), biotite and K-feldspar ^{40}Ar/^{39}Ar ages obtained from them are generally 10–20 Ma younger than the time of intrusion (Axen et al., 2000; M. Grove, unpubl. data). Hence, even in these areas, there is no compelling case for ^{40}Ar loss due to transient heating effects associated with pluton emplacement. Consequently, we consider that biotite and K-feldspar Ar isotopic ages from the eastern Peninsular Ranges batholith at 33°N represent regional cooling patterns in a deep crustal setting (see also Krummenacher et al., 1975).

Krummenacher et al.'s (1975; see also Ortega-Rivera, 2003) 85 Ma biotite K-Ar isochron provides a convenient way of delineating rocks with contrasting cooling histories within the east-central Peninsular Ranges batholith near 33°N. Krummenacher et al. (1975) demonstrated that the positions of the 90 Ma and older biotite K-Ar isochrons were strongly influenced by the boundaries of plutons of the La Posta TTG suite while younger contours (including 85 Ma) are superposed across these intrusions (Fig. 1). Both Silver et al. (1979) and Grove (1994) have described a sharp eastward decrease in K-Ar cooling ages that occurs across the 85-Ma biotite K-Ar isochron. We will use the position of the 85 Ma contour as a geographic reference line throughout this paper.

West of the 85-Ma biotite K-Ar isochron, biotite records early Late Cretaceous (Turonain–Cenomanian) cooling in comparatively shallow rocks (2–3 kbar or 7–11 km; see Berggreen and Wallawender, 1977; Detterman, 1984; Todd et al., 1988; Germinario, 1993; Grove, 1994; Rothstein and Manning, this volume, Chapter 12). Development of poorly dated (Cenomanian–Turonian?), west-flowing, steep gradient, short-length streams and numerous alluvial fan/submarine fan masses (Flynn, 1970; Nordstrom, 1970; Kennedy and Moore, 1971; Peterson and Nordstrom, 1970; Sundberg and Cooper, 1978; Bottjer and Link, 1984) appear to record contemporaneous erosional denudation. Results of Lovera et al. (1999) from the Peninsular Ranges batholith forearc confirm that significant erosional denudation overlapped with final intrusion of the La Posta TTG (see also Kimbrough et al., 2001). Walawender et al. (1991), George and Dokka (1994), and others have proposed that tectonic denudation via normal faulting accompanied emplacement of the La Posta TTG. However, the extent to which the upper crust was extended in response to emplacement of the La Posta TTG may never be known since most of the affected rocks have been eroded away.

Biotite from positions east of the 85 Ma biotite K-Ar isochron record a second phase of rapid cooling with a distinctive early Laramide timing (i.e., beginning after ca. 80 Ma; Krummenacher et al., 1975). The rocks that record Late Campanian–Maastrictian Ar closure in biotite are structurally deeper than their western counterparts (Theodore, 1970; Gastil, 1979; Anderson, 1983; Engel and Schultejann, 1984; Grove, 1987; Ague and Brimhall, 1988; Todd et al., 1988; Rothstein and Manning, 2003). Major sedimentation is recorded along the western margin of the batholith at this time (Kennedy and Moore, 1971; Sundberg and Cooper, 1978; Nilsen and Abbott, 1981; Bottjer et al., 1982; Bottjer and Link, 1984; Fry et al., 1985; Girty, 1987; Bannon et al., 1989).

There is limited evidence for a final pulse of rapid cooling at the very end of the Cretaceous that appears to be restricted to the structurally deepest domains in the eastern batholith. These deep rocks, which cooled rapidly between 60 and 70 Ma (Krummenacher et al., 1975; Goodwin and Renne, 1991; Axen et al., 2000; Wenk et al., 2000), owe their exposure to either late cataclastic deformation along the eastern Peninsular Ranges mylonite zone and/or late Cenozoic extensional faulting within the Salton Trough and northern Gulf (Dokka and Merriam, 1982; Schultejann, 1984; Siem and Gastil, 1994; Axen and Fletcher, 1998). In any case, a mature erosion surface traversed by extra-regional depositional systems was developed throughout the northern Peninsular Ranges batholith by early Tertiary time (Kennedy and Moore, 1971; Peterson and Nordstrom, 1970; Abbott and Smith, 1978; Minch, 1979; Kies and Abbott, 1983; Abbott and Smith, 1989). Apatite fission track and (U-Th)–He data indicate that rocks of the eastern batholith still resided at several kilometers depth during the early Tertiary (Cerveny et al., 1991; Naeser et al., 1996; Wolf et al., 1997; Premo et al., 1998).

Structural Breaks Coincident with the 85 Ma Biotite K-Ar Isochron

Krummenacher et al. (1975) have argued for a significant (~5 km) structural break between eastern and western lobes of the La Posta pluton on the basis of a sharp decrease of K-Ar biotite ages in the vicinity of Carrizo Gorge (Fig. 2). While this proposed structure would coincide approximately with the 85 Ma biotite K-Ar age contour, mapping in the Sweeny Pass 7.5′ quadrangle failed to reveal an important fault (Hoggatt, 1979). Less ambiguous evidence for a significant structural break near the surface expression along the 85 Ma isochron occurs between Cuyamaca Reservoir and Granite Mountain (Fig. 3).

The Chariot Canyon fault zone (Fig. 2; Kofron, 1984; Germinario, 1993; Grove, 1994) is a Late Cretaceous zone of top-to-the-west, ductile-to-brittle shear that coincides with a region of significant Late Cretaceous gold mineralization (Kofron, 1984). West of the Chariot Canyon fault zone, metamorphic grade recorded in pelitic intervals within Julian schist wallrocks increases from upper staurolite zone to sillimanite+K-feldspar zone toward the intrusive contact with the granodiorite of Cuyamaca Reservoir (Fig. 3; Grove, 1994). This metamorphic zonation was clearly developed in response to intrusion of the Cuyamaca Reservoir granodiorite during the middle Jurassic (Thomson and Girty, 1994; Murray and Girty, 1996). Shaw et al. (2003) report U-Pb zircon crystallization ages of 162–168 Ma for two samples of the Cuyamaca Reservoir granodiorite within the area of Figure 3. Garnet + biotite + andalusite + plagioclase + quartz thermobarometry performed with phyllitic rocks west of the Chariot Canyon fault zone indicate ~2.5–3.0 kbar conditions (Grove, 1994; see also Germinario, 1993). Steeply plunging isoclinal folds and lineations related to the Cuyamaca–Laguna Mountains shear zone are prevalent west of the Chariot Canyon fault zone.

The comparatively low-pressure western wallrocks are abruptly truncated against the Chariot Canyon fault zone (Fig. 3, Germinario, 1993; Grove, 1994). East of the Chariot Canyon fault zone, metamorphic grade is uniformly upper amphibolite facies (sillimanite-biotite-orthoclase gneiss) with migmatitic fabrics developed in appropriate bulk compositions regardless of proximity to intrusive contacts (Grove, 1987; see also Lampe, 1988). Petrogenetic relationships and garnet-biotite-Al_2SiO_5-plagioclase-quartz thermobarometry indicate 4.0–5.5 kbar conditions (Grove, 1987). Th-Pb dating of garnet-hosted monazite from one of the thermobarometry samples indicates that the upper amphibolite fabrics of the Granite Mountain area formed contemporaneously with intrusion of the La Posta pluton (M. Grove, unpublished ion microprobe data). While the ~1.5 kbar or roughly 5 km depth increase across the Chariot Canyon fault zone cannot be rigorously interpreted in terms of fault offset since peak grade assemblages on either side differ in age by ~70 m.y., the significant contrast in structural level across the zone (Fig. 3) can only be explained by large fault displacements.

Dominantly cataclastic deformation within the Chariot Canyon fault zone is superposed upon all synbatholithic structures, including well-developed secondary schistosity related to the Cuyamaca–Laguna Mountains shear zone that are developed in the granodiorite of Pine Valley in Oriflamme Canyon (Fig. 3; Thomson and Girty, 1994). The late deformation is characterized by lower greenschist facies recrystallization and ductile shearing along discrete, discontinuous zones of intense, brittle cataclasis (Fig. 3). Prehnite-actinolite facies assemblages (prehnite-epidote-chlorite-actinolite-albite-quartz in sheared tonalite) stable at 250–325 °C (Liou et al., 1987) are associated with the ductile fabrics: prehnite ± chlorite veins formed during the brittle cataclasis. Shallowly inclined, predominantly northwest-striking, northeast-dipping shear planes of cataclastic deformation (dark gray flinty gouge and ultracataclasite) are abundant both within and east of the Chariot Canyon fault zone. Indications of shear sense along these planes are conflicting but generally imply that the Chariot Canyon fault is a west-directed thrust or high-angle reverse fault (Grove, 1994). Although similar shear planes are present throughout the northeast region of Figure 2, they often exhibit normal geometry and appear to be related to Miocene extension (e.g., Schultejann, 1984). The Chariot Canyon fault zone was reactivated as an east-side down normal fault during the Late Cenozoic (Lampe, 1988).

Cenozoic Deformation

Physiographically, the southwestern portion of the batholith shown in Figure 2 is an essentially intact structural block (Todd and Shaw, 1979). In contrast, the region northeast of the Elsinore fault has been noticeably affected by Late Cenozoic deformation. Vestigial, early Tertiary(?) erosion surfaces (e.g., Minch, 1979) preserved within topographically lower desert ranges northeast of the Elsinore fault generally occur at lower elevations than their southwestern counterparts. Because this implies net down-dropping of the desert ranges toward the Salton Trough, the regional northeastward increase in structural depth (e.g., Todd et al., 1988; Ague and Brimhall, 1988) cannot be an artifact of Cenozoic deformation. Middle Tertiary normal faults (Schultejann, 1984; Siem and Gastil, 1994; Axen and Fletcher, 1998; Lough and Stinson, 1991) have been described in a number of ranges northeast of the Elsinore fault. Many of these structures were formerly considered part of the Late Cretaceous eastern Peninsular Ranges mylonite zone (Sharp, 1979). With the exception of the normal fault system within the Sierra Cucupah and Sierra El Major (Siem and Gastil, 1994; Axen et al., 2000) ranges, Late Cenozoic normal faulting within the east-central Peninsular Ranges batholith near 33°N has produced a contrast in structural level that is barely resolved by apatite (U-Th)-He thermochronometry (Kairouz et al., 2003).

Late Cenozoic strike-slip faulting also does not appear to have greatly complicated the Late Cretaceous distribution of cooling ages within the area of Figure 2. While late Cenozoic strike-slip faulting has collectively displaced the main structural block of the batholith 25–40 km northwest relative to the easternmost desert ranges, most of this displacement has occurred along the San Jacinto fault zone to the northeast of the area of Figure 2 (Sharp, 1967; Dorsey, 2002). While the Elsinore fault has accommodated up to 30 km of dextral offset at the northern end of the Peninsular Ranges batholith, much less displacement occurs within the area of Figure 2. Some of the displacement has been transferred further east to the Earthquake Valley fault and San Felipe faults (Magistrale and Rockwell, 1996). Along the Elsinore fault, displaced intrusive contacts limit right-lateral slip to less than ~2 km in the vicinity of Granite Mountain and Vallecito Valley (Fig. 3; Lampe, 1988; Todd et al., 1977a).

METHODS

Granitic rocks were sampled at locations indicated in Figure 2. Sample selection was guided by the results of previous studies (e.g., Krummenacher et al., 1975). For reasons outlined in the introduction, a disproportionately large proportion of the samples were selected from positions that were anticipated to lie in close proximity to the estimated location of the 85 Ma biotite K-Ar isochron. Thermal history information was obtained from $^{40}Ar/^{39}Ar$ step-heating experiments performed using K-feldspar. Complementary $^{40}Ar/^{39}Ar$ results were also generated from coexisting hornblende, muscovite, and biotite. Argon isotopic analysis was performed at the University of California at Los Angeles using techniques and instrumentation discussed by Grove and Harrison (1996) and Quidelleur et al. (1997). Further details are provided in Appendix 1. Loss of radiogenic argon from hornblende and micas (i.e., total fusion ages) has been interpreted only in terms of bulk closure (e.g., Dodson, 1973) as constrained by hydrothermal argon diffusion experiments and/or well-constrained field settings (~525 °C for hornblende, ~400 °C for muscovite, and ~350 °C for biotite; see McDougall and Harrison, 1999). K-feldspar $^{40}Ar/^{39}Ar$ step-heating experiments, on the other hand, have been interpreted using the multi-diffusion domain approach (Lovera et al., 1989).

Our multi-diffusion domain (MDD) approach for recovering crustal thermal histories from $^{40}Ar/^{39}Ar$ step-heating experiments performed with K-feldspar is outlined in Lovera et al. (1997, 2002). The multi-diffusion domain character of K-feldspar appears to be related in poorly understood ways to the typically rich array of microstructures that typify basement feldspar (Lovera et al., 1993). Regardless of the nature of the intracrystalline controls, K-feldspar has been empirically demonstrated to be capable of recording continuous thermal history information from ~350 °C to ~150 °C (Lovera et al., 1997, 2002, and references cited therein; Parsons et al., 1999, offer a dissenting opinion). The extent to which samples we have examined are suitable for thermal history analysis is considered in greater detail in Appendix 2.

The age spectrum and Arrhenius data from a typical Peninsular Ranges batholith K-feldspar (AC) are shown in Figure 4A and Figure 4B respectively. One of the most significant sources of uncertainty involved in the estimation of thermal histories using the MDD approach is determination of activation energy (E). While E is generally estimated from the slope defined by the

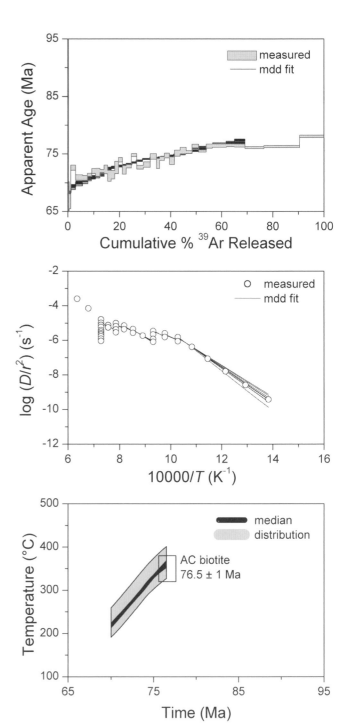

Figure 4. Representative K-feldspar thermal history results from AC K-feldspar. A: Age spectrum. Solid lines are 50 best-fit solutions from multi-diffusion domain thermal history analysis. B: Arrhenius plot. Solid lines represent 10 best-fit domain distributions obtained by allowing activation energy to vary by ± 3 kcal/mol about a mean imposed value of 46 kcal/mol (C) calculated thermal history. Envelopes indicate 90% confidence bounds for mean and overall distribution. See Table 2 for calculation limits and Appendix 2 for additional details of thermal history calculations.

initial low-temperature (~450–650 °C) data in the Arrhenius plot, the limited range over which the regression is performed may give rise to serious error in estimating this parameter. We have found the Arrhenius properties of Peninsular Ranges batholith K-feldspars to be remarkably similar. This has motivated us to employ a single fixed activation energy (46 kcal/mol) indicated by systematic analysis of several hundred K-feldspars (Lovera et al., 1997, 2002). To fit the Arrhenius data, we varied E within ± 3 kcal/mol limits to calculate 10 equivalent domain distributions (Fig. 4B). For each of these, five best-fit monotonically decreasing cooling histories were calculated to yield a total of 50 solutions. From these we determined the 90% confidence limits of the thermal history (Fig. 4C) that correspond to the interpreted portion of the age spectrum (i.e., the fraction of the ^{39}Ar released below melting; see Fig. 4A).

RESULTS

Our ^{40}Ar/^{39}Ar results have been summarized in Table 1. Complete data tables are available in GSA's Data Repository[1]. Biotite and K-feldspar total gas ages have been plotted as a function of distance normal to the 85 Ma biotite age contour in Figure 5. Also included are previously published results from Armstrong and Suppe (1973) and Krummenacher et al. (1975) and new results from Ortega-Rivera (2003). These data define four domains (Fig. 5). Domain A in the southwest region of Figure 2 is characterized by variable biotite K-Ar ages between 95 and 110 Ma. An intermediate domain (= B) is defined by an ~15-km-wide expanse of comparatively constant 89–93 Ma ages. A second intermediate domain (= C) is defined by abruptly decreasing (1.3 m.y./km) biotite and K-feldspar total gas ages within ±5 km of the 85 Ma biotite isochron. The age gradient appears to be steepest (>2.5 m.y./km) across the Chariot Canyon fault zone (Chariot Canyon–Granite Mountain area; Fig. 3). Further north (Lake Henshaw-Ranchita area), the gradient is subtle (ca. 0.5 m.y./km). To the south, near the international border with México (Carrizo Gorge area; Fig. 2), the age gradient is about 1 m.y./km. The last northeastern domain (= D) defined by homogenous biotite K-Ar ages between ca. 80 and 75 Ma extends for at least 45 km to the east of the 85 Ma biotite isochron. Note that the youngest result (CM1) comes from the lower plate of a detachment fault system within the Coyote Mountains (Miller and Kato, 1991). In carrying out the more detailed K-feldspar thermal history analysis described below, we focus upon domains B, C, and D.

K-Feldspar Thermal History Analysis

Details of the multi-diffusion domain modeling process are summarized in Table 2. Specific comments related to our ability to interpret the K-feldspar results are provided in Appendix 2.

Interested readers are also directed to individual sample plots and model results that are available in GSA Data Repository (see footnote 1). Results of our thermal history calculations are summarized in Figure 6. We have also plotted (when available) thermal history estimates from biotite and muscovite total gas ages that assume 350 ± 30 °C and 400 ± 30 °C conditions for bulk closure respectively. To the extent possible, we have attempted to display the results in Figure 6 according to geographic position (see Fig. 2). Results obtained along the top row are from the northernmost samples arranged west to east and so forth. As stated in the "Methods" section, we represent the 50 best-fit thermal histories that were obtained for each sample by 90% confidence intervals for the median (black) and overall (gray) distribution. Note that we have permitted only monotonically decreasing cooling histories in our calculations.

From inspection of Figure 6 it is clear that all but the northernmost of samples (BM) from domain B had cooled to below ~200 °C by ca. 86 Ma. Apatite fission track results from domain B generally record cooling through ~125 °C by ca. 80 Ma (Naeser et al., 1996). Hence, slower cooling seems to be required for these samples subsequent to 86 Ma. Slower cooling subsequent to 86 Ma is recorded by most samples from domain C. Again, the northernmost samples from domain C record somewhat faster cooling between 86 and 78 Ma. Finally, all samples from domain D record rapid cooling from >350 °C conditions subsequent to about 78 Ma. Samples from the most easterly positions record the most rapid cooling. Apatite fission track results from this domain tend to yield 40–60 Ma ages (Naeser et al., 1996). This requires post–70 Ma deceleration of cooling rates throughout domain D.

Median cooling histories are plotted together in Figure 7A (note that we have omitted dissimilar results from the three northwest K-feldspar samples BM, SFH, and TS). These results are differentiated in Figure 7B to yield the Late Cretaceous cooling rates indicated by our calculations. The single bold curve in Figure 7B represents the mean cooling rate that is valid over most of the region outlined in Figure 2. Strictly speaking, the 95–86 Ma portion of this curve applies mainly to the western domain B rocks while the 78–65 Ma segment pertains primarily to the eastern domain D rocks. However, both independent apatite fission track data (Naeser et al., 1996) and results from domain C demonstrate that the later (78–68 Ma) phase of rapid cooling also must apply to domain B. Similarly, petrologic considerations and limited higher temperature thermochronologic constraints from hornblende and muscovite suggest that cooling rates in domain D must have been much slower prior to 78 Ma (i.e., these rocks existed at ~350–450 °C or about 11–15 km prior to 78 Ma, assuming a 30 °C/km geotherm).

In summary, it is clear that the east-central Peninsular Ranges batholith at 33°N experienced two phases of regionally extensive, Late Cretaceous rapid cooling. The first persisted from ca. 91 to 86 Ma and peaked at 88 Ma with a mean rate of ~25 °C/Ma. It overlapped with, and closely followed, emplacement of the La Posta TTG suite. We infer that it was primarily an expression of denudation that was triggered by massive

[1]GSA Data Repository Item 2003176, tabulated ^{40}Ar/^{39}Ar analytical data, is available on request from Documents Secretary, GSA, P.O. Box 9140, Boulder, CO 80301-9140, USA, editing@geosociety.org, at www.geosociety.org/pubs/ft2003.htm, or on the CD-ROM accompanying this volume.

TABLE 1. ^{40}Ar/^{39}Ar RESULTS

	Sample	Total gas age*				Sample locations†		Distance§ (km)	Description
		Hbd (Ma)	Mus (Ma)	Bio. (Ma)	Ksp. (Ma)	Latitude	Longitude		
1	409-B	–	–	78.6	–	32° 59.566'	116° 25.630'	+8.56	Gar-bio tonalite
2	1110-I	–	82.6	–	76.1	33° 02.396'	116° 24.426'	+11.6	Granite pegmatite
3	AC	–	–	76.6	74.9	32° 55.896'	116° 16.450'	+10.8	La Posta tonalite
4	BM	–	–	86.3	87.4	33° 12.901'	116° 42.477'	−5.74	Ranchita tonalite
5	BS	–	–	87.1	–	32° 38.391'	116° 13.786'	−3.6	La Posta tonalite
6	BW-2	–	–	–	75.1	32° 50.222'	116° 14.765'	+7.6	La Posta tonalite
7	CAR	–	–	81.0	–	32° 37.948'	116° 07.038'	+6.6	La Posta tonalite
8	CM-1	–	–	–	72.1	32° 47.266'	116° 01.231'	+21.4	Granodiorite
9	CP-175	100.1	–	84.8	80.8	32° 57.408'	116° 30.390'	−1.2	Gd. of Pine Valley
10	CP-128	101.1	–	91.9	92.7	32° 58.320'	116° 34.554'	−5.9	Gd. of Chiquita Peak
11	CS	–	–	91.9	92.9	32° 56.401'	116° 33.880'	−6.8	Gd. of Chiquita Peak
12	CV		–	79.6	75.2	33° 12.689'	116° 26.592'	+13.1	Ranchita tonalite
13	DIA	–	–	90.7	–	32° 42.290'	116° 21.983'	−9.78	La Posta tonalite
14	DSC	–	–	–	91.1	32° 51.122'	116° 35.622'	−15.0	Granodiorite
15	EM	–	–	–	91.9	32° 54.268'	116° 34.145'	−9.7	Granodiorite
16	EQ	104.6	–	78.3	–	33° 00.827'	116° 23.467'	+11.4	Granite Mountain tonalite
17	EV	101.9	–	84.0	–	33° 05.328'	116° 29.903'	+4.1	Granite Mountain tonalite
18	FCM	–	–	78.4	76.4	33° 01.152'	116° 05.858'	+29.6	La Posta TTG tonalite
19	GM	101.3	–	82.7	81.2	33° 03.061'	116° 28.784'	+6.1	Granite Mountain tonalite
20	IKPG	–	–	77.6	–	32° 41.314'	116° 03.840'	+13.0	La Posta tonalite
21	IRR	–	–	78.1	–	32° 38.948'	116° 06.616'	+7.3	La Posta tonalite
22	JU	103.9	–	84.3	78.0	33° 01.105'	116° 31.694'	+0.43	Granite Mountain tonalite
23	JUCH	–	–	89.4	91.5	33° 04.227'	116° 32.633'	−4.7	Granodiorite
24	KCR	–	–	91.8	–	32° 43.272'	116° 28.315	−17.5	La Posta tonalite
25	KP	–	–	–	91.6	32° 56.004'	116° 28.939'	−1.2	Harper Creek gneiss
26	LOS	–	–	89.5	–	32° 40.617'	116° 19.852'	−9.3	La Posta tonalite
27	LBV-2B	–	84.6	–	75.4	33° 01.521'	116° 25.964'	+8.7	Granite pegmatite
28	LP-80	–	–	90.3	–	32° 43.462'	116° 25.544'	−13.6	La Posta tonalite
29	LPRR	–	–	92.5	–	32° 38.686'	116° 25.481'	−18.5	La Posta tonalite
30	MMVT	–	–	–	79.1	33° 04.292'	116° 32.455'	+0.4	Granite Mtn. tonalite
31	MP-17	108.8	–	90.2	85.7	32° 52.428'	116° 24.765'	0	Las Blancas tonalite
32	MP	–	–	80.7	76.5	32° 58.574'	116° 24.947'	+6.3	La Posta tonalite
33	MSR	–	82.5	79.1	–	32° 40.225'	116° 06.352'	+3.6	La Posta tonalite
34	MVY	–	–	91.2	–	32° 42.586'	116° 23.063'	−10.8	La Posta tonalite
35	PV	–	–	89.9	91.1	32° 48.798'	116° 30.644'	−14.3	Gd. of Pine Valley
36	RR	–	–	88.9	–	33° 05.948'	116° 35.739'	−4.6	Cuyamaca Res. gd.
37	RG	–	–	–	86.4	32° 58.322'	116° 30.956'	−1.1	Rattlesnake Gr.
38	SFH	94.1	–	89.4	85.7	33° 13.082'	116° 34.673'	+2.56	Ranchita tonalite
39	SP	–	–	76.7	75.3	33° 10.675'	116° 06.752'	+41.3	La Posta TTG tonalite
40	SY	98.2	–	90.8	91.9	33° 05.895'	116° 39.850'	−10.8	granodiorite
41	TG	–	–	79.4	78.9	33° 08.138'	116° 22.681'	+15.9	La Posta TTG tonalite
42	TS			87.3	84.3	33° 12.554'	116° 36.251'	+1.5	Ranchita tonalite
43	WC	–	–	83.7	–	32° 39.785'	116° 14.413'	−3.2	La Posta tonalite
44	WYN	–	–	90.3	–	33° 05.670'	116° 37.851'	−7.8	granodiorite
45	YS	–	–	78.9	74.8	33° 00.118'	116° 23.141'	+10.9	La Posta tonalite

*Integrated ^{40}Ar/^{39}Ar ages. Internal precision is generally much smaller (<0.3%) than accuracy (±1.5%) based upon literature uncertainty in age of irradiation standard Fish Canyon sanidine (FCT-1). See text, Appendix 1, and the GSA Data Repository (see footnote 1) for additional details.

†Estimated from map positions.

§Distance normal 85 Ma biotite K-Ar age contour. The position of this curve was established from data in this study, Armstrong and Suppe (1972); Krummenacher et al. (1975); and Ortega-Rivera (this study).

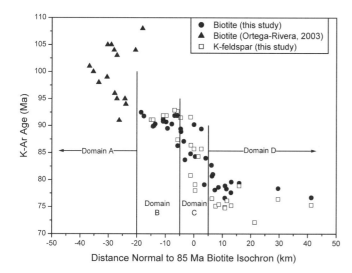

Figure 5. Biotite and K-feldspar total gas (= K-Ar) ages as function of distance (measured perpendicular to 85 Ma biotite K-Ar age isochron). Includes data from Armstrong and Suppe (1973), Krummenacher et al. (1975), and Ortega-Rivera (2003). Domains A, B, C, and D are discussed in text.

TABLE 2. SUMMARY OF K-FELDSPAR MULTIDIFFUSION DOMAIN THERMAL HISTORY ANALYSIS

Sample	MDD*	Interval of ^{39}Ar release modeled		Comments
		Low-T limit[†]	High-T limit[§]	
1110-I	N	-	-	No MDD analysis
AC	Y	3.0	69.2	Inadequate temperature control
BM	Y	0.6	31.3	Interval of ^{39}Ar release not extended with multiple 1100°C steps
BW-2	Y	1.3	76.1	—
CM-1	Y	0.6	64.6	Low-T misfit of age spectrum
CP-175	Y	0.2	85.0	—
CP-128	Y	4.4	95.0	—
CS	Y	0.4	34.4	Minor IAM resolved at 34.4% ^{39}Ar
CV	Y	0.4	77.3	—
DSC	Y	0.4	93.8	—
EM	N	-	-	Inadequate temperature control
FCM	Y	0.3	76.8	—
GM	Y	7.3	32.6	Minor IAM resolved at 32.6% ^{39}Ar
JU	Y	1.4	90.6	High uncertainties in age spectrum
JUCH	N	-	-	Inadequate temperature control
KP	Y	0.6	42.2	Appreciable IAM resolved at 42.2% ^{39}Ar
LBV-2B	Y	0.3	84.0	—
MMVT	Y	1.0	45.2	Minor IAM resolved above 45.2% ^{39}Ar
MP-17	Y	3.4	74.4	-
MP	Y	0.8	81.2	High uncertainties in age spectrum
PV	N	-	-	Inadequate temperature control
RG	Y	0.5	42.0	Inordinately high low-Temp. ^{40}Ar$_E$
SFH	Y	0.1	79.5	—
SP	Y	1.5	76.3	—
SY	Y	3.3	93.2	—
TG	Y	4.1	32.1	Minor IAM resolved above 32.1% ^{39}Ar
TS	Y	0.1	44.6	Interval of ^{39}Ar release not extended with multiple 1100°C steps
YS	Y	0.0	36.7	Interval of ^{39}Ar release not extended with multiple 1100°C steps

Note: IAM refers to intermediate age maxima (see Appendix 2).
*MDD—multidiffusion domain thermal history analysis. Some samples were not analyzed due to poor temperature control.
[†]Lower limit of age spectrum modeled was determined by when evidence for fluid-inclusion hosted excess radiogenic ^{40}Ar was no longer manifested.
[§]Upper limit established at final 1100 °C step (melting behavior is evident at high temperatures). Some samples exhibited problematic behavior at lower temperatures that further limited the interval of ^{39}Ar release that we interpreted (see comment in adjoining column).

intrusion of the La Posta plutonic suite. In addition, the huge heat flux from the La Posta plutons must have significantly outgassed biotite and K-feldspar thermochronometers in their wallrocks. The second episode primarily affected the La Posta rocks ~20 m.y. after they were emplaced. While Krummenacher et al. (1975) originally recognized this relationship on the basis of mica bulk closure ages, the K-feldspar thermal history results presented here demonstrate it quantitatively with stark clarity. The event persisted from ca. 78 to 68 Ma and peaked at a mean rate of ~30 °C/Ma at ca. 73 Ma. More easterly rocks recorded far higher rates of cooling (to 80 °C/m.y.) over a shorter interval (76–72 Ma) centered around 74 Ma.

DISCUSSION

Although the regional eastward decrease in K-Ar ages across the Peninsular Ranges batholith has been known for more than three decades (Everden and Kistler, 1970; Armstrong and Suppe, 1973; Krummenacher et al., 1975; Silver et al., 1979; Ortega-Rivera, 2003), the nature of the control(s) responsible for producing the pattern have remained uncertain. There is little doubt that when viewed at the scale of the entire southwestern margin of North America (e.g., fig. 17 in Ortega-Rivera, 2003), the regular eastward decline in age of K-Ar thermochronometers reflects multiple interacting processes such as magmatism, denudation, and subduction geometry that operate over broad regions (see also Ortega-Rivera, 2003). Hence, it is inescapable that conclusions drawn from small areas (e.g., Fig. 2) will fail to adequately explain complex behavior over the entire orogen. Nevertheless, it is only through such detailed studies that the true nature of the phenomenon can be incrementally understood.

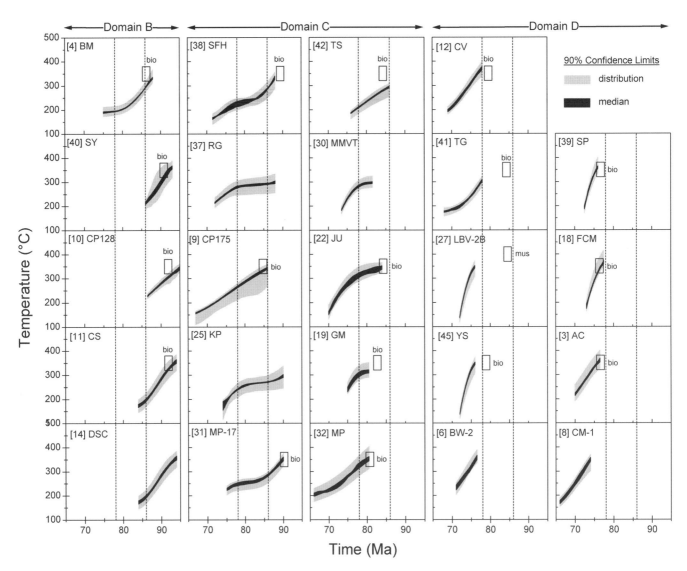

Figure 6: K-feldspar multi-diffusion domain (MDD) thermal history results from 24 K-feldspars from east-central Peninsular Ranges batholith. Confidence limits and age interval modeled for thermal history is same as in Figure 4. Domains B, C, and D are same as in Figure 5. Biotite (~350 ± 30 °C) and muscovite (400 ± 30 °C) bulk K-Ar closure after McDougall and Harrison (1999). See Table 2 and Appendix 2 for additional details.

Three principal schools of thought have arisen regarding the significance of the regional pattern of cooling ages within the northeastern Peninsular Ranges batholith. The first, articulated by Ortega-Rivera (2003), interprets the eastward decrease of cooling ages to fundamentally reflect a progressive eastward shift in magmatism. While such a view may be generally true for the western Cordillera (e.g., Coney and Reynolds, 1977), it provides an inadequate explanation for the east-central Peninsular Ranges batholith near 33°N where K-Ar ages postdate intrusion by as much as 10–20 m.y. The second view, favored by Krummenacher et al. (1975), among others, is motivated by the strong correlation between K-Ar age and erosion depth within the batholith and the coincidence of Late Campanian–Maastrictian rapid cooling with significant forearc

sedimentation of this age. Proponents of this view interpret the timing of Late Cretaceous cooling as being dictated by when significant erosion denudation occurred. While this linkage is tantalizing, the late timing of the Late Campanian–Maastrictian cooling (10–20 m.y. after emplacement of the La Posta TTG suite) seems inconsistent with the expectation that arc crust should strengthen and increasingly resist deformation after intrusion has ceased.

A third school of thought, voiced by Dumitru et al. (1991), attributes the delayed cooling to the onset of Laramide shallow subduction (see Grove et al., 2003b). Late Cretaceous migration of the magmatic arc into the formerly adjacent regions of Sonora (Gastil and Krummenacher, 1977; Silver and Chappell, 1988; Staude and Barton, 2001; McDowell et al., 2001; Ortega-

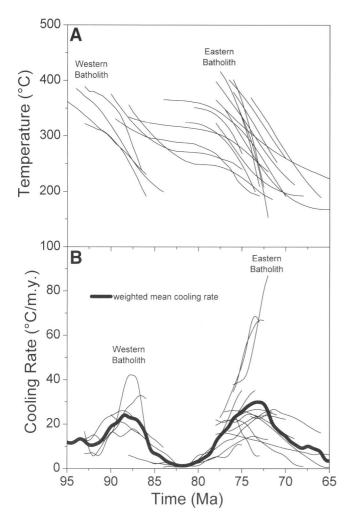

Figure 7. A: Summary plot of median K-feldspar cooling histories from Figure 6. Note that three results (BM, SFH, and TS from Lake Henshaw–Ranchita area) are somewhat different and have been omitted for clarity. B: Calculated cooling rates as function of time obtained by differentiating results in part A, above. Mean cooling history is indicated by bold curve.

that are constrained by thermal history results from the east-central Peninsular Ranges batholith at 33°N.

Numerical Analysis of Cooling Age Patterns

Cooling in Response to Erosional Denudation

The simplest denudation process capable of reproducing the gross characteristics of the observed K-Ar cooling age patterns within the Peninsular Ranges batholith is hinged uplift or regional tilting (Ague and Brimhall, 1988; Butler et al., 1991; Ague and Brandon, 1992; Dickinson and Butler, 1998). Lovera et al. (1999) modeled the effects of overlapping intrusion (120–90 Ma) and hinged denudation (100–65 Ma) to interpret results from detrital thermochronometers from the Peninsular Ranges batholith forearc (see Appendix 3). Thermal histories of "eroded" material were determined by monitoring temperature variation throughout the grid. This information was used to calculate closure age distributions of detrital materials that could be compared with those measured from strata of known depositional age. Forward modeling of the detrital mineral age distributions permitted estimation of the batholiths erosional denudation history (see "Relationship to Forearc Sedimentation" section below).

While reasonably successful in reproducing detrital age distributions from the Peninsular Ranges batholith forearc, Lovera et al.'s (1999) model relied upon an overly simplistic representation of the distribution of erosion depth. The hinged denudation geometry they employed implied extremely deep erosion levels (>30 km) for formerly adjacent rocks in Sonora, México. In fact, available data indicate that erosion depth is more likely to have been symmetrically distributed about the east-central Peninsular Ranges batholith (Gastil, 1979). While Late Cenozoic formation of the Salton Trough and Gulf of California has obliterated key relationships, the Sonoran batholith on mainland México is known to expose shallow crustal levels and greenschist facies metamorphism (~1–3 km; Gastil and Krummenacher, 1977; Staude and Barton, 2001). Moreover, greenschist facies wallrocks that crop out in the Salton Trough (Coyote Mountains; Fig. 1) and Gulf of California (Sierra Pintas; Fig. 1) indicate that shallow erosion levels may also have been prevalent along the eastern margin of the Peninsular Ranges batholith (Gastil, 1979). Clearly, the deep (> 30 km) erosion depths implied for these regions by hinged denudation or regional tilting are inaccurate.

In this paper, we have modified the basement denudation model of Lovera et al. (1999) to permit a more symmetric distribution of erosion depth. We have constrained the denudation history with the new thermal history constraints presented in this paper and have extended the grid so that the shallow crustal levels and younger intrusion (90–65 Ma; see Anderson and Silver, 1974; Gastil and Krummenacher, 1977; Staude and Barton, 2001) within the Sonoran batholith on mainland México are represented. Interested readers are referred to Appendix 3 and Lovera et al. (1999) for further details.

Figure 8A illustrates the relevant portion of our calculation grid and shows the final distribution of K-feldspar bulk closure

Rivera, 2003) seemingly require that Laramide shallow subduction occurred beneath the Peninsular Ranges. Several previous studies have suggested that accelerated Late Cretaceous denudation was triggered by the onset of shallow Laramide subduction beneath the Peninsular Ranges batholith (George and Dokka, 1994; Grove, 1994). Dumitru (1990) and Dumitru et al. (1991) have interpreted Laramide timing of the cooling within the Sierra Nevada batholith in terms of heat flow from the overlying crust and mantle lithosphere into a shallowly subducting slab (subduction refrigeration). In their interpretation, K-Ar cooling ages directly reflect the timing of Laramide subduction and have little, if any, implication for when denudation occurred. In the following section, we evaluate both erosional denudation and subduction refrigeration with numerical models

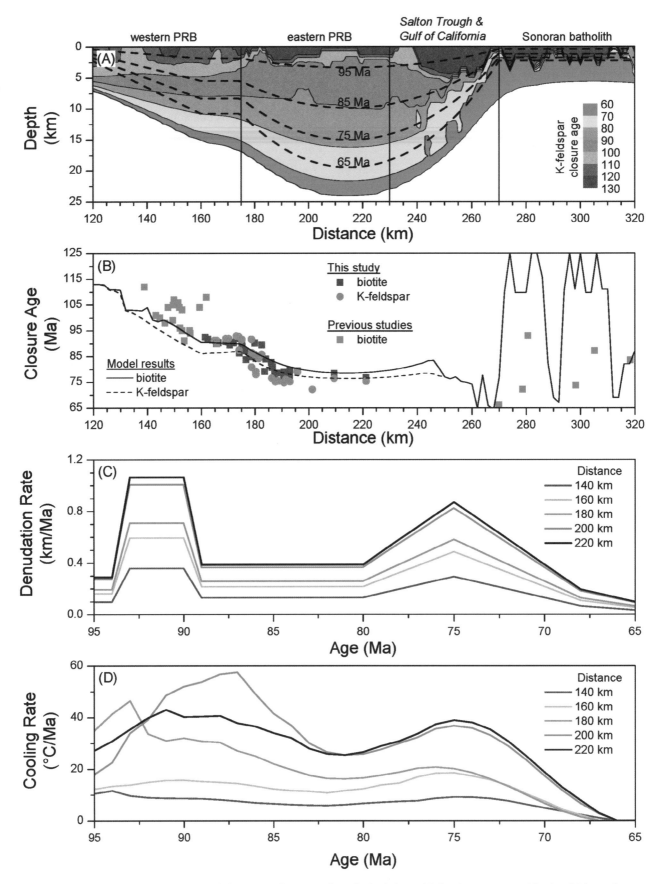

Figure 8. Results of erosional denudation calculations. A: Relevant portion of calculation grid showing contours of final K-feldspar closure ages used to calculate detrital age distributions in Figure 10. Positions of 95, 85, 75, and 65 Ma erosion surfaces are indicated by dashed lines. Irregularities in age contours are produced by transient heating during pluton emplacement. B: Predicted variation of biotite and K-feldspar K-Ar closure ages along final erosion surface in part A. C: Imposed denudation histories for different horizontal positions in grid as function of time. D: Calculated cooling rates for positions in C as function of time. Note that significant lag (~2–3 m.y.) exists between denudation and cooling during early denudation phase. PRB—Peninsular Ranges batholith.

ages produced by our numerical simulation of overlapping intrusion and erosional denudation (see Appendix 3 for additional details). It is important to realize that we simulated erosion by propagating surface temperatures (25 °C) into the grid. The position of the erosion surface at 95, 85, 75, and 65 Ma is represented by dashed lines in Figure 8A. Transient heating effects related to intrusion will be recorded at depths shallower than ~5–7 km beneath the active erosion surface (Lovera et al., 1999). At greater depths, K-feldspar is open to Ar diffusion, and the age distribution within the grid is controlled by denudation. For example, reheating effects related to 90–65 Ma pluton emplacement that are evident for grid positions between $240 < x < 260$ km at 10–20 km depths (i.e., within the region representing the Salton Trough and Gulf of California) were developed after significant denudation had occurred in this region.

The geometry of the final (i.e., 65 Ma) erosion surface in Figure 8A was constrained by the requirement that the available biotite and K-feldspar total gas ages of Figure 5 be reproduced (Fig. 8B). Several points bear mentioning. First, in the case of the Sonoran batholith and the western Peninsular Ranges batholith, we sought only to impose intrusive and denudation histories that simulated the first-order characteristics of these regions. We did not attempt to explicitly fit available data from these areas (Krummenacher et al., 1975; Gastil and Krummenacher, 1977; Ortega-Rivera, 2003) since they appear to reflect specific relationships to plutons that are not easily dealt with in our model. Second, the amount of denudation imposed upon the Salton Trough and Gulf of California region was based primarily upon extrapolation only. The intrusion history we imposed for this region was motivated in part by unpublished U-Pb zircon results from the Sierra El Major and Sierra Cucapa (Premo et al., 1998; Grove et al., 2003a). Finally, we failed to obtain a good numerical match to measured K-feldspar total gas ages from domain B from the east-central Peninsular Ranges batholith. K-feldspars from this domain tend to be slightly older than coexisting biotite; see Table 1 and Appendix 2). We believe this reflects a minor problem with excess ^{40}Ar in the K-feldspars that we were unable to correct for (see Appendix 2). K-feldspars from domains C and D generally lack low-temperature excess ^{40}Ar and yield total gas ages that are generally younger than those determined from biotite (Table 1).

Our calculations indicate erosion depths of about 11 km for domain B (Fig. 8A). An abrupt eastward increase in erosion depth beginning at $x = 175$ km coincides with the western limit of the eastern batholith. Note that the 85 Ma biotite age contour occurs at about $x = 185$ or 10 km east of the western boundary of the eastern batholith. This is compatible with the observed field relationship (Fig. 2). To the east of the 85 Ma contour, model K-feldspar total gas ages remain relatively constant. The maximum erosion depth we calculated for domain D occurs at $x = 215$ km. This value (20 km) corresponds to the upper limit of metamorphic pressures determined for the eastern Peninsular Ranges batholith (12–20 km; Grove, 1987; Ague and Brimhall, 1988; Todd et al., 1988; Rothstein and Manning, 2003). The assumed

background geothermal gradient in our grid was 30 °C/km. Use of higher ambient geotherms would have reduced our estimates of erosion depth and vice versa.

In our calculations, we adjusted the denudation history (Fig. 8C) to obtain cooling histories (Fig. 8D) that were compatible with what we determined from our K-feldspar thermal history results (Fig. 7B). Comparison of Figure 8, parts C and D, reveals that there is a significant lag between the time of rapid denudation and the cooling that is produced by advective heat transport toward the surface. Because the early phase of denudation (93–90 Ma) overlaps with intrusion that ends at 90 Ma, transient heating has an impact on some of the solutions. In general, maxima in the cooling rates appear to be shifted by 2–3 m.y. relative to denudation rates with the greatest delays being recorded by the deepest rocks. An implication of this is that the cooling rate maxima we observe for our K-feldspars (at 88 Ma; see Fig. 7B) actually corresponds to maximum denudation at about 91 Ma. This adjustment would cause the timing of the earlier denudation pulse to be in better agreement with the 98–93 Ma interval for main phase emplacement of the La Posta TTG suite. In contrast, maxima in denudation and cooling both occur ca. 75 Ma during the later 80–68 Ma phase. Better agreement occurs because the rocks are closer to the surface when denudation starts.

Cooling in Response to Shallow Subduction

Simple heat flow calculations performed by Dumitru (1990) involving a horizontal, instantaneously emplaced oceanic plate at very shallow (30–50 km) depths clearly demonstrated that large-scale crustal cooling of a former magmatic arc terrane would result from such a process. Here we present calculations of hanging wall thermal effects produced by a shallowly inclined, noninstantaneously emplaced oceanic plate using a geometry that we believe is more appropriate to the Peninsular Ranges batholith than the one adopted by Dumitru (1990). In our calculations, subduction initiates at 80 Ma (Fig. 9A). The slab has a 15° dip and is underthrust at 4 cm/yr. After 25 Ma of subduction, the thermal gradient within the calculation region has been reduced by a factor of two and a steady-state thermal structure has been approached (Fig. 9B). Cooling histories are presented for crustal depths of 5, 10, 15, and 20 km for both the left and right boundaries of the calculation region (Fig. 9C). Cooling produced by the shallowly subducting slab begins to occur within 1–3 m.y. after shallow subduction is initiated, depending upon its position within the area outlined in Figure 9B. Grid points near the left boundary experienced the most rapid cooling (10–50 °C/m.y., depending upon depth) with a peak at 76 Ma (Fig. 9D). Cooling rates are a factor of two lower (6–27 °C/Ma) along the right boundary with the peak occurring somewhat later (70–72 Ma, depending upon depth). Although we have not yet explored the effects of varying subduction parameters, it is clear that the timing and rates of cooling are comparable to those recorded by the eastern Peninsular Ranges batholith near 33°N (Fig. 7).

The calculated distribution of K-feldspar bulk closure ages for positions within the area outlined in Figure 9B is shown in

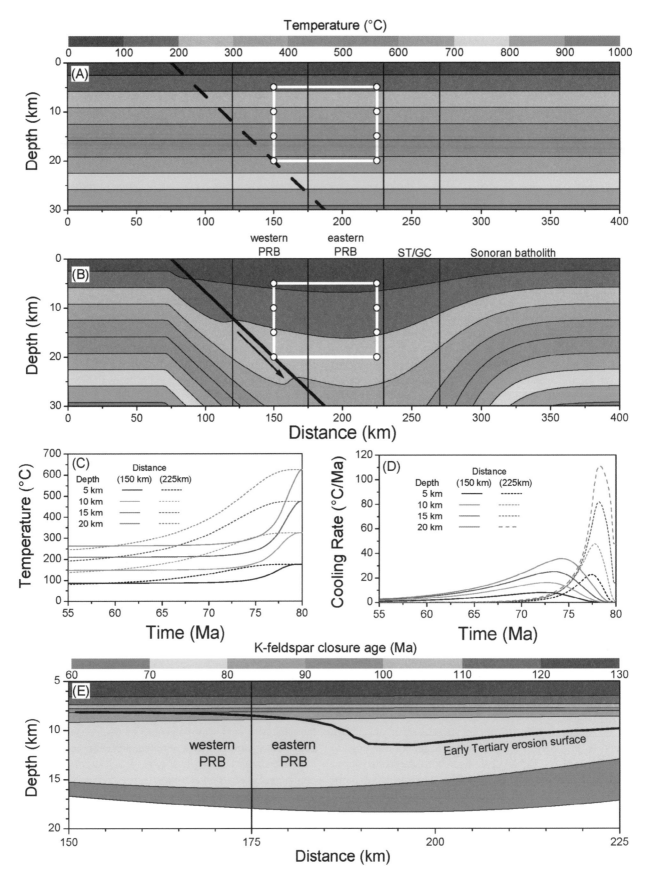

Figure 9. Calculation of hanging wall cooling (= subduction refrigeration) during shallow subduction. A: Initial geothermal structure at 80 Ma. Dashed bold line is future position of subducting oceanic plate (15° dip angle). Rectangular region indicates calculation points for panels C and D, below. Vertical lines indicate positions of western Peninsular Ranges batholith, eastern Peninsular Ranges batholith, Salton Trough/Gulf of California (ST/GC), and Sonoran batholith, respectively. B: Thermal structure developed by 55 Ma after 25 m.y. of 4 cm/yr underthrusting. C: Temperature-time histories at eight calculation points illustrated in panel A, above. D: Cooling histories for same calculation points. E: Distribution of K-feldspar bulk closure ages at 55 Ma. Bold line indicates position of erosion surface that corresponds to closure age profile of Figure 5. Erosion down to this surface must take place during continued steady-state underthrusting. Under these conditions, magnitude of erosional denudation is roughly half of that required in absence of subduction refrigeration (compare with Fig. 8A).

Figure 9E. Note that the subnormal geothermal gradients produced by subduction refrigeration cause K-feldspar to experience bulk Ar closure over a much broader region (and to significantly greater depths) than is the case when rocks cool solely due to proximity to the surface. As a result, the depths implied by the data in Figure 5 are only half as deep as those deduced from the erosion denudation model (compare Figs. 8A and 9E). Calculation of this surface assumes erosion occurs under near steady-state conditions of shallow subduction. However, note that the maximum depth attained (11 km) is at the lower limit estimates of metamorphic depth for portions of the eastern batholith that lie east of the 85 Ma biotite age contour (12–20 km; Grove, 1987; Ague and Brimhall, 1988; Todd et al., 1988; Rothstein and Manning, 2003). In addition, because lateral thermal gradients are greatly subdued, the steep age gradient that occurs in the vicinity of the 85 Ma biotite isochron requires a more abrupt depth increase than is the case for erosion denudation (compare Fig. 8A with Fig. 9D). Such an abrupt transition can only be explained by east-side-up faulting.

Relationship to Forearc Sedimentation

As demonstrated in the previous section, erosional denudation and subduction refrigeration mechanisms imply significantly different erosion depths for the east-central Peninsular Ranges batholith at 33°N. In the subduction refrigeration model, erosion can significantly postdate cooling without leaving a signal resolvable by biotite and K-feldspar thermochronometers provided that the low geothermal gradient produced by shallow subduction is maintained indefinitely. Alternatively, the erosion denudation model predicts that significant sedimentation will correlate strongly with major cooling events in the batholith. Because of this, the erosion denudation model can be further tested by determining how well it accounts for the age distributions of detrital thermochronometers in the forearc (Fig. 10).

Lovera et al. (1999) determined that it was possible to deduce important aspects of basement denudation histories from detrital thermochronometers. In their analysis of materials from the northern Peninsular Ranges batholith forearc (Santa Ana mountains and La Jolla; see locations in Fig. 1), they determined that a constant mean denudation rate of 0.5 mm/yr from 100–65 Ma fit the detrital K-feldspar K-Ar results reasonably well and that a 0.25 mm/yr rate was too slow. In refining their calculations, Lovera et al. (1999) concluded that the best fit to the data involved initially rapid erosion following emplacement of the La Posta TTG during the Cenomanian–Turonian (92–89 Ma; 1.25 mm/yr; Fig. 11A). They also found that erosion rates had to decrease during the Santonian–early Campanian (0.15 mm/yr from 89 to 78 Ma;) to match detrital results from this interval. Finally, a second accelerated phase was indicated by results from Late Campanian–Early Maastrictian sediments (0.45 mm/yr beginning at 78 Ma). We find it remarkable that the erosion denudation history implied by our basement thermal history results so well matches that determined independently from forearc strata

(Fig. 11B). Note that the constant denudation shown after 73 Ma in the Lovera et al. (1999) result is not constrained by data they obtained from the forearc sediments.

To further explore the compatibility of the two independent data sets, we have calculated detrital K-feldspar closure age distributions from our revised erosion denudation model. Because we consider it unlikely that drainage systems would have extended completely across the actively denuding region of Figure 8A, we established a drainage divide at the inferred eastern limit of the Peninsular Ranges batholith (at $x = 230$ km). The broad similarity in the measured and calculated age distributions that we obtained indicates that the overall magnitude and timing of basement denudation that we imposed in the calculations is well calibrated (Fig. 10, Table 3). This is an important point since denudation that is too fast or too slow can dramatically shift calculated detrital age distributions (see Lovera et al., 1999). Even so, Kolmogorov-Smirnoff analysis of the measured and synthetic data sets (see Lovera et al., 1999) reveals that the model age distributions are not as similar to the measured distributions as they could be (Table 3). Specifically, the well-defined maxima evident in our calculated spectra are not expressed in the detrital K-feldspar age distributions measured from the forearc (Fig. 10). Our model produces strong peaks in the detrital age distributions when material from the deeply denuded east-central Peninsular Ranges batholith predominates over other sources (Fig. 8A). As indicated in Figure 8A, western zone rocks yield a more dispersed distribution of bulk closure ages by virtue of their derivation from shallower levels. Based upon these results it seems clear that while our erosion denudation model for the Peninsular Ranges batholith is in reasonable agreement with the forearc results, there is need for further revision.

There are at least two potentially important factors that could cause shallower western zone rocks to be overrepresented in detrital age distributions sampled from the forearc. First, K-feldspar is less prevalent in the eastern batholith because it is dominated by the La Posta TTG suite (Baird and Miesch, 1984). Hence, K-feldspar will be disproportionately derived from the western batholith. If we were to weight our model results with respect to the abundance of K_2O, as constrained by Baird and Miesch's (1984) data set, K-feldspar yields from western zone rocks would be significantly increased. Second, the sedimentary sequences sampled by Lovera et al. (1999) onlap the batholith and could therefore receive an enhanced sediment load from local western sources. More distal forearc strata to be sampled in the future should be more representative of the batholith as a whole.

Extent of Faulting

Our present calculations do not explicitly consider the role of faulting in denuding the Peninsular Ranges batholith. A number of authors (Krummenacher et al., 1975; Goodwin and Renne, 1991; George and Dokka, 1994; Grove, 1994; Ortega-Rivera et al., 1997; Ortega-Rivera, 2003) have argued that faults with large

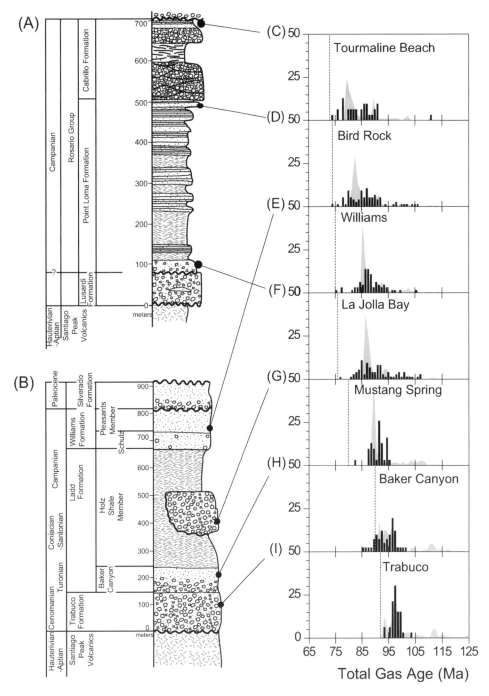

Figure 10. Stratigraphy of Peninsular Ranges batholith forearc strata (localities shown in Fig. 1A). A: La Jolla area. B: Northern Santa Ana Mountains. C–I: Solid (black) histograms represent measured detrital K-feldspar $^{40}Ar/^{39}Ar$ closure age distributions for stratigraphic positions located in panels A and B. Dashed vertical lines indicate depositional age (see Lovera et al., 1999). Gray curves represent calculated detrital K-feldspar distributions from data of Figure 8A. Overall similarity of measured and calculated age distributions indicates that Peninsular Ranges batholith denudation history is well calibrated. Additional details about good fit are presented in Table 3.

vertical components of displacement were locally significant in influencing K-Ar age gradients within the northeastern Peninsular Ranges batholith. The most obvious Late Cretaceous structure to affect the batholith is the west-vergent eastern Peninsular Ranges mylonite zone (Sharp, 1979; Simpson, 1984; Engel and Schultejann, 1984). While high-temperature mylonitization along this zone (e.g., Theodore, 1970) appears to have been largely coeval with emplacement of the La Posta TTG (Todd et al., 1988), continued, Late Cretaceous cataclastic deformation along the

eastern Peninsular Ranges mylonite zone has been documented by Goodwin and Renne (1991) and Wenk et al. (2000) in the San Jacinto and Santa Rosa mountains.

The rocks that we are considering from the east-central Peninsular Ranges batholith near 33°N are all situated southwest of the eastern Peninsular Ranges mylonite zone (Fig. 1). Cataclastic deformation similar to that developed within the eastern Peninsular Ranges mylonite zone occurs but is associated with more subtle structures. Of these, the Chariot Canyon fault zone is the most

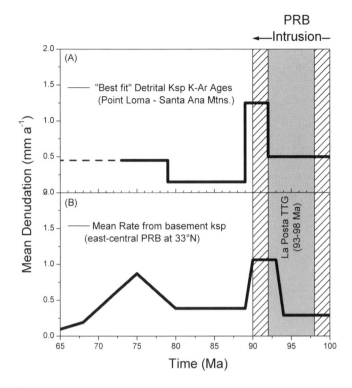

TABLE 3. COMPARISON OF MEASURED AND MODEL DETRITAL K-FELDSPAR CLOSURE AGE DISTRIBUTIONS

Sample	Lovera et al. (1999)		This study	
	Δt^* (Ma)	Log $PROB_{MAX}$	Δt^* (Ma)	Log $PROB_{MAX}$
Trabuco	0	−2.3	0	−2.3
Baker Canyon	0	−1.8	0	−2.1
Mustang Spring	0	−1.8	0	−1.7
La Jolla Bay	0	−2.6	0	−4.3
Williams	+1	−0.8	0	−2.0
Bird Rock	+1	−0.2	−1	−3.2
Tourmaline Beach	+1	−0.2	−2	−1.0

*Δt—difference between the best fit erosion surface predicted by the model and the independently determined depositional age for the sample (see Lovera et al., 1999).

[†]$PROB_{MAX}$—Kolmogorov-Smirnov statistic (Press et al., 1986) obtained from comparison of best-fit closure age distribution obtained from model and detrital K-feldspar $^{40}Ar/^{39}Ar$ closure age distribution measured from sample (see Lovera et al., 1999).

Figure 11. A: Erosional denudation histories deduced from detrital K-feldspars from northern Peninsular Ranges batholith forearc (Lovera et al., 1999). Dashed segment is not constrained by detrital data. B: Erosional denudation histories deduced from basement K-feldspars from east-central Peninsular Ranges batholith near 33°N. Gray region in A and B above represents emplacement interval for La Posta TTG (tonalite-trondhjemite-granodiorite) suite after Kimbrough et al. (2001). Hatched region indicates limits of meaningful intrusion within east-central Peninsular Ranges batholith at 33°N after Silver and Chappell (1988), Walawender et al. (1990, 1991), and Lovera et al. (1999).

obvious example (Fig. 3). Our K-feldspar thermal history results record a 150°C temperature difference across it prior to 78 Ma (Fig. 7). This corresponds to ~5 km vertical separation assuming a 30 °C/km geotherm. By about 68 Ma, rocks positioned east of the Chariot Canyon fault zone had cooled substantially (Fig. 7). Well-developed, early Tertiary(?) erosion surfaces developed on either side of the Chariot Canyon fault zone differ in elevation by less than 650 m. Because the erosion surfaces are actually dropped down to the east in a sense that is compatible with extensional faulting associated with Late Tertiary formation of the Salton Trough (Lampe, 1988), it seems clear that formation of the erosion surfaces post-dated earlier east-side-up displacement along the Chariot Canyon fault zone.

Elsewhere, it is difficult to prove that faulting has influenced cooling age patterns. Krummenacher et al. (1975) argued that the La Posta pluton was offset by an east-side-up fault displacement on the basis of a sharp decrease of K-Ar biotite ages in the vicinity of Carrizo Gorge (Fig. 2). While this proposed structure would coincide approximately with the 85 Ma biotite K-Ar age

contour, mapping in the Sweeny Pass 7.5′ quadrangle failed to reveal an important fault (Hoggatt, 1979).

Conclusions based upon our numerical analysis of erosional denudation would need to be re-evaluated if additional work demonstrated that Late Cretaceous tectonic denudation via normal faulting played a major role in the Late Cretaceous evolution of the east-central Peninsular Ranges batholith near 33°N. Currently there is little evidence that this was an important factor. Normal faults of potential Late Cretaceous age were mapped by Erskine and Wenk (1985) in the San Jacinto and Santa Rosa Mountains. Cataclastic deformation along minor low-angle fault surface occurs sporadically throughout the east-central Peninsular Ranges batholith near 33°N (Ratschbacher and Grove, unpubl. mapping). Some or all of these structures may be late Tertiary in age and similar to normal fault systems developed at Yaqui Ridge and the Vallecito Mountains (Schutejann, 1984; Lough and Stinson, 1991), the Coyote Mountains (Miller and Kato, 1991), and the Sierra El Major and Sierra Cucapá (Siem and Gastil, 1994; Axen and Fletcher, 1998; Axen et al., 2000). The only normal faults that are known to have exposed significantly deeper rocks in their footwalls are Late Miocene–Pliocene structures that occur within the Sierra El Major and Sierra Cucapá (Axen et al., 2000).

Driving Mechanisms for Late Cretaceous Denudation

It seems clear that early Late Cretaceous denudation (centered around 91 Ma; see Figs. 7B and 8C and D) is most simply explained as a consequence of massive intrusion of the eastern Peninsular Ranges batholith by the 96 ± 3 Ma La Posta TTG suite (see also Kimbrough et al., 2001). While more ambiguous, we believe that the driving mechanism(s) behind the later (i.e.,

< 78 Ma) cooling was most likely erosional denudation related to Laramide shallow subduction (Coney and Reynolds, 1977; Dickinson and Snyder, 1978; Bird, 1988). Subduction refrigeration may also have played a role. We speculate that the eastern Peninsular Ranges batholith lithosphere was eroded up to the base of the lower crust during Laramide shallow subduction beginning ca. 80 Ma and that removal of this material destabilized the overlying crust and triggered erosion and localized shortening along features such as the Chariot Canyon fault zone to restore the density balance between the crust and the upper mantle. Regional development of erosional surfaces and extraregional depositional systems across the Peninsular Ranges batholith (Minch, 1979; Kies and Abbott, 1983) signal that these processes had run their course by latest Cretaceous–early Tertiary time.

CONCLUSIONS

We draw the following conclusions for the thermal evolution of the east-central Peninsular Ranges batholith near 33°N:

1. The 85 Ma biotite K-Ar age isochron delineates rocks that have experienced contrasting cooling histories. We recognize four domains. At >20 km southwest of the 85 Ma contour, biotite $^{40}Ar/^{39}Ar$ total gas (= K-Ar) ages are old (>95 Ma) and variable. From 5 to 20 km southwest of the 85 Ma isochron, biotite and K-feldspar ages are 90–93 Ma. Within ± 5 km of the 85 Ma isochron, biotite and K-feldspar ages drop sharply to the east by about 1.3 m.y./km. As much as 40 km northeast of the 85 Ma isochron, biotite and K-feldspar ages are 74–77 Ma over a broad area (>3000 km²).

2. Multi-diffusion domain thermal history analysis of K-feldspar permits us to clearly resolve two phases of rapid cooling. The first episode is recorded primarily by rocks positioned west of the 85 Ma isochron. It persisted from ca. 91–86 Ma and peaked at 88 Ma with a mean rate of ~25 °C/m.y. Our numerical model of erosional denudation indicates a lag between denudation and cooling of ~3 m.y. Hence, peak cooling at 88 Ma corresponds to peak denudation at 91 Ma.

3. The second episode of rapid cooling was recorded mainly by rocks east of the 85 Ma contour. The event persisted from ca. 78–68 Ma and peaked at a mean rate of ~30 °C/Ma ca. 73 Ma. More easterly rocks recorded far higher rates of cooling (to 80 °C/Ma) over a shorter interval (76–72 Ma) centered around 74 Ma. Rocks in close proximity to the 85 Ma contour tend to record both cooling phases as well as slower cooling between 86–78 Ma.

4. A locally sharp K-Ar age gradient across the 85 Ma isochron (up to 2.5 m.y./km) in the vicinity of the Chariot Canyon fault reflects ductile to brittle, east-side-up shearing along it. Elsewhere along the 85 Ma isochron, field evidence for Late Cretaceous faulting is lacking.

5. Numerical analysis of erosion denudation indicates that our thermal history results can be largely explained by rapid denudation (up to 1 mm/yr) in two phases from 93 to 90 Ma and from 80 to 68 Ma. Erosion depths predicted from this model (up to 20 km) are near the upper limits of thermobarometric estimates from the eastern batholith.

6. Numerical analysis of cooling effects produced immediately above shallowly inclined (~15°) subduction zones (subduction refrigeration) indicates that the latter can also account for the later rapid cooling in the eastern Peninsular Ranges batholith if 4 cm/yr underthrusting is initiated at 80 Ma. Erosion depths predicted by this model (up to 11 km) are a factor of two lower than those estimated in the absence of subduction refrigeration and are at the lower limit of thermobarometric estimates for the eastern batholith.

7. Erosional denudation histories deduced for the northern Peninsular Ranges batholith using independent means (basement sampling and analysis of forearc sediments) yield remarkably similar results and indicate that erosional denudation was the most important factor in producing both episodes of Late Cretaceous cooling.

8. We believe that the early denudation phase (~93–90 Ma) was related to massive intrusion of the batholith by the La Posta TTG suite. Later 80–68 Ma denudation may have occurred in response to removal of the lower crust and underlying lithospheric mantle during Laramide shallow subduction.

APPENDIX 1: $^{40}AR/^{39}AR$ ANALYSIS DETAILS

Hand-selected muscovite and biotite (~5 mg) and hornblende (~25 mg) were wrapped in Sn foil and packed along with Al-wrapped Fish Canyon sanidine (FCT-1: 27.8 ± 0.3 Ma) flux monitors in 6 mm ID quartz tubes that were evacuated and sealed. Fe–mica biotite (307.3 Ma) was used to monitor K-feldspar samples (~200 mg) that were similarly packaged for irradiation. See Grove and Harrison (1996) for additional information regarding the Fe–mica biotite standard. Samples were irradiated at the University of Michigan's Ford reactor (L67 position). See McDougall and Harrison (1999) for more information regarding this facility and $^{40}Ar/^{39}Ar$ irradiation procedures. Correction factors for reactor-produced K- and Ca-derived argon were determined by measuring K_2SO_4 and CaF_2 salts included with each irradiation. Because several irradiations were performed, it is inefficient to include further information here. Instead, we provide data reduction parameters relevant to each sample in their respective data tables in the data repository. This information includes irradiation history, the date of $^{40}Ar/^{39}Ar$ analysis, and all irradiation parameters (J, $^{40}Ar/^{39}Ar_K$, $^{38}Ar/^{39}Ar_K$, $^{36}Ar/^{37}Ar_{Ca}$, and $^{39}Ar/^{37}Ar_{Ca}$).

Most K-feldspar samples were incrementally heated in a double vacuum Ta furnace (see Lovera et al., 1997, for additional information regarding K-feldspar step-heating experiments). Evolved gas was transferred with the aid of liquid nitrogen-activated charcoal and purified with a SAES ST-101 50 l/s getter pump. $^{40}Ar/^{39}Ar$ measurements were performed with an automated Nuclide 4.5–60-RSS mass spectrometer that was equipped with a Nier source and a Faraday detector, and it was typically operated at an Ar sensitivity of 1.5×10^{-15} mol/mV. Gas was admitted to the mass spectrometer with the aid of a leak valve to ensure that the quantity of gas analyzed did not exceed the linear range of the detection system. Orifice corrections were performed where necessary. Values of ^{39}Ar have been normalized to 100% gas delivery to the mass spectrometer. Fe–mica biotite flux monitors packed with the K-feldspars were analyzed following similar techniques. Additional experimental details are presented in Grove and Harrison (1996).

Three of the K-feldspars and all of the hornblende and muscovite samples were analyzed with a VG1200S mass spectrometer. This instrument is equipped with a Baur-Signer source and an electron multiplier and was operated at a sensitivity of 4×10^{-17} mol/mV. Evolved gas was transferred by expansion and purified with a SAES ST-101 10 l/s getter pump. Gas delivery to the mass spectrometer was governed by splitting using calibrated procedures.

Biotite samples were fused with a Coherent 5 W Argon ion laser and analyzed with a VG3600 mass spectrometer. This instrument was equipped with a "bright" Nier source and a Daly photomultiplier and was operated at an Ar sensitivity of 2×10^{-17} mol/mV. Generally each analysis entailed fusion of ~five 40–60 mesh grains. The extraction line and procedures were similar to those of the VG1200S. Additional details about both of these extraction line/mass spectrometer systems are provided in Quidelleur et al. (1997).

Intercalibration studies of different splits of FCT-1 sanidine involving the VG1200S and VG3600 extraction lines revealed no statistically significant (<0.3%) difference between either instrument (their pipette systems were filled simultaneously from a common aliquot of atmospheric Ar). J-factors calculated from Fe–mica biotite on the nuclide were generally agreed with splits run on the VG1200S to less than 0.5%. When J-factors calculated from adjacently packaged FCT-1 sanidine and Fe–Mica biotite were run on the VG1200S, they also agreed to within 0.5% (Fe–Mica biotite was systematically high). The error in age implied by this potential miscalibration of FCT-1 sanidine and Fe–Mica biotite for Late Cretaceous samples is less than 0.5 Ma.

Values $^{40}Ar/^{39}Ar$, $^{38}Ar/^{39}Ar$, $^{37}Ar/^{39}Ar$, and $^{36}Ar/^{39}Ar$ listed in data tables were corrected for total system backgrounds, mass discrimination (monitored by measurement of atmospheric Ar introduced by a pipette system), abundance sensitivity, and radioactive decay. Correction of $^{40}Ar/^{39}Ar$ ratios for nuclear interferences and atmospheric argon and calculation of apparent ages was carried out as described in McDougall and Harrison (1999) using conventional decay constants and isotopic abundances Steiger and Jäger (1977). Additional information is provided in footnote form with the data tables.

APPENDIX 2: DETAILS OF K-FELDSPAR MDD THERMAL HISTORY ANALYSIS

The principal requirement for successful application of the MDD approach is that thermally activated release of reactor-induced ^{39}Ar during laboratory step-heating experiments adequately portrays diffusion of radiogenic ^{40}Ar in nature (i.e., the same diffusion mechanisms and boundaries control Ar transport in both cases). This requirement cannot be met if volumetrically significant recrystallization occurs subsequent to the onset of radiogenic argon ($^{40}Ar^*$) retention at ~300–350 °C (Lovera et al., 2002; see also Parsons et al., 1999). Such an assessment can be made by assessing the extent of correlation between ^{39}Ar and $^{40}Ar/^{39}Ar$ spectra (Lovera et al., 2002). The age spectrum reflects millions of years of Ar transport in nature while the ^{39}Ar diffusion properties image this behavior over minutes to hours at elevated temperatures in the laboratory. If both reflect the same intrinsic process and boundaries for Ar transport, then the two spectra should be highly correlated (Lovera et al., 2002). The fact that K-feldspars examined in this study exhibit generally high correlation between nucleogenic (^{39}Ar) and radiogenic (^{40}Ar) argon (typically >0.9) is strong evidence that natural diffusion properties have been reproduced in the laboratory with reasonably high fidelity.

Calculations were performed for 24 of 28 K-feldspars. Step-heating results from samples 1110-I, EM, PV, and JUCH were not interpreted due to highly imprecise temperature control. Based upon the similarity of their age spectra to those of adjacent samples, however, we believe it is likely that they would have yielded compatible results. A first-order expectation of volume diffusion in a multi-diffusion domain sample is that the age spectrum should increase monotonically. Nearly all K-feldspars we have examined (e.g., Lovera et al., 2002)

exhibited evidence of minor low-temperature $^{40}Ar_E$ contamination that we believe to have originated from decrepitating fluid inclusions (Harrison et al., 1994). Evidence for this generally subsided between 600 and 700 °C. For samples that also exhibit signs of high-temperature $^{40}Ar_E$ contamination, evidence for this (i.e., erratic age variation that is uncorrelated with ^{39}Ar diffusion properties; see McDougall and Harrison, 1999) generally is first manifested above 850–900 °C.

The interpreted interval of ^{39}Ar release for each of the samples is listed in Table 2. Generally, the low-temperature limit (f_{min}) corresponds to the point in the age spectrum where indications of fluid-inclusion hosted low-temperature $^{40}Ar_E$ disappear (i.e., the $^{40}Ar/^{39}Ar$ ages begin to increase systematically). In the case of RG K-feldspar, evidence of $^{40}Ar_E$ continued to higher temperature. To improve our ability to interpret results from this sample, we performed isothermal duplicates up to 1100 °C and modeled only the ages yielded by the second, less affected steps (see Mahon et al., 1998, for additional details about this strategy). For most samples, we established the upper limit for analysis (f_{max}) at the fraction of ^{39}Ar release that corresponded to the final 1100 °C step. At higher temperatures evidence for melting in the form of dramatically elevated ^{39}Ar diffusivities is manifested. For samples with problematic high temperature behavior (i.e., intermediate age maxima or high-temperature $^{40}Ar_E$), we set f_{max} to immediately below the point where these anomalies first became resolved.

Three-quarters of the 24 interpreted samples (AC, BM, BW2, CM1, CP128, CP175, CV, DSC, FCM, JU, LBV2B, MP, MP17, SFH, SP, SY, TS, and YS) yielded sufficiently well behaved age and Arrhenius properties that we felt confident in modeling all steps unaffected by low-temperature $^{40}Ar_E$ up to the point of melting (Table 2). K-feldspars from domain B had generally higher levels of low-temperature excess $^{40}Ar^*$. Unfortunately, we could not correct the age spectra of these samples for this effect since ^{38}Ar was not measured (see Harrison et al., 1994). In domain C, several K-feldspars (JU and MP) had relatively high age uncertainties that we attribute to small sample size due to the very K-feldspar–poor nature of the host rocks. Another sample (RG) had an inordinately large amount of low-temperature $^{40}Ar_E$. Additional samples in domains B, C, and D (CS, GM, KS, MMVT, and TG) exhibited intermediate maxima that significantly limited the interval of ^{39}Ar release ($f_{max} - f_{min}$) that we were able to model (Table 2).

A subset of the samples discussed in this paper (AC, CP128, CP175, CV, FCM, JU, LBV2B, MP, MP17, SFH, SP, and SY) were previously analyzed in Grove (1993). Temperature cycling measurements (e.g., Lovera et al., 1989) that were performed with these samples are not presented here. While such experiments are very useful in elucidating the multi-diffusion character of ^{39}Ar release from K-feldspar, in practice it is difficult to adequately constrain temperature in the crucible during conditions of declining temperature. Specifically, sluggish thermal equilibration in the crucible (particularly at low temperature) causes more ^{39}Ar release than should have been the case if the sample temperature corresponded to the set temperature. Because diffusivities fall exponentially with temperature, the effect can be significant and lead to misleading results (i.e., dramatic and geologically unreasonable variation in activation energy is often implied; see Lovera et al., 1997). Because the fraction of ^{39}Ar release associated with these steps is generally very small (typically <0.1%), neglecting them has no discernible impact (relative to the ± 0.05 uncertainties in D/r²) upon ^{39}Ar diffusivities of adjacent steps.

APPENDIX 3 NUMERICAL SIMULATIONS

Erosion Denudation

Our calculations of erosion denudation effects are modified after Lovera et al. (1999). We employed a two-dimensional, Crank-Nicholson finite-difference algorithm to solve the diffusion equation (Press et al.,

1986, p. 638). A constant thermal diffusivity (10^{-6} m^2/s), no radioactive internal heating, and a fixed basal heat flux appropriate to maintain a 30 °C/km thermal gradient (= the initial geothermal gradient) were used. Calculations were performed in a 400-km-wide by 60-km-deep grid (1 km × 1 km resolution). Surface temperature was maintained constant at 25 °C and zero-flux conditions were established at the lateral boundaries. Calculations began at 125 Ma and terminated at 40 Ma. Random distributions of circular plutons with randomly specified radius and emplacement temperature were intruded at 2 m.y. intervals (see Lovera et al., 1999, for further details). Three distinct regions of intrusion were set up. For the region bounded by 120 < x < 175 km (western Peninsular Ranges batholith), we intruded plutons at depths between 2 and 20 km from 120–100 Ma. For the region bounded by 175 < x < 230 km (eastern Peninsular Ranges batholith), we intruded plutons at depths between 2 and 20 km from 105–90 Ma. Finally, for the region bounded by 230 < x < 400 km (Salton Trough/Gulf of California and western Sonora), we intruded plutons at depths between 2 and 10 km from 90 to 65 Ma. Denudation was initiated at 108 Ma. In the model, progressively developed Late Cretaceous erosion surfaces were represented by propagating a 25 °C surface through the finite difference grid in successive time steps. This was accomplished by setting the temperature of all grid points situated at or above the defined surface to a constant value (25°C). The applied denudation history at x = 215 km shown in Figure 8C produced a final erosion depth of 20 km at 65 Ma. A proportionally scaled denudation history was applied at other lateral positions in the grid to produce the final erosion surface shown in Figure 8A. Temperature-time histories were recorded for each grid node in the model. Bulk closure ages were calculated assuming a single domain diffusion model and experimentally determined Arrhenius parameters for K-feldspar (activation energy or E = 46 kcal/mol; frequency factor or log D_o/r_o^2 = 3.5 s^{-1}; Lovera et al., 1997). To calculate detrital age distributions, cooling ages were randomly sampled at 1 m.y. intervals between surfaces separated by ± 0.5 Ma. This procedure is described in more detail in Lovera et al. (1999).

Subduction Refrigeration

We employed a two-dimensional, Crank-Nicholson finite-difference algorithm to solve the diffusion equation (Press, et al., 1986, p. 638). Heat conduction, boundary conditions, and the calculation grid dimensions and resolution were described in the same manner as for erosion denudation. In our calculations, shallow subduction was initiated at 80 Ma at x = 75 km. The slab dipped at 15° and was underthrust to the right at 4 cm/yr. We implemented a 2-km-thick shear zone characterized by a constant friction equivalent to 30 MPa. Temperature-time (T-t) histories were monitored continuously for all points within the 15 × 75 km region outlined in Figure 9A. K-feldspar closure ages were calculated as described above.

ACKNOWLEDGMENTS

We acknowledge support from Department of Energy grant DE-FG-03-89ER14049. Grove received support from National Science Foundation grant EAR-0113563 as well as from additional NSF grants to Harrison. Vicki Todd was an invaluable source of information regarding local geology and contributed several key samples. Important information and insight regarding various aspects of the study have also come from discussions with Pat Abbott, Gary Axen, Brad Erskine, John Fletcher, Gordon Gastil, Peter George, Gary Girty, Laurel Goodwin, Dave Kimbrough, Cynthia Lampe, Harold Lang, Charlie Lough, Craig Manning, Doug Morton, Lothar Ratschbacher, Dave Rothstein, Keegan Schmidt, Lee Silver, Amy Stinson, Rich Wolf, and Mike Walawender. Dave Rothstein performed some of the ^{40}Ar/^{39}Ar measurements included in this study. Matt Heizler was extremely helpful in supporting the ^{40}Ar/^{39}Ar analysis and also helped in our initial attempts to interpret the results through the use of the MDD model. Reviews of this paper by Dave Kimbrough and Gary Girty helped to improve it considerably.

REFERENCES CITED

Abbott, P.L., and Smith, T.E., 1978, Trace element comparison of clasts in Eocene conglomerates, southwestern California and northwestern Mexico: Journal of Geology, v. 86, p. 753–762.

Abbott, P.L., and Smith, T.E., 1989, Sonora, Mexico, source for the Eocene Poway Conglomerate of southern California: Geology, v. 17, p. 329–332.

Ague, J.J., Brandon, M.T., 1992, Tilt and northward offset of Cordilleran batholiths resolved using igneous barometry: Nature, v. 360, p. 146 152.

Ague, J.J., and Brimhall, G.H., 1988, Magmatic arc asymmetry and distribution of anomalous plutonic belts in the batholiths of California; effects of assimilation, crustal thickness, and depth of crystallization: Geological Society of America Bulletin, v. 100, p. 912–927.

Allmendinger, R.W., Figueroa, D., Snyder, D., Beer, J., Mpodozis, C., and Isacks, B.L., 1990, Foreland shortening and crustal balancing in the Andes at 30°S latitude: Tectonics, v. 9, p. 789–809.

Anderson, J.R., 1983, Petrology of a portion of the eastern Peninsular Ranges mylonite zone, Southern California: Contributions to Mineralogy and Petrology, v. 84, p. 253–271.

Anderson, T.H., and Silver, L.T., 1974, Late Cretaceous plutonism in Sonora, Mexico and its relationship to circum-Pacific magmatism: Geological Society of America Abstracts with programs, v. 6, no. 6, p. 484.

Armstrong, R.L., and Suppe, J., 1973, Potassium-argon geochronometry of Mesozoic igneous rocks in Nevada, Utah, and southern California: Geological Society of America Bulletin, v. 84, p. 1375–1392.

Axen, G.J., and Fletcher, J.M., 1998, Late Miocene—Pleistocene extensional faulting, northern Gulf of California, Mexico and Salton Trough, California: International Geology Review, v. 40, p. 217 244.

Axen, G.J., Grove, M., Stockli, D., Lovera, O.M., Rothstein, D.A., Fletcher, J.M., Farley, K., and Abbott, P.L., 2000, Thermal evolution of Monte Blanco Dome; low-angle normal faulting during Gulf of California rifting and late Eocene denudation of the eastern Peninsular Ranges: Tectonics, v. 19, p. 197–212.

Baird, A.K., and Miesch, A.T., 1984, Batholithic rocks of southern California—a model for the petrogenesis of their source materials: U.S. Geological Survey Professional Paper 1284, p. 42.

Bannon, J.L., Bottjer, D.J., Lund S.P., and Saul, L.R., 1989, Campanian/Maastrichtian stage boundary in southern California: Resolution and implications for large-scale depositional patterns: Geology, v. 17, p. 80–83.

Barth, A.P., Wooden, J.L., Grove, M., Jacobson, C.E., and Dawson, J.P., 2003, U-Pb zircon geochronology of rocks in the Salinas Valley region of California: A reevaluation of the crustal structure and origin of the Salinian Block: Geology, v. 31, p. 517–520.

Barton, M.D., Battles, D.A., Bebout, G.E., Capo, R.C., Christensen, J.N., Davis, S.R., Hanson, R.B., Michelsen, C.J., and Trim, H.E., 1988, Mesozoic contact metamorphism in the Western United States, in Ernst, W.G., ed., Metamorphism and crustal evolution of the Western United States, Rubey Volume VII: Englewood Cliffs, New Jersey, Prentice-Hall, p. 110–178.

Bateman, P.C., and Chappell, B.W., 1979, Crystallization, fractionation, and solidification of the Tuolumne Intrusive Series, Yosemite National Park, California: Geological Society of America Bulletin, v. 90, p. I 465-I 482.

Berggreen, R.G., and Walawender, M.J., 1977, Petrography and metamorphism of the Morena Reservoir roof pendant, southern California: California Division of Mines and Geology Special Report 129, p. 61–65.

Bird, P., 1988, Formation of the Rocky Mountains, western United States: A continuum computer model: Science, v. 239, p. 1501–1507.

Bottjer, D.J., and Link, M.H., 1984, A synthesis of Late Cretaceous southern California and northern Baja California paleogeography, in Crouch, J.K., and Bachman, S.B., eds., Tectonics and sedimentation along the California margin: Los Angeles, Society for Ecomonic Paleontologists and Mineralogists Pacific Section, p. 171–188.

Bottjer, D.J., Colburn, I.P., and Cooper, J.D., 1982, Late Cretaceous depositional environments and paleogeography, Santa Ana Mountains, Southern California: Los Angeles, Society for Ecomonic Paleontologists and Mineralogists Pacific Section, p. 121.

Burchfiel, B.C., and Davis, G.A., 1981, Mojave Desert and environs, *in* Ernst, W.G., ed., The geotectonic development of California, Rubey Vol. I: Englewood Cliffs, New Jersey, Prentice Hall, p. 217–252.

Butler, R.H., Dickinson, W.R., and Gerhels, G.E., 1991, Paleomagnetism of coastal California and Baja California: Alternatives to large-scale northward transport: Tectonics, v. 10, p. 561–576.

Cerveny, P.F., Dorsey, R.J., and Burns, B.A., 1991, Apatite and zircon fission-track ages from the Sierra San Pedro Martir, eastern Peninsular Range, Baja California, Mexico: Geological Society of America Abstracts with Programs, v. 23, no. 2, p. 12.

Chase, C.G., and Wallace, T.C., 1986, Uplift of the Sierra Nevada of California: Geology, v. 14, p. 730–733.

Clinkenbeard, J.P., and Walawender, M.J., 1989, Mineralogy of the La Posta pluton: Implications for the origin of zoned plutons in the eastern Peninsular Ranges batholith, southern and Baja California: American Mineralogist, v. 74, p. 1258–1269.

Coleman, D.S., and Glazner, A.F., 1998, The Sierra Crest magmatic event: Rapid formation of juvenile crust during the Late Cretaceous in California, *in* Ernst, W.G., and Nelson, C.A., eds., Integrated Earth and environmental evolution of the southwestern United States: Columbia, Maryland, Bellwether Publishing for the Geological Society of America, p. 253–272.

Coney, P.J., and Reynolds, S.J., 1977, Cordilleran Benioff zones: Nature, v. 270, p. 403–405.

Crowell, J.C., 1981, An outline of the tectonic history of southeastern California, *in* Ernst, W.G., ed., The geotectonic development of California (Rubey Vol. I): Englewood Cliffs, New Jersey, Prentice Hall, p. 583–600.

DePaolo, D.J., 1981, A neodymium and strontium isotopic study of the Mesozoic calc-alkaline granitic batholiths of the Sierra Nevada and Peninsular Ranges, California: Journal of Geophysical Research, v. 86, p. 10470–10488.

Detterman, M.E., 1984, Geology of the Metal Mountain district, In-ko-pah Mountains, San Diego County, California [M.S. thesis]: San Diego, San Diego State University, 216 p.

Dickinson, W.R., and Butler, R.F., 1998, Coastal and Baja California paleomagnetism reconsidered: Geological Society of America Bulletin, v. 110, p. 1268 1280.

Dickinson, W.R., and Snyder, W.S., 1978, Plate tectonics of the Laramide orogeny: Boulder, Colorado, Geological Society of America Memoir 151, p. 355–366.

Dodson, M.H., 1973, Closure temperature in cooling geochronological and petrological systems: Contributions to Mineralogy and Petrology, v. 40, p. 259–274.

Dokka, R.K., and Merriam, R.H., 1982, Late Cenozoic extension of northeastern Baja California, Mexico: Geological Society of America Bulletin, v. 93, p. 371–378.

Dorsey, R.J., 2002, Stratigraphic record of Pleistocene initiation and slip on the Coyote Creek fault, lower Coyote Creek, southern California, in Barth, A., ed., Contributions to crustal evolution of the southwestern United States: Boulder, Colorado, Geological Society of America Special Paper 365, p. 251 269.

Ducea, M., 2001, The California arc: Thick granitic batholiths, eclogite residues, lithospheric-scale thrusting, and magmatic flare-ups: GSA Today, v. 11, no. 11, p. 4–10.

Dumitru, T.A., 1990, Subnormal Cenozoic geothermal gradients in the extinct Sierra Nevada magmatic arc; consequences of Laramide and post-Laramide shallow-angle subduction: Journal of Geophysical Research, v. 95, p. 4925–4941.

Dumitru, T.A., Gans, P.B., Foster, D.A., and Miller, E.L., 1991, Refrigeration of the western Cordilleran lithosphere during Laramide shallow-angle subduction: Geology, v. 19, p. 1145–1148.

Engel, A.E.J., and Schultejann, P.A., 1984, Late Mesozoic and Cenozoic tectonic history of south central California: Tectonics, v. 3, no. 6, p. 659–675.

Erskine, B.G., and Wenk, H.R., 1985, Evidence for Late Cretaceous crustal thinning in the Santa Rosa mylonite zone, southern California: Geology, v. 18, p. 1173–1177.

Everden, J.F., and Kistler, R.W., 1970, Chronology of emplacement of Mesozoic batholithic complexes in California and western Nevada: U.S. Geological Survey Professional Paper 623, 42 p.

Everhart, D.L., 1951, Geology of the Cuyamaca Peak quadrangle, San Diego County, California: California Division of Mines Bulletin 159, p. 51–115.

Flynn, C.J., 1970, Post-batholithic geology of La Gloria Presa Rodriguez area, Baja California, Mexico: Geological Society of America Bulletin, v. 81, p. 1789–1806.

Fry, J.G., Bottjer, D.J., and Lund, S.P., 1985, Magnetostratigraphy of displaced Upper Cretaceous strata in southern California: Geology, v. 13, p. 648–651.

Gastil, R.G., 1979, A conceptual hypothesis for the relation of differing tectonic terranes to plutonic emplacement: Geology, v. 7, p. 542–544.

Gastil, R.G., 1983, Mesozoic and Cenozoic granitic rocks of southern California and western Mexico: Boulder, Colorado, Geological Society of America Memoir 159, p. 265–275.

Gastil, R.G., 1993, Prebatholithic history of peninsular California, *in* The prebatholithic stratigraphy of peninsular California: Geological Society of America Special Paper 279, p. 145–156.

Gastil, R.G., and Krummenacher, D., 1977, Reconnaissance geologic map of coastal Sonora between Puerto Lobos and Bahia Kino: Geological Society of America Bulletin, v. 88, p. 189–198.

Gastil, R.G., Phillips, R.P., and Allison, E.C., 1975, Reconnaissance geology of the state of Baja California: Boulder, Colorado, Geological Society of America Memoir 140, 170 p.

Gastil, R.G., Morgan, G.J., and Krummenacher, D., 1981, The tectonic history of peninsular California and adjacent Mexico, *in* Ernst, W.G., ed., The Geotectonic Development of California, Rubey Volume I: Englewood Cliffs, New Jersey, Prentice-Hall, p. 284–306.

Gastil, G., Diamond, J., Knaack, C., Walawender, M., Marshall, M., Boyles, C., and Chadwick, B., 1990, The problem of the magnetite/ilmenite boundary in southern and Baja California, *in* Anderson, J.L., ed., The nature and origin of Cordilleran magmatism: Boulder, Colorado, Geological Society of America Memoir 174, p. 19–32.

Gastil, G., Kimbrough, J., Tainosho, Y., Shimizu, M., and Gunn, S.H., 1991, Plutons of the eastern Peninsular Ranges, southern California, USA and Baja California, Mexico, in Walawender, M.J., and Hanan Hanan, B., Geological excursions in Southern California and Mexico: San Diego, San Diego State University Publication, p. 319 331.

George, P.G., and Dokka, R.K., 1994, Major Late Cretaceous cooling events in the eastern Peninsular Ranges, California, and their implications for Cordilleran tectonics: Geological Society of America Bulletin, v. 106, p. 903–914.

Germinario, M.P., 1993, The early Mesozoic Julian Schist, Julian, California: Boulder, Colorado, Geological Society of America Special Paper 279, p. 107–118, 1993.

Girty, G.H., 1987, Sandstone provenance, Point Loma formation, San Diego, California: Evidence for uplift of the Peninsular Ranges during the Laramide orogeny: Journal of Sedimentary Petrology, v. 57, p. 839–844.

Goodwin, L.B., and Renne, P.R., 1991, Effects of progressive mylonitization on Ar retention in biotites from the Santa Rosa mylonite zone, California, and thermochronologic implications: Contributions to Mineralogy and Petrology, v. 108, p. 283–297.

Gromet, L.P., and Silver, L.T., 1987, REE variations across the Peninsular Ranges batholith: Implications for batholithic petrogenesis and crustal growth in magmatic arcs: Journal of Petrology, v. 28, p. 75–125.

Grove, M., 1987, Metamorphism and deformation in the Box Canyon area, eastern Peninsular Ranges, San Diego County, California [M.S. thesis]: Los Angeles, University of California, 174 p.

Grove, M., 1993, Thermal histories of southern California basement terranes [PhD. thesis]: Los Angeles, University of California, 451 p.

Grove, M., 1994, Contrasting denudation histories within the east-central Peninsular Ranges batholith (33°N), *in* McGill, S.F., and Ross, T.M., Geological investigations of an active margin: Redlands, California, San Bernardino County Museum Association, p. 235–240.

Grove, M., Fletcher, J., Axen, G., and Stockli, D., 2003a, U-Pb zircon crystallization ages for plutonic rocks within the Sierra el Mayor and Sierra Cucapá, northwestern Baja California, México, Geological Society of America Abstracts with Programs, v. 35, no. 6, p. 27.

Grove, M., Jacobson, C.E., Barth, A.P., and Vucic, A., 2003b, Temporal and spatial trends of Late Cretaceous–early Tertiary underplating of Pelona and related schist beneath southern California and southwestern Arizona, *in* Johnson, S.E., Paterson, S.R., Fletcher, J.M., Girty, G.H., Kimbrough, D.L., and Martín-Barajas, A., eds., Tectonic evolution of northwestern México and the southwestern USA: Boulder, Colorado, Geological Society of America Special Paper 374, p. 381–406 (this volume).

Grove, M., and Harrison, T.M., 1996, $^{40}Ar^*$ diffusion in Fe-rich biotite: American Mineralogist, v. 81, p. 940–951.

Hamilton, W., 1988, Tectonic setting and variations with depth of some Cretaceous and Cenozoic structural and magmatic systems of the western United States, *in* Ernst, W.G., ed., Metamorphism and crustal evolution of the western United States, Rubey Vol. VII: Englewood Cliffs, New Jersey, Prentice Hall, p. 1–40.

Hanson R.B., and Barton, M.D., 1989, Thermal development of low-pressure metamorphic belts: Results from two-dimensional numerical models: Journal of Geophysical Research, v. 94, p. 10363–10377.

Harrison, T.M., Heizler, M.T., Lovera, O.M., Chen, W., and Grove, M., 1994, A chlorine disinfectant for excess argon released from K-feldspar during step-heating: Earth and Planetary Science Letters, v. 123, p. 95–104.

Hill, R.I., 1988, San Jacinto intrusive complex 1. Geology and mineral chemistry, and a model for intermittent recharge of tonalitic magma chambers: Journal of Geophysical Research, v. 93, p. 10,325–10,348.

Hill, R.I., Silver, L.T., and Taylor, H.P., 1986, Coupled Sr-O isotope variations as an indicator of source heterogeneity for the northern Peninsular Ranges batholith: Contributions to Mineralogy and Petrology, v. 92, p. 351–361.

Hoggatt, W.C., 1979, Geologic map of Sweeney Pass Quadrangle, San Diego County, California: U.S. Geological Survey Open-File Report OF 79-754, 37 p.

Jachens, R.C., Simpson, R.W., Griscom, A., and Mariano, J., 1986, Plutonic belts in southern California defined by gravity and magnetic anomalies: Geological Society of America Abstracts with Programs, v. 18, p. 120.

Jennings, C.W., compiler, 1977, Geologic map of California: California Division of Mines and Geology, 1 sheet, scale: 1:750,000.

Kairouz, M., Axen, G.J., Grove, M., Lovera, O., and Stockli, D., 2003, Late Cenozoic $^{40}Ar/^{39}Ar$ ages of fault rocks formed along the west Salton detachment systems, southern California, Geological Society of America Abstracts with Programs, v. 35, no. 6, p. 629.

Kennedy, M.P., and Moore, G.W., 1971, Stratigraphic relations of Upper Cretaceous and Eocene formations, San Diego coastal area, California: American Association of Petroleum Geologists Bulletin, v. 55, p. 709–722.

Kies, R.P., and Abbott, P.L., 1983, Rhyolite clast populations and tectonics in the California continental borderland: Journal of Sedimentary Petrology, v. 53, p. 461–475.

Kimbrough, D.L., Magistrale, H., and Gastil, R.G., 1998, Enormous Late Cretaceous magma flux associated with TTG batholith emplacement; eastern Peninsular Ranges batholith of Southern and Baja California: Geological Society of America Abstracts with Programs, v. 30, no. 7, p. 258.

Kimbrough, D.L., Smith, D.P., Mahoney, J.B., Moore, T.E., Grove, M., Gastil, R.G., and Ortega-Rivera, A., 2001, Forearc-basin sedimentary response to rapid Late Cretaceous batholith emplacement in the Peninsular Ranges of southern and Baja California: Geology, v. 29, p. 491–494.

Kofron, R.J., 1984, Age and origin of gold mineralization in the southern portion of the Julian mining district, Southern California [M.S. thesis]: San Diego, San Diego State University, 75 p.

Krummenacher, D., Gastil, R.G., Bushee, J., and Doupont, J., 1975, K-Ar apparent ages, Peninsular Ranges batholith, southern California: Geological Society of America Bulletin, v. 86, p. 760–768.

Lampe, C.M., 1988, Geology of the Granite Mountain area; Implications of the extent and style of deformation along the southeast portion of the Elsinore fault [M.S. thesis]: San Diego, San Diego State University, 150 p.

Lewis, J.L., Day, S.M., Magistrale, H.C., Raul, R.A., Luciana, R., Cecilio, J., Eakins, J., Vernon, F.L., and Brune, J.N., 2001, Crustal thickness of the Peninsular Ranges and Gulf extensional province in the Californias: Journal of Geophysical Research, v. 106, p. 13,599–13,611.

Liou, J.G., Maruyama, S., and Cho, M., 1987, Very low-grade metamorphism of volcanic and volcaniclastic rocks—mineral assemblages and mineral facies, *in* Frey, M., ed., Low Temperature Metamorphism: Glasgow, Blackie, p. 59–113.

Lough, C.F., and Stinson, A.L., 1991, Structural evolution of the Vallecito Mountains, SW California: Geological Society of America Abstracts with Programs, v. 23, no. 5, p. 246.

Lovera, O.M., Richter, F.M., and Harrison, T.M., 1989, $^{40}Ar/^{39}Ar$ thermochronometry for slowly cooled samples having a distribution of diffusion domain sizes: Journal of Geophysical Research, v. 94, p. 17917–17935.

Lovera, O.M., Heizler, M.T., and Harrison, T.M., 1993, Argon diffusion domains in K-feldspar II: Kinetic properties of MH-10: Contributions to Mineralogy and Petrology, v. 113, p. 381–393.

Lovera, O.M., Grove, M., Harrison, T.M., and Mahon, K.I., 1997, Systematic analysis of K-feldspar $^{40}Ar/^{39}Ar$ step-heating experiments I: Significance

of activation energy determinations: Geochimica et Cosmochimica Acta, v. 61, p. 3171–3192.

Lovera, O.M., Grove, M., Kimbrough, D.L., and Abbott, P.L., 1999, A method for evaluating basement exhumation histories from closure age distributions of detrital minerals: Journal of Geophysical Research, v. 104, p. 29,419–29,438.

Lovera, O.M., Grove, M., and Harrison, T.M., 2002, Systematic analysis of K-feldspar $^{40}Ar/^{39}Ar$ step-heating experiments II: Relevance of laboratory K-feldspar argon diffusion properties to nature: Geochimica Cosmochimica Acta, v. 66, p. 1237–1255.

Magistrale, H., and Rockwell, T.K., 1996, The central and southern Elsinore fault zone, southern California: Bulletin of the Seismological Society of America, v. 86, p. 1793–1803.

Mahon, K.I., Harrison, T.M., and Grove, M., 1998, The thermal and cementation histories of a sandstone petroleum reservoir, Elk Hills, California. 1: $^{40}Ar/^{39}Ar$ thermal history results, Chemical Geology, v. 152, p. 227–256.

Martin, H., 1993, The mechanics of petrogenesis of the Archean continental crust-comparison with modern processes: Lithos, v. 30, p. 373–388.

McDougall, I., and Harrison, T.M., 1999, Geochronology and thermochronology by the $^{40}Ar/^{39}Ar$ method: New York, Oxford University Press, 212 p.

McDowell, F.W., Roldán-Quintana, J., and Connelly, J.N., 2001, Duration of Late Cretaceous–early Tertiary magmatism in east-central Sonora, Mexico: Geological Society of America Bulletin, v. 113, p. 521–531.

Merriam, R.H., 1958, Geology of the Santa Ysabel quadrangle, San Diego County, California: California Division of Mines Bulletin, v. 177, p. 7–20.

Miller, W.J., 1935, A geologic cross section across the southern Peninsular Range of California: California Journal of Mines and Geology, v. 31, p. 115–242.

Miller, D.E., and Kato, T., 1991, Mid-Tertiary continental extension, SW Salton Trough, California: Geological Society of America Abstracts with Programs, v. 23, no. 2, p. 79.

Minch, J.A., 1979, The late-Mesozoic—Early Tertiary framework of continental sedimentation, northern Peninsular Ranges, Baja California, Mexico, *in* Abbott, P.L., ed., Eocene depositional systems: Los Angeles, Society for Sedimentary Geology Pacific Section, p. 43–68.

Murray, G.T., and Girty, G.H., 1996, Pre-Jurassic deformation and metamorphism within the Julian Schist, Peninsular Ranges, Southern California: Geological Society of America Abstracts with Programs, v. 28, no. 5, p. 95.

Naeser, C.W., Naeser, N.D., Todd, V.R., Bohannon, R.G., 1996, Thermochrononology of the Peninsular Ranges Batholith, Southern California, from fission-track and $^{40}Ar/^{39}Ar$ analysis, Geological Society of America Abstracts with Programs, v. 28, no. 7, p. 513.

Nilsen, T.H., and Abbott, P.L., 1981, Paleogeography and sedimentology of Upper Cretaceous turbidites, San Diego, California: American Association of Petroleum Geologists Bulletin, v. 65, p. 1256–1284.

Nordstrom, C.E., 1970, Lusardi Formation—A post-batholithic Cretaceous conglomerate north of San Diego, California: Geological Society of America Bulletin, v. 81, p. 601–605.

Ortega-Gutierrez, F., Mitre-Salazar, L.M., Roldán-Quintana, J., Aranda-Gómez, J., Morán-Zenteno, D.J., Alanís-Alvarez, S., and Nieto-Samaniego, A., 1992, Carta geológica de la República Mexicana, escala 1:2,000,000, quinta edición: Secretaría de Minas e Industria Par-aestatal, Consejo de Recursos Minerales y Universidad Nacional Autónoma de México, Instituto de Geología.

Ortega-Rivera, A., 2003, Geochronological constraints on the tectonic history of the Peninsular Ranges batholith of Alta and Baja California: Tectonic Implications for western México, *in* Johnson, S.E., Paterson, S.R., Fletcher, J.M., Girty, G.H., Kimbrough, D.L., and Martín-Barajas, A., eds., Tectonic evolution of northwestern México and the southwestern USA: Boulder, Colorado, Geological Society of America Special Paper 374, p. 297–335 (this volume).

Ortega-Rivera, A., Farrar, E., Hanes, J.A., Archibald, D.A., Gastil, R.G., Kimbrough, D.L., Zentilli, M., Lopez, M.M., Feraud, G., and Ruffet, G., 1997, Chronological constraints on the thermal and tilting history of the Sierra San Pedro Martir Pluton, Baja California, Mexico, from U/Pb, $^{40}Ar/^{39}Ar$, and fission-track geochronology: Geological Society of America Bulletin, v. 109, p. 728–745.

Parrish, K.E., 1990, Geology, petrology, geochemistry, and isotopic character of the Indian Hill pluton, Jacumba Mountains [M.S. thesis]: San Diego, San Diego State University, 136 p.

Parsons, I., Brown, W.L., and Smith J.V., 1999, $^{40}Ar/^{39}Ar$ thermochronology using alkali feldspars: Real thermal history or mathematical mirage of microtexture?: Contributions to Mineralogy and Petrology, v. 136, p. 92–110.

Peterson, G.L., and Nordstrom, C.E., 1970, Sub-La Jolla unconformity in vicinity of San Diego, California: American Association of Petroleum Geologists Bulletin, v. 54, p. 265–274.

Phillips, E., 1964, Coaxial refolding in southeast San Diego County, California [M.S. thesis], San Diego, San Diego State College, 30 p.

Premo, W., Morton, D.M., Snee, L., Naeser, N.D., and Fanning, C.M., 1998, Isotopic ages, cooling histories, and magmatic origins for Mesozoic tonalitic plutons from the N. Peninsular Ranges batholith, S. California: Geological Society of America Abstracts with Programs, v. 30, no. 5, p. 59–60.

Press, W.H., Flannery, B.P., Teukolsky, S.A., and Vetterling, W.T., 1986, Numerical recipes: The art of scientific computing: Cambridge University Press, New York, 818 p.

Quidelleur, X., Grove, M., Harrison, T.M., and Yin, A., 1997, Thermal evolution and slip history of the Renbu Zedong thrust, southeastern Tibet: Journal of Geophysical Research, v. 102, p. 2659–2679.

Rothstein, D.A., and Manning, C.E., 2003. Geothermal gradients in continental magmatic arcs: Constraints from the eastern Peninsular Ranges batholith, Baja California, México, in Johnson, S.E., Paterson, S.R., Fletcher, J.M., Girty, G.H., Kimbrough, D.L., and Martín-Barajas, A., eds., Tectonic evolution of northwestern México and the southwestern USA: Boulder, Colorado, Geological Society of America Special Paper 374, p. 337–354 (this volume).

Schultejann, P.A., 1984, The Yaqui Ridge antiform and detachment fault: Mid-Cenozoic extensional terrane west of the San Andreas fault: Tectonics, v. 3, no. 6, p. 677–691.

Sharp, R.V., 1967, San Jacinto fault zone in the Peninsular Ranges of Southern California: Geological Society of America Bulletin, v. 78, p. 705 729.

Sharp, R.V., 1979, Some characteristics of the eastern Peninsular Ranges mylonite zone, in Proceedings, Conference VIII, Analysis of actual fault zones in bedrock: U.S. Geological Survey Open File Report 79-1239, p. 258–267.

Shaw, S.E., Todd, V.R., and Grove, M., 2003, Jurassic peraluminous gneissic granites in the axial zone of the Peninsular Ranges, southern California, in Johnson, S.E., Paterson, S.R., Fletcher, J.M., Girty, G.H., Kimbrough, D.L., and Martín-Barajas, A., eds., Tectonic evolution of northwestern México and the southwestern USA: Boulder, Colorado, Geological Society of America Special Paper 374, p. 157–183(this volume).

Siem, M.E., and Gastil, R.G., 1994, Mid-Tertiary to Holocene extension associated with the development of the Sierra El Mayor metamorphic core complex, northeastern Baja California, Mexico, in McGill, S.F., and Ross, T.M., Geological investigations of an active margin, GSA Cordilleran Section Guidebook: Redlands, California, San Bernardino County Museum Association, p. 107–119.

Silver, L.T., and Chappell, B., 1988, The Peninsular Ranges batholith: An insight into the evolution of the Cordilleran batholiths of southwestern North America: Transactions of the Royal Society of Edinburgh: Earth Sciences, v. 79, p. 105–121.

Silver, L. T, Taylor, H.P., Jr., and Chappell, B., 1979, Some petrological, geochemical and geochronological observations of the Peninsular Ranges batholith near the International Border of the U.S.A., and Mexico, in Abbott, P.L., and Todd, V.R., eds., Mesozoic Crystalline Rocks—Peninsular Ranges Batholith and Pegmatites, Point Sal Ophiolite: Geological Society of America Annual Meeting Guidebook: San Diego, San Diego State University, p. 83–110.

Silver, L.T., Chappell, B.C., and Anderson, T.H., 1996, Petrogenetic implications of petrographic and geochemical zonations in the integrated Peninsular Ranges and Sonoran batholiths: Geological Society of America Abstracts with Programs, v. 28, no. 5, p. 112.

Simpson, C., 1984, Borrego Springs–Santa Rosa mylonite zone—A Late Cretaceous west-directed thrust in southern California: Geology, v. 12, p. 8–11.

Staude, J.M.G., and Barton, M.D., 2001, Jurassic to Holocene tectonics, magmatism, and metallogeny of northwestern Mexico: Geological Society of America Bulletin, v. 113, p. 1357–1374.

Steiger, R.H., and Jäger, E., 1977, Subcommission on geochronology: Convention on the use of decay constants in geo- and cosmochronology: Earth and Planetary Science Letters, v. 36, p. 359–362.

Strand, R.G., 1962, Geologic map of California, San Diego–El Centro sheet, Olaf P. Jenkins edition: Sacremento, California Division of Mines and Geology, scale 1:250,000.

Sundberg, F.A., and Cooper, J.D., 1978, Late Cretaceous depositional environments, northern Santa Ana Mountains, Southern California, in Howell, D.G., and McDougall, K.A., eds., Mesozoic paleogeography of the western United States: Pacific Coast Paleogeography Symposium 2,: Los Angeles, Society of Economic Paleontologists and Mineralogists Pacific Section, , p. 535–546.

Taylor, H.P., 1986, Igneous rocks; 2, Isotopic case studies of Circum-Pacific magmatism, in Valley, J.W., Taylor, H.P., Jr., and O'Neil, J.R., eds., Stable isotopes in high-temperature geological processes: Reviews in Mineralogy, v. 16 p. 273–317.

Theodore, T.G., 1970, Petrogenesis of mylonites of high metamorphic grade in the Peninsular Ranges of southern California: Geological Society of America Bulletin, v. 81, p. 435–450.

Thomson, C.N., and Girty, G.H., 1994, Early Cretaceous intra-arc ductile strain in Triassic-Jurassic and Cretaceous continental margin arc rocks, Peninsular Ranges, California: Tectonics, v. 13, p. 1108 1119.

Tobisch, O.T., Saleeby, J., Renne, P.R., McNulty, B.A., and Tong, W., 1995, Variations in deformation fields during emplacement of a large-volume magmatic arc, central Sierra Nevada, California: Geological Society of America Bulletin, v. 107, p. 148–166.

Todd, V.R., 1977a, Geologic map of the Agua Caliente Springs Quadrangle, San Diego County, California: U.S. Geological Survey Open-File Report OF 77-0742, 20 p.

Todd, V.R., 1977b, Geologic map of the Cuyamaca Peak quadrangle, San Diego County, California: U.S. Geological Survey Open File Report 77-405, 13 p.

Todd, V.R., 1978, Geologic map of the Monument Peak quadrangle, San Diego County, California: U.S. Geological Survey Open File Report 78-697, 47 p.

Todd, V.R., 1979, Geologic map of the Mount Laguna quadrangle, San Diego County, California: U.S. Geological Survey Open File Report 79-862, 49 p.

Todd, V.R., and Shaw, S.E., 1979, Structural, metamorphic and intrusive framework of the Peninsular Ranges batholith in southern San Diego County, California, in Abbott, P.L., and Todd, V.R., eds., Mesozoic crystalline rocks—Peninsular Ranges batholith and pegmatites, Point Sal Ophiolite, Geological Society of America Annual Meeting Guidebook: San Diego, San Diego State University, p. 177–231.

Todd, V.R., and Shaw, S.E., 1985, S-type granitoids and an I-S line in the Peninsular Ranges batholith, southern California: Geology, v. 13, p. 231–233.

Todd, V.R., Shaw, S.E., and Hammarstrom, J.M., 2003, Cretaceous plutons of the Peninsular Ranges batholith, San Diego and westernmost Imperial Counties, California: Intrusion across a Late Jurassic continental margin, in Johnson, S.E., Paterson, S.R., Fletcher, J.M., Girty, G.H., Kimbrough, D.L., and Martín-Barajas, A., eds., Tectonic evolution of northwestern México and the southwestern USA: Boulder, Colorado, Geological Society of America Special Paper 374, p. 185–235 (this volume).

Todd, V.R., Erskine, B.G., and Morton, D.M., 1988, Metamorphic and tectonic evolution of the northern Peninsular Ranges batholith, southern California, in Ernst, W.G., ed., Metamorphism and Crustal Evolution of the Western United States, Rubey Volume VII: Englewood Cliffs, New Jersey, Prentice-Hall, p. 894–937.

Walawender, M.J., Gastil, R.G., Clinkenbeard, J.P., McCormick, W.V., Eastman, B.G., Wernicke, R.S., Wardlaw, M.S., Gunn, S.H., and Smith, B.M., 1990, Origin and evolution of the zoned La Posta-type plutons, eastern Peninsular Ranges batholith, southern and Baja California, in Anderson, J.L., ed., The nature and origin of Cordilleran magmatism: Geological Society of America Memoir 174, p. 1–18.

Walawender, M.J., Girty, G.H., Lombardi, M.R., Kimbrough, D., Girty, M.S., and Anderson, C., 1991, A synthesis of recent work in the Peninsular Ranges batholith: Geological Society of America Annual Meeting Guidebook for Field Trips, p. 297–312.

Wenk, H.-R., Johnson, L.R., and Ratschbacher, L., 2000, Pseudotachylites in the eastern Peninsular Ranges of California: Tectonophysics, v. 321, p. 253 277.

Wolf, R.A., Farley, K.A., and Silver, L.T., 1997, Assessment of (U-Th)/He thermochronometry; the low-temperature history of the San Jacinto Mountains, California: Geology, v. 25, p. 65 68.

MANUSCRIPT ACCEPTED BY THE SOCIETY JUNE 2, 2003

Geological Society of America
Special Paper 374
2003

Temporal and spatial trends of Late Cretaceous–early Tertiary underplating of Pelona and related schist beneath southern California and southwestern Arizona

Marty Grove
*Department of Earth and Space Sciences, University of California, 595 Charles Young Drive E,
Los Angeles, California 90095-1567, USA*

Carl E. Jacobson
Department of Geological and Atmospheric Sciences, Iowa State University, Ames, Iowa 50011-3212, USA

Andrew P. Barth
Department of Geology, Indiana University–Purdue University, Indianapolis, Indiana 46202-5132, USA

Ana Vucic
Department of Geological and Atmospheric Sciences, Iowa State University, Ames, Iowa 50011-3212, USA

ABSTRACT

The Pelona, Orocopia, and Rand Schists and the schists of Portal Ridge and Sierra de Salinas constitute a high–pressure/temperature terrane that was accreted beneath North American basement in Late Cretaceous–earliest Tertiary time. The schists crop out in a belt extending from the southern Coast Ranges through the Mojave Desert, central Transverse Ranges, southeastern California, and southwestern Arizona. Ion microprobe U-Pb results from 850 detrital zircons from 40 metagraywackes demonstrates a Late Cretaceous to earliest Tertiary depositional age for the sedimentary part of the schist's protolith. About 40% of the $^{206}Pb/^{238}U$ spot ages are Late Cretaceous. The youngest detrital zircon ages and post-metamorphic mica $^{40}Ar/^{39}Ar$ cooling ages bracket when the schist's graywacke protolith was eroded from its source region, deposited, underthrust, accreted, and metamorphosed. This interval averages 13 ± 10 m.y. but locally is too short (<~3 m.y.) to be resolved with our methods. The timing of accretion decreases systematically (in palinspastically restored coordinates) from about 91 ± 1 Ma in the southwesternmost Sierra Nevada (San Emigdio Mountains) to 48 ± 5 Ma in southwest Arizona (Neversweat Ridge). Our results indicate two distinct source regions: (1) The Rand Schist and schists of Portal Ridge and Sierra de Salinas were derived from material eroded from Early to early Late Cretaceous basement (like the Sierra Nevada batholith); and (2) The Orocopia Schist was derived from a heterogeneous assemblage of Proterozoic, Triassic, Jurassic, and latest Cretaceous to earliest Tertiary crystalline rocks (such as basement in the Mojave/Transverse Ranges/southwest Arizona/northern Sonora). The Pelona Schist is transitional between the two.

Keywords: Pelona Schist, Orocopia Schist, Rand Schist, detrital mineral, zircon, U-Pb.

Grove, M., Jacobson, C.E., Barth, A.P., and Vucic, A., 2003, Temporal and spatial trends of Late Cretaceous–early Tertiary underplating of Pelona and related schist beneath southern California and southwestern Arizona, *in* Johnson, S.E., Paterson, S.R., Fletcher, J.M., Girty, G.H., Kimbrough, D.L., and Martín-Barajas, A., eds., Tectonic evolution of northwestern México and the southwestern USA: Boulder, Colorado, Geological Society of America Special Paper 374, p. 381–406.

INTRODUCTION

The dramatic Laramide craton-ward shift of arc magmatism and contractional deformation was arguably the most important development in the Late Cretaceous–early Tertiary evolution of western North America (Coney and Reynolds, 1977; Dickinson and Snyder, 1978; Bird, 1988). In southern California and adjacent areas, the Pelona, Orocopia, and Rand Schists and the schists of Portal Ridge and Sierra de Salinas have been widely considered to be a product of this event (e.g., Burchfiel and Davis, 1981; Crowell, 1981; Hamilton, 1987, 1988). Referred to collectively in this paper as the "schists," these rocks comprise a distinctive eugeoclinal assemblage that was metamorphosed from epidote-blueschist to oligoclase–amphibolite facies at pressures of ~700–1000 MPa (Fig. 1; Ehlig, 1958, 1968, 1981; Crowell, 1968, 1981; Yeats, 1968; Haxel and Dillon, 1978; Jacobson, 1983a, 1995; Graham and Powell, 1984; Jacobson et al., 1988, 1996, 2002a; Dillon et al., 1990; Malin et al., 1995; Wood and Saleeby, 1997; Haxel et al., 2002). The schists are juxtaposed beneath an upper plate of Precambrian to Mesozoic igneous and metamorphic rocks, with essentially no evidence of intervening lithospheric mantle. They underlie a series of low-angle faults known as the Vincent–Chocolate Mountains fault system. A few of these faults may be remnants of the thrust(s) beneath which the schists underwent prograde metamorphism (Ehlig, 1981; Jacobson, 1983b, 1997), but most are retrograde or postmetamorphic features responsible for exhumation of the schists (Haxel et al., 1985, 2002; Silver and Nourse, 1986; Jacobson et al., 1988, 1996, 2002a; Malin et al., 1995; Wood and Saleeby, 1997; Yin, 2002).

Many Cordilleran geologists have viewed the schists as accretionary rocks analogous to the Franciscan Complex that were underthrust beneath southern California during low-angle, northeast-dipping subduction of the Farallon plate (Fig. 2A; "subduction" model; Crowell, 1968, 1981; Yeats, 1968; Burchfiel and Davis, 1981; Hamilton, 1987, 1988; Jacobson et al., 1996; Malin et al., 1995; Wood and Saleeby, 1997; Yin, 2002). In fact, a correlation between the Pelona Schist and the Franciscan Complex was first suggested on the basis of lithologic and petrologic similarities even before the advent of plate tectonic theory (Ehlig, 1958; Woodford, 1960). The subduction model requires that lowermost North American continental crust and underlying mantle lithosphere were tectonically eroded by shallow subduction of the Farallon slab prior to accretion of the schist.

Hall (1991), Barth and Schneiderman (1996), and Saleeby (1997) have presented an alternative hypothesis in which the schist is considered to be a correlative of the Great Valley Group that was underthrust beneath a northeast-dipping fault between arc and forearc (Fig. 2B; "forearc" model). As with the subduction model, the forearc interpretation calls upon low-angle subduction of the northeast-dipping Farallon plate during the Laramide orogeny as the ultimate driving force for emplacement of the schists. However, in the forearc model, the Vincent–Chocolate Mountains thrust is viewed as a structure *within* the overriding North American plate rather than comprising the

plate boundary. As such, it does not require wholesale removal of lithospheric mantle, consistent with the modern example of the Andes, where South American mantle lithosphere is only partly thinned above the flat Nazca plate (Smalley and Isacks, 1987; Isacks, 1988; Allmendinger et al., 1990).

Additional models for the origin of the schist do not attempt to relate it to Laramide deformation. These models were motivated by early structural studies that indicated top-to-northeast sense of shear along the Chocolate Mountains fault in southeastern California (Dillon, 1976; Haxel, 1977). While it is now generally accepted that this shear sense was imparted during exhumation of the schist (Simpson, 1990; Jacobson et al., 1996, 2002a; Oyarzabal et al., 1997; Haxel et al., 2002; Yin, 2002), two distinct models involving southwest-dipping thrusts were born out of the early work. One of these that held that the schists formed during collision between North America and an exotic microcontinent (Fig. 2C; "Collision" model; Haxel and Dillon, 1978; Vedder et al., 1983) quickly lost favor as it became apparent that the proposed allochthon exhibited clear North American affinities (Haxel et al., 1985; Barth, 1990; Tosdal, 1990; Bender et al., 1993). A second interpretation has persisted, however. It postulates that the schist's protolith was deposited during rifting in a backarc setting within the North American craton (Haxel and Dillon, 1978; Ehlig, 1981). Haxel and Tosdal (1986) went a step further in suggesting that the backarc basin formed during activity along the proposed Mojave–Sonora Megashear of Silver and Anderson (1974). Overall, the backarc model has proven difficult to reconcile with the geology of southern California and western Arizona in that independent evidence for either the opening of the necessary rift basin or the suture that would have formed when it was closed is lacking (Burchfiel and Davis, 1981; Crowell, 1981; Hamilton, 1987, 1988). Very recently, Haxel et al. (2002) dealt with the apparent absence of a suture by proposing that all remnants of the closed backarc basin and the suture itself were restricted to a tectonic window beneath counterposed thrust faults (Fig. 2D). This model is similar to one that is used to explain concealment of the suture between India and Asia in southern Tibet (Yin et al., 1994; Harrison et al., 2000).

A major consideration in evaluating the validity of the models described above is provided by reports of pre-Late Cretaceous U-Pb ages from igneous bodies inferred to crosscut the schist (Mukasa et al., 1984; James, 1986; James and Mattinson, 1988). In particular, Mukasa et al.'s (1984) determination of a 163 Ma crystallization age for a mafic metadiorite within Orocopia Schist of the Chocolate Mountains has been widely cited as indicating a pre–Late Jurassic depositional age for the schist's protolith (Mattinson and James, 1985; Drobeck et al., 1986; Reynolds, 1988; Dillon et al., 1990; Grubensky and Bagby, 1990; Simpson, 1990; Powell, 1993; Richard, 1993; Sherrod and Hughes, 1993; Barth and Schneiderman, 1996; Miller et al., 1996; Robinson and Frost, 1996; Richard et al., 2000). An entirely pre–Late Jurassic depositional age for the schist's protolith would strongly disfavor the subduction and forearc models since they identify either the Late Cretaceous subduction trench or forearc basin as the source of graywacke sediment that was underthrust during Laramide shal-

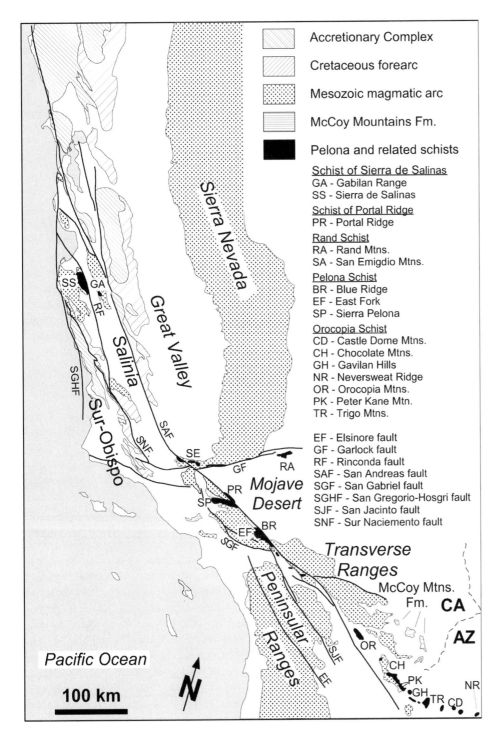

Accretionary Complex

Cretaceous forearc

Mesozoic magmatic arc

McCoy Mountains Fm.

Pelona and related schists

<u>Schist of Sierra de Salinas</u>
GA - Gabilan Range
SS - Sierra de Salinas
<u>Schist of Portal Ridge</u>
PR - Portal Ridge
<u>Rand Schist</u>
RA - Rand Mtns.
SA - San Emigdio Mtns.
<u>Pelona Schist</u>
BR - Blue Ridge
EF - East Fork
SP - Sierra Pelona
<u>Orocopia Schist</u>
CD - Castle Dome Mtns.
CH - Chocolate Mtns.
GH - Gavilan Hills
NR - Neversweat Ridge
OR - Orocopia Mtns.
PK - Peter Kane Mtn.
TR - Trigo Mtns.

EF - Elsinore fault
GF - Garlock fault
RF - Rinconda fault
SAF - San Andreas fault
SGF - San Gabriel fault
SGHF - San Gregorio-Hosgri fault
SJF - San Jacinto fault
SNF - Sur Naciemento fault

Figure 1. Schematic geologic map of central and southern California and southwesternmost Arizona showing distribution of Pelona and related schists. Note that simple geometry of medial Cretaceous Sierran arc, Great Valley forearc basin, and subduction complex in central California is highly disrupted within southern Coast Ranges and Transverse Ranges of southern California; AZ—Arizona, CA—California. Simplified from Jennings (1977).

low subduction. Alternatively, Haxel and Tosdal (1986) pointed out that the backarc model could agree well with a pre–Late Jurassic depositional age if the rift basin formed in response to activity along the proposed Mojave–Sonora Megashear of Silver and Anderson (1974).

To test the validity of Mukasa et al.'s (1984), James' (1986), and James and Mattinson's (1988) findings for the schist terrane in its entirety (Fig. 1), we undertook a regionally comprehensive U-Pb age analysis of detrital zircons from the schist (see also Jacobson et al. [2000] and Barth et al. [2003a]). Provided that U-Pb systematics in the detrital zircons were not disturbed by metamorphism, the youngest U-Pb zircon ages place an upper limit upon the depositional age while the overall age distribution yields a measure of provenance. Earlier reconnaissance

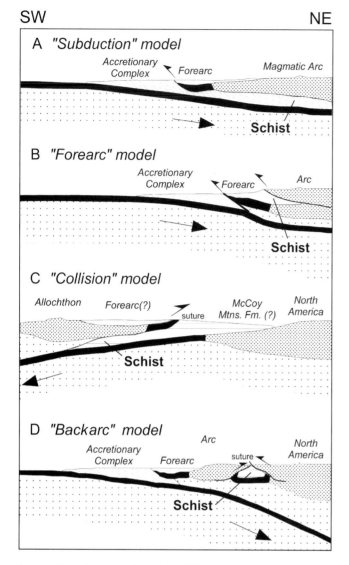

SW   NE

A *"Subduction" model*

Accretionary
Complex   Forearc   Magmatic Arc

Schist

B *"Forearc" model*

Accretionary
Complex   Forearc   Arc

Schist

C *"Collision" model*

Allochthon   Forearc(?)   McCoy   North
                            suture   Mtns. Fm. (?) America

Schist

D *"Backarc" model*

Accretionary   Arc   North
Complex   Forearc   suture   America

Schist

Figure 2 Tectonic models for origin of Pelona and related schists. A: In subduction model, schist's graywacke protolith is deposited in trench, subducted, and accreted beneath tectonically eroded cratonal rocks. Lithospheric mantle has been completely removed by this process. B: In forearc model, sediments in forearc basin are overthrust by magmatic arc; C: In collision model, schist's protolith is overthrust by a northeast-directed allochthon. D: In backarc model, protolith of schist originates in a backarc basin that is subsequently overthrust and closed.

multigrain U-Pb analysis of zircons from the schist performed by Silver et al. (1984) and James and Mattinson (1988) were equivocal with respect to the depositional age of the graywacke protolith. In contrast, our initial ion microprobe results clearly indicated the presence of Late Cretaceous detritus (Jacobson et al., 2000; Barth et al., 2003a). In this paper, we present a much more comprehensive data set involving 40 samples that represent virtually all of the major schist exposures. This compilation, which includes six samples from Jacobson et al. (2000) and Barth et al. (2003a), confirms a Late Cretaceous to earliest Tertiary

depositional age for the graywacke protolith for all of the schists and reveal systematic regional variations in the depositional age and provenance of the protolith.

When combined with $^{40}Ar/^{39}Ar$ cooling ages, our detrital zircon U-Pb data can be used to constrain the "cycling interval" during which the schist's graywacke protolith was eroded from its source region, deposited, underthrust, accreted, and metamorphosed beneath crystalline rocks of southwestern North America. Combined with previously obtained data, new muscovite and biotite $^{40}Ar/^{39}Ar$ cooling ages from our detrital zircon samples clearly indicate that the cycling interval for erosion, deposition, underthrusting, accretion, and metamorphism varied systematically across the schist terrane. Schist located outboard of the craton in the southern Coast Ranges and northwestern Mojave Desert was emplaced well before schists were underplated beneath more cratonal (and progressively more southeasterly) positions in the Transverse Ranges and southeastern California and southwestern Arizona. This spatial-temporal relationship and the overall Laramide depositional age for the schist's graywacke protolith must be explained by any model that seeks to describe the underplating of the schists beneath southwest North America.

CORRELATION OF THE SCHISTS

Early attempts to correlate the Pelona, Orocopia, Rand, Portal Ridge, and Sierra de Salinas Schists (Fig. 1) were based on their field and petrographic characteristics (Ehlig, 1958, 1968). A key observation is that relative proportions of the major rock types are broadly similar in all areas (~90–99% metagraywacke, 1–10% mid-ocean ridge basalt (MORB) metabasite, and trace amounts of Fe-Mn metachert, marble, and serpentinite; Haxel and Dillon, 1978; Haxel et al., 1987, 2002; Dawson and Jacobson, 1989). Moreover, many of the schist bodies exhibit inverted metamorphic zonation; i.e., peak temperature of metamorphism decreases structurally downward (Ehlig, 1958; Graham and England, 1976; Jacobson, 1983b, 1995). In addition, the schists share a number of diagnostic mineralogic features (Haxel and Dillon, 1978): (1) poikiloblasts of sodic plagioclase in metagraywacke that appear gray to black due to inclusions of graphite; (2) local centimeter-scale clots of fuchsite in metagraywacke; (3) centimeter- to meter-scale lenses of talc-actinolite rock in metagraywacke; (4) sodic-calcic amphibole (winchite to barroisite) in low-grade metabasites; and (5) spessartine garnet, piemontite, stipnomelane, and sodic amphibole in Fe-Mn metachert.

The approximate early Tertiary positions of schist outcrops are indicated in Figure 3. Correlation among the various schists has been most readily accepted for the Pelona and Orocopia Schists, which were in close proximity prior to offset on the San Andreas system (Fig. 3; James and Mattinson, 1988; Dillon et al., 1990). Less well established is the relationship of the Pelona–Orocopia schists to the Rand Schist, schist of Portal Ridge, and schist of Sierra de Salinas.

While the Rand Schist exhibits all the key lithologic attributes of the Pelona and Orocopia Schists (Jacobson, 1995), it

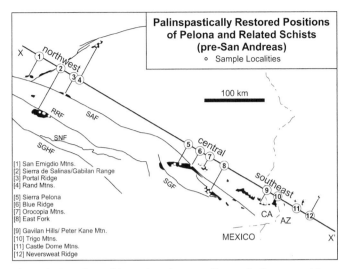

Figure 3. Sample positions plotted on a palinspastically restored base map showing pre–Late Tertiary positions of Pelona and related schist exposures of southern California and southwestern Arizona. Appendix 1 describes how this base map was constructed. Distances along projection line are used in Figure 7. Numbers (1–12) refer to geographic areas sampled in this study (see Table 1). Northwest, central, and southeast refer to three major groupings discussed in text. Abbreviations of fault names are same as in Figure 1.

differs in two respects. Some Rand metagraywackes are more aluminous than any bulk compositions observed in the Pelona and Orocopia Schists (Jacobson et al., 1988). Moreover, as indicated by initial radiometric dating (Kistler and Peterman, 1978; Silver and Nourse, 1986; Jacobson, 1990) and confirmed here, the Rand Schist is older than the Pelona and Orocopia Schists. This age difference and the spatial separation prior to offset on the San Andreas system (Fig. 3) make it difficult to know the exact paleogeographic relationships between the northern and southern schist bodies.

The schists of Sierra de Salinas and Portal Ridge have been correlated with each other on the basis of lithologic similarity and close proximity prior to slip on the San Andreas system (Ross, 1976). While a correlation to the Pelona, Orocopia, and Rand Schists was deemed unlikely by James and Mattinson (1988), other workers (e.g., Haxel and Dillon, 1978) concluded that the schists of Sierra de Salinas and Portal Ridge were simply higher-grade lithologic equivalents of the other schists. Both the schists of Sierra de Salinas and Portal Ridge were metamorphosed in the middle to upper amphibolite facies (Evans, 1966; Ross, 1976). Because they generally lack graphitic plagioclase poikiloblasts and possess a high ratio of biotite to muscovite, the schists of Sierra de Salinas and Portal Ridge typically appear quite different in outcrop than lower-grade schists found elsewhere. While we have not examined the schists of Sierra de Salinas in detail, our work in the Portal Ridge area has revealed the same talc-actinolite lenses, piemontite-bearing metachert, and other features indicative of the schist's eugeoclinal protolith. Hence, we see no reason that the schists of

Sierra de Salinas and Portal Ridge should not be lithologically correlated with the Pelona, Orocopia, and Rand Schists.

METHODS

We sampled metagraywacke from most of the significant schist exposures in southern California and southwestern Arizona (Fig. 3, Table 1; see also Jacobson et al., 2000; Barth et al., 2003a). Brief descriptions of each of the 12 sampling areas are provided in Appendix 2. The sample suite represent the range of geologic attributes (metamorphic grade, structural position, lithology) exhibited by the metagraywacke throughout the region (Table 1). In addition, many samples were selected from outcrops containing metabasite. Contacts between metagraywacke and metabasite could originally have been either tectonic or depositional. In the latter case, graywacke spatially associated with metabasite potentially represents the lowermost and oldest part of the stratigraphic section. Details of the U-Pb and $^{40}Ar/^{39}Ar$ analysis we have employed are provided in Appendices 3 and 4, respectively.

Ion Microprobe U-Pb

The efficiency of the ion microprobe (high analysis throughput with no wet chemical separation of Pb, U, and Th required) and its ability to provide high spatial resolution have resulted in the extensive use of this instrument in detrital zircon U-Pb studies over the past 20 years (Froude et al., 1983; Ireland, 1992; Maas et al., 1992; Lee et al., 1997; Miller et al., 1998; DeGraaff Surpless et al., 2002). Because small amounts of zircon are analyzed (< 1 ng), ion microprobe results are less prone (relative to isotope dilution methods) to be biased by Pb-enriched restitic regions that are likely to be present within some of our grains (e.g., Silver et al., 1984; James and Mattinson, 1988). Reduced precision accompanies smaller sample size, however, and it is generally difficult or impossible to convincingly demonstrate U-Pb concordance of spot analyses from relatively U-poor and/or young zircons. This difficulty applies to the majority of U-Pb results we produced in this study. Only Cretaceous grains with uncharacteristically high U abundances (higher than about 1500 ppm) yield sufficiently precise $^{207}Pb/^{235}U$ results that U-Pb concordance can be convincingly assessed. In most instances, our interpretations are based solely upon the more precisely determined $^{206}Pb/^{238}U$ ages.

Interpretation of detrital zircon U-Pb results from metamorphosed rocks depend significantly upon the extent to which grains examined are affected by Pb loss and/or recrystallization effects. Recent high-temperature experimental diffusion studies have indicated that Pb should be essentially immobile in crystalline zircon under crustal conditions (e.g., Cherniak and Watson, 2000). Problems arise when significant radiation damage or metamorphic zircon growth has occurred. Poorly crystalline zircon that has interacted with fluids produced at metamorphic conditions relevant to the schists is known to be prone to significant Pb loss and/or recrystallization (Pidgeon et al., 1966; Sinha et al.,

TABLE 1. SAMPLE INFORMATION

Sample	Locality*	Type[†]	Area[§]	Group[#]	UTM-x[****]	UTM-y[****]	Grade[††]	Level[§§]	Mafic[##]	Plutons[***]
JM-80-102[†††]	Gablian Range	SS	1	NW	-	-	A	-	-	X
DR-1656	Sierra de Salinas	SS	1	NW	-	-	A	-	-	X
02-342	Sierra de Salinas	SS	1	NW	640.964	4028.026	A	-	-	X
02-345	Sierra de Salinas	SS	1	NW	640.833	4027.773	A	-	-	X
02-358	Sierra de Salinas	SS	1	NW	637.827	4030.154	A	-	-	X
99-57[§§§]	San Emigdio Mtns.	RS	2	NW	307.173	3860.678	A	-	-	-
PR150	Portal Ridge	PRS	3	NW	386.955	3830.870	A	-	-	-
RA169	Rand Mtns.	RS	4	NW	435.621	3912.289	EB	low	-	-
RA170	Rand Mtns.	RS	4	NW	431.438	3909.618	AEA	high	-	X
98-241	Sierra Pelona	PS	5	C	367.073	3821.369	GS	Int.	-	-
SP25D	Sierra Pelona	PS	5	C	371.584	3825.297	GS	low	X	-
SP10	Sierra Pelona	PS	5	C	369.567	3822.959	GS	low	X	-
98-240	Blue Ridge	PS	6	C	431.487	3803.663	AEA	-	-	-
OR307	Orocopia Mtns.	OS	7	C	610.924	3717.529	LA	high	X	-
OR77B	Orocopia Mtns.	OS	7	C	613.409	3714.761	LA	high	X	-
OR312A	Orocopia Mtns.	OS	7	C	610.099	3716.481	AEA	low	-	-
OR113	Orocopia Mtns.	OS	7	C	610.846	3713.131	AEA	low	-	-
OR314	Orocopia Mtns.	OS	7	C	609.938	3716.171	AEA	low	-	-
OR15A	Orocopia Mtns.	OS	7	C	605.971	3717.226	AEA	low	-	-
OR337	Orocopia Mtns.	OS	7	C	600.624	3720.286	AEA	low	-	-
KE224031	Orocopia Mtns.	OS	7	C	614.684	3708.983	AEA	-	-	-
KE24033	Orocopia Mtns.	OS	7	C	616.865	3711.985	AEA	-	-	-
KE127021	Orocopia Mtns.	OS	7	C	617.253	3712.830	AEA	-	-	-
SG69	East Fork	PS	8	C	431.551	3793.948	UGS	high	-	-
98-237	East Fork	PS	8	C	431.641	3802.224	UGS	high	X	-
SG532	East Fork	PS	8	C	433.398	3800.745	GS	Int.	-	-
SG533	East Fork	PS	8	C	444.292	3795.208	LGS	low	-	-
PK114B	Peter Kane Mtns.	OS	9	SE	703.878	3659.663	LA	high	-	X
UG1417A	Gavilan Hills	OS	9	SE	710.524	3653.248	LA	low	-	-
UG-1500	Gavilan Hills	OS	9	SE	712.678	3653.280	LA	low	X	-
YN17	Yellow Narrows	OS	10	SE	720.800	3653.598	AEA	-	-	-
MW10	Marcus Wash	OS	10	SE	725.790	3654.909	AEA	-	-	-
TR23[###]	Trigo Mtns.	OS	10	SE	726.889	3658.508	AEA	-	-	-
TR24A[###]	Trigo Mtns.	OS	10	SE	729.928	3658.508	AEA	-	-	-
CD7	Castle Dome Mtns.	OS	11	SE	765.922	3658.934	LA	-	-	X
CD9	Castle Dome Mtns.	OS	11	SE	765.636	3659.319	LA	-	-	X
KE4	Castle Dome Mtns.	OS	11	SE	776.537	3656.621	LA	-	-	-
KE6	Castle Dome Mtns.	OS	11	SE	775.745	3655.262	LA	-	-	-
NR1	Neversweat Ridge	OS	12	SE	240.274	3662.994	LA	-	-	X
NR2	Neversweat Ridge	OS	12	SE	240.443	3663.413	LA	-	-	X

*See Appendix 1 for additional information.
[†] SS—schist of Sierra de Salinas; RS—Rand Schist; PRS—schist of Portal Ridge; PS—Pelona Schist; OS—Orocopia Schist.
[§] 12 area subdivisions employed in text.
[#] NW—northwest, C—central, SE—southeast.
[††] Metamorphic grade: EB—epidote blueschist; GS—greenschist, AEA—albite epidote amphibolite, A—amphibolite (L—lower, U—upper).
[§§] Structural level: low—structurally deep; int.—structurally intermediate; high—near upper low-angle fault contact.
[##] "x" denotes sample collected adjacent to mafic schist and/or metachert.
[***] "x" denotes potential contact metamorphism due to nearby igneous bodies.
[†††] Sample studied by James and Mattinson (1988).
[§§§] Locality studied by James (1986).
[###] Field relations of metadiorite are similar to body studied by Mukasa et al. (1984).
[****] UTM coordinates are zone 11 (except Gablian Range and Sierra de Salinas which are zone 10 and Neversweat Rdge which is zone 12).

1992; Mezger and Krogstad, 1997; Geisler et al., 2001; Högdahl et al., 2001). In general, the extent of radiation damage depends most strongly upon uranium content and age. Hence, there are obvious tests for such problematic behavior. If, for example, zircons yielding the youngest U-Pb ages were systematically found to possess the highest U-contents, then Pb loss would be identified as a potentially serious problem. Another useful test is to examine the relationship between U-Pb age and Th/U. Metamorphic overgrowths on igneous zircons are often characterized by very low Th/U (Kröner et al., 1994; Carson et al., 2002; Mojzsis and Harrison, 2002; Rubatto, 2002). Consequently, if the youngest detrital zircon U-Pb ages were highly correlated with anomalously low Th/U values, then growth of metamorphic zircon should be suspected to have played a prominent role in producing these young ages.

^{40}Ar/^{39}Ar

Muscovite and/or biotite ages were obtained from most of the detrital zircon samples (Table 1). Because the schist was so extensively recrystallized at upper greenschist to amphibolite facies conditions, it is extremely unlikely that any detrital micas survived metamorphism. Available petrographic and compositional data from the phengitic white mica and biotite present within the schist confirm that they are metamorphic phases (Jacobson, 1983b, 1990, 1995, 1997; Jacobson et al., 1988, 2002a). When transient heating is unimportant and incorporation of excess radiogenic ^{40}Ar is not a factor, ^{40}Ar/^{39}Ar ages from metamorphic micas will record post-metamorphic cooling through their bulk closure temperature (McDougall and Harrison, 1999). Under these conditions, the cycling interval during which the schist's graywacke protolith was eroded from its source region, deposited, underthrust, accreted, and metamorphosed will be bracketed between the youngest detrital U-Pb zircon ages and post-metamorphic ^{40}Ar/^{39}Ar cooling ages recorded by the micas.

Because age spectra yielded by biotite and, to a lesser extent, muscovite are degraded by the dehydroxylation processes during *in vacuo* step-heating (McDougall and Harrison, 1999), thermal history information from micas is most reliably obtained from total gas ^{40}Ar/^{39}Ar ages that can be related to bulk closure temperatures. Although the latter are not well determined, available empirical and experimental data indicate bulk closure temperatures of ~400 °C for muscovite and 350 °C for biotite (McDougall and Harrison, 1999) for geologically reasonable cooling rates and diffusive length scales.

RESULTS

U-Pb zircon

Analytical Considerations

Analytical uncertainties in ^{206}Pb/^{238}U age for unknowns averaged 2.7% throughout our study[1]. This quantity is similar in magnitude to the spot-to-spot reproducibility predicted by the scatter

of calibration data defining the calibration (3.1%, N = 272; see Appendix 3). Average radiogenic yields for ^{206}Pb and ^{207}Pb were 95.9% and 72.1%, respectively. The far more significant common lead corrections that were applied to measured ^{207}Pb caused errors associated with ^{207}Pb/^{235}U ages to be, on average, six times greater than those of corresponding ^{206}Pb/^{238}U ages (Fig. 4; see Appendix 3). The large uncertainties in ^{207}Pb/^{235}U age generally compromised our ability to evaluate results for U-Pb discordance and required our interpretations to be based primarily upon ^{206}Pb/^{238}U ages. In 14% of our measurements, ^{206}Pb/^{238}U and ^{207}Pb/^{235}U ages were distinguishable at the 1σ level. In most of these cases (11% of total), the U-Pb ages were sufficiently old (generally greater than 1 Ga) that well determined ^{207}Pb/^{206}Pb ages were calculated. Since ^{207}Pb/^{206}Pb ages are more reliable than U-Pb ages for approximating the time of crystallization of zircons that have experienced Pb loss, we used them in lieu of ^{206}Pb/^{238}U ages in the cases where they could be precisely measured. While our use of ^{206}Pb/^{238}U ages for the remaining discordant analyses (3% of total) is potentially problematic, it does not bias the remaining data in any systematic way. In fact, we will show that the overall U-Pb age distributions we obtained from the schist are extremely coherent on a regional scale and similar to those exhibited by the likely protolith (DeGraaff Surpless et al., 2002; Jacobson et al., 2002b).

Abundance of Late Cretaceous Zircons

Our detrital zircon U-Pb results are summarized in Table 2. The most fundamental outcome of this study is that Late Cretaceous ^{206}Pb/^{238}U ages were measured in at least one grain in 37 of the samples examined, with two of the remaining samples giving ages as young as late Early Cretaceous (Table 2). This occurred in spite of the fact that fewer than 15 grains were measured for 60% of the samples. In fact, 42% of the grains yielded ^{206}Pb/^{238}U ages less than 100 Ma while an additional 20% of the results were Early Cretaceous (Table 2).

The relevance of the Late Cretaceous zircon U-Pb ages for the age of the schist's sedimentary protolith could be questioned if the detrital zircons experienced Pb loss or were overgrown by metamorphic zircon during upper greenschist to amphibolite facies recrystallization that attended underthrusting. Petrographic inspection and limited cathodoluminescence (CL) imaging (Jacobson et al., 2000) indicate that most samples we examined contain abundant euhedral to partially rounded zircons with characteristically simple oscillatory zoning patterns that we interpret as magmatic features. These characteristics make it seem most reasonable to us that the vast majority of our U-Pb ages represent igneous recrystallization ages.

Barth et al. (2003a) carefully integrated CL imaging and ion probe U-Pb study of detrital zircons from the comparatively high

---

[1]GSA Data Repository Item 2003177, data for palinspastic reconstruction and tabulated analytical data (U-Pb and ^{40}Ar/^{39}Ar) and derivative plots, is available on request from Documents Secretary, GSA, P.O. Box 9140, Boulder, CO 80301-9140, USA, editing@geosociety.org, at www.geosociety.org/pubs/ft2003.htm, or on the CD-ROM accompanying this volume. All errors are ± 1σ values that reflect analytical uncertainties only (see Appendix 3).

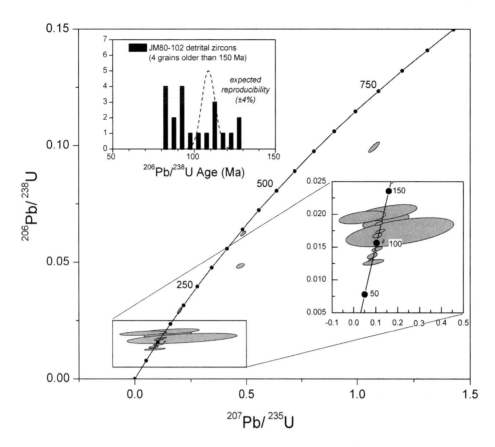

Figure 4. Representative concordia plot from JM80-102. Upper inset is histogram of results from 50 to 150 Ma. Four of 23 results plot outside this range. Gaussian curve represents expected spot-to-spot reproducibility based upon measurements of AS-3 standard zircon. Lower inset is a vertically expanded view of concordia.

grade schist of Sierra de Salinas (see Table 1), which locally contains sillimanite (Ross, 1976) and hence indicates significantly higher peak metamorphic temperatures than were attained in most other bodies of the schist. They determined that thin (1–15 μm), CL bright, low Th/U overgrowths of potential metamorphic origin were present on a number of grains. While these have proven difficult to measure directly in conventionally sectioned and polished grains, we cannot preclude the possibility that some of our zircon U-Pb analyses are anomalously young because the beam overlapped metamorphic zircon.

Because metamorphic overgrowths on igneous zircons generally have very low Th/U (Kröner et al., 1994; Rubatto, 2002; Carson et al., 2002; Mojzsis and Harrison, 2002), we can make some inferences about the extent to which anomalously young ages produced by metamorphic overgrowths influence our results by examining the U-content and Th/U versus U-Pb age for our zircons (Fig. 5). As indicated, there is no clear relationship between U-Pb age and either U or Th/U (note that Th was not determined in 25% of the analyses). To illustrate this, we have used a filled symbol to represent the youngest age measured in each of the samples in Figure 5. As shown, spot analyses from most of these grains define essentially the same range of U concentrations (50–5000 ppm) and Th/U (0.05–1.5) as those exhibited by Proterozoic grains that are the most susceptible to Pb loss by virtue of protracted exposure to radiation damage. Roughly 88% of our results yield Th/U ratios greater than the values (<~0.1) that are

believed suggestive of metamorphic zircon (Kröner et al., 1994; Rubatto, 2002; Carson et al., 2002; Mojzsis and Harrison, 2002). In five of our samples, the youngest zircon ages also have Th/U values below 0.1. (see also Barth et al., 2003a). These included an 80 Ma result with Th/U = 0.009 from JM80–102 (Gabilan Range) and a 59 Ma result with Th/U = 0.028 from TR23 (Trigo Mountains). In both cases, however, the next oldest grains were only slightly older (81 Ma in JM80–102 and 67 Ma in TR23) and had Th/U in excess of 0.1. Moreover, the youngest U-Pb age we measured (a 55 Ma result from Neversweat Ridge) had Th/U = 0.25 and less than 200 ppm U. Hence, while it is likely that limited metamorphic zircon growth has influenced our results, there is no obvious evidence that it impacts them in a serious way.

Provenance Analysis

In the initial stages of the study, we analyzed 30–50 grains per sample. Subsequently, our efforts to detect whether or not Late Cretaceous grains were present in as many samples as possible motivated us to greatly reduce the number of grains analyzed per sample (to as few as 10). Such a limited sampling rate is significantly less than that required for a statistically meaningful age distribution (e.g., Dodson et al., 1988). Inspection of Table 2 reveals that closely spaced samples from a single area can differ quite dramatically in their distribution of detrital ages. Hence, it is clearly desirable to combine data from adjacent samples to ensure that the overall variability from a given

TABLE 2. SUMMARY OF U-Pb AGE* RESULTS FROM DETRITAL ZIRCONS

Sample	Area[§]	Group[§]	N[#]	Late K[††]	Early K[††]	55–70 (Ma)	71–80 (Ma)	81–90 (Ma)	91–145 (Ma)	146–248 (Ma)	>249 (Ma)
9957	1	NW	35	8	22	-	-	-	30	3	2
JM80103	2	NW	23	9	7	-	-	6	10	2	5
DR1656	2	NW	28	11	7	-	51	5	12	2	8
02-342[§§]	2	NW	35	9	14	-	-	7	7	10	11
02-34[§§]	2	NW	24	12	24	-	1	2	21	-	-
02-358[§§]	2	NW	35	12	17	-	-	6	11	5	14
PR150	3	NW	40	19	9	-	6	6	22	5	7
RA169	4	NW	46	17	16	-	1	5	27	7	6
RA170	4	NW	39	17	15	-	1	2	29	1	6
98241[†]	5	C	50	28	-	2	10	14	2	5	17
SP25D	5	C	10	7	3	-	1	1	8	-	-
SP10	5	C	10	8	2	-	-	-	10	-	-
98240[†]	6	C	43	17	22	-	-	5	34	3	1
OR307	7	C	14	4	2	-	-	-	6	3	5
OR312A	7	C	10	2	-	-	-	2	-	3	5
OR113	7	C	9	1	1	-	-	-	2	2	5
OR314	7	C	13	3	-	-	-	2	1	3	7
OR77B	7	C	11	4	3	-	-	2	5	-	4
OR315A	7	C	10	6	8	-	4	1	4	-	1
OR337	7	C	10	8	9	1	4	1	2	-	2
KE224031	7	C	10	3	4	-	3	-	1	1	5
KE24033	7	C	9	-	-	-	-	-	-	5	4
KE127021	7	C	10	2	2	-	1	-	1	2	6
98237	8	C	31	20	5	-	1	13	11	4	2
SG532	8	C	10	6	-	-	5	-	1	-	4
SG533	8	C	32	20	4	1	9	7	7	4	4
SG69	8	C	10	3	1	-	-	2	2	4	2
PK114B	9	SE	11	8	3	-	4	3	4	-	-
UG1417A[†]	9	SE	45	32	2	-	18	11	5	1	10
UG1500	9	SE	10	-	5	-	-	-	5	1	4
YN17	10	SE	14	1	-	-	-	-	1	5	8
MW10	10	SE	13	3	-	-	1	1	1	4	6

*Statistically concordant results (86% of 853 analyses) represented by ^{206}Pb/^{238}U ages. Discordant analyses that were sufficiently old (i.e., >1 Ga; 11%) have been represented by ^{207}Pb/^{206}Pb ages. Remainder (3%) are generally Cretaceous and are represented by ^{206}Pb/^{238}U ages.

[†]Results from Jacobson et al. (2000).

[§]See Table 1 and Appendix 1 for further details.

[#]Total spot analyses obtained per sample (one spot analysis per grain).

**Numbers represent quantity of analyses that fall within indicated age bins.

[††]Early Cretaceous (Early K) is 100–145 Ma, Late Cretaceous (Late K) is 6–99 Ma.

[§§]Results from Barth et al (2003a).

region is adequately represented. To accomplish this, we pooled results at two different scales. At the local level, we combined samples from individual schist bodies or adjacent schist bodies as indicated in Figure 3 (see Table 2). At a more regional scale, we pooled our results into three groups based upon their palinspastically restored positions: (1) samples from the Rand Schist and the schists of Sierra de Salinas and Portal Ridge defined a "northwest" group; (2) samples of the Pelona Schist and the Orocopia Schist of the Orocopia Mountains defined a "central" group; and (3) all bodies of Orocopia Schist southeast of the Orocopia Mountains defined a "southeast" group.

We employed the following age bins in our analysis: 55–70 Ma, 71–80 Ma, 81–90 Ma, 91–145 Ma, 146–248 Ma (i.e., Jurassic and Triassic), and 249–2000 Ma (Paleozoic to middle Early Proterozoic). The comparatively fine subdivision of the Cretaceous is justified since nearly two-thirds of our detrital

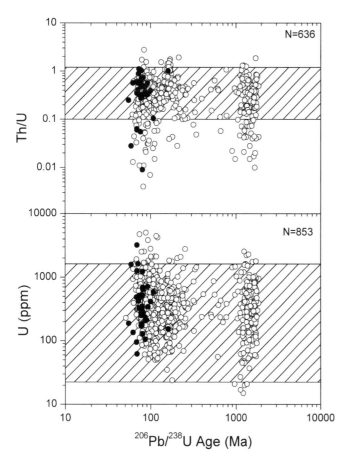

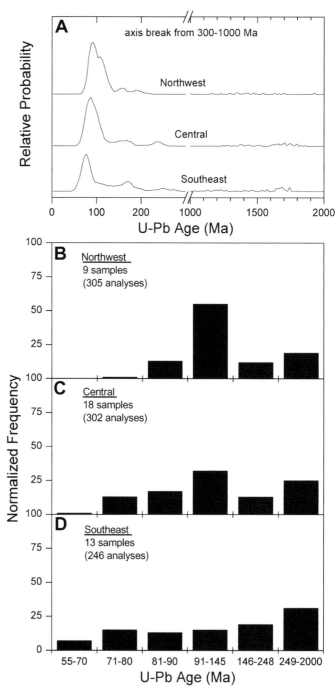

Figure 5. A: U-content versus U-Pb age; B: Th/U versus U-Pb age. Filled symbols indicate youngest ^{206}Pb/^{238}U age obtained each of 40 samples (includes data from Jacobson et al., [2000] and Barth et al., [2003a]). Note that Cretaceous results exhibit same range of U and Th/U as grains that have preserved Proterozoic U-Pb ages. This is evidence against preferential Pb loss by Cretaceous grains. Hatched regions indicate 98% confidence bands calculated from over 950 ion probe U-Pb analyses from detrital zircons from greywackes of the Great Valley Group (DeGraaff-Surpless et al., 2002). By this standard, metagraywacke zircon Th/U values less than 0.1 are anomalously low and may signal metamorphic zircon growth.

zircon U-Pb data ages are Cretaceous. Moreover, as will be evident below, it facilitates comparison to potential source regions of southwestern North America. We combined all Jurassic and Triassic ages into the same bin, since crystalline rocks of this age tend to occur in the same areas. Also note that, while we put all pre-Mesozoic results into a single bin, only about 4% of these are between 249 and 1100 Ma. Moreover, most of the 249–1100 Ma ^{206}Pb/^{238}U ages obtained were yielded by grains that exhibit U-Pb discordance at the 1σ level. Hence the 249–2000 Ma bin is populated primarily by late Early Proterozoic to earliest Late Proterozoic zircons.

Zircon U-Pb age distributions from the three major geographic groups discussed above are plotted in Figure 6. The relative probability plots of Figure 6A clearly demonstrate the abun-

Figure 6. A: U-Pb age frequency spectra for northwest schists (Rand Schist and schists of Portal Ridge and Sierra de Salinas), central schists (Pelona Schist and Orocopia Schist of Orocopia Mountains), and southeastern schists (Orocopia Schist excluding that exposed in Orocopia Mountains). All three geographic groups contain subequal proportions of Mesozoic (dominantly Cretaceous) and late Early to late Middle Proterozoic grains. Normalized histograms of U-Pb ages from (B) northwest, (C) central, and (D) southeast schists shown in (B), (C), and (D), show that progressive decrease of maximum possible depositional age of graywacke protolith from northwest to southeast (in palinspastically restored coordinates) is accompanied by prominent change in provenance.

dance of Cretaceous ages. When the age bins defined above are employed, it becomes evident that Early to early Late Cretaceous (91–145 Ma) zircons overwhelmingly populate samples from the northwest schist exposures (Fig. 6B), whereas subequal proportions of Proterozoic, Triassic/Jurassic, and latest Cretaceous to earliest Tertiary zircons typify detrital zircon populations within the southeast schist exposures (Fig. 6D). The central group is transitional between the other two (Fig. 6C). Note that the proportion of Cretaceous zircons decreases sharply from 69% in the northwest to 48% in the southeast. This falloff is primarily an expression of the proportion of 91–145 Ma zircons in the northwest present (compare parts B and D in Fig. 6). Zircons in this age range constitute 55% of the population in the northwest but only 16% of the population in the southeast. At the same time, the proportion of Jurassic and Triassic grains increases significantly in the southeast. Paleozoic and latest Proterozoic grains are very sparse in all samples (Fig. 6A).

A second important trend is the systematic southeastward decrease in U-Pb ages of the youngest zircons (Fig. 6, B–D). For example, 13% of the grains from the Schist of Sierra de Salinas and the Rand Schist fall in the 80–90 Ma range while an additional 1.3% of the results are as young as 77 Ma (Fig. 6B). In the central region, 13% of the zircon analyses of Pelona Schist fall in the 71–80 Ma range while an additional 1.3% of the results are as young as 68 Ma (Fig. 6C). Finally, 7% of the zircon ages from the schists of the southeastern region are in the 55–70 Ma range (1% are less than 60 Ma) (Fig. 6D). Because the accuracy of our ^{206}Pb/^{238}U ages is on the order of 3% (see Appendix 3), the very youngest ages in each of the three groupings could reasonably be ascribed to analytical scatter. However, the large numbers of grains with slightly older ages cannot be easily dismissed as measurement artifacts and strongly imply that systematically younger detrital zircons are present to the southeast.

^{40}Ar/^{39}Ar Thermal History Results

Muscovite and biotite ^{40}Ar/^{39}Ar age spectra measured in this study (complete ^{40}Ar/^{39}Ar data tables and derivative plots are available from the GSA Data Repository; see footnote 1) all revealed monotonically increasing age gradients at low-temperature gas release. This pattern typifies metamorphic micas from slowly cooled, deep-seated subduction settings (e.g., Grove and Bebout, 1995). In general, there was no indication of excess ^{40}Ar. One sample (SG532A muscovite) did yield an initially old step indicative of excess ^{40}Ar that constituted less than 1% ^{39}Ar release in the age spectrum. In addition, the age spectra of several samples that were run sequentially yielded zero ages during the initial 1% of ^{39}Ar release (KE1 biotite, OR312A biotite, OR314 biotite, PR36A muscovite, and TR18 muscovite). We attribute these initial zero ages to unresolved hydrocarbon interferences at mass 36 that resulted from an under-performing getter pump. Neglecting these problematic initial steps, we calculated bulk closure ages and interpreted the results as indicating the time of post-metamorphic cooling. Results are summarized in Table 3.

Muscovite post-metamorphic cooling ages decrease systematically from 90 Ma in the northwest (San Emigdio Mountains) to 40 Ma in the southeast (Neversweat Ridge; Fig. 7). Although data are less abundant for biotite, a second pattern is also evident in Figure 7. Specifically, biotite is only 2–3 m.y. younger than muscovite in the San Emigdio Mountains and the western Mojave region but up to 20 m.y. younger than muscovite in southeastern California and southwestern Arizona. While we cannot absolutely rule out reheating effects related to Tertiary volcanism or hypabyssal plutonism in the Castle Dome Mountains or Neversweat Ridge area (see Appendix 2), virtually identical muscovite and biotite ages were obtained from the Trigo Mountains, which show relatively little Tertiary intrusion. Consequently, we believe

TABLE 3. ^{40}Ar/^{39}Ar MICA TOTAL GAS AGES*

Sample	Locality	Mus	Bio
9957	San Emigdio Mtns.	89.8 ± 0.2	86.1 ± 0.2
RA138	Rand Mtns.	67.4 ± 0.2	-
RA85	Rand Mtns.	67.8 ± 0.2	-
RA169	Rand Mtns.	68.8 ± 0.2	
RA170	Rand Mtns.	71.7 ± 0.2	
PR36	Portal Ridge	69.6 ± 0.1	67.1 ± 0.6
PR150	Portal Ridge	-	65.5 ± 0.2
98241	Sierra Pelona	58.5 ± 0.1	-
SP25D	Sierra Pelona	51.8 ± 0.2	-
SP10	Sierra Pelona	57.8 ± 0.1	-
98240	San Gabriel Mtns. (BR)	48.8 ± 0.2	-
SG43	San Gabriel Mtns. (EF)	51.4 ± 0.2	-
SG69	San Gabriel Mtns. (EF)	57.8 ± 0.1	-
SG81	San Gabriel Mtns. (EF)	52.8 ± 0.1	-
98237	San Gabriel Mnts. (EF)	55.5 ± 0.3	-
SG530	San Gabriel Mtns. (EF)	42.2 ± 0.1	-
SG531	San Gabriel Mtns. (EF)	43.4 ± 0.2	-
SG532	San Gabriel Mtns. (EF)	34.2 ± 0.2	-
SG533	San Gabriel Mtns. (EF)	31.7 ± 0.2	-
OR77B	Orocopia Mtns.	44.6 ± 0.1	-
OR312A	Orocopia Mtns.	34.3 ± 0.6	25.3 ± 0.1
OR314	Orocopia Mtns.	42.1 ± 0.1	-
PK114B	Peter Kane Mtns.	49.0 ± 0.2	-
YN17	Yellow Narrows	44.3 ± 0.1	22.6 ± 0.4
MW10	Marcus Wash	45.6 ± 0.1	-
TR18	Trigo Mtns.	39.8 ± 0.1	-
TR24A	Trigo Mtns.	44.9 ± 0.2	-
CD9	W. Castle Dome Mtns.	41.1 ± 0.1	-
KE1	E. Castle Dome Mtns.	44.2 ± 0.1	35.8 ± 0.1
KE3	E. Castle Dome Mtns.	42.2 ± 0.1	18.9 ± 0.3
KE6	E. Castle Dome Mtns.	42.7 ± 0.1	-
NR1	Neversweat Ridge	-	19.3 ± 0.3
NR2	Neversweat Ridge	40.4 ± 0.1	-

Note: BR—Blue Ridge; EF—East Fork.
*Integrated from entire age spectrum (= K-Ar age).

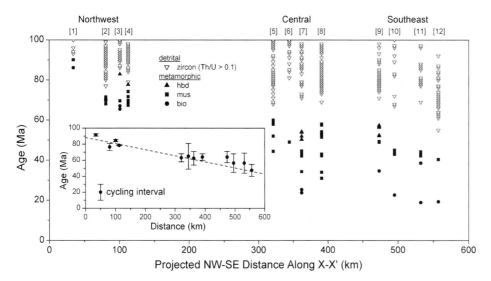

Figure 7. Summary of ^{40}Ar/^{39}Ar cooling ages and zircon U-Pb ages as function of relative northwest-southeast distance for each of 12 areas studied (see Fig. 3 for locations). Additional ^{40}Ar/^{39}Ar data from Jacobson (1990), Jacobson et al. (2002), Vucic et al. (2002), and Barth et al. (2003a) have been included in this plot. Zircons with low Th/U suggestive of metamorphic growth not been used to calculate minimum detrital age. Average time difference between youngest U-Pb detrital zircon age and oldest cooling age is 13 ± 10 Ma. During this time graywacke protolith of schist was eroded from its basement source region, deposited, underthrust, accreted, and metamorphosed.

the results are best interpreted as indicating that cooling from peak metamorphic conditions occurred earlier and more rapidly in the northwest relative to the southeast.

DISCUSSION

Time-Space Constraints upon the Evolution of the Pelona and Related Schists

Late Cretaceous–Early Tertiary Depositional Age for the Schist

The depositional age of the graywacke protolith of the schist is necessarily younger than the youngest detritus contained within it. On the basis of textural and compositional characteristics, we believe that the vast majority of detrital zircons we examined from the schist are igneous in origin. The available evidence (Fig. 5) also strongly indicates that the U-Pb ages yielded by ion probe analysis of the detrital zircons overwhelmingly represent igneous crystallization ages unaffected by Pb loss or metamorphic zircon formation during peak-grade recrystallization of the schist. Hence, the preponderance of ^{206}Pb/^{238}U ages less than 100 Ma (Table 2) requires that the depositional age of the graywacke protolith of all schist samples we examined was also Late Cretaceous or, in the case of the easternmost bodies, as young as early Tertiary. Because our sampling adequately represents the known geographic extent of the schist terrane and all variation manifested within it (Table 1), we conclude that it is highly unlikely that any portion of the schist terrane (Pelona-Orocopia-Rand-Portal Ridge-Sierra de Salinas) was formed from sedimentary sequences deposited prior to Late Cretaceous time.

Our results are in direct conflict with the conclusion of James (1986) and James and Mattinson (1988) that the Rand Schist of the San Emigdio Mountains is intruded by the 131 Ma tonalite of Antimony Peak. We analyzed 35 detrital zircon grains from one sample of the schist in this area (99–57). All ages but

five were less than 131 Ma, and all but 11 were less than 110 Ma. Based on the detrital zircon ages, our limited field observations, and the mapping of Ross (1989), we conclude that the contact between schist and tonalite in this area is a fault and that the age of the tonalite does not constrain either the depositional or metamorphic age of the schist.

Our results also contradict the inference of Mukasa et al. (1984) that a 163 Ma mafic diorite (now metamorphosed) found within the Chocolate Mountains intruded the protolith of the schist. While we were unable to analyze any samples from this locality, which lies within an active bombing range, we did examine two samples of schist adjacent to possible correlatives of the metadiorite in the Trigo Mountains (samples TR23 and TR24A; Haxel et al., 2002). The proportions of Late Cretaceous U-Pb ages yielded by detrital zircons from these samples were 41% and 18%, respectively. This result leads us to doubt the intrusive relation inferred by Mukasa et al. (1984). Instead, we suggest that the metadiorite was incorporated into the schist protolith as either a sedimentary or tectonic fragment. The 163 Ma igneous age of the metadiorite and its arc-like geochemical signature (Haxel et al., 1987, 2002) suggest a potential correlation with the Coast Range ophiolite (Dickinson et al., 1996). The Coast Range ophiolite forms the basement of the western Great Valley forearc basin. Exotic blocks of Coast Range ophiolite are inferred to be present in the Franciscan Complex (MacPherson et al., 1990) and may provide an analog for late Middle Jurassic metadiorite within the schist.

Finally, all previous reports of post-metamorphic, Late Cretaceous intrusion of the schist also appear to be incorrect. The Randsburg granodiorite within the Rand Mountains clearly intrudes the Rand Schist and has produced contact metamorphism within it (Silver and Nourse, 1986). Conventional zircon U-Pb data reported by Silver and Nourse (1986) from the pluton were interpreted as indicating a 79 ± 1 Ma crystallization age.

However, new ion microprobe U-Pb zircon and $^{40}Ar/^{39}Ar$ biotite and hornblende results from the Randsburg granodiorite demonstrate that it was actually emplaced during the early Miocene (Barth et al., 2003b). Hence, the intrusive relationship between the Randsburg granodiorite and Rand Schist does not place any significant constraints on the emplacement history of the schist.

Similarly, past assessments that the schist of Sierra de Salinas was intruded and contact metamorphosed during the Late Cretaceous appear to have been based upon misinterpreted geologic relationships. Granitic rocks spatially associated with the schist of Sierra de Salinas in the southern Coast Ranges comprise a diverse calc-alkalic suite (Ross, 1984) with U-Pb ages of 82–94 Ma (Mattinson, 1978, 1990; Mattinson and James, 1985; Barth et al., 2003a). While Ross and Brabb (1973) originally mapped the contacts between schist and granitic rocks as Cenozoic faults based on their generally straight trace and the lack of intrusive bodies in the schist, Ross (1976) subsequently concluded that the schist bodies in the Sierra de Salinas and Gabilan Range had been contact metamorphosed by the adjacent granitoids. The U-Pb ages of detrital zircons and the contrasting biotite $^{40}Ar/^{39}Ar$ cooling ages exhibited by the schist of Sierra de Salinas and adjacent granitoids (Barth et al., 2003a) clearly demonstrate that deposition of the schist's protolith postdated granitoid intrusion in the area.

Regional Variation in the Timing of Deposition, Underthrusting, and Metamorphism

Results shown in Figure 7 require that the cycling interval for the graywacke protolith of the schist (i.e., the interval during which the schist's protolith was eroded from its source region, deposited, underthrust, accreted, and metamorphosed) was generally short. As indicated in Figure 7, the upper bound of the interval is defined by the youngest detrital zircon U-Pb age while the lower bound is established by the oldest $^{40}Ar/^{39}Ar$ cooling age. The timing of this interval decreases from 93 to 90 Ma in the San Emigdio Mountains (southernmost Sierra Nevada) to 55–43 Ma in the Neversweat Ridge area (southwestern Arizona). Hence, the entire cycle of processes occurred earlier for the Rand Schist and the schists of Portal Ridge and Sierra de Salinas in the northwest than it did for the Pelona and Orocopia Schists situated progressively farther southeast in palinspastically restored coordinates (Fig. 3).

Our data indicate an average duration for the cycling interval of 13 ± 10 m.y. along the entire belt (see inset in Fig. 7). For the Rand Schist and schist of Portal Ridge, the cycling interval is too small to be resolved with the methods we have employed (~3 m.y.). Since our results provide only maximum and minimum bounds, the gap is likely to have been smaller than that depicted in Figure 7. This is due partly to the statistical likelihood of finding detrital zircons just older than the age of deposition. Moreover, to the extent to which the source region was plutonic, nearly synchronous intrusion and rapid exhumation are required to supply materials of appropriate age. In addition, hornblende is available from only a few localities. Muscovite ages defining

the lower bound may significantly postdate the time of peak-grade metamorphism since its closure temperature (~400 °C; McDougall and Harrison, 1999) is significantly lower than the peak metamorphic temperature in many of the areas investigated (~450–>550 °C; Graham and Powell, 1984; Jacobson, 1995). Hornblende, which is more retentive of ^{40}Ar, clearly constrains the time of peak-grade metamorphism more tightly. For example, hornblendes from Portal Ridge (83 Ma) and the Rand Mountains (78 Ma) are respectively 13 m.y. and 4 m.y. older than muscovites sampled from adjacent exposures (Jacobson, 1990). In the Gavilan Hills, hornblende total gas ages range from 52 to 58 Ma and are 3–10 m.y. older than muscovites from adjacent rocks (Jacobson et al., 2002a). These observations and the fact that terminal ages in the muscovite $^{40}Ar/^{39}Ar$ age spectra can be several million years older than the total gas ages reported in Table 3 indicate that the interval for deposition and accretion is likely to be well under 10 m.y. for the entire schist terrane. We are currently working to improve the database of hornblende $^{40}Ar/^{39}Ar$ results to confirm this.

The $^{40}Ar/^{39}Ar$ data collected here also have implications for the exhumation history of the schists. For example, recent work in the Gavilan Hills (Jacobson et al., 2002a) and Orocopia Mountains (Vucic et al., 2002) of southeastern California has recognized two important phases of Tertiary cooling attributed to two discrete denudation events. Closure of hornblende and muscovite occurred in the early Tertiary (between 58 and 44 Ma). Subsequent slow cooling in the temperature range corresponding to biotite closure produced a spread of biotite total gas ages from 44 to 30 Ma in the Gavilan Hills and down to 20 Ma in the Orocopia Mountains. Rapid cooling in the middle Tertiary (28–20 Ma) caused K-feldspar to close (Jacobson et al., 2002a; Vucic et al., 2002). Results shown in Figure 7 suggest that a broadly similar history may apply to rocks east of the Gavilan Hills in southwestern Arizona.

Spatial/Temporal Variation of the Provenance of the Source Region

The northwest to southeast decrease of protolith age evident in our data is accompanied by a dramatic shift in sediment provenance that closely mimics the crystallization ages of basement rocks that structurally overlie the schists. Figure 8 is a first-order palinspastic restoration showing the pre–San Andreas distribution of basement rocks of southern California and western Arizona. Figure 9 indicates the distribution of detrital zircon U-Pb ages from the same region. Comparison of Figures 8 and 9 makes it clear that the dominantly 81–145 Ma detrital zircon population from the Rand Schist and the schist of Portal Ridge and Sierra de Salinas in the northwest forms a good match to the known intrusive ages from the medial Cretaceous Sierran batholith (Stern et al., 1981; Chen and Moore, 1982; Saleeby et al., 1987; Barth et al., 2003a). Similar detrital zircon U-Pb age distributions have been measured by DeGraaff Surpless et al. (2002) from the Upper Cretaceous part of the Great Valley Group in the San Joaquin Valley and by Jacobson et al. (2002b) from Upper Cretaceous

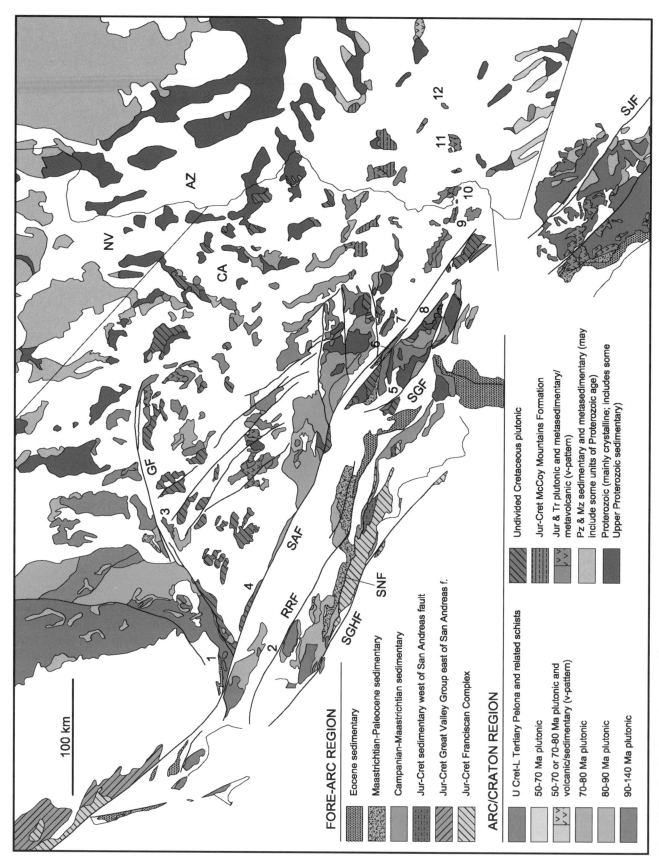

FORE-ARC REGION

- Eocene sedimentary
- Maastrichtian-Paleocene sedimentary
- Campanian-Maastrichtian sedimentary
- Jur-Cret sedimentary west of San Andreas fault
- Jur-Cret Great Valley Group east of San Andreas f.
- Jur-Cret Franciscan Complex

ARC/CRATON REGION

- U Cret-L Tertiary Pelona and related schists
- 50-70 Ma plutonic
- 50-70 or 70-80 Ma plutonic and volcanic/sedimentary (v-pattern)
- 70-80 Ma plutonic
- 80-90 Ma plutonic
- 90-140 Ma plutonic

- Undivided Cretaceous plutonic
- Jur-Cret McCoy Mountains Formation
- Jur & Tr plutonic and metasedimentary/metavolcanic (v-pattern)
- Pz & Mz sedimentary and metasedimentary (may include some units of Proterozoic age)
- Proterozoic (mainly crystalline; includes some Upper Proterozoic sedimentary)

100 km

Figure 8. Pre–late Tertiary paleogeographic reconstruction of southern California, southwestern Arizona, and adjacent areas. Details regarding data sources employed in construction of this map are provided in Appendix 1. Abbreviations of fault names as in Figure 1.

strata within the northern portion of the Central California Coast Ranges. Crystallization ages of basement rocks of southeastern California and southwestern Arizona (Fig. 8) likewise mimic the detrital zircon U-Pb age systematics that we have measured for the Orocopia Schist from this area (Fig. 9; i.e., these schists contain high proportions of zircons yielding Proterozoic, Early Mesozoic, Late Cretaceous ages less than 80 Ma, and even earliest Tertiary ages; Jacobson et al., 2000). Maastrichtian to lower Eocene strata from southern Salinia and the Transverse Ranges contain detrital zircon suites similar to those from the southern schists (Fig. 8; Jacobson et al., 2002b).

Along-Strike Versus Across-Strike Control on Protolith Age and Provenance

Recognition of the northwest to southeast variation in age and provenance of the schist's protolith is one of the most conspicuous outcomes of this study. However, there is a significant ambiguity in interpreting the significance of this trend because the schist crops out along a relatively linear northwest-southeast array that transects the grain of the Mesozoic continental margin. Specifically, the Rand Schist and the schists of Portal Ridge and Sierra de Salinas tend to underlie older, more central to western parts of the Cretaceous arc (Fig. 8; Mattinson, 1978; Mattinson and James, 1985; Silver and Nourse, 1986; James and Mattinson, 1988; Pickett and Saleeby, 1993; Wood and Saleeby, 1997). In contrast, the Pelona and Orocopia Schists are located beneath the youngest (ca. 75 Ma), most inboard part of the main axis of voluminous Cretaceous magmatism (Sierra Pelona, San Gabriel Mountains, Orocopia Mountains; Fig. 8; Ehlig, 1981; Barth et al., 1995). In fact, the easternmost schist exposures in southwest Arizona are positioned east of the limit of extensive Cretaceous magmatism (Fig. 8; Powell, 1993; Tosdal et al., 1989; Haxel et al., 2002).

In light of this relationship, we cannot unambiguously determine the relative roles of position *along* versus *across* strike of the Late Cretaceous–early Tertiary continental margin in controlling detrital zircon age distributions and $^{40}Ar/^{39}Ar$ thermal histories (Fig. 7). Moreover, the highly schematic nature of Figure 8 (see Appendix 1) prevents us from determining the relative inboard-outboard or along-strike location of individual schist bodies on too fine a scale (e.g., the positions of the northern schists are not corrected for oroclinal bending of the southern Sierra Nevada). In any case, the above ambiguity must be kept in mind when considering the implications of our results for tectonic models to explain the genesis of the schists.

Evaluation of Models for Underplating of the Schists

Subduction Model

In the subduction model, graywacke supplied to the trench during the Late Cretaceous–earliest Tertiary was subducted and accreted to the base of the eroded continental lithosphere during shallow Laramide subduction of the Farallon slab (Fig. 2A; Crowell, 1968, 1981; Yeats, 1968; Burchfiel and Davis, 1981; Hamilton, 1987, 1988; Jacobson et al., 1996; Malin et al., 1995;

Wood and Saleeby, 1997). Such a model can explain the key observation of this study. Namely, erosion of basement source regions, trench deposition, underthrusting, accretion, and metamorphism are all predicted to occur in a very short time interval by the subduction model. Moreover, juxtaposition of Upper Cretaceous sediments and eugeoclinal lithologies is easily explained since these materials are expected to mix as they are subducted beneath the margin. Finally, because the model postulates that the lithosphere is eroded upwards to the base of the crust during low-angle Laramide subduction, any subsequently accreted trench sediments would have to post-date this erosion. Hence, the depositional age of the oldest accreted sediment in the subduction model can be no older than Late Cretaceous in age.

In its current form, the subduction model does not explicitly predict the systematic northwest to southeast younging of both protolith and metamorphic age that we observe in the schist (Fig. 7). We believe this pattern makes sense if it is primarily a function of distance normal to, rather than along strike of, the margin. This concept is illustrated in Figure 10 (Jacobson et al., 2002b). Before initiation of the Laramide orogeny, subduction of the Farallon slab at a moderate angle would have been associated with the "classic" belts of Franciscan Complex, Great Valley Group, and Sierran batholith (Fig. 10A). Beginning in the Late Cretaceous, gradual flattening of the subducting plate would result in progressive eastward removal of North American lower crust and mantle lithosphere and concomitant underplating of trench-derived graywacke. In the earliest stages of flat subduction (Fig. 10B), graywacke could be accreted only west of, and beneath, the medial Cretaceous arc. This material would be equivalent to the Rand Schist and schists of Portal Ridge and Sierra de Salinas. We postulate that equivalent age material was underplated to the south beneath the Peninsular Ranges. Post-intrusive rapid cooling initiated within the central and eastern Peninsular Ranges after 80 Ma support this hypothesis (George and Dokka, 1994; Grove et al., this volume, Chapter 13).

As the locus of tectonic erosion shifted inboard, graywacke precursors of the Pelona and Orocopia schists would have been accreted at positions that were ultimately farther east than the medial Cretaceous arc in southeastern California and southwestern Arizona (Fig. 10C). Also inherent in this model is that younger underplated material will have passed beneath previously accreted schist. The resultant thickening of the crust from below would have caused the older accreted material to be driven toward the surface, either by tectonic collapse at high structural levels (e.g., Platt, 1986) or by erosion (Yin, 2002). Thus, older schists accreted in the west are expected to have cooled through Ar closure while underplating continued to carry new material to the east. This provides a ready explanation for the parallel younging of detrital zircon and $^{40}Ar/^{39}Ar$ metamorphic cooling ages evident in Figure 7. It can also explain why cooling was more gradual in the southeast (i.e., larger gap between muscovite and biotite $^{40}Ar/^{39}Ar$ cooling ages). The southeasternmost schists would have been emplaced toward the end of Laramide low-angle subduction. Cessation of further underplating upon return

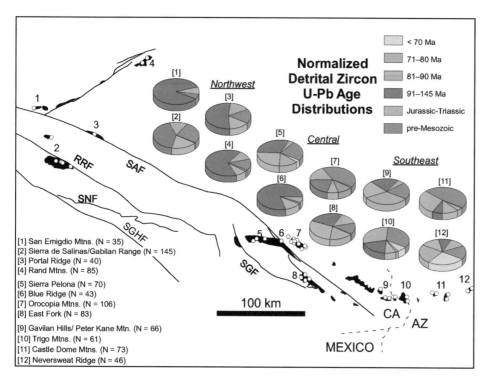

Figure 9. Normalized histograms illustrating northwest to southeast change in provenance for each of 12 major exposure areas of schist defined in Table 1. Age bins in histograms are the same as in Figure 6B–6D and correspond to age groupings used to subdivide crystalline rocks of equivalent age in Figure 8.

Late Cretaceous (ca. 90 Ma)

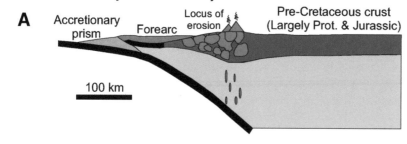

Late Cretaceous (ca. 75 Ma)

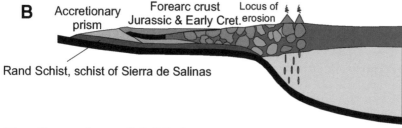

Early Tertiary (ca. 60 Ma)

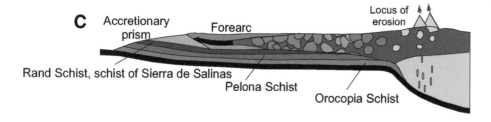

Figure 10. Tectonic model for underplating of schist. A: Before initiation of Laramide orogeny (i.e., subduction of Farallon slab proceeds at moderate angle). Beginning in Late Cretaceous, gradual flattening of subducting plate results in progressive eastward removal of North American lower crust and mantle lithosphere and concomitant underplating of trench-derived graywacke. B: In early stages of Laramide orogeny, graywacke is accreted west of, and beneath, medial Cretaceous arc (Rand Schist and schists of Portal Ridge and Sierra de Salinas). C: As locus of tectonic erosion shifted inboard, graywacke precursors of Pelona and Orocopia Schists are ultimately accreted east of medial Cretaceous arc in southeastern California and southwestern Arizona.

to a normal plate geometry would have removed the driving force for rapid exhumation.

The modified subduction model of Figure 10 may also help explain the northwest to southeast variation in provenance of the schist protolith. To some degree, this trend may simply reflect along-strike differences of bedrock geology; i.e., the dominance of middle to early Late Cretaceous intrusions in the Sierra Nevada compared to an abundance of Proterozoic, Jurassic, and latest Cretaceous crystalline rocks in the Mojave Desert and Transverse Ranges (Fig. 8). If continental drainages were orthogonal to the coast and there was little along-strike transport of sediment within the trench, then it is expected that northern (Rand-Portal Ridge-Sierra de Salinas) and southern (Pelona-Orocopia) schists would exhibit different sediment source characteristics. However, it is not clear that this process can explain the west to east variation in provenance exhibited just within the Pelona and Orocopia Schists (Fig. 9), which are restricted to a relatively small range of latitude. In this context, the model of Figure 10 predicts that sediment composition at any given point along the continental margin should evolve with time. This is consistent with previous studies of forearc sediments in central and southern California (Minch, 1979; Kies and Abbott, 1983; Linn et al., 1992; DeGraaff Surpless et al., 2002), which have suggested that drainage networks tapped progressively more easterly source areas as the magmatic front, the leading edge of the flat slab, and the region of underplating all shifted eastward during the Laramide orogeny (Fig. 10; Coney and Reynolds, 1977; Dickinson and Snyder, 1978; Bird, 1988). Because the Cretaceous arc is most voluminous in the west, the earliest sediment accreted (i.e., that represented by the Rand Schist and the schists of Portal Ridge and Sierra de Salinas) would have been dominated by this component. Sediment accreted during later stages of the Laramide orogeny would have been derived from more cratonal source regions farther east. As indicated in Figure 8, these would include mostly Proterozoic and Early Mesozoic arc detritus, with a smaller component of Laramide age (latest Cretaceous to early Tertiary) material. The net outcome is that the schist would appear to have a provenance that is broadly similar to the composition of the overlying crystalline rocks, exactly as we have observed.

Forearc Model

In this model (Fig. 2B), forearc strata are underthrust and accreted deep beneath the recently emplaced batholithic source region (Hall, 1991; Barth and Schneiderman, 1996; Saleeby, 1997). While the force driving this convergence is considered to be Laramide shallow subduction, underthrusting and metamorphism of the schist are considered to occur entirely within an intraplate setting. The forearc model is motivated to a large degree by the inferred difficulty of completely removing mantle lithosphere during shallow subduction. For example, low-angle subduction of the Nazca plate beneath northern Chile, which is commonly taken as an analog for the Laramide orogeny, has not resulted in substantial thinning of the South American mantle

lithosphere (Smalley and Isacks, 1987; Isacks, 1988; Allmendinger et al., 1990).

Previous discussion of the forearc model has focused upon the northwestern schist exposures (Rand Schist and the schists of Portal Ridge and Sierra de Salinas). The model provides a convenient explanation for the juxtaposition of Franciscan and arc rocks along the Sur–Nacimiento fault zone (Fig. 1; Hall, 1991; Barth and Schneiderman, 1996) and relationships involving exposure of deep crustal rocks in the southernmost Sierra Nevada (Saleeby, 1997). According to Barth and Schneiderman (1996), thrusting of the arc over the forearc was triggered by subduction of a buoyant oceanic plateau. In their view, the intersection between plateau and trench propagated from north to south, thus explaining the younging of the schist in that direction. Thus, the forearc model is sensible if the observed age progression among the schists (Fig. 7) is primarily a function of distance along strike of, rather than normal to, the margin.

It is clear that the forearc model can account for the short timescale for deposition and accretion observed for the northwestern schist exposures (Fig. 7). Moreover, Late Cretaceous strata are prominent in all of the forearc sequences depicted in Figure 8. The main difficulty we see with the model is that the youngest and generally more proximal parts of the forearc that are the most likely to have been overthrust are also the least likely to be spatially associated with eugeoclinal lithologies. Chert and basaltic rocks with mid-oceanic ridge compositional affinities constitute up to 10% of the schist (see Haxel et al., 1987, 2002; Dawson and Jacobson, 1989). These lithologies are most typically associated with distal Late Jurassic to Early Cretaceous forearc strata (Ingersoll, 1983, 1997) that we have not detected in our sampling of the schist. However, we cannot discount the possibility that other sources of eugeoclinal lithologies (such as the Sierran Foothills belt) were present along the western margin of the Cretaceous arc and were underthrust and mixed with forearc strata to produce the observed mix of rock types within the schist.

When the forearc model is extended to apply to all of the schist exposures, potential problems related to scale arise. For example, the schist bodies of southwestern Arizona lie on the cratonal side of the Cretaceous arc. If derived from the Great Valley Group, then the Arizona schists have been carried completely from one side of the arc to the other. This scale of intraplate thrusting (i.e., complete detachment of the arc) has not been recognized in the potentially analogous tectonic setting of low-angle subduction of the Nazca plate beneath northern Chile (Allmendinger et al., 1990).

Finally, the Andean example used to argue in favor of the forearc model may not be entirely relevant to areas proximal to the trench, i.e., the region where underplating of the schist would have occurred. This ambiguity is evident in the interpretation of Livaccari and Perry (1993) for the U.S. Cordillera. These authors argued, by analogy to the modern example from South America, that low-angle subduction during the Laramide orogeny did not result in *widespread* excision of North American lithosphere (e.g., beneath the Colorado Plateau and Rocky Mountains). Nonetheless, Livaccari and Perry (1993) specifically cited the

Pelona and related schists as accretionary wedge deposits that indicated the removal of lowermost North American crust and underlying mantle lithosphere within several hundred kilometers of the trench.

Backarc Model

The backarc model is distinct from the subduction and forearc models in that it predicts that the schist's graywacke protolith was deposited not along the convergent margin but in a basin positioned behind the Cretaceous arc (Fig. 2D; Haxel and Dillon, 1978; Ehlig, 1981; Haxel et al., 2002). The model explicitly addresses the origin of the central and southeastern schist bodies. While originators of the model have not specified the age of the proposed rift basin, our results clearly require that it would have formed in the latest Cretaceous and/or earliest Tertiary. Hence, the possibility raised by Haxel and Tosdal (1986) that the backarc basin was somehow related to the Jurassic Mojave–Sonora Megashear of Silver and Anderson (1974) is precluded.

The principal prediction of the backarc model that is relevant to our study is the expectation that the provenance of basin-filling sediment should closely reflect the local basement (Haxel et al., 2002). Our results appear to uphold this prediction (see also Haxel et al., 2002). Specifically, the schist from the southeastern area contains high proportions of zircons yielding Proterozoic, Early Mesozoic, and Late Cretaceous ages <80 Ma, and even earliest Tertiary ages (Fig. 9) that strongly mimic the local basement geology (Fig. 8). It is somewhat unexpected that 14% of the zircons measured from the southeast group yield U-Pb ages that correspond to the emplacement interval (90–145 Ma) of the Early to early Late Cretaceous arc (Fig. 6). Grains of this age range occur in nearly all samples (Table 2). For the central group (Pelona Schist plus Orocopia Schist from the Orocopia Mountains), the proportion of 91–145 Ma zircons is even higher (38%). Inspection of Figure 8 reveals that the abundance of 91–145 Ma basement rocks from the Transverse Ranges, southeastern California, and southwestern Arizona is nowhere near the abundance implied by our detrital zircon results (particularly for the central schists). Nevertheless, recent characterization of the U-Pb age population of zircons from the McCoy Mountains Formation has clearly documented that 91–145 Ma arc detritus, of probable Sierran-Peninsular provenance, was supplied to basins in southeastern California (Barth et al., 2004). Hence, the existence of 91–145 Ma zircons from the southeastern schist cannot be considered unfavorable to the backarc model.

The main argument against the backarc model remains the lack of obvious evidence for either rift facies or a suture (see also Burchfiel and Davis, 1981; Crowell, 1981; Hamilton, 1987, 1988). These problems become monumental if the backarc model is applied to the northwestern schists. Haxel et al. (2002) acknowledge this and have argued that the origins of the northern and southern schists should be considered independently. However, despite the fact that the northern and southern schists differ in age, their similarities in rock types, geochemistry of the basalt protolith, structural style, and metamorphic petrology are

so striking (Haxel et al., 1987; Jacobson et al., 1988; Dawson and Jacobson, 1989; Jacobson, 1995) that we find it difficult to accept that the different parts of the schist terrane could have been produced by such fundamentally dissimilar tectonic processes.

CONCLUSIONS

More than 850 ion microprobe U-Pb ages of detrital zircons from 40 schist metagraywacke samples that represent virtually all of the important exposures of the Pelona, Orocopia, and Rand Schists and the schists of Portal Ridge and Sierra de Salinas in southern California and southwestern Arizona indicate that the depositional age of their graywacke protolith is Late Cretaceous to earliest Tertiary (see also Jacobson et al., 2000; Barth et al., 2003a). Previous assessments that the schist's protolith was deposited prior to Late Jurassic or Early Cretaceous time appear to have been based upon interpretations of contact relationships that are no longer tenable.

Combined consideration of the youngest detrital zircon U-Pb ages and mica $^{40}Ar/^{39}Ar$ cooling ages indicate that the interval during which the schist's graywacke protolith was eroded from its source region, deposited, underthrust, accreted, and metamorphosed decreases systematically (in palinspastically restored coordinates) from about 92–90 Ma in the southwesternmost Sierra Nevada (San Emigdio mountains) to 55–40 Ma in the southwest Arizona (Neversweat Ridge). The average timescale for all these processes to have occurred is 13 ± 10 Ma and is locally too small (<~ 3 m.y.) to be resolved with the methods we have employed.

Two distinct source regions are indicated for the schists; the Rand Schist and schists of Portal Ridge and Sierra de Salinas were derived from material eroded from Early to early Late Cretaceous basement (like the Sierra Nevada batholith); the Orocopia Schist was derived from a heterogeneous assemblage of Proterozoic, Triassic, Jurassic, and latest Cretaceous to earliest Tertiary crystalline rocks (like basement in the Mojave/Transverse Ranges/southwest Arizona/northern Sonora). The Pelona Schist is transitional between the two.

Our U-Pb detrital zircon and $^{40}Ar/^{39}Ar$ mica age results do not appear to permit us to exclude any of the models proposed for the origin of the schist. The backarc model appears untenable on the basis of independent evidence. We prefer the subduction model but recognize that the forearc model is also capable of explaining key aspects of our data set. Additional tests are required to select between the two.

ACKNOWLEDGMENTS

The University of California at Los Angeles Keck Center for Isotope Geochemistry and Cosmochemistry is supported by a grant from the National Science Foundation's Instrumentation and Facilities program (EAR0113563). Grove acknowledges support from Department of Energy grant DE-FG-03–89ER14049. Jacobson and Barth were supported by National Science Foundation Grants EAR9902788 and EAR0106123. Jane Dawson, Kim Rodgers, and John Uselding are thanked for their help with mineral separations. Jane Dawson also helped with various technical aspects of preparation of the manuscript. We appreciate Jim Mattinson's and Kristen Ebert's willingness to contribute samples from the schist of Sierra de Salinas and southeastern Orocopia mountains respectively. Important insights into the nature of the schist's graywacke protolith came from discussions with Gary Axen, Jane Dawson, Ray Ingersoll, Eric James, Jon Nourse, Stephen Richard, Jason Saleeby, Dick Tosdal, Joe Wooden, and An Yin. This study would not have been possible without the generosity of Gordon Haxel, who spent countless days in the field over many years sharing with us his knowledge of the schists of southeastern California and southwestern Arizona. Finally, Jim Mattinson and Erwin Melis are thanked for reviews that helped to significantly improve this paper.

APPENDIX 1: PALINSPASTIC RESTORATION OF SOUTHWEST NORTH AMERICA

To compare detrital zircon populations in the schist with potential source terranes, we developed a simplified palinspastic reconstruction of southern California and adjoining areas starting with a present-day geologic base compiled from a variety of sources. Primary references were Jennings (1977) and Powell (1993) for California, Richard et al. (2000) for Arizona, and Burchfiel et al. (1992) for Nevada. (Additional sources and a detailed listing of the sequence of steps utilized to generate the reconstruction can be obtained from the GSA Data Repository; see footnote 1.) An ideal palinspastic restoration for interpreting detrital zircon suites within the Pelona and related schists would take into account all compressional, extensional, translational, and rotational deformation since deposition of the schist protolith. Construction of such a map, however, is hindered by insufficient, ambiguous, and/or contradictory evidence regarding Cenozoic deformational history. Furthermore, geologic uncertainty is compounded by geometric complexities associated with drafting an accurate, balanced reconstruction for such a large area of inhomogeneous, multiphase deformation. Easiest, and what we concentrated on here, was to restore those strike-slip faults oriented sub-parallel to the continental margin (e.g., San Andreas, San Gabriel, San Jacinto, Punchbowl, San Gregorio-Hosgri, Rinconada-Reliz, etc.). Partitioning of slip among some of these faults is controversial (Powell, 1993; Dickinson, 1996). We followed Crowell (1962), Ehlig (1981), and Dillon and Ehlig (1993) in utilizing 240 km of displacement on the San Andreas fault and 60 km on the San Gabriel fault. Further discussion of this issue (including additional sources and a detailed listing of the sequence of steps utilized to generate the reconstruction) is presented in the GSA Data Repository (see footnote 1). We also utilized paleomagnetic data of Luyendyk et al. (1985; see also Dickinson, 1996) to back-rotate the western Transverse Ranges. The position of this region along the western edge of the map minimizes the problem of maintaining continuity between units. In contrast, we did not restore any east-west–trending left-lateral faults, such as the Garlock fault or those of the eastern Transverse Ranges. Balancing the complex gaps and overlaps that would result from removing the slip on these faults is beyond the scope of this study. For similar reasons, we did not correct for inferred rotations in the eastern Transverse Ranges or southern Sierra Nevada and Mojave Desert (Kanter and McWilliams, 1982; Luyendyk et al., 1985; Dokka and Ross, 1995). Nor did we take into account middle to late Tertiary extension, which affected the region from Death Valley southeastward to the corridor of the lower Colorado River (e.g., Davis

and Coney, 1979), despite the fact that extension has greatly altered the surface geology compared to Late Cretaceous–early Tertiary time. Notwithstanding these limitations, we consider that Figure 8 accurately portrays the *first-order* distribution of basement terranes within southern California and adjacent areas during the time of deposition of the schist protolith.

APPENDIX 2: SAMPLE LOCATION DETAILS

Sierra de Salinas and Gabilan Range

The Sierra de Salinas is underlain by one of the largest bodies of schist, yet it is not well studied because of poor exposure and lack of accessibility to private land. We include with the Sierra de Salinas body a smaller outcropping in the Gabilan Range to the northeast across the Rinconada-Reliz fault (Ross, 1976, 1984; James and Mattinson, 1988). We analyzed one sample from the Sierra de Salinas (DR-1656) and one from the Gabilan Range (JM80–102). Both were supplied by J.M. Mattinson of the University of California, Santa Barbara, with the sample from the Sierra de Salinas having originally been collected by D.C. Ross of the U.S. Geological Survey. We also incorporated U-Pb ion probe results from the three biotite schists of Barth et al. (2003a) from the Sierra de Salinas (02-242, 02-245, and 02-258) measured into our database. The schist of Sierra de Salinas is metamorphosed to the middle to upper amphibolite facies (Ross, 1976), which represents a higher peak temperature of metamorphism than was attained in most other bodies of the schist.

The schist bodies in the Sierra de Salinas and Gabilan Range are bordered by Cretaceous granitoid intrusions, some as young as 82 Ma (Barth et al., 2003a), although the contact relations have long been a matter of debate. Ross and Brabb (1973) and Ross (1974) interpreted the contacts as faults based on their straightness and the striking paucity of intrusions within the schists. Ross (1976), in contrast, concluded that the granitoids do intrude the schist along its margins, a view also taken by James and Mattinson (1988). However, detrital zircon ages for the schist demonstrate that the schist protolith is younger than the crystallization age of the adjacent pluton (see also Barth et al., 2003a). This, and the fact that the schists exhibit younger biotite $^{40}Ar/^{39}Ar$ ages than the plutons (70–71 Ma versus 75–76 Ma), requires that the contacts are faults.

Portal Ridge

The schist of Portal Ridge forms a sliver northeast of the San Andreas fault on the southwest edge of the Mojave Desert (Evans, 1966; Ross, 1976). This body likely correlates with the schist of Sierra de Salinas and shows similar high-grade metamorphism (middle to upper amphibolite facies). Nonetheless, we are impressed by the likeness in composition and proportion of rock types to the more typical parts of the schist. We analyzed one sample from this area (PR150).

San Emigdio Mountains

The Rand Schist in the San Emigdio Mountains is the most recently recognized of the schist bodies (James, 1986; James and Mattinson, 1988; Ross, 1989). It is not well exposed and has not received detailed study. Nonetheless, descriptions by the above workers and our own limited observations give us no reason to doubt that this body is correlative with the type Rand Schist. James (1986) and James and Mattinson (1988) concluded that the schist in this area was intruded after metamorphism by the 131 Ma tonalite of Antimony Peak. This implies an Early Cretaceous or older age for both the protolith and metamorphism. Ross (1989), however, mapped the contact between the schist and tonalite as a fault. Our one sample (99–57) was collected near the contact and exhibits a protomylonitic texture. Based on this deformational fabric, the fact that we observed no igneous intrusions within the schist, and our finding that detrital zircons in the schist are younger than the age of the tonalite, we concur with Ross (1989) that the contact is a fault.

Rand Mountains

The Rand Mountains comprise the type locality of the Rand Schist (Hulin, 1925; Silver et al., 1984; Silver and Nourse, 1986; Postlethwaite and Jacobson, 1987). Approximately 3 km of structural section is exposed in a post-metamorphic antiform. We analyzed one sample (RA169) from the structurally deepest, epidote-blueschist part of the section in the core of the antiform and one sample (RA170) belonging to the albite-epidote amphibolite facies from the top of the section in the southwestern part of the range. Total structural thickness in the area is about 3 km.

Sierra Pelona

The Sierra Pelona is the type locality of the Pelona Schist (Hershey, 1912; Harvill, 1969). Metamorphism ranges from greenschist to lower amphibolite facies and exhibits a clear inverted zonation (Graham and Powell, 1984). Our samples come from the middle (98–241, SP10) to lower (SP25D) parts of the section. Samples SP10 and SP25D were both collected adjacent to bodies of metabasite. The locality of SP25D is relatively well known because it is the only place in the entire schist terrane where relict pillow structure is evident in the metabasite (Haxel et al., 1987; Dawson and Jacobson, 1989). Absence of primary igneous structures elsewhere is presumably due to the intense deformation and recrystallization.

Blue Ridge (San Gabriel Mountains)

Pelona Schist in the Blue Ridge area of the San Gabriel Mountains occurs as a sliver between the Punchbowl and San Andreas faults. Although the Blue Ridge body currently sits immediately adjacent to Pelona Schist on the southwest side of the Punchbowl fault (East Fork body), this juxtaposition is thought to be coincidental. Most workers consider that the Blue Ridge body has been offset from the east end of the Sierra Pelona (Dibblee, 1967, 1968; Ehlig, 1968). One sample (98–240) was analyzed from this area and lies within the albite-epidote amphibolite facies.

East Fork (San Gabriel Mountains)

Pelona Schist exposed along the East Fork of the San Gabriel River was first studied by Ehlig (1958), who recognized most of the relations still considered key to understanding the schists. These include the fact that the schist originated in the lower plate of a major regional thrust fault, the presence of inverted metamorphism related to underthrusting, the relatively high-P/low-T style of metamorphism, and the eugeoclinal nature of the protolith and its likely Mesozoic age. The East Fork area continues to draw attention because it preserves a 1-km-thick zone of mylonite at the base of the upper plate that appears to be a remnant of the original underthrust (Ehlig, 1958, 1968, 1981; Jacobson, 1983a, 1983b, 1997). In most other areas, the contact between schist and upper plate appears to be an exhumation

structure. The East Fork area exposes a structural thickness of 3–4 km. Metamorphism ranges from lower greenschist facies at the base to uppermost greenschist facies at the contact with the upper plate. We analyzed four samples, two from near the top of the section (98–237, SG69), one from the middle (SG532), and one from the base (SG533). Sample SG69 was collected adjacent to metabasite. Sample 98–237 was collected near metachert, which, in turn, is usually associated with metabasite.

Orocopia Mountains

The Orocopia Mountains form the type locality of the Orocopia Schist. However, according to the pre–San Andreas fault reconstructions of Crowell (1962), Ehlig (1981), and Dillon and Ehlig (1993), the Orocopia Mountains formerly occupied a position adjacent to the Pelona Schist of Blue Ridge and the Sierra Pelona (Fig. 3). A different reconstruction is presented by Powell (1993) and Matti and Morton (1993), who place the Orocopia Mountains opposite the East Fork body of Pelona schist prior to offset on the San Andreas fault. In any case, the Orocopia Schist of the Orocopia Mountains is relatively far removed from the other bodies of Orocopia Schist we analyzed in southeasternmost California and southwestern Arizona (Fig. 3). Consequently, in pooling our results, we combined analyzes from the Orocopia Mountains with those from the Pelona Schist of the Sierra Pelona, Blue Ridge, and East Fork areas, rather than with those from the other bodies of Orocopia Schist.

Orocopia Schist in the Orocopia Mountains is exposed in a broad, northwest-trending post metamorphic arch (Crowell, 1975; Jacobson and Dawson, 1995; Robinson and Frost, 1996). Metamorphism is mostly in the albite-epidote amphibolite facies but locally reaches lowermost amphibolite facies. In the northwest Orocopia Mountains, we analyzed two samples from near the top of the section (OR77B, OR307; both associated with metabasalt) and five near the base (OR312A, OR113, OR314, OR337, and OR15A). In the southeastern Orocopia Mountains, we analyzed three albite-epidote amphibolite facies specimens (KE224031, KE24033, and KE127021) that were supplied to us by Kristen Ebert.

Peter Kane Mountain and Gavilan Hills

These are two closely spaced bodies of Orocopia Schist that lie within the southernmost part of the Chocolate Mountains. The Peter Kane Mountain area is distinctive in that the lower part of the section is intruded by widespread porphyry dikes and granodiorite plutons of latest Oligocene to earliest Miocene age (Haxel, 1977; Miller and Morton, 1977; Uselding et al., 2001). Hornfels texture is characteristic of the schist in this part of the section. The uppermost schist, however, is typical of that in other areas. The one sample from the Peter Kane Mountain body comes from the top of the section.

The Gavilan Hills are underlain by a relatively small outcropping of schist with an exposed structural thickness of only about 300 m (Haxel, 1977; Oyarzabal et al., 1997; Jacobson et al., 2002a). Both samples from this area (UG1417A, UG1500) come from the lower part of the section and were metamorphosed in the lowermost amphibolite facies. Sample UG1500 is associated with metabasite. It is an atypical variety rich in spessartine garnet and transitional to metachert.

Trigo Mountains and Marcus Wash

This area comprises one body of Orocopia Schist divided approximately in two by the Colorado River. The northern half, in Arizona, is part of the Trigo Mountains. The part south of the Colorado River in California will be referred to as the "Marcus Wash schist area." We also include in this grouping a small (1 × 2 km) outcropping of schist to the west of the Marcus Wash body located between Picacho Wash and White Wash (Haxel et al., 1985). The Trigo Mountains include a number of lenses of metabasite tens to hundreds of meters in length which Haxel et al. (2002) correlated with the 163 Ma metadiorite of the Chocolate Mountains (Mukasa et al., 1984) on the basis of field and petrographic character. None of the bodies in the Trigo Mountains have been dated, however. Nor have they been chemically analyzed to determine whether or not they show an arc geochemical signature as is the case for the metadiorite (Haxel et al., 1987, 2002). We analyzed detrital zircons for two samples from the Trigo Mountains (TR23, TR24A), both collected from metagraywacke immediately adjacent to the above mafic bodies. A third sample was analyzed from the Marcus Wash body (MW10) and yet another from the body between Picacho Wash and White Wash (YN17).

Castle Dome Mountains

The Castle Dome Mountains are important because of their position far inboard of the continental margin. Schist is exposed as separate bodies in the southwestern and southeastern parts of the range (Haxel et al., 2002). Appearance is identical to that of Orocopia Schist in ranges to the west; i.e., we are confident that these bodies should be included in the schist terrane. The schist in the southwest Castle Dome Mountains is heavily intruded by dikes and small stocks of early Miocene rhyolite porphyry. Two samples (CD7, CD9) were analyzed from this area. Schist in the southeastern part of the range is much less intruded than that in the southwest. Two samples (KE4 and KE6) were analyzed from the southeast. Cursory examination of both areas suggests metamorphism in the lowermost amphibolite facies.

Neversweat Ridge

Neversweat Ridge includes the easternmost outcrop of Orocopia Schist (Haxel et al., 2002). As is the case in the southwest Castle Dome Mountains, schist at Neversweat Ridge is strongly intruded by early Miocene rhyolite porphyry, and much of it is converted to hornfels or altered to chlorite and clay minerals (Haxel et al., 2002). Nonetheless, abundance of graphitic poikiloblasts of sodic plagioclase in metagraywacke and local presence of metachert and metabasite provide convincing evidence that this body is correlative with the type Orocopia Schist. Two samples, NR-1 and NR-2, were analyzed from this area.

APPENDIX 3: U-Pb Analytical Techniques

Zircon crystals were hand-selected from heavy mineral concentrates ($\rho > 3.30$) obtained from the <250 μm size fraction. Grains were selected to be representative of the range of size, color, and morphology, with the exception that those with inclusions were avoided. Zircon grains mounted on double-sided tape were potted in epoxy, sectioned with 4000 grit SiC paper, and polished to 0.3 μm with polycrystalline diamond. Sample mounts were then ultrasonically cleaned and coated with Au. With the exception of Barth et al. (2003a) samples 02-342, 02-345, and 02-358, U-Pb ages were obtained using the University of California at Los Angeles Cameca ims 1270 ion microprobe (Dalrymple et al., 1999). A mass-filtered 10–20 nA $^{16}O^-$ beam with 22.5 kV impact energy was focused to a 30–35 μm spot. The region immediately above the sample surface was flooded with O_2 at a pressure of ~4 10^{-3} Pa. This has the effect of increasing Pb^+ yields by a factor of ~1.7. Secondary ions were extracted at 10 kV with an energy

band-pass of 50 eV. The mass spectrometer was tuned to obtain a mass resolution (~5000) that was sufficient to resolve the most troublesome molecular interferences (i.e., those adjacent to the Pb peaks). For most samples we took 15 cycles of measurements at $^{94}Zr_2^{16}O$, ^{204}Pb, ^{206}Pb, ^{207}Pb, ^{208}Pb, ^{232}Th, ^{238}U, and $^{238}U^{16}O$. Characteristics of the energy spectra of Th+, U+, and UO+ (relative to the other ions) lead us to apply small energy offsets so that these species could be measured at their maximum intensities.

For most samples we used ^{208}Pb as a proxy for common Pb (Compston et al., 1984). While this worked well for Th-poor samples, our initial analyses were forced to rely upon measurement of ^{204}Pb since ^{208}Pb and ^{232}Th were not determined (Jacobson et al., 2000). Nearly all of our samples yielded initially high common Pb signals derived from surficial contamination that rapidly decayed over the first several minutes of sputtering. Because of this behavior, we pre-sputtered the sample surface for about four minutes before taking measurements. During this interval, we performed ion imaging to align the analysis pit with the optical axis of the instrument and peak centering to fine-tune the mass calibration. Because sputtering proceeds more slowly at the edges of the crater than in the central region, resolvable surface contamination continued to contribute to the measured Pb signal after pre-sputtering had been completed. To further enhance the radiogenic lead yield, we cropped the ion beam with the field aperture to permit only ions sputtered from the central portion of the crater to be transmitted through the mass spectrometer.

The relative sensitivities for Pb and U were determined on reference zircon AS-3 (Paces and Miller, 1993) using a modified calibration technique described in Compston et al. (1984). Calibration data obtained from all of our standard measurements are displayed in Figure A1A and listed in Table A1. Assuming that analytical error alone is the main source of the observed variations, we estimate that the reproducibility of our $^{206}Pb/^{238}U$ apparent ages for such non-radiogenic samples is 3–5 % (Fig. A1B). This is reflected by the spread of $^{206}Pb/^{238}U$ determined for our AS-3 zircon standard data (Fig. A1C). More precise results (~2%) are anticipated for Proterozoic $^{207}Pb/^{206}Pb$ ages that are independent of the Pb-U relative sensitivity factor (Fig. A1D).

APPENDIX 4: $^{40}Ar/^{39}Ar$ Analytical Techniques

Hand-selected muscovite and biotite samples (~5 mg) were wrapped in copper foil and packed along with Fish Canyon sanidine flux monitors in quartz tubes that were evacuated and sealed. Sample results reported in this paper originated from three separate irradiations at the University of Michigan's Ford reactor (L67 position). In addition, several samples were irradiated in the McMaster Reactor (Ontario, Canada). See McDougall and Harrison (1999) for more information regarding these facilities and $^{40}Ar/^{39}Ar$ irradiation procedures. In all instances, ^{39}Ar production from ^{39}K (J-factor) was monitored with Fish Canyon sanidine (27.8 ± 0.3 Ma; Cebula et al., 1986) that had been interspersed with samples in each tube (1 cm spacing). Correction factors for nucleogenic K- and Ca-derived argon were determined by measuring K_2SO_4 and CaF_2 salts that were included with each irradiation. Because several irradiations were performed, it is inefficient to include further information here. Instead, we provide data reduction parameters relevant to each sample in the GSA Data Repository (see footnote 1). This information includes irradiation history and the date of $^{40}Ar/^{39}Ar$ analysis, all irradiation parameters (J, $^{40}Ar/^{39}Ar_K$, $^{38}Ar/^{39}Ar_K$, $^{36}Ar/^{37}Ar_{Ca}$, and $^{39}Ar/^{37}Ar_{Ca}$) instrumental backgrounds (m/e 40, 39,38,37,and 36), and mass discrimination (based upon blank-corrected measurement of atmospheric $^{40}Ar/^{36}Ar$).

Incremental heating was conducted with a double vacuum Ta furnace (details provided in Lovera et al., 1997). Temperature was generally increased from 500 to 1350 °C in 15-min intervals. Evolved gas was transferred by expansion and purified with an SAES ST-101 50 l/s

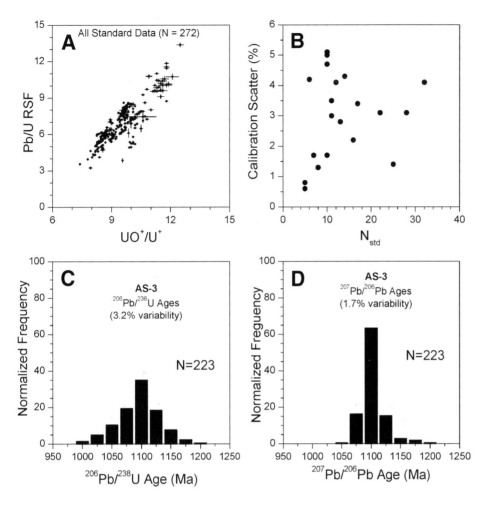

Figure A1. Summary of U-Pb calibration data (see Appendix 3). A: Plot of Pb/U relative sensitivity factor (RSF) vs. UO$^+$/U$^+$ for determined from measurements of AS-3 standard zircon. B: Mean reproducibility of calibration array used to define U-Pb relative sensitivity factor is 3%. C: Histogram of ^{206}Pb/^{238}U ages calculated for AS-3 standard zircon. D: Histogram of ^{207}Pb/^{206}Pb ages calculated for AS-3 standard zircon.

getter pump in a LABVIEW automated, all-stainless steel extraction line. Note that values quoted for absolute quantities of ^{39}Ar have been normalized to 100% gas delivery to the mass spectrometer. Although 66% of the gas was generally transferred to the mass spectrometer, quantities of gas that exceeded the linear range of the detection system were split statically according to previously calibrated procedures. Argon isotopic measurements were performed using an automated VG1200S mass spectrometer equipped with a Baur-Signar ion source and an axially fitted electron multiplier (Quidelleur et al., 1997). The instrument is typically operated at an Ar sensitivity of 4×10^{-17} mol/mV. Apparent ages were calculated using conventional decay constants and isotopic abundances (Steiger and Jäger, 1977).

REFERENCES CITED

Allmendinger, R.W., Figueroa, D., Snyder, D., Beer, J., Mpodozis, C., and Isacks, B.L., 1990, Foreland shortening and crustal balancing in the Andes at 30°S latitude: Tectonics, v. 9, p. 789–809.

Armstrong, R.L., and Suppe, J., 1973, Potassium-argon geochronometry of Mesozoic igneous rocks in Nevada, Utah, and southern California: Geological Society of America Bulletin, v. 84, p. 1375–1392.

Barth, A.P., 1990, Mid-crustal emplacement of Mesozoic plutons, San Gabriel Mountains, California, and implications for the geologic history of the San Gabriel terrane: Geological Society of America Memoir 174, p. 33–45.

Barth, A.P., and Schneiderman, J.S., 1996, A comparison of structures in the Andean orogen of northern Chile and exhumed midcrustal structures in southern California, USA: An analogy in tectonic style: International Geology Review, v. 38, p. 1075–1085.

Barth, A.P., Wooden, J.L., Tosdal, R.M., and Morrison, J., 1995, Crustal contamination in the petrogenesis of a calc-alkalic rock series: Josephine Mountain intrusion, California: Geological Society of America Bulletin, v. 107, p. 201–211.

Barth, A.P., Wooden, J.L., Grove, M., Jacobson, C.E., and Dawson, J.P., 2003a, U-Pb zircon geochronology of rocks in the Salinas Valley region of California: A reevaluation of the crustal structure and origin of the Salinian Block: Geology, v. 31, p. 517–520.

Barth, A.P., Coleman, D.S., Grove, M., Jacobson, C.E., Miller, B.V., and Wooden, J.L., 2003b, Geochronology of the Randsburg Granodiorite: Reevaluation of the tectonics of the southern Sierra Nevada and western Mojave Desert: Geological Society of America Abstracts with Programs, v. 35, no. 4, p. 70.

Barth, A.P., Wooden, J.L., Jacobson, C.E., and Probst, K., 2004, U-Pb geochronology and geochemistry of the McCoy Mountains Formation, southeastern California: A Cretaceous retro-arc foreland basin: Geological Society of America Bulletin (in press).

Bender, E.E., Morrison, J., Anderson, J.L., and Wooden, J.L., 1993, Early Proterozoic ties between two suspect terranes and the Mojave crustal block of the southwestern U.S.: Journal of Geology, v. 101, p. 715–728.

Bird, P., 1988, Formation of the Rocky Mountains, western United States: A continuum computer model: Science, v. 239, p. 1501–1507.

Burchfiel, B.C., and Davis, G.A., 1981, Mojave Desert and environs, *in* Ernst, W.G., ed., The geotectonic development of California (Rubey Vol. I): Englewood Cliffs, New Jersey, Prentice Hall, p. 217–252.

Burchfiel, B.C., Cowan, D.S., and Davis, G.A., 1992, Tectonic overview of the Cordilleran orogen in the western United States, *in* Burchfiel, B.C.,

TABLE A1. SUMMARY OF U-Pb CALIBRATION DATA

Samples Analyzed	Session	Calibration*[†] scatter (%)	UO⁺/U⁺[§] (Minimum)	UO⁺/U⁺[§] (Maximum)	Slope[†]	Intercept[†]	N_{std}	N_{unk}
98-240	09/18/98	1.3	7.38	8.70	1.5	−7.3	8	26
UG1417A	11/13/98	1.7	9.53	10.00	2.3	−14.4	7	27
UG1417A	11/28/98	0.6	10.90	11.30	0.8	1.9	5	18
98-241	02/06/99	2.8	8.98	9.34	2.5	−16.8	13	23
98-241	04/02/99	4.2	8.70	9.00	2.5	−16.1	6	27
98-240	04/04/99	3.4	8.44	10.00	1.6	−8.4	17	21
PR150, RA170	08/26/99	1.4	7.93	8.85	1.8	−9.7	25	79
99-57, RA169	11/29/99	4.1	8.19	8.97	1.7	−8.2	32	61
98-237, SG533	03/23/00	4.7	11.30	12.30	2.0	−14.6	10	42
SG533	03/23/00	3.5	10.90	12.30	1.0	−1.7	11	21
CD7, NR2	03/25/00	5.1	11.40	12.10	2.5	−18.8	10	68
CD9, MW10, OR30, PK114B, YN17	07/25/00	3.0	9.35	9.78	2.5	−16.1	11	50
SG69, SG532, SP10, SP25D	07/26/00	3.1	9.32	9.91	2.5	−16.3	22	40
OR-77B, OR307, PK114B	07/27/00	3.1	9.32	9.91	2.5	−16.1	28	23
KE4, KE6, NR1	03/17/01	4.1	8.93	10.10	2.5	−16.2	12	30
JM80-102, NR1	03/18/01	4.3	8.46	10.10	2.4	−15.9	14	29
TR23, TR24C	03/19/01	5.0	9.13	9.86	2.3	−14.9	10	12
TR23, TR24C	04/01/01	1.7	9.25	9.85	2.4	−16.6	10	22
DR-1656, UG1500	06/11/01	2.2	9.21	10.50	1.9	−11.5	16	38
JM80-102	06/13/01	0.8	10.00	10.60	1.4	−6.3	5	5
OR312A, OR313	01/30/02	2.9	7.76	8.25	2.0	−11.4	9	19
99-57, OR314	02/03/02	3.3	9.28	9.74	2.0	−12.6	8	20

*Calibration scatter defined as 100 x [1σ standard deviation of AS-3 standard measurements]/[mean ²⁰⁶Pb/²³⁸U age of AS-3 standard measurements].

[†]Pb-U relative sensitivity obtained from: RSF_{Pb-U} = slope x UO/U + intercept.

[§]Range of UO⁺/U⁺ values defined by the standard data.

Lipman, P.W., and Zoback, M.L., eds., The Cordilleran orogen: Conterminous U.S.: Boulder, Colorado, Geological Society of America, Geology of North America, v. G-3, p. 407–479.

Carson, C.J., Ague, J.J., Grove, M., Coath, C.D., and Harrison, T.M., 2002, U-Pb isotopic behavior of zircon during upper-amphibolite facies fluid infiltration in the Napier Complex, east Antarctica: Earth and Planetary Science Letters, v. 199, p. 287–310.

Cebula, G.T., Kunk, M.J., Mehnert, H.H., Naeser, C.W., Obradovich, J.D., and Sutter, J.F., 1986, The Fish Canyon Tuff, a potential standard for the ⁴⁰Ar/³⁹Ar and fission-track dating methods: TERRA Cognita, v. 6, p. 139–140.

Chen, J.H., and Moore, J.G., 1982, Uranium-lead isotopic ages from the Sierra Nevada batholith, California: Journal of Geophysical Research, v. 87, p. 4761–4784.

Cherniak, D.J., and Watson, E.B., 2000, Pb diffusion in zircon: Chemical Geology, v. 172, p. 5–24.

Compston, W., Williams, I.S., and Meyer, C.E., 1984, U-Pb geochronology of zircons from lunar breccia 73217 using a sensitive high mass-resolution ion microprobe: Journal of Geophysical Research B Supplement, v. 89, p. 525–534.

Coney, P.J., and Reynolds, S.J., 1977, Cordilleran Benioff zones: Nature, v. 270, p. 403–406.

Crowell, J.C., 1962, Displacement along the San Andreas fault, California: Boulder, Colorado, Geological Society of America Special Paper 71, 61 p.

Crowell, J.C., 1968, Movement histories of faults in the Transverse Ranges and speculations on the tectonic history of California, *in* Dickinson, W.R., and Grantz, A. eds., Proceedings of conference on geologic problems of San Andreas fault system: Stanford University Publications in the Geological Sciences, v. 11, p. 323–341.

Crowell, J.C., 1975, Geologic sketch of the Orocopia Mountains, southeastern California, *in* Crowell, J.C., ed., San Andreas fault in southern California: California Division of Mines and Geology Special Report 118, p. 99–110.

Crowell, J.C., 1981, An outline of the tectonic history of southeastern California, *in* Ernst, W.G., ed., The geotectonic development of California (Rubey Vol. I): Englewood Cliffs, New Jersey, Prentice Hall, p. 583–600.

Dalrymple, G.B., Grove, M., Lovera, O.M., Harrison, T.M., Hulen, J.B., and Lanphere, M.A., 1999, Age and thermal history of The Geysers plutonic complex (felsite unit), Geysers geothermal field, California: A ⁴⁰Ar/³⁹Ar and U-Pb study: Earth and Planetary Science Letters, v. 173, p. 285–298.

Davis, G.H., and Coney, P.J., 1979, Geologic development of the Cordilleran metamorphic core complexes: Geology, v. 7, p. 120–124.

Dawson, M.R., and Jacobson, C.E., 1989, Geochemistry and origin of mafic rocks from the Pelona, Orocopia, and Rand Schists, southern California: Earth and Planetary Science Letters, v. 92, p. 371–385.

DeGraaff Surpless, K., Graham, S.A., Wooden, J.L., and McWilliams, M.O., 2002, Detrital zircon provenance analysis of the Great Valley Group,

California: Evolution of an arc-forearc system: Geological Society of America Bulletin, v. 114, p. 1564–1580.

Dibblee, T.W., Jr., 1967, Areal geology of the western Mojave Desert, California: U.S. Geological Survey Professional Paper 522, 153 p.

Dibblee, T.W., Jr., 1968, Displacements on the San Andreas fault system in the San Gabriel, San Bernardino, and San Jacinto Mountains, southern California, *in* Dickinson, W.R., and Grantz, A., eds., Proceedings of Conference on Geologic Problems of San Andreas Fault System: Stanford, California, Stanford University Publications in the Geological Sciences, v. 11, p. 260–278.

Dickinson, W.R., 1996, Kinematics of transrotational tectonism in the California Transverse Ranges and its contribution to cumulative slip along the San Andreas transform fault system: Boulder, Colorado, Geological Society of America Special Paper 305, 46 p.

Dickinson, W.R., and Snyder, W.S., 1978, Plate tectonics of the Laramide orogeny, *in* Matthews, V., III, ed., Laramide folding associated with basement block faulting in the western United States: Boulder, Colorado, Geological Society of America Memoir 151, p. 355–366.

Dickinson, W.R., Hopson, C.A., and Saleeby, J.B., 1996, Alternate origins of the Coast Range Ophiolite (California): Introduction and implications: GSA Today, v. 6, no. 2, p. 1–10.

Dillon, J.T., 1976, Geology of the Chocolate and Cargo Muchacho Mountains, southeasternmost California, California [Ph.D. thesis]: Santa Barbara, University of California, 405 p.

Dillon, J.T., and Ehlig, P.L., 1993, Displacement on the southern San Andreas fault, *in* Powell, R.E., Weldon, R.J., II, and Matti, J.C., eds., The San Andreas fault system: Displacement, palinspastic reconstruction, and geologic evolution: Boulder, Colorado, Geological Society of America Memoir 178, p. 199–216.

Dillon, J.T., Haxel, G.B., and Tosdal, R.M., 1990, Structural evidence for northeastward movement on the Chocolate Mountains thrust, southeasternmost California: Journal of Geophysical Research, v. 95, p. 19953–19971.

Dodson, M.H., Compston, W., Williams, I.S., and Wilson, J.F., 1988, A search for ancient detrital zircons in Zimbabwean sediments: Journal of the Geological Society [London], v. 145, p. 977–983.

Dokka, R.K., and Ross, T.M., 1995, Collapse of southwestern North America and the evolution of early Miocene detachment faults, metamorphic core complexes, the Sierra Nevada orocline, and the San Andreas fault system: Geology, v. 23, p. 1075–1078.

Drobeck, P.A., Hillemeyer, F.L., Frost, E.G., and Liebler, G.S., 1986, The Picacho mine: A gold mineralized detachment in southeastern California, *in* Beatty, B., and Wilkinson, P.A.K., eds., Frontiers in geology and ore deposits of Arizona and the southwest: Arizona Geological Society Digest, v. 16, p. 187–221.

Ehlig, P.L., 1958, The geology of the Mount Baldy region of the San Gabriel Mountains, California, California [Ph.D. thesis]: Los Angeles, University of California, 195 p.

Ehlig, P.L., 1968, Causes of distribution of Pelona, Rand, and Orocopia Schists along the San Andreas and Garlock faults, *in* Dickinson, W.R., and Grantz, A., eds., Proceedings of Conference on Geologic Problems of San Andreas Fault System: Stanford University Publications in the Geological Sciences, v. 11, p. 294–306.

Ehlig, P.L., 1981, Origin and tectonic history of the basement terrane of the San Gabriel Mountains, central Transverse Ranges, *in* Ernst, W.G., ed., The geotectonic development of California (Rubey Volume I): Englewood Cliffs, New Jersey, Prentice Hall, p. 253–283.

Evans, J.G., 1966, Structural analysis and movements of the San Andreas fault zone near Palmdale, southern California [Ph.D. thesis]: Los Angeles, University of California, 186 p.

Froude, D.O., Ireland, T.R., Kinny, P.D., Williams, I.S., Compston, W., Williams, I.R., and Myers, J.S., 1983, Ion microprobe identification of 4,100–4,200-Myr-old terrestrial zircons: Nature, v. 304, p. 616–618.

Geisler, T., Ulonska, M., Schleicher, H., Pidgeon, R.T., and van Bronswijk, W., 2001, Leaching and differential recrystallization of metamict zircon under experimental hydrothermal conditions: Contributions to Mineralogy and Petrology, v. 141, p. 53–65.

George, P.G., and Dokka, R.K., 1994, Major Late Cretaceous cooling events in the eastern Peninsular Ranges, California, and their implications for Cordilleran tectonics: Geological Society of America Bulletin, v. 106, p. 903–914.

Graham, C.M., and England, P.C., 1976, Thermal regimes and regional metamorphism in the vicinity of overthrust faults: An example of shear heating and inverted metamorphic zonation from southern California: Earth and Planetary Science Letters, v. 31, p. 142–152.

Graham, C.M., and Powell, R., 1984, A garnet-hornblende geothermometer: Calibration, testing, and application to the Pelona Schist, southern California: Journal of Metamorphic Geology, v. 2, p. 13–31.

Grove, M., and Bebout, G.E., 1995, Cretaceous tectonic evolution of coastal southern California: Insights from the Catalina Schist: Tectonics, v.14, p.1290–1308.

Grubensky, M.J., and Bagby, W.C., 1990, Miocene calc-alkaline magmatism, calderas, and crustal evolution in the Kofa and Castle Dome Mountains, southwestern Arizona: Journal of Geophysical Research, v. 95, p. 19989–20003.

Hall, C.A., Jr., 1991, Geology of the Point Sur-Lopez Point region, Coast Ranges, California: A part of the southern California allochthon: Boulder, Colorado, Geological Society of America Special Paper 266, 40 p.

Hamilton, W., 1987, Mesozoic geology and tectonics of the Big Maria Mountains region, southeastern California, *in* Dickinson, W.R., and Klute, M.A., eds., Mesozoic rocks of southern Arizona and adjacent areas: Arizona Geological Society Digest, v. 18, p. 33–47.

Hamilton, W., 1988, Tectonic setting and variations with depth of some Cretaceous and Cenozoic structural and magmatic systems of the western United States, *in* Ernst, W.G., ed., Metamorphism and crustal evolution of the western United States (Rubey Vol. VII): Englewood Cliffs, New Jersey, Prentice Hall, p. 1–40.

Harrison, T.M., Yin, A., Grove, M., Lovera, O.M., and Ryerson, F.J., 2000, The Zedong Window: A record of superposed Tertiary convergence in southeastern Tibet: Journal of Geophysical Research, v. 105, p. 19211–19230.

Harvill, L.L., 1969, Deformational history of the Pelona Schist, northwestern Los Angeles County, California [Ph.D. thesis]: Los Angeles, University of California, 117 p.

Haxel, G.B., 1977, The Orocopia Schist and the Chocolate Mountain Thrust, Picacho-Peter Kane mountain area, southeasternmost California [Ph.D. thesis]: Santa Barbara, University of California, 277 p.

Haxel, G.B., and Dillon, J.T., 1978, The Pelona-Orocopia Schist and Vincent–Chocolate Mountain thrust system, southern California, *in* Howell, D.G., and McDougall, K.A., eds., Mesozoic paleogeography of the western United States: Los Angeles, Society of Economic Paleontologists and Mineralogists Pacific Section Pacific Coast Paleogeography Symposium 2, p. 453–469.

Haxel, G.B., and Tosdal, R.M., 1986, Significance of the Orocopia Schist and Chocolate Mountains thrust in the late Mesozoic tectonic evolution of the southeastern California-southwestern Arizona region, *in* Beatty, B., and Wilkinson, P.A.K., eds., Frontiers in geology and ore deposits of Arizona and the Southwest: Arizona Geological Society Digest, v. 16, p. 52–61.

Haxel, G.B., Tosdal, R.M., and Dillon, J.T., 1985, Tectonic setting and lithology of the Winterhaven Formation: A new Mesozoic stratigraphic unit in southeasternmost California and southwestern Arizona: U.S. Geological Survey Bulletin 1599, 19 p.

Haxel, G.B., Budahn, J.R., Fries, T.L., King, B.W., White, L.D., and Aruscavage, P.J., 1987, Geochemistry of the Orocopia Schist, southeastern California: Summary, *in* Dickinson, W.R., and Klute, M.A., eds., Mesozoic rocks of southern Arizona and adjacent areas: Arizona Geological Society Digest, v. 18, p. 49–64.

Haxel, G.B, Jacobson, C.E., Richard, S.M., Tosdal, R.M., and Grubensky, M.J., 2002, The Orocopia Schist in southwest Arizona: Early Tertiary oceanic rocks trapped or transported far inland, *in* Barth, A.P., ed., Contributions to crustal evolution of the southwestern United States: Boulder, Colorado, Geological Society of America Special Paper 365, p. 99–128.

Hershey, O.H., 1912, The Belt and Pelona series: American Journal of Science, v. 34, p. 263–273.

Högdahl, K., Gromet, L.P., and Broman, C., 2001, Low P-T Caledonian resetting of U-rich Paleoproterozoic zircons, central Sweden: American Mineralogist, v. 86, p. 534–546.

Hulin, C.D., 1925, Geology and ore deposits of the Randsburg quadrangle, California: California State Mining Bureau Bulletin 95, 152 p.

Ingersoll, R.V., 1983, Petrofacies and provenance of Late Mesozoic forearc basin, northern and central California: American Association of Petroleum Geologists Bulletin, v. 67, p. 1125–1142.

Ingersoll, R.V., 1997, Phanerozoic tectonic evolution of central California and environs: International Geology Review, v. 39, p. 957–972.

Ireland, T.R., 1992, Crustal evolution of New Zealand—Evidence from age distributions of detrital zircons in western province paragneisses and Torlesse greywacke: Geochimica et Cosmochimica Acta, v. 56, p. 911–920.

Isacks, B.L., 1988, Uplift of the central Andean plateau and bending of the Bolivian orocline: Journal of Geophysical Research, v. 93, p. 3211–3231.

Jacobson, C.E., 1983a, Structural geology of the Pelona Schist and Vincent thrust, San Gabriel Mountains, California: Geological Society of America Bulletin, v. 94, p. 753–767.

Jacobson, C.E., 1983b, Relationship of deformation and metamorphism of the Pelona Schist to movement on the Vincent thrust, San Gabriel Mountains, southern California: American Journal of Science, v. 283, p. 587–604.

Jacobson, C.E., 1990, The ^{40}Ar/^{39}Ar geochronology of the Pelona Schist and related rocks, southern California: Journal of Geophysical Research, v. 95, p. 509–528.

Jacobson, C.E., 1995, Qualitative thermobarometry of inverted metamorphism in the Pelona and Rand Schists, southern California using calciferous amphibole in mafic schist: Journal of Metamorphic Geology, v. 13, p. 79–92.

Jacobson, C.E., 1997, Metamorphic convergence of the upper and lower plates of the Vincent thrust, San Gabriel Mountains, southern California: Journal of Metamorphic Geology, v. 15, p. 155–165.

Jacobson, C.E., and Dawson, M.R., 1995, Structural and metamorphic evolution of the Orocopia Schist and related rocks, southern California: Evidence for late movement on the Orocopia fault: Tectonics, v. 14, p. 933–944.

Jacobson, C.E., Dawson, M.R., and Postlethwaite, C.E., 1988, Structure, metamorphism, and tectonic significance of the Pelona, Orocopia, and Rand Schists, southern California, *in* Ernst, W.G., ed., Metamorphism and crustal evolution of the western United States (Rubey Vol. VII): Englewood Cliffs, New Jersey, Prentice Hall, p. 976–997.

Jacobson, C.E., Oyarzabal, F.R., and Haxel, G.B., 1996, Subduction and exhumation of the Pelona-Orocopia-Rand schists, southern California: Geology, v. 24, p. 547–550.

Jacobson, C.E., Barth, A.P., and Grove, M., 2000, Late Cretaceous protolith age and provenance of the Pelona and Orocopia Schists, southern California: Implications for evolution of the Cordilleran margin: Geology, v. 28, p. 219–222.

Jacobson, C.E., Grove, M., Stamp, M.M., Vucic, A., Oyarzabal, F.R., Haxel, G.B., Tosdal, R.M., and Sherrod, D.R., 2002a, Exhumation history of the Orocopia Schist and related rocks in the Gavilan Hills area of southeasternmost California, *in* Barth, A.P., ed., Contributions to crustal evolution of the southwestern United States: Boulder, Colorado, Geological Society of America Special Paper 365, p. 129–154.

Jacobson, C.E., Grove, M., Barth, A.P., Pedrick, J.N., and Vucic, A., 2002b, "Salmon tectonics" as a possible explanation for Laramide sedimentation and underplating of schist in southern California and southwestern Arizona: Geological Society of America Abstracts with Programs, v. 34, no. 6, p. 510.

James, E.W., 1986, U/Pb age of the Antimony Peak tonalite and its relation to Rand Schist in the San Emigdio Mountains, California: Geological Society of America Abstracts with Programs, v. 18, no. 2, p. 121.

James, E.W., and Mattinson, J.M., 1988, Metamorphic history of the Salinian block: An isotopic reconnaissance, *in* Ernst, W.G., ed., Metamorphism and crustal evolution of the western United States (Rubey Vol. VII): Englewood Cliffs, New Jersey, Prentice Hall, p. 938–952.

Jennings, C.W., compiler, 1977, Geologic map of California: California Division of Mines and Geology, scale 1:750,000, 1 sheet.

Kanter, L.R., and McWilliams, M.O., 1982, Rotation of the southernmost Sierra Nevada, California: Journal of Geophysical Research, v. 87, p. 3819–3830.

Kies, R.P., and Abbott, P.L., 1983, Rhyolite clast populations and tectonics in the California continental borderland: Journal of Sedimentary Petrology, v. 53, p. 461–475.

Kistler, R.W., and Champion, D.E., 2001, Rb-Sr whole-rock and mineral ages, K-Ar, ^{40}Ar/^{39}Ar, and U-Pb mineral ages, and Strontium, Lead, Neodymium, and Oxygen isotopic compositions for granitic rocks from the Salinian composite terrane, California: U.S. Geological Survey Open File Report 01-453, 84 p.

Kistler, R.W., and Peterman, Z.E., 1978, Reconstruction of crustal blocks in California on the basis of initial strontium isotopic compositions of Mesozoic granitic rocks: U.S. Geological Survey Professional Paper 1071, 17 p.

Kröner, A., Jaeckel, P., and Williams, I.S., 1994, Pb-loss patterns in zircons from a high-grade metamorphic terrain as revealed by different dating methods: U–Pb and Pb–Pb ages for igneous and metamorphic zircons from northern Sri Lanka: Precambrian Research, v. 66, p. 151–181.

Lee, J.K.W., Williams, I.S., and Ellis, D.J., 1997, Pb, U and Th diffusion in natural zircon: Nature, v. 390, p. 159–161.

Linn, A.M., DePaolo, D.J., and Ingersoll, R.V., 1992, Nd-Sr isotopic, geochemical, and petrographic stratigraphy and paleotectonic analysis: Mesozoic Great Valley forearc sedimentary rocks of California: Geological Society of America Bulletin, v. 104, p. 1264–1279.

Livaccari, R.F., and Perry, F.V., 1993, Isotopic evidence for preservation of Cordilleran lithospheric mantle during the Sevier-Laramide orogeny, Western United States: Geology, v. 21, p. 719–722.

Lovera, O.M., Grove, M., Harrison, T.M., and Mahon, K.I., 1997, Systematic analysis of K-feldspar ^{40}Ar/^{39}Ar step-heating experiments I: Significance of activation energy determinations: Geochimica et Cosmochimica Acta, v. 61, p. 3171–3192.

Luyendyk, B.P., Kamerling, M.J., Terres, R.R., and Hornafius, J.S., 1985, Simple shear of southern California during Neogene time suggested by paleomagnetic declinations: Journal of Geophysical Research, v. 90, p. 12454–12466.

Maas, R., Kinny, P.D., Williams, I.S., Froude, D.O., and Compston, W., 1992, The earth's oldest known crust—A geochronological and geochemical study of 3900–4200 Ma old detrital zircons from Mt. Narryer and Jack Hills, Western Australia: Geochimica et Cosmochimica Acta, v. 56, p. 1281–1300.

MacPherson, G.J., Phipps, S.P., and Grossman, J.N., 1990, Diverse sources for igneous blocks in Franciscan mélanges, California Coast Ranges: Journal of Geology, v. 98, p. 845–862.

Malin, P.E., Goodman, E.D., Henyey, T.L., Li, Y.G., Okaya, D.A., and Saleeby, J.B., 1995, Significance of seismic reflections beneath a tilted exposure of deep continental crust, Tehachapi Mountains, California: Journal of Geophysical Research, v. 100, p. 2069–2087.

Matti, J.C., and Morton, D.M., 1993, Paleogeographic evolution of the San Andreas fault in southern California: A reconstruction based on a new cross-fault correlation, *in* Powell, R.E., Weldon, R.J., and Matti, J.C., eds., The San Andreas fault system: Displacement, palinspastic reconstruction and geologic evolution: Boulder, Colorado, Geological Society of America Memoir 178, p. 107–159.

Mattinson, J.M., 1978, Age, origin, and thermal histories of some plutonic rocks from the Salinian block of California: Contributions to Mineralogy and Petrology, v. 67, p. 233–245.

Mattinson, J.M., 1990, Petrogenesis and evolution of the Salinian magmatic arc, *in* Anderson, J.L., ed., The nature and origin of Cordilleran magmatism: Boulder, Colorado, Geological Society of America Memoir 174, p. 237–250.

Mattinson, J.M., and James, E.W., 1985, Salinian block U/Pb age and isotopic variations: Implications for origin and emplacement of the Salinian terrane, *in* Howell, D.G., ed., Tectonostratigraphic terranes of the circum-Pacific region: Houston, Texas, Circum-Pacific Council for Energy and Mineral Resources Earth Science Series, no. 1, p. 215–226.

McDougall, I., and Harrison, T.M., 1999, Geochronology and thermochronology by the ^{40}Ar/^{39}Ar method, second edition: New York, Oxford University Press, 269 p.

Mezger, K., and Krogstad, E.J., 1997, Interpretation of discordant U-Pb zircon ages: An evaluation: Journal of Metamorphic Geology, v. 15, p. 127–140.

Miller, F.K., and Morton, D.M., 1977, Comparison of granitic intrusions in the Pelona and Orocopia Schists, southern California: U.S. Geological Survey Journal of Research, v. 5, p. 643–649.

Miller, J.S., Glazner, A.F., and Crowe, D.E., 1996, Muscovite-garnet granites in the Mojave Desert: Relation to crustal structure of the Cretaceous arc: Geology, v. 24, p. 335–338.

Miller, C.F., Hatcher, R.D., Harrison, T.M., Coath, C.D., and Gorisch, E.B., 1998, Cryptic crustal events elucidated through zone imaging and ion microprobe studies of zircon, southern Appalachian Blue-Ridge, North Carolina-Georgia: Geology, v. 26, p. 419–422.

Minch, J.A., 1979, The late-Mesozoic–early Tertiary framework of continental sedimentation, northern Peninsular Ranges, Baja California, Mexico, *in* Abbott, P.L., ed., Eocene depositional systems, San Diego: San Diego, Pacific Section, Society of Economic Paleontologists and Mineralogists, p. 43–68.

Mojzsis, S.J., and Harrison, T.M., 2002, Establishment of a 3.83-Ga magmatic age for the Akilia tonalite (southern West Greenland): Earth and Planetary Science Letters, v. 202, p. 563–576.

Mukasa, S.B., Dillon, J.T., and Tosdal, R.M., 1984, A Late Jurassic minimum age for the Pelona-Orocopia Schist protolith, southern California: Geological Society of America Abstracts with Programs, v. 16, p. 323.

Oyarzabal, F.R., Jacobson, C.E., and Haxel, G.B., 1997, Extensional reactivation of the Chocolate Mountains subduction thrust in the Gavilan Hills of southeastern California: Tectonics, v. 16, p. 650–661.

Paces, J.B., and Miller, J.D., 1993, Precise U-Pb age of Duluth Complex and related mafic intrusions, northeastern Minnesota: Geochronological insights into physical, petrogenetic, paleomagnetic, and tectonomagmatic processes associated with the 1.1 Ga midcontinent rift system: Journal of Geophysical Research, v. 98, p. 13997–14013.

Pickett, D.A., and Saleeby, J.B, 1993, Thermobarometric constraints on the depth of exposure and conditions of plutonism and metamorphism at deep levels of the Sierra Nevada batholith, Tehachapi Mountains, California: Journal of Geophysical Research, v. 98, p. 609–629.

Pidgeon, R.T., O'Neil, J.R., and Silver, L.T., 1966, Uranium and lead isotopic stability in a metamict zircon under experimental hydrothermal conditions: Science, v. 154, p. 1538–1540.

Platt, J.P., 1986, Dynamics of orogenic wedges and the uplift of high-pressure metamorphic rocks: Geological Society of America Bulletin, v. 97, p. 1037–1053.

Postlethwaite, C.E., and Jacobson, C.E., 1987, Early history and reactivation of the Rand thrust, southern California: Journal of Structural Geology, v. 9, p. 195–205.

Powell, R.E., 1993, Balanced palinspastic reconstruction of pre-late Cenozoic paleogeology, southern California, *in* Powell, R.E., Weldon, R.J., and Matti, J.C., eds., The San Andreas fault system: Displacement, palinspastic reconstruction, and geologic evolution: Boulder, Colorado, Geological Society of America Memoir 178, p. 1–106.

Quidelleur, X., Grove, M., Lovera, O.M., Harrison, T.M., Yin, A., and Ryerson, F.J., 1997, The thermal evolution and slip history of the Renbu Zedong thrust, southeastern Tibet: Journal of Geophysical Research, v. 102, p. 2659–2679.

Reynolds, S.J., 1988, Geologic map of Arizona: Arizona Geological Survey Map 26, scale 1:1,000,000, 1 sheet.

Richard, S.M., 1993, Palinspastic reconstruction of southeastern California and southwestern Arizona for the middle Miocene: Tectonics, v. 12, p. 830–854.

Richard, S.M., Reynolds, S.J., Spencer, J.E., and Pearthree, P.A., 2000, compilers, Geologic Map of Arizona: Arizona Geological Survey Map 35, scale 1:1,000,000, 1 sheet.

Robinson, K.L., and Frost, E.G., 1996, Orocopia Mountains detachment system: Progressive development of a tilted crustal slab and a half-graben sedimentary basin during regional extension, *in* Abbott, P.L., and Cooper, J.D., eds., American Association of Petroleum Geologists Field Conference Guide 73: Bakersfield, California, Pacific Section American Association of Petroleum Geologists, p. 277–284.

Ross, D.C., 1974, Map showing basement geology and location of wells drilled to basement, Salinian block, central and southern Coast Ranges, California: U.S. Geological Survey Miscellaneous Field Studies Map MF-588, scale 1:500,000, 1 sheet.

Ross, D.C., 1976, Metagraywacke in the Salinian block, central Coast Ranges, California—and a possible correlative across the San Andreas fault: U.S. Geological Survey Journal of Research, v. 4, p. 683–696.

Ross, D.C., 1984, Possible correlations of basement rocks across the San Andreas, San Gregorio-Hosgri, and Rinconada-Reliz-King City faults, California: U.S. Geological Survey Professional Paper 1317, 37 p.

Ross, D.C., 1989, The metamorphic and plutonic rocks of the southernmost Sierra Nevada, California, and their tectonic framework: U.S. Geological Survey Professional Paper 1381, 159 p.

Ross, D.C., and Brabb, E.E., 1973, Petrography and structural relations of granitic basement rocks in the Monterey Bay area, California: U.S. Geological Survey Journal of Research, v. 1, p. 273–282.

Rubatto, D., 2002, Zircon trace element geochemistry, partitioning with garnet and the link between U-Pb ages and metamorphism: Chemical Geology, v. 184, p. 123–138.

Saleeby, J.B., 1997, What happens to the California Great Valley province (GVP) at its northern and southern ends?: Geological Society of America Abstracts with Programs, v. 29, no. 5, p. 62.

Saleeby, J.B., Sams, D.B., and Kistler, R.W., 1987, U/Pb zircon, strontium, and oxygen isotopic and geochronological study of the southernmost Sierra Nevada batholith, California: Journal of Geophysical Research, v. 92, p. 10443–10466.

Sherrod, D.R., and Hughes, K.M., 1993, Tertiary stratigraphy of the southern Trigo Mountains, Ariz., and eastern Chocolate Mountains, Calif.: Picacho State Park area, *in* Sherrod, D.R., and Nielsen, J.E., eds., Tertiary stratigraphy of highly extended terranes, California, Arizona, and Nevada: U.S. Geological Survey Bulletin 2053, p. 189–191.

Silver, L.T., and Anderson, T.H., 1974, Possible left-lateral Early to Middle Mesozoic disruption of the southwestern North American craton margin: Geological Society of America Abstracts with Programs, v. 6, p. 955–956.

Silver, L.T., and Nourse, J.A., 1986, The Rand Mountains "thrust" complex in comparison with the Vincent thrust-Pelona Schist relationship, southern California: Geological Society of America Abstracts with Programs, v. 18, no. 2, p. 185.

Silver, L.T., Sams, D.B., Bursik, M.I., Graymer, R.W., Nourse, J.A., Richards, M.A., and Salyards, S.L., 1984, Some observations on the tectonic history of the Rand Mountains, Mojave Desert, California: Geological Society of America Abstracts with Programs, v. 16, no. 5, p. 333.

Simpson, C., 1990, Microstructural evidence for northeastward movement on the Chocolate Mountains fault zone, southeastern California: Journal of Geophysical Research, v. 95, p. 529–537.

Sinha, A.K., Wayne, D.M., and Hewitt, D.A., 1992, The hydrothermal stability of zircon: Preliminary experimental and isotopic studies: Geochimica Cosmochimica Acta, v. 56, p. 3551–3560.

Smalley, R.F., and Isacks, B.L., 1987, A high-resolution local network study of the Nazca plate Wadati-Benioff zone under western Argentina: Journal of Geophysical Research, v. 92, p. 13,903–13,912.

Steiger, R.H., and Jäger, E., 1977, Subcommission on geochronology: Convention on the use of decay constants in geo- and cosmochronology: Earth and Planetary Science Letters, v. 36, p. 359–362.

Stern, T.W., Bateman, P.C., Morgan, B.A., Newell, M.F., and Peck, D.L., 1981, Isotopic U-Pb ages of zircon from the granitoids of the central Sierra Nevada, California: U.S. Geological Survey Professional Paper 1185, 17 p.

Tosdal, R.M., 1984, Pelona-Orocopia schist protolith: Accumulation in a Middle Jurassic intra-arc basin: Geological Society of America Abstracts with Programs, v. 16, no. 5, p. 338.

Tosdal, R.M., 1990, Tectonics of the Mule Mountains thrust system southeast California and southwest Arizona: Journal of Geophysical Research, v. 95, p. 20035–20048.

Tosdal, R.M., Haxel, G.B., and Wright, J.E., 1989, Jurassic geology of the Sonoran Desert region, southern Arizona, southeastern California, and northernmost Sonora: Construction of a continental-margin magmatic arc, *in* Jenney, J.P., and Reynolds, S.J., eds., Geologic evolution of Arizona: Arizona Geological Society Digest, v. 17, p. 397–434.

Uselding, J.E., Jacobson, C.E., and Haxel, G.B., 2001, Tertiary structural development of Orocopia Schist and related lithologies in the Peter Kane Mountain area of SE California: Geological Society of America Abstracts with Programs, v. 33, no. 6, p. 74.

Vedder, J.G., Howell, D.G., and McLean, H., 1983, Stratigraphy, sedimentation, and tectonic accretion of exotic terranes, southern Coast Ranges, California, *in* Watkins, J.S., and Drake, C.L., eds., Studies in continental margin geology: American Association of Petroleum Geologists Memoir 34, p. 471–496.

Vucic, A., Grove, M., Jacobson, C.E., and Pedrick, J.N., 2002, Multi-stage exhumation history of the Orocopia Schist in the Orocopia Mountains of southeast California: Geological Society of America Abstracts with Programs, v. 34, no. 6, p. 83.

Wood, D.J., and Saleeby, J.B., 1997, Late Cretaceous-Paleocene extensional collapse and disaggregation of the southernmost Sierra Nevada batholith: International Geology Review, v. 39, p. 973–1009.

Woodford, A.O., 1960, Bedrock patterns and strike-slip faulting in southwestern California: American Journal of Science, v. 258-A, p. 400–417.

Yeats, R.S., 1968, Southern California structure, seafloor spreading, and history of the Pacific Basin: Geological Society of America Bulletin, v. 79, p. 1693–1702.

Yin, A., 2002, Passive-roof thrust model for the emplacement of the Pelona-Orocopia schist in southern California, United States: Geology, v. 30, p. 183–186.

Yin, A., Harrison, T.M., Ryerson, F.J., Chen, W., Kidd, W.S.F., and Copeland, P., 1994, Tertiary structural evolution of the Gangdese thrust system, southeastern Tibet: Journal of Geophysical Research, v. 99, p. 18,175–18,201.

MANUSCRIPT ACCEPTED BY THE SOCIETY JUNE 2, 2003

Geological Society of America
Special Paper 374
2003

Offset of Pliocene ramp facies at El Mangle by El Coloradito fault, Baja California Sur: Implications for transtensional tectonics

Markes E. Johnson
David H. Backus
Department of Geosciences, Williams College, Williamstown, Massachusetts 01267, USA

Jorge Ledesma-Vázquez
Facultad de Ciencias Marinas, Universidad Autónoma de Baja California, Ensenada, Baja California, México 22800

ABSTRACT

Near Loreto in Baja California Sur, México, the Pliocene–Pleistocene Cerro Mencenares volcanic complex is cut by north-south–trending faults compatible with an extensional tectonic history. In contrast, El Coloradito fault strikes onshore from the Gulf of California toward Cerro Mencenares on an azimuth of N55°W oblique to the dominant north-south structural pattern on land. A 30-m-high scarp traces part of this fault and defines the south side of the uplifted "El Mangle" block formed by a headland of Miocene volcanic rocks in the Comondú Group. El Coloradito fault is oriented parallel to active transform faults in the Gulf of California.

Development of the coast on the east flank of the Cerro Mencenares volcanic complex is recorded by a sequence of terrestrial and marine facies exposed in a unified sedimentary ramp near the mouth of Arroyo El Mangle. With a seaward inclination of 6°, the ramp succession sits unconformably on a 310-m-wide adesitic shelf. Several distinctive units extend 8 km across Ensenada El Mangle along the shoreface of the Gulf of California. The sequence represents a transgressive-regressive package that includes a succession of red claystones and tuffs, silicified debris from salt-tolerant land plants, conglomerate beds with an intertidal marine biota, and limestone dominated by pectens and echinoids. The ramp correlates with the middle Pliocene Piacenzian Stage based on tuffs near the base of the succession that yield a K/Ar age of 3.3 ± 0.5 Ma. Fossils from the limestone, including the echinoid *Clypeaster marquerensis* and sand dollar *Encope shepherdi*, confirm a middle- to upper-Pliocene position. The same limestone crops out on the opposite side of El Coloradito fault from the main ramp, which is buttressed against a sequence of columnar basalts and volcanic breccia. In this case, however, the ramp has a seaward dip of 12°. The elevation difference between adjacent limestone beds indicates a minimum uplift of 70 m for the upthrown block on the north side of El Coloradito fault. The exaggerated dip of the limestone also indicates tectonic over-steepening compared to the undisturbed ramp on the south side of El Coloradito fault. The abrupt westward termination of El Coloradito fault on north-south–trending normal faults that define El Mangle block suggest the geometry of a fracture zone reactivated by compression against the flanks of the Cerro Mencenares volcanic complex sometime after 3.3 Ma. Carbonate ramps are well documented elsewhere in the Pliocene of Baja California, but at El Mangle a major ramp that postdates orthogonal extension was disrupted by a transtensional oblique fault.

Keywords: Baja California Sur, Mexico; Pliocene strata; carbonate ramps; paleotopography; tectonics.

Johnson, M.E., Backus, D.H., and Ledesma-Vázquez, J., 2003, Offset of Pliocene ramp facies at El Mangle by El Coloradito Fault, Baja California Sur: Implications for transtensional tectonics, *in* Johnson, S.E., Paterson, S.R., Fletcher, J.M., Girty, G.H., Kimbrough, D.L., and Martín-Barajas, A., eds., Tectonic evolution of northwestern México and the southwestern USA: Boulder, Colorado, Geological Society of America Special Paper 374, p. 407–420. For permission to copy, contact editing@geosociety.org. © 2003 Geological Society of America.

INTRODUCTION

Rifting of Baja California from the mainland and the opening of a proto-gulf inland from the peninsula began 14–10 Ma through extensional tectonics distally associated with the Basin and Range province of western North America (Karig and Jensky, 1972; Stock and Hodges, 1989; Lyle and Ness, 1991). Changeover to the present tectonic regime of transform-rift faults in the Gulf of California is widely agreed to have occurred by middle Pliocene time ca. 3.5 Ma (Lonsdale, 1989; Umhoefer et al. 1994; Zanchi, 1994), although there is some evidence this changeover may have occurred earlier ca. 5.1–6.1 Ma (Oskin et al., 2001). Normal faulting also took place, in addition to strike-slip faulting, after this time of transition (Fletcher and Munguía, 2000). The changeover from dominantly extensional to transtensional tectonics resulted in the complete transfer of the Baja California peninsula from the North American plate to the Pacific plate through oblique plate divergence.

Rich in igneous and sedimentary rocks of Pliocene age, the Loreto area on the Gulf of California in Baja California Sur has attracted the attention of geologists interested in traditional stratigraphy (Anderson, 1950; McLean, 1989), paleontology (Durham, 1950; Piazza and Robba, 1998), geochemistry (Bigioggero et al., 1995), structural geology (Zanchi, 1994), and the relationship between regional tectonics and basin development (Umhoefer et al., 1994; Dorsey et al., 1995; Dorsey, 1997). The origin of the Loreto Basin is attributed to a transtensional stress regime (Zanchi, 1994; Umhoefer et al., 1994; Dorsey et al., 1995). Likewise, the Cerro Mencenares volcanic complex on the north edge of the Loreto Basin is a stratovolcano that displays geochemical changes associated with the middle Pliocene transition from extensional to transtensional tectonics in the Gulf of California (Bigioggero et al., 1995). The Loreto fault, which runs along the western side of the Loreto Basin, may merge into the Concepción fault, which runs parallel to Bahía Concepción in the neighboring Mulegé region 70–100 km northwest of Loreto, but the latter has a different origin because it bounds an extensional basin of middle to late Miocene age (Ledesma-Vázquez and Johnson, 2001). Similarly, all faults mapped on the Cerro Mencenares volcanic complex by Bigioggero et al. (1995) are oriented north-south, consistent with extensional tectonics and tilt blocks of a horst-and-graben style.

It is difficult to link any previously known fault systems onshore in the central part of the Baja California peninsula between the latitudes of Isla San Luis in the north to Isla San José in the south with fracture zones now active in the Gulf of California (Fig. 1). The purpose of this contribution is to describe coastal development, particularly the interactions of fault structures and sedimentary patterns found in a small part of the North Loreto Basin along the east flank of the Cerro Mencenares volcanic complex. Emphasis is given to the origin and timing of a distinctive fault scarp that follows an onshore-offshore trace at odds with the pattern of north-south extensional faults that are dominant through the central Baja California peninsula.

The concept of a carbonate ramp as "a gently sloping surface from shallow to deeper water, underlain by pellet grainstones and allied carbonate sediments" (Bates and Jackson, 1987) is essential to this study. Our only enlargement on this definition is that ramps may incorporate terrigenous sediments and fossils at the landward end, where they abut against shorelines. Carbonate ramps that accrued during early Pliocene time as a result of relative changes in sea level in the Gulf of California are well defined around paleoislands at Punta Chivato and San Nicolas (Johnson and Ledesma-Vázquez, 2001). These ramps sit on basal conglomerates derived from eroded andesite and show a radial pattern of offshore dip, on average 6°, around former islands. Preservation of such symmetry provides evidence for the interpretation of synsedimentary deposition and local tectonic quiescence. In contrast, the succession of ramp stratigraphy found at Punta Bajo in the southeastern part of the Loreto Basin is interpreted to have been tectonically over-steepened to the extent that the oldest ramp at the base of the sequence exhibits an inclination twice as great as the youngest ramp (Dorsey, 1997). Ramp buildups on paleoshores are worthy of study for additional reasons. First, they provide precise pinning points to changing sea level through geologic time (Goldstein and Franseen, 1995). Second, they may show marked variations in facies indicative of exposed as opposed to sheltered environments of deposition (Simian and Johnson, 1997). These paleoenvironmental factors and others related to the changing tectonic setting of El Mangle paleoshore on the Cerro Mencenares volcanic complex are treated in this study.

LOCATION AND GEOLOGICAL SETTING

Centered 30 km north of Loreto between Mexican Federal Highway 1 and the Gulf of California, the Cerro Mencenares volcanic complex covers a semicircular area more than 150 km^2 and reaches a maximum elevation of 790 m above sea level (Fig. 2). Zanchi (1994) mapped associated faults and Bigioggero et al. (1995) described the igneous petrology and stratigraphy of this large Pliocene-Pleistocene stratovolcano in detail. Along the Gulf coast, the volcanic complex is bounded by Ensenada San Basilio to the north and Ensenada El Mangle to the south. According to the map by Bigioggero et al. (1995), a large horst block of Miocene rocks from the Comondú Group borders Cerro Mencenares between ensenadas San Basilio and El Mangle. This narrow horst, herein designated as El Mangle block, is separated from Cerro Mencenares by a pair of aligned and deeply incised faults that lead inland from the opposing embayments on a north-south strike. The mouth of each valley is filled by Pliocene marine sediments left behind in the San Basilio and El Mangle embayments during a high stand in sea-level (Bigioggero et al., 1995). An earlier study by Zanchi (fig. 14 in Zanchi, 1994) correctly identifies these strata as limestone but mistakenly attributes them to the Lower Quaternary San Juan Formation.

This paper focuses on the stratigraphy and facies of a carbonate ramp south of Punta El Mangle (Fig. 2). Arroyo El Mangle is the formal name of the faulted stream valley that descends

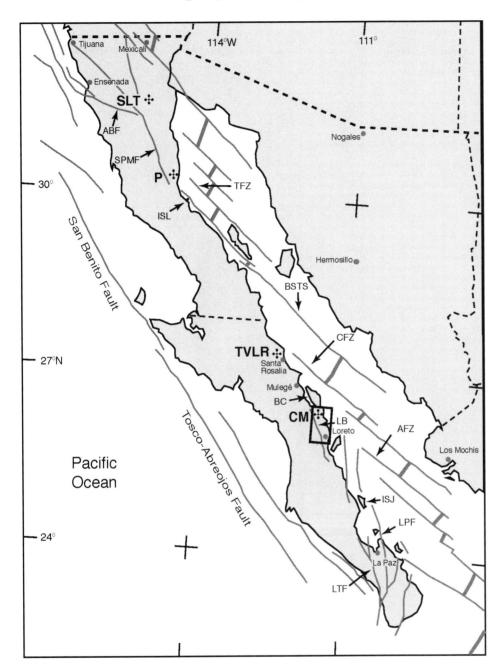

Figure 1. Map of Baja California and Gulf of California showing volcanic centers, major faults, and other features. From north to south, volcanic centers are Sierra Las Tinajas (SLT), Puertecitos (P), Tres Virgenes and La Reforma (TVLR), Cerro Mencenares (CM); Peninsular faults include Agua Blanca fault (ABF), San Pedro Mártir fault (SPMF), Los Tules fault (LTF), La Paz fault (LPF); faults associated with rifting in the Gulf of California include the Tiburon Fracture Zone (TFZ), Ballenas–Salsipuedes Transform System (BSTS), Carmen Fracture Zone (CFZ), Atl Fracture Zone (AFZ); islands include Isla San Luis (ISL), Isla San José (ISJ); Other features include Bahía Concepción (BC) and Loreto Basin (LB). Box marks location of Cerro Mencenares area detailed in Figure 2.

from the north between Cerro Mencenares and El Mangle block (topographic sheet Santa Rosa G12A78, Instituto Nacional de Estadistíca Geografía e Informática [INEGI]). Thick outcrops of Pliocene limestone are exposed on the shoreline both north and south of Arroyo El Mangle, but those to the south are more extensive and continuous (Fig. 3). The limestone caps a ramp sequence buttressed against volcanic hills. Contact between conglomerate beds and the volcanic rocks represents a distinct paleoshore. We follow the pioneering geological and paleontological studies by Anderson (1950) and Durham (1950) to correlate the local stratigraphic succession at El Mangle with other Pliocene sequences

around the Gulf of California. The Pliocene time scale used in this study, however, conforms to the global standards erected by Castradori et al. (1998) and Rio et al. (1998).

FACIES COMPOSITION OF EL MANGLE RAMP

Pliocene strata around El Mangle clearly overlap older andesitic basement rocks (Fig. 4). The onlapping beds reach nearly to the 40 m topographic level. We measured stratigraphic sections at four sites: three where the thicker, distal end of the ramp is truncated by the modern shoreline (Fig. 4, sites 1, 2a, and

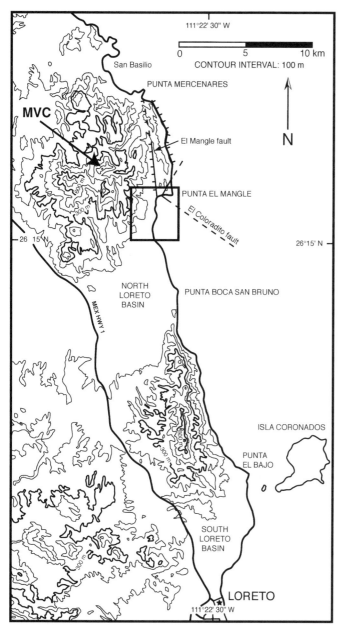

Figure 2. Map showing detailed topography of Cerro Mencenares north of Loreto, Baja California Sur, México. Arrow points to Mencenares volcanic center (MVC). Box outlines area of map detailed in Figure 4.

cation, sedimentary structures, and fossils. The unconformity at the base of this unit is poorly known, and the maximum exposed thickness is 5 m. The unit is interpreted as a hardpan deposit consisting of weathered andesite washed down slope from the surrounding volcanic hills. Tuff deposits occur locally within this unit. They presumably derive from the Cerro Mencenares volcanic complex, which is the closest possible source. A modern hardpan deposit with a brown-red crust occurs close to sea level immediately adjacent to Arroyo El Mangle (Fig. 4) and provides a similar environmental model for this unit.

Coastal Plain Deposit

Overlying the brown-red siltstone at localities 2b and 3 is a distinctive deposit up to 2 m in thickness and composed mainly of a friable mixture of silicified clay-size through sand-size particles with small clots of opal. Thin lenses of solid, gray-white opal occur throughout the deposit. The friable material yields abundant silicified plant twigs that account for as much as 20% of a given sediment sample. Most of the twigs, which are preserved well enough to retain the texture of their cortex cells, bear a striking resemblance to living *Atriplex* salt bushes growing at the mouth of Arroyo El Mangle (Mitchell et al., 2001). Less common are fragments of fossil wood large enough to retain limbs having recognizable knots. The deposit also contains tiny gastropods and salt-water diatoms identified as the extant species *Campylodiscus clypeus* (E. Ruck, 2001, personal commun., 2001).

We interpret this unit as a coastal-plain deposit that encroached shoreward over the basal hardpan deposit. This is compatible with the marine and non-marine fossils as well as the low-energy character of the underlying mud. We attribute opal replacement to circulation of silica-charged waters emanating from local hot springs. Contemporary hot springs still occur in the vicinity of El Mangle.

Intertidal Conglomerate Deposits

At locality 2b is a 2-m-thick unit that verges on a clast-supported conglomerate with rounded cobbles and small boulders of andesite. Virtually all of the igneous clasts are thickly encrusted with coralline red algae. Small pectens and gastropods occur within the carbonate matrix. Poorly preserved echinoid spines and barnacles are less common. A stratigraphically higher cobble-boulder bed is separated from the lower by limestone beds devoid of igneous clasts. Clasts in the upper conglomerate bed also are thickly coated by rinds of coralline red algae. Large oysters encrust boulders where conglomerate beds buttress against the basement andesite. We interpret the conglomerate beds as intertidal deposits, where large clasts eroded by wave action from the volcanic shoreface were rounded and sufficiently jostled to be colonized by coralline red algae. A conglomerate with similar encrustations of coralline red algae is known from the Upper Pleistocene of Bahía San Antonio near San Nicolas, where intertidal zonation is confirmed by the occurrence of fossil mollusks representing extant species in the Gulf of California (Johnson and Ledesma-Vázquez, 1999).

3) and one inland, where the proximal edge of the ramp nears the paleoshore (Fig. 4, site 2b). Appendix 1 contains the stratigraphic columns and unit descriptions for these sites, including general fossil content for all sections. We identified several distinct facies ubiquitous throughout the succession, as summarized below.

Hardpan and Tuff Deposits

The base of the sedimentary succession is represented by a brown-red siltstone with a variable clay fraction. It lacks stratifi-

Figure 3. View looking southwest over Pliocene ramp and paleoshoreline on flanks of Cerro Mencenares volcanic complex at El Mangle, taken from vantage point on uplifted El Mangle block.

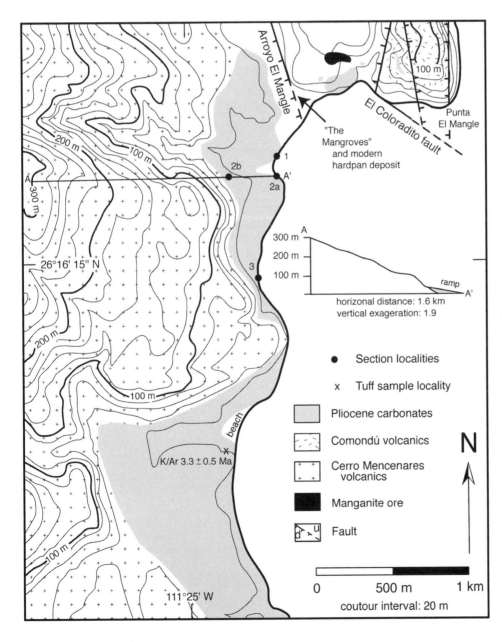

Figure 4. Detailed topographical and geological map of area around El Mangle. See inset in Figure 1 for regional location.

Limestone Deposit (Proximal and Distal Facies)

The most extensive component of the depositional suite at El Mangle is a limestone unit that caps most of the sedimentary package. Stratigraphic correlation of the sections at localities 2a and 2b demonstrates that this unit is wedge-shaped: the limestone unit thins to zero from east to west as it approaches the paleoshore (Fig. 5). Therefore, the conglomerate beds that sit stratigraphically above and below this unit (as described above) are contiguous and wrap around the landward terminus of the limestone wedge tracking the paleoshore through time. In limestone beds near the shoreline, the fossil assemblage includes jack-knife clams (*Tagelus californiaus*), ark shells (*Anadara multicostata*), and strombid gastropods (*Strombus subgracilior, S. granulatus*); clusters of the coral *Porites panamensis* also occur in these nearshore beds, particularly in the most sheltered part of the embayment on the north side of Arroyo El Mangle. We interpret these fossil assemblages as nearshore biofacies and assign to them a coastal setting.

The wedge-shaped limestone facies reach a maximum thickness of about 12 m at sites 1 and 2a, where the most distal beds are

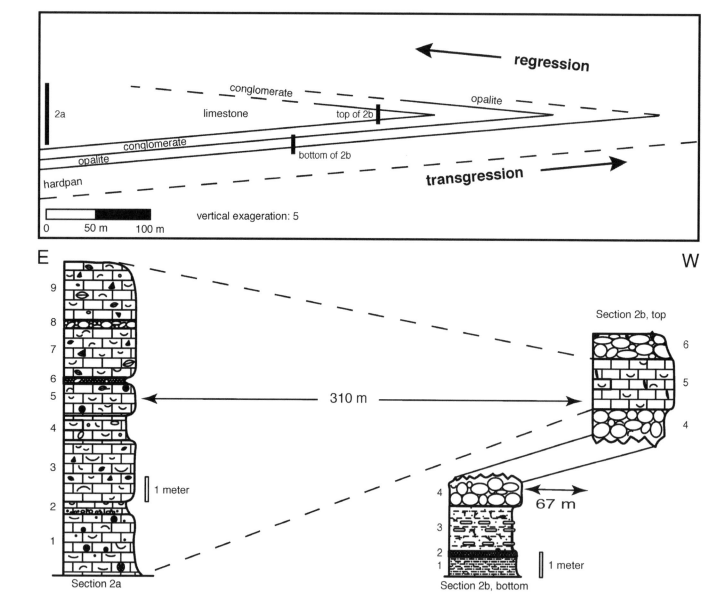

Figure 5. Stratigraphic columns for sections 2a and 2b (see transect in Fig. 4 for orientation). Numbers on left side of each stratigraphic column represent individual units described in full in Appendix 1. Lithological symbols follow standard patterns for limestone, conglomerate, and siltstone. Fossil content is represented symbolically: single curved line—disarticulated bivalve; two curved lines co-joined—articulated bivalve; circle with five-fold division—echinoids; spiral—gastropod. Box projects correlation of a stratigraphic wedge with respect to the interpretation of transgressive and regressive facies. Dashed lines indicate possible extent of regressive facies now removed by erosion.

exposed in coastal cliffs. Layers range from medium-bedded to very thickly bedded. The most abundant fossils are pectens, with *Aequipecten abietis* accounting for as much as 95% of sampled pecten specimens (R. Doménech, 1999, pers. commun.). *Nodipecten subnodosus* occurs less commonly. Oysters (*Myrakeena angelica*) are sparse but significant. Barnacles encrust some of the fossil pectens, and echinoderms are less common than molluscs. Two species occur, including the echinoid *Clypeaster marquerensis* and the sand dollar *Encope shepherdi*. These typically are well preserved, suggesting minimal transport. Found closely related to one another in the South Loreto Basin, *M. angelica* and *A. abietis* are interpreted by Piazza and Robba (1998, p. 245) as pointing "toward a water depth of approximately 5 m."

A distinct feature of the distal limestone deposit is the common inter-fingering of clast-supported conglomerate beds consisting principally of andesitic pebbles and cobbles. These tend to occur in profile as long stringers less than 0.5 m thick. Fossils are rare to absent in these beds. None of the clasts is encrusted by coralline red algae or barnacles. Conglomerate associated with these limestone beds presumably bypassed the nearshore limey deposits via channels that extended offshore from arroyos. Overall, the distal limestone deposit is regarded as having accumulated in a subtidal setting that was episodically affected by land-derived sediments flushed out of the volcanic hills by violent rainstorms. Such events are known as *chubascos* in Baja California. An unusually massive conglomerate bed more than 4 m thick occurs near the top of the section at site 1 (see Appendix 1) and probably represents a zone directly offshore a feeder arroyo. Given the geographic location of site 1 (Fig. 4), this conglomerate bed was derived from alluvial deposits scoured from the nearby fault valley of Arroyo El Mangle.

AGE AND INTERNAL CONFIGURATION OF THE RAMP SUCCESSION

A tuff bed immediately below the ramp succession near El Mangle is known to date from 3.3 ± 0.5 Ma (Berry et al., 2002). The site where the interval was sampled for K/Ar age analysis is marked on the map in Figure 4. There is good lateral continuity in the sedimentary package that constitutes the ramp succession exposed along the shores of Ensenada El Mangle. Thus, the age of the overlying terrestrial and marine facies in the package postdates 3.3 ± 0.5 Ma. McLean (1987) reported a similar age for the lowermost pyroclastic deposit in the southern end of the Loreto basin.

According to current standards established by the Commission on Stratigraphy under the International Union of Geological Sciences, the Pliocene Epoch is divided into three stages: the Zanclean Stage (lower Pliocene), Piacenzian Stage (middle Pliocene), and the Gelasian Stage (upper Pliocene). The base of the Piacenzian Stage is formally defined as correlative with a precessional excursion estimated to date from 3.6 Ma (Castradori et al., 1998). The base of the Gelasian Stage is formally defined in a similar way as dating from 2.5 Ma (Rio et al., 1998). Thus, the stratigraphic position of the ramp package at El Mangle is best accommodated by an assignment to the middle Pliocene Piacenzian Stage.

Traditional biostratigraphy as practiced in the Gulf of California region is in urgent need of better integration with current international standards for Neogene epochs and stages. The divisions of lower, middle, and upper Pliocene used in the Gulf region do not correspond to the boundaries now defined by global stratotype sections in Italy. According to Durham (1950, p. 41 and 48), *Clypeaster marquerensis* and *Encope shepherdi* are restricted in range to the upper Pliocene. In his stratigraphic arrangement, the Marquer Formation is equated with the upper Pliocene and the two previously mentioned echinoderms have a history of co-occurrence at localities such as Marquer Bay on Carmen Island and the Arroyo de Arce and Arroyo de Gua areas near Loreto (table 9 in Durham, 1950). *Aequipecten abietis* may have a wider Pliocene range, but it co-occurs with the forementioned echinoderms on Carmen Island and the arroyos north of Loreto (table 9 in Durham, 1950). The oyster *Myrakeena angelica* has a known range from the upper Miocene through the Pliocene (Smith, 1991). Taken together, the biostratigraphy of the common macrofossils from strata around El Mangle clearly indicate a late Pliocene age.

The relationship of the marine ramp with adjoining land is illustrated through cross section A–A' (Fig. 4). Having a seaward inclination of 6°, the entire ramp package sits unconformably on a 310-m-wide shelf eroded in andesitic rocks that forms the underlying basement of the Mencenares volcanic complex. These same basement rocks form uplands that fringe the sedimentary basin at El Mangle and rise steeply with a maximum 44% gradient (Fig. 4). Internal configuration of the ramp succession at El Mangle is best shown by the hinged nature of lower and upper conglomerate beds that stratigraphically wrap around the landward termination of limestone layers in the ramp core (box, Fig. 5). This arrangement clearly defines a flooding surface that follows a transgressive-regressive pathway. Particularly south of Arroyo El Mangle, the Pliocene ramp is preserved in pristine condition (Fig. 3).

EL MANGLE BLOCK AND EL COLORADITO FAULT

The photo in Figure 6 represents the southwestern corner of a tilted block near the southwestern corner of El Mangle block. This view shows the westward termination of a new fault we name El Coloradito. The fault is notable for a regionally uncharacteristic strike of N55°W with a dip of 54° to the south. Highly polished and bright red in color, El Coloradito fault was not identified by previous workers who described the geology of the area (Zanchi, 1994; Bigioggero et al., 1995). A fault breccia approximately 0.75 m thick underlies the fault surface and is primarily composed of andesitic clasts entrained from the Miocene volcaniclastic units offset by the fault. The Miocene units that compose the footwall of the fault lose their primary bedding characteristics and become more fractured as they are traced into the fault zone. Next to the fault zone, the dominant fracture becomes oriented parallel to the fault surface. The footwall rocks also lose their original coloration, becoming bright red as well as highly indurate as they are incorporated into the fault breccia.

Figure 6. View of southwestern corner of El Mangle block showing fault scarp at El Coloradito and profile of an uplifted Pliocene ramp.

El Coloradito fault is one of several faults that define the southwestern corner of a tilted block. A second fault with an initial orientation of N15°E and dip of 42°W cuts across the southern end of El Mangle block to the northeast with an overall trend of N40°E. A third fault surface oriented N24°W with a dip of 44°S, which we informally name the "Nose" fault, cuts off the corner where El Coloradito and the northeast-trending fault meet (see Fig. 6). Both of these fault surfaces show a red coloration and polished surfaces similar to El Coloradito fault, but they do not have similarly well-developed fault breccias. Fault striae are found on both the "Nose" fault (counter-clockwise 72° from horizontal) and the northeast-trending fault (clockwise 72° from horizontal). In contrast, no fault striations occur on any part of the El Coloradito fault surface we were able to reach.

Extending to the southeast from the corner of the block and along the trend of El Coloradito fault, the lithology of El Mangle block passes through volcanic breccia, followed by interbedded sandstone and mudstone strata, succeeded by columnar basalt and topped by an uppermost volcanic breccia. All units in this sequence dip 25–35°E on a strike of N34–40°E; the total exposed thickness of the package is approximately 115 m. A number of minor north-south–trending faults mapped by Zanchi (1994) extend northward from Punta El Mangle and offset this package of Comondú Group rocks.

Looking north across the Ensenada El Mangle, a 30-m-high fault scarp cuts through the succession as described above. Found close below the skyline in the same view (Fig. 6), a single 1-m-thick limestone ledge sits unconformably on eroded volcanic breccia beds. The limestone has the appearance of a carbonate ramp, but its dip is over-steepened at 12°, twice the average value for the development of a synsedimentary ramp.

The unit contains large oysters and distinct fragments of the coral *Porites panamensis*. Given the same faunal composition as strata exposed at sea level around Arroyo El Mangle, we conclude that the narrow limestone deposit represents a fringe of the same Pliocene ramp so fully exposed along much of the Ensenada El Mangle to the immediate south.

El Coloradito fault, therefore, offsets a small part of the original ramp and postdates deposition of the ramp sequence. Based on the elevation of the limestone ledge, which sits well above the 30-m-fault scarp on a slope eroded back from the scarp, a minimum uplift of 70 m for the southwestern corner of El Mangle block is estimated. Other Comondú rocks rise behind the limestone ledge to the north, and they are interpreted as representing the original paleoshore for this part of the Pliocene basin. The fact that different kinds of basement rocks underlie the same Pliocene limestone on opposite sides of El Coloradito fault implies that the limestone was deposited over a preexisting fracture zone that was reactivated sometime after ramp formation and lithification.

The southeastern corner of El Mangle block (out of view in Fig. 6 to the right) ends at Punta El Mangle. Two minor faults branch away from El Coloradito fault and pass northward through the block (Fig. 3). The faults cut through Comondú rocks composed of volcaniclastic strata that dip 28°E on a strike of N10°W. These Comondú rocks comprise a second and younger stratigraphic package, because they sit unconformably on the shoulder of the uppermost volcanic breccia that has a different strike and dip. The younger (upper) sequence is ~115 m in total exposed thickness. Based on the two distinct packages of stratified rocks exposed on the south side of El Mangle block, it is clear that the horst mapped by Bigioggero et al. (1995, Fig. 3) is not a stratigraphically unified Comondú block.

A sizable ore body of manganite along strike with El Colo-radito fault is located <300 m beyond the southwestern corner of El Mangle block. This ore body is adjacent to Pliocene carbonate deposits east of the normal fault that projects into Arroyo El Mangle (Fig. 4), although the association with El Coloradito fault is not clear. Geothermal alteration and discoloration of basalt columns inboard from El Coloradito fault on El Mangle block also suggest fault-related hydrothermal circulation.

DISCUSSION

Coherence of the Baja California Central Domain

The change in plate boundary from the Pacific coast to the nascent Gulf of California brings into question the tectonic stability of the Baja California Central Domain. Srike-slip faults that reach from the active gulf to the Pacific side of the Baja California peninsula are absent from the Baja California Central Domain over a distance of ~700 km between the latitudes of Isla San Luis in the north to Isla San José in the south (Lonsdale, 1989; Ness and Lyle, 1991). By contrast, extensive fault systems such as the Agua Blanca and San Pedro Mártir in the north or the La Paz and Los Tules in the south help define the Baja California northern and southern domains (Fig. 1). This implies a fundamentally different tectonic role for the Baja California Central Domain in contrast to domains to the north and south.

Termination of El Coloradito fault against the southeastern flank of the Cerro Mencenares volcanic complex (Fig. 2) suggests a linked geological and tectonic history between the emplacement of this Pliocene-Pleistocene stratovolcano and a failed breach of the Baja California Central Domain by faults activated or reactivated by transtensional stress in the Gulf of California. Axen (1995, fig. 4) divided the Gulf of California escarpment along the eastern edge of the Baja California peninsula into a series of several rift segments separated by accommodation zones. Although he noted some examples of accommodation zones that apparently control the location of syn- or post-tectonic volcanic centers at Sierra Las Tinajas, Puertecitos, and Tres Virgenes, no explicit causal relationship between oceanic fracture zones and accommodation zones was recognized (Axen, 1995).

Comparison of volcanic and tectonic elements in the El Mangle area with other volcanic centers along the gulf escarpment suggests that a reexamination of a possible causal relationship between these elements may be warranted. One of the best-studied accommodation zones–volcanic centers is found at Puertecitos (Fig. 1), where a connection between the volcanic province and the Ballenas–Salsipuedes Transform System, as well as the Tiburon Fracture Zone, is inferred (Lonsdale, 1989; Stock, 2000). Of particular interest is the potentially close relationship between the southern end of the San Pedro Mártir fault, which represents the main gulf escarpment northwest of the Puertecitos volcanic province (Gastil et al., 1975) and the Ballenas–Salsipuedes Transform System at the north end of the Baja California Central Domain (Fig. 1). In Baja Cali-

fornia Sur, Fabriol et al. (1999) collected seismic, gravity, and magnetic data that show a complex relationship between the Tortuga Volcanic Ridge and the Rosalia Volcanic Ridge, which sit offshore the La Reforma caldera. They further suggest that the Rosalia Volcanic Ridge lies along the projection of the Carmen Fracture Zone (Fig. 1). Though the ridge does not project directly into the La Reforma caldera, recent bathymetric studies of the region offshore of Santa Rosalia show a trench next to the Rosalia Volcanic Ridge that does project into the caldera region (Nava-Sánchez et al., 2001). In addition, the orientation of the trench changes abruptly where it is crossed by a north-south–trending fault, suggesting this region of the gulf has a more complex tectonic history than previously realized.

Despite the similarities between each of these volcanic centers, significant differences exist with respect to: (1) the age of erupted volcanics, (2) the extent to which the centers have been dissected by faults, and (3) how directly the centers and their related master faults can be connected to active oceanic fracture zones in the Gulf of California. Each center has a distinct history. The Puertecitos volcanic province is highly dissected by faults (Gastil et al., 1975; Martín-Barajas et al., 1997) but has the most obvious connection to active oceanic fracture zones. In contrast, the La Reforma volcano is young, relatively undissected by faults, and exhibits less obvious connections to tectonic features offshore (Fabriol et al., 1999). We speculate that each accommodation zone along the gulf escarpment had a connection with oceanic fracture zones in the Gulf of California at some point during its development, although the connection may no longer be obvious. Detailed studies are required that involve the stratigraphy and mapping of Pliocene basins as well as the detection and mapping of anomalous faults that diverge from north-south trends associated with extensional tectonics.

The smallest of fault lineaments along the margins of the Gulf coast, such as El Coloradito fault, are significant but easy to overlook. The Loreto-Mencenares area is a case in point. Umhofer et al. (1994) used Ar/Ar dating on tuffs to constrain the age of formation and subsequent episodes of faulting on unnamed units within the southern Loreto Basin to an interval ca. 2.5 Ma near early late Pliocene (Gelasian) time, when a minor change in the plate boundary was inferred to have occurred as an ongoing consequence of transtensional tectonics. The biostratigraphic evidence from around El Mangle, however, does not preclude the possibility that El Coloradito fault may already have been active during middle Pliocene (Piacenzian) time. A more careful analysis must be conducted with respect to the submarine bathymetry offshore El Mangle, but some preliminary data are available. Figure 8 in Nava-Sánchez et al. (2001) provides two profiles that define the San Bruno Trough bounded by the Mangle Bank in a position roughly parallel to the shoreface of Ensenada El Mangle. La Giganta Basin is the next basin offshore that impinges south of the Mangle Bank (Nava-Sánchez et al., 2001). Part of this basin has an appropriate shape to accommodate a fault trace on a direct heading to Punta El Mangle.

Integration with Pliocene Standards for Chronostratigraphy and Eustasy

The ramp stratigraphy so well preserved on the east margin of the Cerro Mencenares volcanic complex is a perfect match for the classic model of coastal onlap with marine transgression and regression promoted by Vail et al. (1977, fig. 4) for the evaluation of global sea-level changes from seismic patterns reflected by strata on continental shelves. In particular, the nested arrangement of nonmarine, littoral, and inner neritic facies is easy to recognize in the outcrops at El Mangle (Fig. 5). The stratigraphic pattern that results from multiple facies that advance shoreward and retreat seaward in a unified but diachronous fashion also is known as an "Israelsky wedge" (Prothero and Schwab, 1996, p. 355). This term reflects the pioneering contribution of the petroleum geologist who recognized subsurface facies patterns of transgression and regression using electric logs (Israelsky, 1949). Deposition of such facies in time and space along a well-defined and continuous paleoshore required gradual changes in global sea level and/or gentle tectonic uplift and subsidence of a local to regional nature. To sort out competing factors, it is important to reach a better integration of local stratigraphy with respect to global standards for Pliocene chronostratigraphy and eustasy. The pristine preservation of the succession, not withstanding the faulted northern edge of the ramp by El Coloradito fault, also necessitates the same considerations.

The local history and relationship of rocks and tectonic events is clarified by reference to Figure 7, which summarizes two independent sources of information. The first represents the global standard for the chronostratigraphic definition of stages within the Pliocene series by Rio et al. (1998) and Castradori et al. (1998). The second entails the global standard for Pliocene eustasy, as promulgated by Haq et al. (1988).

The most prominent transgressive-regressive excursions of the Pliocene occurred during the early part of that epoch and are recorded by strata from the Zanclean Stage. In Baja California Sur, these events are represented by ramped strata in the San Marcos Formation arrayed around paleoislands formed by tilted fault blocks of Miocene Comondú volcanics at Punta Chivato and San Nicolas (Johnson and Ledesma-Vázquez, 2001). One of the diagnostic index fossils for the lower Pliocene of Baja California Sur is the echinoid *Clypeaster bowersi* (Durham, 1950). The tilting of the fault blocks occurred prior to ramp formation at Punta Chivato and San Nicolas, due to extensional tectonics prevalent in the region during Miocene time (Ledesma-Vázquez and Johnson, 2001). In post Zanclean time, the ramp structures in northern Baja California Sur were left untouched by subsequent tectonic events associated with the continued opening of the Gulf of California.

Clypeaster bowersi is not found in the strata around El Mangle. Instead, its ecological place is filled by the index fossils *Clypeaster marquerensis* and *Encope shepherdi*, said to be restricted in range to the upper Pliocene of Durham (1950). Two eustatic transgresive-regressive events occurred during Piacenzian time, and one is believed to have taken place during subsequent Gelasian time (Fig. 7). Given the age of 3.3 ± 0.5 Ma for tuff near

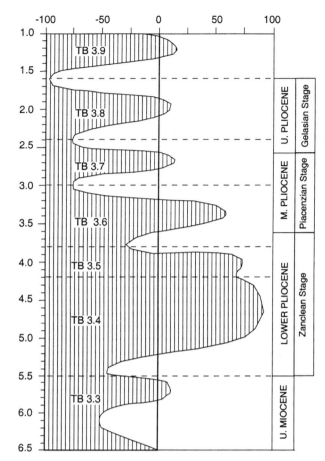

Figure 7. Pliocene chronostratigraphy (after Rio et al., 1998, and Castradori et al., 1998) integrated with Pliocene eustatic sea-level curve (after Haq et al., 1988). Abbreviation TB is a prefix (meaning Tejas, part B) to a numerical scheme that identifies each highstand in sea level recorded in strata from Upper Oligocene through Pleistocene.

the base of the ramp succession at El Mangle (Berry et al., 2002), it is possible that the transgressive-regressive event recorded by the ramp succession at El Mangle corresponds to event TB 3.7 of Haq et al. (1988). What is more certain, however, is that a small part of El Mangle ramp was offset by El Coloradito fault. If initiation of transtensional tectonics in the Gulf of California dates from 3.5 Ma and no earlier, then the circumstances fit the facts based on geological relationships found around El Mangle. That means the timing of events on the Loreto fault proposed by Umhoefer et al. (1994) and Dorsey et al. (1995) reflects more a continuation of transtensional tectonics as opposed to their onset in the greater Loreto area. It also could signify that the Loreto fault was locally reactivated by transtensional tectonics after the failure of El Coloradito fault to penetrate the Cerro Mencenares volcanic center.

CONCLUSIONS

The Pliocene ramp succession at El Mangle on the east flank of the Cerro Mencenares volcanic center is one of the best exposed and best preserved of its kind in Baja California Sur. Composition of the ramp facies includes a range of terrestrial, intertidal, and shallow marine units that indicates a relatively sheltered shoreline consistent with the present local geography. Key facies in the ramp may be traced laterally for at least 8 km along much of the Ensenada El Mangle. The northern edge of the ramp was offset by El Coloradito fault, leaving a fault scarp oblique to the dominant north-south fault pattern expressed elsewhere on the Pliocene-Pleistocene Cerro Mencenares volcanic complex. Projection of transform faults and fracture zones from the spreading ridge in the Farallon Basin, for example, are crudely parallel to El Coloradito fault on a bearing of N55°W. Ramp-related sequences that were deposited along the margins of Baja California help to document the transition from orthogonal extension to oblique extension as the Gulf of California evolved through Pliocene-Pleistocene time. El Coloradito fault may be the landward terminus of a fracture zone against the Mencenares volcanic complex within the Baja California Central Domain.

Future attention to the detailed bathymetry of the Gulf of California with respect to fault traces and the adjacent accommodation zones recognized on land along the Baja California Central Domain is needed. This will require better coordination between marine geologists and traditional land-based geologists. Attention to ramp stratigraphy throughout the Pliocene of the Baja California Central Domain must play an important role in this process.

APPENDIX 1

Hotel section at El Mangle, 28.5 km north of Loreto, Baja California Sur. Elevation 1 m above sea level.

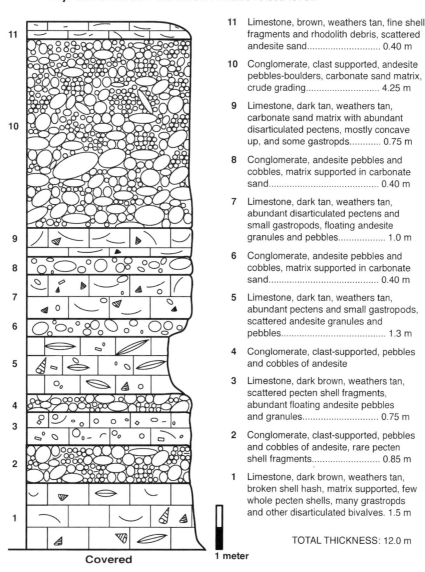

11 Limestone, brown, weathers tan, fine shell fragments and rhodolith debris, scattered andesite sand............................ 0.40 m

10 Conglomerate, clast supported, andesite pebbles-boulders, carbonate sand matrix, crude grading............................ 4.25 m

9 Limestone, dark tan, weathers tan, carbonate sand matrix with abundant disarticulated pectens, mostly concave up, and some gastropods............ 0.75 m

8 Conglomerate, andesite pebbles and cobbles, matrix supported in carbonate sand.. 0.40 m

7 Limestone, dark tan, weathers tan, abundant disarticulated pectens and small gastropods, floating andesite granules and pebbles.................. 1.0 m

6 Conglomerate, andesite pebbles and cobbles, matrix supported in carbonate sand.. 0.40 m

5 Limestone, dark tan, weathers tan, abundant pectens and small gastropods, scattered andesite granules and pebbles.. 1.3 m

4 Conglomerate, clast-supported, pebbles and cobbles of andesite

3 Limestone, dark brown, weathers tan, scattered pecten shell fragments, abundant floating andesite pebbles and granules............................ 0.75 m

2 Conglomerate, clast-supported, pebbles and cobbles of andesite, rare pecten shell fragments.......................... 0.85 m

1 Limestone, dark brown, weathers tan, broken shell hash, matrix supported, few whole pecten shells, many grastropods and other disarticulated bivalves. 1.5 m

TOTAL THICKNESS: 12.0 m

Covered

1 meter

B: Road cut above caretaker's house at El Mangle, 28 km north of Loreto, Baja California Sur. Elevation 20 m (65 ft) above sea level.

6 Cobble-boulder conglomerate, clast supported, andesite clasts (3-40 cm diameter), completely encrusted with red algae (bleached white), large oysters associated with upper boulders..... 1.0 m

5 Limestone, yellow-tan, carbonate sand matrix, pectin coquina consisting of disarticulated valves (mostly concave upward), razor clam, gastropods, and barnacles.................... 2.0 m

4 Cobble-boulder conglomerate, clast supported, andesite clasts (2-50 cm diameter), completely encrusted with red algae, carbonate matrix includes occasional small pectens, gastropds, echinoid spines, trace worm tubes and barnacles... 2.0 m

3 Opalite deposit, white weathers to olive-white, soft layers with 30% clay-size fraction include abundant twig-size woody debris and scattered small branches interbedded with co-extensive lenses of solid opal up to 3 cm thick, vertical orientation of twigs in some layers suggestive of root zone............ 1.8 m

2 Conglomerate lag deposit, clast supported with pebble-size andesite bits (1-3 cm diameter), black weathers brown-red.......... 0.2 m

1 Siltstone, unlithified, fine grained (30% clay fraction), light brown-red, admixture of scattered quartz and carbonate grains, no internal stratification, salty to taste, basal contact covered.........0.8 m

TOTAL THICKNESS: 7.8 m

1 meter

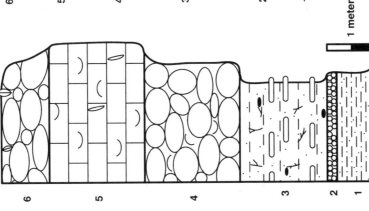

Covered

A: Main ramp outcrop below caretake's house at El Mangle, 28 km north of Loreto, Baja California Sur. Elevation at sea level.

9 Limestone, light gray, massive, high diversity molluscan coquina, some large articulated shells, rare andesite pebbles............ 2.4 m

8 Pebble-boulder conglomerate, andesite clasts in carbonate sand matrix........ 0.3 m

7 Limestone, light gray, massive, high diversity molluscan coquina, some large articulated shells, rare andesite pebbles............ 2.0 m

6 Conglomerate, clast-supported andesite pebbles, and few cobbles in a carbonate sand matrix...................... 0.25 m

5 Limestone, light gray, very abundant mollusks verging on a coquina, high diversity with some bivalves, gastropods, and a few echinoids, rare andesite pebbles................ 1.3 m

4 Limestone, gray-tan, weathers pinkish-gray, grain-supported carbonate sand matrix with disarticulated pectens and associated broken shells mixed with pebbles and cobbles.... 1.0 m

3 Limestone, flat gray, grain-supported carbonate sand with a few large disarticulated bivalves and abundant small bivalves and gastropods.. 2.5 m

2 Limestone, gray-tan, weathers pinkish-gray, carbonate sand matrix, basal conglomerate of rounded andesite clasts with graded upper boundary, pectens near top of unit......... 0.5 m

1 Limestone, tan, weathers pinkish-gray, carbonate sand with disarticulated pectens, 75% of shells concave up, common whole echinoids of Clypeaster marquerensis, scattered small pebbles of andesite.... 2.5 m

TOTAL THICKNESS: 12.75 m

1 meter

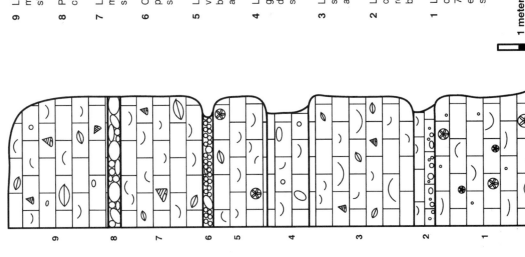

Covered

APPENDIX 1 (*continued*)

Shore-side cliffs below road immediately north of gated entrance to El Mangle, 27.5 km north of Loreto, Baja California Sur. Elevation 1 m above sea level.

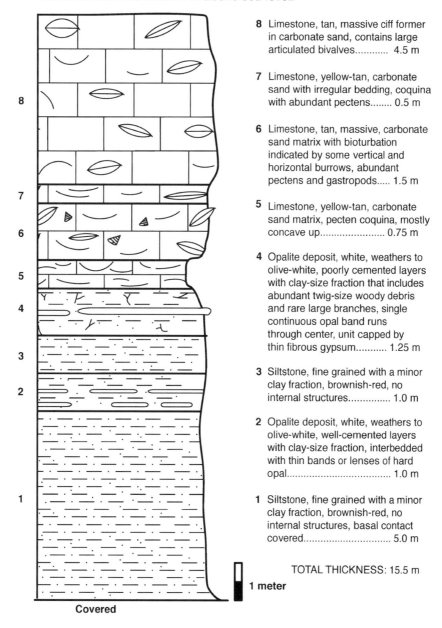

8 Limestone, tan, massive ciff former in carbonate sand, contains large articulated bivalves............ 4.5 m

7 Limestone, yellow-tan, carbonate sand with irregular bedding, coquina with abundant pectens........ 0.5 m

6 Limestone, tan, massive, carbonate sand matrix with bioturbation indicated by some vertical and horizontal burrows, abundant pectens and gastropods..... 1.5 m

5 Limestone, yellow-tan, carbonate sand matrix, pecten coquina, mostly concave up........................ 0.75 m

4 Opalite deposit, white, weathers to olive-white, poorly cemented layers with clay-size fraction that includes abundant twig-size woody debris and rare large branches, single continuous opal band runs through center, unit capped by thin fibrous gypsum........... 1.25 m

3 Siltstone, fine grained with a minor clay fraction, brownish-red, no internal structures.............. 1.0 m

2 Opalite deposit, white, weathers to olive-white, well-cemented layers with clay-size fraction, interbedded with thin bands or lenses of hard opal.................................... 1.0 m

1 Siltstone, fine grained with a minor clay fraction, brownish-red, no internal structures, basal contact covered.............................. 5.0 m

TOTAL THICKNESS: 15.5 m

1 meter

Covered

ACKNOWLEDGMENTS

We thank the donors of the Petroleum Research Fund (American Chemical Society) for full support of this project through grant 35475-B8 to M.E. Johnson at Williams College. We thank Rosa Doménech (Departament d'Estratigrafia i Paleontologia, Facultat de Geologia, Universitat de Barcelona) for identification of fossil pectens and oysters; Andres Lopez Perez (Department of Geoscience, University of Iowa) for identification of the fossil coral; and Elizabeth Ruck (San Francisco State University) for identification of the salt-water diatom. Williams College students Nathan Cardoos, Wei-Li Deng, Jamon Fronstenson, Lisa Marco, Elizabeth Mygatt, and Nicholas Nelson participated in a mapping exercise on the geology of the El Mangle area in January 2001. The maps and stratigraphic columns compiled for this report were drafted by Darius E. Mitchell and Gudveig Baarli. Reviews that helped substantially to improve the quality of this paper were contributed by Jochen Halfar (Stuttgart University) and Jonathan C. Matti (U.S. Geological Survey, Southwest Field Office, Tucson, Arizona).

REFERENCES CITED

Anderson, C.A., 1950, 1940 E.W Scripps cruise to the Gulf of California; Part I, Geology of islands and neighboring land areas: Geological Society of America Memoir 43, p. 1–53.

Axen, G., 1995, Extensional segmentation of the main gulf escarpment, Mexico and United States: Geology, v. 23, p. 515–518.

Bates, R.L., and Jackson, J.A., eds., 1987, Glossary of geology (third edition): Alexandria, Virginia, American Geological Institute, 788 p.

Berry, R.W., Ledesma-Vázquez, J., Johnson, M.E., and Backus, D.H., 2002, 3.3 Ma sea water incursion of San Nicolás and El Mangle basins during opening of the Proto-Gulf of California, Mexico: Geological Society of America Abstracts with Programs, v. 34, no. 6, p. 84.

Bigioggero, B., Chiesa, S., Zanchi, A., Montrasio, A., and Vezzoli, L., 1995, The Cerro Mencenares volcanic center, Baja California Sur: Source and tectonic control on postsubduction magmatism within the gulf rift: Geological Society of America Bulletin, v. 107, p. 1108–1122.

Castradori, D., Rio, D., Hilgen, Fl.J., and Lourens, L., 1998, The global standard stratotype-section and point (GSSP) of the Piacenzian Stage (middle Pliocene): Episodes, v. 21, p. 88–93.

Dorsey, R.J., 1997, Origin and significance of rhodolith-rich strata in the Punta El Bajo section, southeastern Pliocene Loreto basin, *in* Johnson, M.E., and Ledesma-Vázquez, J., eds., Pliocene carbonates and related facies flanking the Gulf of California, Baja California, Mexico: Geological Society of America Special Paper 318, p.119–126.

Dorsey, R.J., Umhoefer, P.J., and Renne, P.R., 1995, Rapid subsidence and stacked Gilbert-type fan deltas, Pliocene Loreto basin, Baja California Sur, Mexico: Sedimentary Geology, v. 98, p. 181–204.

Durham, J.W., 1950, 1940 E.W. Scripps cruise to the Gulf of California; Part II, Megascopic paleontology and marine stratigraphy: Geological Society of America Memoir 43, p. 1–216.

Fabriol, H., Delgado-Argote, L.A., Danobeitia, J.J., Cordoba, D., Gonzalez, A., Garcia-Abdeslem, J., Bartolome, R., Martin-Atienza, B., and Frias-Camacho, V., 1999, Backscattering and geophysical features of volcanic ridges offshore Santa Rosalia, Baja California Sur, Gulf of California, Mexico: Journal of Volcanology and Geothermal Research, v. 93, p. 75–92.

Fletcher, J.M., and Munguía, L., 2000, Active continental rifting in southern Baja California, Mexico: Implications for plate motion partitioning and the transition to seafloor spreading in the Gulf of California: Tectonics, v. 19, p. 1107–1123.

Gastil, R.G., Phillips, R.P., and Allison, E.C., 1975, Reconnaissance geology of the state of Baja California: Geological Society of America Memoir 140, 170 p.

Goldstein, R.H., and Franseen, E.K., 1995, Pinning points: A method providing quantitative constraints on relative sea-level history: Sedimentary Geology, v. 95, p. 1–10.

Haq, B.U., Hardenbol, J. and Vail, P.R., 1988, Mesozoic and Cenozoic chronostratigraphy and cycles of sea-level change, *in* Wilgus, C.K., Hastings, B.S., Kendall, C.G., Posamentier, H.W., Ross, C.A., and Van Wagoner, J.C., eds., Sea-level changes: An integrated approach: Society of Economic Paleontologists and Mineralogists Special Publication 42, p. 72–108.

Israelsky, M.C., 1949, Oscillation chart: American Association of Petroleum Geologists Bulletin, v. 33, p. 92–98.

Johnson, M.E., and Ledesma-Vázquez, J., 1999, Biological zonation on a rocky-shore boulder deposit: Upper Pleistocene Bahía San Antonio (Baja California Sur, Mexico): Palaios, v. 14, p. 569–584.

Johnson, M.E., and Ledesma-Vázquez, J., 2001, Pliocene-Pleistocene rocky shorelines trace coastal development of Bahía Concepción, gulf coast of Baja California Sur (Mexico): Palaeogeography, Palaeoclimatology, Palaeoecology, v. 166, p. 65–88.

Karig, D.E., and Jensky, W., 1972, The protogulf of California: Earth and Planetary Science Letters, v. 17, p. 169–174.

Ledesma-Vázquez, J., and Johnson, M.E., 2001, Miocene-Pleistocene tectono-sedimentary evolution of Bahía Concepción region, Baja California Sur (México): Sedimentary Geology, v. 144, p. 183–196.

Lonsdale, P., 1989, Geology and tectonic history of the Gulf of California, *in* Winterer, E.L., et al., eds., The eastern Pacific Ocean and Hawaii: Boulder, Colorado, Geological Society of America, Geology of North America, v. N, p. 499–521.

Lyle, M., and Ness, G.E., 1991, The opening of the southern Gulf of California, *in* Dauphin, J.P., and Simoneit, B.R.T., eds., The gulf and peninsular provinces of the Californias: American Association of Petroleum Geologists Memoir, 47, p. 403–423.

Martín-Barajas, A., Téllez-Durarte, M., and Stock, J.M., 1997, Pliocene volcanogenic sedimentation along an accommodation zone in northeastern Baja California: The Puertecitos Formation, *in* Johnson, M.E., and Ledesma-Vázquez, J., eds., Pliocene carbonates and related facies flanking the Gulf of California, Baja California, Mexico: Geological Society of America Special Paper 318, p. 1–24.

McLean, H., 1987, K-Ar age confirms Pliocene age for the oldest Neogene marine strata near Loreto, Baja California Sur, Mexico: American Association of Petroleum Geologists Bulletin, v. 71, p. 591.

McLean, H., 1989, Reconnaissance geology of a Pliocene marine embayment near Loreto, Baja California Sur, Mexico, *in* Abbott, P.L., ed., Geologic studies in Baja California: Los Angeles, Society of Economic Paleontologists and Mineralogists Pacific section Book 63, p. 17–25.

Mitchell, D.E., Backus, D.H., and Johnson, M.E., 2001, Opalized wood from the upper Pliocene of Baja California Sur: A costal-plain deposit on the Gulf of California: Geological Society America Abstracts with Program, v. 33, no. 6, p. A197-A198.

Nava-Sánchez, E.H., Gorslin, D.S., and Molina-Cruz, A., 2001, The Baja California peninsula borderland: Structural and sedimentological characteristics: Sedimentary Geology, v. 144, p. 63–82

Ness, G.E., and Lyle, M.W., 1991, A seismo-tectonic map of the Gulf and Peninsular Province of the Californias, *in* Dauphin, J.P., and Simoneit, B.R.T., eds., The gulf and peninsular provinces of the Californias: American Association of Petroleum Geologists Memoir, 47, p. 71–77, and Plates 7 and 8.

Oskin, M., Stock, J., and Martín-Barajas, A., 2001, Rapid localization of Pacific-North America plate motion in the Gulf of California: Geology, v. 29, p. 459–462.

Piazza, M., and Robba, E., 1998, Autochthonous biofacies in the Pliocene Loreto Basin, Baja California Sur, Mexico: Rivista Italiana di Paleontologia e Stratigrafia, v. 104, p. 227–262.

Prothero, D.R., and Schwab, F., 1996, Sedimentary geology: An introduction to sedimentary rocks and stratigraphy: W.H., Freeman and Company, New York, 575 p.

Rio, D., Sprovieri, R., Castradori, D., and Di Stefano, E., 1998, The Glasian State (Upper Plliocene): A new unit of the global standard chronostratigraphic scale: Episodes, v. 21, p. 82–87.

Simian, M.E., and Johnson, M.E., 1997, Development and foundering of the Pliocene Santa Ines archipelago in the Gulf of California: Baja California Sur, Mexico, *in* Johnson, M.E., and Ledesma-Vázquez, J., eds., Pliocene carbonates and related facies flanking the Gulf of California, Baja California, Mexico: Boulder, Colorado, Geological Society of America Special Paper 318, p. 25–38.

Smith, J.T., 1991, Cenozoic marine mollusks and paleogeography of the Gulf of California, *in* Dauphin, J.P., and Simoneit, B.R.T., eds., The gulf and peninsular provinces of the Californias: American Association Petroleum Geologists Memoir 47, p. 637–666.

Stock, J.M., 2000, Relation of the Puertecitos volcanic province, Baja California, Mexico, to development of the plate boundary in the Gulf of California, *in* Delgado-Granados, H., Aguirre-Díaz, G., and Stock, J.M., eds., Cenozoic tectonics and volcanism of Mexico: Boulder, Colorado, Geological Society of America Special Paper 334, p. 143–156.

Stock, J.M., and Hodges, K.V., 1989, Pre-Pliocene extension around the Gulf of California and the transfer of Baja California to the Pacific plate: Tectonics, v. 8, p. 99–115.

Umhoefer, P.J., Dorsey, R.J., and Renne, P., 1994, Tectonics of the Pliocene Loreto basin, Baja California Sur, Mexico, and evolution of the Gulf of California: Geology, v. 22, p. 649–652.

Vail, P.R., Mitchum, Jr., R.M., and Thompson, S., 1977, Seismic stratigraphy and global changes of sea level, Part 3: Relative changes of sea level from coastal onlap, *in* Playton, C.E., ed., Seismic stratigraphy—Applications to hydrocarbon exploration: American Association of Petroleum Geologists Memoir 26, p. 63–81.

Zanchi, A., 1994, The opening of the Gulf of California near Loreto, Baja California, México: from basin and range extension to transtensional tectonics: Journal Structural Geology, v. 16, p. 1619–1639.

MANUSCRIPT ACCEPTED BY THE SOCIETY JUNE 2, 2003

Geological Society of America
Special Paper 374
2003

Cenozoic volcanism and tectonics of the continental margins of the Upper Delfín basin, northeastern Baja California and western Sonora

Michael Oskin

Institute for Crustal Studies, University of California, Santa Barbara, California 93106, USA

Joann Stock

Division of Geological and Planetary Sciences, California Institute of Technology, Pasadena, California 91125, USA

ABSTRACT

Pre- and syn-rift stratigraphy of conjugate rifted margins of the Upper Delfín basin provides a rare opportunity to explore proximal relationships between the loci of volcanism and rifting during formation of new ocean basin. The Upper Delfín basin is one of a series of youthful, en-echelon ocean basins that accommodate spreading between the Pacific and North American plates in the Gulf of California. Four groups of volcaniclastic stratigraphy are described from the Baja California rifted margin from the Puertecitos Volcanic Province to the Sierra San Felipe and from the conjugate Sonora rifted margin from Isla Tiburón to the adjacent mainland coastal region. Excluding the uppermost post–6 Ma group, these strata predate opening of the Upper Delfín basin and thus similar facies relationships occur on both conjugate rift margins. Pre-rift, mostly arc-related volcanism from 21 to 12 Ma built isolated volcanic centers over a regional Eocene(?) erosion surface cut onto pre-Tertiary basement rocks. The Puertecitos Volcanic Province formed as a concentration of arc-related volcanism with a peak of activity at 18–15 Ma. Rift-related faulting and basin formation initiated after 12.6 Ma to the north and east of the Puertecitos Volcanic Province, but extension largely bypassed the center of the volcanic province. Rather, rifting stepped eastward along the Matomí Accommodation Zone, perhaps by taking advantage of crust weakened by prior arc-related volcanism in the Puertecitos Volcanic Province. Later rift-related volcanism localized along intersections of the Matomí Accommodation Zone and north-striking extensional faults. The crustal break that opened the Upper Delfín basin ca. 6 Ma coincides with the most voluminous exposures of late Miocene syn-rift volcanism and the vent area for 6.7–6.1 Ma rhyolite ignimbrites that blanketed the region just prior to the onset of marine sedimentation. Overall, the pattern of arc- and rift-related volcanism, sedimentation, and faulting on the margins of the Upper Delfín basin indicates a close association between continental extension and volcanism. These relationships support that magmatism and crustal rift structure evolved as a coupled system to localize Pacific–North America plate motion into the Gulf of California.

Keywords: Gulf of California, rifting, volcanism, continental margin.

Oskin, M., and Stock, J., 2003, Cenozoic volcanism and tectonics of the continental margins of the Upper Delfín basin, northeastern Baja California and western Sonora, *in* Johnson, S.E., Paterson, S.R., Fletcher, J.M., Girty, G.H., Kimbrough, D.L., and Martín-Barajas, A., eds., Tectonic evolution of northwestern México and the southwestern USA: Boulder, Colorado, Geological Society of America Special Paper 374, p. 421–438. For permission to copy, contact editing@geosociety.org. © 2003 Geological Society of America.

INTRODUCTION

The role of volcanism in the transition from continental rifting to continental rupture and seafloor spreading is a fundamental yet poorly understood process in plate tectonics. Examples range from (1) wide rifts where volcanism plays an insignificant role in complex break-up of a rheologically stratified lithosphere (Louden and Chian, 1999) to (2) narrow rifts where localized, runaway lithospheric necking was accompanied by large-volume volcanic activity in some regions (Cochran, 1983) and (3) rifts accompanied by immense volumes of flood basalt volcanism (Tegner et al., 1998). Ultimately, volcanism and rifting are intertwined in seafloor spreading. However, the role of volcanism in facilitating continental rupture remains unclear. Workers in east Africa and the Red Sea have suggested a close association of localization of volcanism and crustal break-up (Cochran, 1983; Ebinger and Casey, 2001). Additional examples from continental margin settings where both structure and associated volcanism are well-exposed and well-timed are required to characterize the role of volcanism in continental break-up and formation of new ocean basins.

In this paper, we document the relationships between volcanism and rift structure from the continental margins of the Upper Delfín basin of the Gulf of California (Fig. 1). The Gulf of California is one of a few examples of a youthful ocean basin undergoing the transition from continental rifting to seafloor spreading (Curray et al., 1982; Lonsdale, 1989). Major episodes of the Cenozoic tectonic evolution of the Upper Delfín basin are well represented by the volcanic and sedimentary rocks preserved on its conjugate rifted margins (Fig. 2). Present understanding of this record stems from the pioneering regional geochronologic study by Gastil et al. (1979) and reconnaissance geologic mapping of Baja California Norte by Gastil et al. (1975) and of coastal western Sonora and Isla Tiburón by Gastil et al. (1974). Although many aspects of the Cenozoic tectonic history of northwestern México have been significantly advanced since this early work, most of the regional stratigraphic relationships remain as they were originally mapped by Gastil, Allison, Krummenacher, and their army of students from San Diego State University (Gastil et al., 1974, 1975).

Subsequent studies of Neogene rocks on margins of the Upper Delfín basin have focused on the Puertecitos Volcanic Province of Baja California (Figs. 1 and 3). The stratigraphic record documented in the Puertecitos Volcanic Province has yielded important insight into the geochemical evolution of rift volcanism (Martín-Barajas et al., 1995) and a regionally correlative volcaniclastic stratigraphy (Stock, 1989; Martín-Barajas and Stock, 1993; Lewis, 1996; Nagy et al., 1999; Stock et al., 1999) from which substantial progress has been made in understanding the structural evolution of the Baja California rift margin (Stock and Hodges, 1989, 1990; Stock et al., 1991; Lewis and Stock, 1998a, 1998b; Nagy, 2000; Nagy and Stock, 2000; Oskin et al., 2001). Building on the well-documented stratigraphic record and regionally correlative marker horizons of the Puertecitos Volcanic Province (Fig. 2) we present an updated regional-scale view

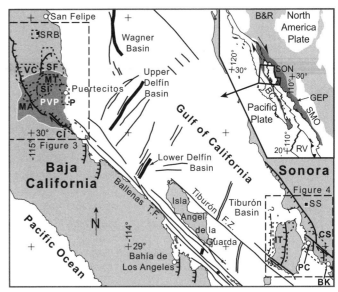

Figure 1. Index map of northern Gulf of California. Marine faults from Fenby and Gastil (1991). Selected faults in Baja California and western Sonora from Gastil et al. (1974, 1975). Puertecitos Volcanic Province (PVP) is shown in darker gray. Areas of reconnaissance mapping for this study are shown by large dashed boxes. Areas of detailed mapping shown by smaller dashed lines: SRB—Santa Rosa Basin (Bryant, 1986); SF—Sierra San Fermín (Lewis, 1994); VC—Valle Chico (Stock, 1993); MT—Arroyo Matomí Transect (Stock et al., 1991, and unpubl. mapping); SI—Santa Isabel Wash (Nagy, 1997); P—Puertecitos (Martín-Barajas and Stock, 1993); MA—Mesa El Avion (Stock, unpublished mapping); CI—Cinco Islas (Oskin, 2002); IT—Isla Tiburón (Oskin, 2002); CS—Coastal Sonora (Oskin, 2002); BK—Bahía Kino; PC—Punta Chueca; SS— Sierra Seri. Inset map of southwestern North America and the Pacific–North America plate boundary: B&R—Basin and Range Province; GEP—Gulf Extensional Province, SMO—Sierra Madre Occidental; SON—Sonora, BC, Baja California; RV—Rivera Plate.

of the Cenozoic stratigraphy of the continental margins of the Upper Delfín basin of the Gulf of California. The scope of this study is intermediate between the reconnaissance maps of Gastil et al. (1974, 1975) and many of the detailed studies that have followed this work (Fig. 1). This new synthesis supports spatial and temporal correlation of arc- and rift-related volcanism with major structural elements of the Upper Delfín basin segment of the Gulf of California rift.

RIFT STRUCTURE

The Upper Delfín basin is one of a series of en echelon rift basins in the Gulf of California that opened via spreading between the Pacific and North America plates (Lonsdale, 1989; Fig. 1). In the northern Gulf of California, rift basins are defined by north- to northeast-trending bathymetric lows separated by northwest-striking transform faults (Fenby and Gastil, 1991). Crustal spreading in these northern basins occurs by diffuse

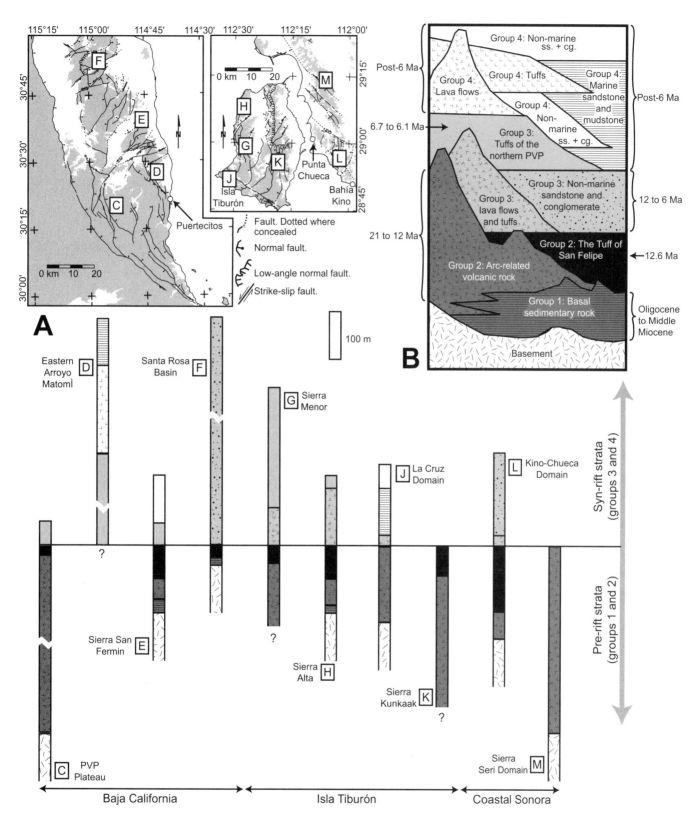

Figure 2. A: Index map of Baja California and Sonora conjugate rifted margins of Upper Delfin basin. Labels correspond to stratigraphic columns below. B: Schematic stratigraphic column of conjugate rifted margins of Upper Delfin basin. See text for discussion of stratigraphic groups depicted here. Tuff of San Felipe and Tuffs of the northern Puertecitos Volcanic Province (PVP) form principal regional markers between groups. C–M: Stratigraphic columns and correlation of localities in northeastern Baja California, Isla Tiburón, and coastal Sonora. Localities are as listed beside each column and shown in part A.

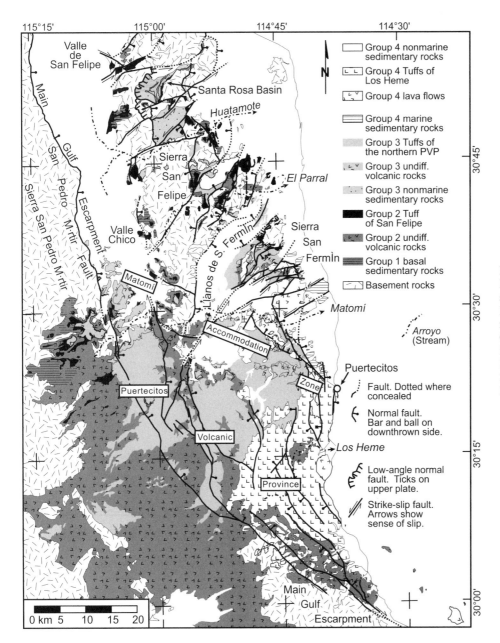

Figure 3. Geologic map of northeastern Baja California. Stratigraphic relationships of these groups are shown in Figure 2. Geology based on detailed mapping of areas shown in Figure 1, Gastil et al. (1975), field reconnaissance in Sierra San Felipe, interpretation of a Landsat Thematic Mapper image (path 38 row 39), and aerial photography of Puertecitos Volcanic Province.

extension and volcanism under rapid sedimentation (Henyey and Bischoff, 1973; Persaud et al., 2003). Presently, the locus of active spreading is asymmetrically positioned near the Baja California rifted margin, northwest of Isla Angel de La Guarda. Lonsdale (1989) proposed that spreading jumped northwestward from the Tiburón basin to the Delfín basin ca. 3 Ma and detached Isla Angel de La Guarda from Baja California. Opening of the Upper Delfín basin is defined here to include this earlier period of opening of the Tiburón basin. The Upper Delfín basin segment of the Gulf of California is flanked by rifted continental margins in northeastern Baja California and western Sonora. The alignment of these margins was first proposed by correlation of a pre-rift conglomerate by Gastil et al. (1973) and later confirmed by stud-

ies of basement rocks (Gastil et al., 1991; Silver and Chappell, 1988) and pre- and syn-rift ignimbrites (Oskin et al., 2001).

On the Baja California rift shoulder, the major structural features are the Sierra San Pedro Mártir, Main Gulf Escarpment, and southwest Puertecitos Volcanic Province (Fig. 3). The Sierra San Pedro Mártir and the southwest Puertecitos Volcanic Province lie outside of the Gulf extensional province as part of the tectonically stable central Baja California peninsula. Down-to-the-east normal and oblique-normal faults form the Main Gulf Escarpment and the western boundary of the Gulf Extensional Province (Gastil et al., 1975). The north-northwest–striking San Pedro Mártir fault is the most dramatic of these rift shoulder structures, with at least 5 km of structural relief between the Sierra San Pedro

Mártir and the floor of the Valle de San Felipe–Valle Chico basin (Slyker, 1970). Slip on the San Pedro Mártir fault diminishes as it enters the Puertecitos Volcanic Province (Stock and Hodges, 1990). The rift shoulder also corresponds to a change in crustal thickness from 40 km beneath the Sierra San Pedro Mártir to less than 20 km beneath the rifted margin (Lewis et al., 2001).

Within the Baja California rifted margin, the principal structural features are the Sierra San Felipe, Sierra San Fermín, northeast Puertecitos Volcanic Province, and the Matomí Accommodation Zone (Fig. 3). Although the northeast Puertecitos Volcanic Province is cut by numerous normal faults, the total crustal extension here does not exceed 5% (Stock and Hodges, 1990; Nagy, 1997). Significantly greater extension of 10% to over 40% occurred in the Sierra San Felipe and Sierra San Fermín (Stock and Hodges, 1990; Lewis and Stock, 1998b), and a low-angle normal fault in the Santa Rosa basin of the Sierra San Felipe may have accommodated even greater extension (Bryant, 1986). The west-northwest–trending Matomí Accommodation Zone transfers differential extension and shear between the Puertecitos Volcanic Province plateau and the Sierra San Felipe by complex, diffuse right-lateral and extensional faulting (Dokka and Merriam, 1982; Stock and Hodges, 1990; Stock et al., 1991; Nagy, 2000). At a more regional scale, the Matomí Accommodation Zone probably also accommodates a vergence reversal in the Baja California rift

shoulder from a footwall position in the Sierra San Pedro Mártir to a hanging wall position from the Puertecitos Volcanic Province to Bahía de Los Angeles (Axen, 1995; Stock, 2000; Fig. 1). A series of northeast-striking, left-lateral faults transect the northernmost Puertecitos Volcanic Province, Sierra San Felipe, and Sierra San Fermín (Fig. 3). These faults have accommodated 30 ± 15° of clockwise rotation of multiple crustal blocks east of the Main Gulf Escarpment and north of the Puertecitos Volcanic Province (Lewis and Stock, 1998a). This rotational deformation overprinted older rift structures (Lewis and Stock, 1998b) and probably fragmented and rotated older rift basins.

The Sonora rifted margin is a continuation of the Basin and Range province of northern Mexico and southern Arizona (Henry and Aranda-Gómez, 1992). Principal structural features adjacent to the Upper Delfín segment of the Gulf of California rift are the Sierra Menor, Sierra Kunkaak, and La Cruz fault in Isla Tiburón, and the Sierra Seri and Sacrificio faults in coastal Sonora (Fig. 4). These structures define five structural domains in the Sonora margin that are shown in the next section to each have distinct stratigraphy (Fig. 5). The Sierra Menor and Sierra Kunkaak define two structural domains separated by down-to-the-west normal slip on the complexly segmented Tecomate fault (Fig. 4). An older strand of this fault system has been uplifted and tilted to a low angle within the northern Sierra Kunkaak. The La Cruz

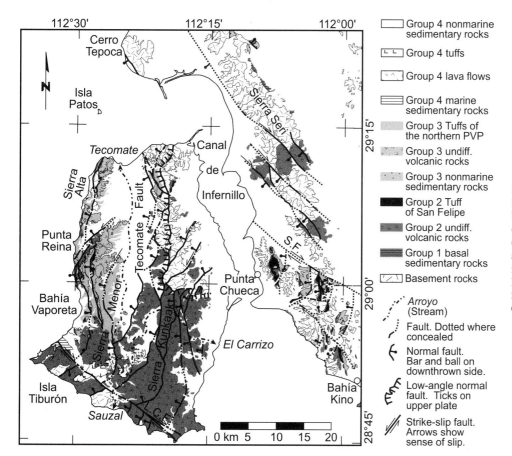

Figure 4. Geologic map of coastal Sonora and Isla Tiburón. Geology based on detailed mapping of areas shown in Figure 1, Gastil et al. (1974) and Gastil et al. (1999) with additional field reconnaissance and interpretation of a Landsat Thematic Mapper image (path 36, row 40). S.F.—Sacrificio fault; L.C.F.—La Cruz fault.

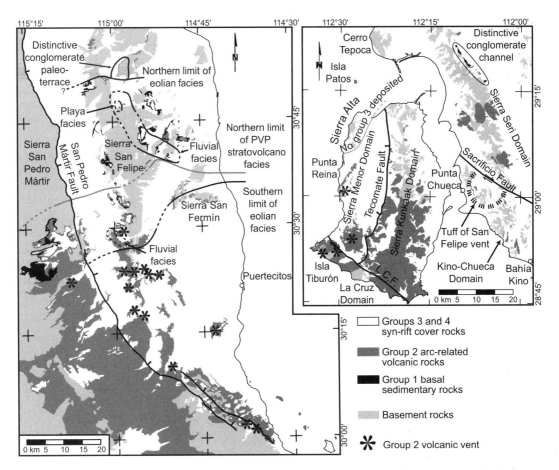

Figure 5. Outcrops of pre-rift stratigraphy on conjugate rift margins of Upper Delfín basin. Facies distribution of group 1 in Baja California is divided into eolian, fluvial, and playa-dominated depositional environments. A distinctive fluvial conglomerate containing clasts of extra-regional lithologies (Bryant, 1986) is shown as a paleo-terrrace isolated from younger group 1 deposits. Correlative distinctive conglomerate outcrops of Gastil et al. (1973) are circled in northeast corner of Sonora map. Facies distribution of group 2 deposits is shown as distribution of known group 2 vents (asterisks) and northern limit of stratovolcano volcaniclastic facies of Puertecitos Volcanic Province. Five domains of group 2 volcanic arc strata are shown with boundary structures for Isla Tiburón and coastal Sonora. On Isla Tiburón, Sierra Menor domain is separated by Tecomate fault from Sierra Kunkaak. This fault is shown in simplified form here. La Cruz domain spans southwest shoreline of Isla Tiburón and is separated from adjacent domains by La Cruz fault (L.C.F.). On coastal Sonora, Kino–Chueca domain and Sierra Seri domain are separated by Sacrificio fault. Hatched line surrounds vent region for Tuff of San Felipe (Oskin et al., 2001). See text for discussion of stratigraphy of these domains.

strike-slip fault through southern Isla Tiburon juxtaposes a third structural domain against the southern edge of the Sierra Menor and Sierra Kunkaak (Gastil et al., 1974). Inferred strike-slip motion of the Sacrificio fault through coastal Sonora separates the intensively faulted Kino-Chueca domain on the southwest from the less deformed Sierra Seri domain to the northeast. This fault is one of a series of northwest-striking strike-slip faults first proposed for the Sonora coastal region by Gastil and Krummenacher (1977). The offshore continuation of the Sacrificio fault, its interaction with the Tecomate fault, and the relationship of basement exposures at Cerro Tepoca to the structural domains described here are uncertain.

Connections between offshore and onshore rift structure are not straightforward. Offshore extensional structures of the Upper Delfín basin are aligned orthogonal to the Pacific–North America spreading direction but oblique to the coastline and major onshore extensional faults (Henyey and Bischoff, 1973; Fenby and Gastil, 1991; Persaud et al., 2003). Major continental boundary structures probably reside close to the present-day shoreline since pre-6-Ma pyroclastic flow sheets restore to close proximity (Oskin et al., 2001; Fig. 1). It is likely that these boundary structures were aligned with the north-striking structural grain of both continental margins and that the present northeast-striking offshore rift basins have evolved with crustal spreading in the Upper Delfín basin.

STRATIGRAPHY

The strata of conjugate rifted margins of the Upper Delfín basin are divided into four groups overlying basement rocks. These groups encompass four partially interfingering packages corresponding to time periods prior to and during Gulf rifting (Fig. 2). Groups 1 through 3 predate opening of the Upper Delfín basin and display a close association when restored, whereas group 4 evolved independently as the Upper Delfín basin opened. This stratigraphic framework is based on previous groupings defined by Stock (1989), Lewis (1996), and Nagy et al. (1999). These groups are described briefly below and in more detail in the following sections.

Basement and groups 1 and 2 contain rocks that predate Neogene rifting in the Gulf of California region. Basement rocks are dominantly plutons of the Peninsular Ranges batholith with lesser amounts of metamorphosed sedimentary rocks. Group 1 rocks nonconformably overlie the basement as a thin, discontinuous veneer of Eocene through middle Miocene fluvial, lacustrine, and eolian sedimentary rock. These deposits grade upward and laterally into lava flows, intrusions, and volcaniclastic deposits of the middle Miocene volcanic arc that comprise most of group 2. Group 2 is capped by the Tuff of San Felipe, a 12.6 Ma ignimbrite of regional extent in northeastern Baja California and western Sonora (Stock et al., 1999; Oskin et al., 2001). Where preserved, this distinctive ignimbrite marks the transition from the middle to late Miocene arc volcanism to rifting, volcanism, and sedimentation related to formation of the Gulf Extensional Province (Stock, 1993; Lewis, 1994).

Lithologic groups 3 and 4 were deposited synchronously with rifting in the Gulf of California Extensional Province. Group 3 is an assemblage of volcanic, volcaniclastic, and fluvial deposits. This group is capped by an extensive series of 6.7–6.1 Ma ignimbrites that form much of the volcanic cover of the northern and western Puertecitos Volcanic Province (Fig. 3). These ignimbrites also crop out on western and northern Isla Tiburón (Oskin et al. 2001; Oskin, 2002; Fig. 4). Group 4 contains a younger volcanic and fluvial assemblage similar to group 3 and also includes marine deposits associated with opening of the Upper Delfín basin of the northern Gulf of California.

Basement Complex (Pre-Cenozoic)

Pre-Cenozoic rocks of northeastern Baja California are predominantly intrusions of the Peninsular Ranges batholith (Gastil et al., 1975). Matching basement rocks east of the Gulf of California are known as the Sonoran Coastal batholith (Gastil and Krummenacher, 1977; Silver and Chappell, 1988; Mártin-Valencia et al., 2001). The combined Peninsular Ranges–Sonoran Coastal batholith was emplaced during Jurassic through Cretaceous time as the roots of a volcanic arc at the western margin of North America (Gastil et al., 1975; Silver and Chappell, 1988). Pluton emplacement continued into the early Cenozoic east of the study area (Silver and Chappell, 1988). Isotopic studies of the batholith

intrusions and the provenance of pre-batholithic sedimentary strata indicate that the basement beneath the study area is probably rifted Proterozoic North American continental crust (DePaolo, 1981; Silver and Chappell, 1988; Gastil et al., 1991; Rothstein, 1997; Valencia-Moreno et al., 2001). The oldest exposed host rocks for the batholithic intrusions consist of Neoproterozoic through Jurassic siliciclastic and carbonate sequences (Gastil et al., 1991). Within the Gulf extensional province these strata are generally preserved as steeply dipping, isoclinally folded intrabatholithic screens metamorphosed to amphibolite grade (Gastil et al., 1975; Rothstein, 1997). The youngest metasedimentary basement of northern Baja California comprises syn-intrusive volcaniclastic deposits of the Cretaceous Alisitos Formation, exposed west of the Main Gulf Escarpment (Gastil et al., 1975). Within the study region, some metasedimentary deposits may record a Jurassic and Cretaceous marine backarc basin (Radelli, 1989; Gastil, 1993; Phillips, 1993) or forearc basin (Johnson et al., 1999; Dickinson and Lawton, 2001) that was closed in middle Cretaceous time.

Basement is exposed heterogeneously in the study area (Figs. 3 and 4). Within the Puertecitos Volcanic Province, a thick Cenozoic volcaniclastic section covers basement (Gastil et al., 1975). North of the Puertecitos Volcanic Province, basement is extensively exposed in the Sierra San Pedro Mártir and Sierra San Felipe. In the Sierra San Fermín, adjacent to the Gulf of California, exposures are about equally divided between basement and volcanic cover (Lewis, 1994). The transition between thick Cenozoic volcanic cover and basement exposures east of the Main Gulf Escarpment corresponds approximately to the position of Arroyo Matomí (Fig. 3). Outcrops of basement dominate the coastal Sonora study area and northern Isla Tiburón (Gastil et al., 1974). A separate belt of basement crops out south of the La Cruz fault on southern Isla Tiburón.

Group 1: Basal Sedimentary Rocks (Oligocene–middle Miocene)

Group 1 sedimentary deposits nonconformably and unconformably overlie the basement complex. Strata of group 1 comprise a variety of fine- to coarse-grained colluvial, fluvial, lacustrine, and eolian deposits and rarely exceed 100 m in total thickness. Clast composition dominantly reflects the local basement substrate, with the exception of rare fluvial conglomerates containing exotic lithologies. Several of these exotic conglomerates are correlated to sources east of the Gulf of California (Gastil et al., 1973; Minch, 1979; Bryant, 1986; Abbott and Smith, 1989). Group 1 basal sedimentary rocks are recognized regionally across northeastern Baja California and the Peninsular Ranges. Gabb (1882) first proposed the name Mesa Formation to describe the combined basal sedimentary rocks and overlying volcaniclastic strata in the Sierra San Pedro Mártir. Dorsey and Burns (1994) studied group 1 rocks west of the Main Gulf Escarpment and restricted the Mesa Formation to the dominantly locally derived basal sedimentary rocks.

Within the study area, group 1 forms a thin sedimentary veneer (Fig. 2). The contact on basement is an irregular erosional surface and weathering horizon. Discontinuous colluvial and fluvial facies characterize the basal deposits of group 1 strata and are commonly buttressed against local paleo-relief. Tabular-bedded sandstones, mudstones, and cross-bedded eolian sandstones are more evenly distributed in the upper sections of group 1 in Baja California but are absent from Isla Tiburón and coastal Sonora. The upper contact of group 1 sedimentary rocks with group 2 volcaniclastic strata is gradational and laterally interfingering. At the northern end of Isla Tiburón and the northern half of the Baja California study area, most of group 2 is absent, and group 1 sedimentary rocks are overlain directly by the Tuff of San Felipe or younger deposits.

The lithologies of group 1 strata define four distinct facies (Fig. 5). The most common facies are (1) the basal basement weathering profile, present in all exposures, and (2) eolian sandstone, present over most of the Baja California study area. (3) Fluvial facies are present to a lesser extent in all areas and (4) lacustrine facies are recognized in one area of Baja California (Fig. 5). The basal weathering profile facies are recognized by immature, poorly bedded sandstone, grus, and basement breccia overlying a 5–10 m thickness of weathered basement. Where the overlying sedimentary rocks are absent, the weathered basement surface may be recognized by rounded corestones exhumed from the weathering profile. Medium- to fine-grained, poorly to moderately indurated eolian deposits form the most common facies of upper group 1 strata in Baja California and are present everywhere from southern Valle Chico (Stock, 1993) and the Sierra San Fermín (Lewis, 1994) to the Santa Rosa Basin (Bryant, 1986). Medium- to coarse-grained, moderately indurate tabular sandstone and conglomerate (fluvial facies) interfinger locally throughout these eolian sand deposits. Between Arroyo Huatamote and the Santa Rosa Basin, fluvial deposits are the dominant facies, with local deposits of eolian sandstone. Mudstone with authigenic gypsum, siltstone with mud cracks, and laminated fine-grained sandstone comprise a lacustrine/playa facies exposed within the central Sierra San Felipe adjacent to Arroyo Huatamote.

Fluvial conglomerate containing a distinctive, exotic clast assemblage comprises group 1 strata at the base of the Santa Rosa basin of Baja California (Figs. 5 and 6A) and the Sierra Seri of Sonora (Gastil et al., 1973; Bryant, 1986; Fig. 5). From the presence of distinctive limestone clasts containing Permian fusilinid fossils, Gastil et al. (1973) correlated this conglomerate and estimated 300 km of dextral offset across the Gulf of California. Locally derived basement clasts and basalt boulders are also present in this deposit in the Santa Rosa basin. Thickness changes in the overlying Tuff of San Felipe at the southern end of the basin indicate that the conglomerate formed an older, incised terrace deposit during deposition of adjacent, locally derived fluvial and eolian deposits. Exotic clasts from the Santa Rosa basin conglomerate are recycled into these younger group 1 deposits, although the distinctive limestone clasts are not present in the younger strata.

The age of group 1 strata is best constrained from Baja California. Most likely, these strata are Oligocene (?) to middle-Miocene age (Dorsey and Burns, 1994), although similar strata northwest of the study area are as old as Late Cretaceous (Minch, 1979). The oldest group 1 deposits must postdate Cretaceous intrusions of the Peninsular Ranges batholith. Apatite fission-track cooling ages as young as 57 ± 15 Ma are reported for the Sierra San Pedro Mártir pluton, located in the northwestern part of Figure 3 and west of the area of Cenozoic extension (Ortega-Rivera et al., 1997). In northern Baja California, known Eocene and Paleocene marine sedimentary deposits are restricted to the western slope of the Peninsula (Minch, 1979; Gastil et al., 1975). Together, these data suggest that the basal unconformity/nonconformity between basement and group 1 in the study area formed during early Tertiary time and that an Eocene age is probably the oldest possible age for group 1 strata. Upper group 1 strata are overlain by, and laterally equivalent to, Miocene volcaniclastic strata of group 2. The oldest ages of intercalated and overlying volcanic flows are 20–21 Ma (Lewis, 1994; Stock, 1989). The youngest conformable overlying unit observed is the ca. 12.6 Ma Tuff of San Felipe (Stock et al., 1999). These relationships support the Oligocene to middle Miocene age assignment by Dorsey and Burns (1994) and indicate that group 1 depositional systems were active synchronously with deposition of group 2 volcanic strata in other parts of the study area.

Group 2: Pre-Rift Volcanic Rocks (21–12 Ma)

Extensive volcanic and volcaniclastic deposits of dominantly andesitic composition comprise the first volumetrically significant lithologic group overlying the basement nonconformity in northeastern Baja California and western Sonora (Fig. 2). These deposits form part of a belt of calc-alkaline volcanic rocks located adjacent to the eastern shoreline of the Baja California peninsula (Gastil et al., 1979; Sawlan and Smith, 1984). In southern Baja California, equivalent deposits comprise a continuous volcanic pile up to 2 km thick deposited between 24 and 12 Ma (Hausback, 1984). Unlike the continuous belt of arc-related volcanic rocks present in Baja California Sur, volcanic rocks of this age in northern Baja California are preserved in discrete volcanic provinces such as at Puertecitos. Gastil et al. (1975) first defined these provinces in Baja California and recognized that these areas also concentrated later rift-related volcanism. The limited distribution of the andesitic rocks surrounding the northern Gulf of California may reflect a shorter residence time of the arc here (Sawlan, 1991). Within the study area, group 2 deposits overlie a veneer of group 1 sedimentary rocks or basement. Over much of the area, the distinctive Tuff of San Felipe caps group 2 deposits. Elsewhere, unconformable contacts with overlying syn-rift group 3 and 4 strata mark the top of group 2.

Within the Baja California margin, group 2 is divided into (1) an andesitic stratovolcano facies centered at the Puertecitos Volcanic Province, (2) a smaller-volume but more evenly distributed basaltic facies, and (3) the Tuff of San Felipe. Deposits

Figure 6. Field photographs. A: Group 1 distinctive conglomerate of Gastil et al. (1973) and Bryant (1986), Santa Rosa basin. B: Proximal group 2 volcaniclastic debris flows and andesite lava flows, southern Puertecitos Volcanic Province at Cinco Islas (Oskin, 2002). Cliff is ~200 m high. C: Outcrops of Tuff of San Felipe capping andesite on basement in Sierra San Felipe, west of Santa Rosa basin. D: Proximal group 2 pyroclastic flow deposits, Sierra Kunkaak. E: Early syn-rift boulder conglomerate, Santa Rosa basin. F: Outcrops of Tuffs of northern Puertecitos Volcanic Province capped by Pico Los Heme andesite volcano, Santa Isabel Wash, northern Puertecitos Volcanic Province. G: Outcrops of Tuffs of northern Puertecitos Volcanic Province at Punta Reina, western Isla Tiburón. Tuffs overlie basalt in lower right corner of outcrop. Cliff on horizon is ~120 m high. H: Cerro Colorado Volcano (group 4) and mesas formed of related pyroclastic deposits, southern Isla Tiburón.

of the stratovolcano facies reach a composite thickness of up to 500 m within the Puertecitos Volcanic Province (Nagy et al., 1999), where eroded volcanic edifices with up to 500 m of additional relief are present locally (Gastil et al., 1975). Individual flow fields and dome complexes of the stratovolcano facies range in volume up to ~20 km³ (e.g., the Tombstone Dacite of Nagy et al., 1999). Separation of stratovolcano facies into vent, proximal-medial, and distal facies is incomplete in the study region, although some vent localities are known from areas of detailed mapping (Figs. 5 and 6B). West of the Puertecitos Volcanic Province, a westward-fining apron of volcaniclastic conglomerate and sandstone conformably overlies group 1 strata (Dorsey and Burns, 1994). Volcaniclastic deposits of the stratovolcano facies thin abruptly north of the Puertecitos Volcanic Province and pinch out completely in the northern Sierra San Fermín and adjacent parts of the Sierra San Felipe (Fig. 5). The basaltic facies of group 2 consists of small-volume basalt and basaltic andesite intrusions and flows distributed throughout the Baja California margin. Flows of the basaltic facies seldom exceed 10 m thickness. Few vent areas have been identified for these deposits, but the distribution of individual flows suggests a similarly widespread distribution of sources. Intercalation of the basaltic facies and the stratovolcano facies of group 2 with group 1 sedimentary strata indicates that these groups represent depositional systems that coexisted in the study area.

The densely welded, rhyolitic, crystal-rich Tuff of San Felipe caps group 2 and forms an important pre-rift marker horizon across the Baja California margin. The significance of the Tuff of San Felipe as a stratigraphic marker was recognized by Stock (1993) and Lewis (1994), who correlated the deposit from the Main Gulf Escarpment eastward to the coastal plain (Fig. 3). Stock et al. (1999) compiled lithologic, petrographic, geochemical, paleomagnetic, and geochronologic data for the Tuff of San Felipe and correlated this unique, ca. 12.6 Ma deposit across an 1800 km² area of northeastern Baja California (Figs. 3 and 6C). The Tuff of San Felipe is preserved discontinuously within the Puertecitos Volcanic Province, where it occupies canyons cut westward across group 2 andesitic rocks (Fig. 3). North of the Puertecitos Volcanic Province, the Tuff of San Felipe was deposited as a continuous sheet on group 1 sedimentary rocks and basement. Oskin et al. (2001) extended correlation of the Tuff of San Felipe across the northern Gulf of California to Isla Tiburón and coastal Sonora, where this distinctive unit also caps group 2 deposits (Fig. 4).

Group 2 deposits of Isla Tiburón are divided into three separate domains, each with a unique volcanic stratigraphy (Fig. 5). These domains have been separated and juxtaposed by rift-related faulting that postdates deposition of group 2. Group 2 deposits of the Sierra Menor domain are andesite lava and proximal andesitic debris-flow deposits capped by discontinuous outcrops of the Tuff of San Felipe. The overall pattern of deposition of andesitic rocks suggests the existence of a volcanic center in the area of the southern Sierra Menor. At this center, massive andesite bodies make up the bulk of exposures. Group 2 andesitic

rocks of the Sierra Menor domain become dominantly volcaniclastic in the northern Sierra Menor and pinch out in the Sierra Alta (located north of the Sierra Menor on Fig. 5), where younger syn-rift volcanic rocks rest directly upon basement and the Tuff of San Felipe (Fig. 4). Very little is known about the geology of the Sierra Kunkaak domain, as this area has not been mapped in detail. Reconnaissance investigation of the southern part of the Sierra Kunkaak domain at Arroyo El Carrizo (Figs. 4 and 6D) revealed a variety of volcaniclastic deposits intercalated with lava flows. The base of the section here is a sequence of andesite lava flows and volcaniclastic debris-flow deposits similar to the Sierra Menor domain. Above these deposits lies a welded tuff with a red glassy matrix and 50% angular aphanitic orange volcanic lithic fragments. This distinctive tuff is exposed throughout the Sierra Kunkaak and may exceed 200 m thickness in the central part of the range. An isolated outcrop of the lithic tuff is overlain by the Tuff of San Felipe in the northern part of the Sierra Kunkaak.

The La Cruz domain of group 2 deposits (Fig. 5) defines a distinctive terrane on the southern edge of Isla Tiburón. This domain is separated from the rest of Isla Tiburón by the La Cruz fault (Fig. 4). Gastil et al. (1999) subdivided group 2 strata of the western part of the La Cruz domain. The base of this section consists of thinly bedded lacustrine chert and carbonate rocks interbedded with fluvial arkose and conglomerate. Basaltic to andesitic flows and breccias with red volcanic sandstone and tephra overlie these basal sedimentary rocks and a dacite lava flow caps the group 2 section here. Reconnaissance investigations of the central and eastern areas of the La Cruz domain reveal additional silicic lava flows and tuffs not present on the western part of the island. These silicic volcanic deposits may have been emplaced during rifting. However, most of these rocks were deposited conformably on older strata and are confined to the La Cruz domain. None of the volcanic rocks of the La Cruz domain except the most recent group 4 syn-rift strata can be correlated across the La Cruz fault.

Coastal Sonora is divided into 2 additional domains of distinct stratigraphy (Fig. 5). The Kino–Chueca domain forms a belt of north- to northeast–trending ridges on the Sonoran mainland adjacent to the Gulf of California. In the southern and central parts of this region, a 20–80-m-thick section of andesitic lava flows and volcaniclastic deposits nonconformably overlies basement and is conformably overlain by thick, densely welded deposits of the Tuff of San Felipe. In the northern part of the Kino–Chueca domain, high-grade, densely welded tuff deposits characterized by rheomorphic flow texture and an abundance of dark, crystal-rich rhyolite lithic inclusions reside within the probable vent for the Tuff of San Felipe (Oskin et al., 2001; Fig. 5).

The Sierra Seri domain of group 2 deposits forms a northwest-trending mountain range adjacent to the Sonoran coastline separated from the Kino–Chueca domain by the northwest-striking Sacrificio fault (Fig. 4). The base of group 2 in the Sierra Seri consists of 20 m of volcaniclastic sandstones and conglomerate overlain by 50 m of andesitic lava flows. The upper part of the volcanic section is formed of an additional 100–200 m of the rhyolite ignimbrites and lava flows. The lithology of these deposits is distinct from pyroclastic flow deposits elsewhere in the study area.

Cenozoic pre–rift volcanism on the continental margins of the Delfín basin occurred during a short interval of Early to middle Miocene time (Gastil et al., 1975; Sawlan, 1991). Isotopic ages of group 2 strata range from 21 to 12 Ma (Gastil et al., 1979; Stock, 1989; Martín-Barajas et al., 1995; Lewis, 1996; Nagy et al., 1999; Gastil et al., 1999). Age determinations for probable arc-related andesitic to dacitic rocks span a slightly more restricted range, from 14 to 20 Ma. Deposits of this type in the Puertecitos Volcanic Province cluster even more closely at 15–18 Ma, suggesting a pulse of activity in this region during this time interval. The top of group 2 is capped by the ca. 12.6 Ma Tuff of San Felipe (Stock et al., 1999; Oskin et al., 2001).

Group 3: Early Syn-Rift Deposits (12–6 Ma)

Rifting that ultimately led to formation of the Gulf of California caused a pronounced change in sedimentary and volcanic deposition in northeastern Baja California and western Sonora. Group 3 comprises an early syn-rift sequence of volcanic and non-marine sedimentary rocks deposited between 12 Ma and 6 Ma. Over much of the study area, group 3 deposits overlie group 2 and are capped by ca. 6.7–6.1 Ma ignimbrites erupted from the northern Puertecitos Volcanic Province (Nagy et al., 1999). Rift-related bimodal volcanism in the northern Puertecitos Volcanic Province and western Isla Tiburón generated an abundance and variety of flows and pyroclastic deposits that covered most of the preexisting arc strata (Martín-Barajas et al., 1995; Figs. 3, 4, and 7). North of the Puertecitos Volcanic Province erosion of fault-generated topographic relief produced locally derived conglomerate and sandstone that filled adjacent rift basins (Gastil et al., 1974, 1975; Lewis, 1994; Bryant, 1986). Group 3 strata may be most easily differentiated from older deposits by the presence of extensive rhyolitic ash-flow tuffs and proximal rift basin conglomerates and breccias. Often, the stratigraphic succession of group 3 deposits shows evidence for rifting such as internal angular unconformities and angular contacts with older strata.

It is likely that rift basin formation has continued to the present over much of the study area, burying older syn-rift strata. However, a chain of early syn-rift basins is presently exposed within the Sierra San Felipe of Baja California where faulting has uplifted older syn-rift strata. The most studied of these is the Santa Rosa basin (Bryant, 1986, Figs. 3, 6E, and 7A). Here, early syn-rift conglomeratic strata were deposited on a conformable section of older fluvial conglomerate (group 1), basalt and andesite flows, and the Tuff of San Felipe (group 2). Group 3 strata in the Santa Rosa basin contain boulder conglomerate interstratified with conglomeratic sandstone (Bryant, 1986). South of the Santa Rosa basin, the Huatamote, Cañon El Parral, and Llanos de San Fermín basins occupy similar structural positions within and adjacent to the Sierra San Felipe (Fig. 7A). The Huatamote

basin fill preserves no volcanic units useful for confirming the timing of deposition here. The Cañon el Parral and Llanos de San Fermín basins contain outcrops of the ca. 6.1–6.7 Ma rhyolite ignimbrites that mark the upper part of group 3.

Similar to syn-rift deposits of Baja California, group 3 sedimentary rocks of the Sonoran margin record formation of rift basins that filled with fluvial deposits. The principal rift basins of group 3 age in the Sonoran margin study area are the Valle de Tecomate, which forms the central valley of Isla Tiburón, and the Infernillo channel between Isla Tiburón and mainland coastal Sonora (Fig. 7A). Early rift-related deposits from these two basins have been uplifted and exposed on the basin margins in the Sierra Menor, the northern Sierra Kunkaak, and in the Kino–Chueca domain of coastal Sonora. Gastil and Krummenacher (1977), Smith et al. (1985), and Gastil et al. (1999) document middle Miocene proto-Gulf marine deposits on southwest Isla Tiburón. Reevaluation of the volcanic rocks intercalated with these strata indicates a latest Miocene to early Pliocene age (Oskin and Stock, 2003). These deposits are classified as group 4 syn-rift strata and will be described further in the next section.

Group 3 volcanic strata form the majority of exposures in the northern Puertecitos Volcanic Province and western Isla Tiburón (Figs. 3, 4, and 7). Rhyolitic pyroclastic flow deposits comprise the greatest volume of group 3 rocks mapped in the Puertecitos Volcanic Province, where sections of ash-flow tuff may exceed 300 m thickness and cap a ~2,000 km^2 area of mesas and plateaus. These tuffs, which are subdivided and described in detail by Nagy et al. (1999), are grouped together here as the Tuffs of the northern Puertecitos Volcanic Province (Fig. 7A). The vent(s) for the Tuffs of the northern Puertecitos Volcanic Province probably resides within a ~25 km in diameter collapse caldera centered on the eastern end of Arroyo Matomí (Fig. 7). In outcrops surrounding this area, abrupt thickness changes that occur within the ignimbrite section are interpreted here to mark caldera-collapse structures (Stock et al., 1991; Lewis, 1994; Nagy, 1997; Figs. 6F and 7). Individual units of the Tuffs of the northern Puertecitos Volcanic Province form regionally correlative marker horizons (Stock and Hodges, 1990; Lewis, 1996; Nagy et al., 1999). Several of these units are thickest adjacent to the west coast of the Gulf of California (Lewis, 1994; Nagy, 1997) and correlate across the Gulf of California to western Isla Tiburón (Oskin et al., 2001; Oskin, 2002; Fig. 6G). Deposits of the Tuffs of the northern Puertecitos Volcanic Province on Isla Tiburón filled an east-tilted, half-graben basin and locally exceed 200 m thickness.

A variety of group 3 lava flows and local pyroclastic flow deposits preceded eruption of the Tuffs of the northern Puertecitos Volcanic Province (Fig. 7B). In southern Valle Chico, 5.8 ± 0.5 Ma rhyolitic and 6.47 ± 0.5 Ma andesitic lava flows and related local pyroclastic deposits underlie the Tuffs of the northern Puertecitos Volcanic Province (Stock, 1993). Undated rhyolite lava flows near Puertecitos (Martín-Barajas et al., 1995) and 11.7 ± 0.6 Ma andesite in the Sierra San Fermín (Lewis, 1996) also underlie the Tuffs of the northern Puertecitos Volcanic

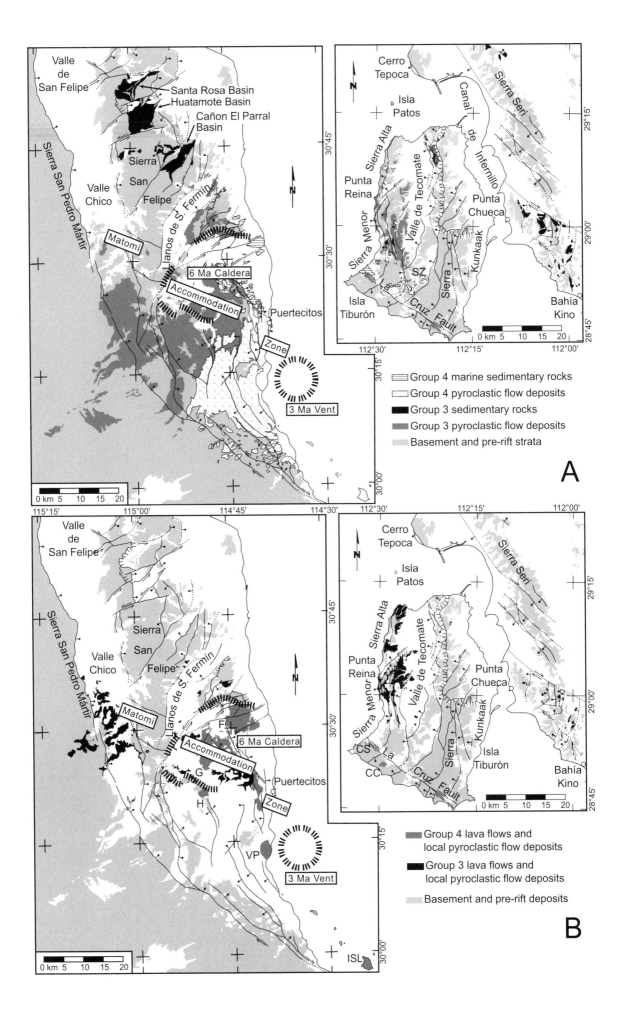

Group 4 marine sedimentary rocks
Group 4 pyroclastic flow deposits
Group 3 sedimentary rocks
Group 3 pyroclastic flow deposits
Basement and pre-rift strata

A

Group 4 lava flows and local pyroclastic flow deposits
Group 3 lava flows and local pyroclastic flow deposits
Basement and pre-rift deposits

B

Figure 7. A: Outcrop map of syn-rift sedimentary rocks and regionally extensive pyroclastic flow deposits. SZ marks outcrop of four cooling units of Tuffs of Arroyo Sauzal. Hatched lines indicate vents for ash-flow tuffs: onshore caldera associated with ca. 6.1–6.7 Ma Tuffs of northern Puertecitos Volcanic Province; proposed offshore vent for 3.3–2.6 Ma Tuffs of Los Heme (Martín-Barajas et al., 1995). B: Outcrop map of syn-rift proximal volcanic rocks and local pyroclastic flow deposits. Volcanism is localized along Matomí Accommodation Zone and western shore of Isla Tiburón. Group 4 andesitic to dacitic volcanoes: F—Pico San Fermín (Lewis, 1994); G—Pico de los Gemelos (Nagy et al. 1999); H—Pico Los Heme (Nagy et al. 1999); VP—Volcan Prieto (Martín-Barajas et al., 1995); ISL—Isla San Luis (Paz-Moreno and Demant, 1999); CS—Cerro Starship (Gastil et al., 1999; Oskin and Stock, 2003); and CC—Cerro Colorado (Oskin, 2002). Hatched lines as in A.

◄────────────────────────

Province. On western Isla Tiburón, undated basaltic, andesitic, and rhyolitic lava flows and breccias occur above the Tuff of San Felipe (group 2) and beneath correlative deposits of the Tuffs of the northern Puertecitos Volcanic Province (Oskin et al., 2001). Vents in the northern Sierra Menor and adjacent parts of the Sierra Alta are a likely source of basalts, which appear to have flowed into a basin in the central and southern Sierra Menor. Andesite and rhyolite lava flows occur above the basalt flows in the northern Sierra Menor and southern Sierra Alta, culminating in a ~200-m-thick accumulation of ash deposits, lava flows, and breccias that crop out in the northern Sierra Alta. Undated group 3 lava and pyroclastic flow deposits also overlie the Tuff of San Felipe (group 2) in the Sierra Kunkaak and in mainland coastal Sonora. In the central and southern parts of the Kino–Chueca domain, rhyolite lava flows and non-welded pyroclastic deposits conformably overlie the Tuff of San Felipe and are unconformably overlain by syn-rift fluvial conglomerates. Basalt flows crop out at this same stratigraphic position in the northern part of the Kino–Chueca domain, and rhyolite lava flows capped by a basalt flow overlie outcrops of the Tuff of San Felipe in the northern Sierra Kunkaak.

Group 4: Late Syn-Rift Deposits (Post–6 Ma)

Group 4 comprises a continuation of group 3 syn-rift sedimentary deposits but with the addition of marine strata of the Gulf of California and a shift in the loci of volcanism to the eastern Puertecitos Volcanic Province and southern Isla Tiburón. The base of group 4 is defined by the top of the extensive Tuffs of the Northern Puertecitos Volcanic Province that covered much of the study area ca. 6.7–6.1 Ma. Where these tuffs are not present, the base of group 4 may be difficult to differentiate from underlying group 3. Group 4 syn-rift strata include present-day sedimentary and volcanic rocks, and thus the age of group 4 spans from latest Miocene/earliest Pliocene to Late Quaternary time. Volcanic strata of group 4 (Fig. 7) are similar to those of group 3 and include flows of a range of compositions erupted in the Puertecitos Volcanic Province and on southwest Isla Tiburón. Group 4

also includes a regionally extensive Late Pliocene (3.3–2.6 Ma) sequence of welded rhyolitic ignimbrites in the Puertecitos Volcanic Province known as the Tuffs of Los Heme (Martín-Barajas and Stock, 1993; Fig. 7B) and the correlative Tuffs of Mesa El Tábano (Stock et al., 1991; Lewis, 1996). Sedimentary strata of group 4 include non-marine breccia, conglomerate, and sandstone as well as marine sedimentary rocks.

Clastic sedimentary deposits of group 4 are common throughout the study area. Pliocene conglomerate crops out in the Cañon El Parral basin and adjacent to the Sierra San Fermín (Lewis, 1994). However, most group 4 basins continue to subside and collect sediment, such as at Valle Chico–Valle de San Felipe, Valle Tecomate, and most of the Llanos de San Fermín basin (Fig. 7). Marine sedimentary deposits are exposed in an uplifted marginal marine sequence adjacent to the east side of the Sierra San Fermín extending southward to Puertecitos (Lewis, 1994; Martín-Barajas and Stock, 1993; Fig. 3). Martín-Barajas et al. (1997) summarized the stratigraphy here and designated these deposits as the Puertecitos Formation. This formation records two Pliocene marine transgressions that postdate the Tuffs of the northern Puertecitos Volcanic Province (Lewis, 1994).

Exposures of pre-Quaternary marine rocks also occur on southwestern Isla Tiburón (Gastil and Krummenacher, 1974; Gastil et al., 1999; Oskin and Stock, 2003). These outcrops were previously identified as the oldest proto-gulf of California marine rocks, based upon a 12.5 Ma intercalated volcanic breccia and a capping 11.2 Ma rhyodacite flow (Smith et al., 1985; Gastil et al., 1999). The outcrops on southwest Isla Tiburón are primarily shallow-marine to littoral conglomeratic delta-fan deposits (Cassidy, 1988; Neuhaus, 1989) that fill in an embayment formed by right-lateral displacement on the La Cruz fault (Fig. 4). New mapping and reevaluation of existing geochronology indicates that these marine rocks are of latest Miocene to early Pliocene rather than middle Miocene age (Oskin and Stock, 2003).

Volcanic deposits of group 4 are concentrated in the eastern Puertecitos Volcanic Province and southern Isla Tiburón (Fig. 7). The northeastern area of the Puertecitos Volcanic Province hosted waning volcanism ca. 6 Ma within the collapse caldera at Arroyo Matomí and formation of a new eruptive center for the Tuffs of Arroyo Los Heme to the southeast. Volcanism at the eastern end of Arroyo Matomí formed a thick accumulation of coalesced rhyolite domes (Martín-Barajas and Stock, 1993) and three andesitic volcanoes (Figs. 6F and 7B, F–H). The Los Heme eruptive center, located to the southeast offshore of Puertecitos (Martín-Barajas et al., 1997; Fig. 7), started to erupt during middle Pliocene time. The 3.3–2.6 Ma Tuffs of Los Heme form a stack of up to 28 densely welded cooling units increasing in thickness and number toward this eruptive center (Martín-Barajas et al., 1995). Volcán Prieto, a 10 km^2, 2.6 ± 0.1 Ma basaltic andesite volcano is interstratified at an unconformity within the Tuffs of Los Heme (Martín-Barajas and Stock, 1993; Martín-Barajas et al., 1995; VP on Fig. 7B). The youngest volcanism in the Puertecitos Volcanic Province is a series of Late Pleistocene and Holocene rhyolite lava domes and airfall deposits erupted from Isla San Luis (Paz-Moreno and Demant, 1999; Hausback et al., 2000; Fig. 7B).

Group 4 volcanic deposits on Isla Tiburón are concentrated on the southern part of the island, adjacent to the onshore La Cruz fault and offshore Tiburón Fracture Zone (Figs. 1 and 4). The oldest group 4 volcanic rocks on Isla Tiburón are four thin pyroclastic flow deposits, designated here as the Tuffs of Arroyo Sauzal, which are exposed within a small half-graben basin in the southern part of the island (SZ in Fig. 7A). The Tuffs of Arroyo Sauzal range from 1 to 3 m thickness each and the upper three units are densely welded. The lowermost non-welded unit of the Tuffs of Arroyo Sauzal is correlated to a pyroclastic flow deposit exposed beneath the marine rocks of southwest Isla Tiburón (Oskin and Stock, 2003). Younger volcanic rocks on southwest Isla Tiburón include the Rhyodacite of Cerro Starship (Gastil et al., 1999) and the Tuffs of Bahía Vaporeta (Oskin and Stock, 2003), both of which are present along the modern shoreline and overlie the marine rocks at southwest Isla Tiburón. Other young volcanic rocks of southern Isla Tiburón are a dacitic volcano that forms Cerro Colorado (Fig. 6H; CC on Fig. 7B) and related flat-lying, partially welded pyroclastic deposits that cover the interior area northeast of Cerro Colorado (Fig. 7A).

DISCUSSION

Correlation, subdivision, and mapped relationships of the stratigraphy of the conjugate margins of the Upper Delfin basin support links between magmatism and rifting. Four volcanic and structural associations stand out from this analysis: (1) Pre-rift volcanism, concentrated at the Puertecitos Volcanic Province, formed a discontinuity that localized formation of an extensional accommodation zone between rift segments of Gulf Extensional Province (Stock and Hodges, 1990; Axen, 1995; Stock, 2000). (2) Early syn-rift volcanism localized at the intersections of north-striking extensional faults and this accommodation zone. (3) Enhanced rift-related volcanism, culminating with eruption of the Tuffs of the northern Puertecitos Volcanic Province, preceded formation of the crustal break that opened the Upper Delfin ocean basin. (4) Late syn-rift volcanism on the Baja California margin migrated to the eastern Puertecitos Volcanic Province, probably contemporaneous with opening of the Lower Delfin ocean basin (Stock, 2000) while late syn-rift volcanism on the Sonora margin localized adjacent to the Tiburón fracture zone. Each of these associations and its implications for connections between magmatism and rifting processes is discussed in the following sections.

Pre-Rift Volcanism and Initial Rift Geometry

At the scale of the entire Gulf Extensional Province, the axis of the Early to Middle Miocene volcanic arc correlates to the position of the Gulf of California (Sawlan, 1991). The stratigraphy preserved on conjugate margins of the Upper Delfin basin permits analysis of this correlation and its relationship to the

structural development of the northern Gulf of California. The locus of voluminous pre-rift andesitic volcanism at the Puertecitos Volcanic Province appears to have been split from similarly concentrated volcanism across central Isla Tiburón. Arc-related volcanic rocks thin in all directions away from all parts of the Puertecitos Volcanic Province and central Isla Tiburón that have been mapped thus far. Only the southeast direction, which should lie on northwest Isla Angel de La Guarda or which has subsided into the Lower Delfin basin, has not yet been mapped and correlated to the rest of the province. This result implies that the greater Puertecitos Volcanic Province, including volcanic rocks on Isla Tiburón, probably developed as either a termination point or an isolated center of arc volcanism and that continental break-up localized within this axis of arc volcanism.

North of the Puertecitos Volcanic Province and central Isla Tiburón, pre-rift volcanic rocks are sparse on both conjugate rift margins of the Upper Delfin basin (Fig. 5). Even in this region, however, an indirect relationship exists between arc volcanism and the geometry of continental break-up. Accommodation zones between rift segments of the northern Gulf Extensional Province correspond to the prior locations of concentrated arc-related volcanic activity (Stock and Hodges, 1990; Stock et al., 1991; Axen, 1995; Lee et al., 1996). The structural development of the Matomí Accommodation Zone in Baja California and the La Cruz fault of Isla Tiburón confirm that the greater Puertecitos Volcanic Province formed a discontinuity during initial rift development. Several significant north to northeast–trending rift basins lie north of the Matomí Accommodation Zone, whereas the Puertecitos Volcanic Province south of this zone underwent less extension and did not develop significant basins (Figs. 3 and 7A). Similar to the pattern in northeastern Baja California, north-trending rift basins on Isla Tiburón diminish and then terminate as they approach the La Cruz fault (Figs. 4 and 7A).

By restoring post–6 Ma opening of the Upper Delfin basin (Oskin et al., 2001) the Matomí Accommodation Zone aligns with the La Cruz fault in southern Isla Tiburón. The apparent continuity of these accommodation zones, together with separation of concentrated arc volcanism in central Isla Tiburón from the Puertecitos Volcanic Province, further supports that crust weakened by arc volcanism may have localized an extensional segment boundary during initial rift development. As proposed by Axen (1995), this rift segmentation is preserved on both conjugate rift margins. However, the vergence of extensional faulting is opposite on these conjugate margins (Figs. 1 and 7) and does not support that vergence reversals correlate in a straightforward manner across the Gulf of California.

Early Syn-Rift Volcanism and Continental Rift Structure

Volcanism during early continental extension, prior to opening of the Upper Delfin ocean basin, was concentrated along the intersections of north-striking extensional faults and the Matomí Accommodation Zone (Fig. 7). This relationship is well expressed at southern Valle Chico (Stock and Hodges, 1990; Stock, 1993) and may also occur in the southern Sierra San Felipe adjacent to Arroyo Matomí (Stock et al., 1991). More detailed mapping and additional geochronology of rocks in southern Isla Tiburón are required to test for a similar correlation of volcanism along the La Cruz fault. Existing data suggest that >20 km of strike-slip displacement occurred on this fault sometime prior to marine deposition on southwest Isla Tiburón (Oskin and Stock, 2003). This significant strike slip motion complicates interpretation of the initial configuration of the La Cruz fault and probably translated any correlative pre- and early syn-rift strata away from Isla Tiburón. The most voluminous early syn-rift volcanism occurred near the intersection of the Matomí Accommodation Zone and the present-day coastlines of Baja California and western Isla Tiburón (Fig. 7). This volcanism culminated with eruption of the Tuffs of the northern Puertecitos Volcanic Province ca. 6.7–6.1 Ma (Nagy et al., 1999) and formation of a collapse caldera near the present-day outlet of Arroyo Matomí (Fig. 7).

Syn-rift volcanism localized at the Matomí Accommodation Zone suggests a genetic relationship between this volcanism and the termini of major continental extensional faults. Magma intrusion may have accommodated extension as slip diminished on brittle extensional faults, similar to relationships described from the northern and central Basin and Range Province (Bursik and Sieh, 1989; Faulds and Varga, 1998; Fig. 1). In a related manner, early east-directed extension (Stock and Hodges, 1990; Lewis and Stock, 1998b) would have been transtensional relative to the west-northwest orientation of the Matomí Accommodation Zone. This transtensional relative motion may have promoted volcanism along "leaky transform faults" within the accommodation zone. In either case, early syn-rift volcanism along the Matomí Accommodation Zone appears most likely to have been a response to extensional fault geometry and extension magnitude.

Opening of the Upper Delfín Basin

The Upper Delfín basin opened as a narrow, north-trending zone of oblique extension near the present day coastlines of northeastern Baja California and Isla Tiburón (Oskin et al., 2001). The location of this zone coincides with a concentration of early syn-rift volcanism exposed along the western shoreline of Isla Tiburón and probably intersected the vent area for the voluminous ca. 6.7–6.1 Ma Tuffs of the northern Puertecitos Volcanic Province (Fig. 7). It is unlikely that more than a 10–25 km width of the continental margins has subsided out of view, beneath the Gulf of California, because the conjugate rifted margins of the Upper Delfín basin fit closely together at ca. 12.6 and 6.3 Ma and the northward-thinning distribution of pre-rift volcanic rocks matches on either side (Oskin et al., 2001; Fig. 5). Although some part of the volcanic record that accompanied initial opening of the Upper Delfín basin has undoubtedly been submerged, the pulse of volcanism preserved on western Isla Tiburón records significant, localized volcanic activity leading up to crustal rupture and opening of the Gulf of California. These rocks are capped by the Tuffs of the northern Puertecitos Vol-

canic Province, which correlate closely across the Upper Delfín basin (Oskin et al., 2001; Oskin, 2002). In contrast to the abundant volcanism that predated eruption of these tuffs and opening of the Upper Delfín basin, volcanic rocks that immediately postdate the Tuffs of the northern Puertecitos Volcanic Province are preserved mostly within the related collapse caldera in Baja California (Fig. 7). Elsewhere on the margins of the Upper Delfín basin, marine sedimentary rocks overlie the Tuffs of the northern Puertecitos Volcanic Province and record opening of the Gulf of California ocean basin (Lewis, 1994; Martín-Barajas and Stock, 1997; Oskin and Stock, 2003; Fig. 7A).

The pulse of volcanic activity preserved on western Isla Tiburón, combined with the location of the collapse caldera formed by eruption of the Tuffs of the northern Puertecitos Volcanic Province, supports that localized volcanism preceded the crustal break that opened the Upper Delfín ocean basin. This phenomenon is similar to observations of localized volcanism during break-up in the Red Sea and East Africa region (Cochran, 1983; Ebinger and Casey, 2001). However, in contrast to the southern Red Sea and East Africa, the continental margins of the Upper Delfín basin did not host widespread rift-related volcanism prior to the pulse of localized volcanism at the nascent crustal break. Though concentrations of rift-related volcanism did occur along the Matomí Accommodation Zone (Fig. 7B) and along other accommodation zones in the Gulf Extensional Province (Axen, 1995), rifting of the margins of the Gulf of California has been generally amagmatic. For example, exposures of pre-rift volcanic rocks and basement dominate both margins of the Upper Delfín basin (Figs. 3 and 4). Detailed mapping elsewhere in the Gulf Extensional Province has yielded similar limited volumes of syn-rift volcanism (Gastil et al., 1974, 1975; Lee et al., 1996; Axen and Fletcher, 1998; Martín-Barajas et al., 2000; Dorsey and Umhoefer, 2000; Delgado-Argote et al., 2000; Mora-Alvarez and McDowell, 2000). Concentrated rift-related volcanism at the nascent axis of the Upper Delfín basin strongly supports that volcanism and crustal break-up are directly related even within a dominantly amagmatic rift. Bi-modal volcanism, culminating with large-volume high-silica rhyolitic ignimbrite eruptions just prior to break-up, further suggests that crustal heating and weakening through input of mafic melt and crustal anatexis preceded localization of rifting into the nascent Upper Delfín ocean basin.

Late Syn-Rift Volcanism and Oceanic Rift Structure

Volcanism after opening of the Upper Delfín basin followed development of rift structure within the Gulf of California. Post-6 Ma volcanism on Isla Tiburón is confined to the southernmost part of the island, adjacent to the Tiburón fracture zone. This fracture zone and the Ballenas fracture zone west of Isla Angel de la Guarda have together accommodated over 250 km of strike-slip displacement (Lonsdale, 1989; Oskin et al., 2001; Fig. 1). The La Cruz fault probably formed as an early strand of the Tiburón fracture zone, and the block south of the La Cruz fault was probably transferred from Isla Angel de la Guarda to Isla Tiburón during opening of the Gulf of California. Post-6 Ma volcanism in Baja California began by infilling the ca. 6.7–6.1 Ma collapse caldera surrounding eastern Arroyo Matomí and later migrated to the eastern coastal zone of the Puertecitos Volcanic Province (Martín-Barajas et al., 1995; Fig. 7). Volcanism here culminated with eruption of the ca. 3.3–2.6 Ma Tuffs of Los Heme (Martín-Barajas and Stock, 1993) that probably accompanied opening of the Lower Delfin ocean basin (Stock, 2000).

Late syn-rift volcanism on the margins of the Upper Delfin basin further supports correlation of magmatism with active rifting. Low-volume volcanism associated with transform faulting on southern Isla Tiburón shares a similar structural setting to volcanism localized along the Matomí Accommodation Zone in Baja California. Although correlation of the ca. 3.3–2.6 Ma Tuffs of Los Heme to opening of the Lower Delfin basin is speculative at present, it is consistent with the pattern of volcanism that preceded opening of the Upper Delfin basin ca. 6 Ma.

CONCLUSIONS

The Upper Delfin basin opened ca. 6 Ma as a narrow zone of oblique divergence between the Pacific and North American plates in the northern Gulf of California. Prior to localization of plate motion into the Upper Delfin basin, a system of north-northwest–striking continental extensional faults and basins opened in northeastern Baja California and western Sonora. To analyze relationships between early extensional faulting, crustal rupture, and pre- and syn-rift magmatism in the Gulf Extensional Province, the stratigraphy of the continental margins of the Upper Delfin basin is divided into four groups overlying a pre-Cenozoic basement complex: (1) Pre-rift Oligocene–middle Miocene basal sedimentary rocks, (2) pre-rift, 21–12 Ma mostly arc-related volcanic rocks, (3) early syn-rift, 12–6 Ma volcanic and sedimentary rocks, and (4) late syn-rift, 6–0 Ma volcanic and sedimentary rock. Excluding the uppermost post–6 Ma group, this stratigraphy was emplaced prior to opening of the Upper Delfin basin and thus similar facies relationships are expressed on conjugate rifted margins in northeastern Baja California and western Sonora.

Repeated patterns of volcanism and rifting preserved on the conjugate rifted margins of the Upper Delfin basin support that magmatism and crustal rift structure evolve as a coupled system. Arc-related volcanism guided later development of rift structure in the Gulf Extensional Province both at province-wide scale and at local structural scale. Although rifting proceeded amagmatically over most of the Gulf Extensional Province, limited early rift-related volcanism localized at structural accommodation zones and particularly at the termini of major extensional faults. Enhanced rift-related volcanism, culminating with emplacement of large-volume silicic ignimbrites, accompanied formation of the crustal break that opened the Upper Delfin basin. Although this enhanced volcanism was localized structurally at the axis of the nascent ocean basin, field relationships support that most of the volcanic output preceded localization of the Pacific–North

America plate boundary into the Upper Delfin basin. A similar pattern of volcanism and rifting may also have preceded opening of the Lower Delfin basin during Late Pliocene time. Taken together, these relationships support that crustal weakening through magmatism, both prior to and concurrent with crustal extension, played a critical role in localizing rifting into the Gulf of California.

ACKNOWLEDGMENTS

This work was supported by National Science Foundation grants EAR-9614674 and EAR-0001248. We also appreciate the support of C. González-León of the Universidad Nacional Autónoma de México. Permission to enter Isla Tiburón was granted by the Secretaría de Medio Ambiente y Recursos Naturales and the Cumcaác (Seri) Indian Tribe. E. Molina, C. Lewis, S. Dobner, R. Houston, N. Marks, J. Wise, and L. Perg assisted with field studies. Discussions with C. Lewis, E. Nagy, A. Martín-Barajas, and R. Dorsey contributed to development of this paper, and reviews by D. Kimbrough and R. Dorsey substantially improved the manuscript and figures. California Institute of Technology, Division of Geological and Planetary Sciences contribution #8868.

REFERENCES CITED

Abbott, P.L., and Smith, T.E., 1989, Sonora, Mexico, Source for the Eocene Poway Conglomerate of southern California: Geology, v. 17, p. 329–332.
Axen, G.J., 1995, Extensional segmentation of the Main Gulf Escarpment, Mexico and the United States: Geology, v. 23, p. 515–518.
Axen, G.J., and Fletcher, J.M., 1998, Late Miocene–Pliocene extensional faulting, northern Gulf of California and Salton Trough, California: International Geological Review, v. 40, p. 217–244.
Bryant, B.A., 1986, Geology of the Sierra Santa Rosa Basin, Baja California, Mexico [M.S. thesis]: San Diego, San Diego State University, 75 p.
Bursik, M., and Sieh, K., 1989, Range front faulting and volcanism in the Mono Basin, eastern California: Journal of Geophysical Research, v. 94, p. 15587–15609.
Cassidy, M.E., 1988, Marine stratigraphy and paleontology of southwestern Isla Tiburón [M.S. thesis]: San Diego, San Diego State University, 177 p.
Cochran, J.T., 1983, A model for development of the Red Sea: American Association of Petroleum Geologists Bulletin, v. 67, p. 41–69.
Curray, J.R., Moore, D.G., Kelts, K., and Einsele, G., 1982, Tectonics and geological history of the passive continental margin at the tip of Baja California, in Curray, J.R., and Moore, D.G., eds., Initial Reports of the Deep Sea Drilling Project, Leg 64: Washington, D.C., U.S. Government Printing Office, p. 1089–1116.
Delgado-Argote, L.A., López-Martínez, M., and Perrilliat, M.C., 2000, Geologic reconnaissance and age of volcanism and associated fauna from sediments of Bahia de Los Angeles Basin, central Gulf of California, in Stock, J., Aguirre, G., and Delgado, H., eds., Cenozoic tectonics and volcanism of Mexico: Boulder, Colorado, Geological Society of America Special Paper 334, p. 111–121.
DePaolo, D.J., 1981, A neodymium and strontium isotopic study of the Mesozoic calc-alkaline granitic batholiths of the Sierra Nevada and Peninsular Ranges, California: Journal of Geophysical Research, v. 86, p. 10470–10488.
Dickinson, W.R., and Lawton, T.F., 2001, Carboniferous to Cretaceous assembly and fragmentation of Mexico: Geological Society of America Bulletin, v. 13, p. 1142–1160.
Dokka, R.K., and Merriam, R.H., 1982, Late Cenozoic extension of northeastern Baja California, Mexico: Geological Society of America Bulletin, v. 93, p. 371–378.

Dorsey, R.J., and Burns, B., 1994, Regional stratigraphy, sedimentology, and tectonic significance of Oligocene-Miocene sedimentary and volcanic rocks, northern Baja California, Mexico: Sedimentary Geology, v. 88, p. 231–251.

Dorsey, R.J., and Umhoefer, P.J., 2000, Tectonic and eustatic controls on sequence stratigraphy of the Pliocene Loreto basin, Baja California Sur, Mexico: Geological Society of America Bulletin, v. 112, p. 177–199.

Ebinger, C.J., and Casey, M., 2001, Continental breakup in magmatic province: An Ethiopian example: Geology, v. 29, p. 527–530.

Faulds, J.E., and Varga, R.J., 1998, The role of accommodation zones and transfer zones in the regional segmentation of extended terranes, *in* Faulds, J.E., and Stewart, J.H., eds., Accommodation zones and transfer zones: The regional segmentation of the Basin and Range Province: Boulder, Colorado, Geological Society of America Special Paper 323, p. 1–46.

Fenby, S.S., and Gastil, R.G., 1991, A seismo-tectonic map of the Gulf of California and surrounding areas, *in* Dauphin, J.P., and Simoneit, B.R., eds., The Gulf and Peninsular Provinces of the Californias: American Association of Petroleum Geologists Memoir 47, p. 79–83.

Gabb, W.M., 1882, Notes on the geology of Lower California, *in* Whitney, J., ed., Geology: Sacramento, California, California Geological Survey, p. 137–148.

Gastil, R.G., 1993, Prebatholithic history of peninsular California, *in* Gastil, R.G., and Miller, R.H., eds., The prebatholithic stratigraphy of peninsular California: Boulder, Colorado, Geological Society of America Special Paper 279, p. 145–156.

Gastil, R.G., and Krummenacher, D., 1977, Reconnaissance geology of coastal Sonora between Puerto Lobos and Bahia Kino: Geological Society of America Bulletin, v. 88, p. 189–198.

Gastil, R.G., Lemone, D.V., and Stewart, W.J., 1973, Permian fusulinids from near San Felipe, Baja California: American Association of Petroleum Geologists Bulletin, v. 57, p. 746–747.

Gastil, R.G., Krummenacher, D., and students, 1974, Reconnaissance geologic map of coastal Sonora between Puerto Lobos and Bahia Kino: Geological Society of America Map and Chart Series MC-16, scale 1:150,000, 1 sheet.

Gastil, R.G., Phillips, R.P., and Allison, E.C., 1975, Reconnaissance geology of the State of Baja California: Boulder, Colorado, Geological Society of America Memoir 140, 170 p.

Gastil, R.G., Krummenacher, D., and Minch, J.A., 1979, The record of Cenozoic volcanism around the Gulf of California: Geological Society of America Bulletin, v. 90, p. 839–857.

Gastil, R.G., Miller, R.H., Anderson, P., Crocker, J., Campbell, M., Buch, P., Lothringer, C., Leier-Engeldardt, P., DeLattre, M., Hoobs, J., and Roldán-Quintana, J., 1991, The relation between the Paleozoic strata on opposite sides of the Gulf of California, *in* Perez-Segura, E., and Jacques-Ayala, C., eds., Studies of Sonoran geology: Boulder, Colorado, Geological Society of America Special Paper 254, p. 7–16.

Gastil, G.R., Neuhaus, J., Cassidy, M., Smith, J.T., Ingle, J.C., and Krummenacher, D., 1999, Geology and paleontology of southwestern Isla Tiburón, Sonora, Mexico: Revista Mexicana de Ciencias Geológicas, v. 16, p. 1–34.

Hausback, B.P., 1984, Cenozoic volcanic and tectonic evolution of Baja California Sur, Mexico, *in* Frizzell, V.A., Jr., ed., Geology of the Baja California peninsula: Los Angeles, Pacific Section of the Economic Paleontologists and Mineralogists, p. 219–236.

Hausback, B., Cook, A., Farrar, C., Giambastiani, M., Martín-Barajas, A., Paz-Moreno, F., Stock, J., Dmochowski, J., Fowler, S., Sutter, K., Verke, P., and Winant, C., 2000, Isla San Luis Volcano, Baja California, Mexico—Timing of volcanic activity [abs.]: Eos (Transactions, American Geophysical Union), p. 1359.

Henry, C.D., and Aranda-Gómez, J.J., 1992, The real southern basin and range: Mid Cenozoic to Late Cenozoic extension in Mexico: Geology, v. 20, p. 701–704.

Henyey, T.L., and Bischoff, J.L., 1973, Tectonic elements of the northern part of the Gulf of California: Geological Society of America Bulletin, v. 84, p. 315–330.

Johnson, S.E., Tate, M.C., and Fanning, C.M., 1999, New geologic mapping and SHRIMP U-Pb zircon data in the Peninsular Ranges batholith, Baja California, Mexico: Evidence for a suture?: Geology, v. 27, p. 743–756.

Lee, J., Miller, M., Crippen, R., Hacker, B., and Ledesma-Vazquez, J., 1996, Middle Miocene extension in the Gulf extensional province, Baja California: Evidence from the southern Sierra Juárez: Geological Society of America Bulletin, v. 108, p. 505–525.

Lewis, C.J., 1994, Constraints on extension in the Gulf extensional province from the Sierra San Fermín, northeastern Baja California, Mexico [Ph.D. thesis]: Cambridge, Massachusetts, Harvard University, 361 p.

Lewis, C.J., 1996, Stratigraphy and geochronology of Miocene and Pliocene volcanic rocks in the Sierra San Fermín and southern Sierra San Felipe, Baja California, Mexico: Geofisica Internacional, v. 35, p. 1–31.

Lewis, C.J., and Stock, J.M., 1998a, Paleomagnetic evidence of localized vertical-axis rotation during Neogene extension of the Sierra San Fermín, northeastern Baja California, Mexico: Journal of Geophysical Research, v. 103, p. 2455–2470.

Lewis, C.J., and Stock, J.M., 1998b, Paleomagnetic evidence of localized vertical-axis rotation during Neogene extension of the Sierra San Fermín, northeastern Baja California, Mexico: Journal of Geophysical Research, v. 103, p. 2455–2470.

Lewis, J.L., Day, S.M., Magistrale, H., Castro, R.R., Astiz, L., Rebollar, C., Eakins, J., Vernon, F.L., and Brune, J.N., 2001, Crustal thickness of the Peninsular Ranges and Gulf Extensional Province in the Californias: Journal of Geophysical Research, v. 106, p. 13,599–13,611.

Lonsdale, P., 1989, Geology and tectonic history of the Gulf of California, *in* Winterer, E.L., Hussong, D.M., and Decker, R.W., eds., The eastern Pacific Ocean and Hawaii: Boulder, Colorado, Geological Society of America, Geology of North America, v. n, p. 499–521.

Louden, K.E., and Chian, D., 1999, The deep structure of non-volcanic rifted continental margins: Philosophical Transactions of the Royal Society of London, v. 357, p. 767–804.

Martín-Barajas, A., and Stock, J.M., 1993, Estratigrafía y petrología de la secuencia volcánica de Puertecitos, Noreste de Baja California. Transición de un arco volcánico a rift, *in* Delgado-Argote, L.A., and Martín-Barajas, A., eds., Contribuciones a la tectónica del occidente de México. Monografía 1 de la Unión Geofísica Mexican: Mexico City, Mexico, Mexican Geophysical Union, p. 66–89.

Martín-Barajas, A., Stock, J.M., Layer, P., Hausback, B., Renné, P., and Martínez-López, M., 1995, Arc-rift transition volcanism in the Puertecitos Volcanic Province, northeastern Baja California, Mexico: Geological Society of America Bulletin, v. 107, p. 407–424.

Martín-Barajas, A., Tellez-Duarte, M., and Stock, J.M., 1997, The Puertecitos Formation: Pliocene volcaniclastic sedimentation along an accommodation zone in northeastern Baja California, *in* Johnson, M.E., and Ledesma-Vasquez, J., eds., Pliocene carbonate and related facies flanking the Gulf of California, Baja California, Mexico: Boulder, Colorado, Geological Society of America Special Paper 318, p. 1–24.

Martín-Barajas, A., Fletcher, J.M., and López-Martínez, M., 2000, Waning Miocene subduction and arc volcanism in Baja California: The San Luis Gonzaga volcanic field: Tectonophysics, v. 318, p. 27–52.

Minch, J.A., 1979, The Late Mesozoic–Early Tertiary framework of continental sedimentation, northern Peninsular Ranges, Baja California, Mexico, *in* Minch, J.A., ed., Eocene Depositional Systems: Los Angeles, Pacific Section, Society of Economic Mineralogists and Paleontologists, p. 43–68.

Mora-Alvarez, G., and McDowell, F.W., 2000, Miocene volcanism during late subduction and early rifting in the Sierra Santa Ursula of western Sonora, Mexico, *in* Delgado-Granados, H., Aguirre-Díaz, G., and Stock, J.M., eds., Cenozoic tectonics and volcanism of Mexico: Boulder, Colorado, Geological Society of America Special Paper 334, p. 123–141.

Nagy, E.A., 1997, Extensional deformation and volcanism within the northern Puertecitos volcanic province, Sierra Santa Isabel, Baja California, Mexico [Ph.D. thesis]: Pasadena, California Institute of Technology, 333 p.

Nagy, E.A., 2000, Extensional deformation and paleomagnetism at the western margin of the Gulf Extensional Province, Puertecitos Volcanic Province, northeastern Baja California, Mexico: Geological Society of America Bulletin, v. 112, p. 857–870.

Nagy, E.A., Grove, M., and Stock, J.M., 1999, Age and stratigraphic relationships of pre- and syn-rift volcanic deposits in the northern Puertecitos Volcanic Province, Baja California, Mexico: Journal of Volcanology and Geothermal Research, v. 93, p. 1–30.

Neuhaus, J.R., 1989, Volcanic and nonmarine stratigraphy of southwest Isla Tiburón, Gulf of California, Mexico [Master's thesis]: San Diego, San Diego State University, 170 p.

Ortega-Rivera, A., Farrar, E., Hanes, J.A., Archibald, D.A., Gastil, R.G., Kimbrough, D.L., Zentilli, M., López-Martínez, M., Féraud, G., and Ruffet, G., 1997, Chronological constraints on the thermal and tilting history of the Sierra San Pedro Mártir Pluton, Baja California, Mexico, from U/Pb, $^{40}Ar/^{39}Ar$, and fission-track geochronology: Geological Society of America Bulletin, v. 109, p. 728–745.

Oskin, M., 2002, Tectonic evolution of the northern Gulf of California, Mexico, Deduced from conjugate rifted margins of the upper Delfín Basin [Ph.D. thesis]: Pasadena, California Institute of Technology, California, 481 p.

Oskin, M., Stock, J., and Martín-Barajas, A., 2001, Rapid localization of Pacific–North America plate motion in the Gulf of California: Geology, v. 29, p. 459–462.

Oskin, M., and Stock, J., 2003, Marine incursion synchronous with plate boundary localization in the Gulf of California: Geology, v. 31, p. 23–26.

Paz-Moreno, F.A., and Demant, A., 1999, The recent Isla San Luis volcanic centre: Petrology of a rift-related suite in the northern Gulf of California, Mexico: Journal of Volcanology and Geothermal Research, v. 93, p. 31–52.

Persaud, P., Stock, J., Steckler, M., Martín-Barajas, A., Diebold, J., Gónzalez-Fernández, A., and Mountain, G., 2003, Active deformation and shallow structure of the Wagner, Consag and Delfín Basins, Northern Gulf of California, Mexico: Journal of Geophysical Research. v. 108, DOI: 10.1029/2002JB001937.

Phillips, J.R., 1993, Stratigraphy and structural setting of the mid-Cretaceous Olvidada Formation, Baja California Norte, Mexico, *in* Gastil, R.G., and Miller, R.H., eds., The prebatholithic stratigraphy of peninsular California: Boulder, Colorado, Geological Society of America Special Paper 279, p. 97–106.

Radelli, L., 1989, The ophiolites of Caimalli and the olvidada nappe of northeastern Baja California and west-central Sonora, Mexico, *in* Abbott, P.L., ed., Geologic studies in Baja California: Los Angeles, California, Society of Economic Paleontologists and Mineralogists Pacific Section, p. 79–85.

Rothstein, D., 1997, Metamorphism and denudation of the eastern Peninsular Ranges batholith, Baja California Norte, Mexico [Ph.D. thesis]: Los Angeles, University of California, 445 p.

Sawlan, M.G., 1991, Magmatic evolution of the Gulf of California rift, *in* Dauphin, J.P., and Simoneit, B.R.T., eds., The Gulf and Peninsular Province of the Californias: Tulsa, Oklahoma, The American Association of Petroleum Geologists, p. 301–370.

Sawlan, M.G., and Smith, J.G., 1984, Petrologic characteristics, age, and tectonic setting of Neogene volcanic rocks in northern Baja California, *in* Frizzell, V.A., ed., Geology of the Baja California peninsula: Los Angeles, Pacific Section of the Society of Economic Paleontologists and Mineralogists, p. 237–251.

Silver, L.T., and Chappell, B.W., 1988, The Peninsular Ranges batholith: An insight into the evolution of the cordilleran batholiths of southwestern North America: Transactions of the Royal Society of Edinburgh, Earth Sciences, v. 79, p. 105–121.

Slyker, R.G., 1970, Geologic and geophysical reconnaissance of the Valle De San Felipe region, Baja California, Mexico [M.S. thesis]: San Diego, California State University, 97 p.

Smith, J.T., Smith, J.G., Ingle, J.C.J., Gastil, R.G., Boehm, M.C., Roldán, Q.J., and Casey, R.E., 1985, Fossil and K-Ar age constraints on upper middle Miocene conglomerate, SW Isla Tiburon, Gulf of California, *in* Proceedings, The Geological Society of America Cordilleran Section 81st Annual Meeting, Volume 17, p. 409.

Stock, J.M., 1989, Sequence and geochronology of Miocene rocks adjacent to the main gulf escarpment: Southern Valle Chico, Baja California Norte, Mexico: Geofisica Internacional, v. 28, p. 851–896.

Stock, J.M., 1993, Geologic map of southern Valle Chico and adjacent regions, Baja California, Mexico: Geological Society of America Map and Chart Series 76, 2 sheets.

Stock, J.M., 2000, Relation of the Puertecitos Volcanic Province, Baja California, to development of the plate boundary in the Gulf of California, *in* Delgado-Granados, H., Aguirre-Diaz, G., and Stock, J.M., eds., Cenozoic tectonics and volcanism of Mexico: Boulder, Colorado, Geological Society of America Special Paper 334, p. 143–155.

Stock, J.M., and Hodges, K.V., 1989, Pre-Pliocene extension around the Gulf of California and the transfer of Baja California to the Pacific Plate: Tectonics, v. 8, p. 99–115.

Stock, J.M., and Hodges, K.V., 1990, Miocene to recent structural development of an extensional accommodation zone, northeastern Baja California, Mexico: Journal of Structural Geology, v. 12, p. 315–328.

Stock, J.M., Martín-Barajas, A., Suárez-Vidal, F., and Miller, M.M., 1991, Miocene to Holocene extensional tectonics and volcanic stratigraphy of northeastern Baja California, Mexico, *in* Walawender, M., and Hanan, B., eds., Guidebook for the 1991 Annual Meeting of the Geological Society of America: Boulder, Colorado, Geological Society of America, p. 44–67.

Stock, J.M., Lewis, C.J., and Nagy, E.A., 1999, The Tuff of San Felipe: An extensive middle Miocene pyroclastic flow deposit in Baja California, Mexico: Journal of Volcanology and Geothermal Research, v. 93, p. 53–74.

Tegner, C., Duncan, R., Bernstein, S., Brooks, C., Bird, D., and Storey, M., 1998, ^{40}Ar/^{39}Ar geochronology of Tertiary mafic intrusions along the East Greenland rifted margin: Relation to flood basalts and the Iceland hotspot track: Earth and Planetary Science Letters, v. 156, p. 75–88.

Valencia-Moreno, M., Ruiz, J., Barton, M., Patchett, P., Zürcher, L., Hodkinson, D., and Roldán-Quintana, J., 2001, A chemical and isotopic study of the Laramide granitic belt of northwestern Mexico: Identification of the southern edge of the North American Precambrian basement: Geological Society of America Bulletin, v. 113, p. 1409–1422.

MANUSCRIPT ACCEPTED BY THE SOCIETY JUNE 2, 2003

Geological Society of America
Special Paper 374
2003

The Quaternary Moctezuma volcanic field:
A tholeiitic to alkali basaltic episode in the central Sonoran
Basin and Range Province, México

Francisco A. Paz Moreno
Universidad de Sonora, Departamento de Geología, Apdo. Postal 847, 83000 Hermosillo, Sonora, México

Alain Demant
Jean-Jacques Cochemé
Laboratoire de Pétrologie Magmatique, Université Aix-Marseille, Case courrier 441, 13397 Marseille Cedex 20, France

Jaroslav Dostal
Department of Geology, Saint Mary's University, Halifax, Nova Scotia B3H 3C3, Canada

Raymond Montigny
Institut de Physique du Globe de Strasbourg, 5 rue Descartes, 67084 Strasbourg Cedex, France

ABSTRACT

The Quaternary Moctezuma volcanic field at the foothills of the Sierra Madre Occidental, México, is characterized by a close association of tholeiitic and alkaline magmas. The vents of the tholeiitic magmatism (1.7 Ma), which forms the overall mesa morphology of the volcanic field, lie along the major faults that define the eastern margin of the Moctezuma basin. This half-graben formed during the early Miocene as indicated by the emplacement of basaltic flows (22.3 Ma), which were intercalated with fanconglomerates of the Báucarit Formation. The youngest volcanics are alkaline lavas (0.53 Ma), which erupted from scoria cones located in the center of the basin. The main mineralogical difference between the Quaternary tholeiitic and alkaline lavas is the composition of the clinopyroxenes: they are Ca-rich in the alkaline lavas, whereas they are subcalcic with orthopyroxene or pigeonite in the tholeiitic lavas. The geochemical data show a gradual change in the characteristics of the lavas. The Sr and Nd isotopic compositions, and the shape of the mid-oceanic-ridge basalt (MORB)–normalized patterns, are similar to oceanic island basalt suites, suggesting that the source of the Moctezuma volcanic field lavas was dominated by asthenospheric mantle.

The absence of mantle or granulite xenoliths in the Moctezuma volcanic field lavas, compared to the neighboring Geronimo volcanic field, reflects slower ascent rates. The Southern Cordillera Basaltic Andesite (SCORBA)–type geochemical signature of the early Miocene synextensional basalts of Moctezuma compared to the Quaternary Moctezuma volcanic field lavas can be correlated with a progressive thinning of the mantle lithosphere during the Neogene.

Keywords: Quaternary volcanism, México, mineralogy, geochemistry, isotopes.

---

*paz@geologia.uson.mx

Paz Moreno, F.A., Demant, A., Cochemé, J.-J., Dostal, J., and Montigny, R., 2003, The Quaternary Moctezuma volcanic field: A tholeiitic to alkali basaltic episode in the central Sonoran Basin and Range Province, México, *in* Johnson, S.E., Paterson, S.R., Fletcher, J.M., Girty, G.H., Kimbrough, D.L., and Martín-Barajas, A., eds., Tectonic evolution of northwestern México and the southwestern USA: Boulder, Colorado, Geological Society of America Special Paper 374, p. 439–455.

INTRODUCTION

The transition from tholeiitic to alkali basalt volcanism has been observed in many oceanic intraplate volcanoes (Feigenson et al., 1983; Chen et al., 1991; Frey et al., 1991; Fodor et al., 1992; Bardintzeff et al., 1994). It is far less commonly observed in an intracontinental environment with the exception of some rift systems. The processes invoked for this change include: (a) a decrease of the degree of partial melting, (b) a change from a lithospheric to an asthenospheric mantle source, and (c) interaction with the continental crust. However, these processes have remained controversial for intracontinental volcanism. The Quaternary Moctezuma volcanic field in central Sonora, at the western margin of the Sierra Madre Occidental, displays such an evolution from tholeiitic to alkali basalts and represents a unique example of this transition in the Basin and Range province of northwestern México.

Previous investigations in the Moctezuma area were only of a reconnaissance nature and defined two volcanic episodes (Paz-Moreno, 1985; Paz et al., 1986). The principal goals of this paper are: (1) to establish the chronostratigraphy of the basaltic successions and to relate the volcanic activity to the tectonic evolution of western Sonora; (2) to document the petrography, mineral chemistry, and geochemistry of the Moctezuma volcanic field lavas; and (3) to infer the mantle sources and the petrogenetic evolution of this magmatic suite.

GEOLOGICAL SETTING

The geology of northwestern México is characterized by a widespread late Eocene–Oligocene volcanic episode that built the Sierra Madre Occidental, one of the largest ignimbritic provinces in the world. Older rocks capped by the ignimbrites included andesites and volcaniclastic rocks intruded and hydrothermally altered by Laramide plutons. The huge volume of rhyolitic tuffs of the Sierra Madre Occidental, commonly referred to as the Upper volcanic sequence (McDowell and Keizer, 1977), erupted in a brief interval between 34 and 27 Ma (McDowell and Clabaugh, 1979; Cocheme and Demant, 1991). The petrology and geochemistry of the lavas as well as the associated chalcophile mineralizations are typical of calc-alkaline lavas. The Oligocene sequence is in turn overlain by a bimodal sequence composed of (1) poorly welded ash-flow tuffs and perlitic dacitic to rhyolitic volcanic domes, and (2) trap-like (locally >300 m thick) basaltic rocks. The basalts, abundant in the northern part of the Sierra Madre Occidental, have ages ranging between 30 and 22 Ma (Swanson and McDowell, 1984; Montigny et al., 1987; Paz-Moreno, 1992). Their chemical characteristics are those of continental tholeiites (Cameron et al., 1989; Demant et al., 1989). The emplacement of this bimodal sequence was related to an early episode of crustal extension (Cocheme and Demant, 1991).

A further middle Cenozoic extension disrupted the Sierra Madre Occidental volcanic plateau both to the east and west,

giving rise to the actual Basin and Range morphology of the states of Sonora and Chihuahua (Fig. 1). This period of large-scale northeast-southwest extension (Gans, 1997) was responsible for the development of basins bordered on the east by strike-slip faults (half-grabens). Different stages in the evolution of the basins are recorded by changes in the sedimentary accumulations. The oldest sediments are sandy to conglomeratic alluvial deposits. These deposits, known locally as the Báucarit Formation (Dumble, 1900; King, 1939), are frequently tilted and strongly cemented by low-temperature zeolitization processes (Cocheme et al., 1988; Münch et al., 1996). Intercalations of basalts in the lower part of the Báucarit Formation are relatively frequent. A characteristic feature of these lavas is the complete alteration of olivine phenocrysts to reddish clay minerals and vesicles filled by zeolites or calcite (Cocheme et al., 1994). Basalts from the Báucarit Formation yielded ages in the range 23–18 Ma (Demant et al., 1989; Paz-Moreno, 1992; Bartolini et al., 1994; Gans, 1997; McDowell et al., 1997), whereas post-Báucarit volcanism started at 17 Ma (Damon et al., 1981; Cocheme and Demant, 1991; Gans, 1997; McDowell et al., 1997). The main episode of Neogene extension in this part of Sonora is early Miocene, coeval with the deposition of the Báucarit Formation.

Unconsolidated alluvial fan or talus deposits occurred throughout the Basin and Range province in central Sonora (Paz-Moreno, 1992; McDowell et al., 1997). These deposits result from west-east extension due to transtensional dextral deformation related to the coupling of the Baja California peninsula with the Pacific plate and the progressive opening of the Gulf of California rift system since ca. 12 Ma (Atwater, 1970; Dickinson and Snyder, 1979; Spencer and Normark, 1979; Angelier et al., 1981; Atwater, 1989; Martín-Barajas et al., 1995; Nagy et al., 1999). Gordon Gastil has done pioneering work on the mechanism and age of the opening of the Gulf of California. His paper on the chronology of the volcanic formations in coastal Sonora (Gastil and Krummenacher, 1977) has inspired most of the further works on the stratigraphy of the volcanic successions in Sonora.

Mafic volcanism associated with Late Cenozoic extension is relatively scarce in Sonora, with the exception of the Pinacate area (Lynch, 1981). The volcanism produced small volcanic fields, such as Sierra Santa Ursula (Mora-Alvarez and McDowell, 2000), the Comedores Formation (Salas, 1970; Bartolini et al., 1993), and Pozo de Leyva (McDowell et al., 1997). Nevertheless, a gradual change of the source of the basalts from lithospheric to asthenospheric has been documented since the Neogene (Paz-Moreno, 1992) and can be compared to the model inferred for southwestern U.S. magmatism (Fitton et al., 1988, 1991; Kempton et al., 1991). This evolution in magma composition has been linked to changes in the regional tectonic setting, and in particular to the development of a slab window (Hole et al., 1995) and a new transform plate boundary between the Pacific and the North America plates after subduction ceased (Stock and Molnar, 1988; Martín-Barajas et al., 1995; Stock, 2000).

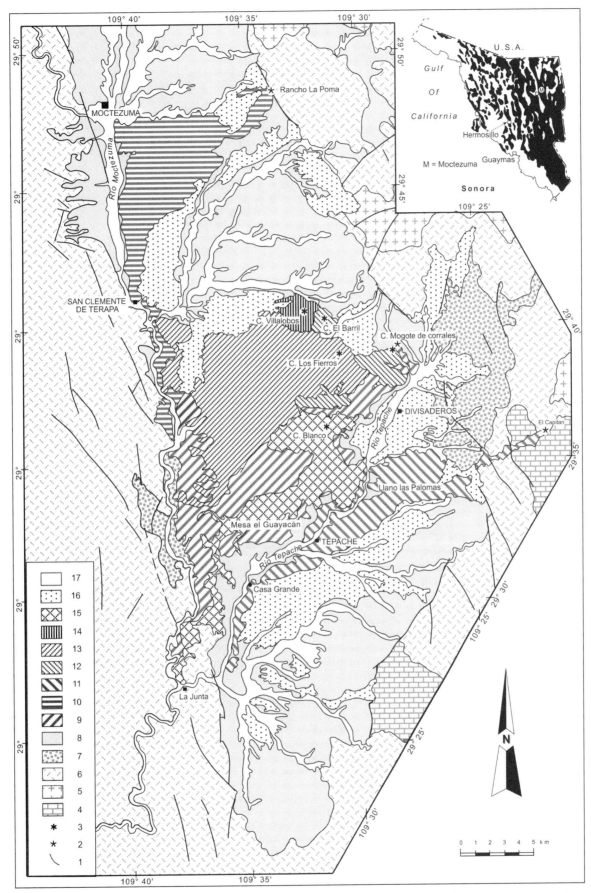

Figure 1. Geologic map of Moctezuma volcanic field, Sonora, México. As an insert, location of Moctezuma graben and aspect of Basin and Range extension in Sonora State. Legend: 1—principal faults; 2—vents of the fissure eruptions; 3—scoria cone vents of the alkaline lavas; 4— Cretaceous limestones; 5—granite intrusions; 6—upper volcanic complex of Sierra Madre Occidental (mainly ignimbrites); 7—sandstones and conglomerates of Báucarit Formation sensu stricto; 8—unconsolidated alluvial fan deposits; 9–11—different tholeiitic lava flows in chronological order; 12–15—alkaline lavas from monogenetic cones in chronological order; 16—late Quaternary alluvial and terrace deposits; 17—Recent fluvial deposits.

THE MOCTEZUMA VOLCANIC FIELD

Geology and Volcanic Stratigraphy

The Quaternary Moctezuma volcanic field, located in the Moctezuma valley, covers the central part of a half-graben. This graben is bounded on the west by the middle Tertiary felsic volcanic sequences of Sierra El Coyote and to the east by Sierra La Madera, which represents the footwall of this half-graben. Sierra La Madera comprises a granitoid batholithic body that intruded Lower Cretaceous limestones and is in turn covered by Tertiary volcanics (Fig. 1). Miocene basin infilling is limited in the western side of the Moctezuma valley and is only preserved near Río Moctezuma. Báucarit-type sedimentary deposits with their unique morphological features outcrop more extensively east of Divisaderos. They are unconformably overlain by nonconsolidated sandy to silty reddish or greenish sediments. Gypsum lenses and white subhorizontal zeolite layers occur within the Pliocene sediments. The presence of alluvial fan deposits, overlying the Quaternary basalts (Fig. 1), demonstrates that vertical fault movements are still active in the Moctezuma region (Paz-Moreno, 1992). Therefore, the Moctezuma graben, like the San Bernardino basin (Suter, 2000), represents the tectonically active western edge of the Sierra Madre Occidental.

The Quaternary Moctezuma volcanic field can be subdivided into (1) an earlier fissural tholeiitic event followed by (2) alkaline lavas erupted from small monogenetic cones (Figs. 1 and 2). The tholeiitic lavas, which erupted from fissure vents along the faults on the eastern boundary of the Moctezuma graben, form planar mesas. The lavas flowed mainly toward the southwest and then toward the south, thickening in the same direction from a few meters near the vents to ~40 m in the paleovalley of Río Moctezuma. Columnar jointings are common and particularly visible in the cliffs, which dominate the Río Tepache and Moctezuma valleys. The presence of lava tubes, pressure ridges, and smooth pāhoehoe-type flow surfaces with tree molds indicate that lavas erupted during the fissural episode were fluid and voluminous. Spectacular exposures of pillow lavas occur at the base of the volcanic mesa, 4.5 km south of the village of Moctezuma along the road to Tepache (Figs. 3 and 4). Three and a half kilometers west of the Casa Grande ranch, in the eastern cliff of Río Moctezuma (Fig. 1), the pillow package is more than 15 m thick. The surface features of the pillow lobes (Yamagashi, 1985) and their glassy rims are well preserved. Pillows and vesicle pipes (Williams and McBirney, 1979; Goff, 1996), are evidence that tholeiitic lavas flooded wet zones in the paleovalley of Río Moctezuma. A relative chronology of the different flows has been established during the mapping (Fig. 1).

The five alkaline monogenetic vents are concentrated in the central part of the graben. Here also lava flows have been mapped individually (Fig. 1). Cerro Blanco is the youngest volcano in the Moctezuma volcanic field. Volcanic bombs (from 6 cm up to 1.5 m in size) occur on the flanks of this 90-m-high strombolian cone that has been smoothed by the erosion. A'a-type lava flows

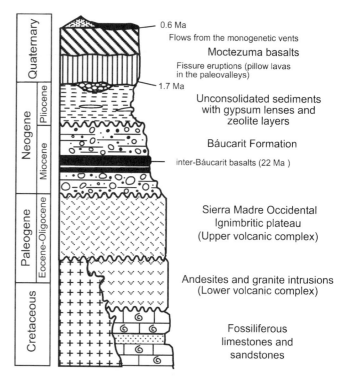

Figure 2. Schematic stratigraphic column of Moctezuma basin (no vertical scale).

Figure 3. Tholeiitic flow in northern part of Moctezuma volcanic field (vent of Rancho La Poma). This outcrop is located 4.5 km southeast of Moctezuma, along Road Moctezuma–Tepache. Limit between pillows and 2-m-thick massive upper part is clearly seen.

that issued from the breached cone form the Mesa Grande malpaís. Toward the south, the Cerro Blanco flows cascaded down the central mesa into the Río Tepache and Río Moctezuma valleys. This morphological feature is consistent with a recent Quaternary age for this volcanic episode.

Figure 4. Detail of pillow package showing glassy margins and radial cracks (hammer gives scale, 32 cm.). Same locality as Figure 3.

The Moctezuma volcanic field scattered an area of ~380 km², but the total volume of the erupted magma is estimated to be <2 km³, about two-thirds of which corresponds to the tholeiitic lavas.

Geochronology

Three radiometric ages have been determined (Table 1 and Appendix 1). Plagioclase feldspars from a basaltic lava flow intercalated in the Báucarit sequence (sample 9-88), on the eastern side of Río Moctezuma (Fig. 1), were dated by the $^{40}Ar/^{39}Ar$ method. The age for the second step, representing 90.8% of ^{39}Ar released, was 22.3 ± 0.13 Ma. This early Miocene age is consistent with the generally accepted age of the Báucarit Formation in

the region (Cochemé et al., 1988). This eruptive stage predated the emplacement of the Moctezuma volcanic field lavas.

A whole rock $^{40}Ar/^{39}Ar$ age of 1.7 ± 0.74 Ma was obtained on a tholeiitic lava flow (sample 78–82) from the Mogote de Corrales vent (Fig. 1). This age is not excellent as it was obtained from the second step at 1300 °C, while 44.5% of the ^{39}Ar has degassed at 550 °C, probably due to the presence of a K-rich phase that does not retain the argon well. It is nevertheless in good agreement with the field stratigraphy and gives the age of the Moctezuma volcanic field tholeiitic eruptions.

Lavas from the monogenetic vents are younger than the tholeiitic volcanism. This is confirmed by a K/Ar age of 0.53 ± 0.2 Ma obtained on a basaltic sample from Cerro Blanco (sample 8–82), representing the most recent volcanic event in the Moctezuma volcanic field (Paz et al., 1986; Montigny et al., 1987).

In summary, the Moctezuma half-graben formed during an early Miocene extensional phase. Volcanic activity in the Moctezuma basin began during the early Quaternary with eruptions of tholeiitic lavas fed by vents located along the main fault, which borders the graben to the east. This first volcanic episode (~1.7 Ma) produced the overall planar morphology of the Moctezuma volcanic field. After a gap in the volcanic activity of ca. 1 Ma, during which erosion carved the Moctezuma and Tepache valleys, renewed activity migrated toward the center of the graben, building scoriaceous cones and associated alkaline lava flows.

PETROGRAPHY AND MINERAL CHEMISTRY

Petrography

The texture of lavas reflects their mode of emplacement and cooling conditions. However, differences are apparent between the lavas from the fissural event and those from the monogenetic

TABLE 1. K/AR AND $^{39}AR/^{40}AR$ AGES										
Sample	Rock type	Material used	Long (°W)	Lat (°N)	Fraction (microns)	K (wt %)	$^{40}Ar^*$ 10^{11} mol/g	$^{40}Ar^*$ (%)		Age ± 1σ
8-82	Basalt	Wr	109°31'	29°36'	100-125	1.120	0.09846	5.46		0.61 ± 0.08
		Wr			125-160	1.222	0.08276	5.90		0.47 ± 0.06

Sample	Rock type	Material used	Long (°W)	Lat (°N)	Temp. (°C)	$^{40}Ar^*$ (%)	^{39}Ar (%)	$^{37}Ca/^{39}K$	$^{40}Ar^*/^{39}K$	Age ± 1σ
9-88	Basalt	Pl	109°37'	29°30'	550	4.2	3.3	0.8398	0.9455	20.8 ± 2.6
					1300	77.1	90.8	0.9206	1.0140	22.3 ± 0.13
					1500	6.4	5.8	1.0200	0.6837	15.1 ± 1.9
		J-value = 0.012264							Integrated age	21.8 ± 0.2
78-82	Basalt	Wr	109°28'	29°39'	550		44.5	1.0953		
					1300	4.5	43.0	14.45	0.07706	1.7 ± 0.74
					1500	0.004	12.4	29.69	0.01344	0.3 ± 1.10

Notes: $\lambda_\beta = 4.962 \times 10^{-10}$ yr⁻¹; $\lambda_\epsilon = 0.581 \times 10^{-10}$ yr⁻¹; $^{40}K/K = 1.167 \times 10^{-4}$ moles/mole. Material used for datation: Wr—whole rock; pl—plagioclase. Analytical techniques in Appendix.
*radiogenic argon in percent of total ^{40}Ar.

cones. The tholeiitic basalts are aphyric to slightly porphyritic. They have dominantly doleritic intersertal to intergranular textures; less commonly hyalophitic to hyaloporphyritic textures are seen in the quenched facies. The tholeiitic basalts contain less than 10% phenocrysts (Table 2), typically with olivine as the only phenocryst phase. Alkalic lavas from the monogenetic cones are more porphyritic (>30% phenocrysts) and exhibit an intergranular groundmass. In the alkalic lavas, olivine, clinopyroxene, and plagioclase phenocrysts frequently form glomeroporphyritic clusters. Plagioclase is abundant and ranges in size from phenocrystic laths (>> 1 mm) to microlites in the matrix of the seriate-textured lavas. Groundmass assemblages include glass, plagioclase, clinopyroxene, oxides, and subordinate olivine (< 6%).

In contrast with the Quaternary alkali basaltic fields from northwestern México [El Pinacate, Sonora, and San Quintín, Baja California (Aranda-Gómez and Ortega-Gutiérrez, 1987)] and the southwestern United States (Menzies et al., 1987), mantle xenoliths are lacking in the Moctezuma volcanic field.

Mineral Chemistry

Mineral compositions were determined on 10 polished thin sections. These analyses were obtained on a Cameca-Camebax electron microprobe at the University of Montpellier, France, using wavelength-dispersive spectrometers and natural material as standards. The analytical conditions were typically 15 kV accelerating voltage, 10 nA beam current, 1μm spot size, and integrated counting times ranging from 6 to 20 s, depending on the analyzed elements. For feldspar analyses, sodium was determined first to minimize the effect of its tendency to volatilize under the electron beam. Representative microprobe analyses of the different mineral phases are reported in Table 3.

Olivine

In the tholeiitic basalts, olivine phenocrysts are skeletal, which is consistent with rapid cooling as indicated by the glassy matrix. They are poorly zoned (Fo_{84-76}). In the alkalic lavas, the cores of olivine phenocrysts have about the same composition, but as olivine also crystallizes in the matrix, the total compositional range is greater (up to Fo_{48}). Olivine phenocrysts in both types of basalts contain sparse inclusions (< 30 μm in diameter) of spinel. These crystals are Cr- and Mg-rich (chromian spinel) and have relatively low Ti and Fe^{3+}. They plot in the oceanic-island basalt (OIB) field of the TiO_2 versus Al_2O_3 binary diagram of Kamenetsky et al. (2001).

Pyroxenes

Clinopyroxenes are present only as a groundmass phase (< 0.5 mm) in the tholeiites. These late crystallizing colorless crystals have compositions evolving from Ca-poor augite ($Wo_{40}En_{44}Fs_{16}$) to pigeonite ($Wo_7En_{72}Fs_{21}$). Some tholeiitic basalts also contain subcalcic augite (Fig. 5). Such a trend, characterized by a decrease of the Wo component during crystallization, is typical of the tholeiitic series (Mellini et al., 1988). Orthopyroxene phenocrysts (>1 mm), with Mg-rich compositions (En_{80-72}) and clinopyroxene rims, were found in one tholeiite (sample 78–82); they represent the early crystallizing phases.

Clinopyroxenes from the alkali lavas have a purplish color and crystallize either as phenocrysts or microphenocrysts in the groundmass. They have a distinct compositional range compared to the tholeiites as they plot in the fields of Ca-rich augites and diopsides (Fig. 5). Augite cores are Al_2O_3- (up to 3.6 wt %) and Cr_2O_3-rich (up to 0.70 wt %) while microphenocrysts tend to have low Al_2O_3 and Cr_2O_3 and slightly higher Ca with increasing iron.

TABLE 2. MODAL COMPOSITIONS OF THOLEIITIC AND ALKALINE LAVAS FROM THE MOCTEZUMA VOLCANIC FIELD

Sample no.	Tholeiitic lavas				Alkaline lavas					
	78-82	8-81	6-88	54-82	71-82	28-81	27-81	29-81	1-81	8-82
Phenocrysts										
Olivine	6.4	7.1	6.4	2.0	8.3	7.2	7.6	8.0	7.5	7.3
Clinopyroxene	0.5*				3.2	2.3	0.5	0.4		
Plagioclase				0.2*	26.5	21.6	24.2	23.4	29.6	
Microcrystals and microliths										
Olivine			6.5	13.9	3.4	5.7	3.0	5.1	3.8	3.8
Clinopyroxene		32.9			15.5	14.7	17.6	20.6		14.5
Plagioclase		34.8	25.4	42.1	32.3	37.4	33.3	27.1		16.1
Oxides		11.0			7.1	7.8	8.2	8.8		13.9
Glass	82.9	4.9	55.4	38.0					57.2	40.0
Vesicles	10.3	9.3	6.3	3.8	3.7	3.3	5.6	6.6	1.9	4.4

Note: 1 500 counted points per sample (thin section).
*In sample 78-82, orthopyroxene phenocrysts; in sample 54-82, sieve textured plagioclases.

TABLE 3. REPRESENTATIVE SELECTION OF MINERAL COMPOSITIONS FROM THE THOLEIITIC AND ALKALINE LAVAS FROM THE MOCTEZUMA VOLCANIC FIELD

Olivines

	Tholeiitic lavas						Alkaline lavas								
Sample no.:	25-82	25-82	25-82	71-82	71-82	71-82	29-81	29-81	29-81	27-81	27-81	27-81	1-81	1-81	1-81
Analysis:	89	106	124	5	2	24	48	57	64	18	15	28	160	161	170
Mineral size:	ph c	Mph	mc	ph	mph	mc	ph	ph	mc	ph	mph	mc	ph c	ph r	mc
SiO_2	38.83	38.45	37.26	38.97	37.99	36.16	39.07	38.24	37.29	39.65	38.54	36.83	39.24	37.02	33.46
FeO	14.87	18.51	22.74	17.26	25.31	33.97	15.95	19.03	26.04	16.94	21.47	31.26	17.71	26.56	42.76
MnO	0.19	0.27	0.34	0.27	0.36	0.51	0.23	0.28	0.28	0.22	0.43	0.35	0.27	0.49	0.88
MgO	44.99	41.78	38.42	42.84	36.55	29.32	44.25	41.73	35.31	42.97	39.54	31.02	43.64	36.09	21.71
CaO	0.19	0.28	0.28	0.20	0.27	0.31	0.21	0.22	0.30	0.20	0.23	0.33	0.19	0.26	0.33
Total	99.09	99.29	99.04	99.54	100.48	100.27	99.71	99.50	99.22	99.98	100.21	99.79	101.05	100.42	99.14
Fo	84.35	80.09	75.07	81.56	72.02	60.60	83.18	79.63	70.73	81.89	76.65	63.88	81.45	70.77	47.50

Pyroxenes

	Tholeiitic lavas						Alkaline lavas								
Sample no.:	78-82	78-82	78-82	78-82	78-82	25-82	28-81	28-81	27-81	27-81	29-81	29-81	1-81	1-81	1-81
Analysis:	177	178	44	39	67	119	128	125	52	49	153	154	152	164	167
Mineral size:	ph (*)	ph (*)	ph (+)	mc	mc	mc	ph	ph	ph	mc	mc	ph	mc	mc	mc
SiO_2	53.23	53.29	52.92	50.37	52.12	51.22	50.31	50.78	51.05	48.55	50.55	51.76	51.94	51.97	48.58
Al_2O_3	2.64	0.39	0.49	1.48	1.03	2.56	3.36	1.92	2.54	3.43	1.47	1.91	1.21	1.02	4.42
FeO	10.46	14.43	17.55	13.71	15.15	7.95	6.36	10.61	7.06	10.35	11.50	6.60	11.68	13.62	8.50
MnO	0.24	0.32	0.35	0.25	0.39	0.21	0.19	0.25	0.17	0.30	0.29	0.20	0.38	0.33	0.26
MgO	29.43	27.12	23.21	15.18	16.02	17.34	15.68	13.30	16.31	13.13	13.63	16.55	16.16	14.95	15.06
CaO	1.96	2.29	4.01	16.58	14.61	18.27	21.12	21.43	20.46	20.51	20.56	20.73	17.85	17.16	20.04
Na_2O	0.07	0.04	0.09	0.29	0.25	0.34	0.40	0.54	0.36	0.62	0.39	0.37	0.34	0.31	0.43
TiO_2	0.26	0.37	0.40	1.06	0.95	1.09	1.22	1.36	1.27	2.67	1.28	1.15	0.90	0.77	2.15
Cr_2O_3	0.27	0.16				0.71	0.65		0.39			0.19	0.01	0.03	0.46
Total	98.56	98.41	99.02	98.92	100.42	98.69	99.29	100.19	99.61	99.56	99.67	99.46	100.47	100.16	99.90
Wo	3.82	4.43	7.98	34.12	29.85	37.46	43.96	44.27	41.93	43.55	42.16	42.25	35.88	35.13	41.91
En	79.88	73.22	64.22	43.44	45.53	49.48	45.40	38.22	46.50	38.79	38.93	46.93	45.19	42.57	43.80
Fs	16.30	22.34	27.80	22.44	24.62	13.06	10.64	17.51	11.57	17.66	18.90	10.82	18.93	22.29	14.29

Plagioclases

	Tholeiitic lavas						Alkaline lavas								
Sample no.:	78-82	78-82	78-82	78-82	78-82	25-82	28-81	28-81	27-81	27-81	27-81	27-81	27-81	1-81	1-81
Analysis:	25	27	50	72	64	80	88	92	119	33	132	118	145	183	173
Mineral size:	mph	Mph	mc	mph	mph	mc	ph	mc	ph	mc	mc	mc	ph	mc	mc
SiO_2	52.51	54.03	54.96	52.77	53.91	55.81	53.13	58.11	55.00	57.16	65.45	63.63	54.34	53.17	64.57
Al_2O_3	29.04	28.29	27.32	29.70	28.52	27.33	29.25	25.45	28.29	26.12	19.01	21.32	29.23	29.03	20.72
FeO	0.66	0.75	0.85	0.35	0.36	0.55	0.46	0.63	0.35	0.37	0.46	0.45	0.43	0.45	0.62
CaO	12.22	11.00	10.12	12.80	11.49	9.71	11.73	6.81	10.59	8.10	0.27	2.54	11.49	12.22	2.44
Na_2O	4.37	5.07	5.39	4.38	4.81	5.57	4.34	6.81	5.27	6.32	6.66	7.91	4.77	4.42	8.31
K_2O	0.10	0.21	0.28	0.15	0.19	0.33	0.15	0.60	0.43	0.54	7.32	3.58	0.26	0.21	2.54
Total	98.90	99.35	98.92	100.15	99.28	99.30	99.06	98.41	99.93	98.61	99.17	99.43	100.52	99.50	99.20
An	60.33	53.88	50.10	61.26	56.27	48.08	59.36	34.31	51.31	40.14	1.28	12.03	56.24	59.73	11.93
Ab	39.07	44.88	48.26	37.90	42.62	49.96	39.74	62.09	46.21	56.67	57.29	67.79	42.26	39.07	73.34
Or	0.60	1.25	1.65	0.84	1.11	1.96	0.90	3.60	2.48	3.19	41.43	20.18	1.50	1.20	14.73

Note: Symbols used for the crystal size: ph—phenocryst (>1 mm); ph c—core of the phenocryst, ph r—rim of the phenocryst; mph—microphenocryst (between 1 and 0.1 mm); mc—microcrystal (<0.1 mm); for pyroxenes, (*)—represent orthopyroxene analyses, (+)—pigeonite analysis.

(continued)

TABLE 3. REPRESENTATIVE SELECTION OF MINERAL COMPOSITIONS FROM THE THOLEIITIC AND
ALKALINE LAVAS FROM THE MOCTEZUMA VOLCANIC FIELD (*continued*)

Fe-Ti oxides	Tholeiitic lavas				Alkaline lavas			
Sample no.:	78-82	78-82	25-82	25-82	28-81	27-81	27-81	1-81
Analysis:	31	52	85	129	143	110	139	150
Mineral size:	mc	mc	mc	mc	mc	mph	mc	mc
SiO_2	0.24	0.19	0.35	0.22	0.37	0.21	0.33	0.27
FeO	42.13	42.18	43.27	43.08	41.37	40.86	43.13	44.49
Fe_2O_3	3.02	6.84	5.21	4.11	5.79	6.22	2.33	3.07
MnO	0.30	0.36	0.49	0.64	0.55	0.63	0.73	0.64
MgO	2.39	1.10	0.68	1.14	1.69	1.99	1.60	0.81
TiO_2	51.82	49.44	50.01	50.89	49.92	50.07	51.92	51.79
Total	99.90	100.11	100.01	100.08	99.69	99.98	100.04	101.07

Spinels	Tholeiitic lavas	
Sample no.:	25-82	25-82
Analysis:	108	110
Mineral size:	mc	mc
SiO_2	0.27	0.28
Al_2O_3	16.02	19.94
FeO	32.06	28.72
MnO	0.31	0.21
MgO	9.79	11.02
TiO_2	3.06	2.72
Cr_2O_3	36.82	37.60
Total	98.33	100.49

Note: Symbols used for the crystal size: ph—phenocryst (>1 mm); ph c—core of the phenocryst;
ph r—rim of the phenocryst; mph—microphenocryst (between 1 and 0.1 mm); mc—microcrystal (<0.1 mm); for
pyroxenes, (*)—represent orthopyroxene analyses, (+)—pigeonite analysis.

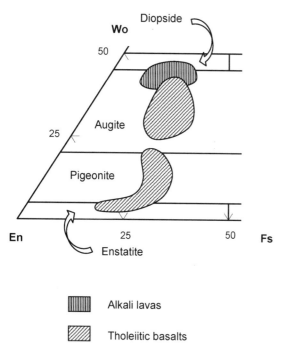

Figure 5. Ca-Fe-Mg (mol%) plot of pyroxenes (Morimoto et al., 1988) from Moctezuma lavas.

Plagioclase

Plagioclase is a conspicuous phase in the Moctezuma basalts. In the tholeiites, plagioclases range in size from microphenocrysts (less than 1 mm) to microlites and have a rather homogeneous composition (An_{61} to An_{47}). Sample 54–82 contains scarce, clouded plagioclase phenocrysts (Macdonald and Katsura, 1965). Their sieve-textured central parts (An_{50-47}) are surrounded by clear outer rims with more calcic compositions (An_{64}). These phenocrysts likely represent partly melted xenocrysts that reacted with a hotter, more primitive basaltic liquid.

Plagioclase phenocrysts in the alkali lavas have a labradorite composition (An_{66-50}) and normal zoning. Feldspar laths and microlites from the groundmass are more sodic plagioclase (up to An_{30}), anorthoclase ($An_{12}Ab_{68}Or_{20}$), or sanidine ($An_1Ab_{57}Or_{42}$). Plagioclase has an overall higher Or component in the alkaline lavas than in the tholeiites.

Fe-Ti Oxides

Ilmenite, the most frequent iron-titanium oxide in the Moctezuma lavas, is typically a late crystallizing phase in the tholeiites. Ilmenite in the alkaline lavas has a similar composition (1–8% hematite). Titanomagnetite is scarce and only occurs in some alkaline lavas.

GEOCHEMISTRY

Major and Trace Elements

Whole-rock major-oxide analyses and trace-element concentrations were determined on 23 samples representative of the different flows. Analyses were performed by inductively coupled plasma–atomic emission spectrometry (ICP-AES) at the University of Aix-Marseille (France), except for Na, K, and Rb, which were determined by atomic absorption spectrophotometry and Fe^{2+} by titration. Rare-earth elements (REE) were analyzed by ICP-AES at Centre de Recherches Pétrographiques et Géochimiques Nancy (France). All the rock samples were ground in an agate mill.

The Moctezuma lavas were classified based on their major-element chemistry and CIPW norms (calculated on an anhydrous basis). The Quaternary tholeiites are hypersthene- to quartz-normative whereas alkali lavas generally contain less than 5% normative nepheline. The total alkali-silica diagram (Le Bas et al., 1986; Le Maitre et al., 1989) discriminates the two series. The tholeiites plot into the basalt and basaltic andesite fields, below the Irvine and Baragar (1971) subalkaline-alkaline divide, while the alkalic magmas are classified as hawaiites (Fig. 6). Some lavas have transitional character as they plot in the alkali field but do not contain normative nepheline (Table 4). The Moctezuma lavas have mg numbers [$100*(Mg/Mg+Fe^{2+})$] ranging from 57 to 66, but most of the values are between 60 and 62. Higher TiO_2 and alkalis, and lower silica contents, are distinctive features between tholeiitic and alkaline lavas.

The tholeiites, like the alkaline rocks, do not show any distinct elongated fractionation trends on the major- or trace-element variation diagrams (Fig. 7). The exceptions are TiO_2 and V, which decrease with increasing SiO_2. In the tholeiitic series, alkalis are scattered but tend to decrease with increasing silica (Fig. 7). Mg numbers and Ni and Cr contents in both the Quaternary suites (Table 4) suggest that the lavas did not undergo extensive fractional crystallization. However, the rocks are evolved compared to primary magmas, which are in equilibrium with mantle olivine (Fo_{92-88}). Such primary magmas should have mg number > 68 and Ni contents of ~300 ppm (e.g., Frey et al., 1978). The relatively low mg numbers (mostly 60–62), low Ni contents (100–220 ppm), high SiO_2 (between 50%–54%) of the lavas, as well as the Mg content of olivine phenocrysts (Fo_{84-78}), demonstrate that the Moctezuma volcanic field rocks are not primary magmas derived from simple partial melting of normal mantle peridotite. The Moctezuma volcanic field lavas have high Ti/V ratios like many continental alkali basalts (Frey et al., 1978).

The abundances of incompatible elements (Table 5)—Th, Zr, Hf, Nb, Ta, light REE (LREE)—are correlated and increase with the degree of silica-undersaturation. The Th/La ratios (~0.1) in both suites are similar to primitive mantle values. Since this ratio is a sensitive indicator of crustal contamination (Taylor and McLennan, 1985), it appears that the rocks were not significantly modified by this process. The Nb/La and Nb/Y ratios are, however, significantly higher than those of primitive mantle. An Nb enrichment relative to LREE has also been reported from continental alkali basaltic suites that were not affected by crustal contamination.

The REE patterns strongly differentiate the two volcanic series. The alkaline rocks have La concentrations ~70 times chondrite values and the patterns have a linear steep slope decreasing from La to Yb (Fig. 8) with $(La/Yb)_n$ ~7–9 and $(La/Sm)_n$ ~2. The tholeiitic lavas exhibit a smaller enrichment in LREE (La ~15–30 times chondrite) and have relatively flat LREE patterns [$(La/Sm)_n$ ratios ~0.75–1.0]. The LREE abundances in the Moctezuma volcanic field lavas are significantly more variable than heavy REE (HREE). HREE are approximately the same in both suites, suggesting the presence of garnet in the source. Such differences between various rock types have been attributed, in part, to the varying degree of partial melting of the source accompanied by the variation of the garnet/clinopyroxene ratio (Kay and Gast, 1973).

On the MORB-normalized plots (Fig. 9), hawaiites have a humped shape with positive anomalies in K, Rb, Ba, and Nb, similar to the average of recent (<5 Ma) basalts from the southwestern U.S. Basin and Range Province (Fitton et al., 1991). This shape of pattern is characteristic of OIB (Sun and McDonough, 1989) and similar to those of other intraplate basaltic lavas from both continental and oceanic environments. The patterns increase with increasing element incompatibility from Yb to Ce and peak at Ta-Nb-Th unlike most continental tholeiites, which usually display negative Nb-Ta anomalies. The Moctezuma volcanic field lavas were probably derived from a source with an incompatible element composition similar to those of the mantle

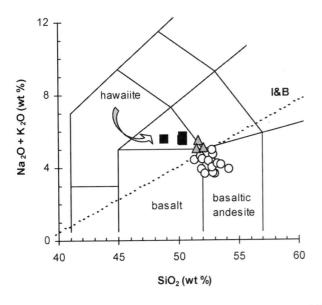

Figure 6. Total alkali-silica (TAS) diagram (Le Maitre et al., 1989) for Moctezuma lavas. I&B line separates alkaline and subalkaline fields (Irvine and Baragar, 1971). Filled squares show hawaiites, gray triangles show transitional basalts, and empty circles show basalts and basaltic andesites.

TABLE 4. MAJOR- AND TRACE-ELEMENT COMPOSITIONS OF THE MOCTEZUMA LAVAS

Tholeiitic lavas

Sample number:	8-88	54-82	40-82	78-82	26-81	6-88	5-88	25-82	12-81	8-81	28-82	29-82	24-82	21-82	1-88	14-88
SiO$_2$ (wt %)	50.53	50.92	51.09	51.36	51.70	51.88	52.03	52.04	52.20	52.31	52.43	52.46	52.53	53.02	53.40	54.01
TiO$_2$	2.05	1.98	1.99	2.00	1.90	2.13	2.10	2.21	1.95	2.20	2.11	2.27	1.93	2.12	2.08	2.01
Al$_2$O$_3$	14.83	15.38	15.08	14.80	15.47	14.73	14.30	15.15	15.08	16.05	14.86	15.66	15.06	15.00	14.77	14.16
Fe$_2$O$_3$	2.18	1.55	0.74	0.41	1.96	2.39	1.38	1.10	2.03	1.48	0.70	0.28	0.14	1.49	2.69	0.48
FeO	8.28	8.50	9.49	9.80	8.29	7.55	9.00	7.18	7.45	7.73	9.20	8.39	9.87	8.05	7.53	9.11
MnO	0.14	0.13	0.16	0.16	0.16	0.13	0.13	0.12	0.16	0.15	0.15	0.15	0.16	0.15	0.14	0.12
MgO	7.67	7.24	7.35	7.35	6.69	7.15	7.41	7.48	6.84	6.60	7.30	6.35	6.76	6.90	6.93	7.69
CaO	8.31	7.84	8.43	8.65	8.60	8.12	8.06	8.24	8.50	8.40	8.55	8.39	8.85	8.28	7.76	7.95
Na$_2$O	3.72	3.92	3.33	3.29	3.82	3.30	3.15	3.80	3.59	3.58	3.21	3.92	3.67	3.42	3.62	3.47
K$_2$O	0.61	0.61	0.51	0.32	0.60	0.70	0.42	0.81	0.58	0.78	0.45	0.99	0.31	0.78	0.54	0.42
P$_2$O$_5$	0.19	0.20	0.26	0.20	0.27	0.25	0.18	0.30	0.31	0.41	0.28	0.48	0.21	0.33	0.25	0.19
LOI	0.82	0.74	0.65	0.86	0.96	0.80	0.94	0.69	1.00	0.90	0.68	0.60	0.80	0.95	1.15	0.82
Total	99.33	99.01	99.08	99.20	100.42	99.13	99.10	99.12	99.69	100.59	99.92	99.94	100.29	100.49	100.86	100.43
mg number	61.33	60.78	60.51	60.48	58.49	60.95	60.51	65.97	60.96	60.67	61.13	60.88	58.88	60.88	59.59	63.05
CIPW norm																
Qtz						1.09	2.04		.42	0.29	1.03			1.71	2.04	1.74
Hy	10.55	11.89	20.48	22.77	13.88	22.65	24.93	15.45	20.91	20.64	25.03	13.10	19.40	21.22	22.60	25.84
Ol	9.64	8.49	4.19	2.70	5.36			4.03				5.74	3.61			
Rb (ppm)	16	17	14	5	10	19	15	21	10	13	9	16	7	13	15	14
Sr	319	331	319	259	321	368	271	412	352	408	275	470	212	354	326	286
Ba	131	159	163	103	142	185	112	194	187	194	99	226	80	158	182	99
Cr	224	197	246	317	209	210	254	197	249	205	322	190	272	218	225	304
V	153	141	156	147	143	144	147	135	134	130	138	130	147	129	138	133
Ni	200	178	182	220	172	181	196	168	185	174	215	154	182	191	195	218
Co	44	42	35	41	35	41	43	40	34	34	38	34	41	33	42	43

Note: mg number = [100*Mg/(Mg + Fe^{2+})]. For CIPW norm: Qtz—quartz; Ne—nepheline; Ol—olivine; Hy—hypersthene.

(continued)

TABLE 4. MAJOR- AND TRACE-ELEMENT COMPOSITIONS
OF THE MOCTEZUMA LAVAS (*continued*)

	Alkaline lavas						
Sample number:	8-82	28-81	29-81	27-81	71-82	11-81	1-81
SiO_2 (wt %)	48.64	50.20	50.22	50.27	51.18	51.34	51.60
TiO_2	2.74	2.60	2.48	2.76	2.54	2.42	2.18
Al_2O_3	15.49	16.13	15.56	15.61	16.21	13.67	15.06
Fe_2O_3	1.34	2.35	2.16	1.55	0.58	1.57	1.21
FeO	9.07	7.58	6.83	7.93	7.69	9.36	7.98
MnO	0.17	0.16	0.16	0.16	0.15	0.17	0.16
MgO	7.31	6.15	6.97	6.95	6.31	7.31	6.80
CaO	8.92	8.59	9.34	8.34	8.48	8.55	8.66
Na_2O	4.28	4.17	4.16	4.31	4.20	4.05	4.05
K_2O	1.25	1.32	1.26	1.39	1.27	0.98	1.02
P_2O_5	0.52	0.56	0.54	0.58	0.48	0.30	0.46
LOI	0.60	0.60	0.60	0.58	0.61	0.88	0.46
Total	100.27	100.41	100.28	100.43	99.70	100.60	99.64
mg number	60.24	57.33	62.72	61.21	61.94	58.96	61.36
CIPW norm							
Ne	5.31	1.19	2.55	2.07	0.09		
Hy						3.26	5.63
Ol	15.13	13.19	12.40	13.80	13.52	12.70	9.96
Rb (ppm)	21	20	21	23	20	10	17
Sr	589	584	588	643	590	310	463
Ba	308	312	349	333	290	146	259
Cr	272	183	225	212	171	243	197
V	197	159	160	160	134	177	151
Ni	171	101	122	152	133	203	131
Co	39	36	33	36	34	39	32

Notes: mg number = $[100*Mg/(Mg + Fe^{2+})]$. For CIPW norm: Qtz—quartz; Ne—nepheline; Ol—olivine; Hy—hypersthene.

sources of continental and oceanic island basalts without a significant contribution of a subduction component. The tholeiitic basalts have lower concentrations in many of these elements and, on the whole, the patterns of the Moctezuma tholeiites resemble those of enriched-MORB or OIB tholeiites (Wilson, 1989).

The position of the Moctezuma basalts on the La/10-Y/15-Nb/8 diagram (Cabanis and Lecolle, 1989), as well as their low Th/Ta ratios, indicates that these lavas are akin to magmatism associated with an extensional tectonic regime. Alkaline rocks are enriched in Nb compared to the transitional and tholeiitic lavas (Fig. 10). In contrast, SCORBA basalts (Cameron et al., 1989) emplaced during the early Miocene northeast-southwest extensional event on the Sierra Madre Occidental plateau plot in a transitional domain near the field of the orogenic lavas, suggesting the contribution of a subduction component.

Isotope Geochemistry

The Sr and Nd isotope compositions of four representative samples from the Moctezuma volcanic field are given in Table 6 and plotted in Figure 11. Also reported for comparison are the isotopic compositions of Quaternary basalts from the Pinacate (Lynch et al., 1993), San Quintín (Luhr et al., 1995), Geronimo (Menzies et al., 1985), and San Luis Potosi (Pier et al., 1989) volcanic fields, as well as recent volcanics from the southwestern U.S. Basin and Range province (Kempton et al., 1991; Bradshaw et al., 1993). The Quaternary Moctezuma lavas have high εNd (between +8 and +11) and low Sr-isotopic values (in the range 0.7028–0.7036). These lavas have about the same range of Sr-isotopic values as alkali basalts from the Pinacate, San Quintín, San Luis Potosi, and Geronimo volcanic fields but have higher εNd values. The high initial εNd values for the Moctezuma volcanic field rocks (Table 6) suggest a derivation from a depleted mantle source and an absence of contamination by older continental crust during their evolution (DePaolo, 1988a, 1988b).

The mantle source for the alkali basalts was probably enriched in strongly incompatible trace elements including LREE. Such an enrichment cannot be readily generated from a LREE-depleted source with a high $^{143}Nd/^{144}Nd$ ratio. Thus, more likely, the alkali basaltic lavas were derived from a LREE-enriched mantle source. However, the Nd isotopic systematics suggest that on a time-integrated basis their source was depleted with respect to chondritic mantle. The apparent discrepancy, frequently observed in alkali basaltic suites, implies that metasoma-

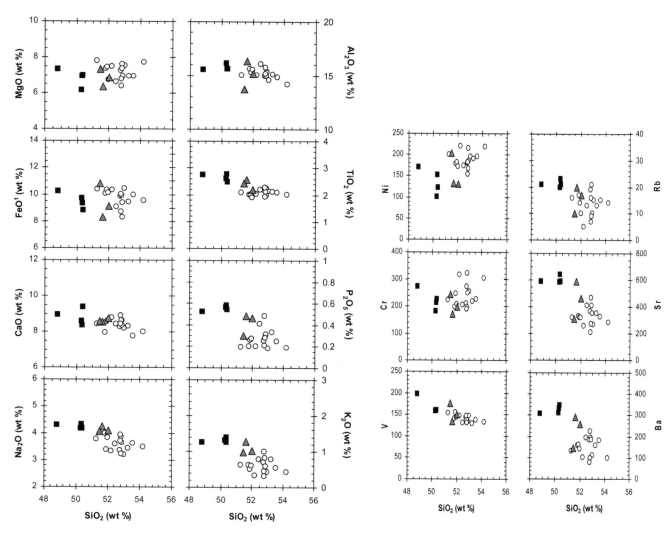

Figure 7: Major- and trace-element variation diagrams of Moctezuma lavas, using SiO$_2$ as a differentiation index. Same symbols as in Figure 6.

TABLE 5. TRACE-ELEMENT COMPOSITIONS
OF SELECTED SAMPLES FROM THE MVF

Sample number:	Tholeiitic lavas				Alkaline lavas	
	8-88	78-82	28-82	24-82	8-82	27-81
Zr (ppm)	110	99	111	95	183	178
Nb	14	8	14	7	35	46
Y	25	24	22	23	29	27
U		0.19			0.7	
Th		0.64			2.42	
Ta		0.6			2.19	
Hf		3.11			4.74	
Pb		2.2			2.1	
La	8.6	6.6	7.5	5.0	22.4	25.1
Ce	28.3	23.1	21.0	10.8	47.1	50.2
Nd	14.2	12.7	12.7	10.0	25.9	27.1
Sm	4.7	4.4	4.4	3.7	6.5	6.8
Eu	1.64	1.70	1.56	1.34	2.21	2.23
Gd	5.20	5.70	5.42	4.84	6.99	6.71
Dy	4.71	4.67	4.41	4.40	5.61	5.34
Er	2.59	2.39	1.94	2.21	2.65	2.56
Yb	1.84	1.64	1.43	1.55	2.00	1.79
Lu	0.36	0.30	0.27	0.25	0.33	0.30

Note: Zr, Nb, Y and REE concentrations were determined by ICP-AES at CRPG, Nancy (France). U, Th, Ta, Hf and Pb determined by ICP-MS at Montpellier (France).

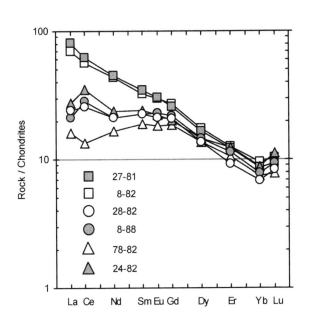

Figure 8. Chondrite-normalized REE patterns of representative Moctezuma lavas. Normalizing values from Sun and McDonough (1989). Squares show hawaiites; triangles and circles show tholeiitic lavas.

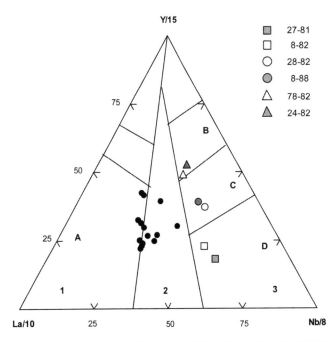

Figure 10: La/10-Y/15- Nb/8 diagram of Cabanis and Lecolle (1989). Legend: 1—domain of orogenic lavas; 2—domain of tardi- to post-orogenic lavas; 3—domain of nonorogenic lavas; A—calc-alkaline lavas; B—MORB-N; C—MORB-E; D—alkali basalts. Black dots correspond to the SCORBA lavas of Cameron et al. (1989).

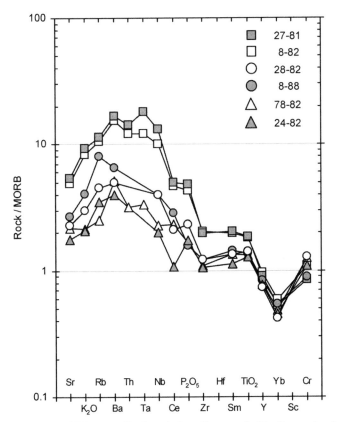

Figure 9: MORB-normalized variation diagram (spiderdiagram) of lavas from Moctezuma basin. Normalizing values from Pearce (1983). Same symbols are used as in Figure 8.

tism took place shortly before the melting event. Thus, the differences between the Moctezuma volcanic field alkali and tholeiitic suites can, in part, be attributed to this process.

The Sr and Nd isotopic compositions of the Moctezuma lavas and the shape of the MORB-normalized patterns are similar to many oceanic island basalt suites, the source of which did not contain a significant contribution of an old crust and/or subduction component but was probably dominated by asthenospheric mantle. By contrast, the Tertiary SCORBA lavas from the Sierra Madre Occidental have higher Sr and lower Nd isotopic compositions that are suggestive of a major contribution from a lithospheric mantle source (Cameron et al., 1989).

DISCUSSION

The occurrence of late-extensional Quaternary magmatism with OIB trace-element and Sr-Nd isotope signatures is a good indicator of the tectonic evolution of the Moctezuma half-graben. This region is located along the western edge of the Sierra Madre Occidental, at the limit with the zone of major extension corresponding to the main Basin and Range Province to the west. Extensional processes at the western margin of the Sierra Madre Occidental were initiated in the early Miocene (22 Ma) as indicated by the ages of: (1) a mafic flow intercalated in the Báucarit Formation in the Moctezuma basin; (2) a lava flow from the 350 m thick volcanic pile of Sierra Basómari, located at the

TABLE 6. SR AND ND ISOTOPIC COMPOSITIONS OF THE MOCTEZUMA
LAVAS (SEE APPENDIX FOR ANALYTICAL TECHNIQUES)

Sample:	27-81	8-82	24-82	78-82
Rb	23	20	7	5
Sr	651	594	213	273
$^{87}Sr/^{86}Sr$	0.703258	0.703692	0.703591	0.702828
2σ	0.000025	0.000033	0.000049	0.000068
E Sr	−17.63	−11.47	−12.90	−23.73
$^{87}Rb/^{86}Sr$	0.09991	0.09325	0.09120	0.04883
2σ	0.00023	0.00004	0.00014	0.00003
Nd	27.11	25.78	9.98	13.29
Sm	6.87	6.48	3.86	4.77
$^{143}Nd/^{144}Nd$	0.513105	0.513059	0.513223	0.513078
2σ	0.000009	0.000011	0.000019	0.000013
$^{147}Nd/^{144}Nd$	0.15656	0.15511	0.23902	0.22128
2σ	0.00001	0.00001	0.00008	0.00028
εNd	9.11	8.22	11.41	8.58

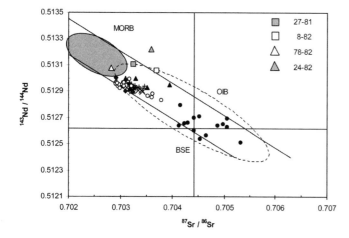

Figure 11. $^{143}Nd/^{144}Nd$ versus $^{87}Sr/^{86}Sr$ diagram for Moctezuma volcanic rocks. Legend: X symbol—Pinacate volcanic field (Lynch et al., 1993); crosses—San Quintín volcanic field (Luhr et al., 1995); open circles—Quaternary basalts from San Luis Potosi (Pier et al., 1989); black triangles—recent basalts from Basin and Range Province (Kempton et al., 1991; Bradshaw et al., 1993); black diamonds—Geronimo volcanic field (Menzies et al., 1985); black dots— Southern Cordillera Basaltic Andesite (SCORBA) lavas (Cameron et al., 1989).

western side of the San Bernardino valley (Paz-Moreno, 1992). Tectonic extension generated tilt-block/half-graben structures (Leeder and Gawthorpe, 1987) bounded on the east by north-northwest–trending high-angle faults. The asymmetric basin that developed as a result of the rotation of the block was filled by

sandy to conglomeratic alluvial-fan deposits. The dip of the tilted blocks indicates a relatively minor amount of extension during that time, far smaller than the amount commonly considered for the Basin and Range Province to the west. The high heat flow that accompanied this initial extensional phase may account for the zeolite cementation of the sediments of the Báucarit Formation (Cochemé et al., 1988).

After this early Miocene west-southwest–east-northeast extensional phase, the extension was limited and characterized by vertical movements rather than by the rotation of the blocks. In the active San Bernardino fault system, the direction of extension is east-west, perpendicular to the fault trace, and it produced vertical surface ruptures (Suter, 2000). In the Moctezuma graben, the upper Miocene to Quaternary sedimentary deposits are alluvial fans produced by the erosion of Sierra La Madera and, in the central part of the depression, they are lacustrine sediments. Continental infilling progressively flattened the paleotopography so that the early Quaternary lavas flowed down a relatively smooth, gently dipping slope toward the west, giving rise to the mesa morphology of the volcanic field.

The geochemistry of the lavas from the Moctezuma basin provides further constraints on the changes that have occurred in the underlying mantle during tectonic evolution (DePaolo and Daley, 2000). Pre-Moctezuma volcanic field intra-Báucarit synextensional basalts of Moctezuma and basaltic andesites from Sierra Basómari have the chemical characteristics of SCORBA lavas (Paz-Moreno, 1992), which are thought to have been derived from the subcontinental lithospheric mantle (Cameron et al., 1989; Fitton et al., 1991; Bradshaw et al., 1993). In the Moctezuma basin, the early Miocene basalts are poorly represented, in contrast to the high lava volume in the nearby San Bernardino valley.

Magmatic activity was renewed in the Quaternary after a long period of quiescence between 22 and 2 Ma. The geochemical characteristics indicate that these Quaternary basalts were derived from the asthenospheric mantle. The Moctezuma lavas like the Geronimo volcanic field in Arizona (Menzies et al., 1985) have the characteristics of magmas derived from an OIB-type mantle source. An important difference between the two volcanic fields is the abundance of mantle and granulite xenoliths in the Geronimo volcanic field (Kempton et al., 1984, 1990). This confirms a higher production rate and a more rapid ascent of the basaltic magmas in the San Bernardino valley. The slower melt ascent rate of the Moctezuma magmas accounts for the absence of xenoliths and the slightly more evolved character of the lavas. Finally, the major change in mantle signature between the early Miocene and Quaternary basalts in the Moctezuma valley is also inferred to record the progress in the thinning of the continental lithosphere. As for the southwestern Basin and Range Province (DePaolo and Daley, 2000), the geochemical signatures indicate the involvement of subcontinental lithospheric mantle as a major source of basalt magma during the early stages of extension, giving way to asthenospheric sources as extension proceeds. In the region located at the foothills of the Sierra Madre Occidental, asthenospheric melts could reach the surface only in recent time.

CONCLUSIONS

The Quaternary Moctezuma volcanic field is a unique example in Sonora of a close association of tholeiitic and alkaline magmas in an intracontinental setting. The Moctezuma volcanic field is located at the southern end of a valley that formed in response to west-southwest–east-northeast extension. The rotation of the blocks gave rise to half-graben structures that were later filled by fanconglomerates of the Báucarit Formation. This extensional phase is of early Miocene age (22 Ma). The topography of the basin was subsequently flattened by Neogene unconsolidated alluvial fans so that the first Quaternary basaltic lava flows, erupted from vents located along the major faults on the eastern border of the Moctezuma basin, formed subhorizontal mesas. Alkaline lavas erupted later from scoria cones located in the center of the basin.

Mineral chemistry and geochemistry show a change in the characteristics of the lavas from tholeiitic to alkaline. The Moctezuma lavas were derived from a metasomatized garnet-bearing upper mantle source. Relative to primordial mantle, the source was depleted in incompatible elements. Shortly before the melting events, the source was metasomatically enriched in incompatible elements. The similarities of incompatible element compositions between the Quaternary Moctezuma volcanic field basaltic rocks and oceanic island basalts imply generation from sources with comparable compositional characteristics, suggesting that metasomatic agents, which superimposed incompatible element enrichment on the mantle parents, were derived from the same source: asthenospheric mantle. The absence of mantle or granulite xenoliths in the Moctezuma volcanic field lavas, compared to the neighboring Geronimo volcanic field, shows an easy access to the surface for the latter. Finally, the strong difference in geochemistry between the early Miocene SCORBA-type synextensional lavas and the Quaternary Moctezuma volcanic field lavas is indicative of a progressive thinning of the mantle lithosphere during the Neogene.

APPENDIX 1. ANALYTICAL METHODS

Geochronology

The K-Ar investigation was carried out at the university of Strasbourg. Whole-rock samples were obtained from the 100–160 μm sieve fraction. The analytical procedure, more extensively described elsewhere (Bonhomme et al., 1975; Montigny et al., 1988), is as follows. Potassium was measured by flame photometry with a lithium internal standard. Argon was extracted in a heat-resistant glass vacuum apparatus and determined by isotopic dilution (^{38}Ar as a tracer) using a MS 20 mass spectrometer.

The set of constants recommended by Steiger and Jäger (1977) was used for age calculation. Quoted uncertainties represent estimates of analytical precision at two standard deviations. They were calculated following the procedure given by Cox and Dalrymple (1967).

The $^{40}Ar/^{39}Ar$ experimental techniques have been described elsewhere (Montigny et al., 1988). The samples are sealed in quartz vials and irradiated for 30 h in the Osiris reactor at the Centre d'Etudes Nucléaires de Saclay, France. CaF_2 and K_2SO_4 are included in the irradiation package to determine the correction factors for interfering isotopes produced by nuclear reactions during irradiation. Standard biotite LP6 (129 Ma) was utilized as flux monitor with aliquots of at least 100 mg. The errors, quoted as 2σ, were obtained following the procedure given by Albarède (1976).

Isotopes

Sm and Nd abundance and Sr- and Nd-isotope ratios were determined by isotope dilution mass spectrometry in the Atlantic University Radiogenic Isotope Facility (AURIF) laboratory of Memorial University of Newfoundland. A description of the analytical technique was given by Kerr et al. (1995). Measured $^{143}Nd/^{144}Nd$ values were normalized to a $^{146}Nd/^{144}Nd$ ratio of 0.7219. An average value for $^{143}Nd/^{144}Nd = 0.511849 \pm 9$ resulted from replicate analyses of the La Jolla standard, which was analyzed repeatedly throughout. The $^{87}Sr/^{86}Sr$ was corrected using $^{86}Sr/^{88}Sr = 0.1194$. The NBS 987 Sr standard gave an average value of $^{87}Sr/^{86}Sr = 0.710250 \pm 11$ on replicate analyses. Depleted mantle model ages (T_{DM}) were calculated assuming values of $^{143}Nd/^{144}Nd = 0.513114$ and $^{147}Sm/^{144}Nd = 0.213$ for modern depleted mantle.

ACKNOWLEDGMENTS

This research was supported by an interuniversity convention between Universidad de Sonora, México, and Université d'Aix-Marseille, France, as well as by a research grant from NSERC of Canada. Fieldwork benefited from the logistical support of the Departamento de Geología de la Universidad de Sonora. The authors are grateful to Sr. José Cruz, "El Compita" (Divisaderos, Sonora), for his hospitality and introduction to the field knowledge, to Saúl Herrera U. (UniSon) for the field assistance, to Marie Odile Trensz (Marseille) for the chemical data, and to Claude Merlet for technical assistance during microprobe work. Reviews by Todd Housh, Gary Girty, and an anonymous reviewer provided useful comments that improved the manuscript.

REFERENCES CITED

Albarède, F., 1976, Géochronologie comparée par la méthode ^{39}Ar-^{40}Ar de deux régions d'histoire post-hercynienne différente: la Montagne Noire et les Pyrénées orientales: Thèse d'Etat, Université Paris 7, 116 p.

Angelier, J., Colletta, B., Chorowicz, J., Ortlieb, L., and Rangin, C., 1981, Fault tectonics of the Baja California peninsula and the opening of the Sea of Cortez, Mexico: Journal of Structural Geology, v. 3, p. 347–357.

Aranda-Gómez, J.J., and Ortega-Gutiérrez, F., 1987, Mantle xenoliths of Mexico, in Nixon, P., ed., Mantle xenoliths: New York, USA, J. Wiley & Sons, p. 75–84.

Atwater, T.A., 1970, Implications of plate tectonics for the Cenozoic evolution of western North America: Geological Society of America Bulletin, v. 81, p. 3513–3536.

Atwater, T.A., 1989, Plate tectonic history of northeast Pacific and western North America, in Winterer, E.L., Hussong, D.M., and Decker, R.W., eds., The eastern Pacific Ocean and Hawaii: Boulder, Colorado, Geological Society of America, Geology of North America, v. N, p. 21–71.

Bardintzeff, J-M., Leyrit, H., Guillou, H., Guille, G., Bonin, B., Giret, A., and Brousse, R., 1994, Transition between tholeiitic and alkali basalts: petrographical and geochemical evidence from Fangataufa, Pacific Ocean, and Kerguelen, Indian Ocean: Geochemical Journal, v. 28, p. 489–515.

Bartolini, C., Shafiqullah, M., Damon, P.E., and Morales-Montaño, M., 1993, Preliminary geochronology (K-Ar) and chemistry of the Comedores Formation (Volcanic Field), Northern Sonora, Mexico [abs.]: Eos (Transactions, American Geophysical Union), v. 74, p. 43.

Bartolini, C., Damon, P.E., Shafiqullah, M., and Morales-Montaño, M., 1994, Geochronologic contribution to the Tertiary sedimentary-volcanic sequences (Báucarit Formation) in Sonora, Mexico: Geofísica Internacional, v. 33, p. 67–77.

Bonhomme, M.G., Thuizat, R., Pinault, Y., Clauer, N., Wendling, A., and Winckler, R., 1975, Méthode de datation potassium-argon, Appareillage et technique: Notes techniques de l'Institut de Géologie, Université Louis Pasteur, Strasbourg, 53 p.

Bradshaw, T.K., Hawkesworth, C.J., and Gallagher, K., 1993, Basaltic volcanism in the Southern Basin and Range: no role for a mantle plume: Earth and Planetary Science Letters, v. 116, p. 45–62.

Cabanis, B., and Lecolle, M., 1989, Le diagramme La/10-Y/15-Nb/8 : un outil pour la discrimination des séries volcaniques et la mise en évidence des processus de mélange et/ou de contamination crustale: Comptes Rendus de l'Académie des Sciences Paris, t. 309, p. 2023–2029.

Cameron, K.L., Nimz, G.J., Kuentz, D., Niemeyer, S., and Gunn, S., 1989, Southern Cordilleran basaltic andesite suite, southern Chihuahua, Mexico: A link between tertiary continental arc and flood basalt magmatism in northern America: Journal of Geophysical Research, v. 94, p. 7817–7840.

Chen, C.Y., Frey, F.A., Garcia, M.O., Dalrymple, G.B., and Hart, S.R., 1991, The tholeiitic to alkalic basalt transition at Haleakala volcano, Maui, Hawaii: Contributions to Mineralogy and Petrology, v. 106, p. 183–200.

Cochemé, J–J., and Demant, A., 1991, Geology of the Yécora area, northern Sierra Madre Occidental, Mexico, in Pérez-Segura, E., and Jacques-Ayala, C., eds., Studies of Sonoran geology: Boulder, Colorado, Geological Society of America Special Paper 254, p. 81–94.

Cochemé, J–J., Demant, A., Aguirre, L., and Hermitte, D., 1988, Présence de heulandite dans les remplissages sédimentaires liés au "Basin and Range" (formation Báucarit) du Nord de la Sierra Madre Occidental (Mexique): Comptes Rendus de l'Académie des Sciences Paris, t. 307, p. 643–649.

Cochemé, J–J., Aguirre, L., Bevins, R.E., Robinson, D., and Münch, P., 1994, Zeolitization processes in basic lavas of the Báucarit Formation, northwestern México: Revista Geológica de Chile, v. 21, p. 217–232.

Cox, A., and Dalrymple, G.B., 1967, Statistical analysis of geomagnetic reversal data and the precision of potassium-argon dating: Journal of Geophysical Research, v. 72, p. 2603–2614.

Damon, P.E., Clark, K.F., Shafiqullah, M., and Roldán-Quintana, J., 1981, Geology and mineral deposits of southern Sonora and the Sonoran Sierra Madre Occidental, in Ortlieb, L., and Roldán-Quintana, J., eds., Geology of northwestern Mexico and southern Arizona: Hermosillo, Sonora, Geological Society of America, Cordilleran Section Field Guide and Papers, p. 369–426.

Demant, A., Cochemé, J–J., Delpretti, P., and Piguet, P., 1989, Geology and petrology of the Tertiary volcanics of the northwestern Sierra Madre Occidental, Mexico: Bulletin de la Société Géologique de France, (8), t. V, p. 737–748.

DePaolo, D.J., 1988a, Age dependence of the composition of continental crust as determined from Nd isotopic variations in igneous rocks: Earth and Planetary Science Letters, v. 59, p. 263–271.

DePaolo, D.J., 1988b, Neodymium Isotope Geochemistry: An Introduction, Springer-Verlag, Heidelberg, 187 p.

DePaolo, D.J., and Daley, E.E., 2000, Neodynium isotopes in basalts of the southwest Basin and Range and lithospheric thinning during continental extension: Chemical Geology, v. 169, p. 157–185.

Dickinson, W.R., and Snyder, W.S., 1979, Geometry of subducted slabs related to San Andreas transform: Journal of Geology, v. 87, p. 609–627.

Dumble, E.T., 1900, Notes on the geology of Sonora, Mexico: American Institute of Mining Engineers, v. 29, p. 122–152.

Feigenson, M.D., Hofmann, A.W., and Spera, F.J., 1983, Case studies on the origin of basalt II: The transition from tholeiitic to alkalic volcanism at Kohala volcano, Hawaii: Contributions to Mineralogy and Petrology, v. 84, p. 390–405.

Fitton, J.G., James, D., Kempton, P.D., Ormerod, D.S., and Leeman, W.P., 1988, The role of lithospheric mantle in the generation of Late Cenozoic basic magmas in the western United States: Journal of Petrology, Special lithosphere issue, p. 331–349.

Fitton, J.G., James, D., and Leeman, W.P., 1991, Basic magmatism associated with late Cenozoic extension in the western United States: compositional variations in time and space: Journal of Geophysical Research, v. 96, p. 13693–13711.

Fodor, R.V., Frey, F.A., Baeur, G.R., and Clague, D.A., 1992, Ages, rare-earth element enrichment, and petrogenesis of tholeiitic and alkalic basalts from Kahoolawe Island, Hawaii: Contributions to Mineralogy and Petrology, v. 110, p. 442–462.

Frey, F.A., Green, D.H., and Roy, S.D., 1978, Integrated models of basaltic petrogenesis: a study of quartz tholeiites to olivine melilites from southeastern Australia utilizing geochemical and experimental petrological data: Journal of Petrology, v. 19, p. 463–513.

Frey, F.A., Garcia, M.O., Wise, W.S., Kennedy, A., Gurriet, P., and Albarède, F., 1991, The evolution of Mauna Kea volcano, Hawaii: petrogenesis of tholeiitic and alkalic basalts: Journal of Geophysical Research, v. 96, p. 14347–14375.

Gans, P.B., 1997, Large-magnitude Oligo-Miocene extension in southern Sonora: Implications for the tectonic evolution of northwest Mexico: Tectonics, v. 16, p. 388–408.

Gastil, R.G., and Krummenacher, D., 1977, Reconnaissance geology of coastal Sonora between Puerto Lobos and Bahia Kino: Geological Society of America Bulletin, v. 88, p. 189–198.

Goff, F., 1996, Vesicle cylinders in vapor-differentiated basalt flows: Journal of Volcanology and Geothermal Research, v. 71, p. 167–185.

Hole, M.J., Saunders, A.D., Rogers, G., and Sykes, M.A., 1995, The relationship between alkaline magmatism, lithospheric extension and slab window formation along continental destructive plate margins, in Smellie, J.L., ed., Volcanism associated with extension at consuming plate margins: Geological Society of London Special Publication 81, p. 265–285.

Irvine, T.N., and Baragar, W.R., 1971, A guide to the chemical classification of the common volcanic rocks: Canadian Journal of Earth Sciences, v. 8, p. 523–548.

Kamenetsky, V.S., Crawford, A.J., and Mefre, S., 2001, Factors controlling chemistry of magmatic spinels: an empirical study of associated olivine, Cr-spinels and melt inclusions from primitive rocks: Journal of Petrology, v. 42, p. 655–671.

Kay, R.W., and Gast, P.W., 1973, The rare earth content and origin of alkali-rich basalts: Journal of Geology, v. 81, p. 653–682.

Kempton, P.D., Menzies, M.A., and Dungan, M.A., 1984, Petrography, petrology and geochemistry of xenoliths and megacrysts from the Geronimo volcanic field, southeastern Arizona, in Kornprobst, J., ed., Kimberlites II: the mantle and crust-mantle relationship: Amsterdam, Elsevier, p. 71–83.

Kempton, P.D., Harmon, R.S., Hawkesworth, C.J., and Moorbath, S., 1990, Petrology and geochemistry of lower crustal granulites from the Geronimo Volcanic Field, southeastern Arizona: Geochimica et Cosmochimica Acta, v. 54, p. 3401–3426.

Kempton, P.D., Fitton, J.G., Hawkesworth, C.J., and Ormerod, D.S., 1991, Isotopic and trace elements constraints on the composition and evolution of the lithosphere beneath the southwestern United States: Journal of Geophysical Research, v. 96, p. 13713–13735.

Kerr, A., Jenner, G.A., and Fryer, B.J., 1995, Sm-Nd isotopic geochemistry of Precambrian to Paleozoic granitoid suites and the deep-crustal structure of the southeast margin of the Newfoundland Appalachians: Canadian Journal of Earth Sciences, v. 32, p. 224–245.

King, R.E., 1939, Geological reconnaissance in northern Sierra Madre Occidental: Geological Society of America Bulletin, v. 50, p. 1625–1722.

Le Bas, M.J., Le Maitre, R.W., Streckeisen, A., and Zanettin, B., 1986, A chemical classification of volcanic rocks based on the total alkali-silica diagram: Journal of Petrology, v. 27, p. 745–750.

Leeder, M.R., and Gawthorpe, R.L., 1987, Sedimentary models for extensional tilt-block/half-graben basins, *in* Coward, M.P., Dewey, J.F., and Hancock, P.L., eds., Continental extensional tectonics: Geological Society of London Special Publication 28, p. 139–152.

Le Maitre, R.W., Bateman, P., Dudek, A., Keller, J., Lameyre, J., Le Bas, M.J., Sabine, P.A., Schmid, R., Sorensen, H., Streckeisen, A., Wooley, A.R., and Zanettin, B., 1989, A classification of igneous rocks and glossary of terms: Oxford, United Kingdom, Blackwell, 193 p.

Luhr, J.F., Aranda-Gómez, J.J., and Housh, T.B., 1995, San Quintín volcanic field, Baja California norte, México: Geology, petrology and geochemistry: Journal of Geophysical Research, v. 100, p. 10353–10380.

Lynch, D.J., 1981, Genesis and geochronology of alkaline volcanism in the Pinacate volcanic field of northwestern Sonora, Mexico [Ph.D. thesis]: Tucson, University of Arizona, 215 p.

Lynch, D.J., Musselman, T.E., Gutmann, J.T., and Patchett, P.J., 1993, Isotopic Evidence for the origin of Cenozoic volcanic rocks in the Pinacate volcanic field, northwestern Mexico: Lithos, v. 29, p. 295–302.

Macdonald, G.A., and Katsura, T., 1965, Chemical composition of Hawaiian lavas: Journal of Petrology, v. 5, p. 82–133.

Martín-Barajas, A., Stock, J.M., Layer, P., Hausback, B., Renne, P., and López-Martínez, M., 1995, Arc-rift transition volcanism in the Puertecitos Volcanic Province, northeastern Baja California, Mexico: Geological Society of America Bulletin, v. 107, p. 407–424.

McDowell, F.W., and Clabaugh, S.E., 1979, Ignimbrites of the Sierra Madre Occidental and their relation to the tectonic history of western Mexico, *in* Chapin, C.E., and Elston, W.E., eds., Ash-flow tuffs: Boulder, Colorado, Geological Society of America Special Paper 180, p. 113–124.

McDowell, F.W., and Keizer, R.P., 1977, Timing of mid-Tertiary volcanism in the Sierra Madre Occidental between Durango city and Mazatlan, Mexico: Geological Society of America Bulletin, v. 88, p. 1479–1487.

McDowell, F.W., Roldán-Quintana, J., and Amaya-Martínez, R., 1997, Interrelationship of sedimentary and volcanic deposits associated with Tertiary extension in Sonora, Mexico: Geological Society of America Bulletin, v. 109, p. 1349–1360.

Mellini, M., Carbonin, S., Dal Negro, A., and Piccirillo, E.M., 1988, Tholeiitic hypabyssal dykes: how many clinopyroxenes?: Lithos, v. 22, p. 127–134.

Menzies, M.A., Kempton, P.D., and Dungan, M., 1985, Interaction of continental lithosphere and asthenospheric melts below the Geronimo volcanic field, Arizona, USA: Journal of Petrology, v. 26, p. 663–693.

Menzies, M.A., Arculus, R.J., Best, M.G., Bergman, S.C., Ehrenberg, S.N., Irving, A.J., Roden, M.F., and Schulze, D.J., 1987, A record of subduction processes and within-plate volcanism in lithospheric xenoliths of the southwestern USA, *in* Nixon, P., ed., Mantle xenoliths: New York, USA, J. Wiley & Sons, p. 59–74.

Montigny, R., Demant, A., Delpretti, P., Piguet, P., and Cochemé, J–J., 1987, Chronologie K/A des séquences volcaniques tertiaires du Nord de la Sierra Madre Occidental (Mexique): Comptes Rendus de l'Académie des Sciences Paris, t. 304, p. 987–992.

Montigny, R., Le Mer, O., Thuizat, R., and Whitechurch, H., 1988, K/Ar and ^{40}Ar/ ^{39}Ar study of metamorphic rocks associated with the Oman ophiolite; tectonic implications: Tectonophysics, v. 151, p. 345–352.

Mora-Alvarez, G., and McDowell, F.W., 2000, Miocene volcanism during late subduction and early rifting in the Sierra Santa Ursula of western Sonora, Mexico, *in* Delgado-Granados, H., Aguirre-Díaz, G., and Stock, J.M., eds., Cenozoic tectonics and volcanism of Mexico: Boulder, Colorado, Geological Society of America Special Paper 334, p. 123–141.

Morimoto, N., Fabries, J., Ferguson, A., Ginzburg, I., Roos, M., Seifert, F., and Zussman, J., 1988, Nomenclature of pyroxenes: Bulletin de Minéralogie, v. 111, p. 535–550.

Münch, P., Duplay, J., and Cochemé, J–J., 1996, Alteration of acidic vitric tuffs interbedded in the Báucarit Formation, Sonora State, Mexico. Contribution of transmission and analytical electron microscopy: Clays and Clay Minerals, v. 44, p. 49–67.

Nagy, E.A., Grove, M., and Stock, J.M., 1999, Age and stratigraphic relationships of pre- and syn-rift volcanic deposits in the northern Puertecitos Volcanic Province, Baja California, Mexico: Journal of Volcanology and Geothermal Research, v. 93, p. 1–30.

Paz-Moreno, F.A., 1985, Composición y origen de los basaltos (Malpaís) Plio-Cuaternarios de Moctezuma, Sonora, México: Boletín del Departamento de Geología Uni-Son., v. 2, no. 1/2, p. 9–15.

Paz-Moreno, F.A., 1992, Le volcanisme mio-plio-quaternaire de l'Etat du Sonora (nord-ouest du Mexique): évolution spatiale et chronologique; implications pétrogénétiques [Ph.D. thesis]: Marseille, Université Aix-Marseille, 282 p.

Paz, F., Cochemé, J–J., Demant, A., and Piguet, P., 1986, Le champ basaltique quaternaire de Moctezuma, Sonora (Mexique): Comptes Rendus de l'Académie des Sciences Paris, t. 303, p. 701–706.

Pearce, J.A., 1983, The role of sub-continental lithosphere in magma genesis at destructive plate margins, *in* Hawkesworth, C.J., and Norry, M.J., eds., Continental basalts and mantle xenoliths: Nantwich, United Kingdom, Shiva Publishing, p. 230–249.

Pier, J.G., Podosek, F.A., Luhr, J.F., and Brannon, J.C., 1989, Spinel-lherzolite bearing Quaternary volcanic centers in San Luis Potosi, Mexico, Sr and Nd isotopic systematics: Journal of Geophysical Research, v. 94, p. 7941–7951.

Salas, G.A., 1970, Areal geology and petrology of the igneous rocks of the Santa Ana region, northwestern Sonora, Mexico: Boletín de la Sociedad Geológica Mexicana, v. 31, p. 11–63.

Spencer, J.E., and Normark, W.R., 1979, Tosco-Abreojos fault zone: A Neogene transform plate boundary within the Pacific margin of southern Baja California, Mexico: Geology, v. 7, p. 554–557.

Steiger, R.H., and Jäger, E., 1977, Subcommission on Geochronology: convention of the use of decay constants in geo- and cosmochronology: Earth and Planetary Science Letters, v. 36, p. 359–362.

Stock, J.M., 2000, Relation of the Puertecitos Volcanic Province, Baja California, Mexico, to development of the plate boundary in the Gulf of California, *in* Delgado-Granados, H., Aguirre-Díaz, G., and Stock, J.M., eds., Cenozoic tectonics and volcanism of Mexico: Boulder, Colorado, Geological Society of America Special Paper 334, p. 123–141.

Stock, J.M., and Molnar, P., 1988, Uncertainties and implications of the Late Cretaceous and Tertiary position of North America relative to the Farallon, Kula, and Pacific Plates: Tectonics, v. 7, p. 1339–1384.

Sun, S.S., and McDonough, J.D., 1989, Chemical and isotopic systematics of oceanic basalts: implications for mantle compositions and process, *in* Saunders, A.D., and Norry, M.J., eds., Magmatism in the ocean basins: Geological Society of London Special Publication 42, p. 313–345.

Suter, M., 2000, Seismotectonics of Northeastern Sonora, *in* Calmus, T., and Pérez-Segura, E., eds., Cuarta Reunión sobre la Geología del Noroeste de México y Areas adyacentes, v. 2, p. 132–133.

Swanson, E.R., and McDowell, F.R., 1984, Calderas of the Sierra Madre Occidental Volcanic Field, western Mexico: Journal of Geophysical Research, v. 89, p. 8787–8799.

Taylor, S.R., and McLennan, S.M., 1985, The Continental Crust; Its Composition and Evolution: Blackwell Scientific, Oxford, U.K., 312 p.

Williams, H., and McBirney, A.R., 1979, Volcanology: San Francisco, USA, Freeman, 397 p.

Wilson, M., 1989, Igneous petrogenesis; a global tectonic approach: London, United Kingdom, Harper Collins Academic, 466 p.

Yamagashi, H., 1985, Growth of pillow lobes. Evidence from pillow lavas of Hokkaido, Japan, and North Island, New Zealand: Geology, v. 13, p. 499–502.

MANUSCRIPT ACCEPTED BY THE SOCIETY JUNE 2, 2003

Index

Caborca terrane. (*continued*)
 transport of: 15, 20–21, 23–26
Calamajué
 Arroyo Calamajué fault: 11, 14
 batholith. *see* Peninsular Ranges batholith
 Cañon Calamajué Unit: 103, 104
 Sierra Calamajué: 103–104, 110
calcite: 440
California
 Bedford Canyon Formation. *see* Bedford Can-
  yon Formation
 Caborca terrane: 5
 Catalina Island: 66, 287, 288
 Central Valley: 66
 Coast Ranges. *see* Coast Ranges
 Colorado River: 3
 compaction studies in: 17–18
 French Valley Formation: 6, 96, 98
 geobarometric studies of: 19
 geotherm, background: 12
 Inglewood fault: 3
 Julian Schist. *see* Julian Schist
 Klamath Mountains. *see* Klamath Mountains
 Laramide orogeny. *see* Laramide orogeny
 Mojave-Snow Lake fault: 122, 131
 Newport fault: 3
 Orocopia Schist: 21–22, 382–385, 393,
  395–399
 Pelona Schist. *see* Pelona Schist
 Peninsular Ranges. *see* Peninsular Ranges
 Peninsular Ranges batholith. *see* Peninsular
  Ranges batholith
 Rand Schist. *see* Rand Schist
 Rose Canyon fault: 3
 San Andreas fault. *see* San Andreas fault
 Sierra Nevada. *see* Sierra Nevada
 Sierra Nevada batholith. *see* Sierra Nevada
  batholith
 Yuma terrane. *see* Yuma terrane
Campylodiscus clypeus: 410
Canada
 Alberta: 122
 British Columbia. *see* Baja British Columbia;
  British Columbia
 Canada basin: 121, 122
 Rocky Mountains: 122, 128–131, 397
 Yukon: 128
Candelero plutons: 251, 254–255, 265
Cañon Calamajué Unit: 103, 104
Cañon El Parral: 431, 433
Cantwell Basin: 121
Capnodoce: 48, 57
carbonate
 geochronology of: 125, 252
 petrology of: 47, 97, 103, 300, 430
 structures, formation of: 408
Caribbean: 124, 126–128
Carlsbad twins: 168
Carmen Fracture Zone: 415
Carmen Island: 33, 413
Carnian: 57
Carrizo Gorge: 361, 363, 373
Cascade Mountains: 338
Cascadia: 106
Castle Dome Mountains: 391, 401

Catalina Island: 66, 287, 288
Caucasus Mountains: 131–132
Cedros Island. *See* Isla Cedros
Cedros Island Ophiolite. *See also* Vizcaíno
  Peninsula Ophiolite
 correlations to: 65–66
 depositional setting of: 48, 65–68
 geochemical analysis of: 65–66
 geochronology of: 52, 63, 66, 88
 petrology of: 44, 48
Central America: 19, 67, 266. *See also* specific
  locations in
Central Valley: 66
Cerralvo Island: 8
Cerro Blanco: 442, 443
Cerro Colorado: 434
Cerro El Calvario: 48, 62
Cerro Mencenares volcanic complex: 33, 34,
  408–417
Cerro Starship: 434
Cerro Tepoca: 426
Chariot Canyon fault zone: 229–230, 361, 363,
  372–373
chert
 geochronology of: 17, 57
 paleomagnetic studies of: 17
 petrology of
  Bedford Canyon Formation: 96
  Cochimí terrane: 17
  Coloradito Formation: 63
  El Marmol: 103
  Eugenia Formation: 63
  French Valley Formation: 96
  Gran Cañon Formation: 48
  Isla Tiburón: 430
  La Cruz: 430
  Morro Hermoso: 47
  Olvidada Formation: 103
  Orocopia Schist: 384
  Pelona Schist: 384
  Peninsular Ranges: 298
  Portal Ridge schist: 384, 397
  Punta Quebrada-Rompiente coastline: 47
  Punta San Hipolito: 48
  Rand Schist: 384, 397
  San Hipolito Formation: 44, 60
  Santiago Peak Volcanics: 99
  Sierra Calamajue: 103
  Sierra de Salinas schist: 384, 397
  Sierra de San Andres: 47
  Sierra de San Andres Ophiolite: 79
  Sinaloa pre-batholithic strata: 252
  Tangaliote faults: 85
Chihuahua: 31, 265, 266
Chile
 Andes Mountains. *see* Andes Mountains
 isotopic age pattern in: 326
 Nazca plate: 382, 397
Chiquito Peak suite
 deformation in: 204, 205
 depositional setting of: 227–228
 elements in, major and trace: 218–223, 227–228
 geobarometry of: 194
 geochemical analysis of: 209–218
 geological setting of: 204–208, 227–228

 Harker diagrams for: 217
 isotopic analysis of: 223–225
 petrology of: 205, 227–228
chlorite: 49, 75, 260, 341–342
Chocolate Mountains
 correlations to Coast Ranges: 392
 faults in: 382
 geochronology of: 392, 401
 geological setting of: 400
Chocolate Mountains fault: 382
Chortis block: 19, 266
Choyal terrane: 44–49, 63
chromite: 75
chrysotile: 75, 81
chubascos: 413
clams: 412
clinopyroxene
 in Cuyamaca Reservoir suite: 169
 in Moctezuma volcanic field: 444
 in San Ignacio plutons: 254
 in Santiago Peak Volcanics: 98
 in Sierra de San Andres Ophiolite: 47
 in Sinaloa batholith: 246, 252
 in Tayoltita mines: 260
 in Vizcaíno Peninsula Ophiolite: 47
Clypeaster bowersi: 416
Clypeaster marquerensis: 413, 416
Coast Plutonic complex: 339
Coast Range Ophiolite
 correlations to: 61–62, 65–66, 392
 depositional setting of: 44, 64–67
 geochemical analysis of: 65–66
 geochronology of: 52, 66
 geological setting of: 66
 paleomagnetic studies of: 67
Coast Ranges
 faults in: 126
 geochronology of
  amphibolite: 66
  eclogite: 66
  schists: 384, 393
  underplating of: 356
 geological setting of: 44
 reconstructive modeling of: 128
 Sierra de Salinas schist: 356, 382–385, 388,
  393, 395–399
Coastal batholith (BC): 326
Coastal batholith (Peru): 201, 288–290
Cochimí terrane. *See also* Choyal terrane; Vizcaíno
  terrane
 boundaries of: 20
 definition of: 5
 depositional setting of: 8–9, 16
 faults in: 9, 75
 field relations with
  Alisitos Formation: 8
  Valle Group: 16, 44
 geobarometry of: 8
 geochronology of: 8–9, 17, 74–75
 geological setting of: 74
 geothermometry of: 8
 metamorphism in: 8–9
 paleomagnetic studies of: 17, 20
 petrology of: 8–9, 16–17, 74–75
 transport of: 17, 20

New England batholith
  correlations to: 165
  depositional setting of: 177
  elements in, major and trace: 176
  geochemical analysis of: 217
  isotopic analysis of: 172–173
New Zealand
  Alpine fault: 278, 279, 288
  Fiordland block: 278–279, 287
  geochronology of
    magmatism in: 279–281
    rift from Australia: 286
  Gondwana margin in: 278
  Median batholith. *see* Median batholith
  Nelson block: 278–279
  oil wells in: 278
  Rahu suite: 278–286, 291
  Separation Point suite. *see* Separation Point suite
New Zealand Eastern Province: 279
Newport fault: 3
Nodipecten subnodosus: 413
Norian: 57
norite: 75, 207
North America. *See also* specific nations; specific provinces; specific states
  Basin and Range extension: 29
  continental shelf: 20
  crust/mantle structure: 29
  Laramide orogeny. *see* Laramide orogeny
  terrane accretion process: 118
North America–Kula–Farallon triple junction: 28, 121, 126
North American craton
  boundaries of: 5
  continental-margin arc: 67, 266
  field relations with
    Peninsular Ranges batholith: 427
    Yaqui Ridge plutons: 358
  geochronology of: 5
  geological setting of: 74
  reference poles for: 148
North American Magnetic Anomaly Map: 3–4
North America plate
  accretionary prism
    modeling of: 66
    removal of: 14, 111
    turbidite deformation in: 96
  continental-margin arc
    arc-arc collision: 124
    formation of: 64
    magma: 23, 275–276
    modeling of: 66, 126–128
  crust/mantle structure
    lithospheric thinning: 453
    necking in: 422
    rifting vs. rupture: 422
    below Sonora: 3
    TTG suites from: 275–276
    underplating of: 13
    underthrusting in: 119
  faults in
    dextral: 22, 118
    normal: 3
    northward transport along: 20–21, 23, 118
    sinistral: 22, 118, 177

    slip rate: 3
    southward transport along: 23
    Tosco-Abreojos: 20, 30
    transform: 177
  geochronology of
    accretion of: 13–14, 120, 124
    arc-arc collision: 124
    continental arc magmatism: 177
    island arc evolution: 177
    magmatism, migration of: 28–29
    pluton emplacement: 13
    subduction zone: 13, 28, 29, 122
    triple junction formation: 28, 121
  geographic setting of: 3
  motion of
    Atlantic Ocean impact on: 131–132
    convergence velocity: 326
    correlations to Pacific plates: 22, 120
    direction of: 118
    margin-parallel transport: 6
    polar wander path: 64–65, 148
    reconstruction of: 23–27
    speed of: 120
  subduction zone
    arc magmatism along: 5, 13, 266, 326
    batholith depositional setting of: 13, 395–397
    conversion of, to transform margin: 28
    crust/mantle structure: 382, 395
    geochronology of: 118
    rollback in, of Farallon: 29
    schist formation from: 21–22, 382, 395
    shallowing of: 13, 326
    structure of: 328
    underplating in: 395
    westward transport of: 28
  suture zones: 13
North Victoria Land: 288

O

olistostrome
  geochronology of: 52–54
  petrology of: 49, 63, 96, 97
olivine
  geochemical analysis of: 444
  petrology of
    Báucarit Formation: 440
    Moctezuma volcanic field: 444
    Santiago Peak Volcanics: 98
    Sierra de San Andres Ophiolite: 47, 75
    Vizcaíno Peninsula Ophiolite: 47
Olvidada Formation: 103
opal: 410
ophiolites
  description of: 73
  geochronology of: 8–9
  paleomagnetic studies of: 17
  petrology of: 7, 8–9
Oppelia sp.: 48
Oregon
  Blue Mountains: 126–128
  Cascade Mountains: 338
  Coast Ranges. *see* Coast Ranges
  Columbia River: 10
  geochronology of bivalves: 49
  Klamath Mountains. *see* Klamath Mountains

Oriflamme Canyon pluton: 228, 361
Orocopia Mountains: 393, 400
Orocopia Schist: 21–22, 382–385, 393, 395–399
orthogneiss: 5–6, 10, 11, 251
orthopyroxene: 47, 60–61, 227
orthopyroxenite: 75
Otago Schist: 286, 287
Owens Mountain: 64–65
oysters: 410, 413–414

P

Pacific Ocean
  paleocirculation: 28
  paleomagnetic studies in: 22, 151
  spreading center basins: 177
  transition zone in: 3
Pacific plate. *See also* Pacific–North America boundary
  faults along: 3, 20, 30
  geographic setting of: 3
  modeling of: 20, 22–27, 30–31
Pacific-Australian boundary: 278
Pacific-Farallon boundary: 22
Pacific–Farallon–North America junction: 28
Pacific-Kula boundary: 22
Pacific–North America boundary
  faults in
    Agua Blanca fault. *see* Agua Blanca fault
    Sierra Placer mélange: 89
    Tangaliote: 89
    Tosco-Abreojos: 20, 30
    transform: 440
  geochemical analysis of: 440
  geochronology of: 22, 28–29
  in Gulf of California. *see* Gulf of California
  modeling of: 23–27, 30–31, 89
Panthalassan lithosphere: 180
Panuco volcanics: 260
Paparoa Range: 286
Pasayten fault: 122, 128
Peacock Index: 57
pectens: 410, 413
pegmatite dikes: 12, 204, 207, 360
Pelona Schist
  correlations to: 382, 384–385
  cycling interval: 393
  depositional setting of: 382, 395–399
  evolution of: 21–22
  geobarometry of: 382
  geochronology of: 22, 382–384
  geological setting of: 382, 395
  metamorphism in: 384, 400
  petrology of: 384
  underplating of: 356
Penasquitos Canyon: 64
Peninsular Ranges. *See also* Peninsular Ranges batholith
  Agua Blanca fault. *see* Agua Blanca fault
  Alisitos Formation. *see* Alisitos Formation
  Bedford Canyon Complex. *see* Bedford Canyon Complex
  bivalves in: 64
  Calamajué. *see* Calamajué
  correlations to: 64, 327
  crust/mantle structure: 3